DIE PHYSIK
DER HOCHPOLYMEREN

HERAUSGEGEBEN

VON

H. A. STUART

O. PROFESSOR FÜR CHEMISCHE PHYSIK AN DER UNIVERSITÄT MAINZ

VIERTER BAND

THEORIE
UND MOLEKULARE DEUTUNG
TECHNOLOGISCHER EIGENSCHAFTEN
VON HOCHPOLYMEREN WERKSTOFFEN

Springer-Verlag Berlin Heidelberg GmbH

THEORIE UND MOLEKULARE DEUTUNG TECHNOLOGISCHER EIGENSCHAFTEN VON HOCHPOLYMEREN WERKSTOFFEN

BEARBEITET

VON

J. P. BERRY · J. D. FERRY · E. JENCKEL · W. KAST · H. F. MARK

W. MESKAT · O. ROSENBERG · F. SCHWARZL · A. J. STAVERMAN

R. S. STEIN · H. A. STUART · H. THURN · L. R. G. TRELOAR

A. K. VAN DER VEGT · F. WÜRSTLIN

MIT 367 TEXTABBILDUNGEN

Springer-Verlag Berlin Heidelberg GmbH

ISBN 978-3-662-01379-3 ISBN 978-3-662-01378-6 (eBook)
DOI 10.1007/978-3-662-01378-6

Ursprünglich erschienen bei Springer-Verlag OHG., Berlin . Göttingen . Heidelberg 1956.
Softcover reprint of the hardcover 1st edition 1956

Vorwort des Herausgebers zum vierten Band.

Der vierte und letzte Band der „Physik der Hochpolymeren" soll einen Überblick über die charakteristischen mechanischen und elektrischen Eigenschaften hochpolymerer Körper sowie über die Möglichkeiten geben, diese Eigenschaften auf die Elemente der molekularen Struktur zurückzuführen. Es ist daher bei der Herausarbeitung der allgemeinen Zusammenhänge sowohl von der phänomenologischen als auch von der molekularen Betrachtungsweise Gebrauch gemacht worden.

Auch da, wo die Entwicklung einer quantitativen molekularen Theorie wenig aussichtsvoll erscheint, würde es eine Einschränkung der Forschungs- und Erkenntnismöglichkeiten bedeuten, wenn man neben der phänomenologischen Untersuchung nicht etwa auch die direkt molekular interpretierbaren röntgenographischen und optischen Methoden möglichst weit ausschöpfen würde. Viele Erscheinungen werden in ihrem Kern viel schneller verständlich und technisch beherrschbar, und mancher Umweg und Fehlschluß ist dabei vermeidbar, wenn man sie von vornherein auch im molekularen Bilde durchdenkt und dabei auf die wichtigsten Elemente des molekularen Bewegungs- und Ordnungszustandes zurückführt. Die Erforschung der Kaltverstreckung, die typisch hochmolekulare Merkmale zeigt und in ihrem Ablauf wesentlich auch von der molekularen Textur des Anfangszustandes abhängt, vgl. Band III, ist ein Beispiel dafür.

Die Denk- und Arbeitsweise von Wissenschaftlern oder Praktikern ist verschieden. Der Herausgeber bekennt sich zu denjenigen, für die das molekulare Bild nicht nur das Verständnis erleichtert, sondern auch die unentbehrliche und oft bewährte Arbeitshypothese liefert, deren Grenzen man sich natürlich immer bewußt bleiben muß. Er sieht geradezu eine Aufgabe darin, dem Lernenden wie auch dem Praktiker vereinfachte, aber die wesentlichen Punkte treffende molekulare Vorstellungen an die Hand zu geben. Wie sollte man ohne molekulare Betrachtungen das so komplexe mechanische Verhalten hochpolymerer Körper, die Abhängigkeit ihrer Eigenschaften von der Temperatur und der Frequenz sowie von der mechanischen und thermischen Vorgeschichte verstehen und übersehen?

Diese Gesichtspunkte bestimmten die Disposition dieses Bandes. Die drei ersten allgemeinen Kapitel behandeln vorwiegend die phänomenologischen Theorien des linearen und nichtlinearen Deformationsverhaltens sowie der Bruchspannung und Festigkeit, immer wieder ergänzt durch molekulare und andere Modellvorstellungen. Es folgt dann ein Kapitel mit kritischen Betrachtungen über die technologischen Prüfmethoden. In den folgenden drei Kapiteln werden die Zusammenhänge zwischen der Struktur und den mechanischen Eigenschaften von kautschukartigen und plastischen Körpern sowie von Faserstoffen behandelt. Dabei sind nur solche Untersuchungen berücksichtigt, die für die Entwicklung unserer Kenntnisse der Eigenschaften hochpolymerer Körper und für das Verständnis ihrer Beziehungen zur molekularen Struktur von allgemeiner Bedeutung sind. Derartige Untersuchungen sind aber nur an relativ wenigen Stoffen durchgeführt worden.

Den Zusammenhängen zwischen Struktur und elektrischen Eigenschaften ist, ebenso wie der Weichmacherwirkung und ihrer molekularen Deutung, ein besonderes Kapitel gewidmet. Schließlich wurden im letzten Kapitel einige allgemeine Zusammenhänge zwischen der molekularen Struktur und den makroskopischen Eigenschaften von hochpolymeren Körpern herausgestellt. Hierbei kann es sich nur um einen ersten Versuch handeln, der zu weiteren Arbeiten anregen möge.

Es ist mir eine Freude, allen Mitarbeitern für ihre Mitwirkung und für ihr Eingehen auf die Wünsche des Herausgebers vielmals zu danken, insbesondere den Herren STAVERMAN und SCHWARZL, die wesentlich dazu beigetragen haben, daß dieser Band in der gewünschten Form und rechtzeitig erscheinen konnte.

Herr O. FUCHS, Höchst, hat auch bei diesem Bande die gesamte Korrektur mitgelesen und manchen Verbesserungsvorschlag gemacht, wofür ihm auch im Namen der Mitarbeiter herzlich gedankt sei.

Schließlich sei auch dem Springer-Verlag für sein Verständnis für alle Wünsche des Herausgebers, nicht nur bei diesem Bande, sowie für die vorzügliche Ausstattung auch dieses letzten Bandes, bestens gedankt.

Mainz, Dezember 1955 H. A. STUART.

Mitarbeiter des vierten Bandes.

Dr. J. P. Berry, The British Rubber Producers' Research Association, London, England.

Professor J. D. Ferry, University of Wisconsin, Department of Chemistry, Madison Wisconsin, USA.

Professor Dr. E. Jenckel, Direktor des Instituts für theoretische Hüttenkunde und physikalische Chemie der Rhein.-Westf. Technischen Hochschule, Aachen.

Professor Dr. W. Kast, Honorarprofessor an der Universität Köln, früher Direktor des Instituts für experim. Physik der Universität Halle.

Director H. F. Mark, Polymer Research Institute, Polytechnic Institute of Brooklyn, Brooklyn, NY., USA.

Dr. phil. W. Meskat, Farbenfabriken Bayer AG., Leverkusen.

Dr. phil. O. Rosenberg, Farbenfabriken Bayer AG., Leverkusen.

Dr. phil. F. Schwarzl, Centraal-Laboratorium-T.N.O., Delft, Holland.

Dr. A. J. Staverman, Centraal-Laboratorium-T.N.O., Delft, Holland.

Professor R. S. Stein, Department of Chemistry, University of Massachusetts, Amherst, Mass., USA.

Professor Dr. H. A. Stuart, o. Professor für chemische Physik an der Universität Mainz.

Dr. rer. nat. H. Thurn, Meß- und Prüfabteilung der Badischen Anilin- u. Soda-Fabrik AG., Ludwigshafen/Rh.

Dr. L. R. G. Treloar, British Rayon Research Association, Heald Green Laboratories, Wythenshawe, Manchester, England.

Dr. ir. A. K. van der Vegt, Centraal-Laboratorium-T.N.O., Delft, Holland.

Dr. F. Würstlin, Meß- und Prüfabteilung der Badischen Anilin- u. Soda-Fabrik AG. Ludwigshafen/Rh.

Inhaltsverzeichnis.

First Chapter:

Linear Deformation Behaviour of High Polymers.

By A. J. STAVERMAN and F. SCHWARZL.

With 59 figures.

Second Chapter:

Non Linear Deformation Behaviour of High Polymers.

By A. J. STAVERMAN and F. SCHWARZL.

With 32 figures.

Drittes Kapitel:

Bruchspannung und Festigkeit von Hochpolymeren.

Von F. SCHWARZL und A. J. STAVERMAN.

Mit 51 Abbildungen.

Viertes Kapitel:

Kritischer Überblick über die technologischen Prüfmethoden.

Von W. MESKAT, O. ROSENBERG, F. SCHWARZL und A. J. STAVERMAN.

Mit 15 Abbildungen.

Fifth Chapter:

The Structure and Mechanical Properties of Rubberlike Materials.

By L. R. G. Treloar.

With 59 figures.

Sixth Chapter:

Structure and Mechanical Properties of Plastics.

By John D. FERRY.

With 35 figures.

Siebentes Kapitel:

Struktur und mechanische Eigenschaften von Faserstoffen.

Von W. Kast, W. Meskat, O. Rosenberg und A. K. van der Vegt.

Mit 80 Abbildungen.

Achtes Kapitel:

Struktur und elektrische Eigenschaften.

Von F. WÜRSTLIN und H. THURN.

Mit 17 Abbildungen.

Neuntes Kapitel:

Die Wirkung von Weichmachern und ihre molekulare Deutung.

Von Ernst Jenckel.

Mit 17 Abbildungen.

Zehntes Kapitel:

Vergleichende Betrachtungen zum Zusammenhang zwischen der molekularen Struktur und den makroskopischen Eigenschaften von Körpern aus Fadenmolekülen.

Von H. Mark und H. A. Stuart.

Mit 2 Abbildungen.

Zusammenstellung der am häufigsten benutzten Symbole.

Die Seitenzahl gibt an, wo die betreffende Größe eingeführt wird.

			Seite
α	Oberflächenenergie, surface energy	§ 17,	86
c	Fortpflanzungsgeschwindigkeit, velocity of propagation	§ 8,	90
$\gamma_{xy}, \gamma_{yz}, \gamma_{zx}$	Scherungsdeformationen, shear strains	§ 2,	6
$\dot{\gamma}_{xy}, \dot{\gamma}_{yz}, \dot{\gamma}_{zx}$	Scherungsgeschwindigkeit, rate of shear	§ 2,	6
δ	Verlustwinkel, loss angle	§ 3,	24
$\operatorname{tg}\delta$	Dämpfung, damping, Verlust, loss	§ 3, § 67,	24, 28 531
Δ	logarithmisches Dekrement, logarithmic decrement	§ 8,	86
E	Elastizitätsmodul, Youngs modulus	§ 2,	8
$E(t)$	Spannungsrelaxationsmodul bei Dehnung, extension stress relaxation modulus	§ 6,	72
$\varepsilon_x, \varepsilon_y, \varepsilon_z$	Normaldeformationen, normal strains	§ 2,	6
$\dot{\varepsilon}_x, \dot{\varepsilon}_y\ \dot{\varepsilon}_z$	Normal-Deformationsgeschwindigkeiten, rate of normal strains	§ 2,	6
η	Viskosität, viscosity	§ 2,	10, 142
$[\eta]$	Grenz-Viskositätszahl, intrinsic viscosity	§ 12,	146
f_1, f_2, f_3	konventionelle Spannungen, bezogen auf den nichtdeformierten Körper, conventional stresses	§ 11,	131
f_t	Zugfestigkeit, tensile strength (bezogen auf den nichtdeformierten Querschnitt)	§ 15,	169
f_c	Druckfestigkeit, compressive strength (bezogen auf den nichtdeformierten Querschnitt)	§ 15,	169
f_b	Biegefestigkeit, bending strength (bezogen auf den nichtdeformierten Querschnitt)	§ 15,	169
f_s	Scherfestigkeit, shear strength (bezogen auf den nichtdeformierten Querschnitt)	§ 15,	169
$f(\tau)$	Retardationsspektrum, retardation spectrum	§ 4,	41
$f(T)$	Zeit-Temperatur-Verschiebungsfunktion, time temperature shift function	§ 5,	57
F	Freie Energie, free energy	§ 5,	54
$F(t)$	Kriechfunktion bei Dehnung, extension compliance	§ 6,	72
$g(\tau)$	Relaxationsspektrum, relaxation spectrum	§ 4,	40
G	Scherungsmodul, shear modulus	§ 2,	8
$G(t)$	Spannungsrelaxationsmodul bei Scherung, stress relaxation modulus in shear	§ 3,	16
$G_1(\omega)$ oder G'	Realteil des komplexen Moduls, storage (shear) modulus	§ 3,	25, 28
$G_2(\omega)$ oder G''	Imaginärteil des komplexen Moduls, loss (shear) modulus	§ 3,	25, 28
$G^*(\omega)$	komplexer Modul, complex shear modulus	§ 3,	25, 28
$\lvert G\rvert$	absoluter dynamischer Modul, absolute dynamic modulus	§ 3,	24, 28
H	Härte, hardness	§ 17,	209
$H(\ln\tau)$	logarithmisches Relaxationsspektrum, logarithmic relaxation spectrum	§ 4,	42
ΔH_a	Aktivierungsenergie, activation energy	§ 5,	61
I_1, I_2, I_3	Invarianten der Hauptdehnungen, invariants of principal stretches	§ 11,	131

J	Nachgiebigkeit, allgemein wie besonders bei Scherung, compliance and shear compliance	§ 3, 12
$J(t)$	Kriechfunktion, Nachgiebigkeitsfunktion, creep compliance, compliance function	§ 3, 12
$J_1(\omega)$	Realteil der komplexen Nachgiebigkeit, storage (shear) compliance	§ 3, 24, 28
$J_2(\omega)$	Imaginärteil der komplexen Nachgiebigkeit, loss (shear) compliance	§ 3, 24, 28
$J^*(\omega)$	komplexe Nachgiebigkeit, complex (shear) compliance	§ 3, 25
$\|J\|$	absolute dynamische Nachgiebigkeit, absolute dynamic compliance	§ 3, 24, 28
J_1, J_2, J_3	Invarianten des Deformationstensors, invariants of the strain tensor	§ 2, 7
k	Kerbzahl, stress concentration factor	§ 17, 188
k	BOLTZMANN-Konstante, BOLTZMANN constant	§ 17, 194
K	Kompressionsmodul, bulk modulus	§ 2, 8
$K(t)$	Spannungsrelaxationsmodul bei isotroper Kompression, bulk stress relaxation modulus	§ 6, 70
K_1, K_2, K_3	Invarianten des Deformationsgeschwindigkeitstensors, invariants of the rate of strain tensor	§ 2, 7
$\varkappa(t)$	Kriechfunktion bei isotroper Kompression, creep-function	§ 6, 72
$L(\ln\tau)$	logarithmisches Retardationsspektrum, logarithmic retardation spectrum	§ 4, 42
λ	Länge des Griffith Risses, length of Griffith crack	§ 17, 189
λ	Wellenlänge, wave length	§ 8, 90
$\lambda_1, \lambda_2, \lambda_3$	Hauptdehnungen, principal stretches	§ 11, 131
N	Anzahl Lastwechsel, number of loading cycles	§ 15, 172
ν	POISSONsche Verhältnis, POISSON's ratio	§ 2, 8
$\nu(t)$	POISSONsche Verhältnis, POISSON's ratio	§ 6, 73
ω	Kreisfrequenz, angular frequency	§ 3, 23
ω_c	Eigenfrequenz, Eigenfrequency	§ 8, 86
ω_0	Resonanzfrequenz, resonance frequency	§ 8, 87
$\Delta\omega$	Halbwertsbreite der Resonanzkurve, half width of resonance curve	§ 8, 84
p	hydrostatischer Druck, hydrostatic pressure	§ 2, 4
$p(\sigma)$	Differentielle Bruchspannungsverteilungskurve, differential distribution curve of breaking stresses	§ 16, 176
$P(\sigma)$	Kumulative oder integrale Bruchspannungsverteilungskurve, cumulative distribution curve of breaking stresses	§ 16, 177
q	Deformationsgeschwindigkeit (Scherungsgeschwindigkeit), rate of deformation (shear)	§ 12, 141
r	Dämpfung per Wellenlänge, attenuation per wavelength	§ 8, 90
R	Rückprall-Elastizität, rebound resilience	§ 8, 94
ϱ	Dichte, density	§ 5, 58
S	Entropie, entropy	§ 5, 54
S_1, S_2, S_3	Invarianten des Spannungstensors, invariants of the stress tensor	§ 2, 6
σ_M	Molekulare Zerreißfestigkeit, molecular strength	§ 16, 179
$\sigma_x, \sigma_y, \sigma_z$	Normalspannungen, normal stresses	§ 2, 4
$\sigma_{xy}, \sigma_{yz}, \sigma_{zx}, \sigma$	Scherungsspannungen, Schubspannung, shearing stresses	§ 2, 4
$\sigma_1, \sigma_2, \sigma_3$	Hauptspannungen, principal stresses	§ 2, 4
t	Zeit, time	§ 3, 12
t_b	Bruchzeit, rupture time	§ 15, 172
T	absolute Temperatur, absolute temperature	§ 5, 54
T_b	Sprödigkeitstemperatur, brittle temperature	§ 3, 17
T_E	Einfriertemperatur, transition temperature	§ 3, 17

τ	Retardationszeit, retardation time, auch Relaxationszeit, relaxation time	§ 4, 31
τ'	Relaxationszeit, relaxation time	§ 4, 32
ξ, η, ζ	Verschiebungsvektor, displacement vector	§ 2, 5
u, v, w	Geschwindigkeitsvektor, velocity vector	§12, 141
U	innere Energie, internal energy	§ 5, 54
$\Delta V/V$	relative Volumenänderung, relative change in volume	§ 2, 7
W	Deformationsenergie, deformation energy	§ 1, 7
x	reduzierte logarithmische Zeit, reduced logarithmic time	§ 5, 58
y	reduzierte logarithmische Frequenz, reduced logarithmic frequency	§ 5, 58

First Chapter.

Linear Deformation Behaviour of High Polymers.

By

A. J. Staverman and F. Schwarzl.

With 59 figures.

§ 1. Introduction.

A study of the physics of high polymers would be incomplete without a thorough treatment of the "mechanical behaviour" of these substances. This is not only because mechanics is one chapter of general physics, but particularly because the immense development of practical applications of high polymers is due to a large extent to their mechanical properties.

To study the mechanical behaviour of a material we make a sample out of it, a "body", and study the deformations of that body under the influence of various external forces. The ultimate purpose of these studies is to find a certain number of constants, such as elasticity moduli, Poissons' ratio and the like which are characteristic for the material and with the aid of which we can predict the deformation of arbitrary bodies of that material if subjected to arbitrary forces.

We see immediately that this problem is much more complicated than that of many other properties like thermal, electrical or optical properties. In investigations of these latter properties, also a body is made from the material under consideration, but the shape of this body is chosen in such a way that its dimensions enter into the measurable relationships in a simple well-defined way. However in the study of mechanical properties the deformation or the change of shape of the body is the object of the study and therefore even if the initial shape of the body is simple it may change in a complicated way in the course of the experiment.

The first major problem encountered in the study of mechanical behaviour is, therefore, the mathematical problem of calculating local deformations from a given change of shape of a given body. This general problem, well known to engineers, who meet the problem of predicting the local deformations of complicated pieces of equipment subjected to complicated forces, is still unsolved in classical elasticity theory and is even more intricate if the body under consideration is made from a visco-elastic high polymer. As we are interested in the properties of *materials* and not of *bodies* we will as much as possible avoid this problem by considering bodies of the simplest shape only. However, in all interpretation of measurements this problem will complicate matters to some extent.

The second major problem in our study is due to the fact that neither "change of shape" nor "local deformation" are variables of state like the volume or the pressure of a gas. Classical elasticity theory is based upon the conception of a material of which the deformation is indeed a variable of state, which means that the internal state of the material of which the deformed body is made does not depend on the history by which this deformation has been reached. Without going into the question whether such materials exist at all, we must realize from the start that high polymers are not among them. An important question which is basic to the whole of our study will be the question of which variables of state should be used to describe the local deformation in a polymer and how these variables make themselves perceptible in experiments. This is the problem of the thermodynamics of viscoelastic behaviour, which will be treated in § 5.

The variables of state are not immediately measurable. In fact, their disentanglement from experimental data is a problem in itself. Therefore in § 3 we give an account of how experimental data on linear viscoelastic behaviour can be described in a purely phenomenological manner, while § 4 is intended to approach the somewhat abstract variables of state as perspicuously as possible by the treatment of models.

In the first four sections of chapter I the dependence of deformation and of the conforming force on time and temperature are considered but no account is given of the geometry of the deformation. In § 6 the complications arising from the fact that generally three-dimensional deformations are considered. Finally in § 7 an account is given of experimental methods for investigations of mechanical behaviour of polymers.

While chapter I deals exclusively with deformations which are so small, that quadratic terms in deformations and forces can be neglected, in chapter II non-linear behaviour is treated. Whereas linear behaviour can be treated with a closed physical theory, although determination of all the material constants needed for an complete description of a particular material is nearly beyond the experimental possibilities, non-linear behaviour is not yet accessible to such a treatment. Here understanding the phenomena has to proceed in a more haphazard and descriptive way.

§ 2. General considerations about stress and strain.

a) Preliminary remarks.

As pointed out in the introduction, a description of the mechanical behaviour of deformable substances consists of relations between the mechanical forces applied to the material and the resulting changes in shape, including the occurrence of rupture. The task of finding such relations meets with three problems. The first problem is to describe the way in which the external forces affect the substance under consideration, i.e. to define the state of stress in a body and to extend NEWTONS law of motion to derive the equations of motion for the stressed body. The second problem is to define the change of shape in a stressed material, or, more accurately, to describe the state of strain in a deformed body.

The solution of these problems is well established at least in the region of small strains and is shortly reviewed in sections b) and c). The third and most difficult problem is to derive the relations between stress and strain in different materials. This will concern us throughout this book and is to be considered as the proper problem of *rheology*.

This first chapter deals entirely with linear deformation properties, which means that the forces are assumed to be so small that the deformations are proportional to them. This has the great advantage that a consistent theory can be built up, which describes the properties common to solids, liquids and high polymers. All mechanical properties to be dealt with can be described by some simple fundamental methods of a completely phenomenological character, which are independent of the special nature of the material under consideration. A vast number of experimental properties of high polymers can be understood and classified in this way.

In classical theory the deformation behaviour of two particular types of materials has been investigated, the *elastic solid* and the *viscous liquid*. A hard solid is deformed under external forces into a new equilibrium state and goes back to its original form after release of the forces. It is characterised by *elasticity*. A liquid has no definite shape but flows under the action of external forces under generation of heat due to internal friction. It is characterised by a *viscosity*. The theory of hard solids and of viscous liquids will be reviewed in sections e) and f).

In recent times, especially as a result of the appearance of high polymers, it has turned out that neither the ideal elastic solid nor the ideal viscous liquid are general types of materials existing in nature. These model substances are to be considered as limiting cases of materials, the one showing only phenomena generally observed in very short time behaviour, the other in very long time behaviour. High polymers can show all intermediate properties between elastic and fluid, depending on the temperature and the experimentally chosen time scale.

b) The state of stress.

The theory of stress and strain and the classical theories of elasticity and of viscosity are assumed to be familiar to the reader. We give a very short account of these topics and refer the reader for extended studies to the literature cited at the end of this chapter.

A body which is in equilibrium under the action of external forces, transmits the action of these forces through its interior. The state of stress in such a body can be described in the following way. At the point under consideration we choose a right handed coordinate system x, y, z and cut out a small cube with edges parallel to

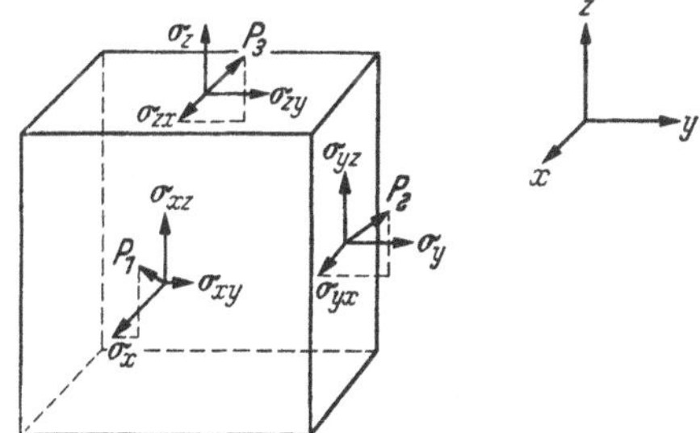

Fig. I, 1. Components of stress tensor.

the coordinate axes (fig. I,1). The forces to which the cube is subjected act on the three pairs of coordinate planes. These forces, divided by the area of the cube faces, are

P_1 acting on the yz plane
P_2 acting on the zx plane
P_3 acting to the xy plane.

These three tractions (forces per unit area) can each be resolved into three components

force	component in x direction	component in y direction	component in z direction
P_1	σ_x	σ_{xy}	σ_{xz}
P_2	σ_{yx}	σ_y	σ_{yz}
P_3	σ_{zx}	σ_{zy}	σ_z

The three stresses σ_x, σ_y, σ_z are called *normal stresses* and the six stresses with mixed indices are called *shearing stresses*. As a result of a consideration of the moments of the forces, it is found that only three of the six shearing stresses are independent the other three follow from the relations

$$\sigma_{xy} = \sigma_{yx}, \quad \sigma_{xz} = \sigma_{zx}, \quad \sigma_{yz} = \sigma_{zy}. \tag{I,1}$$

The state of stress is therefore known, if in each point of the body six stress components are known or as we say, if the components of the *stress tensor* are known.

If we choose another orientation of the cubes, say a new coordinate system x', y', z' the new stress components can be calculated from the old ones by

$$\sigma'_{i'k'} = \sum_{l,m} \sigma_{lm} \cos(l\,i') \cos(m\,k'). \tag{I,2}$$

Here the letters i, k, l, m stand for one of the coordinate directions x, y, z and $\cos(x, y')$ e.g. means the cosine of the angle detween the old x and the new y direction.

There is one special orientation of the coordinate system at the point considered for each state of stress, in which only the normal stresses are different from zero. The shearing stresses vanish. The normal stresses in this special coordinate system are called the *principal stresses* σ_1, σ_2, σ_3 ($\sigma_1 \geq \sigma_2 \geq \sigma_3$) and can be obtained as the roots of the equation

$$\begin{vmatrix} \sigma_x - \sigma & \sigma_{xy} & \sigma_{xz} \\ \sigma_{xy} & \sigma_y - \sigma & \sigma_{yz} \\ \sigma_{xz} & \sigma_{yz} & \sigma_z - \sigma \end{vmatrix} = 0. \tag{I,3}$$

It follows that, apart from a rotation of the coordinate axis, any state of stress can be charaterised by its three principal stresses.

Another simplification which is often convenient is to subtract from each of the normal stresses one third of their sum which quantity is called the negative *hydrostatic pressure* p:

$$p = -[\sigma_1 + \sigma_2 + \sigma_3]/3. \tag{I,4}$$

The stress left after this subtraction is called the *deviatoric component* of the stress, while the stress described by three normal stresses equal to p and shear stresses equal to zero is called the *isotropic component*. The total stress is considered as the sum of these two components.

The maximum shearing stress existing in a particular state of stress is given by

$$\tau_{\max} = [\sigma_1 - \sigma_3]/2 . \tag{I,5}$$

In the following table we give some simple states of stress and the corresponding hydrostatic pressure and maximum shearing stress contained in them.

Table I,1. *Simple states of stress.*

State of stress	principal stresses	hydrostatic pressure	maximum shearing stress
Simple tension in one direction	$\sigma_1 > 0 \quad \sigma_2 = \sigma_3 = 0$	$\sigma_1/3$	$\sigma_1/2$
Pure shear	$\sigma_1 = \tau \quad \sigma_2 = 0 \quad \sigma_3 = -\tau$	0	τ
Hydrostatic pressure	$\sigma_1 = \sigma_2 = \sigma_3 = -p$	p	0

c) The state of strain.

A convenient method of describing the changes of shape in a stressed body starts from the definition of the strain tensor. Each material point of the body in the undeformed state can be fixed by a coordinate vector x, y, z relative to an arbitrarily chosen origin 0. In the deformed state the position of the material point will be displaced, the point will have the new coordinates $x + \xi$, $y + \eta$, $z + \zeta$. The *displacement vector* ξ, η, ζ of the point considered will depend on the original position of this point in the unstrained state

$$\xi = \xi(x, y, z), \quad \eta = \eta(x, y, z), \quad \zeta = \zeta(x, y, z) . \tag{I,6}$$

If the displacement vector ξ, η, ζ is known for all positions x, y, z the state of strain in the body is described. But if we are interested in the properties of the material, it is not the displacement but the deformation which is of primary importance. The deformation, then, depends upon differences in displacement between neighbouring elements and can be described by the *strain tensor* (*deformation tensor*). Like the stress tensor of the foregoing section, the deformation tensor consists of six components[1]:

$$\left.\begin{aligned}
&\varepsilon_x = \partial\xi/\partial x \qquad\qquad\qquad\qquad \varepsilon_y = \partial\eta/\partial y\\
&\frac{1}{2}\gamma_{xy} = \frac{1}{2}\left(\frac{\partial\xi}{\partial y} + \frac{\partial\eta}{\partial x}\right) = \frac{1}{2}\gamma_{yx}; \quad \frac{1}{2}\gamma_{yz} = \frac{1}{2}\left(\frac{\partial\zeta}{\partial y} + \frac{\partial\xi}{\partial z}\right) = \frac{1}{2}\gamma_{zy};\\
&\varepsilon_z = \partial\zeta/\partial z\\
&\frac{1}{2}\gamma_{xz} = \frac{1}{2}\left(\frac{\partial\zeta}{\partial x} + \frac{\partial\xi}{\partial z}\right) = \frac{1}{2}\gamma_{zx}.
\end{aligned}\right\} \tag{I,7}$$

[1] For many purposes it is convenient to define the shear strains γ_{xy}, γ_{yz} and γ_{zx} as twice the tensor components of the deformation tensor. The proper tensor components of the deformation tensor are the quantities ε_x, ε_y, ε_z, $^1/_2\,\gamma_{xy}$, $^1/_2\,\gamma_{yz}$, $^1/_2\,\gamma_{xz}$.

The geometrical meaning of the strain tensor is as follows: ε_x, called the *normal strain* in the x direction, is the relative change in length $\Delta l/l_0$ of a line parallel to the x coordinate. The normal strains in y and z directions, ε_y and ε_z have a similar meaning. The *shear strain* γ_{xy} describes the change of shape caused by the deformation: the right angle formed by the two planes of a cube, perpendicular to the x and y axis before deformation is changed to an angle $\frac{\pi}{2} - \gamma_{xy}$ by the deformation. The *relative change in volume*, resulting from the deformation, is given by the sum of the normal strains

$$\frac{\Delta V}{V} = \varepsilon_x + \varepsilon_y + \varepsilon_z . \tag{I,8}$$

Some simple examples of strain will be discussed in the following section.

It must be mentioned that the strain tensor as defined in the above form is the appropriate mathematical tool if we deal with small deformations only. ($\varepsilon_{xx} \ll 1$, $\gamma_{xy} \ll 1$.) For homogeneous deformations that implies that the *changes in dimensions of the sample during deformation are always small compared to the original dimensions of the sample.* With large deformations the appropriate way of describing the strain will have to be considered again.

In viscous liquids and also in viscoelastic solids it is not only the strain itself which is important for the behaviour of the material but also the *rate of strain.* The general description of the rate of strain involves no further difficulties. A new tensor, the rate of strain tensor, can be formed quite simply from the time derivatives of the strain components. Generally time derivatives will be denoted by an upper dot

$$\dot{\varepsilon}_x = \frac{d\,\varepsilon_x}{d\,t} .$$

The rate of strain tensor consists of $\dot{\varepsilon}_x$, $\dot{\varepsilon}_y$, $\dot{\varepsilon}_z$, $^1/_2\,\dot{\gamma}_{xy}$, $^1/_2\,\dot{\gamma}_{yz}$ and $^1/_2\,\dot{\gamma}_{xz}$.

d) Invariants of the tensors of stress, strain and rate of strain.

As already indicated in formula (I,2), a component of the stress tensor, σ_x say, depends on the coordinate system chosen. For if a different coordinate system is chosen, the new component $\sigma'_{x'}$, is different from σ_x.

However, there exist certain functions of the components of the stress tensor, which do not change by a rotation of the coordinate system. For instance the sum of the three normal stresses has this property:

$$\sigma_x + \sigma_y + \sigma_z = \sigma'_{x'} + \sigma'_{y'} + \sigma'_{z'} . \tag{I,9}$$

Each function of the stress components with a property similar to equation (I,9) is called an *invariant* of *the stress tensor.* There are three independent invariant functions, the first, second and third invariants of the stress tensor, which are of the form

$$\left.\begin{aligned} S_1 &= \sigma_x + \sigma_y + \sigma_z \\ S_2 &= \sigma_x\sigma_y + \sigma_y\sigma_z + \sigma_z\sigma_x - \sigma_{xy}^2 - \sigma_{yz}^2 - \sigma_{zx}^2 \\ S_3 &= \sigma_x\sigma_y\sigma_z + 2\,\sigma_{xy}\sigma_{yz}\sigma_{zx} - \sigma_x\sigma_{yz}^2 - \sigma_y\sigma_{zx}^2 - \sigma_z\sigma_{xy}^2 \end{aligned}\right\} \tag{I,10}$$

All other functions of the stress components, which are invariant with respect to rotation of the coordinate system, can be expressed as functions of S_1, S_2 and S_3. In the coordinate system of the principal stresses, the invariants take the simple form

$$\left.\begin{aligned} S_1 &= \sigma_1 + \sigma_2 + \sigma_3 \\ S_2 &= \sigma_1\sigma_2 + \sigma_2\sigma_3 + \sigma_3\sigma_1 \\ S_3 &= \sigma_1\sigma_2\sigma_3 \end{aligned}\right\} \qquad (\mathrm{I}, 11)$$

In a similar way we can define the three *invariants of the strain tensor*

$$\left.\begin{aligned} J_1 &= \varepsilon_x + \varepsilon_y + \varepsilon_z \\ J_2 &= \varepsilon_x\varepsilon_y + \varepsilon_y\varepsilon_z + \varepsilon_z\varepsilon_x - \frac{1}{4}\gamma_{xy}^2 - \frac{1}{4}\gamma_{yz}^2 - \frac{1}{4}\gamma_{zx}^2 \\ J_3 &= \varepsilon_x\varepsilon_y\varepsilon_z + \frac{1}{4}\gamma_{xy}\gamma_{yz}\gamma_{zx} - \frac{1}{4}\varepsilon_x\gamma_{yz}^2 - \frac{1}{4}\varepsilon_y\gamma_{zx}^2 - \frac{1}{4}\varepsilon_z\gamma_{xy}^2 \end{aligned}\right\} \qquad (\mathrm{I}, 12)$$

and the *invariants of the rate-of-strain tensor*

$$\left.\begin{aligned} K_1 &= \dot\varepsilon_x + \dot\varepsilon_y + \dot\varepsilon_z \\ K_2 &= \dot\varepsilon_x\dot\varepsilon_y + \dot\varepsilon_y\dot\varepsilon_z + \dot\varepsilon_z\dot\varepsilon_x - \frac{1}{4}\dot\gamma_{xy}^2 - \frac{1}{4}\dot\gamma_{yz}^2 - \frac{1}{4}\dot\gamma_{zx}^2 \\ K_3 &= \dot\varepsilon_x\dot\varepsilon_y\dot\varepsilon_z + \frac{1}{4}\dot\gamma_{xy}\dot\gamma_{yz}\dot\gamma_{zx} - \frac{1}{4}\dot\varepsilon_x\dot\gamma_{yz}^2 - \frac{1}{4}\dot\varepsilon_y\dot\gamma_{zx}^2 - \frac{1}{4}\dot\varepsilon_z\dot\gamma_{xy}^2 \end{aligned}\right\} \qquad (\mathrm{I}, 13)$$

The importance of the invariants lies in the following consideration: Assume a certain property of the material e. g. the elastic energy W stored in the deformation, to be known to be a unique function of the state of deformation:

$$W = W(\varepsilon_x, \varepsilon_y, \varepsilon_z, \gamma_{xy}, \gamma_{yz}, \gamma_{zx}). \qquad (\mathrm{I}, 14)$$

Assume further the material to be isotropic, then the elastic energy can not depend on the direction of the coordinate system chosen, i. e. the elastic energy must be an invariant function. Therefore W can depend on the strain components only through the strain invariants, i. e.

$$W = W(J_1, J_2, J_3). \qquad (\mathrm{I}, 15)$$

As (I, 15) is considerably more simple than (I, 14), it can be used as starting point for general phenomenological theories of the deformation of isotropic elastic bodies. Considerations of this kind will be used especially in the formulation of the theory of rubber elasticity [§ 11] and in the theory of non-linear flow phenomena [§ 12].

e) The classical theory of elasticity.

In order to define the behaviour of a given material, we have to find relationships between the stress tensor and the strain tensor.

The *classical theory of elasticity*, which within a certain range of temperature describes very closely the mechanical behaviour of crystalline solids, is based upon the assumption of HOOKE's *law:* the components of the stress tensor are proportional to the components of the strain tensor. If we confine ourselves to an isotropic solid, *Hooke's law* leads to the following equations between stress and strain

$$\varepsilon_x = \frac{1}{E}\sigma_x - \frac{\nu}{E}(\sigma_y + \sigma_z) \qquad \gamma_{xy} = \frac{1}{G}\sigma_{xy} \tag{I,16}$$

and two analogue pairs by cyclic interchange of x, y, z. The behaviour of an Hookean solid is therefore known, if two material constants, YOUNG's *modulus* E and POISSON's *ratio* ν are known. The constant G, the *shear modulus*, can be derived from these two

$$G = E/2(1 + \nu). \tag{I,17}$$

Of course it is also possible to calculate the stresses, if the strains are known, with the aid of the equations

$$\sigma_x = 2\mathrm{G}\,\varepsilon_x + [K - 2\mathrm{G}/3]\,(\varepsilon_x + \varepsilon_y + \varepsilon_z) \qquad \sigma_{xy} = \mathrm{G}\,\gamma_{xy}. \tag{I,18}$$

One further elastic constant is of importance in the theory of elasticity, the *bulk modulus* K, (in German: Kompressionsmodul) defined by

$$K = E/3(1 - 2\nu) \tag{I,19}$$

YOUNG's modulus, shear modulus and bulk modulus have the dimension force per area[1], POISSON's ratio is dimensionless and can vary between 0 and 1/2. The bulk modulus varies between $E/3$ and infinity, the shear modulus between $E/2$ and $E/3$. It is clear that if two of the above constants are known, the others can be calculated at once, a consequence of the isotropy of the solid.

It will further be observed that time does not enter explicitly into HOOKE's law. The stresses and strains are independent of stress rates and strain rates. The stress at a certain time depends only on the strain at the same moment and not on the strain history.

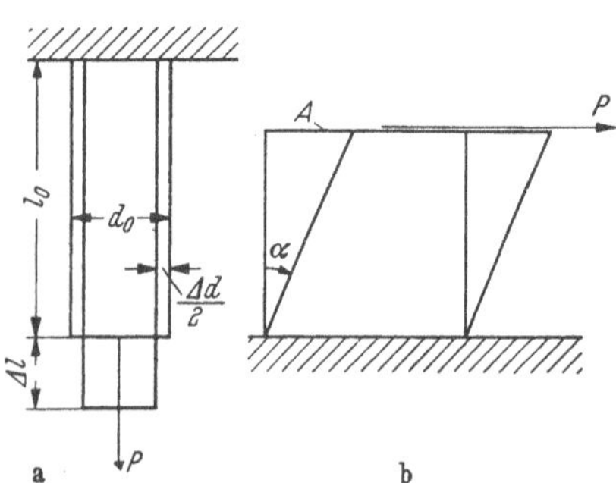

Fig. I, 2. a) Linear extension, b) Simple shear.

To make clear the meaning of the different elastic constants, we consider briefly the most simple forms of deformation.

Linear extension in one direction. Assume a rod-shaped sample (fig. I, 2a) fixed at its upper end and loaded with a force P at the lower end. Let l_0 and d_0 be the longitudinal and lateral dimensions of the sample before loading, and A its cross sectional area. The only stress acting on the sample is $\sigma_x = P/A$; the other stresses vanish. Due to the load the

[1] Moduli are given in dynes/cm², in kg/cm², in pounds per square inch or in NEWTON per square meter. – 1 kg/cm² $\simeq$ 9,81 · 10⁵ dynes/cm², 1 P.S.I. $\simeq$ 6,897 · 10⁴ dynes/cm², 1 N/m² = 10 dynes/cm².

sample changes its longitudinal dimension to $l_0 + \Delta l$ and its lateral dimension to $d_0 - \Delta d$. The deformations in the x and y directions are $\varepsilon_x = \Delta l / l_0$, $\varepsilon_y = -\Delta d / d_0$, and from (I,16) we have

$$\frac{\Delta l}{l_0} = \frac{\sigma_x}{E}, \quad \frac{\Delta d}{d_0} = \nu \frac{\sigma_x}{E}. \tag{I,20}$$

YOUNG's modulus gives the force necessary for unit elongation in linear tension and POISSON's ratio gives the ratio between lateral contraction and longitudinal extension in linear tension.

Simple shear. A rectangular sample of lateral cross section A is deformed by a tangential force P as indicated in fig. I,2b. The tangential force results in a shear stress $\sigma_{xy} = P/A$ and the original right angle of the sample is deformed to an angle $\frac{\pi}{2} - \alpha$. The shear deformation is measured by the angle $\gamma_{xy} = \operatorname{tg} \alpha$. According to (I,16) we have

$$\operatorname{tg} \alpha = \sigma_{xy} / G. \tag{I,21}$$

The shear modulus is the shearing stress necessary to produce a unit shear deformation.

Simple shear is the type of deformation on which we will base our considerations in § 3 and § 4. We shall denote there the shear strain by γ instead of γ_{xy} and the shear stress by σ instead of σ_{xy}. This will produce no confusion as simple shear is the only type of deformation considered in these two sections.

Isotropic compression. Finally we consider a sample under a uniform pressure p. All three principal stresses are equal, $\sigma_x = \sigma_y = \sigma_z = -p$, and all shearing stresses vanish. The deformation consists in a uniform compression which results in a decrease of the volume of the sample ΔV given by

$$\frac{\Delta V}{V_0} = p/K. \tag{I,22}$$

The bulk modulus is therefore equal to the ratio of uniform pressure and relative decrease in volume in isotropic compression.

A solid which does not change its volume under uniform pressure is called incompressible. For an incompressible solid we have $K = \infty$, $\nu = 1/2$ and $G = E/3$.

Table I, 2.

Relations between elastic moduli for an isotropic Hookean elastic solid.

	(G, E)	(G, K)	(E, K)	(G, ν)	(E, ν)	(K, ν)
E	—	$\frac{9GK}{3K+G}$	—	$2G(1+\nu)$	—	$3K(1-2\nu)$
G	—	—	$\frac{3EK}{9K-E}$	—	$\frac{E}{2(1+\nu)}$	$\frac{3K(1-2\nu)}{2(1+\nu)}$
K	$\frac{EG}{9G-3E}$	—	—	$G\frac{2(1+\nu)}{3(1-2\nu)}$	$\frac{E}{3(1-2\nu)}$	—
ν	$\frac{E}{2G} - 1$	$\frac{3K-2G}{2[3K+G]}$	$\frac{1}{2}\left[1 - \frac{E}{3K}\right]$	—	—	—

f) The classical theory of hydrodynamics.

The second simple relation between stress and strain is NEWTON's *viscosity law*. It describes the mechanical behaviour of simple liquids and gases and forms the basis of classical hydrodynamics.

NEWTON's viscosity law states that the components of stress are proportional to the components of rate of strain rather than of strain itself. For an incompressible liquid it leads to the equations

$$\sigma_x = 2\eta\dot{\varepsilon}_x - p, \quad \sigma_{xy} = \eta\dot{\gamma}_{xy}. \tag{I,23}$$

For the viscous liquid one material constant, *the shear viscosity* η, is sufficient to characterise the behaviour. This is a consequence of the fact that liquids generally have a very low compressibility. Therefore NEWTON assumed them to be incompressible, which is expressed by the condition

$$\dot{\varepsilon}_x + \dot{\varepsilon}_y + \dot{\varepsilon}_z = 0. \tag{I,24}$$

The shear viscosity has the dimension of force times time per unit area and is measured in Poises. (1 Poise = 1 dyne sec/cm².)

We shall not discuss the general equations (I,23) here but we mention briefly three examples of viscous flow which are of importance for experimental measurement.

Laminar flow between parallel plates. Assume a *Newtonian liquid* between two parallel plates of area A, the one kept fixed in space, the other being moved by a tangential force F (fig. I,3). Due to the tangential force, which gives rise to the shear stress $\sigma_{xy} = F/A$, a laminar velocity distribution is set up in the liquid. The liquid layer at a distance y from the stationary plate will have a velocity $v = v(y)$ in the x direction which is proportional to y and to the shear stress

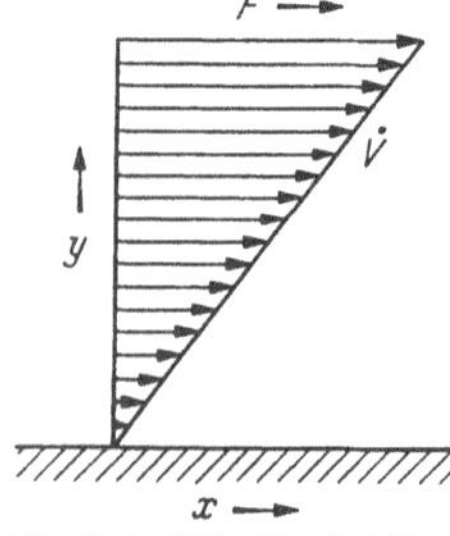

Fig. I, 3. Velocity distribution in laminar flow.

$$v = \frac{1}{\eta}\sigma_{xy}\cdot y. \tag{I,25}$$

The *velocity grandient* equals the *rate of shear* (compare § 12) an dwill be proportional to shear stress and reciprocal viscosity

$$\dot{\gamma}_{xy} = \frac{\partial v}{\partial y} = \frac{1}{\eta}\sigma_{xy}. \tag{I,26}$$

Flow through a capillary tube. If a viscous liquid is forced through a capillary tube of radius R and length l by a pressure difference P, a parabolic velocity distribution is set up in each cross section of the tube. The volume of liquid flowing through a cross section per unit of time is given by *Poiseuilles law*

$$Q = \frac{\pi R^4}{8l}\cdot\frac{P}{\eta}. \tag{I,27}$$

This law is widely used for the measurement of viscosity (capillary tube viscometers).

Flow between concentric cylinders. We consider a viscous liquid contained in the interspace between two concentric cylinders of radii R_1 and R_2 ($R_2 > R_1$) and of length l. If a constant torque M is applied to the inner cylinder while the outer cylinder is kept fixed in position, the inner cylinder will rotate with a constant angular velocity

$$\omega = \frac{M}{4\pi\eta l}(1/R_1^2 - 1/R_2^2). \qquad (I,28)$$

This type of experiment too is widely used for the determination of viscosity (cylinder or rotation viscometer).

Literature to § 2.

Theory of Elasticity:

GEIGER, H. u. KARL SCHEEL: Mechanik der elastischen Körper. Bd. 6: Handbuch der Physik. Springer, Berlin 1928.

LOVE, E. A.: A treatise on the mathematical theory of elasticity. Cambridge University Press, London 1927.

SOKOLNIKOFF, I. S.: Mathematical theory of Elasticity. McGraw Hill Book Comp. Inc., New York 1946.

SOUTHWELL, R. V.: An Introduction to the Theory of Elasticitiy. Oxford University Press, New York 1936.

TIMOSHENKO, S.: Theory of Elasticity, 1934; Theory of Elastic Stability, 1936; Theory of Plates and Shells, 1940; Strength of Materials, 1941. McGraw Hill Book Comp. Inc., New York.

Hydrodynamics:

AUERBACH-HORT: Hydrodynamik. Bd. 5: Handbuch der physikalischen und technischen Mechanik. J. A. Barth, Leipzig 1931.

DURAND, W. F.: Aerodynamics Theory, Vol. 1, 2, 3. Springer, Berlin 1928.

GEIGER, H. u. KARL SCHEEL: Mechanik der flüssigen und gasförmigen Körper. Bd. 7: Handbuch der Physik. Springer, Berlin 1927.

LAMB, SIR HORACE: Hydrodynamics. Dover Publications, New York 1945.

MÜLLER, W.: Einführung in die Theorie der zähen Flüssigkeiten. Akademische Verlagsges. m. b. H., Leipzig 1932.

PRANDTL, L.: Führer durch die Strömungslehre. Friedrich Vieweg u. Sohn, Braunschweig 1949.

For the Theory of Stress and Strain:

MURNAGHAN, F. D.: Finite Deformations of an Elastic Solid. Amer. J. Math. **59**, 235 (1937).

NADAI, A.: Theory of Flow and Fracture of Solids. McGraw Hill Book Comp. Inc., New York 1950, Part I.

§ 3. Phenomenological treatment of linear viscoelastic behaviour.

a) Introduction.

The deformation behaviour of high polymers differs considerably from both ideal Hookean behaviour and Newtonian flow.

For an ideal elastic material in simple shear a linear relation between the force and the displacement is obeyed

$$\sigma = G_0 \cdot \gamma \quad \text{or} \quad \gamma = J_0 \cdot \sigma \quad \text{with} \quad G_0 = 1/J_0. \qquad (I,29)$$

Here σ is the shearing stress, γ the shear strain and G_0 represents a material constant, the shear modulus, with the dimension of a force per unit area (dynes/cm^2). The reciprocal shear modulus has become an important quantity in high polymer physics and has been given a special name, the *shear compliance* J_0 (in German: Nachgiebigkeit) (dimension cm^2/dyne)[1].

The type of deformation considered here and throughout the following two chapters is simple shear. The moduli are shear moduli, the compliances are shear compliances, the viscosity is a shear viscosity, and so on. Though the whole theory will be developed for simple shear only, it will remain valid for every other simple type of deformation such as for instance linear extension. All quantities must then be appropriately chosen for extension, modulus and viscosity become tensile modulus and tensile viscosity and so on. More complicated types of deformation as well as the complete theory of three-dimensional viscoelastic behaviour will be treated in § 6.

We call equation (I,29) a time-independent relation between stress and strain, since according to (I,29) a time-independent stress causes a time-independent strain and vice versa. If for instance a constant shear stress σ_0 is applied at time t_1, a constant shear deformation is produced at the same moment, which remains constant as long as the stress remains constant. In removing the stress at time t_2 no after-effects are observed, but the strain vanishes at the same moment (see fig. I,4).

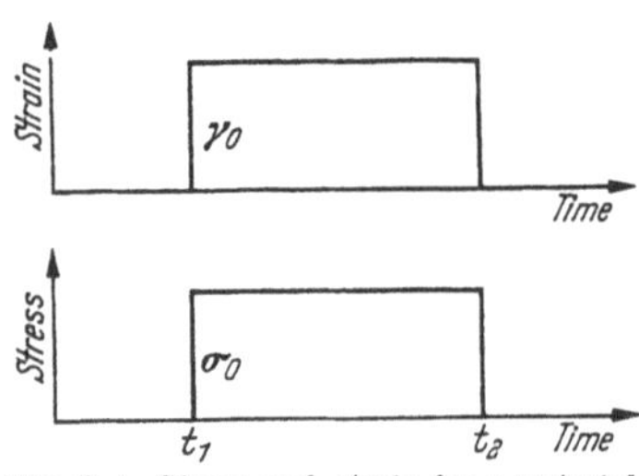

Fig. I, 4. Stress and strain for a material obeying relation (I, 29).

A completely different deformation behaviour is shown by high polymers. Here a time-independent constant shear stress produces a deformation which increases with time (creep experiment) and a time-independent deformation gives rise to a force which decreases with time (stress relaxation experiment). These two basic experiments can be used to characterise the deformation behaviour of high polymers.

b) The creep experiment.

As the simplest rheological experiment we can consider simple *creep* (in German *Kriechexperiment*). At time zero the material is given a constant stress σ_0 and its deformation is measured as function of the time. It is assumed that no forces have been applied to the sample before the time zero.

A pure Hookean material would respond with a constant deformation of magnitude $J_0\sigma_0$ according to equation (I,29). The time dependence of high polymers manifests itself in a monotonic increase of the deformation with time

$$\gamma(t) = \sigma_0 J(t)\,. \qquad \text{(I,30)}$$

The monotonically increasing function $J(t)$ is called the *creep compliance* (in German: *Kriechfunktion*) and its time dependence determines the

[1] In German the term: *Nachgiebigkeit* has been proposed.

deviation from Hookean behaviour. At time zero we have $J(0) = J_0$, and as the deformation is positive and can only increase with time

$$J(t) > 0, \quad J(t) > 0. \tag{I,31}$$

The special form (I,30) is subject to the restriction that the deformation behaviour must be linear. It implies that the magnitude of the stress influences the deformation as a constant factor only, so that $\gamma(t)/\sigma_0$ is a pure function of time only, independent of the stress. A simple criterion can be given to decide experimentally whether the creep behaviour can be represented by (I,30) in any particular case. The experiment has to be repeated using the double (threefold) stress and must yield then exactly the double (threefold) creep curves at all times fig. (I,5). This is what we call linearity in time-dependent materials. One generally finds, in accordance with expectation, that high polymers behave linearly as long as the stresses and strains do not become too high. Some examples of such limits of linearity will de discussed in the beginning of section d.

We now turn to a closer investigation of the creep behaviour of high polymers[1]. We can split the deformation under constant stress into three parts according to their different nature with respect to their dependence on time and temperature

$$\gamma(t) = \gamma_1 + \gamma_2 + \gamma_3. \tag{I,32}$$

All three terms are assumed to be proportional to the constant stress σ_0. The quantity γ_1 is the *instantaneous* Hookean *deformation*, independent of time and stress history

$$\gamma_1 = \sigma \cdot J_0 \tag{I,33}$$

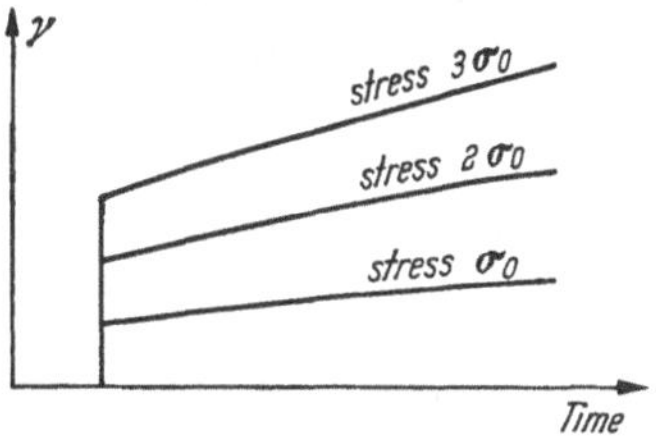

Fig. I, 5. Creep curves at constant stress for linear creep behaviour.

and can be found by measurement of the creep after very short times of loading.

The ordinary elastic compliance J_0 is very low for all polymers, in the order of 10^{-12} to 10^{-10} cm²/dynes. The corresponding deformation γ_1 is therefore very small and never exceeds a few percent. A material exhibiting only the deformation γ_1 is said to be in the *glassy or brittle state* (in German: *Glasartiger Zustand*). It was first realised by Kuhn[2] that the shear moduli of all solid materials whether polymers or not, lie in the region between 10^{10} to 10^{12} dynes/cm² if the temperature is chosen low enough to exclude deformation mechanisms γ_2 and γ_3.

Let us consider next the deformation γ_3 which is called NEWTONian *flow* (in German: NEWTONscher *Fluß*) and which is proportional to the time of loading

$$\gamma_3 = \sigma_0 t/\eta. \tag{I,34}$$

The material constant η is the *shear viscosity* (in German: *Scherungsviskosität*), its reciprocal is sometimes denoted by the term *fluidity*.

[1] JENCKEL, E.: Angew. Chem. **51**, 563 (1938); Kunststoffe **31**, 209 (1941). – W. KUHN: Z. physik. Chem. **B 42**, 1 (1939); Angew. Chem. **52**, 289 (1939). – A. P ALEKSANDROV u. Y. LAZURKIN: Z. techn. Physik (USSR) **9**, 1249, 1261, 1267 (1939). – R. F. TUCKETT: Trans. Faraday Soc. **38**, 310 (1942); **39**, 158 (1943); **40**, 448 (1944); Chem. and. Ind. **62**, 430 (1943). – F. H. MÜLLER: Kolloid-Z. **120**, 119 (1951).

[2] KUHN, W.: Z. physik. Chem. **B 42**, 1 (1939).

γ_3 is the type of deformation generally occurring in viscous low-molecular liquids. The difference between the flow behaviour of high polymers and low-molecular liquids lies in the magnitude of the viscosity, which is very much higher for polymers. (High polymers show viscosities between 10^5 and 10^{15} poises at ordinary temperatures, whilst low-molecular liquids have viscosities between 10^{-2} and 10^{+4} poises.)

Linear amorphous polymers like raw rubbers or thermoplastic materials (consisting of linear unbranched long chain molecules) show a finite but high viscosity, which is very temperature-dependent. At room temperatures the viscosities are generally so high that the deformation γ_3 is completely dominated by the other deformation mechanisms γ_1 and γ_2. Only after extremely long time or at elevated temperatures can γ_3 become the dominating term in deformation behaviour.

Crosslinked polymers with a three-dimensional structure (for instance vulcanised rubbers or thermosetting plastics) do not show any Newtonian flow at all. Except for very high temperatures where chemical degradation may take place, crosslinked polymers show infinite viscosity and therefore no flow, i.e. $\gamma_3 = 0$.

Neither γ_1, which is exhibited by solids, nor γ_3, which is shown by liquids, is characteristic for high polymers. The characteristic behaviour of high polymers is associated with in the deformation γ_2, which is called the *highly elastic deformation* (in German: *hochelastische Deformation*). It will be assumed proportional to the stress and a function of time only

$$\gamma_2 = \sigma_0 J \Psi(t). \qquad (I, 35)$$

We have expressed the highly elastic deformation by a normalised function $\Psi(t)$, the *retarded elasticity function*, with $\Psi(0) = 0$ and $\Psi(\infty) = 1$, whilst the normalisation constant J, the *equilibrium shear compliance* gives the constant value reached by γ_2/σ_0 after very long times.

The equilibrium shear compliance J is very much higher than the instantaneous compliance J_0, namely of the order of 10^{-8} to 10^{-6} cm²/dynes. Therefore, after the time needed to reach equilibrium, the highly elastic deformation γ_2 surpasses the instantaneous elastic deformation γ_1 by a factor of 100 to 10.000. A material exhibiting highly elastic deformation in times of the order of seconds is said to be in the *rubbery state* (in German: *kautschukelastisch*).

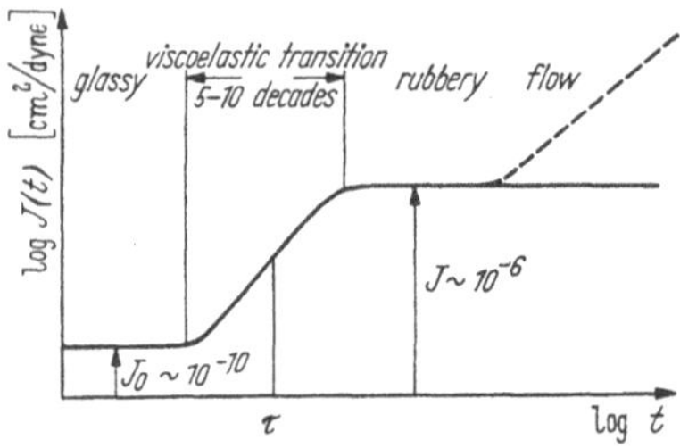

Fig. I, 6. Creep behaviour of linear polymers (dotted curve) and of network polymers (drawn line) on a double logarithmic diagram.

We have decomposed the creep compliance into three parts

$$J(t) = J_0 + J\Psi(t) + t/\eta \qquad (I, 36)$$

and we may now survey viscoelastic behaviour of high polymers graphically (fig. I, 6). The logarithm of creep compliance has been plotted against the logarithm of time. This double logarithmic plot shows clearly the characteristic features of *viscoelastic behaviour*.

At very short times the deformation consists only of the instantaneous elastic deformation γ_1. The material behaves like a glass or hard solid. In the vicinity of the *transition time* τ the highly-elastic deformation γ_2 develops and becomes the dominating term in the deformation. During this period, extending over at least 5 to 10 decades of time, the material becomes soft *(softening region,* in German: *Erweichungszone).* After the highly-elastic equilibrium has been reached the creep compliance remains constant, the material shows high elasticity and is very soft, but most of the deformation is reversible (rubbery state). After still longer times linear polymers show a further increase in deformation due to Newtonian flow γ_3. The flow term becomes finally dominant and the polymer behaves like a very viscous liquid. Cross linked polymers do not show this second transition but remain in the rubbery state.

It is seen that a high polymer can behave like an elastic solid, like a rubber or like a liquid according to the position of the *transition time* τ with respect to the experimental timescale. Indeed the difference between rubbers and thermoplastic materials lies chiefly in the position of τ. For a rubber at room temperature the transition time is very small compared to the time of experiment, for a plastic it is very long.

A further characteristic feature of high polymers is the strong temperature dependence of the transition time τ. Generally for each 10 degrees centigrade increase in temperature, the transition time decreases by between a half and two decades on a logarithmic timescale. So a rubber will always behave as a hard solid at very low temperatures, at which the transition time τ is shifted beyond the experimental time. On the other hand a plastic softens at high temperatures and becomes rubberlike.

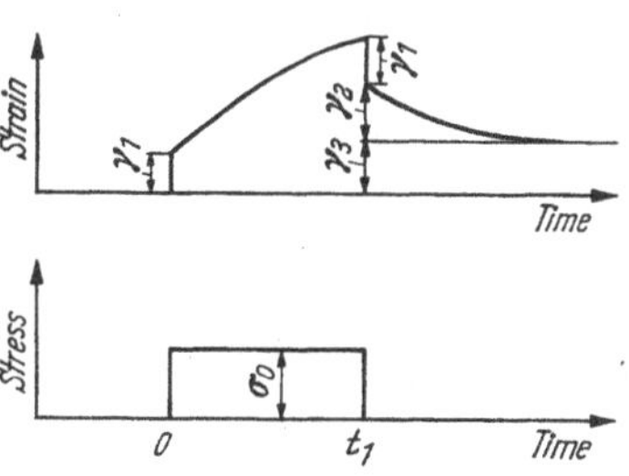

Fig. I, 7. Stress and strain as function of time in a creep recovery experiment.

In a *recovery experiment*[1] the sample is subjected to constant stress σ_0 during the time t_1 (between $t = 0$ and $t = t_1$). Then the force is removed and the deformation is observed as function of time (see fig. I,7).

The recovery is important for the reason (among others) that it allows an unambiguous separation of γ_3 and γ_2. The viscoelastic deformation γ_2 is completely recovered after a sufficient lapse of time whereas γ_3 does not recover at all. So the deformation left in a recovery experiment conducted until the deformation has become constant is γ_3. By dividing this quantity by the product of the stress σ_0 and the time t_1 *the fluidity* or *reciprocal viscosity* is obtained.

Mathematically, the recovery experiment can be described as follows: Between time zero and t_1 when the stress σ_0 is applied, the usual creep equations holds:

$$\gamma(t)/\sigma_0 = J(t); \quad 0 < t < t_1. \qquad \text{(I, 30)}$$

For times greater t_1, the recovery equation applies

$$\gamma(t)/\sigma_0 = J(t) - J(t - t_1) = J[\Psi(t) - \Psi(t - t_1)] + t_1/\eta; \; t_1 < t \qquad \text{(I, 37)}$$

which is the difference between two creep curves, the one beginning at $t = 0$ and the other at $t = t_1$. This follows immediately from the superposition principle (section d).

[1] JENCKEL, E. u. K. UEBERREITER: Z. physik. Chem. A 182, 361 (1938).

c) Stress relaxation experiment.

The counterpart of the creep experiment (deformation under constant stress) is the ***stress relaxation*** experiment (in German: *Spannungsrelaxation*). The sample is subjected to a constant shear strain γ_0 at time zero. The reactive stress necessary to maintain this constant deformation decreases with time and can be measured by suitable means

$$\sigma(t) = \gamma_0 G(t)\,. \tag{I,38}$$

Equation (I,38) for the stress relaxation has much in common with equation (I,30) for the deformation: The reactive stress must be proportional to γ_0 and a pure time function $G(t)$, the ***stress relaxation modulus in shear*** (in German: *Spannungsrelaxationsmodul*).

$G(t)$ is a monotonically decreasing function of time and can be written in the form

$$G(t) = G + G_0[1 - \Phi(t)]\,. \tag{I,39}$$

The function $\Phi(t)$ is the *relaxation function* (in German: *Relaxationsfunktion*). It is normalised to $\Phi(0) = 0$ and $\Phi(\infty) = 1$ and since the relaxing stress decreases with time, we must have $\Phi(t) > 0$ and $\dot{\Phi}(t) > 0$ or $G(t) > 0$ and $\dot{G}(t) < 0$.

The reactive stress (per unit deformation) after very short times is given by

$$G(0) = G + G_0 \simeq G_0\,. \tag{I,40}$$

G_0 determines the stress after very short times and is called *ordinary* (elastic) *shear modulus*. For long times we have

$$G(\infty) = G \tag{I,41}$$

G determines the equilibrium stress (per unit deformation) and is called *equilibrium shear modulus.*

In discussing the stress relaxation behaviour of high polymers we must distinguish between linear polymers and network polymers (more accurately between polymers showing finite Newtonian flow and those which do not flow at all). The ordinary elastic shear modulus is of the order 10^{10} to 10^{12} dynes/cm² ($\simeq 1/J_0$) for all polymers and solid materials. The equilibrium elastic shear modulus is of the order 10^6 to 10^8 dynes/cm² for network polymers and is zero for linear polymers.

$G = 1/J$ for polymers showing no Newtonian flow,

$G = 0$ for polymers showing finite flow.

Comparing the order of magnitudes of G_0 and G, we see that the latter is always negligible compared with the first. This fact has been used in equation (I,40).

We draw attention to the fact that polymers which show a finite Newtonian flow, show always total stress relaxation (the stress vanishes after very long time). Polymers which do not show flow, exhibit only partial stress relaxation (the stress is reduced to a finite equilibrium value G).

The stress relaxation behaviour of high polymers is surveyed in fig. I, 8 in a double logarithmic plot. For very short times the modulus is very high, the material behaving as a hard solid. Around the characteristic transition time τ' a transition extending over several decades takes place, in which the material softens. After this transition the rubbery state is obtained. Cross linked polymers remain in the rubbery state, whilst linear polymers show a second transition which is connected with flow.

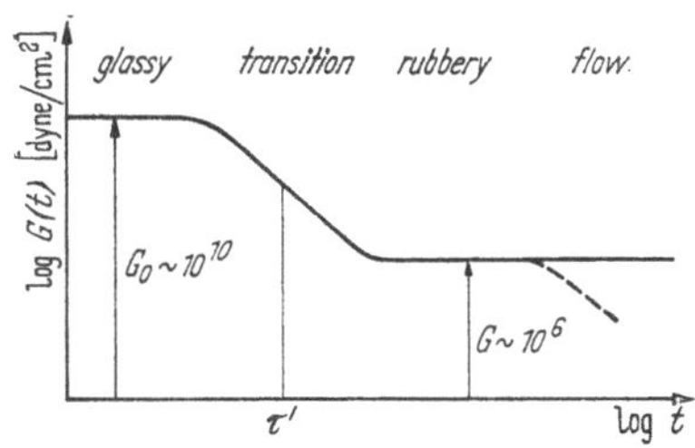

Fig. I, 8. Relaxation behaviour of linear polymers (dotted curve) and of network polymers (drawn line) on a double logarithmic plot.

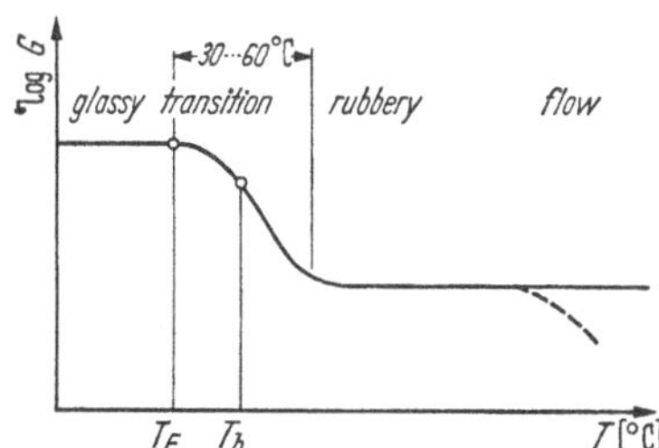

Fig. I, 9. Shear modulus as a function of temperature for linear polymers (dotted curve) and for network polymers (drawn curve).

The viscoelastic transition times τ and τ' of highly-elastic deformation and of stress relaxation respectively are of course closely connected, but they are not identical. For practical purposes both transition regions can be said to coincide and the remarks about the temperature dependence, which have been made for the compliance, remain valid for the stress relaxation modulus. Increase of temperature shifts the transition region to shorter times.

The equivalence of logarithmic time and temperature suggest a plot of modulus as a function of temperature. We determine the modulus after a certain fixed time (for instance a "ten seconds modulus") as a function of the temperature of the experiment (fig. I,9). We get a picture which resembles very much the picture of fig. I,8. At low temperatures we have a high modulus and glasslike behaviour. At higher temperatures there is a transition extending over a temperature range of between 30 and 60 degrees centigrade. Then the rubbery state is reached and at still higher temperatures the second transition turns the material into a viscous liquid. The effect of time in this picture is obvious: Increase of time makes the curve shift towards lower temperatures, and conversely.

Two important characteristic temperatures may be marked in our plot, the second order *transition temperature* T_E (in German: *Einfriertemperatur*) and the *brittle temperature* T_b (in German: *Sprödigkeitstemperatur*). The transition temperature T_E is the temperature[1] at which

[1] The relation between the transition temperature and mechanical properties has been considered by various authors: A. Smekal: Ergebn. exakt. Naturwiss. **15**, 174 (1936); Glastechn. Ber. **16**, 198 (1938); Z. physik. Chem. **B 44**, 286 (1939); **B 48**, 114 (1940). – E. Jenckel u. K. Ueberreiter: Z. physik. Chem. **A 182**, 361 (1938). – E. Jenckel: Z. Elektrochem. **45**, 202 (1939). – K. H. Meyer: Z. Elektrochem. **45**, 156 (1939). – K. Ueberreiter: Z. physik.Chem. **B 45**, 361 (1940); **B 46**, 157 (1940); **B 48**, 197 (1941); Kunststoffe **30**, 170 (1940); Kautschuk **19**, 11 (1943). – E. Jenckel: Kunststoffe **31**, 209 (1941); Kolloid-Z. **100**, 163 (1942). – F. H. Müller: Kolloid-Z. **95**, 138, 306 (1941); **108**, 66 (1944). – R. F. Tuckett: Trans. Faraday Soc. **38**, 310 (1942). – R. F. Boyer u. R. S. Spencer: Advances in Colloid Science Vol. 2, New York 1946. – E. Jenckel: Kolloid-Z. **120**, 160 (1951). – K. Ueberreiter u. H. J. Orthman: Kolloid-Z. **123**, 84 (1951).

Table I, 3. *Properties of different Rubbers and Plastics at the transition region.*

Material	Time (sec.) or period (sec.)	Shear modulus		Transition region °C		T_E °C	T_b °C	Length of transition, decades (log time)	References		
		glassy state 10^9 dynes/cm² G_0	rubbery state 10^6 dynes/cm² G						Mod.	T_E	T_b
Natural Rubber Vulc. N-2 (S-2%)	$0{,}3 \cdot 10^{-2}$	4	7,1	−60°	−20°	−73°	−57°		1	8	1
Neoprene Vulc. GN-50	$0{,}5 \cdot 10^{-2}$	6,7	11	−40°	0°		−40°		1		1
Buna-N-carbon-filled Vulc. B 5	$0{,}4 \cdot 10^{-2}$	12	33	−20°	30°		−19°	7 — 8	1		1
GR-S gumstock (Butadiene-styrene copolymer 75/25)	$3{,}6 \cdot 10^{+3}$	5,7	8,4	−70°	−45°	−61°	−49°	7 — 8	6	8	10
Polyisobutylene	$3{,}6 \cdot 10^{+3}$	13	2,5	−75°	−10°	−74°	−50°	7 — 8	4, 5	8	8
Polyvinylchloride	1	10	32	80°	110°	75°	81°		2	8	8
Nitrocellulose	1	13	—			53°			2	8	
Polystyrene	1	13	2,0	85°	125°	81°	80°		3	11	8
Phenolformaldehyde (lightly crosslinked)	1	21	32	60°	125°				3		
Plasticized Cellulose Acetate	1	5,0	—	75°	—	69°			3	8	
Polymethylmethacrylate	$3{,}6 \cdot 10^{+3}$	4,2	7,1	90°	120°	105°		8	7	9	

[1] NOLLE, A. W., J. Polymer Sci. **5**, 1 (1950). — [2] SCHMIEDER, K. u. K. WOLF, Kolloid-Z. **127**, 65 (1952). — [3] NIELSEN, L. E. u. R. BUCHDAHL, S. P. E. Journal **16** (May 1953). — [4] LEADERMAN, H. u. R. S. MARVIN, Report Nat. Bur. of Standards, Washington D. C., Febr. 1953, Proc. Sec. Int. Congr. Rheology, Oxford 1953. Page 165, 203. — [5] TOBOLSKY, A. V. u. J. R. MC LOUGHLIN, J. Polymer Sci. **8**, 543 (1952). — [6] BISCHOFF, J., E. CATSIFF u. A. V. TOBOLSKY, J. Am. Chem. Soc. **74**, 3378 (1952). — [7] MC LOUGHLIN, J. R. u. A. V. TOBOLSKY, J. Colloid Sci. **7**, 555 (1952). — [8] BOYER, R. F. u. R. S. SPENCER, Adv. Colloid Sci. Vol. II (Rubber). — [9] HEYBOER, J., Chem. Weekblad **48**, 264 (1952). — [10] KEMP, A. R., J. H. INGMANSON, J. B. HOWARD u. V. T. WALDER, Ind. Eng. Chem. **36**, 361 (1944). — [11] PATNODE, W. u. W. J. SCHEIBER, J. Am. Chem. Soc. **61**, 3449 (1939).

the volume-temperature curve of the polymer shows a sudden change of slope. It is situated just before the beginning of the fall of the elasticity modulus in the transition region. The brittle point T_b is the temperature at which the polymer can no longer resist a sudden blow without breaking. It is situated somewhere in the upper region of the transition and lies generally 10° to 25° C higher than T_m.

We have collected in table I,3 some examples of recent experimental work on modulus-temperature curves for high polymers. The second column contains the times at which the shear moduli are taken. The values of the shear moduli in the glassy state G_0 and in the rubbery state G depend slightly on the temperature. The values given refer to respectively the beginning and the end of the transition region. The beginning and the end of the transition region (column four) have been selected as these temperatures at which the steep decrease in the modulus-temperature curves starts or ends.

d) The superposition principle.

We have discussed the deformation behaviour of polymers by considering two simple basic experiments, behaviour under constant stress and behaviour under constant strain. We defined two characteristic material functions, the creep compliance and the stress relaxation modulus describing these experiments. The only underlying assumption was linearity: Strain should be proportional to stress.

In practice it is assumed that linear behaviour is obeyed provided that deformation and stresses do not become too large. For each material there will be a certain *linearity limit* either in stress or in strain, beyond which equations (I,30) or (I,38) cease to hold[1].

For instance it has been shown[2] that Polymethylmethacrylate behaves linearly in torsion creep at room temperatures up to stresses of $7{,}5 \cdot 10^7$ dynes/cm². On the other hand non-linear effects for this material in tension, torsion, bending and compression have been reported[3] at stresses of $14 \cdot 10^7$ dynes/cm². We conclude therefore that the limit of linearity for this material lies in the region of 10^8 dynes/cm². Similar considerations lead to the linearity limits of table I,4.

Table I,4. *Linearity limits of plastics and rubbers**.

Material Plastics:	characteristic stress limit in creep in 10^7 dynes/cm²	Reference
Polymethylmethacrylate	10	2, 3
Polystyrene	5	3, 4
Plasticized Polyvinyl chloride	> 1	5
Phenolic resins	10	6
Polythene	12	7
Rubbers:	**characteristic strain limit[8] in relaxation in % (linear tension)**	
Polyisobutylene	50	9
Natural rubber	100	10, 11
G R S	100	11

[1] Compare also: F. H. MÜLLER: Kolloid-Z. **120**, 119 (1951).
[2] LETHERSICH, W.: Brit. J. appl. Physics **1**, 294 (1950).
* These limits are based on a accuracy of measurement of about 5%.

Fußnoten 3–11 auf S. 20.

The stress limits of linearity for plastics may be converted into strain limits using elastic moduli of the order of 10^{10} d/cm² (compare table I,3) and the strain limits for rubbers may be converted into stress limits using a shear modulus of the order of 10^7 d/cm². We arrive at

	plastics	rubbers
stress limits of linearity	$10^7 - 10^8$ d/cm²	$10^7 - 10^8$ d/cm²
strain limits of linearity	0,1 – 1%	10 – 100%

Whilst the stress limits of rubbers and plastics are of the same order of magnitude, the strain limits differ by a factor hundred.

For classical elastic bodies the entire mechanical behaviour can be predicted from the measurement of two moduli and from the experimental determination of the limit of linearity. For one particular type of deformation (tensile elongation, pure shear etc.) one modulus suffices. No physical assumption is needed except that a linear region exists, and there is good reason for confidence in this assumption in nearly all circumstances.

With viscoelastic bodies matters are distinctly more complicated. Not only is one modulus insufficient to describe the behaviour of a material in a very simple experiment like the relaxation experiment. Already for this case instead of the modulus a continuous function of time $G(t)$ is needed. Besides, in another simple experiment like the creep experiment another time function, the creep function, $J(t)$ describes the behaviour. The only difference between a relaxation experiment and a creep experiment is that the time dependence, the history of strain and of stress, is different. Besides a constant stress (creep experiment) or a constant strain (relaxation experiment), one may impose any other stress-strain history on the material; without further assumptions no predictions about the behaviour of the material can then be made from a knowledge of only the creep and relaxation functions. Moreover, it is probable that some relation exists between the creep function and the relaxation function and one may ask how much freedom there is for the behaviour of a material once the creep behaviour is given. These various examples all point

Fortsetzung der Fußnoten von S. 19.

[3] MARIN, J. u. Y. PAO: Trans. A. S. M. E. (October 1952) 1231; Proc. A. S. T. M. **51**, 1277 (1951). – J. MARIN u. G. CUFF: Proc. A. S. T. M. **49**, 1158 (1949). – J. A. SAUER, J. MARIN u. C. C. HSIAO: J. appl. Physics **20**, 507 (1949).

[4] BUCHDAHL, R., L. E. NIELSON u. E. H. MERZ: J. Polymer Sci. **6**, 403 (1951).

[5] LEADERMAN, H.: Ind. Engng. Chem. **35**, 374 (1943).

[6] TELFAIR, D., T. S. CARSWELL u. H. K. NASON: Mod. Plastics **21** (Febr. 1944) 137.

[7] HILLIER, K. W. u. H. KOLSKY: Proc. physic. Soc. (London) **B 62**, 111 (1949).

[8] These strain limits are valid, if the quantity $(1 + \varepsilon) - 1/(1 + \varepsilon)^2$ is taken as the measure of tensile strain, rather than ε itself. (Compare § 10, § 11.)

[9] ANDREWS, R. D., N. HOFMAN-BANG u. A. V. TOBOLSKY: J. Polymer Sci. **3**, 669 (1948). – G. M. BROWN u. A. V. TOBOLSKY: J. Polymer Sci. **6**, 165 (1951).

[10] MOONEY, M., W. E. WOLSTENHOLME u. D. S. WILLIAMS: J. appl. Physics **15**, 324 (1944).

[11] TOBOLSKY, A. V. u. R. D. ANDREWS: J. chem. Physics **13**, 3 (1945). – A. V. TOBOLSKY, I. B. PRETTYMAN u. J. H. DILLON: J. appl. Physics **15**, 380 (1944).

to the need for a further physical principle which allows an arbitrary history of strain to be related to the corresponding history of stress, and conversely. This new principle is the *superposition principle* (in German: *Superpositionsprinzip*) which was put forward as early as 1876 by BOLTZMANN[1].

The superposition principle is a logical extension of the principle of linearity. In fact, it is identical with the latter if we define the concept of linearity as the principle that the total "effect" of a sum of "causes" is equal to the sum of the effects of each of these causes. For "effects" and "causes" we may substitute strains and stresses or the reverse.

According to the superposition principle the effects of different stresses are superimposed by simple addition: If a certain stress $\sigma_1(t)$ produces a deformation $\gamma_1(t)$, and if the stress $\sigma_2(t)$ results in the deformation $\gamma_2(t)$, then the sum of the two stresses $\sigma_1(t) + \sigma_2(t)$ just produces the sum of the two deformations $\gamma_1(t) + \gamma_2(t)$. Also: if a stress $\sigma_1(t)$ is necessary to produce a deformation $\gamma_1(t)$ and if a stress $\sigma_2(t)$ is necessary to produce a deformation $\gamma_2(t)$, then the sum of the two stresses $\sigma_1(t) + \sigma_2(t)$ will be necessary to produce the sum of the deformations.

The superposition principle has far-reaching consequences. First, if we have determined the creep compliance $J(t)$ for all values of t from zero to infinite by means of a creep experiment, we can predict the deformation which is produced by an arbitrary stress prescribed as function of time $\sigma(t)$ between $t = 0$ and $t = \infty$. This deformation is given by[2]

$$\gamma(t) = \int_{-\infty}^{t} \dot{\sigma}(\xi) J(t-\xi)\, d\xi \,. \qquad \text{(I, 42a)}$$

Conversely, knowing the stress-relaxation modulus $G(t)$, we can predict the stress necessary to produce an arbitrary deformation behaviour $\gamma(t)$. This stress is[2]

$$\sigma(t) = \int_{-\infty}^{t} \dot{\gamma}(\xi) G(t-\xi)\, d\xi \,. \qquad \text{(I, 42b)}$$

Both in (42a) and (42b) we see the general idea underlying the superposition principle: The total effect [$\gamma(t)$ in (42a) and $\sigma(t)$ in (42b)] is calculated as the sum [integral] of the effects of a great number of causes, each of the causes being increase of σ [in (42a)] or γ [in (42b)] during the time interval between ξ and $\xi + d\xi$.

The form of the superposition principle as written in (I,42a) and (I,42b) contains the time derivatives of the stress history $\dot{\sigma}(t)$ or the strain history $\dot{\gamma}(t)$ under the sign of integration. It may be used for calculation of all cases in which the prescribed stress or strain histories are continuous functions of time without jumps.

However we often meet strain or stress histories which are not continuous and of which no derivatives can be defined. An example of such

[1] BOLTZMANN, L.: Pogg. Ann. Physik 7, 624 (1876). See further H. LEADERMAN: Elastic and creep properties of filamentous materials, Textile foundation 1943, page 20; C. ZENER: Elasticity and anelasticity of metals, Chicago (1948), p. 49.

[2] Here the dot means: derivative with respect to the argument: $\dot{\sigma}(\xi) = d\sigma(\xi)/d\xi$.

a discontinuous stress history is the creep experiment, where the stress is assumed to be zero for times smaller zero and constant for times larger zero, which implies a discontinuity at the point $t = 0$. For discontinuous stress or strain histories the superposition principle may be used in the form[1]

$$\gamma(t) = J(0)\,\sigma(t) + \int_{-\infty}^{t} \sigma(\xi)\,\dot{J}(t-\xi)\,d\xi\,, \tag{I,42c}$$

$$\sigma(t) = G(0)\,\gamma(t) + \int_{-\infty}^{t} \gamma(\xi)\,\dot{G}(t-\xi)\,d\xi\,, \tag{I,42d}$$

$J(0)$ and $G(0)$ mean the values of the creep compliance or the stress relaxation modulus at time zero.

The superposition principle is of the utmost importance for the description of viscoelastic behaviour under small stresses and strains. It tells how many and what kind of measurements are necessary to characterise this behaviour completely. It tells further how the results of different experimental techniques can be related to each other, for instance how we can predict the result of a creep measurement from a stress relaxation measurement, and conversely. The great power of the superposition principle lies in its generality as it is not restricted to any special material: all its consequences remain valid independently of the microrheological structure of the material under consideration. Therefore the superposition principle may be used for inorganic glasses as well as for organic high polymers or metals.

Equation (I,42a) shows how the course of strain with time can be calculated under an arbitrary course of stress, once the creep function, that is the course of strain under constant stress, is given. If, then, a stress is prescribed which decreases exactly as the relaxation function, the corresponding strain must be constant as is the case in a relaxation experiment. Mathematically this leads to[2]

$$G(0)\,J(t) + \int_{0}^{t} \dot{G}(\xi)\,J(t-\xi)\,d\xi = 1\,. \tag{I,43}$$

Equation (I,43) can be considered as determining $G(t)$ if $J(t)$ is known or as an equation for $J(t)$ if $G(t)$ is known. Equation (I,43) may be written in a more symmetrical form

$$\int_{0}^{t} G(t-\tau)\,J(\tau)\,d\tau = t\,. \tag{I,43a}$$

By differentiation of (I,43a) with respect to time, equ. (I,43) is obtained.

Thus we find as a further consequence of the superposition principle that if the stress-relaxation function is known, the creep function can be calculated, and vice versa. Together with our earlier statement about the superposition principle we find that, when the creep function is known, the deformation behaviour can be predicted for all arbitrary stress or strain histories. The same is true when the relaxation function is known.

---

[1] These equations may be derived by partial integrations of (I,42a) or (I,42b).

[2] GROSS, B., J. appl. Physics 18, 212 (1947). — R. SIPS: J. Polymer Sci. 5, 69 (1950).

Any function, such as the creep function or the relaxation function, by the aid of which the entire deformation behaviour can be characterised will be called a *characteristic function:* The stress-relaxation function and the creep function if measured for all times between zero and infinity are characteristic functions, each characterising the deformation behaviour of the material in the region of validity of the superposition principle in a simple type of deformation.

The relation (I,43) written in appropriate form[1] tells us that the product of the compliance and the modulus is always equal to or less than unity

$$J(t)\,G(t) \leq 1 \quad \text{for all values of } t. \tag{I,44}$$

This product deviates from unity only in the transition region and, for polymers showing flow, in the liquid region. As a first crude approximation one can consider the relaxation modulus and the creep compliance as inverse functions. This approximation will hold in all regions where the modulus remains approximately constant, but will fail in transition regions.

An experimental examination of the validity of the superposition principle has been carried out by LEADERMAN[2]. He concludes that the superposition principle holds for deformations which do not exceed a few percent.

e) Dynamic measurements.

In creep and relaxation experiments the lower limit of time for which information can be obtained lies in the region of some seconds. There is of course no fundamental long time limit of observation in creep or relaxation. We call such experiments static[3] experiments.

If information about the behaviour of a material at very short times after applying the load is required, vibration experiments have to be performed. By vibration experiments we mean experiments with sinusoidal stress or strain. We give below a short account of the theoretical background of dynamic viscoelastic response of high polymers.

In static experiments the inertial forces do not play a considerable role. But if the motions within the viscoelastic body become faster and more accelerated as in dynamic experiments, the inert masses will play a more and more important role. However in most practical cases inertial forces are small compared to elastic and frictional forces and by neglecting the inertial forces the theoretical treatment is simplified very much. We will therefore in the following neglect inertial forces, either by assuming the density of the sample to be sufficiently low, or by assuming the frequency of the experiment to be sufficiently small. We shall investigate the influence of inertia in more detail in § 7.

A polymer subjected to a sinusoidal shearing stress will reach the steady-state condition after a number of cycles. Let the stress be written

$$\sigma = \sigma_0 \sin \omega t\,, \tag{I,45}$$

---

[1] ZENER, C.: loc. cit. page 21.

[2] LEADERMAN, H.: J. appl. Mechan. **6 A** 79 (1939), Bakelite; Textile Res. **11**, 171 (1941), Silk filament; Ind. Engng. Chem. **35**, 374 (1943), Plasticized Polyvinylchloride.

[3] FERRY (chapter 6) calls such experiments: transient experiments.

where σ_0 is the amplitude of stress, ω is the angular frequency of vibration. As a consequence of the superposition principle, in the steady state the strain will vary sinusoidally with the same frequency ω, with an amplitude γ_0 and with a certain phase lag δ (see fig. I,10).

$$\gamma = \gamma_0 \sin(\omega t - \delta)\,. \tag{I,46}$$

The strain lags behind the stress, there exists a phase angle, δ, between the two which is called the *loss angle* (in German: *Verlustwinkel*). As a further consequence of the superposition principle, the ratio of the amplitudes of stress and strain and the loss angle are independent of the stress or strain amplitude, but depend on the frequency ω only. The ratio between strain and stress amplitudes is called the *absolute dynamic compliance*

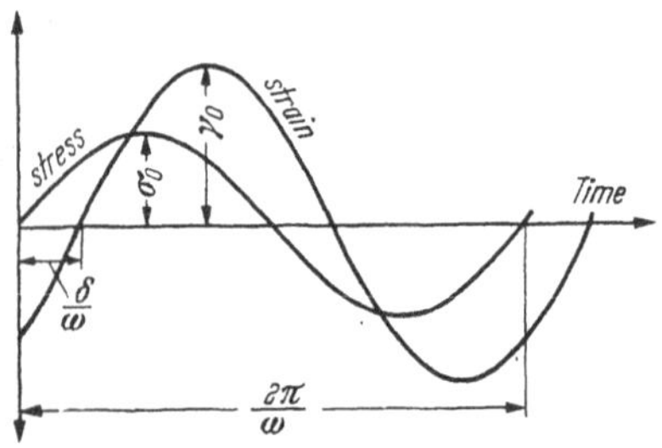

Fig. I,10. Steady state response of a viscoelastic material to sinusoidal shear stress or strain.

$$\gamma_0/\sigma_0 = |J|\,. \tag{I,47}$$

With these two functions of frequency, the absolute dynamic compliance $|J|$ and the loss angle δ, the dynamical properties of a material can be described completely. It is of course not essential that the sinusoidal stress be given and the strain be considered as the response. A simple shift of the origin of the time scale $t' = t - \delta/\omega$ gives for (I,45) and (I,46) the equations

$$\gamma = \gamma_0 \sin \omega t'\,. \tag{I,48}$$

$$\sigma = \sigma_0 \sin(\omega t' + \delta)\,. \tag{I,49}$$

Here the strain can be considered to be given, whilst the resultant stress then precedes the strain by the angle δ. The ratio between the amplitudes of stress and strain is called the *absolute dynamic modulus* $|G|$ and is of course the reciprocal of the absolute shear compliance

$$\sigma_0/\gamma_0 = |G| = 1/|J|\,. \tag{I,50}$$

The dynamic modulus together with the loss angle or the absolute compliance together with the loss angle are sufficient to describe dynamic behaviour. But sometimes it is convenient to introduce two other pairs of characteristic functions for the description of vibration experiments. The deformation under given stress (I,46) can be split into two parts

$$\gamma = \sigma_0 [J_1(\omega) \sin \omega t - J_2(\omega) \cos \omega t]\,. \tag{I,46a}$$

One part is in phase with stress and has the relative magnitude (per unit stress amplitude) $J_1(\omega)$, the other part differs in phase by $\pi/2$ with stress and has the amplitude $J_2(\omega)$. J_1, the *storage compliance* is a measure of the elastic part of the deformation, J_2, the *loss compliance* determines the internal frictional loss. The storage compliance and the loss compliance are connected with the absolute compliance and with the loss angle

$$\left.\begin{aligned} J_1 &= |J| \cos\delta \quad & |J|^2 &= J_1^2 + J_2^2 \\ J_2 &= |J| \sin\delta \quad & J_2/J_1 &= \operatorname{tg}\delta . \end{aligned}\right\} \tag{I,51}$$

The meaning of these formulas is illustrated in fig. I,11. A right-angled triangle with δ as angle and with $|J|$ as hypotenuse possesses J_1 and J_2 as cathetes.

If we consider the strain to be given, the resulting stress (I,49) can be split into two parts

$$\sigma = \gamma_0 [G_1(\omega) \sin\omega t' + G_2(\omega) \cos\omega t'] . \tag{I,49 a}$$

The part in phase with strain determines the elastic reactive forces and is called the *storage modulus* G_1 and the part out of phase with strain gives the dissipative forces and is called the *loss modulus* G_2. The relationships between the absolute modulus, the loss angle and the storage and loss moduli are similar to those for the corresponding compliances (I,51). Therefore fig. I,11 remains valid for the corresponding moduli.

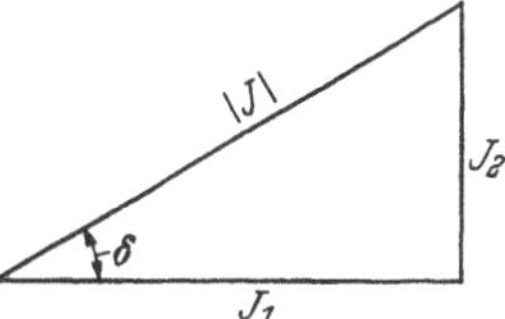

Fig. I, 11. The relationship between absolute, storage and loss compliances and the phase angle.

The interrelationships between the seven characteristic quantities J_1, J_2, $|J|$, G_1, G_2, $|G|$, δ are represented by equ. (I,52). If two of these quantities are known, all the others can be calculated.

$$\left.\begin{aligned} J_1 &= G_1/|G|^2 \qquad & J_2 &= G_2/|G|^2 \\ G_1 &= J_1/|J|^2 \qquad & G_2 &= J_2/|J|^2 \\ |G|^2 &= G_1^2 + G_2^2 \qquad & |J|^2 &= J_1^2 + J_2^2 \end{aligned}\right\} \tag{I,52}$$

$$\operatorname{tg}\delta = G_2/G_1 = J_2/J_1 .$$

The meaning of the loss modulus and the loss compliance can be illustrated by the calculation of the dissipation of energy due to viscoelastic losses in the material. The energy dissipated during one cycle is

$$\Delta W = \int \sigma\, d\gamma = \int_0^{2\pi/\omega} \sigma \frac{d\gamma}{dt}\, dt ,$$

or with (I,48) and (I,46a)

$$\Delta W = \sigma_0^2 J_2(\omega) \cdot \omega \int_0^{2\pi/\omega} \sin^2 \omega t\, dt .$$

Therefore we obtain

$$\Delta W = \pi J_2(\omega)\sigma_0^2 = \pi G_2(\omega)\gamma_0^2 . \tag{I,53}$$

These formulas give the generation of heat per cycle due to viscoelastic losses.

The storage and the loss compliance are sometimes expressed as a single complex quantity $J^* = J_1 - iJ_2$, the *complex shear compliance* and the moduli are written as a *complex shear modulus* $G^* = G_1 + i G_2$. The compliances are the real and the negative imaginary parts of the complex compliance, the moduli are the real and the imaginary parts of the complex modulus. In this notation the whole system (I,52) becomes simply $J^* = 1/G^*$. The absolute compliance and the absolute modulus become the absolute values of the corresponding complex quantities and the loss angle is the argument (negative argument) of the complex compliance (complex modulus). Stress and strain can be represented in a complex vector-diagram as complex vectors σ^* and γ^*. At zero time the stress is represented by a vector of length σ_0 coinciding with the real axis and the strain as a vector of length γ_0 making an angle δ, in a clockwise direction, with the stress. With proceeding time the entire picture rotates counterclockwise around the origin with angular velocity ω. The complex stress and strain vectors are connected by the equations $\gamma^* = J^*\sigma^*$ and $\sigma^* = G^*\gamma^*$.

We turn now to the consideration of dynamic behaviour of high polymers beginning with the elastic moduli. In static experiments the time t has been the independent variable, in dynamic experiments it is the angular frequency ω. We plot the logarithm of the moduli G_1, G_2, and $|G|$ and the tangent of the loss angle, tg δ, (the *mechanical loss*) against the logarithm of frequency (fig. I,12). A polymer showing no flow (network polymer) has a low equilibrium modulus G_1 at low frequencies in the rubbery state. In a certain frequency range a transition occurs characterised by a sharp increase in modulus to the high value of the glassy state. A similar behaviour is shown by the absolute modulus $|G|$ which coincides with the storage modulus G_1 in all regions where no considerable loss takes place. Only in the transition region the absolute modulus is always larger than the storage modulus. The loss modulus G_2 differs from zero only in the transition region. It shows a maximum at approximately the frequency at which the derivative of the storage modulus with respect to logarithmic frequency has a maximum. The transition region is also characterised by a maximum of the dynamic loss tg δ, but this maximum is found at lower frequencies than the maximum in G_2.

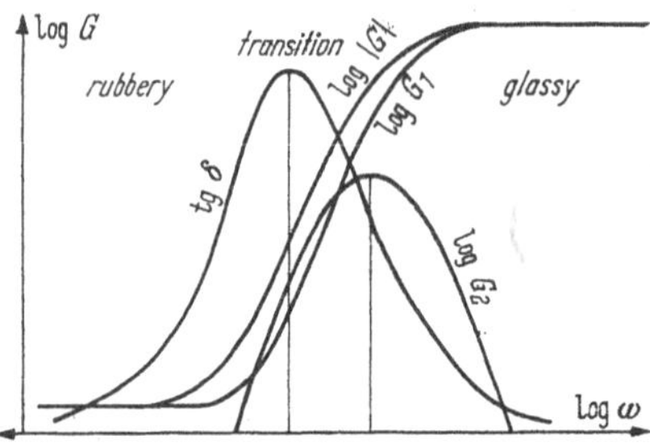

Fig. I, 12. The logarithm of the dynamic moduli G_1, G_2, $|G|$ and the dynamic loss tg δ as function of the logarithmic frequency for a polymer showing no flow.

So far we have discussed network polymers only. A polymer showing NEWTONian flow exhibits a further decrease of the absolute and of the storage moduli to zero at very low frequencies. A second strong increase occurs in tg δ in this frequency region and a second maximum in the loss modulus.

In considering compliances we find the reverse picture: J_1 and $|J|$ are high at low frequencies and show a transition to low values. In this transition region the loss compliance shows a maximum, which appears at lower frequency than the maximum of tg δ.

The picture given here is very similar to that obtained in considering the stress relaxation modulus or the creep compliance as functions of logarithmic time. We can make the two pictures coincide by identifying the frequency in dynamic measurement with the reciprocal time in a static measurement. In fig. I,13 a logarithmic time axis is drawn, which is at the same time the logarithmic frequency and time axis by means of the definition

$$t = 1/\omega \quad \text{or} \quad \log t = -\log \omega . \qquad \text{(I,54)}$$

The logarithm of the relaxation modulus is then compared with the logarithm of the storage and loss moduli. The transition regions in relaxation and dynamic measurements coincide and the relaxation modulus appears to be equal to the storage modulus in all loss-free regions. In the transition region the storage modulus is always somewhat larger than the stress-relaxation modulus. The same equivalence of course exists between the storage compliance and the creep compliance by virtue of relation (I, 54).

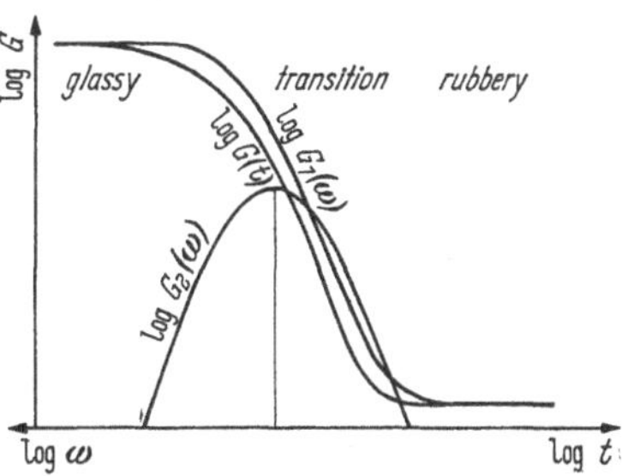

Fig. I, 13. Illustrating the connection between static and vibration experiments performed in the same time scale ($\omega = 1/t$).

In the above discussion we have tacitly assumed that stress relaxation could be measured in the same time region as the dynamic modulus. In reality the two types of measurement cover different time scales so that the storage modulus represents a continuation of the stress relaxation modulus rather than a duplication of it. In practice static experiments yield the mechanical behaviour for times ranging from about one second to infinity. *Dynamic* measurements complete this picture down to 10^{-8} seconds (frequency region 10^{-1} to 10^{+8} cycles per second). There is no fundamental objection to performing dynamic measurements at frequencies lower than 10^{-1} or 10^{-3} per sec. corresponding to times of 10 to 1000 sec. So it is possible in principle to perform static and dynamic experiments in the same range of time.

In fact, as stated above, once the relaxation or the creep function of a material is known, its viscoelastic behaviour can be predicted for any stress-strain history, and hence for a sinusoidal stress or strain. Therefore, it is obvious from the start that it must be possible to express the dynamic characteristic functions in terms of the static ones and conversely.

Gross[1] has given the following expressions[2]

$$\left.\begin{aligned} J_1(\omega) &= J_0 + J\int_0^\infty \cos\omega y\,\dot{\Psi}(y)\,dy = J(0) + \int_0^\infty \dot{J}(t)\cos\omega t\,dt \\ J_2(\omega) &= J\int_0^\infty \sin\omega y\,\dot{\Psi}(y)\,dy + \frac{1}{\omega\eta} = \int_0^\infty \dot{J}(t)\sin\omega t\,dt, \end{aligned}\right\} \quad (\mathrm{I}, 55)$$

$$\left.\begin{aligned} G_1(\omega) &= G + G_0\left[1 - \int_0^\infty \cos\omega y\,\dot{\Phi}(y)\,dy\right] = G(0) + \int_0^\infty \dot{G}(t)\cos\omega t\,dt \\ G_2(\omega) &= G_0\int_0^\infty \sin\omega y\,\dot{\Phi}(y)\,dy = -\int_0^\infty \dot{G}(t)\sin\omega t\,dt. \end{aligned}\right\} \quad (\mathrm{I}, 56)$$

The dynamic quantities are the Fourier transforms of the static quantities and conversely. There are also relations between the two components of the dynamic quantities, as shown by the Kramers-Kronig relations[3].

---

[1] Gross, B.: J. appl. Physics **19**, 257 (1948).

[2] $\Psi(t)$ and $\Phi(t)$ are retarded elasticity function and relaxation function as defined in (I, 35) and (I, 39).

[3] Kronig, R. de L.: J. opt. Soc. America **12**, 547 (1926). – H. A. Kramers: Atti Congr. fis. Como 1927, 545. – B. Gross: Amer. mathem. Monthly **50**, 89 (1943); Phys. Rev. **59**, 748 (1941). – R. H. Cole: Phys. Rev. **60**, 172 (1941). – E. Hiedemann u. R. D. Spence: Z. Physik **133**, 109 (1952). – H. Eisenlohr: Kolloid-Z. **138**, 57 (1954).

Summarising, we can say that each of the following material functions is a characteristic function of a linear viscoelastic material, determining the stress strain behaviour completely:

creep compliance	$J(t)$
stress relaxation modulus	$G(t)$
storage compliance	$J_1(\omega)$
loss compliance	$J_2(\omega)$
absolute dynamic compliance	$\lvert J\rvert(\omega)$
storage modulus	$G_1(\omega)$
loss modulus	$G_2(\omega)$
absolute modulus	$\lvert G\rvert(\omega)$
mechanical loss	$\operatorname{tg}\delta(\omega)$

If one of these nine functions is known for all times (for all frequencies), all the others can be calculated, though these calculations are far from simple in practice.

For practical purposes it is sometimes advantageous to make use of the approximate relationships[1]:

$$G(t) \simeq [G_1(\omega)] \qquad \omega = 1/t. \tag{I, 57}$$

The static modulus and the storage modulus form approximately two branches of the same function, defined in different time regions.

$$G_2(\omega) \simeq \frac{\pi}{2}\, d\, G_1(\omega)/d \ln \omega; \quad \operatorname{tg}\delta \simeq \frac{\pi}{2}\, d \ln G_1(\omega)/d \ln \omega. \tag{I, 58}$$

The loss modulus is approximately proportional to the slope of the storage modulus on a logarithmic frequency scale and the mechanical loss is approximately proportional to the slope of the logarithm of the storage modulus on a logarithmic frequency scale. The same approximate relationships hold if the moduli are replaced by the corresponding compliances.

Integrating equation (I, 58) with respect to logarithmic frequency, we get the two approximate relations

$$\left.\begin{aligned} \frac{2}{\pi}\int_{\omega_1}^{\omega_2} G_2(\omega)\, d \ln \omega &\simeq G_1(\omega_2) - G_1(\omega_1) \\ \frac{2}{\pi}\int_{\omega_1}^{\omega_2} \operatorname{tg}\delta\, d \ln \omega &\simeq \ln G_1(\omega_2)/G_1(\omega_1)\,. \end{aligned}\right\} \tag{I, 59}$$

The areas under the curves $G_2(\omega)$ or $\operatorname{tg}\delta$ on a logarithmic frequency scale are approximately connected with the difference or the ratio respectively of the modulus values at the initial and final points of the area considered[2].

---

[1] SCHWARZL, F.: Physica **17**, 830, 923 (1951). – J. HEYBOER, P. DEKKING u. A. J. STAVERMAN: Proc. 2nd. Int. Congr. on Rheology, Oxford 1953, page 123.

[2] Relations (I, 57) to (I, 59) may be derived by approximation methods which are treated in § 4f. They must be considered as rather crude approximations, the value of which lies more in their qualitative than in their quantitative information.

Literature to § 3.

Books:

ALFREY, T., JR.: Mechanical behaviour of high polymers. High Polymers VI, New York 1948, Interscience Publishers.

First and Second Report on Viscosity and Plasticity. North Holland Publ. Comp., Amsterdam 1935, 1938.

GROSS, B.: Mathematical structure of the theories of viscoelasticity. Hermann & Cie., Editeurs, Paris 1953.

HAWARD, R. N.: The strength of plastics and glass. Interscience. New York 1949.

KOLSKY, H.: Stress waves in solids. Clarendon Press, Oxford 1953.

LEADERMAN, H.: Elastic and Creep Properties of filamentous Materials. Textile Foundation, Washington 1943; Physics of high polymers. University Press, Utrecht 1951.

REINER, M.: Twelve lectures on theoretical rheology, Amsterdam 1949; Deformation and flow. New York 1949.

SPENCER, R. S. and R. F. BOYER: Adv. Colloid. Sci. Vol. II, Interscience Publ., New York 1946.

ZENER, CL.: Elasticity and Anelasticity of Metals. Univ. Chicago Press, 1948.

Articles:

ALFREY, T., JR. and P. DOTY: J. appl. Physics **16**, 700 (1945).

ANDREWS, R. P.: Ind. Engng. Chem. **44**, 707 (1952).

BOYER, R. F. and R. S. SPENCER: J. appl. Physics **15**, 398 (1944); **16**, 594 (1945).

FERRY, J. D.: Annu. Rev. physic. Chem. **4**, 345 (1953).

GROSS, B.: J. appl. Physics **19**, 257 (1948); **18**, 212 (1947); Proc. 2nd. Int. Congr. Rheology, Oxford 1953, page 221.

GROSS, B. and H. PELZER: J. appl. Physics **22**, 1035 (1951).

HIEDEMANN, E. u. R. D. SPENCE: Z. Physik **133**, 109 (1952).

JENCKEL, E.: Kunststoffe **31**, 209 (1941); Kolloid-Z. **120**, 160 (1951); **130**, 64 (1953).

JENCKEL, E. u. K. UEBERREITER: Z. physik. Chem. **A 182**, 361 (1938).

KNESER, H. O.: Kolloid-Z. **134**, 20 (1953).

KUHN, W.: Z. physik. Chem. **B 42**, 1 (1939); Helv. chim. Acta **30**, 487 (1947); Die makromol. Chem. **6**, 224 (1951).

KUHN, W., O. KÜNZLE u. H. PREISMANN: Helv. chim. Acta **30**, 464 (1947); **30**, 307 (1947).

MARK, H.: Trans. Faraday Soc. **43**, 447 (1947).

MARVIN, R. S.: Ind. Engng. Chem. **44**, 696 (1952); Phys. Rev. **86**, 644 (1952).

MÜLLER, F. H.: Kolloid-Z. **120**, 119 (1951); **122**, 109 (1951); **134**, 77 (1953); **123**, 65 (1951).

SIMHA, R.: J. appl. Physics **13**, 201 (1942); J. physic. Chem. **47**, 348 (1943).

SIPS, R.: J. Polymer Sci. **5**, 69 (1950); **6**, 285 (1951); **7**, 191 (1951).

TUCKETT, R. F.: Trans. Faraday Soc. **38**, 310 (1942); **39**, 158 (1943); **40**, 448 (1944).

UMSTÄTTER, H.: Schweiz. Arch. angew. Wiss. Techn. **19**, 184 (1953).

§ 4. Model theory of linear viscoelastic behaviour.

a) General meaning of mechanical models.

In the introduction to this chapter we have mentioned a fundamental difficulty underlying all study of viscoelastic behaviour, namely the difficulty of describing the *state* of a viscoelastic body completely and unambiguously by specifying a number of suitable parameters of state. The fact that stress varies at constant strain and that strain varies at constant stress shows that neither stress nor strain alone describe the state of the material completely and that "internal parameters" may change when these measurable "external parameters" are kept constant.

A complete elucidation of the nature of the internal parameters would require a study of all configurations of the atoms contained in the system. This, in turn, would require quantitative knowledge of the valence forces and the VAN DER WAALS forces between different atoms, of the dependence of these forces on atom distances and valence angles and of the change of the number of configurations on changing the valence angles and atom distances. Certainly this very detailed information is not nearly obtainable with our present knowledge of interatomic forces and of configurational statistics and it does not appear probable that it will become available in the near future. On the other hand, viscoelastic substances generally have a common chemical feature in that they contain macromolecules and so much information about macromolecules has been obtained recently that it is certainly possible to have some qualitative understanding of the nature of these internal parameters. Qualitative pictures will be discussed in § 5.

On the other hand, already in the nineteenth century, when virtually nothing was known about macromolecules, the need was felt of a "picture" and for that purpose mechanical models have been drawn up consisting of springs and dashpots. In order to serve as a model of a particular viscoelastic body the mechanical system must show the same relations between force, elongation and time as those between stress, strain and time for the viscoelastic body.

Although mechanical models do not add anything really new to our knowledge of viscoelastic substances, since everything that is derived from these models can be derived from a suitable choice of the characteristic phenomenological constants and the superposition principle, nevertheless they are extremely useful. First they give a perspicuous picture of the meaning and the importance of internal parameters of state, these being represented by the elongation of each of the springs and dashpots. Moreover, springs and dashpots embody the essential features of the real molecular parameters in a pure form; a spring is an element which stores energy on deformation, a dashpot dissipates energy on deformation. Today not much more can be said about the internal parameters of real substances than that on deformation some of them store energy in a reversible way whereas others dissipate it irreversibly.

We feel justified therefore, in considering mechanical models in some detail.

b) Simple mechanical models.

The first basic element of mechanical models, the *spring*, represents a HOOKEan deformation mechanism, characterised by proportionality between force σ, and elongation, γ.

$$\sigma = G\gamma, \quad (I, 60)$$

where G is the spring constant (see fig. I, 14).

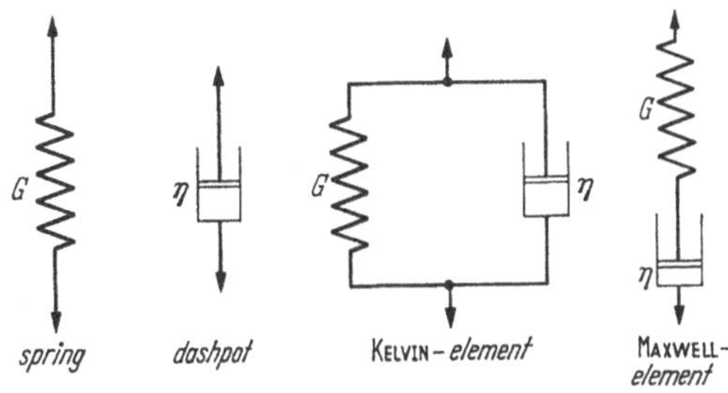

Fig. I, 14. Simple mechanical models.

The second basic element is the *dashpot* representing a molecular process in which the whole deformation energy is dissipated as heat. The mechanical model consists of a piston which moves in a viscous liquid with the equation of motion

$$\sigma = \eta\dot{\gamma}, \quad (I, 61)$$

η is a friction constant which may be called the viscosity of the dashpot.

HOOKEan deformation can be represented by a spring alone, NEWTONian flow can be represented by a dashpot alone, but for the description of the more complicated viscoelastic deformations, springs and dashpots must be combined in various ways. The composition of basic elements can be performed by connecting them either in series or parallel. If two basic units, governed by the equations $\gamma_1 = f(\sigma_1)$, $\gamma_2 = f(\sigma_2)$ are combined in series, the resulting model is obtained by

$$\gamma = \gamma_1 + \gamma_2, \quad \sigma = \sigma_1 = \sigma_2 \quad (I, 62)$$

and if they are combined parallel, by

$$\gamma = \gamma_1 = \gamma_2, \quad \sigma = \sigma_1 + \sigma_2. \quad (I, 63)$$

If a spring and a dashpot are linked in parallel, we obtain the KELVIN (VOIGT)-*element*[1]. Its equation of motion is easily seen to be

$$\dot{\gamma} + \frac{1}{\tau}\gamma = \frac{1}{\eta}\sigma \quad [\text{or } \sigma = \eta\dot{\gamma} + G\gamma] \quad (I, 64)$$

where η is the viscosity of the dashpot and G is the elasticity of the spring. The constant $\tau = \eta/G$ has the dimension of a time and is called the *retardation time* (in German: *Retardationszeit*) of the element. The KELVIN element is a good example of a mechanical model of a viscoelastic body.

---

[1] This model has been proposed independently by Lord KELVIN: Encyclopaedia Britannica (1875), and by W. VOIGT: Abh. K. Ges. Wiss. Göttingen **36** (1890). See further: J. H. C. THOMPSON: Phil. Trans Roy. Soc. London (A) **231**, 339 (1933); H. v. RÖTGER: Glastechn. Ber. **19**, 192 (1941).

If the model is stressed, one part of the deformation energy is stored in the spring while the remainder is dissipated in the dashpot. After release of the force, the deformation recovers due to the stressed spring. The dissipative force of the dashpot effects the retardation of the deformation, and the retardation time τ is a measure of the time needed to approach equilibrium.

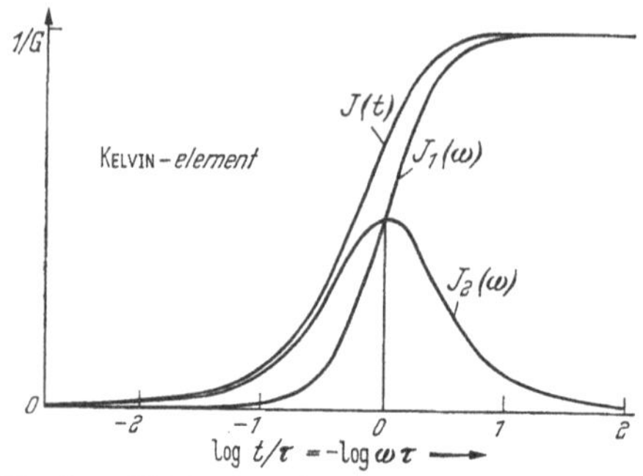

Under a constant stress the KELVIN element shows creep, the deformation being of the form

$$J(t) = \frac{1}{G}\left[1 - e^{-t/\tau}\right]. \qquad (I, 65)$$

1/G
J(t)
$J_1(\omega)$
KELVIN-element
$J_2(\omega)$
0
-2
-1
0
1
2
log t/τ = -log $\omega\tau$ →

Fig. I, 15. Kelvin element, creep compliance, storage compliance and loss compliance.

This function shows the characteristic features of viscoelastic deformation. On the logarithmic time diagram there is a transition region, in the neighbourhood of the retardation time τ. In the case of vibration measurements, the storage compliance and the loss compliance are

$$J_1(\omega) = \frac{1}{G}\,\frac{1}{1+\omega^2\tau^2}, \quad J_2(\omega) = \frac{1}{G}\,\frac{\omega\tau}{1+\omega^2\tau^2}. \qquad (I, 66)$$

The time dependence of these quantities is represented in fig. I,15 on a logarithmic scale of time and frequency, where the retardation time τ has been chosen as unit of the time scale.

The KELVIN element does not account for stress relaxation phenomena occurring under constant strain. To represent this behaviour, we have to consider another basic element, the MAXWELL *element*[1], which consists of a spring and a dashpot connected in series. Its equation is

$$\dot{\sigma} + \frac{1}{\tau'}\sigma = G\dot{\gamma} \quad \left[\text{or } \dot{\gamma} = \frac{1}{G}\dot{\sigma} + \frac{1}{\eta}\sigma\right], \qquad (I, 67)$$

where the ratio $\tau' = \eta/G$ is called the *relaxation time* (in German: ***Relaxationszeit***). If the element undergoes a constant deformation at time zero, only the spring is stretched in the first moment. In the beginning the whole deformation energy is stored in the spring; with increasing time the dashpot flows and the energy is transmitted to the dashpot and dissipated there (the stress relaxes).

The stress relaxation modulus of the MAXWELL element decays exponentially

$$G(t) = G e^{-t/\tau'}. \qquad (I, 68)$$

In logarithmic timescale a transition in stress relaxation modulus

[1] MAXWELL, J. C.: Phil. Mag. J. Sci. **35**, 134 (1868); Phil. Trans. Roy. Soc. London (A) **157**, 49 (1867). Compare further: H. UMSTÄTTER: Kolloid-Z. **107**, 81 (1943); **105**, 182 (1943). – H. UMSTÄTTER u. H. FLASCHKA: Kolloid-Z. **110**, 208 (1948).

occurs at the relaxation time τ'. A vibration experiment under sinusoidal strain is described by the dynamic modulus quantities

$$G_1(\omega) = G\frac{\omega^2\tau'^2}{1+\omega^2\tau'^2}, \quad G_2(\omega) = G\frac{\omega\tau'}{1+\omega^2\tau'^2}. \tag{I, 69}$$

The time dependence of the moduli is represented by fig. I, 16. (Here the relaxation time τ' has been chosen as unit of the time scale.)

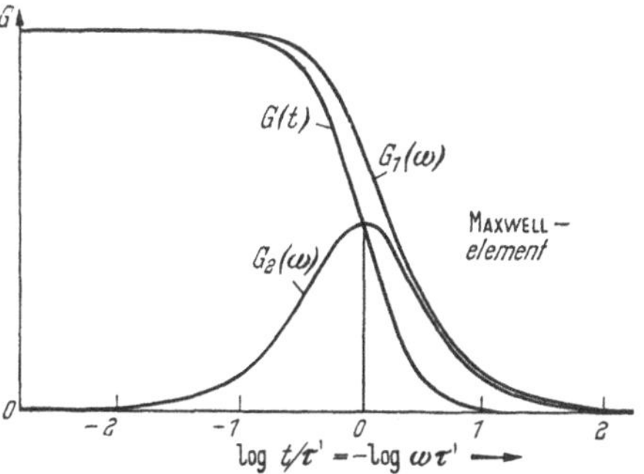

Fig. I, 16. MAXWELL element, relaxation modulus storage modulus and loss modulus.

c) The four-parameter model.

The two most simple models, the KELVIN and MAXWELL unit are insufficient to describe the viscoelastic behaviour of a real polymer completely since the KELVIN unit only represents retarded deformation, the MAXWELL unit only represents stress relaxation. If we want to have a model which behaves like a real polymer, we need a combination of at least three or four basic units.

Models consisting of four units have been treated in the literature frequently[1]. Model A of fig. I, 17 shows a spring of high modulus G_0, a KELVIN unit with the constants G and $G\tau$ and a dashpot with viscosity η linked in series. The deformation under constant stress of this model is the sum of the HOOKEan deformation of the single spring, of the highly-elastic deformation of the VOIGT (KELVIN) element and of the flow of the single dashpot. The creep compliance yielded by model A becomes therefore

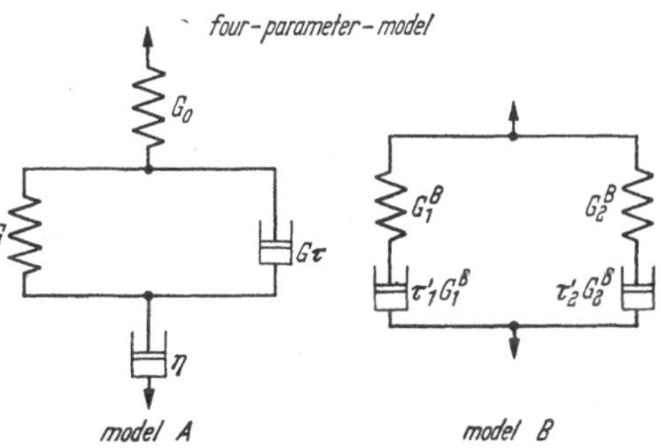

Fig. I, 17. Four-parameter models, A and B.

$$J(t) = \frac{1}{G_0} + \frac{1}{G}[1 - e^{-t/\tau}] + t/\eta. \tag{I, 70}$$

[1] BURGERS, J. M.: First report on Viscosity and Plasticity. Amsterdam 1935. – K. BENNEWITZ u. H. RÖTGER: Physik. Z. **37**, 578 (1936); **40**, 416 (1939). – A. P. ALEKSANDROV u. Y. LAZURKIN: Acta physicochim. USSR **12**, 647 (1940). – H. RÖTGER: Glastechn. Ber. **19**, 192 (1941). – R. F. TUCKETT: Trans. Faraday Soc. **39**, 158 (1943); **40**, 448 (1944). – T. ALFREY: Mechanical behaviour of high polymers, New York (1948), page 104. – E. JENCKEL: Kunststoffe **40**, 98 (1950). – R. S. MARVIN: Ind. Engng. Chem. **44**, 696 (1952). – E. KLEIN u. E. JENCKEL: Z. Naturforsch. **7 a**, 305 (1952). – P. L. KIRKBY: Trans. Soc. Glass. Techn. **37**, 7 (1953).

This compliance represents exactly the behaviour we found to be characteristic for high polymers [equation (I,36)]. The retarded elasticity function is given by the expression in square bracketts.

Under a sinusoidal force the behaviour of the model is given by the dynamic compliances

$$J_1(\omega) = \frac{1}{G_0} + \frac{1}{G}\frac{1}{1+\omega^2\tau^2}, \quad J_2(\omega) = \frac{1}{\omega\eta} + \frac{1}{G}\frac{\omega\tau}{1+\omega^2\tau^2}. \quad \text{(I, 71)}$$

Fig. I,18 contains a survey of the deformation behaviour of model A in a double logarithmic plot of the compliance against time or frequency.

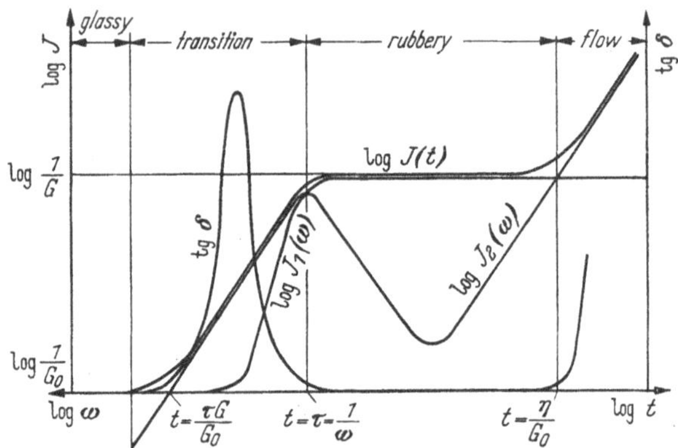

Fig. I, 18. Deformation behaviour of model A, logarithm of J, J_1, J_2 and $\operatorname{tg}\delta$ as functions of $\log t$ or $\log \omega$.

In the glassy state the compliances are of magnitude $1/G_0$. The viscoelastic transition begins at the time $t = \tau G/G_0$ and ends at time τ. The compliances in the rubbery state are of the order $1/G$ and the rubbery state reaches from $t = \tau$ to $t = \eta/G_0$, where the flow region begins. Attention is drawn to the fact that the transition in the storage compliance J_1 is considerably shorter than the transition in the creep compliance J. Whilst the maximum in loss compliance J_2 occurs at time τ, the maximum in dynamic loss $\operatorname{tg}\delta$ occurs at the shorter time $\tau\sqrt{G/G_0}$.

The general behaviour of model A can be determined by the differential equation between stress and strain, which can be derived in the following way:

The deformation of model A is composed of the strain γ_1 of the single spring, the deformation γ_2 of the Kelvin element and the deformation γ_3 of the single dashpot

$$\gamma = \gamma_1 + \gamma_2 + \gamma_3.$$

The deformation of the spring, the Kelvin element and the single dashpot conform respectively to the equations

$$\sigma = G_0\gamma_1$$

$$\frac{1}{G\tau}\sigma = \dot{\gamma}_2 + \frac{1}{\tau}\gamma_2$$

$$\sigma = \eta\dot{\gamma}_3.$$

From these four equations we can eliminate the quantities $\gamma_1, \gamma_2, \gamma_3$ replacing them by the deformation of the four parameter model γ and its time derivatives. In this way we end up with the differential equation:

$$\gamma + \frac{1}{\tau}\dot{\gamma} = \frac{1}{G_0}\left[\ddot{\sigma} + \left(\frac{1}{\tau_1'} + \frac{1}{\tau_2'}\right)\dot{\sigma} + \frac{1}{\tau_1' \tau_2'}\sigma\right] \qquad \text{(I, 72)}$$

$\left(\frac{1}{\tau_1'} + \frac{1}{\tau_2'}\right)$ and $\frac{1}{\tau_1' \tau_2'}$ are abbrevations for the expressions

$$\left.\begin{aligned} \frac{1}{\tau_1' \tau_2'} &= \frac{G_0}{\tau \eta} \\ \frac{1}{\tau_1'} + \frac{1}{\tau_2'} &= \frac{G_0}{G\tau} + \frac{G_0}{\eta} + \frac{1}{\tau}. \end{aligned}\right\} \qquad \text{(I, 72 a)}$$

If the stress is prescribed, equation (I, 72) forms a differential equation for the strain while if the strain is prescribed, it determines the stress. The two new constants τ_1' and τ_2' are called the relaxation times of the four parameter model and are connected with the constants G_0, G, η, τ by equation (I, 72 a) which can be solved with respect to the relaxation times:

$$\frac{1}{\tau_{1\,2}'} = \frac{1}{2}\left[\frac{G_0}{G\tau} + \frac{G_0}{\eta} + \frac{1}{\tau}\right] \pm \frac{1}{2}\left\{\left[\frac{G_0}{G\tau} + \frac{G_0}{\eta} + \frac{1}{\tau}\right]^2 - \frac{4 G_0}{\tau\eta}\right\}^{\frac{1}{2}}. \qquad \text{(I, 72 b)}$$

We can use equation (I, 72) to calculate the stress relaxation behaviour of our model. We put γ = unity viz. $\dot{\gamma} = \ddot{\gamma} = 0$ and solve the differential equation for the stress under the initial conditions

$$\sigma(0) = G_0, \quad \dot{\sigma}(0) = -G_0^2\left[\frac{1}{\eta} + \frac{1}{G\tau}\right].$$

We get for the relaxing stress under unit strain amplitude, i.e. for the stress relaxation modulus

$$G(t) = G_1^B e^{-t/\tau_1'} + G_2^B e^{-t/\tau_2'}; \qquad \text{(I, 73)}$$

τ_1' and τ_2' are the two relaxation times defined in (I, 72 b) and G_1^B and G_2^B are abbrevations for the following expressions

$$\left.\begin{aligned} G_1^B &= G_0\left[\frac{1}{\tau_1'} - \frac{1}{\tau}\right] \Big/ \left[\frac{1}{\tau_1'} - \frac{1}{\tau_2'}\right] \\ G_2^B &= G_0\left[\frac{1}{\tau} - \frac{1}{\tau_2'}\right] \Big/ \left[\frac{1}{\tau_1'} - \frac{1}{\tau_2'}\right]. \end{aligned}\right\} \qquad \text{(I, 74)}$$

The stress relaxation behaviour (I, 73) of our model is interesting because it shows two transitions of the modulus. The first, the viscoelastic transition takes place at a time of the order of the relaxation time τ_1' and reduces the modulus from the value $G_1^B + G_2^B = G_0$ in the glassy region to a value G_2^B in the rubbery region. The second transition is connected with flow and takes place around the time τ_2'.

This relaxation behaviour cannot be read off immediately from the form of model A, but it can be represented by a model consisting of two Maxwell units linked parallel with the spring constants G_1^B and G_2^B and with the viscosities $\tau_1' G_1^B$ and $\tau_2' G_2^B$ (fig. I, 17 model B). Each of the relaxation processes of (I, 73) is then represented by one Maxwell unit. Model B is nothing but an other representation of the mechanical behaviour of model A; both models represent the same stress strain behaviour.

We encounter here a very important result: One particular viscoelastic material can be represented by two different mechanical models, the one consisting of a number of Voigt elements in series plus a single spring and a single dashpot, the other consisting of a number of Maxwell units linked in parallel. The constants of these two models are interrelated, model B can be calculated if model A is known and vice versa [equation (I,72) and (I,74) in our case]. Two such models, representing the same stress strain behaviour are called *mechanically equivalent.* Of course the choice of one of these equivalent models is only a matter of convenience. It appears that the model of the VOIGT type is to be preferred, if the stress is given and the strain is to be calculated. If the strain is given and the stress is to be determined, the MAXWELL type is the more convenient.

Model A is characterised by one retardation time τ, model B is characterised by two relaxation times τ_1' and τ_2'. These times must always satisfy the inequality

$$\tau_1' < \tau < \tau_2'. \tag{I,75}$$

The dynamic moduli of the four-parameter model can be read off directly from model B

$$G_1(\omega) = G_1^B \frac{\omega^2 \tau_1'^2}{1+\omega^2 \tau_1'^2} + G_2^B \frac{\omega^2 \tau_2'^2}{1+\omega^2 \tau_2'^2}, \quad G_2(\omega) = G_1^B \frac{\omega \tau_1'}{1+\omega^2 \tau_1'^2} + G_2^B \frac{\omega \tau_2'}{1+\omega^2 \tau_2'^2}. \tag{I,76}$$

Before concluding this section we want to say a few words about the three-parameter model. If the material to be represented does not show NEWTONian flow, the viscosity η becomes infinite and the flow term in equation (I,70) and (I,71) has to be dropped. Considering model A this means that the single dashpot must be dropped and we are left with a three-parameter model of one VOIGT element and one spring in series. The transition from the four-parameter model to the three-parameter model can be performed in all formulas of this section by putting $\eta = \infty$. At the same time model B degenerates into a three-parameter model, consisting of one MAXWELL unit and one single spring linked in parallel (the dashpot in the second MAXWELL unit has to be dropped). This means that the relaxation time τ_2' becomes infinite, whilst the only relaxation time τ_1' of the model is connected with the retardation time by

$$\tau = [1 + G_0/G]\tau_1'.$$

d) Composite finite models.

We have seen in the last section that the mechanical behaviour of high polymers can be represented roughly by a four-parameter model, showing one viscoelastic transition in creep or relaxation and one transition connected with flow.

If the mechanical behaviour of real polymers is observed in more detail, it turns out that the above picture is too simple in many cases. The creep curve does not show one, but a number of viscoelastic transitions, the large transition (softening) being preceded by one or more smaller transitions characterised by much smaller increases in compliance than the transitions at the softening point.

Similar observations have been made in measurements of the damping of high polymers. The loss modulus G_2 and the mechanical loss $\operatorname{tg}\delta$ show a large and broad maximum, corresponding to the softening point, and a number of smaller secondary peaks[1].

In order to account for such series of transitions one has to generalise the models treated so far. One may use a generalisation of model A (if one deals with creep behaviour) or of model B (if one deals with relaxation). Generalisations of the MAXWELL model were first used in this con-

[1] Examples of secondary transitions in organic polymers are considered in chapter VI, § 52 (see for instance table VI,4). Also in inorganic glasses the transitions are stepwise: compare A. SMEKAL: Glastechn. Ber. **16**, 198 (1938); Z. physik. Chem. (B) **44**, 286 (1939); (B) **48**, 114 (1940).

nection independently by KUHN and BENNEWITZ and RÖTGER (page 32 loc. cit.).

Generalised VOIGT model[1] (model A). A mechanical model which shows instantaneous elasticity G_0, NEWTONian flow with viscosity η and n different retardation times $\tau_1, \tau_2, \ldots \tau_n$ is shown in fig. I, 19. It consists of a single spring G_0, a single dashpot η and of n VOIGT elements all connected in series. The VOIGT element with retardation time τ_i has an elasticity G_i and a viscosity $\eta_i = G_i \tau_i$. It contributes to the equilibrium compliance with a term $J_i = 1/G_i$.

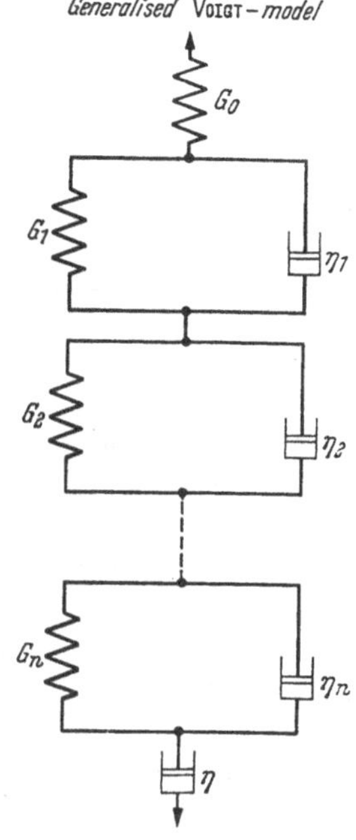

Fig. I, 19. Generalised VOIGT model.

From model A the creep behaviour can be read off at once. Besides instantaneous elasticity and flow there are n different transitions in the creep curve corresponding to the elastic compliance

$$\left.\begin{aligned} J(t) &= \frac{1}{G_0} + J\Psi(t) + t/\eta \\ \text{with} \qquad J\Psi(t) &= \sum_{i=1}^{n} J_i (1 - e^{-t/\tau_i}) \end{aligned}\right\} \qquad \text{I, (77)}$$

In the same way the storage and loss compliance are written down at once

$$\left.\begin{aligned} J_1(\omega) &= \frac{1}{G_0} + \sum_i J_i \frac{1}{1+\omega^2 \tau_i^2}; \\ J_2(\omega) &= \sum_i J_i \frac{\omega \tau_i}{1+\omega^2 \tau_i^2} + \frac{1}{\omega \eta}. \end{aligned}\right\} \qquad \text{(I, 78)}$$

It is not necessary to give a figure of the characteristic quantities, since all conclusions from fig. I, 15 and I, 18 remain valid, the only difference being the occurrence of more than one transition. Our model is determined by $2n + 2$ quantities, the $n + 1$ spring constants and the $n + 1$ dashpot viscosities.

Generalised MAXWELL model[2] (model B). In considering stress relaxation phenomena it is more convenient to use a generalisation of model B which consists of $n + 1$ MAXWELL elements connected parallel (fig. I, 20). The i-th MAX-

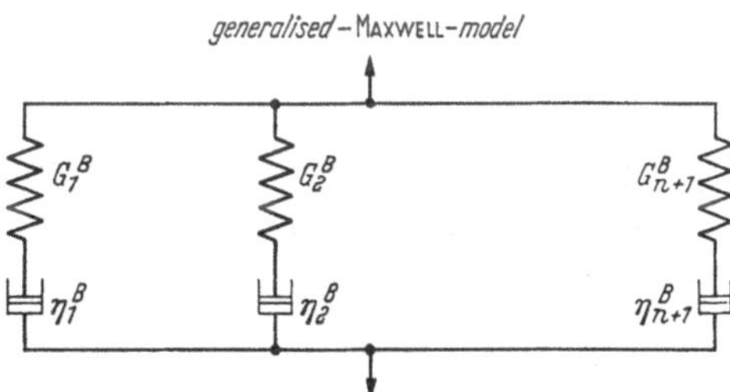

Fig. I, 20. Generalised MAXWELL model.

---

[1] ALFREY, T. u. P. DOTY: J. appl. Physics **16**, 700 (1945). – See also E. MEWES: Kolloid-Z. **131**, 84 (1953).

[2] KUHN, W.: Z. physik. Chem. (B) **42**, 1 (1939); Z. angew. Chem. **52**, 289 (1939); Helv. chim. Acta **30**, 487 (1947); Die makromol. Chem. **6**, 224 (1951). – K. BENNEWITZ u. H. RÖTGER: Physik. Z. **40**, 416 (1939); Z. physik. Chem. B **48**, 108 (1940). – W. HOLZMÜLLER u. E. JENCKEL: Z. physik. Chem. A **186**, 359 (1940). – D. D. ELEY: Trans. Faraday Soc. **38**, 299 (1942). – E. JENCKEL: Kautschuk **19**, 25 (1943).

WELL unit will consist of a spring G_i^B and a dashpot η_i^B, showing a relaxation time $\tau_1' = \eta_i^B/G_i^B$. The model shows $n+1$ different relaxation times, n of which correspond to the n retardation times of model A and one additional relaxation time which represents the flow. The constants of model B are indicated by an upper index B.

The stress relaxation behaviour of model B is found by simple addition of the behaviour of the different MAXWELL units. For instance the stress relaxation modulus shows $n+1$ transitions at the characteristic relaxation times ξ_i

$$G(t) = \sum_{i=1}^{n+1} G_i^B e^{-t/\tau_i'} \tag{I, 79}$$

and the dynamic quantities show transitions or maxima at the characteristic frequencies $1/\xi_i$

$$G_1(\omega) = \sum_i G_i^B \frac{\omega^2 \tau_i'^2}{1+\omega^2 \tau_i'^2} \qquad G_2(\omega) = \sum_i G_i^B \frac{\omega \tau_i'}{1+\omega^2 \tau_i'^2}. \tag{I, 80}$$

The differential equation[1]. The mechanical behaviour of either model A or model B can be represented by a differential equation which is of the order $n+1$ in stress and strain

$$\sigma^{(n+1)} + p_n \sigma^{(n)} + \cdots + p_1 \dot{\sigma} + p_0 \sigma = q_{n+1} \gamma^{(n+1)} + \cdots + q_1 \dot{\gamma}. \tag{I, 81}$$

This equation determines the stress, if the strain is prescribed, and the strain if the stress is prescribed. It is known, if the $2n+2$ constants $\{p, q\}$ are known. The set of constants $\{p, q\}$ of the operator equation can be calculated from either the $2n+2$ constants of model A $\{G, \tau\}$ or from the $2n+2$ constants of model B $\{G^B, \tau'\}$. If therefore one of these three ways of representation of mechanical behaviour is given, the other two can be calculated. It follows that the constants of the general VOIGT model (the creep behaviour) yield the constants of the general MAXWELL model as well as the differential equation, though the algebraic equations between the sets $\{p, q\}$, $\{G, \tau\}$, $\{G^B, \tau'\}$ are not simple[2]. We restrict ourselves to the following remarks: The negative retardation frequencies $-1/\tau_i$ are the characteristic roots of the right hand side of equation (I, 81) and the negative relaxation frequencies $-1/\tau_i'$ are the characteristic roots of the left hand side of equation (I, 81). Further the constant q_{n+1} equals the single spring constant G_0 in model A and the sum of all elasticities in model B. The quantity q_1/p_0 equals the viscosity of the single dashpot in model A and the sum of all viscosities in model B.

This short account shows that model A and B always represent the same mechanical behaviour, being *mechanically equivalent models*. Which of them is chosen for a representation of the material is a matter of convenience rather than a fundamental question. Again model A is preferable, if the stress is prescribed, and model B is preferable, if the strain is prescribed. For each general VOIGT model there exists a general MAXWELL model and vice versa. The bridge between these two ways of representation is the differential equation (I, 81). The special case $n = 1$ has been considered already in the four-parameter model.

The arrangement of springs and dashpots as series of KELVIN units or rows of MAXWELL units is not the only possible way of arranging a given number of these elements, but many other ways are possible. For instance networks containing 3, 4, 5, 6 elements can be arranged in 2, 4, 6, 10 independent ways. Any network obtained by whatever arrangement of springs and dashpots is equivalent to one generalised VOIGT model and to one generalised MAXWELL model. The values of the characteristic constants of the springs and dashpots of the various equivalent net-

---

[1] ALFREY, T. u. P. DOTY: J. appl. Physics **16**, 700 (1945).

[2] For a further account of this subject we refer the reader to the article by ALFREY and DOTY and to ALFREY's book.

works are generally completely different. These values for one arrangement can always be calculated from the values of an equivalent arrangement but the calculations may be laborious.

Electro-mechanical analogies[1]. The differential equation which governs the theory of mechanical networks (I,81) is of the same type as the equation which governs the theory of electrical networks consisting of resistances and condensors. Therefore an analogy between mechanical and electrical networks exists.

If the mechanical stress corresponds to voltage and mechanical deformation velocity to electric current, the behaviour of a mechanical spring G can be described by an electrical condensor of capacity $C = 1/G$ and the behaviour of a dashpot η corresponds to the behaviour of a resistance $R = \eta$. Connection in series (parallel) of two mechanical units then corresponds to connection in parallel (series) of the electric units. This analogy has the advantage that the mechanical energy which is stored in the springs equals the electrical energy stored in the capacitors and that the mechanical energy dissipated in the dashpots equals the electrical energy disipated in the resistances.

The electro-mechanical analogy has been used frequently to calculate mechanical networks with the help of the better known electrical ones. But in recent times mechanical networks become gradually better known so this indirect method of treating mechanical models no longer fulfils any purpose. But there is one important application of the analogy which should be mentioned. To analyse an experimental relaxation curve as a sum of simple exponential terms as formulated in (I,79) is a very tedious job. Therefore, recently it has been proposed to analyse mechanical behaviour with the help of electrical network systems, which can be more easily varied to fit experimental curves[2].

e) Continuous models, the spectra.

So far we have considered mechanical models composed of a finite number $(2n + 2)$ of springs and dashpots. An obvious extension of our generalisation of the Voigt model and the Maxwell model is to take up the consideration of models of an infinite number of units with characteristic constants varying continuously with the relaxation or retardation time. Such infinite networks are not characterised by a finite number of constants, but by a continuous function of one independent variable. These functions defining continuous models are called distribution functions, the *distribution function of retardation times* in the case of a generalised continuous Voigt model and the *distribution function of relaxation times* in the case of a generalised continuous Maxwell model.

The extension to continuous models is not merely a matter of mathematical elegance. Restriction to a finite number of units each with a definite characteristic time implies the assumption that in the real viscoelastic material processes always proceed with a very definite characteristic time. This state of affairs may be acceptable in crystalline materials — which indeed sometimes show viscoelastic behaviour due to well defined atomic rearrangements[3] — but it is not expected to exist in polymers. On the contrary the amorphous or liquid structure which is to some extent always present in polymers makes us believe that a definite

[1] An extended discussion of all possible electro-mechanical analogies and their physical meaning has been published recently: Kegel, G.: Kolloid-Z. **135**, 125 (1954)

[2] Stambaugh, R. B.: Ind. Engng. Chem. **44**, 1590 (1952).

[3] Fast, J. D. u. L. J. Dijkstra: Philips' techn. Rev. **13**, 172 (1951). — J. D. Fast u. J. L. Meyering: Philips Res. Rep. **8**, 1 (1953).

molecular rearrangement responsible for viscoelastic behaviour may proceed in a fairly large range of times depending on the accidental state of the environment of the atoms involved.

A further reason for the use of distribution functions lies in the shapes of the transition curves observed in experiment. Fig. I,21 shows as example a transition in creep as measured on Polyisobutylene[1], compared with the theoretical transition due to one VOIGT element according to equation (I,65). The actual transition cannot be described by the theoretical curve, as the real transition increases more gradually and extends over a longer time region than the theoretical transition. Although one can approach the experimental curve by the theoretical one by gradually increasing the — finite — number of elements from which the theoretical curve is calculated, the limitation to a finite number appears to be rather artificial.

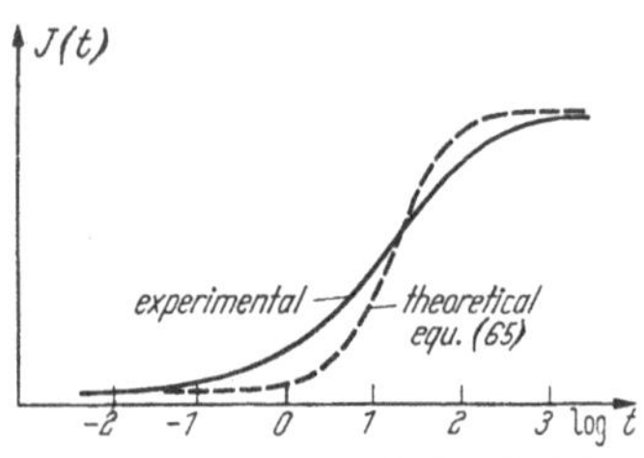

Fig. I, 21. Transition in Polyisobutylene, drawn line experimental curve, dotted line theoretical curve according to equation (I, 65).

The distribution of relaxation times[2]. We consider a general MAXWELL model (fig. I,20) and we assume that the different MAXWELL units are arranged according to increasing relaxation times[3]

$$\tau_1 < \tau_2 < \tau_3 < \cdots.$$

The model is determined completely, if the corresponding spring constant G_i corresponding to each relaxation time τ_i is given. In order to generalise the model more and more, we refine the division into different relaxation times more and more, so that at last the variable τ becomes continuous and takes all values between zero and infinity. The "spring constant" which belongs to the relaxation processes with relaxation times between τ and $\tau + d\tau$ has the magnitude

$$g(\tau)\, d\tau \quad 0 < \tau < \infty$$

where $g(\tau)$ is a continuous function of the relaxation time, the distribution function of relaxation times or shortly the *relaxation spectrum* (in German: *Relaxationsspektrum*).

If the relaxation spectrum $g(\tau)$ is known for all times, the mechanical behaviour of the model is determined completely. Particularly the stress relaxation modulus is given by the generalisation of equation (I,79), if

---

[1] LEADERMAN, H., R. G. SMITH u. R. W. JONES: J. Polymer Sci. **14**, 47 (1954). — H. LEADERMAN: Proc. 2nd. Int. Congr. Rheology, Oxford 1953, page 203.

[2] SMEKAL, A.: Z. physik. Chem. B **44**, 286 (1939). — W. KUHN: Angew. Chem. **52**, 289 (1939). — E. JENCKEL u. W. HOLZMÜLLER: Z. physik. Chem. **A 186**, 359 (1940). — R. SIMHA: J. appl. Physics **13**, 201 (1942). — E. JENCKEL u. J. FUEHLES: Makromol. Chem. **1**, 203 (1943). — W. BRENSCHEDE: Kolloid-Z. **104**, **1** (1943). — R. F. TUCKETT: Trans. Faraday Soc. **40**, 448 (1944). — W. KUHN, O. KÜNZLE u. A. PREISSMANN: Helv. chim. Acta **30**, 307, 464 (1947). — R. SIPS: J. Polymer Sci. **5**, 69 (1950). — D. TER HAAR: Physica **16**, 719, 738, 839 (1950). — F. H. MÜLLER: Kolloid-Z. **122**, 109 (1951). — W. KUHN: Makromol. Chem. **6**, 224 (1951).

[3] We use further for the relaxation- and retardationtime the same symbol τ.

summation over different MAXWELL units is replaced by integration over all relaxation times:

$$G(t) = \int_0^\infty g(\tau)\, e^{-t/\tau}\, d\tau + G. \tag{I,82}$$

G is the constant value which is attained by the modulus after infinite time. In our model the constant G has the meaning of the elasticity of a single spring, which must be connected in parallel to the model of fig. I,20 so as to assure that the stress after infinite time remains finite.

Now from (I,82) the physical meaning of the relaxation spectrum can be understood. The atomic rearrangements occurring in a stress relaxation experiment can be considered as composed of an infinite number of relaxation processes (representable by MAXWELL units). The relaxation spectrum $g(\tau)$ gives the contribution per unit interval of the time scale of the relaxation processes with relaxation times around τ to the instantaneous elasticity modulus.

The distribution of retardation times[1]. In complete analogy a continuous model can be constructed starting from the generalised VOIGT model (fig. I,19). This model consists of a single spring of compliance J_0, a single dashpot of viscosity η and of an infinite number of VOIGT units. Again the VOIGT units can be arranged according to increasing retardation times and the compliance J_i can be considered as a function of the retardation time τ_i. In the continuous model the VOIGT units with retardation times between τ and $\tau + d\tau$ will contribute to the compliance

$$f(\tau)\, d\tau \quad 0 < \tau < \infty;$$

$f(\tau)$ is called the distribution function of retardation times or shortly the *retardation spectrum* (in German: *Retardationsspektrum*). The creep behaviour of such a model is given by the creep compliance

$$J(t) = J_0 + \int_0^\infty f(\tau)\, [1 - e^{-t/\tau}]\, d\tau + t/\eta. \tag{I,83}$$

Equation (I,83) describes the physical meaning of the retardation spectrum: the highly-elastic deformation $J\Psi(t)$ can be considered to be caused by an infinite number of retardation processes. The retardation spectrum $f(\tau)$ gives the contribution per unit interval of the time scale of the retardation processes with retardation time around τ to the equilibrium compliance of highly-elastic deformation.

Spectra in logarithmic time scale. The relaxation and retardation spectra as given above are the appropriate quantities, if we deal with material behaviour on the linear time scale. As we have already seen in § 3 the time or frequency dependence becomes more pronounced if we consider the curves in logarithmic time or frequency diagrams. In log-

[1] WIECHERT, E.: Ann. Physik **50**, 335, 546 (1893). – H. LEADERMAN: Elastic and creep properties of filamentous materials and other high polymers. Textile Foundation, Washington 1943.

arithmic scales two other distributions must be used, which are defined by

$$\left.\begin{aligned} L(\ln\tau) &= \tau f(\tau) \\ H(\ln\tau) &= \tau g(\tau). \end{aligned}\right\} \qquad (I, 84)$$

The *logarithmic retardation spectrum* $L(\ln\tau)$ has the dimension of a compliance, the *logarithmic relaxation spectrum* $H(\ln\tau)$ has the dimension of a modulus. They are usually used in the literature instead of f or g. The meaning of the logarithmic distributions $L d\ln\tau$ and $H d\ln\tau$ is the contribution of retardation or relaxation processes with characteristic times between $\ln\tau$ and $\ln\tau + d\ln\tau$ to creep or stress relaxation.

From the logarithmic distribution functions stress relaxation and creep behaviour can be calculated by means of the equations

$$G(t) = \int_{-\infty}^{\infty} H(\ln\tau)\, e^{-t/\tau}\, d\ln\tau + G \qquad (I, 85)$$

$$J(t) = J_0 + \int_{-\infty}^{\infty} L(\ln\tau)\,[1 - e^{-t/\tau}]\, d\ln\tau + t/\eta\,. \qquad (I, 86)$$

The instantaneous modulus and the equilibrium compliance can be calculated by integration of the distribution functions in logarithmic time scale

$$\int_{-\infty}^{\infty} H(\ln\tau)\, d\ln\tau = G_0 - G,\ \int_{-\infty}^{\infty} \tau H(\ln\tau)\, d\ln\tau = \eta,\ \int_{-\infty}^{\infty} L(\ln\tau)\, d\ln\tau = J - J_0. \qquad (I, 87)$$

Also the dynamic quantities such as storage and loss modulus and storage and loss compliance can be expressed in the distribution functions[1]

$$\left.\begin{aligned} G_1(\omega) &= \int_{-\infty}^{\infty} H(\ln\tau)\, \frac{\omega^2\tau^2}{1+\omega^2\tau^2}\, d\ln\tau + G \\ G_2(\omega) &= \int_{-\infty}^{\infty} H(\ln\tau)\, \frac{\omega\tau}{1+\omega^2\tau^2}\, d\ln\tau\,. \end{aligned}\right\} \qquad (I, 88)$$

$$\left.\begin{aligned} J_1(\omega) &= J_0 + \int_{-\infty}^{\infty} L(\ln\tau)\, \frac{1}{1+\omega^2\tau^2}\, d\ln\tau \\ J_2(\omega) &= \int_{-\infty}^{\infty} L(\ln\tau)\, \frac{\omega\tau}{1+\omega^2\tau^2}\, d\ln\tau + \frac{1}{\omega\eta}\,. \end{aligned}\right\} \qquad (I, 89)$$

For polymers showing NEWTONian flow the constant G vanishes whilst the flow term $1/\eta$ is finite. Polymers showing no flow have a finite equilibrium modulus G and a vanishing flow term. The above formulas show that the spectra are characteristic functions describing the material properties completely. Besides the retardation spectrum $L(\ln\tau)$ the viscosity η has to be given to determine viscoelastic behaviour. Similarly, besides the relaxation spectrum $H(\ln\tau)$ the equilibrium modulus G has to be given.

A great advantage in using the spectra rather than one of the experimentally accessible quantities for characterising viscoelastic behaviour, lies in a simple relation between stress relaxation and creep yielded by the spectra. As can be shown

[1] TOBOLSKY, A. u. H. EYRING: J. chem. Physics **11**, 125 (1943). – T. ALFREY u. P. DOTY: J. appl. Physics **16**, 700 (1945). – W. KUHN, O. KÜNZLE u. A. PREISSMANN: Helv. chim. Acta **30**, 307, 464 (1947).

by rather intricate calculations[1] a simple relation between the two spectra exists by which they can be converted into each other

$$\left.\begin{aligned} L(\ln\tau) &= H(\ln\tau)\Bigg/\left\{\left[G_0 - \int_{-\infty}^{\infty}\frac{H(\ln\xi)}{1-\xi/\tau}\,d\ln\xi\right]^2 + \pi^2 H^2(\ln\tau)\right\}\\ H(\ln\tau) &= L(\ln\tau)\Bigg/\left\{\left[J_0 + \int_{-\infty}^{\infty}\frac{L(\ln\xi)}{1-\xi/\tau}\,d\ln\xi - \tau/\eta\right]^2 + \pi^2 L^2(\ln\tau)\right\}\end{aligned}\right\}. \tag{I, 90}$$

These equations allow some interesting remarks: apart from discontinuities, both spectra are zero or finite always in the same time region. If the retardation spectrum for instance is different from zero only in a certain time region, the relaxation spectrum can only differ (and must differ) from zero in the same time region. The equations show further that only the relaxation spectrum is affected by a change in NEWTONian viscosity. The retardation spectrum is not affected by flow properties.

f) Approximation methods for determination of the spectra.

In the foregoing section we have discussed a fairly large number of functions of time or frequency each of which is sufficient to describe completely linear viscoelastic behaviour of a material. Some of these functions, such as the stress relaxation function or the creep function can be measured immediately. Others such as the relaxation and retardation spectra cannot be measured immediately and their physical meaning is related to mechanical models of the viscoelastic material rather than to this material itself.

One might be inclined to consider the spectra as somewhat obscure functions of secondary importance and generally try to describe the viscoelastic behaviour by means of functions which are immediately accessible to measurement. However, it will be shown in § 5 that of all the various functions the spectra give the closest representation of the molecular processes occurring in viscoelastic deformation. This is understandable as we know that the spectra give a picture of the springs and dashpots of the model and these units can be thought of as representing molecular processes.

This fact, that the spectra give an insight into the molecular processes is, of course, of great importance to the chemist who studies viscoelastic behaviour with the object of improving or changing it by changing the chemical constitution of the material.

Moreover, the apparent disadvantage that the spectra cannot be measured directly cannot be considered to be very serious because for a complete characterisation of a particular material measurements have to be made over such a large range of the time and frequency scale that one single type of measurement (relaxation, creep a.s.o.) never suffices. Therefore, in order to characterise the material completely by one single function over the entire accessible range of time, functions have to be transformed into each other anyhow. However, by means of equations like

[1] GROSS, B.: J. appl. Physics **19**, 257 (1948); **18**, 212 (1947). – B. GROSS u. H. PELZER: J. appl. Physics **22**, 1035 (1951).

(I, 90), the spectra are more easily transformed into a number of the phenomenological functions than each of the functions themselves.

We see that there are several reasons for considering the spectra as the most convenient characteristic functions describing the behaviour of a material and we will discuss, therefore, how these spectra can be obtained from experimental measurements. One could try to invert equations like

$$G(t) - G = \int_{-\infty}^{\infty} H(\ln\tau)\, e^{-t/\tau}\, d\ln\tau \qquad \text{(I, 85)}$$

in order to obtain $H(\ln t)$ in terms of the function $G(t) - G$. This method has been followed by GROSS[1], who formulated the inverse equations of (I, 85), (I, 88) and (I, 89). Unfortunately his equations cannot be applied immediately to experimental results, as they contain the measured function $G(t)$ as a function of a complex variable. The spectrum can be calculated if the analytical expression for $G(t)$ together with the complex continuation of this expression is known, which is hardly the case with an experimentally measured function. Therefore the usual inversion equations cannot be used in dealing with experimental results, and a number of approximation methods have been proposed. We want to deal shortly with these approximation methods and begin with the determination of the spectrum from static measurements.

Equation (I, 85) is in principle a LAPLACE transform, which can be inverted by an equation first proposed by WIDDER[2]. If WIDDER's formula is applied to our problem, we get the spectrum in the following form[3]

$$H(t) = \lim_{k\to\infty} \frac{(-1)^k}{(k-1)!}\, t^k\, d^k G(k\,t)/d\,t^k\,. \qquad \text{(I, 91)}$$

Here the spectrum is assumed to be given as a function of t rather than of $\ln t$.

In order to obtain the spectrum with the aid of (I, 91) one has to differentiate the relaxation function k times and to multiply it with t^k and then let k approach infinity. It is impossible to do this rigorously not only because of the work involved but also because of the insufficient accuracy with which the relaxation function is known. However, by assigning to k values of 1, 2 and so on, approximations to the spectrum of increasing accuracy are obtained.

$k = 1$ gives an approximation of first order, which has been proposed by ALFREY[4] and which represents the spectrum by the slope of the stress relaxation modulus in logarithmic time scale

$$H(\ln t) \simeq H_1(\ln t) = -\,dG/d\ln t\,. \qquad \text{(I, 92)}$$

This approximation method has been widely used in literature[5] and gives a fairly good picture of the real distribution, if this extends over several

---

[1] GROSS, B.: J. appl. Physics **18**, 212 (1947); **19**, 257 (1948). — B. GROSS u. H. PELZER: J. appl. Physics **22**, 1035 (1951).

[2] WIDDER, D. V.: Trans. Amer. math. Soc. **36**, 107 (1934).

[3] SCHWARZL, F.: Physica **17**, 830, 923 (1951). — F. SCHWARZL u. A. J. STAVERMAN: Physica **18**, 791 (1952).

[4] ALFREY, T. u. P. DOTY: J. appl. Physics **16**, 700 (1945). — T. ALFREY: Mechanical behaviour of high polymers. New York 1948, page 551.

[5] HAAR, D. TER: Physica **16**, 719, 738, 839 (1950). — J. D. FERRY, W. M. SAWYER, G. V. BROWNING u. A. H. GROTH: J. appl. Physics **21**, 513 (1950). — J. D. FERRY u. M. L. WILLIAMS: J. Colloid Sci. **7**, 347 (1952).

decades and is rather flat. In the case of narrow distributions and of spectra with different neighbouring peaks the approximation (I, 92) fails[1]. Therefore we can not expect that (I, 92) would reproduce the details in a complicated spectrum with sufficient resolving power.

To get better approximations we take $k = 2$ and $k = 3$ and obtain thus second and third order approximations

$$H_2(\ln t/2) = -\, dG/d\ln t + d^2 G/d\ln t^2 \tag{I,93}$$

$$H_3(\ln t/3) = -dG/d\ln t + \frac{3}{2} d^2 G/d\ln t^2 - \frac{1}{2} d^3 G/d\ln t^3 . \tag{I,94}$$

Here we have adopted the following notation: H without index indicates the exact spectrum, H_k indicates an approximation to it, derived from the stress relaxation modulus. The index k gives the order of the approximation, that is the highest logarithmic derivative occurring in the formula. It is expected that the higher the order of the approximation, the closer the corresponding expression H_k approaches to the exact spectrum. To prove this assumption various approximations of a large number of fictitious spectra have been calculated [1,2]. In all cases the quality of the approximations and their resolving power with respect to the details in the real spectrum improved considerably by increase of the order k.

If we want to calculate the retardation spectrum from the creep compliance exactly the same formula hold. We only have to replace the relaxation spectrum H by the retardation spectrum L and the stress relaxation modulus $G(t)$ by the negative creep compliance without NEWTONian flow $-[J(t) - t/\eta]$.

Next we consider the calculation of the spectrum H from dynamic measurements, which comes down to an inversion of equation (I, 88). One can proceed in the same way as above so we restrict ourselves to giving the results[3]. From the storage modulus a first approximation is given by the logarithmic derivative[2].

$$H(\ln 1/\omega) \simeq H_1'(\ln 1/\omega) = dG_1(\omega)/d\ln\omega . \tag{I,95}$$

A higher approximation turns out to be of the third order and runs

$$H_3'(\ln 1/\omega) = dG_1/d\ln\omega - \frac{1}{4} d^3 G_1/d\ln\omega^3 . \tag{I,96}$$

Similarly from the loss modulus an approximation of zeroth order is[4] given by

$$H_0''(\ln 1/\omega) = \frac{2}{\pi} G_2(\omega) \tag{I,97}$$

[1] LEADERMAN, H.: Rheology on Polyisobutylene: Rep. natur. Bur. Stand., Washington D.C., March 1953. Proc. 2nd Int. Congr. Rheology, Oxford 1953, page 203. — H. LEADERMAN, R. G. SMITH u. R. W. JONES: J. Polymer Sci. **14**, 47 (1954).

[2] SCHWARZL, F.: loc. cit. page 44.

[3] See F. SCHWARZL: Proc. 2nd Int. Congr. Rheol., Oxford, England 1953, page 197. — F. SCHWARZL u. A. J. STAVERMAN: Appl. Sci. Res. **A 4**, 127 (1953). — H. LEADERMAN: N. B. S. Rep. 2672, Washington (July 1953).

[4] IVEY, D. G., B. A. MROWCA u. E. GUTH: J. appl. Physics **20**, 486 (1949). — A. W. NOLLE: J. Polymer Sci. **5**, 1 (1950). — D. J. FERRY, W. M. SAWYER, G. V. BROWNING u. A. H. GROTH: J. appl. Physics **21**, 513 (1950). — R. D. ANDREWS: Ind. Engng. Chem. **44**, 707 (1952).

and a higher approximation is of the form

$$H_2''(\ln 1/\omega) = \frac{2}{\pi}\left[G_2(\omega) - d^2 G_2/d\ln\omega^2\right]. \qquad (\text{I},98)$$

Here we indicate an approximation derived from the storage modulus by one dash and an approximation derived from the loss modulus by two dashes. The lower index gives again the order of the approximation.

A quantitative criterion for the closeness of an approximation method can be found in the following way. The simplest possible spectrum consists of one infinitely sharp peak, one "line". If this line is situated at time τ, the spectrum can be written as a delta-function

$$H(\ln t) = \tau\,\delta(t-\tau) = \delta\,[\ln t/\tau]\,. \qquad (\text{I},99)$$

This is the spectrum of a material with a simple exponential relaxation function and of a model consisting of one Maxwell unit.

We assume a spectrum consisting of two such sharp lines, the one at the time $t=\tau$, the other at the time $t=\alpha\tau$, having a distance $\log\alpha$ in logarithmic time scale. It will be clear that if the distance $\log\alpha$ in the original spectrum is very small, the approximate spectrum calculated with one of the foregoing methods will not show two separate peaks but one broad maximum. The minimum distance between the peaks in the original spectrum at which in the approximate spectrum the separation of the peaks can be observed, is a good measure of the resolving power of the approximation method. The smaller α, the higher this resolving power.

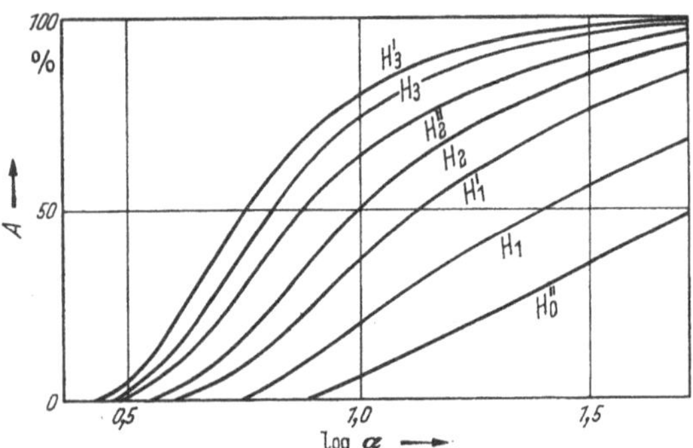

Fig. I, 22. The resolving power $\log\alpha$ as a function of the relative experimental error A for different approximations [eq. I, 92–I, 98].

The resolving power of a certain approximation method depends of course on the accuracy with which the experimental function, from which the calculation starts, can be obtained. Fig. I,22 shows the resolving power as function of the relative experimental error A (measured in %).

Fig. I,22 leads to interesting conclusions. First, if we were able to determine experimentally all derivatives with the same relative accuracy, the approximations would increase in resolving power in the order

$$H_0'' < H_1 < H_1' < H_2 < H_2'' < H_3 < H_3'.$$

Secondly if we should desire more and more knowledge about the details of the spectrum, we should increase the relative accuracy of our experiments more and more. But even if we could measure damping with infinite accuracy, we should not be able to reach a resolving limit smaller than $\log\alpha = 0{,}87$. For each approximation there is a theoretical limit to the resolving power, attained when the accuracy of the experiment is absolute.

The quest of higher approximation methods is not merely a matter of theoretical interest. It will be very difficult indeed to measure relaxation functions with such high accuracy that derivatives higher than the first could be obtained from them with any accuracy at all. But it is cei tainly possible to develop subtle experimental methods directed towards measuring higher derivatives immediately. For this purpose recovery measurements after one or two loading cycles appear particularly suitable. So the possibility of obtaining spectra with a much higher accuracy has now become a matter of experimental skill.

A slightly different system of approximations has been proposed by FERRY and coworkers[1]. They assume as first order approximations from the storage and loss modulus

$$H(\ln 1/\omega) \simeq H_1'(\ln 1/\omega) = G_1\, d\ln G_1/d\ln\omega \qquad \text{(I, 100 a)}$$

$$H(\ln 1/\omega) \simeq H_1''(\ln 1/\omega) = G_2[1 - d\ln G_2/d\ln\omega] \qquad \text{(I, 100 b)}$$

The first of these equations is identical with the first order approximation (I,95), the second equation is in between formulae (I,97) and (I,98), as it is a first order approximation from the loss modulus.

As second order approximation FERRY proposes expressions like (I,100), but multiplied by a factor which itself depends on the slope of the first order approximations

$$H(\ln 1/\omega) \simeq A\, G_1\, d\ln G_1/d\ln\omega \qquad \text{(I, 101 a)}$$

$$H(\ln 1/\omega) \simeq B\, G_2[1 - d\ln G_2/d\ln\omega], \qquad \text{(I, 101 b)}$$

where A and B are defined by

$$A = (2-m)/2\,\Gamma\left(2-\frac{m}{2}\right)\Gamma\left(2+\frac{m}{2}\right)$$

$$B = (1+m)/2\,\Gamma\left(\frac{3}{2}-\frac{m}{2}\right)\Gamma\left(\frac{3}{2}+\frac{m}{2}\right).$$

m contains the slope of the first order approximations (I,100) in logarithmic frequency scale

$$m = d\ln H_1/d\ln\omega$$

and is therefore in fact a second order derivate of the experimental curve with respect to frequency.

A completely different method for the numerical determination of the spectrum from experimental functions has been proposed by ROESLER and PEARSON[2] who assume the loss modulus G_2 given as function of the logarithmic frequency

$$G_2 = G_2(x), \quad x = (\pi/l)\ln\omega.$$

The loss modulus is assumed to be measured in the whole logarithmic frequency range, where it differs considerably from zero and is assumed to vanish for $x < 0$ and $x > \pi$ (the latter condition can be realised by choosing the normalizing constant l large enough). The loss modulus is an odd function of logarithmic frequency

$$G_2(-x) = -G_2(x)$$

and is defined outside the range $0 < x < \pi$ by periodic continuation

$$G_2(x + 2\pi) = G_2(x).$$

Under these rather restrictive conditions G_2 can be developed into a FOURIER series

$$G_2(x) = a_1 \sin x + a_2 \sin 2x + \cdots + a_n \sin n x + \cdots \qquad \text{(I, 102)}$$

---

[1] FERRY, J. D., E. R. FITZGERALD, L. D. GRANDINE JR. u. M. L. WILLIAMS: Ind. Engng. Chem. **44**, 703 (1952). – J. D. FERRY u. M. L. WILLIAMS: J. Colloid Sci. **7**, 347 (1952). – M. L. WILLIAMS u. J. D. FERRY: J. Polymer Sci. **11**, 169 (1953).

[2] ROESLER, F. C. u. J. R. A. PEARSON, Proc. physic. Soc. **B 67**, 338 (1954).

and ROESLER and PEARSON have shown that the relaxation spectrum can be written

$$H(x) = b_1 \sin x + b_2 \sin 2x + \cdots + b_n \sin nx + \cdots \quad (I,103)$$

where the coefficients b_k can be calculated from the coefficients a_k

$$b_k = a_k \cdot \frac{2}{\pi} \cosh(k\pi^2/2l). \quad (I,104)$$

By calculating the series (I,102) and (I,103) up to terms with $n = 1, 2, 3$ and so on, a number of approximations to the relaxation spectrum of increasing accuracy is obtained. This system closely resembles the approximations treated earlier. In one system the resolution of the method is limited by the accuracy of higher order frequency derivatives, in the other system by the accuracy of higher order FOURIER coefficients.

Literature to § 4.

Books:

ALFREY JR, T.: Mechanical behaviour of high polymers. High Polymers VI, Interscience Publ. New York 1948.

Articles:

ANDREWS, R. D.: Ind. Engng. Chem. **44**, 707 (1952).

FERRY, J. D. and M. L. WILLIAMS: J. Colloid Sci. **7**, 347 (1952).

FERRY, J. D., E. R. FITZGERALD, L. D. GRANDINE and M. L. WILLIAMS: Ind. Engng. Chem. **44**, 703 (1952).

GROSS, B.: J. appl. Physics **18**, 212 (1947); **19**, 257 (1948); Kolloid-Z. **131**, 161 (1953).

HAAR, D. TER: Physica **16**, 719, 738, 839 (1950).

HOLZMÜLLER, W. and E. JENCKEL: Z. physik. Chem. **A 186**, 359 (1940).

JENCKEL, E.: Z. Elektrochem. **45**, 202 (1939); Kolloid-Z. **134**, 47 (1953).

KEGEL, G.: Kolloid-Z. **135**, 125 (1954).

KIRBY, P. L.: Trans. Soc. Glass Techn. **37**, 7 (1953).

KUHN, W.: Makromol. Chemie **6**, 224 (1951).

LEADERMAN, H. and R. S. MARVIN: J. appl. Physics **24**, 812 (1953).

MEWES, E.: Kolloid-Z. **131**, 84 (1953).

MÜLLER, F. H.: Kolloid-Z. **134**, 77 (1953); Kolloid-Z. **122**, 109 (1951).

SCHWARZL, F.: Physica **17**, 830, 923 (1951); Proc. 2nd Int. Congr. Rheology, Oxford 1953, page 197.

SCHWARZL, F. and A. J. STAVERMAN: Physica **18**, 791 (1952); Appl. Sci. Res. **A 4**, 127 (1953).

STAMBAUGH, R. B.: Ind. Engng. Chem. **44**, 1590 (1952).

UEBERREITER, K.: Angew. Chem. **65**, 121 (1953).

WILLIAMS, M. L. and J. D. FERRY: J. Polymer Sci. **11**, 169 (1953).

§ 5. Temperature dependence and thermodynamics of viscoelastic behaviour.

a) The molecular processes.

There are several reasons why the influence of temperature on viscoelastic behaviour is of great interest. The first is of a purely practical nature. Viscoelastic substances like rubbers, fibres and plastics are used at many different temperatures. Rubber in car tyres is used at temperatures down tó $-20°C$ and up to $+60°C$ or more. Fibres are subjected to high temperatures in cleaning and pressing operations, plastics and coatings meet with a great variety of temperatures in practical use. So a complete specification of the properties of these materials cannot be restricted to the properties at room temperature but must include higher and lower temperatures.

Another reason for studying the temperature dependence of viscoelastic behaviour is the difficulty of covering experimentally a sufficient part of the time scale at one single temperature. This generally involves a great number of experimental techniques, many of them of an intricate nature, so it is customary to shift processes of interest into the most convenient time range simply by changing the temperature. This, of course, implies the rather crude assumption that by changing the temperature only the rate of the processes is changed and not their sequence and also not the structure of the substance as a whole. We will see in the following to what extent this assumption is justified.

The most important reason for studying the influence of temperature is due to the thermodynamic information, it gives about the molecular processes occurring on deformation. This information is not only interesting from an academic point of view but also from the point of view of the chemist who aims at improving particular properties of a particular material.

As we shall see thermodynamics alone cannot completely elucidate the molecular mechanism behind macroscopic viscoelastic processes. In order to reach a complete understanding of this mechanism one must necessarily do many experiments with subtle variations in chemical structure. However a thermodynamical study gives important information about the molecular processes. It reveals unambiguously, in an isothermal deformation, what part of the stored elastic energy is increased internal energy and what part is decreased entropy. Moreover, this division between energy and entropy can be studied more or less independently for various processes occurring in different ranges of the time scale, provided these ranges are sufficiently separated.

The signifiance of this information should not be underrated. If the internal energy appears to dominate the elastic response of a material we have to look for causes of increased potential energy in deformation. These may consist of distortion of valence angles and of change of interatomic distances. If, however, the decrease of entropy is dominant we have to look for a decrease of the number of possible configurations for the molecules in the material (establishment of order). In metals and crystalline bodies it is difficult to think of mechanisms which would involve an appreciable change of the number of configurations but in polymers there exists the possibility of coiling of macromolecules which implies a large number of configurations all with about the same potential energy.

Seen in retrospect it is easy to understand that already in the nineteenth century[1] the discovery of decrease of entropy as a source of elastic

[1] Gough: Mem. lit. Phil. Soc. Manchester **1**, 288 (1805). – J. P. Joule: Phil. Trans. **149**, 91 (1859). – For more accurate experiments see: H. M. James u. E. Guth: J. chem. Physics **11**, 455 (1943). – S. L. Dart, R. L. Anthony u. E. Guth: Ind. Engng. Chem. **34**, 1340 (1942).

forces has been found for rubbers. Rubbers are extreme cases in that by far the greater part of their elastic response is due to a change of entropy. At the other end of the scale materials like steel and glass are examples of materials which store potential energy on deformation without much change of entropy. Generally the instantaneous elastic response of polymers — as distinct from the rubberlike elasticity — is found to be mainly due to storage of potential energy.

However, as more detailed information on the viscoelastic behaviour of substances of a subtly varied chemical structure becomes available more intermediate cases will be found. There the thermodynamic study answers the important question whether a particular chemical substitution affects the viscoelastic properties mainly by changing the interatomic forces (potential energy dominates) or by changing the flexibility of the molecules (entropy dominates).

Besides elastic response of materials we have to take viscous response into account. We have discussed already that irreversible flow may be responsible for a large part of the deformation particularly at high temperatures and long times. Thermodynamically flow implies the appearance of a large number of states with different configurations all of equal free energy. It is easily understood that these different states do not differ in average interatomic distances nor in average coiling of the macromolecules but only in the relative positions of the centres (for instance centres of gravity) of these molecules. It is also understood that this state of affairs will occur only in non-crosslinked polymers where no molecular network extends through the entire material.

Summarising, we see three mechanisms by which a material may respond to deformation: increase of potential energy, decrease of number of configurations (decrease of entropy — establishment of order) and flow. These three mechanisms conform roughly to the classification of the deformation behaviour into three types, as introduced in formula (I, 32).

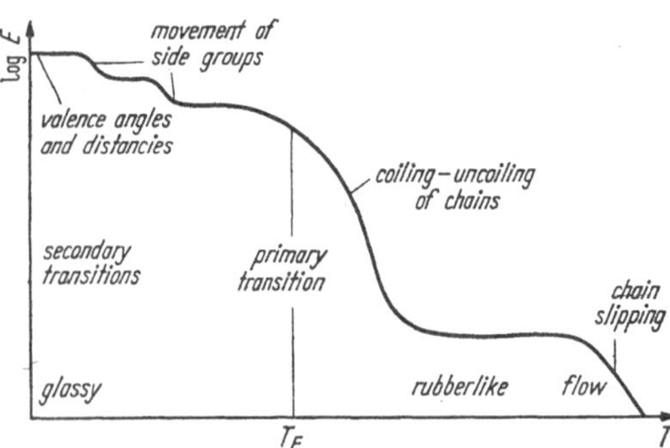

Fig. I, 23. The course of modulus with temperature for an amorphous uncrosslinked polymer, and the molecular movements.

A short summary of the molecular movements underlying the deformation behaviour of high polymers may be given with reference to fig. I, 23. There the modulus of an amorphous uncrosslinked polymer is plotted schematically as a function of temperature.

The modulus may show a large number of transitions, each of them being connected with a certain molecular movement: At temperatures below the corresponding *'transition temperature'* the molecular process in question is frozen in. In the vicinity of the transition temperature the molecular movement begins to contribute to the deformation mechanism and lowers the resistance against deformation, i.e. the modulus. At temperatures higher than the transition temperature the molecular process participates completely in the deformation process. Similarly each transition in logarithmic time or frequency can be attributed to a molecular process, increasing temperatures correspond to increasing time or decreasing frequency. In this way one may build up a *mechanical spetroscopy* by studying molecular processes with the aid of the transitions, which may occur either as a lowering of the modulus or as maxima in the damping and the spectra.

At very low temperatures the only deformation occurring is HOOKEan elastic deformation, which is time independent and therefore *mechanically reversible* (deformation recovers completely after removal of the force) and *thermodynamically reversible* (no energy is dissipated during a stress-strain cycle)[1]. HOOKEan deformation is attributed to the slight change in valence angles and valence distances, which takes place against the high binding forces leading to a high modulus [$10^{11} - 10^{12}$ dynes/cm^2].

The primary and secondary transitions occurring in fig. I,23 are examples of linear viscoelastic deformation behaviour [as expressed in the second term of formula (I,32), § 3 section b], which is time dependent, reaching its equilibrium value after a certain time and recovering after a time of the same magnitude. It is mechanically reversible, but owing to its time dependence, thermodynamically irreversible. One part of the deformation energy is stored as free energy — this part takes care of the mechanical reversibility — another part of the deformation energy is dissipated as heat.

Secondary transitions which may occur in the hard region (more accurately below the transition to the rubbery state) have recently[2] been found. They lead to a relative decrease in modulus and to damping maxima, which are much smaller than the corresponding quantities in the primary transition. Up to the primary transition the modulus remains larger than 10^{10}dyn/cm^2. The secondary transitions are attributed to the exitation of the movements of side groups of the chain. The free rotation of the $C-C$ bonds is still strongly hindered and the chain as a whole is stiff.

The *primary transition* — characterised by the "*brittle point*" or the "*softening temperature*" —leads from the hard region (moduli larger than 10^{10} d/cm^2) to the rubber-elastic region (moduli $10^6 - 10^8$ dynes/cm)^2. It is caused by the exitation of the free rotation of the primary bonds of the long-chain molecules, by which the coiling-uncoiling process of the molecules becomes possible (in German: freie *mikrobrownsche* Bewegung).

---

[1] There is a fundamental difference between thermodynamical and mechanical reversibility. Thermodynamical reversibility implies mechanical reversibility, but the opposite is not true. Each time dependent process, which may be mechanically reversible, is accompanied by energy dissipation and is therefore thermodynamically irreversible.

[2] SCHMIEDER, K. u. K. WOLF: Kolloid-Z. **127**, 65 (1952); **134**, 149 (1953). — J. HEYBOER, P. DEKKING u. A. J. STAVERMAN: Proc. 2nd. Int. Congr. Rheology, Oxford 1953, page 123. — K. DEUTSCH, E. A. W. HOFF u. W. REDDISH: J. Polymer Sci. **13**, 565 (1954).

This coiling-uncoiling process is the elementary mechanism underlying highly-elastic deformation in the rubbery region. In the unstressed state the long-chain molecules are not extended but are coiled as this configuration is the most probable. If a stress is applied to the material the most probable state will be different, with the molecules more uncoiled and oriented. There occurs orientation in the direction of the principal stresses and the uncoiling of the molecules results in a large deformation, orders of magnitude larger than those produced by the HOOKEan mechanism or by the movement of sidegroups. Though part of the work of deformation is dissipated, another part of this work is stored as configurational free energy (the stretched molecule possesses a larger free energy than the coiled one). This configurational free energy results in mechanical reversibility of the highly-elastic deformation.

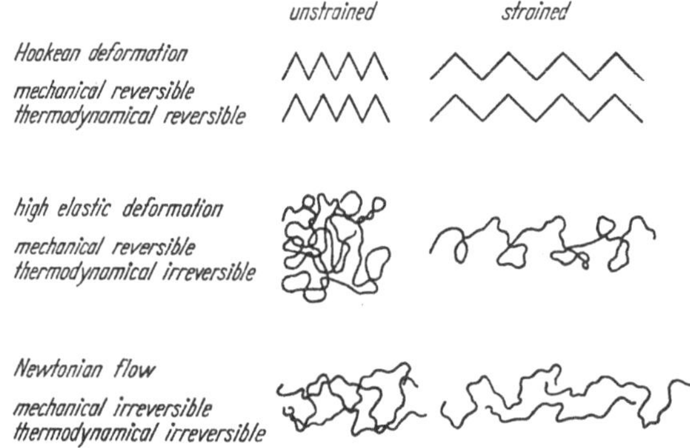

Fig. I, 24. Hookean deformation, highly-elastic deformation and Newtonian flow: molecular processes.

At still higher temperatures NEWTONian flow, which is totally irreversible, both mechanically and thermodynamically, takes place. The whole deformation energy is dissipated as heat; there is no configurational free energy left after deformation, and no recovery. The deformation mechanism responsible for flow is the slipping of chains relative to each other. This mechanism can only take place if the chain molecules are not interconnected by strong bonds. Network polymers are therefore not expected to show NEWTONian flow phenomena, which assumption is confirmed by experiment. Fig. I, 24 illustrates the three main types of deformation mechanism.

Crystalline polymers show a modulus temperature dependence, which differs in some respects from the behaviour of amorphous polymers (fig. I, 25). At low temperatures we have again the glassy region with HOOKEan deformation behaviour followed by an even larger number of secondary transitions. At the primary transition (second order transition temperature T_E), however, the decrease in modulus is much smaller than it is for amorphous polymers. In the "rubbery region" between the transition temperature T_E and the melting temperature of crystallites T_s, the modulus has values of about 10^9 d/cm^2 and the crystalline polymer behaves as a tough rather than a rubberlike material. The molecular reason for this "undeveloped" high elasticity are the strong intermolecular forces, which do not allow the full development of the coiling-uncoiling process. The highly-elastic deformation mechanism can take place only in the amorphous regions which connect the different crystallites. Therefore these polymers do not show the typical rubberlike behaviour.

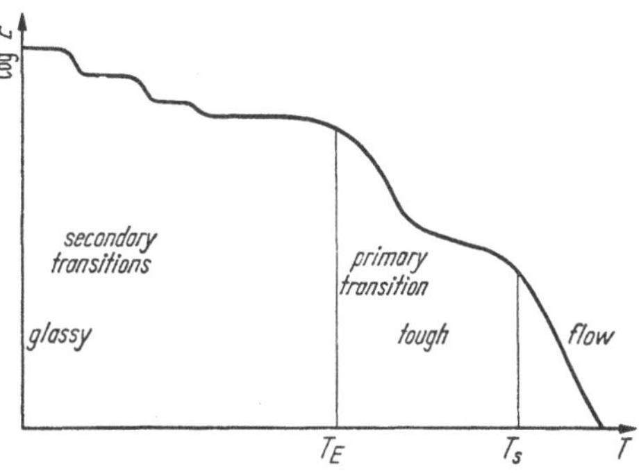

Fig. I, 25. The course of modulus with temperature for crystalline polymers.

At the melting temperature of crystallites T_s, where the crystallites melt and the strong intermolecular forces cease to exist, molecular slip occurs together with rubberlike elasticity and the polymer changes rather abruptly into the flow region.

Molecular theories of the different deformation mechanism are surveyed in table I, 5 compare also § 82.

Table I,5. *Molecular theories of deformation mechanism.*

Type of deformation	Molecular movement	Reference
brittle deformation	valence angle and distances	—
secondary transitions	sidegroup movements	—
primary transition	exitation of free rotation	volume 4 chapter 1 § 8
rubberlike elasticity	coiling-uncoiling of long chains	volume 4 chapter V
flow	chain slipping	volume 2 chapter 11

In the following we consider the temperature dependence and thermodynamics of viscoelastic behaviour. To this end we distinguish between the equilibrium rubberlike region and the "quasi-equilibrium" brittle region on the one hand and the non equilibrium transition region on the other hand (compare fig. I,27).

Section b) deals with the temperature dependence and thermo-dynamics of the equilibrium and quasi-equilibrium states, where the classical thermodynamics may be applied, section c) treats the temperature dependence of the viscoelastic transition and in section d) viscoelastic behaviour in the transition region is discussed by means of non equilibrium thermodynamics.

b) Rubberlike elasticity[1].

When isothermal change of a parameter Y, of a system is accompanied by change of free energy, it is usual to denote the increase of free energy, F, per unit change of Y as a "force" X.

$$X = \left(\frac{\partial F}{\partial Y}\right)_T. \tag{I,105}$$

[1] See also: This Vol., chapter V A and L.R.G. TRELOAR: The Physics of rubber elasticity, Oxford 1949.

Generally the change of free energy may be due to change of the internal energy U or of the entropy S of the system

$$F = U - T S \tag{I,106}$$

and

$$X = \left(\frac{\partial U}{\partial Y}\right)_T - T\left(\frac{\partial S}{\partial Y}\right)_T. \tag{I,107}$$

There are two common methods to find out the magnitude of the two terms in the right hand side of (I, 107). One is to measure X as a function of temperature at constant Y, the other is to compare the heat developed or absorbed on isothermal change of X with the work done.

In order to understand the first method we remember that

$$\left(\frac{\partial F}{\partial T}\right)_Y = -S \quad \text{or with (I, 105)}$$

$$\left(\frac{\partial X}{\partial T}\right)_Y = -\left(\frac{\partial S}{\partial Y}\right)_T. \tag{I,108}$$

The increase of X with temperature is $1/T$ times the second term of the right hand side of (I,107).

When X is exactly proportional to T, which means that the graph of X versus T at constant Y extrapolates towards $X = 0$ at $T = -273°$C, the first term of (I,107) is exactly zero. This has been found to occur in dilute gasses (with Y for the volume, X for the negative pressure) and with good approximation in rubbers (with Y for the strain and X for the stress). In these cases the entire free energy increase on isothermal change of Y is due to change of entropy.

When, on the other hand, X at constant Y is independent of T, the second term of (I,107) is zero and the increase of free energy is due to change of internal energy. This has been found to be approximately true for steel, glass and the instantaneous elasticity of plastics.

The second method for investigating the nature of the free energy change is less convenient because it involves calorimetric measurement. However, it is sensitive to small deviations from pure energy effects because in an isothermal change involving energy change only, no heat is developed or absorbed at all.

$$dQ = d\,U - X\,d\,Y \tag{I,109}$$

is the amount of heat developed. This is exactly zero in an isothermal process if

$$X = \left(\frac{\partial U}{\partial Y}\right)_T$$

or if the second term in (I,107) vanishes.

An experimental investigation of the shear stress as a function of temperature at constant strain for rubber[1] shows that the stresses are proportional to absolute temperature even at the smallest shear strains.

[1] MEYER, K. H. u. A. J. A. VAN DER WIJK: Helv. chim. Acta **29**, 1842 (1946).

Generally the determination of the stress-temperature relation is performed in extension where an additional complication arises, the thermoelastic inversion phenomenon: In the highly-elastic state the tensile stress increases proportionally to temperature only at deformations larger than about 10% (thermoelastic inversion point). At smaller elongations the stresses decrease with temperature. This phenomenon shows that in the highly-elastic state in the beginning of the elongation the internal energy also contributes to the tensile forces and that it is only beyond 10% elongation that the decrease in entropy becomes the dominating part in tensile behaviour. This is connected with small volume changes occurring in extension.

The increase of the highly-elastic modulus in extension with temperature has been studied for various natural and synthetic rubbers[1] and special attention has been drawn to the thermoelastic inversion phenomen. We show as an example the stress-temperature relations of a natural rubber according to ANTHONY, CASTON and GUTH[1] for various elongations (fig. I, 26). The change in slope of the stress temperature curves (i.e. the thermoelastic inversion) between 6% and 13% elongation is clearly shown.

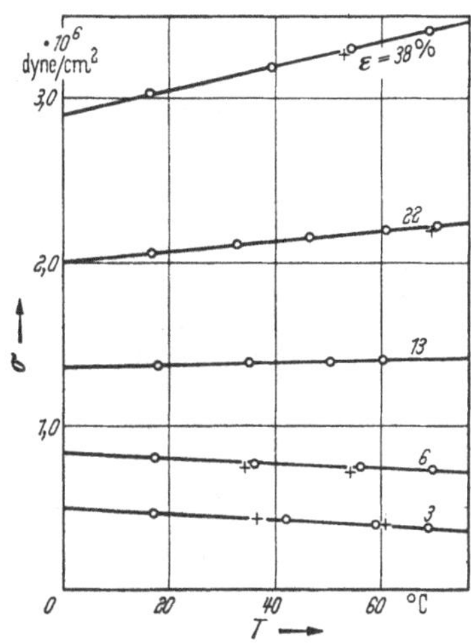

Fig. I, 26. Stress temperature curves for a natural vulcanised rubber (ANTHONY, CASTON and GUTH).

It has further been shown[2] that the thermoelastic inversion does not occur if the extension is performed at constant volume rather than at constant pressure. This proves that the *thermoelastic inversion* is caused by volume changes and explains at the same time why thermoelastic inversion does not occur in shear.

Of course we would expect the equilibrium modulus to be proportional to absolute temperature not only for rubbers but also for thermoplastic materials in the highly-elastic region. This has been proved by investigations on cross linked Polystyrene[3] at temperatures between 120 and 220° C.

Concluding this section we should point out, that, strictly speaking, classical thermodynamics may be applied only to true equilibrium states. This condition is apparently fulfilled in the extension of rubberlike materials after sufficient time and also for steel after deformation but not for the instantaneous deformation of plastics. However, at low enough temperatures the viscoelastic changes occurring after the instantaneous deformation are so slow that the state of the material can be treated as an equilibrium in the thermodynamic sense. However, this simplification is not allowed in the transition region where classical thermodynamics fails completely.

---

[1] MEYER, K. H. u. C. FERRI: Helv. chim. Acta **18**, 570 (1935). – V. HAUK u. W. NEUMANN: Z. physik. Chem. **A 182**, 285 (1938). – K. H. MEYER u. H. MARK: Hochpolymere Chemie, Bd. 2, Akad. Verlagsges. Becker & Erler, Leipzig 1940. – R. L. ANTHONY, R. H. CASTON u. E. GUTH: J. physik. Chem. **46**, 826 (1942). – L. E. PETERSON, R. L. ANTHONY u. E. GUTH: Ind. Engng. Chem. **34**, 1349 (1942). – E. JENCKEL u. J. FÜHLES: Makromol. Chem. **1**, 203 (1943). – F. L. ROTH u. L. A. WOOD: J. appl. Physics **15**, 781, 749 (1944). – R. S. WITTE u. R. L. ANTHONY: Rubber Chem. Technol. **25**, 468 (1952).

[2] GEE, G.: Trans. Faraday Soc. **42**, 585 (1946).

[3] KLEIN, E. u. E. JENCKEL: Z. Naturforsch. **7**, 800 (1952).

We are now able to give a more accurate picture of the shear modulus-temperature curve than has been possible in § 3 (fig. I,27). At low temperatures we have the glassy region, the quasi equilibrium state in which the modulus decreases slightly with temperature[1]. At higher temperature the non-equilibrium transition region follows with a steep decrease in modulus. At still higher temperatures the rubbery state appears with a low modulus which increases proportionally to absolute temperature.

c) Temperature dependence of viscoelastic deformation in the transition region.

We turn now to the consideration of the temperature dependence of viscoelastic behaviour in the transition region, where the thermodynamic considerations of section b no longer apply.

Consider the stress relaxation modulus $G(t)$ in the transition region on a logarithmic time scale at a constant temperature T_2 (fig. I,28). If

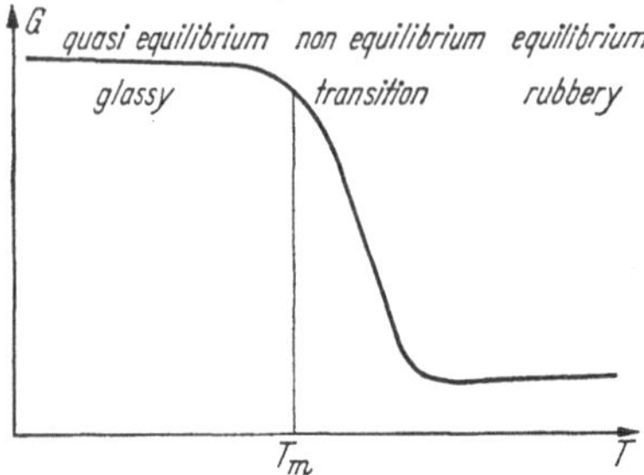

Fig. I, 27. The shear modulus as a function of temperature, for a vulcanised rubber.

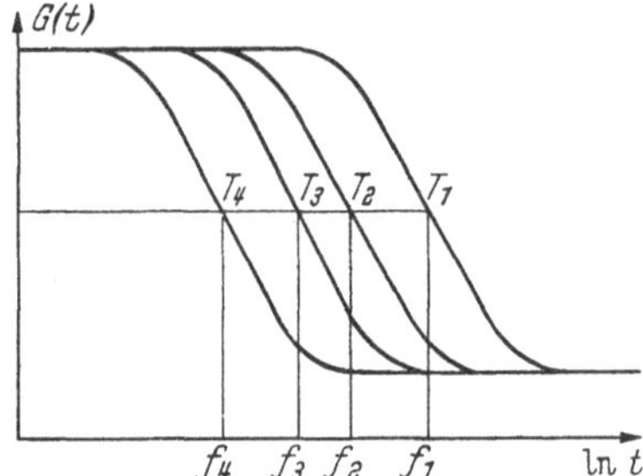

Fig. I, 28. Stress relaxation modulus as a function of time for different temperatures.

the same curve is measured at a higher temperature T_3 we get a similar picture shifted to shorter times. If the modulus is observed at lower temperature T_1, the curve is shifted towards longer times.

Figures of this kind have been obtained by numerous authors, of whom we only cite a few examples[2]. Extensive examples of *temperature reductions* will be treated in Chapter VI. The figure gives the impression

[1] Secondary transitions are not considered here for simplicity.

[2] Kobeko, P., E. Kuvshinsky u. G. Guvevitch: J. techn. Physics (USSR) **4**, 622 (1937). – G. Guvevitch u. P. Kobeko: J. techn. Physics (USSR) **9**, 1267 (1939); Rubber Chem. Technol. **13**, 904 (1940), Creep of natural rubber in torsion. – E. Jenckel: Z. Elektrochem. **45**, 202 (1939), General considerations over the time dependence of the softening temperature. – A. P. Aleksandrov u. Y. S. Lazurkin: J. techn. Physics (USSR) **9**, 1249, 1261, (1939); Rubber Chem. Technol. **13**, 886 (1940), Dynamic measurements on natural rubber, Chloroprene, Polymethylmethacrylate. – J. R. Scott: Trans. Faraday Soc. **38**, 284 (1942), Torsion creep of ebonite. – F. H. Müller: Kolloid-Z. **114**, 2 (1949); Kunststoffe **39**, 215 (1949), Dynamic measurements on natural rubber, bitumina, Buna S., Polystyrene, Polyethylene, etc. – D. G. Ivey, B. A. Mrowca u. E. Guth: J. appl. Physics **20**, 486 (1949), Ultrasonic pulse propagation in Butylrubber and natural rubber. – A. W. Nolle: J. Polymer Sci. **5**, 1 (1950), Dynamic measurements on various rubbers. – K. Schmieder u. K. Wolf: Kolloid-Z. **127**, 65 (1952), Torsional vibrations on plasticized Polyvinylchloride, Polymethylmethacrylate.

that temperature and logarithmic time are equivalent parameters. One of the first to realise this equivalence between time and temperature was LEADERMAN[1], who also proposed to use this equivalence to extend the experimental time scale. Recently it has become common practice to change the temperature in order to extend the experimentally accessible time region.

If by a temperature change only the position of the modulus curve on a logarithmic time scale, but not its shape, is altered, we are confronted with a simple material, which has been called *thermorheologically simple*[2]. The stress relaxation modulus, which generally depends on time and temperature $G(\ln t, T)$, depends in this case only on the combined quantity x (the reduced time)

$$G(\ln t, T) = G(x), \quad x = \ln t - f(T). \tag{I,110}$$

The function $G(x)$ is called the *master curve in stress relaxation* and determines the shape common to all curves in fig. I,28, whilst the function $f(T)$ is a function of temperature only, fixing the position of the modulus curve on a logarithmic time scale.

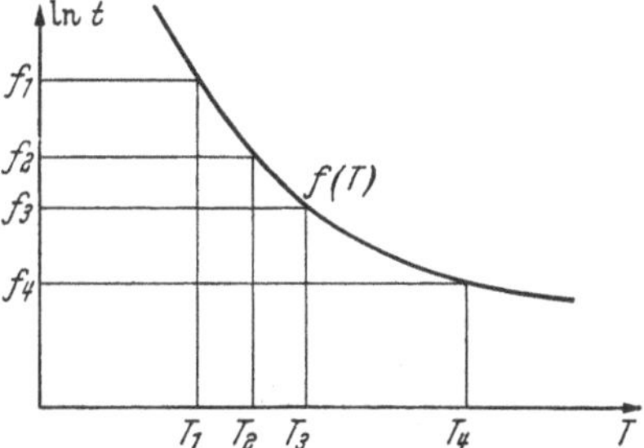

Fig. I, 29. Construction of the time-temperature relation (from fig. I, 28).

$f(T)$ is a decreasing function of temperature and can be constructed by plotting the relative shifts $f_1 - f_2$, $f_1 - f_3$, $f_1 - f_4$ versus the temperature differences $T_2 - T_1$, $T_3 - T_1$, $T_4 - T_1$ (fig. I, 29). $f(T)$ is of course indeterminate to the extent of an additive constant (the chosen value of x), but this is sufficient as only the time differences matter in practice.

If one characteristic quantity (e.g. the relaxation modulus) obeys the relations (I,110), then all the other quantities obey the same time-temperature relation[2]. The creep compliance is a function of reduced time x, whilst the dynamic moduli are functions of the reduced frequency $y = -x$, where

$$G_1(\ln \omega, T) = G_1(y), \quad y = \ln \omega + f(T). \tag{I,111}$$

If the material under consideration shows NEWTONian flow and obeys relations (I,110), then these relations can be derived from the temperature dependence of viscosity

$$f(T) = \ln \eta(T). \tag{I,112}$$

Instead of the reduced time or the reduced frequency, the following expressions can be used

$$x = \ln t/\eta, \quad y = \ln(\omega \eta). \tag{I,113}$$

A consequence of the validity of (I,113) is that creep compliances and stress relaxation moduli at different temperatures fall upon one single curve (the master curve), if they are plotted versus t/η and that loss and storage moduli at different temperatures fall upon one single curve if plotted against the reduced frequency $\omega \eta$.

---

[1] LEADERMAN, H.: Elastic and creep properties of filamentous materials, Textile Foundation, Washington, D.C., 1943, page 175. — In dielectric measurements the temperature reduction has been realised already by K. W. WAGNER in 1914 [Discussion remark by B. GROSS: Kolloid-Z. **134**, 197 (1953)].

[2] SCHWARZL, F. u. A. J. STAVERMAN: J. appl. Physics **23**, 838 (1952).

This reduction scheme was proposed originally by FERRY[1] and TOBOLSKY[2] in a slightly altered form which is theoretically somewhat more satisfactory. As the theory of rubber elasticity shows that the equilibrium modulus is proportional to absolute temperature, the quantity G/T rather than G itself should obey the *time temperature relation*. FERRY therefore assumes that

$$\frac{1}{\varrho T} G(\ln t, T)$$

is a function of reduced time x only (ϱ is the density of the polymer, depending slightly on temperature). The factors $1/T$ and $1/\varrho$ lead to small corrections and are introduced for theoretical reasons. So far it has not been possible to decide which assumption (G = function of x or G/T = function of x) gives the better fit to experimental data in the reduced curves[3].

FERRY reduces all curves to a standard temperature T_0 (the *master curve*). For the moduli in stress relaxation and dynamic measurements he puts

$$\left.\begin{aligned} \frac{\varrho_0 T_0}{\varrho T} G(a\cdot t, T) &= G_{T_0}(t) \\ \frac{\varrho_0 T_0}{\varrho T} G_1(\omega/a, T) &= G_{1,T_0}(\omega) \\ \frac{\varrho_0 T_0}{\varrho T} G_2(\omega/a, T) &= G_{2,T_0}(\omega) \end{aligned}\right\} \qquad \text{(I, 114)}$$

$G_{T_0}(t)$, $G_{1,T_0}(\omega)$ and $G_{2,T_0}(\omega)$ are the *master curves* of the stress relaxation function, the storage and the loss modulus respectivily. They are functions of time or frequency only and describe the viscoelastic properties of the material at a single reference temperature T_0. The effect of temperature change consists in multiplying all relaxation times by a common factor $a(T)$, which is a function of temperature only and which embodies the equivalence of time and temperature by

$$\ln a = f(T) - f(T_0). \qquad \text{(I, 115)}$$

The *reduced logarithmic time* or *frequency* can be expressed by

$$\left.\begin{aligned} x &= \ln t/a - f(T_0) \\ y &= \ln(a\omega) + f(T_0) \end{aligned}\right\} \qquad \text{(I, 115a)}$$

and the equations (I, 114) become identical with the statement that $\frac{1}{\varrho T} G(t, T)$ is a unique function of x and that $\frac{1}{T\varrho} G_1(\omega, T)$ is a unique function of y.

[1] FERRY, J. D.: J. Amer. chem. Soc. **72**, 3746 (1950). – J. D. FERRY, E. R. FITZGERALD, L. D. GRANDINE u. M. L. WILLIAMS: Ind. Engng. Chem. **44**, 703 (1952).

[2] TOBOLSKY, A. V. u. R. D. ANDREWS: J. chem. Physics **13**, 3 (1945). – R.D. ANDREWS, N. HOFMAN-BANG u. A. V. TOBOLSKY: J. Polymer Sci. **3**, 669 (1948).

[3] Recently, two investigators advocated the factor $1/T$ for the treatment of their data: W. PHILIPPOFF: J. appl. Physics **25**, 1102 (1954). – I. L. HOPKINS: J. appl. Physics **24**, 1300 (1953).

If compliances are to be reduced, we have to take the formulas

$$\left.\begin{aligned} \frac{\varrho T}{\varrho_0 T_0} J(a\cdot t, T) &= J_{T_0}(t) \\ \frac{\varrho T}{\varrho_0 T_0} J_1(\omega/a, T) &= J_{1,T_0}(\omega) \\ \frac{\varrho T}{\varrho_0 T_0} J_2(\omega/a, T) &= J_{2,T_0}(\omega) \end{aligned}\right\} \qquad \text{(I, 116)}$$

which means that $\varrho T\, J(t, T)$ and $\varrho T\, J_1(\omega, T)$ are unique functions of the reduced time or frequency. The tangent of loss angle does not depend explicitly on temperature after reduction

$$\operatorname{tg}\delta(\omega/a, T) = \operatorname{tg}\delta(\omega, T_0) \qquad \text{(I, 117)}$$

and for the logarithmic relaxation spectra and retardation spectra equations similar to (1,114) and (I,116) hold.

If the polymer shows NEWTONian flow, then the value of the "shift function" $a(T)$ can be obtained immediately from the dependence of the viscosity η on temperature

$$a = \frac{\varrho_0 T_0}{\varrho T} \frac{\eta}{\eta_0} \qquad \text{(I, 118)}$$

(η = viscosity at temperature T, η_0 = viscosity at temperature T_0).

From equation (I,118) together with equations (I,114) or (I,116) we obtain a further method for determining the master curves. If the moduli $(T_0\varrho_0/T\varrho)\cdot G$ or $(T_0\varrho_0/T\varrho)\cdot G_1$ are plotted against the reduced frequency $\omega\,\eta\, T_0\varrho_0/\eta_0 T\varrho$ all points must fall on a single line, the master curve. Similarly if $(T\varrho/T_0\varrho_0)\cdot J$ or $(T\varrho/T_0\varrho_0)\cdot J_1$ are plotted against $\omega\,\eta T_0\varrho_0/\eta_0 T\varrho$, we get the master curves of the compliances.

TOBOLSKY has proposed[1] that the same relationships should be valid for moduli and compliances in linear extension.

The time-temperature equivalence has been found to exist at least approximately in a large number of cases. Some examples may be cited here:

CONANT and coworkers[2] have determined the time temperature shift in creep measurements on various rubbers. They shifted the creep curves obtained at different temperatures to fit onto one single master curve and obtained for the shift function the equation

$$\log a = c/(T - b)$$

where T is the absolute temperature and the constants c and b are given in the following table

	c	b	Temperature region
Hevea	1780	122	− 60 to 31°C
GRS	3121	100	− 50 30
Neoprene GN	480	193	− 40 40
Butylrubber	5040	33	− 60 40
Butaprene	720	183	− 30 30

[1] TOBOLSKY, A. V.: loc. cit. page 58.
[2] CONANT, F. S., G. L. HALL u. W. L. LYONS: J. appl. Physics **21**, 499 (1950).

TOBOLSKY and coworkers[1,2] obtained the shift function for Hevea, GRS and polyisobutylene from stress relaxation data in linear tension. A careful test of the time temperature shift was made for polyisobutylene and showed its approximate but not accurate validity in the temperature region between 30 and 100° C. Besides the time factor $a(T)$, a small correction in the form of a stress factor had to be used. $a(T)$ has been found to be independent of molecular weight. Further, the validity of equation (I,118) has been checked[3] by calculating flow viscosities from stress relaxation data. The predicted viscosities agreed quite well with the measured ones.

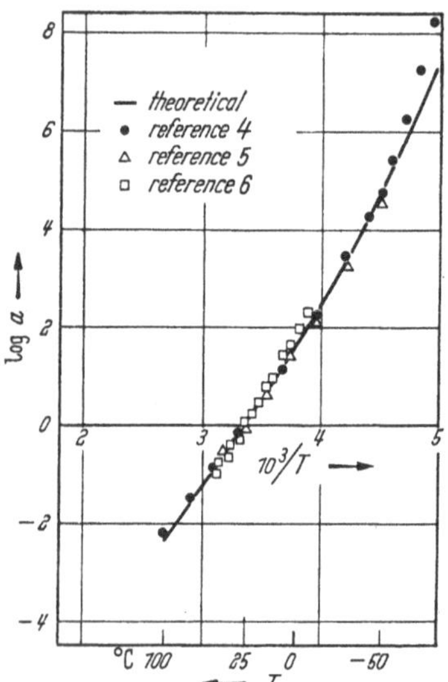

Fig. I, 30. Time-temperature relation for polyisobutylene.

The experimental equality of the shift function in creep and relaxation was established for GRS[3] and a full account of the temperature shift from stress relaxation data for polyisobutylene has been given[4].

DAHLQUIST and HATFIELD[5] made tensile creep measurements on polyisobutylene and GRS and determined the time-temperature shift. They found in both cases activation energies of about 20 kcal/mole.

The function $a(T)$ has also been obtained from dynamic experiments on high molecular weight polyisobutylene[6].

The time temperature relation for polyisobutylene determined by various techniques is shown in fig. I, 30. From viscosity measurements by FOX and FLORY[7] a theoretical expression can be derived for the shift function

$$\log a = 5{,}6 \cdot 10^5 \left[\frac{1}{T^2} - \frac{1}{(298)^2}\right] + \log \frac{208}{T}.$$

As will be seen from fig. I, 30 excellent agreement exists between this theoretical relation and the shifts obtained by stress relaxation[4], by creep[5] or by dynamic measurements[6].

The time temperature relation for a low molecular weight polyisobutylene ($M_n = 800$) has been determined by viscosity measurements[8]. Between −30 and 63°C approximately constant activation energies of about 15,8 kcal/mole are found. Dynamic quantities could be reduced by plotting them against $\omega\eta$.

LEADERMAN and SMITH[9] have reported that creep curves for polyisobutylene at different temperatures fall upon one master curve, if plotted as a function of t/η. MARVIN and LEADERMAN[10] found that stress relaxation and dynamic measurements on polyisobutylene could be reduced to master curves by plotting them against t/η or $\omega\eta$.

---

[1] TOBOLSKY, A. V. u. R. D. ANDREWS: loc. cit. page 58.

[2] ANDREWS, R. D., N. HOFMANN-BANG u. A. V. TOBOLSKY: loc. cit. page 58.

[3] ANDREWS, R. D. u. A. V. TOBOLSKY: J. Polymer Sci. **7**, 221 (1951).

[4] TOBOLSKY, A. V. u. J. R. MCLOUGHLIN: J. Polymer Sci. **8**, 543 (1952).

[5] DAHLQUIST, C. A. u. M. R. HATFIELD: J. Colloid Sci. **7**, 253 (1952).

[6] MARVIN, R. S., E. R. FITZGERALD u. J. D. FERRY: J. appl. Physics **21**, 197 (1950). – J. D. FERRY, E. R. FITZGERALD, M. F. JOHNSON u. L. D. GRANDINE: J. appl. Physics **22**, 717 (1951).

[7] FOX, T. G. u. P. J. FLORY: J. Amer. chem. Soc. **70**, 2384 (1948).

[8] HARPER, R. C., H. MARKOVITZ u. T. W. DEWITT: J. Polymer Sci. **8**, 435 (1952).

[9] LEADERMAN, H. u. R. G. SMITH: Phys. Rev. **81**, 303 (1951). – H. LEADERMAN: Proc. 2nd Int. Congr. Rheology, Oxford 1953, page 203. — H. LEADERMAN, R. G. SMITH u. R. W. JONES: J. Polymer Sci. **14**, 47 (1954).

[10] MARVIN, R. S.: Proc. 2nd Int. Congr. Rheology, Oxford 1953, page 156. – H. LEADERMAN u. R. S. MARVIN: J. appl. Physics **24**, 812 (1953).

Having thus established that change of temperature and a parallel shift on the logarithmic time scale are at least approximately equivalent for a number of substances we will discuss the implications of this equivalence and also the restrictions to which it is subject.

For the engineer who wants to test his materials for a large range of temperatures and times the equivalence, even when it is approximate only, means a tremendous simplification. Instead of a three dimensional representation of deformation in $\ln t$, T space he can give an approximate description of the behaviour in two simple functions, the master curve (of deformation or stress, or one of the spectra versus the logarithm of time) and the shift function of temperature.

For the chemist the existence of time temperature equivalence means that all the molecular processes occurring on deformation are accelerated or retarded to the same extent by change of temperature. Putting this a little more quantitatively one may say that it means that the *activation energy* of all these molecular processes, including flow if present, is the same. For this activation energy can always be calculated from the increase of rate with increase of temperature from equations of the type

$$\Delta H_a = R \frac{d f(T)}{d(1/T)} = R \frac{d \ln a}{d(1/T)}. \qquad \text{(I, 119)}$$

For instance, the activation energy of polyisobutylene can be constructed from fig. I, 30 (as the slope of the shift function plotted versus $1/T$). One obtains activation energies which decrease from 40 kcal at $-70°$ C to 16 kcal at high temperatures[1].

Equality of the activation energy of all the rate processes occurring in deformation means that the interatomic forces involved in change of configuration are all of the same nature. Apparently such a situation can be expected if only non-polar forces are present. Polar forces and forces involved in crystallisation generally depend on temperature in a different manner from non-polar forces, so if these forces operate to an appreciable extent in the changes of atomic configurations occurring in viscoelastic processes one would not expect to find equivalence of time and temperature over a large range.

It is, therefore, understandable that equivalence has been found particularly in non-polar substances such as polyisobutylene and rubber and one must expect it to be absent in polymers with polar groups and crystalline regions.

An example of a polymer with a *secondary viscoelastic transition* is polymethylmethacrylate. Tobolsky and coworkers[2] have tried to apply their reduction scheme to this material. This results in a time temperature shift shown in fig. I, 31, which shows a quite irregular form. Moreover, accurate measurements on dynamic modulus and damping[3] showed that polymethylmethacrylate has a secondary transition

---

[1] Brown, G. M. u. A. V. Tobolsky: J. Polymer Sci. **6**, 165 (1951).

[2] McLoughlin, J. R. u. A. V. Tobolsky: J. Colloid Sci. **7**, 555 (1952). J. Bischoff, E. Catsiff u. A. V. Tobolsky: J. Amer. chem. Soc. **74**, 3378 (1952).

[3] Schmieder, K. u. K. Wolf: Kolloid-Z. **127**, 65 (1952). — J. Heyboer, P. Dekking u. A. J. Staverman: Proc. 2nd Int. Congr. Rheology, Oxford (Engl.), July 1953, page 123. — Further: J. Heyboer; private communications.

at a somewhat smaller frequency (larger time) than the primary transition and that these two transitions shift at different rates on change of temperature. Whilst the primary transition (softening) has an activation energy of the order of 100 kcal/mole, the activation energy for the secondary process is only 20 kcal. Therefore the time temperature equivalence does not exist in this material. This can be understood from the presence of polar groups in the ester linkages.

Finally it should be mentioned that FERRY and coworkers have reported similar time temperature relationships for polymer solutions[1]. Also in a polymer solution viscoelastic effects are present leading to a distribution of relaxation times, but here we have three independent variables instead of two, namely time, temperature and the concentration of solute polymer c. FERRY[2] has reduced the viscoelastic behaviour of polymer solutions to one standard temperature and to one standard concentration by means of an extended reduction scheme.

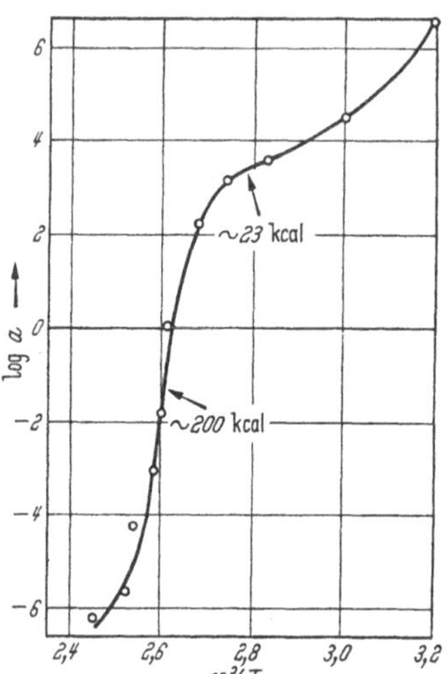

Fig. I, 31. The time-temperature shift of polymethylmethacrylate according to TOBOLSKY.

According to FERRY the following relations should be valid for a polymer solution: time — temperature relation (at constant concentration an increase of temperature shifts the dynamic modulus curves to shorter times on a logarithmic time scale).

Time — concentration relation (at constant temperature an increase of concentration shifts the curves to longer times on a logarithmic time scale).

It follows that an increase of temperature is equivalent to a decrease in concentration and vice versa. This is reasonable as the relaxation times in a polymer solution will be the shorter the higher the temperature and the more dilute the polymer.

FERRY was able to prove these general features by dynamic measurements on polymer solutions and was able to show that master curves could be obtained by superposition at different concentrations and temperatures[3].

FERRY[4] has further compared the dynamic properties of pure polyisobutylene with those of solutions in xylene. In taking the concentration of solute equal to $c = 100\%$ in the time-concentration relation, one should obtain the results for the bulk polymer. The obtained distribution functions for pure polymer and solutions could not be reduced to the same master curves, showing that the concentration-

[1] Compare chapter VI.

[2] FERRY, J. D.: J. Amer. chem. Soc. **72**, 3746 (1950).

[3] PHILIPPOFF, W.: Physik. Z. **35**, 900 (1934); Cellulose acetate in Dioxane. — J. D. FERRY: J. Amer. chem. Soc. **64**, 1323, 1330 (1942); Polystyrene in Xylene. — T. L. SMITH, J. D. FERRY u. F. W. SCHREMP: J. appl. Physics **20**, 144 (1949); Polyisobutylene in Xylene. — ASHWORTH, J. N. u. J. D. FERRY: J. Amer. chem. Soc. **71**, 622 (1949); Polyisobutylene in Xylene and n-Heptan. — F. T. ADLER, M. W. SAWYER u. J. D. FERRY: J. appl. Physics **20**, 1036 (1949); Polyvinylacetate in 1,2,3,-Trichlorpropane. — J. D. FERRY: J. Amer. chem. Soc. **72**, 3746 (1950); Polystyrene and Polyisobutylene in Xylene. — J. D. FERRY, W. M. SAWYER, G. V. BROWNING u. A. H. GROTH JR.: J. appl. Physics **21**, 513 (1950); Polyvinylacetate in 1,2,3,-Trichlorpropane. — W. M. SAWYER u. J. D. FERRY: J. Amer. chem. Soc. **72**, 5030 (1950); Polyvinylacetate in different solvents. — F. W. SCHREMP, J. D. FERRY u. W. W. EVANS: J. appl. Physics **22**, 711 (1951); Polyisobutylene and Polystyrene in Dekalin.

[4] FERRY, J. D., E. R. FITZGERALD, M. F. JOHNSON u. L. D. GRANDINE JR.: J. appl. Physics **22**, 717 (1951).

time relation possesses only restricted validity in a region of medium concentrations. An extrapolation to $c = 100\%$ leads to wrong results. This apparently means that the molecular processes occurring on deformation which are dominant in solution are different from those predominating in the pure polymer. Apparently there is a difference between the activation energy in the two cases.

d) Thermodynamics of linear viscoelastic behaviour.

It has been pointed out earlier in this chapter that a study of the thermodynamics of viscoelastic behaviour is an indispensable link in the chain of understanding of the mechanical properties of high polymers from the knowledge of their chemical constitution. First a thermodynamical study will reveal what parameters of state are necessary and sufficient to describe the state of a material involved in a viscoelastic process. The next step will be to discover how the value of these parameters can be measured quantitatively as a function of time from a macroscopic study of the behaviour. Next one should try to relate these parameters of state quantitatively to molecular configurations of the substance under consideration. This would afford a solution to the problem of the influence of change of chemical structure on the mechanical properties of a substance. Automatically a complete thermodynamic treatment will account for the temperature dependence of the mechanical properties which is so very important in practice.

A first attempt to treat viscoelastic behaviour thermodynamically has been made by BRIDGMAN[1]. However his treatment has not been developed sufficiently to allow quantitative application to specific experimental data. The main point made by BRIDGMAN was, that application of thermodynamics to viscoelastic properties requires an extension of classical thermodynamics since classical thermodynamics is essentially confined to states of equilibrium whereas it is essential for viscoelasticity that the material passes through a series of non-equilibrium states.

Apparently at the time of his publication BRIDGMAN was not sufficiently familiar with the development of an extension of thermodynamics particularly suitable for application to viscoelasticity. This branch of thermodynamics, generally known as the *thermodynamics of irreversible processes* has been summarised in a recent monography[2] after the foundations had been laid by de DONDER[3], PRIGOGINE[4], MEIXNER[5] and several others.

The feature of this new thermodynamics of irreversible processes which makes it particularly suitable for the treatment of viscoelastic behaviour is that it starts from the basic assumption that the non-equilibrium state is not far from equilibrium. In fact, the deviation from equilibrium is supposed to be so small, that all the "effects" of the deviation are proportional to the "causes" or, to put it in more physical

[1] BRIDGMAN, P. W.: Rev. mod. Physics **22**, 56 (1950).

[2] GROOT, S. R. DE: Thermodynamics of irreversible processes, Amsterdam 1951.

[3] DONDER, TH. DE: L'Affinité (Paris, 1936).

[4] PRIGOGINE, J.: Etude Thermodynamique de Phénomènes irréversibles (Liège, 1947).

[5] MEIXNER, J.: Z. Naturforsch. **4a**, 594 (1949).

language, that all the "currents" are proportional to the "forces". Exactly as in the treatment of linear viscoelastic behaviour it is always very reasonable that such a region of proportionality, i.e. such a linear region, exists and it is a matter of experiment to find out how far it extends.

It would lead us outside the scope of this book to give a full treatment of the fundamentals of non-equilibrium thermodynamics so here we must refer to the quoted literature, particularly de GROOT's book. We shall, however, put forward some of the principles used in such a treatment.

The deviation from equilibrium of a particular system can always be described completely and unambiguously by specifying a sufficient number of parameters $\xi_1 \ldots \xi_i \ldots$. One can arbitrarily chose the zero-point of these parameters such that in equilibrium

$$\xi_i \equiv 0 \quad \text{for all } i\,. \tag{I, 120}$$

If we consider processes at constant temperature throughout, equilibrium is characterised by a minimum of free energy. If in a non-equilibrium state we denote the free energy in excess over the minimum as F we have in equilibrium, besides $F = 0$ and $\xi_i = 0$,

$$\left(\frac{\partial F}{\partial \xi_i}\right)_{\xi_k} = 0\,. \tag{I, 121}$$

So outside equilibrium we can develop F with respect to the ξ's and obtain as the first term

$$F = \frac{1}{2} \Sigma_i \Sigma_k b_{ik} \xi_i \xi_k \tag{I, 122}$$

with $b_{ik} = \left(\frac{\partial^2 F}{\partial \xi_i \partial \xi_k}\right)_{\xi_l}$.

The basic assumption of linearity, then, implies that higher terms in the development of F with respect to ξ can be neglected in comparison to (I, 122).

By differentiating F with respect to the ξ's one obtains a new set of quantities, the so-called "forces" X.

$$X_i = \left(\frac{\partial F}{\partial \xi_i}\right)_{\xi_k} = \Sigma_k b_{ik} \xi_k\,. \tag{I, 123}$$

A set of forces X is just as useful for specifying a particular non-equilibrium state as a set of ξ's.

Another consequence of the concept of linearity is that all rates of change are proportional to the ξ's or, what amounts to the same thing, to the X's.

$$\dot{\xi}_i = \Sigma_k a_{ik} X_k\,. \tag{I, 124}$$

We have here, in fact, a formalism which allows a thermodynamical description of systems not in equilibrium in the linear region. The sets of coefficients b_{ik} and a_{ik} characterise the properties of the system, a set of ξ's *or* a set of X's characterises a particular state of that system. With (I, 124) we can calculate the series of states passed through by the system on its path from a given non-equilibrium state towards equilibrium.

This formalism can be applied to viscoelastic behaviour without any change[1]. The ξ's represent parameters describing such configurational features as discussed in sec. a); coiling of macromolecules, orientation of macromolecules, degree of crystallisation, orientation of crystallites and the like.

When trying to relate this thermodynamical treatment with the phenomenological treatment of viscoelastic behaviour we encounter the difficulty that in the free energy as shown by (I,122), as well as in the rate of change of any parameter as shown by (I,124), all individual parameters are mixed. The rate of change of one parameter and the effect of change of one parameter on the free energy depends on the value of all the other parameters. This is not a formal result from the assumptions made but it definitely represents the physical situation. If we select configurational parameters as discussed above (coiling, orientation ect.) we must expect that in fact the rate of change and the effect on the free energy of one individual parameter will depend upon the values of all the others.

However, it has been proved by MEIXNER (loc. cit. page 63) that owing to certain symmetry properties of the sets of coefficients, it must always be possible to replace the set of parameters ξ_i by another set γ_μ with the characteristic property that the rate of change and the effect on the free energy of one individual parameter depend on the value of that particular parameter only.

Mathematically this is formulated as follows. Since the set of ξ's is an arbitrarily chosen set of parameters any linear combination is equally good for describing the state of the system, so we can always form a new set by the equations

$$\gamma_\mu = \Sigma_i A_{\mu i} \xi_i . \tag{I,125}$$

With this new set of parameters we obtain instead of (I, 122)

$$F = \frac{1}{2} \Sigma_\mu \Sigma_\nu b_{\mu\nu} \gamma_\mu \gamma_\nu \tag{I,126}$$

and instead of (I,124)

$$\dot{\gamma}_\mu = \Sigma_\nu a_{\mu\nu} \gamma_\nu . \tag{I,127}$$

MEIXNER has shown that one can always construct a set of coefficients $A_{\mu i}$ in such a way that the coefficients $b_{\mu\nu}$ and $a_{\mu\nu}$ obey very special conditions, namely:

$$b_{\mu\nu} = 0 \quad \text{for} \quad \mu \neq \nu \tag{I,128}$$

and

$$a_{\mu\nu} = 0 \quad \text{for} \quad \mu \neq \nu . \tag{I,129}$$

If our γ's are selected in this way the equations (I,126) and (I,127) become much simpler because we have to take into account only the terms with coefficients $a_{\mu\mu}$ and $b_{\mu\mu}$. Calling

$$a_{\mu\mu} = -\frac{1}{\tau_\mu} \tag{I,130}$$

[1] STAVERMAN, A. J. u. F. SCHWARZL: Proc. Acad. Sci. A'dam **B 55**, 486 (1952).

and

$$b_{\mu\mu} = c_\mu \tag{I,131}$$

we obtain from (I,126) and (I,127)

$$F = \frac{1}{2} \Sigma_\mu c_\mu \gamma_\mu^2 \tag{I,132}$$

and

$$\dot{\gamma}_\mu = -\frac{1}{\tau_\mu} \gamma_\mu . \tag{I,133}$$

We see that molecular processes represented by the γ's behave as if they were completely independent of each other. We call these processes *quasi-independent*. The addition "quasi" serves to stress that the processes are not really independent physically. They interact in the same way as the normal vibrations of a set of coupled oscillators. In this respect it is interesting to pursue our model considerations of § 4 a little further. In these models the *parameters of state* are clearly the extensions of the springs and the dashpots. We have pointed out in § 4 that any experimental relaxation function $G(t)$ can be represented generally by a large number of models, one of them consisting of a row of parallel MAXWELL-elements with spring constants $G(\tau)$ and viscosities $G \cdot \tau$, the others with differently built networks of springs and dashpots with completely different spring constants and viscosities. In a relaxation experiment of such a network according to the above the quasi-independent internal processes are always the same. If the network is the row of MAXWELL elements these quasi-independent processes are identical with the elongation of the individual MAXWELL elements. However in one of the other, equivalent networks, the quasi-independent processes do not coincide with the movement of any single element or pair of elements.

Without proof[1] we mention that the coefficients c_μ and τ_μ are immediately related to phenomenological quantities. Indeed the τ_μ's, the relaxation times of the quasi-independent molecular processes are identical with the relaxation times found in an isothermal relaxation experiment. And the value of c_μ represents the value of the "intensity" of the relaxation spectrum for the time τ_μ.

So far, the meaning of the value of the *relaxation spectrum* was intimately connected with the spring constants for a mechanical model of the material rather than with anything in the material itself. We now see, that the value of the relaxation spectrum has a definite physical meaning. The value of the spectrum at τ_μ is the amount of free energy stored instantaneously on unit deformation in the quasi-independent molecular processes with relaxation time τ_μ.

The first purpose of our thermodynamical treatment has been attained. The parameters of state of a substance are the quasi-independent parameters. Their value can be calculated from (I,133) in a relaxation experiment starting at $t = 0$ with all γ_μ equal to the instantaneous deformation. In other experiments they can be calculated from (I,133) and the superposition principle.

[1] STAVERMAN, A. J. u. F. SCHWARZL: loc. cit. p. 65.

Once the values of the γ's are known the free energy as a function of time can be found with (I,132). It is interesting to find that in a relaxation experiment the free energy per unit deformation $F(t)$ does not follow the relaxation function $G(t)$ closely. $F(t)$ decays faster than $G(t)$ since it runs like $G(2t)$ (see fig. I,32). The analogy between viscoelastic substances and mechanical models is more than purely formal also in this connection. It has been shown by STAVERMAN and SCHWARZL[1] that the total energy stored in all springs together in a relaxation experiment runs parallel with $G(2t)$ independently of the construction of the mechanical model provided it shows the relaxation function $G(t)$.

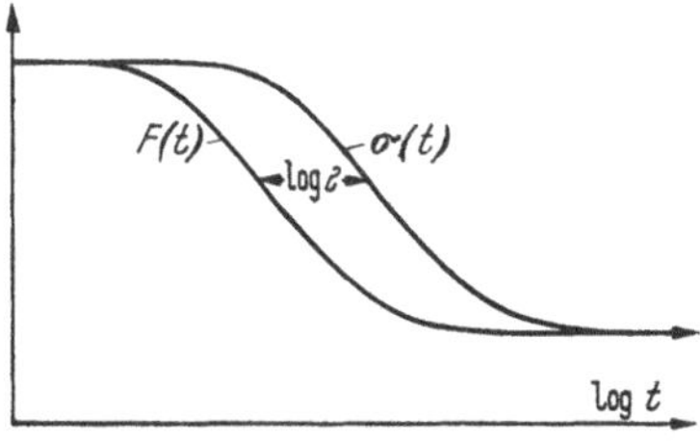

Fig. I, 32. Dependence of stress (σ) and free energy F on time in a relaxation experiment.

This can be proved without resorting to any thermodynamics but some physical properties of the links between the elements of the model play a part. Apparently and understandably the total energy stored in the springs is equivalent to the total free energy of the system although there exists no physical method of turning this free energy reversibly into work, as is required of free energy in classical thermodynamics.

Having thus acquired confidence in the physical meaning of the spectra our next intention should be to relate the molecular processes which build up the spectrum to specific changes of molecular configurations. However, our knowledge of interatomic forces is far too scanty to allow of even a very rough theoretical calculation of the spectrum from the chemical constitution of a substance. The only way to proceed here is to investigate the changes of the spectrum with changing chemical constitution, as has been done with infrared and ultraviolet spectra.

However somewhat more information about the molecular processes can be obtained without change of the chemical constitution by changing the temperature. If the value of the spectrum at a certain time τ is a measure of the contribution to the free energy of the system from molecular processes with relaxation time τ, then we can apply to this quantity the equations of thermodynamics. Writing f for the free energy change of a particular process μ, u for the internal energy and s for the entropy associated with this process we have

$$f = u - T s \tag{I,134}$$

or

$$f = u + T \frac{df}{dT} . \tag{I,135}$$

From the spectrum the values of f of a large number of processes can be determined. If from the temperature dependence of the spectrum $\frac{df}{dT}$ could be determined it would be possible to specify for each molecular

[1] STAVERMAN, A. J. u. F. SCHWARZL: Proc. Acad. Sci. A'dam **B 55**, 474 (1952).

process the part of the free energy due to internal energy and the part due to entropy. This would certainly aid in the molecular interpretation of the behaviour of the material.

However, except in simple cases serious difficulties arise here. The first is that on change of temperature not only is the value of f of a certain process changed but the relaxation time is affected much more strongly. So there is some difficulty in finding out which relaxation time τ_2 at temperature T_2 must be attributed to a particular molecular process characterised by τ_1 and f_1 at temperature T_1. Even, if such an assignment of different relaxation times to identical processes at different temperatures could be made without ambiguity by virtue of the shape of the spectrum, another difficulty of a more fundamental nature arises. The molecular processes showing up in the spectrum are not really independent but only behave quasi-independently and, therefore on change of temperature the processes themselves which behave quasi-independently may be different. In mathematical language: not only may the coefficients f_μ and τ_μ depend on the temperature but also the coefficients $A_{i\mu}$ of eq. (I, 125). This means that a process which belongs to the set of quasi-independent processes at one temperature may no longer belong to this set at a different temperature. So far there is no proof that changes of f and τ would be expected to be more important than changes of the A's[1].

Concluding this section on thermodynamics we may say that progress in the molecular interpretation of viscoelastic behaviour can be expected in three directions:

1. experimental methods to measure spectra with higher accuracy should be developed;

2. the influence of chemical structure on the spectra should be investigated systematically;

3. the influence of temperature on the spectra should be investigated both theoretically and experimentally.

As to 3., one interesting point is to obtain information about the coupling between quasi-independent processes from the temperature dependence of the spectra. Another point is that in the slow processes occurring in the measurement of time dependent transition points (see Vol. III of this work) apparently the same parameters of state are involved as in viscoelastic behaviour and it would be very interesting to see whether data from one type of experiments tally with those from the other type.

Finally the temperature dependence of the spectra is connected intimately with the heat development on deformation. In fact this heat development depends on time, not in the same way as the free energy but rather as the difference between free energy and internal energy. This heat development is accessible to experiment since in dynamic experiments the rate of diffusion of heat affects the results and this effect depends on the dimensions of the sample.

[1] An attempt to prove this, A. J. STAVERMAN: Proc. 2nd Int. Congr. Rheology, Oxford 1953, page 134, has turned out to be erroneous.

Literature to § 5.

Books:

GROOT, S. R. DE: Thermodynamics of irreversible processes. Amsterdam 1951, North Holl. Publ. Cy.

TRELOAR, L. R. G.: The physices of rubber elasticity. Oxford 1949, Clarendon Press.

WILDSCHUT, A. J.: Technological and physical investigations on natural and synthetic rubbers. Elsevier, Amsterdam 1946.

Articles:

BISCHOFF, J., E. CATSIFF and A. V. TOBOLSKY: J. Amer. chem. Soc. **74**, 3378 (1952).

DAVIES, R. O.: Proc. 2nd Int. Congr. Rheology, Oxford 1953, page 171.

DAVIES, R. O. and G. O. JONES: Proc. Roy. Soc. [London] **A 217**, 26 (1953).

FERRY, J. D.: J. Amer. chem. Soc. **72**, 3746 (1950).

FERRY, J. D., E. R. FITZGERALD, M. F. JOHNSON and L. D. GRANDINE JR.: J. appl. Physics **22**, 717 (1951).

FERRY, J. D., E. R. FITZGERALD, L. D. GRANDINE and M. L. WILLIAMS: Ind. Engng. Chem. **44**, 703 (1952).

FERRY, J. D. and E. R. FITZGERALD: J. Colloid Sci. **8**, 224 (1953).

FERRY, J. D. and E. R. FITZGERALD: Proc. 2nd Int. Congr. Rheclogy, Oxford 1953, page 140.

FERRY, J. D., L. D. GRANDINE and E. R. FITZGERALD: J. appl. Physics **24**, 911 (1953).

HOPKINS, I. L.: J. appl. Physics **24**, 1300 (1953).

KUHN, W.: Angew. Chem. **49**, 858 (1936); **51**, 640 (1938); **52**, 289 (1939).

LYONS, W. J.: J. appl. Physics **24**, 217 (1953).

MARVIN, R. S., E. R. FITZGERALD and J. D. FERRY: J. appl. Physics **21**, 197 (1950).

MARVIN, R. S.: Proc. 2nd Int. Congr. Rheology, Oxford 1953, page 156.

MCLOUGHLIN, J. R. and A. V. TOBOLSKY: J. Colloid Sci. **7**, 555 (1952).

MEIXNER, J.: Kolloid-Z. **134**, 3 (1953); Z. f. Naturf. **9 a**, 654 (1954); Z. f. Phys. **139**, 30 (1954); Proc. Roy. Soc. **A 226**, 51 (1954).

MOONEY, M.: J. appl. Physics **19**, 434 (1948).

MÜLLER, F. H.: Kollcid-Z. **134**, 207 (1953).

SCHWARZL, F.: Kolloid.-Z. **139**, 52 (1954).

SCHWARZL, F. and A. J. STAVERMAN: J. appl. Physics **23**, 838 (1952).

STAVERMAN, A. J. and F. SCHWARZL: Proc. Acad. Sci. A'dam **B 55**, 474, 486 (1952).

STAVERMAN, A. J.: Kolloid-Z. **134**, 189 (1953).

TOBOLSKY, A. V. and R. D. ANDREWS: J. chem. Physics **13**, 3 (1945).

TOBOLSKY, A. V. and J. R. MCLOUGHLIN: J. Polymer Sci. **8**, 543 (1952).

WALL, W. and P. J. FLORY: J. chem. Physics **19**, 1435 (1951).

§ 6. Viscoelastic behaviour under three dimensional stresses.

a) Introduction.

So far our considerations about the time-dependent stress-strain behaviour of high polymers have been restricted in one respect. We have only considered simple deformations as for instance simple shear; stress and strain had the meaning of shear stress and shear strain, moduli and compliances of shear moduli and shear compliances.

As measurements are often done with other types of deformation such as linear extension, torsion and bending, it is necessary to generalise our considerations to viscoelastic behaviour under arbitrary three-dimensional stress distributions.

We shall give a short survey in the following, chiefly based on a publication by SIPS[1]. Though much theoretical work has been done on shear, comparatively little attention has been paid to general viscoelastic deformation.

b) The general stress-strain behaviour of viscoelastic polymers.

We base our treatment of three-dimensional viscoelastic behaviour on two assumptions:

(i) The material under consideration is homogenous and isotropic.

(ii) The material obeys BOLTZMANN's superposition principle (is linear).

In order to find general relations between stress and strain for a material obeying the conditions (i) and (ii), we remember that an isotropic HOOKEan elastic solid can be described by the stress-strain relations

$$\sigma_x = 2G\,\varepsilon_x + [K - 2G/3]\,\Delta \qquad \text{(I, 18a)}$$

$$\sigma_{xy} = G\,\gamma_{xy} \qquad \text{(I, 18b)}$$

where Δ means the relative volume dilatation

$$\Delta = \varepsilon_x + \varepsilon_y + \varepsilon_z\,. \qquad \text{(I, 18c)}$$

G and K are the shear and bulk modulus, the two constants describing the isotropic HOOKEan solid.

Turning to the time dependent viscoelastic behaviour in simple shear we have to replace equ. (I,18b) by

$$\sigma_{xy}(t) = \int_{-\infty}^{t} G(t-\xi)\,\dot{\gamma}_{xy}(\xi)\,d\xi\,. \qquad \text{(I, 42b)}$$

Instead of a material constant G defining the shear behaviour, we have a function of time $G(t)$, the stress relaxation modulus in shear. By means of the superposition principle (I,42b) the general stress strain behaviour in shear is known, if the stress relaxation modulus is determined.

The generalisation to the three dimensional behaviour of the isotropic viscoelastic solid is straightforward. Two characteristic material properties have to be known, the stress relaxation modulus in shear $G(t)$, which is identical with the quantity dealt with in § 3 and § 4, and the stress relaxation modulus in compression $K(t)$. Then an arbitrary prescribed strain condition $\varepsilon_x = \varepsilon_x(t)$, $\gamma_{xy} = \gamma_{xy}(t)$ and so on, allows the calculation

[1] SIPS, R.: J. Polymer Sci. **7**, 191 (1951). — See further: E. SKUDRZYK: Österr. Ing.-Archiv **3**, 356 (1949). — T. ALFREY: Mechanical behaviour of high polymers Interscience Publ. New York 1948, page 204, 557. — W. KUHN and S. VIELHAUER: Z. physik. Chem. **202**, 124, 161 (1953).

of the resulting time dependent stress distribution by means of the equations

$$\left.\begin{aligned}\sigma_x(t) &= 2\int_{-\infty}^{t} G(t-\xi)\,\dot{\varepsilon}_x(\xi)\,d\xi + \int_{-\infty}^{t}\left[K(t-\xi) - \frac{2}{3}G(t-\xi)\right]\dot{\Delta}(\xi)\,d\xi \\ \sigma_{xy}(t) &= \int_{-\infty}^{t} G(t-\xi)\,\dot{\gamma}_{xy}(\xi)\,d\xi \qquad \Delta(t) = \varepsilon_x(t) + \varepsilon_y(t) + \varepsilon_z(t)\,.\end{aligned}\right\} \quad \text{(I,136)}$$

These equations give all the information obtainable about the linear mechanical properties of a viscoelastic material. The state of stress (the components of the stress tensor) at time t depend on the whole strain history. As only two material functions G and K are involved we have the statement: A linear viscoelastic isotropic material is completely characterised, if two time dependent functions, the stress relaxation modulus in shear $G(t)$ and the stress relaxation modulus in compression $K(t)$, are known.

Equ. (I,136) can, of course be reversed to express the strains as functions of the stresses, but we reserve this problem to a later stage. We will first consider some simple types of deformation.

Simple shear. In simple shear we have the condition that all stresses except $\sigma_{xy}(t)$ vanish, resulting in a strain system, where only the shear strain $\gamma_{xy}(t)$ is different from zero. If the shear strain is prescribed as function of time, the resulting shear stress is given by

$$\sigma_{xy}(t) = \int_{-\infty}^{t} G(t-\xi)\,\dot{\gamma}_{xy}(\xi)\,d\xi\,. \qquad \text{(I,137)}$$

If on the other hand the shear stress is prescribed, the resulting strain is given by

$$\gamma_{xy}(t) = \int_{-\infty}^{t} J(t-\xi)\,\dot{\sigma}_{xy}(\xi)\,d\xi. \qquad \text{(I,138)}$$

$G(t)$ is the ***stress relaxation modulus in shear*** and $J(t)$ the ***creep compliance in shear***, which are connected by the equation

$$\int_{0}^{t} G(t-\tau)\,J(\tau)\,d\tau = t\,. \qquad \text{(I,139)}$$

Of course, these equations are identical with those of § 3d [equations (I,42a), (I,42b) and (I,43a)], since in this case no isotropic compression is involved. The torsion of a rod with circular cross section can be treated by means of the above equations. Equ. (I,137), (I,138) and (I,139) remain valid, if we replace σ_{xy} by $M(t)/I$ and γ_{xy} by $\vartheta(t)$. Here $M(t)$ is the time dependent torque on the rod, I the polar moment of the cross section and $\vartheta(t)$ the time dependent angle of torsion per unit length.

Isotropic Compression. Next we consider the case of isotropic compression, where the shearing stresses vanish and the tensile stresses are all equal to $-P(t)$. The hydrostatic pressure $P(t)$ results in a volume contraction $-\Delta(t)$, both quantities being connected by

$$P(t) = -\int_{-\infty}^{t} K(t-\xi)\,\dot{\Delta}(\xi)\,d\xi \tag{I,140}$$

$$\Delta(t) = \frac{\delta V}{V_0}$$

and inversely

$$\Delta(t) = -\int_{-\infty}^{t} \varkappa(t-\xi)\,\dot{P}(\xi)\,d\xi\,. \tag{I,141}$$

The negative sign means that the volume is decreased under a positive pressure. Equ. (I,140) and (I,141) are precisely of the same form as (I,137) and (I,138) and define the meaning of the *bulk modulus* $K(t)$ and the creepfunction(t)[1]. The bulk modulus $K(t)$ is the hydrostatic pressure necessary to maintain a constant unit decrease of volume in isotropic compression. The bulk compliance gives the decrease of volume as function of time resulting from a constant unit hydrostatic pressure. Between bulk modulus and bulk compliance the relationship (I,139) is valid.

Linear extension. Finally we consider the most important case, linear extension. Linear extension means the elongation of a rod-shaped sample by forces acting in the longitudinal direction only. The lateral dimensions are considered to be stress free. Therefore only $\sigma_x(t)$ is different from zero, all the other stresses vanishing. The tensile stress in the longitudinal direction results in a relative elongation of the longitudinal dimension $+\varepsilon_x(t)$ and in a lateral contraction of the sample $-\varepsilon_y(t)$.

Longitudinal stress and strain are connected by two equations, whether we express the tensile stress in terms of the tensile strain, or vice versa.

$$\sigma_x(t) = \int_{-\infty}^{t} E(t-\xi)\,\dot{\varepsilon}_x(\xi)\,d\xi \tag{I,142}$$

$$\varepsilon_x(t) = \int_{-\infty}^{t} F(t-\xi)\,\dot{\sigma}_x(\xi)\,d\xi\,. \tag{I,143}$$

The meaning of the characteristic functions E and F will be quite clear. The *stress relaxation modulus in extension* $E(t)$ gives the stress necessary to maintain a constant unit elongation. The *extension compliance* $F(t)$ gives the extension as function of time under constant unit stress. Between modulus and compliance again a relationship of the form (I,139) is valid.

The compliance in extension is connected with the compliances in shear and compression by the simple relation

$$F(t) = \frac{1}{3}J(t) + \frac{1}{9}\varkappa(t)\,. \tag{I,144}$$

[1] $\varkappa$ is generally known under the name compressibility, and turns out to be a time dependent function.

Of course there exists a relationship between the moduli E, G and K too, but it is of a much more complicated form.

Considering linear extension we can ask how the lateral contraction $-\varepsilon_y(t)$ is connected with the longitudinal extension. This question defines a new material function, POISSON's *ratio*, $\nu(t)$, which is considered to be a function of time in the general case

$$-\varepsilon_y(t) = \int_{-\infty}^{t} \nu(t-\xi)\,\dot{\varepsilon}_x(\xi)\,d\xi\,. \tag{I,145}$$

The meaning of POISSON's ratio becomes clear, if we consider a linear extension experiment, where the elongation $\varepsilon_x(t)$ is held constant, say ε_0. The lateral contraction is then determined by the function

$$-\varepsilon_y(t) = \nu(t)\cdot\varepsilon_0\,. \tag{I,146}$$

Equ. (I,146) gives a way of measuring the function $\nu(t)$ as the lateral contraction in a stress relaxation experiment under constant unit elongation.

If POISSON's ratio together with any arbitrary characteristic function is known, the viscoelastic behaviour is determined. If we consider for instance POISSON's ratio and the extension compliance to be known, we have

$$\left.\begin{aligned} J(t) &= 2F(t)\,[1+\nu_0] + 2\int_0^t \dot{\nu}(t-\xi)\,F(\xi)\,d\xi \\ \varkappa(t) &= 3F(t)\,[1-2\nu_0] - 6\int_0^t \dot{\nu}(t-\xi)\,F(\xi)\,d\xi \end{aligned}\right\} \tag{I,147}$$

where ν_0 means $\nu(0)$.

If the shear modulus together with POISSON's ratio are known, we get

$$E(t) = 2G(t)\,[1+\nu_0] + 2\int_0^t \dot{\nu}(t-\xi)\,G(\xi)\,d\xi \tag{I,148}$$

It may be mentioned that there are more possibilities of defining POISSON's ratio for a viscoelastic material. Amongst others two of these possibilities are:

1. The ratio of lateral contraction to longitudinal elongation $(-\varepsilon_y/\varepsilon_x)$ in a creep experiment $[\sigma_x = \text{const.},\ \sigma_y = 0]$.

2. The ratio of lateral contraction to longitudinal elongation $(-\varepsilon_x/\varepsilon_x)$ in a stress relaxation experiment $[\varepsilon_x = \text{const.},\ \sigma_y = 0]$.

For a purely elastic material both definitions are identical. For a viscoelastic material the second possibility has been chosen rather arbitrarily.

The characteristic functions. We have seen that in considering the viscoelastic behaviour of an isotropic material, we have a series of characteristic material properties. The relaxation moduli in shear, compression and extension $G(t)$, $K(t)$ and $E(t)$ describe the stress relaxation properties in the corresponding experiments and are monotonically decreasing functions of time. The compliances $J(t)$, $\varkappa(t)$ and $F(t)$ describe the creep properties in these experiments and are increasing functions of time. Moduli and compliances are always connected by an equation of the form (I,139). Finally POISSON's ratio $\nu(t)$ is dimension less and can vary between 0 and 1/2. The characteristic functions and their properties are summarised in table (I,6).

Table I, 6. *The characteristic functions.*

Stress relaxation moduli		Creep compliances		Experiment
$G(t)$	$\dot{G}(t) < 0$	$J(t)$	$\dot{J}(t) > 0$	simple shear, torsion
$K(t)$	$\dot{K}(t) < 0$	$\varkappa(t)$	$\dot{\varkappa}(t) > 0$	isotropic compression
$E(t)$	$\dot{E}(t) < 0$	$F(t)$	$\dot{F}(t) > 0$	linear extension, bending
Poisson's ratio $\nu(t)$		$0 < \nu(t) \leq \frac{1}{2}$		

In conclusion we repeat that two time-dependent quantities have to be measured to specify the three-dimensional viscoelastic behaviour of an isotropic polymer. The following pairs of independent characteristic functions may be chosen:

Two different stress relaxation moduli.

Two different creep compliances.

One stress relaxation modulus together with one creep compliance of another kind.

One stress relaxation modulus and POISSON's ratio.

One creep compliance and POISSON's ratio.

c) The stationary state of vibration.

In view of the practical importance of vibration experiments it is worth while to say a few words about them. Our considerations can be applied to free vibrations, forced standing vibrations and the propagation of sinusoidal waves. In these experiments stresses and strains posses a sinusoidal time dependence

$$\sigma_x = \sigma_x^0 e^{i\omega t} \qquad \varepsilon_x = \varepsilon_x^0 e^{i\omega t} \tag{I, 149}$$

σ_x^0 and ε_x^0 are the amplitude factors of stress and strain and may depend on the coordinates and on the angular frequency ω.

The equations between stress and strain have a simple form completely analogous to the stress strain equations for the HOOKEan solid

$$\left.\begin{aligned} \sigma_x &= 2G^* \varepsilon_x + \left[K^* - \frac{2}{3} G^*\right] (\varepsilon_x + \varepsilon_y + \varepsilon_z) \\ \varepsilon_x &= \frac{1}{2} J^* \sigma_x + \frac{1}{3}\left[\frac{1}{3}\varkappa^* - \frac{1}{2} J^*\right] (\sigma_x + \sigma_y + \sigma_z) \\ \sigma_{xy} &= G^* \gamma_{xy} \qquad \gamma_{xy} = J^* \sigma_{xy}. \end{aligned}\right\} \tag{I, 150}$$

The characteristic material functions appear here in the form of dynamic moduli or compliances, depending on the frequency ω, but not on the stress and strain amplitudes.

The definition of the dynamic quantities is the same as we have already used in § 3, equ. (I,55) and (I,56). For instance, the *complex shear modulus* and the *complex shear compliance* are defined by

$$\left.\begin{aligned} &G^* = G_1 + iG_2 && J^* = J_1 - iJ_2 \\ &G_1 = G_0 + \int_0^\infty \dot{G}(t)\cos\omega t\,dt && J_1 = J_0 + \int_0^\infty \dot{J}(t)\cos\omega t\,dt \\ &G_2 = -\int_0^\infty \dot{G}(t)\sin\omega t\,dt && J_2 = \int_0^\infty \dot{J}(t)\sin\omega t\,dt\,. \end{aligned}\right\} \quad (\mathrm{I},151)$$

The definitions of the *complex bulk modulus* K^* and of the *complex extension modulus* E^* correspond to the definition of G^*, the definitions of the *complex bulk compliance* $\varkappa^*$ and the *complex extension compliance* F^* correspond to those of J^*. G_0 and J_0 are the values of stress relaxation modulus in shear and of shear compliance at time zero.

Further we define a *complex* POISSON's *ratio* by

$$\left.\begin{aligned} &\nu^* = \nu_1 - i\nu_2 \\ &\nu_1 = \nu_0 + \int_0^\infty \dot{\nu}(t)\cos\omega t\,dt; \quad \nu_2 = +\int_0^\infty \dot{\nu}(t)\sin\omega t\,dt\,. \end{aligned}\right\} \quad (\mathrm{I},152)$$

We have therefore the following characteristic quantities specifying vibration properties:

Table I, 7. *The dynamic characteristic functions.*

Storage modulus	Loss modulus	Storage compliance	Loss compliance	Damping	in
G_1	G_2	J_1	J_2	$J_2/J_1 = G_2/G_1 = (\mathrm{tg}\,\delta)_s$	shear
K_1	K_2	$\varkappa_1$	$\varkappa_2$	$\varkappa_2/\varkappa_1 = K_2/K_1 = (\mathrm{tg}\,\delta)_c$	compression
E_1	E_2	F_1	F_2	$F_2/F_1 = E_2/E_1 = (\mathrm{tg}\,\delta)_e$	extension

The whole theory which has been developed for complex shear modulus and shear compliance remains valid for all moduli and compliances. The storage moduli G_1, K_1, E_1 represent those parts of the stress, which are in phase with a sinusoidal prescribed strain of unit amplitude, the loss moduli G_2, K_2, E_2 are those parts of stress, which are 90° out of phase with the strain. The storage compliances J_1, $\varkappa_1$, F_1 are those parts of strain, which are in phase with the prescribed sinusoidal stress of unit amplitude, the loss compliances J_2, $\varkappa_2$, F_2 are those parts of strain, which are 90° out of phase with the stress. The storage POISSON's ratio ν_1 is that part of the lateral contraction which is in phase with the sinusoidal longitudinal extension of unit amplitude in linear extension, the loss POISSON's ratio ν_2 is that part of lateral contraction, which is 90° out of phase with the extension.

All storage moduli are monotonically increasing functions of frequency and all storage compliances are monotonically decreasing functions of frequency. The shape of an arbitrary storage modulus or storage compliance taken as functions of frequency is very similar to the shape of the corresponding static modulus or compliance taken as a function of time, with $\omega = 1/t$. The loss moduli or compliances are positive functions and show maxima in the transition regions of the corresponding storage moduli or compliances. There exist relationships between each storage modulus (compliance) and the corresponding loss modulus (compliance) of the form of the KRONIG-KRAMERS relations (see § 3e page 27).

The definition of the storage and loss parts of POISSON's ratio (equ. I,152) has been chosen in such a way, that we may expect ν_1 to be a monotonically decreasing function of frequency and ν_2 a positive function of frequency. This assumption cannot be proved theoretically, but may be expected as POISSON's ratio $\nu(t)$ has been experimentally shown to be an increasing function of time.

The utmost usefulness of the complex quantities lies in the fact that all relationships which are valid for the HOOKEan elastic solid remain valid for the viscoelastic solid if expressed in complex quantities. For instance, each complex modulus is the reciprocal of its compliance

$$G^* J^* = K^* \varkappa^* = E^* F^* = 1 \qquad (I, 153)$$

and between the different types of moduli and compliances we have the equations summarised in tables I,8 and I,9.

Table I, 8. *Relations between complex moduli.*

	(G^*, E^*)	(G^*, K^*)	(E^*, K^*)	(G^*, ν^*)	(E^*, ν^*)	(K^*, ν^*)
E^*	—	$\frac{9G^* K^*}{3K^* + G^*}$	—	$2G^*(1+\nu^*)$	—	$3K^*(1-2\nu^*)$
G^*	—	—	$\frac{3E^* K^*}{9K^* - E^*}$	—	$\frac{E^*}{2(1+\nu^*)}$	$\frac{3K^*(1-2\nu^*)}{2(1+\nu^*)}$
K^*	$\frac{E^* G^*}{9G^* - 3E^*}$	—	—	$G^* \frac{2(1+\nu^*)}{3(1-2\nu^*)}$	$\frac{E^*}{3(1-2\nu^*)}$	—
ν^*	$\frac{E^*}{2G^*} - 1$	$\frac{3K^* - 2G^*}{2[3K^* + G^*]}$	$\frac{1}{2}\left[1 - \frac{E^*}{3K^*}\right]$	—	—	—

Table I, 9. *Relations between complex compliances.*

	(J^*, F^*)	$(J^*, \varkappa^*)$	$(F^*, \varkappa^*)$	(J^*, ν^*)	(F^*, ν^*)	$(\varkappa^*, \nu^*)$
F^*	—	$\frac{1}{3}J^* + \frac{1}{9}\varkappa^*$	—	$J^*/2(1+\nu^*)$	—	$\varkappa^*/3(1-2\nu^*)$
J^*	—	—	$3F^* - \frac{1}{3}\varkappa^*$	—	$2F^*(1+\nu^*)$	$\varkappa^* \frac{2(1+\nu^*)}{3(1-2\nu^*)}$
$\varkappa^*$	$9F^* - 3J^*$	—	—	$J^* \frac{3(1-2\nu^*)}{2(1+\nu^*)}$	$3F^*(1-2\nu^*)$	—
ν^*	$\frac{J^*}{2F^*} - 1$	$\frac{3J^* - 2\varkappa^*}{2[3J^* + \varkappa^*]}$	$\frac{1}{2}\left[1 - \frac{\varkappa^*}{3F^*}\right]$	—	—	—

From the above tables a large number of very useful relations between different storage moduli and compliances, loss moduli and compliances and various types of damping may be derived. For a survey of these relations we refer to an article which is to be published soon[1].

d) Examples of three-dimensional viscoelastic behaviour.

We have seen that the mechanical behaviour of a viscoelastic substance can be described by one time-dependent function, only if experiments are restricted to one type of deformation. However, in order to cover a sufficient range of the time scale it is necessary to use different experimental techniques generally involving different types of deformation. Also the deformations undergone by the material in practical use are generally of different types.

A rigorous description of the complete behaviour requires knowledge of two time dependent functions. In order to avoid the work involved in measuring these two functions many investigations resort to more or less arbitrary assumptions, all reducible to the statement that one characteristic function does not depend on time but is a simple constant.

Of this kind of assumptions three have gained some confidence:

a) POISSON's ratio is a time independent constant; $\nu_1 = \nu_0$, $\nu_2 = 0$

b) The compressibility is time independent; $\varkappa_1 = \varkappa_0$, $\varkappa_2 = 0$

c) The material is incompressible; $\nu = 1/2$, $\varkappa = 0$.

a) If POISSON's ratio is constant ν_0, all characteristic functions are connected in a simple way, for instance,

$$\left.\begin{aligned} E(t) &= 2[1+\nu_0]\,G(t) = 3[1-2\nu_0]\,K(t) \\ \varkappa(t) &= 3[1-2\nu_0]\,F(t) \quad J(t) = 2[1+\nu_0]\,F(t). \end{aligned}\right\} \qquad (\text{I}, 154)$$

The same relationships are valid for storage and loss moduli and compliances with respect to frequency dependence. It is seen that all stress relaxation moduli (in shear, compression or tension) have the same time dependence and differ only by a common factor.

There is not very much experimental material concerning POISSON's ratio for high polymers. Generally we can say that rubbers (at ordinary temperatures and times) have a POISSON's ratio very near to 1/2, while plastics have a POISSON's ratio lower than 1/2. Though the shear modulus of rubbers is low (order of magnitude 10^6 to 10^8 dynes/cm^2), the bulk modulus is much higher (order of 10^{10} dynes/cm^2)[2].

We quote some results[3] from static measurements at 30° C.

	ν
Polystyrene	0,305
Polymethylmethacrylate	0,351
Cellulose Acetate	0,435
Cellulose Acetate Butyrate	0,459

[1] SCHWARZL, F. and A. BURGERS to be published.

[2] See for instance: L. E. COPELAND: J. appl. Physics **19**, 445 (1948).

[3] SHARP, E. B. and B. MAXWELL: Technical report Princeton University, August 1951.

It seems that POISSON's ratio is the higher, the softer the material. HOFF[1] has reported static measurements on polymethylmethacrylate, where POISSON's ratio equals 0,35 $\pm$ 0,02 over a temperature region from $-25°$C to $+50°$C. Similarly HUGHES and coworkers[2] report stress propagation methods on polystyrene, where POISSON's ratio does not vary considerably with temperature.

Polystyrene[2].

$T°$ C	E_1 in 10^{11} d/cm²	G_1 in 10^{11} d/cm²	K_1 in 10^{11} d/cm²	ν
31	0,363	0,136	0,370	0,336
51	0,350	0,131	0,357	0,337
72	0,342	0,128	0,353	0,339
82	0,333	0,124	0,339	0,339
92	0,324	0,121	0,335	0,340

These results are very remarkable as they emphazise that POISSON's ratio is constant over a wide temperature and therefore probably over a wide time range, provided we do not enter the transition region. Combining this with the results on various rubbers, we may assume: POISSON's ratio is constant (very near to 1/2) in the rubbery region and is also constant (with values considerably lower than 1/2), in the brittle region of polymers. In the transition region we must assume ν to increase monotonically with increasing temperature towards the value 1/2. Therefore the special case a) will apply to the rubbery and to the brittle state, *but certainly not to the transition region itself*. In this region assumption b) will give a better description of the behaviour.

b) We assume that there is no stress relaxation in isotropic compression. Then the bulk compliance is constant and equal to $\varkappa_0$ and the bulk modulus is constant and equal to $K_0 = 1/\varkappa_0$. It can be shown that creep compliances in shear and tension differ by an additive constant rather than by a multiplicative factor

$$F(t) = \frac{1}{3} J(t) + \frac{1}{9} \varkappa_0 . \tag{I, 155}$$

POISSON's ratio will be constant in regions where the stress relaxation modulus is constant (in the brittle and in the rubbery state), but will depend on time in each transition region.

$$\nu(t) = \frac{1}{2} [1 - E(t)/3 K_0]. \tag{I, 156}$$

Equation (I,156) describes exactly the behaviour outlined above. Taking polymethylmethacrylate as an example (fig. I,33) we see that in the brittle region ν equals 0,35, $E/K_0 = 0{,}90$ and $G/K_0 = 0{,}33$. Assuming a transition of two decades in E, we get for the rubbery state $E/K_0 = 0{,}009$ and $G/K_0 = 0{,}003$. ν will then be equal to 0,498 which is practically the value for an incompressible material.

[1] HOFF E. W. A.: J. appl. Chem. [London] **2**, 441 (1952).

[2] HUGHES, D. S., E. B. BLANKENSHIP and R. L. MIMS: J. appl. Physics **21**, 294 (1950).

A further interesting consequence of b) is that the damping in shear and the damping in tension are not equal, but are connected by

$$\frac{G_2}{G_1} = \frac{E_2}{E_1}\left[1 + G_1/3K_0\right]. \tag{I,157}$$

In the rubbery state the factor in brackets will be practically equal to unity, but in the brittle state it has a magnitude of 1,1 which means that the damping in shear is larger than the damping in tension.

An investigation of the damping in shear and compression on the same material has been performed by NOLLE and SIECK[1]. They measured the wave propagation of compressional and shear waves in a vulcanized Buna N Rubber and report the complex moduli in shear and bulk at frequencies of 10 megacycles per second:

$T°C$	−65	−50	−30	−20	−10	0	10	15
G_1 in 10^{10} d/cm²	2,06	2,00	1,81	1,72	1,56	1,36	1,14	1,02
K_1 in 10^{10} d/cm²	4,84	4,79	4,61	4,46	4,23	4,03	3,70	3,46
G_2/G_1	0,053	0,056	0,061	0,074	0,11	0,16	0,25	—
K_2/K_1	0,0085	0,0093	0,0085	0,011	0,019	0,014	—	0,0054

The damping in compression, K_2/K_1, is in general lower than that in shear, G_2/G_1, by a factor of 10 and can be neglected in comparison to G_2/G_1. On the other hand, the bulk modulus K_1 depends clearly on temperature, which means that the assumption b) of a constant bulk modulus is not rigorously valid.

Similar results on polyisobutylene are reported by MARVIN[2], who found that K_1 increases slightly with increasing frequency and that K_2 shows a maximum at about 10^7 cps at 25°C. These results show that assumption b) will only be approximately valid and none of the possibilities a), b) or c) will be rigorously true for high polymers in the whole time and temperature range.

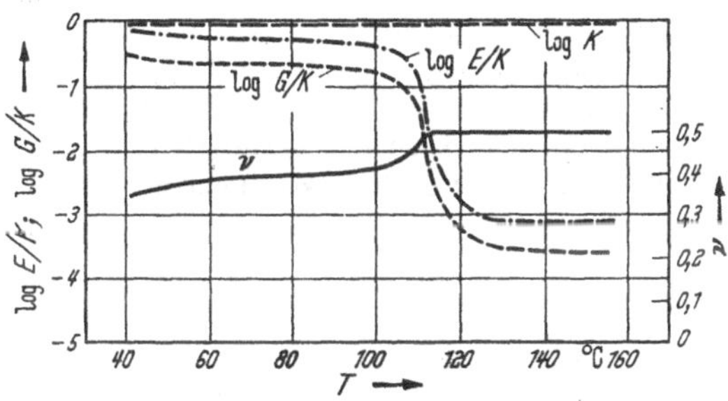

Fig. I, 33. The moduli and POISSON's ratio of polymethylmethacrylate as a function of temperature in the transition region according to assumption b.

c) To assumption c), which applies only to the rubbery state we need not give much attention, as it is a special case of a) as well as of b) POISSON's ratio is constant and equal to 1/2 and the bulk modulus is constant and infinite. The tension modulus is exactly three times the shear modulus and the tension compliance is exactly one third of the shear compliance.

[1] NOLLE, A. W. u. P. W. SIECK: J. appl. Physics **23**, 888 (1952).

[2] MARVIN, R. S.: Proc. 2nd Int. Congr. Rheology, Oxford 1953, page 156.

Concluding, we may summarise the three dimensional viscoelastic behaviour of polymers in the following way: The polymer will be practically incompressible in the rubbery state with POISSON's ratio equal to 1/2 and with extension moduli equal to three times the shear moduli. In the hard region the polymer will show a POISSON's ratio lower than 1/2, ν being the lower, the higher the shear modulus. A transition in shear will generally be accompanied by an increase in POISSON's ratio whereas a relaxation of the bulk modulus may or may not occur at the same frequencies. We may expect POISSON's ratio to be a monotonically increasing function of time, of which the main increase takes place in the transition region.

Literature to § 6.

Book:

ALFREY JR., T.: Mechanical behaviour of high polymers, page 204, 557.

Articles:

COPELAND, L. E.: J. appl. Physics **19**, 445 (1948).

HOFF, E. A. W.: J. appl. Chem. [London] **2**, 441 (1952).

HUGHES, D. S., E. B. BLANKENSHIP and R. L. MIMS: J. appl. Physics **21**, 294 (1950).

KUHL, W. and E. MEYER: Noise & Sound Transmission 1948 (Physical Society, London, 1949) 181.

KUHN, W. and S. VIELHAUER: Z. physik. Chem. **202**, 124, 161 (1953).

MARVIN, R. S.: Proc. 2nd Int. Congr. Rheology, Oxford 1953, page 156.

MEYER, E. and K. TAMM: Akust. Z. **7**, 45 (1942).

NOLLE, A. W. and P. W. SIECK: J. appl. Physics **23**, 888 (1952).

REINER, M.: Appl. Sci. Res. **A 1**, 475 (1949).

REINER, M.: Amer. J. Math. **68**, 672 (1946).

SHARP, E. B. and B. MAXWELL: Techn. Rep. Princeton Univ. Aug. 1951.

SIPS, R.: J. Polymer Sci. **7**, 191 (1951).

TAYLOR, G.: Proc. Roy. Soc. [London] **A 226**, 38 (1954).

§ 7. Experimental techniques.

a) Introduction.

We close the first chapter with a general survey of experimental techniques for the measurement of viscoelastic and flow behaviour. Our short survey will not enlarge upon experimental details but aims at a comparison of the different techniques available. For experimental details we refer to the literature cited at the end of each section.

Further we shall not discuss here the results of measurements on high polymers, as these are treated in other parts of the book. For a treatment of flow properties of polymer solutions and molten polymers we refer to volume II, chapter V and chapter XI; for a treatment of the viscoelastic properties of polymeric liquids, polymer solutions, rubberlike polymers and hard plastics to this volume, chapter VI and for the treatment of viscoelastic properties of fibers to this volume, chapter VII.

Rheological instruments for studying viscoelastic mechanical properties may be devided into three groups:

(i) Viscometers for the investigation of flow properties.

(ii) Static experiments for the investigation of viscoelastic effects showing up in times longer than 10 seconds.

(iii) Dynamic measurements for the investigation of viscoelastic effects in times shorter than 10 seconds.

There is not always a clear distinction between these three groups; they partly overlap, and in some cases it is doubtful whether a certain technique belongs to the one or the other. Our classification is not of a fundamental nature but is introduced for convenience only.

Fig. I,34 demonstrates the wide time range which is covered by the different experimental techniques. Though this region ranges from 10^{-8} seconds up to many years (10^8 seconds), each special technique covers only a small part of this region (hardly more than two or three decades).

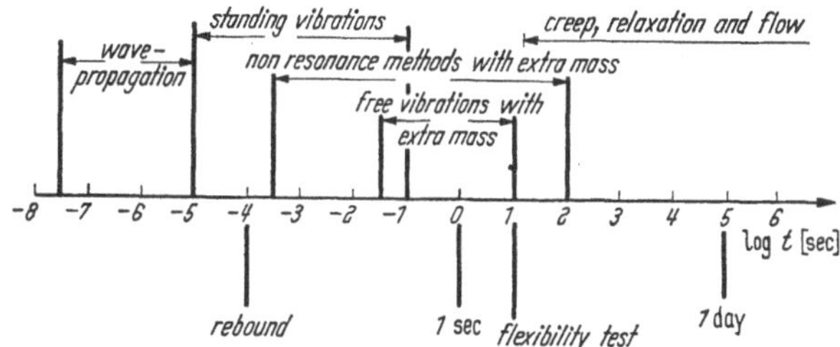

Fig. I, 34. The position of different experimental techniques in logarithmic time scale.

Therefore a large number of different techniques, which supplement each other rather than overlap, is needed. It is hardly necessary to say, that much work is involved in determining the behaviour of the material in the whole time region available. If we did not have the possibility of changing temperature besides time (compare § 5), we would not expect to have such complete information about rheological behaviour as we have to-day.

b) Viscometers.

If we are interested in the flow properties of high polymers, viscometers are the appropriate tool. In viscometers the steady state viscosity η [defined in formula (I,34)] is determined under simple types of stress geometry, in which shearing stresses play the dominant role.

The choice of the viscometer depends on the order of magnitude of the viscosity to be measured. Table I,10 taken from LEADERMAN[1], lists different viscometers and their range of applicability.

The *falling ball viscometer* measures the STOKES' resistance which a ball falling in a viscous liquid undergoes. Calling d the diameter of the

[1] LEADERMAN, H.: The physics of high polymers. Utrecht Univ. Press 1951, page 60.

Table I, 10. *Different viscometers and their range of applicability.*

Instrument	Viscosity range, Poises	Applicable to Non-Newtonian flow	Applicable to viscoelastic behaviour
Falling ball	10^{-2} — 10^{5}	—	—
Capillary viscometer (Ostwald, Ubbelohde, Bingham, Nason)	10^{-2} — 150	+	—
Coaxial cylinder, Couette type	10^{-2} — 10^{6}*	+	+
Coaxial cylinder, Pochettino type	10^{2} — 10^{6}	+	+
Parallel plate	10^{5} — 10^{11}	—	+

sphere, v its velocity relative to the liquid, ϱ and ϱ' the densities of the sphere and of the liquid, then the viscosity of the liquid is given by

$$\eta = \frac{(\varrho - \varrho')\, g d^2}{18\, v} \tag{I, 158}$$

(g = gravitation constant).

In the *capillary flow viscometer* the liquid is forced through a capillary tube by a pressure difference P. The volume Q of the liquid flowing out of the tube per unit of time is inversely proportional to the viscosity

$$Q = \frac{\pi R^4}{8 l} \cdot \frac{P}{\eta} \tag{I, 159}$$

where R is the radius of the capillary and l its length.

In *rotational viscometers* the material is placed in the annular space between two concentric cylinders (with height h and radii R_1 and R_2). If the outer cylinder is fixed in space and the inner cylinder is subjected to a constant torque T, this results in a uniform rotation of the inner cylinder with a constant angular velocity equal to

$$\Omega = \frac{T}{4 \pi h} \frac{1}{\eta} \left[\frac{1}{R_1^2} - \frac{1}{R_2^2} \right]. \tag{I, 160}$$

Rotational viscometers are constructed either with rotating outer cylinder or with rotating inner cylinder. The latter is simpler to construct but the former is more satisfactory from a theoretical point of view because of absence of turbulence up to a higher spead of rotation.

In the *Pochettino viscosimeter* the sample is pressed between two coaxial cylinders, the outer being fixed. The inner cylinder is pulled in axial direction and the sample is subjected to shear stress. The velocity v of the cylinder under constant force P is measured. From this velocity we get the viscosity by the equation

$$v = \frac{P}{2 \pi h} \frac{1}{\eta} \cdot \ln (R_2/R_1). \tag{I, 161}$$

* depending upon design.

In the *parallel plate viscometer* the sample is held between two parallel plates, which are pressed together by a constant pressure. The distance between the plates as a function of time gives information about the viscosity.

Some of these viscometers may also be used to measure non-NEWTONian flow or linear viscoelastic behaviour. For this purpose the formulas have to be changed in the appropriate way (LEADERMAN loc. cit. page 81).

Literature to § 7b.

Books:

British Rheologist's Club, The principles of rheological measurement. London 1949.

First and Second report on viscosity and plasticity. Amsterdam 1935, 1938.

GREEN H.: Industrial rheology and rheological structures. New York 1949.

LEADERMAN, H.: The physics of high polymers. Utrecht 1951.

MERRINGTON, A. C.: Viscometry. London 1949.

PHILIPPOFF, W.: Viscosität der Kolloide. Dresden-Leipzig 1942.

SCOTT BLAIR, G. W.: A survey of general and applied rheology. New York 1944.

UMSTÄTTER, H.: 1953. Einführung in die Viskometrie und Rheometrie, Springer 1952.

Articles:

ASBECK, W. K., P. D. LAIDERMAN and M. VON LOO: J. Colloid Sci. **7**, 306 (1952).

BRAUNBECK, W.: Z. Physik **57**, 501 (1929).

COUETTE, M.: Ann. Chim. Phys. **21**, 433 (1890).

DIENES, G. J.: J. Colloid Sci. **2**, 131 (1947).

DIENES, G. J. and H. F. KLEMM: J. appl. Physics **17**, 458 (1946).

FERRY, J. D.: Physics **6**, 356 (1935).

GREEN, H.: Ind. Engng. Chem., analyt. Edit. **14**, 576 (1942).

HULL, H. H.: J. Colloid Sci. **7**, 316 (1952).

IBRAHIM, A. K. and A. M. KABIEL: J. appl. Physics **23**, 754 (1952).

KRIEGER, I. M. a.o.: J. appl. Physics **23**, 147 (1952); **24**, 134 (1953); **25**, 72 (1954).

MOONEY, M.: Physics **7**, 413 (1936); Rubber Chem. Technol. **10**, 214 (1937); J. appl. Physics **11**, 582 (1940).

OSTWALD, W.: Kolloid-Z. **60**, 159 (1932); **63**, 61 (1933); **67**, 211 (1934).

PAWLOWSKY, J.: Kolloid-Z. **130**, 129 (1953).

PHILIPPOFF, W.: Arch. techn. Mess. 9122 T, 147 (1936); Kautschuk **15**, 191 (1939).

SCHULTZ-GRUNOW, F. and H. WEYMANN: Kolloid-Z. **131**, 61 (1953).

SCOTT, J. R.: Trans. I.R.I. **7**, 1 (1931).

SAECHTLING, H.: Kunststoffe **33**, 127 (1943).

SINGER, R. and K. BERNEIS: Makromol. Chem. **8**, 268 (1952).

TILLMANN, W.: Kolloid-Z. **131**, 66 (1953).

UMSTÄTTER, H.: Makromol. Chem. **10**, 30 (1953).

c) Static (transient) measurements.

The most important static measurements are those in which either the strain is observed under constant stress (creep measurement) or the stress is observed under constant strain (relaxation measurement).

The *creep measurement* consists usually in loading the sample with a constant weight and observing the deformation as function of time. Creep instruments have been constructed for linear tension as well as for other types of deformation such as simple shear, bending, torsion and axial compression. The creep measurement in shear or torsion yields the creep compliance in shear $J(t)$; the measurement in linear tension, bending or axial compression yield the extension compliance $F(t)$.

For polymers which show large elongations in a creep test, applying a constant weight does not ensure constant stress because of the diminution of the cross section of the sample during elongation. To correct for this increase of stress, instruments have been designed in which the loading force diminishes during elongation so as to obtain constant stress. These constructions use either an appropriately shaped weight immersed partially in a liquid[1] or means by which the load is reduced mechanically[2].

In *stress relaxation measurements* the load necessary to maintain a constant deformation is determined as a function of time. This gives either the stress relaxation modulus in shear $G(t)$ or the stress relaxation modulus in extension $E(t)$. These measurements are usually made with a modulus balance, which reduces the weight automatically by a servomechanism so as to maintain constant strain. Such relaxometers have been constructed for linear tension and axial compression.

Besides creep and relaxation measurements a large number of routine tests are performed which allow the determination of something like a "10 seconds modulus". Such *flexibility tests* yield only one arbitrary point of the whole time curve of rheological behaviour, but they have the advantage of their rapidity. Such tests have been developed for tensile elongation, bending, torsion and axial compression.

All the measurements treated above are performed at constant temperature. It may be mentioned that a technique in which, after cooling of the sample, the temperature is raised continuously, hasbeen developed in rubber technology as the *temperature retraction* test. This test allows the direct determination of the deformation-temperature curve of the material, though its interpretation is certainly much more complicated than the results of experimental techniques at constant temperature.

Literature to § 7c.

Creep.

Brenschede, W.: Kolloid-Z. **104**, 1 (1943).

Buchdahl, R., L. E. Nielsen and E. H. Merz: J. Polymer Sci. **6**, 403 (1951).

Clash, R. F. and R. M. Berg: Mod. Plastics **21**, 119 (1944).

Conant, F. S., G. L. Hall and G. R. Thurman: J. appl. Physics **20**, 526 (1949).

Dyson, A.: J. Polymer Sci. **7**, 147 (1951).

Gehman, S. D.: J. appl. Physics **19**, 456 (1948).

[1] Andrade, E. N.: Proc. Roy. Soc. [London] **A 84**, 1 (1910). — C. A. Dahlquist, J. O. Hendricks u. N. W. Taylor: Ind. Engng. Chem. **43**, 1404 (1951).

[2] Andrade, E. N.: Proc. Roy. Soc. [London] **A 138**, 348 (1932) — W. L. Holt, E. O. Knox u. F. L. Roth: J. Res. nat. Bur. Standards **41**, 95 (1948). — F. L. Roth u. R. V. Stiehler: J. Res. nat. Bur. Standards **41**, 87 (1948).

LETHERSICH, W.: Brit. J. appl. Physics **1**, 294 (1950).

MARIN, J. and G. CUFF: Proc. A.S.T.M. (1949), 1158.

UMSTÄTTER, H.: Schweiz. Arch. **19**, 184 (Juni 1953).

Stress Relaxation.

BEATTY, J. R. and A. E. JUVE: India Rubber World (Febr. 1950), 537.

BOSTRÖM, S.: Kautschuk u. Gummi **3**, 276 (1950).

FARNAM, R. G.: India Rubber World (March 1951), 679.

MCLOUGHLIN, J. R.: Rev. Sci. Instr. **23**, 459 (1952).

MOONEY, M., W. E. WOLSTENHOLME and D. S. WILLIAMS: J. appl. Physics **15**, 324 (1944).

PEDERSEN, H. L. and B. NIELSEN: J. Polymer Sci. **7**, 97 (1951).

SCHREMP, F. W., J. D. FERRY and W. W. EVANS: J. appl. Physics **22**, 711 (1951).

TOBOLSKY, A. V., I. B. PRETTYMAN and J. H. DILLON: J. appl. Physics **15**, 380 (1944).

WATSON, M. T., W. D. KENNEDY and G. M. ARMSTRONG: Physic. Rev. **82**, 301 (1951).

Flexibility tests.

CLASH, R. F. and R. M. BERG: Mod. Plastics **21**, 119 (1944).

CONANT, F. S. and J. W. LISKA: J. appl. Physics **15**, 767 (1944).

GEHMAN, S. D., D. E. WOODFORD and C. S. WILKINSON: Ind. Engng. Chem. **39**, 1108 (1947).

GREENE, H. L. and D. L. LOUGHBOROUGH: J. appl. Physics **16**, 3 (1945).

LISKA, J. W.: Ind. Engng. Chem. **36**, 40 (1944).

STECHERT, D. G.: A.S.T.M. Bull. **157**, 61 (1949).

Temperature retraction test.

GEHMAN, S. D., D. E. WOODFORD and C. S. WILKINSON: Ind. Engng. Chem. **39**, 1108 (1947).

GIBBONS, W. A., R. H. GERKE and H. C. TINGEY: Ind. Engng. Chem., analyt. Edit. **5**, 279 (1933).

SMITH, O. H., W. A. HERMONAT, H. E. HAXO and A. W. MEYER: Rubber Chem. Technol. **24**, 684 (1951).

STAMBAUGH, R. B., M. ROHNER and S. D. GEHMAN: J. appl. Physics **15**, 740 (1944).

YERZLEY, F. L. and D. F. FRASER: Ind. Engng. Chem. **34**, 332 (1942).

d) Dynamic measurements.

The majority of dynamic measurements are based upon vibrating systems in which stress and strain vary sinusoidally with time[1]. The time scale involves ranges from about 10 seconds[2] to 10^{-8} seconds (frequency range $10^{-1} < \omega < 10^{8}$) and a classification can be given into three groups based upon the ratio of the wave length of the elastic waves to the dimensions of the sample.

[1] There are however exceptions such as the free damped vibrations and the rebound test, which do not have simple sinusoidal time dependence. This leads to complications in the theoretical treatment of these experiments which will be discussed below.

[2] In exceptional cases vibrations with very low frequencies have been performed corresponding to time ranges up to 10^{4} seconds: W. PHILIPPOFF: J. appl. Physics **24**, 685 (1953). — W. LETHERSICH: Brit. J. appl. Physics **1**, 294 (1950).

Vibrations with additional mass: The dimensions of the sample are small compared to the wave length, the sample shows elasticity and viscosity but no effects due to inertia associated with the mass of the sample. In order to set up vibrations an additional mass may be fixed to the sample. Frequency range $10^{-1} < \omega < 10$.

Vibrations without additional mass: The dimension of the sample is of the same order of magnitude as the wave length. Under vibrating forces standing waves are built up in the sample which are dominated by the mass of the sample itself. Frequency range $10 < \omega < 10^4$.

Propagation of elastic waves: If the dimension of the sample is very large compared with the wave length, elastic waves are propagated through the medium. Frequency range $10^4 < \omega < 10^8$.

Free vibrations with additional mass. A mass M is attached to the sample and the system mass-sample is deformed at time zero and then released. It returns to its equilibrium position by means of *free damped vibrations*, from which the "*Eigenfrequency*" ω_e and the *logarithmic decrement* Δ (the natural logarithm of the ratio of two subsequent amplitudes) are determined.

Free damped vibrations of a linear viscoelastic material are governed by the equation

$$m\ddot{\gamma} + \frac{E_2}{\omega}\dot{\gamma} + E_1\gamma = 0. \qquad \text{(I,162)}$$

If the experiment is performed in tension or bending, γ is the displacement and E_1 and E_2 are dynamic storage and loss modulus in tension; m is a constant which equals the additional mass M times a dimensional factor A containing the dimensions of the sample. The dimension of A is a reciprocal length. If the experiment is performed in torsion the shear moduli G_1 and G_2 have to be used and M is the moment of inertia. A becomes the dimension of a reciprocal volume.

Equ. (I,162) is simply the equation for the equilibrium of forces during the vibration. The first term represents D'ALEMBERT's inertia forces, the second term the viscous forces and the third term the elastic forces of the sample.

A solution of equ. (I,162) which describes free vibrations is of the form

$$\gamma = \gamma_0 e^{-\frac{\omega_e \Delta}{2\pi}\cdot t} \cos \omega_e t \quad t > 0. \qquad \text{(I,163)}$$

An example of such an oscillation is shown in fig. I,35, from which the meaning of the Eigenfrequency ω_e and the logarithmic decrement Δ may be read off.

The experimental quantities Δ and ω_e can be related to the physical quantities m, E_1 and E_2. We insert (I,163) into (I,162) and obtain the two equations (by equating cosine and sine terms seperately).

$$E_1/m = \omega_e^2 [1 + \Delta^2/4\pi^2] \qquad \text{(I,164)}$$

$$E_2/m = \omega_e^2 \cdot \Delta/\pi.$$

For the loss angle we obtain

$$(\operatorname{tg}\delta)_e = E_2/E_1 = \frac{\Delta/\pi}{1+\Delta^2/4\pi^2}\,. \tag{I,165}$$

We are interested chiefly in the case of small damping, where

$$\Delta \ll 2\pi \quad \text{or} \quad (\operatorname{tg}\delta) \ll 1\,.$$

In this case we can neglect the terms of second order in (I,164) and (I,165). The Eigenfrequency yields the storage modulus, and the logarithmic decrement yields the loss angle:

$$\omega_0^2 \equiv E_1/m \simeq \omega_e^2 \tag{I,166}$$

$$(\operatorname{tg}\delta)_e = E_2/E_1 \simeq \Delta/\pi\,. \tag{I,167}$$

If the experiment is performed in torsion, we get the storage shear modulus G_1 and the damping in shear $(\operatorname{tg}\delta)_s$.

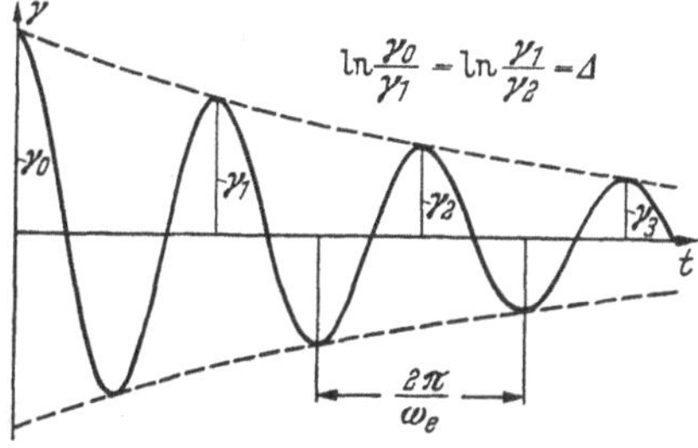

Fig. I, 35. Free damped vibration.

We have distinguished between two different frequencies, the Eigenfrequency ω_e and the resonance frequency ω_0. The Eigenfrequency is the angular frequency of the damped, free vibration, the resonance frequency is the angular frequency at which the amplitude in forced vibrations has its maximum. Generally ω_e and ω_0 are different, the difference being of second order in the logarithmic decrement. Only for small damping ($\Delta^2 \ll 4\pi^2$) ω_e and ω_0 may be put equal.

Equ. (I,164) and (I,165) do not contain the restriction to small damping and one would be inclined to use them for the calculation of the moduli and the damping in the case of large damping values, but there is a serious objection to this procedure. For it may be shown that the initial equation (I,162) itself, on which all considerations are based, will cease to be valid for large damping values.

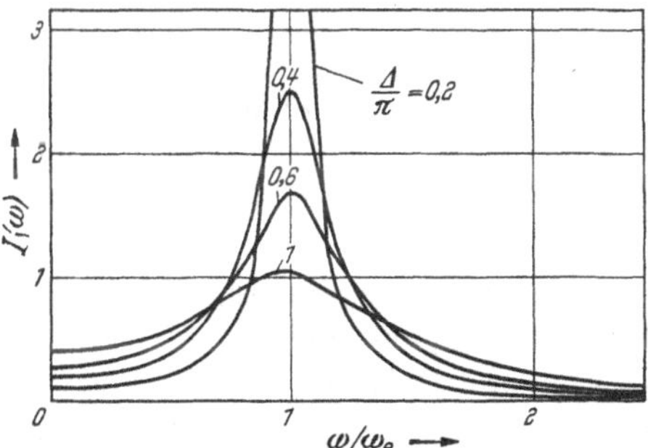

Fig. I, 36. Intensity functions I (ω) of the cosine terms contained in I, 163 for different values of the logarithmic decrement.

Indeed equ. (I,162) uses the concept of dynamic moduli E_1 and E_2 which is justified only when dealing with a pure sinusoidal deformation containing one single frequency (compare § 6c). On the other hand, the damped vibration (I,163) does not contain one frequency only, but — due to the damping factor — a whole spectrum of frequencies. The result of a Fourier analysis of equ. (I,163) is shown in fig. I,36, where the intensity functions $I(\omega)$ of the cosine terms[1] with frequency ω contained in (I,163) are

[1] There are also sine terms contained in (I,163), but these can be eliminated by an appropriate symmetrisation of the experiment in time scale and are not considered in the following.

plotted against ω for different values of the logarithmic decrement ($\Delta/\pi = 0{,}2$, 0,4, 0,6 and 1). For small damping these intensity curves show sharp maxima at the frequency $\omega = \omega_e$ and (I,163) may be approximately treated as a vibration with frequency ω_e, i.e. equ. (I,162) may be used for the description of the free vibration. For higher damping however the frequencies different from ω_e become more and more important and equ. (I,162) will fail.

The theory of free vibrations of viscoelastic materials with high damping has been treated very recently[1].

Forced vibrations with additional mass. The system mass-sample is driven by a sinusoidal force with amplitude a and an angular frequency ω different from the *resonance frequency* ω_0. In the case of forced vibrations of a viscoelastic material we have the equation

$$m\ddot{\gamma} + \frac{E_2}{\omega}\dot{\gamma} + E_1\gamma = a\cos\omega\, t. \tag{I,168}$$

The meaning of the physical constants is the same as in equ. (I,162). After some periods initial effects have been damped out and in stationary state the system is forced to vibrations with the imposed frequency ω and an amplitude $f(\omega)$, which depends on ω:

$$\gamma = a\cdot f(\omega)\cdot\cos(\omega t - \varphi). \tag{I,169}$$

The displacement is not in phase with the imposed stress but precedes it by the phase angle φ.

Inserting (I,169) into equation (I,168) and equating sine and cosine terms seperately, we get the relations between the amplitude and phase angle and the physical quantities E_1, E_2 and m.

$$f(\omega) = \frac{1}{\sqrt{[E_1 - m\omega^2]^2 + E_2^2}} \tag{I,170}$$

$$\operatorname{tg}\varphi = \frac{(E_2/E_1)}{1 - m\omega^2/E_1} = \frac{(\operatorname{tg}\delta)_e}{1 - m\omega^2/E_1}. \tag{I,171}$$

In the discussion of these equations it is advantageous to distinguish the two cases, when the imposed frequency ω is in the neighbourhood of the Eigenfrequency (resonance frequency) ω_0 and when it is very small compared with the resonance frequency.

Resonance case, small damping: In the case of small damping $(\operatorname{tg}\delta) \ll 1$, the moduli E_1 and E_2 vary only slowly with frequency and may be taken as approximately constant in equ. (I,170) and (I,171). Further we can introduce the resonance frequency ω_0 from equ. (I,166) and obtain for amplitude and phase angle

$$f(\omega)/f_0 = \frac{E_2}{\sqrt{m^2(\omega_0^2 - \omega^2)^2 + E_2^2}} \qquad f_0 = f(\omega_0) = 1/E_2 \tag{I,172}$$

$$\operatorname{tg}\varphi = (\operatorname{tg}\delta)/[1 - \omega^2/\omega_0^2]. \tag{I,173}$$

Vibration methods working in the resonance region *(resonance vibrators)* measure the amplitude as a function of the frequency ω in the neighbourhood of the resonance frequency ω_0.

[1] BRINKMAN, H. C.: Central Lab. T. N. O., Delft. Holland; to be published.

This resonance curve [equ. (I,172)] is shown in fig. I,37. It has a maximum of height $f_0 = 1/E_2$ at the resonance frequency $\omega = \omega_0$. From the position of this maximum the storage modulus can be derived [equ. (I,166)]. From the *half width* $\Delta\omega$ of the resonance curve ($2 \cdot \Delta\omega$ equals the width of the resonance curve at the amplitude $f_0/2$) the damping can be calculated[1]

$$\left(\frac{\Delta\omega}{\omega_0}\right) \simeq \frac{1}{2}\sqrt{3}\,E_2/E_1 = \frac{1}{2}\sqrt{3}\,(\mathrm{tg}\,\delta)_e\,. \tag{I,174}$$

It may be mentioned that the phase lag φ between stress and strain is not equal to the loss angle δ, but is given by equ. (I,173). For instance at resonance ($\omega = \omega_0$) the phase lag between stress and strain is 90° independently from the magnitude of the loss angle.

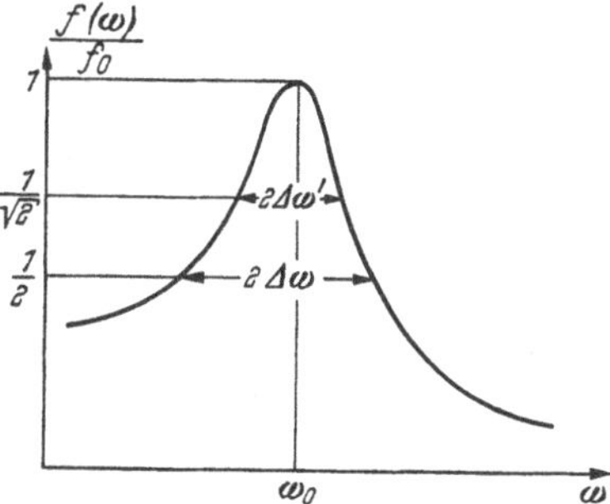

Fig. I, 37. The resonance curve as function of frequency (eq. I, 172).

Resonance case, large damping: There is no objection to the use of equation (I,168) in the case of large damping. Indeed the deformation (I,169) is now of a pure sinusoidal type with the frequency ω. However the evaluation of those measurements would imply the comparison of the resonance curve with the expression (I,170), in which E_1 and E_2 are now to be considered as functions of frequency ω. Apart from the fact that such a comparison is not unique (both E_1 and E_2 have to be determined as function of frequency from the same curve) it would call for an experimental accuracy which is not easily reached.

Non-resonance case: Very often measurements are made with imposed frequencies ω, which are very low compared to the resonance frequency *(non-resonance vibrators)*. In this case force and deformation are compared with respect to amplitude and phase. In the non-resonance case we may neglect $m\omega^2$ against E_1 in (I,170) and (I,171) and get the equations

$$f(\omega) = \frac{1}{\sqrt{E_1^2 + E_2^2}} = \frac{1}{|E|}, \quad \mathrm{tg}\,\varphi = \mathrm{tg}\,\delta = E_2/E_1\,. \tag{I,175}$$

The ratio of the amplitudes of stress and strain gives the absolute dynamic modulus $|E|$, and the phase lag between force and deformation yields the mechanical loss.

Equ. (I,175) are not restricted to small damping, but are valid always if the frequency is small compared with the resonance frequency. Therefore the non-resonance vibration is the most appropriate method for measuring large losses. In fact, our entire discussion on vibration measurements in § 3d, including the definitions of storage and loss modulus and damping [equ. (I,45) to (I,53)] has been based upon non-resonance forced vibrations.

---

[1] Sometimes the quantity $2\Delta\omega'$ is used, which equals the width of the resonance curve at the amplitude $f_0/\sqrt{2}$. $\Delta\omega'$ is related to $\Delta\omega$ by $\Delta\omega = \Delta\omega' \cdot \sqrt{3}$.

Non-resonance methods have a further advantage compared with the resonance method. Whilst the resonance method is restricted to a series of single resonance frequencies of the system, the non-resonance method allows a larger frequency range to be measured and over this range frequency can be varied continuously.

Vibrations without additional mass, or standing vibrations, are governed by the same equations as vibrations with additional masses, the role of the mass is taken here by the inert forces of the sample itself. Equ. (I,162) and (I,168), with their consequences for free and forced vibrations remain valid. The only difference lies in the meaning of the constant m, which is now related to the density ϱ of the material: $m = B \cdot \varrho$, where B is a factor of the dimension of a length squared, containing the geometrical dimensions of the sample.

The standing wave technique has been developed for different shapes of samples and for different modes of vibration as the transversal vibrations of reeds and thin strips, bending vibrations of rods, longitudinal vibrations of rods, and torsional vibrations of rods.

A sample shows a resonance frequency ω_0 which depends – apart from its dependence on moduli and density – on the dimensions of the sample and the geometry of the vibration. This dependence is contained in the magnitude of the factor B. Besides ω_0 there is a discrete number of higher resonance frequencies – corresponding to higher harmonic forms of vibration – which are simply related to ω_0.

In free vibration the lowest resonance frequency is measured together with the logarithmic decrement of the amplitude. In forced resonance vibrations the position and halfwidth of the resonance amplitudes are determined at the resonance frequencies corresponding to several of the lowest harmonics. In forced non-resonance vibrations force and deformation are compared with respect to magnitude and phase.

Wave-propagation methods. In considering *wave-propagation methods* in viscoelastic media we restrict ourselves, to plane harmonic waves propagated along the x direction. Such a plane wave is of the form

$$s = s(x, t) = s_0 e^{-\alpha x} \cos \omega [t - x/c] \qquad \text{(I,176)}$$

in which the displacement s depends on time and space and the meaning of the constants is the following:

ω angular frequency of the propagated wave
c velocity of propagation
α attenuation per unit of distance[1].

Further we define:

$\lambda = 2\pi c/\omega$ wave length
$2\pi r = \alpha\lambda$ attenuation per wave length.

[1] The quantity α has the dimension of a reciprocal length and is called "attenuation in Nepers per cm". However, the dimensionless quantity r is a more appropriate measure of the attenuation.

The physical meaning of these constants becomes clear if we consider the wave motion (I,176) at one constant time ($t = 0$) as a function of the x coordinate. Such an "instantaneous photograph" of the motion is shown in fig. I,38, where the amplitude $s(x, 0)$ is plotted against x:

$$s(x,0) = s_0 e^{-\alpha x} \cos \omega x/c = s_0 e^{-2\pi r x/\lambda} \cos 2\pi x/\lambda . \qquad (I,177)$$

The wave is taken to be generated at the point $x = 0$, where the maximum amplitude s_0 is reached at time zero. At a distance of one *wave length*, at $x = \lambda$, the wave motion is again in phase with the motion at $x = 0$, but the amplitude has decreased to s_1. The *attenuation per wave length* is formed by the logarithm of the ratio of two consecutive amplitudes

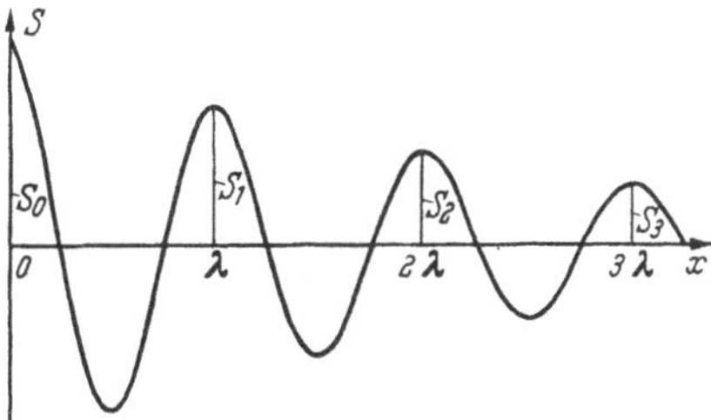

Fig. I, 38. Dependence of the amplitude of the propagated wave on distance x at time zero.

$$2\pi r = \ln s_0/s_1 = \ln s_1/s_2 . \qquad (I,178)$$

Wave propagation in viscoelastic materials is governed by an equation of the form[1]

$$M_1 \frac{\partial^2 s}{\partial x^2} + \frac{M_2}{\omega} \frac{\partial^3 s}{\partial t\, \partial x^2} = \varrho \frac{\partial^2 s}{\partial t^2} , \qquad (I,179)$$

M_1 and M_2 are the storage and loss parts of an effective stiffness. The meaning of these quantities depends on the special form of the wave considered. In wave propagation two quantities are measured, namely the velocity of propagation c and the attenuation per wave length $2\pi r$.

Inserting (I,176) into (I,179), the measured quantities r and c may be related to the physical quantities occurring in equ. (I,179). We get the equations

$$\left.\begin{aligned} M_1 &= \varrho c^2 \frac{1 - r^2}{[1 + r^2]^2} \\ M_2 &= 2 \varrho c^2 \frac{r}{[1 + r^2]^2} \\ M_2/M_1 &= 2r/[1 - r^2] . \end{aligned}\right\} \qquad (I,180)$$

Equations (I,180) are not restricted by the magnitude of the damping and may be used for large damping too. However, if the damping is small, $M_2/M_1 \ll 1$, the equations simplify considerably. The *velocity of propagation* yields an effective storage modulus and the attenuation per wave length yields the loss angle:

$$c \simeq \sqrt{M_1/\varrho} \qquad 2\pi r \simeq \pi M_2/M_1 . \qquad (I,181)$$

For propagation of shear waves, the effective stiffness has the meaning of the complex shear modulus $M_1 = G_1$, $M_2 = G_2$ and for propagation

[1] This may be shown by inserting (I,176) into the general equations of motion of a linear isotropic viscoelastic solid.

of compressional waves in bulk, the effective stiffness is the sum of complex bulk modulus and shear modulus $M_1 = K_1 + 4/3\,G_1$, $M_2 = K_2 + 4/3\,G_2$. Another technique is the propagation of longitudinal waves along a thin filament, in which case the effective stiffness is equal to YOUNG's modulus $M_1 = E_1$, $M_2 = E_2$.

We conclude the consideration of vibration measurements with a survey of the methods of determination of damping and modulus at small damping (table I,11).

Table I,11. *Methods of determination of damping and dynamic modulus.*

Method	Damping (determined by)	Modulus (determined by)	Formulae
Free vibrations	logarithmic decrement	Eigenfrequency	(I,166), (I,167)
Forced vibrations, resonance	half width of resonance curve	resonance frequency	(I,166), (I,174)
Forced vibrations, non-resonance	phase lag	ratio stress-strain amplitudes	(I,175)
Wave propagation	attenuation per wave length	velocity of propagation	(I,181)

Literature to § 7d.

General.

DILLON, J. H. and J. D. GEHMAN: India Rubber World **115**, 61, 76, 217 (1946); Rubber Chem. Technol. **20**, 827 (1947).

FERRY, J. D., W. M. SAWYER and J. N. ASWORTH: J. Polymer Sci. **2**, 593 (1947).

MARVIN, R. S.: Ind. Engng. Chem. **44**, 696 (1952).

NOLLE, A. W.: J. appl. Physics **19**, 753 (1948).

SPÄTH, W.: Kautschuk u. Gummi **2**, 311 (1949).

Free vibrations.

BALLOU, J. W. and J. C. SMITH: J. appl. Physics **20**, 493 (1949).

BENBOW, J. J.: J. Sci. Instr. **30**, 412 (1953).

CASSIE, A. B. D., M. JONES and W. J. S. NAUNTON: Trans. Inst. Rubber Ind. **12**, 49 (1936).

HOPKINS, I.: Trans. Amer. Soc. mechan. Engr. **73**, 195 (1951).

IWAYANAGI, S. and T. HIDESHIMA: J. physic. Soc. Japan **8**, 365 (1953).

JENCKEL, E.: Kunststoffe **40**, 98 (1950).

JENCKEL, E., H. HERTOG and E. KLEIN: Z. Naturforsch. **8a**, 255 (1953).

KIRBY, P. L.: J. Soc. Glass. Technol. **37**, 7T (1953).

KUHN, W. and O. KÜNZLE: Helv. chim. Acta **30**, 839 (1947).

NIELSEN, L. E.: Rev. Sci. Instr. **22**, 690 (1951).

NIELSEN, L. E., R. BUCHDAHL and R. LEVRAULT: J. appl. Physics **21**, 607 (1950).

NOLLE, A. W.: J. appl. Physics **19**, 753 (1948).

RORDEN, H. C. and A. GRIECO: J. appl. Physics **22**, 842 (1951).

SCHMIEDER, K. and K. WOLF: Kolloid-Z. **127**, 65 (1952).

SCHREUER, E.: Kolloid-Z. **132**, 75 (1953).

WOLF, K.: Kunststoffe **41**, 89 (1951).

YERZLEY, F. L.: Ind. Engng. Chem., analyt. Edit. **9**, 392 (1937).

Resonance vibrations.

DILLON, J. H., I. B. PRETTYMAN and G. I. HALL: J. appl. Physics **15**, 304 (1944).

FLETCHER, W. P. and J. R. SCHOFIELD: J. Sci. Instr. **21**, 193 (1944).

GEHMAN, S. D.: J. appl. Physics **13**, 402 (1942).

GEHMAN, S. D., D. E. WOODFORD and R. B. STAMBAUGH: Ind. Engng. Chem. **33**, 1032 (1941); **35**, 964 (1943).

HOFF, E. A. W.: J. Polymer Sci. **9**, 41 (1952).

KOSTEN, C. W. and C. ZWIKKER: Physica **4**, 221 (1937).

MEIJ, S. DE and G. J. VAN AMERONGEN: Kautschuk u. Gummi, in Press.

NAUNTON, W. J. S. and J. R. WAVIN: Rubber Chem. Technol. **12**, 332 (1939).

Non-resonance vibrations.

ALEKSANDROV, A. P. and Y. S. LAZURKIN: J. techn. Physics (USSR) **9**, 1249 (1939).

BLIZARD, R. B.: J. appl. Physics **22**, 730 (1951).

COOPER, L. V.: Ind. Engng. Chem., analyt. Edit. **5**, 350 (1933).

DAVIES, D. M.: Brit. J. appl. Physics **3**, 285 (1952).

FITZGERALD, E. R. and J. D. FERRY: J. Colloid Sci. **8**, 1 (1953).

GEHMAN, S. D., P. J. JONES and D. E. WOODFORD: Ind. Engng. Chem. **35**, 964 (1943).

HAVENHILL, R. S.: Physics **7**, 179 (1936).

HOLZMÜLLER, W.: Chem. Techn. (Berlin) **4**, 541 (1942).

LESSIG, E. T.: Ind. Engng. Chem., analyt. Edit. **9**, 582 (1937).

MARVIN, R. S., E. R. FITZGERALD and J. D. FERRY: J. appl. Physics **21**, 197 (1950).

MORRIS, R. E., R. R. JAMES and H. L. SNIJDER: Ind. Engng. Chem. **43**, 2540 (1951).

PHILIPPOFF, W.: J. appl. Physics **24**, 685 (1953).

ROELIG, H.: Rubber Chem. Technol. **12**, 395 (1939); **18**, 62 (1945).

ROELIG, H. and G. FROMANDI: Kautschuk u. Gummi **5**, 157 (1952).

Standing vibrations.

BALLOU, J. W. and J. C. SMITH: J. appl. Physics **20**, 493 (1949).

DEUTSCH, K., E. A. W. HOFF and W. REDDISH: J. Polymer Sci. **13**, 565 (1954).

ECKER, R.: Kautschuk u. Gummi **6**, 127 (1953).

HEYBOER, J., P. DEKKING and A. J. STAVERMAN: Proc. 2nd Int. Congr. Rheology, Oxford 1953, page 123.

HORIO, M., S. ONOGI, C. NAKAYAMA and K. YAMAMOTO: J. appl Physics **22**, 966, 977 (1951).

LABBE, B. G.: India Rubber World **128**, 193 (1953).

MÜLLER, F. H.: Kolloid-Z. **114**, 2 (1949); Kunststoffe **39**, 215 (1949).

NOLLE, A. W.: J. appl. Physics **19**, 753 (1948).

OBERST, H.: Acustica **2** (1952), Beih. 4, 181.

OBERST, H.: Kunststoffe **43**, 446 (1953).

OBERST, H. and G. W. BECKER: Akust. Beih., H. 1 (1954), 433.

POEL, J. VAN DER: Proc. 2nd Int. Congr. Rheology, Oxford 1953, page 331.

SACK, H. S., J. MOTZ, H. L. RAUB and R. N. WORK: J. appl. Physics **18**, 450 (1947).

WEGEL, R. L. and H. WALTHER: Physics **6**, 141 (1935).

Wave propagation methods.

ALTENBURG, K.: Z. physik. Chem. **202**, 14 (1953).

BALLOU, J. W. and J. C. SMITH: J. appl. Physics **20**, 493 (1949).

HATFIELD, P.: Brit. J. appl. Physics **1**, 252 (1950).

HILLIER, W.: Trans. I.R.I. **26**, 64 (1950).

IVEY, D. G., B. A. MROWCA and E. J. GUTH: J. appl. Physics **20**, 486 (1949).

KOLSKY, H.: Proc. 2nd Int. Congr. Rheology, Oxford 1953, page 79.

KUHN, W. and S. VIELHAUER: Z. physik. Chem. **202**, 124, 161 (1953).

LEE, E. H. and I. KANTER: J. appl. Physics **24**, 1115 (1953).

MARVIN, R. S., R. ALDRICH and H. S. SACK: J. appl. Phys. **25**, 1213 (1954).

MCSKIMIN, H. V.: J. acoust. Soc. Amer. **23**, 429 (1951).

NOLLE, A. W.: J. acoust. Soc. Amer. **19**, 194 (1947).

NOLLE, A. W. and S. C. MOWRY: J. acoust. Soc. Amer. **20**, 432 (1948).

NOLLE, A. W. and WESTERVELT: J. appl. Physics **21**, 304 (1950).

NOLLE, A. W. and P. W. SIECK: J. appl. Physics **23**, 288 (1952).

ROTH, W. and S. R. RICH: J. appl. Physics **24**, 940 (1953).

SOFER, G. A. and E. A. HAUSER: J. Polymer Sci. **8**, 611 (1952).

SACK, H. S. and R. W. ALDRICH: Physic. Rev. **75**, 1285 (1949).

WITTE, R. S., B. A. MROWCA and E. GUTH: J. appl. Physics **20**, 481 (1949).

e) Rebound tests.

Apart from the large number of vibration experiments, there are also some measurements involving rheological properties at very short times, in which stress and strain are not sinusoidal. An important example is the *rebound test* or *impact resilience* (in German: *Rückprallelastizität*), in which the sample under consideration is struck by a falling ball or pendulum. The energy dissipated in the sample during impact ΔW is measured as the difference of the kinetic energy of the striker before and after the blow

$$\Delta W = W_0 - W_1 .$$

The ratio of the kinetic energy after impact, W_1, to the kinetic energy before impact W_0 is called the impact resilience R:

$$R = W_1/W_0 = 1 - \Delta W/W_0 . \tag{I,182}$$

Resilience measurements are usually made by means of a ball dropped on the sample and by measuring the heigth of the ball h_0 before dropping, h_1 after bouncing once, h_2 after bouncing twice and so on. The ratio of these heights is equal to the ratio of the corresponding kinetic energies of the ball before the first, second and third rebound. To a first approximation the resilience (I,182) will be independent of the original height of dropping (as the dissipated energy will depend on amplitude in the same way as the energy stored elastically in the sample) and we have the equations

$$\left.\begin{aligned} R &= W_1/W_0 = W_2/W_1 = W_3/W_2 = \cdots \\ &= h_1/h_0 = h_2/h_1 = h_3/h_2 = \cdots. \end{aligned}\right\} \tag{I,183}$$

A physical interpretation of the rebound resilience has been given by MARVIN[1] who treats the rebound measurement approximately as a damped free vibration. The impact resilience R is related to the mechanical loss

$$R \simeq e^{-\pi \operatorname{tg} \delta} \simeq 1 - \pi \operatorname{tg} \delta \tag{I,184}$$

which equation will be true for small damping only. It follows that the ratio of the energy dissipated during the impact to the maximum elastic energy stored in the sample is equal to

$$\Delta W/W_0 \simeq \pi \operatorname{tg} \delta. \tag{I,185}$$

Also a more profound investigation of the rebound test[2] shows that the disipated energy may be related to the damping by means of equ. (I,185.) The value of the damping obtained in this way corresponds to an angular frequency $\omega = \pi/T$, where T is the time of contact between the sphere and the sample.

In this way the rebound test could be considered as a convenient, rapid method of determining the mechanical damping. The experimental times T involved in this experiment depend on the elasticity of the material and on geometrical factors. Measurements by TILLET[3] show the order of magnitude for the contact time T to be 10^{-4} seconds. The frequency at which the mechanical loss will be obtained lies therefore in the region of 10^4 cps.

Literature to § 7e.

BOONSTRA, B. B. S. T.: J. Rubber Ind. **119**, 4 (1950).

DILLON, J. H., I. B. PRETTYMAN and G. L. HALL: J. appl Physics **15**, 309 (1944).

FRIEDLANDER, G.: Anal. Chem. **22**, 1545 (1950).

JENCKEL, E. and E. KLEIN: Z. Naturforsch. **7**a, 619 (1952).

JONES, H. J.: Soc. chem. Ind. **67**, 415 (1948).

KLEIN, E. and E. JENCKEL: Z. Naturforsch. **7**a, 305 (1952).

LABBE, R. G.: India Rubber World **121**, 547 (1950).

MULLINS, L.: Trans. Inst. Rubber Ind. **22**, 235 (1947).

MULLINS, L.: J. Rubber Res. **16**, 180 (1947).

SCHREUER, E.: Kolloid-Z. **129**, 123 (1952).

SHAW, R. F.: India Rubber World **118**, 796, 868 (1948).

SPÄTH, W.: Gummi u. Asbest **2**, 180 (1949), 57 (1949).

STÖCKLIN, P.: Kautschuk **15**, 118 (1938); **18**, 151 (1942); **19**, 3 (1943).

[1] MARVIN, R. S.: Ind. Engng. Chem. **44**, 696 (1952).
[2] ZENER, C.: Physic. Rev. **59**, 669 (1941).
[3] TILLETT, J. P. A.: Proc. physic. Soc. **B 67**, 677 (1954).

§ 8. The relation of viscoelastic behaviour to molecular structure[1].

By JOHN D. FERRY.

a) Introduction.

The proceding Sections have shown how the viscoelastic properties of polymers and their solutions can be visualized in terms of mechanical models. This phenomenological approach is of great value in understanding the interrelation of different properties and gives some insight into the effects of temperature and concentration. However, the interpretation is not complete until the relation of viscoelastic properties to molecular structure is established.

It is clear from the relationships given in the preceding Sections that, insofar as the viscoelastic behaviour is linear, a molecular interpretation of one time-dependent mechanical property (such as stress relaxation) will serve as a molecular interpretation of all other such properties. Various molecular theories have been proposed, some to interpret creep, others stress relaxation, and others dynamic mechanical properties. For comparing different theories and testing them against experimental data, it is convenient to use the spectrum of relaxation times or retardation times as a common denominator. The results of certain theories are expressible in the form of discrete spectra and those of others in the form of continuous spectra. The following definitions, which include both types, are for the case of shear (see also § 4):

Relaxation Spectrum:

Discrete: G_i = contribution to instantaneous shear modulus associated with relaxation time τ_i.

Continuous: $H d \ln \tau$ = contribution to instantaneous shear modulus associated with relaxation times whose logarithms lie between $\ln \tau$ and $\ln \tau + d \ln \tau$.

Retardation Spectrum:

Discrete: J_i = contribution to equilibrium or steady-state compliance associated with retardation time τ_i.

Continuous: $L\, d \ln \tau$ = contribution to equilibrium or steady-state compliance associated with retardation times whose logarithms lie between $\ln \tau$ and $\ln \tau + d \ln \tau$.

The relaxation spectrum may, of course, be visualized as a MAXWELL model, with springs G_i in parallel, each relaxed by a dashpot of viscosity $G_i \tau_i$; while the retardation spectrum may be visualized as a VOIGT model,

[1] The writer is much indebted to Drs. P. E. ROUSE JR., F. BUECHE and W. G. HAMMERLE for the opportunity of seeing their respective theoretical calculations in advance of publication.

with springs of stiffness $1/J_i$ in series, each retarded by a dashpot of viscosity τ_i/J_i.

A complete description of the viscoelastic properties may require, in addition to a continuous spectrum H or L, certain supplementary constants. Thus in a cross-linked system H does not include the equilibrium modulus, which is an integration constant for the integral of $H d \ln\tau$; whereas in an uncross-linked system L does not include the steady-flow viscosity, which must be specified separately. The function L must also be supplemented by the instantaneous modulus, whether the system is cross-linked or not. The continuous spectra H and L are interconvertible if these supplementary constants are known[1].

Examination of these spectra as determined by experimental measurements on polymers reveals certain characteristic regions of time scale, which differ so widely that it is scarcely to be expected that one simple molecular theory could interpret the entire panorama. For example, the continuous distributions H and L for a sample of polyisobutylene (viscosity-average molecular weight $1{,}35 \cdot 10^6$), compiled from various sources[2] and reduced to 25°C, are reproduced as double logarithmic plots in fig. I, 39. The H curve shows a maximum at very short times, a gradual descent, a plateau, and a steep descent. The mechanical consistencies which would be revealed by experiments within the corresponding regions of time scale are indicated. For some polymers in the glassy region, to the left of the principal maximum, there may be subsidiary maxima associated with motions of side groups of the polymer molecules[3]; these are beyond the scope of present theories, which deal only with motions of the central molecular chain. The function H as determined for various other polymers and polymer solutions resembles that in fig. I, 39 except that the plateau is absent under certain conditions, for example in very dilute solutions and also for low molecular weights (less than the order of 10^4). Numerous examples of experimentally determined spectra are given in Chapter VI.

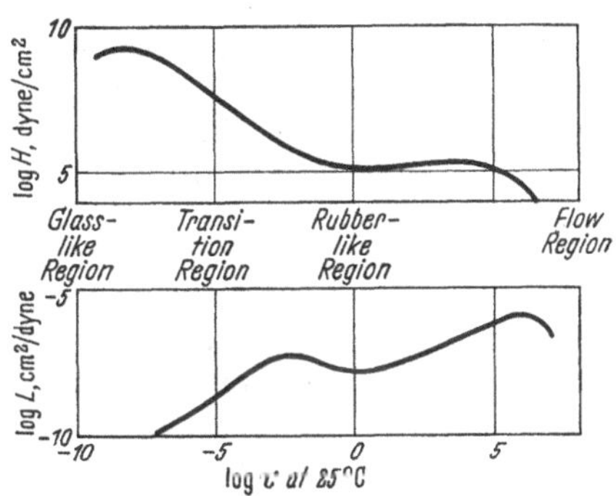

Fig. I, 39. Experimental relaxation spectrum H and retardation spectrum L, plotted logarithmically, for polyisobutylene (viscosity average molecular weight $1{,}35 \cdot 10^6$) at 25°C.

[1] GROSS, B. and H. PELZER: J. appl. Physics **22**, 1035 (1951).

[2] MARVIN, R. S.: Proc. Sec. Int. Congr. Rheology, page 156, London: Butterworths Ltd. (1954). — E. R. FITZGERALD, L. D. GRANDINE JR. and J. D. FERRY: J. appl. Physics **24**, 650 (1953). — R. D. ANDREWS and A. V. TOBOLSKY: Unpublished work.

[3] SCHMIEDER, K. and K. WOLF: Kolloid-Z. **134**, 149 (1953). — K. DEUTSCH, E. A. W. HOFF and W. REDDISH: J. Polymer Sci. **13**, 565 (1954). — J. HEYBOER, P. DEKKING and A. J. STAVERMAN: Proc. Sec. Int. Congr. Rheology, page 123, London: Butterworths Ltd. (1954).

Most molecular theories of polymer viscoelasticity are more or less related to the kinetic theory of equilibrium rubberlike elasticity, in which the storage of elastic energy is attributed to the entropy decrease associated with changes in the configurational distributions of flexible chain molecules. In the equilibrium theory for a cross-linked network, only the equilibrium distribution has to be treated, as modified by the applied stress which can be considered as transmitted to individual strands through the network junctions. To describe time-dependent mechanical properties, it is necessary to treat rates of configurational changes.

Qualitatively, short relaxation and retardation times reflect configurational changes requiring short-range cooperation of segments of the molecular chain, and long times reflect long-range cooperation. Near the maximum of H in fig. I,39 the cooperation is so local that it depends on steric features and intermolecular forces which are peculiar to the detailed chemical structure. In the region of gradual descent, the cooperation involves segments farther removed, and is less specifically related to chemical structure, so that when H has dropped to 10^7 to 10^6 dyn/cm^2 the curves are similarly shaped for all polymers and all their solutions (Chapter VI). In the plateau region the cooperation involves more than one polymer molecule, apparently because occasional entanglements provide rather tight coupling between a given molecule and some of its neighbors. This effect appears only at high molecular weights, and the plateau width depends on the molecular weight (and, in solutions, on the polymer concentration). Finally, the longest relaxation time represents cooperation at the longest range, and beyond this there are no elastic contributions. As the maximum relaxation time is approached, H drops steeply, with a slope which depends primarily on molecular weight distribution. Similar qualitative statements can be made concerning the retardation spectrum L.

A fundamental parameter which enters most of the molecular theories is the force required to move a monomer unit of a polymer chain through its surroundings at unit velocity. This friction coefficient, ζ_0, determines the time scale where the characteristic regions of the relaxation spectrum fall; and the effect of temperature on H, which is essentially a shift of the logarithmic time scale, reflects primarily the temperature dependence of ζ_0. The magnitude of ζ_0 cannot be predicted, however, by the theories which are reviewed here.

The position of the relaxation spectrum at short times, corresponding for example to about $\tau < 10^{-5}$ sec. in fig. I, 39, is characteristic of the chemical structure of a polymer but does not depend much on the molecular weight or the presence of a mild degree of cross-linking. This reflects the qualitative expectation that motions involving short-range cooperation of polymer segments will involve mostly portions of molecules far removed from free ends or from cross-links, and will be oblivious of any influence from the latter. For most of the segments participating in such motions, the friction coefficient ζ_0 should be practically independent of molecular weight and cross-linking.

The molecular theories of viscoelastic behaviour discussed in the following paragraphs are introduced in the chronological order of their development.

b) Early theories.

Theory of ALFREY. The first attempt[1] to relate viscoelastic behaviour quantitatively to molecular structure pictured the response of an uncrosslinked polymer in creep as the sum of configurational changes of molecular segments of various lengths. The original derivation for compliance in extension may be converted to compliance in shear by multiplying by 3; then the compliance contribution associated with any type of segment, regardless of length, is calculated to be $J_i = 3/5nkT$, where n is the number of molecules per cc., and the associated relaxation time is assumed to be proportional to $e^{B\sqrt{jq}}$ where jq is the number of bonds in the segment and B is an undefined constant. (The number of monomer units in the segment is q, and j is the number of bonds per monomer unit, usually 2.) The latter relation was taken by analogy with the dependence of steady flow viscosity on degree of polymerization found in early studies of polymers of rather low molecular weight[2], a dependence which, however, was later found to be without general applicability[3]. The assumed exponential square root relation led to the following expression for the continuous retardation spectrum:

$$L = 3(C\ln\tau - D), \quad \alpha > \ln\tau > \alpha + B\sqrt{jZ} \qquad (I,186)$$
$$C = 2/5B^2 nkT$$

where k is BOLTZMANN's constant and D and α are constants which depend on temperature but are not predictable by the theory; jZ is the number of chain bonds per molecule, Z being the degree of polymerization. The polymer is assumed to be homogeneous with respect to molecular weight.

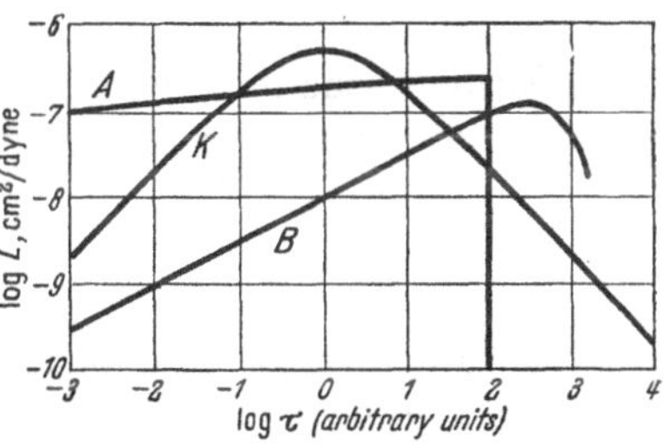

Fig. I, 40. Retardation spectra as predicted by theories of ALFREY (A), KIRKWOOD (K), and BLIZARD (B). The parameters for A are chosen to correspond to a total spectrum width of 8 cycles of logarithmic time; the height of K corresponds to a molecular weight between cross-links of 50,000, at 25° C; the remaining parameters are all arbitrary.

ALFREY's function L, with parameters chosen to correspond to a total width on the time scale of 8 powers of ten, is plotted with arbitrary scales as curve A in fig. I,40. Its shape is superficially similar to that of the second peak in the experimental curve for polyisobutylene (fig. I,39), but it rises too slowly and falls too abruptly. The theory correctly predicts that at short τ the value of L is independent of molecular weight; that L vanishes at a critical value of τ which increases with increasing molecular weight; and

[1] ALFREY, T.: J. chem. Physics **12**, 374 (1944).
[2] FLORY, P. J.: J. Amer. chem. Soc. **62**, 1057 (1940).
[3] FOX JR., T. G. and P. J. FLORY: J. Amer. chem. Soc. **70**, 2384 (1948).

that the effect of temperature is primarily a shift of the logarithmic time scale.

Theory of KIRKWOOD. In the first derivation of the retardation spectrum of a cross-linked polymer, KIRKWOOD[1] considered the response of a flexible polymer chain to a sinusoidally varying force applied to its ends, while the chain undergoes Brownian motion in accordance with the diffusion tensor previously calculated by KIRKWOOD and FUOSS[2] in their theory of dielectric dispersion. The bond angles are assumed to be 90°. The calculation is made for extension; when converted to shear it yields the following continuous distribution:

$$\left.\begin{aligned} L &= J\,(\tau/\tau_m)/(1+\tau/\tau_m)^2 \\ \tau_m &= 2\,j\,Z_c\,b^2\,\zeta_0/k\,T \end{aligned}\right\} \qquad (\mathrm{I}, 187)$$

where J is the equilibrium compliance (given by the kinetic theory of rubberlike elasticity as $Z_c M_0/\varrho R T$, Z_c being the number of monomer units between cross-links, M_0 the molecular weight of a monomer, and ϱ the density); j is again the number of bonds per monomer unit (usually 2); b the bond length, and ζ_0 the friction coefficient per monomer unit. KIRKWOOD's function is also plotted logarithmically as curve K in fig. I,40; the time scale is arbitrary, but the magnitude of L is that predicted by theory for a molecular weight between cross-links ($Z_c M_0$) of 50,000, an appropriate value for a lightly cross-linked rubber[3]. It is symmetrical about a maximum at $\tau = \tau_m$, and at this point $L = J/4$.

At short times, the function reduces to $L = (M_0/2\varrho N_0 j b^2 \zeta_0)\,\tau$, where N_0 is AVOGADRO's number; here it is directly proportional to τ, and it is independent, as it should be, of the concentration of cross-links; the magnitude is determined essentially by the friction coefficient ζ_0 (the bond length b being the same for most polymers). At intermediate times, the position of the maximum on the time scale, τ_m, is proportional to ζ_0 and to the molecular weight between cross-links, measured by Z_c. At long times L is inversely proportional to τ. The effect of temperature is, again, primarily a shift in the time scale through the temperature dependence of ζ_0.

Theory of BLIZARD. In another theory, due to BLIZARD[4], the polymer molecules are represented as springs moving in a viscous medium. The detailed molecular structure is ignored, but more attention is paid to the lengths of the chains and the topology of their interconnections, so that the treatment is applicable to both cross-linked and uncross-linked systems; and in the cross-linked case it includes chains bound at only one end as well as at both.

The important parameters are the compliance per unit length of a molecular spring and the viscous coupling constant per unit length of a

[1] KIRKWOOD, J. G.: J. chem. Physics **14**, 51 (1946).
[2] KIRKWOOD, J. G. and R. M. FUOSS: J. chem. Physics **9**, 329 (1941).
[3] FLORY, P. J.: Ind. Engng. Chem. **38**, 417 (1946).
[4] BLIZARD, R. B.: J. appl. Physics **22**, 730 (1951).

spring with the surrounding medium. These remain unspecified, and no attempt is made to define the absolute time scale nor the absolute values of mechanical properties. The relative frequency dependence of the real and imaginary parts of the complex modulus, G' and G'', is obtained in a very complicated form for the general case of an arbitrary degree of cross-linking.

The result is considerably simpler for the case of a linear polymer, homogeneous with respect to molecular weight, with no cross-linking at all. Here $G' + iG'' = n[(iK\omega)^{1/2} \coth (iK\omega)^{1/2} - 1]$, where n is the number of molecules per cc., K is a constant and ω the circular frequency. The corresponding relaxation and retardation spectra at *short* times are simply proportional to $\tau^{-1/2}$ and $\tau^{1/2}$ respectively, and are independent of molecular length. At *long* times the behaviour approaches that of a single MAXWELL element with a relaxation time equal to $K/45$. Since K is proportional to Z^2, where Z is the number of monomer units in a chain, it is to be expected that the relaxation and retardation spectra will vanish above a value of τ which is proportional to the square of the molecular weight.

The BLIZARD function L for an uncross-linked polymer is also plotted schematically as curve B in fig. I, 40. In shape it is intermediate between the other two, and it resembles the predictions of the more recent theories of ROUSE and BUECHE, to be described below.

Theory for dilute solutions of KIRKWOOD. In very dilute solution, a completely general formulation of the motion of a macromolecule has been given by KIRKWOOD[1], embodying the interaction of Brownian motion with hydrodynamic forces exerted by the solvent when the system is subjected to alternating shear. The general case is not readily expressible in terms of measurable quantities; but for the specific case of a rigid rod, represented by a rectilinear series of structural elements each with the same friction coefficient, KIRKWOOD and AUER[2] have obtained explicit equations for the complex rigidity $G' + iG''$. They correspond to a single MAXWELL element with a relaxation time approximately proportional to be cube of the molecular length, and a maximum rigidity $G = (3/5)nkT$ where n is again the number of molecules per cc.

c) Theory of Rouse.

A less detailed description of macromolecular motions in dilute solution than KIRKWOOD's is provided by the theory of ROUSE[3]. This is the first successful detailed quantitative interpretation of viscoelastic properties of flexible polymers. Although designed for very dilute solutions only, it can be applied with some success to more concentrated systems also.

The polymer molecule is divided into equal segments whose length is arbitrary but long enough so that the distance between the ends of a

[1] KIRKWOOD, J. G.: Recueil Trav. chim. Pays-Bas **68**, 649 (1949).
[2] KIRKWOOD, J. G. u. P. L. AUER: J. chem. Physics **19**, 281 (1951).
[3] ROUSE, P. E. JR.: J. chem. Physics **21**, 1272 (1953).

segment follows the usual Gaussian distribution function of the theory of rubberlike elasticity. There are N segments per molecule and $q(=Z/N)$ monomer units per segment. The friction coefficient of a segment is $f_0 = q\,\zeta_0$ and an effective average of f_0 over all segments is supposed to be applicable to describe the resistance encountered in various configurational changes; f_0 depends on the viscosity of the solvent and is found empirically to be roughly proportional to it, but it also depends on the density of polymer within a single molecular coil. When the solution is subjected to an applied stress, each junction between two segments is assumed to move with a velocity equal to the vector sum of the solvent velocity at that point and the velocity due to the Brownian motion of the chain.

The simultaneous motions of all segment junctions can be described as the sum of a series of cooperative modes. It turns out that the response of the system associated with each mode of cooperative motion corresponds to a MAXWELL element; that is, energy is stored proportional to a modulus contribution G_i and dissipated by relaxation with a time constant τ_i. This leads to the following discrete relaxation spectrum:

$$\left.\begin{aligned} G_p &= n\,k\,T \\ \tau_p &= \sigma^2 f_0/24\,k\,T \sin^2[p\,\pi/2(N+1)]\,, \end{aligned}\right\} \qquad \text{(I, 188)}$$

where n is, as usual, the number of polymer molecules per cc., σ is the root mean square end-to-end separation of a segment, and the index p goes from 1 to N. As $p \to N$, the theory breaks down because it ignores configurational changes *within* segments, which have shorter relaxation times than the minimum value of τ_N given by equations I, 188; these would involve local cooperation which would be influenced by intramolecular steric effects and interactions and could not be treated so simply anyway. If the modes are limited to those with $p < N/5$, a minor additional restriction, and if $N \gg 1$, the second of equations I, 188 reduces to

$$\tau_p = \sigma^2 N^2 f_0/6\,\pi^2 p^2 k\,T = a^2 Z^2 \zeta_0/6\,\pi^2 p^2 k\,T \qquad \text{(I, 189)}$$

where a is a length[1] of the order of a chain bond, defined by the property of a Gaussian chain that $\sigma^2 = a^2 q$. Thus the arbitrary choice of segment length does not influence the final result, which depends only on ζ_0 and a, properties of the monomer unit, and on Z. The choice of segment length influences only the maximum value of p (minimum τ) at which the treatment is applicable.

Evaluation in terms of steady flow viscosity. Equation I, 189 still contains one unknown parameter, ζ_0. This can, however, be eliminated by the relation that the contribution of the polymer to steady flow viscosity η is $\sum G_p \tau_p$. Thus $\eta - \eta_s = n a^2 Z^2 \zeta_0/36$, and τ_p may be specified by

$$\tau_p = 6\,(\eta - \eta_s)/\pi^2 p^2 n\,k\,T \qquad \text{(I, 190)}$$

[1] The relation of the length a to the familiar statistical segment length A_m of KUHN is given by the equation $a^2 = b\,A_m$, where b is the length of a fully extended monomer unit [W. KUHN u. H. KUHN: Helv. chim. Acta **26**, 1394 (1943)].

where η_s is the solvent viscosity; so the entire linear viscoelastic behaviour can be predicted from knowledge of the molecular weight and the steady flow viscosities of solvent and solution. For very dilute solutions this prediction is remarkably successful[1] (see Chapter VI). It should be remarked, however, that the viscosity increment $\eta - \eta_s$ is predicted here to be proportional to the product of concentration (c) and molecular weight (M), since $nZ^2 \alpha c M$; thus the intrinsic viscosity is supposed to be directly proportional to molecular weight. This is in fact the steady-state behaviour of an unshielded or free-draining molecular coil as treated by DEBYE[2], but the observed steady-state behaviour is different and the intrinsic viscosity increases with a lower power of M, owing to partial hydrodynamic shielding (see Volume II of this Series). The effect which such partial shielding might have on the relaxation spectrum is not clear*.

Relaxation and retardation spectra. At short times ($\tau < \tau_3$, as specified by equation I,189 with $p = 3$), the discrete spectrum given by equations I,188 and I,189 can be replaced within 0,1% error by a continuous relaxation spectrum, $H\,d\ln\tau = -nkT\,(dp/d\tau)\,d\tau$; differentiation of equation I,189 yields

$$H = (aZn/2\pi)\,(\zeta_0 kT/6)^{1/2}\tau^{-1/2} \tag{I,191}$$

To derive the corresponding retardation spectrum, an exact interconversion formula of GROSS and PELZER[3] may be used. When H is proportional to $\tau^{-1/2}$ and $\tau < \tau_3$, but τ is not so small that H approaches the maximum value indicated in fig. I,39, this formula reduces[4] to a very simple relation $H = 1/\pi^2 L$. Hence

$$L = (2/\pi a Z n)\,(6/\zeta_0 kT)^{1/2}\tau^{1/2}. \tag{I,192}$$

Thus at short times H and L are proportional to $\tau^{-1/2}$ and $\tau^{1/2}$ respectively. Here the ROUSE theory resembles that of BLIZARD; the two are similar also in specifying that the maximum relaxation time τ_1 is proportional to the square of the molecular length (equation I,189 with $p = 1$). They differ, of course, in that ROUSE's, unlike BLIZARD's, provides all constants in terms of molecular parameters.

In terms of the steady flow viscosity, equations I,191 and I,192 become

$$H = \left(\sqrt{6}/2\pi\right)[nkT\,(\eta - \eta_s)]^{1/2}\tau^{1/2} \tag{I,193}$$

$$L = \left(2/\pi\sqrt{6}\right)[nkT\,(\eta - \eta_s)]^{-1/2}\tau^{1/2}. \tag{I,194}$$

At long times, the spectra are not expressible by continuous functions if the polymer is homogeneous with respect to molecular weight, since τ_1,

---

[1] ROUSE, P. E., JR. u. K. SITTEL: J. appl. Physics 24, 690 (1953).

[2] DEBYE, P.: J. chem. Physics 14, 636 (1946).

* This question has been re.ently treated theoretically by B. H. ZIMM (Meeting Amer. Chem. So ., Sept. 1954).

[3] GROSS, B. u. H. PELZER: J. appl. Physics 22, 1035 (1951).

[4] FERRY, J. D., R. F. LANDEL u. M. L. WILLIAMS: J. appl. Physics 26, 359 1955).

τ_2, and τ_3 are spaced too far apart. Nevertheless, a heterogeneous polymer should give continuous spectra which would be related to the molecular weight distribution.

Deduction of reduced variables. The dependence of H and related quantities on temperature and concentration, as expressed in the foregoing equations, can be very conveniently formulated in terms of certain reduced variables. The following definitions of the reduced distribution function H_r and the reduced relaxation time τ_r are also inherent in some of the earlier theories; they can alternatively be derived from some simple assumptions[1] quite independently of any prescribed functional form of H, and they have been extensively used to combine data at different temperatures and concentrations (see also § 5).

$$\left.\begin{aligned} \tau_r &= \tau\, T\, c/(\eta - \eta_s)\, T_0 \\ H_r &= H\, T_0/Tc \end{aligned}\right\} \qquad \text{(I, 195)}$$

where T is the absolute temperature of measurement, T_0 a standard temperature (often chosen as 298° K), and c the polymer concentration in g./cc. In these terms, noting that $nk = cR/Z M_0$, where M_0 is the molecular weight of a monomer unit, equation I, 193 becomes

$$H_r = \left(\sqrt{6}/2\pi\right) (R\, T_0/Z\, M_0)^{1/2}\, \tau_r^{-1/2}. \qquad \text{(I, 196)}$$

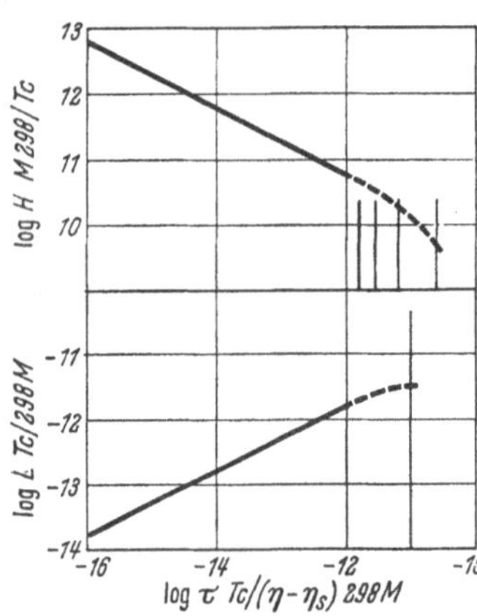

Fig. I, 41. Above: relaxation spectrum of ROUSE (discrete spectrum at right, equivalent continuous spectrum at left), reduced to a universal function independent of temperature, concentration, and molecular weight. Below: corresponding continuous retardation spectrum, together with terminal discrete contribution. The standard temperature T_0 is indicated as 298° K.

This is now a master function independent of temperature and concentration, and also of the choice of solvent, but still dependent on the degree of polymerization Z. Although equation I, 196 applies only to the continuous part of the relaxation spectrum, the reduced variables are of course equally applicable to the discrete part of the spectrum, where the corresponding reduced relaxation time τ_{rp} is $6\, Z M_0/\pi^2 p^2 R\, T_0$.

A further reduction could be made to eliminate dependence on Z by plotting $H_r M$ against τ_r/M, where $M (= Z M_0)$ is the polymer molecular weight. It can be seen from equation I, 196 that this yields a universal function for all systems subject to the assumptions underlying the ROUSE theory. The continuous and discrete parts of this function are plotted logarithmically in fig. I, 41, together with a plot of $L T c/T_0 M$ which shows the continuous region and the terminal retardation time. The universal curve for the relaxation spectrum can, of course, be converted to $\log H$ *vs.* $\log \tau$

[1] FERRY, J. D.: J. Amer. chem. Soc. **72**, 3746 (1950).

for any specific dilute solution by shifting vertically, adding $\log c\,T/T_0 M$, and horizontally, adding $\log(\eta - \eta_s) M T_0/cT$. The range of concentrations for which the theory is supposed to be applicable is of the order of $c = 10^{-3}$ g./cc.

Application to concentrated solutions. In concentrated solutions, the polymer molecules interpenetrate and a moving segment must push aside not only solvent molecules and segments belonging to its own molecule but also segments from foreign molecules. As long as neither concentration nor molecular weight is too high, an effective average value of f_0 can again apparently be used for all modes of motion, and the shape of fig. I, 41 is unchanged; equation (I, 193) can be used even for undiluted polymers with some success at quite low molecular weights (see Chapter VI). In the definition of reduced time, η is used instead of $\eta - \eta_s$, the solvent viscosity being negligible by comparison.

For high molecular weights, however, apparently coupling to neighboring molecules causes translation of a molecule, together with motions involving long-range segment cooperation, to be abnormally slow. In BUECHE's treatment of steady flow viscosity, f_0 is replaced by f, a much larger effective average frictional coefficient[1]. Since the slow relaxation mechanisms and the steady flow viscosity are affected in a similar way by this coupling, the theory in terms of reduced variables, or with τ_p defined in terms of viscosity as in equation (I, 190), is still applicable at the long-time end of the time scale; f cancels out in terms of viscosity just as f_0 (or ζ_0) did before[2].

But the short-range cooperative motions are oblivious of coupling and their rate is still governed by f_0 rather than f. Hence the linear portion of $\log H$ *vs.* $\log \tau$, at short times, appears far to the left of the point which would be predicted by equation (I, 190). In between the region where $H \propto \tau^{-1/2}$, now displaced to the left, and the discrete region near τ_1, still in its proper place, there appears a flatter portion of the H function, or a plateau as seen in fig. I, 39. To the right of the plateau, the theory correctly predicts the behaviour with τ defined by equation (I, 190); to the left, the theory correctly predicts the behaviour with τ defined by equation (I, 189), but with ζ_0 no longer determinable from η.

Application to undiluted polymers. In undiluted polymers, as in concentrated solutions, the relaxation spectrum includes a plateau except for samples of quite low molecular weight. To the left of the plateau there is usually a region with a slope of $-1/2$ on a log-log plot, corresponding to a range where equation (I, 191) is applicable and from which ζ_0 can be calculated. This region is narrower, however, than in concentrated solutions (where it may extend over several decades); with decreasing τ, H rises more sharply and passes through a maximum. The detailed shape of the spectrum here depends on specific chemical structure, which is not embodied in the theory.

---

[1] BUECHE, F.: J. chem. Physics **20**, 1979 (1952).

[2] FERRY, J. D., M. L. WILLIAMS and D. M. STERN: J. physic. Chem. **58**, 987 1954).

Application to Heterogeneous Polymers. At short times, the relaxation and retardation spectra are independent of molecular length as well as molecular weight distribution; this can be seen from equations (I, 191) and (I, 192), noting that nZ measures the weight concentration of polymer and is independent of molecular weight. At long times, the behaviour of heterogeneous polymers could in principle be described in terms of molecular weight distribution by equation (I, 188) or related forms. For one particular case the calculation is quite simple — the steady-state compliance of a uncross-linked polymer, corresponding to the maximum elastic energy storage in steady-state flow[1], as follows:

$$J = \sum G_p \tau_p^2/\eta^2 = (2/5)\, \overline{M}_{z+1}\, \overline{M}_z/\overline{M}_w\, c\, R\, T \qquad \text{(I, 197)}$$

where the molecular weight averages have their usual significance. In an undiluted polymer c would be replaced by ϱ, the density. The derivation assumes that the average friction coefficient f is the same for the slow relaxation mechanisms of all molecules, which should be true unless some of the molecules are quite short. The result reflects qualitatively the experimental observation that J increases with breadth of molecular weight distribution.

d) Theory of Bueche.

A less realistic, but more versatile molecular model is used by Bueche[2] to develop a theory of viscoelastic properties. The molecule is represented as a 3-dimensional zigzag with links at 90° angles. Each set of 3 mutually perpendicular links constitutes a segment which acts like a spring with the properties of a rubberlike Gaussian coil. The normal coordinates of motion are then described in terms of masses, spring constants, and the viscous reaction of the surrounding medium which is, as usual, measured by a segmental friction coefficient f. The calculations are made for extension, but for comparison with the preceding theories they may be converted to shear by the usual factor of 3. Forces are assumed to be applied to the molecules in different ways according to the type of time-dependent loading pattern (creep, stress relaxation, etc.). Comparison of the predicted responses indicates that the principles of linear viscoelasticity will not be followed exactly by this molecular model.

Uncross-linked polymers. The response of an undiluted, uncross-linked polymer in creep is expressed by a discrete retardation spectrum

$$\left.\begin{aligned} J_p &= 32/\pi^4 n\, k\, T\, (2p-1)^4 \\ \tau_p &= N^2 \sigma^2 f/3\pi^2 k\, T\, (2p-1)^2 \end{aligned}\right\} \qquad \text{(I, 198)}$$

where the index p goes from 1 to $(N+1)/2$. The symbols have the same significance as before. The coefficient f may be expected to have the value f_0 (characteristic of no coupling) at short τ, and the higher value f (which depends on the degree of coupling) at long τ. The form is reminiscent of Rouse's relaxation spectrum (equation (I, 188)) and it can be treated in

[1] Fitry, J. D., M. L. Williams and D. M. Stern: J. physic. Chem. 58, 987 (1954).
[2] Bueche, F.: J. chem. Physics 22, 603 (1954).

an analogous manner. First, the retardation time can be redefined to eliminate the arbitrary segment length σ:

$$\tau_p = a^2 Z^2 \zeta_0 / 3\pi^2 k T (2p-1)^2 \qquad \text{(I, 199)}$$

and then alternatively in term of the steady flow viscosity

$$\tau_p = 12\eta/\pi^2 (2p-1)^2 n k T . \qquad \text{(I, 200)}$$

It will be noted that the maximum retardation time in the BUECHE theory is $\tau_1 = 12\,\eta/\pi^2 nkT$. The maximum relaxation time of ROUSE is $\tau_1 = 6\eta/\pi^2 nkT$ and his corresponding maximum retardation time would be $6(\pi^2 - 6)\,\eta/\pi^4 nkT$ which is smaller than BUECHE's by about a factor of 5. Further comparison of the two theories is difficult, however, since the treatment of BUECHE is not intended to be applicable to short times where L can be approximated by a continuous spectrum.

The response of an undiluted polymer, or a polymer solution, to sinusoidally varying stresses is expressed in terms of a discrete series of retardation times defined by equation (I, 190) (with η replaced by $\eta - \eta_s$ for solutions). However, the contributions to the dynamic viscosity do not correspond to elements of a simple mechanical model, and the real part of the complex dynamic viscosity is represented by a rather complicated summation:

$$\left.\begin{aligned} \eta' = \eta_s + (\eta_0 - \eta_s) \Bigg[1 &- \frac{12}{\pi^2} \sum_1^N \frac{\omega^2 \tau_1^2}{p^2 (p^4 + \omega^2 \tau_1^2)} \\ &- \frac{6}{\pi^2} \sum_1^N \frac{\omega^2 \tau_1^2 (p^4 - \omega^2 \tau_1^2)}{p^2 (p^4 + \omega^2 \tau_1^2)^2} \Bigg] . \end{aligned}\right\} \qquad \text{(I, 201)}$$

The numerical values of η' obtained from this expression are somewhat smaller than those derived from the theory of ROUSE by equations (I, 188) and (I, 190).

BUECHE's theory provides a value for the steady-state compliance which is identical with that derived from ROUSE's theory (equation (I, 197) except for a minor difference in the numerical factor, 2/5 being replaced by 1/3.

Cross-linked polymers. The response of a cross-linked polymer, in which all network junctions are equally spaced, to a sinusoidally varying stress is expressed by a discrete relaxation spectrum

$$\left.\begin{aligned} J_p &= 8/\pi^2 n_c k T (2p-1)^2 \\ \tau_p &= N^2 \sigma^2 f_0 / 3\pi^2 k T (2p-1)^2 \end{aligned}\right\} \qquad \text{(I, 202)}$$

where n_c is now the number of *network strands* per cc. and the other symbols have their usual significance. The retardation time can be redefined in terms of ζ_0 to give an expression identical with equation (I, 199) except that Z is replaced by Z_c, the number of monomer units between cross-links. It cannot of course be expressed in terms of η, which is infinite for a cross-linked system. The sum of J_p from $p = 0$ to $p = \infty$ gives the equilibrium value of the compliance as $J = 1/nkT = Z_c M_0/\varrho RT$, as specified by the equilibrium theory of rubberlike elasticity.

In any actual network there must be a distribution of Z_c values, and the associated spectrum must be broader than equation (I, 202) at long times. At short times, however, it can be shown that the spectrum becomes independent of Z_c, just as the ROUSE spectrum for linear polymers becomes independent of molecular weight. Thus at short times equation (I, 202) can be replaced by the continuous function

$$L = (1/\pi a Z_c n)(12/\zeta_0 kT)^{1/2}\tau^{1/2}. \tag{I, 203}$$

The corresponding relaxation distribution function derived from the interconversion formula of GROSS and PELZER[1] is

$$H = (aZ_c n/\pi)(\zeta_0 kT/12)^{1/2}\tau^{-1/2}. \tag{I, 204}$$

The BUECHE spectra for cross-linked polymers are seen to have the same form as the ROUSE spectra for uncross-linked polymers [equations (I, 191) and (I,192)], and in fact they are identical except that L is smaller and H larger by a factor of $\sqrt{2}$. Since Zn or $Z_c n = \varrho N_0/M_0$, the two relaxation distributions can be alternatively written

$$\left.\begin{array}{c} H = \left(1/2\pi\sqrt{6}\right)(a\varrho N_0/M_0)(\zeta_0 kT)^{1/2}\tau^{-1/2} \\ \text{(Rouse, uncross-linked)} \end{array}\right\} \tag{I, 205}$$

$$\left.\begin{array}{c} H = \left(1/\pi\sqrt{12}\right)(a\varrho N_0/M_0)(\zeta_0 kT)^{1/2}\tau^{-1/2} \\ \text{(Bueche, cross-linked)} \end{array}\right\} \tag{I, 206}$$

According to these equations, the behaviour is not influenced by molecular weight or the disposition of cross-links except insofar as these may affect the magnitude of the monomeric friction coefficient ζ_0. This conclusion is in general agreement with experiment, as illustrated in Chapter VI.

Of course, at extremely short times, the BUECHE theory breaks down as does that of ROUSE, because the molecular segments can no longer be considered as having the elastic properties of a rubberlike GAUSSian coil.

e) Theory of Hammerle.

None of the preceding theories has been able to predict the plateau region of the relaxation spectrum seen in fig. I, 39, although we have seen qualitatively how this phenomenon can arise when the effective segmental friction coefficient changes from f_0 to the larger value f as we progress from short range cooperative motions with small τ (large p in the ROUSE and BUECHE theories) to long range cooperative motions with large τ (small p).

A calculation by HAMMERLE[2] introduces a variable friction coefficient which depends on position along the polymer chain. It is expressed as the force per unit velocity for a single chain atom (ordinarily, half a monomer unit), thus resembling KIRKWOOD's treatment rather than those of ROUSE

[1] GROSS, B. u. H. PELZER: J. appl. Physics 22, 1035 (1951).
[2] HAMMERLE, W. G.: Ph. D. Thesis, Princeton Univ., 1954.

and BUECHE which employ arbitrary segments. In terms of the effective frictional coefficient per monomer unit ζ_0, the assumed dependence is equivalent to

$$\left.\begin{aligned} \zeta_i &= \zeta_0 A\, i^P & i &\leq Z/2 \\ &= \zeta_0 A\,(Z-i)^P & i &\geq Z/2 \end{aligned}\right\} \qquad \text{(I, 207)}$$

where ζ_i is the friction coefficient of the i'th monomer unit, Z the degree of polymerization, and A and P unspecified constants. At each chain end ζ_i has a minimum value of $\zeta_0 A$, and ζ_i increases rapidly with progression toward the center of the chain where its maximum value is $\zeta_0 A\,(Z/2)^P$. The potential of viscous forces acting on a molecule when an undiluted, uncross-linked polymer is elongated, as well as the diffusion tensor analogous to that of KIRKWOOD and FUOSS[1], are evaluated under these conditions, leading to a continuous relaxation spectrum (for deformation in extension).

The analytical function is too complicated to reproduce here; a log-log plot for the case of $P = 5$ is given in fig. I, 42. It shows a slow increase at short times [where the slope is $1/(P+1)$], a shallow maximum, and a sharp drop at long times. This is not unlike the rubbery zone of polyisobutylene as shown in fig. I, 39. However, the details are not in complete accordance with experiment. The position of the shallow maximum on the time scale, τ_m, is calculated to be proportional to Z^{P+1}, which is undoubtedly too sharp a dependence on molecular weight. The height of the shallow maximum, H_m, which is essentially the level of the plateau, is calculated to be $CZ\varrho RT/M_0$, where C is a numerical constant of the order of 10^{-3}. This incorrectly predicts the height of the plateau level to be proportional to molecular weight; actually it is nearly independent of molecular weight. The width of the plateau cannot be derived, because the theory does not predict the rise in H at the left of the plateau.

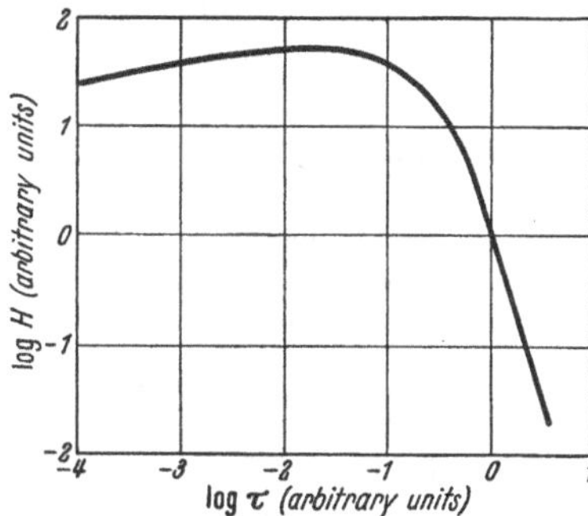

Fig. I, 42. Relaxation spectrum of HAMMERLE (assuming spectrum for shear is proportional to that for extension).

An alternative description of the properties of H in the plateau region may be derived from considerations of localized entanglement points which change the effective segment friction coefficient from f_0 to f as reflected by the dependence of steady flow viscosity on molecular weight[2]. BUECHE[3] has introduced these directly into the equations of motion of his molecular model, while FERRY, LANDEL, and WILLIAMS[4] have modified the ROUSE theory in an *ad hoc* manner to take them into account.

[1] KIRKWOOD, J. G. u. R. M. FUOSS: J. chem. Physics **9**, 329 (1941).

[2] BUECHE, F.: J. chem. Physics **20**, 1979 (1952).

[3] BUECHE, F.: J. appl. Physics **26**, 738 (1955).

[4] FERRY, J. D., R. F. LANDEL u. M. L. WILLIAMS: J. appl. Physics **26**, 359 (1955).

f) Conclusions.

The interpretation of viscoelastic properties of polymers in terms of energy storage associated with configurational entropy, and energy dissipation associated with diffusion processes, is well established by the qualitative success of several different theories. For quantitative relations between measured properties and molecular characteristics, the theories of ROUSE and BUECHE are the most useful, providing in some cases explicit predictions of time-dependent mechanical properties with no unknown parameters.

For uncross-linked systems, the theory of ROUSE is most successful; for very dilute solutions it agrees closely with experiment, and in more concentrated systems it predicts the general shape of the time spectra in the transition zone between glasslike and rubberlike consistency (fig. I, 39). The theory of BUECHE describes the properties of lightly cross-linked systems in this same zone, where as a matter of fact the behaviour is almost independent of the presence of cross-links or of the magnitude of the molecular weight.

Neither of these theories as originally developed describes the plateau region of the relaxation spectrum, though both can be modified to account for its presence. The shape of the plateau is approximately predicted by the theory of HAMMERLE, but without specification of its width nor correct evaluation of its height. No existing theory describes the principal maximum of the relaxation spectrum as shown in fig. I, 39.

The greatest need for further understanding appears to be in interpreting the shape, height and position of the principal maximum as shown in fig. I, 39; the magnitude of the friction coefficient ζ_0 which determines the position of the transition zone; and the height and detailed shape of the plateau zone.

Many additional examples comparing theory with experiment are given in Chapter VI, where experimental data are presented both in the form of the spectra H and L and as direct measurements of creep, stress relaxation, and dynamic mechanical properties.

§ 9. Strain-Birefringence measurements as a means for investigating viscoelastic properties.

By R. S. STEIN.

a) General considerations.

If a high-polymer is stretched there is a restoring force which originates from two types of molecular processes (compare Chapter X). One of these is an entropy component which is a result of the orientation produced by the deformation. The Brownian motion of each segment of the distorted polymer chain operates to return the chain to its more probable configurations. The entropy force becomes greater at higher temperatures because of the greater BROWNian motion. There is also a potential energy component of the force which is a result of the attraction and repulsion between molecules which are pulled apart or squeezed together during the deformation.

These two contributions to the force, f, may be resolved by the thermodynamic equation[1, 2]

$$f=\left(\frac{\partial W}{\partial l}\right)_T=\left(\frac{\partial U}{\partial l}\right)_T-T\left(\frac{\partial S}{\partial l}\right)_T \qquad \text{(I, 208)}$$

the first term representing the change in potential energy, U, with length, l, and the second, the change in entropy, S compare § 25.

An ideal rubber has been defined as one in which the potential energy is independent of length and for which the stress arises entirely from the entropy contribution.

The birefringence, Δn, is a measure of the orientation of the polymer chains. For a crosslinked ideal rubber, where both stress and birefringence arise from orientation, KUHN[3] and TRELOAR[4] have related the stress and birefringence by the following equation of Chapter V of Vol. III

$$\Delta n=\frac{2\pi(n^2+2)^2}{45\,k\,T\,n}(\alpha_{01}-\alpha_{02})\sigma \qquad \text{(I, 209)}$$

where α_{01} and α_{02} are the principal polarizabilities of the statistically independent link, σ is the stress per actual cross-section and T is the absolute temperature. For the derivation of this equation compare Chapter V, § 35 of this Volume.

This equation does not apply to non-ideal rubber type polymers where there is potential energy contribution to the force. In these cases the birefringence still serves as an indication of the degree of orientation of the chains.

If a polymer is stretched and held at constant length, the stress relaxes as a result of a rearrangement of the chains from their initial deformed configurations to more probable configurations. This takes place by such processes as (1) chemical reactions, (2) viscous flow, (3) rearrangement to relieve local strains, and (4) recrystallization or rearrangement of crystallites. In general, some sort of combination of these processes occurs, but the slowest will be the rate-determining one. The nature of the rate-determining process in a particular case depends upon the nature of the sample, the temperature, the amount of strain, and in some cases (hydrogen bonded polymers such as cellulose, nylon) the humidity.

The measurement of the birefringence as well as the stress during a viscoelastic experiment serves as an additional means for studying the nature of the processes by means of which the relaxation occurs.

---

[1] WIEGAND, W. B. and J.W. SNYDER: Trans. Inst. Rubber Ind. **10**, 239 (1934).
[2] ELLIOT, D. R. and S. LIPPMANN: J. appl. Physics **16**, 50 (1945).
[3] KUHN, W. and F. GRÜN: Kolloid-Z. **101**, 248 (1942).
[4] TRELOAR, L. R. G.: Trans. Faraday Soc. **43**, Pt. 5, 277, 289 (1947).

The change in birefringence occurring during the relaxation of a sample may be studied by means of an apparatus of the type shown in fig. I,43[1]. The retardation of a beam of monochromatic polarized light which passes through the stretched sample is measured by means of a Babinet Compensator[2]. The stress is measured simultaneously by determining the force between the clamps holding the sample with a mechanical balance[3,4] or an electric strain gauge[5,6].

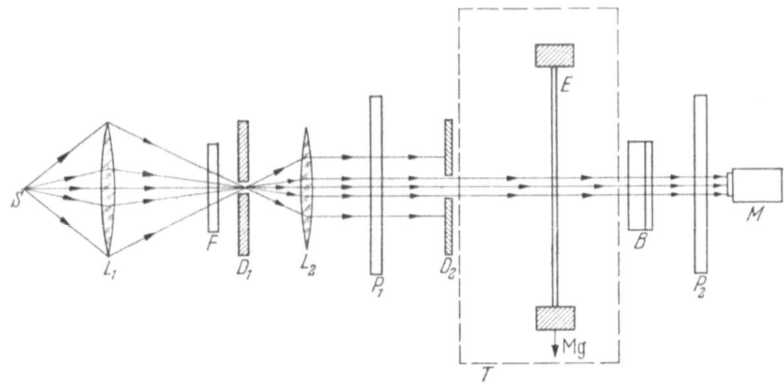

Fig. I, 43. Experimental arrangement. S = source, $L_1\,L_2$ = lenses, F = filter, $D_1\,D_2$ = diaphragms, $P_1\,P_2$ = polaroid screens, E = specimen, B = Babinet compensator, M = travelling microscope, T = constant temperature enclosure.

b) Rubber-like materials.

Crosslinked or vulcanized rubbers.

For these materials the stress-optical properties are adequately described by equation (I,210). The combination of terms

$$\frac{\sigma}{T\,\Delta n} = \frac{45\,k\,n}{2\pi\,(n^2+2)^2\,(\alpha_{01}-\alpha_{02})} \tag{I,210}$$

should be a constant which is independent of stress, elongation, degree of crosslinking, and temperature, but which depends only upon the molecular constitution of the chain. One might expect that in a process in which the number of crosslinks changes with time, this equation would still be obeyed, providing the relaxation process occurs at a rate which is sufficiently slow so that the structure is always in equilibrium with its external constraints. The average time for a crosslink to form or to be lost must be long compared with the time required for a chain to diffuse to its equilibrium configuration.

---

[1] Kolsky, H. and A. C. Shearman: Proc. physical. Soc. **55**, 383 (1943).

[2] Born, M.: "Optik" page 244. Springer, Berlin (1933).

[3] Tobolsky, A. V., I. B. Prettyman and J. H. Dillon: J. appl. Physics **15**, 380 (1944).

[4] Stein, R. S. and A. V. Tobolsky: Text. Res. J. **18**, 302 (1948).

[5] Stein, R. S. and H. Schaevitz: Rev. sci. Instr. **19**, 835 (1948).

[6] For further literature see Vol. III, § 28 d.

The rate-determining processes for the long-time decay of the stress on vulcanized rubbers are believed to be chemical reactions such as oxidation and depolymerization. The viscoelastic behaviour induced by such chemical reactions is called "*chemorheology*" by TOBOLSKY. One finds that during such processes, $\sigma/T\Delta n$ remains constant. Fig. I, 44 and I, 45 illustrate the changes in stress, birefringence and their ratio occurring during the relaxation of vulcanized Hevea rubber held at constant elongation[1]. Similar results are obtained for the relaxation of polysulfide or "Thiokol" rubber. The former process is believed to occur as a result of oxidation of the polyisoprene chain[2] while the latter occurs as the result of an exchange reaction between the S—S and S—H bonds[3].

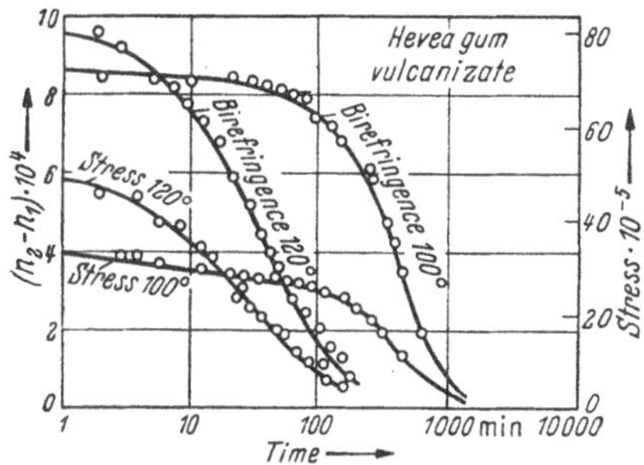

Fig. I, 44. Relaxation of the stess and birefringence of a sample of Hevea gum vulcanizate stretched and held at 48% elongation at several temperatures.

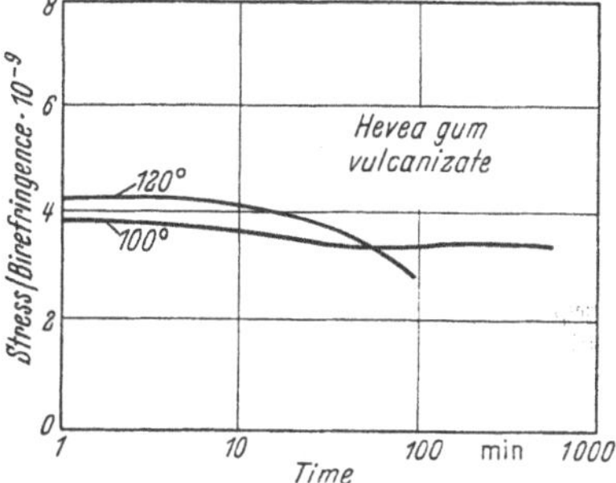

Fig. I, 45. Variation of the ratio of stress to birefringence during the Hevea gum vulcanizate relaxation.

Linear amorphous rubbers.

The relaxation of the stress of unvulcanized or uncross linked rubbers occurs by a physical process involving the flow or diffusion of chains and the unsnarling of chain entanglements. Experiments indicate that equation (I, 210) adequately describes the relaxation of the stress and birefringence of these materials. For example, MISCH and PICKEN[4] find that at temperatures above 50° C., the stress and birefringence of polyvinyl acetate decay in a corresponding manner. STEIN and TOBOLSKY[5] have obtained similar results for unvulcanized Hevea rubber. They found that the value of $(\sigma/T\Delta n)$ obtained by using equation (I,210) remains essentially constant during the relaxation and has a value of $1{,}4 \cdot 10^7$ dynes/cm^2/°K in good agreement with the values of $1{,}2$ to $1{,}7 \cdot 10^7$ reported for vulcanized Hevea rubber[6, 7].

[1] STEIN, R. S. and A. V. TOBOLSKY: see page 304.
[2] WOLSTENHOLME, W. E. and D. S. VILLARS: J. appl. Physics **15**, 324 (1944). — A. V. TOBOLSKY, I. B. PRETTYMAN and J. H. DILLON: see page 112.
[3] STERN, M. D. and A. V. TOBOLSKY: J. chem. Physics **14**, 93 (1946).
[4] MISCH, L. and L. PICKEN: Z. physik. Chem. **B 36**, 400 (1937).
[5] STEIN, R. S. and A. V. TOBOLSKY: see page 305.
[6] TRELOAR, L. R. G.: Trans Faraday Soc. **42**, 83 (1946).
[7] THIBODEAU, W. E. and A. T. MCPHERSON: J. Res. nat. Bur. Standards **21**, 257 (1938).

Similar behaviour has been found[1] for polyisobutylene over a wide range of temperature and molecular weight giving a value of $(\sigma/T\Delta n)$ of about $2{,}3 \cdot 10^7$ dynes/cm²/°K (figs. I, 46 and I, 47).

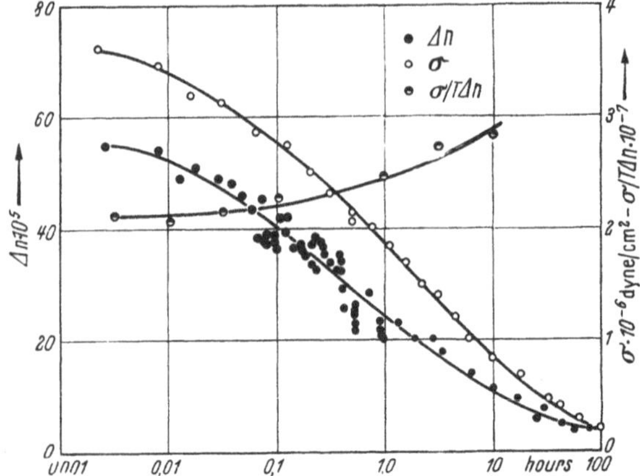

Fig. I, 46. The Variation in the Stress, Birefringence and the Ratio, $(\sigma/T\,\Delta n)$ during the Relaxation of a Sample of polyisobutylene of Molecular Weight 1,500,000 at 30° C and 47% Elongation.

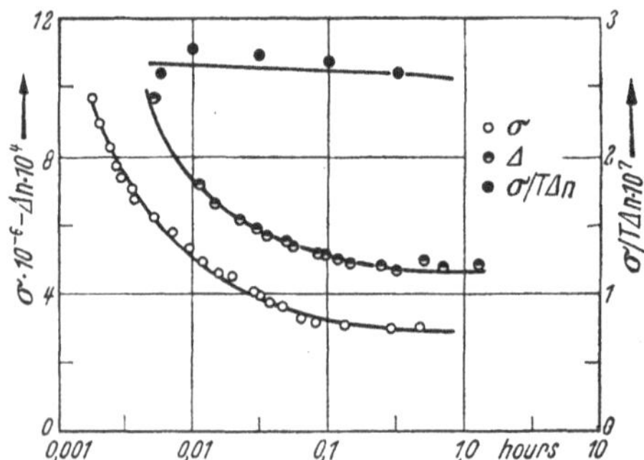

Fig. I, 47. The Variation in the Stress, Birefringence and the Ratio, $(\sigma/T\,\Delta n)$ during the Relaxation of a Sample of Polyisobutylene of Molecular Weight 1,500,00 at −35°C and 48% Elongation.

A complete relaxation curve for polyisobutylene has the form in dicated in fig. I, 48. It is difficult to carry out an experiment over a wide enough range of time to observe the complete relaxation curve with a single sample at a single temperature. However, by judiciously choosing the molecular weight of the sample and the temperature, it is possible to observe a particular region of the generalized curve. Fig. I, 46 is obtained with a sample of molecular weight 1,500,000 at a temperature of 30°C. and is typical of relaxations in regions II and III of fig. I, 48, while fig. I, 47 is obtained with the same sample at −35°C and is characteristic of region I. The relaxation mechanisms in these regions are different, as may be indicated by the relaxation rate in region III being dependent upon the molecular weight, while that in region I being independent of molecular weight[2,3]. The fact that the stress-birefringence ratio is essentially the same in these two regions in-

[1] Stein, R. S., F. H. Holmes and A. V. Tobolsky: in publication.

[2] Andrews, R. D., N. Hofman-Bang and A. V. Tobolsky: J. Polymer Sci. **3**, 669 (1948).

[3] Andrews, R. D. and A. V. Tobolsky: J. Polymer Sci. **6**, 221 (1951).

dicates that the stress has the same molecular origin within these two regions, but that the mechanism by which the chains rearrange to relieve the stress is different.

It is expected that at temperatures below −40° C polyisobutylene is no longer rubbery and that the stress-birefringence behaviour is more characteristic of that of the "glassy state" described in the next section.

The stress and birefringence of polyisobutylene have been shown[1] to vary with temperature in the manner predicted by theory (fig. I, 49). The birefringence at constant length independent of temperature, while the stress is proportional to absolute temperature. The molecular weight of the sample was sufficiently high (1,500,000) and the temperature sufficiently low (−35° to 20° C) so that little relaxation occurred during the course of this experiment.

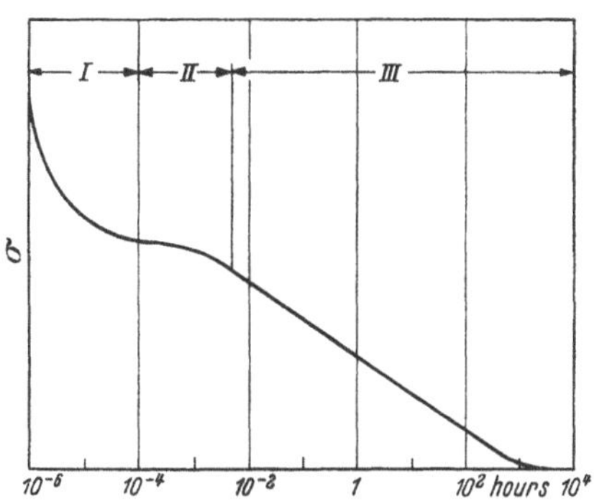

Fig. I, 48. A Generalized Relaxation Curve for Polyisobutylene at Constant Length and Temperature.

NIELSON and BUCHDAHL[2] have measured the birefringence of polystyrene films at 110° C (higher than the softening point) during both creep and relaxation experiments. They have observed that the ratio of stress to birefringence remains constant during these experiments at about −9,0 pounds per square inch per Å retardation per mil of sample thickness. They find that this value is independent of the molecular weight of the sample. The ratio of strain to birefringence does not remain constant. On the basis of these experiments and on the basis of the observation that polystyrene films stretched 95% at 110° C return to practically their original length, they conclude that some type of semi-permanent network structure must exist. Similar results have been found for these and other polymers by KOLSKY and SHEARMAN[3], KUHN[4], and HERMANS[5].

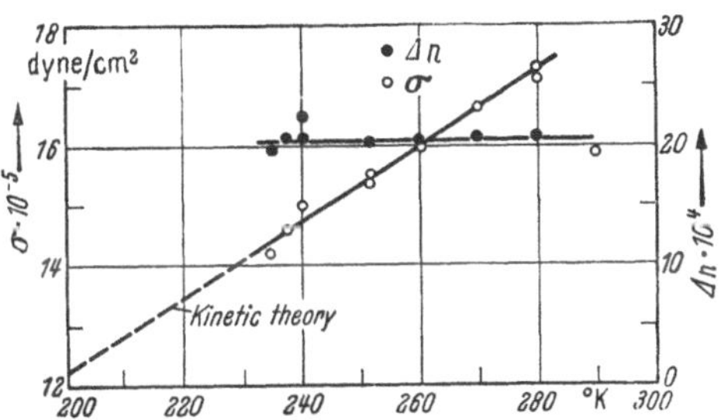

Fig. I, 49. The Variation of Stress and Birefringence with Temperature at Constant Length for a Sample of Polyisobutylene Previously Stretched 100% at 20° C.

MÜLLER[6] has treated the calculation of orientation birefringence in a manner which is independent of the assumption of the existence of chemical crosslinks and

[1] STEIN, R. S., F. H. HOLMES and A. V. TOBOLSKY: see page 114.
[2] NIELSON, L. E. and R. BUCHDAHL: J. chem. Physics **17**, 839 (1949).
[3] KOLSKY, H. and A. C. SHEARMAN: see page 112.
[4] KUHN, W. and F. GRÜN: see page 111.
[5] HERMANS, J. J.: Kolloid-Z. **103**, 210 (1943).
[6] MÜLLER, F. H.: Kolloid-Z. **95**, 172, 307 (1941).

is therefore applicable to the description of unvulcanized rubbers. He describes the orientation of the molecular segments by an orientation function, $f(\Theta)$ which is equal to the number of segments oriented between angles Θ and $\Theta + d\Theta$ from the direction of stretching. In terms of this function, the birefringence is given by

$$\Delta n = \frac{\pi (n^2 + 2)^2}{9n} (\alpha_1 - \alpha_2) \int_0^\pi F(\Theta)(3\cos^2\Theta - 1)\, d\Theta \qquad \text{(I, 211)}$$

where n is the average refractive index, and α_1 and α_2 are the principal polarizabilities of the statistical segment.

The orientation function may be expanded in terms of even spherical harmonics[1]

$$F(\Theta) = F_0(1 + A_2 P_2 + A_4 P_4 + \cdots) \qquad \text{(I, 212)}$$

where F_0 is a normalizing constant, A_2, A_4 are constants and P_2, P_4, etc. are the second, fourth, etc. spherical harmonics. If this is substituted into equation (I, 211), all of the terms except the one containing P_2 drop out and one finds that

$$\Delta n = \frac{2}{45} \pi \frac{(n^2 + 2)^2}{n} (\alpha_{01} - \alpha_{02}) A_2. \qquad \text{(I, 213)}$$

Thus the birefringence is a measure of the coefficient A_2. This equation is general for orientation birefringence and not restricted to the rubbery state.

If A_2 is expanded as a power series in strain, γ,

$$A_2 = a_{21}\gamma + a_{22}\gamma^2 + \cdots \qquad \text{(I, 214)}$$

in which for small strains only the first term is important, then

$$\Delta n = \frac{2}{45} \pi \frac{(n^2 + 2)^2}{n} (\alpha_1 - \alpha_2) a_{21} \gamma. \qquad \text{(I, 215)}$$

The entropy may be evaluated in terms of this same orientation function for an ideal rubber, and the stress may be calculated from the derivate of entropy with respect to strain. This proves to be (neglecting higher spherical harmonics than P_2 and neglecting higher than linear terms in strain),

$$\sigma = \frac{1}{5} k T a_{21}^2 \gamma. \qquad \text{(I, 216)}$$

Thus the ratio

$$\frac{\sigma}{T \Delta n} = \frac{9 k n a_{21}}{2\pi (n^2 + 2)^2 (\alpha_{01} - \alpha_{02})} \qquad \text{(I, 217)}$$

should be a constant for small elongations from which the parameter a_{21} may be determined. For the special case of a crosslinked network, comparison with the network theory shows $a_{21} = 5$. For an uncrosslinked rubber, one finds experimentally that $\sigma/T\Delta n$ is the same as for the same rubber when it is crosslinked. Thus, to the approximation of the second spherical harmonic term, the distribution of orientations produced by stretching an uncrosslinked rubber is the same as that for a crosslinked one.

It is apparent that for rubbery materials, both stress and birefringence vary in a similar fashion during relaxation or creep experiments. A simultaneous measurement of these two quantities during such experiments permits the calculation of $(\alpha_{01} - \alpha_{02})$, the apparent optical anisotropy of the statistical link, which is a fundamental property of the polymer chain, see Vol. III § 28b and Vol. IV, § 35d. Deviations will be found at low temperatures or high elongation where the rubbers are no longer ideal.

[1] Odd spherical harmonics need not be included if the distribution has cylindrical symmetry.

c) The glassy state.

If a rubber-like polymer is cooled, a temperature (or narrow range of temperatures) is reached at which the mechanical properties of the polymer drastically change. Below this transition temperature, (the "softening point", "glass transition temperature") the polymer is no longer rubbery, and the stress arises principally from the $\left(\frac{\partial U}{\partial l}\right)_{V,T}$ term of equation (I, 208). For example, figs. I, 50 and I, 51 show how both the force and $\left(\frac{\partial S}{\partial l}\right)_{V,T}$ on an extended sample of polyvinyl chloride change with temperature[1]. This extensive region between the temperatures of the transition point and absolute zero is called the "glassy state" of a polymer; compare also Vol. III, Chapter X.

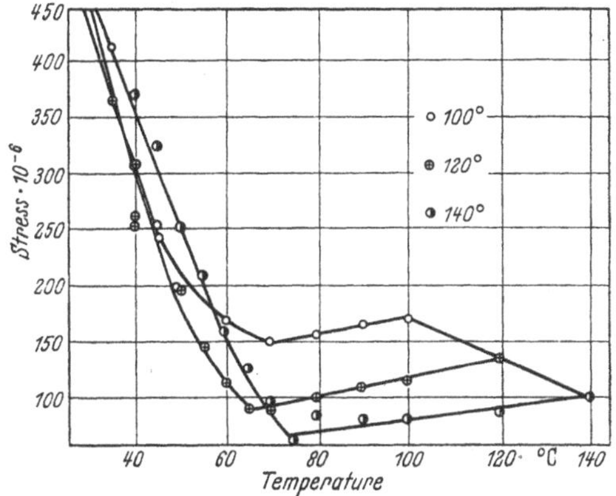

Fig. I, 50. The variation of stress with temperature of a sample of unplasticized polyvinyl chloride which was stretched and held at 100% elongation at 100° C.

It has been postulated that the mechanism of deformation of a polymer is different above and below the transition temperature. Above the transition temperature a polymer chain may diffuse freely subject only to the constraints that it meet other chains at certain points of interaction or crosslinking. Below the transition temperature the thermal energy is not great enough to overcome the attractive forces between portions of the chain. Interaction occurs all along the chain rather than at only a few widely separated points. Consequently, the mechanism of stretching is not the straightening-out of coiled chains, as in the rubbery state, but rather the squeezing together and pulling apart of chains against the forces of VAN DER WAALS interaction. The YOUNG's Modulus in the glassy state is of the order of 10^9

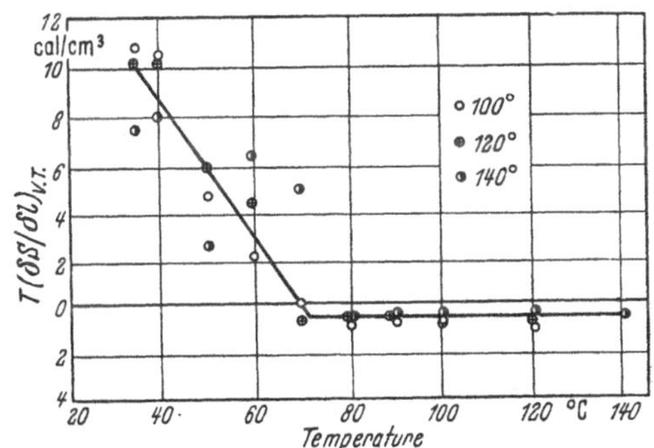

Fig. I, 51. The variation of $T(\delta S/\delta L)_{V,T}$ with temperature for a sample of unplasticized polyvinyl chloride which has been stretched 100% and relaxed at 100° C.

[1] STEIN, R. S., S. KRIMM and A. V. TOBOLSKY: Text. Res. J. 19, No. 1 (1949).

to 10^{12} dynes/cm² as compared with about 10^7 dynes/cm² in the rubbery state.

The two states may be readily distinguished on the basis of the birefringence behaviour. If a sample of "Lactoprene"[1] rubber is stretched and allowed to relax, the usual rubber-like relaxation curves are obtained at temperatures above −17° C[2,3].

If a sample of this material is stretched (30% at 30° C), allowed to relax, and cooled at constant length while the change in birefringence accompanying a small change in stress is measured as a function of temperature, the data of fig. I, 52 are obtained. The quantity (Δ Birefringence/Δ Stress) is plotted. Some time-dependency of this quantity is observed, and the two sets of points represent initial values and values obtained after waiting about a half hour.

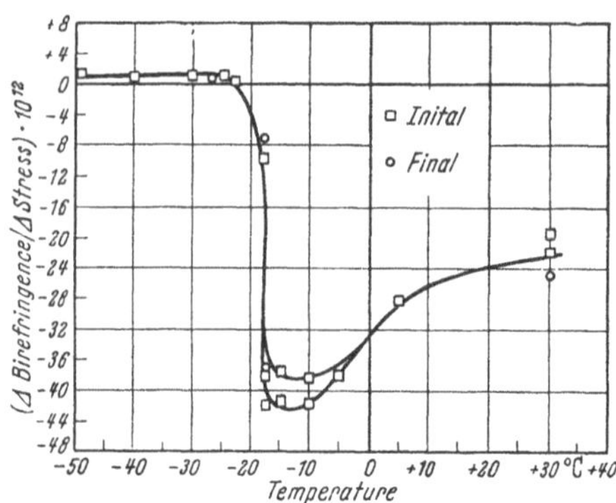

Fig. I, 52. The variation of the ratio $\frac{(\Delta \text{ birefringence})}{(\Delta \text{ stress}}$ with temperature for lactoprene gum-stock.

It is noted that the behaviour of "Lactoprene" is very different above and below −20° C. Above −20° C, the birefringence is negative and characteristic of rubber-like birefringence. Below −20° C, the birefringence is positive and fairly constant. The ratio of birefringence to stress in this region is about 1/20 of that in the rubbery region. We shall call the birefringence in this region "distortional birefringence".

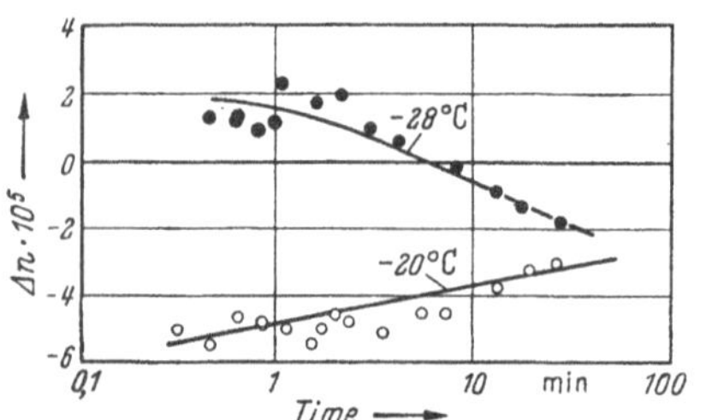

Fig. I, 53. The Relaxation of the Birefringence of Lactoprene in the Transition region.

The time-dependency of birefringence in this transition region is apparent from fig. I, 53[2] in which the change in birefringence with time is compared for samples of "Lactoprene" stretched 10% at −20° C and −28° C. At −20° C, the stretching produces orientation of the chains, and the resulting negative birefringence decreases with time as the chains return to less oriented configurations. At −28° C, the segments do not have sufficient kinetic energy to rearrange while the polymer is being stretched. Consequently, the structure be-

---

[1] Crosslinked polyethylacrylate.

[2] Stein, R. S., S. Krimm and A. V. Tobolsky: see page 117.

[3] It is noted that the birefringence is negative for this polymer. Therefore $\alpha_1 - \alpha_2$) for the statistical segment is negative. This is a result of the high polarizability of the acrylate side group perpendicular to the chain.

comes distorted, and chains are no longer at their equilibrium VAN DER WAALS' distances apart. The resulting positive birefringence decreases with time as these local strains are relieved by orientation of the chains. The birefringence eventually becomes negative as the orientation effect becomes dominant. Presumably, at longer times the birefringence would again increase toward zero as the orientation is lost.

Similar behaviour has been described for polystyrene by KOLSKY[1] (fig. I, 54) and for polymethyl methacrylate by KOLSKY and SHERMAN[2].

McLOUGLIN[3] has investigated the change in birefringence during the relaxation of polymethyl methacrylate. He finds that during the relaxation the ratio of birefringence to strain remains fairly constant (at about .01) but that the ratio of birefringence to stress increases with time (fig. I, 55) in the glassy region. The stress-optical coefficient appears to be of opposite sign above and below the transition temperature (about 120°C) (fig. I, 56).

These findings are confirmed by the observation of KOLSKY that at

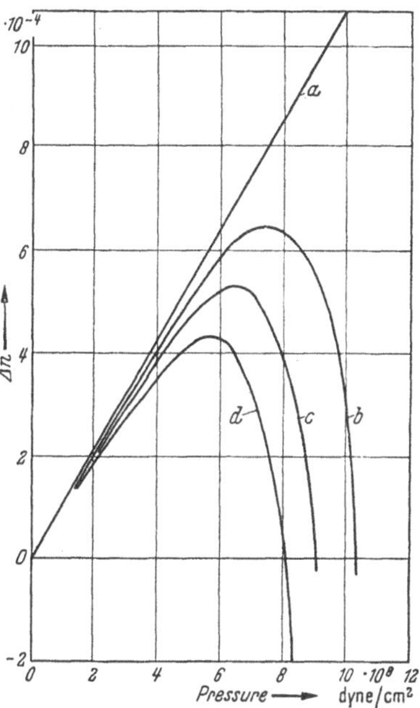

Fig. I, 54. The Variation of Birefringence with Stress for Polystyrene (a) immediately after applying stress, (b) after being stressed for 1 minute, (c) 10 minutes, (d) 1 hour.

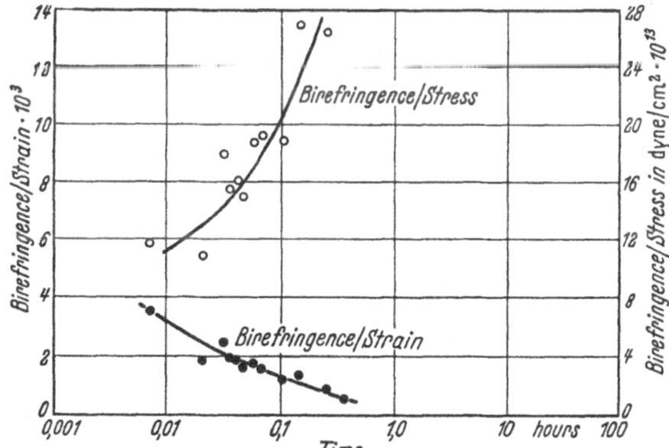

Fig. I, 55. The Variation of the Stress-Optical Coefficient and Strain-Optical Coefficient with Time for a Sample of Polymethyl Methacrylate which has been Stretched at 60° C.

[1] KOLSKY, H.: Nature **166**, 235 (1950).

[2] KOLSKY, H. and A. C. SHEARMAN: see page 392.

[3] McLOUGHLIN, J. R.: „Physical Studies of some High Polymers", Princeton Univ. Doctoral Diss. (1951).

20°C the birefringence is a uni-valued function of strain even at strains (.06) which far exceed the elastic limit (occurring at a strain of about (.015) where considerable plastic distortion takes place. For strains that are below the elastic limit, the birefringence is a uni-valued function of stress. For example, ROBINSON, RUGGY and SLANTZ[1] find that the birefringence of polymethyl methacrylate varies linealy with stress up to stresses of 5000 lbs/sq. in with a stress-optical coefficient of −3,8 BREWSTERS at 20°C. PEUKERT[2] has recently made similar observations.

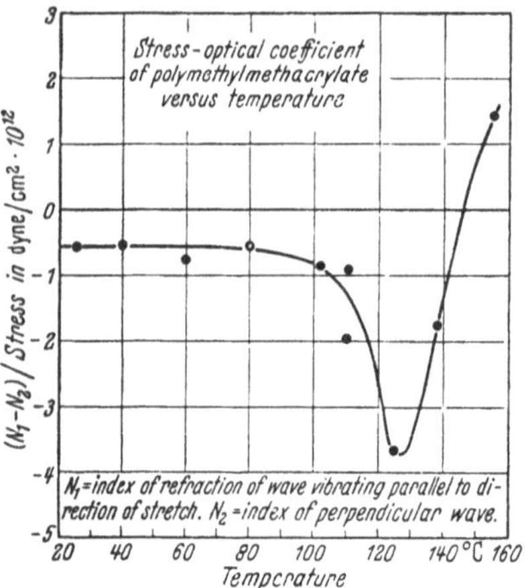

Fig. I, 56. The Variation of the Stress-Optical Coefficient of Polymethyl Methacrylate with Temperature.

PEUKERT[2] has studied samples of polymethyl methacrylate that have been heated to 150°C in an oven, removed and stretched (109%) as the sample begins to cool. He has measured the birefringence during the elongation and during the subsequent cooling to 20°C as the sample was held at constant length, and he found that the initial positive birefringence produced by the elongation at high temperature rapidly decreases and becomes negative. The stress on the sample increases during the cooling as a consequence of the tendency for thermal contraction. This result is to be expected on the basis of MCLOUGHLIN's observations.

To summarize, one finds that birefringence of polymers in the glassy region is different from that in the rubbery region. The ratio of stress to birefringence in the glassy region is not independent of stress except at small stresses. The ratio may have a different sign and is usually much

Table I, 12.

Substance Rubbers	T (°C)	Δn/stress (dynes/cm²)	Δn/strain
Hevea Rubber[3, 4]	30°	2,2 · 10^{-10}	2,6 · 10^{-3}
Polyisobutylene[5]	30°	1,5 · 10^{-10}	1,7 · 10^{-3}
Polystyrene	115°	− 5,2 · 10^{-10} [6]	− 3 · 10^{-3} (100°)[7]
Lactoprene[8]	30°	− ,22 · 10^{-10}	,17 · 10^{-3}
Polysulfide Rubber[9]	60°	1,4 · 10^{-10}	3,2 · 10^{-3}

(1) ROBINSON, H. A., R. RUGGY and E. SLANTZ: J. appl. Physics **15**, 343 (1944).
(2) PEUKERT, H.: Kunststoffe **41**, 154 (1951).
(3) TRELOAR, L. R. G.: Trans. Faraday Soc. **42**, 83 (1946).
(4) THIBODEAU, W. E. and A. T. MCPHEASON: see page 113.
(5) STEIN, R. S., F. H. HOLMES and A. V. TOBOLSKY: see page 114.
(6) NIELSON, L. E. and R. BUCHDAHL: see page 115.
(7) MÜLLER, F. H.: see page 115.
(8) STEIN, R. S., S. KRIMM and A. V. TOBOLSKY: see page 117.
(9) STEIN, R. S. and A. V. TOBOLSKY: see page 113.

Table I, 12. *(continued.)*

Substance Glasses	T (° C)	Δn/stress (dynes/cm²)	Δn/strain
Polystyren[1]	16°	$10 \cdot 10^{-13}$	—
Polymethyl methacrylate	40°	$-4 \cdot 10^{-13}$ [2]	$12 \cdot 10^{-3}$ [2]
	20°	$-3{,}8 \cdot 10^{-13}$ [3]	$8 \cdot 10^{-3}$ [3]
Lactoprene[4]	25°	$10 \cdot 10^{-13}$	
Flint Glass[5]	25°	$1{,}4 \cdot 10^{-13}$	

greater in magnitude in the glassy region. The glassy state birefringence is proportional to strain over a wide range of strains. The optical properties of some polymers in the rubbery and glassy regions are summarized in table I,12. The value of the strain optical coefficient for crosslinked rubbers is dependent upon the degree of crosslinking. Those reported here are merely typical values.

The order of magnitude of the distortional strain-optical coefficients in the glassy region may be accounted for by considering the interaction of polarizabilities of adjacent molecules and the effect of changing the distances between them as the structure is distorted when the polymer is stretched. For example, for a simple cubic lattice of atoms having a polarizability of 10^{-24} cm³ separated by 3 Å, one would calculate a strain-optical coefficient of about 10^{-2} [6].

In the glassy state as in the rubbery state, the birefringence will return to zero if the force on the sample is released, providing the elastic limit of the material is not exceeded. When the elastic limit is exceeded, the force is sufficiently great to produce orientation as well as distortion. In this case, the birefringence is the sum of the rubber-like orientational birefringence and the distortional birefringence. The relative contributions of these may change during the relaxation. When the stress is released, the internale restoring forces may not be sufficient to produce a corresponding disorientation. The orientation birefringence is then "frozen in" but may be removed by annealing the sample at a temperature above its softening point.

The orientation contribution to the birefringence in the glassy state may be resolved by Müller's treatment, and the strain-optical coefficient may be interpreted in terms of the parameter A_2 of equation (I, 213).

d) Crystalline polymers.

For crystalline polymers, the contribution to the birefringence resulting from the orientation of anisotropic crystallites must be considered.

[1] Kolsky, H.: see page 119.
[2] McLoughlin, J. R.: see page 119.
[3] Robinson, H. A., R. Rugg and E. Slantz: see page 120.
[4] Stein, R. S., S. Krimm and A. V. Tobolsky, see page 117.
[5] Coker and Filon: „Treatise on Photoelasticity", Cambridge Univ. Press, Cambridge (England) 1931.
[6] This is calculated by assuming that polarizability of a pair of atoms may be calculated by Silberstein's equation. It is assumed that all of the interatomic distances change by an affine transformation and that Poisson's ratio is 1/3.

The degree of orientation of crystallites in a stretched polymer is usually greater than that of the amorphous portions, and consequently their contribution to the birefringence is greater. The change in birefringence accompanying the relaxation of stress in a crystalline polymer results from (1) change in the *orientation* of the crystallites, (2) change in the *amount* of crystalline material, (3) change in the orientation of the amorphous portions of the polymer, and (4) change in the distortion of the amorphous portions[1]; compare Vol. III, § 28d.

THIESSEN and WITTSTADT[2] have noticed that if soft vulcanized rubber is stretched and held at constant length, the birefringence continues to increase, even though the stress is decreasing during this period. The rate and amount of increase in the birefringence are greater at greater elongations. At higher temperatures, the amount of birefringence increase is less but the rate at which it occurs is greater. If a stretched sample which has attained constant birefringence is cooled while being held at constant length, the birefringence increases. Upon subsequent heating and cooling, the birefringence decreases and increases in a reversible manner providing the highest temperature does not exceed the temperature at which the sample was initially elongated. These changes have been interpreted as resulting from the growth and melting of crystallites. The crystallization which occurs when the rubber is stretched is a slow process which continues as the polymer is held elongated[3]. As these crystallites which grow are oriented and anisotropic, the birefringence continues to increase during the relaxation. Since crystallization is a process in which coiled chains straighten out in the direction of stretching, it is accompanied by a decrease in stress.

TRELOAR[4] has compared the change in birefringence with time with the change in density for samples of unvulcanized rubber which have been stretched various amounts at 0° C. He has found that the birefringence is almost exactly proportional to the density change between 100% and 700% elongation. Since the density is believed to increase as a result of the formation of more dense crystallites, this would indicate that, within this range of elongations, the birefringence changes arise almost entirely from the growth of oriented crystallites. At elongations lower than 100%, the contribution of the oriented amorphous regions must be considered, while at elongations higher than 700%, the distortional birefringence of the amorphous material becomes significant.

TRELOAR has also shown that at low elongations (less than 360% at 25° C) the birefringence actually decreases rather than increases with time. This decrease is presumably the result of the loss of orientation of the amorphous material in a process of the type occurring during the relaxation of unvulcanized rubber. Above 360%, there is an inversion and the

---

[1] There is also the possibility of birefringence resulting from the distortion of crystallites. This contribution, however, is probably negligible for polymers.

[2] THIESSEN, P. A. and W. WITTSTADT: Z. physik. Chem. **B 41**, 33 (1938).

[3] BEKKEDAHL, N.: J. Res. nat. Bur. Standards **13**, 410 (1934).

[4] TRELOAR, L. R. G.: Trans. Faraday Soc. **37**, 84 (1941).

birefringence increases with time due to crystallization. This inversion elongation is higher at higher temperatures (680% at 50° C).

Somewhat similar behaviour is found for partially crystalline substituted polyamides[1], the birefringence increasing during the relaxation at constant strain when the stress decreases[2]. There is a considerable decrease in stress during the relaxation, probably resulting from loss of distortional stresses. If a relaxed, stretched sample of one of these polyamides is cooled below the temperature of the relaxation, the stress and birefringence change in a reversible manner (fig. I, 57 and I, 58). This sample, after being stretched 100% at 60° C and relaxed at constant length, was cooled to 30° C, and the stress was observed to decrease and birefringence increase along portions *BA* of the figures. Upon heating back to 60° C, the stress and birefringence returned reversibly to their original values. The changes in stress and birefringence along this portion of the curve are believed to result from the growth and melting of oriented crystallites as the sample is cooled and heated. The crystallites do not completely melt but only increase and decrease in size.

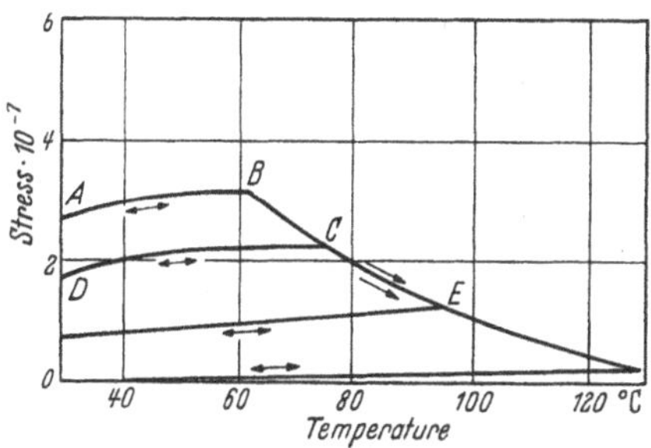

Fig. I, 57. The Variation of Stress with Temperature for a Substituted Polyamide. The sample was stretched and held at 100% elongation at 60° C and allowed to relax for a day before these curves were determined. The double headed arrows indicate reversible portions of the curve, while the single headed arrows indicate irreversible portions.

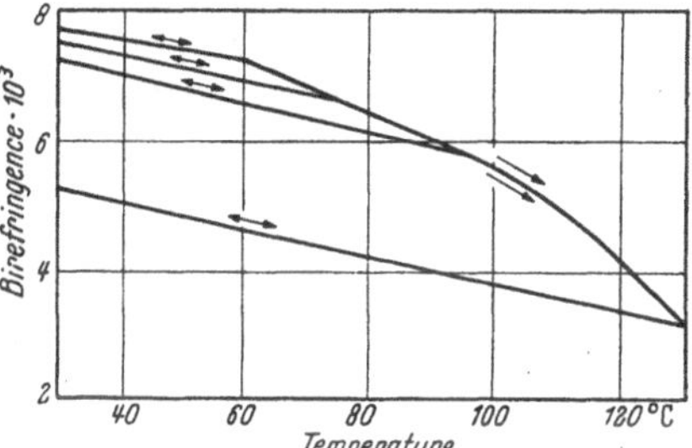

Fig. I, 58. The Variation of Birefringence with Temperature measured Simultaneously with the Stress Determination of fig. I, 57.

If the temperature of the relaxation (60° C) is exceeded and the sample is heated to 80° C, the stress and birefringence both decrease irreversibly along *BC*, resulting apparently from the complete melting of crystallites which act as crosslinks in the amorphous network. This allows a partial relaxation of stress with a partial loss of orientation of the remaining crystallites. If the material is now cooled, the remaining crystallites grow and the stress and birefringence change along the new reversible curve *CD*. This may be repeated at higher temperatures.

Thus, this material behaves as a rubber-like network which is crosslinked by crystallites, the number of which depends upon temperature.

[1] Copolymers of N,N'-diisobutyl hexamethylene diamine, mono—N—isobutyl hexamethylene diamine and unsubstituted hexamethylene diamine with sebacic acid.

[2] STEIN, R. S. and A. V. TOBOLSKY: see page 306.

It has been shown[1] that the material attains the same state if it is stretched at a low temperature and then heated to a higher temperature while at constant length, as it does when it is stretched directly at the higher temperature.

Polyvinyl chloride, which is poorly crystalline[2] and on the border of being a rubbery polymer, behaves in much the same manner. There is little birefringence change during the relaxation of this polymer[3]. Although the birefringence produced upon stretching is positive, there is a reversible decrease of birefringence on cooling the stretched, relaxed polymer at constant length. This is contrary to what would be expected if cooling is accompanied by a growth of oriented crystals, and it may be a result of decreasing vibration of the chlorine atoms or of increasing distortion as the sample is cooled.

One observes that if the temperature of relaxation is exceeded in this experiment, there is an irreversible increase in birefringence rather than a decrease as with the polyamides. This indicates that when the polyvinyl chloride is heated, distortion is relieved by producing orientation. Similar behaviour is observed with the polyamides at low temperatures.

Polyethylene is another case of a crystalline polymer in which there is little change in birefringence during relaxation at constant length[1, 2]. The birefringence appears to depend only upon the strain of this polymer and is independent of stress. The birefringence appears to result entirely from the orientation of crystallites. The crystallinity of the polythene appears to depend only upon temperature and not upon the degree to which the polymer has been stretched.

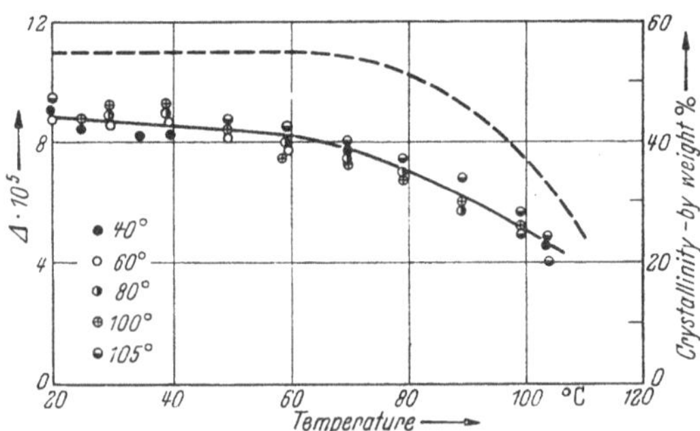

Fig. I, 59. The variation of birefringence with temperature for a sample of polyethylene initially stretched and relaxed at 40° C. The dotted line shows the variation of per cent crystallinity with temperature for polyethylene.

If a sample of polythene which has been stretched is cooled at constant length, the birefringence increases reversibly, presumably because of the growth of oriented crystals (fig. I,59)[4]. If the sample is then heated to a temperature exceeding the temperature of the original relaxation,

---

[1] Stein, R. S. and A. V. Tobolsky: see page 112.
[2] Fuller, C. S.: Chem. Rev. **26**, 143 (1940).
[3] Stein R. S. and A. V. Tobolsky: see page 112.
[4] Stein, R. S., S. Krimm and A. V. Tobolsky: see page 117.

the crystals melt and the birefringence drops. If the sample is again cooled the birefringence returns to its original value. There appears to be no irreversible loss of birefringence even if the sample is heated (while elongated) to 100° C (within 15° of its melting point). The stress, however, behaves differently. There is an irreversible decrease in stress every time the temperature exceeds the highest temperature to which the sample has been heated. At 100° C most of the stress has been lost. When the sample is cooled to 20 C, the stress is only about one-sixth of its original value, while the birefringence returns to the original value. It is clear the stress must be associated with a different molecular process from the birefringence and probably arises from potential energy factors.

Second Chapter.

Non Linear Deformation Behaviour of High Polymers.

By

A. J. STAVERMAN and F. SCHWARZL.

With 32 figures.

§ 10. On the nature of large deformations.

In this section we deal with deformation behaviour of polymers at higher forces and deformations, with *non-linear deformation behaviour*, in contradistinction to what we called linear deformation behaviour. In the first part we called the (time dependent) deformation behaviour linear, if it satisfied the superposition principle of BOLTZMANN. We have pointed out that polymers behave linearly up to certain critical stresses and strains, the limits of linearity. These limits may be very different in different cases ranging from less than one percent deformation for brittle plastics up to more than ten percent for rubbers. In the present part we are therefore concerned with force-deformation laws in the region of stresses and strains higher than the linearity limits, where the superposition principle is no longer obeyed.

The treatment of linear deformation behaviour has been based upon one single general law, the superposition principle which is obeyed at small stresses by a large number of materials, such as plastics, rubbers, polymer melts and solutions and anorganic glasses. Therefore one common theory could be given for all these materials, completely general and independent of the physical or chemical constitution of the matter under consideration. In the linear region therefore, common methods of investigation, description and interpretation could be applied and a logically closed and complete description of mechanical properties could be given.

Beyond the limits of linearity it is no longer possible to formulate such a general theory of mechanical properties common to various materials. A principle similar to but more general than the superposition principle, applicable to non-linear behaviour, has not been found so far and is not likely to exist.

Therefore we are forced to treat non-linear properties in a more descriptive way. This involves a different treatment for different groups of polymers and for different aspects of the mechanical behaviour. Accordingly, a classification of polymers with respect to chemical or structural features and also of the mechanical properties under consideration becomes essential.

An important structural characteristic of polymers affecting their mechanical behaviour is presence or absence of cristallinity. Another one, particularly important in rupture phenomena, is their homogeneity as will be elaborated in chapter III.

Typical non-cristalline polymers are glasslike at temperatures below the transition temperature T_E, rubberlike between the transition temperature T_E and the flow temperature T_f and – if not crosslinked – fluid above T_f. Typical crystalline polymers are glasslike below the transition temperature T_m, tough between the transition temperature and the melting point of crystallites T_s and fluid above T_s. Therefore a first distinction for the classification of mechanical properties can be made between properties below the transition temperature T_E, properties above the flow temperature T_f (above T_s for crystalline polymers) and properties in between T_E and T_f. Below T_m polymers are generally hard and rigid (they show brittle behaviour), above T_E they are rubberlike and above T_f they behave as liquids.

A problem appearing in the treatment of non-linear behaviour of an even more elementary nature than that of structure and temperature is that of the mathematical description of deformations of finite magnitude. Independent of the question whether a deformation exceeds the linearity limit or not, we can distinguish "infinitesimal" and finite deformations. A deformation is called infinitesimal (or sometimes simply small), if the changes in dimensions of the sample are very small in comparison with the original dimensions, and it is called finite, if changes and original dimensions are of the same order of magnitude. Of course the concept of infinitesimal deformation, and therefore the practical decision whether a deformation is to be considered as finite or infinitesimal, depends on the accuracy of the experiment.

Consider the uniform extension of a rod in the longitudinal direction (fig. II,1). We call L_0 the length of the rod in the undeformed state and L the length in the deformed state. The *stretch* or elongation ratio λ is defined as the ratio of the stretched length to the unstretched length

$$\lambda = L/L_0. \qquad \text{(II,1)}$$

The meaning of the stretch λ is seen in fig. II,1. A straight line (parallel to the direction of stretch) of length unity in the undeformed state is changed by the deformation into a line of length λ, whilst a line of length unity in the deformed state originates from a line of length $1/\lambda$ in the undeformed state.

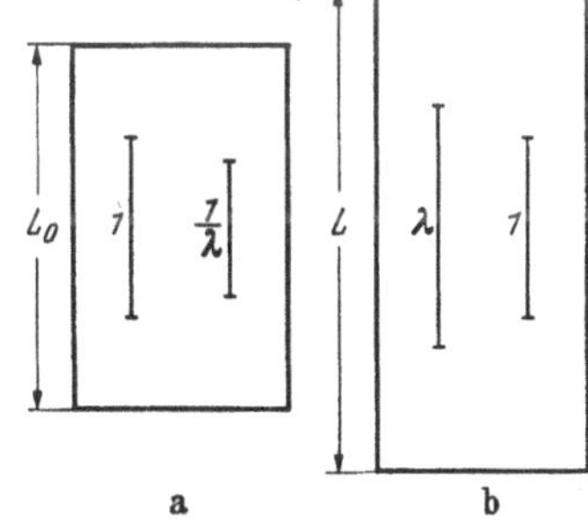

Fig. II,1. Uniform extension of a rod: a) undeformed, b) deformed.

The relative increase ε in length of a line of original length unity is called the elongation, or the CAUCHY *measure* of strain

$$\varepsilon = \lambda - 1 = (L - L_0)/L_0 \quad \text{CAUCHY}. \qquad \text{(II,2)}$$

The CAUCHY definition of strain is not the only possible definition nor is it preferable to other possible definitions for any physical reason. In fact every function of the shape of a body which vanishes when the body under consideration is undeformed and which increases with increasing deformation may be used as a measure of strain[1]. Obviously any such quantity can be developed into powers of ε and by a suitable choice of the units the coefficient of the first power can be reduced to unity. So for small deformations for which $\varepsilon \ll 1$ every definition of the strain leads to the same value, ε, as the CAUCHY definition.

However for "finite" strains (ε comparable with unity) different values are obtained for different definitions. Besides the CAUCHY definition the following definitions of strain have been proposed:

Kirchhoff measure $\quad \varepsilon^K = {}^1/_2[\lambda^2 - 1]$

Murnaghan measure $\quad \varepsilon^M = {}^1/_2[1 - 1/\lambda^2]$

Hencky measure $\quad \varepsilon^H = \ln \lambda$.

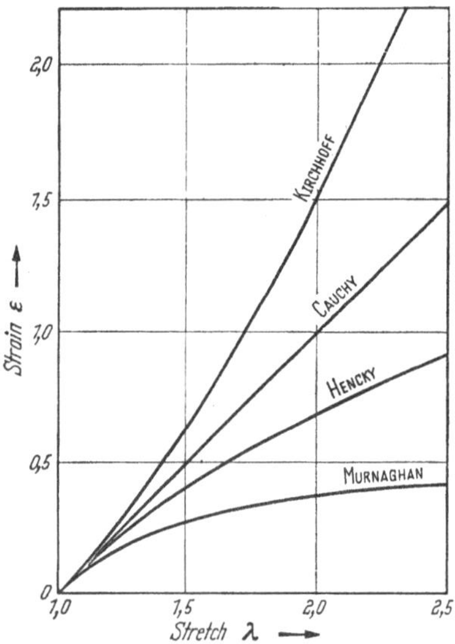

Fig. II,2. The different definitions of strain as a function of stretch.

The difference between these definitions is illustrated in fig. II,2, where the different measures of strain are plotted against the stretch λ (between 0 and 150% elongation). With the CAUCHY definition the strain is a linear function of stretch[2].

As stated before there is no preferential choice of the measure of strain with regards to the formulation of rheological laws[3], each of the above-mentioned forms may be used equally well. Of course the form of the relation between stress and strain will depend on the method of representation of the strain. In particular, if the stress strain relation is linear, when based on one of the strain definitions, it will become nonlinear when based on a different strain definition. The concept of linearity

[1] WEISSENBERG, K.: British Rheologist's Club Conference, London 1946. Compare further: M. REINER: Amer. J. Mathem. **70**, 433 (1948).

[2] Note that in the unstrained state the strain definitions ε, ε^K, ε^M and ε^H take the value zero, whilst the stretch λ takes the value unity. Therefore a CAUCHY strain of 100% corresponds to a stretch $\lambda = 2$.

[3] There may exist appropriate measures of strain, in which rheological laws become as simple as possible, but generally we do not know them. Two articles must be mentioned here: J. G. OLDROYD [Proc. Roy. Soc. (London) **A 200**, 523 (1950)] and A. S. LODGE [Proc. Camb. Phil. Soc. **47**, 575 (1951)], which advise the use of convected coordinate systems as advantagous in formulating rheological laws with correct invariance properties.

therefore looses its meaning if we proceed into the region of finite strains. It is always possible to choose a measure of strain such as to linearise an arbitrary stress strain relationship. Many of the definitions of strain found in the literature originated in this way.

§ 11. Large deformations of rubberlike materials[1].

a) Introduction.

We begin with the discussion of *rubberlike elasticity* (in German: *Kautschukelastizität*), which is characteristic of high polymers in a temperature region above the softening temperature T_m and which is shown at normal temperature especially by vulcanized natural and synthetic rubbers.

Rubberlike elasticity shows certain characteristic general features:

(i) The deformations obtainable are large[2].

(ii) The deformations are reversible.

(iii) The compressibility of the material is low.

(i) Polymers in the highly-elastic state possess a low shear modulus ($10^6 - 10^7$ dynes/cm²) and high extensibility and therefore show large deformations even under medium forces. Fig. II, 3 represents the example of a force elongation diagram for a natural vulcanized rubber in simple tension after MOONEY[3]. The deformations occurring are as high as 600% ($\lambda = 7$) and the S-shaped diagram is characteristic for rubberlike behaviour.

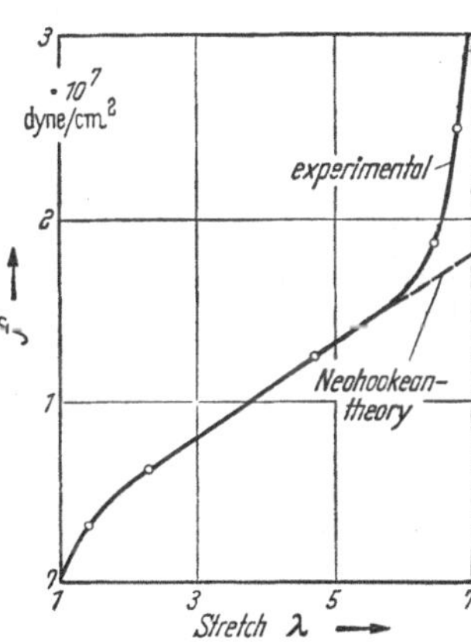

Fig. II, 3. Tensile force f against elongation ratio λ for natural rubber, after MOONEY.

(ii) The highly-elastic deformation behaviour of well vulcanized rubbers at temperatures sufficiently far above the softening temperature T_m, is to a great extent free from time effects. The amount of energy involved in hysteresis, stress relaxation and internal damping is generally very small in comparison to the energy stored in the large elastic reversible deformations occurring[4]. The reversibility of rubberlike elasticity guarantees that the

[1] For further information compare chapter V.

[2] The mechanical behaviour of rubberlike polymers at small deformations has already been treated in the first chapter of this volume.

[3] MOONEY, M.: J. appl. Physics **11**, 582 (1940).

[4] In many cases disturbing hysteresis effects or relaxation can be eliminated by appropriate preliminary treatment and conditioning (relaxation at higher temperatures before the beginning of the measurement). In this way reversible stress strain relations can be obtained. Of course also the time effects play a certain role in high elastic deformation and we shall deal with these problems shortly in the last section.

state of deformation will be a unique function of the forces or of the state of stress and yields the possibility of developing a general theory of rubberlike elasticity.

(iii) Polymers in the rubberlike state are practically incompressible, i.e. the changes in volume which occur are generally smaller by a factor of 100 to 1000 than the corresponding changes in shape (shear modulus 10^7 d/cm^2, bulk modulus 10^{10} d/cm^2). The incompressibility of rubbers has a remarkable consequence: For a given state of deformation the applied forces are indetermined to the extent of an additive component p in the nature of a hydrostatic pressure. If a state of deformation can be produced by the three principal stresses σ_1, σ_2, σ_3, the same deformation can also be realised by the principal stresses $\sigma_1 + p$, $\sigma_2 + p$, $\sigma_3 + p$, p being an arbitrary additional pressure.

We shall develop the theory of rubberlike behaviour in three steps:

a) The *Neo*-Hookean[1] or *statistical theory* of *rubberlike elasticity*.

b) The *Superelastic theory*[2] of *rubberlike elasticity*.

c) The *general theory* of *rubberlike elasticity*.

In cases a) and b) a certain expression for the stored energy under deformation is assumed [equ. (II,4) and (II,14) respectively], whilst in case c) the form of the stored energy function is left undetermined and can be adapted to the experiment. Both the Neo-Hookean and the Superelastic theory yield a linear relation between the shear stress and the shear strain in simple shear, in the general theory this relationship may have a nonlinear form.

The Neo-Hookean theory is the most simple. It describes rubberlike elasticity by means of one material constant, the shear modulus G. It further agrees with the statistical theory of rubberlike elasticity, which is able to explain Neo-Hookean behaviour by the movements of the long chain molecules forming a three dimensional molecular network. Neo-Hookean theory may be used as a good approximation of the mechanical behaviour of rubbers up to medium elongations. For instance the force elongation diagram of fig. II,3 can be explained by Neo-Hookean theory up to 400% elongation.

The Superelastic theory is slightly more general than the Neo-Hookean theory, as it describes rubberelasticity by two material constants C_1 and C_2. The sum of these constants conforms to the shear modulus G, whilst the ratio C_2/C_1 determines the deviations of Superelastic behaviour from Neo-Hookean behaviour. The Superelastic theory has been introduced for experimental reasons. It describes the mechanical behaviour of rubbers at small and medium elongations more accurately than the Neo-Hookean theory does.

---

[1] The term Neo-Hookean has been proposed by Rivlin loc. cit., page 131. It should be indicate that the Neo-Hookean theory is the simplest one which can be used to treat rubberlike elasticity, just as the Hookean theory is the simplest for treating the elasticity of hard solids at small deformations.

[2] The term Superelastic has been proposed by Mooney, loc. cit., page 135.

For very large deformations neither the Neo-Hookean nor the Superelastic theory are sufficient to describe the mechanical behaviour of rubbers. For instance the deviation of the force-elongation curve to higher values of f at about 500% deformation in fig. II, 3 (the upper part of the S-shape), which is connected with the finite length of the chain molecules, cannot be explained by Neo-Hookean or Superelastic theory. For the range of very large elongations a general theory of rubberlike elasticity has been developed, based on only two assumptions, the reversibility of deformation and the isotropy of material[1].

b) The Neo-Hookean theory of rubberelasticity.

The Neo-Hookean theory represents a very good approximation of rubberlike elasticity in the deformation range of medium magnitude. The theory originated in two independent ways: 1. From purely phenomenological considerations as the most simple form of description of ideal elastic behaviour. 2. From the statistical molecular theory of networks. We start with the phenomenological treatment, which was developed after the statistical theory in a series of publications by MOONEY and RIVLIN[2].

Assume a cube of the material of unit dimensions deformed by the three forces f_1, f_2, f_3 perpendicular to its surfaces, into a rectangular parallelepiped having edge-lengths λ_1, λ_2, λ_3 (see fig. II, 4). The *principal stretches* λ_1, λ_2, λ_3 will obey an equation expressing the incompressibility of the rubber [§ 11 a), condition (iii)]

$$\lambda_1 \lambda_2 \lambda_3 = 1. \qquad \text{(II,3)}$$

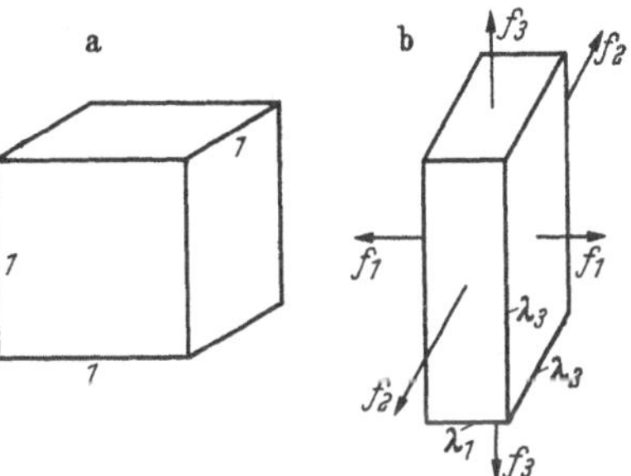

Fig. II, 4. Homogeneous deformation a) unstrained, b) strained state.

The deformation energy W performed by the external forces (the *stored energy function*) will be a unique function of the principal stretches because of condition (ii). Neo-Hookean theory now assumes that this stored energy is given by

$$W = {}^1/_2 G[\lambda_1^2 + \lambda_2^2 + \lambda_3^2 - 3] = {}^1/_2 G[I_1 - 3]. \qquad \text{(II, 4)}$$

The stored energy function vanishes in the unstrained state ($\lambda_1 = \lambda_2 = \lambda_3 = 1$) and rubberlike elasticity is characterised by one material constant, the

[1] For a more thorough treatment we refer to the book of L. R. G. TRELOAR: The physics of rubber elasticity, Oxford, Clarendon Press 1949, where an excellent survey is given of the different phenomenological and molecular theories and where also the experimental contributions are treated much more extensively.

[2] MOONEY, M.: J. appl. Physics **11**, 582 (1940). - R. S. RIVLIN: J. appl. Physics **18**, 444 (1947); Philos. Trans. Roy. Soc. London, **A 240**, 459, 491, 509 (1948); **A 241**, 379 (1948); Proc. Roy. Soc. [London] **A 195**, 463 (1949).

shear modulus G. The stored energy contains the principal stretches only in the symmetrical form I_1, where

$$I_1 = \lambda_1^2 + \lambda_2^2 + \lambda_3^2 \tag{II, 5}$$

represents the first invariant of the deformation tensor[1].

From the deformation energy (W) the stress strain equations have been deduced by MOONEY[2] and TRELOAR[3]. If f_1, f_2, f_3 are the tension forces per unit of cross section in the undeformed state (sometimes called *conventional stresses* or nominal stresses) and if σ_1, σ_2, σ_3 are the tensions per unit of cross section in the deformed state (the *true* principal stresses), we have the equations

$$\sigma_1 = \lambda_1 f_1, \quad \sigma_2 = \lambda_2 f_2, \quad \sigma_3 = \lambda_3 f_3.$$

An increase of the three principal stretches ($d\lambda_1$, $d\lambda_2$, $d\lambda_3$) leads to an increase in deformation energy

$$dW = G[\lambda_1 d\lambda_1 + \lambda_2 d\lambda_2 + \lambda_3 d\lambda_3]$$

which must be equal to the expression

$$dW = f_1 d\lambda_1 + f_2 d\lambda_2 + f_3 d\lambda_3.$$

Comparison of both expressions leads to

$$(G\lambda_1 - f_1)d\lambda_1 + (G\lambda_2 - f_2)d\lambda_2 + (G\lambda_3 - f_3)d\lambda_3 = 0.$$

However the increases $d\lambda_1$, $d\lambda_2$, $d\lambda_3$ are not independent from each other, but must obey the condition of incompressibility (II, 3), which can be written

$$d\lambda_1/\lambda_1 + d\lambda_2/\lambda_2 + d\lambda_3/\lambda_3 = 0.$$

From the last two equations we obtain, by the method of undetermined multipliers, the stress strain equations.

The stress strain equations for a Neo-Hookean solid are of the form

$$\left.\begin{aligned} \lambda_1 f_1 &= \sigma_1 = G\lambda_1^2 + p \\ \lambda_2 f_2 &= \sigma_2 = G\lambda_2^2 + p \\ \lambda_3 f_3 &= \sigma_3 = G\lambda_3^2 + p \end{aligned}\right\} \tag{II, 6}$$

where p is the undetermined isotropic component of stress.

The Neo-Hookean behaviour has a number of remarkable consequences, which we shall illustrate by consideration of some experiments. In *unidirectional elongation* (or compression) the sample is stretched by a longitudinal force f (measured per unit area in undeformed state) to λ times its original length. Because of incompressibility the lateral dimensions contract in the ratio $1 : \sqrt{\lambda}$. (fig. II, 5.)

---

[1] An invariant is a function of the extension ratios, which does not change if the coordinate system is rotated. Compare § 2d.

[2] MOONCY, M., loc. cit., page 131.

[3] TRELOAR, L. R. G.: Trans. Faraday Soc. **39**, 241 (1943).

The relation between tensile force f and stretch λ can be derived from equation (II,6) inserting $\lambda_1 = \lambda$, $\lambda_2 = \lambda_3 = \lambda^{-1/2}$. The first of the equations (II,6) gives the relation between f and λ, which however still contains the undetermined hydrostatic component p. The latter is determined from the second of the equations (II,6) by putting the transverse force f_2 equal to zero. In this way we get:

$$f = G[\lambda - 1/\lambda^2]. \quad \text{(II,7)}$$

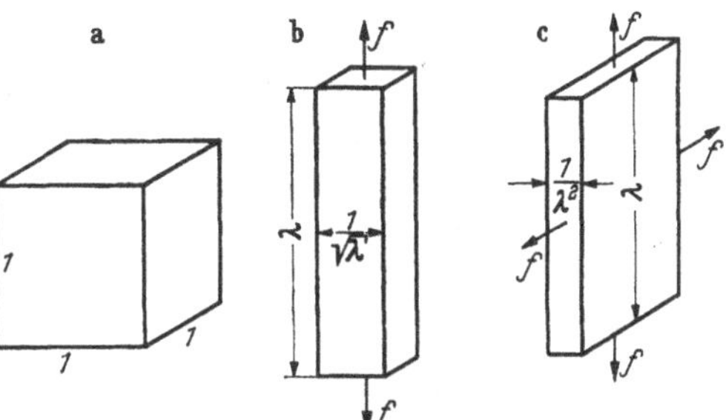

Fig. II, 5. (a) undeformed, (b) simple elongation, (c) uniform two-dimensional elongation.

The shape of this force elongation curve is illustrated by fig. II,6. Equation (II,7) describes both the elongation curve (f positive, $\lambda > 1$) and the compression curve (f negative, $\lambda < 1$). In the vicinity of the point $\lambda = 1$, i.e. for very small elongations, the curve can be approximated by its tangent having the slope $3G$. (II,7) contains therefore as a special case the Hookean law of an incompressible elastic material with a Young's modulus $E = 3G$.

In *uniform two-dimensional extension*, a cube of the material is stretched by two equal forces f in two perpendicular directions ($\lambda_1 = \lambda_2 = \lambda$, $\lambda_3 = 1/\lambda^2$). The relation between force and extension becomes in this case

$$f = G\lambda[1 - 1/\lambda^6]. \quad \text{(II,8)}$$

For large values of λ the force increases linearly with the extension. (II, 8) has been plotted together with (II,7) in fig. II,6. This type of deformation can be realized by inflating a balloon with gas[1].

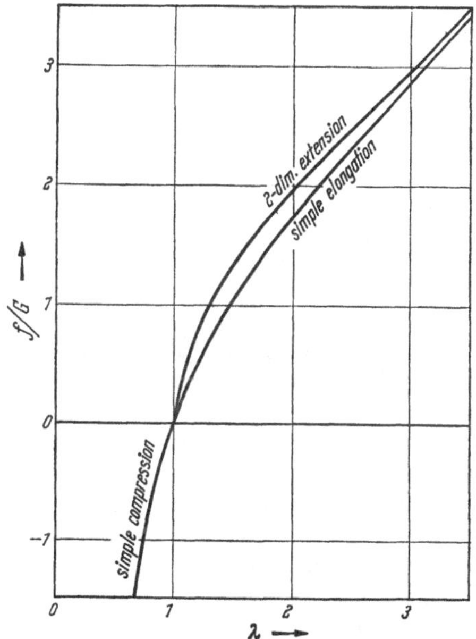

Fig. II, 6. Force-elongation curve for simple elongation or compression and for uniform two dimensional elongation.

The stress strain equations (II,6) describe only the case of homogeneous deformation as characterised by the three constant principal stretches λ_1, λ_2, λ_3. Rivlin[2] has generalised the Neo-Hookean theory to make it applicable to an arbitrary non-

[1] Boonstra, B. B. S. T.: J. appl. Physics **21**, 1098 (1950).

[2] Rivlin, R. S.: Philos. Trans. Roy. Soc. London **A 240**, 459 (1948).

homogeneous deformation. An arbitrary non-homogeneous deformation can be defined by the components of the displacement vector

$$\xi = \xi(x, y, z), \quad \eta = \eta(x, y, z), \quad \zeta = \zeta(x, y, z)$$

defined in a coordinate system x, y, z, fixed in space.

The Neo-Hookean behaviour is then characterised by the following equations between the stress components and the partial derivatives of the displacements:

$$\left.\begin{aligned} \sigma_{xx} &= G\left[(1+\xi_x)^2 + \xi_y^2 + \xi_z^2\right] + p \\ \sigma_{yz} &= G\left[\eta_x \zeta_x + (1+\eta_y)\zeta_y + \eta_z(1+\zeta_z)\right] \end{aligned}\right\} \qquad \text{(II, 9)}$$

and corresponding equations for the other stress components. From equations (II, 9) we may deduce again equations (II, 6) by putting

$$\xi = (\lambda_1 - 1)x, \quad \eta = (\lambda_2 - 1)y, \quad \zeta = (\lambda_3 - 1)z \qquad \text{(II, 10)}$$

which is the displacement vector belonging to the simple state of homogeneous deformation as described in fig. II, 4.

Consider next the case of *simple shear*. Each plane parallel to the x-y-plane is deformed as given in fig. II, 7. The rectangle $ABCD$ becomes a trapezoid $A'B'CD$ of equal height and base. The shear strain γ is measured by the tangent of the angle φ, $\gamma = \operatorname{tg}\varphi$. Simple shear is characterised by a displacement vector of the form

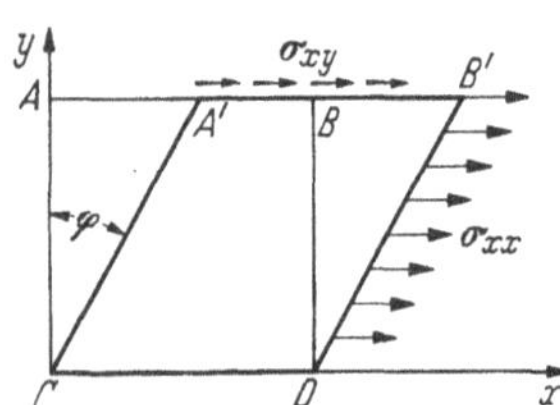

Fig. II, 7. Simple shear: $ABCD$ undeformed, $A'B'CD$ deformed and distribution of the shearing and tensile forces.

$$\xi = \gamma y, \quad \eta = \zeta = 0.$$

It follows from Neo-Hookean theory [i. e. from equations (II, 9)], that simple shear can be realized only by the application of at least two types of stress

$$\sigma_{xy} = G\gamma, \quad \sigma_{xx} = G\gamma^2. \qquad \text{(II, 11)}$$

The shear stress σ_{xy} acts on the x–z plane in direction of the x axis and is proportional to the amount of shear γ, the proportionality factor being the shear modulus G. Therefore the relation between shear stress and shear strain is just the same in Neo-Hookean theory as it is in Hookean theory. But apart from the shear stress, a tensile stress in the x direction is necessary, proportional to the square of γ, which vanishes at low shear. The fact that a simple shear deformation necessitates a tensile stress represents an example of a so called *cross effect* (in German: *Wechselwirkungseffekt*) and has no analogy in the theory of small deformations.

Similar considerations apply to the *pure torsion* of a circular cylinder. A cylinder of height l and radius a is twisted about an angle $l\psi$, ψ being the twisting angle per unit length (fig. II, 8). On the basis of Neo-Hookean theory a shear and a compressive force are necessary for this type of deformation. These are given by

$$\left.\begin{aligned} \sigma_{z\vartheta} &= r \cdot G\psi \\ \sigma_{zz} &= -{}^1/_2 G\left[a^2 - r^2\right]\psi^2. \end{aligned}\right\} \qquad \text{(II, 12)}$$

The torque

$$M = G\pi a^4/2 \cdot \psi \qquad \text{(II, 13)}$$

is proportional to the twist, just as in Hookean theory. A completely new fact is the requirement of a pressure in the longitudinal direction σ_{zz} having a parabolic distribution as shown in fig. II,8a. If only the torque M, but no pressure is applied, we do not get a pure torsion, but a torsion combined with a change in shape as shown in fig. II,8b (constriction in the middle and bulging out at the ends).

The Neo-Hookean theory degenerates into the common Hookean theory of an incompressible material in the limit of small deformations. This may be shown by neglecting all second order terms like ξ_x^2, $\eta_x \zeta_x$ etc. in equ. (II,9) and introducing the CAUCHY definitions of strain. We get the equations:

$$\sigma_{xx} = 2G\,\varepsilon_x + p'$$
$$\sigma_{yz} = G\gamma_{yz}$$

which are the stress strain equations of a Hookean incompressible material with shear modulus G.

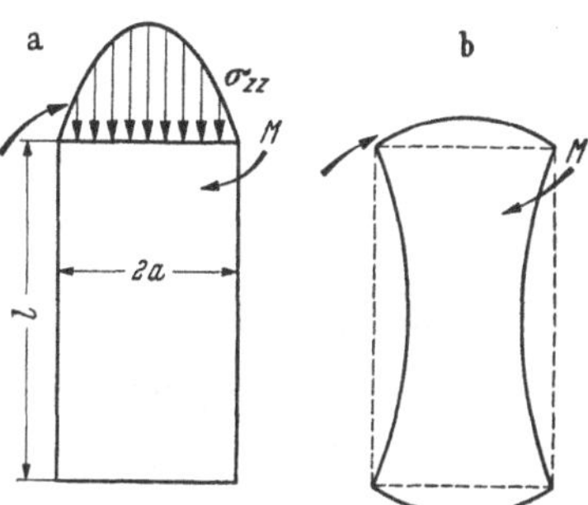

Fig. II,8a. Pure torsion of a circular cylinder and distribution of compressive force.
Fig. II,8b. Shape of cylinder twisted without longitudinal compressive force.

The Neo-Hookean theory of rubber elasticity may be explained by molecular considerations[1]. The long chain molecules in rubber form a molecular network. It can be shown that by a deformation the entropy of this network is decreased. The negative decrease in entropy times absolute temperature gives the energy stored during deformation:

$$W = -T(S - S_0) = \tfrac{1}{2} G[\lambda_1^2 + \lambda_2^2 + \lambda_3^2 - 3]$$

which is equal to equ. (II,4). The *statistical theory* of *rubberlike elasticity* is therefore identical with the Neo-Hookean theory as for as macroscopic results are concerned. Besides it yields a shear modulus G which is proportional to absolute temperature.

c) The superelastic theory of rubberelasticity[2].

In the preceeding section we stated that the Neo-Hookean theory describes to a good approximation the mechanical behaviour of vulcanized rubbers in the range of medium deformations. However, more accurate experiments have proved that in the region of medium deformations a better description of rubber elasticity is yielded by a slightly more general theory, the Superelastic theory.

The Superelastic theory has been developed by MOONEY[3] and is based on the following expression for the stored energy

$$W = C_1[I_1 - 3] + C_2[I_2 - 3] \tag{II,14}$$

[1] For a detailed discussion of the molecular theory of rubberlike elasticity see this volume, chapter V, § 27. — [2] For further information see chapter V, § 33.

[3] MOONEY, M.: J. appl. Physics **11**, 582 (1940); J. Colloid. Sci. **6**, 96 (1951); further references 2) on page 131.

where I_2 is the second invariant of the deformation tensor

$$I_2 = 1/\lambda_1^2 + 1/\lambda_2^2 + 1/\lambda_3^2. \tag{II, 15}$$

C_1 and C_2 are two material constants, describing completely the Superelastic behaviour. In the special case $C_1 = G/2$, $C_2 = 0$ Superelastic theory reduces to the Neo-Hookean form. Therefore the differences between these two theories will be the smaller, the smaller the ratio C_2/C_1.

Generally the differences between the Superelastic and the Neo-Hookean theory are small, and have more the character of a quantitative correction than of a qualitative difference. For instance in simple elongation or compression we get instead of (II, 7) the force extension relation

$$f = 2C_1[\lambda - 1/\lambda^2] + 2C_2[1 - 1/\lambda^3]. \tag{II, 16}$$

In order to obtain a deformation of simple shearing type, we have to apply a shear stress proportional to shear, a tensile stress in the x direction and a compressive stress in the y direction, both proportional to γ^2

$$\sigma_{xy} = 2[C_1 + C_2]\gamma, \quad \sigma_{xx} = 2C_1\gamma^2, \quad \sigma_{yy} = -2C_2\gamma^2. \tag{II, 17}$$

In simple torsion we have to apply a torque M and a normal pressure in the longitudinal direction

$$M = \pi a^4 (C_1 + C_2)\psi, \quad \sigma_{zz} = -[2a^2 C_2 + (a^2 - r^2)(C_1 - 2C_2)]\psi^2. \tag{II, 18}$$

Experimental verifications of the Neo-Hookean or Superelastic behaviour in the region of medium deformation have been given by several authors, chiefly on vulcanised natural rubber.

TRELOAR[1] found excellent agreement between experiment and Neo-Hookean theory up to 100% deformation and unimportant deviations between 100% and 400% deformation in linear extension. The curve of simple compression could be explained completely by Neo-Hookean theory. The theory yielded further agreement with experiment in uniform two dimensional extension up to 200% deformation and in pure shear up to 100% shear. The same author reported[2] that in an investigation of two-dimensional states of stress he found that Neo-Hookean theory did not describe the experimental results satisfactorily and that the Superelastic theory had to be used with a value of C_2/C_1 of 0,10.

MOONEY[3] compared experimental stress strain curves of soft rubber with the Superelastic theory and got agreement up to 500% elongation with the choice $C_2/C_1 = 0{,}38$.

RIVLIN[4] actually demonstrated the existence of compressive stresses in the longitudinal direction during the torsion of a rubber cylinder and measured them. He proved that the compressive stresses had a parabolic distribution over the cross section and were proportional to the square of the twisting angle. He obtained a value $C_2/C_1 = 0{,}14$ in good agreement with TRELOAR.

[1] TRELOAR, L. R. G.: Trans. Faraday Soc. **40**, 49 (1944).
[2] TRELOAR, L. R. G.: Proc. physic. Soc. **60**, 135 (1948).
[3] MOONEY, M.: J. appl. Physics **11**, 582 (1940). – L. E. COPELAND u. M. MOONEY: J. appl. Physics **9**, 450 (1948).
[4] RIVLIN, R. S.: J. appl. Physics **18**, 444 (1947).

Very comprehensive experiments on vulcanised rubber are reported by RIVLIN and SAUNDERS[1], who investigated pure shear, pure shear with superimposed extension, simple extension and compression, pure torsion and torsion with superimposed extension. They concluded that within certain limits, the behaviour of the material could be described by the Superelastic theory. They proved that C_1 is a material constant independent of I_1 and I_2 and that C_2 is independent of I_1 but slightly dependent on I_2. The ratio C_2/C_1 decreased from 0,25 to 0,04 with increasing I_2.

Concluding, we may say that the behaviour of vulcanised natural rubber at deformations of medium magnitude can be excellently described by the Superelastic theory with a ratio C_2/C_1 lower than 0,2. It follows that Neo-Hookean theory too describes to a very good approximation the mechanical behaviour of natural rubber for deformations smaller than about half the ultimate extensions.

d) The general equilibrium theory of rubberelasticity.

In the region of very large deformations neither the Neo-Hookean nor the Superelastic theory remain valid. The forces always increase strongly above the theoretical values, because of the finite extensibility of the network. For these cases a general theory of a purely descriptive character has been developed.

This theory of large elastic deformations appropriate for the phenomenological description of rubberlike elasticity over the entire deformation range has been given independently by MOONEY[2] and RIVLIN[3]. These theories [compare also Chapter V, C] are in principle equivalent and differ only in their starting point: MOONEY starts from the stress strain relation in simple shear while RIVLIN starts from a certain form for the stored energy function.

The MOONEY-RIVLIN theory is based on the postulates (i) to (iii) on page 129 and on the fourth assumption:

(iv) The material is isotropic, its elastic properties are the same in all directions.

As the deformations are reversible (i) and the material is isotropic (iv), the stored energy function W must be a unique function of the three strain invariants I_1, I_2 and I_3 defined by (II,5), (II,15) and

$$I_3 = \lambda_1^2 \lambda_2^2 \lambda_3^2 .$$

In virtue of the incompressibility condition (iii), the third invariant I_3 is equal to unity and the stored energy function can only depend on the two invariants I_1 and I_2:

$$W = W(I_1, I_2) \qquad \text{(II,19)}.$$

---

[1] RIVLIN, R. S. u. D. W. SAUNDERS: Philos. Trans. Roy. Soc. London **A 243**, 251 (1951).

[2] MOONEY, M.: J. appl. Physics **19**, 434 (1948).

[3] RIVLIN, R. S.: Philos. Trans. Roy. Soc. London **A 241**, 379 (1948). – See further M. REINER: Amer. J. Mathem. **70**, 433 (1948).

The most simple forms of the deformation energy would involve only first order terms in the invariants

$$W = C_1[I_1 - 3]$$

or

$$W = C_1[I_1 - 3] + C_2[I_2 - 3]$$

which represent the cases of Neo-Hookean and Superelastic behaviour.

RIVLIN however did not assume any special form for the expression (II, 19) but treated it quite generally. He succeeded in giving the form of the stress strain equations, the equations of motion and the boundary conditions in terms of W and the partial derivatives of W with respect to I_1 and I_2. RIVLIN treated further a number of simple types of strain such as simple extension or compression, simple shear, pure shear[1], problems of cylinder symmetry[2] and pure bending[3].

The MOONEY-RIVLIN theory investigates the interconnections which exist between different types of experiments due to the isotropy of the material. For instance if we are concerned with a Neo-Hookean material one measurement is sufficient to determine the one material constant G. The results of all other measurements are then predetermined. This one measurement can be for instance the measurements of the stress-strain relation in simple shear.

If the material obeys the more complicated Superelastic theory, two measurements are necessary to determine the two material constants. The stress-strain relation in simple shear gives us the sum $C_1 + C_2$ and the relation between tensile stress (in the x direction) and shear strain gives us C_1. Then again all other experiments are predetermined.

The question arises how to know whether an elastic material obeys the simple Neo-Hookean or the Superelastic law or neither the one nor the other. This question has been answered by MOONEY: If and only if the stress strain relation in simple shear is linear, the material obeys either the Neo-Hookean or the Superelastic law. To decide between these possibilities a further measurement has to be made. If the relationship is non-linear, none of these two theories can describe the material behaviour and the general MOONEY-RIVLIN theory is to be used.

If the form of the stress strain relation in simple shear is non-linear, the experiment in simple shear or in simple extension does not specify the material behaviour completely. Indeed if the form of the stored energy function (II, 19) is known, the behaviour of the rubber in simple elongation, simple shear, torsion or bending may be predicted. However, neither a measurement of the shear stress-shear strain relation in simple shear, nor a measurement of the force elongation curve in simple elongation is sufficient to specify the form of the stored energy function. For this purpose a sample has to be stretched simultaneously in two directions and both forces and stretches have to be measured.

e) Time dependence of large deformations of rubbers.

In the foregoing sections we have treated the highly elastic deformation behaviour of rubbers on the assumption of condition (ii): All deformations are reversible and are without time effects. This assumption enabled us to consider the deformation energy W as a unique function

---

[1] RIVLIN, R. S.: Philos. Trans. Roy. Soc. London **A 241**, 379 (1948).
[2] RIVLIN, R. S.: Philos. Trans. Roy. Soc. London **A 242**, 173 (1949).
[3] RIVLIN, R. S.: Proc. Roy. Soc. [London] **A 195**, 463 (1949).

of the state of deformation. This is of course an oversimplification and in many cases time effects will be of considerable importance.

We have seen in chapter I that in the case of small deformations, time effects can be treated quite generally by the superposition principle without any special assumptions. On the other hand in the case of large deformations without time effects, the behaviour of isotropic rubbers can be treated quite as well generally. But if we consider large deformations *and* time effects no general theory is available and each theoretical consideration has to start with certain simplifying assumptions, which have to be taken from experimental evidence.

A simple assumption based on experimental evidence has been given by GUTH and coworkers[1], who investigated the relaxation behaviour of natural and synthetic rubbers in simple elongation. They showed that the stress relaxation curves, which will be generally a function of time t and deformation λ, could be decomposed into a product

$$\sigma(\lambda, t) = F(\lambda) \cdot G(t). \qquad \text{(II,20)}$$

This factorisation separates the influence of time and elongation in a simple way. Relaxation curves at different elongations are similar, as they can be derived by simple magnification from another. The stress relaxation curves at different elongations fall on a single curve, the stress relaxation modulus $G(t)$, if the quantities $\sigma(\lambda, t)/F(\lambda)$ are plotted against time.

The function $F(\lambda)$ will depend generally on the type of deformation considered and on the law of elasticity which is obeyed by the rubber in equilibrium. If we assume that the rubber follows Neo-Hookean behaviour in equilibrium (which implies that the deformations do not approach the breaking strains), we have

$$\text{for simple elongation} \quad F(\lambda) = \lambda^2 - 1/\lambda$$
$$\text{for simple shear} \quad F(\gamma) = \gamma.$$

The stress relaxation function (in simple shear) $G(t)$ is a monotonically decreasing function of time and tends to the equilibrium value G of Neo-Hookean theory for infinite times.

In agreement with GUTH and coworkers, STERN and TOBOLSKY[2] found that the factorisation in stress relaxation fitted quite well the experimental data on polysulfide rubbers in simple elongation. GREEN and TOBOLSKY[3] gave a molecular theory, explaining stress relaxation by the breaking and reforming of cross links. The macroscopic result of this theory agrees with the form (II,20).

These experimental facts could be used to set up a theory of time-dependent rubberlike behaviour, based on the assumptions:

(i) Rubberlike behaviour at equilibrium is given by the Neo-Hookean law. This implies that the deformation does not come into the range of ultimate elongations and that the rubber is incompressible and isotropic.

(ii) In stress relaxation a simple factorisation of the form (II,20) is valid.

Consider first the stress relaxation experiment under constant stretch λ_0 in simple elongation or compression. The stress can be written

$$\sigma(\lambda_0, t) = F(\lambda_0) \cdot G(t) \quad \text{where} \quad F(\lambda_0) = \lambda_0^2 - 1/\lambda_0. \qquad \text{(II,21)}$$

---

[1] GUTH, E., P. E. WACK u. R. L. ANTHONY: J. appl. Physics **17**, 347 (1946).
[2] STERN, M. D. u. A. V. TOBOLSKY: J. chem. Physics **14**, 93 (1946).
[3] GREEN, M. S. u. A. V. TOBOLSKY: J. chem. Physics **14**, 80 (1946).

(II,21) may be generalised to a superposition formula which is quite similar to BOLTZMANN's principle. Under an elongation arbitrarily prescribed as a function of time, $\lambda = \lambda(t)$, the resulting stress can be found

$$\sigma(t) = G(0)\,F(t) + \int_{-\infty}^{t} \dot{G}(t-t')\,F(t')\,dt'. \tag{II,22}$$

The function $F(t)$ is defined by inserting the prescribed time dependence of λ into $F(\lambda)$. This superposition principle (given already by GUTH and coworkers) can be interpreted in a simple way: We can use each arbitrarily measure to fix the amount of elongation. Take especially $F = \lambda^2 - 1/\lambda$ as a measure of strain, then BOLTZMANN's superposition principle remains valid between the stress and the appropriately chosen strain F. We can for instance invert equation (II,22). Under a tensile stress, arbitrarily prescribed as a function of time, $\sigma = \sigma(t)$, the strain $F(t)$ is given by

$$F(t) = J(0)\,\sigma(t) + \int_{-\infty}^{t} \dot{J}(t-t')\,\sigma(t')\,dt'. \tag{II,23}$$

The function $J(t)$, monotonically increasing with time, corresponds to the creep compliance. If the stress relaxation modulus is known, the creep compliance can be calculated and vice versa, by means of the equation

$$\int_{0}^{t} G(t-t')\,J(t')\,dt' = t. \tag{II,24}$$

In the special case of creep under constant stress σ_0, we have

$$F(t) = \sigma_0 J(t). \tag{II,25}$$

Formula (II,25) shows that in the case of creep the strain $F(\sigma, t)$ can be factorised into the stress and into the creep compliance. The elongation λ has to be calculated from F by means of equation (II,21) and cannot be factorised into a stress and a time factor.

If we consider a vibration experiment with the stress prescribed in sinusoidal form, the "strain" F has sinusoidal time dependence with the same angular frequency and a phase lag α. The elongation will be a periodic but non-sinusoidal function with the same period as the stress.

In the case of simple shear, the appropriate measure of strain is the shear $F(\gamma) = \gamma$ and the superposition principle holds between shear stress and shear strain up to moderately large strains.

§ 12. Non-linear flow of polymers and polymer solutions.

a) Introduction.

Next we consider the non-linear mechanical behaviour of polymers at very high temperatures. In a temperature region, above the rubberlike range, linear uncrosslinked polymers become soft and behave like highly viscous liquids showing flow as their main characteristic mechanical property. This temperature region is identical with the processing range and therefore the flow properties of polymers are not only interesting from a purely scientific point of view but also because of their influence on the processing behaviour of the material.

Similar behaviour is shown by solutions of high polymers in low molecular solvents at room temperatures.

The kinematics of flow will be treated with reference to a Cartesian coordinate system fixed in space $(x\,y\,z)$. The description of a steady state of flow involves the knowledge of the velocity vector of flow $\vec{V} = (u, v, w)$ at each point $P = (x\,y\,z)$ of space. Assume the Cartesian components of the velocity $\vec{V}$ to be given

$$u = u(x, y, z), \quad v = v(x, y, z), \quad w = w(x, y, z).$$

The state of flow is generally described by the *flow tensor* (*deformation velocity tensor*) with the components[1]

$$e_{11} = \partial u/\partial x \ldots, \quad e_{12} = e_{21} = 1/2\,[\partial v/\partial x + \partial u/\partial y]. \qquad \text{(II, 26)}$$

The simplest state of flow is laminar shearing motion (fig. II, 9), in which all velocity vectors are parallel to the x-direction and where the magnitude of the velocity is proportional to the distance from the x–z plane: $u = q\,y$, $v = 0$, $w = 0$. In this case the only non-vanishing components of the flow tensor are $e_{12} = e_{21} = q/2$. q is called the *"rate of shear"*.

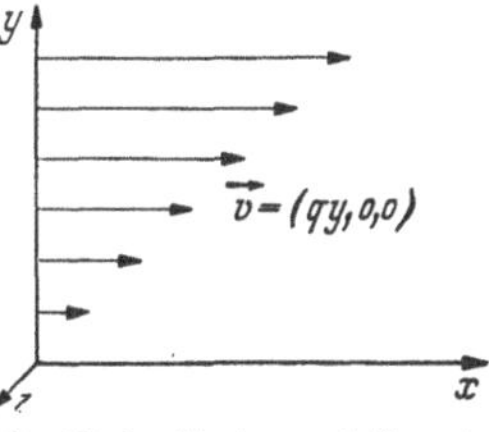

Fig. II, 9. Vectors of flow in laminar shearing motion.

Viscous flow without elastic effects will be characterised by a relation ship between the stresses and the components of the flow tensor. The simplest case is Newtonian *flow*, in which the stress components are proportional to the corresponding flow components

$$\left.\begin{aligned} \sigma_{11} &= 2\eta_0 e_{11} + p \ldots, \quad \sigma_{12} = 2\eta_0 e_{12} \\ \operatorname{div} \vec{V} &= e_{11} + e_{22} + e_{33} = 0. \end{aligned}\right\} \qquad \text{(II, 27)}$$

The last equation expresses the incompressibility of the liquid. As a consequence of this incompressibility, the normal stresses $\sigma_{11}, \ldots$ are in determinate to the extent of a pressure p. The only physical constant of the incompressible Newtonian liquid is the viscosity η_0.

In laminar shearing motion (characterised by the rate of shear q) the normal stresses become identical with the isotropic pressure and the only non-vanishing stress is[2]

$$\sigma \equiv \sigma_{12} = \eta_0 q. \qquad \text{(II, 28)}$$

Newtonian behaviour is therefore characterised by two restrictions:

1. The occurring shear stresses are proportional to the rate of shear.

[1] The components of the flow tensor may be related to the components of the deformation velocity tensor as introduced in § 2, c by means of the equations

$$e_{11} = \frac{\partial u}{\partial x} = \frac{\partial}{\partial t}\left(\frac{\partial \xi}{\partial x}\right) = \dot{\varepsilon}_x, \quad e_{12} = \frac{1}{2}\dot{\gamma}_{xy}.$$

[2] Here and throughout this section (§ 12) we denote the shear stress shortly as σ.

2. In a state of laminar steady flow – apart from a purely hydrostatic pressure – only shear stresses occur.

The simple viscosity laws (II,27) and (II,28) are valid for low-molecular liquids. High-molecular liquids (polymer melts and polymer solutions) show Newtonian behaviour only at low shear stresses[1]. At higher shear stresses there are deviations from the simple Newtonian behaviour, which can occur in two ways, depending on whether statement 1 or 2 is violated:

(i) The shear stresses in laminar flow are non-linear functions of the rate of shear.

(ii) Normal stresses occur besides the shear stresses in a state of laminar steady flow.

(i) If the shear stress is no longer proportional to the rate of shear, equation (II,28) is to be replaced by a more general law (for the case of laminar flow)

$$q = F(\sigma) = \sigma/\eta. \tag{II,29}$$

The quantity η depends on shear stress and may be called a "viscosity", which is not a constant, but decreases with increasing shear stress. (For low shear stresses the viscosity tends to a limit, η_0, which is measured as the Newtonian viscosity at low rates of shear.)

(ii) A new feature is the occurrence of normal stresses as secondary phenomena, the occurrence of cross effects. These effects, which are similar to those occurring in rubberlike behaviour, have been demonstrated in polymer solutions by WEISSENBERG[2] and others.

b) Non-Newtonian flow of polymer melts.

We consider first the *non*-Newtonian *flow* behaviour of polymer melts, i. e. the form of the relation between stress and rate of shear (II, 29). Non-Newtonian flow behaviour of polymer melts may be illustrated by fig. II, 10 where the rate of shear is plotted against shear stress for molten polystyrene, after BUCHDAHL[3]. Validity of NEWTON's law would imply straight lines for these plots. It is seen that the flow curves deviate from straight lines with a concave curvature in the direction of higher shearing velocities. This is the general picture for polymer melts.

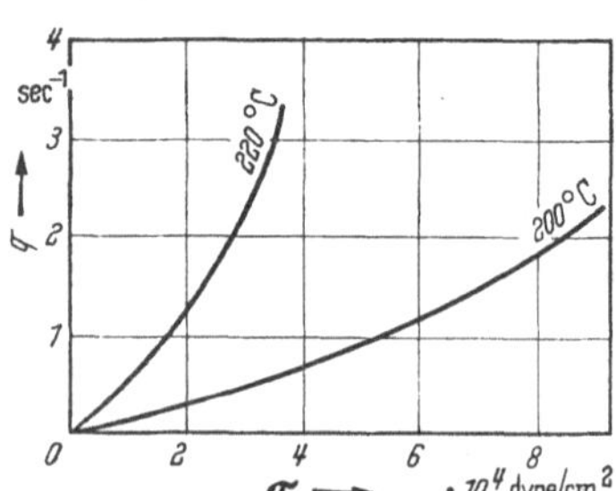

Fig. II, 10. Non-Newtonian flow of polystyrene at 200° and 220° C after BUCHDAHL[3].

This behaviour can be approximated by a relation originally proposed by FERRY[4]

$$\frac{1}{\eta} = \frac{1}{\eta_0}[1 + a\sigma]. \tag{II, 30}$$

---

[1] A survey on Newtonian flow behaviour of polymer solutions and melts is given in volume II, chapter V. — [2] WEISSENBERG, K.: loc. cit., page 147.

[3] BUCHDAHL, R.: J. Colloid Sci. **3**, 87 (1948).

[4] FERRY, J. D.: J. Amer. chem. Soc. **64**, 1330 (1942).

The material constant a has the dimensions of a reciprocal shear stress and its value has been found to be between 10^{-4} and 10^{-6} cm²/dynes. Non-linear effects become considerable therefore in a region of shear stresses between 10^4 and 10^6 dynes/cm². Relation (II,30) has been used to describe non-Newtonian flow behaviour of polyisobutylene[1] and a has been shown to be approximately proportional to molecular weight and independent of temperature.

FERRY's relation (II,30) is not general, but should be considered as a first approximation which does not remain valid at higher shear stresses. As has been shown by SPENCER[2] in the case of polystyrene, the viscosity – shear stress relation takes the form

$$\frac{1}{\eta} = \frac{1}{\eta_0}\left[1 + a\sigma + 0{,}48\,a^2\sigma^2 + 0{,}15\,a^3\sigma^3 + 0{,}034\,a^4\sigma^4\right]. \qquad \text{(II,31)}$$

This represents the development of the reciprocal viscosity into a power series of the stress, the coefficients being all positive and determined by the first constant a, which has been found independent of temperature and proportional to molecular weight. SPENCER's equation (II,31) has been further used to calculate the extrusion properties of polystyrene[3] and to describe the non-Newtonian flow of mastificated rubber[4].

SCHEELE and coworkers[5] have used the OSTWALD-DE WAELE law[6] as a rheological description of non-Newtonian flow

$$q = K\sigma^n \quad \text{or} \quad \frac{1}{\eta} = K\sigma^{n-1}. \qquad \text{(II,32)}$$

Both the exponent n and the constant K depend strongly on temperature. n tends to unity (NEWTONS law) for high and low temperatures and has a maximum in the transition region. However, it seems to us that at temperatures below the transition temperature, the observed deformation is partially reversible. It includes the beginning of a very slow retarded viscoelastic mechanism[7].

A decision about the question, which of the two equations (II,31) or (II,32) is preferable, must be delayed, until experiments of sufficient

---

[1] FERRY, J. D.: Physics **6**, 356 (1935). – T. G. FOX u. P. J. FLORY: J. Amer. chem. Soc. **70**, 2384 (1948). – H. LEADERMAN: Physics of high polymers, Utrecht, Univ. Press 1951, page 26.

[2] SPENCER, R. S. u. R. E. DILLON: J. Colloid Sci. **4**, 241 (1949); **3**, 163 (1948). – R. S. SPENCER: J. Polymer Sci. **5**, 591 (1950).

[3] SPENCER, R. S.: Proc. Sec. Int. Congr. Rheology, Oxford 1953, page 20. – C. E. BEYER u. F. E. TOWSLEY: J. Colloid Sci. **7**, 236 (1952).

[4] TRELOAR, L. R. G. u. D. W. SAUNDERS: Trans. Instn. Rubber Ind. **24**, 92 (1948).

[5] SCHEELE, W., M. ALFEIS u. L. LAHAYE: Kolloid-Z. **103**, 1 (1943). – W. SCHEELE Kolloid-Z. **105**, 209 (1943). – W. SCHEELE, M. ALFEIS u. I. FRIEDRICH: Kolloid-Z. **108**, 44 (1944). – W. SCHEELE: Z. Naturforsch. **4a**, 433 (1949); Kolloid-Z. **115**, 113 (1949). – W. SCHEELE u. TH. TIMM: Kolloid-Z. **116**, 131 (1950); **120**, 103 (1951); **122**, 129 (1951). – Compare further L. R. G. TRELOAR: Trans. Instn. Rubber Ind. **25**, 167 (1949).

[6] WAELE, A. DE: J. Oil Col. Chem. Assoc. **4**, 33 (1923). – WO. OSTWALD: Kolloid-Z. **36**, 99 (1925).

[7] Compare J. M. BURGERS: Proc. Roy. Soc. Amsterdam **51**, No. 7 (1948).

accuracy and over a sufficiently large range of shear stresses have become available. But there is an essential difference between these two equations: If the equation of FERRY or SPENCER holds, the polymer behaves in a Newtonian manner at small shear stresses (more accurately: in the region of shear stress $\sigma \ll 1/a$). But if the OSTWALD-DE WAELE law were to hold exactly, there would be no stress region, however small, in which the flow behaviour could be described by NEWTON's law.

However, it has been experimentally observed and theoretically explained in many domains of physics that the velocity of rate phenomena occurring as a consequence of a deviation from equilibrium is proportional to the amount of deviation, provided that this deviation is small. Therefore, we believe[1] that the equation of FERRY or SPENCER, or, more generally, a development into a power series with respect to the stress, is preferable to an exponential equation.

We did not use here the term *plastic flow*, which is sometimes used in the literature for either the non-Newtonian flow or for the cold drawing of polymers. For plasticity has a very definite meaning in the deformation behaviour of metals, where it describes the flow under shear stresses beyond the yield point. So far we have insufficient insight to judge whether the deformation mechanism in non-Newtonian flow or cold drawing of polymers resembles that of plastic flow of metals to such an extent as to justify this nomenclature.

c) Non-Newtonian flow of polymer solutions.

For many years it has been recognised[2] that high polymer solutions may behave as non-Newtonian liquids. The *non*-Newtonian *behaviour* of polymer solutions at low shear stress is very similar to the behaviour of polymer melts: The shearing velocity-shear stress relation begins to deviate from a straight line in the direction of higher shearing velocities.

For instance equation (II,30) has been used to describe the flow behaviour of polystyrene solutions in xylene by FERRY[3]. The parameter a turned out to be independent of temperature and inversely proportional to concentration. BESTUL and BELCHER[4] used FERRY's equation for the description of the flow behaviour of GRS dissolved in o-dichlorobenzene. MARON and coworkers[5] on the other hand described the flow behaviour of GRS latex at small shear stresses by the OSTWALD-DE WAELE equation (II,32) and found exponents n from 1 at very high dilution up to 2 at higher concentrations.

However, at medium and higher shear stresses the flow behaviour of polymer solutions is entirely different from that of polymer melts.

---

[1] A critism of the OSTWALD-DE WAELE law based on other physical grounds has been given by A. NADAI u. P. G. McVETTY: Proc. ASTM **43**, 735 (1943).

[2] STAUDINGER, H.: Die hochmolekularen organischen Verbindungen, Springer Berlin 1932, page 189.

[3] FERRY, J. D.: J. Amer. chem. Soc. **64**, 1330 (1942).

[4] BESTUL, A. B. u. H. V. BELCHER: J. Colloid Sci. **5**, 303 (1950).

[5] MARON, S. H., B. P. MADOW u. I. M. KRIEGER: J. Colloid Sci. **6**, 584 (1951). — S. H. MARON u. B. P. MADOW: J. Colloid Sci. **8**, 130 (1953).

The general flow curves of polymer solutions show the character of fig. II, 11. At low shear stress there is an initial Newtonian region (shear velocity proportional to stress) with a high viscosity η_0. Then deviations from Newtonian behaviour to higher shear velocities occur as in the case of polymer melts. Up to the inflection point P the flow curve can be described approximately by FERRY's equation, but then the curvature changes sign and at still higher rates of shear a second Newtonian region with a much smaller viscosity η_∞ is found. The second part, from the point P upwards, is not observed in molten polymers. The slopes of the two dotted lines give the reciprocal viscosities $1/\eta_0$ and $1/\eta_\infty$.

Fig. II, 11. The flow curve (rate of shear against shear stress) for polymer solutions.

This behaviour can be interpreted in terms of an appearent viscosity η which decreases strongly with increasing shear stress from a high initial value η_0 to a much lower final value η_∞. Such *structural viscosity* (*Strukturviscosität*) is attributed by OSTWALD[1] to change in the structure of the solution during laminar flow. At low shear stresses molecules are oriented at random giving a high resistance to flow (high viscosity η_0). At higher shear stresses the molecules attain a fixed orientation with respect to the velocity gradient, the resistance against flow is lowered, and a strong decrease in viscosity is observed[2].

Flow curves showing the general character of fig. II, 12 have been reported by REINER[3] for rubber solutions in toluene and by PHILIPPOFF and coworkers[4] for cellulose nitrate solutions. PHILIPPOFF recognized the existence of two limits for the appa-

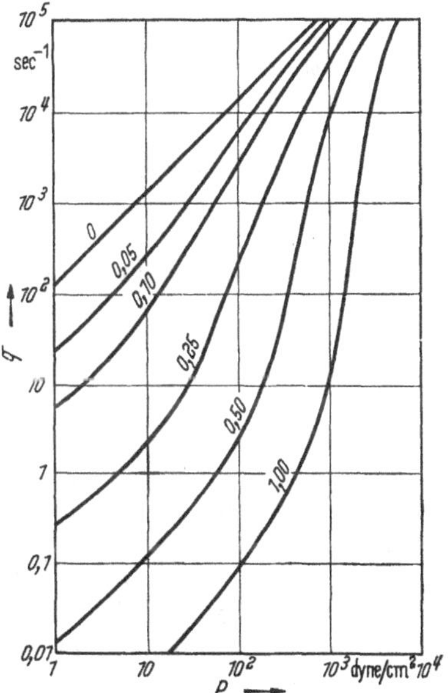

Fig. II, 12. Flow diagram of trinitro cellulose in butyl acetate, after PHILIPPOFF and HESS[4]; concentrations as indicated in g/100 cm[3].

[1] OSTWALD, WO.: Z. physik. Chem. **111**, 62 (1924).

[2] SIMHA, R.: J. physic. Chem. **44**, 25 (1940). — J. J. HERMANS: Physica **10**, 777 (1943); Kolloid-Z. **106**, 28 (1944). — W. KUHN u. H. KUHN: Helv. chim. Acta **28**, 1572 (1945).

[3] REINER, M. u. R. SCHOENFELD-REINER: Kolloid-Z. **65**, 44 (1953). — M. REINER: Physics **5**, 343 (1934).

[4] PHILIPPOFF, W.: Kolloid-Z. **71**, 1 (1935); Angew. Chem. **49**, 855 (1936). — W. PHILIPPOFF u. K. HESS: Z. physik. Chem. **B 31**, 237 (1936); Ber. **70**, 639 (1937).

rent viscosity, η_0 and η_∞, and tried to describe the flow curve by an equation

$$\eta = \eta_\infty + \frac{\eta_0 - \eta_\infty}{1 + \sigma^2/b^2}\,. \tag{II,33}$$

The constant b, having the dimensions of a shear stress, gives the position of the non-Newtonian region: The inflection point P in fig. II,11 is determined by the abscissa $\sigma = b\sqrt{\eta_0/3\eta_\infty}$.

PHILIPPOFF plotted the logarithm of rate of shear against the logarithm of stress. Fig. II,12 shows double logarithmic diagrams of rate of shear against pressure (proportional to shear stress) from capillary viscometer experiments on solutions of trinitro-cellulose in butyl acetate at different concentrations. The pure solvent ($c = 0$) behaves like a Newtonian liquid yielding straight lines of slope unity. The higher the concentration of the solute, the larger the deviations from Newtonian behaviour. All curves have a limiting slope of unity at very low shear stress (first Newtonian region) and at very high shear stress (second Newtonian region). The concentration c affects chiefly the ratio of the limiting viscosities η_0/η_∞, which varies between unity (for zero concentration) and 10^5.

Similar condition diagrams have been observed by UMSTÄTTER[1] for lubricating oils and by EDELMANN[2] for nitrated rayon, cotton and polyacrylonitrile. The latter pointed out that the position of the inflection points with respect to the logarithmic q axis is independent of concentration and depends on the molecular weight of the dissolved polymer. With increasing molecular weight the position of the inflection point is lowered and non-Newtonian behaviour occurs at lower rate of shear.

In the case of very dilute solutions special attention has been drawn to the dependence of the *intrinsic viscosity* (*Viscositätszahl*)

$$[\eta] = \lim_{c \to \infty} [\eta - \eta_c]/\eta_0 c \tag{II,34}$$

on the shear stress (η is the viscosity of the solution, η_0 the viscosity of the pure solvent). The question whether this intrinsic viscosity depends on shear stress or not has been investigated by many authors[3] and generally a decrease of intrinsic viscosity with increasing rate of shear is found. SHARMAN and coworkers[4] found that the intrinsic viscosity of polystyrene fractions in various solvents decreases linearly with increasing shear rate. For a given temperature the change of the viscosity in-

---

[1] UMSTÄTTER, H.: Kolloid-Z. **116**, 18 (1950).

[2] EDELMANN, K.: Proc. Sec. Int. Congr. Rheology, Oxford 1953, page 107.

[3] LYONS, W. J.: J. chem. Physics **13**, 43 (1945). — FOX, T. G. JR., J. C. FOX u. P. J. FLORY: J. Amer. chem. Soc. **73**, 1901 (1951). — C. M. CONRAD, V. W. TRIPP u. T. MARES: J. physic. Colloid Chem. **55**, 1474 (1951). — R. M. FUOSS u. W. N. MACLAY: J. Polymer Sci. **6**, 305 (1951). — F. AKKERMANS, D. T. F. PALS u. J. J. HERMANS: Recueil Trav. chim. Pays-Bas **71**, 56 (1952). — G. DE WIND u. J. J. HERMANS: Recueil Trav. chim. Pays-Bas **70**, 521 (1951). — V. W. TRIPP, C. M. CONRAD u. T. MARES: J. physic. Chem. **56**, 693 (1953).

[4] SHARMAN, L. J., R. H. SONES u. L. H. CRAGG: J. appl. Physics **24**, 703 (1953).

creases with increasing molecular weight of the polymer and with increasing solvent power of the solvent. With increasing temperature the effect of shear decreases in a good solvent and increases in a poor solvent[1].

d) Viscous cross effects and theories of non-Newtonian flow.

Apart from the non-linearity of the relation between shear stress and rate of shear, the occurrence of cross effects gives rise to some remarkable phenomena, first reported by WEISSENBERG[2].

WEISSENBERG describes experiments with rotating cylinders containing concentrated polymer solutions and he distinguishes "general" and "special" liquids, according to whether the liquid under consideration does or does not show the typical *cross effect*. Fig. II, 13 shows three examples of cross effects. In all cases the outer cup contains the liquid and rotates with constant angular velocity. If there is no inner gap or cylinder (case *a*), the surface of the rotating liquid shows the usual parabolic form and no difference between a general and a special liquid is observed. If there is a fixed inner cylinder immersed in the liquid (case *b*), the level of a general liquid rises at the inner, fixed cylinder and falls on the outer, rotating cylinder. This centripetal pump effect is of considerable importance in cylinder viscometry. A special liquid does not show this effect. If there is a fixed open tube immersed in the rotating liquid (case *c*), the liquid level in the tube is higher than in the outer cylinder in the case of a general liquid and equal in the case of a special liquid.

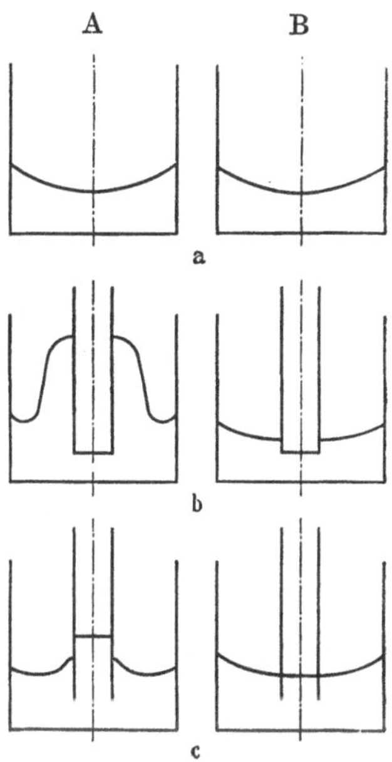

Fig. II, 13. Behaviour of general and special liquids in rotating cups, after WEISSENBERG. A general liquid, B special liquid.

Further cross effects are described by GARNER, NISSAN and WOOD[3], who observed the "general" behaviour of soap-solutions, solutions of glue in water, perspex and rubber in benzene. They report the appearance of secondary flow phenomena from regions of higher shear velocity to regions of lower shear velocity. If, for instance, a stationary disk is immersed in a liquid rotating with the outer cylinder

[1] For further information over the dependence of intrinsic viscosity on velocity gradient and concentration we refer to volume II, chapters V and XI.

[2] WEISSENBERG, K.: Nature **159**, 310 (1947); Proc. Int. Congr. Rheology, Holland 1948; Conf. British Rheologist's Club, London 1948; Proc. Roy. Soc. [London] **A 200**, 183 (1950).

[3] GARNER, F. H. u. A. H. NISSAN: Nature **158**, 634 (1946). – G. F. WOOD, A. H. NISSAN u. F. H. GARNER: J. Inst. Petrol **33**, 71 (1947). – F. H. GARNER, A. H. NISSAN u. G. F. WOOD: Philos Trans. Roy. Soc. London **A 234**, 37 (1950).

(fig. II, 14), secondary flow takes place in a general liquid from the periphery of the disk towards its centre and counter flow is observed in the relatively unsheared material above and below the disk.

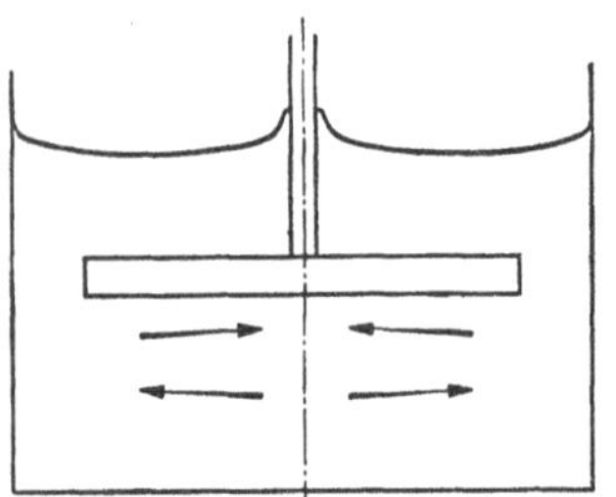
Fig. II, 14. Stationary disk in rotating liquid, directions of secondary flow.

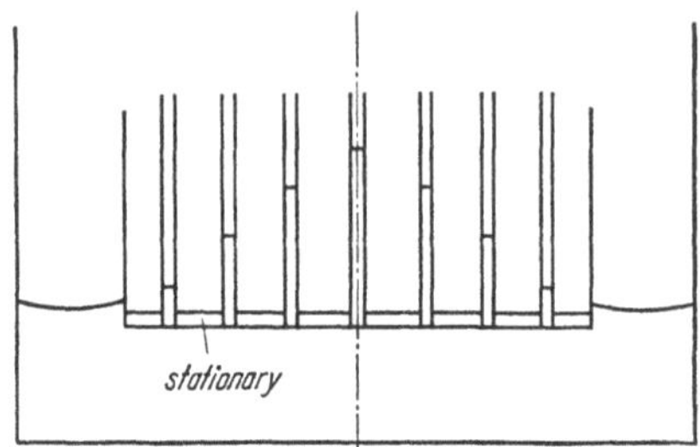

Fig. II, 15. Measurement of normal pressure during shearing.

To measure the pressure differences arising in a sheared liquid, GARNER, NISSAN and WOOD, and GREENSMITH and RIVLIN[1] used an experimental arrangement shown in fig. II, 15. The liquid under investigation was subjected to a torsional motion between the parallel bases of two coaxial cylindrical cups, the inner cup being stationary, the outer cup rotating with constant speed. The inner cup carried a number of pressure gauges, by which the pressure could be read off as function of the position.

WARD and LORD[2] made a geometrical study of the WEISSENBERG *effect*. The liquid was sheared between a stationary inner cylinder and a rotating outer cylinder and the height h (see fig. II, 16) was measured as a function of several geometrical factors such as the radii r_1 and r_2 of the cylinders, the depth of immersion of the inner cylinder d and the space s between the bases of the cylinders. WARD and LORD found that the height h increases with increasing angular velocity ω at constant s and decreases with increasing s at constant ω. The torque has been shown to be proportional to $r_1^2 r_2^2/(r_1^2 - r_2^2)$.

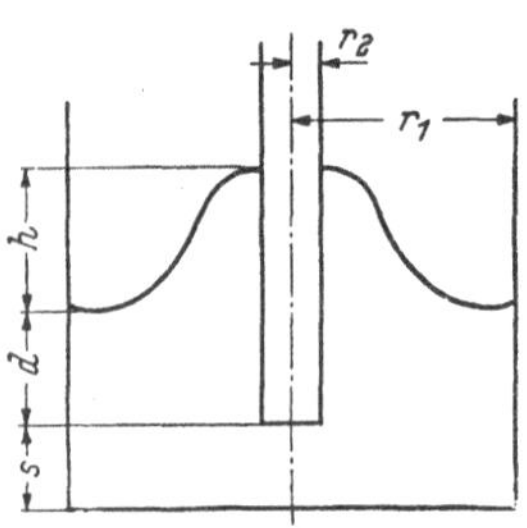

Fig. II, 16. Geometry of the WEISSENBERG effect.

Cross effects as treated above can be understood with the present theories of non-Newtonian liquids. The existing theories of non-Newtonian behaviour can be divided into two classes, the prototypes of which are

A. The theory of REINER-RIVLIN,

B. The theory of MOONEY.

The main difference between A and B lies in the assumption of absence

[1] GREENSMITH, H. W. u. R. S. RIVLIN: Nature **168**, 664 (1951); Trans. Roy. Soc. [London] **A 245**, 399 (1953).

[2] WARD, A. F. H. u. P. LORD: Proc. 2nd Int. Congr. Rheology, Oxford 1953, page 214.

or presence of elasticity in the liquid. A investigates a viscous liquid, which does not exhibit elastic effects. Such a liquid may exhibit viscous cross effects of the character of WEISSENBERG's phenomena. On the other hand B describes a liquid of a still more general character in that it shows viscous resistance combined with elastic effects. In a steady-state motion this liquid will store elastic energy and the elastic structure can give rise to cross effects, which are in this case a phenomenon of cross elasticity.

Both theories A and B are able to explain the existence of cross phenomena, though they explain it in different ways (as cross viscosity and cross elasticity respectively). It can hardly be said today which of the two theories is the more correct. Also it should be mentioned, that these two theories do not exclude one another, the existence of cross elasticity does not exclude cross viscosity or vice versa. Possibly in high polymeric liquids both effects may be present at the same time.

A. *The Theory of* REINER[1] *and of* RIVLIN[2]. RIVLIN restricted his considerations to an incompressible liquid obeying the following conditions

(i) The liquid is incompressible.
(ii) The liquid does not show viscoelastic effects (is inelastic).
(iii) The liquid is isotropic.

The REINER-RIVLIN theory of a general viscous liquid without elastic effects is the counterpart of the RIVLIN-MOONEY theory of a general elastic solid without viscous (time) effects. As the latter starts with a stored energy function W as a function of the state of deformation, the REINER-RIVLIN theory starts with the dissipation function Φ as a function of the state of flow. In a steady state of flow the rate of dissipation of energy $\Phi(xyz)$ is defined as the energy dissipated per unit time and per unit volume in the neighbourhood of the point (xyz). Conditions (ii) and (iii) imply that Φ can depend on the three invariants K_1, K_2, K_3 of the deformation velocity tensor (II,26) only[3]. As, further, the first invariant K_1 must vanish identically, because of the condition of incompressibility, the general isotropic viscous liquid will be characterised by a dissipation function

$$\Phi = \Phi(K_2, K_3). \tag{II,35}$$

The equation of state of an inelastic viscous liquid can involve only the stresses and the components of the flow tensor (II,26), but not the time derivatives of these quantities. As has been shown by REINER[1], for an incompressible and isotropic liquid the rheological equations must have the form

$$\left.\begin{aligned} \sigma_{11} &= F_1 e_{11} + F_2 [e_{11}^2 + e_{12}^2 + e_{13}^2] + p \\ \sigma_{12} &= F_1 e_{12} + F_2 [e_{12} e_{13} + e_{22} e_{23} + e_{23} e_{33}] \end{aligned}\right\} \tag{II,36}$$

F_1 and F_2 are arbitrary coefficients which may be functions of both invariants K_2 and K_3 and which determine the properties of the liquid

[1] REINER, M.: Amer. J. Mathem. **67**, 350 (1945).
[2] RIVLIN, R. S.: Proc. Roy. Soc. [London] **A 193**, 260 (1948); Nature **160**, 611 (1947). — [3] For the definition of the invariants K_1, K_2, K_3 see § 2d, formulae (I,13).

and the dissipation function Φ. The liquid is Newtonian only if F_2 vanishes and if F_1 degenerates into a constant (equal to twice the viscosity). Under a state of simple laminar shearing motion (only e_{12} different from zero) the occurrence of normal stresses depends on the quantity F_2: liquids with a vanishing normal stress coefficient F_2 are special liquids showing no cross effects, liquids with non-vanishing F_2 are general liquids showing cross effects.

RIVLIN[1] as well as BRAUN and REINER[2] have calculated the behaviour of general liquids in a rotational viscometer, a capillary viscometer and a parallel plate viscometer. It is shown that cross effects of the correct sign are to be expected if F_2 is taken positive.

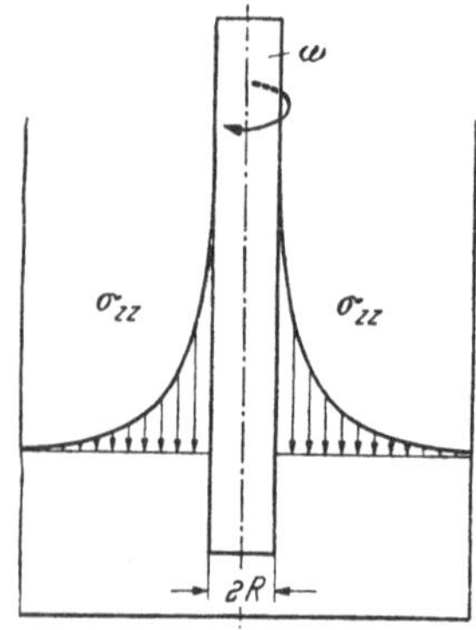

Fig. II, 17. Vertical pressure distribution necessary for pure shearing motion.

Consider as an example a non-Newtonian liquid contained in a large stationary cup of which the radius is taken as infinite for convenience. At the centre a cylindrical rod of radius R is rotating with a constant angular velocity ω. In order to produce a pure shearing motion in the liquid without any motion in the vertical direction, we must apply a pressure in the vertical direction whose magnitude depends on the distance r from the centre

$$\sigma_{zz} = -F_2 \omega^2 R^4/r^4$$

If F_2 is assumed to be a positive quantity, the stress to be applied is a pressure distributed as indicated in fig. II, 17. If this pressure is *not* applied, it is quite clear that the liquid will move upwards in the neighbourhood of the centre and will show the WEISSENBERG phenomenon.

B. *The Theory of* MOONEY[3]. MOONEY makes the following assumptions

(i) The viscous properties of the liquid are given by NEWTON's viscosity law.

(ii) The elastic properties of the liquid are given by the Superelastic theory [compare § 11 c)].

(iii) The viscous and elastic properties are connected by MAXWELL's hypothesis: The rate of relaxation of shear stress is proportional to the instantaneous stress.

MOONEY calculated the behaviour of such liquids in different viscometers and showed that they exhibit cross effects similar to WEISSENBERG's phenomena.

Recapitulating, we may say that the occurrence of a cross effect is always connected with a relationship between a strain of one type and a stress of an other geometrical type.

---

[1] RIVLIN, R. S.: Proc. Cambridge philos. Soc. **45**, 88 (1949).

[2] BRAUN, I. u. M. REINER: Quart. J. Mech. and appl. Mathem. **5**, 42 (1952).

[3] MOONEY, M.: J. Colloid Sci. **6**, 96 (1951); J. appl. Physics **24**, 675 (1953).

We may distinguish:

1. elastic cross effects (as occurring in the theory of MOONEY-RIVLIN for rubber). The stresses are unique functions of the strains. A shear strain gives rise to a compressive stress.

2. viscous cross effects (as occurring in the theory of REINER-RIVLIN for non-Newtonian liquids). The stresses are unique functions of the rate of strain. A rate of shear gives rise to a compressive stress.

3. viscoelastic cross effects (as occurring in the theory of MOONEY for non-Newtonian liquids). The stresses are functions of the strain and of the rate of strain. A shear strain or a rate of shear gives rise to a compressive stress.

§13. Non-linear deformation behaviour of plastic materials.

a) Introduction.

In this chapter we consider the mechanical behaviour of polymers at relatively low temperatures, below the softening temperature, and at high stresses, i. e. of those materials which are generally called "plastics". Plastic materials can behave either in a *brittle* or a *ductile* manner, depending on the temperature, the rate of strain and the geometrical conditions of stress.

As an example fig. II,18 shows the tensile stress-strain curves of polystyrene at room temperature at different constant rates of deformation[1]. The diagram is a typical example of brittle behaviour, the curves being nearly straight lines up to the breaking point. Rupture occurs at very small deformations (1,5%). The apparent modulus (the initial slope of the stress-strain curves) increases with increasing rate of deformation.

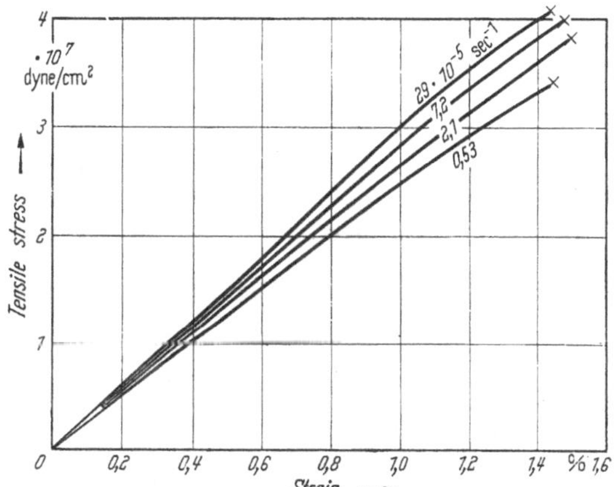

Fig. II, 18. Stress-strain curves in tension of polystyrene at various rates of strain.

The same material behaves completely different when subjected to unidirectional compression. Fig. II,19 shows the ductile behaviour of polystyrene under compression at various constant strain rates. The stress-strain diagram shows a maximum, the yield point, after which large deformations occur without further increase in nominal stress. Both, the yield stress and the yield elongation increase with increasing rate of

[1] HSIAO, C. C. u. J. A. Sauer: ASTM-Bulletin No. 127 (Febr. 1951).

deformation. Fig. II, 18 and II, 19 illustrate the influence of the geometry of the stress on the brittleness: Tensile stresses favour brittle behaviour, whilst compressive or shear stresses favour ductile behaviour.

In order to demonstrate the large differences in mechanical behaviour between hard plastics and rubbers we compare the stress-strain diagrams of polystyrene in tension (fig. II, 18), polystyrene in compression (fig. II, 19)

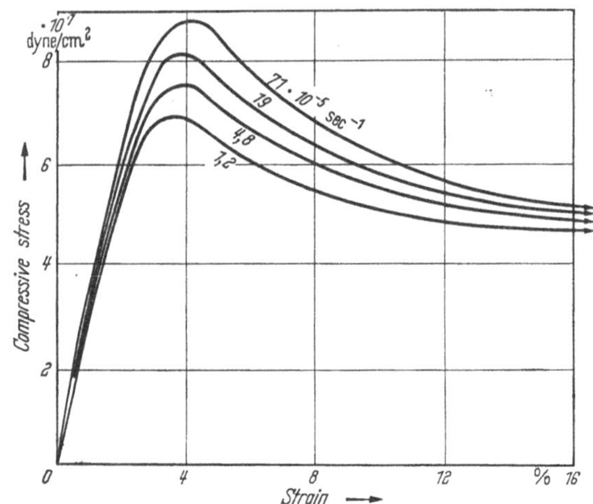

Fig. II, 19. Stress-strain curves in compression for polystyrene at various rates of strain.

and natural rubber (fig. II, 3). In table II, 1 the modulus (the initial slope of the stress-strain diagrams), the elongation at rupture and the ultimate stress have been listed.

Table II, 1. *Comparison of brittle, ductile and rubberlike behaviour.*

	modulus dyn/cm²	elongation at rupture	stress at rupture or yield dyn/cm²
brittle behaviour polystyrene (fig. II, 18)	$3{,}0 \cdot 10^{10}$	1,4%	$4{,}2 \cdot 10^{8}$
ductile behaviour polystyrene (fig. II, 19)	$3{,}4 \cdot 10^{10}$	$>$ 16 %	$9{,}1 \cdot 10^{8}$
rubberlike behaviour natural rubber (fig. II, 3)	$1{,}0 \cdot 10^{7}$	$>$ 600 %	$2{,}1 \cdot 10^{8}$

In all three cases the true stress at rupture is of the same order of magnitude (10^8 to 10^9 dynes/cm²). However large differences may be seen in the initial slopes of the stress-strain diagrams and the ultimate elongations. Plastics show a modulus 300 times as high as rubber and the ultimate elongations vary between one percent (brittle behaviour) up to almost thousand percent (for rubbers).

A further important fact is the influence of temperature on the stress-strain diagram, which is demonstrated by fig. II, 20 showing the stress-strain curves of polymethyl methacrylate at different temperatures. At low temperatures the material is essentially brittle whereas it becomes more and more ductile at higher temperatures. At 60° C the ultimate elongation has increased to 130%.

It is quite clear that the speed of testing has a similar influence on the stress-strain curves as the temperature: a slow rate of straining favours ductile behaviour, a high rate of straining favours brittle behaviour.

In the chapter on linear behaviour we have seen that a single quantity from classical elasticity theory like the modulus is, in the theory of visco-elastic behaviour replaced by a continuous spectrum at one temperature and by a series of such spectra if different temperatures are considered.

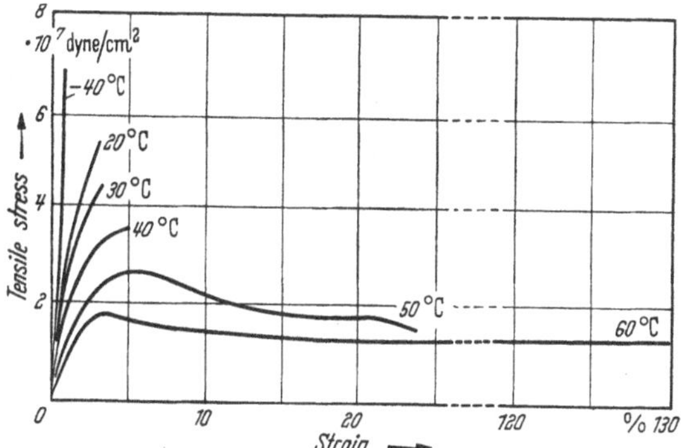

Fig. II, 20. The effect of temperature on the tensile stress strain curves of polymethyl methacrylate, after WALL[1].

For the description of larger deformations classical theory needs a multitude of coefficients, partly to account for higher terms in the development with respect to strain or stress, partly to describe cross effects. Each of these coefficients is or can be dependent on time and temperature in the case of viscoelastic materials, so that a complete description of the behaviour of such materials at high deformation, even in geometrically simple cases is not possible. However, for practical purposes one is interested in the behaviour of these materials under various conditions, very often including large deformations.

At the present moment one cannot do much more than measure more or less systematically relationships between measurable quantities which may be expected to be of interest in practice. The experiments quoted above are examples of such measurements. In the beginning of the mechanical testing of plastics considerable importance was attributed to stress-strain diagrams in investigating mechanical behaviour. However in the last decennium it has become more and more clear that the time dependence of the properties of plastics complicates the mechanical behaviour considerably, and no simple method such as a stress-strain measurement will be sufficient to describe all the mechanical properties. More intricate techniques such as creep, stress relaxation, dynamic experiments etc. have to be used. Today we should say that the value of a stress-strain diagram lies in a quick survey of a variety of mechanical

[1] WALL, W. C.: Aero Digest **41**, 122 (Oct. 1942).

properties. From the stress-strain diagram a "modulus of elasticity" can be calculated from the initial slope, it can be judged whether the behaviour is brittle, ductile or rubberlike and an estimate can be made of the magnitudes of the yield or breaking stress and strain.

In general not very much is known today about the deformation behaviour of plastics at large stresses. For our subsequent considerations we shall pick out two subjects, which are of considerable interest: the creep of hard plastics in the region of high stresses and the phenomenon of cold drawing.

b) Non-linear creep of hard plastics.

Investigations on the creep behaviour of hard plastics at large stresses have been performed by MARIN and coworkers. They describe testing machines[1] to measure the creep properties in linear tension, unidirectional compression, in torsion and in bending and give data on the creep behaviour of polystyrene[2], polymethyl methacrylate[3] and several laminates[4].

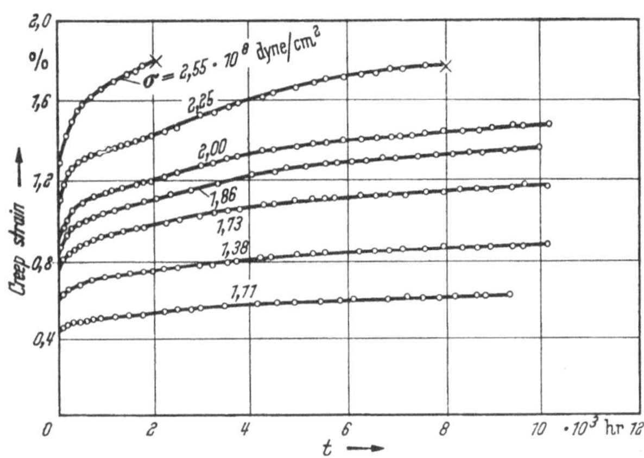

Fig. II, 21. Long time tension creep of polymethyl methacrylate, after MARIN[3].

Fig. II, 21, II, 22 and II, 23 represent the creep curves of polymethyl methacrylate in tension and in compression and the torsional creep of polystyrene. All the creep diagrams show a similar shape, whether we deal with tensile, compressional or with torsional creep. In some cases fracture occurs, showing that the stresses used are of the magnitude of the breaking stress. However, the deformation at fracture is lower than 2% in all cases; this is a characteristic feature of brittle behaviour.

---

[1] MARIN, J., Y. PAO u. G. CUFF: Trans. A. S. M. E. **73**, 705 (1951).

[2] MARIN, J. u. G. CUFF: Proc. A. S. T. M. **49**, 1158 (1949). – J. A. SAUER, J. MARIN u. C. C. HSIAO: J. appl. Physics **20**, 507 (1949). – J. MARIN u. Y. PAO: Proc. A. S. T. M. **51**, 1277 (1951).

[3] MARIN, J., Y. PAO u. G. CUFF: Trans. A. S. M. E. **73**, 705 (1951). – J. MARIN u. Y. PAO: Trans. A. S. M. E. **74**, 1231 (1952).

[4] MARIN, J. u. D. E. HARDENBERG: Mod. Plastics **26**, 101 (Febr. 1949).

Though the deformations occurring remain very small, the creep behaviour is non-linear. Generally deformation depends on both time t and stress σ

$$\gamma = \gamma(t, \sigma), \qquad (\text{II},37)$$

which, in case of linearity, can be written

$$\gamma(t, \sigma) = \sigma \cdot F(t). \qquad (\text{II},38)$$
(for linear creep)

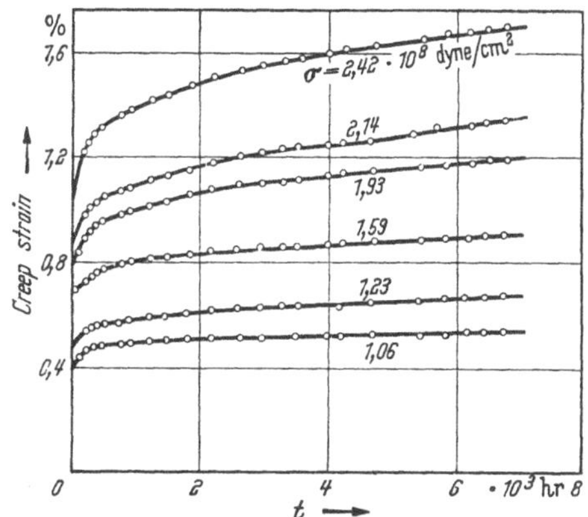

Fig. II, 22. Long time compression creep of polymethyl methacrylate, after MARIN[1].

Now the torsional creep data of polystyrene of fig. II, 23 have been replotted in the form $\gamma(t, \sigma)/\sigma$ for different stresses as a function of time in fig. II, 24. The deviations from linear behaviour can be clearly seen: The creep increases with increasing stress much more strongly than proportionately to the stress. The same property is shown by the creep curves of polymethylmethacrylate (fig. II, 21 and II, 22).

MARIN and coworkers interprete their creep data by decomposing the creep curves into three parts[2]

$$\gamma(t, \sigma) = \gamma_1(\sigma) + \gamma_2(t, \sigma) + \gamma_3(t, \sigma). \qquad (\text{II},39)$$

The first term $\gamma_1(\sigma)$ is the instantaneous elastic strain, independent of time and completely recoverable. Generally it cannot be represented by HOOKE's simple law, as it deviates from linearity at high stress.

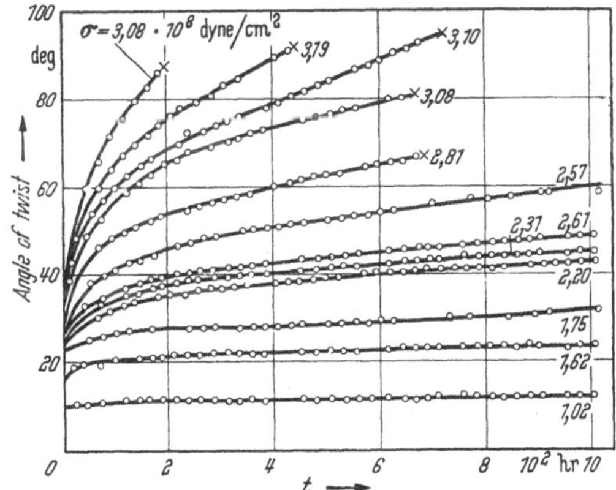

Fig. II, 23. Long time torsion creep of polystyrene, after MARIN[1].

The second term $\gamma_2(t, \sigma)$ is the transient creep strain, which increases with increasing time, being zero at time zero and reaching a finite, equilibrium value after in-

---

[1] MARIN u. CUFF: see page 154, note [2].

[2] This decomposition of creep is similar to that used in the case of linear creep. In the case of linear creep the instantaneous strain corresponds to HOOKEan deformation, the transient creep strain corresponds to retarded elastic deformation and the steady state creep corresponds to NEWTONian flow.

finite time. MARIN and coworkers propose for the transient creep strain the form

$$\gamma_2(t,\sigma) = K\sigma^m[1 - e^{-qt}].$$

After longer times creep reaches a steady state velocity and the total creep consists chiefly of the third part, the steady state creep $\gamma_3(t,\sigma)$ which is proportional to time. MARIN describes the steady state creep by the OSTWALD-DE WAELE law

$$\gamma_3(t,\sigma) = Bt\sigma^n.$$

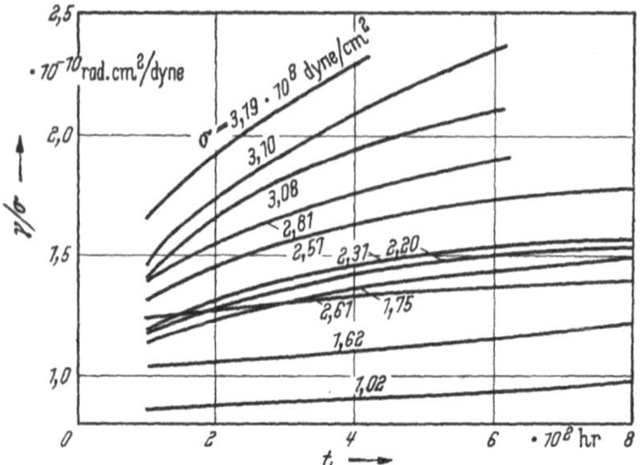

Fig. II,24. Plot of $\gamma(t,\sigma)/\sigma$ as function of time for various stresses, torsional creep of polystyrene. Stress in 10^8 dyn/cm².

Not very much general information is available about non-linear time dependent deformation behaviour. Lacking more extensive experimental information, we can give here a very general discussion of the subject only. Creep properties can be described by means of an empirical equation of the form (II,37), which permits two alternatives:

(i) The independent variables stress and time can be separated in their action on creep, so that the deformation can be factorized thus

$$\gamma(t,\sigma) = f(\sigma)\cdot F(t). \tag{II,40}$$

(ii) Time and stress are interrelated with respect to their action on creep, so that the deformation cannot be written in the simple form (II,40).

(i). If a factorization (II,40) is possible, we can reduce creep curves at different stresses to one characteristic time function $F(t)$ by plotting $\gamma(t,\sigma)/f(\sigma)$ against time. All creep curves can be constructed by simple magnification and the position of transitions occurring in creep (the characteristic retardation times occurring) are not altered by a change in stress. The choice of the stress may influence therefore the relative amplitudes of creep, but not the retardation times.

MARIN's theory has been an example of such behaviour, another example is represented by an equation originally proposed by NUTTING[1] for the creep of metals. This equation describes the strain as a power function of stress and time

$$\gamma = \sigma^\beta t^k \cdot 1/\Psi \tag{II,41}$$

where β, k and Ψ are constants. The NUTTING *equation* has been used

[1] NUTTING, P. G.: Proc. A. S. T. M. **21**, 1112 (1921).

by SCOTT BLAIR and coworkers[1] to describe the creep properties of different materials. A special use of equation (II, 41) with the choice $\beta = 1$ has been made by BUCHDAHL and NIELSEN[2] to describe the creep properties of polyvinylchloride, GRS and polystyrene. However, putting β equal to unity in the NUTTING equation is equivalent to linear creep behaviour, therefore the results of BUCHDAHL and NIELSEN could equally well be considered as special cases of linear viscoelastic behaviour.

It might be felt that an equation of such a particular form as the NUTTING equation (II, 41) can only represent a rough approximation to real creep time dependence. Indeed as long as creep curves are plotted on the ordinary time scale and as long as observations are extended over a short time interval only, the NUTTING equation will give a good fit to either linear or non-linear creep data. If, however, the creep data are taken logarithmically and if observations are extended over many decades, the NUTTING equation, and also the more general form (II, 40) would be expected to become insufficient.

(ii). Generally the magnitude of the applied stress will influence the amplitudes and the retardation times of deformation processes: the higher the applied stress, the shorter the retardation times. The stress thus becomes a variable which can shift retardation positions and "activate" a retardation process just as the temperature activates retardation processes. In fig. II, 25 the creep curves of cellulose acetate are plotted for various stresses on a logarithmic timescale comprising several decades[3]. It is seen that the retardation time of the creep process is strongly influenced (shortened) by an increase of stress[4].

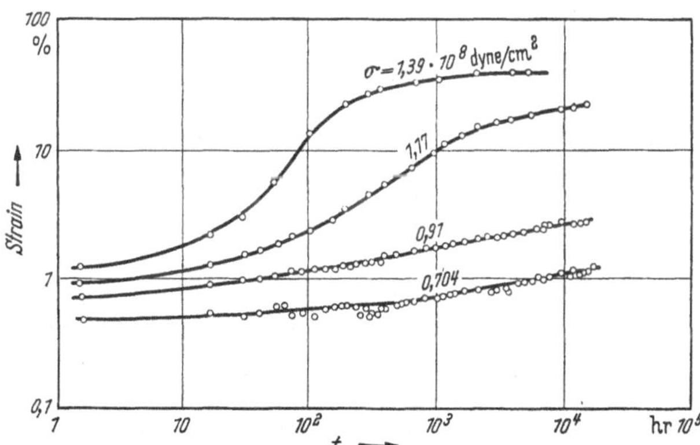

Fig. II, 25. Creep at various stresses for cellulose acetate, after FINDLEY[3].

[1] SCOTT BLAIR, G. W. u. B. C. VEINOGLOU: J. sci. Instr. **21**, 149 (1944). — G. W. SCOTT BLAIR, B. C. VEINOGLOU u. J. E. CAFFYN: Proc. Roy. Soc. [London] **A 189**, 69 (1947). — G. W. SCOTT BLAIR, u. J. E. CAFFYN: Philos. Mag. J. Sci. **40**, 80 (1949). — G.W. SCOTT BLAIR u. M. REINER: Appl. Sci. Research **A 2**, 225 (1951).

[2] BUCHDAHL, R. u. L. E. NIELSEN: J. appl. Physics **22**, 1344 (1951).

[3] FINDLEY, W. N.: "Creep properties of plastics" in: Symposium on Plastics A. S. T. M. Philadelphia (Febr. 1944).

[4] More examples of this phenomenon can be found in chapter VII dealing with the mechanical behaviour of fibers.

One has tried to represent this type of non-linear behaviour, where the stress influences the characteristic times as well as the equilibrium values of the retardation processes, by means of mechanical models, similar to those used in § 4, but consisting of non-linear units.

Certain types of three and four parameter models have been calculated by HALSEY[1] and HOGAN[2], who assume all springs of the model to be Hookean, characterised by a spring constant G, and all dashpots of the model to be non-Newtonian, obeying the hyperbolic sine law proposed by EYRING[3]

$$d\gamma/dt = \frac{\alpha}{\eta} \sinh(\sigma/\alpha).$$

A dashpot is therefore characterised by two constants, η and α. HALSEY and HOGAN studied creep, relaxation and the behaviour under constant stress and strain rates of such models. However such models can not account for a mechanical behaviour as represented in fig. II, 25[4]. Furthermore the benefit derived from mechanical models in the description of linear deformation behaviour cannot be obtained in the description of non-linear deformation behaviour. For so far non-linear models cannot be handled mathematically as conveniently and generally as linear models and, therefore, no general descriptive theory of non-linear viscoelastic behaviour can be formulated.

c) Cold drawing of polymers[5].

We conclude with a consideration of the *cold drawing* of plastics (*Kaltverstreckung*), which is of great practical importance, as it is the basic rheological process underlying the drawing of fibres and films.

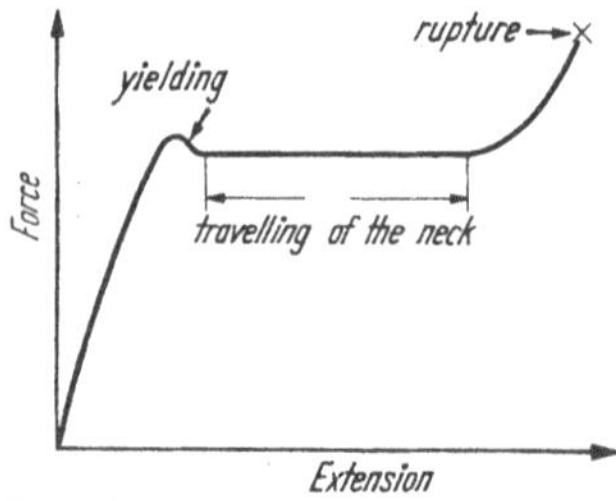

Fig. II, 26. Force extension diagram of a polymer which shows cold drawing (schematically).

In 1947 MIKLOWITZ[6] described the drawing process in Nylon strips and filaments strained at constant rate. Nylon, like all other materials which can be cold drawn, shows a typical force-extension diagram, when extended at a constant rate (see fig. II, 26). In the beginning the sample is uniformly extended over its whole length and shows a high modulus of elasticity. When the yield stress is reached, a sudden change in the force elongation curve occurs. The force often decreases somewhat and then remains constant during further stretching. At the yield point a constriction is formed in the

[1] HALSEY, G.: J. appl. Physics **18**, 1072 (1947).

[2] HOGAN, M. B.: The engineering application of the absolute rate theory to plastics. Report No. 39 (May 1953), Univ. of Utah, Inst. for the study of rate processes. — [3] EYRING, H.: J. chem, Physics **4**. 263 (1936).

[4] Compare the extensive treatment given by VAN DER VEGT in chapter VII, B.

[5] For the molecular interpretation compare also the third volume, § 32. — An excellent survey has been published recently K. JÄCKEL: Kolloid-Z. **137**, 130 (1954).

[6] MIKLOWITZ, J.: J. Colloid Sci. **2**, 193, 217 (1947).

sample at the weakest point. During further stretching the transition between the stretched and unstretched part, the neck, travels along the sample, and the constriction extends more and more (fig. II, 27).

Beyond the yield point the extension of the specimen does not take place uniformly, but the sample consists of two regions, an undrawn region and a drawn region. The transition boundary between the two forms is well defined but not abrupt and consists of a curved shoulder connecting the undrawn and drawn material. More and more material is changed from the undrawn into the drawn state and an extension of several hundred percent can be reached in this way. When the constriction has extended through the whole sample, the force again begins to rise and generally the sample breaks soon afterwards.

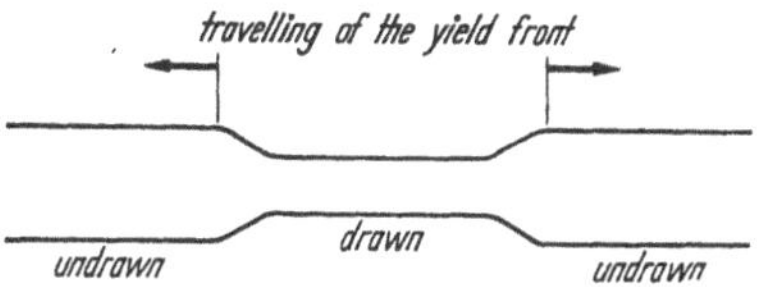

Fig. II, 27. The neck in the specimen during cold drawing.

The change of the polymer from the undrawn to the drawn state is accompanied by a change in optical and other physical properties. Polyvinylchloride becomes faint and dull in the drawn state whilst polythene and polyamides become more transparent. Fig. II, 28 shows a sample of polyethylene which has been partially cold drawn. The vast difference in transparency between undrawn and drawn parts is clearly illustrated.

To illustrate this change in optical properties in a more quantitative way, transmissibility measurements of visible light have been made[1] with samples of drawn and undrawn Nylon and polyethylene. The result is expressed as extinction coefficient α, defined by the equation $J = J_0 (10)^{-\alpha d}$. (J ... intensity of transmitted light, J_0 ... intensity of incident light, d thickness of sample.) The result of these measurements and the corresponding densities are given in the following table.

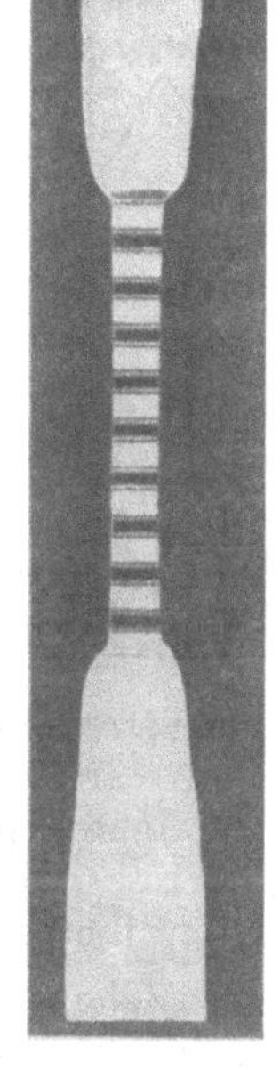
Fig. II, 28. Partially drawn sample of polyethylene.

Table II, 2. *Extinction coefficients and specific volume of undrawn and drawn polymers.*

material	extinction coeff · α	specific volume cm^3/g
Nylon undrawn	$5{,}7\ cm^{-1}$ ($\lambda = 5600$ Å)	0,870
Nylon drawn	$3{,}7\ cm^{-1}$ ($\lambda = 5600$ Å)	0,872
polyethylene undrawn	$5{,}9\ cm^{-1}$ ($\lambda = 5600$ Å)	1,079
polyethylene drawn	$2{,}8\ cm^{-1}$ ($\lambda = 5600$ Å)	1,079

We see that the density does not change appreciably in the drawing process, but a considerable change in optical properties takes place. Samples of undrawn Nylon or polyethylene of 1 cm thickness both transmit

[1] The transmissibility measurements have been performed by D. T. F. Pals, Centraal Laboratorium T. N. O., Delft.

approximately 10^{-6} times the incident light intensity. Drawn Nylon transmits 100 times and drawn polyethylene 1000 times as much light.

Typical cold drawing phenomena are shown by a number of mostly crystalline polymers like polythene, polyamides, polyvinyl- and polyvinylidene chloride, polyvinyl alcohol, polyethylene terephtalate, and cellulose fibres. Cold drawing is further restricted to a certain temperature range and to a certain range of rate of straining. If the rate of straining is too high or the temperature is too low, the sample shows brittle rupture. If the temperature is too high, no necking occurs and the sample stretches continuously over its whole length just as a rubber. The temperature range of cold drawing reaches from a critical temperature in the plastic range up to about the softening temperature.

HOFF[1] has reported that also a completely uncrystalline material like polymethyl methacrylate shows necking at temperatures higher than 45° C. If the partially stretched sample is unloaded at the drawing temperature no considerable recovery occurs, so cold drawing would seem te be an irreversible process. But if the drawn sample is heated up to the softening temperature (in the present case 140° C), complete recovery occurs.

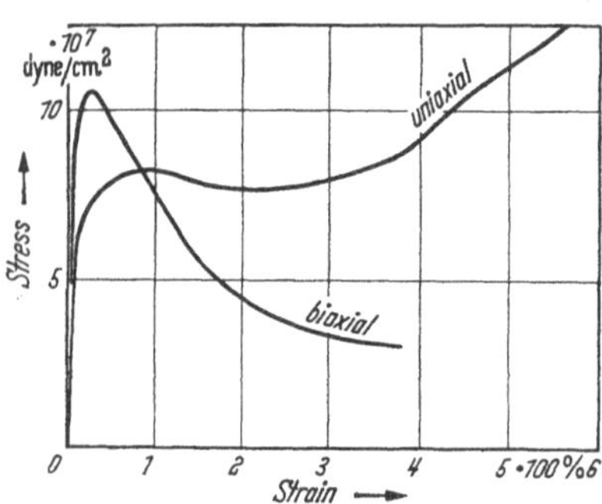

Fig. II, 29. Nominal stress strain curves under uniaxial and biaxial tension for polythene, after HOPKINS and al[3].

A comparison between the stress-strain behaviour under uniaxial and biaxial tensile stresses has been made by MIKLOWITZ[2] for Nylon and by HOPKINS, BAKER and HOWARD[3] for polythene. Nekking occurs also under biaxial stresses, but generally, other conditions being equal, the sample behaves more brittle when subjected to biaxial stresses. Fig. II, 29 shows the lower elongations and the higher yield stress under biaxial tension compared with uniaxial tension.

A careful investigation of the cold drawing of fibres and strips of chlorinated polyvinyl chloride has been performed by MÜLLER[4], who describes the characteristic features of the stress-strain diagram at different temperatures as shown in fig. II, 30.

a) The yield stress (the stress corresponding to cold drawing) decreases with increasing temperature.

b) The difference between the undrawn and the drawn material, especially the difference in length, decreases with increasing temperature.

---

[1] HOFF, E. A. W.: J. appl. Chem. **2**, 441 (1952).

[2] MIKLOWITZ, J.: loc. cit. page 158.

[3] HOPKINS, I. L., W. O. BAKER u. J. B. HOWARD: J. appl. Physics **21**, 206 (1950).

[4] MÜLLER, F. H.: Kolloid-Z. **114**, 59 (1949); **115**, 118 (1949); **126**, 65 (1952). – F. H. MÜLLER u. K. JÄCKEL: Kolloid-Z. **129**, 145 (1952); Makromol. Chem. **9**, 97 (1953). – F. H. MÜLLER: Kolloid-Z. **135**, 65, 188 (1954).

Therefore the region of cold drawing (the hatched zone in fig. II, 30) becomes smaller with increasing temperature.

c) At a certain "critical" temperature in the vicinity of the softening temperature (in this case 77° C) the phenomenon of necking ceases to exist and the sample stretches continuously over its whole length.

The stress-strain curves of fig. II, 30 at constant temperature are very similar to the pressure volume isotherms of a VAN DER WAALS gas, if we identify the stress σ with the pressure p, and the inverse deformation $1/\varepsilon$ with the volume v. The hatched coexistence region of drawn and undrawn material corresponds to the liquid vapor coexistence region in the VAN DER WAALS diagram. Indeed MÜLLER could describe purely empirically his stress-strain diagrams by VAN DER WAALS' equation

$$\sigma = E_0\, \varepsilon/(1 - b\, \varepsilon) - a\, \varepsilon^2$$

where E_0 is the initial slope of the stress-strain diagram and a and b are two constants connected with the critical point (ε_k, σ_k) by the equations $b = \varepsilon_k/3$, $a = 3\sigma_k/\varepsilon_k^2$.

Insofar as MÜLLER draws a mathematical analogy between the p–V-curves of a VAN DER WAALS gas and the σ–ε-curves of a cold-drawn polymer, no objections need be raised. But MÜLLER assumes that the analogy is not only mathematical but also physical. He considers the state of a partly drawn polymer as a state of equilibrium between drawn and undrawn material to which CLAPEYRON's equation may be applied. This equation he writes

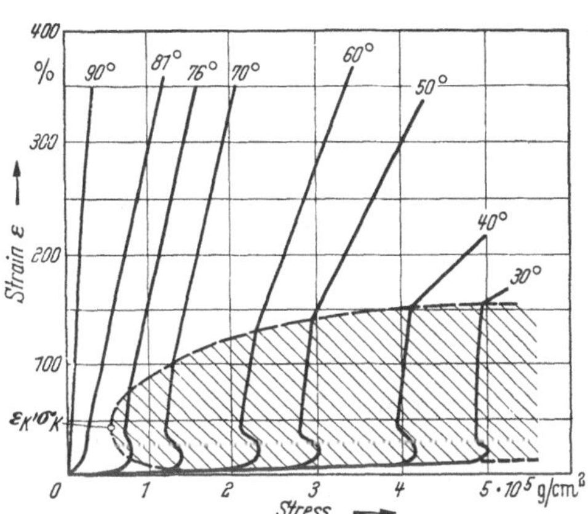

Fig. II, 30. Deformation-stress diagram in cold drawing of chlorinated polyvinyl chloride, after MÜLLER.

$$q = (v_{\text{undrawn}} - v_{\text{drawn}})\, T \left(\frac{\partial \sigma}{\partial T}\right)$$

in which the v's represent the specific volumes, σ is the stress at drawing and q is a heat of transformation from the undrawn into the drawn state. The total energy Q necessary for the drawing process is of course much higher and can be calculated from the area of the stress- strain diagram.

MÜLLER determined the quantities q and Q experimentally and found for polyvinyl chloride at 20° C the values $q = 2{,}5$ cal/g and $Q = 17{,}8$ cal/g.

MÜLLERS' derivation of the heat of transformation appears open, to criticism. For in this derivation σ must be regarded as the tension needed to keep the drawn and undrawn state in equilibrium. However, it is clear that there is no real equilibrium between these two states at all since the process of drawing is irreversible and the stress σ depends on the rate of strain.

Therefore, little meaning can be attached to MÜLLERS q. On the other hand his quantity Q follows immediately from the load-extension curve and may well have a physical meaning. MÜLLER observes that the energy difference is just enough to warm up the sample to the softening temperature. Upon this fact MÜLLER bases his assumption about the mechanism of cold drawing: The sample is warmed up locally at the necking shoulder to temperatures above the softening temperature, all other portions of the specimen remaining at a lower temperature. MÜLLER could support this assumption by experimental measurement of the temperature in the sample. At regions, where the neck passes, the temperature rises considerably.[1]

Apparently this part of MÜLLER's treatment can be considered as a good approximation to a physical description of the cold drawing process. A better approximation could be obtained if the heat developed at drawing could be measured accurately, for instance in a calorimeter. By subtracting this heat from MÜLLERS' Q, which is, in fact, the total work done in the drawing process, an energy of transformation from the undrawn into the drawn state could be calculated. In view of the stability of the drawn state and the "irreversibility" of the drawing process, one would expect that this quantity q will turn out to be negative as the change in entropy will be positive due to the orientation effects. This means that the entire work of drawing is converted into heat and besides this also a quantity $-q$ is develop as heat during the drawing process.

MARSHALL and THOMPSON[2] made an effort to arrive at a quantitative description of the drawing process. They investigated the behaviour of terylene fibres during continuous drawing in spinning machines. The natural draw ratio (cross section of undrawn fibre divided by the cross section of drawn fibre) r is approximately equal to 4,5 at room temperature and decreases strongly with increasing temperature. At 80° C it has fallen to unity, the sample stretches continuously without necking.

MARSHALL and THOMPSON measured the natural draw ratio together with the tension as a function of temperature and speed. In this way they were able to deduce the isothermal load extension curves of terylene filaments which were amorphous in the beginning as reproduced in fig. II,31. From these they calculated the hypothetical adiabatic load extension curve, neglecting crystallinity changes and elastic potential energy. This adiabatic extension curve is also shown in fig. II,31. It starts at 20° C but as adiabatic extension causes a rise in temperature, the sample is heated to about 80° C. There occurs an intermediate region of a negative slope. MARSHALL and THOMPSON suggest that a load extension curve with negative slope will be unstable. In such a case an alternative

---

[1] Very recently MÜLLER gave a theory in which the formation of the neck is explained by general thermodynamic considerations very simular to these of MARSHALL and THOMPSON. [F. H. MÜLLER private comunication.]

[2] MARSHALL, I. u. A. B. THOMPSON: Nature **171**, 38 (1953); Proc. Roy. Soc. [London] **A 221**, 541 (1954).

process of extension appears to be possible, which takes place at constant tension in a shoulder and which is coupled with heat conduction along the specimen. Heat is conducted through the shoulder from the warmer stretched region to the colder unstretched region. In this way the unstretched zone is "preheated" and the extension process can take place at a lower tension than that needed for the adiabatic process without necking. MARSHALL and THOMPSON conclude that necking is not intimately connected with crystallinity or orientation, but will always occur where a negative slope in adiabatic extension exists.

The fact that terylene can be drawn also in the non-crystalline state[1] and that polymethylmethacrylate in which no crystalline state is known, shows the phenomenon of neck formation, is sufficient proof that crystallinity is not essential for the orientation process which takes place in cold drawing. Apparently in a polymer two kinds of order can be established more or less independently, viz. crystallinity and orientation. For crystallinity periodicity of atoms is essential whereas for orientation the sequence of atoms or other structural elements, whether periodic or not, must differ in different directions.

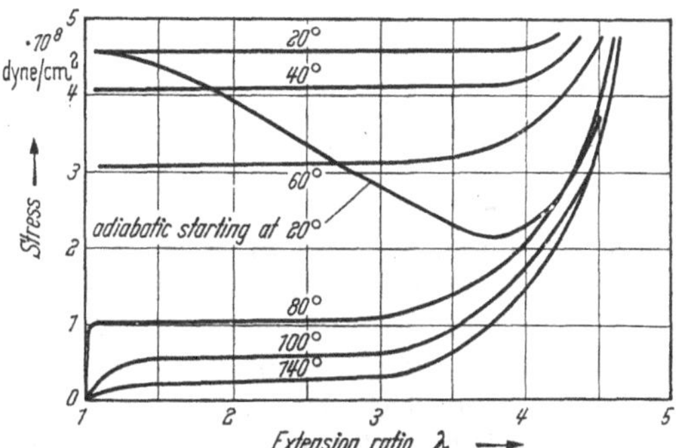

Fig. II, 31. Isothermal load extension curves at different temperatures for terylene and the hypothetical adiabatic load-extension curve, after MARSHALL and THOMPSON[1].

On the other hand crystallinity has such a strong influence on all mechanical properties of polymers that it must be expected to have a marked effect upon orientation processes such as those which occur in cold drawing. Apparently, if no crystallites are present the orientation is performed by separate macromolecules, but if the polymer is crystalline the main orientation effects will originate from orientation of the crystallites.

This is already borne out by the fact that most polymers which show necking are crystalline. As, however, every crystalline polymer possesses amorphous parts, the average orientation of the crystallites alone does not completely describe the state of the polymer. Moreover, KRATKY[2] has pointed out that the "average orientation" of crystallites is fixed by one parameter only in the special case in which they have cylindrical

---

[1] MARSHALL, I. u. A. B. THOMPSON: Proc. Roy. Soc. [London] **A 221**, 541 (1954).

[2] KRATKY, O.: Kolloid-Z. **64**, 213 (1933); **84**, 149 (1938); **120**, 24 (1951). — O. KRATKY, G. POROD u. E. TREIBER: Kolloid-Z. **121**, 1 (1951). — J. J. HERMANS: J. Colloid Sci. **1**, 235 (1946); Trans. Faraday Soc. **B 42**, 160 (1946). — P. H. HERMANS, J. J. HERMANS, D. VERMAAS u. A. WEIDINGER: J. Polymer Sci. **1**, 389, 393 (1946); **2**, 632 (1947); **3**, 1 (1948). — P. H. HERMANS u. W. KAST: Kolloid-Z. **121**, 21 (1951). — W. KAST u. A. PRIETZSCHK: Kolloid-Z. **114**, 23 (1949). — W. KAST: Kolloid-Z. **120**, 40 (1951); **125**, 45 (1952).

symmetry. If the crystallites have three unequal edges two parameters are needed for the description of the orientation, namely one for the average orientation of the largest plane and another for the orientation of the longest edge (see fig. II, 32).

It must therefore be concluded, that a physical description of the state of a polymer drawn in one direction (we will not complicate matters by considering different degress of drawing in different directions) involves at least 4 parameters in the general case: one to represent the degree of crystallinity, two to define the orientation of the crystallites and one to define the orientation of the amorphous component.

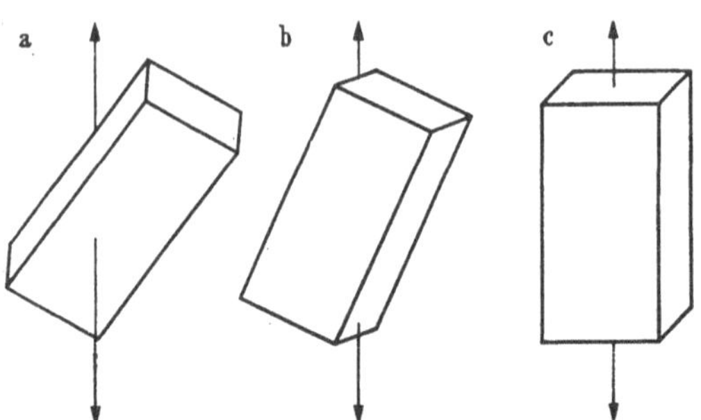

Fig. II, 32. Sheet orientation of crystallites: a) completely unoriented, b) plane orientation, c) plane and edge orientation.

At present experimental technique is not sufficiently advanced to measure these 4 parameters independently and accurately[1]. Particularly HERMANS and KRATKY[2] have studied the orientation of cellulose fibres carefully.

Experimental results on X-ray scattering show that the orientation generally occurs in two steps: the orientation of the sheet plane precedes the orientation of the edge. FANKUCHEN and MARK[3] studied the X-ray pattern of Nylon at short intervals through the shoulder region and found a gradual increase in orientation from undrawn to drawn material. BRYANT[4] was able to demonstrate the increase in orientation during drawing of polythene. On heating the cold drawn material to a temperature near the melting point, it recovers completely to the undrawn form and at the same time the orientation vanishes. BROWN[5] reported that the orientation of polythene during drawing occurs in two steps just as in the case of cellulose. In the early state of drawing, from 20 to 200% elongation, the long chain axes of the crystallites are inclined at an angle of about 64° to the stretching direction. At later stages between 200% and 600% elongation the crystallites are turned completely parallel to the stretching direction. HOPKINS and coworkers[6] investigated the X-ray diagram of polythene under biaxial tension and found orientation of the crystallites in the plane of the principal stresses.

---

[1] For a detailed investigation of crystallinity and structure of high polymers we refer the reader to the third volume of this series.

[2] KRATKY: loc. cit. page 163.

[3] FANKUCHEN, I. u. H. MARK: J. appl. Physics **15**, 364 (1944).

[4] BRYANT, W. M. D.: J. Polymer Sci. **2**, 547 (1947).

[5] BROWN, A.: J. appl. Physics **20**, 552 (1949); compare also C. W. BUNN: J. appl. Chem. **1**, 266 (1951).

[6] HOPKINS, I. L., W. O. BAKER u. J. B. HOWARD: J. appl. Physics **21**, 206 (1950).

Drittes Kapitel.

Bruchspannung und Festigkeit von Hochpolymeren.

Von

F. SCHWARZL und A. J. STAVERMAN.

Mit 51 Abbildungen.

§ 14. Allgemeines über das Studium von Bruchvorgängen.

Beim Studium von Brucherscheinungen von Polymeren stößt der Untersucher auf eine Reihe besonderer Schwierigkeiten. Sowohl das lineare, wie auch das nichtlineare Deformationsverhalten lassen sich auf mathematisch-physikalische Weise beschreiben. Mit Hilfe der Begriffe Deformation und Spannung kann man den Zustand des Probekörpers in jedem Zeitaugenblick charakterisieren und die Beschreibung des Deformationsverhaltens besteht stets aus der Angabe von Zusammenhängen zwischen Deformation, Spannung und Zeit. Durch physikalische Theorien sucht man dann allgemeine Gesetze aufzustellen – die rheologischen Gesetze –, denen bestimmte Deformationstypen gehorchen.

Ein analoger Versuch bei der Untersuchung der Bruchvorgänge scheitert an der unstetigen Natur des Bruchvorganges selbst. Wir haben ein Stück Material vor dem Bruch und zwei oder mehrere Stücke nach dem Bruch. Der eigentliche Prozeß der Stofftrennung beim Bruche geht so schnell vor sich, daß eine direkte Beobachtung dieses Vorganges im allgemeinen nicht möglich ist. Letztere wird obendrein noch dadurch erschwert, daß der größte Teil des Bruchvorganges im Inneren des Materials vor sich geht.

Man hat sich daher bis vor kurzem damit begnügt, auf eine Beschreibung des Bruchvorganges selbst zu verzichten und die Bedingungen zu studieren, unter welchen Bruch eintritt. Die markanteste dieser Bedingungen ist das Überschreiten eines gewissen kritischen Spannungszustandes, der *Bruchspannung*. Die Existenz einer Bruchspannung ist der Ausgangspunkt aller älteren Bruchtheorien gewesen und ihre Bestimmung der Hauptzweck aller technologischen Untersuchungsmethoden des Bruchverhaltens. Die Bruchfestigkeiten geben auch heute noch dem Ingenieur allgemeine Richtlinien für den Entwurf von technisch sicheren Konstruktionen.

Die sogenannten phänomenologischen Bruchtheorien befassen sich mit der Untersuchung der Bedingungen, die zum Auftreten eines Bruches führen. Verschiedene Gesichtspunkte hiervon, wie der Einfluß der Geometrie des Spannungszustandes, der Einfluß der Belastungszeit oder der Deformationsgeschwindigkeit sollen in § 15 behandelt werden.

Bei der Bestimmung der Bruchspannung stößt man auf eine besonders große experimentelle Streuung der Resultate, die sich nicht durch die Fehlerstreuung der Apparate erklären läßt, sondern zu der Schlußfolgerung zwingt, daß dem Bruchvorgang Gesetze besonderer statistischer Natur zugrunde liegen. Die statistische Auffassung der dem Bruchvorgange unterliegenden Gesetze (§ 16) führt zu wertvollen Schlüssen über die Natur von spröden Brüchen und erklärt außerdem zwanglos zwei bis dahin unbegreifliche Tatsachen: Die niederen Werte der experimentell ermittelten Bruchspannungen und die Abhängigkeit der Festigkeitseigenschaften vom Volumen und der Form des Probestückes.

Will man die physikalische Natur von Bruchvorgängen eingehend studieren, so kann man sich nicht auf den Zeitpunkt des Bruchbeginnes beschränken, sondern muß den Verlauf der Bruchfortpflanzung und Stofftrennung zu erforschen suchen. Hier hat besonders SMEKAL in den letzten Jahren gezeigt, daß auf indirektem Wege – durch mikroskopische Beobachtung und richtige Interpretierung der resultierenden Bruchflächen – weitgehende Schlüsse über den Vorgang des Bruchprozesses gezogen werden können (§ 17). So konnte SMEKAL beweisen, daß das Auftreten von sprödem Bruchverhalten stets an die Wirksamkeit von *schwachen Stellen* im Material gebunden ist, während in kleinsten, fehlerfreien Bereichen niemals spröder Bruch, sondern lediglich plastische Verformung auftreten kann. Weiterhin ist es auch möglich, auf indirektem Wege die Fortpflanzungsgeschwindigkeit von Bruchvorgängen zu messen.

§ 15. Phänomenologische Bruchtheorien.

a) Abhängigkeit von der Spannungsgeometrie.

Wie bereits in der Einleitung erwähnt, wird der Eintritt des Bruches charakterisiert durch Überschreiten einer bestimmten Spannung, der *Bruchspannung*. Der Spannungszustand in einem Punkte eines deformierten Körpers ist jedoch nicht eine reine Zahl, sondern ist bestimmt durch die sechs Komponenten des Spannungstensors[1]. Eine Scherungsspannung wird im allgemeinen einen anderen Einfluß auf das Bruchverhalten zeigen wie eine Zugspannung oder ein hydrostatischer Druck. Das erste Problem ist daher der Einfluß der *Geometrie* des *Spannungszustandes* auf das Festigkeitsverhalten. Wir beschäftigen uns in diesem Abschnitt mit dem geometrischen Bruchproblem, ohne vorläufig näher auf die Frage einzugehen, ob die Größe des Spannungszustandes auch der einzige Parameter ist, der das Auftreten des Bruches bestimmt.

Wir beschränken die nachfolgende Diskussion auf Materialien, die isotrop sind und bis zum Eintritt des Bruches isotrop bleiben. Bei solchen Stoffen können die Festigkeitseigenschaften nicht von der Richtung abhängen. Ein Spannungszustand kann, abgesehen von der Richtung des

---

[1] Vgl. hierzu die Ausführungen in § 2b.

Hauptachsensystems, vollständig festgelegt werden durch die drei Hauptspannungen $(\sigma_1, \sigma_2, \sigma_3)$[1]. Ein Spannungszustand, der gerade zum Bruche führt, ist bestimmt durch die drei Hauptspannungen

$$F(\sigma_1, \sigma_2, \sigma_3) = 0. \qquad \text{(III,1)}$$

Jeder Spannungszustand $(\sigma_1, \sigma_2, \sigma_3)$ kann in einem kartesischen Koordinatensystem als ein Punkt $P = (\sigma_1, \sigma_2, \sigma_3)$ festgelegt werden (vgl. Abb. III,1). Gleichung (III,1) ist dann die Gleichung einer teilweise geschlossenen Fläche im $(\sigma_1, \sigma_2, \sigma_3)$-Raum mit der folgenden physikalischen Bedeutung: Ein Spannungszustand, der im $(\sigma_1, \sigma_2, \sigma_3)$-Raum innerhalb der Fläche F liegt, führt nicht zum Bruche, ein Spannungszustand, der auf der Fläche oder außerhalb der Fläche liegt, verursacht Brucheintritt. Die Fläche F heißt *Grenzfläche des Bruchеintrittes* (in englisch: *limiting surface of failure*).

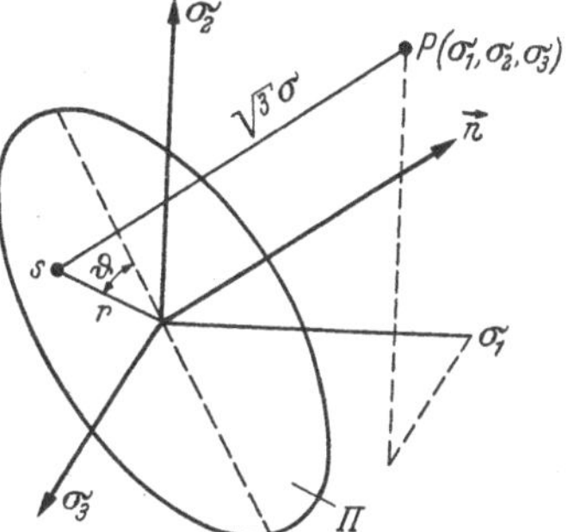

Abb. III, 1. Die Darstellung des Spannungspunktes P durch kartesische Koordinaten $(\sigma_1\ \sigma_2, \sigma_3)$ und Zylinderkoordinaten $(\Theta, r, \sqrt{3}\,\sigma)$.

Infolge der Isotropie des betrachteten Stoffes besitzt die Grenzfläche noch eine Anzahl von Symmetrieeigenschaften. Um die Gestalt der Grenzfläche diskutieren zu können, wollen wir erst noch eine andere Darstellung des Spannungspunktes $(\sigma_1, \sigma_2, \sigma_3)$ besprechen.

Jeder Spannungszustand läßt sich schreiben als die Summe eines hydrostatischen Spannungszustandes und eines Spannungsdeviators:

$$(\sigma_1, \sigma_2, \sigma_3) = (\sigma, \sigma, \sigma) + (\sigma_1 - \sigma, \sigma_2 - \sigma, \sigma_3 - \sigma), \qquad \text{(III,2)}$$

wobei

$$\sigma = [\sigma_1 + \sigma_2 + \sigma_3]/3. \qquad \text{(III,3)}$$

Der isotrope Spannungszustand (σ, σ, σ) ist ein allseitiger hydrostatischer Zug [für $\sigma > 0$] oder ein allseitiger Druck [für $\sigma < 0$, in welchem Falle man $p = -\sigma$ als hydrostatischen Druck bezeichnet] und enthält keine Scherungsspannungen. Der Spannungsdeviator $(\sigma_1 - \sigma, \sigma_2 - \sigma, \sigma_3 - \sigma)$ enthält die maximalen Scherspannungen

$$\tau_1 = (\sigma_2 - \sigma_3)/2, \quad \tau_2 = (\sigma_3 - \sigma_1)/2, \quad \tau_3 = (\sigma_1 - \sigma_2)/2, \qquad \text{(III,4)}$$

aber keine hydrostatische Komponente [die Summe der Spannungsdeviatorkomponenten im Hauptachsensystem ist gleich Null]. Es ist einleuchtend, daß isotrope Komponente und Spannungsdeviator in anderer Weise auf das Festigkeitsverhalten einwirken werden, da die isotrope Komponente eine Volumenänderung, der Deviator eine Formänderung des Stoffes bewirkt.

Die geometrische Bedeutung von Gleichung (III,2) ersehen wir aus Abb. III,1. Ein hydrostatischer Spannungszustand (σ, σ, σ) liegt stets

[1] Siehe S. 166, Fußnote 1.

auf der Raumdiagonalen $\vec{n}$ mit den Richtungskosinussen $\left(\frac{1}{\sqrt{3}}, \frac{1}{\sqrt{3}}, \frac{1}{\sqrt{3}}\right)$, während alle Spannungsdeviatoren $(\sigma_1 - \sigma, \sigma_2 - \sigma, \sigma_3 - \sigma)$ in der Ebene Π liegen müssen [Π ist die Ebene durch den Ursprung senkrecht zu $\vec{n}$].

Nun bestimmen wir zu jedem Spannungszustand P den zugehörigen Deviator S durch Fällung des Lotes von P auf die Ebene Π. Der Abstand $\overline{PS}$ beträgt dann $\sqrt{3}\sigma$. Der Deviatorpunkt S kann festgelegt werden durch seine Polarkoordinaten (r, θ) in der Ebene Π.

Nun läßt sich zeigen[1], daß die Koordinaten r und θ gegeben sind durch die Gleichungen:

$$\left.\begin{aligned} r &= \frac{2}{\sqrt{3}} \sqrt{\tau_1^2 + \tau_2^2 + \tau_3^2} = \sqrt{3}\,\tau_{\text{okt}} \\ \operatorname{tg}\theta &= \frac{1}{\sqrt{3}} \frac{2\sigma_3 - \sigma_1 - \sigma_2}{\sigma_2 - \sigma_1}, \end{aligned}\right\} \qquad \text{(III, 5)}$$

wenn der Anfangspunkt in der Zählung von θ geeignet gewählt wird. Aus Gleichung (III, 4) ersehen wir, daß der Radiusvektor r ein Maß ist für die Quadratsumme der maximalen Scherspannungen, $r/\sqrt{3}$ wird manchmal[2] als Oktaederspannung bezeichnet.

Wir können nun jeden Spannungszustand auch durch seine Zylinderkoordinaten $\left(r, \theta, \sqrt{3}\sigma\right) = \left(\sqrt{3}\tau_{\text{okt}}, \theta, \sqrt{3}\sigma\right)$ festlegen. Hieraus folgt, daß sich auch die Grenzfläche des Brucheintrittes in Zylinderkoordinaten schreiben läßt, und wir erhalten an Stelle von Gleichung (III, 1):

$$F\left(\sqrt{3}\,\tau_{\text{okt}}, \theta, \sqrt{3}\,\sigma\right) = 0. \qquad \text{(III, 6)}$$

Die Isotropie des Materials beschränkt die Abhängigkeit der Grenzfläche vom Winkel[2] θ. Diese Isotropie verlangt nämlich, daß der Charakter des Materialverhaltens nicht verändert wird, wenn man die Werte der Komponenten σ_1, σ_2 und σ_3 oder auch τ_1, τ_2 und τ_3 untereinander vertauscht. Das heißt, der Radiusvektor r – und damit die Gestalt der Grenzfläche – muß, wenn überhaupt, von θ periodisch abhängen mit einer Periode von 60 Graden. Die Gestalt der Grenzfläche ist bis jetzt ausschließlich für Metalle eingehender untersucht worden. Dort wurde gefunden, daß Bruch und Fließbedingungen von θ überhaupt nicht abhängen[2]. Übertragen wir diese Annahme auch auf das Gebiet von Hochpolymeren, so erhalten wir für die Grenzfläche eine Gleichung der Form

$$F\left(\sqrt{3}\,\tau_{\text{okt}}, \sqrt{3}\,\sigma\right) = 0. \qquad \text{(III, 7)}$$

Hieraus können wir uns auch den Radiusvektor gegeben denken als

---

[1] LODE, W.: Z. Physik **36**, 913 (1926). – R. HILL: The mathematical theory of plasticity, Oxford, Clarendon press, 1950, page 17.

[2] MISES, H. v.: Göttinger Nachr. 1913. – A. NADAI: Theory of flow and fracture of solids, Mac Graw Hill Book Comp. New York 1950, page 103.

Funktion der hydrostatischen Spannung, so daß sich die Gleichung der Grenzfläche auch schreiben läßt

$$\sqrt{3}\,\tau_{\text{okt}} = f(\sqrt{3}\,\sigma). \qquad \text{(III, 8)}$$

Die geometrische Bedeutung von Gl. (III, 7) oder (III, 8) ersehen wir aus den Abb. III, 2 und III, 4. Wenn die Grenzfläche nicht vom Winkel θ abhängt, so ist sie eine Umdrehungsfläche um die Achse $\vec{n}$. Der Schnitt der Grenzfläche mit der Ebene Π – oder mit einer Ebene parallel zu Π, gekennzeichnet durch eine konstante hydrostatische Komponente σ – ist ein Kreis vom Radius r. Der Radius dieses Schnittkreises hängt von σ ab und ist gegeben durch Gl. (III, 8).

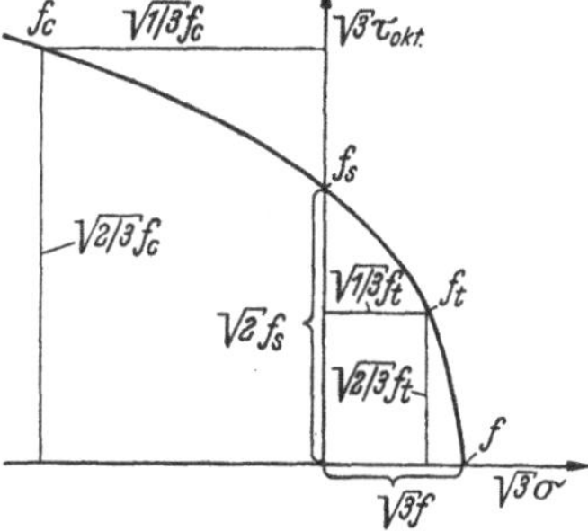

Abb. III, 2. Die Grenzkurve des Brucheintrittes und die Lage von Scherfestigkeit f_s, Zugfestigkeit f_t und Druckfestigkeit f_c.

Gl. (III, 8) ist die mathematisch einfachste Annahme für die Bruchbedingung. Vom physikalischen Standpunkt würde man vielleicht eine andere Annahme für wahrscheinlicher halten. Man könnte sich denken, daß, bei gegebenem hydrostatischen Druck oder Zug, statt der Oktaederspannung die maximale Scherspannung, also der größte der drei Werte τ_1, τ_2 und τ_3 maßgebend für den Bruch sei. Dies würde bedeuten, daß der Schnitt der kritischen Grenzfläche mit der Ebene Π kein Kreis, sondern ein regelmäßiges Sechseck wäre. Prinzipiell würde eine solche Situation sich von der in Gl. (III, 8) gegebenen nur wenig unterscheiden, so daß wir uns weiter auf diese einfache Bedingung beschränken werden.

Eine Bruchbedingung in der Form Gl. (III, 8) wurde zuerst von SCHLEICHER[1] vorgeschlagen. Diese Form scheidet den Einfluß der Scherspannungen und der hydrostatischen Komponente auf das Bruchverhalten. Bei jedem Spannungszustand mit einer hydrostatischen Komponente σ ist eine bestimmte Scherspannung $\sqrt{3}\,\tau_{\text{okt}}$ notwendig, um Bruch zu erzielen (vgl. Abb. III, 2). Die Kurve $\sqrt{3}\,\tau_{\text{okt}}$ als Funktion von $\sqrt{3}\,\sigma$ nennen wir die *Grenzkurve des Brucheintrittes* [in englisch: *limiting curve of failure*]. Ein Spannungszustand, der innerhalb der Grenzkurve liegt, führt nicht zum Bruch, ein Spannungszustand, der auf oder außerhalb der Grenzkurve liegt, verursacht Brucheintritt.

Um das geometrische Bruchverhalten vollständig zu charakterisieren, muß man die gesamte τ_{okt}—σ-Kurve messen. In vielen Fällen wird sich diese Kurve jedoch abschätzen lassen durch eine Anzahl einfacher konventioneller Prüfungsmethoden, wie mit Hilfe der Zugfestigkeit f_t, der Scherfestigkeit f_s und der Druckfestigkeit f_c, die bestimmte charakteristische Punkte der Grenzkurve vermitteln. Die Werte von $\sqrt{3}\,\tau_{\text{okt}}$ und $\sqrt{3}\,\sigma$ für diese Versuche sind in Tabelle III, 1 zusammengefaßt.

Der Schnittpunkt der Grenzkurve mit der σ-Achse gibt den (hypothetischen) Versuch unter triaxialer Zugspannung f. Die Abszisse dieses

[1] SCHLEICHER, F.: Z. angew. Math. Mechan. **6**, 216 (1926).

Tabelle III, 1.
Oktaederspannungen und hydrostatische Komponente für einfache Experimente.

Experiment	Hauptspannungen $\sigma_1, \sigma_2, \sigma_3$	$\sqrt{3}\,\tau_{okt}$	$\sqrt{3}\,\sigma$
hydrostatischer Zug in drei Richtungen f	f, f, f	0	$\sqrt{3}\,f$
hydrostatischer Druck p	$-p, -p, -p$	0	$-\sqrt{3}\,p$
Scherfestigkeit f_s	$+f_s, -f_s, 0$	$\sqrt{2}\,f_s$	0
Zugfestigkeit f_t	$f_t, 0, 0$	$\sqrt{2}\,f_t/\sqrt{3}$	$f_t/\sqrt{3}$
Druckfestigkeit f_c	$-f_c, 0, 0$	$\sqrt{2}\,f_c/\sqrt{3}$	$-f_c/\sqrt{3}$

Punktes ist gleich $\sqrt{3}f$. Der Schnittpunkt der Grenzkurve mit der τ_{okt}-Achse stellt den Scherversuch dar. Seine Ordinate ist gleich $\sqrt{2}f_s$. Der allgemeine Verlauf der Grenzkurve ist physikalisch einleuchtend. Je niedriger der hydrostatische Zug (je größer der hydrostatische Druck), desto größer ist die zum Brucheintritt notwendige Scherkomponente. Auf der Zugseite der Grenzkurve liegt ein Punkt, dessen Ordinate und Abszisse sich verhalten wie $\sqrt{2}:1$. Dieser Punkt entspricht der Zugfestigkeit. Ein ebensolcher Punkt auf der Druckseite entspricht der Druckfestigkeit.

Auf diese Weise sind in Abb. III, 3 zwei Grenzkurven konstruiert aus der Messung von f_s, f_t und f_c für Polystyrol[1] und für eine Mischung von Stearinsäure mit Gips[2].

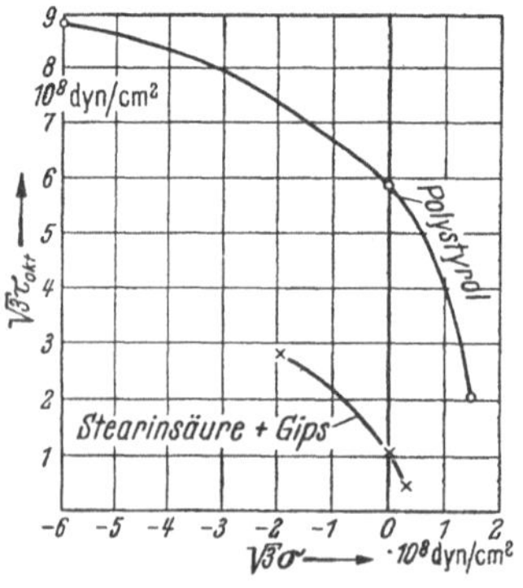

Abb. III, 3. Konstruktion der Grenzkurven für Polystyrol und eine Mischung von Stearinsäure mit Gips aus Zugfestigkeit, Druckfestigkeit und Scherfestigkeit.

Wir müssen hier erwähnen, daß nach obiger einfacher Theorie die Zugfestigkeit f_t und die Biegefestigkeit f_b übereinstimmen sollten, da bei beiden Experimenten Bruch unter reiner Zugspannung auftritt.

Im Gegensatz dazu findet man stets eine Biegefestigkeit, die wesentlich höher liegt als die Zugfestigkeit. So gilt nach KUNTZE[2] für die Stearinsäure-Gips-Mischung

$$f_b/f_t \sim 1{,}5\,.$$

Nach HAWARD[3] liegt das Verhältnis f_b/f_t für verschiedene Fraktionen von Vinyl-Chlorid-Acetat-Copolymeren zwischen 1,3 und 1,5. Eine Erklärung dieser Tatsache ist mit Hilfe der phänomenologischen Bruchtheorie nicht mög-

---

[1] SCHWARTZ, R. T. u. E. DUGGER: Mod. Plastics (March 1944), 117.
[2] KUNTZE, G.: Ann. Physik **11**, 1020 (1903).
[3] HAWARD, R. N.: The strength of plastics and glass, Cleaver Hume Press London 1949, page 14.

lich, sondern erfordert eine statistische Betrachtungsweise des Bruchproblems (vgl. § 16).

Aus der Grenzkurve von Abb. III,2 läßt sich die Grenzfläche sofort konstruieren. Wir lassen die Grenzkurve um die σ-Achse rotieren und erhalten so eine Umdrehungsfläche mit der σ-Achse als Symmetrieachse. Diese Fläche stellen wir mit der Symmetrieachse als Raumdiagonale $\vec{n}$ in den $\sigma_1, \sigma_2, \sigma_3$-Raum. Ein Beispiel einer solchen Grenzfläche ist in Abb. III, 4 wiedergegeben.

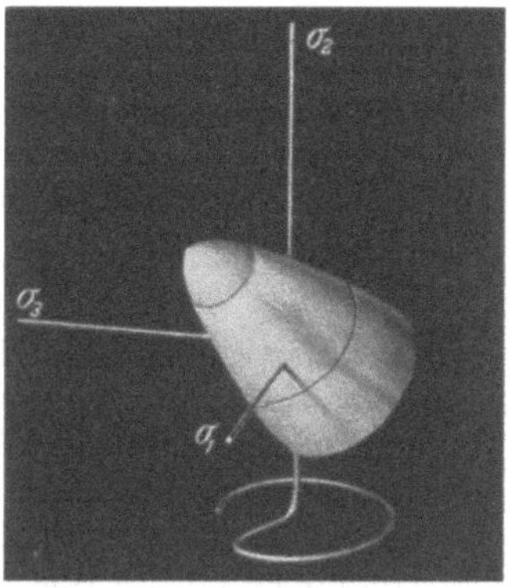

Außer dem Bruchkriterium in der Form Gl. (III, 8) sind noch eine große Anzahl anderer Bruchkriteria vorgeschlagen worden, die sich teilweise als Spezialform von Gl. (III, 8) schreiben lassen, teilweise sich in ihren Folgerungen nicht wesentlich von Gl. (III, 8) unterscheiden. Wir gehen hierauf nicht näher ein, verweisen den Leser jedoch auf einen Übersichtsartikel, in dem alle Original-Zitate zu finden sind[1].

Abb. III, 4. Hypothetische Form der Grenzfläche des Brucheintrittes für ein isotropes Material nach der Theorie von SCHLEICHER.

b) Zeiteffekte.

Wir haben im letzten Abschnitt die Abhängigkeit des Bruchbeginns von der Spannungsgeometrie behandelt, ohne auf die Frage einzugehen, ob die Höhe des Spannungszustandes auch die einzige Variable ist, die das Auftreten des Bruches bestimmt. Nun ist es gerade eine der wesentlichen Schwierigkeiten des Festigkeitsproblems, daß wir kein physikaliches Kriterium für das Eintreten des Bruchbeginns besitzen. Sicher ist heutzutage, daß der Brucheintritt in sehr komplizierter Weise von der gesamten Spannungs-Deformations-Vorgeschichte des Materials abhängt. Weder die Größe der Spannung allein, noch die der Deformation, noch die der Verformungsenergie oder ähnliche einfache Kriteria kennzeichnen eindeutig den Eintritt des Bruches.

Um in dieser unbefriedigenden Situation doch zu einer physikalisch-eindeutigen Formulierung des Bruchproblems zu kommen, muß man eine Anzahl Experimente mit wohldefinierter Spannungs-Deformations-Vorgeschichte betrachten. Die Vorgeschichte darf nur von einem Parameter abhängen und die resultierende gemessene Bruchspannung ist dann eine Funktion dieses Parameters. Beispiele hiervon sind:

1. Der *Zerreißversuch unter konstanter Deformationsgeschwindigkeit* $v = d\gamma/dt$; der Parameter ist hier die Deformationsgeschwindigkeit v und die Bruchspannung hängt ab von der Deformationsgeschwindigkeit

$$\sigma = \sigma(v). \tag{III,9}$$

[1] NADAI, A.: Theory of flow and fracture of solids, Mac Graw Hill Book Comp., New York 1950, page 207.

2. Der *Standzeitversuch*, wobei das Probestück mit einer konstanten Spannung belastet und die vom Beginn der Belastung bis zum Eintritt des Bruches verstreichende Bruchzeit t_b gemessen wird. Zu jeder angelegten Spannung erhält man eine Bruchzeit t_b (eventuell $t_b = 0$, wenn Bruch augenblicklich eintritt, oder $t_b = \infty$, wenn die Spannung nicht zum Bruche führt). Umgekehrt können wir die zu einer bestimmten Bruchzeit gehörende Spannung ausdrücken in der Standzeitkurve

$$\sigma = \sigma(t_b). \tag{III,10}$$

3. Das *dynamische Ermüdungsexperiment*, wobei das Probestück an eine Wechselspannung mit der Amplitude σ unterworfen wird. Man mißt dann die Anzahl Lastwechsel N bis zum Brucheintritt und erhält die WÖHLER-*Kurve*

$$\sigma = \sigma(N). \tag{III,11}$$

Jedes der obigen Experimente definiert so eine Bruchspannung σ, wobei die Bedeutung von σ abhängt von der geometrischen Form des Versuches. Bei einfachen Versuchen, wie axialer Zug, axialer Kompression, Biegung oder Scherung hat σ die Bedeutung von f_t, f_c, f_b oder f_s, bei mehrdimensionalen Spannungszuständen ist σ eine Abkürzung für die Komponenten des Spannungszustandes. Da wir nicht über ein allgemeines Bruchkriterium verfügen, ist es nicht möglich, die in Gleichungen (III,9), (III,10) und (III,11) definierten Bruchspannungen miteinander in Beziehung zu setzen.

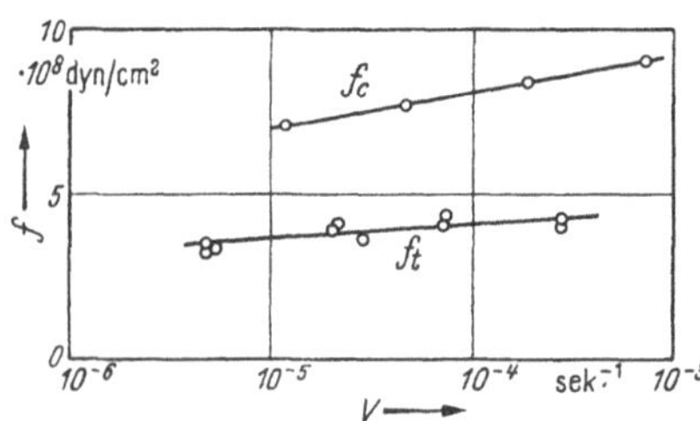

Abb. III, 5. Abhängigkeit der Zugfestigkeit f_t und Druckfestigkeit f_c von der Deformationsgeschwindigkeit v für Polystyrol. (Nach HSIAO und SAUER.) [HSIAO C. C., SAUER J. A., ASTM Bulletin No. 172 (Febr. 1951)].

Zu 1. Mit zunehmender Deformationsgeschwindigkeit wechselt das Festigkeitsverhalten langsam von zähem nach sprödem Bruch, wobei die Bruchspannung σ zunimmt und die Bruchdeformation abnimmt. Ein Beispiel dafür haben wir bereits in Abb. II,18 und II,19 kennen gelernt (vgl. auch § 15c).

Die Abhängigkeit von Zugstärke und Druckstärke von der Deformationsgeschwindigkeit für Polystyrol zeigt Abb. III,5 [1]. Wie man sieht, ist die Bruchspannung in beiden Fällen proportional zum Logarithmus der Deformationsgeschwindigkeit

$$\left.\begin{aligned} f_t &= a_t + b_t \log v/v_0 \\ f_c &= a_c + b_c \log v/v_0 \end{aligned}\right\} \tag{III,12}$$

Die Konstante b bestimmt den Einfluß der Deformationsgeschwindigkeit auf die Festigkeit. Im vorliegenden Fall ist $b_t = 5{,}5 \cdot 10^7\ \mathrm{d/cm^2}$, die Zugstärke nimmt ungefähr 11% zu, wenn die Deformationsgeschwindigkeit um eine Zehnerpotenz größer wird.

[1] HSIAO, C. C. u. J. A. SAUER: A. S. T. M. Bulletin No. 172 (Febr. 1951).

Zu 2. Ein Beispiel für die Ermüdungskurve von Polystyrol finden wir in Abb. IV, 10. Messungen über ein langes Zeitintervall an verschiedenen Glas- und Porzellansorten sind von BAKER und PRESTON[1] ausgeführt worden. Abb. III, 6 zeigt die Standzeitkurve von trockenem Pyrexglas über einen Zeitbereich von mehr als 6 Zehnerpotenzen. Die kurzzeitige Festigkeit ist fast doppelt so hoch wie die Bruchspannung nach 24 Stunden.

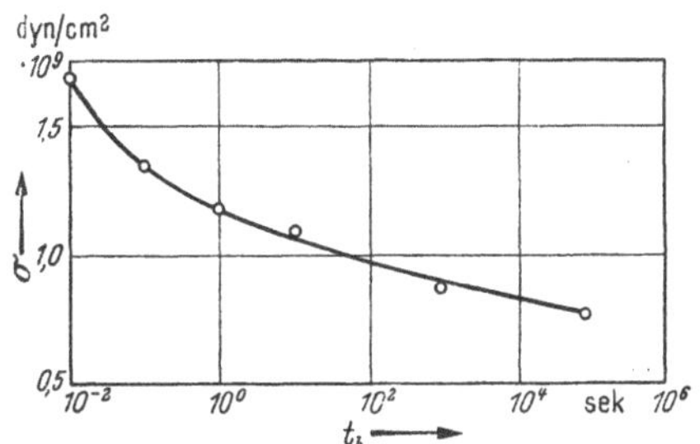

Abb. III, 6. Die Standzeitkurve von trockenem Pyrexglas. (Nach BAKER und PRESTON.) [BAKER, T. C., F. W. PRESTON: J. Applied Phys. 17, 170 (1946)].

Wie wir bereits in § 5 gesehen haben, sind alle zeitabhängigen mechanischen Eigenschaften von Polymeren auch sehr stark von der Temperatur abhängig, wobei eine Erhöhung der Temperatur ungefähr gleichwertig ist einer Parallelverschiebung in der logarithmischen Zeitachse nach kürzeren Zeiten. Einen ähnlichen Effekt einer „Zeit-Temperatur-Relation" sollten wir auch bei der Standzeitkurve von Polymeren erwarten.

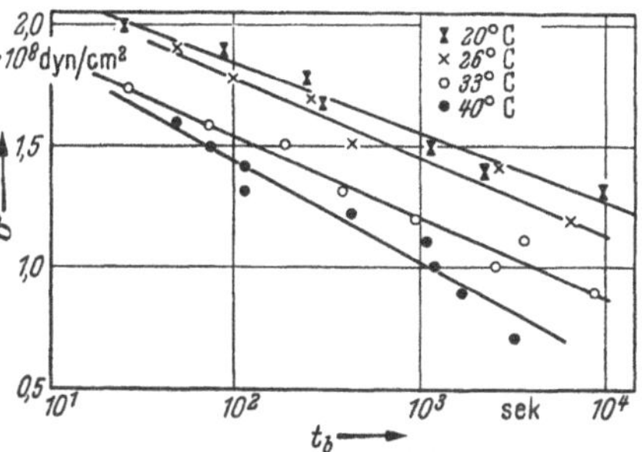

Abb. III, 7. Standzeitkurven an Zellulose-Azetat bei verschiedenen Temperaturen. (Nach HAWARD.) [HAWARD, R. N.: Trans. Farad. Soc. 39, 267 (1943)].

Einen solchen Effekt zeigen die Standzeitkurven an Celluloseacetat bei vier verschiedenen Temperaturen in Abb. III, 7[2]. Obwohl die Temperaturunterschiede sehr gering sind, geben sie doch beträchtliche Verschiebungen in der Zeitachse.

Die Temperaturabhängigkeit der Lebensdauer (Bruchzeit unter einer konstanten Spannung) für Baumwollgarn nach BUSSE[3] ist in Abb. III, 8 wiedergegeben. Hieraus läßt sich eine Aktivierungsenergie berechnen von 12,5 kcal/Mol.

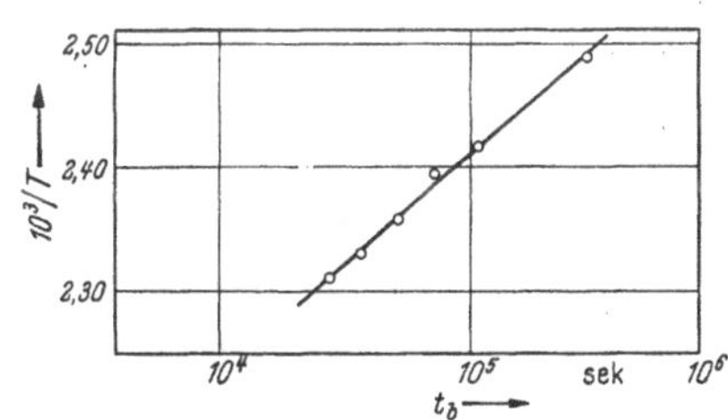

Abb. III, 8. Bruchzeit unter konstanter Spannung, t_b als Funktion der Temperatur T. (Nach BUSSE.) [BUSSE, W. F., E. T. LESSING, D. L. LOUGHBOROUGH, L. LARRIK: J. Applied Phys. 13, 715 (1942)].

Zu 3. Als Beispiel für dynamische Ermüdungsexperimente zeigen wir Torsionsermüdungsversuche von VAN DER VEGT[4].

---

[1] BAKER, T. C. u. F. W. PRESTON: J. appl. Physics **17**, 170 (1946).

[2] HAWARD, R. N.: Trans. Faraday Soc. **39**, 267 (1943).

[3] BUSSE, W. F., E. T. LESSING, D. L. LOUGHBOROUGH u. L. LARRIK: J. appl. Physics **13**, 715 (1942).

[4] VEGT, A. K. VAN DER: Centraal Laboratorium T. N. O., Delft.

Eine große Anzahl (100) Nylonfasern wurden in einer Versuchsreihe unter gleichen Bedingungen mit einer konstanten maximalen Amplitude tordiert und die nach jedem Deformationscyclus gebrochenen Exemplare gezählt. Man erhält so die kumulative Verteilungsfunktion

$$P(N),$$

das ist die relative Anzahl Fasern, die nach N Lastwechseln bereits gebrochen ist. Abb. III, 9 zeigt die kumulativen Verteilungsfunktionen der Lebensdauer bei verschiedenen konstanten Maximalamplituden entsprechend den maximal auftretenden Scherdeformationen von $\gamma = 0{,}60$ bis $\gamma = 1{,}10$.

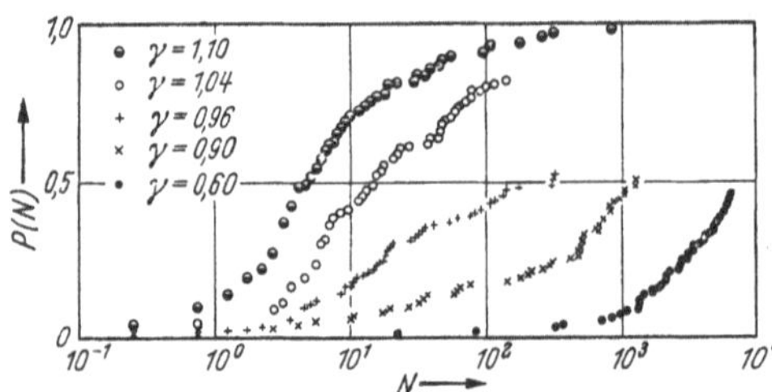

Abb. III, 9. Kumulative Verteilungsfunktionen $P(N)$ der Lebensdauer N von je 100 Nylonfasern bei den konstanten Maximalamplituden $\gamma = 1{,}10$; 1,04; 0,96; 0,90 und 0,60. (Nach v. D. VEGT.) (A. K. v. D. VEGT, Centraal Laboratorium T. N. O., Delft).

Die Verteilungsfunktion für die größte Torsion $\gamma = 1{,}10$ ist voll durchgemessen und zeigt in erster Näherung einen logarithmisch normalen Verlauf, wobei der Logarithmus der Lebensdauer N normal verteilt ist. Ist einmal bekannt, daß die Lebensdauer nahezu logarithmisch normal verteilt ist, so genügt es, die restlichen Verteilungskurven nur bis zum Median $N_{.5}$ durchzumessen, d.h. bis zu der Anzahl Lastwechsel, bei der die Hälfte aller Fasern gebrochen ist. Abb. III, 10 gibt die dynamische Ermüdungskurve, wobei der Median $N_{.5}$ der Lebensdauer gezeigt ist als Funktion der Maximalamplitude γ.

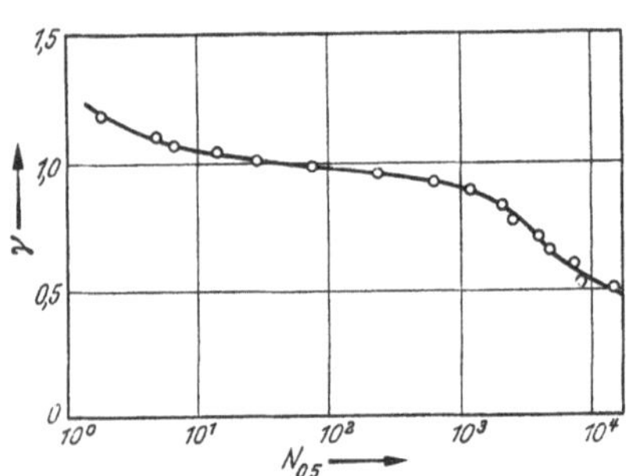

Abb. III, 10. Die Maximalamplitude γ als Funktion des Medians der Lebensdauer $N_{.5}$ für Torsionsermüdung von je 100 Nylonfasern. (Nach v. D. VEGT.) (A. K. v. D. VEGT, Centraal Laboratorium T. N. O., Delft).

c) Sprödes und zähes Bruchverhalten.

Vom Gebiete der Metallprüfung her ist man gewöhnt, bei auftretenden Brüchen zähe und spröde Bruchtypen zu unterscheiden. Diese Terminologie wird heute auch allgemein auf Polymere angewandt. Streng genommen müßte man die folgenden Bruchtypen unterscheiden

spröder Bruch	zäher (plastischer) Bruch
Zugbruch	Gleitungsbruch.

Man spricht von zähem Bruch oder sprödem Bruch, je nachdem das Auftreten des Bruches von großen irreversiblen Deformationen eingeleitet wird oder nicht. Ein ideal spröder Stoff ist Glas bei niederen Temperaturen, ein ideal zähes Material bilden gut verformbare Metalle. Während die Unterscheidung zähe – spröde nach der Größe der auftretenden Deformationen oder Bruchenergien geschieht, bezieht sich die Einteilung in Zugbruch oder Gleitungsbruch auf den Bruchvorgang und die Morphologie der Bruchfläche selbst. Wir sprechen von Zugbruch, wenn der eigentliche Vorgang der Stofftrennung hauptsächlich durch Zugspannungen vor sich geht und wenn der Bruch durch senkrechtes Auseinanderziehen der beiden Bruchflächen entsteht. Wir sprechen von Scherbruch, wenn der Bruchvorgang durch Scherkräfte bewirkt wird und die Bruchflächen durch einen Gleitungsprozeß entstehen. Die Unterscheidung in Zugbruch und Gleitungsbruch läßt sich meistens nach dem Aussehen der Bruchfläche treffen.

Bei Metallen fallen die Begriffe spröder und Zugbruch bzw. zäher und Gleitungsbruch in vielen Fällen zusammen. Der Grund hierfür ist, daß bei Metallen große Deformationen stets in der Form von plastischem Fließen auftreten, und zwar ausschließlich unter dem Einfluß von Scherkräften. Bei Polymeren müssen diese Begriffe nicht notwendig zusammenfallen, da hier große Deformationen auch unter Zugspannungen auftreten können.

Das Auftreten von Zug- oder Gleitungsbruch – und damit bis zu einem gewissen Grad auch das Auftreten von sprödem oder zähem Bruch – hängt natürlich stark zusammen mit der Geometrie des betrachteten Spannungszustandes. Spannungszustände mit hohen positiven hydrostatischen Spannungen σ zeigen vorzugsweise Zugbruch, dessen Eintreten bestimmt wird durch die Größe der maximalen auftretenden Zugspannung und wobei die Bruchfläche senkrecht zur Zugspannungsrichtung gebildet wird (*Normalspannungsgesetz* des spröden Bruches). Spannungszustände mit negativen hydrostatischen Spannungen neigen zu Gleitungsbruch, dessen Eintreten hauptsächlich von der maximalen Scherspannung bestimmt wird und wobei die Bruchflächen mit den Flächen maximaler Scherung zusammenfallen. Im Lichte von Abb. III,2 gesehen: Spannungszustände rechts vom Punkte f_t neigen zu Zugbruch, Spannungszustände links vom Punkte f_s neigen zu Gleitungsbruch. Dazwischen liegt natürlich eine Übergangszone, wo die beiden Bruchkriteria allmählich ineinander übergehen. Ein Beispiel für sprödes Verhalten bei axialem Zug und zähes Verhalten bei axialer Kompression unter sonst gleichen Umständen beim selben Stoff haben wir bereits in Abb. II,18 und II,19 kennen gelernt[1].

Einen wesentlichen Einfluß auf das Bruchverhalten besitzt die Temperatur. Bei hohen Temperaturen zeigen die meisten Stoffe zähes Bruchverhalten, bei sehr niederen Temperaturen, wenn die Wärmeenergie keinen Anteil am Bruchprozeß hat, ist das ideal spröde Verhalten verwirklicht.

Der Großteil der experimentellen und theoretischen Betrachtungen von § 16 und § 17 werden sich auf das spröde Bruchverhalten und den Zugbruch beziehen, da diese Bruchtypen vom theoretischen Standpunkt aus

---

[1] Vgl. auch P. W. Bridgeman: Studies in large plastic flow and fracture, Mac Graw Hill Book Comp., Inc. New York 1952.

die einfachsten sind. Wenn nämlich das Eintreten des Bruches nur von kleinen reversiblen Formänderungen begleitet wird, können wir von einer Struktur des Stoffes sprechen und diese im unbelasteten Zustand untersuchen. Bei großen bleibenden Deformationen wird sich jedoch die Struktur des Stoffes selbst während der Deformation ändern und die Untersuchung der Brucheigenschaften wird wesentlich erschwert.

Bei harten Kunststoffen haben wir es bei normaler Temperatur größtenteils mit einem Übergangstypus zu tun, der zwar nicht ideal spröde ist aber in vielen Fällen vom spröden Verhalten nicht so weit abweicht, daß die nachfolgenden Betrachtungen nicht anzuwenden wären.

§ 16. Statistische Aspekte des Bruchproblems.

a) Die Bruchspannungsverteilung.

Wir haben bis jetzt die Bruchspannung σ – abgesehen von den Einflüssen der Spannungsgeometrie und der Zeit – als eine Materialkonstante angesehen in dem Sinne, daß eine große Anzahl identischer Probestücke unter gleichen Bedingungen stets denselben Festigkeitswert zeigt. Nun ist es gerade ein typischer Charakterzug aller Festigkeitseigenschaften, daß sie eine sehr große experimentelle Streuung aufweisen. Diese Streuung kann nicht den Fehlergrenzen der Versuchsanordnung zugeschrieben werden, sondern ist als eine Materialeigenschaft anzusehen. Die Gründe hierfür sind, daß die auftretende Streuung viel größer ist, als sich nach den Fehlergrenzen der Versuchsanordnung erwarten läßt und daß sie unabhängig von der Untersuchungsmethode bei allen Festigkeitsmessungen immer wieder zutage tritt. Wenn also die Streuung der Bruchwerte als eine Materialeigenschaft angesehen werden soll, so ist es das erste Problem, diese Eigenschaft zu *messen*, und das zweite, sie molekular oder strukturell zu *deuten*. Die Frage, wie man diese Materialeigenschaft messen soll, ist nicht schwer zu beantworten. Statt in der üblichen Weise Messungen an drei oder fünf Probestücken vorzunehmen und von den Resultaten einen Mittelwert zu bilden, soll man die ganze Bruchspannungsverteilung aus Messungen an sehr vielen Probestücken bestimmen. Die exakte Form dieser Bruchspannungsverteilung ist genau so eine Materialeigenschaft wie ihr Mittelwert, und die Aufgabe des Experimentators wird dann auch die Bestimmung der ganzen Verteilungskurve, nicht etwa des Mittelwertes allein.

Die Bruchspannungsverteilung definieren wir in der üblichen Weise durch eine Verteilungsfunktion

$$p = p(\sigma), \qquad \text{(III,13)}$$

wobei $p(\sigma)\,d\sigma$ der Bruchteil aller Probestücke ist, die eine Bruchspannung zwischen σ und $\sigma + d\sigma$ aufweisen. Die Verteilungsfunktion zeigt im allgemeinen zwei charakteristische Abszissenpunkte, den Modul und die mittlere Bruchspannung. Der Modul M ist die Spannung, bei der das Maximum der Verteilungskurve liegt, M stellt also die am häufigsten vorkommende Bruchspannung dar. Die mittlere Bruchspannung $\bar{\sigma}$ ist definiert durch die Gleichung

$$\bar{\sigma} = \int_0^\infty \sigma\, p(\sigma)\, d\sigma \tag{III,14}$$

und ist angenähert gleich dem Mittelwert einer Anzahl von Messungen. Mittelwert und Modul sind nur bei besonderer, symmetrischer Form der Verteilungskurve identisch, wie z. B. bei der GAUSS-Verteilung.

Ein Maß für die Streuung ist die Standardabweichung s, definiert durch die Gleichung

$$s^2 = \int_0^\infty [\sigma - \bar{\sigma}]^2\, p(\sigma)\, d\sigma. \tag{III,15}$$

Außer der Verteilungskurve (III,13) verwendet man auch häufig die kumulative Verteilungskurve

$$P(\sigma) = \int_0^\sigma p(\sigma)\, d\sigma. \tag{III,16}$$

Die kumulative Verteilung $P(\sigma)$ stellt die Wahrscheinlichkeit dar, daß ein Probestück unter einer Spannung bricht, die kleiner oder gleich σ ist. Anders ausgedrückt, $P(\sigma)$ ist der Bruchteil aller Probestücke mit einer Bruchspannung zwischen 0 und σ.

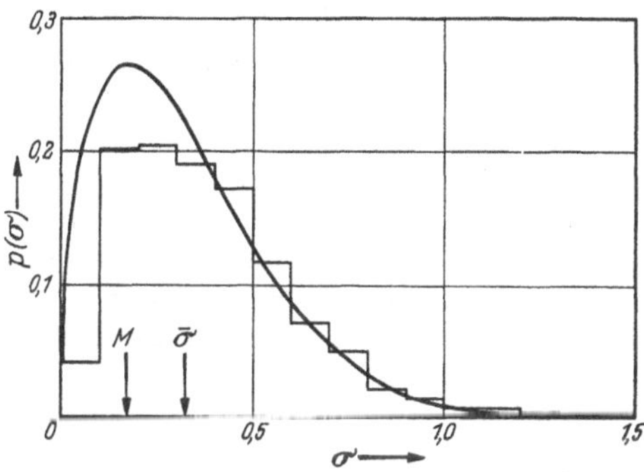

Abb. III,11. Experimentelle und theoretische Bruchspannungsverteilungskurve von 1000 Baumwollfasern im Zugversuch. (Nach KOSHAL und TURNER.) [KOSHAL, R. S., A. J. TURNER: J. Text. Inst. 21, 325 (1930).]

Wir betrachten eine Bruchspannungsverteilung, gemessen an 1000 Baumwollfasern im Zugversuch[1]. Abb. III,11 gibt die Bruchspannungsverteilung. Die Stufenkurve stellt die experimentelle Verteilungsfunktion dar, wobei der Flächeninhalt jeder Stufe gleich ist der relativen Anzahl Fasern mit Bruchspannungen zwischen den Abszissengrenzen der Stufe. Als Stufengruppen sind gewählt Bruchspannungen zwischen 0,0 und 0,1, zwischen 0,1 und 0,2 usw. Die glatte Kurve ist die theoretische Verteilungskurve, die im vorliegenden Fall die von WEIBULL vorgeschlagene Form hat (vgl. Abschnitt c):

$$p(\sigma) \sim \left[\frac{\sigma}{3{,}77}\right]^{0{,}53} e^{-[\sigma/3{,}77]^{1{,}53}}.$$

Wie man sieht, ist die Bruchspannungsverteilung schief. Der größere Teil der gemessenen Bruchspannungen liegt rechts vom Modul. Daraus folgt, daß auch der Modul M und der Mittelwert $\bar{\sigma}$ nicht zusammenfallen, der Modul ist kleiner als der Mittelwert.

[1] KOSHAL, R. S. u. A. J. TURNER: J. Textile Inst. 21, 325 (1930).

Abb. III,12 zeigt die kumulative Verteilungsfunktion, wobei die Stufenhöhe die relative Anzahl Fasern mit einer Bruchspannung kleiner als die mittlere Abszisse der Stufe wiedergibt. Die theoretische kumulative Verteilungskurve hat die Form

$$P(\sigma) = 1 - e^{-[\sigma/3{,}77]^{1{,}53}}.$$

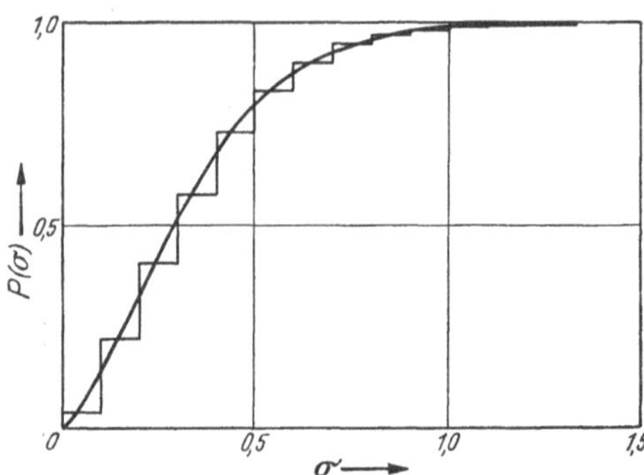

Abb. III,12. Experimentelle und theoretische kumulative Bruchspannungsverteilungskurve von 1000 Baumwollfasern im Zugversuch. (Nach KOSHAL und TURNER.) [KOSHAL, R. S., A. J. TURNER: J. Text. Inst. 21, 325 (1930).]

Ein anderes Beispiel einer Verteilungskurve findet man in Abb. IV,9 für die Zugfestigkeiten von Naturkautschuk. Auch die Kurve ist schief, aber nach der entgegengesetzten Richtung. Der größere Teil der gemessenen Bruchspannungen liegt dort links vom Modul. Dementsprechend ist dort der Modul größer als der Mittelwert.

b) Form- und Volumenabhängigkeit der Bruchspannung.

Nachdem die „Bruchspannung" bei tiefergehender Betrachtung fast alle Züge einer makroskopischen Materialeigenschaft verloren hat (Abhängigkeit von der Spannungsgeometrie, von der Deformationsvorgeschichte, Darstellung durch eine Verteilungsfunktion), verliert sie nun auch den Charakter einer spezifischen Größe, nämlich eine Konstante zu sein, die man auf die Volumeneinheit des betrachteten Probestückes beziehen kann. Die Bruchspannung hängt ab von den Dimensionen des betrachteten Probestückes im einzelnen, sowie auch vom totalen Volumen des Probestückes.

Bei Probestücken, deren Dimensionen alle von derselben Größenordnung sind, ist die Bruchspannung vom Volumen V abhängig. HIGUSHI und Mitarbeiter[1] konnten die Volumenabhängigkeit der Zugfestigkeit für natürlichen Kautschuk und Buna zeigen. Der Mittelwert der Bruchspannung $\bar{\sigma}$ (Mittel aus 16 Versuchen) wurde mit zunehmendem Volumen kleiner nach der Gleichung

$$\bar{\sigma} = a - b \log V, \qquad \text{(III,17)}$$

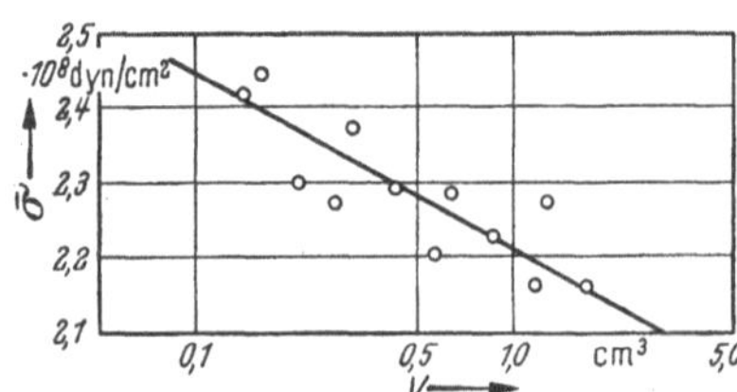

Abb. III,13. Abhängigkeit der mittleren Bruchspannung $\bar{\sigma}$ vom Volumen V für Buna. (Nach HIGUSHI.) [HIGUSHI, T., H. M. LEEPER u. D. S. DAVIS: Analyt. Chemistry 20, 1029 (148).]

während die Standardabweichung ungefähr unabhängig vom Volumen war. Abb. III,13 zeigt ihre Daten an Buna.

Am stärksten wird die Abhängigkeit der Bruchspannung von den Dimensionen, wenn eine der Dimensionen des Probestückes sehr klein

[1] HIGUSHI, T., H. M. LEEPER u. D. S. DAVIS: Analyt. Chemistry 20, 1029 (1948).

wird gegen die übrigen, wie das z. B. bei Fasern der Fall ist. Bei Textilfasern ist allgemein bekannt, daß die Festigkeit mit abnehmendem Faserdurchmesser zunimmt und mit zunehmender Faserlänge abnimmt.

Ein besonders eindrucksvolles Beispiel bilden die kombinierten Versuche von GRIFFITH[1] und GOODING[2] an gezogenen Glasfasern von sehr stark variierenden Durchmessern (Abb. III, 14). Während GOODING mit Durchmessern von 0,1 bis 1 mm arbeitet, untersuchte GRIFFITH Fasern bis zu wenigen μ dick.

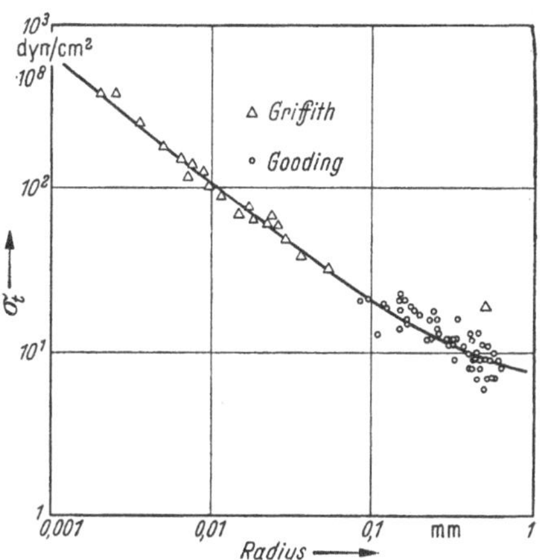

Abb. III, 14. Die Abhängigkeit der Bruchspannung σ_t beim Zugversuch von Glasfasern vom Faserhalbmesser: Dreiecke [GRIFFITH, A. A.: Phil. Trans. London A **221**, 163 (1920)], Punkte [GOODING, E. J.: J. Soc. Glass. Tech. **16**, 145 (1932)].

Die Bruchspannung steigt von der technischen Zerreißfestigkeit von weniger als 10^9 dyn/cm² bis zu $4 \cdot 10^{10}$ dyn/cm² bei sehr dünnen Fasern. Bei sehr kleinen Durchmessern erreicht die Bruchspannung die Größenordnung der „molekularen Zerreißfestigkeit" σ_M (vgl. § 17a). Daraus muß man schließen, daß, mit verursacht durch die Wirkung des Warmziehens, Glasfasern von wenigen μ Dicke praktisch keine Kerbstellen mehr enthalten. (Vergleiche die Ausführungen in § 17.) Der steile Abfall zu den gewöhnlichen technischen Zerreißfestigkeiten ist bei etwa 100-mal größeren Durchmessern beendet. Diese Versuche bilden eine der stärksten experimentellen Stützen für die Schwachstellenhypothese (§ 17a).

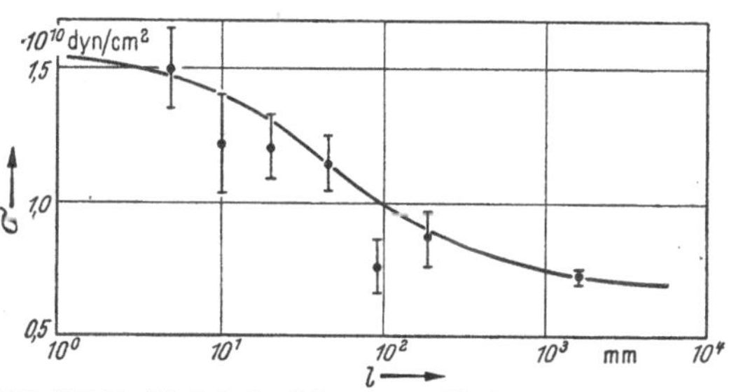

Abb. III, 15. Einfluß der Länge l von Glasfasern auf die Zugfestigkeit σ. (Nach ANDEREGG.) [ANDEREGG, F.O.: Ind. Eng. Chem. **31**, 290 (1939).]

ANDEREGG[3] demonstrierte den Einfluß der Faserlänge auf die Festigkeit von Glasfasern (Dicke 13 μ). Wie man aus Abb. III, 15 ersieht, nimmt hier die Festigkeit mit dem Logarithmus der Länge ab.

c) Statistische Bruchtheorien.

Haben wir also oben gezeigt, wie man die Materialeigenschaft der Bruchspannungsverteilung messen kann, so müssen wir uns jetzt noch mit der Frage nach der Deutung dieser Verteilung beschäftigen. Schon

[1] GRIFFITH, A. A.: Philos. Trans. Roy. Soc. London **A 221**, 163 (1920).
[2] GOODING, E. J.: J. Soc. Glass. Techn. **16**, 145 (1932).
[3] ANDEREGG, F. O.: Ind. Engng. Chem. **31**, 290 (1939).

mit Hinsicht auf die Tatsache, daß die Resultate von Bruchspannungsmessungen in ganz eigenartiger Weise vom „Zufall" beherrscht werden, ist man geneigt, die Bruchspannung zu den strukturempfindlichen Eigenschaften im SMEKALschen Sinne zu rechnen. Das sind Eigenschaften, die ganz oder größtenteils von kleinen lokalen Einzelheiten der sog. *Mikrostruktur*, wie Fehl- und Lockerstellen sowie Inhomogenitäten (s. S. 186), abhängen. Jedoch genügt diese Annahme nicht, um die eigenartige Statistik der Resultate von Bruchspannungsmessungen zu erklären. Denn es gibt viele Stoffeigenschaften, deren Strukturempfindlichkeit nicht angezweifelt werden kann, wie z. B. Fluorescenz, Halbleitung und ähnliche, deren Messung aber nicht die obenerwähnten statistischen Schwierigkeiten aufweist. Es muß also zu dem Merkmal der Strukturempfindlichkeit im Falle der Bruchspannung noch etwas Besonderes hinzu kommen. Dieses Besondere ist, wie wir unten sehen werden, die Tatsache, daß die Bruchspannung nicht von der Summe oder von einem Mittelwert von Eigenschaften der strukturellen Sonderstellen abhängt, sondern ausschließlich von einer einzigen Sonderstelle und dann noch von derjenigen, die in der ganzen Reihe der Sonderstellen das Extremum darstellt. Es ist von vornherein deutlich, daß ein solcher Umstand die Statistik der Meßresultate ganz beträchtlich beeinflussen muß.

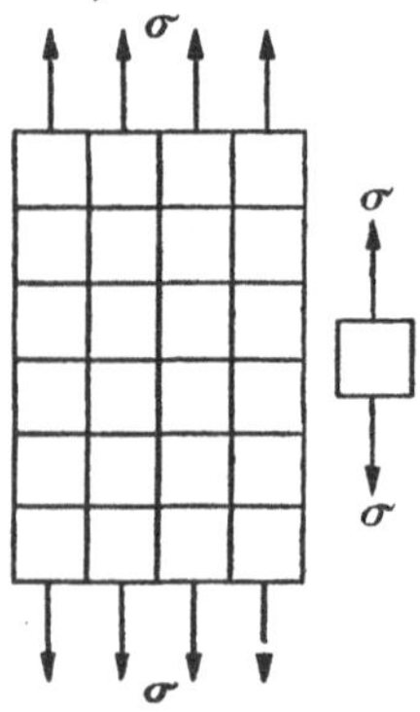

Abb. III, 16. Zur Definition der Mikrostruktur mit Hilfe der Elementarvolumina.

Die *statistische Bruchtheorie*[1] wurde hauptsächlich von WEIBULL[2] entwickelt und von anderen Autoren weiter ausgearbeitet[3]. Sie nimmt an, daß die lokale Festigkeit eines Probestückes von Ort zu Ort verschieden ist und behandelt das Problem, wie man aus der lokalen Festigkeitsverteilung auf die Festigkeit des Probestückes schließen kann.

Wir denken uns das Probestück zerlegt in eine große Anzahl von kleinen Volumelementen, die alle das gleiche Volumen V_0 besitzen (Abb. III, 16). Ein Probestück vom Volumen V enthält dann V/V_0 dieser Volumelemente. Die Größe der Volumelemente, V_0, muß geeignet gewählt sein. V_0 muß groß genug sein, so daß jeder der auftretenden Risse der Mikrostruktur vollständig in einem Volumelement enthalten ist. Andrerseits muß V_0 klein genug sein, um für makroskopische Betrachtungen als infinitesimales Volumelement dienen zu können.

Wir unterwerfen nun nicht ein Probestück, sondern eine sehr große Anzahl Probestücke vom gleichen Material der gleichen Zerlegung und erhalten eine sehr große (theoretisch unendlich große) Zahl von Volumelementen. Wir prüfen jedes der Volumelemente unter einem einfachen

[1] PEIRCE, F. T.: J. Textile Inst. **17**, 355 (1926).

[2] WEIBULL, W.: Ing. Vetenskaps Akad. Handl. No. 151 (1939), No. 153 (1939); Kungl. Tekniska Högskolans Handl. No. 27 (1949); Appl. Mech. Rev. (Nov. 1952), **449**.

[3] KONTOROVA, T.A.: J. techn. Physics USSR **10**, 886 (1940). – J. I. FRENKEL u. T. A. KONTOROVA: J. Phys. USSR **7**, 108 (1943). – N. DAVIDENKOW, E. SHEVANDIN u. F. WITTMAN: J. appl. Mechan. **14**, A 63 (1947). – B. EPSTEIN: J. appl. Physics **19**, 140 (1948). – W. GEORGE: Text. Res. J. **21**, 847 (1951).

Spannungszustand – z. B. einer Zugspannung σ – auf seine Festigkeit und erhalten eine Verteilung von Bruchspannungen

$$w = w(\sigma), \tag{III,18}$$

wobei $w(\sigma)d\sigma$ die relative Anzahl der Volumelemente ist mit einer Stärke zwischen σ und $\sigma + d\sigma$. Gleichzeitig ist $w(\sigma)d\sigma$ auch die Wahrscheinlichkeit, daß ein willkürlich herausgegriffenes Elementarvolumen V_0 eine Festigkeit zwischen σ und $\sigma + d\sigma$ besitzt. Die kumulative Verteilungsfunktion

$$W(\sigma) = \int_0^\sigma w(\sigma)\, d\sigma \tag{III,19}$$

gibt die Wahrscheinlichkeit, daß ein willkürlich herausgegriffenes Elementarvolumen eine Festigkeit kleiner oder gleich σ besitzt.

Die Verteilungsfunktionen $w(\sigma)$ oder $W(\sigma)$ beschreiben die Festigkeitseigenschaften des betrachteten Stoffes vollständig und wir können daher $w(\sigma)$ ansehen als die Definition der Mikrostruktur.

Für viele Zwecke ist es nützlich, Gl. (III,18) und Gl. (III,19) noch in einer etwas anderen Form zu schreiben, nämlich

$$W(\sigma) = 1 - e^{-n(\sigma)} \tag{III, 19a}$$

oder

$$w(\sigma) = \frac{d\,n(\sigma)}{d\sigma}\, e^{-n(\sigma)}. \tag{III, 18a}$$

Die Mikrostruktur läßt sich dann ebensogut beschreiben durch die Funktion $n(\sigma)$ zusammen mit der Größe des Elementarvolumens V_0. (Durchläuft σ den Bereich von 0 nach ∞, so durchläuft $n(\sigma)$ denselben Bereich.)

Wir denken uns nun mit Weibull jedes makroskopische Probestück vom Volumen V zusammengesetzt aus V/V_0 willkürlich herausgegriffenen Elementarvolumina. Wesentlich für die ganze Theorie ist nun die Annahme, daß die Festigkeit des Probestückes bestimmt wird von der Festigkeit des schwächsten in ihm enthaltenen Elementarvolumens. Dann läßt sich die Bruchspannungsverteilung der Probestücke sofort angeben.

Wir beginnen mit der einfachen Annahme, daß auch das makroskopische Probestück unter einer einfachen, überall im Probestück konstanten, Spannung σ geprüft wird. Dann ist die kumulative Bruchspannungsverteilung der Probestücke gegeben durch

$$P(\sigma) = 1 - [1 - W(\sigma)]^{V/V_0} = 1 - e^{-(V/V_0)\,n(\sigma)} \tag{III, 20}$$

und die differentielle Bruchspannungsverteilung der Probestücke wird

$$p(\sigma) = \left(\frac{V}{V_0}\right)\frac{d\,n(\sigma)}{d\sigma}\, e^{-(V/V_0)\,n(\sigma)}. \tag{III, 21}$$

Die Gültigkeit der Theorie läßt sich mit Hilfe von Formel (III,20) prüfen. Wir messen die kumulativen Bruchspannungsverteilungen für Probestücke verschiedener Volumina V_1, V_2 usw. und erhalten so[1]

$$P(\sigma,V) = 1 - \exp\left[-\frac{V}{V_0}\, n(\sigma)\right]. \tag{III, 20}$$

[1] Insbesondere sieht man, daß $P(\sigma, V_0) = W(\sigma)$, d. h. $W(\sigma)$ ist die kumulative Bruchspannungsverteilung von Probestücken der Größe des Elementarvolumens V_0 und $w(\sigma)$ ist die entsprechende differentielle Bruchspannungsverteilung: $p(\sigma, V_0) = w(\sigma)$.

Die kumulativen Bruchwahrscheinlichkeiten, bei konstanter Spannung als Funktion des Volumens V, müssen die einfache Form haben

$$-\ln[1 - P(\sigma, V)] = \frac{1}{V_0} n(\sigma) \cdot V. \qquad \text{(III, 22)}$$

Aus Gl. (III, 22) läßt sich dann für jedes σ der Wert von $n(\sigma)/V_0$ bestimmen, nicht jedoch die Werte von $1/V_0$ und $n(\sigma)$ getrennt. Will man die Bruchspannungsverteilung für beliebige Probestücke berechnen, so ist die Größe $n(\sigma)/V_0$ ausreichend. Will man jedoch die Mikrostruktur charakterisieren, so muß man $1/V_0$ auf unabhängige Weise bestimmen.

Gl. (III, 20) erklärt:

1. Die Existenz einer Bruchspannungsverteilung für Probestücke konstanten Volumens V.

2. Die Abhängigkeit der Bruchspannungsverteilung vom Volumen der Probestücke. Eine Abhängigkeit von den Dimensionen, die sich nicht als Volumenabhängigkeit ausdrücken läßt, muß auf Oberflächeneffekte zurückgeführt werden und ist mit Hilfe von Gl. (III, 20) nicht zu verstehen.

Epstein[1] hat bestimmte Verteilungen für die Mikrostruktur – d.h. bestimmte Formen von $w(\sigma)$ – angenommen und die daraus resultierenden Bruchspannungsverteilungen $p(\sigma)$ berechnet. Eine rechtwinkelige oder Cauchy-Verteilung für $w(\sigma)$ geben physikalisch unwahrscheinliche Resultate. Tabelle III, 2 enthält drei mögliche Verteilungsfunktionen $w(\sigma)$, nämlich die Verteilung von Laplace, die Verteilung von Gauss und eine von Weibull vorgeschlagene Verteilungsfunktion, sowie den Modul M, den Mittelwert $\bar{\sigma}$ und die Standardabweichung s der resultierenden Bruchspannungsverteilungen für Probestücke des Volumens V.

Tabelle III, 2.
Verteilungen für die Mikrostruktur und die Parameter der resultierenden Bruchspannungsverteilung.

Verteilung der Mikrostruktur	Resultierende Bruchspannungsverteilung*		
	Modul M	Mittelwert $\bar{\sigma}$	Standardabweichung s
1. Laplace $w(\sigma) = \frac{1}{2\sigma_0} e^{-\lvert\sigma/\sigma_0 - \alpha\rvert}$	$\sigma_0[\alpha - \log(V/2V_0)]$		$\frac{\pi}{\sqrt{6}}\sigma_0$
2. Gauss $w(\sigma) = \frac{1}{\sqrt{2\pi}\sigma_0} e^{-[(\sigma/\sigma_0)-\alpha]^2}$	$\sigma_0\left[\alpha - \sqrt{2\log(V/V_0)} + \frac{\log\log(V/V_0) + \log 4\pi}{2\sqrt{2\log(V/V_0)}}\right]$		$\frac{\pi\sigma_0}{2\sqrt{3}} \frac{1}{\sqrt{\log(V/V_0)}}$
3. Weibull $w(\sigma) = \frac{\beta}{\sigma_0}(\sigma/\sigma_0)^{\beta-1} e^{-(\sigma/\sigma_0)^\beta}$	$[1 - 1/\beta]^{1/\beta}\sigma_0\left(\frac{V_0}{V}\right)^{1/\beta}$	$\sigma_0\left(\frac{V_0}{V}\right)^{1/\beta} I_\beta$	$\sigma_0\left(\frac{V_0}{V}\right)^{1/2\beta}[I_{\beta/2} - I_\beta^2]$

[1] Siehe S. 180.

* Nur gültig für $V/V_0 \gg 1$.

I_β ist eine numerische Konstante mit den Werten:

β	1	2	3	4	8	16	∞
I_β	1,000	0,886	0,896	0,908	0,940	0,965	1,000

Abb. III, 17 zeigt die Verteilung der Mikrostruktur nach LAPLACE mit einem Parameter $\alpha = 20$ (die Kurve mit $V/V_0 = 1$) und die zugehörigen Bruchspannungsverteilungen für Probestücke des Volumens $V/V_0 = 10$, 100, 1000. Die Verteilungsfunktion der Mikrostruktur ist symmetrisch und hat ihr Maximum bei $\sigma/\sigma_0 = \alpha$. Die Bruchspannungsverteilung eines Probestückes mit einem Volumen größer als V_0 ist stark asymmetrisch nach links. Für zunehmende Volumina der Probestücke verschiebt sich die ganze Verteilungsfunktion nach kleineren Spannungen, ohne ihre Form weiter zu ändern. Der Modul und der Mittelwert der Verteilung nehmen daher mit zunehmendem Volumen ab, die Standardabweichung und die „Schiefe" bleiben konstant.

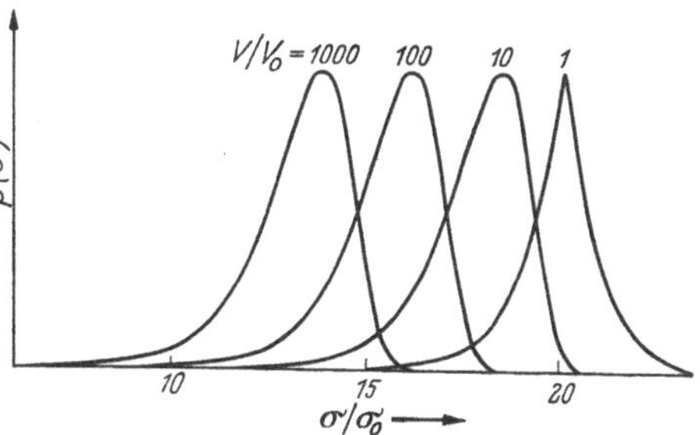

Abb. III, 17. Die Verteilungsfunktion der Mikrostruktur nach LAPLACE ($V/V_0 = 1$) und die Bruchspannungsverteilungen von Probestücken des relativen Volumens $V/V_0 = 10$, 100, 1000.

Abb. III, 18 zeigt die Verteilung der Mikrostruktur nach GAUSS mit dem Parameter $\alpha = 20$ und die zugehörigen Bruchspannungsverteilungen für Probestücke des Volumens $V/V_0 = 10$, 100, 1000. Die Verteilungsfunktion der Mikrostruktur ist wieder symmetrisch mit einem Maximum bei $\sigma/\sigma_0 = \alpha$. Die Bruchspannungsverteilungen der Probestücke sind asymmetrisch nach links und werden mit zunehmendem Volumen immer stärker asymmetrisch. Die Verteilungen verschieben sich nach links und werden schmäler. Daher werden sowohl der Modul als auch die Standardabweichung mit zunehmendem Volumen der Probestücke kleiner.

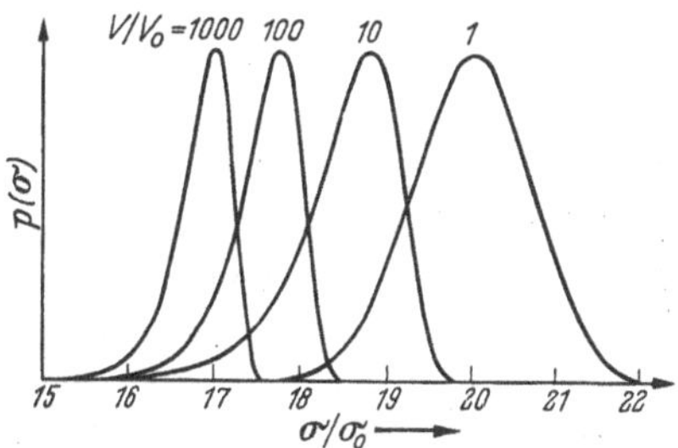

Abb. III, 18. Die Verteilungsfunktion der Mikrostruktur nach GAUSS ($V/V_0 = 1$) und die Bruchspannungsverteilungen von Probestücken des relativen Volumens $V/V_0 = 10, 100, 1000$.

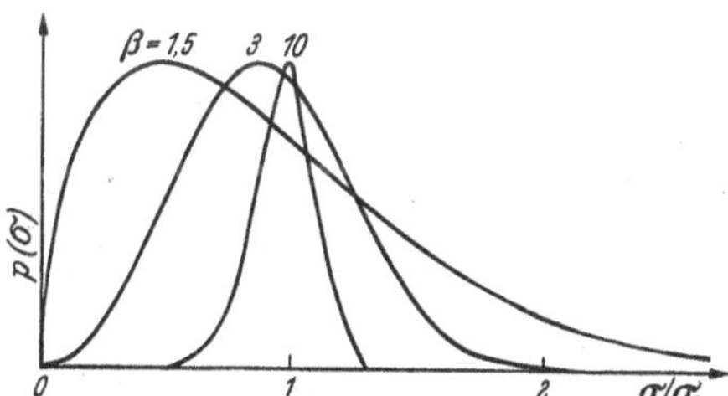

Abb. III, 19. Die Verteilungsfunktion der Mikrostruktur nach WEIBULL für verschiedene Werte des Parameters β: $\beta = 10$ asymmetrisch nach links, $\beta = 3$ annähernd symmetrisch, $\beta = 1,5$ asymmetrisch nach rechts.

Betrachten wir schließlich die WEIBULL-Verteilung. Diese kann sowohl asymmetrisch nach links, wie auch asymmetrisch nach rechts, wie auch symmetrisch sein, je nach dem Werte des Parameters β (vgl. Abb. III, 19).

In Abb. III,20 ist die Verteilung der Mikrostruktur nach WEIBULL mit einem Parameter $\beta = 3$ zusammen mit den zugehörigen Bruchspannungsverteilungen wiedergegeben. Man sieht, daß die Verteilungsfunktion der Mikrostruktur schwach asymmetrisch nach rechts ist. Mit zunehmendem Volumen wird diese Asymmetrie schwächer, die Kurve verschiebt sich nach links und wird sehr schnell schmäler. Daraus folgt wieder, daß der Modul und die Standardabweichung mit zunehmendem Volumen sehr schnell kleiner werden.

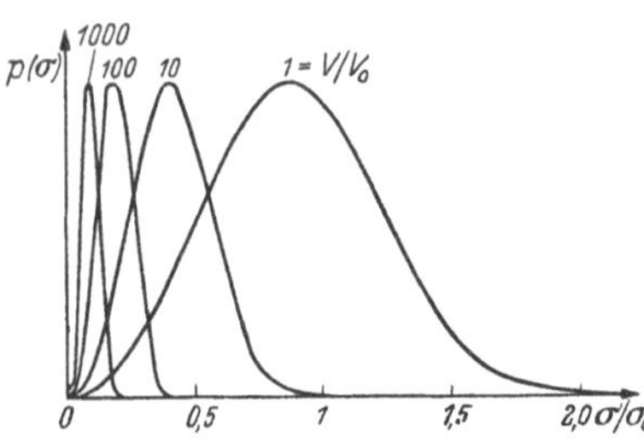

Abb. III,20. Die Verteilungsfunktion der Mikrostruktur nach WEIBULL mit $\beta = 3$ ($V/V_0 = 1$) und die zugehörigen Bruchspannungsverteilungen von Probestücken des relativen Volumens $V/V_0 = 10, 100, 1000$.

Die Wirkung zunehmenden Volumens der Probestücke hängt in den Einzelheiten ab von der Form der ursprünglichen Verteilung. Als allgemeine Tendenz können wir jedoch erwähnen: Der Modul und der Mittelwert nehmen mit zunehmendem Volumen ab und die Verteilungskurven werden mit zunehmendem Volumen mehr und mehr asymmetrisch nach kleinen Spannungen zu.

WEIBULL hat gezeigt, daß sich die Zugspannungsverteilungen von Textilfasern in vielen Fällen durch die einfache Form 3 in Tabelle III,2 darstellen lassen. So ist z.B. auch die von uns in Abb. III,11 gezeigte Verteilung eine WEIBULL-Verteilung. KASE[1] gibt die Bruchspannungsverteilung von natürlichem Kautschuk und Buna. Seine Data lassen sich beschreiben mit einer WEIBULL-Verteilung mit Parameter $\beta = 16{,}6$. (Die Bruchspannungsverteilung ist in Abb. IV,9 wiedergegeben.)

Wir gehen nun über auf den allgemeineren Fall, daß die Spannungsverteilung σ in den makroskopischen Probestücken nicht konstant ist, sondern auf bekannte Weise von den Koordinaten (xyz) abhängt: $\sigma = \sigma(x, y, z)$. WEIBULL hat gezeigt, daß die kumulative Bruchwahrscheinlichkeit dann gegeben ist durch den Ausdruck

$$P(\sigma) = 1 - \exp\left\{-\frac{1}{V_0}\int_V n[\sigma(x,y,z)]\,dx\,dy\,dz\right\}, \qquad \text{(III,23)}$$

wobei das Integral zu erstrecken ist über das Volumen des Probestückes.

Die Gl. (III,23) erklärt, warum die Biegefestigkeit f_b eines Materials im allgemeinen höher liegt als die Zugfestigkeit f_t und die Torsionsfestigkeit f_{tor} höher liegt als die Scherfestigkeit f_s. Beim Zugversuch wird das gesamte Volumen des Probestückes der maximalen Spannung σ ausgesetzt, während bei der Biegung nur ein kleiner Teil des Volumens die maximalen Zugspannungen aufweist. Daher ist die Bruchwahrscheinlichkeit beim Zugversuch wesentlich höher als beim Biegeversuch, und ähnliches gilt für Scherung und Torsion:

$$f_b > f_t \quad \text{und} \quad f_{tor} > f_s. \qquad \text{(III, 24)}$$

---

[1] KASE, S.: J. Polymer Sci. 11, 425 (1953).

Allgemein läßt sich aussagen, daß die Bruchspannung bei einer inhomogenen Spannungsverteilung stets höher liegen muß als bei der entsprechenden homogenen Spannungsverteilung.

Unter der speziellen Annahme, daß die Mikrostruktur die Verteilungsfunktion von WEIBULL besitzt (Tabelle III,2, Zeile 3), hat WEIBULL den Zusammenhang zwischen Zug und Biegungsfestigkeit[1] einerseits und zwischen Torsions- und Scherfestigkeit andererseits abgeleitet:

$$f_b/f_t = [2\,\beta + 2]^{1/\beta} \tag{III, 25}$$

$$f_{\text{tor}}/f_s = [\beta/2 + 1]^{1/\beta}; \tag{III, 26}$$

β ist der Parameter der WEIBULL-Verteilung.

Wir haben bis jetzt ausschließlich den Fall behandelt, daß die Probestücke einem eindimensionalen Spannungszustand σ unterworfen sind (σ war entweder eine reine Zugspannung oder eine reine Scherspannung). Die Theorie läßt sich verallgemeinern auf den Fall, daß die isotropen Probestücke unter einem dreidimensionalen Spannungszustand mit den Hauptspannungen σ_1, σ_2, σ_3 geprüft werden. Wir definieren dann eine kumulative Verteilungsfunktion der Mikrostruktur an Stelle von Gl. (III,19) in der Form

$$W(\sigma_1, \sigma_2, \sigma_3) = 1 - e^{-n(\sigma_1, \sigma_2, \sigma_3)}, \tag{III, 27}$$

wobei $W(\sigma_1, \sigma_2, \sigma_3)$ die Wahrscheinlichkeit ist, daß ein Elementarvolumen unter einem Spannungszustand bricht, dessen Hauptspannungen kleiner oder gleich als die drei Werte σ_1, σ_2, σ_3 sind. Dann ist die kumulative Bruchspannungsverteilung der Probestücke gegeben durch die Gleichung

$$P(\sigma_1, \sigma_2, \sigma_3) = 1 - \exp\left[-\frac{1}{V_0}\int_V n(\sigma_1, \sigma_2, \sigma_3)\,d\,x\,d\,y\,d\,z\right]. \tag{III, 28}$$

Wir wollen zum Abschluß dieses Abschnitts auf einen schwachen Punkt der statistischen Bruchtheorie hinweisen. Bei der Berechnung der Bruchwahrscheinlichkeit unter homogener Spannung wurde stets angenommen, daß alle Punkte des Probekörpers als Ausgangspunkt des Bruches gleichwahrscheinlich sind. Nun wird gerade bei Zugversuchen häufig festgestellt (siehe § 17), daß der Bruchbeginn vielfach am Rande des Probestückes gelegen ist. Für solche Fälle muß die Theorie dahingehend modifiziert werden, daß das Volumenintegral in Gl. (III,23) nicht über den ganzen Probekörper, sondern lediglich über eine gewisse Oberflächenzone der Dicke h zu erstrecken ist,

$$\int_V n(\sigma)\,d\,x\,d\,y\,d\,z = h\int_A n(\sigma)\,d\,o, \tag{III, 29}$$

A = Oberfläche des Probekörpers.

§ 17. Morphologie der Brucherscheinungen.

a) Die Schwachstellenhypothese.

Die Schwachstellenhypothese besagt, daß ein scheinbar homogener, spröder Stoff eine große Anzahl kleiner Inhomogenitäten enthält, die die Festigkeit des Materials örtlich sehr stark herabsetzen. Die Gesamtheit

[1] Angenommen ist dabei eine reine Biegung eines Probestückes mit rechteckigem Querschnitt.

dieser Inhomogenitäten, die entweder Fehler im regelmäßigen Bau des Stoffes – worunter auch wirklich Risse und Kerbstellen oder auch Fremdeinschlüsse sein können – nennen wir die *Mikrostruktur* des Stoffes. Jede Inhomogenität bewirkt eine starke Spannungskonzentration in ihrer unmittelbaren Umgebung, die sowohl von der Geometrie und Lage der Inhomogenität, wie auch von dem makroskopischen Spannungszustand abhängig ist (vgl. Abschn. b). Die „gefährlichste" Inhomogenität wird zum Ausgangspunkt des spröden Bruches.

Indirekte Beweise für die Existenz der Mikrostruktur haben wir bereits in § 16 kennengelernt. Sowohl die Existenz einer Bruchspannungsverteilung wie auch die Form und Volumenabhängigkeit dieser Verteilung lassen sich mit Hilfe der Schwachstellenhypothese erklären (§ **16c**).

Der direkte Ausgangspunkt für die Schwachstellenhypothese[1] war jedoch der große Unterschied zwischen theoretischen Festigkeiten und dem wirklich gemessenen Bruchverhalten. Die theoretische Festigkeit eines vollständig homogenen fehlerfreien harten Stoffes, die sogenannte „molekulare Zerreißfestigkeit" läßt sich abschätzen zu

$$\sigma_M \sim 10^{10} \text{ bis } 2 \cdot 10^{11} \text{ dyn/cm}, \qquad \text{(III, 30)}$$

während die Bruchspannung makroskopischer Probestücke eines spröden Materials 2 bis 3 Zehnerpotenzen niedriger liegt. Dieser große Unterschied läßt sich nur verstehen durch die Annahme, daß durch die Mikrostruktur die Festigkeit stark vermindert wird.

Smekal[2] hat die folgende Abschätzung für die Größenordnung der molekularen Zerreißfestigkeit gegeben. Die Arbeit, die zur Erzeugung oder Vergrößerung einer Oberfläche je Flächeneinheit erforderlich ist, die spezifische Oberflächenenergie α, ist für kristallisierte und amorphe Festkörper von der Größenordnung

$$\alpha \sim 10^2 \text{ bis } 10^3 \text{ erg/cm}^2. \qquad \text{(III, 31)}$$

Smekal nimmt an, daß eine Bruchfläche gebildet wird, wenn die elastische Deformationsenergie, aufgespeichert in einer Schicht von der Dicke der doppelten Reichweite der Molekularkräfte, $2\,r_0$, gleich wird der Oberflächenenergie der zu bildenden Bruchfläche, d.h. wenn[3]

$$\alpha = r_0 \cdot \sigma^2/2\,E$$

oder

$$\sigma_M \sim \sqrt{2\,E\,\alpha/r_0}\,. \qquad \text{(III, 32)}$$

Mit $r_0 \sim 5 \cdot 10^{-8}$ cm und $E \sim 10^{11}-10^{12}$ dyn/cm² ergibt sich die Größenordnung von Gl. (III, 30).

Eine andere Abschätzung für die theoretische Festigkeit eines idealisierten Polymers, bestehend aus einer Anzahl paralleler Cellulosemoleküle, hat Mark gegeben[4]. Nimmt man als Bruchursache das Zerreißen von Hauptvalenzbindungen an, so erhält man für die Größenordnung der Bruchspannung (III, 30). Nimmt man als Bruchursache das Auseinanderziehen von Molekülketten gegen die van der Waals-Kräfte, so erhält man

$$\sigma_M \sim 3 \cdot 10^9 \text{ dyn/cm}^2.$$

---

[1] Griffith, A. A.: Philos. Trans. Roy. Soc. London **A 221**, 163 (1920).

[2] Smekal, A.: Die Festigkeitseigenschaften spröder Körper, Ergebn. exakt. Naturwiss., Bd. 15, Berlin: Springer 1936.

[3] E ist der Elastizitätsmodul des betrachteten Materials.

[4] Mark, H.: High Polymers V: Cellulose and its derivatives, Interscience Publ. New York 1943, page 1001.

Die Realität der molekularen Zerreißfestigkeit σ_M konnte durch GRIFFITH experimentell bewiesen werden. Die Zugstärke sehr dünner Glas- und Quarzglasfäden kann die Werte $4 \cdot 10^{10}$ dyn/cm² aufweisen (vgl. Abb. III, 14). Das weist darauf hin, daß Glasfäden vom Durchmesser weniger μ praktisch frei von Inhomogenitäten sind.

Eine andere Bestätigung der Größenordnung von Gl. (III, 30) fand SMEKAL im Mikroritz-Versuch (siehe Abschn. f).

Schließlich ergibt sich ein direkter Beweis für die Existenz der Mikrostruktur aus der mikroskopischen Untersuchung von Bruchflächen (Abschn. d). Dabei stellt sich heraus, daß die Inhomogenitäten nicht nur als Beginnpunkt des Bruches, sondern auch bei der Ausbreitung der Bruchfront eine wesentliche Rolle spielen.

Über die Einzelheiten der Mikrostruktur für Polymere können wir heute noch kaum Aussagen machen. SMEKAL[1] hat jedoch für die Mikrostruktur von Glas die folgenden Schlüsse ziehen können:

„Der mittlere Abstand der Kerbstellen beträgt höchstens $5 \cdot 10^{-5}$ cm. Die Kerbstellendichte in massiven Glaskörpern ist groß, wobei mit einer gegenseitigen Beeinflussung der Spannungshöfe benachbarter Kerbstellen zu rechnen ist. Die untere Grenze der Abmessungen der Kerbstellen beträgt $5 \cdot 10^{-6}$ cm. Die Kerbstellenverteilung ist quasi-isotrop, wird jedoch durch starke bleibende Verformungen beeinflußt (Ziehen von Fäden). In gezogenen Fäden von ungefähr 10^{-4} cm Durchmesser findet man praktisch keine Kerbstellen mehr."

b) Die Kerbwirkung der Mikrostruktur.

Wir untersuchen nun die Kerbwirkung eines Fehlers auf ein sonst homogenes Probestück, d.h. wir betrachten ein Elementarvolumen V_0 im Sinne des § 16c, das *eine* Inhomogenität enthält. Die Wirkung der gesamten Mikrostruktur auf die Festigkeit eines Probestückes läßt sich dann aus einer großen Anzahl der hier betrachteten Elementarvolumina zusammensetzen (statistische Bruchtheorie). Die Voraussetzung hierfür ist jedoch, daß die verschiedenen Kerbstellen soweit voneinander entfernt sind, daß ihre Spannungshöfe sich gegenseitig nicht beeinflussen. Die Kerbstellendichte im Material muß also relativ klein sein. Nach dem oben Gesagten, trifft diese Voraussetzung für Gläser nur in beschränktem Maße zu.

Eine allgemeine Übersicht über die Kerbwirkung verschiedener geometrischer Typen von Kerbstellen hat SMEKAL[1] gegeben. Wir können uns daher hier auf den wichtigsten Fall beschränken, daß die Inhomogenität die Form eines langen schmalen Spaltrisses hat. Aus mathematischen Gründen geschieht die Behandlung des Problems zweidimensional, das in Wirklichkeit vorliegende dreidimensionale Problem führt jedoch zu Resultaten, die qualitativ überhaupt nicht, quantitativ nur sehr wenig von denen des zweidimensionalen Problems abweichen.

Wir betrachten ein elliptisches Loch in der unendlich ausgedehnten Ebene (Abb. III, 21). In großer Entfernung von der Kerbstelle sei die äußere

[1] Siehe S. 186, Fußnote 2.

Beanspruchung durch die zwei zueinander senkrechten Hauptspannungen σ_1 und σ_2 gegeben, das elliptische Loch mit den Halbachsen a und b habe die Form eines schmalen Risses ($b/a \ll 1$) und seine Längsachse sei um einen Winkel θ gegen die Hauptspannungsrichtung σ_1 geneigt.

Die maximalen Zugspannungen treten am Rande der Ellipse auf[1], und zwar bei langen, schmalen Ellipsen sehr nahe an den beiden Rißenden. Am Rißende selbst ist die Zugspannung R tangential an den Scheitel und hat die Größe

$$R = (\sigma_2 - \sigma_1) \cos 2\theta + \frac{a}{b} [(\sigma_1 + \sigma_2) + (\sigma_2 - \sigma_1) \cos 2\theta]. \qquad \text{(III, 33)}$$

Die Normalspannung σ_n auf die Rißebene in großer Entfernung von der Inhomogenität ist gleich dem halben Ausdruck in der eckigen Klammer wir können daher Gl. (III, 33) auch schreiben

$$R = (\sigma_2 - \sigma_1) \cos 2\theta + 2 \frac{a}{b} \sigma_n. \qquad \text{(III, 33a)}$$

In Spannungszuständen, bei denen σ_n nicht verschwindet, wird das zweite Glied in Gl. (III, 33 a) das erste in den meisten Fällen weit überwiegen ($a/b \gg 1$) und die maximale Spannungskonzentration wird im wesentlichen bestimmt durch die Größe der zur Rißebene senkrechten Normalspannung (Normalspannungsgesetze des spröden Bruches).

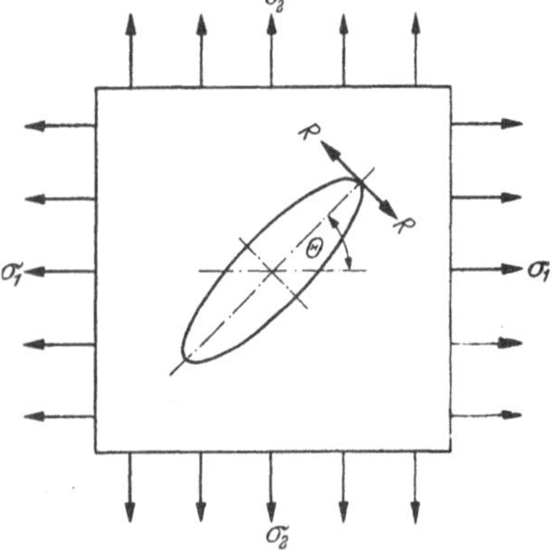

Abb. III, 21. Zur Theorie von GRIFFITH: Elliptisches Loch in der unendlich ausgedehnten Ebene.

Der wichtigste Spezialfall ist, daß das Elementarvolumen unter einer axialen Zugspannung steht ($\sigma_2 = \sigma$, $\sigma_1 = 0$). Dann wird die Randspannung als Funktion der Orientierung θ des Risses

$$R = \sigma [(1 + a/b) \cos 2\theta + a/b]. \qquad \text{(III, 34)}$$

Die maximale Spannungskonzentration wird erreicht, wenn der Riß senkrecht zur Zugspannungsrichtung orientiert ist ($\theta = 0$):

$$R_{\max} = \sigma [1 + 2a/b] \simeq \sigma \cdot 2a/b. \qquad \text{(III, 35)}$$

Bei axialer Zugspannung nennt man das Verhältnis von Maximalspannung R zur makroskopischen Zugspannung σ die *Kerbzahl* k des Risses. Diese hängt ab von der Orientierung der Rißebene und vom Achsenverhältnis

$$\frac{R_{\max}}{\sigma} = k = 1 + 2a/b \simeq 2a/b. \qquad \text{(III, 36)}$$

Obige Betrachtungen gelten auch für den geraden Spaltriß, der nicht genau die Form einer Ellipse besitzt. Dann muß man in den Formeln

[1] GRIFFITH, A. A.: Proc. Int. Congr. appl. Mechanics, Delft 1924, page 55.

(III, 33) bis (III, 36) das doppelte Achsenverhältnis $2a/b$ ersetzen durch das Verhältnis $\sqrt{2\lambda/\varrho}$, wobei $\lambda = 2a$ die Länge des Risses und $\varrho = b^2/a$ der Krümmungsradius an den Rißenden ist. Die Kerbzahl eines geraden Risses senkrecht zur Zugrichtung wird daher:

$$k = \sqrt{2\lambda/\varrho}\,. \tag{III, 37}$$

SMEKAL[1] hat nun aus Gl. (III, 34) die Bruchspannung des Elementarvolumens mit einem ellipsenförmigen Riß durch folgende Überlegung abgeleitet: Bruch tritt ein, wenn die maximale Spannung R die Größenordnung der molekularen Zerreißfestigkeit σ_M erreicht[2]:

$$R = \sqrt{\frac{4}{\pi}}\,\sigma_M\,. \tag{III, 38}$$

Man erhält für das Elementarvolumen unter axialer Zugspannung mit einem Riß senkrecht zur Zugspannungsrichtung aus Gl. (III, 36), (III, 37) und (III, 38) die Bruchspannung

$$\sigma = \frac{\sigma_M}{k}\sqrt{\frac{4}{\pi}} = \sqrt{\frac{4E\alpha}{\pi\lambda}}\sqrt{\frac{\varrho}{r_0}} \tag{III, 39}$$

oder mit Gl. (III, 37)

$$\sigma_M = \sigma\sqrt{\frac{\pi\lambda}{2\varrho}}\,. \tag{III, 40}$$

Das Verhältnis von molekularer Zerreißfestigkeit σ_M zur makroskopischen Bruchspannung σ wird gemäß Gl. (III, 39) allein bestimmt durch die Lage und Gestalt des Risses.

Diese Gleichung gilt auch noch für einen Riß, dessen Ebene gegen die Zugspannungsebene in einem Winkel θ geneigt ist. In diesem Fall muß man für k den aus Gl. (III, 34) folgenden Wert einsetzen.

In der Praxis kennt man die Gestalt und Größe der Risse der Mikrostruktur im einzelnen nicht. Trotzdem kann man durch Messung der Zugstärke das Verhältnis σ_M/σ als sogenannte *empirische Kerbzahl* experimentell bestimmen.

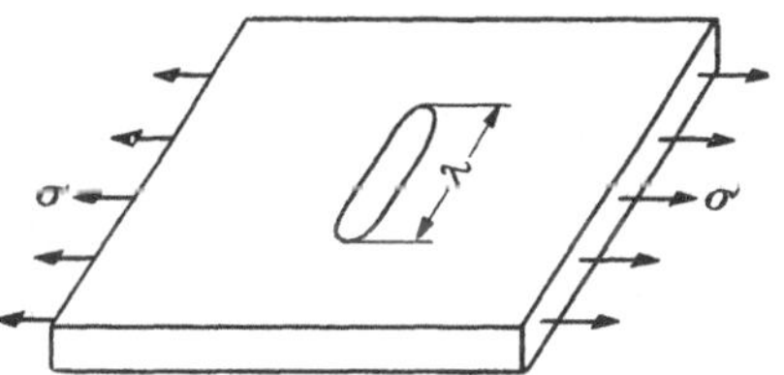

Abb. III, 22. Zur Theorie von GRIFFITH: Unendlich ausgedehnte Platte mit geradem Spaltriß unter axialer Zugspannung.

Wie GRIFFITH[3] gezeigt hat, läßt sich das vorliegende Bruchproblem für einen etwas spezielleren Fall streng lösen. GRIFFITH betrachtet eine unendlich ausgedehnte Platte der Dicke 1 unter einer axialen Zugspannung σ. Ein gerader Spaltriß der Länge λ befinde sich senkrecht zur Normalspannung (Abb. III, 22). SMEKAL hat darauf hingewiesen, daß die Breite des Spaltrisses mit $2r_0$ und sein Krümmungsradius an den Enden $\varrho = r$ anzusetzen ist.

GRIFFITH konnte die elastische Deformationsenergie bei konstanter Spannung σ und bei konstanter Rißlänge λ berechnen. Die Energiediffe-

[1] SMEKAL, A.: Z. Physik **103**, 495 (1936).

[2] Der Faktor $\sqrt{4/\pi} \cong 1{,}13$ ist eingeführt, um quantitative Übereinstimmung mit der von GRIFFITH vorgeschlagenen Theorie zu erzielen [vgl. Gl. (III, 45)].

[3] GRIFFITH, A. A.: Philos. Trans. Roy. Soc. London A **221**, 163 (1920).

renz W zwischen der elastischen Deformationsenergie einer Platte ohne Riß und einer Platte unter den gleichen äußeren Bedingungen mit Riß, ist gleich

$$W = \frac{\pi \lambda^2 \sigma^2}{4E}. \tag{III, 41}$$

Die Oberflächenenergie des Risses beträgt

$$\Omega = 2\lambda\alpha. \tag{III, 42}$$

Die Energiezufuhr, die notwendig ist, um einen Riß der Länge λ um die Längeneinheit zu verlängern, wird

$$\frac{d}{d\lambda}[\Omega - W] = 2\alpha - \frac{\pi\sigma^2}{2E}\lambda. \tag{III, 43}$$

Diese Energiezufuhr ist bei festem σ für kleine Rißlängen positiv, nimmt mit zunehmender Rißlänge linear ab und wird bei einer kritischen Rißlänge Null. Bei dieser kritischen Rißlänge wird der Riß instabil und breitet sich von selbst in seiner Ebene weiter aus, d.h. er wird zum Ausgangspunkt der Bruchfläche. Für die kritische Rißlänge bei fester Spannung σ finden wir

$$\lambda = \frac{4E\alpha}{\pi\sigma^2}. \tag{III, 44}$$

Umgekehrt existiert bei gegebener Rißlänge λ eine kritische Zugspannung σ, bei der der Riß instabil wird; diese Spannung läßt sich aus Gl. (III,44) berechnen und gibt die Bruchspannung

$$\sigma = \sqrt{4E\alpha/\pi\lambda}. \tag{III, 45}$$

Zieht man in Betracht, daß die Rechnung von GRIFFITH für den Fall $\varrho = r$ gilt, so wird für Spaltrisse der Dicke $2r_0$ die Formel (III,45) identisch mit der Formel von SMEKAL (III,39). Das rechtfertigt zugleich den Ansatz (III,38) von SMEKAL.

Nehmen wir an, daß die Risse der Mikrostruktur allgemein eine Breite $2r_0$ haben, so gilt für die empirische Kerbzahl

$$\sigma_M/\sigma = \sqrt{\frac{\pi\lambda}{2r_0}}. \tag{III, 46}$$

Die empirische Kerbzahl findet man experimentell zu 10–10^2 (vgl. § 17a). Um die Größenordnung der empirischen Kerbzahl zu erklären, müßten die größten in der Mikrostruktur auftretenden Rißlängen die Größenordnung besitzen:

$$\lambda_{\max} \sim 10^2 r_0 - 10^4 r_0 \sim 10^{-3} - 10^{-4}\,\text{cm}.$$

SMEKAL[1] hat jedoch darauf hingewiesen, daß bei übermikroskopischen Untersuchungen an Glasoberflächen und Bruchflächen keine den GRIFFITHschen Abmessungen entsprechende Risse gefunden werden. Man kommt daher zu dem Schluß, daß die Mikrostruktur im unbelasteten Zu-

[1] SMEKAL, A.: Österr. Ing.-Archiv 7, 49 (1953).

stand noch feiner sein muß, als GRIFFITH ursprünglich angenommen hat. Erst unter Belastung können die Kerbstellen der Mikrostruktur unter Mitwirkung der Wärmeenergie auf die kritische Größe des GRIFFITH-Risses anwachsen und zum Ausgangspunkt der Bruchfläche werden.

c) Thermischer Mechanismus des Bruchbeginnes.

Wir haben bis jetzt angenommen, daß der Bruchvorgang bei der Kerbstelle mit optimaler Kerbwirkung beginnt, sobald die elastische lokale Energiekonzentration größenordnungsmäßig gleich wird der Oberflächenenergie α [Gl. (III,39) oder (III,46)]. Bei dieser Energiebilanz haben wir die Wärmebewegung der Molekularbausteine vollständig außer acht gelassen. Nun muß die Wärmebewegung beim Bruchvorgang wesentlich beteiligt sein, wie man aus der starken Abhängigkeit der Bruchzeit von der Temperatur bei Dauerstandversuchen schließen kann (vgl. § 15b, Abb. III,7 und III,8).

SMEKAL hat vorgeschlagen, beim Bruchvorgang zwei Phasen zu unterscheiden, eine *thermische Beginnphase* und eine *athermische Endphase* des Bruchprozesses. Wir erläutern den Gedankengang SMEKALs an Hand von Abb. III,23, die schematisch die Gestalt der Bruchfläche eines Zugstabes und die Fortpflanzungsgeschwindigkeit des Bruchvorganges wiedergibt. Auf der Bruchfläche unterscheidet man die primäre Kerbstelle K, von der aus die Bruchfront nach allen Richtungen radial fortschreitet. Die primäre Kerbstelle ist umgeben von einer halbkreisförmig begrenzten glatten Fläche, dem Spiegel, der seinerseits in die rauhe Furchungsfläche mündet (vgl. dazu die ausführlichen Erörterungen in Abschn. d). Über der Bruchfläche ist die von SMEKAL gemessene Fortpflanzungsgeschwindigkeit der Bruchfront aufgetragen als Funktion des Bruchweges, d.h. des Abstandes, den die Bruchfront von K aus bereits zurückgelegt hat.

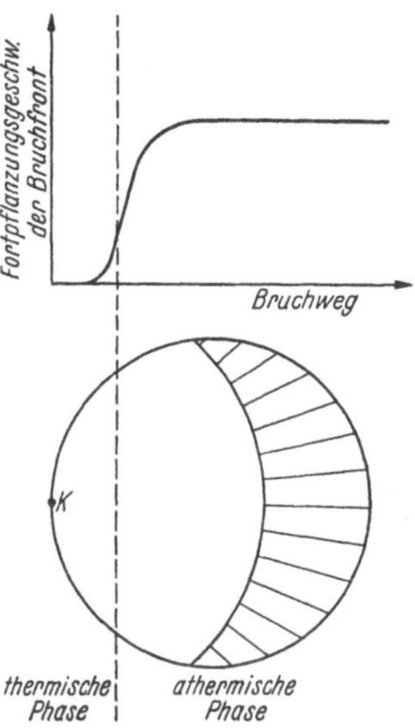

Abb. III,23. Schematische Darstellung der Ausbreitung des Bruchvorganges nach SMEKAL: Gestalt der Bruchfläche und Ausbreitungsgeschwindigkeit der Bruchfront.

Nach SMEKAL ist die Wärmebewegung gerade beim ersten Beginn des Bruches wesentlich beteiligt, während sie bei der weiteren Fortpflanzung der Bruchfront keine Rolle mehr spielt. SMEKAL nennt daher den mit kleiner Geschwindigkeit durchlaufenen Teil des Bruchweges die thermische Bruchphase, den Rest des Bruchvorganges die athermische Bruchphase.

Betrachten wir die Gestalt der Bruchfläche, so müssen wir schließen, daß nur der Teil, der sich in der unmittelbaren Umgebung der optimalen Kerbstelle befindet, mit dem thermischen Teil des Bruchvorganges korrespondiert. Der Rest der Bruchfläche entsteht mit sehr großer Bruchgeschwindigkeit, d.h. athermisch. Daraus erklärt sich, daß die Gestalt der Bruchfläche in weitem Maße unabhängig ist von der Temperatur und der Belastungsgeschwindigkeit.

Betrachten wir jedoch den Bruchvorgang in der Zeit, so nimmt der thermische Bruchbeginn praktisch die ganze erforderliche Bruchzeit in Anspruch. Ist der Bruchbeginn einmal eingeleitet, so pflanzt sich die weitere athermische Bruchwelle mit sehr großer Geschwindigkeit (mit der Größenordnung der Schallgeschwindigkeit) durch den Restquerschnitt fort. Hieraus folgt, daß die Bruchzeit bestimmt wird von der Dauer des Bruchbeginnes. Da andererseits die Bruchzeit stark von der Temperatur abhängig ist, folgt, daß auch der Bruchbeginn stark temperaturabhängig ist, d.h. durch einen thermischen Mechanismus dargestellt werden muß.

Der Mechanismus des thermischen Bruchbeginns sollte eine Erklärung geben für die folgenden Erscheinungen:

1. Die Abhängigkeit der Bruchzeit von der Spannung bei Dauerstandversuchen (§ 15b).

2. Die Temperaturabhängigkeit des Brucheintrittes und die Existenz einer Aktivierungsenergie für Brucherscheinungen.

Für den Mechanismus des Bruchbeginns wurden verschiedene Theorien vorgeschlagen, die auf verschiedene Formen der Bruchspannungs-Zeit-Relation führen. Da alle diese Theorien jedoch mehrere anpassungsfähige Konstante enthalten, ist eine experimentelle Entscheidung zwischen diesen Theorien noch nicht möglich gewesen[1].

Wir wollen hier lediglich auf ein schematisiertes Bild von STUART und ANDERSON[2] hinweisen, das uns in großen Zügen ein Verständnis der Zeitabhängigkeit der Brucherscheinungen liefern kann.

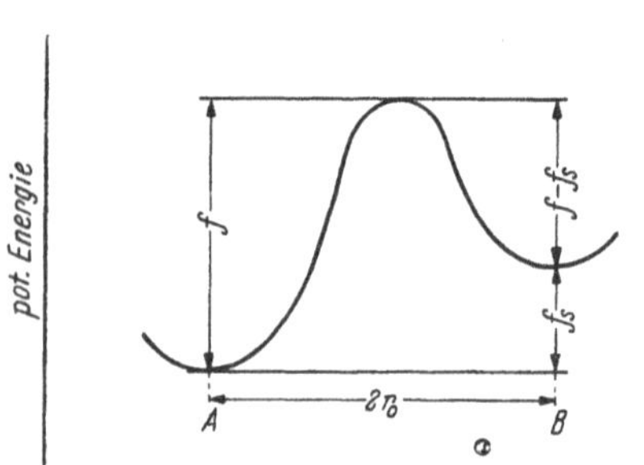

Abb. III,24. Potentialverlauf zweier Molekülbausteine als Funktion ihres Abstandes im spannungsfreien Fall. (Nach STUART und ANDERSON.)

Wir betrachten zwei Molekularbausteine an den Enden einer optimalen Kerbstelle. Diese Molekularbausteine können gegeneinander eine relative Bewegung ausführen, wodurch die Kerbstelle um ein Stück erweitert wird. In Stand A (siehe Abb. III,24) bilden die Bausteine eine normale Bindung und die Kerbstelle endet gerade vor den betrachteten Molekülteilen. In Stand B wird die Bindung auseinandergerissen und die beiden Bausteine sind die äußersten Teile der freien Oberfläche des Risses, der nun um eine Bindung länger geworden ist. Abb. III,24 gibt schematisch die potentielle Energie eines dieser Bausteine als Funktion der relativen Abstandes, wenn keine äußere Span-

[1] Für eine gute Übersicht über diese Theorien verweisen wir auf einen zusammenfassenden Artikel: O. L. ANDERSON u. D. A. STUART: Ind. Engng. Chem. **46**, 154 (1954).

[2] STUART, D. A. u. O. L. ANDERSON: J. Amer. ceram. Soc. **36**, 416 (1953).

nung wirksam ist. Die beiden Potentialmulden A und B sind um einen Abstand voneinander entfernt, der gleichgesetzt ist $2r_0$, der Breite des GRIFFITH-Risses. Lage A stellt ein absolutes Minimum dar, während die Energie in Lage B um einen Betrag f_s höher ist, der gleich ist der Oberflächenenergie je Bindung

$$f_s = \alpha/N.$$

N ist die Anzahl Bindungen je Flächeneinheit. Zwischen Lage A und B befindet sich ein Potentialberg der Höhe f (Aktivierungsenergie des Bruchprozesses je Bindung). Im spannungsfreien Fall ist die Aktivierungsenergie für den Sprung $B \to A$ niedriger (nämlich $f - f_s$) als für den Sprung $A \to B$. Somit wird die Wärmebewegung eher die Tendenz haben, den Riß zu schließen, als ihn zu erweitern.

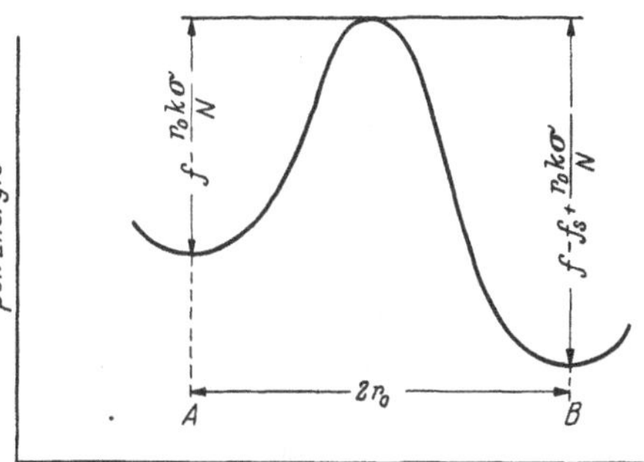

Abb. III, 25. Potentialverlauf zweier Molekülbausteine als Funktion ihres Abstandes bei Anwesenheit einer makroskopischen Spannung σ. (Nach STUART und ANDERSON.)

Bei Anlegen einer makroskopischen Spannung σ wird jedoch die Form der Potentialkurve verändert (Abb. III, 25). STUART und ANDERSON nehmen an, daß die linke Potentialmulde um einen Energiebetrag $r_0 k \sigma/N$ gehoben, die rechte Potentialmulde um denselben Betrag gesenkt wird. $k\sigma/N$ ist die maximale Spannung bei der Kerbstelle je molekulare Bindung. Wenn die rechte Potentialmulde tiefer liegt als die linke, so tritt gemittelt ein Wachsen des Risses ein, bis nach einer gewissen Zeit t_b der Bruchprozeß übergeht in den athermischen Mechanismus. Die Bruchzeit t_b wird dabei bestimmt von der Aktivierungsenergie f, der Temperatur T und der Spannung σ.

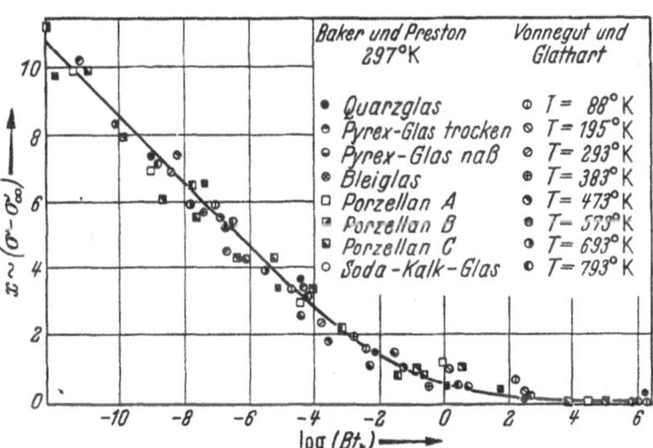

Abb. III, 26. Die theoretische Bruchspannungs-Zeit Relation nach der Theorie von STUART und ANDERSON und die Lage von Meßpunkten an verschiedenen Gläsern. Experimentelle Data: BAKER und PRESTON: J. Applied Phys. **17**, 174 (1946). VONNEGUT und GLATHART: J. Applied Phys. **17**, 1084 (1946).

Die Theorie erklärt die Existenz einer Langzeitgrenze der Dauerstandfestigkeit σ_∞. Ist die Spannung σ kleiner als σ_∞, so tritt kein Bruch auf. σ_∞ ist dabei die Spannung, bei der die beiden Potentialmulden gerade gleich tief sind:

$$\sigma_\infty = \alpha/2r_0 k. \tag{III,47}$$

Ist die Spannung größer als σ_∞, so tritt nach einer gewissen Zeit t_b Bruch auf. Abb. III, 26 zeigt die theoretische Bruchspannungs-Zeit-Relation zusammen mit experimentellen Messungen an verschiedenen Gläsern nach STUART und ANDERSON. Auf der Ordinate ist die Differenz $(\sigma - \sigma_\infty)$ in geeigneten Einheiten aufgetragen, auf der Abszisse die Bruchzeit t_b multipliziert mit einem temperaturabhängigen Faktor B

$$B \sim \exp\left[-(f - 2f_s)/k_B T\right]$$

(k_B ist hier die BOLTZMANN-Konstante).

Es muß darauf hingewiesen werden, daß die Theorie von STUART und ANDERSON im Widerspruch steht zur Bruchtheorie von GRIFFITH. Nimmt man in Abb. III,25 den Potentialberg, der die beiden Zustände A und B trennt, sehr niedrig, so verlaufen alle Prozesse sehr schnell, d.h. athermisch. Gl. (III,47) gibt dann die Grenze des athermischen Brucheintritts, die identisch sein sollte mit der Bruchspannungsformel von GRIFFITH (III,45). Wie man sich durch Einsetzen der Kerbzahl für den GRIFFITH-Riß $k = \sqrt{2\lambda/r_0}$ in (III,47) leicht überzeugt, steht (III,47) im Widerspruch zu der Formel von GRIFFITH.

Der Grund hierfür liegt darin, daß STUART und ANDERSON annehmen, daß die Erniedrigung bzw. Erhöhung der potentiellen Energie der Mulden A und B proportional ist zur Spannung σ. Eine bessere Annahme wäre, daß diese Erniedrigung proportional ist zur elastischen Deformationsenergie $\sigma^2/2E$. Nehmen wir an, daß bei Anwesenheit der Spannung die Potentialmulde A um den Betrag

$$\frac{\pi}{4}\,\frac{r_0}{2N}\,\frac{k^2\sigma^2}{2E}$$

gehoben und die Mulde B um den gleichen Betrag gesenkt wird, so erhält man an Stelle von Gl. (III,47) für die Dauerstandfestigkeit den Wert

$$\sigma_\infty = \frac{1}{k}\sqrt{\frac{4}{\pi}\,\frac{2E\alpha}{r_0}} = \sqrt{\frac{4}{\pi}\,\frac{E\alpha}{\lambda}}\,, \qquad \text{(III,48)}$$

in Einklang mit der Theorie von GRIFFITH.

d) Athermische Bruchausbreitung und die Gestalt der Bruchflächen.

Bei der weiteren Diskussion der Ausbreitung des Bruches und der Gestalt der Bruchflächen wollen wir von dem Einfluß der Wärmebewegung absehen. Der Bruch beginnt bei der Kerbstelle, die bezüglich Beanspruchung und Form optimal ist, d.h. die größte Kerbzahl k besitzt. In vielen Fällen liegt der Bruchbeginn beim Zugversuch in der Nähe des Randes der Bruchfläche, was im allgemeinen zwei Ursachen haben kann. Wie SMEKAL[1] gezeigt hat, haben Kerbstellen in der Nähe des Randes eine größere Kerbzahl als solche gleicher Lage und Form im Innern des Probestückes. Obendrein ist es denkbar, daß bei der Bildung von Kerbstellen unter atmosphärischen Bedingungen die randnahen Gebiete des Probestückes bevorzugt werden.

Das makroskopische Probestück wird eine sehr große Anzahl Kerbstellen verschiedener Abmessungen und verschiedener Orientierung ent-

[1] SMEKAL, A.: Ergebn. exakt. Naturwiss. **15**, 106 (1936).

halten, ist also, mikroskopisch gesehen, inhomogen und anisotrop. Wenn die Ebenen der Kerbrisse jedoch vollkommen willkürlich über alle Richtungen verteilt sind, so ist das Probestück trotzdem *makroskopisch isotrop*[1].

Unter einer äußeren Beanspruchung, die positive Zugspannungen enthält (die rechte Hälfte von Abb. III,2), werden die Kerbstellen die gefährlichsten sein, die senkrecht zur Richtung der maximalen Zugspannung σ_n orientiert sind [Gl. (III,33a)]. Für den Bruchbeginn ist dann im wesentlichen die maximale Zugspannung charakteristisch und die Bruchfläche liegt senkrecht zu dieser Richtung.

Die Hypothese von Griffith ist jedoch nicht imstande, den Mechanismus des Gleitungsbruches oder zähen Bruches unter Spannungszuständen mit negativer hydrostatischer Komponente zu erklären. Smekal[2] hat z. B. aus der Bruchtheorie von Griffith abgeleitet, daß die Scherfestigkeit übereinstimmen sollte mit der Zugfestigkeit, was im Widerspruch zu experimentellen Resultaten steht.

Bei der Ausbreitung des Bruchvorganges von der optimalen Kerbstelle aus können wir uns drei Fälle denken[2] (Abb. III,27):

A. Die Kerbstelle kann sich unbeschränkt weiter ausbreiten und liefert die Bruchfläche, ohne daß die Bruchfront auf ihrem Wege in die Nähe anderer Kerbstellen kommt.

B. Die Kerbstelle besitzt eine endliche Ausbreitungszone (sie mündet z. B. in eine andere Kerbstelle, die eine sehr ungünstige Lage mit niedriger Kerbzahl besitzt).

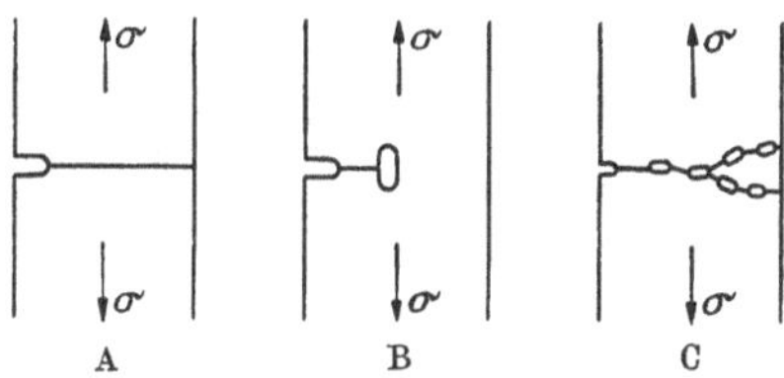

Abb. III,27. Schematische Darstellung der drei Fälle A, B und C.

C. Die Kerbstelle führt zum Bruch, aber die Bruchfront trifft auf ihrem Wege noch viele andere Kerbstellen und pflanzt sich über diese fort.

Fall A wird im allgemeinen auftreten, wenn die Kerbstellendichte $1/V_0$ im betrachteten Probestück äußerst gering ist. Man sollte dann vollkommen glatte Bruchflächen erhalten, vollkommene Spiegel, die mit der Ebene der Kerbstelle zusammenfallen. Man kann diesen Fall nur angenähert realisieren durch Anbringen einer künstlichen Inhomogenität, nämlich durch Anritzen des Probestabes vor dem Versuch.

Im Falle B mündet die primäre Kerbstelle in eine andere, die durch ihre kleine Kerbzahl den Bruchvorgang vorläufig zum Stehen bringt. Die Spannung muß dann weiter erhöht werden, bis eine andere Kerbstelle instabil wird und dieser Vorgang kann sich mehrere Male wiederholen, bevor eine Bruchfront auftritt. Der Mechanismus *B* bewirkt daher eine irreversible Veränderung der Mikrostruktur ohne direkt zum Bruche zu führen. Dieser Mechanismus wird, stark verknüpft mit thermischen Prozessen, in der ersten Bruchphase eine wesentliche Rolle spielen.

---

[1] Es handelt sich also um jene makroskopische Isotropie, die wir in § 15a stillschweigend vorausgesetzt haben. — [2] Siehe S. 194, Fußnote 1.

Wegen der großen Kerbstellendichtheit wird Fall C den Mechanismus der athermischen Bruchausbreitung am besten beschreiben. Im Beginn des Bruches werden nur sehr dicht bei der Bruchfront gelegene Kerbstellen mit hoher Kerbzahl (parallel zur Bruchebene) am Bruchvorgang teilnehmen können. Je weiter die Bruchfront fortschreitet, desto kleiner wird der noch tragende Querschnitt und desto höher wird die mittlere Spannung im Restquerschnitt. Daher werden mit Fortschreiten des Bruches mehr und mehr Kerbstellen „kritisch", darunter auch solche, die außerhalb der Bruchebene gelegen sind oder die schräg zur Bruchebene orientiert sind. Diese Kerbstellen werden Ausgangspunkte von Sekundärbrüchen, die gegenüber der primären Bruchfront geneigt sind und sich mit ihr vereinigen. Daher wird die Bruchfläche um so verzweigter und rauher, je mehr der Bruch fortgeschritten ist.

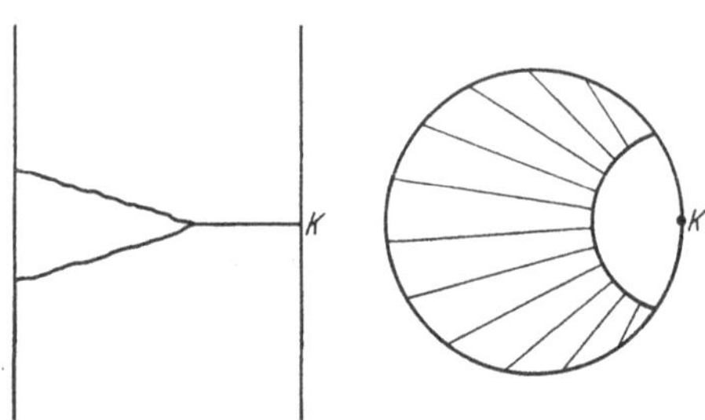

Abb. III,28. Morphologisches Bild des Bruches von Rundstäben im Zugversuch: Primäre Kerbstelle K, umgeben vom Spiegel, der in die Furchungsfläche ausmündet.

Ein Zugversuch an runden Probestäben gibt daher das folgende morphologische Bild (Abb. III,28)[1]. Ausgehend von der primären Kerbstelle K breitet sich die Bruchfront erst sehr glatt, mit fortschreitendem Bruche immer rauher werdend, aus und kann sich schließlich auch teilen, so daß das Probestück in drei Bruchstücke zerrissen wird. In diesem Fall wird das zwischen den Reißstücken fehlende Keilstück abgeschleudert.

Die Bruchfläche selbst setzt sich zusammen aus verschiedenen Teilen, der primären Kerbstelle, dem Spiegel und der Furchungsfläche. Der Spiegel ist ein kreisförmig begrenztes, glattes ebenes Gebiet, das um die primäre Kerbstelle senkrecht zur Zugrichtung orientiert ist. Der Spiegel liegt häufig am Rande und ist dann halbkreisförmig begrenzt, seltener sind im Innern der Bruchfläche liegende vollkreisförmig begrenzte Spiegel[2]. Im Mittelpunkt des Spiegels findet man manchmal eine Unebenheit, die sich als die primäre Kerbstelle deuten läßt.

Das optisch glatte Gebiet der Spiegelfläche besitzt keine scharfe Grenze gegen die Furchungsfläche, sondern einen Übergang, der durch eine kreisringartige Zone gebildet wird und von innen nach außen an Rauhigkeit zunimmt. Diese Rauhigkeit wird nach außen hin aus optischen Gründen erst allmählich sichtbar und ist wahrscheinlich bereits innerhalb der eigentlichen Spiegelfläche vorhanden. Dort bleiben die Unebenheiten jedoch unterhalb der Wellenlänge des Lichtes und darum

[1] Dieses Bruchbild ist von SMEKAL (siehe Seite 194) für Zugversuche an Glasstäben eingehend mit Beispielen belegt worden, es gilt in großen Zügen auch für organische Gläser, wie z. B. Plexiglas.

[2] Nach SMEKAL kommen bei Glasstäben Randspiegel viel häufiger vor als Innenspiegel, was darauf hinweist, daß bei Glas die Oberflächenkerbstellen wirksamer sind als die inneren Kerbstellen. Ähnliches findet man für Bruchflächen an Plexiglas.

unsichtbar. (Bei Bruchflächen an Plexiglas sind auch die Spiegel nicht optisch glatt, sondern zeigen eine feine Zeichnung, wie aus Abb. III,32 ersichtlich ist.)

Die Furchungsfläche setzt sich zusammen aus mehr oder weniger glatten Bruchflächen sekundärer Brüche, die andere Richtungen aufweisen wie die primäre Bruchfläche und scharfkantig aufeinanderstoßen. Dadurch erhalten die Furchungsflächen ein stufenförmig zerklüftetes, rauhes Aussehen. Beim Zugbruch von Glasstäben streben die Furchungsflächen der beiden Reißflächen auseinander, und zwar um so mehr, je kleiner die relative Spiegelgröße ist. Das weist auf die in Abb. III,28 angedeutete Verzweigung des Bruchvorganges hin.

Abb. III,29 zeigt die Bruchfläche eines runden Glasstabes[1], die deutlich alle oben beschriebenen Züge erkennen läßt. Der glatte Spiegel und die Furchungsfläche sind beide sehr deutlich ausgebildet, die Übergangszone zwischen Spiegel und Furchungsfläche ist schmal. Diese Gestalt der Bruchfläche wird nach SMEKAL[2] in der überwiegend großen Mehrzahl aller einwandfrei ausgeführten Zugversuche an Glas angetroffen.

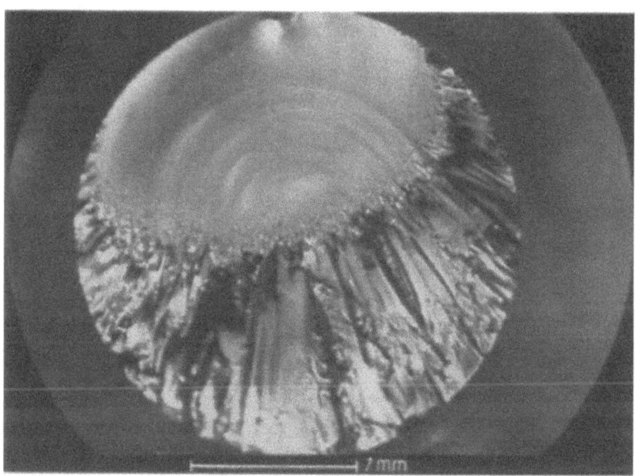

Abb. III,29. Zugbruchfläche eines runden Glasstabes, Raumtemperatur, Zuggeschwindigkeit 10mm/min; Leitz Ultropak, Auflicht Hellfeld, Objektiv 3,8 ×, Vergrößerung 20 ×.

Eine Ausnahme einer sogenannten „muscheligen" Bruchfläche ist in Abb. III,30 wiedergegeben. Diese Form der Bruchfläche beruht nach SMEKAL auf einer mangelhaften Versuchseinrichtung, wobei der Zugspannung noch eine Biege- oder Torsionsbeanspruchung überlagert ist, oder auf sehr groben Fehlern im Bau des Materials. An Stelle der ebenen Spiegel erscheinen in der Mitte muschelige Bruchflächenteile. Der gezeigte

[1] Die Verfasser sind den Herren J. LEEUWERIK und A. BURGERS, Centraal Laboratorium T.N.O., Delft, die die in Abb. III,29 bis III,40 gezeigten Aufnahmen zur Verfügung stellten, zu großem Dank verpflichtet.

[2] Wir wollen hier ausdrücklich auf den auf Seite 194 zitierten Übersichtsartikel von SMEKAL hinweisen, der eine eingehende Diskussion einer großen Anzahl Bruchflächen von Glasstäben enthält.

Muschelbruch entstand an einem Plexiglasstab durch schiefes Einklemmen in der Zugbank. Das Bruchflächenbild von Muschelbrüchen an Glas und Plexiglas zeigt eine überraschend große Ähnlichkeit[1].

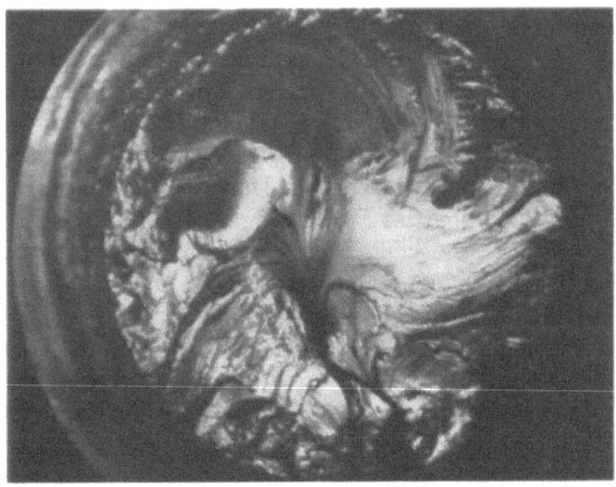

Abb. III, 30. Muschelbruch an einem runden Plexiglasstab, Zugbeanspruchung überlagert von Biegung und Torsion. Raumtemperatur, Zuggeschwindigkeit 10 mm/min; Leitz Ultropak, Auflicht Hellfeld, Objektiv 3,8 ×, Vergrößerung 20 ×.

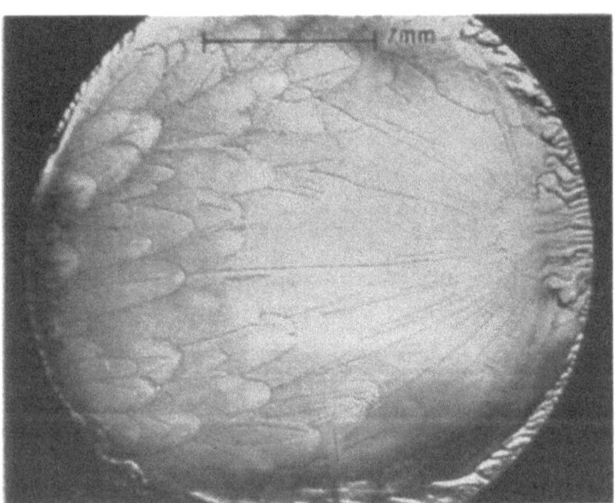

Abb. III, 31. Zugbruchfläche eines runden Plexiglasstabes. Raumtemperatur, Zuggeschwindigkeit 10 mm/min; Leitz Ultropak, Durchlicht Hellfeld, Objektiv 3,8 ×, Vergrößerung 20 ×.

Vergleichen wir nun die Bruchfläche eines Glasstabes (Abb. III, 29) mit einer solchen eines organischen Glases (Abb. III, 31). Abb. III, 31 zeigt die Zugbruchfläche eines langsam gezogenen runden Plexiglasstabes. Im Gegensatz zur Bruchfläche an Glasstäben sind hier die charakteristischen Unterschiede zwischen Spiegel, Übergangsgebiet und Fur-

[1] Ein Beispiel eines Muschelbruches an einem Glasstab, dessen morphologisches Bild fast mit dem von Abb. III, 30 identisch ist, gibt SMEKAL (siehe Seite 194).

chungsfläche viel weniger deutlich ausgebildet. Der Spiegel (rechts) ist nicht optisch glatt, sondern zeigt eine feine fiederförmig radiale Zeichnung, wie man aus Abb. III,32 entnehmen kann. Außerhalb des kleinen

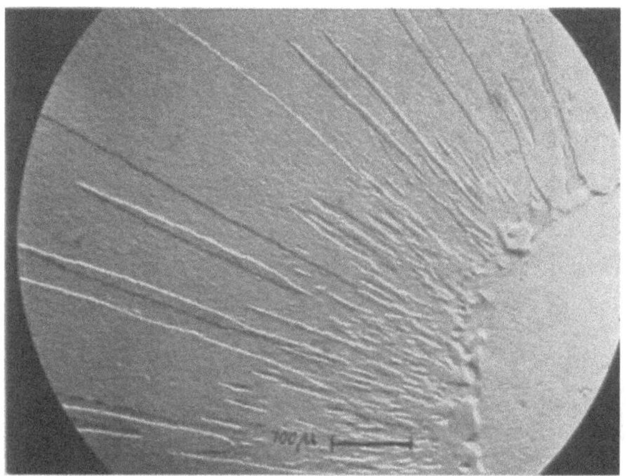

Abb. III, 32. Vergrößerte Wiedergabe des Spiegels aus Abb. (III, 31), Leitz Ultropak, Durchlicht Hellfeld, Objektiv 50 ×, Vergrößerung 96 ×.

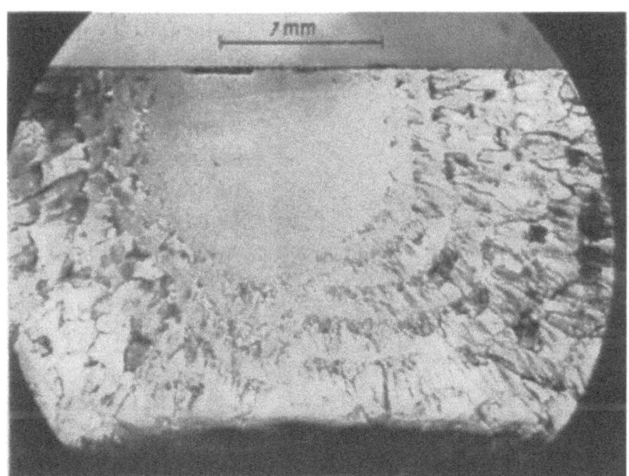

Abb. III, 33. Bruchfläche eines rechteckigen Plexiglasstabes, gebrochen unter Biegungsschlag. Schlagrichtung von oben nach unten, Raumtemperatur, Schlaggeschwindigkeit 5,2 m/sek; Leitz Ultropak, Durchlicht Hellfeld, Objektiv 3,8 ×, Vergrößerung 20 ×.

Doppelspiegels erstreckt sich eine sehr große Übergangszone, die man noch nicht als Furchungsfläche ansprechen kann. Diese Übergangszone wird allmählich gegen den Rand zu rauher und enthält eine große Anzahl Bruchfiguren, auf deren Besprechung wir noch zurückkommen. Erst ganz am Rande der Bruchfläche findet man eine Andeutung eines Furchungsgebietes.

Diese wesentlichen Unterschiede der Bruchflächen zwischen Glas und Plexiglasstäben weisen darauf hin, daß die ziemlich starken auftretenden Fließerscheinungen bei Plexiglas kein rein sprödes Bruchverhalten mehr

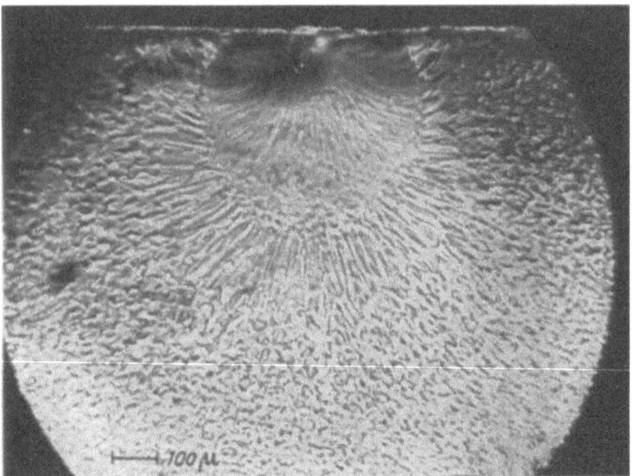

Abb. III,34. Vergrößerte Wiedergabe von Übergangszone und Spiegel aus Abb. (III,33); Leitz Ortholux, Phasenkontrast, Durchlicht, Objektiv 10 ×, Vergrößerung 55 ×.

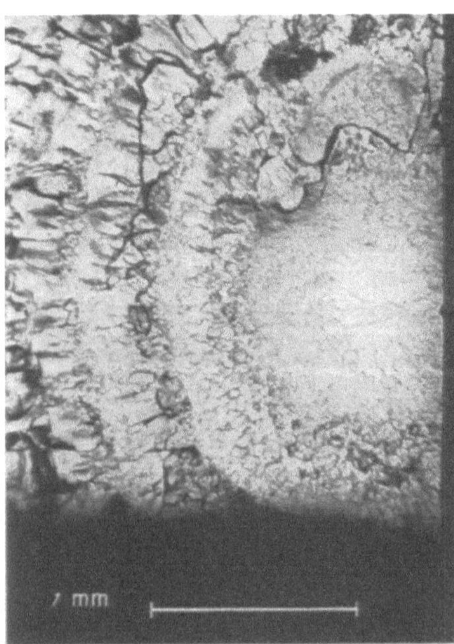

Abb. III,35. Bruchfläche eines rechteckigen Plexiglasstabes, gebrochen unter Biegungsschlag. Schlagrichtung von rechts nach links, Raumtemperatur, Schlaggeschwindigkeit 2,1 m/sek, Leitz Ultropak, Durchlicht Hellfeld, Objektiv 3,8 ×, Vergrößerung 27 ×.

zulassen. Trotzdem kann man auch bei Plexiglas typisch sprödes Bruchverhalten erzwingen, wenn man dafür sorgt, daß die Spannungszunahme sehr schnell erfolgt. Bei großen Deformationsgeschwindigkeiten zeigt Plexiglas dann wieder ein sprödes Bruchbild, wie man aus Abb. III,33 und III,35 entnehmen kann.

Abb. III,33 zeigt die Bruchfläche eines rechteckigen Plexiglasstabes, der durch Biegungsschlag unter großen Deformationsgeschwindigkeiten gebrochen wurde. Man sieht einen kleinen Randspiegel, eingebettet in eine große Übergangszone, die ihrerseits von sehr gut ausgebildeten Furchungsflächen begrenzt wird. Die Furchungsflächen liegen in der Form konzentrischer

Ringe um das glatte Gebiet. Abb. III, 34 zeigt eine Vergrößerung der Übergangszone und des Spiegels aus Abb. III, 33. Deutlich hebt sich der eigentliche Spiegel mit einer feinen fiederförmigen Zeichnung ab vom

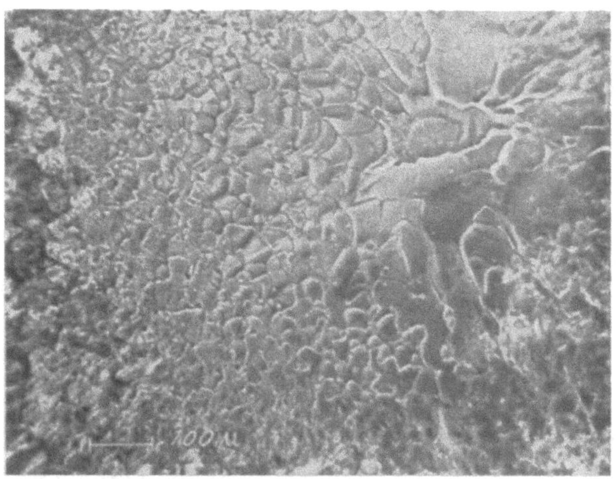

Abb. III, 36. Vergrößerte Wiedergabe von Übergangszone und Spiegel aus Abb. (III, 35); Leitz Ortholux, Phasenkontrast, Durchlicht, Objektiv 10 ×, Vergrößerung 82 ×.

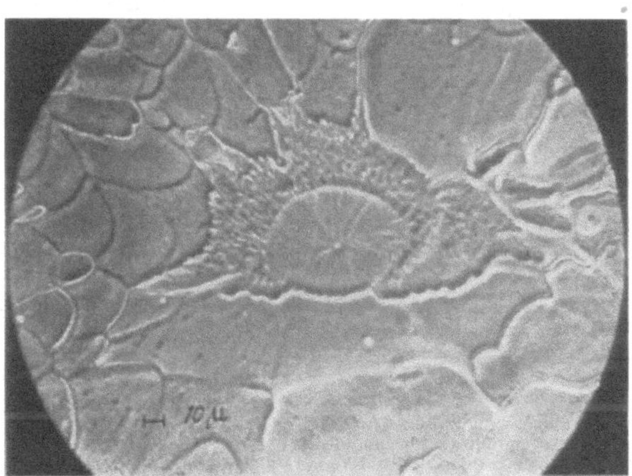

Abb. III, 37. Vergrößerte Wiedergabe des Spiegels aus Abb. (III, 36); Leitz Ortholux, Phasenkontrast, Durchlicht, Objektiv 40 ×, Vergrößerung 240 ×.

Übergangsgebiet, das hyperbelförmige Bruchfiguren in großer Anzahl enthält.

Schließlich zeigen Abb. III, 35 die Bruchfläche eines Biegeschlagstükkes mit Innenspiegel und Abb. III, 36 und III, 37 zwei aufeinanderfolgende Vergrößerungen von Übergangszone mit Spiegel. Der eigentliche Innenspiegel liegt in Abb. III, 37 etwa in der Mitte und hat ellipsenförmige Gestalt mit Halbachsen von etwa 20 und 30 μ. Der Spiegel ist deutlich radial gezeichnet und man erkennt in seinem Mittelpunkt die primäre Kerbstelle.

SMEKAL[1] hat für Bruchflächen an runden Glasstäben das Auftreten und die Größe des Spiegels in Zusammenhang gebracht mit der Kerbzahl der primären Kerbstelle. Der Spiegel entsteht jedenfalls unter relativ kleinen makroskopischen Spannungen und unter Mitwirkung von sekundären Kerbstellen, die ungefähr in der primären Bruchebene liegen. Je kleiner daher die mittlere Spannung im Beginn des Bruches sein wird, desto langsamer ist die Bruchausbreitung und desto größer wird die Spiegelfläche. Andererseits sind kleine Beginnspannungen bei der Bruchausbreitung gleichbedeutend mit einer großen Kerbzahl k der primären Kerbstelle. SMEKAL fordert daher einen Zusammenhang zwischen der relativen Spiegelgröße s/q (s = Fläche des Spiegels, q = Querschnittsfläche des Stabes) und der primären Kerbzahl k. Ein Beispiel dafür bilden die besonders großen Spiegelflächen (s/q bis zu 95%), die sich bei Glasstäben durch Anritzen der Oberfläche künstlich erzwingen lassen, da durch den Vorgang des Anritzens eine Kerbstelle mit besonders großer Kerbzahl geschaffen wird.

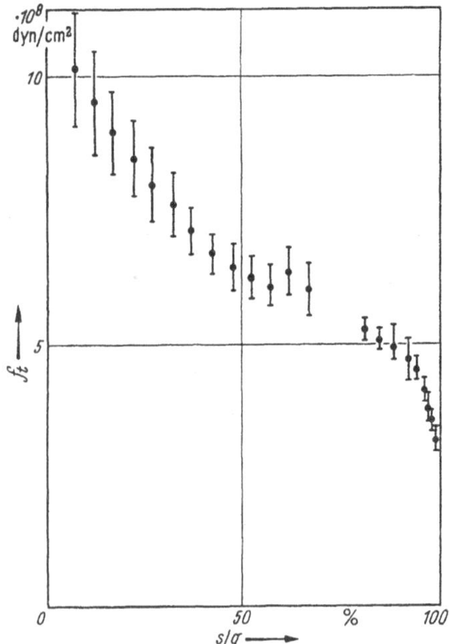

Abb. III, 38. Der Zusammenhang zwischen Zugfestigkeit f_t und relativer Spiegelgröße s/q für runde Glasstäbe im Zugversuch. (Nach SMEKAL.) [SMEKAL, A.: Ergebn. exakt. Naturwiss. **15**, 106 (1936).]

SMEKAL konnte den erwarteten Zusammenhang zwischen Spiegelgröße und Kerbzahl an feuerpolierten ungeritzten und angeritzten Glasstäben zeigen, wobei Belastungsgeschwindigkeit und Temperatur in weitem Bereiche variiert wurden. Als Maß für die primäre Kerbzahl wurde die Zugfestigkeit gewählt, die nach Gl. (III,39) umgekehrt proportional zu k ist. Abb. III,38 zeigt den erwarteten Zusammenhang zwischen relativer Spiegelgröße und Zugfestigkeit nach SMEKAL.

Im Übergangsgebiet zwischen Spiegel und Furchungsfläche, teilweise auch noch am Rande des Spiegels, sieht man eine große Anzahl charakteristischer Bruchfiguren, die besonders bei Perspex sehr deutlich ausgebildet sind (Abb. III,31 und III,36). Diese Bruchfiguren können die mannigfaltigsten offenen und geschlossenen Formen besitzen, erscheinen sehr häufig jedoch als Ast einer Hyperbel, deren Scheitel alle nach dem Ort des Bruchbeginns hin gerichtet sind.

---

[1] SMEKAL, A.: siehe Seite 194.

Abb. III,39 zeigt die Vergrößerung eines solchen Hyperbelfeldes und Abb. III,40 die Vergrößerung einer einzelnen Hyperbel. Man erkennt in jeder Hyperbel den Brennpunkt als eine Inhomogenität, von der aus feine Linien radial nach der Hyperbelkontur zu verlaufen. Dieser Brennpunkt ist der Ausgangspunkt einer sekundären Bruchfront, die beim Herannahen der primären Bruchfront ausgelöst wird. Das Zusammentreffen

Abb. III,39. Ausschnitt aus dem Hyperbelfeld der Zugbruchfläche eines Plexiglasstabes. Raumtemperatur, Zuggeschwindigkeit 10 mm/min; Leitz, Opak, Durchlicht Hellfeld, Objektiv 16 ×, Vergrößerung 70 ×.

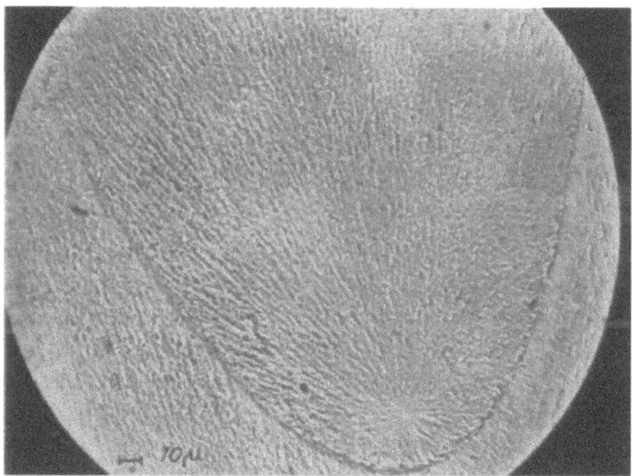

Abb. III,40. Vergrößerung einer Bruchhyperbel aus dem Hyperbelfeld der Zugbruchfläche eines Plexiglasstabes. Raumtemperatur, Zuggeschwindigkeit 10mm/min; Leitz Ultropak, Durchlicht Hellfeld, Objektiv 50 ×, Vergrößerung 270 ×.

der primären und sekundären Bruchfront veranlaßt die Entstehung der Hyperbelkontur. Aus Abb. III,40 erkennt man, daß die feine Zeichnung im Innern der Hyperbel radial gerichtet ist und etwa die Richtung der Trajektorien der sekundären Bruchfront besitzt. Im Äußeren der Hyperbel besitzt die Zeichnung die Richtung der Trajektorien der primären Bruchfront. Weiter ersieht man aus Abb. III,39, daß zwei benachbarte Hyperbeln sich bei ihrer Ausbreitung gegenseitig stören. Überall, wo zwei benachbarte Bruchhyperbeln sich schneiden, laufen die beiden Hyperbelkonturen nach dem Schnittpunkt nicht durch, sondern werden durch eine, beiden Figuren gemeinsame, Grenzlinie fortgesetzt. Diese Grenzlinie entsteht beim Zusammentreffen der beiden sekundären Bruchfronten (hinter dem Schnittpunkt erfolgt das Zusammentreffen der beiden sekundären Bruchfronten noch bevor eine der sekundären Bruchfronten von der primären Bruchfront eingeholt wird).

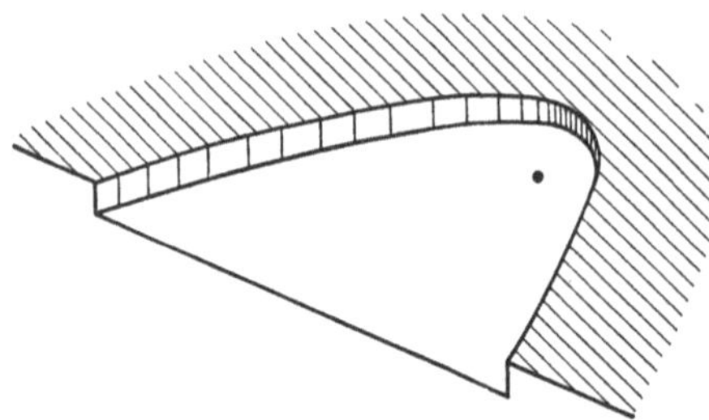

Abb. III,41. Schematische Darstellung einer Bruchhyperbel an Zugbruchflächen von Glasstäben. Das Hyperbelinnere besitzt ein tieferes Niveau als die Umgebung.

Nach SMEKAL[1] besitzt bei Glas das Hyperbelinnere ein anderes Niveau als die Umgebung. Beide Bruchflächen passen genau zusammen, auf einer Fläche liegt das Hyperbelniveau tiefer, auf der Gegenfläche höher als der Rest der Bruchfläche (vgl. Abb. III,41). Ähnliches gilt für die Bruchhyperbeln an Plexiglas.

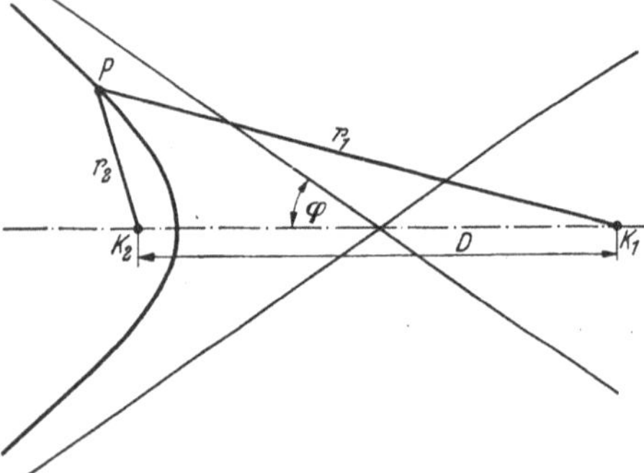

Abb. III,42. Schematische Darstellung der Entstehung der Bruchhyperbeln: K_1 und K_2 sind die Ausgangspunkte der primären und sekundären Bruchfront, P ein laufender Punkt der Bruchhyperbel in dem sich die beiden Bruchfronten schneiden.

Der Mechanismus der Entstehung dieser Bruchhyperbeln ist in Abb. III,42 veranschaulicht. Von der primären Kerbstelle des Bruchbeginns, K_1, breitet sich die Bruchfront in Form einer Kreiswelle aus. Noch bevor die Bruchfront K_2 erreicht hat, kann infolge der Spannungskonzentration von einer inneren Kerbstelle, K_2, aus ein sekundärer Bruch entstehen, der sich auch in der Form einer Kreiswelle ausbreitet und teilweise der primären Bruchfront entgegenläuft. An den Orten, wo sich beide Bruchfronten schneiden, entsteht eine Bruchkontur, die in gewissen Fällen die Form einer Hyperbel haben kann.

Nehmen wir an, daß beide Bruchfronten die konstante hohe Ausbreitungsgeschwindigkeit c besitzen[2] und daß die primäre Bruchfront im Zeitpunkt $t = 0$ von K_1 aus, die sekundäre Bruchfront im späteren Zeitpunkt t_0 von K_2 aus fortschreiten. Zum Zeitpunkt t hat die primäre Bruchfront den Weg $r_1 = ct$ und die

[1] SMEKAL, A.: Glastechn. Ber. **23**, 57 (1950).

[2] Nachdem die Hyperbeln am Rande des Spiegels auftreten, muß man annehmen, daß die primäre Bruchfront bereits ungefähr die hohe konstante Endgeschwindigkeit c der Bruchausbreitung besitzt (vgl. Abschn. e). Die sekundäre Bruchfront muß sich dann auch mit derselben hohen Geschwindigkeit ausbreiten.

sekundäre Bruchfront den Weg $r_2 = ct - ct_0$ zurückgelegt. Beide Fronten treffen sich dann in einem Punkt P der Bruchfigur. Es gilt daher für die Bruchfigur die Gleichung

$$r_1 - r_2 = ct_0 = \text{konstant} \tag{III,49}$$

und das ist die Gleichung eines Hyperbelastes. Die Hyperbel läßt sich in einem geeigneten Koordinatensystem (die x-Achse geht durch K_1 und K_2, die y-Achse teilt den Abstand $\overline{K_1 K_2}$ in gleiche Teile) durch die Gleichung darstellen

$$\frac{x^2}{a^2} - \frac{y^2}{b^2} = 1 .$$

Die beiden Halbachsen a und b der Hyperbel hängen mit dem doppelten Brennpunktabstand D zusammen durch

$$2a = ct_0 \qquad b^2 = D^2/4 - a^2 .$$

Die Orte des primären und sekundären Bruchbeginns K_1 und K_2 bilden die beiden Brennpunkte der Hyperbel mit einem Abstand D. Der halbe Asymptotenwinkel φ der Hyperbel läßt sich in Zusammenhang bringen mit der Zeit des Entstehens des sekundären Bruches

$$ct_0/D = \cos\varphi , \tag{III,50}$$

ct_0 ist dabei der Abstand, den der primäre Bruch bis zum Beginn des sekundären Bruches zurücklegt.

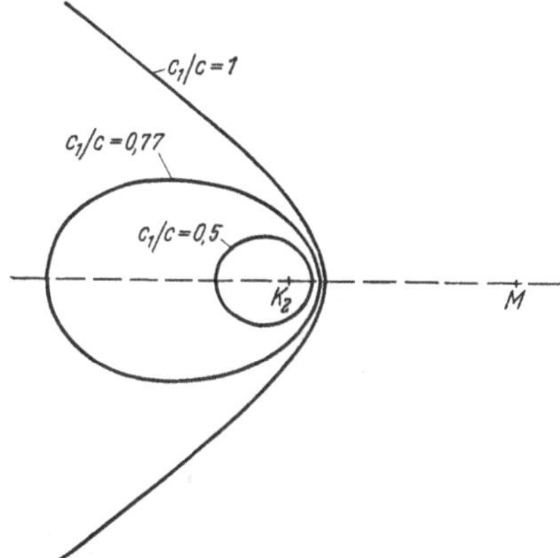

Abb. III,43. Bruchfiguren, die beim Aufeinandertreffen von primären und sekundären Bruchfronten verschiedener Ausbreitungsgeschwindigkeiten entstehen. K_1 und K_2 sind die Ausgangspunkte der primären und sekundären Bruchfront mit einer Ausbreitungsgeschwindigkeit c bzw. c_1. K_1 liegt rechts außerhalb der Abb., M ist der Mittelpunkt der Strecke $\overline{K_1 K_2}$.

Ist die Geschwindigkeit des sekundären Innenbruches kleiner als die der primären Bruchfront, dann wird der Innenbruch von Primärbruch überholt und von der weiteren Ausbreitung abgeschnitten[1]. Nehmen wir als Beispiel, daß c die konstante Fortpflanzungsgeschwindigkeit des Primärbruches, c_1 die konstante Fortpflanzungsgeschwindigkeit des Sekundärbruches und $c_1/c = \alpha$ kleiner als Eins ist. Dann ist die Bruchfigur an Stelle von Gl. (III,49) bestimmt durch die Gleichung

$$r_1 - \frac{1}{\alpha} r_2 = ct_0 . \tag{III, 51}$$

In Abb. III,43 haben wir eine Reihe dieser Bruchfiguren konstruiert, wobei das Verhältnis von Sekundär- und Primärgeschwindigkeit gewählt ist zu 1, 0,77 und 0,50. Sobald das Verhältnis α kleiner ist als Eins, sind die entstehenden Bruchfiguren geschlossene Kurven.

e) Messung der Bruchgeschwindigkeit.

Aus der Gestalt von Zugbruchflächen (Abschn. d) folgt, daß der Bruchvorgang an der primären Kerbstelle K beginnt und sich von da aus radial durch den Querschnitt des Probestückes fortpflanzt. Aus allgemeinen Gründen haben wir bereits angenommen, daß die Ausbreitungsgeschwindigkeit der Bruchfront im Anfang in der Nähe von K sehr klein ist und mit zunehmender Größe des Bruchweges stark anwächst, um schließlich

[1] Smekal, A.: Österr. Ing.-Archiv 7, 49 (1953).

einen sehr hohen konstanten Endwert zu erreichen. SMEKAL[1] hat einen Weg angegeben, um diese Ausbreitungsgeschwindigkeit an Hand der Gestalt von Bruchflächen von Glasstäben zu messen.

Die Methode beruht auf der Entdeckung[2] von Ultraschallspuren auf der Bruchfläche von Glasstäben. Auf den glatten Teilen von Biege- und Zugbruchflächen von Glasstäben sieht man zwei sich durchkreuzende Kurvenscharen, die symmetrisch zur primären Kerbstelle K liegen. Diese WALLNER-*Linien* können in großer Anzahl hervorgebracht werden durch Schmirgeln oder Anätzen der Oberfläche des Zugstabes vor dem Versuch.

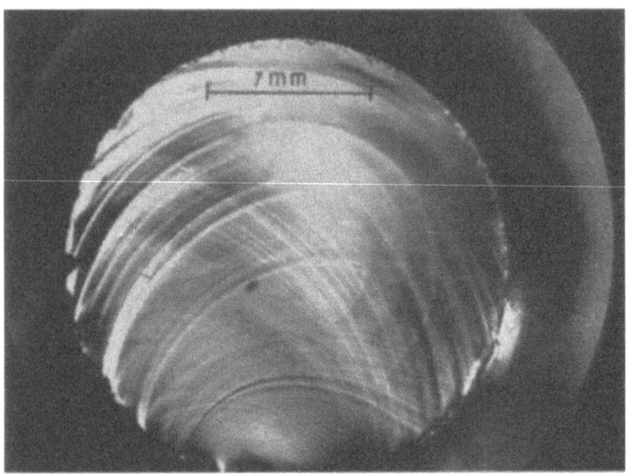

Abb. III,44. WALLNER-Linien an einer Zugbruchfläche von Glas. Raumtemperatur, Zuggeschwindigkeit 10 mm/min; Leitz Ultropak, Auflicht Hellfeld, Objektiv 3,8 ×, Vergrößerung 20 ×.

WALLNER-Linien treten hauptsächlich im Spiegel auf – jedoch nicht in der unmittelbaren Umgebung der primären Kerbstelle – und in schwächerem Maße auch in der den Spiegel umgebenden Zone feiner Rauhigkeit. Ein Beispiel von WALLNER-Linien an einer Zugbruchfläche von Glas zeigt Abb. III,44.

Die WALLNER-Linien sind äußerst flache Erhebungen oder Vertiefungen von etwa 1 μ Breite, die auf der Bruchfläche und Gegenfläche genau aufeinanderpassen. Der Beginn der WALLNER-Linien liegt am Rande des Probestabes an den durch die Oberflächenbehandlung künstlich angebrachten Kerbstellen. SMEKAL[1] konnte die WALLNER-Linien als Spuren von transversalen Ultraschallwellen sehr hoher Frequenzen identifizieren. Während die Bruchfront von der Kerbstelle K aus fortschreitet, wird am Randpunkt der WALLNER-Linie eine Ultraschallwelle ausgelöst.

[1] SMEKAL, A.: Glastechn. Ber. **23**, 57, 186 (1950); Acta physic. austr. **7**, 110 (1953). – F. KERKHOF [Naturwiss. 8, 478 (1953)] konnte durch Bestrahlung der Probestücke mit transversalen Ultraschallwellen (Frequenz 9 MHz) während des Zerreißens von Glasstäben künstlich eine dritte Schar von WALLNER-Linien auf den Bruchflächen erhalten. Dadurch wurde bewiesen, daß transversale Ultraschallwellen hoher Frequenzen WALLNER-Linien hervorbringen können.

[2] WALLNER, H.: Z. Phys. **114**, 368 (1939).

An den Stellen, wo sich die Ultraschallwelle und die Bruchfront treffen, entsteht eine Kontur, die WALLNER-Linie.

Geht man von der Annahme aus, daß die Ultraschallgeschwindigkeit überall den konstanten Wert c_t besitzt, so lassen sich die WALLNER-Linien zur Messung der Fortpflanzungsgeschwindigkeit der Bruchfront verwenden. Abb. III,45 zeigt eine schematische Darstellung der beiden Scharen der WALLNER-Linien. Die linke Schar habe die Randpunkte P_0, P_1, P_2 usw., die rechte Schar die Randpunkte Q_0, Q_1, Q_2 usw. Die WALLNER-Linien sind keine Kreise, sondern Kurven mit positiver Krümmung, die vom Rande nach innen zu durchgehend kleiner wird. Betrachten wir eine beliebige WALLNER-Linie der linken Schar, z. B. die von P_2 ausgehende. Diese WALLNER-Linie wird von den Linien der rechten Schar in den Punkten M_0, M_1, M_2 und M_3 geschnitten. Die Auslösung der Schallwelle von P_2 aus erfolgt zu dem noch unbekannten Zeitpunkt t_0. Die von P_2 aus gezogenen Sehnen nach den Punkten M_0, M_1, M_2, M_3 entsprechen dem von der Schallwelle zurückgelegten Weg und geben daher, dividiert durch die Schallgeschwindigkeit c_t, die Zeitpunkte, zu denen die Bruchfront bis zu M_0, M_1, M_2, M_3 fortgeschritten ist, gerechnet von t_0 aus. So erhalten wir zu den Punkten M_0, M_1, M_2, M_3 Zeitmarken, die die Ankunft der Bruchfront in diesen Punkten angeben. Weiter lassen sich mit Hilfe der Sehnen $\overline{M_0 Q_0}$, $\overline{M_1 Q_1}$, $\overline{M_2 Q_2}$, $\overline{M_3 Q_3}$ die Anfangspunkte der rechten Schar konstruieren. Haben wir alle Beginnpunkte der rechten Schar mit einer Zeitmarke versehen, so können wir ebenso mit allen Beginnpunkten der linken Schar verfahren und erhalten schließlich für alle Flächenpunkte, durch die eine WALLNER-Linie läuft, eine Zeitmarkierung.

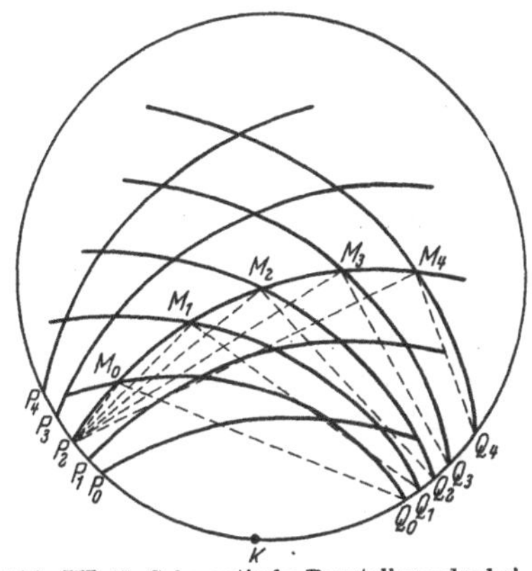

Abb. III, 45. Schematische Darstellung der beiden Scharen von WALLNER-Linien an Zugbruchflächen von Glasstäben und der Konstruktion der Zeitmarken in den Schnittpunkten M_0, M_1, M_2 usw.

Verbinden wir alle Punkte mit gleicher Zeitmarke, so erhalten wir die Bruchfronten zu den entsprechenden Zeitpunkten. Diese Bruchfronten sind für größere Abstände von K Kreise, was aus Symmetriegründen auch zu erwarten war, und eine Rechtfertigung der der Methode zugrunde liegenden Annahmen darstellt. Die radiale Ausbreitungsgeschwindigkeit der Bruchfront ergibt sich dann aus den Zeitmarken, wobei der Beginnpunkt der Zeitzählung noch willkürlich bleibt. Die nähere Umgebung des Bruchbeginns enthält keine WALLNER-Linien, so daß sich der unmittelbare Beginn der Bruchfront auf diese Weise nicht erfassen läßt.

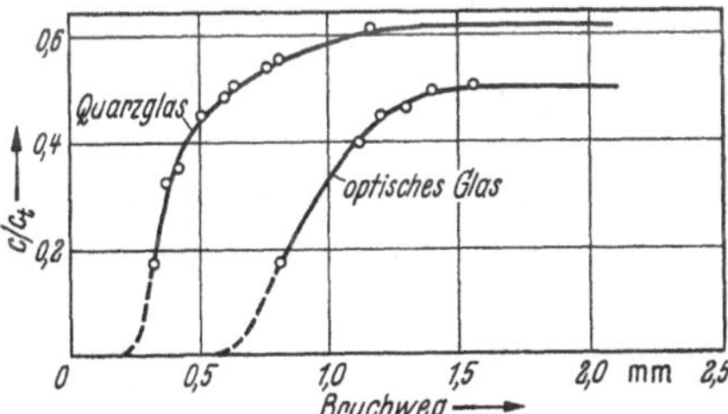

Abb. III, 46. Relative Fortpflanzungsgeschwindigkeit der Bruchfront in Abhängigkeit vom Bruchweg für Quarzglas und für optisches Glas. (Nach SMEKAL.) Zugbruchstäbe mit 6 mm Durchmesser. [SMEKAL, A.: Glastechn. Ber. 23, 57, 186 (1950).]

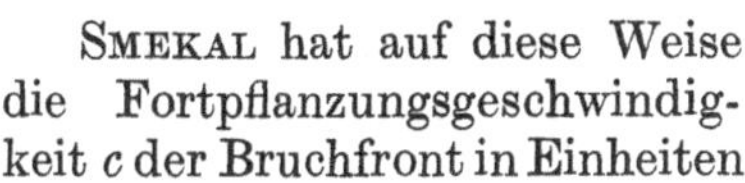

SMEKAL hat auf diese Weise die Fortpflanzungsgeschwindigkeit c der Bruchfront in Einheiten der Ultraschallgeschwindigkeit c_t an verschiedenen Glassorten gemessen (Abb. III,46). Im Bereiche der unteren Grenze der Meßbarkeit von c/c_t (die bei SMEKAL ungefähr 0,2 beträgt) steigt die Bruchgeschwindigkeit steil an und erreicht einen hohen konstanten Endwert von 50 bzw.

62 % der transversalen Schallgeschwindigkeit. Eine Extrapolation nach dem Bruchweg Null zeigt, daß die Kurven einen Wendepunkt besitzen müssen. Deutlich lassen sich in Abb. III,46 die beiden Bruchphasen erkennen. Im Beginn erfolgt ein sehr langsames Anwachsen der Bruchgeschwindigkeit, das stark zeit- und temperaturabhängig ist und den thermischen Auslösungsmechanismus wiedergibt. Die andere, athermische Phase, beginnt mit dem starken Anstieg der Bruchgeschwindigkeit. Der konstante Endwert der Bruchgeschwindigkeit beträgt stets 50–60% der transversalen Ultraschallgeschwindigkeit und ist augenscheinlich als eine Stoffkonstante anzusehen. Während der Bruchweg, der in den ersten 30 Sekunden zurückgelegt wird, nur 0,3 mm beträgt, werden die restlichen 5,7 mm des Bruchweges in $3 \cdot 10^{-6}$ Sekunden zurückgelegt. Die stationären Endwerte der Bruchfortpflanzungsgeschwindigkeit liegen für Quarzglas bei 2100 m/sek und für optisches Glas bei 1800 m/sek.

Die stationäre Endgeschwindigkeit der Bruchausbreitung ist noch auf eine andere unabhängige Weise gemessen worden[1], nämlich durch die funkenkinematische Aufnahme der Ausbreitungsgeschwindigkeit von Kerben. Dieses Verfahren besitzt ein hohes zeitliches, aber nur ein geringes räumliches Auflösungsvermögen und kann daher nur zur Messung der schnellen Endphase des Bruches dienen. Die hier gemessenen Werte betragen in guter Übereinstimmung mit SMEKAL für Quarzglas 2200 m/sek und für optisches Glas 1700 m/sek. Für Plexiglas fand SCHARDIN[2] Bruchgeschwindigkeiten zwischen 500 und 700 m/sek.

f) Der Mikroritzversuch und das Stoffverhalten in kerbfreien Bereichen.

In den vorhergehenden Abschnitten ist gezeigt worden, daß das spröde Stoffverhalten verknüpft ist mit der Wirksamkeit von Inhomogenitätsstellen. Es bleibt noch zu beweisen, daß bei Abwesenheit von Kerbstellen kein sprödes Bruchverhalten auftritt. Auch hier hat SMEKAL den geeigneten Weg angegeben.

Werden die mechanischen Beanspruchungen auf Raumgebiete beschränkt, die klein genug sind, um keine Kerbstellen zu enthalten, so trifft man auch bei typisch spröden Stoffen auf vollkommen andere Verhältnisse als im Makroversuch. Es werden keine Bruchflächen mehr gebildet, sondern es tritt eine rein plastische Formänderung auf.

Mechanische Beanspruchungen in sehr kleinen Gebieten hat SMEKAL[3] im Mikroritzversuch mit Hilfe von Diamanten unter sehr kleinen Belastungen realisiert. Man findet dann feine Ritzspuren, die – bei genügend kleiner Belastung und Kontaktfläche – keine Spur von Bruchbildung mehr aufweisen. Abb. III,47 zeigt eine mehrfache Ritzspur an Plexiglas. SMEKAL[4] konnte durch Mikrointerferometrie und Phasenkontrastverfahren auch das Profil der Ritzbahnen ausmessen. Diese haben die Form von seichten Gräben, flankiert von seitlichen Wällen (Abb. III,48).

[1] SCHARDIN, H. u. W. STRUTH: Glastechn. Ber. **16**, 219 (1938). – H. M. DIMMICK: J. Soc. Glass. Techn. **35**, 318 (1951).

[2] SCHARDIN, H.: Kunststoffe **44**, 48 (1954).

[3] KLEMM, W. u. A. SMEKAL: Naturwiss. **29**, 688, 710, 769 (1941).

[4] SMEKAL, A. u. W. KLEMM: Mh. Chem. **82**, 411 (1951).

Bruchfreie Ritzspuren treten nach SMEKAL nur bei sehr kleinen Kontaktflächen auf (Größenordnung $1\,\mu$[1]). Vergrößert man die Ritzbreite auf $10\,\mu$, dann entstehen Ritzspuren mit lokalen Bruchstellen. Das beweist direkt die Existenz von Kerbstellen: Nur wenn die Kontaktfläche klein genug ist, um auf ihrem Weg ausschließlich homogenes Material anzutreffen, erhält man bruchfreie Ritzbahnen.

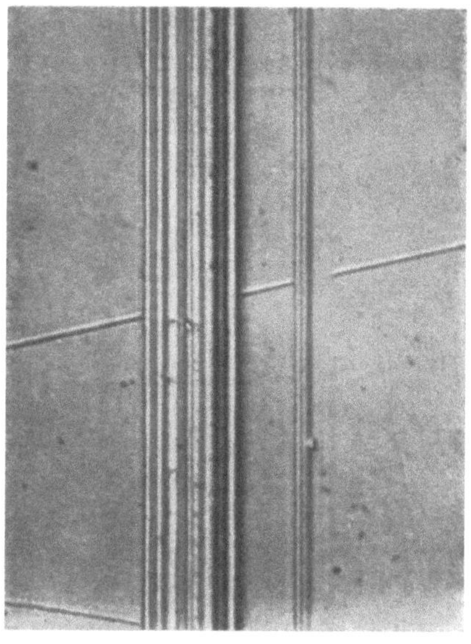

Abb. III, 47. Mehrfache Mikroritzspur auf Plexiglas. (Nach SMEKAL.) Diamantbelastung 8 g. Gesamtbreite der Spur etwa $10\,\mu$. [SMEKAL, A., W. KLEMM: Mh. Chemie **82**, 411 (1951).]

Ein etwas abweichendes Verhalten zeigen Ritzspuren auf Plexiglas[2] gezogen mit einem Langspielsaphir von der Form eines Kegels mit einem Abrundungsradius von $25\,\mu$. Abb. III, 49, III, 50 und III, 51 zeigen drei Ritzbahnen von etwa $30\,\mu$ Breite gezogen unter verschiedenen Belastungen (25 g, 30 g, 35 g). Wie man sieht, treten die ersten lokalen Brüche bei einer kritischen Last von etwa 30 g auf.

Nach SMEKAL beruht der Mikroritzversuch auf der unmittelbaren Überwindung der chemischen Bindungskräfte und ist so ein Maß für die Kohäsion des Stoffes im homogenen Zustand. Darauf weisen der athermische Charakter und die überaus großen Werte der erhaltenen Mikrohärte. SMEKAL erhält für Plexiglas unter 8 g Belastung Ritzspuren von $6{,}5\,\mu$ Breite, was einer Mikrohärte von

$$H \sim 2 \cdot 10^{10}\ \mathrm{d/cm^2} \tag{III,52}$$

entspricht. Für Silikatglas liegt die Härte etwa 5mal höher, was darauf zurückzuführen ist, daß bei Plexiglas die zu überwindenden chemischen Bindungen VAN DER WAALS-Bindungen, bei Glas jedoch Hauptvalenzbindungen sind. Beide Härten liegen in der Größenordnung der molekularen Zerreißfestigkeit [vgl. Gl. (III, 30)].

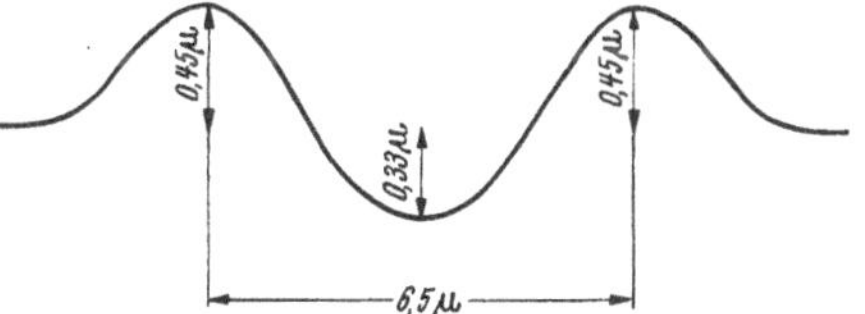

Abb. III, 48. Schematische Zeichnung des Querschnittes einer Ritzspur an Plexiglas. (Nach SMEKAL.) [SMEKAL, A., W. KLEMM: Mh. Chemie **82**, 411 (1951).]

Ein Vergleich der Mikroritzhärte mit der Makrohärte, gemessen durch die Kugeleindringung, zeigt, daß die Mikrohärte etwa 1 Zehnerpotenz höher liegt. Jedoch muß man sich bei einem solchen Vergleich realisieren, daß Mikrohärte und

[1] Siehe S. 208, Fußnote 4.
[2] Versuche von P. DEKKING: Centraal Laboratorium T. N. O., Delft.

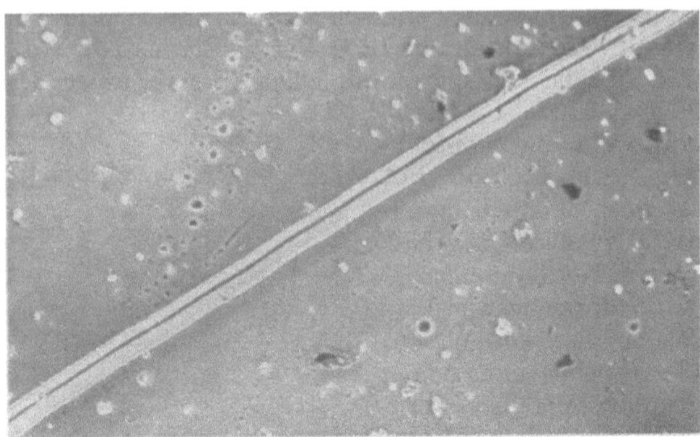

Abb. III, 49. Ritzbahn auf Plexiglas, gezogen mit einem Langspielsaphir unter 25 g Belastung. Breite der Spur etwa 30 μ. Leitz Ortholux, Phasenkontrast, Durchlicht, Objektiv 10 ×, Vergrößerung 94 ×. (P. DEKKING, Centraal Laboratorium T.N.O., Delft.)

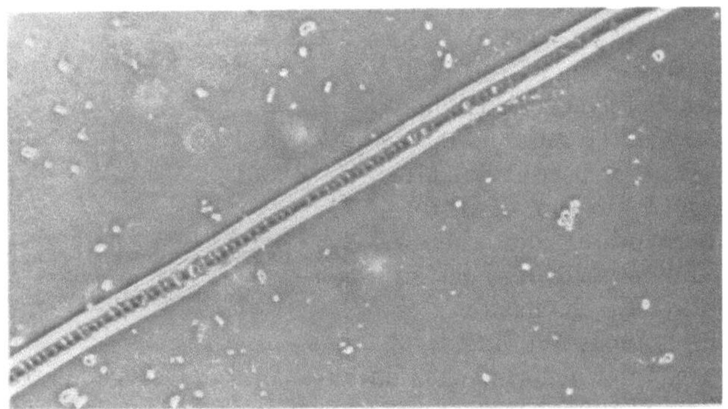

Abb. III, 50. Ritzbahn auf Plexiglas, gezogen mit einem Langspielsaphir unter 30 g Belastung. Breite der Spur etwa 30 μ. Leitz Ortholux, Phasenkontrast, Durchlicht, Objektiv 10 ×, Vergrößerung 94 ×. (P. DEKKING, Centraal Laboratorium T.N.O., Delft.)

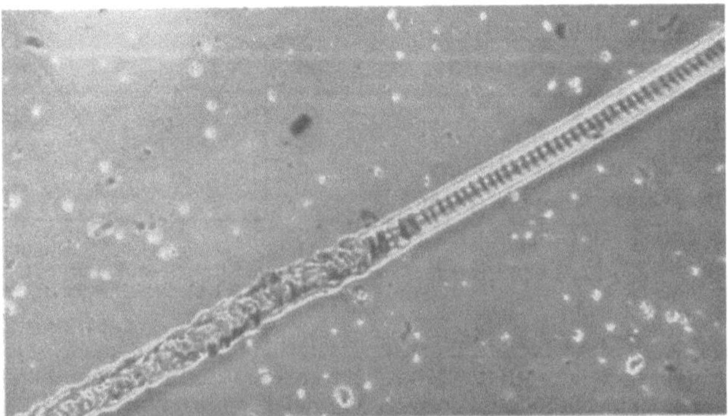

Abb. III, 51. Ritzbahn auf Plexiglas, gezogen mit einem Langspielsaphir unter 35 g Belastung. Breite der Spur etwa 30 μ. Leitz Ortholux, Phasenkontrast, Durchlicht, Objektiv 10 ×, Vergrößerung 94 ×. (P. DEKKING, Centraal Laboratorium T.N.O., Delft.)

Makrohärte die Ausdrücke verschiedener physikalischer Eigenschaften sind. Die Makrohärte ist ein Maß für den Widerstand gegen elastische Deformation[1] und kann zum Elastizitätsmodul in Beziehung gesetzt werden, die Mikroritzhärte ist ein Maß für die Kohäsionsenergie des Materials.

g) Zeitlicher Verlauf des Bruchvorganges.

Fassen wir die Ergebnisse von § 17 kurz zusammen, so kommen wir zu dem folgenden Bild für den zeitlichen Ablauf des Bruchgeschehens:

Das Auftreten von sprödem Stoffverhalten ist stets an die Wirksamkeit von Kerbstellen gebunden. In kleinsten, fehlerfreien Bereichen tritt niemals sprödes Bruchverhalten, sondern lediglich plastische Verformung auf.

Spröde Stoffe besitzen eine Mikrostruktur, bestehend aus einer großen Anzahl von Kerbstellen großer räumlicher Dichte, die eine starke Spannungskonzentration bewirken. Die Spannungshöfe dieser Kerbstellen können einander gegenseitig beeinflussen. Die Kerbstellen sind kleiner als die GRIFFITHschen Risse und können in ihrer ursprünglichen Form das Auftreten eines Bruches noch nicht veranlassen.

Bei Anlegen einer mechanischen Belastung wird die Mikrostruktur unter Mitwirkung der Wärmeenergie irreversibel verändert. Eine große Anzahl Risse beginnt zu „wachsen“. Dieses Wachsen ist ein Prozeß, der die thermische Überwindung von Energieschranken erfordert (thermische Phase des Bruchprozesses). Das Wachsen geht mit sehr kleiner Geschwindigkeit vor sich, die stark von der Temperatur und von der makroskopischen Spannung abhängig ist.

Die erste der Kerbstellen, die die kritische Größe des GRIFFITH-Risses erreicht, wird zur primären Kerbstelle und leitet die athermische Phase des Bruchvorganges ein. Die Ausbreitungsgeschwindigkeit der Bruchfront steigt steil an und erreicht einen stationären Endwert von etwa der halben transversalen Ultraschallgeschwindigkeit. Die Bruchfront breitet sich über die anderen Kerbstellen der Mikrostruktur aus und besteht aus einer Aufeinanderfolge von sekundären Brüchen, die die Bruchfläche bilden. Überall, wo sich zwei Bruchfronten oder eine Bruchfront und eine Ultraschallwelle treffen, entstehen Bruchfiguren auf der Bruchfläche.

h) Ausblick.

Übersehen wir den heutigen Stand unserer Kenntnis auf dem Gebiete der Brucherscheinungen, so müssen wir feststellen, daß wir noch weit entfernt sind von einer exakten quantitativen Beschreibung des Bruchverhaltens. Eine solche Beschreibung fordert die Aufstellung einer Gleichung oder eines Satzes von Gleichungen für die *äußeren Bedingungen*, die zu bestimmten Brucherscheinungen führen. Die Koeffizienten in diesen Gleichungen können für jedes Material verschieden sein und sind als *Materialkonstanten* zu betrachten.

Ein vollständiges Verständnis der Physik der Brucherscheinungen können wir erst gewinnen, wenn es gelingt, diese phänomenologischen Materialkonstanten quantitativ zu berechnen aus strukturellen oder chemischen Eigenschaften des betreffenden Materials.

---

[1] Vgl. dazu die Ausführungen über die Härteprüfung in Kap. IV, § 19.

Wir wollen uns überlegen, wie weit wir auf diesem Wege vorgeschritten sind und welche weiteren Schritte noch vorgenommen werden sollten.

Die wichtigste äußere Bedingung ist die äußere Kraft, aus der sich makroskopisch eine innere Spannung berechnen läßt. Bereits hier besteht Unsicherheit im Hinblick auf die Frage, welche Kombinationen der Spannungskomponenten als kritische Größen angesehen werden müssen: die Oktaederspannung τ_{okt} oder die maximale Scherspannung $\tau_{\max}$. Sicher ist außerdem auch die hydrostatische Spannungskomponente σ von Einfluß auf das Bruchverhalten, so daß der kritische Spannungszustand von der Scherspannung τ und der hydrostatischen Spannung σ abhängig sein wird (Kriterium von SCHLEICHER). Für jeden Spannungszustand lassen sich außerdem noch die Belastungszeit und die Temperatur variieren, wodurch das Bruchverhalten von mindestens drei äußeren Bedingungen abhängig wird.

Ein wirkliches Verständnis der Brucherscheinungen kann jedoch durch rein makroskopische Betrachtungen nicht erhalten werden. Sowohl die statistische Streuung der Meßresultate und die Abhängigkeit der Brucherscheinungen von den Dimensionen des Probestückes, wie auch die Morphologie der Bruchflächen führen zwangsläufig zu dem Schluß, daß beim Bruchverhalten bestimmte Inhomogenitäten oder Kerbstellen (die *Mikrostruktur*) eine entscheidende Rolle spielen.

Als die geeignete Materialkonstante zur Beschreibung des Bruchverhaltens ist daher die Verteilung der im Material vorhandenen Kerbstellen bezüglich ihrer Stärke anzusehen: Die Anzahl der Kerbstellen je Volumeinheit, die bei einem makroskopischen Spannungszustand mit den Komponenten zwischen τ, σ und $\tau + d\tau$, $\sigma + d\sigma$ zum Bruche führen.

Ist die Verteilung der Mikrostruktur bezüglich ihrer Stärke bekannt, so lassen sich hieraus mit Hilfe der statistischen Bruchtheorie auch die makroskopischen Brucheigenschaften ableiten.

Von direkten Untersuchungen der Mikrostruktur wird man aber kaum hoffen können, die Verteilung der Stärken der Kerbstellen zu erfahren, sondern bestenfalls die Verteilung ihrer geometrischen Beschaffenheit: Die Anzahl Kerbstellen je Volumeinheit, mit einer gewissen Länge, gewissen Breite und einer bestimmten Orientierung im Material. Wir stehen dann vor der Aufgabe, aus dieser geometrischen Verteilungsfunktion auf die Stärkeverteilung zu schließen. Im Falle kleiner Kerbstellendichte – wenn die Kerbstellen sich gegenseitig in ihrer Wirkung nicht beeinflussen – läßt sich diese Aufgabe lösen durch zuerst die kritische Spannung für die einzelne Kerbstelle bestimmter Abmessungen und Orientierung zu berechnen. Für den einfachsten Fall einer Kerbstelle von der Form eines langen schmalen Risses ist diese kritische Spannung durch die Theorie von GRIFFITH bekannt.

Die Kenntnis der im ursprünglichen Material vorhandenen Verteilung der Schwachstellen ist jedoch nicht ausreichend. Während das Material unter Spannung steht, verändert sich die Form der Mikrostruktur, die Kerbstellen beginnen zu „wachsen“, bis eine die kritische Größe erreicht. Der nächste Schritt muß daher das Studium der Veränderung der Kerbstellenverteilung unter äußerer Spannung mit der Zeit sein.

In einem besonders einfachen Fall, nämlich wenn die Kerbstellendichte klein ist, alle Kerbstellen von derselben Art sind und ihre Stärke durch einen einzigen Parameter charakterisiert werden kann, könnte der Wachstumsvorgang makroskopisch auf besonders einfache Weise beschrieben werden als eine Veränderung der Verteilungsfunktion dieses Parameters.

In diesem Spezialfall, der zur Zeit wohl als der einzige lösbare Fall betrachtet werden muß, wären also die Materialkonstanten: Die Verteilung der Kerbstellen hinsichtlich ihrer Stärke im ursprünglichen Material und die Funktion, nach der die Stärke einer einzelnen Kerbstelle unter einer äußeren Spannung als Funktion der Zeit zunimmt. Als äußere Bedingungen wären in diesem Fall anzusehen: die makroskopische Spannung, die Zeit und die Temperatur.

Es ist deutlich, daß wir noch weit davon entfernt sind – sei es auch nur in einem Einzelfall – alle diese Materialkonstanten bestimmen und für die quantitative Deutung des Bruchverhaltens verwenden zu können. Jedoch kann man die weitere Entwicklung der Forschung auf dem Gebiete der Brucherscheinungen in großen Zügen übersehen. Erstens wird man Methoden suchen, um die Kerbstellenverteilung auf unabhängigem Wege experimentell zu bestimmen. Eine große Schwierigkeit ist hierbei die Bestimmung einzelner Kerbstellen. Weiter wird man trachten, die zeitliche Veränderung der Stärke einer einzelnen Kerbstelle als Funktion der lokalen Spannung und Temperatur zu verfolgen.

Außer den experimentellen Schwierigkeiten sind es jedoch auch solche theoretischer und mehr prinzipieller Natur, die bei der Untersuchung von Brucherscheinungen auftauchen. Zum Beispiel ist die Frage „Was ist eine Kerbstelle?" nicht eindeutig zu beantworten. Ein Strukturfehler, der sich im homogenen Material als eine Kerbstelle auswirkt, kann in der Nähe einer größeren – gefährlicheren – Kerbstelle als homogenes Material betrachtet werden. Besonders wird hierbei die Schwierigkeit, die bei größerer Kerbstellendichtheit auftritt, merklich: Sobald verschiedene Kerbstellen mit einander in Wechselwirkung stehen, kann man von einzelnen Kerbstellen eigentlich gar nicht mehr sprechen. Auch das Auftreten von Kerbstellen ganz verschiedener Art mit ganz verschiedener Spannungsempfindlichkeit, wie z. B. Kerbstellen herrührend von Lockerstellen, Kristallkeimen oder Verunreinigungen, wird die Untersuchung wesentlich erschweren.

Abschließend können wir sagen, daß sich allmählich ein befriedigendes Bild der Brucherscheinungen herauszuschälen beginnt, aber daß es noch viel Zeit und Mühe kosten wird, bevor dieses qualitative Bild in eine quantitative Theorie verwandelt werden kann.

Allgemeine Literatur zum Kapitel III.

Bücher:

FREUDENTHAL, A. M.: The inelastic behaviour of engineering materials and structures, John Wiley & Sons, New York 1950.

HAWARD, R. N.: The strength of plastics and glass, Cleaver Hume Press, London 1949.

KOLSKY, H.: Stress Waves in solids, Oxford Clarendon Press 1953.

LESSELLS, J. M.: Strength and Resistence of metals, John Wiley & Sons, New York 1954.

NADAI, A.: Theory of flow and fracture of solids, Mac Graw Hill Book Comp., New York 1950.

OROWAN, E.: Fracture and strength of solids, Report on Progress in Physics **12**, 186 (1948). Physical Soc., London.

PETCH, N. J.: The fracture of metals, Pergamon Press Ltd., London 1954.

SMEKAL, A.: Die Festigkeitseigenschaften spröder Körper. Ergebn. exakt. Naturwiss. **15**, 106, Springer 1936. Berlin.

STANWORTH, J. E.: Physical properties of glass, Oxford, Clarendon Press 1950.

Zeitschriften:

ANDERSON, O. L. u. D. A. STUART: Ind. Engng. Chem. **46**, 154 (1954).

APELT, G.: Z. Physik **91**, 336 (1934).

CONDON, E. U.: Physics of the glassy state, Amer. J. Physics **22**, 43, 132, 224, 310 (1954).

EPSTEIN, B.: J. appl. Physics **19**, 140 (1948).

FISHER, J. C. u. J. H. HOLLOMON: Amer. Inst. min. metallurg. Engr., techn. Publ. No. 2218 (August 1947).

GRIFFITH, A. A.: Philos. Trans. Roy. Soc. London **A 221**, 163 (1920); Proc. Int. Congr. appl. Mechan. Delft 1924.

HOLLAND, A. J. u. W. E. S. TURNER: J. Soc. Glass Technol. **20**, 279 (1936); **21**, 383 (1937).

HSIAO, C. C. u. J. A. SAUER: J. appl. Physics **21**, 1071 (1950).

IRWIN, G. R. u. J. A. KIES: Welding J. **31**, 95-S (1952).

JONES, G. O.: J. Soc. Glass Technol. **33**, 120 (1949).

KAINRADL, P. u. F. HÄNDLER: Kautschuk und Gummi **7**, 34 (1954).

KIES, J. A., A. M. SULLIVAN u. G. R. IRWIN: J. Appl. Phys. **21**, 716 (1950).

KÜCH, W.: Luftfahrt-Forsch. **19**, 111 (1942).

ROBERTS, D. K. u. A. A. WELLS: Engineering **178**, 820 (1954).

ROŠ, M. u. A. EICHINGER: EMPA-Bericht 172, 173 Zürich (1949, 1950).

SCHARDIN, H.: Kunststoffe **44**, 48 (1954); Glastechn. Ber. **23**, 1, 67, 325 (1950).

SHAND, E. B.: J. Am. Ceram. Soc. **37**, 52, 559 (Febr. 1954).

SMEKAL, A.: Z. Physik **103**, 495 (1936).

SMEKAL, A.: Glastechn. Ber. **23**, 57, 186 (1950); Österr. Ing.-Arch. **7**, 49 (1953).

SMEKAL, A. u. W. KLEMM: Mh. Chem. **82**, 411 (1951).

Sprödbruchkolloquium Leoben: Radex Rundschau **4-5** (1953).

STUART, D. A. u. O. L. ANDERSON: J. Am. Ceram. Soc. **36**, 416 (1953).

TAYLOR, N. W.: J. appl. Physics **18**, 943 (1947).

WALLER, G.: Ind. Eng. Chem. **44**, 1328 (1952).

WAPLER, D.: Dissertation, Forschungsinstitut für Pigmente und Lacke, Stuttgart.

WECK, R.: Trans. of the inst. of welding **41** (April 1950).

WEIBULL, W.: Ing. Vetensk. Akad., Handl. 151 (1939).

YOFFE, E. H.: Phil. Mag. **42**, 739 (1951).

Viertes Kapitel.

Kritischer Überblick über die technologischen Prüfmethoden.

Von

W. MESKAT, O. ROSENBERG, F. SCHWARZL und A. J. STAVERMAN.

Mit 15 Abbildungen.

A. Kautschukartige und plastische Werkstoffe.

Von F. SCHWARZL und A. J. STAVERMAN.

§18. Einleitung.

Der vorliegende Abschnitt A dieses Kapitels soll eine Übersicht geben über die bei Kunststoffen und kautschukartigen Materialien verwendeten Prüfmethoden. Die Methoden für Faserstoffe werden im Abschnitt B behandelt. Der Zweck einer solchen Darstellung liegt jedoch nicht in der Aufzählung und Beschreibung einer großen Anzahl von Prüfmethoden, hierfür verweisen wir den Leser auf die Standardliteratur[1]. Wir wollen vielmehr an Hand von einigen typischen Beispielen den Zusammenhang erläutern zwischen Prüfmethoden und den in den vorangehenden drei Kapiteln behandelten physikalischen Gesichtspunkten. Wegen weiterer grundsätzlicher Betrachtungen verweisen wir auf den Abschnitt B, vor allem auf § 21.

Ein logisches und in sich abgeschlossenes System von mechanisch-technologischen Prüfmethoden – das jedoch heutzutage noch lange nicht realisierbar ist – sollte eine eindeutige Charakterisierung aller mechanisch-

---

[1] Für eine ausführliche Beschreibung von Prüfmethoden verweisen wir auf: Für die deutschen Vorschriften die Normblätter DIN 53452 (Biegefestigkeit), DIN 53453 (Schlagzähigkeit und Kerbschlagzähigkeit), DIN 53454 (Druckfestigkeit), DIN 53455 (Zugfestigkeit), DIN 53456 (Eindruckhärte) und DIN 53462 (Martenstemperatur).

Für die amerikanischen Normen: A. S. T. M. Standards 1952, part 6, Amer. Soc. Test. Mat. Philadelphia 1953.

Für die englischen Vorschriften: British Standard 1948 (Synthetic Resins) British Standard Institution London SW 1.

Für die französischen Vorschriften z.B.: Union technique des syndicats de l'électricité, Méthodes d'essais des matières plastiques utilisées dans la construction électrique, Paris 1946.

Für die niederländischen Vorschriften die Normblätter: Hoofd commissie voor de Normalisatie in Nederland V 933, V 1509, V 1512.

rheologischen Eigenschaften eines Stoffes möglich machen. Es sollten in einem solchen System keine zwei verschiedenen Experimente vorkommen, die dieselbe mechanische Eigenschaft messen und es sollte auch keine mechanische Eigenschaft unbestimmt bleiben. Daher erhebt sich als erstes die Frage, welches sind die voneinander unabhängigen Materialeigenschaften, die das mechanische Verhalten festlegen?

Eine Einteilung des mechanischen Verhaltens in großen Zügen, gültig bei den verschiedenartigsten Stoffen, wurde bereits bei der Einteilung des vorliegenden Bandes benützt. Diese unterscheidet drei Stufen der rheologischen Forschung, fortschreitend von kleinen nach großen Deformationen oder Beanspruchungen:

1. Lineares Deformationsverhalten.
2. Nichtlineares Deformationsverhalten.
3. Bruchverhalten.

Lineares Deformationsverhalten finden wir in einem Bereich bei kleinen Spannungen und Deformationen bei praktisch allen Stoffen, unabhängig von der chemischen oder physikalischen Konstitution des betrachteten Materials. Darum ist es möglich, alle diese Stoffe im Gültigkeitsbereich des Superpositionsprinzipes mechanisch zu charakterisieren durch einen zeitabhängigen Elastizitätsmodul $E(t)$ und eine zeitabhängige Konstante von POISSON $\nu(t)$. Statt dessen können natürlich auch zwei andere unabhängige charakteristische Funktionen gewählt werden wie z.B. der YOUNGsche Modul $E(t)$ und der Schermodul $G(t)$. Sind zwei dieser charakteristischen Funktionen *für alle Zeiten* bekannt, so läßt sich das lineare Deformationsverhalten bei konstanter Temperatur vollständig berechnen[1].

Prüfmethoden, die das lineare Deformationsverhalten – also im wesentlichen den Elastizitätsmodul – messen, sind sehr zahlreich: Der Zugversuch, Biegungs- und Torsionsversuch, die Erweichungstemperatur, die Härtemessung und viele andere. Bei richtiger Interpretation führen alle diese Teste auf eine Messung des Elastizitätsmoduls, sie sind jedoch nicht imstande, eine vollständige Beschreibung des linearen Deformationsverhaltens zu liefern. Ein linearer Test gibt nämlich stets nur einen Punkt der gesamten Modulzeitabhängigkeit wieder, z.B. einen „Modul nach einer Minute", während zur vollständigen Charakterisierung die gesamte Modul-Zeit-Funktion gemessen werden müßte.

Im Falle des nichtlinearen Deformationsverhaltens sind wir heutzutage noch nicht imstande, in voller Allgemeinheit Aussagen über die charakteristischen Funktionen zu machen. Wir konnten lediglich eine Einteilung geben nach den auffallendsten Eigenschaften in Kaltverstrekkung, Kautschukelastizität und nichtlineare Fließerscheinungen, charakteristisch für bzw. Polymere im glasartigen Zustand, im kautschukelastischen Zustand und im flüssigen oder gelösten Zustand. Eine vollständige Beschreibung dieser drei Phänomene ist nicht gelungen, denn wir waren gezwungen, bei der Kautschukelastizität die Zeitabhängigkeit und bei

[1] Das gilt für isotrope Stoffe, im Falle von Anisotropie müssen mehr als zwei unabhängige charakteristische Funktionen bestimmt werden.

den nichtlinearen Fließerscheinungen die elastischen Eigenschaften zu vernachlässigen. Unter den Prüfmethoden finden wir heutzutage noch nicht viele, die diese nichtlinearen Deformationseigenschaften unabhängig von anderen rheologischen Eigenschaften messen[1]. Vielmehr tritt das nichtlineare Deformationsverhalten meistens als ein störender Faktor bei der Messung des linearen Deformationsverhaltens oder der Brucherscheinungen auf.

Brucherscheinungen bieten die meisten Schwierigkeiten bei der physikalischen Beschreibung. Wie bereits im dritten Kapitel gezeigt wurde, ist sprödes Bruchverhalten grundsätzlich an die Wirksamkeit von Inhomogenitätsstellen geknüpft. Ein scheinbar homogener Körper besitzt in Wirklichkeit eine große Anzahl von kleinen Fehlstellen, von denen jeder spröde Bruch ausgeht und über die sich jeder Bruch fortpflanzt. Das hat zur Folge, daß die wirklich auftretenden Festigkeitswerte weit unter den theoretischen Bindungsfestigkeiten des Materials bleiben.

Brucherscheinungen werden daher im allgemeinen weniger empfindlich sein gegenüber Änderungen im chemischen Aufbau des Materials (bei gleichbleibender Mikrostuktur) und werden hauptsächlich abhängen von der Größe und Verteilung der Fehlstellen, der *Mikrostruktur* des Stoffes. Die richtige, für die Festigkeit eines Materials charakteristische „Materialeigenschaft" wäre daher die Angabe der Größe, Verteilung und Anzahl der Fehlerstellen. Bei vollständiger Kenntnis dieser Mikrostruktur sollte diese alle charakteristischen Züge des Bruchverhaltens ebenso beschreiben, wie etwa der Elastizitätsmodul das lineare Deformationsverhalten beschreibt[2].

Beim heutigen Stand der Prüfung von Festigkeitseigenschaften ist es noch nicht möglich, eine Analyse der Mikrostruktur zu geben und man begnügt sich mit der Angabe einer „gemittelten Bruchspannung". Bei der Berechnung dieser Bruchspannung geht man bewußt von der – in Wirklichkeit unrichtigen – Annahme aus, daß das Material sich während des Bruches wie ein homogener Stoff verhält. Die Bruchspannung läßt sich natürlich zur Beschreibung des Bruchverhaltens verwenden, wenn man in Rechnung zieht, daß diese Bruchspannung infolge der Existenz der Mikrostruktur eine Reihe von ungewöhnlichen Eigenschaften zeigt.

Die Bruchspannung ist keine Materialkonstante, sondern eine statistische Eigenschaft in dem Sinne, daß man – auch bei vollkommener Unterdrückung von apparativen und subjektiven Meßfehlern – noch eine Verteilung von Bruchspannungen erhält. Die Bruchspannung wird bestimmt durch den *schwächsten Punkt* in der Mikrostruktur und es ist deutlich, daß eine solche Eigenschaft wesentlich mehr streuen muß als eine Eigenschaft wie der Elastizitätsmodul, der von der *gemittelten* Struktur abhängt.

Die Bruchspannung ist keine spezifische Materialeigenschaft in dem

---

[1] Eine Ausnahme bildet die Messung des Spannungs-Deformations-Diagramme zur Charakterisierung der Kautschukelastizität und der Kaltverstreckung.

[2] Dabei ist natürlich zu beachten, daß die Mikrostruktur außer von der chemischen Zusammensetzung noch von vielen anderen Faktoren abhängen wird, wie Herstellungsweise, thermische Vorbehandlung usw.

Sinne, daß sie von dem Volumen des betrachteten Probestückes unabhängig wäre. Unter sonst gleichen Verhältnissen liegt die Bruchspannung stets um so niedriger, je größer das Volumen des betrachteten Probestückes ist. Dieser Volumeneffekt erklärt sich dadurch, daß die schwächsten Punkte der Mikrostruktur in einem großen Probestück im Mittel schwächer sind als in einem kleinen. Auf ähnliche Weise wirkt sich auch die Form des Probestückes auf die Bruchspannung aus.

Die Bruchspannung ist abhängig von der Belastungsdauer, da nicht nur die Größe der Kraft, sondern auch die Beanspruchungsdauer den Brucheintritt bestimmt. Je kleiner eine Kraft ist, desto länger ist die zugehörige Bruchzeit – die Zeit zwischen Anbringen der Kraft und Eintritt des Bruches.

Schließlich ist das Bruchverhalten abhängig von der Geometrie des Spannungszustandes.

Zusammenfassend können wir sagen, daß es bei dem Stande unserer heutigen Kenntnis noch nicht möglich ist, ein System von Prüfmethoden anzugeben, um das mechanische Verhalten von Hochpolymeren zu charakterisieren. Wir müssen uns daher im folgenden damit begnügen, die bestehenden Prüfmethoden zu analysieren und – wo möglich – auf physikalische Begriffe zurückzuführen.

§19. Zerstörungsfreie Prüfmethoden, die das Deformationsverhalten messen.

Wir beginnen mit der Betrachtung der einfachsten Prüfmethoden, das sind solche, bei denen das Deformationsverhalten unter relativ kleinen Belastungen gemessen wird. Wie bereits in der Einleitung erörtert, mißt man bei diesen Prüfmethoden im wesentlichen den Elastizitätsmodul. Unter günstigen Umständen muß also zwischen diesen Prüfmethoden und dem Elastizitätsmodul ein eindeutiger Zusammenhang bestehen, also auch zwischen den nachfolgenden Prüfmethoden untereinander. In gewissen Fällen kann die vorgeschriebene Belastung jedoch so groß sein, daß außerdcm nichtlineare Effekte auftreten und der Zusammenhang zwischen Prüfmethode und elastischem Deformationsverhalten minder gut wird.

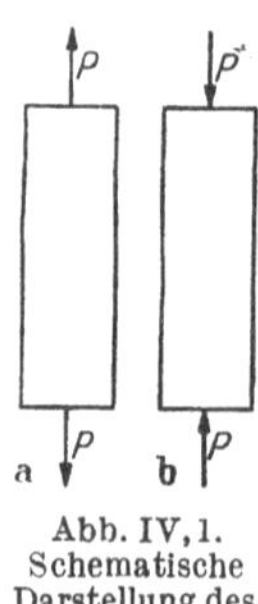

Abb. IV, 1. Schematische Darstellung des Zugversuches (a) und des Druckversuches (b).

Die einfachste Methode zur Bestimmung des (YOUNGschen) Elastizitätsmoduls ist der Zugversuch[1] (Abb. IV, 1). Obwohl dieser meistens bei konstanter Deformationsgeschwindigkeit bis zum Bruche ausgeführt wird, kann man doch aus der Anfangssteigung des Spannungs-Deformations-Diagrammes den Zugmodul berechnen

$$E = d\sigma/d\varepsilon. \tag{IV,1}$$

Der so berechnete Modul ist um so größer, je größer die gewählte Deformationsgeschwindigkeit ist. In ähnlicher Weise kann man auch aus dem

[1] DIN 53455, DIN 53457, A. S. T. M. D 412–51 T, A. S. T. M. D 638–52 T.

Druckversuch[1] einen Modul berechnen unter Benützung der Anfangssteigung der Druck-Deformations-Kurve. Dieser Modul sollte theoretisch übereinstimmen mit dem aus dem Zugversuch bestimmten. Beide entsprechen einem Punkt der Spannungs-Relaxations-Kurve nach etwa wenigen Sekunden bis einer Minute Belastungszeit.

Das Zug-Dehnungs-Diagramm, aufgenommen mit konstanter Dehnungsgeschwindigkeit bis zum Bruchpunkt, enthält wesentlich mehr Information als nur die linearen Eigenschaften. Bei geeigneter Interpretation gibt das Zug-Dehnungs-Diagramm Auskunft über lineares Deformationsverhalten, nicht-lineares Deformationsverhalten und Brucheigenschaften.

1. Die Anfangssteigung ergibt den Zugmodul nach Gl. (IV,1).

2. Die Form des Spannungs-Deformationsverlaufes vermittelt einen Eindruck über die nicht-linearen Deformationseigenschaften. Wir können unterscheiden (siehe Kap. II): sprödes Deformationsverhalten, Abb. II,18; zähes Deformationsverhalten, Abb. II,19; Kaltverstreckung, Abb. II,26; kautschukelastisches Verhalten, Abb. II,3.

3. Die Lage des Bruchpunktes ergibt die Bruchspannung und die Bruchdeformation.

All diese Eigenschaften sind abhängig von der Deformationsgeschwindigkeit und der Temperatur, so daß zur vollständigen Charakterisierung eine Reihe Spannungs-Deformations-Kurven bei verschiedenen Deformationsgeschwindigkeiten und Temperaturen nötig ist.

Eine etwas andere, aber prinzipiell ebenso einfache Methode ist die Bestimmung der *Biegungssteifheit*, wobei das Probestück auf eine der in Abb. IV,2 dargestellten Weisen auf Biegung beansprucht wird. Sowohl aus der Dreipunktsbiegung[2] wie auch aus der Biegung des einseitig eingespannten Probestückes[3] läßt sich der Biegungsmodul E_b berechnen. Es sei bei rechteckigem Querschnitt des Probestückes l die Länge zwischen den Stützen (bzw. die Spanlänge), h die Dicke des Probestückes in der Kraftrichtung und b die Breite, P die angreifende Kraft und δ die erhaltene Durchbiegung. Dann gilt

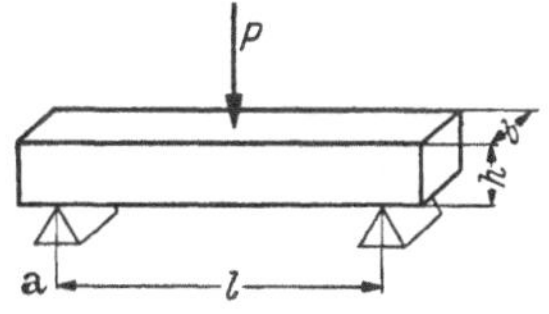

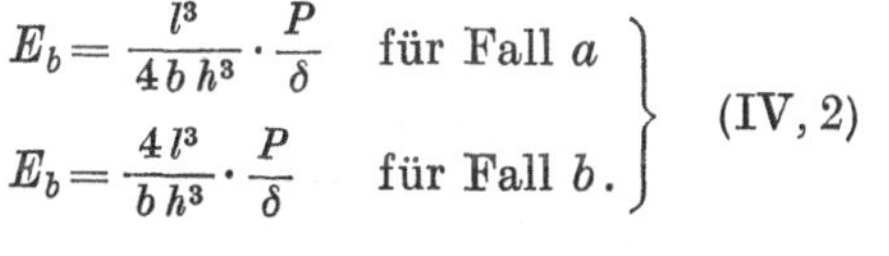

$$\left.\begin{aligned} E_b &= \frac{l^3}{4\,b\,h^3}\cdot\frac{P}{\delta} \quad \text{für Fall } a \\ E_b &= \frac{4\,l^3}{b\,h^3}\cdot\frac{P}{\delta} \quad \text{für Fall } b. \end{aligned}\right\} \qquad \text{(IV, 2)}$$

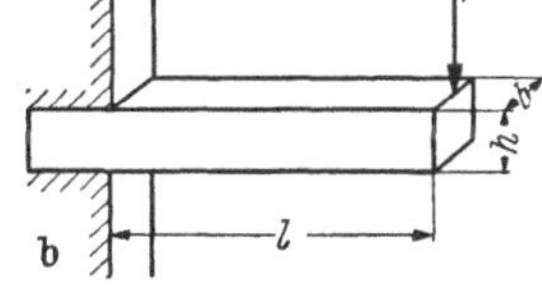

Abb. IV, 2. Biegungssteifheit: (a) Dreipunktsbiegung, (b) einseitig eingespanntes Probestück.

In der Praxis treten manchmal Unterschiede zwischen Zug, Druck und Biegungsmodul auf. Diese sind meistens zurückzuführen auf Anisotropie oder Inhomogenität des Probematerials oder auf Unterschiede im Verformungsgrad, in der Deformationsgeschwindigkeit usw. zwischen den verschiedenen Prüfungsmethoden.

Der resultierende Biegungsmodul sollte bei isotropen Probestücken mit dem Elastizitätsmodul übereinstimmen, den man aus Zug und Kom-

[1] DIN 53454, A. S. T. M. D 575–46, A. S. T. M. D 621–51, A. S. T. M. D 695–52 T.
[2] DIN 53452, A. S. T. M. 797–46. — [3] A. S. T. M. D 747–50.

pressionsversuch erhält. Im allgemeinen ist die Übereinstimmung von Biegungs- und Zugmodul innerhalb der Fehlergrenzen bei nichtkristallinen und nichtorientierten Stoffen gut[1]. Der aus dem Kompressionsversuch bestimmte Modul liegt in vielen Fällen etwas niedriger.

Bei kurzen und dicken Probestücken treten bei der Biegung außer einem reinen Biegungsmoment noch Scherkräfte auf. Der Elastizitätsmodul E ist dann nicht identisch mit dem aus Gl. (IV,2) bestimmten Biegungsmodul E_b, sondern hängt noch von dem Verhältnis Dicke zu Länge h/l und von dem Wert der POISSONschen Konstanten ν ab:

$$\left.\begin{aligned} E &= E_b\,[1 + 3\,(1+\nu)\,(h/l)^2] \quad \text{für Fall } a \\ E &= E_b\,[1 + 3/4\,(1+\nu)\,(h/l)^2] \quad \text{für Fall } b \end{aligned}\right\} \qquad \text{(IV, 3)}$$

Abb. IV,3 zeigt das Verhältnis E/E_b als Funktion von h/l bei inkompressiblem Material ($\nu = 0{,}5$). Man sieht, daß bei kurzen und dicken Probestücken der Elastizitätsmodul wesentlich höher sein kann als der Biegungsmodul. Bei einem Verhältnis $h/l < 1/8$ (für Fall a) und $h/l < 1/4$ (für Fall b) stimmen beide Moduli praktisch überein.

Bei der Bestimmung der Torsionssteifheit[2] wird ein Probestück der Länge l durch ein Moment M tordiert und der auftretende Torsionswinkel φ gemessen. Bei isotropen Stoffen läßt sich daraus der Schermodul G berechnen. Bei kreisförmigem Probenquerschnitt vom Radius r gilt:

$$G = \frac{2\,M\,l}{\pi\,r^4\,\varphi}\,. \qquad \text{(IV, 4a)}$$

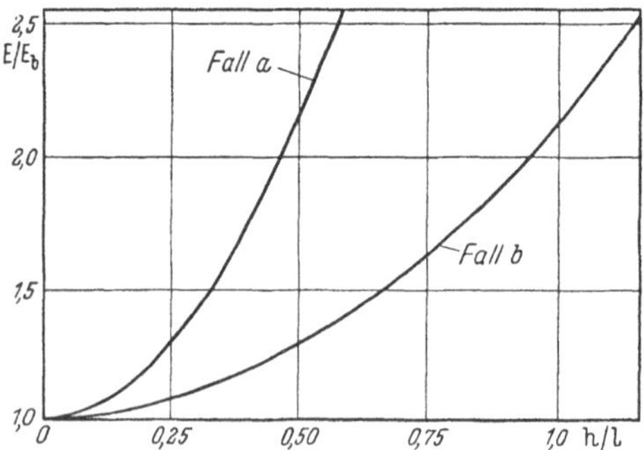

Abb. IV,3. Das Verhältnis E/E_b als Funktion von h/l für ein inkompressibles Material.

Im Falle eines rechteckigen Probenquerschnittes, wobei a die längere und b die kürzere Querschnittsdimension ist, gilt statt dessen:

$$G = \frac{4\,M\,l}{a\,b^3\,\varphi} \cdot \frac{1}{f}\,. \qquad \text{(IV, 4b)}$$

f ist dabei ein Formfaktor, der vom Verhältnis a/b abhängt, siehe Tab. IV,1.

Tabelle IV,1. *Der Formfaktor f.*

a/b	1,00	1,25	1,50	2,00	3,00	4,00	5,00	10,00	∞
f	2,25	2,75	3,13	3,66	4,21	4,49	4,66	5,00	5,33

Der so gemessene Schermodul hängt bei isotropen Stoffen mit dem Zug oder Biegungsmodul zusammen über die POISSONsche Kontante ν

$$E = 2\,G\,[1 + \nu]\,. \qquad \text{(IV, 5)}$$

Nur im Falle von kautschukelastischen Substanzen kann man die POISSONsche Konstante 0,5 setzen, bei harten Kunststoffen variiert sie zwischen 0,3 und 0,5.

[1] Siehe z.B. Technical data of plastics, Plastics Mat. Manuf. Ass. Inc. Washington 5 DC, 1948. — [2] A.S.T.M. D 1053–52 T, A.S.T.M. D 1043–51.

Bei gummielastischen Polymeren verwendet man die Messung der Torsionssteifheit als Funktion der Temperatur zur Bestimmung des Temperaturbereiches, in dem man den Kautschuk noch verwenden kann. Ein Kautschuk hat bei normalen Temperaturen einen niedrigen Schermodul, wird jedoch bei sehr niedrigen Temperaturen hart und verliert seine gummielastischen Eigenschaften, siehe Abb. IV,4. Durch die Torsionsmessung läßt sich so die Übergangstemperatur als die untere Grenze für die Gebrauchstemperatur ermitteln.

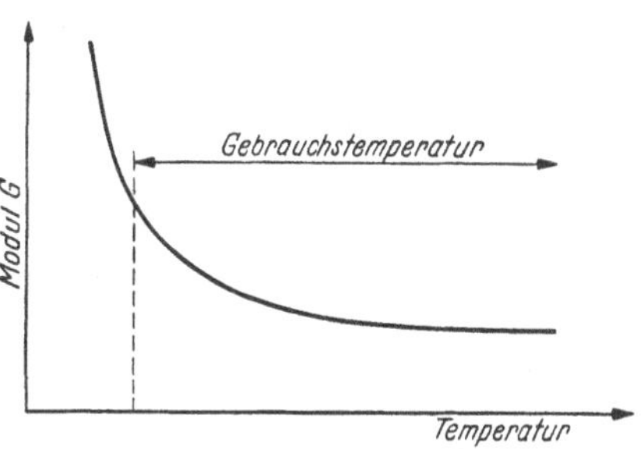

Abb. IV,4. Schematischer Verlauf der Modul-Temperaturkurve und der Gebrauchstemperaturbereich für Kautschuk.

Bei harten Kunststoffen bestimmt man die obere Grenze des Gebrauchstemperaturbereiches als die sog. *Erweichungstemperatur*[1]. Das Probestück wird einer Durchbiegung unter konstanter Kraft unterworfen, einer Dreipunktsbiegung im Falle des amerikanischen Standards (siehe Abb.IV,2a) und einer Vierpunktsbiegung im Falle des deutschen Standards. Gleichzeitig wird die Temperatur des Probestückes in einem Ölbad bzw. einem Luftthermostat kontinuierlich erhöht. Die Temperatur, bei der eine bestimmte Durchbiegung erzielt wird, heißt ‚heat distortion temperature' bzw. ‚Martenstemperatur'. Beide Temperaturen stimmen innerhalb weniger Grade überein, so daß es genügt, die Bedeutung der Methode an der amerikanischen Vorschrift zu erläutern.

Nach der amerikanischen Vorschrift ist die Erweichungstemperatur diejenige, bei der eine Durchbiegung von 0,01 inch erzielt wird. Die vorgeschriebene Belastung entspricht dabei einer maximalen Biegungsspannung von 264 P. S. I. ($l = 5$ inch, $h = 0,5$ inch). Diese Vorschrift ist gleichwertig mit einer Bestimmung des Elastizitätsmoduls: Die Erweichungstemperatur T_w ist die Temperatur, bei der das Probematerial einen Elastizitätsmodul von ungefähr 10^{10} dynes/cm² erreicht. Abb.IV,5 soll die physikalische Bedeutung der Prüfmethode erläutern. Der Gebrauchstemperaturbereich von harten Kunststoffen ist der glasartig harte Bereich, wo der Modul durchweg einen Wert größer als 10^{10} d/cm² besitzt. Dieser Gebrauchstemperaturbereich wird begrenzt durch den Übergang zum kautschukelastischen bzw. Fließbereich, wo der Modul um mehrere Zehnerpotenzen absinkt. Die Erweichungstemperatur charakterisiert

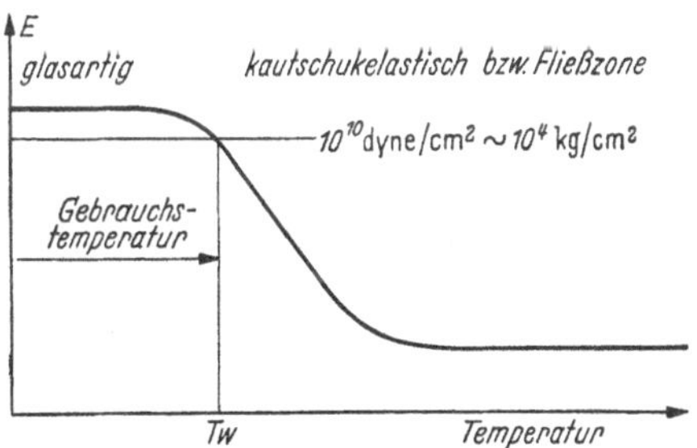

Abb.IV, 5. Der Übergangsbereich glasartig-kautschukelastisch und die Lage der Erweichungstemperatur für harte Kunststoffe.

[1] Deutscher Standard: Martenstemperatur DIN 53462; amerikanischer Standard: Heat distortion temperature A. S. T. M. D 648- 45 T.

daher die untere Temperaturgrenze des Übergangsbereiches bzw. die obere Temperaturgrenze des Gebrauchstemperaturbereiches für Kunststoffe.

Als besonders schönes Beispiel für die physikalische Interpretation von Prüfmethoden erwähnen wir die Messung der *Eindringungshärte*, die sich in vielen Fällen mit dem Elastizitätsmodul des Prüfmaterials in Zusammenhang bringen läßt. Unter Härte versteht man die Fähigkeit der Oberfläche eines Materials, konzentrierten mechanischen Kräften zu widerstehen. Bei jeder Härtemessung wird deshalb ein sehr harter Eindringungskörper (meist aus gehärtetem Stahl) unter bestimmter Druckkraft P gegen die glatte Oberfläche des Probematerials gedrückt. Die Eindringungstiefe ergibt dann ein reziprokes Maß für die Härte.

Der Eindringungskörper kann verschiedene geometrische Formen besitzen, wie die einer Kugel, einer Pyramide, eines Kegels oder eines Zylinders. Die der physikalischen Behandlung am besten zugängliche Form der Härteprüfung ist die Kugeleindringung, auf die wir die folgende Diskussion hauptsächlich beschränken.

Eine gehärtete Stahlkugel vom Radius r und sehr hohem Elastizitätsmodul wird unter einer Druckkraft P in die Oberfläche des Probestückes gepreßt. Die Kugel selbst wird nicht wesentlich deformiert, sondern dringt um eine Tiefe b in das Probestück ein, wobei sich eine Kontaktfläche zwischen Probematerial und Kugel bildet, deren Projektion die Form eines Kreises vom Radius a hat (siehe Abb. IV, 6). Die Eindringtiefe b hängt mit dem Kontaktradius zusammen

$$a^2 = r b. \qquad \text{(IV, 6)}$$

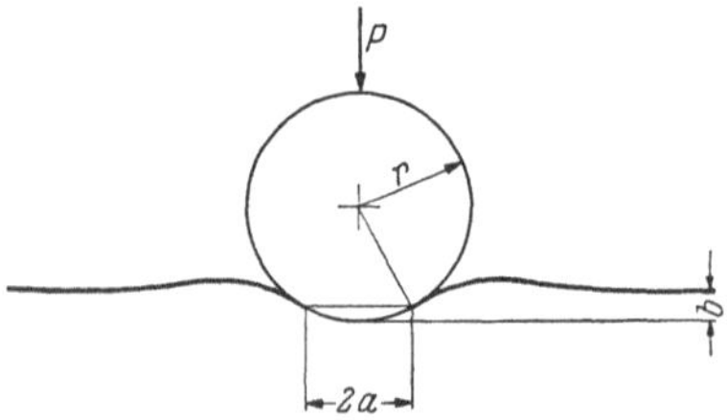

Als Definition der Härte verwendet man in vielen Fällen nach MEYER den Quotienten aus Druckkraft und der Projektion der Kontaktfläche

$$H = \frac{P}{\pi a^2} = P/\pi r b. \qquad \text{(IV, 7)}$$

Abb. IV, 6. Die Kugeleindringung.

Bei der Härteprüfung müssen wir zwei Deformationsmechanismen unterscheiden, die reversible Kugeleindringung und die irreversible Kugeleindringung, aus denen sich die totale Eindringungstiefe b additiv zusammensetzt

$$b = b_{\text{rev}} + b_{\text{irr}}. \qquad \text{(IV, 8)}$$

b_{rev} ist der Anteil, der nach Entlasten wieder augenblicklich oder allmählich verschwindet, und entsteht durch elastische oder viscoelastische Verformung. Der reversible Anteil läßt sich in Beziehung setzen zum Elastizitätsmodul E und zur POISSONschen Konstante ν des Probematerials. Der irreversible Anteil b_{irr} entsteht durch NEWTONsches, Nicht-NEWTONsches oder plastisches Fließen und läßt sich bei Hochpolymeren nicht auf einfache Weise deuten.

Die reversible Kugeleindringung wird durch die HERTZsche Theorie[1]

[1] Siehe z. B. eines der Standardwerke über Elastizitätstheorie (Literaturzitate S. 11).

in Abhängigkeit von Belastung P und Kugelradius r vollständig beschrieben:

$$b_{\text{rev}} = r^{-1/3} \left[\frac{3}{4} \frac{(1-\nu^2)}{E}\right]^{2/3} P^{2/3}. \tag{IV, 9}$$

Die in Gl. (IV, 7) definierte Härte ist also keineswegs eine Materialeigenschaft in dem Sinne, daß sie unabhängig vom Druck P wäre, sie ist vielmehr proportional zur dritten Wurzel aus P

$$H = \frac{1}{\pi} r^{-2/3} \left[\frac{4}{3} \frac{E}{1-\nu^2}\right]^{2/3} P^{1/3} \sim E^{2/3} P^{1/3}. \tag{IV, 10}$$

Um also aus der Eindringungshärte auf den Elastizitätsmodul schließen zu können, muß man den reversiblen Anteil der Kugeleindringung betrachten. Die „reversible Eindruckshärte" läßt sich auffassen als eine schnelle und bequeme Methode zur Bestimmung des Elastizitätsmoduls.

Bei der physikalischen Deutung der bestehenden Härteprüfungen stehen wir jedoch vor der Schwierigkeit, daß hier teilweise die irreversible, teilweise die totale Kugeleindringung zugrunde gelegt wird, in keinem Fall jedoch die reversible Kugeleindringung allein. Härtemessungen, die auf der Bestimmung der Größe b_{irr} beruhen[1], sind einer physikalischen Deutung mit Hilfe des Elastizitätsmoduls nicht zugänglich. Wir schließen diese Methoden ausdrücklich von der nachfolgenden Diskussion aus. Die meisten Härtemessungen beruhen auf der Bestimmung der totalen Eindringung b und enthalten daher sowohl die elastischen und viskoelastischen Eigenschaften, wie auch die Fließeigenschaften des Materials. Diese Methoden sind einer Deutung zugänglich unter der Voraussetzung, daß die reversiblen Deformationstypen den irreversiblen Fluß überwiegen, d.h., daß $b_{\text{rev}} \gg b_{\text{irr}}$ oder daß

$$b \simeq b_{\text{rev}} \tag{IV, 11}$$

ist. Voraussetzung (IV, 11) ist nun gerade bei kautschukelastischen und teilweise auch bei harten Polymeren erfüllt.

Die *deutsche Norm* für die *Härteprüfung*[2] schreibt die Messung der Eindringung b einer Stahlkugel von 5 mm Durchmesser unter einem Druck von 50 kg vor. Die Härte H' wird berechnet nach der Formel $H' = P/2\pi b r = 1/2\,H$ und angegeben in kg/cm². Innerhalb der Genauigkeit der Theorie können wir die Größe $8\,E/9\,(1-\nu^2)$ für harte Kunststoffe gleich E setzen. Wir erhalten so unter Benützung von (IV,10) den Zusammenhang

$$H' \simeq 1{,}9\,E^{2/3}, \tag{IV, 12}$$

wo die Härte H' und der Modul E beide in kg/cm² einzusetzen sind.

Die *Rockwell Härteprüfung*[3] läßt sich nur teilweise mit unserer Theorie beschreiben. Die Rockwell R, L, M und E Härteskalen beruhen auf der Messung der irreversiblen Kugeleindringung und können daher nicht behandelt werden. Die Rockwell α Härteskala schreibt jedoch die Messung der totalen Eindringung vor: Eine Stahlkugel von 1/2 inch Durchmesser wird zuerst unter einer Vorlast von 10 kg auf das Probematerial gedrückt, anschließend wird die Belastung auf 60 kg erhöht. Der Unterschied der beiden Eindringungen dient als Maß für die Härte. Wenden

---

[1] Beispiele solcher Messungen sind die Rockwellhärte (R, L, M oder E-Skala) und die Härteprüfung bei Metallen. — [2] DIN 53456. — [3] A.S.T.M. D 785-51.

wir Gl. (IV, 9) zweimal an, so erhalten wir den Zusammenhang zwischen der Differenz der beiden Eindringungen $b' = b_{60\,\mathrm{kg}} - b_{10\,\mathrm{kg}}$ und dem Elastizitätsmodul:

$$E \simeq 2{,}9 \cdot 10^7 b'^{-3/2}\,. \qquad \text{(IV, 13)}$$

Hier sind b' in μ (= 0,001 mm) und E in kg/cm² einzusetzen.

Als α Rockwellhärtezahl bezeichnet man die Differenz (150 – Eindringung gemessen in Einheiten von $2\,\mu$). Diese Definition kann also für kleine Moduli auch zu negativen Härtezahlen führen. Tabelle IV, 2 gibt den Zusammenhang zwischen der α-Rockwellhärtezahl, der Eindringung und dem Elastizitätsmodul.

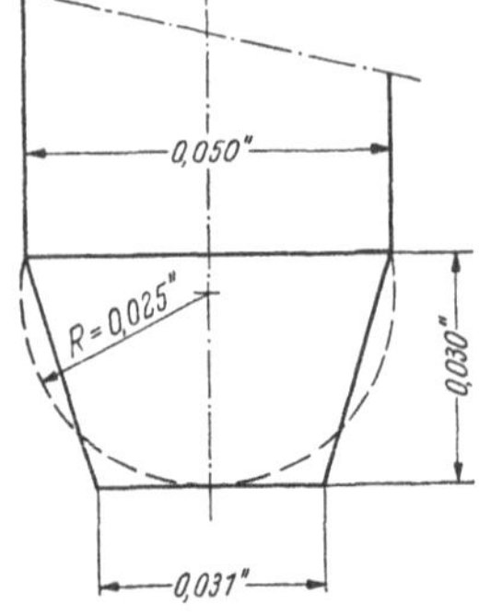

Abb. IV, 7. Der Eindringungskörper des Durometers und die äquivalente Kugel.

Für weiche Polymere mißt man die Härte mit Hilfe eines Durometers[1]. Der Eindringungskörper hat hier die Form eines abgeschnittenen Kreiskegels (siehe Abb. IV, 7). Die *Shore-Härteskala* ist in 100 Einheiten eingeteilt, bei Härte 100 haben wir keine Eindringung unter der maximalen Last, bei Härte 0 die maximale Eindringung unter der minimalen Last. Da die Belastung durch eine Druckfeder hervorgerufen wird, nimmt die Eindringung linear mit der Härteskala ab und gleichzeitig nimmt die Belastung linear mit der Härteskala zu. Tabelle II, 3 zeigt die charakteristischen Daten des Durometers A. Für die Umrechnung der Härteskala in einen Elastizitätsmodul war es notwendig, die komplizierte Form des Eindringungskörpers zu ersetzen durch eine „äquivalente" Kugel, wie in Abb. IV, 7 angegeben.

Die gegebene Umrechnung von der deutschen Härteprüfung, der Rockwellhärte und der Shorehärte in einen Elastizitätsmodul E_k beruht auf der Voraussetzung rein elastischen Deformationsverhaltens. In der Praxis ist diese Voraussetzung nicht immer erfüllt:

1. Auftreten von Fließerscheinungen, die den Spannungszustand verändern.
2. Auftreten von sehr hohen Maximalspannungen, die zu nicht-linearem viscoelastischem Verhalten führen.

Tabelle IV, 2.

α-Rockwellhärtezahl	Eindringung in μ	Modul in kg/cm²
150	0	∞
140	20	$3{,}3 \cdot 10^5$
130	40	1,2
120	60	$6{,}3 \cdot 10^4$
110	80	4,1
100	100	2,9
90	120	2,2
80	140	1,8
70	160	1,5
60	180	1,2
50	200	1,0
40	220	$9{,}0 \cdot 10^3$
30	240	7,9
20	260	7,0
10	280	6,3
0	300	5,6
— 10	320	5,1
— 20	340	4,7
— 30	360	4,3

[1] Shore hardness A. S. T. M. D 676-49 T.

Tabelle IV, 3. *Shorehärte und Elastizitätsmodul.*

Durometer A Shorehärte	Eindringung in mm	Last in g	Modul E_k in kg/cm²
0	2,54	56	0,98
10	2,29	133	2,1
20	2,03	209	5,1
30	1,78	286	8,5
40	1,52	362	13,5
50	1,27	439	22
60	1,02	516	36
70	0,76	592	62
80	0,51	669	130
90	0,25	745	412
100	0,00	822	∞

Beide Faktoren bewirken, daß der aus der Kugeleindringung berechnete Elastizitätsmodul E_k nicht immer übereinstimmt mit dem Zugmodul E. Statt der theoretischen Relation $E_k \cong E$ wurde gefunden[1]

$$E_k \cong 3E/2.$$

Die Abweichungen von der Identität von E_k und E sind im allgemeinen um so stärker, je größer der relative Anteil der irreversiblen Deformation an der totalen Eindringung ist.

Wenn bei der Härteprüfung der irreversible Anteil der Deformation überwiegt, d. h. wenn $b_{\mathrm{irr}} \gg b_{\mathrm{rev}}$ ist, so ist eine physikalische Deutung der Härteprüfung viel schwieriger. Man definiert dann als Härte an Stelle von Gl. (IV, 7) allgemeiner

$$H = \frac{\text{Belastung des Eindruckkörpers}}{\text{Eindruck-Oberfläche}}.$$

Diese Definition ist geeignet für alle möglichen geometrischen Formen von Eindruckskörpern und ist mit (IV, 7) identisch, wenn der Eindruckskörper eine Kugel ist.

Unter den verschiedenen Formen von Eindruckskörpern (Tab. IV, 4) können wir solche unterscheiden, welche gleichförmige oder nicht gleichförmige Eindrücke geben, je nachdem unter verschiedenen Belastungen die entstehenden Eindrücke einander geometrisch ähnlich sind oder nicht. Kegel und Pyramiden geben gleichförmige, Kugeln geben ungleichförmige Eindrücke[2].

Tabelle IV, 4. *Verschiedene Formen der Eindruckshärte.*

	Meyer, Brinell	Rockwell	Ludwig	Vickers, Knoop
Eindruckskörper	Kugel	Kugel	Kegel	4seitige Pyramide
	Ungleichförmige Eindrücke		Gleichförmige Eindrücke	
Besonders geeignet für	Elastisch-reversibles Deformationsverhalten		Plastisch-irreversibles Deformationsverhalten	

[1] Heijboer, J., J. Leeuwerik u. F. Schwarzl: Unveröffentlichte Resultate an Polymethylacrylsäureestern.

[2] Vgl. J. H. Zaat: Metaalinstituut T. N. O., Delft, Publ. Nr. 29 (1955). — E. Kruse: Schweizer Archiv f. angew. Wiss. u. T. **16**, 225 (1950).

Bei plastisch-irreversiblem Deformationsverhalten ist die Härte H innerhalb nicht zu weiter Grenzen von der Größe der Belastung unabhängig, wenn man Eindruckskörper verwendet, die gleichförmige Eindrücke geben[1].

Es ist jedoch bis jetzt bei Kunststoffen noch nicht gelungen, die Härte zu einer physikalisch definierten Deformationseigenschaft in Beziehung zu setzen.

§ 20. Prüfmethoden, die das Bruchverhalten messen.

Prüfmethoden zur Messung der mechanischen Stärke von Polymeren lassen sich einteilen nach der Zeitdauer, die zwischen dem Beginn der Belastung und dem Auftreten des Bruches verstreicht.

Zeitdauer des Bruchversuches	Prüfmethode	gemessene Größe
10^{-3} bis 10^{-2} Sek.	Schlagversuche	Bruchenergie
Minuten	Festigkeitsversuche	Bruchspannung
sehr lang	Ermüdungsversuche	Belastung und Bruchzeit

Diese drei Arten von Stärkemessungen sind entstanden aus einer Nachahmung der Verhältnisse in der Praxis: Plötzlicher Schock, steigende Belastung, lang andauernde konstante oder Wechselbelastung.

a) Festigkeit.

Festigkeitsversuche werden in einer Zugbank ausgeführt, die auf eine gleichmäßige Deformationszunahme – seltener auf eine gleichmäßige Spannungszunahme – eingerichtet ist. Der Bruchpunkt wird innerhalb weniger Minuten erreicht und die größte auftretende Zug- oder Druckbelastung P wird umgerechnet in die größte im Probestück auftretende Zug-, Druck-, Biege- oder Scherspannung.

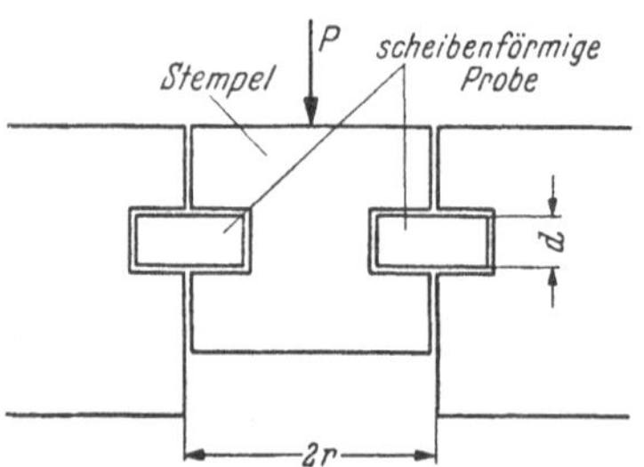

Abb. IV, 8. Bestimmung der Scherfestigkeit.

Nach der geometrischen Form des Versuches unterscheiden wir die *Zugfestigkeit*[2] (Abb. IV, 1 a), die *Druckfestigkeit*[3] (Abb. IV, 1 b), verschiedene Formen der *Biegefestigkeit*[4] (Abb. IV, 2 a und IV, 2 b) und die *Scherfestigkeit*[5] (Abb. IV, 8). Bei der Bestimmung der Scherfestigkeit wird das scheibenförmige Probestück der Dicke d zwischen dem ruhenden Außenblock und einem Stempel vom Radius r abgeschert.

[1] Tabor, D.: „The hardness of metals", Oxford Clarendon Press 1951. – K. V. Shooter, D. Tabor: Proc. Phys. Soc. **B 65**, 661 (1952).

[2] DIN 53455, A. S. T. M. D 412–51 T, A. S. T. M. D 638–52 T.

[3] DIN 53454, A. S. T. M. D 575–46, A. S. T. M. D 621–51, A. S. T. M. D 695–52 T.

[4] DIN 53452, A. S. T. M. D 797–46, A. S. T. M. D 747–50. – [5] A. S. T. M. D 732–46.

Ist P die größte auftretende Zug- oder Druckkraft, dann definiert man als Festigkeiten die im Probestück maximal auftretenden Spannungen:

$$\left.\begin{array}{lll} \text{Zugfestigkeit} & f_t = P/A_0 & \left.\right\} A_0 \text{ Anfangsquerschnitt} \\ \text{Druckfestigkeit} & f_c = P/A_0 & \quad \text{des Probestückes} \\ \text{Biegefestigkeit Fall } a & f_b = 3Pl/2bh^2 & \\ \qquad\qquad\text{Fall } b & f_b = 6Pl/bh^2 & \\ \text{Scherfestigkeit} & f_s = P/2\pi r d & \end{array}\right\} \qquad \text{(IV, 14)}$$

Die in Gl. (IV, 14) definierten Festigkeitswerte sind jedoch keine Materialkonstanten, sondern statistische Eigenschaften, die außerdem noch von den Dimensionen des Probestückes und von der Deformationsgeschwindigkeit abhängen. Bestimmt man die Zugfestigkeit f_t an einer großen Anzahl identischer Probestücke, so erhält man eine ganze Verteilung verschiedener Festigkeitswerte. Eine solche Verteilungskurve[1] zeigt Abb. IV, 9. Die Abszisse gibt die Festigkeitswerte f_t und die Ordinate die Frequenz $p(f_t)$ ihres Auftretens. Zur Festlegung der Brucheigenschaften ist darum die Messung der ganzen Verteilungskurve $p(f_t)$ notwendig.

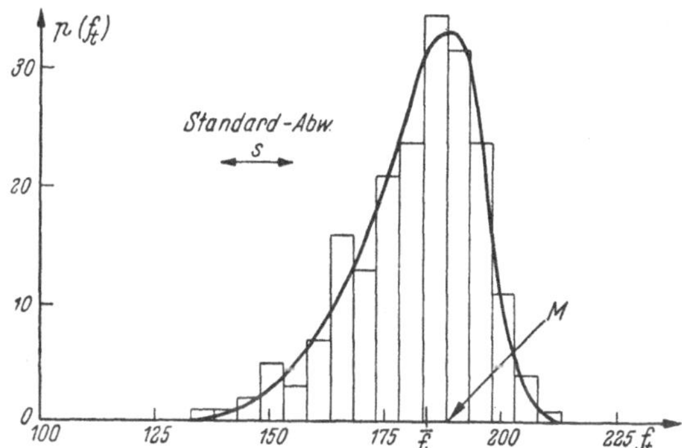

Abb. IV, 9. Verteilung von Zugfestigkeiten von Naturkautschuk[1].

In der Praxis wird man sich aus Zweckmäßigkeitsgründen meistens beschränken auf die Bestimmung einer „gemittelten Bruchspannung“ und der Streuung. Als Maß für die gemittelte Bruchspannung verwendet man entweder den Mittelwert[2]

$$\bar{f}_t = \int_0^\infty f_t\, p(f_t)\, df_t \qquad \text{(IV, 15)}$$

oder den Modul M, definiert als die am häufigsten auftretende Bruchfestigkeit. (M ist die Abszisse des Maximums der Verteilungskurve.) Nur bei symmetrischen Verteilungskurven sind Mittelwert und Modul identisch, bei schiefen Verteilungen

---

[1] Kase, S.: J. Polymer Sci. **11**, 425 (1953).

[2] Die Verteilungskurve $p(f_t)$ wird als normiert angenommen, d.h. das Integral

$$\int_0^\infty p(f_t)\, df_t = 1.$$

sind $\bar{f_t}$ und M verschieden. Ein Beispiel hiervon zeigt Abb. IV,9. Als Maß für die Streuung nimmt man die Standardabweichung s, deren Quadrat definiert ist durch

$$s^2 = \int_0^\infty [f_t - \bar{f_t}]^2 \, p(f_t) \, df_t \,. \qquad \text{(IV, 16)}$$

Aus einer Anzahl gemessener Festigkeitswerte $f_1, f_2, \ldots, f_n$ bestimmt man Mittelwert und Standardabweichung bekanntlich aus den Formeln

$$\bar{f} = \frac{1}{n} \sum_{i=1}^{n} f_i \,, \quad s^2 = \frac{1}{n-1} \sum_{i=1}^{n} (f_i - \bar{f})^2 \,. \qquad \text{(IV, 17)}$$

Der Modul läßt sich nur aus der Verteilungskurve selbst bestimmen.

Die Existenz einer Bruchspannungsverteilung erklärt man durch die Inhomogenität der Probestücke, die eine große Anzahl kleiner Fehlerstellen und Risse enthalten[1]. Nach der statistischen Bruchtheorie[2] wird der Bruchbeginn im Probestück bestimmt durch das schwächste der Elementarvolumina. Die resultierende Verteilung von Bruchspannungen von Probestücken $p(f_t)$ hängt dann ab von der Stärkeverteilung der Elementarvolumina und von der Anzahl der Elementarvolumina, d. h. von dem Volumen des Probestückes. Die Bruchspannung des Probestückes nimmt ab mit zunehmendem Volumen.

Die statistische Bruchtheorie erklärt lediglich qualitativ die Abhängigkeit der gemittelten Bruchspannung vom Volumen. In Wirklichkeit tritt neben der Volumenabhängigkeit auch eine Formabhängigkeit der Bruchspannung auf, so daß nicht nur das gesamte Volumen, sondern auch alle Dimensionen einzeln einen Einfluß auf die Bruchspannung ausüben. Dieser Einfluß tritt besonders zutage, wenn eine der Dimensionen sehr klein wird, nämlich vergleichbar mit der Größe der in der Mikrostruktur auftretenden Fehlerstellen. Als Beispiel erwähnen wir Zugversuche von GRIFFITH an Glasfasern verschiedener Dicke (Abb. IV, 10). Je kleiner der Durchmesser der Fasern, desto größer ist ihre nominale Zugfestigkeit. Die Relation zwischen Festigkeit f_t und Durchmesser d konnte in der Form geschrieben werden[3]

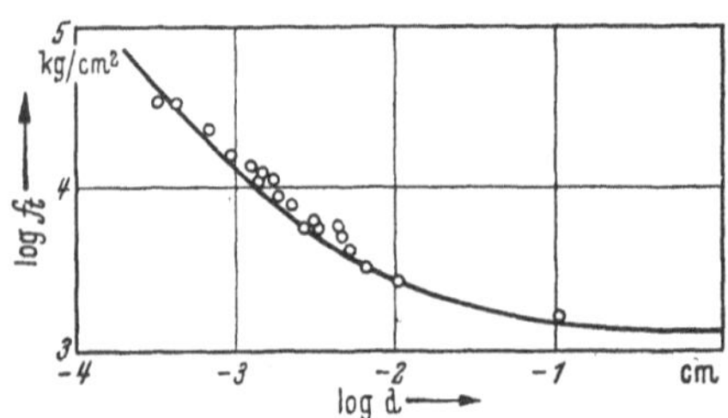

Abb. IV, 10. Abhängigkeit der Zugfestigkeit von Glasfasern vom Durchmesser nach GRIFFITH.

$$f_t = a + b/d \qquad \text{(IV, 18)}$$

(a und b Konstanten). Die totale Bruchkraft ist also die Summe eines gewöhnlichen Spannungsteils (proportional zum Querschnitt) und eines Teiles proportional zum Umfang des Querschnittes. Der letztere Teil bestimmt das Bruchverhalten bei kleinen Durchmessern.

---

[1] Vgl. die ausführliche Behandlung in Kap. 3 dieses Bandes.

[2] WEIBULL, W.: Ing. Vetensk. Akad. Handl. No. 151 (1939); No. 153 (1939). – B. EPSTEIN: J. appl. Physics **19**, 140 (1948).

[3] Vgl. dazu E. U. CONDON: Amer. J. Physics **22**, 224 (1954).

Aus dem Obenstehenden folgt, daß die in Gl. (IV, 14) definierten Festigkeiten abhängig sind von den Dimensionen des betrachteten Probestückes. Es ist daher im allgemeinen nicht möglich, Festigkeiten miteinander zu vergleichen, die auf verschiedenen Standards beruhen. Dazu kommt noch die Abhängigkeit der Festigkeiten von der Deformationsgeschwindigkeit: Je größer die Deformationsgeschwindigkeit, desto höher die Bruchspannungen. Für einen Vergleich verschiedener Materialien untereinander sind wir daher darauf angewiesen, nur Festigkeitsexperimente zu verwenden, die auf derselben Form und Größe des Probestükkes und auf derselben Deformationsgeschwindigkeit basieren.

Über den Zusammenhang der verschiedenen in Gl. (IV, 14) definierten Festigkeiten können wir nur qualitative Aussagen machen. Die Biegefestigkeit f_b ist stets größer als die Zugfestigkeit f_t, was wir an Hand der statistischen Auffassung verstehen können: Bei der Biegung wird lediglich ein kleiner Teil des Probestückes den gefährlichen (maximalen) Zugspannungen ausgesetzt und das wirksame Volumen, in dem der Bruchbeginn auftreten kann, ist dort wesentlich geringer als bei der Dehnung, wo eine homogene Verteilung der maximalen Spannungen vorliegt. Ferner liegt die Druckfestigkeit f_c meist höher als die Zugfestigkeit f_t.

b) Ermüdungsversuche.

Polymere können unter Belastungen brechen, die wesentlich niedriger sind als die Festigkeiten, wenn die Belastungen nur lange genug oder oft genug auf das Material einwirken. Der Widerstand der Materiale gegen langzeitige oder wiederholte Belastungen wird durch *Ermüdungsversuche* bestimmt. Wir unterscheiden statische[1] und dynamische[2] Versuche, je nachdem das Probestück einer konstanten Belastung oder einer Wechsellast mit konstanter Amplitude unterworfen wird.

Beim statischen Ermüdungsexperiment mißt man die zum Brucheintritt notwendige Belastungszeit unter einer bestimmten Spannung und erhält so einen Zusammenhang zwischen Spannung σ und Bruchzeit t_b, die *Standzeitkurve*:

$$\sigma = \sigma(t_b). \qquad \text{(IV, 19)}$$

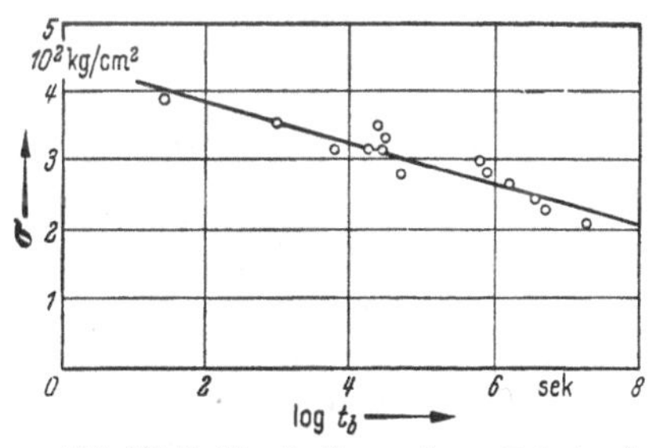

Abb. IV, 11. Standzeitversuche an Polystyrol.

Als Beispiel zeigen wir in Abb. IV, 11 Ermüdungsexperimente an Polystyrol[3]. Man sieht, daß die Bruchspannung bei einer Belastungszeit von einem Jahr nur noch etwa halb so groß ist wie die Bruchspannung bei 10 Sekunden. Die Zeitabhängigkeit der Bruchspannung ist somit ganz beträchtlich und muß bei der Berechnung von Konstruktionen stets in Betracht gezogen werden.

---

[1] A. S. T. M. D 674–51 T. – [2] A. S. T. M. D 623–52 T, A. S. T. M. D 671–51 T.
[3] HSIAO, C. C. u. J. A. SAUER: J. appl. Physics **21**, 1071 (1950).

Ferner sieht man, daß die Bruchspannung etwa linear mit dem Logarithmus der Bruchzeit abnimmt. Auf Grund von statistischen Theorien über die Entstehung und das Anwachsen von Rissen sind verschiedene theoretische Ausdrücke für die Relation Gl. (IV,19) entwickelt worden[1], wie z.B.

$$\left.\begin{aligned} \log t_b &= A - B \log \sigma \\ \text{oder} \quad \log t_b &= A + B/\sigma . \end{aligned}\right\} \qquad \text{(IV, 20)}$$

Bei geeigneter Wahl der Konstanten A und B ergeben diese Relationen in bestimmten Zeitbereichen eine lineare Abnahme der Spannung mit $\log t_b$. Für sehr lange Zeiten jedoch biegt die Standzeitkurve nach Gl. (IV,19) um und nähert sich asymptotisch entweder der Zeitachse oder einem konstanten minimalen Spannungswert. Im letzteren Fall spricht man von der Existenz einer Ermüdungsgrenze.

Schließlich fällt an Abb. IV,11 die ziemlich große Streuung der experimentellen Punkte auf, die eher auf die Existenz eines Bruchspannungs-Zeit-Bandes als einer Bruchspannungs-Zeit-Relation hinweist. Der Grund ist leicht ersichtlich: Auch die Bruchspannungs-Zeit-Relation muß als ein statistischer Zusammenhang aufgefaßt werden und eine korrekte Beschreibung müßte ausgehen von der Beziehung

$$p = p\{\sigma, t_b\}. \qquad \text{(IV,21)}$$

Dabei gibt p die Wahrscheinlichkeit an, um unter einer Spannung σ nach einer Belastungszeit t_b einen Bruch zu erhalten.

Bei dynamischen Ermüdungsexperimenten mißt man die Anzahl Lastwechsel N, die notwendig ist, um bei einer maximalen Spannungsamplitude σ Bruch zu erzielen. Man erhält so die Wöhler-Kurve

$$\sigma = \sigma(N). \qquad \text{(IV,22)}$$

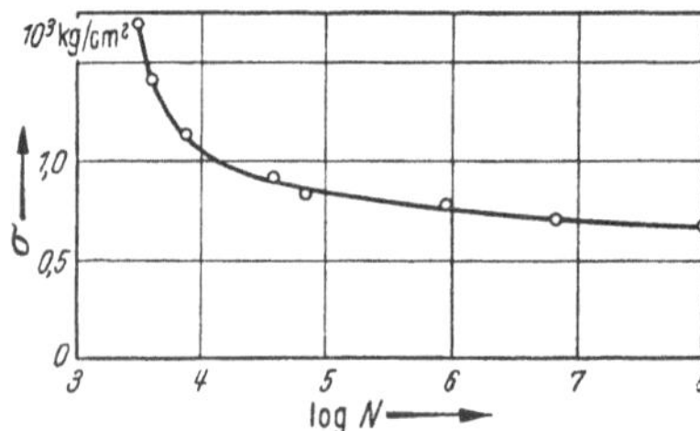

Abb. IV,12. Wöhler-Kurve für Polyester-Glas-Kunststoffe nach Findley[2].

Ein Beispiel einer solchen Wöhler-Kurve für Polyester-Glas-Kunststoffe[2] zeigt Abb. IV, 12. Auch hier tritt eine sehr beträchtliche Erniedrigung der Bruchspannung auf (bis auf ein Drittel des Kurzzeitwertes) und die Abbildung deutet auf die Existenz einer langzeitigen Ermüdungsgrenze hin.

Weibull[3] und Freudenthal[4] haben darauf hingewiesen, daß auch bei der Wöhler-Kurve eine statistische Beschreibung am Platze ist. Diese erfordert die Angabe der Wahrscheinlichkeit p, um unter der Amplitude σ nach N Lastwechsel einen Bruch zu erhalten:

$$p = p\{\sigma, N\}. \qquad \text{(IV,23)}$$

[1] Vgl. O. L. Anderson u. D. A. Stuart: Ind. Engng. Chem. **46**, 154 (1954).

[2] Findley, W. N. u. W. J. Worley: S.P.E.-Journal (April 1951) 9.

[3] Weibull, W.: Trans. Roy. Inst. Techn. Stockholm Nr. 27 (1949).

[4] Freudenthal, A. M.: Planning and interpretation of fatigue tests in: Symposium on statistical aspects of fatigue, A.S.T.M. Techn. Publ. No. 121 (June 1951), Philadelphia.

Bei fester Spannungsamplitude erhalten wir dann nicht einen Wert für N, sondern eine ganze Verteilung von N-Werten. FREUDENTHAL[1] hat vorgeschlagen, die Anzahl Lastwechsel bei festem σ als logarithmisch normal verteilt anzusehen. Diese Auffassung wird durch Torsionsermüdungsversuche an Kunststoffasern bestätigt[2].

c) Schlagversuche.

Bei Schlagexperimenten wird die mechanische Beanspruchung in Form eines Stoßes mit Hilfe eines Schlaghammers im Bruchteil von Sekunden angebracht. Als charakteristische Größe läßt sich hier am einfachsten die Bruchenergie messen als die Energiedifferenz des Schlaghammers vor und nach dem Schlag.

Nach der geometrischen Form können wir wieder unterscheiden den Zugschlagversuch (Abb. IV, 13a), den Biegeschlag – einseitig eingeklemmt – (Abb. IV, 13b) und den Biegeschlag für Dreipunktsbiegung (Abb. IV, 13c). Der Biegeschlag auf ein einseitig eingeklemmtes Probestück[3] und der Biegeschlag in Form einer Dreipunktsbiegung[4] werden als Prüfungsmethoden verwendet, und zwar sowohl mit vorgekerbten wie auch mit ungekerbten Probestücken.

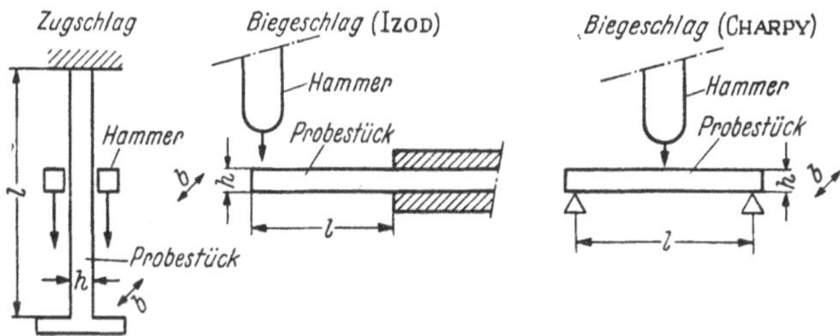

Abb. IV, 13. Verschiedene geometrische Formen von Schlagexperimenten: a) Zugschlagversuch, b) Biegeschlagversuch – einseitig eingeklemmt, c) Biegeschlagversuch – Dreipunktsbiegung.

Der gemessene Energieverlust des Pendels umfaßt im allgemeinen mehr als nur die uns interessierende Bruchenergie, nämlich:

den Energieverlust im Apparat,
den „Toss-Faktor" und
die Bruchenergie W.

Während des Schlages wird sowohl der Schlaghammer wie auch – in minderem Maße – die Einklemmungsvorrichtung in Schwingungen versetzt. Diese Schwingungsenergie ist in der Messung enthalten und bewirkt eine Abhängigkeit der gemessenen Schlagenergien von der Konstruktion des Schlaghammers.

Unter „*Toss-Faktor*" versteht man den Energieverlust, der durch das Wegschleudern des abgebrochenen Teiles des Probestückes entsteht. Er

---

[1] Siehe S. 230, Fußnote 4.

[2] VEGT, A. K. VAN DER: Centraal Laboratorium T.N.O. Delft, private Mitteilung.

[3] DIN 53453, A.S.T.M. D 256–47 T (Izod-Type).

[4] DIN 53453, A.S.T.M. D 256–47 T (Charpy-Type).

besteht aus Translations- und Rotationsenergie des weggeschlagenen Stückes und kann beträchtliche Werte annehmen.

Zu einer Abschätzung der Größenordnung des Toss-Faktors beim Izodversuch[1] wurde das weggeschlagene Stück wieder auf seinen Platz zurückgesetzt und der Schlagversuch wiederholt. Die so gemessene Toss-Energie erwies sich als proportional zur kinetischen Energie des Pendels – proportional zum Quadrat der Schlaggeschwindigkeit – und konnte zwischen 8 und 40% der gesamten gemessenen Brucharbeit ausmachen. Der Toss-Faktor kann also nicht nur einen beträchtlichen Anteil an der Brucharbeit einnehmen, sondern verursacht obendrein noch eine Abhängigkeit der Schlagarbeit von der Schlaggeschwindigkeit.

Die eigentliche Bruchenergie W ist wieder eine statistische Größe, so daß eine vollkommene Bestimmung die Ausmessung der Frequenzkurven erfordern würde. Auch hier begnügt man sich wieder mit einer angenäherten Beschreibung durch Angabe des Mittels und der Standardabweichung einer größeren Anzahl von Versuchen.

Die Abhängigkeit der Bruchenergie W von den Dimensionen des Probestückes l, b und h (vgl. Abb. IV,13) ist wiederum nur sehr unvollständig bekannt. Es ist daher im allgemeinen nicht möglich, Ergebnisse von Schlagexperimenten der gleichen Type mit verschiedenen Dimensionen des Probestückes miteinander zu vergleichen. Im besonderen ist es z.B. noch nicht möglich, die deutschen und amerikanischen Standards miteinander zu vergleichen.

Man könnte die Bruchenergie W in zwei Teile zerlegen:

$$W = W' + W''. \tag{IV,24}$$

W' sei der Teil der Brucharbeit, der im Augenblick des Bruchbeginnes als elastische oder viscoelastische Deformationsenergie im Probestück enthalten ist, W'' sei der Rest der Bruchenergie, notwendig, um den einmal entstandenen Bruch fortzupflanzen, die zwei Stücke voneinander loszureißen usw. Die Existenz einer elastischen Deformationsenergie W' setzt voraus, daß der Schlag langsam genug erfolgt, um vor dem Bruchbeginn im Probestück einen elastischen Spannungszustand aufzubauen. Während die Energie W'' hauptsächlich von der Querschnittsfläche des Probestückes abhängen wird, ist der Teil W' wahrscheinlich vom Volumen V des Probestückes abhängig.

Eine sehr vereinfachte Abschätzung des Anteiles W' läßt sich folgendermaßen geben. Nehmen wir an, daß während des Schlages im Probestück der Deformationszustand analog aufgebaut wird wie im Fall der entsprechenden statischen Belastung[2] und daß der Bruch in dem Augenblick beginnt, in dem die größte auftretende statische Zugspannung den Wert der Zugfestigkeit f_t erreicht, so wird

$$\left.\begin{array}{ll} W' = V \cdot w & \text{Zugschlag} \\ W' = V \cdot w/9 & \text{Biegeschlag Izod (ungekerbt)} \\ W' = V \cdot w/9 & \text{Biegeschlag Charpy (ungekerbt)}, \end{array}\right\} \tag{IV,25}$$

wobei V das Volumen des Probestückes und w eine spezifische Energie ist:

$$w = f_t^2/2E. \tag{IV,26}$$

[1] Unpublizierte Versuche von R. A. J. BOSSCHART u. D. J. VAN WIJK: Kunststoffen Instituut T.N.O., Delft.

[2] Diese Annahme ist ein schwacher Punkt in der Abschätzung, da in Wirklichkeit der Deformationszustand nicht statisch aufgebaut wird, sondern sich Spannungswellen in das Probestück ausbreiten.

Die spezifische Energie w ist proportional zum Quadrat der Zugfestigkeit – im Falle des Biegeschlages der Biegefestigkeiten – und umgekehrt proportional zum Elastizitätsmodul. Für den Elastizitätsmodul E ist ein Wert entsprechend den durch die Schlaggeschwindigkeit erzwungenen Deformationsgeschwindigkeiten einzusetzen.

Im Lichte der Bruchtheorie von GRIFFITH würde in der Zerlegung (IV, 24) W'' die Oberflächenenergie enthalten, die notwendig ist, um die beiden Bruchflächen zu liefern. Da jedoch ein Teil der elastischen Deformationsenergie verwendet wird, um diese Oberflächenenergie zu liefern (vgl. § 17), ist die in Gl. (IV, 24) enthaltene Summe zu groß und eine bessere Deutung der Gleichung wäre: W' ist die Summe von elastischer und plastischer Deformationsenergie; W'' ist der Teil der Oberflächenenergie, welche nicht durch die elastische Deformationsenergie geliefert wird.

Vergleichen wir an Hand eines Zahlenbeispiels die Größenordnung der Bruchenergie W und der Oberflächenenergie, so finden[1] wir für Polymethylmetacrylat $W \simeq 0{,}31$ kg/cm², während die Oberflächenenergie der gebildeten Bruchflächen [berechnet mit Hilfe von Gl. (III, 31)] etwa 1000 mal kleiner ist. Berechnet man die Bruchenergie unter Benützung von Gl. (IV, 25) aus Elastizitätsmodul und Biegefestigkeit, so enthält man $W' \sim 0{,}47$ kg/cm².

Gl. (IV, 25) gibt also größenordnungsmäßig richtige Resultate, die jedoch zu hoch liegen. Bedenkt man all die Näherungsannahme, unter denen Gl. (IV, 25) abgeleitet wurde, so ist das nicht weiter verwunderlich.

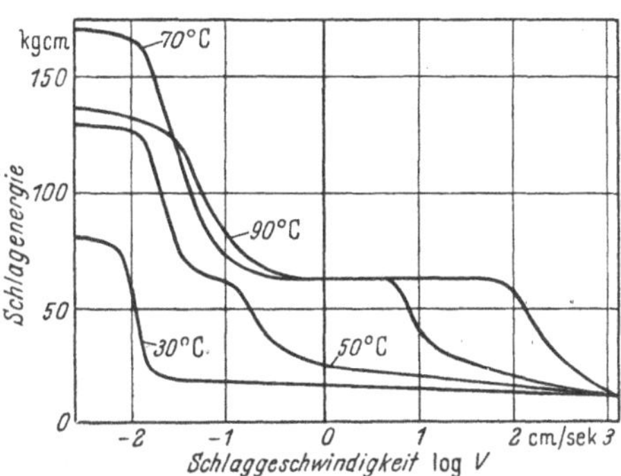

Abb. IV, 14. Abhängigkeit der Schlagenergie von Polymethylmethacrylat von der Schlaggeschwindigkeit v und der Temperatur[2].

Die Schlagenergie von Polymeren hängt weiter stark ab von der Belastungszeit – d.h. der Schlaggeschwindigkeit v – und von der Temperatur. Abb. IV, 14 zeigt diese Abhängigkeit der Schlagenergie von Polymethylmethacrylat[2]. Die Schlagenergie ist groß bei hohen Temperaturen und kleinen Schlaggeschwindigkeiten und niedrig bei niedrigen Temperaturen und großen Schlaggeschwindigkeiten. Charakteristisch ist das Auftreten von „Übergängen" in der Schlagenergie bei bestimmten Geschwindigkeiten. Diese Übergänge zeigen die Zeit-Temperatur-Verschiebung[3] – sie verschieben sich nach höheren Schlaggeschwindigkeiten bei höheren Temperaturen.

Das Auftreten von Übergängen in der Schlagenergie läßt sich verstehen durch die Existenz von Übergängen im Elastizitätsmodul E. Bei kleinen Deformationsgeschwindigkeiten (entsprechend großen Belastungszeiten) ist der Elastizitätsmodul niedrig und es können große Deformationen auftreten, bevor die Bruchspannung erreicht wird und der Bruch beginnt. Der Anteil W' der Bruchenergie ist dann hoch [vgl. Gl. (IV, 26)]. Bei größeren Deformationsgeschwindigkeiten tritt ein Steilanstieg im Elastizitätsmodul auf, der zu einem starken Abfall im Teil W'

[1] Versuche von J. HEYBOER, Centraal Laboratorium T.N.O., Delft: Der Schlagversuch war ein Dynstat-Biegeschlag eines Probestückes aus Polymethylmethacrylat von den Abmessungen $l = 7$ mm, $b = 10$ mm, $h = 3$ mm. Der Elastizitätsmodul dieses Materials ist $E \sim 5{,}5 \cdot 10^4$ kg/cm² und die Biegefestigkeit $f_b \sim 1{,}5 \cdot 10^3$ kg/cm².

[2] MAXWELL, B. u. J. HARRINGTON: Trans. A.S.M.E. 72, 579 (1952). – [3] Vgl. § 5.

der Bruchenergie führt. Auf diese Weise könnte man jedem Übergang im Elastizitätsmodul (als Funktion der Zeit) einen Übergang in der Bruchenergie (als Funktion der Schlaggeschwindigkeit) zuordnen. Die Zeit-Temperatur-Verschiebung in den Übergängen des Elastizitätsmoduls muß sich dann widerspiegeln in einer Schlaggeschwindigkeits-Temperatur-Verschiebung in den Übergängen der Brucharbeit.

Im engen Zusammenhang mit obigen Betrachtungen steht die Bestimmung der *Sprödigkeitstemperatur*[1] (englisch: *brittle point*) · T_b von Kunststoffen und Gummi. Hierbei werden die Probestücke bei verschiedenen Temperaturen einem Biegungsschlag ausgesetzt. Bei hohen Temperaturen ist das Material hochelastisch und vermag dem Schlag auszuweichen ohne zu brechen, bei niedrigen Temperaturen tritt spröder Bruch auf. Die tiefste Temperatur, bei der das Material gerade noch nicht bricht, heißt die Sprödigkeitstemperatur T_b. Die Temperatur T_b bestimmt die untere Grenze des Gebrauchstemperaturbereiches für gummielastische Materialien im praktischen Gebrauch.

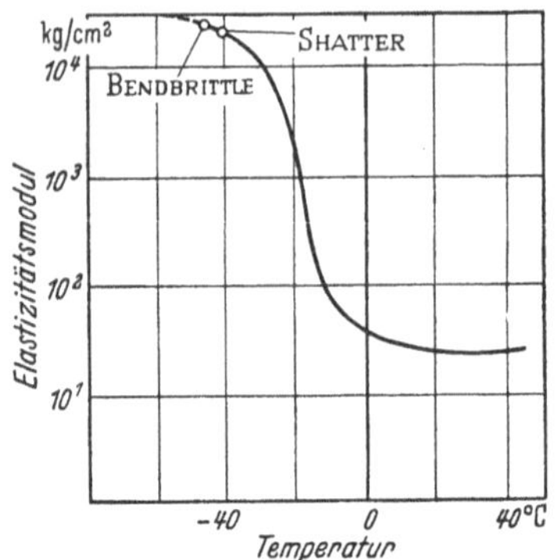

Abb. IV, 15. Der dynamische Elastizitätsmodul (~ 10^{-3} sec) für Neoprene GN 50 und die Lage von Sprödigkeitspunkten.

Die Sprödigkeitstemperatur kann aufgefaßt werden als die Temperatur, bei der der Übergang im E-Modul (oder auch in der Schlagenergie) gerade in die charakteristische Schlaggeschwindigkeit des Versuches fällt[2]. Abb. IV, 15 soll diese Verhältnisse illustrieren. Hier sind die Temperaturabhängigkeit des Elastizitätsmoduls im Übergangsbereich für einen Kautschuk nach Messungen von NOLLE[3] und die Sprödigkeitstemperaturpunkte[4] markiert. Wie man sieht, fallen die Sprödigkeitstemperaturen gerade in den Beginn des Überganges hart–kautschukelastisch.

Die Bestimmung der Sprödigkeitstemperatur könnte man durch Experimente ersetzen, die mehr physikalische Information enthalten, nämlich durch die Bestimmung der Übergänge der Schlagenergie bei verschiedenen Temperaturen. Das erfordert die Messung der Schlagenergie als Funktion der Schlaggeschwindigkeit und Temperatur. Die Sprödigkeitstemperatur ist dann gekennzeichnet durch das Passieren eines Steilabfalls der Schlagenergie durch das Meßgebiet.

Zusammenfassend können wir sagen, daß im Gegensatz zu den Prüfmethoden, die das Verhalten bei kleinen Deformationen messen, die Methoden, die das Verhalten beim Zerbrechen des Materials zu charakteri-

[1] A.S.T.M. D 746–52 T.

[2] Hieraus folgt auch, daß die Sprödigkeitstemperatur mit zunehmender Schlaggeschwindigkeit höher werden muß. Vgl. hierzu Abb. IV, 15, wo die zwei Sprödigkeitstemperaturen (bend brittle point, shatter point) etwas verschiedene Werte besitzen, was auf eine etwas verschiedene Deformationsgeschwindigkeit in den zwei Versuchen hinweist. — [3] NOLLE, W.: J. Polymer Sci. **5**, 1 (1950).

[4] KING, G. E.: Ind. Engng. Chem. **35**, 949 (1943).

sieren suchen, physikalisch sehr unbefriedigend sind. Die üblichen Meßmethoden auf diesem Gebiet stellen keine physikalischen Eigenschaften, charakteristisch für das Material, dar, sondern willkürliche Größen, die in unübersichtlicher Weise von den Dimensionen des Apparates und des Probekörpers abhängen.

Diese unbefriedigende Situation ist nicht die Folge eines Versäumnisses der Prüftechnologen, sondern eher ein Ausdruck des heutigen Standes der Wissenschaft auf diesem Gebiete. Zwar ist unsere Kenntnis bereits so weit fortgeschritten, daß sich die wirklichen *Materialkonstanten* hinter dem Bruchverhalten erkennen lassen, jedoch gibt es noch keine sicheren Laboratoriumsmethoden – noch weniger technologische Prüfmethoden –, um diese Konstanten zu bestimmen.

Als charakteristische Materialeigenschaften zur Beschreibung des Bruchverhaltens wären zu nennen:

1. Die Mikrostruktur, d.h. die Verteilung der Inhomogenitäten nach Anzahl, Größe und „Gefährlichkeit".

2. Die Energie, notwendig zur Vergrößerung eines bestehenden Risses pro Oberflächeneinheit.

Wenn es möglich sein wird, diese Materialkonstanten in eindeutiger Weise zu bestimmen, kann man mit der Entwicklung eines Systems von Prüfmethoden beginnen, das die Bestimmung dieser Konstanten ermöglicht. Ein solches System würde im Vergleich zur heutigen Lage einen wesentlichen Fortschritt bedeuten.

Solange jedoch die Wissenschaft solche Methoden nicht bieten kann, bleibt dem Technologen nichts anderes übrig, als sich mit den existierenden, unbefriedigenden Prüfmethoden zu helfen, um die Probleme der Praxis im groben beantworten zu können, und auf wirkliche „Verbesserung" seiner Methoden zu verzichten.

B. Prüfmethoden an Faserstoffen[1].

Von W. Meskat und O. Rosenberg.

§ 21. Die Rheologie der Faserstoffe und ihre Beziehung zu der Entwicklung der Prüfgeräte

Formal der gegebene Weg der Beschreibung des zu jedem Stoff gehörenden und ihn kennzeichnenden Zusammenhanges zwischen dem Spannungszustand, dem Deformationszustand und der Deformations-

[1] Wir haben in diesem Abschnitt grundsätzlich davon abgesehen, die Probenahme, die Probenvorbereitung sowie die Fragen des Prüfraumes, z.B. Normklima usw., zu behandeln. Es sei auf die Literatur verwiesen: E. Wagner, Mechanisch-technologische Textilprüfungen 1953, Wool Research 1918–1954, Vol. 3, herausgegeben von der Wool Industries Research Association, Leeds 6, England, sowie auf die Normvorschriften DIN 53803 (vgl. auch die gültigen BISFA Liefer- und Prüfvorschriften für Reyon, Zellwolle, Polyamidfäden und -fasern), DIN 50012 und DIN 53802. Auch die Fragen der Konditionierung bei der Verarbeitung werden nicht berührt. Hier sei der Hinweis auf J. Wagget: Reyon, Zellwolle und andere Chemie-Fasern 8, 454 (1953) und Reyon, Zellwolle und andere Chemie-Fasern 9, 511 (1953) sowie auf Ja. Ja. Innolitow: Textil-Praxis 9, 441 (1954) usw. gestattet.

geschwindigkeit ist die Darstellung der Beziehung nach H. FROMM[1], C. TRUESDELL u. a. durch eine allgemeine Gleichung zwischen dem Spannungstensor, dem Deformationstensor und den zeitlichen Ableitungen beider Tensoren.

Der besondere Aufbau der Faserstoffe aus Einzelfasern bzw. Monofilen kommt nun der Idealisierung dieses komplizierten Deformationsverhaltens weitgehend entgegen[2]. Aber auch der vereinfachte konkrete mathematische Ausdruck für den Zusammenhang[3] setzt zu seiner Ermittlung nicht nur das Experiment voraus und damit das Meßgerät, sondern auch die Grundforderung, daß das Prüfgerät mathematisch erfaßbare Deformationen und Deformationsgeschwindigkeiten der zu untersuchenden Faser aufprägt und die zugehörigen Spannungen einwandfrei angibt.

a) Modellvorstellungen über das Deformationsverhalten von Monofilen.

In dem Bestreben, das experimentell ermittelte Deformationsverhalten von Monofilen auf einige allgemeingültige stoffspezifische Einflußfaktoren zurückzuführen und in der Erwartung, damit auch der Entwicklung der Meßgeräte neue Impulse zu geben, ging man dazu über, die Deformationseigenschaften durch Modellvorstellungen in ein anschauliches Gewand zu kleiden; vgl. die ausführlichen Erörterungen in Kap. I, § 4. Das erste derartige mechanische Modell zur Beschreibung des Spannungs-Dehnungs-Zeitverhaltens von J. CL. MAXWELL[4] wird symbolisiert durch eine Feder (spring), kennzeichnend das HOOKEsche Verhalten, und durch ein Dämpfungsglied (dashpot), das Verhalten einer NEWTONschen Flüssigkeit darstellend, wobei diese beiden Glieder in Reihe geschaltet sind. Dieses Modell erlaubt eine gewisse qualitative Beschreibung der Spannungs-Dehnungs-Eigenschaften und der Relaxation bei konstanter Dehnung von Monofilen. Die Kriechvorgänge werden in keiner Weise erfaßt. W. VOIGT[5] verwendet dieselben Glieder des Modells, schaltet sie jedoch nicht in Reihe, sondern parallel und kann damit die Kriechvorgänge bei konstanter Spannung bzw. Belastung im gewissen Umfang darstellen, jedoch sind hier wiederum die Relaxationsvorgänge ausgeschlossen. Die unvollständige Beschreibung des linearen Deformationsverhaltens durch die beiden Modelle legt ihre Kombination nahe. Da auch in diesem Falle für zahlreiche Monofile das Deformationsverhalten nicht befriedigend beschrieben werden kann, macht man von dem BOLTZMANNschen Superpositionsprinzip[6] Gebrauch und schaltet ganze Reihen derartiger Modelle zusammen, wie es bereits von E. WIECHERT[7] und neuerdings auch von

[1] FROMM, H.: Ing.-Arch. **4**, 432 (1933). – W. LODE: Kolloid-Z. **138**, 28 (1954). – C. TRUESDELL: J. Kat. Mech. and Analysis **1**, 125 (1952).

[2] Es wird dabei nur immer wieder übersehen, daß die Querschnittsabmessungen selbst der Einzelfasern noch immer sehr groß im Verhältnis zu den Abmessungen der Kettenmoleküle sind und daher für ein tieferes Eindringen die Vernachlässigung bzw. Idealisierung gar nicht zulässig ist. Auf diesen Idealisierungen beruhen aber die zu schildernden Modellvorstellungen. — [3] Vgl. Kap. I, § 2 und § 3 sowie § 6b.

[4] MAXWELL, J. CL.: Philos. Mag. J. Sci. IV, 35, 134 (1868).

[5] VOIGT, W.: Lehrbuch der Kristall-Physik, Leipzig 1928.

[6] BOLTZMANN, L.: Pogg. Ann. d. Physik, Leipzig **7**, 624 (1876).

[7] WIECHERT, E.: Ann. Physik **50**, 335, 546 (1893).

E. JENCKEL[1] gezeigt worden ist. Eine zusammenfassende Darstellung dieser Modellvorstellungen ist außer in Kap. I von T. ALFREY[2] gegeben worden. Alle diese Modelle, auch in den verschiedensten Kombinationen, sind in der Beschreibung des Deformationsverhaltens adäquat linearen totalen oder partiellen Differentialgleichungen mit konstanten Koeffizienten.

Inwieweit es daher sinnvoll sein soll, ganze Reihen von MAXWELL- und VOIGT-Modellen zu schalten, muß man mit Recht bezweifeln, wenn man überlegt, daß die Aussagen, die dabei gewonnen werden, physikalisch immer inhaltsleerer werden und mit diesen Modellbetrachtungen ein Weg eingeschlagen wird, der mathematisch gesehen, einer Approximation entspricht. Die Modellvorstellungen begünstigen an sich die Entwicklung von Prüfgeräten, die dem Modell entsprechend, Beanspruchungen zu realisieren gestatten. Die Überlagerung mehrerer Modelle zerstört jedoch wieder die einfache physikalische Realisierung der Beanspruchungsart durch ein Prüfgerät. Berücksichtigt man außerdem noch die sehr engen Spannungs- und Dehnungsgrenzen des linearen Deformationsverhaltens, wie sie in Kap. I § 3d angegeben worden sind, in denen diese Überlagerung überhaupt möglich ist, so ist es verständlich, daß, von diesen theoretischen Überlegungen vollkommen unberührt, die konstruktive Entwicklung der Prüfgeräte ihren eigenen Weg gegangen ist und sich einerseits auf die Ausmerzung physikalischer Mängel in der Wirkungsweise der Geräte beschränkte (vgl. §§ 22a, 22c und 22d) und andererseits eine Komplizierung der Prüfgeräte mit unkontrollierter Zunahme der Beanspruchungsmöglichkeiten des zu prüfenden Fadens in Erscheinung getreten ist (vgl. §§ 22d und 23).

Einen Abschluß der geschilderten Modellvorstellungen mit ausführlicher Diskussion der Zusammenhänge zwischen den Modellen brachte die von B. GROSS[3] entwickelte allgemeine Spektraltheorie des linearen viscoelastischen Verhaltens, wobei eingehend die mathematische Struktur dieses Formalismus behandelt und der mathematische Zusammenhang zwischen dem komplexen E-Modul bzw. der komplexen Nachgiebigkeit bei periodischer Wechselbeanspruchung und der Relaxationsfunktion bzw. Kriechfunktion bei Be- und Entlastungsversuchen sowie der Zusammenhang beider Funktionsgruppen zu dem entsprechenden Relaxations- bzw. Retardationsspektrum dargestellt wird. Allerdings bleibt auch hier die Untersuchung auf den Zusammenhang zwischen zwei konjugierten Variablen beschränkt. E. HIEDEMANN und R. D. SPENCE[4] versuchen zunächst, die engen Grenzen zu erweitern, indem sie von zwei konjugierten Tensoren ausgehen, um dann jedoch bei der weiteren Untersuchung wieder auf ein paar konjugierte Variablen zurückzugehen. Erst J. MEIXNER[5] macht sich davon frei und kommt im Rahmen des Gültigkeitsbereiches des BOLTZMANNschen Superpositionsprinzipes zu einer thermodynamischen Erweiterung der Nachwirkungstheorie für beliebige

[1] JENCKEL, E.: Kolloid-Z. **134**, 47 (1953).

[2] ALFREY, T.: Mechanical Behaviour of High Polymers, Interscience Publ. New York 1948.

[3] GROSS, B.: Mathematical Structure of the Theories of Viscoelasticity, Paris 1953. — [4] HIEDEMANN, E. u. R. D. SPENCE: Z. Physik **133**, 109–123 (1952).

[5] MEIXNER, J.: Z. Naturforsch. **9a**, 654–663; Z. Physik **139**, 30 (1954).

anisotrope und isotrope Körper unter Einschluß der visko-elastischen Stoffe[1], die über die Nachwirkungsfunktion hinaus zu einer Nachwirkungsmatrix mit 49 Komponenten führt. Diese Theorie führt auch zu einer engen Analogie zwischen Nachwirkungserscheinungen und elektrischen Netzwerken. Es sei jedoch besonders hervorgehoben, daß wir uns immer noch, wie auch bei der Untersuchung von R. SIPS, J. Polymer Sci. **7**, 191 (1951), in dem erwähnten engen Bereich des linearen Deformationsverhaltens bewegen.

In den Rahmen dieser Skizzierung der phänomenologischen Betrachtungsweise gehört auch der Hinweis auf die Untersuchungen über die Zusammenhänge zwischen dem viscoelastischen Verhalten bzw. der Relaxations- sowie Retardationsspektra und der Molekularstruktur, vgl. Kap. I, § 8c, wobei die Erweiterung der in § 8c dargestellten Theorie des viscoelastischen Verhaltens verdünnter Lösungen von Polymeren von P. E. ROUSE durch J. D. FERRY, R. F. LANDEL und M. L. WILLIAMS: J. appl. Physics **26**, 359 (1955) auf ungelöste lineare Polymeren besonders zu erwähnen ist. Im Hinblick auf die Textilfasern lohnt dieser theoretische Aufwand nicht, denn die zu engen Anwendungsgrenzen der geschilderten Theorien des linearen Deformationsverhaltens ergeben sich bereits bei der Untersuchung der Kriechvorgänge und führten H. LEADERMAN[2] zu einer Modifikation des BOLTZMANNschen Superpositionsprinzips, um auch das beobachtete nicht-lineare Verhalten[3] angenähert beschreiben zu können[4]. G. HALSEY, H. J. WHITE und H. EYRING[5] erweiterten zu diesem Zweck die in § 21a kurz skizzierten Modellvorstellungen, indem sie das Dämpfungsglied (dashpot), das bisher das NEWTONsche Verhalten einer viscosen Flüssigkeit symbolisieren sollte, durch ein die charakteristischen Züge des nicht-NEWTONschen Fließverhaltens tragendes Dämpfungsglied ersetzten. Es lag dabei nahe, diese nicht-NEWTONschen Eigenschaften durch die von H. EYRING[6] einerseits und von L. PRANDTL[7] andererseits unter ganz verschiedenen Gesichtspunkten aus Platzwechselvorstellungen abgeleiteten sin-h-Formeln darzustellen[8]. Das Federglied des Modells, das elastische Verhalten im HOOKEschen Sinne kennzeichnend, wird zunächst beibehalten, um dann im Rahmen der Weiterentwicklung ebenfalls aufgegeben zu werden (vgl. S. KATZ, G. HALSEY und H. EYRING[9]).

[1] Vgl. Kap. I, § 5d.

[2] LEADERMAN, H.: Elastic and Creep Properties of Filamentous Materials and other High Polymers, Text. Foundation, Washington D. C. 1943.

[3] Es sei in diesem Zusammenhang auf die experimentelle Studie von M. T. O'SHAUGNESSY,: Text. Res. J. **18**, 263 (1948) über das Kriechverhalten von Reyon und auf die Untersuchung von E. CATSIFF, T. ALFREY u. M. T. O'SHAUGHNESSY: Text. Res. J. **23**, 808 (1953) über das Kriechen von Nylon hingewiesen.

[4] Vgl. Kap. II, § 13b.

[5] HALSEY, G., H. J. WHITE u. H. EYRING: Text. Res. J. **15**, 295 (1945).

[6] EYRING, H.: J. chem. Physics **4**, 283 (1936). – S. GLASSTONE, K. LAIDLER u. H. EYRING: The Theory of Rate Processes, New York 1941.

[7] PRANDTL, L.: Z.A.M.M. **8**, 85 (1928).

[8] WEYMANN, H.: Kolloid-Z. **138**, 41 (1954) hat zu den Ableitungen der sin-h-Formeln kritisch Stellung genommen und einen neuen Zugang zu dieser Formel angegeben.

[9] KATZ, S., G. HALSEY u. H. EYRING: Text. Res. J. **16**, 378 (1946).

Nachdem auch diese Erweiterung nicht genügte, ging G. HALSEY[1] dazu über, dem Dämpfungsglied sogar thixotrope Eigenschaften zuzuordnen. Damit sind auch hier die Grenzen aufgezeigt, die in der nicht mehr physikalisch auswertbaren Zuordnung der verschiedenen rheologischen Eigenschaften zu den Modellgliedern infolge der Ungültigkeit des BOLTZMANNschen Superpositionsprinzipes liegen[2]. Einen Mittelweg schlägt ICHIRO SAKURADA ein[3].

Trotz dieser unzweifelhaften Schwierigkeiten haben CH. J. GEYER JR., C. H. REICHARDT und G. HALSEY[4], von dem nicht-linearen Drei-Elementen-Modell ausgehend, versucht, eine Theorie der handelsüblichen Garnprüfung mit dem Pendelprüfgerät und der GONSALVESschen Neigungswaage[5] aufzubauen und aus den Meßgrößen die Parameter zur Beschreibung des Deformationsverhaltens abzuleiten. Während man hier jedoch keine weitergehenden Folgerungen aus den physikalischen Mängeln, die gerade diesen Prüfgeräten anhaften[5], zog, hat B. STEENBERG[6] diese Mängel erkannt und speziell für die Papierprüfung das in § 22a im einzelnen beschriebene Gerät entwickelt. Dieses erlaubt das Deformationsverhalten von Papierstreifen, das eine Ähnlichkeit mit dem Verhalten von Viscosereyon zeigt, so genau zu beschreiben, daß B. IVARSSON und B. STEENBERG[7] sowie B. IVARSSON[8] die experimentell gefundenen Kurven nach dem Drei-Elementen-Modell auswerten und die Grenze der Anwendung dieses von H. EYRING und G. HALSEY aufgestellten Modells auf das Deformationsverhalten von Papier angeben konnten[9].

Die besonderen Schwierigkeiten, die sich der mathematischen Erfassung des Deformationsverhaltens nasser Wollfäden entgegenstellen, versuchten H. BURTE und G. HALSEY[10] durch eine neue Modellvorstellung zu überwinden, die als Zwei-Zustands-Modell (Two-State-Model) bezeichnet wird, wobei sie von der Konzeption der α–β-Umwandlung des Keratins ausgehen und von der Annahme Gebrauch machen, daß ein Teil der Molekülketten, die das Material aufbauen, während des Fließvorganges von der Konfiguration A zu der Konfiguration B übergeht und dabei die vorher besessenen Dehnungseigenschaften verliert (vgl. auch W. FUHRMANN[11]). Dieses Modell wurde durch die Einführung eines intermediären Zustandes zu einem Drei-Zustands-Modell (Three-State-Model)

[1] HALSEY, G.: J. appl. Physics **18**, 1072 (1947).

[2] Diese von G. HALSEY angegebenen komplizierten Modelle zur Beschreibung des nichtlinearen Deformationsverhaltens sind mathematisch etwa mit einer Approximation mittels semikonvergenter Reihen vergleichbar.

[3] SAKURADA, ICHIRO: III. Rheology Symposium Japan 1953.

[4] GEYER, CH. J. JR., C. H. REICHARDT u. G. HALSEY: Textile Res. J. **18**, 338 (1948). — [5] Vgl. § 22a.

[6] Aus H. F. RANCE: Some Recent Developments in Rheology (London 1950), S. 84.

[7] IVARSSON, B. u. B. STEENBERG: Svensk Papperstidn. **50**, 419 (1947). — R. MEREDITH: Mechanical Properties of Wood and Paper, Amsterdam 1953, S. 258–262. — [8] IVARSSON, B.: Svensk Papperstidn. **51**, 383 (1948).

[9] Auch hier muß wieder betont werden, daß mit diesen Modellvorstellungen die Eindimensionalität verknüpft ist, die nach der Fußnote 2, S. 236 selbst für Einzelfasern *nicht zutreffend* ist.

[10] BURTE, H. u. G. HALSEY: Textile Res. J. **17**, 465 (1947).

[11] FUHRMANN, W.: Melliand Textilber. **33**, 911 (1952).

noch erweitert und damit ist man bereits wieder auf dem Weg zu einem Mehr-Zustands-Modell (Multi-State-Model). Auch L. PETERS[1] versucht von den üblichen Modellvorstellungen freizukommen und das nichtlineare viscoelastische Verhalten durch eine mehrmolekulare Betrachtung zu erfassen, indem er an Stelle der üblichen Modellglieder valenzmäßige Bindungen zwischen Atomen bzw. Molekülen einführt und annimmt, daß die Aktivierungsenergie für den Bruch einer chemischen Bindung nach einem Exponentialansatz von der Entfernung der Atome abhängig ist. Ebenso sind die molekularen Betrachtungen A. H. NIRSAN[1] über das Spannungsdehnungsverhalten von Papier hervorzuheben. Es ist ein erfolgversprechender Anfang.

Im allgemeinen ist es der Zug, der bei allen Modellvorstellungen von den zwei- bis dreigliedrigen zu den vielgliedrigen Modellen führt und damit den Approximationscharakter der Aussagen über das Deformationsverhalten zum Ausdruck bringt, das die Aufstellung eines Auswahlprinzips zur Ermittlung der für eine vollständige Beschreibung des Deformationsverhaltens notwendige und hinreichende Anzahl von Meßgrößen ausschließt. Daher hat die bereits erwähnte Entwicklung der Prüfgeräte trotz der verschiedenen geschilderten Ansätze von der Theorie eine Brücke zur Prüfungsmethode zu schlagen, nur wenig Impulse empfangen. Auch der zur Zeit in Arbeit befindliche Normblattentwurf DIN 53835[2], der vorschreibt, daß die Belastungsgeschwindigkeit so einzustellen ist, daß die vorgeschriebene Kraftstufe in etwa 20 Sekunden erreicht wird, zeigt die Neigung zu Einpunktsmethoden, die eine Vergleichbarkeit der bei verschiedenen Substanzen ermittelten Werte einer Kenngröße zumindest erschweren und sogar unmöglich machen können.

Durch diese Einschränkungen verlieren die Meßgrößen, die das Deformationsverhalten beschreiben sollen, noch mehr an Bedeutung und sind kaum für eine Aussage über das spätere Verhalten der Faser bei betrieblicher Beanspruchung zu gebrauchen. Diese Erkenntnis führte zu der Entwicklung von Prüfgerätetypen (vgl. §§ 22d und 23b), die eine weitgehende Variation der Beanspruchung des Prüffadens unter den verschiedensten Bedingungen ermöglichen und charakteristische Funktionen zu ermitteln gestatten, die mit größerer Sicherheit das Deformationsverhalten dieser Fasern auf dem Verarbeitungsweg vorauszusagen erlauben. G. SUSICH und ST. BACKER[3] zeigen, wie man mit einem derartigen Prüfgerät, wie den INSTRON-Tester, Modell TT-B 9[4], bereits drei Parameter des Erholungsvorganges ermitteln kann, abgesehen von den zahlreichen Möglichkeiten, die übrigen Größen des Deformationsvorganges zu differenzieren. Allerdings kommt auch hier zum Ausdruck, daß diese Entwicklung infolge Fehlens einer allgemein gültigen Theorie ohne ein leitendes Prinzip vor sich gegangen ist, das aus dieser Vielzahl der Variationsmöglichkeiten die notwendigen und hinreichenden, das Deformationsverhalten beschreibenden Parameter auswählt. Eine Methode, die mit einfachen physikalischen Prüfgeräten auskommt und trotzdem tiefe Einblicke über

---

[1] PETERS, L.: Textile Res. J. **25**, 262 (1955). — A. H. NIRSAN: Nature **175**, 424 (1955). — [2] DIN 53835 — Entwurf August 1954.

[3] SUSICH, G. u. ST. BACKER: Textile Res. J. **21**, 482–509 (1951). — [4] Vgl. S. 276.

den Zusammenhang zwischen dem Deformationsverhalten und dem molekularen Bau gibt, ist von W. SCHEFER[1] bei der Untersuchung der Relaxation von Polyamidborsten bei konstanter Dehnung unter der Einwirkung von wasserfreiem Benzol, Äthanol und Wasser angewendet worden und ergab, daß die unter dem Einfluß dieser Substanzen erfolgende Relaxation auf die Lockerung des zwischenmolekularen Bindungszustandes zurückzuführen ist, wobei vom Wasser anzunehmen ist, daß es sich nur an die polaren Stellen, also an die Säureamidgruppen anlagert und infolgedessen nur die Wirkung der Wasserstoffbrücken aufheben kann, während Alkohole darüber hinaus mit ihrem lipophilen Rest auch die zwischen den benachbarten Methylengruppen vorhandenen VAN DER WAALSschen Kräfte verringern können. Daraus folgt, daß z. B. bei einem Nylon 610 in Äthanol mit seiner größeren Zahl an Methylengruppen eine stärkere Relaxation als im Wasser zu erwarten ist, in Übereinstimmung mit dem Experiment. Auf zahlreiche weitere Folgerungen wollen wir hier nicht eingehen.

Einen weiteren Hinweis, von welchen Einflüssen das Deformationsverhalten der Faserstoffe begleitet ist, und wie man diese Einflüsse zur Erkennung der geschilderten Zusammenhänge ausnutzen kann, gibt uns die Arbeit von C. M. BRYANT und J. WAKEHAM[2], die die Verstreckung von Zellulosefasern in Wasser begleitenden Entropieänderungen durch die Messung der Temperaturabhängigkeit der bei der Längenänderung auftretenden Spannung untersuchen. Bemerkenswert sind auch die Ergebnisse die A. BROWN: Text. Res J. **25**, 891 (1955) bei der Untersuchung des Deformationsverhaltens verschiedener Fasern in Abhängigkeit von der Temperatur erhält.

b) Die statistischen Festigkeitseigenschaften von Fasern und Monofilen.

Mit den entwickelten theoretischen Vorstellungen kann man wohl das Deformationsverhalten von Monofilen vor dem Bruch approximativ beschreiben. Mit der Einleitung des Bruchvorganges sind nun neue Vorstellungen verknüpft; vgl. Kap. III, § 16. Danach wird der Standpunkt der klassischen Festigkeitstheorie, daß die Reiß- bzw. Zugfestigkeit eine Materialkonstante sei und die Streuung der experimentell ermittelten Werte mit der wirklichen Eigenschaft nichts zu tun hat, sondern lediglich eine Folge der unvollkommenen Prüfmethoden ist, verlassen und durch die Annahme ersetzt, daß die Grenzfestigkeit durch irgendwelche im Material statistisch verteilte Schwachstellen bedingt sei[3].

Aus dieser Annahme folgt zwangsläufig, daß die Verteilungskurven der Festigkeitswerte nicht eine zwar unvermeidliche, aber bedeutungslose Begleiterscheinung der wirklichen Werte sind, sondern daß sie das Charakteristische des Festigkeitsverhaltens darstellen.

Damit ist z. B. die Zugfestigkeit selbst von homogenen Fasern keine Konstante mehr, sondern zeigt einen statistischen Charakter, und man

[1] SCHEFER, W.: Textil-Rundschau **10**, 365 (1955), vgl. auch Abschnitt c 4.

[2] BRYANT, C. M. u. J. WAKEHAM: Textile Res. J. **25**, 224 (1955).

[3] Bereits F. T. PEIRCE[4] konnte daraus die Abhängigkeit der Festigkeit des Fadens von der Einspannlänge ableiten.

[4] PEIRCE, F. T.: J. Textile Inst. **17**, T 355/68 (1926).

ordnet daher dieser Zugfestigkeit nach W. WEIBULL[1] eine Wahrscheinlichkeitsdichte $w(\sigma)$ oder die integrale Wahrscheinlichkeit

$$W(\sigma) = \int_0^\sigma w(\sigma)\, d\sigma \tag{IV, 27}$$

zu, in welcher σ die Spannung bedeutet, bei der der Bruch eintritt. Es ist eine Größe, welche die Wahrscheinlichkeit angibt, daß der Prüffaden bei der Beanspruchung durch die Spannung zu Bruch geht[2].

Nunmehr können wir zu der Fragestellung des Zusammenhanges zwischen der so definierten Wahrscheinlichkeitsverteilung und der Statistik der Schwachstellen übergehen, wobei als Schwachstelle eine räumlich eng begrenzte Stelle des Prüffadens angesehen werden soll, die bei Beanspruchung durch die örtliche Spannung σ einen Bruch bewirkt. Weiter sollen die einzelnen Schwachstellen unabhängig voneinander zur Wirkung gelangen und jeder örtlich hervorgerufene Bruch sich über den gesamten Querschnitt ausbreiten. Diese Forderung ist einerseits bei kompakten Materialien immer mit genügender Näherung erfüllt, auch bei Monofilen, nicht jedoch ohne weiteres für Lamellen oder Faserbündel, bei denen der Bruch einer Einzelfaser nicht zwangsläufig zum Zerreißen des ganzen Fadens zu führen braucht. Außerdem sollen die Schwachstellen unabhängig voneinander verteilt sein und ihre mittlere Dichte sei durch die Funktion $n(\sigma)$, die im allgemeinen Fall vom Ort abhängig ist, gegeben.

Unter diesen Voraussetzungen (vgl. Kap. III) erhalten wir den fundamentalen WEIBULLschen Ausdruck

$$W(\sigma) = 1 - e^{-\int n(\sigma)\, dV} \tag{IV, 28}$$

zwischen der Materialfunktion $n(\sigma)$ und der durch Messungen zu ermittelnden Wahrscheinlichkeitsverteilung $W(\sigma)$[3].

Diese Theorie wurde von H. GIESEKUS[4] auf die Zugfestigkeit von realen Perlon-Monofilen angewandt und führte zu dem Ergebnis, daß ein enger Zusammenhang zwischen den Schwankungen des Titers der Zugfestigkeit und der Bruchdehnung besteht, wobei die Ursachen der Schwankungen, die in den entsprechenden Korrelationsgrößen zum Ausdruck kommen, Verstreckungsungleichmäßigkeiten sind; denn starke Verstreckung bedingt kleinere Titer, größere Zugfestigkeit, größere Bruchdehnung und umgekehrt in voller Übereinstimmung mit den aufgetretenen Vorzeichen der verschiedenen Korrelationskoeffizienten.

Die Uneinheitlichkeit der Zugfestigkeitsverteilungskurven kann auf diesem Wege verständlich gemacht werden. Dabei zeigte sich auch, wie empfindlich statistische Untersuchungsmethoden auf scheinbar geringfügige Änderungen der Versuchsbedingungen reagieren. Bereits der Über-

[1] WEIBULL, W.: A Statistical Theory of the Strength of Materials. Ing. Vetensk. Akad., Handl. Nr. **151**, Stockholm 1939.

[2] $w(\sigma)\, d\sigma$ bedeutet demgegenüber die Wahrscheinlichkeit dafür, daß ein Prüffaden, welcher bei der Beanspruchung durch σ noch nicht gerissen ist, bei der Zunahme dieser Belastung um $d\sigma$ nunmehr zu Bruch geht.

[3] Diese Beziehung gilt auch für eine unhomogene Spannungsverteilung, nur muß dann die Größe σ im Ausdruck $W(\sigma)$ durch $\bar{\sigma}$ bzw. durch ein Maß für die Gesamtbelastung und das feste σ in der Funktion $n(\sigma)$ entsprechend durch $\sigma(v)$ ersetzt werden. — [4] GIESEKUS, H.: Nicht veröffentlicht. Vgl. S. 248, Fußnote 2.

gang von einer Klemmenart zu einer anderen am Zerreißapparat bedingte eine wesentliche Verschiebung in der Korrelation. Auf diese Einzelheiten ist besonders zu achten.

Da der grundlegende Gesichtspunkt der WEIBULLschen Theorie darin besteht, daß die experimentell ermittelten Festigkeitswerte in eine quantitative Beziehung zur Struktur des realen Festkörpers, die durch die Schwachstellendichte $n(\sigma)$ (Materialfunktion) gegeben ist, gebracht werden, so hat bereits W. WEIBULL diese Gedankengänge auch auf die Behandlung mehrachsiger Spannungszustände, insbesondere der Torsions- und Biegefestigkeit übertragen.

Gleichfalls kann man auf diesem Wege die Gruppe der Ermüdungserscheinungen, insbesondere die WÖHLER-Kurve, erfassen. Derartige Ansätze stammen von E. OROWAN[1], A. M. FREUDENTHAL[2] und W. WEIBULL[3]. Allen diesen Phänomenen ist gemeinsam, daß nur jeweils eine einzige Schwachstelle zur Wirkung gelangt. Aber auch das elastische und das beschriebene rheologische Verhalten wird von der Schwachstellenverteilung bestimmt, wenn auch in komplizierterer Weise, da hier alle Schwachstellen zusammenwirken[4, 5].

Es sei noch eine elementare Folgerung aus der Existenz der Verteilungskurve $W(\sigma)$ erwähnt, das ist die Abhängigkeit der Zugfestigkeit, verstanden als Durchschnittswert der Verteilung, von der Form bzw. von der Länge des Fadens, und zwar ergibt sich, daß die Zugfestigkeit des Fadens mit wachsender Prüflänge abnimmt, genau das gleiche Ergebnis, wie es bereits von F. T. PEIRCE gefunden worden ist. Damit ist es nunmehr auch möglich geworden, Werte, die mit Prüfgeräten verschiedener Einspannlänge gefunden worden sind, aufeinander zurückzuführen[6].

c) Die statistischen Festigkeitseigenschaften von zusammengesetzten Fasern und Fadengebilden.

Die bisher geschilderten Eigenschaften bezogen sich auf Fasern und Monofile, die als Bauelement für die auf dem Wege der Weiterverarbeitung entstehenden Gewirke und Gewebe dienen. Nun lassen jedoch die Überlegungen des § 21b bereits eindeutig erkennen, daß nicht nur die

[1] OROWAN, E.: Proc. Roy. Soc. [London] **A 171**, 79 (1939).

[2] FREUDENTHAL, A. M.: Proc. Roy. Soc. [London] **A 187**, 416 (1946). — A. M. FREUDENTHAL: Symposium on Statistical Aspects of Fatigue, A.S.T.M. Techn. Publ. Nr. 121 (Juni 1951).

[3] WEIBULL, W.: Trans. Roy. Inst. Technol. Nr. **27**, Stockholm 1949.

[4] Auch die Scheuerfestigkeit kann mit größerem Erfolg als bisher mit Hilfe dieser statistischen Auffassung analysiert werden, ferner kann der Nachweis erbracht werden, wie weit im Innern liegende Schwachstellen neben den Oberflächenschwachstellen den Verschleißvorgang mitbestimmen.

[5] Daraus folgt, daß auch das in den § 21a geschilderte rheologische Verhalten statistische Aspekte trägt und daher statistische Funktionen das Verhalten mit beschreiben.

[6] A. SIPPEL[7] macht mit Recht darauf aufmerksam, daß die Festigkeitswerte nicht nur von der Einspannlänge, sondern auch von der Relativgeschwindigkeit der Einspannklemmen, bezogen auf die Einspannlänge des Prüffadens, abhängen, übersieht jedoch dabei, daß diese Abhängigkeit durchaus nicht im Widerspruch zu der statistischen Betrachtungsweise steht.

[7] SIPPEL, A.: Faserforsch. u. Textiltechn. **4**, 152 (1953).

Festigkeit, sondern das gesamte Deformationsverhalten schon von Einzelfasern und Monofilen durch statistische Verteilungsfunktionen mit beschrieben werden. Es ist daher verständlich, daß beim Übergang zu den zusammengesetzten textilen Gebilden die statistischen Methoden zum ausschlaggebenden Faktor werden.

1. Die statistischen Festigkeitseigenschaften von Faserbündeln und Garnsträngen[1].

Wir beginnen mit dem Vergleich der Prüfverfahren, und zwar mit der Messung der Festigkeit von Einzelfasern und der vorwiegend in den USA gebräuchlichen Bündel- bzw. Strangfestigkeitsprüfung. Wir betrachten beispielsweise einen Prüffaden von der Länge $2l$ von gleichmäßigem Querschnitt und denken uns diesen Faden aus zwei gleich langen Stücken der Länge l zusammengesetzt, deren Summenverteilung $W_l(\sigma)$ gegeben sei. Da die Wahrscheinlichkeit, daß der Gesamtfaden nicht reißt, gleich der Wahrscheinlichkeit ist, daß sowohl im ersten als im zweiten Faden kein Bruch entsteht und durch die Beziehung

$$1 - W_{2l}(\sigma) = [1 - W_l(\sigma)]^2 \qquad \text{(IV,29)}$$

darstellbar ist, so erhalten wir als Ausdruck für die Summenverteilung des ganzen Prüffadens

$$W_{2l}(\sigma) = W_l(\sigma) + W_l(\sigma)\,[1 - W_l(\sigma)]\,. \qquad \text{(IV,30)}$$

Man erkennt hieraus, daß $W_{2l}(\sigma)$, außer für die Grenzwerte null und eins, stets größer als $W_l(\sigma)$ ist. Somit stehen auch die zugeordneten Durchschnittswerte in der gleichen Relation zueinander, d.h., die Festigkeit nimmt mit wachsender Prüflänge ab.

Denken wir uns nun an Stelle eines Prüffadens der Länge $2l$ zwei parallel eingespannte Prüffäden von der Länge l, so erhalten wir für die Wahrscheinlichkeit, daß beide Prüffäden nicht reißen, den gleichen Ausdruck Gl. (IV,29). Allgemein gilt für r Fasern der Länge l oder für eine Faser der Länge $r \cdot l$ die gleiche Beziehung für die Wahrscheinlichkeit, daß ein Bruch nicht eintritt

$$1 - W_{r \cdot l}(\sigma) = [1 - W_l(\sigma)]^r\,. \qquad \text{(IV,31)}$$

Demnach ergibt die WEIBULLsche Theorie, auf die Bündelfestigkeitsprüfung übertragen, daß nur dann die gleichen Festigkeiten zu erwarten sind, wenn das Produkt aus der Anzahl der Fäden und der Einspannlänge konstant ist. Andernfalls wird bei gleicher Einspannlänge eine um so niedrigere Festigkeit zu erwarten sein, je mehr Fäden oder Fasern im Bündel vereinigt sind.

Vergleiche auch F. WINKLER[2], der durch Auswertung zahlreicher experimenteller Arbeiten und Aufstellung der Summenhäufigkeitskurven der Bruchwahrscheinlichkeit von Baumwollfaserbündeln verschiedener Einspannlängen und eines Baumwollgarns bei verschiedenen Einspann-

---

[1] Es besteht in bezug auf die Begriffe Faden und Garn immer noch ein Durcheinander in der Literatur, vgl. E. CUCHE: Textil-Rundschau **10**, 374 (1955) und Textile Terms and Definition II. Ed. Textile Inst. Manchester, September 1955.

[2] WINKLER, F.: Faserforsch. u. Textiltechn. **5**, 398 (1954).

längen und Fadenzahlen sowie von Jutegarn die Gültigkeit der Theorie nachweisen konnte, wobei eine recht gute Übereinstimmung zwischen der Theorie und dem Experiment festzustellen war. Daraus ergibt sich auch eine theoretische Grundlage für das neue Harfenreißgerät von SCHUMACHER[1]. Diese kollektive Garnprüfung auf Festigkeit, Dehnung und Streuung ist gegenüber der eingeführten Einzelzugfestigkeitsmessung, ebenso wie die Gebindeprüfung in Amerika und England, durch einen erheblich niedrigeren Zeitaufwand ausgezeichnet. Eine ausführliche Darstellung der Prüfung mit dem Harfenreißgerät gab die Arbeitsgemeinschaft deutscher Textilingenieure[2], die von H. J. HENNING, der sich ebenfalls mit den statistischen Fragen dieses Gerätes befaßte, ergänzt worden ist. Bei seinen Ausführungen kommt H. J. HENNING vor dem Fachausschuß Messen und Prüfen der ADT/VDI, am 24. 6. 1955, zu dem Schluß, daß damit keine neuen Prüfmöglichkeiten erschlossen werden und auch keine Aufschlüsse, die über die Ergebnisse von den üblichen Zugfestigkeitsprüfungen hinausgehen, zu erwarten sind, daß jedoch der erhebliche Zeitgewinn bei der Prüfung bei einem geringen Informationsverlust von etwa 30% gegenüber dem Ergebnis der Einzelmessungen für die Verwendung des Gerätes spricht[3].

2. *Garneigenschaften und das Problem der Garngleichmäßigkeit.*

Wir haben gesehen, wie man die Bündelfestigkeit statistisch auf die Einzelfaserfestigkeit zurückführen kann.

Nunmehr kommen wir zu der Untersuchung des ersten zusammengesetzten Textilgebildes, des Garns, das, wie z.B. DIN 60305[4] für die Baumwoll- und Zellwollspinnerei erkennen läßt, zu seiner Herstellung mehrere Arbeitsgänge benötigt. Nun erhebt sich auch hier die Frage nach der Zurückführung der Festigkeit dieses zusammengesetzten Gebildes des entstandenen Garns auf die entsprechenden Eigenschaften der einzelnen Fasern[5]. Diese Zusammenhänge sind in zahlreichen Arbeiten behandelt worden. Wir wollen uns auf einige dieser Arbeiten beschränken, um die wesentlichen Faktoren herauszustellen, und erwähnen hierbei die Untersuchung von H. WAKEHAM[6], der bei der Verarbeitung von Baumwollfasern feststellte, daß die Garnfestigkeit eine Funktion der mittleren

---

[1] REINFELD: Zellwoll-Lehrspinnerei, Denkendorf bei Eßlingen, Messen und Prüfen. 1. Folge (1954). — [2] Arbeitsgemeinschaft deutscher Textilingenieure, Arbeitskreis Augsburg: Textil-Praxis **10**, 235 (1955).

[3] Wieweit die Unsicherheit in der Beurteilung bisher gegangen ist, kann man aus der Liefer- und Prüfvorschrift für Zellwolle 1953 der intern. Chemiefasernormen (BISFA) ersehen. Hier wird für die Reißfestigkeitsbestimmung die Einzelfaserprüfung vorgeschrieben und gleichzeitig, ohne auf die Vergleichbarkeit oder Umrechnungsmöglichkeit der Ergebnisse auch nur einzugehen, die Einschränkung gemacht, daß, wenn ein derartiges Gerät nicht zur Verfügung steht, auch ein Faserbündelprüfer verwendet werden kann. — Vgl. auch H. KÖB: Textilprax. **10**, 11 (1955) sowie H. M. BROWN: Textile Res. J. **24**, 251 (1954).

[4] DIN 60305 — Entwurf Januar 1953.

[5] In welchem Umfang die mathematische Statistik angewandt werden kann, sieht man auch aus der Arbeit von E. SCHENKEL: Textilprax. **7**, 875 (1952), sowie aus den einschlägigen Arbeiten von U. GRAF, H. J. HENNING u. R. WARTMANN im Mitteilungsblatt für math. Statistik, Physica Verlag, Würzburg 1952–1955.

[6] WAKEHAM, H.: Text. Wld. **4**, 98 (1953).

Faserfestigkeit und, es sei besonders betont, auch der Schwankungen sowohl der Faserfestigkeit, der Faserlänge wie auch der Feinheit[1] und der Dehnung ist. Außerdem geht die Verdrehung der Baumwollfaser mit ein und ebenso ist zu beachten, daß die Meßlängen zwischen den Klemmen des Prüfgerätes, z. B. beim PRESSLEY-Prüfer[2] die Meßergebnisse beeinflussen, so daß die Garnfestigkeit durch diese Schwankungen der Eigenschaften, wobei auch noch die Lageverteilung der Fasern im Garn eingeht, immer erheblich geringer ist als die Summe der Festigkeiten der Einzelfasern, die das Garn aufbauen. Die Lage bzw. Anordnung der Fasern, insbesondere im Zellwollgarn, wurde von W.E. MORTON und K. C. YEN[3] untersucht und dabei festgestellt, daß die Fasern mehr oder weniger angenähert nach einer harmonisch abklingenden Sinuslinie im Gespinst angeordnet sind (vgl. auch die Arbeit von M. N. BELEZIN[4]). Außer der Lagenverteilung der Fasern ist von Bedeutung die Verteilung der Drehung in einem Garn. Hier ist zu erwähnen die Untersuchung von F. MONFORT[5], der experimentell an verschiedenen Garnen von der Ringspinnmaschine und dem Selfaktor nachwies, daß ein negativer Korrelationskoeffizient zwischen der Garndrehung je Einheitslänge und dem Gewicht dieser Einheitslänge vorhanden ist. Der negative Korrelationskoeffizient bringt zum Ausdruck, daß die Anzahl der Drehungen je Bezugslänge zunimmt, wenn das entsprechende Gewicht dieser Bezugslänge abnimmt oder, mit anderen Worten, daß die Drehung über die Fadenlänge sich so verteilt, daß in die gewichtsmäßig geringere Stelle der größere Draht kommt. J. GREGORY[6] zeigte bei der Untersuchung des Zusammenhanges zwischen den Eigenschaften der Fasern und der Festigkeit gedrehter Garnelemente aus Baumwolle u. a., daß mit zunehmendem Draht die Festigkeit bis zu einem Maximum anwächst und dann zurückgeht, wobei die maximale Garnfestigkeit, wie L. A. FIORY und J. J. BROWN[7] nach kritischer Stellungnahme zu verschiedenen Arbeiten wiederum nachwiesen, sehr wesentlich von der Faserfeinheit abhängt, und zwar verläuft der Festigkeitsverlust von Garnen, die aus groben Fasern versponnen werden, mit abnehmendem Drehungskoeffizienten schneller als bei Garnen aus feinen Fasern. Über die maximale Garnfestigkeit hinaus reagieren mit zunehmender Drehung grobe Garne weniger empfindlich als feine Garne[8–13]. Hier sieht man bereits, daß die besondere Beanspruchung der Fasern auf dem Ver-

---

[1] DIN 53812-Entwurf Juni 1954. — [2] PRESSLEY-Prüfer, A. S. T. M. D 414/47 T.

[3] MORTON, W. E. u. K. C. YEN: J. Textile Inst. **43**, T 60 (1952).

[4] BELEZIN, M. N.: Ref. Textilprax. **4**, 80 (1949). Bezügl. d. Drehung vgl. DIN 53832.

[5] MONFORT, F.: Bull. Inst. France **6**, 55 (1952).

[6] GREGORY, J.: J. Textile Inst. **44**, T 499 (1953).

[7] FIORY, L. A. u. J. J. BROWN: Textile Res. J. **10**, 750 (1951).

[8] Weitere Untersuchungen, die sich mit diesem Problem beschäftigen, sind u. a. von M. M. PLATT[9], S. M. MUNKHERYEE u. M. K. SEN[10] sowie M. M. PLATT, W. G. KLEIN u. W. HAMBURGER[11] und D. M. THORNTON[12] durchgeführt worden.

[9] PLATT, M. M: Textile Res. J. **20**, 665 (1950); **24**, 132 (1954).

[10] MUNKHERYEE, S. M. u. M. K. SEN: Text. Manufacturer **7**, 319 (1949).

[11] PLATT, M. M., W. G. Klein u. W. J. HAMBURGER: Textile Res. J. **22**, 641 (1952). — [12] THORNTON, D. M.: Mod. Textil Mag. **34**, 60 (1953).

[13] Bei Nylonfäden wurde der Einfluß der Drehung von F. J. ALEXANDER u. C. H. STURLEY untersucht: J. Textile Inst. **43**, 1 (1952).

arbeitungswege zu spezifischen Untersuchungsmethoden führt. Die Verhältnisse beim Verspinnen von Fasern, insbesondere der Widerstand der Fasern gegen Torsion, wie er bei der Drahtgebung und Zwirnung zu überwinden ist, werden von R. MEREDITH[1] untersucht. Mittels der Methode der Drehschwingungen wird der Schubelastizitätsmodul bestimmt und das erforderliche Drehmoment, das zur Verdrehung der Faser aufgebracht werden muß, abgeleitet. Zur Ermittlung des Verdrehungswiderstandes wird eine Drehmomenten-Drahtkurve angegeben.

Von anderer Seite, und zwar von W. FRENZEL und H. PERNER[2] wurden die Torsionsfestigkeit und Bruchdehnung von Fäden und Gespinsten ermittelt, um daraus Schlüsse auf das Verhalten bei der Verarbeitung auf Flyer- und Vorspinnmaschinen zu ziehen. Am Verlauf der gewonnenen Kurven wird gezeigt, bis zu welcher Drehzahl die Zugbeanspruchung durch die Drahtgebung für das Gespinst zur Vermeidung von Fehlverzügen und Fadenbrüchen tragbar ist.

Bei allen diesen Untersuchungen zeigt sich eine neue Tendenz, die charakteristisch für die zusammengesetzten Fasergebilde ist. Es ist das Auftreten statistischer Verteilungen geometrischer Größen, wie Faserlänge, Faserdurchmesser und Faserlage im Gespinst[3,6], sowie ihre enge Wechselwirkung mit den statistischen Verteilungen physikalischer Größen, wie sie bisher geschildert wurden. Diese Gruppe der Verteilungen geometrischer Größen ist nun für die Garngleichmäßigkeit maßgebend, und ihre Kenntnis hat unser Wissen über die Spinnprozesse erheblich erweitert.

Um einen Einblick zu erhalten, wie diese für die Garnungleichmäßigkeit charakteristischen Verteilungen zustande kommen, gehen wir folgendermaßen vor. Aus der zu bewertenden Garnmenge wird eine größere Anzahl Stücke der Länge L entnommen und das Gewicht je Längeneinheit bestimmt. Daraus ergibt sich der Variationskoeffizient $V_{ZL}\%$, dessen Quadrat uns die Garnungleichmäßigkeit zwischen den Stücken der Länge L, bezogen auf den Mittelwert, angibt. Außer dieser Ungleichmäßigkeit zwischen Garnstücken besteht noch eine Ungleichmäßigkeit innerhalb der Garnstücke, die entsprechend durch den Variationskoeffizienten gegeben ist. Auf Grund der einfachen Streuungsanalysen (vgl. U. GRAF und H. J. HENNING[7]) setzen sich die Quadrate der Variationskoeffizienten additiv zusammen

$$V_{\text{ges}}^2 = V_{ZL}^2 + V_{iL}^2, \tag{IV,32}$$

[1] MEREDITH, R.: J. Textile Inst. **43**, 755 u. 785 (1952).

[2] FRENZEL, W. u. H. PERNER: Faserforsch. u. Textiltechn. **4**, 1 (1953); **4**, 63 (1953); **6**, 465 (1955).

[3] Noch schwieriger werden die Probleme, wenn man z. B. Zellwollen verschiedener Schnittlängen und Titer in Mischungen verspinnt (L. RUDOLPH[4]), bzw. wenn man Mischungen aus verschiedenen textilen Rohstoffen verarbeitet (H. NUDING[5]).

[4] RUDOLPH, L.: Faserforsch. u. Textiltechn. **5**, 391 (1954).

[5] NUDING, H.: Reyon, Zellwolle u. a. chemische Fasern **9**, 478 (1952).

[6] FRENZEL, W. u. H. PERNER: Faserforsch. u. Textiltechn. **4**, 63 (1953).

[7] GRAF, U. u. H. J. HENNING: Statistische Methoden bei textilen Untersuchungen, Springer-Verlag, Berlin 1952, S. 109, 114, 123. V_{iL} bedeutet anschaulich die Schwankungen, z. B. der Garnnummer auf kürzere Längen, die sich in einem mehr oder minder unruhigen Warenbild äußern kann, und V_{ZL} die Schwankungen auf größere Längen, die zu den Schußbanden Anlaß geben können.

d. h. die gesamte Ungleichmäßigkeit folgt additiv aus der durchschnittlichen Ungleichmäßigkeit eines Garnstückes von der Länge L und der Ungleichmäßigkeit zwischen Garnabschnitten der Länge L. J. G. MARTINDALE[1] hat nun bewiesen, daß V_{ges}^2 bei Vorgarnen und Lunten eine untere Grenze besitzt, die von den Durchmesserschwankungen der versponnenen Fasern und von der mittleren Anzahl der Fasern im Querschnitt abhängt. Sie ist die geringste Ungleichmäßigkeit, die infolge der mit den Spinnverfahren und den Maschinen verbundenen zufälligen Schwankungen mindestens auftritt und wird bedingt durch die Schwankungen in der Faserzahl im Querschnitt und durch die Schwankungen im Faserdurchmesser, die vom Fasermaterial abhängen und durch den Variationskoeffizienten V_D gegeben sind. Nach der POISSONschen Verteilung hängt nun die kleinste Schwankung der Faserzahl von der mittleren Faserzahl $\bar{n}$ im Querschnitt ab. Bei Berücksichtigung der Unabhängigkeit beider Einflußgrößen ergibt sich die Formel von J. G. MARTINDALE für den Grenzwert der Variationskoeffizienten

$$V_{(\text{ges})\,\text{lim}} = \frac{100\sqrt{(1 + 0{,}0004\,V_D^2)}}{\sqrt{\bar{n}}}\,. \qquad \text{(IV, 33)}$$

An Stelle $\bar{n}$ kann man auch $Z_m = \frac{N_m\,(\text{Faser})}{N_m\,(\text{Garn})}$ einführen, wobei N_m = Nummer metrisch[2] bedeutet, und entsprechend umformen. Diese Beziehung erlaubt bereits wesentliche Einblicke in die Spinnverfahren, insbesondere in die Wirkung der Passagen. Von A. HUBERTY[3] ist für Klassifizierungszwecke der Quotient des praktisch ermittelten Variationskoeffizienten V_{ges} und des Variationskoeffizienten $V_{(\text{ges})\,\text{lim}}$ eingeführt und mit $K = \frac{V_{\text{ges}}}{V_{(\text{ges})\,\text{lim}}}$ bezeichnet worden. Kehren wir nun zu der Gl. (IV, 32) zurück, und denken wir uns V_{iL}^2 und V_{ZL}^2 für verschiedene Werte von L ermittelt und in Abhängigkeit von der Länge L dargestellt, so erhalten wir die Streuungs-Längenkurve, wie sie von M. W. TOWNSEND[4] und

[1] MARTINDALE, J. G.: J. Textile Inst. **36**, T 35 (1945).

[2] Bisher gebräuchliche Maße für die Feinheit bzw. der mittleren linearen Dichte eines Faserstoffes wie auch von Garnen und Zwirnen sind bei der Längennummerierung die metrische Nummer

$$N_m = \frac{L\,[\text{m}]}{G\,[\text{g}]}\,,$$

die Anzahl der Längeneinheiten in Metern, je Gewichtseinheit in Gramm dargestellt und bei der Gewichtseinheit der legale Titer

$$T_d = \frac{9000 \cdot G\,[\text{g}]}{L\,[\text{m}]} \quad \text{in Denier [den]}\,,$$

die Anzahl der Gewichtseinheiten in Gramm, je Längeneinheit von 9000 m bezeichnend. Der Übergang zu dem dezimalen Titer in Grex oder Tex, wobei

$$g_x = \frac{1000}{N_m} \quad \text{und 1 Grex} = 10\,\text{Tex}$$

bedeuten, wird angestrebt. – Vgl. LANDOLT-BÖRNSTEIN: Zahlenwerte und Funktionen aus Physik, Chemie, Astronomie, Geophysik, Technik, Springer, Berlin 1955, 6. Aufl., IV. Bd., 1. Teil, 374–384. DIN 53831, sowie ISO/TC 38 Textilien, Unterkommission SC 5, Garnprüfung/Entwurf, Januar 1954. – K. HENTSCHEL: Melliand Textilber. **36**, 1303 (1955). – [3] HUBERTY, A.: Inter. Wool. Text., Organ. Techn. Comm. Proc. **1**, 55 (1947). – [4] TOWNSEND, M. W.: J. Textile Inst. **40**, 566 (1949).

D. R. Cox[1] für verschiedenes Garnmaterial in mehreren Arbeiten ermittelt und angegeben worden ist[2]. Für die praktische Ermittlung der Kurven, die uns weitgehenden Aufschluß über Garnungleichmäßigkeiten geben, ist es von erheblicher Bedeutung, daß die Gesamtstreuung V^2_{ges}, wie man leicht nachweist, unabhängig von der Prüflänge L ist[3]. G. V. LUND hat in einer auch für die Praxis bemerkenswerten Arbeit [Textilprax. **10**, 330 (1955)] diese Gedankengänge von J. G. MARTINDALE auf Mischgarne übertragen und dabei gezeigt, daß feinere Garne weniger gut gemischt erscheinen als gröbere, wenn die Garne aus gleichen proportionalen Mischungen ähnlicher Fasern gesponnen werden, und daß Garne gleicher Nummer streifiger erscheinen, wenn sie aus Fasern gröberen Titers versponnen werden. Ebenfalls erscheinen Garne gleicher Nummer mit zunehmendem Unterschied in den Faseranteilen einer Mischung streifiger. Erwähnt wird auch die Entstehung streifiger Garne, wenn in der Mischung geringe Anteile, farbig stark verschieden, grobe Fasern vorhanden sind. Diese Gedankengänge G. V. LUNDS sind von H. J. HENNING bei der Berechnung der Streuung bei Fasermischungen weitergeführt worden. Vgl. H. J. HENNING: Melliand Textilberichte **36**, 702, 785 (1955) sowie Fortsetzungen, und im besonderen seine interessanten Ausführungen vor dem A. D. T./V. D. J.-Ausschuß am 24. 6. 1955.

Für die Bestimmung dieser statistischen Größen und damit der Garnungleichmäßigkeit, wie sie durch die statistische Verteilung geometrischer Größen gegeben ist, sind große Gruppen von Prüfapparaten entwickelt worden, die gleichberechtigt neben den in den §§ 22–24 beschriebenen physikalischen Meßgeräten stehen, sogar eine notwendige Ergänzung bilden und z. T. ebenfalls eine einwandfreie physikalische Grundlage besitzen. Es sei nur an die kapazitative Meßmethode und darauf beruhende Geräte: Zellweger USTER[6, 7], Textronograph von H. STEIN[8, 9] erinnert. Eine zusammenfassende und ausführliche Darstellung der Methoden und Prüfapparate geben W. WEGENER und W. ZAHN[10] sowie H. SULSER[11], auf die hier verwiesen sei[12].

[1] COX, D. R. u. M. W. TOWNSEND: J. Textile Inst. **42**, 145 (1951). — M. W. TOWNSEND u. D. R. COX: J. Textile Inst. **42**, 107 (1951). — Vgl. auch W. WEGENER u. W. ZAHN: Melliand Textilber. **36**, 686, 776 (1955). Von wesentlicher Bedeutung für die Auswertung ist dabei der enge Zusammenhang zwischen der Autokorrelationsfunktion und der Streuungs-Längenkurve.

[2] Zu erwähnen sind noch die bemerkenswerten theoretischen Ergänzungen der Arbeiten von M. W. TOWNSEND u. D. R. COX durch H. OLERUP[4] u. H. BRENY[5] und von zuletzt genannten außerdem die Auffindung period. Garnfehler mittels harmonis her Analyse, Ann. Sci. textiles belges **2**, 7 (1954).

[3] Vgl. W. MASING: Textilprax. **10**, 1237 (1955).

[4] OLERUP, H.: J. Textile Inst. **43**, 290 (1952). — [5] BRENY, H.: J. Textile Inst. **44**, 1 (1953). — [6] SCHLIEN, K.: Melliand Textilber. **34**, 1061 (1953). — [7] PRINS, I. u. W. H. PETERS: Reyon Revue 11 (1955). — [8] STEIN, H.: Textilprax. **8**, 312 (1953).

[9] FISCHER, B.: Textilprax. **9**, 423 (1954). — [10] WEGENER, W. u. W. ZAHN: Textilprax. **9**, 21 (1954). — [11] SULSER, H.: Schweiz. Arch. angew. Wiss. Techn. **19**, 85 (1953).

[12] Wie weit diese Arbeiten bereits gediehen sind und welche Bedeutung ihnen mit Recht beigemessen wird, ersieht man aus den Normvorschlägen von F. MONFORT[13] für die Bestimmung der Querschnittsgleichmäßigkeit von Kammgarnen, die auf der diskutierten Gleichmäßigkeit des Durchmessers der Fasern und ihrer Verteilung beruhen.

[13] MONFORT, F.: Reyon, Zellwolle und andere Chemiefasern **1**, 6, 24 (1951).

Die Untersuchung von J. E. DOUGHERT, L. H. DANCE und W. H. MARTIN[1] zeigt, wie eine zunehmende Ungleichmäßigkeit im Garn eine Abnahme der Garnfestigkeit hervorruft und wie eine Erholung des Verzuges auf der Spinnmaschine ebenfalls zu einer Verringerung der Garnfestigkeit führt. Verzug und Doublierung beeinflussen gemeinsam die Garnfestigkeit, dagegen wurde keine Korrelation zwischen der Garnfestigkeit und der Vorgarndrehung festgestellt.

An diesem Beispiel sehen wir die enge Wechselwirkung zwischen der Garnungleichmäßigkeit, durch die statistische Verteilung geometrischer Größen bedingt, und den physikalischen Meßgrößen dieses Garns, die zwangsläufig die Untersuchung der Garnungleichmäßigkeit, durch statistische Verteilungen physikalischer Meßgrößen hervorgerufen, zur Folge hat. Daraus entwickelt sich die Prüfung der Festigkeit und anderer mechanischer Größen am laufenden Faden[2]. Auf die Gerätetypen, den USTER-Dynamometer und die Universal-Garnprüfmaschine von FRENZEL-HAHN, wird kurz in dem § 22b eingegangen, wobei beide Geräte auch zur Untersuchung der Haftgleiteigenschaften von Gespinsten nach H. STEIN[4] und nach W. MEYER[5] sowie der Beziehungen zwischen Gewichtshaftfestigkeit und Querschnittsschwankungen bei Kammzugbändern Eingang gefunden haben, vgl. auch G. NITSCHKE[6]. Daß die Beanspruchungen der Garne in der Spinnerei und Weberei dynamischer Natur sind und daher die statisch ermittelten Meßgrößen z. B. Festigkeit und Dehnung kein zuverlässiges Maß für das Verhalten auf dem Webstuhl sind, zeigte M. LEBLANC[7].

Die Ermüdung in Abhängigkeit von der Garngeometrie behandelt M. M. PLATT[8]. Die dynamische Prüfung ergab befriedigende Übereinstimmung mit dem Verhalten des Garns auf dem Webstuhl. Berücksichtigt man noch die große Bedeutung der dynamischen Prüfung für die Cordbeurteilung, so wird man die Anstrengungen bei der Entwicklung einwandfreier Prüfgeräte verstehen. Es wird daher in § 23 ausführlich auf diese Prüfgeräte eingegangen und in § 23b der Vorschlag gemacht, auch die dynamische Prüfung auf den laufenden Faden zu übertragen. Wir erhalten also ein geschlossenes Bild, das uns die mathematische Statistik, insbesondere die Korrelationstheorie in Verbindung mit den beiden großen Meßgerätegruppen für die Ermittlung des Zusammenhanges zwischen den Faser- und den Garneigenschaften liefert. Es ist kein Raum für Modellgeräte; betrachten wir jedoch die Tab. IV, 5, die ein Schema der Meßgrößen in ihrem stochastischen Zusammenhang darstellt, wobei die größeren waagerecht strichliniierten Zwischenräume zwischen den Kästen das Vorhandensein nicht eingezeichneter Prüfgerätetypen, Arbeits-

---

[1] DOUGHERT, J. E., L. H. DANCE u. W. H. MARTIN: Text. Ind. Exporter **10**, 116 (1952).

[2] Bezüglich der Toleranzbestimmung für die Gleichmäßigkeit textiler Materialien vgl. auch U. GRAF u. H. J. HENNING[3].

[3] GRAF, U. u. H. J. HENNING: Melliand Textilber. **31**, 818 (1950).

[4] STEIN, H.: Textilprax. **8**, 312 (1953).

[5] MEYER, W.: Textilprax. **9**, 17 (1954).

[6] NITSCHKE, G.: Faserforsch. u. Textiltechn. **5**, 297 (1954).

[7] LEBLANC, M.: L'Industrie Textile **7**, 361 (1952).

[8] PLATT, M. M.: Textile Res. J. **20**, 1 (1950); **20**, 519 (1950).

Tabelle IV, 5.

Die an Einzelfasern und Fäden mit den verschiedenen Prüfgeräten bei definierten einachsigen Beanspruchungen ermittelten Werte-Verteilungen.

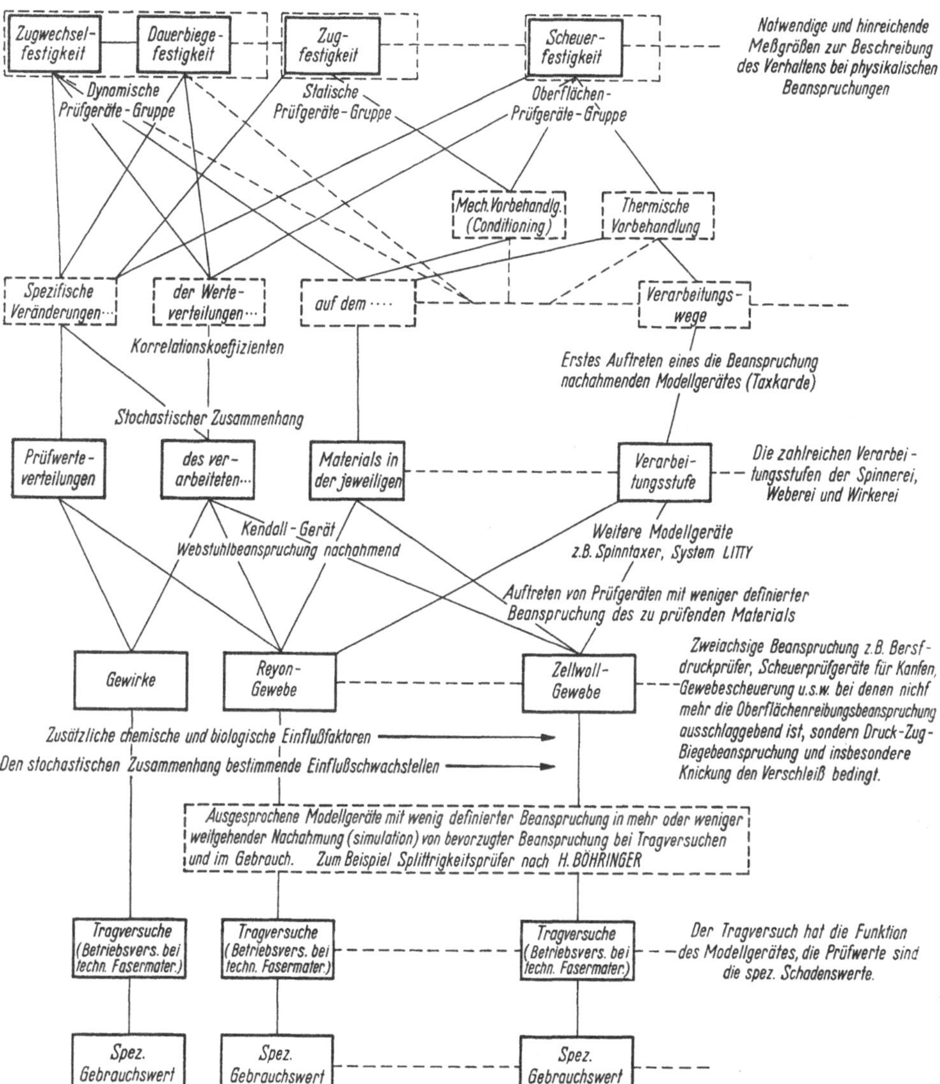

gänge usw. andeuten sollen, so sehen wir, daß mit Auftreten zusammengesetzter Fasergebilde auch die irgendwelche komplexe Beanspruchung nachahmende Modellgeräte auftauchen. Begünstigt wurde die Entwicklung derartiger Modellgeräte durch die bisher noch wenig verbreiteten mathematisch-statistischen Methoden und durch den mit ihrer Anwendung verbundenen großen Arbeitsaufwand.

3. Die Gewebeeigenschaften und ihre Prüfung[1].

Bei dem Übergang vom Garn zum Gewebe nimmt die Zahl der Einflußfaktoren, die den stochastischen Zusammenhang zwischen den Garn- und den Gewebeeigenschaften bestimmen, erheblich zu. G. SATLOW und H. GRIESE[2,3] erläuterten das am Beispiel der Berechnung der Gewebefestigkeit G_w aus der Garnfestigkeit G_g nach der Beziehung $G_w = G_g \cdot A \cdot F$, wobei A die Anzahl Fäden je Prüfbreite bezeichnet und F ein veränderlicher Faktor ist. Der Faktor F hängt von dem Garnmaterial, dem Gewebeaufbau und dem Webprozeß ab. Nun weist allein das Garnmaterial wiederum vier Variationsmöglichkeiten auf, die durch die Höhe des Drehungskoeffizienten, durch die Ungleichmäßigkeit der Garne sowie durch die Glätte[4] des verarbeiteten Materials und schließlich durch die Feinheit des Garns gegeben sind. Die Auswirkungen des Gewebeaufbaus machen sich durch die Gewebedichte und durch die Lage der Bindungsstellen geltend, während nach dem Webprozeß sich noch die Spannungsverhältnisse und die Beschädigung der Garne bemerkbar machen. Man sieht also, daß eine Vielzahl von Einflußfaktoren vorliegt, deren Gewichte schwer abzuschätzen sind. Ein bemerkenswerter Beitrag zu diesen Ausführungen sind die qualitativen Betrachtungen von A. BROWN[5], der glaubt, allein mit 4 Punkten der zugelastischen Kurve der Einzelfaser auszukommen, die die maßgebenden Einflußgrößen der Gewebeeigenschaften darstellen sollen, und diese sind, wie im einzelnen erläutert wird, der sogenannte Fließpunkt, der Steifheitsbereich, der Bruchpunkt und der sogenannte Kräuselungsbereich. Hohe Werte an diesen charakteristischen Stellen der zugelastischen Kurve sollen bei Geweben hohe Reißfestigkeit und Scheuerfestigkeit, sowie Knitterfestigkeit und guten Griff bedeuten. Außerdem wird daraus die Folgerung für eine leichte Verarbeitung des Stapels zum Garn gezogen. Weiter werden diese charakteristischen Stellen der zugelastischen Kurve in Verbindung mit gewissen molekularen Eigenschaften des hochpolymeren Ausgangsmaterials gebracht. Auch hier kommt jedoch zum Ausdruck, daß der Übergang zu undefinierten Beanspruchungsarten und damit zu den sogenannten Modellgeräten zur Prüfung von Geweben nicht notwendig erscheint; vgl. auch A. BAINES: J. Text. Inst. **44**, 645 (1953). Von R. F. DYER, V. G. FAW und R. L. BEARD sowie M. LEBLANC[6] liegt eine Untersuchung über die Fadenbeanspruchung bei dem Zetteln und Scheren in der Webereivorbereitung vor. Über den Einfluß der Fadenspannung während der Vorbereitung der Ketten auf das

---

[1] Die Beziehungen zwischen den mechanischen Charakteristiken von Strickwaren und den Eigenschaften der dazu verwendeten Strickgarne untersucht P. J. DOYLE: J. Textile Inst. **43**, 19 (1952).

[2] SATLOW, G. u. H. GRIESE: Textilprax. **3**, 274 (1948).

[3] BARELLA, A., F. MAILLARD, O. ROEHRICH, E. AMOUROUX, J. M. GARCIA-PLANAS u. S. PERICH: J. Textile Inst. **45**, 82 (1954).

[4] Über das Reibungsverhalten der Fasern existiert eine umfangreiche Literatur, es sei z. B. auf H. L. RÖDER: J. Textile Inst. **46**, 84 (1955) — D. G. LYNE: J. Textile Inst. **46**, 112 (1955) und E. CORD: J. Textile Inst. **46**, 41 (1955) hingewiesen.

[5] BROWN, A: Melliand Textilber. **36**, 48 (1955). Tagungsbericht.

[6] DYER, R. F., V. G. FAW u. R. L. BEARD: Textile Res. J. **22**, 487 (1952). — M. LEBLANC: Reyon, Zellwolle und andere Chemiefaser **2**, 194, 242 (1952).

Verhalten auf dem Webstuhl gibt H. E. ABRUKOW[1] ausführliche Unterlagen. Die Kettfadenbeanspruchung am Webstuhl wird sehr gründlich von H. KELLER[2] sowie von W. FRENZEL und H. MARTIN[3] behandelt. Bezüglich der Abhängigkeit der Kettspannung von der Gewebestruktur erfahren diese Arbeiten durch W. A. WOROBJEW[4] noch eine gewisse Ergänzung. Die Kettfadenspannung, die Anzahl der Kapillarfäden sowie die Anzahl der Schüsse und der Titer von Viscose-Schußgarnen werden von L. H. JAMESON, B. L. WHITTIER und H. SCHIEFER[5] von 32 Webwaren in Beziehung zu dem Gewicht der Konstruktion des Reyongewebes, der Reißfestigkeit, der Dehnung sowie der Weiterreißfestigkeit usw. gebracht. Bei Geweben aus gesponnenen Garnen sind ähnliche Untersuchungen über den Einfluß des Gewebeaufbaus auf die Reißfestigkeit, Tragfähigkeit und Scheuerfestigkeit zur Durchführung gekommen. Von anderen Überlegungen gehen die Arbeiten von J. M. ESSAM[6], I. A. TEN BRUGGENCATE[7] sowie L. HANUSS[8] aus, um den Zusammenhang zwischen der Garn- und der Gewebefestigkeit zu ermitteln[9]. Die Ergebnisse kann man noch nicht als befriedigend bezeichnen, und das ist auch nicht weiter erstaunlich, wenn berücksichtigt wird, daß die Garnprüfung nach DIN 53834[10] bei 500 mm Einspannlänge 20 Sekunden Reißdauer verlangt, während das Technische Komitee 38 der ISO dafür 800 mm/min Klemmengeschwindigkeit vorschreibt, und für den Gewebereißversuch 300 mm Einspannlänge bei 60 Sekunden Reißdauer[11] genormt ist. Die Beanspruchung ist vollkommen verschieden, und es ist daher ein Vergleich nach unserer Auffassung gar nicht zulässig. Die Prüfvorschriften sind reformbedürftig, wobei wieder einmal darauf hingewiesen werden muß, daß die Einpunktsmethoden überwunden werden müssen, um vergleichbare physikalische Beanspruchungen den Prüfstücken aufprägen zu können. Auch die Zugfestigkeitsprüfgeräte, wie sie in den §§ 22a und 22d beschrieben werden, sind für die Gewebeprüfung nicht einwandfrei. C. H. REICHARDT, H. K. WOO und D. J. MONTGOMERY[12] haben diesen Sachverhalt erkannt und durch ein einfaches Zusatzgerät zu dem in dem § 22d diskutierten Instron-Tester eine einwandfreie zweiaxiale Beanspruchung des Gewebes ermöglicht. Weiterhin haben H. K. WOO und D. J. MONTGOMERY[13] den Einfluß der Gewebekonstruktion und der Fasertypen auf das zweiachsige Spannungs-Dehnungs-Verhalten von Geweben

[1] ABRUKOW, H. E.: Tekstiljnaja Promyschlennostj H. **2**, 15 (1954).

[2] KELLER, H.: Diss. E. T. H. Zürich 1943.

[3] FRENZEL, W. u. H. MARTIN: Faserforsch. u. Textiltechn. **8**, 319 (1953).

[4] WOROBJEW, W. A.: Tekstiljnaja Promyschlennostj **2**, 23 (1951). - W. A. WOROBJEW u. T. J. ISTOMINE: Faserforsch. u. Textiltechn. **2**, 497 (1951).

[5] JAMESON, L. H., B. L. WHITTIER u. H. SCHIEFER: Textilprax. **8**, 769 (1953).

[6] ESSAM, J. M.: J. Textile Inst. **19**, T 37 (1928); **20**, T 275 (1929).

[7] BRUGGENCATE, J. A. TEN: Melliand Textilber. **19**, **41** (1938).

[8] HANUSS, L.: Melliand Textilber. **24**, 389 (1943).

[9] Speziell die Beziehungen zwischen der Struktur und den Eigenschaften der Garne und den Schmalgeweben behandelt W. E. MORTON: J. Textile Inst. **42**, 692 (1951). — [10] DIN 53834 - Entwurf August 1954.

[11] J. Textile Inst. **40**, 1 (1949). DIN Entwurf 53857 v. Mai 1955.

[12] REICHARDT, C. H., H. K. WOO u. D. J. MONTGOMERY: Textile Res. J. **23**, 424 (1953). — Vgl. EEG-OLOFSSON: T. Medd. svenska Textilforskningsinst. Nr. 50 (1955).

[13] WOO, H. K. u. D. J. MONTGOMERY: Textile Res. J. **23**, 925 (1953).

untersucht. Der Fortschritt ist unverkennbar, wenn man die in § 24 behandelte Berstdruckprüfung mit ihren Beschränkungen, die bisher bei Flächengebilden wie Geweben und Papier die einzige in kleinen Deformationsbereichen noch einwandfreie zweiachsige Beanspruchungen ergab, betrachtet.

Wie zu erwarten ist, zeigen auch die Versuche von A. KOCHANSKI[1], daß der Webstuhl die Kettfäden während des Webens dynamisch beansprucht und dadurch eine Ermüdung der Kettfäden eintritt, die mit einer Veränderung der elastischen Eigenschaften verbunden ist. Die in § 23 geschilderten Geräte sind für die dynamische Prüfung der Kettfäden anwendbar, für eine zweiachsige Wechselbeanspruchung liegen jedoch noch keine Meßgerätetypen vor. Nur für die bei dem Instron-Tester möglichen, sehr langsamen Wechselbeanspruchungen dürfte das von C. H. REICHARDT, H. K. WOO und D. J. MONTGOMERY entwickelte Gerät auch für derartige Beanspruchungen herrichtbar sein.

Am weitesten sind die Untersuchungen der Scheuerfestigkeit von Geweben gediehen, aus deren großer Zahl wir nur die zusammenfassenden Arbeiten von H. F. SCHIEFER und C. W. WERNTZ[2] sowie von H. SULSER[3] herausgreifen wollen, um sie in bezug auf die Auswertungsmethoden gegenüberzustellen. Beide Arbeiten machen von der Korrelationstheorie Gebrauch; während H. F. SCHIEFER und C. W. WERNTZ den KENDALLschen Rangkorrelationskoeffizienten τ mit nicht befriedigendem Erfolg zur Beurteilung der Wirkungsweise zweier sehr verschiedener Scheuerelemente heranzogen, benutzte H. SULSER die Maßkorrelationskoeffizienten höherer Ordnung, um nunmehr den stochastischen Zusammenhang zwischen den Meßwerten der Scheuerfestigkeit, des Quadratmetergewichts, der Bruchdehnung und der Bruchreißlänge[4] zu prüfen, wobei sich herausstellte, daß bei den geprüften Geweben die Bruchreißlänge keinen Einfluß auf die Scheuerfestigkeit besitzt, daß dagegen ein ziemlich enger Zusammenhang zwischen der Bruchdehnung und der Scheuerfestigkeit vorhanden ist. Ebenso wird ein Einfluß des Quadratmetergewichtes festgestellt. Die Differenz zwischen den berechneten und den gemessenen Scheuerfestigkeitswerten bei den 14 geprüften Geweben ist gering. Die Abweichung von Berechnung und Messung beträgt im Durchschnitt etwa 3%. Bei dem untersuchten Gewebetyp kann auf Grund des Quadratmetergewichtes und der Dehnung in Kette und Schuß zuverlässig auf die Scheuerfestigkeit geschlossen werden. Wie sorgfältig man auch bei der Anwendung der statistischen Methoden vorgehen muß, zeigt die Berechnung des sog. einfachen Korrelationskoeffizienten zwischen der Bruchreißlänge und der Scheuerfestigkeit. Wir finden eine deutlich negative Korrelation $r_{14} = -0{,}650$, während der höhere Korrelationskoeffizient mit Berücksichtigung des Quadratmetergewichtes und der Bruchdehnung zwischen der Bruchreißlänge und der Scheuerfestigkeit nur den Wert $r_{14 \cdot 23} = 0{,}026$ ergibt. Die Untersuchung desselben Zusammen-

---

[1] KOCHANSKI, A.: L'Industrie Textile **10**, 461 (1951).

[2] SCHIEFER, H. F. u. C. W. WERNTZ: Textilprax. **7**, 1, 57 (1952); Textile Res. J. **22**, 1 (1952). — [3] SULSER, H.: Diss. E.T.H. Zürich 1953.

[4] Vgl. Fußnote zu Tab. VII, 5.

hanges zwischen der Bruchreißlänge und der Scheuerfestigkeit führt also einmal ohne Berücksichtigung weiterer Einflußgrößen zu einer ausgeprägten negativen Korrelation, d. h. niedrige Werte der Bruchreißlänge entsprechen hohen Scheuerfestigkeitswerten, und das andere Mal mit Berücksichtigung weiterer Einflußgrößen zu einem nicht wesentlich von Null verschiedenen Korrelationskoeffizienten, mit anderen Worten, es besteht kein Zusammenhang bzw. kein Einfluß der Bruchreißlänge auf die Scheuerfestigkeit. Die Berechnung derartiger Größen, und zwar des einfachen Korrelationskoeffizienten und die damit verknüpfte Bewertung eines stochastischen Zusammenhanges zwischen den Werten zweier Meßreihen x_{1i} und x_{2i} ist durch die Beziehung

$$r_{12} = r_{x_1 x_2} = \frac{\sum_{i=1}^{N} (x_{1i} - \overline{x}_1)(x_{2i} - \overline{x}_2)}{\sqrt{\sum_{i=1}^{N} (x_{1i} - \overline{x}_1)^2 \sum_{i=1}^{N} (x_{2i} - \overline{x}_2)^2}}$$

gegeben, wobei $\overline{x}_1$ und $\overline{x}_2$ die Durchschnittswerte der Meßreihen bedeuten. Für die Korrelationskoeffizienten höherer Ordnung gilt allgemein der folgende Ausdruck

$$r_{12\cdot3\ldots n} = \frac{r_{12\cdot3\ldots(n-1)} - r_{1n\cdot3\ldots(n-1)} \cdot r_{2n\cdot3\ldots(n-1)}}{\left(1 - r^2_{1n\cdot3\ldots(n-1)}\right)^{1/2} \cdot \left(1 - r^2_{2n\cdot3\ldots(n-1)}\right)^{1/2}}.$$

Bezüglich der Ableitung und näheren Begründung dieser Formeln sei auf M. G. KENDALL[1] verwiesen.

Auf die Auswertung soll im einzelnen im Kap. 7 eingegangen werden. Hier wollten wir nur zum Ausdruck bringen, daß mit abnehmender Definiertheit oder zunehmender Komplexität der Beanspruchung, wie es bei der Scheuerprüfung (vgl. § 24b) zutrifft, und sich auch in einer Erhöhung der Streuung der Werte[2] bemerkbar macht, nur noch die Maßkorrelation oder Rangkorrelationskoeffizienten den Weg einer Beurteilungsmöglichkeit zeigen, wobei auch hier, wie wir gesehen haben, mit einer zunehmenden Einschränkung der Aussagefähigkeit dieser Methoden bei ihrer Anwendung zu rechnen ist. Man sieht daraus, zu welchen zweifelhaften Ergebnissen eine weitere Unbestimmtheit der Beanspruchung des zu prüfenden Materials durch den Übergang zu den sog. Modellgeräten führen kann.

4. Die Eigenschaften des Papiers.

Das Zustandekommen der mechanischen Festigkeit von Papier scheint erst in neuester Zeit eine befriedigende Aufklärung zu finden. Bereits L. NORDMAN, CH. GUSTAFSSON und G. OLOFSSON[4] beobachteten, daß der

[1] KENDALL, M. G.: The Advanced Theory of Statistics. London: Charles Griffin and Comp. Ltd. 1948.

[2] Wie ungünstig sich eine nicht genügend definierte Beanspruchung auf die Streuung der Meßwerte auswirkt, haben W. BRECHT u. L. KÖRNER[3] am SCHOPPER-Falzer nachgewiesen. Während für die gleichen Papiere bei der Bestimmung der Bruchlast eine mittlere Streuung der Einzelwerte von $\pm 4{,}1\%$, für die Durchreißarbeit von $\pm 5{,}4\%$ und von $\pm 6{,}6\%$ für den Berstdruck gefunden wurde, wurde für die Falzzahl $\pm 22\%$ bestimmt.

[3] BRECHT, W. u. L. KÖRNER: Papier **6**, 161 (1952).

[4] NORDMAN, L., CH. GUSTAFSSON u. G. OLOFSSON: Paperi ja Pu vom 15. 3. 1952.

größte Teil der von einem Papierstreifen bei seiner Be- und Entlastung aufgenommenen Arbeit für die Trennung von Bindungen verwandt wird; sie benutzten als Maß für die Abnahme der gebundenen Faseroberfläche die Zunahme des Lichtstreuungskoeffizienten, die der Zunahme der freien Faseroberfläche proportional ist[1].

G. BROUGHTON und J. P. WANG[2] führten Festigkeitsuntersuchungen von Papier in Wasser und verschiedenen organischen Flüssigkeiten mit dem Instron-Tester durch und fanden, daß sowohl die Zerreißarbeit als auch die Form der Spannungs-Dehnungs-Kurven von der Fähigkeit der Flüssigkeit abhängt, Wasserstoff-Nebenvalenzbindungen einzugehen vgl. auch W. SCHEFER, § 21a. Noch weitergehend sind die Aussagen von A. H. NIRSAN und insbesonders H. CORTE[3], der experimentelle Untersuchungen über das Gleichgewicht der Reaktion Cellulose mit schwerem Wasser auf Cellulose in Faser- und Papierform bei niedrigen Deuterierungsgraden übertrug und dabei die bei der Herstellung von Papier aus Fasern gebildete Menge an Wasserstoffbrücken abschätzen konnte. Eine in diesem Zusammenhang durchgeführte quantenmechanische Schätzung der Bindungsenergie von Wasserstoffbrücken führte zu Zahlenwerten, deren experimentelle Nachprüfung bestätigte, daß die Festigkeit von Papier der Bildung von Wasserstoffbrücken zwischen den Hydroxylgruppen benachbarter Cellulosefasern zuzuschreiben ist. Daraus ergibt sich auch ein neuer Aspekt für die Festigkeit von Celluloseregeneratfäden und -garnen unter verschiedenen Bedingungen[4].

VAN DEN AKKER[5] nahm das Problem von der statistischen Seite her in Angriff, um den Zusammenhang zwischen den Fasereigenschaften und den Eigenschaften des Papierblattes aufzudecken und um Beziehungen zwischen der Papierstruktur und dem rheologischen Verhalten zu finden. In diesem Zusammenhang sei auch auf eine sorgfältige und gründliche Untersuchung von H. L. COX[6] aufmerksam gemacht, der das mechanische Verhalten von Faservliesen und speziellen papierartigen Flächengebilden aus den Eigenschaften der Fasern ableiten konnte. Die Übertragung dieser Gedankengänge auf textile Faservliese erscheint wertvoll.

Ebenfalls die Verwandtschaft in dem Deformationsverhalten, wie sie in der Verwendung derselben Modellvorstellungen zum Ausdruck kommt (vgl. insbesondere § 21a W. BRECHT und W. VOLK[7] und R. MEREDITH[8]) hat

---

[1] Anwendung der MUNK-KUBELKAschen Theorie. A. S. STENIUS: Svensk Papperstidn. **56**, 607 (1953). — [2] BROUGHTON, G. u. J. P. WANG: Tappi **37**, 72 (1954).

[3] NIRSAN, A. H.: Nature **175**, 424 (1955). — CORTE, H.: Vortrag vor dem Fachausschuß Hochpolymere d. Deutschen Physikalischen Gesellschaften am 17. 9. 1954 in Hamburg. — CORTE, H. u. H. SCHASCHEK: Das Papier **9**, 519 (1955). Experimentelle Befunde, unter anderem von L. M. LYNE u. W. GALLAY: Tappi **37**, 581, 698 (1954), scheinen die Überlegungen von H. CORTE zu bestätigen.

[4] MESKAT, W.: Diskussionsbemerkungen zum Vortrag von H. CORTE am 17. 9. 1954 in Hamburg. Über den Einfluß der Wasserstoffbruchverbindung auf die Festigkeit der Textilfasern, vgl. auch W. HOPPE: Textilprax. **9**, 564 (1954). — D. K. ACHPOOLE: Nature London **169**, Nr. 4288, 37 (1952) u. a.

[5] AKKER, J. A. VAN DEN: Tappi **33**, 398 (1950).

[6] COX, H. L.: Brit. J. appl. Physics **3**, 72 (1952).

[7] BRECHT, W. u. W. VOLK: Das Papier **8**, 365 (1954). — [8] MEREDITH, R.: Mechanical Properties of Wood and Paper, Kap. VII (Amsterdam 1953).

sich in der Entwicklung der Meßgeräte ausgeprägt. Wir finden auch in der Papierprüfung das Pendelgerät für die Zugfestigkeitsprüfung mit seinen Mängeln (vgl. § 22a) und die Entwicklung zu dem Festigkeitsprüfgerät nach dem Prinzip der stetigen Kettenbelastung (vgl. § 22a) von B. STEENBERG sowie die Berstdruck- und Dauerbiegeprüfer (vgl. § 23d) wieder. Die Parallelität in der Entwicklung geht noch weiter, wenn man berücksichtigt, daß auch in der Papierprüfung die Tendenz besteht, die dynamische bzw. Ermüdungsprüfung des Papiers aufzunehmen[1–3], und wenn man dann bedenkt, daß in neuester Zeit nicht nur bei Papier, sondern auch bei Geweben von der Weiterreißfestigkeit gesprochen wird[4].

Selbst den im folgenden Abschnitt § 21d erläuterten spezifischen Gebrauchswert und die Gebrauchswertprüfung findet man in der Papierprüfung wieder. F. BURGSTALLER und R. A. KRAUSS[6] berichten über jahrelange umfangreiche Untersuchungen über die Eignungsbeurteilung von Sackpapier. Eingehende Prüfungen der Papiere auf ihre rheologischen Eigenschaften bei den verschiedenen mechanischen Beanspruchungen und zahlreiche praktische Transportversuche dienten dazu, die Zusammenhänge zwischen der praktischen Beanspruchung von Säcken bei dem Transport und den an Papier meßbaren rheologischen Eigenschaften zu bestimmen und hatten zum Ziel, geeignete Prüfmethoden auszuwählen, um die Voraussetzung für eine zutreffende Eignungsbeurteilung bzw. für eine Gebrauchswertbestimmung zu schaffen.

d) Der Gebrauchswert und die Gebrauchswertprüfung von Faserstoffen.

Unter dem Gebrauchswert werden in der Textilprüfung alle für den Gebrauch maßgeblichen Eigenschaften verstanden. Da nun diese Eigenschaften bei einem Teppich ganz andere sein werden als z. B. beim Damenstrumpf oder beim Monteuranzug, so sieht man bereits, daß es nur einen, auf den spezifischen Verwendungszweck bezogenen spezifischen Gebrauchswert gibt. Je nach dem Verwendungszweck durchläuft jedoch auch das Rohmaterial einen bestimmten Verarbeitungsweg mit entsprechenden Beanspruchungen, so daß sich aus den gesamten Relationen der Tab. IV,5 und IV,6 ein dem Verwendungszweck spezifisches Relationengefüge stochastischer Art heraushebt. Die in den §§ 21c und 22–24 geschilderten Prüfmethoden und Prüfgeräte sind dabei für die Beschreibung des Verhaltens auf dem Verarbeitungsweg bei

[1] LAGALLY, P. u. LANDOLT-BÖRNSTEIN: Stoffwerte, 6. Aufl. Bd. IV, Teil 1, S. 296 ff. Springer-Verlag 1955 mit ausführlichen Hinweisen auf die DIN-, FAK-, Tappi- und PMA-Blätter.

[2] Es sei nur L. RAGOSSNIG[3] erwähnt, der sich eingehend mit den besonderen Problemen der dynamischen Papierprüfverfahren für Sackpapiere befaßt.

[3] RAGOSSNIG, L.: Papier 6, 407 (1952).

[4] Es gibt nur einige Ausnahmefälle, die der Textilprüfung eigentümlich sind, wie z. B. die Ermittlung der Verschiebefestigkeit der Fäden im Gewebe, H. HOLDERER u. H. E. COWLES[5].

[5] HOLDERER, H.: Textilprax. 6, 114 (1951). — H. E. COWLES: J. Textile Inst. 44, T 293 (1953). — [6] BURGSTALLER, F. u. R. A. KRAUSS: Das Papier 9, 237 (1955).

Tabelle IV, 6.

Gebrauchswert
Prüfung der verschiedenen Eigenschaften der Faserstoffe unter verschiedenen Klimabedingungen, lufttrocken, naß, gewaschen u.s.w.
Widerstandsfähigkeit gegen physikalische Beanspruchung
Widerstandsfähigkeit gegen chem. u. biologische Dauerbeanspruchung
Widerstandsfähigkeit gegen physikal. Wechselbeanspruchung
Scheuer-, Zug-, Wechsel-, Dauerbiege-, Dauerknickfestigkeit
Elastizität
Festigkeitseigenschaften
Temperaturbeständigkeit
Alterungsbeständigkeit
Waschbeständigkeit
Feuchtigkeitsbeständigkeit Witterungsbeständigkeit
Beständigkeit gegen Strahlung verschiedener Wellenlänge
Beständigkeit gegen Pilze u. Bakterien
Beständigkeit gegen tierische Schädlinge
Zug-, Biege-, Torsionselastizität
Zugfestigkeit, Dehnbarkeit Knoten-, Schlingen-, Torsions-, Knick-, Scher-, Berst-, Durchstoß-, Schlag-, Schiebe-, Naht-, Einreiß-, Weiterreißfestigkeit
Aussehen
Physikalische Struktur
Verarbeitungstechnische Eigenschaften
Formbeständigkeit
Chemische u. färberische Eigenschaften
Hygienische Eigenschaften
Elektrische Eigenschaften
Schmutzempfindlichkeit
Glanz
Faserstruktur Faserhaftfestigkeit Faserlänge Faserfeinheit Gleichmäßigkeit Glätte Garnfeinheit Garn-, Zwirndrehung Webtechnische Struktur
Weichheit Steifheit
Spinn-, Gleit-, Nähfestigkeit Formbarkeit
Ausbeulung Knittern
Wasch-, Dämpf-, Dekatier-, Filzeingang
Chemische Struktur
Farbaufnahme
Wasser-, Wasch-, Koch-, Schweiß-, Reib-, Meerwasserempfindlichkeit
Walk-, Dekatierempfindlichkeit
Hygroskopische Eigenschaften Quellung Trocknung
Wärmeisolation Wasserdampf-, Wasser-, Luft-, Staubdurchlässigkeit
El. Aufladung El. Leitfähigkeit
El. Isolationsfähigkeit

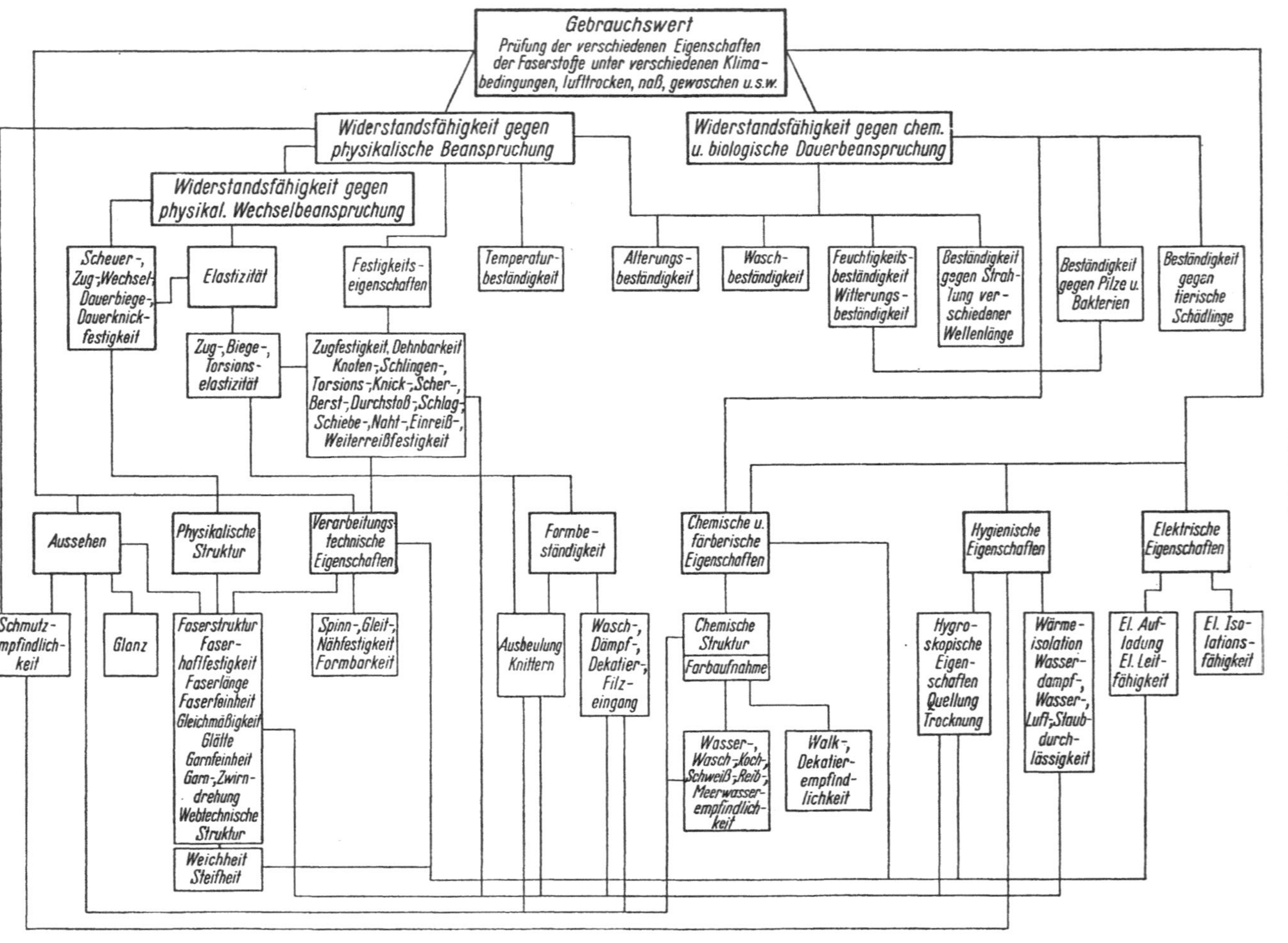

den auftretenden physikalischen Beanspruchungen notwendig und hinreichend[1].

Damit ist jedoch noch nicht die Eignung des Materials unter den Beanspruchungen des Gebrauchs bei einem bestimmten Verwendungszweck gegeben. Um diesen stochastischen Zusammenhang im Rahmen des Relationengefüges zahlenmäßig ausdrücken zu können, muß das Verhalten im praktischen Gebrauch wenigstens halb quantitativ bewertet werden. Zur Schaffung dieser zahlenmäßigen Basis ist der Weg über die Trag- und Betriebsversuche eine zur Zeit noch zwingende Notwendigkeit. Der Tragversuch ist also einem Modellgerät, das die Beanspruchungen der späteren Verwendung weitestgehend nachzuahmen gestattet, gleichzusetzen (vgl. Tab. IV, 5) und damit einer spezifischen Prüfmethode, die der Untersuchung des textilen Materials eigentümlich ist[3].

Ganz analog wie bei den Modellgeräten erfolgt nun auch hier eine ziffernmäßige Auswertung, die zu den Schadenswerten, wie sie von H. BÖHRINGER[4] bezeichnet und in Kap. 7 noch erläutert werden, führt. Auch diese Schadenswerte sind statistischer Natur, da ja die Versuchsstücke von den einzelnen Trägern verschieden stark beansprucht werden. Damit erhalten wir noch das fehlende Endglied, die stochastische Relation zwischen der Beanspruchung im Gebrauch und den Meßwerten des verarbeiteten Materials. H. BÖHRINGER unterscheidet noch den absoluten Schadenswert, der gegeben ist durch:

$$\text{Schadenswert absolut} = \frac{\text{Summe der Schadensziffern}}{\text{Anzahl der Tragperioden} \times \text{Anzahl der Stücke}},$$

wobei die Anzahl der Stücke von dem zu erwartenden Variationskoeffizienten der Schadenswerte abhängt. Unter Schadensziffern werden dabei die Verschleißschäden verstanden, die an den verschiedenen genau gekennzeichneten Stellen auftreten, festgestellt und je nach ihrer Art mit unterschiedlichen Bewertungsziffern belegt werden. Allerdings, was nicht betont wird, ist, daß dieser absolute Schadenswert je nach dem Verwendungszweck verschieden ist und außerdem von den geometrischen Abmessungen der Probe abhängt, so daß die Bezeichnung spezifischer Schadenswert hier angebrachter wäre. Setzt man nach H. BÖHRINGER diesen Wert

$$S_a = S_{sp} = \frac{\alpha \cdot F}{D},$$

[1] Daß allein für die Scheuerprüfung 50 Prüfgeräte (vgl. die Tabelle IV, 6 von H. SULSER[2]) entwickelt und gebaut worden sind mit zum größten Teil undefinierter physikalischer Beanspruchung des zu prüfenden Materials, zeigt nur das Bestreben, in der Praxis unbekannte funktionale Zusammenhänge durch Faustformeln und spezielle Erfahrungswerte zu überbrücken – ein Streben, das nicht nur bei der Scheuerprüfung, sondern auch bei der Gleichmäßigkeitsprüfung, Feinheitsbestimmung usw. zu sehr zahlreichen, jedoch physikalisch meist nicht befriedigenden Geräten geführt hat. Vgl. dagegen J. B QUIG: Reyon, Zellwolle und andere Chemiefasern 5, 851 (1955). – [2] H. SULSER: Diss. E.T.H. Zürich 1953.

[3] Eine statistische Analyse der Schadensfälle würde eine Auswertung von Tragversuchen auf breiter Grundlage und unter natürlichen Bedingungen bedeuten, wobei allerdings nur die im Gebrauch als nicht genügend befundenen Textilien erfaßt werden. Dieser Weg sollte mehr als bisher beschritten werden.

[4] BÖHRINGER, H.: Faserforsch. u. Textiltechn. 5, 55 (1954).

wobei F die Fläche und D die Dicke der Proben bezeichnen, so stellt α einen vom Material abhängigen Proportionalitätsfaktor dar, der nach H. BÖHRINGER mit spezifischem Schadenswert S_{sp}, nach unserer Auffassung richtiger mit spezifischem Gütewert $Gü_{sp}$ zu benennen ist und durch folgende Formel:

$$\alpha = Gü_{sp} = \frac{S_{sp} \cdot \left(1 + \frac{f}{100}\right) \cdot g^2}{G \cdot \gamma} \qquad \text{(IV, 34)}$$

wiedergegeben werden kann, wobei

S_{sp} = spezifischer (absoluter) Schadenswert,
f = hygroskopische Feuchtigkeit in % bei 65% rel. Luftfeuchtigkeit,
g = Quadratmetergewicht in g,
G = Gewicht des Versuchsstückes in g,
γ = Dichte der Fasersubstanz

bedeuten.

Tragen wir nun die im Tragversuch ermittelten Schadenswerte von Beanspruchungsperiode zu -periode in einem rechtwinkligen Koordinatensystem auf, so erhalten wir die Kurve des Güteverlaufs in Abhängigkeit von der Tragzeit, und der spezifische Schadenswert ist die charakteristische Maßzahl für die Beurteilung des Tragversuchsstückes bei der Ausscheidung durch Verschleiß. Der spezifische Schadenswert charakterisiert den spezifischen Gebrauchswert bzw. steht in enger Relation zu ihm. Der spezifische Gütewert läßt dagegen die Qualität des eingesetzten Fasermaterials erkennen. Die halb quantitativen Bewertungsziffern der Tragversuche legen es nahe, zur Untersuchung des Zusammenhanges zwischen den Meßgrößen des textilen Materials vor der Beanspruchung im Gebrauch und den Bewertungsziffern nach dem Gebrauch die Rangkorrelation nach C. SPEARMAN[1] heranzuziehen.

Diese Methode soll an einem Beispiel erläutert werden, das ausführlich von R. G. STOLL[2] und vereinfacht von U. GRAF u. H. J. HENNING[3] behandelt worden ist und den Vergleich von Labor-Meßwerten mit der Bewertung von Tragversuchen zeigt. An 7 Stoffen gleichartiger Konstruktion, aber verschiedener Zusammensetzung, wurde die Rundscheuerprüfung – Zahl der Scheuertouren bis zur Lochbildung –, die Biegescheuerung in Kettrichtung sowie in Schußrichtung – Zahl der Scheuerung bis zum Bruch – und die Prüfung der Weiterreißfestigkeit – Mittel aus Kette und Schuß – durchgeführt und außerdem mit diesen 7 Stoffen Tragversuche unternommen.

In der Tabelle IV, 7 sind links die geprüften Gewebe mit den Buchstaben A bis G bezeichnet und rechts die bei diesen Geweben durch die Auswertung der Tragversuche festgestellte Rangfolge in der Haltbarkeit wiedergegeben. Die Rangziffer 1 bedeutet das in der Haltbarkeit beste Gewebe und die Rangziffer 7 das schlechteste Gewebe. Weiter sind in den Spalten die quantitativen Maßzahlen der Scheuerung und Weiterreißfestigkeit in Prozenten des jeweiligen Mittels wiedergegeben und in Klammern daneben die Rangziffern für jede geprüfte Eigenschaft, wobei auch hier, z.B. bei der Rundscheuerung, die niedrigste Rangziffer dem höchsten Scheuerwert und die höchste Rangziffer dem niedrigsten Scheuerwert zugeordnet werden. Das gleiche gilt für die Weiterreißfestigkeit.

Nun können wir genau so wie bei der Maßkorrelation auch hier die Frage stellen nach dem Zusammenhang zweier Rangfolgen, z.B. die der Rundscheuerungsrang-

[1] SPEARMAN, C.: Amer. J. Physiol. **15**, 79 (1904). – M. G. KENDALL: Rank Correlation Methods. Charles Griffin and Comp. Ltd., London 1948. – U. GRAF u. H. J. HENNING: Melliand Textilber. **32**, 850 (1951). – U. GRAF u. H. J. HENNING: Statistische Methoden bei textilen Untersuchungen. Springer-Verlag, Berlin 1952.

[2] STOLL, R. G.: Textile Res. J. **19**, 394 (1949).

[3] GRAF, U. u. H. J. HENNING: Statistische Methoden bei textilen Untersuchungen. Springer-Verlag, Berlin 1952, S. 201.

Tabelle IV, 7.

Gewebe	Prüfwerte, Rangziffern in Klammern				Tragversuch Rangziffer
	Rundscheuerung	Biegescheuerung		Weiterreiß-festigkeit	
		Kette	Schuß		
A	120 (1)	72 (7)	115 (2)	89 (4)	6
B	86 (6)	85 (4)	78 (6)	75 (6)	5
C	92 (5)	78 (6)	91 (5)	68 (7)	7
D	80 (7)	82 (5)	102 (4)	87 (5)	4
E	107 (3)	91 (3)	67 (7)	125 (2)	3
F	101 (4)	138 (2)	111 (3)	119 (3)	2
G	114 (2)	154 (1)	136 (1)	137 (1)	1

ziffern und die der Tragversuchsrangziffern. Bezeichnen wir allgemein die Rangziffern mit μ_i und v_i und nehmen wir die entsprechenden Ziffern unseres Beispiels aus der Spalte 2 und der Spalte 6 der Tabelle IV, 7, so erhalten wir folgendes Schema:

Tabelle IV, 8.

Geprüftes Gewebe	A	B	C	D	E	F	G
Rangzifferfolge Rundscheuerung	$u_1 = 1$	$u_2 = 6$	$u_3 = 5$	$u_4 = 7$	$u_5 = 3$	$u_6 = 4$	$u_7 = 2$
Rangzifferfolge Tragversuche	$v_1 = 6$	$v_2 = 5$	$v_3 = 7$	$v_4 = 4$	$v_5 = 3$	$v_6 = 2$	$v_7 = 1$

Die Rangziffern u_i und v_i nehmen die aufgeführten ganzzahligen Werte von 1 bis $N = 7$ an. Für den SPEARMANschen Rangkorrelationskoeffizienten ϱ erhält man mit Benutzung dieser Rangziffern den folgenden Ausdruck:

$$\varrho = 1 - 6 \frac{\sum_{i=1}^{N} (u_i - v_i)^2}{N(N^2 - 1)} .$$

Hierbei bedeuten

$\varrho = +1$ volle Übereinstimmung der beiden Rangfolgen

und

$\varrho = 0$ keinen Zusammenhang.

Der erhaltene Rangkorrelationskoeffizient $\varrho = 0{,}21$ zwischen der Rundscheuerungsprüfung und dem Ergebnis des Tragversuches ist gering. Rechnet man nach demselben Schema den Zusammenhang zwischen der Biegescheuerung Kette und Tragversuch, so erhält man für ϱ den Wert 0,96. Ebenso ergeben sich für die Prüfung Biegescheuerung Schuß gegenüber dem Tragversuch ein Rangkorrelationskoeffizient $\varrho = 0{,}32$ und für die Prüfung Weiterreißfestigkeit gegenüber dem Tragversuch $\varrho = 0{,}86$. Daraus folgt, daß die Biegescheuerung Kette und die Weiterreißfestigkeit gegenüber diesen Tragversuchen eine hohe Korrelation zeigen.

Die Rundscheuerung und die Biegescheuerung Schuß haben demgegenüber als Einflußgrößen nur ein geringes Gewicht und können daher vernachlässigt werden. Setzen wir den Einfluß der Biegescheuerung Kette mit 60% und den Einfluß der Weiterreißfestigkeit mit 40% an, so erhalten wir für das Gewebe A eine Meßziffer, die sich zu $0{,}60 \times 72 + 0{,}40 \times 89 = 78{,}8\%$ ergibt. Für die anderen Gewebe ergeben sich entsprechende Meßziffern, und die Rangfolge dieser neuen Meßziffern stimmt bei den untersuchten Gewebearten vorzüglich mit den Ergebnissen des Tragversuches überein.

Dieses Beispiel zeigt uns bereits, daß es Prüfgeräte zur Ermittlung der Gebrauchswerteigenschaften gar nicht geben kann. Es sind immer mehrere Eigenschaften vorhanden mit einem ganz verschiedenen statistischen Gewicht, die sich im Gebrauch entsprechend auswirken[1]. Nur in Ausnahmefällen wird eine Meßgröße allein maßgebend für den Gebrauchswert sein. Wir können daher nicht H. BÖHRINGER[2] folgen, wenn er von Gebrauchswertprüfapparaten spricht. Uns erscheint daher die Kritik an derartigen Geräten, wie sie in § 24 ausgesprochen wird, berechtigt und die Verwendung von Modellgeräten besonders zur Erfassung der physikalischen Zusammenhänge gefährlich, vgl. § 23d.

Die Ausführungen des § 21 lassen die Mannigfaltigkeit der Einflußgrößen, die das rheologische Verhalten der Faserstoffe auch auf ihrem Verarbeitungswege und später im Gebrauch bestimmen, erkennen. Dabei schält sich die Bedeutung der Korrelationsmethoden zur Bestimmung des statistischen Gewichts dieser Einflußgrößen heraus, und sie erlauben damit auch einen Schluß auf die Vollständigkeit der Prüfmethoden bei der Erfassung der Zusammenhänge.

Wir kommen nunmehr zu den Prüfgerätetypen, die uns die physikalischen Größen zu bestimmen gestatten sollen, die als Einflußgrößen für das rheologische Verhalten der Faserstoffe charakteristisch sind. Eine dieser Größen ist z.B. die Zugfestigkeit, und hier werden wir in dem folgenden § 22 sehen, daß die Zugfestigkeitsprüfer, die auch gleichzeitig zur Bestimmung der zugelastischen Eigenschaften der Faserstoffe herangezogen werden, aus der üblichen Waagentechnik entnommen worden sind, wie z.B. die Pendelgewichtswaage, die Neigungswaage, die Torsionswaage usw. Man hat dabei die sogenannte Lastschale ersetzt durch 2 Klemmen, zwischen denen das zu untersuchende Textilmaterial eingespannt wird. Man übernimmt also für die Kraftmessung irgendeine Waagentype, während die Lastschale der Waage durch die Einschaltung des zu untersuchenden Fadens oder der Faser ihre bisherigen Funktionen verliert. Die für die Güte einer Waage maßgebenden 5 Eigenschaften, vgl. J. KROENER: Handbuch der technischen Betriebskontrolle, Bd. 2, Mengenmessung im Betrieb von E. PADELT, 1955, sind in wesentlichen Punkten nicht mehr übertragbar, ja, z.B. die Beweglichkeit der oberen Klemme bei der Pendelgewichtswaage ist sogar, wie wir sehen werden, nachteilig und störend. Erst in neuester Zeit hat man sich von dieser Waagentechnik bei der Kraftmessung mehr oder weniger gelöst und ist, vgl. § 22c, 22d und 23c, zu der weglosen Kraftmessung gekommen[3].

§ 22. Statische Prüfmethoden (Einachsige stetige Zugbeanspruchung).

Die statische Prüfung beschäftigt sich vorwiegend mit der Bestimmung der Zugfestigkeit und Dehnung sowie der zugelastischen Eigen-

[1] WEINER, L. J. u. ST. J. KENNEDY: J. Textile Inst. **44**, 433 (1953).

[2] BÖHRINGER, H.: Faserforsch. u. Textiltechn. **5**, 242 (1954).

[3] Es ist bemerkenswert, daß auch die Waagentechnik diesen Weg beschreitet und von den auf S. 274 und 275 beschriebenen Hilfsmitteln Gebrauch macht, vgl. V. G. KENNEDY: Instruments **28**, 272 (1955).

schaften von Faserstoffen. Dabei geht man so vor, daß man entweder aus dem einmaligen Zerreißvorgang mit vorgegebener konstanter Belastungs- bzw. Dehnungsgeschwindigkeit das Kraft-Dehnungs-(KD)-Diagramm[1] aufnimmt und daraus Zugfestigkeit und Bruchdehnung ermittelt oder aus der stufenweisen Belastung nach DIN 53835 die zugelastischen Eigenschaften ableitet[2]. Das Kraft-Dehnungs-Diagramm bei stufenweiser Belastung eines ideal-elastischen Stoffes versinnbildlicht die Aufeinanderfolge von stationären Gleichgewichtszuständen und ist daher frei von Zeiteffekten, d.h. jedem Kraftwert ist ein bestimmter Dehnungswert eineindeutig zugeordnet. Bei Faserstoffen wird dieses Ideal in keiner Weise erreicht. Dieser Sachverhalt wurde leider oft übersehen und damit sind die zahlreichen Prüfmethoden zur Erfassung des linearen Deformationsverhaltens, die Einpunktsmethoden sind und nicht einmal zur Bestimmung dieser linearen Deformationen ausreichen, hier nur in einem sehr engen Rahmen anwendbar. Zeiteffekte spielen eine große Rolle und sowohl die Geschwindigkeit der Beanspruchung als auch die Reihenfolge der Dehnungs- bzw. Kraftänderungen haben einen wesentlichen Einfluß auf die Gestalt der Kraft-Dehnungs-Kurve. In diesem Falle ist die Zuordnung zwischen Kraft- und Dehnungswerten nicht mehr eindeutig[3–6]. Eine der Größen, welche die Eindeutigkeit der Zuordnung stört und damit die Form des KD-Diagrammes beeinflußt, ist – wie bereits erwähnt – die Zeit. Die Versuchsbedingungen müssen daher so gewählt werden, daß der zeitliche Ablauf definiert und festgelegt ist. Daraus folgt, daß die Prüfgeräte zur Aufnahme des KD-Diagrammes bestimmte konstruktive Voraussetzungen erfüllen müssen. Diese Grundforderungen sind:

1. Konstante Dehnungs- bzw. Deformationsgeschwindigkeit, d.h., die Dehnung muß proportional mit der Zeit zunehmen oder, mit anderen Worten, die eine Klemme des Prüfgerätes muß mit konstanter Geschwindigkeit bewegt werden, während die zweite Klemme ihre Lage im Raum nicht verändern darf.

2. Konstante Belastungsgeschwindigkeit, d.h., die Zugkraft muß proportional mit der Zeit zunehmen. Hierbei ist jedoch besonders zu beachten, daß die Belastungsgeschwindigkeit nicht mehr unabhängig vom

---

[1] In DIN 53816 und 53834 – Entwurf August 1954 – ist diese Bezeichnung durch Kraft-Längenänderungs-Diagramm (P–ΔL-Diagramm) ersetzt worden.

[2] Unter Anlehnung an DIN 53834 wurde der DIN-Entwurf 53836, März 1954, bzw. 53866a und 53867 für den Zugversuch an geknoteten Garnen, Zwirnen und Borsten aufgestellt, der aus physikalisch einwandfrei arbeitenden Geräten Modellgeräte macht, die mehr oder weniger undefiniert zusammengesetzte Größen, die aus dem Zusammenwirken von Längs-, Quer- und Druckkräften usw. hervorgehen, erfassen. Dasselbe gilt, wenn auch in wesentlich geringerem Umfange, für die Prüfung der Schlingenfestigkeit nach DIN-Entwurf 53866.

[3] DIN-Textilnormentwurf: Melliand Textilber. **35**, 9 (1954).

[4] Eisenhut, O. u. W. Grether: Melliand Textilber. **22**, 3 (1941).

[5] Vries, H. de: Ausschuß IVC (3/1954) Enka and Breda Rayon Revue (Nov. 1953).

[6] Brecht, W. u. W. Volk: Papier 8. 365–70 (1954).

Querschnitt ist, d. h. auf den Querschnitt bzw. auf den Titer zu beziehen ist[1].

Diese Grundforderungen werden von den zur Zeit in der Textilprüfung geltenden Normen für die Bestimmung der Zugfestigkeit nicht vollständig erfüllt, weil hier eine konstante Zeit für die Durchführung des Zerreißversuches vorgeschrieben wird[2]. Bei Stoffen mit wesentlich verschiedener Bruchdehnung wird demnach mit entsprechend verschiedenen Dehnungsgeschwindigkeiten gearbeitet. Die praktische Verwirklichung der Forderung (1) stößt besonders bei Faserstoffen mit sehr kleinen Bruchdehnungen auf große technische Schwierigkeiten. Abgesehen von dieser durch die geltende Normung bedingten Einschränkung soll nunmehr untersucht werden, inwieweit die aufgestellten Grundforderungen durch die bekannten Konstruktionen der Festigkeitsprüfgeräte erfüllt werden.

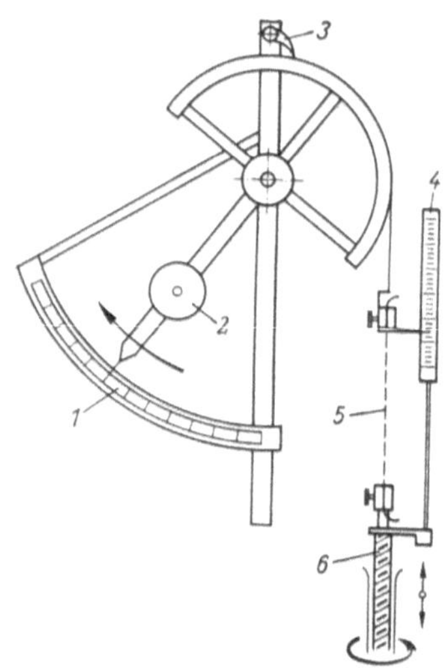

Abb. IV, 16. Schematische Darstellung eines Festigkeitsprüfers nach dem Pendelprinzip. (System L. SCHOPPER.)

1 Skala der Belastungswerte, *2* verschiebbares Gewicht, *3* Sperrklinke, *4* Skala der Dehnungswerte, *5* zu untersuchende Probe, *6* bewegliche Spindel.

a) Festigkeitsprüfgeräte mit nicht wegloser Kraftmessung.

In der Abb. IV, 16 ist ein *Festigkeitsprüfer nach dem Gewichtshebelprinzip* schematisch dargestellt. Wir haben das bekannte System SCHOPPER gewählt. Aus dieser Abbildung kann man sofort entnehmen, daß das zu prüfende Fasermaterial zwischen die Klemmen eingespannt wird. Ein Elektromotor überträgt durch eine Spindel die gewünschte Deformationsgeschwindigkeit auf den Prüffaden. Dabei sieht man, daß mit der Bewegung der unteren Klemme auch eine Verschiebung der oberen Klemme gekoppelt ist, so daß die geforderte konstante Deformationsgeschwindigkeit nicht mehr gewährleistet ist.

---

[1] Die in DIN 53816 und 53834, Abschnitt 5.1, aufgestellten Forderungen, daß für den Zugversuch alle Gerätetypen verwendbar sind, die nach den verschiedenen Prinzipien arbeiten, wie z. B. mit:

a) konstanter Belastungsgeschwindigkeit,

b) konstanter Reckgeschwindigkeit,

c) konstanter Geschwindigkeit der ziehenden Klemme,

sind physikalisch nicht einwandfrei. Vgl. auch Kap. III, § 15b. Die Aufnahme der Forderung c ist als Verlegenheitslösung zu werten, da die bisher gebräuchlichen Prüfgeräte die Bedingungen a und b der Normvorschrift nicht erfüllen, vgl. auch F. STURZMANN: Chem. Ztg. **79**, 664, 704 (1955).

[2] Nach DIN 53816 und 53834, Abschnitt 7,3, soll die mittlere Versuchsdauer etwa 20 Sekunden betragen; sie ist durch Vorversuche einzustellen. – Nach ISO, Techn. Komitee 38, dagegen soll mit 800 mm/min Klemmengeschwindigkeit gearbeitet werden. Diese Klemmengeschwindigkeit wird von dem größten Teil der in Deutschland geführten Prüfgeräte gar nicht erreicht. Vgl. auch BISFA, Liefer- und Prüfvorschriften für Fahrzeugreifen Reyon, Ausgabe 1952, S. 48.

Hinzu kommt, daß als Kraftmesser[1] das bekannte Neigungspendel dient. Dieses ist direkt, wie man aus Abb. IV,16 entnehmen kann, mit dem Prüffaden gekoppelt, so daß mit der Einleitung des Zerreißversuchs mehr oder weniger große Massenwirkungen auftreten.

H. SCHIEFER[3] hat diese Masseneffekte an derartigen Geräten nach dem Pendelprinzip experimentell untersucht, indem er ein elektrisches Dynamometer nach dem Prinzip der Dehnungsmeßstreifen in Temperaturkompensation in Serie mit dem Prüffaden schaltete. Es wurde daher der Kraftverlauf trägheitslos in Abhängigkeit von der Zeit gemessen. Das Oszillogramm des Einschwingvorganges ist in Abb. IV,17 zu erkennen. Deutlich ist das durch die Trägheit der Massen und die Elastizität des Prüffadens zustande kommende Schwingen der aufgeprägten Kraft zu sehen. Ebenso schwanken Belastungsgeschwindigkeit und Trägheitskräfte, wie aus Abb. IV,17 folgt. Aus dem tatsächlichen schwankenden Verlauf der Belastungsgeschwindigkeit ist die Größe der Schwingungsamplitude gegenüber dem Mittelwert der als Versuchsbedingung vorgegebenen konstanten Belastungsgeschwindigkeit (strichliniert) besonders gut zu sehen. Die Frequenz des Einschwingvorganges ist zwischen der 1. und 6. Sekunde einigermaßen konstant. Dies ändert sich erst nach etwa 6,2 Sekunden, zu einem Zeitpunkt, wo der Bruch beginnt. In dieser eigentlichen Bruchzone ändern sich, wie die Abb. IV,18 zeigt, Belastungsgeschwindigkeit und Trägheitskräfte des Pendels noch besonders stark. Der Versuch ergibt einwandfrei, daß die Belastung, die auf die Probe tatsächlich einwirkt, in ihrem zeitlichen Verlauf ganz anders aussieht, als man für gewöhnlich unter der Voraussetzung konstanter Belastung bzw. konstanter Deformationsgeschwindigkeit stillschweigend annimmt.

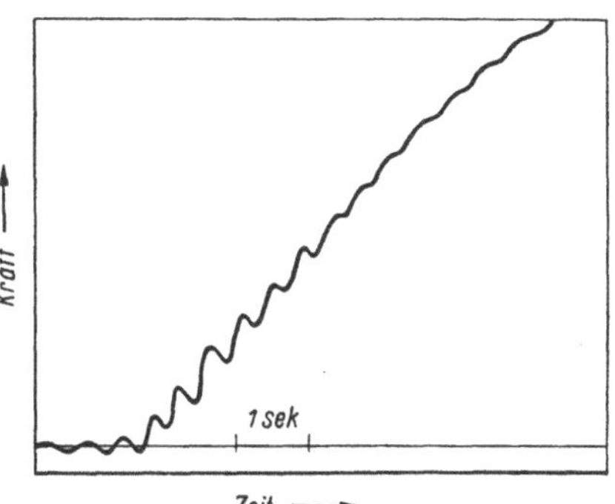

Abb. IV,17. Oszillogramm des Einschwingvorganges bei einem statischen Zerreißapparat nach dem Pendelprinzip. (Nach H. SCHIEFER.)

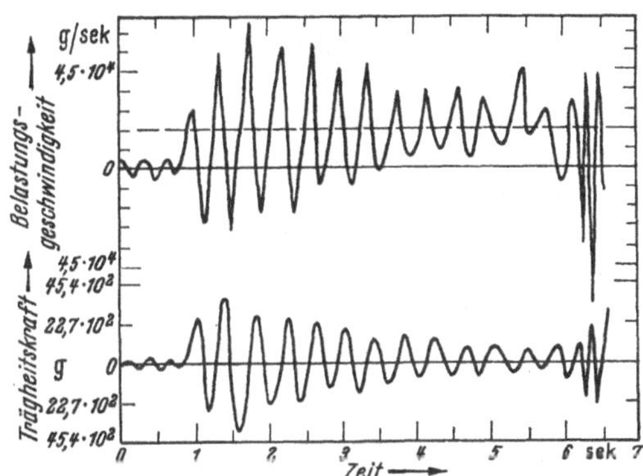

Abb. IV,18. Belastungsgeschwindigkeit und Trägheitskräfte beim Einschwingvorgang eines statischen Zerreißapparates nach dem Pendelprinzip. (Nach H. SCHIEFER.)

Damit ist gezeigt, daß bei diesem Gerätetyp, der weiteste Verbreitung gefunden

---

[1] Die Kraftmessungen mit schwingungsfähigen Systemen wird eingehend von S. MEYER[2] behandelt, wobei die Gleichgewichtsverhältnisse schwingungsfähiger Systeme (Festigkeitsprüfer ohne und mit Dämpfung und Prüffaden) untersucht werden. Der Folgerung, daß die größere Festigkeitsanzeige bei größerer Belastungsgeschwindigkeit allein durch die Eigenschaften des Prüfapparates zu erklären sind, können wir nicht beipflichten, da hier auch die Deformationseigenschaften des Prüffadens eingehen. — [2] MEYER, S.: Faserforsch. u. Textiltechn. 5, 302 (1954).

[3] SCHIEFER, H. F.: Vortrag, gehalten in New York vor dem Committee D–13 on Textile Materials (März 1947). Ein Vorschlag zur Überwindung der Schwierigkeiten wird auch von K. J. KRYSTGAN, D. A. MILLAR und J. WULFF: Rev. Sci. Instrum. 24, 196 (1953) gemacht.

hat, nicht nur die Voraussetzung der konstanten Dehnungsgeschwindigkeit nicht erfüllt ist, sondern durch die geschilderten Masseneffekte noch zusätzliche Störungen auftreten, die sich ganz verschieden je nach dem zu prüfenden Material auswirken können[1].

Auch die in den DIN-Normen 53816 und 53834 angegebenen Verfahren der Überwachung und Kontrolle derartiger Zugprüfgeräte, z. B. mittels des Schwingversuchs, sind der Güteprüfung, wie sie in der Waagentechnik üblich ist, entnommen. Es wird dabei die grundsätzliche Annahme gemacht, daß die Einspannung des zu untersuchenden Fadens die Übertragbarkeit dieser Waagen-Prüfmethoden auf das veränderte System gestattet. Das ist jedoch nicht der Fall (vgl. auch die Überleitung zu § 22) und insbesondere die umfangreichen Untersuchungen von J. JUILFS[2], über die Eichung von Zugprüfgeräten, die auch die Genauigkeit der Kraftanzeige bei den Prüfgeräten nach dem Pendelprinzip umfassen, wobei er zu dem Schluß kommt, daß die Angabe des Fehlers der Kraftanzeige gegenüber der tatsächlichen Bruchlast deshalb mit erheblichen Schwierigkeiten verknüpft ist, weil die Endgeschwindigkeit der oberen Klemme im Zeitpunkt des Bruchs keineswegs einwandfrei definiert ist. Sicher ist, daß bei dem Auseinanderziehen des Prüflings bzw. des zu prüfenden Fadens während des Bruchs die obere Klemme, auf die es bei der Kraftanzeige allein ankommt, ihre Geschwindigkeit wesentlich herabsetzt, da im Augenblick des Bruchs die Dehnung erheblich zunimmt. Nun ist aber der Ausschlag des Kraftanzeigers über die im Zeitpunkt des Bruchs wirkende Kraft hinaus ausschließlich von der dem System in diesem Zeitpunkt innewohnenden kinetischen Energie, also von der innewohnenden Geschwindigkeit, abhängig, so daß die Differenz des Endausschlags gegenüber der wahren Bruchlast, die wir als Anzeigefehler bezeichnen müssen, sicher nicht exakt zu erfassen ist. Der von C. KRAMERS[3] vorgeschlagene Weg der Kontrolle und Eichung von derartigen Textildynamometern arbeitet mit einem ovalen Meßbügel aus Stahl, der zwischen den beiden Klemmen des Dynamometers bzw. des Zugprüfgerätes befestigt wird und an welchem an der unteren Hälfte ein Meßmikroskop festgeklemmt ist. An der Unterseite der oberen Hälfte des Bügels ist ein Spiegel befestigt, auf dem eine vertikale Maßeinteilung angebracht ist. Die von dem Prüfgerät auf den Meßbügel einwirkende Kraft vergrößert den Abstand zwischen dem unteren und oberen Teil des Bügels, so daß das Maßeinteilungsbild im Meßmikroskop sich verschiebt. Wie aus dem Vorstehenden ersichtlich, sagt dieses Verfahren ebenfalls nichts über die wahren Anzeigefehler aus. Wir stellen vielmehr fest, daß hier die Ergebnisse von H. SCHIEFER[4] bestätigt werden, wonach der exakte Zugfestigkeitswert nur über ein elektrisches trägheits- und weglos arbeitendes Dynamometer bestimmbar ist. Diese Schwierigkeiten dürften

[1] Dieser Gerätetyp genügt damit auch nicht den Forderungen des Abschnitts 5.2 der DIN-Normen 53816 und 53834. Leider existiert noch kein DIN-Blatt für die Eichung der Zugfestigkeitsprüfgeräte auch DIN 51211 mit Beiblatt 2 sagen nichts darüber aus, so daß wesentliche Fehler, wie sie hier beschrieben sind, meistens unerkannt bleiben. — [2] Den Verfassern freundlicherweise überlassene Angaben, deren Veröffentlichung durch J. JUILFS vorgesehen ist.

[3] KRAMERS, C.: De Tex 14, 316 (1955). — [4] Siehe S. 265, Fußnote 3.

der Hauptgrund sein, warum immer noch kein DIN-Blatt für die Eichung der üblichen Zugfestigkeitsprüfgeräte vorliegt.

In der Abb. IV,19 ist ein Gerätetyp dargestellt, bei dem durch gleichförmiges Neigen einer Ebene entsprechend einer *Neigungswaage* eine konstante Belastungsgeschwindigkeit dem zu prüfenden Faden aufgeprägt wird. Hier ist die zweite Klemme in ihrer Lage fixiert und genügt damit einer der aufgestellten Forderungen. Da jedoch der schwere Wagen des Neigungsdynamometers direkt mit dem Prüffaden in Verbindung ist, liegt auch hier ein schwingendes System vor, das bei jeder Geschwindigkeitsänderung zu Einschwingvorgängen Anlaß gibt, deren Einfluß auf den Meßvorgang, wie wir gesehen haben, beträchtlich sein kann. Hierzu kommt noch die bereits eingangs erwähnte Einschränkung, daß bei Gerätetypen mit konstanter Belastungsgeschwindigkeit der Querschnitt des Prüffadens bzw. der Titer zu berücksichtigen ist.

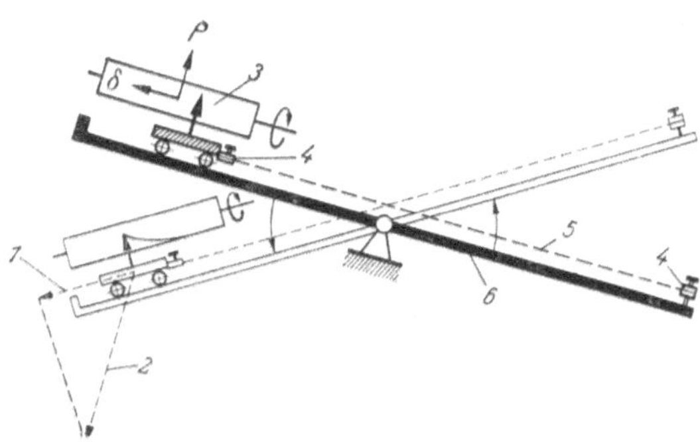

Abb. IV,19. Schematische Darstellung eines Festigkeitsprüfers nach dem Belastungsprinzip der Neigungswaage. (Nach G. W. Scott und Gonsalves/Schultz.)

1 Belastung des zu prüfenden Fadens, *2* Gewicht des Belastungswagens, *3* Schreibtrommel, *4* Klemme, *5* Faden, *6* drehbare Ebene mit Sinusantrieb.

Der Weg, die *stetige Belastung des Prüffadens durch Flüssigkeiten* vorzunehmen, ist von Krais und Keyl beschritten worden. In neuester Zeit ist dieses Prinzip in dem Garnprüfgerät von W. Wegener[1] (siehe Abb. IV,20) verwirklicht worden, das über die Aufnahme des gewonnenen KD-Diagrammes hinaus die Dehnungsänderungen in Abhängigkeit von der Belastung bzw. von der Zeit aufzeichnet. Es ist nämlich zu beachten, daß die während der Verformung auftretenden Veränderungen der inneren Struktur dazu beitragen können, bisher unbekannte Vorgänge im Innern des Fadens sichtbar werden zu lassen. Das Gerät soll zur Erfassung derartiger kleiner und kleinster Änderungen dienen, die in dem gewöhnlichen KD-Diagramm nicht mehr in Erscheinung treten. Es ist dadurch möglich geworden, in einem Garn Aufschluß über die vor sich gehenden inneren Strukturwandlungen, z.B. das abwechselnde Haften und Gleiten der einzelnen Elementarfasern, Aufschluß zu erhalten.

Die Belastung erfolgt, wie erwähnt, durch Wasserzulauf. Registriert wird erstens die Verlängerung des Fadens, ausgehend von der Drehung der Rolle *2* über den Stahldraht *5*, über die Rolle *6* auf den Spiegel *8*. Der Lichtstrahl von der Lichtquelle *9* geht über diesen Spiegel auf die Registriertrommel *10*, die mit photographischem Papier bespannt ist. Die Registrierung der Dehnungsänderung wird zweitens dadurch erreicht, daß auf der Achse der Rolle *2* eine Hülse sitzt, die von dieser bei der Drehung gerade noch mitgenommen wird. Auf der Hülse befinden sich der Spiegel *11* und die beiden aufeinander senkrecht stehenden Metallflügel *12*. Ein

[1] Wegener, W.: Melliand Textilber. **32**, 12 (1951).

Elektromagnet *13* bewirkt über eine Blattfeder durch Kontaktsteuerung vom Antriebssystem her eine Wiederherstellung der Ausgangslage des Spiegels nach bestimmten kleinen vorgegebenen Zeitabständen (z. B. 1 Sekunde). Während dieser

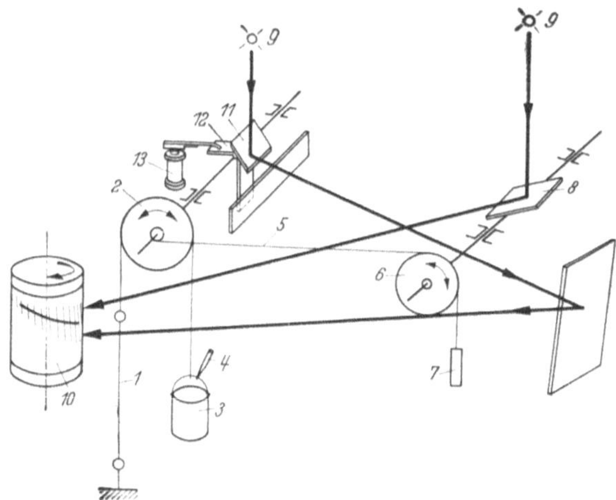

Abb. IV, 20. Schematische Darstellung des Garnprüfgerätes. (Nach W. WEGENER.)
1 Faden, *2* Rolle, *3* Belastungsgefäß, *4* Wasserzulauf, *5* Stahldraht, *6* Rolle, *7* Spanngewicht, *8* Spiegel zur Registrierung der Dehnung, *9* Lichtquellen, *10* Registriertrommel, *11* Spiegel zur Registrierung der Dehnungsänderung, *12* Metallflügel, *13* Elektromagnet.

Zeit folgt der Spiegel *11* jeweils der Drehung der Achse *2* und ein zweiter Lichtstrahl registriert die in diesem Zeitintervall auftretende Dehnung auf der Registriertrommel *10*. Je kleiner das durch Einstellung des Kontaktsystems gewählte Zeitintervall wird, um so mehr entspricht die Begrenzungskurve *2* in Abb. IV, 21 der ersten Differentialkurve der Kraft-Dehnungs-Linie *1*.

Abb. IV, 21. Registrierung einer Dehnungsänderung bei Kupferkunstseide. (Nach W. WEGENER.)

Bei diesem Gerät ist die konstante Energie der zuströmenden Flüssigkeit zu beachten.

Von STEENBERG[1] ist ein Prüfgerät angegeben worden, das wohl nach dem bereits von LEIS angegebenen *Prinzip der Kettenwaage* arbeitet, jedoch durch die Eigenart seiner Konstruktion die Einschränkung, die durch die konstante Belastungsgeschwindigkeit bei diesen Geräten gegeben ist, überwindet.

Das Gerät ist schematisch in der Abb. IV, 22 dargestellt. Diese Abbildung läßt erkennen, daß die untere Klemme durch eine Mikrometerspindel *5* mit Hilfe eines kleinen Motors *4* (Dehnungsmotor) mit konstanter Geschwindigkeit nach unten bewegt wird. Die obere Einspannklemme *10* ist an dem einen Ende des Waagebalkens *11* befestigt. Am

[1] STEENBERG, B.: Svensk Papperstidn. **50**, 127 (1947). In dieser Arbeit verwendet B. STEENBERG noch ein SCHOPPER-Gerät, das jedoch bereits nach dem Prinzip der Kettenwaage umgebaut ist. Die endgültige Konstruktion, wie sie hier beschrieben wird, ist in der folgenden Arbeit[2] angegeben.

[2] STEENBERG, B.: Svensk Papperstidn. **50**, 346 (1947).

anderen Ende ist die Kette *1* angebracht, und die Belastung erfolgt so, daß das zweite Ende der Kette mit Hilfe eines Elektromotors *2* (Belastungsmotor) auf und ab gefahren werden kann. Dieser Motor dreht gleichzeitig die Trommel *3* des Diagrammschreibers, so daß in Laufrichtung des Papiers die Kraft aufgezeichnet wird.

Die Steuerung des Motors erfolgt durch einen Kontakt *13*, der am Waagebalken angebracht ist. Wird durch Betätigung des Dehnungsmotors im Faden eine Kraft erzeugt und schlägt daher der Waagebalken am Kettenende um einen sehr kleinen Betrag nach oben hin aus, wird der Kontakt geschlossen, der den Belastungsmotor in Betrieb setzt und damit durch Absenken des Kettenendes eine Belastung erzeugt, die den Waagebalken wieder in die Normallage zurückführt (Kompensationsprinzip). Bei konstant gehaltener Dehnungsgeschwindigkeit wird die Last in Abhängigkeit von der Zeit auf dem Diagramm registriert. Je nachdem Dehnungs- oder Belastungsmotor primär eingesetzt werden, arbeitet das Gerät mit konstanter Dehnungs- oder Belastungsgeschwindigkeit.

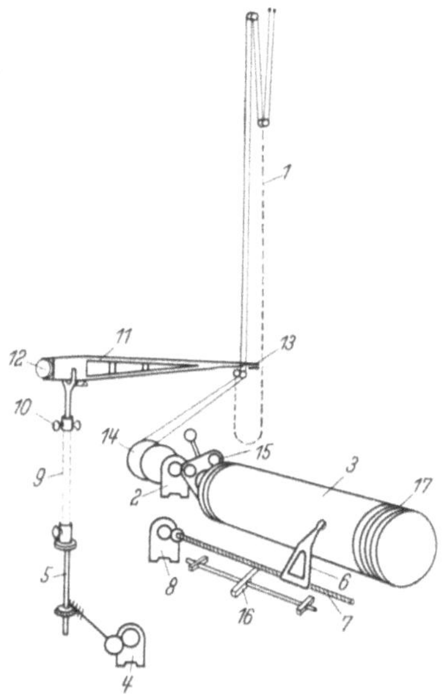

Abb. IV, 22. Schematische Darstellung des Festigkeitsprüfers. (Nach B. STEENBERG.) *1* Kette, *2* Kettenmotor, *3* Schreibtrommel, *4* Dehnungsmotor, *5* Mikrometerspindel, *6* Schreibfeder, *7* Schreibspindel, *8* Federmotor, *9* Probe, *10* obere Klemme, *11* Waagebalken, *12* Justiergewicht, *13* Kontakt, *14* Drahttrommel, *15* Zahnradgetriebe, *16* Dehnungsbegrenzungskontakte, *17* Kraftbegrenzungskontakte.

Die Dehnung wird auf dem Diagrammschreiber durch einen dritten Elektromotor *8* (Federmotor) über eine Mikrometerspindel *7* senkrecht zur Laufrichtung des Papiers aufgezeichnet. Das Verhältnis der Drehzahlen von Dehnungsmotor und Federmotor ist konstant. Durch Stellung von Kontakten ist es möglich, langsame periodische Be- und Entlastungen zwischen zwei vorgegebenen Dehnungs- oder Belastungswerten automatisch zu fahren. Das Diagrammpapier läuft dann gleichfalls vor und zurück. Durch die Kettenbelastung ist das Gerät in den Kräftebereichen und durch die geringe Bewegungsmöglichkeit der unteren Klemme für Textilien nicht ohne weiteres anwendbar und speziell für die Prüfung von Papierfasern geeignet.

Speziell bei Feinfaserprüfgeräten[1] hat man von dem *Belastungsprinzip der gleichmäßig zunehmenden Torsionsfederspannung* Gebrauch gemacht, da Einzelfasern auf Massenwirkungen besonders empfindlich reagieren. Nach der Abb. IV, 23, die uns eine anschauliche Darstellung der Arbeits-

[1] Kraftmeßfedern in weitestem Umfang verwendet CHEVENARD in seiner Mikromaschine mit Lichtschaubildzeichner für Zugvermerke an allen fadenförmigen Stoffen, Bändern, Folien usw. Vgl. A. J. Amslen u. Co. Blatt Mi 4, Schaffh./Schweiz.

weise dieses Gerätetyps gibt, wird die konstante Belastungsgeschwindigkeit des Prüffadens *8* durch gleichmäßiges Spannen der Torsionsfeder *3* hervorgerufen durch einen stufenlos regelbaren Antrieb *1* über die Kegelräder *2* und *2a* erreicht. Auch hier wird die untere Fadenklemme *7* mit Hilfe der Handkurbel *9* so nachgestellt, daß der Zeigerhebel *4* stets horizontal bleibt und damit die Lage der oberen Fadenklemme *7* fixiert.

Das neueste Gerät dieser Art ist zur Prüfung von Feinfasern konstruiert worden. Durch die elektrische Schaltanordnung ist es möglich, auch periodische Be- und Entlastungen an Fasern durchzuführen, wobei der jeweilige Belastungshöchstwert vorher eingestellt werden kann. Die entsprechende Verstellung der Dehnung bei Hin- und Rücklauf erfolgt von Hand aus wieder durch Beobachtung des Zeigers an der Nullmarke. Das Diagrammpapier läuft gleichfalls vor und zurück, wobei die Aufzeichnung des Diagrammes mit einer Schreibfeder so erfolgt, daß keine Rückwirkung der Reibungskräfte auf den Zerreißvorgang der Faser stattfindet[1]. Bei den älteren Geräten dagegen kommt noch der Einfluß der Schreibfederreibung hinzu, der von F. WINKLER[2] untersucht wurde und in der Abb. IV, 24 dargestellt ist. Man erkennt deutlich, daß bei Benutzung des Diagrammschreibers eine zu niedrige Festigkeit, praktisch nur etwa 70% des tatsächlichen Festigkeitswertes, zur Registrierung gelangen – eine Einflußgröße, die oft übersehen wird.

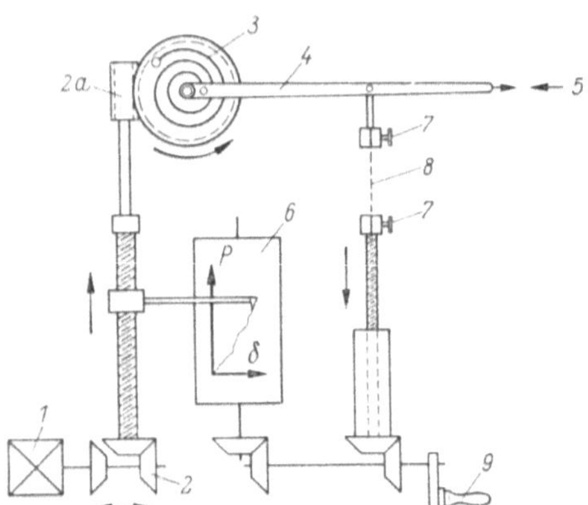

Abb. IV, 23. Schematische Darstellung eines Festigkeitsprüfers nach dem Belastungsprinzip der zunehmenden Torsionsfederspannung. (System AEG-FRANK.)
1 Antriebsmotor, *2* Kegelradantrieb, *2a* Schnecke mit Antriebsrad, *3* Torsionsfeder, *4* Zeigerhebel, *5* Nullmarke, *6* Registriergerät, *7* Klemmen, *8* zu untersuchende Probe, *9* Kurbel zum Ausgleich der Dehnung.

b) Untersuchung der Festigkeit sowie der Gleichmäßigkeit der Fadenspannung und -dehnung von endlosen Fäden und Garnen[3].

Das Prinzip der Neigungswaage bildet auch die Grundlage für Zugfestigkeitsprüfungen am laufenden Faden, wie sie von der Zellweger AG. in dem Garnfestigkeitsprüfer (USTER) verwirklicht worden sind. Das Gerät arbeitet automatisch und zeichnet fortlaufend sowohl die Einzelwerte der Bruchbelastung als auch der Bruchdehnung in Form farbiger Striche auf getrennten Diagrammstreifen auf, summiert dabei gleich-

[1] PREUSSER, H. M., R. A. O'CONNELL, A. S. YEISER u. H. P. LUNDGREN: Textile Res. J. **24**, 2 (1954), haben ein vollautomatisch arbeitendes Einzelfaser-Prüfgerät entwickelt, das die stetige Belastung der Einzelfaser durch Verdrehen eines Torsionsdrahtes erzeugt und damit über die hier geschilderte Gerätetype in der Anwendbarkeit noch hinausgeht. — [2] WINKLER, F.: Faserforsch. u. Textiltechn. **5**, 358 (1954). — [3] Vgl. auch den DIN-Entwurf 53829: Gleichmäßigkeitsprüfung am laufenden Faden bei konstanter Dehnung.

zeitig die Werte und ordnet außerdem die Zugfestigkeitswerte nach Größenklassen in ein Häufigkeitsstaffelbild. Die Dehnungswerte werden dagegen automatisch nicht statistisch erfaßt. Es ist aber möglich, diese Werte abzugreifen und von Hand aus ein derartiges statistisches Schaubild aufzunehmen. Auch für dieses Gerät gelten die aufgeführten Einschränkungen, nicht jedoch für das neue Gerät von Zwick u. Co., Einsingen, mit dem Induktionsspulen-Meßkopf.

Wie bereits im § 21 b ausgeführt, ist die Erkenntnis, daß der arithmetische Mittelwert zur Beurteilung der Eigenschaften nicht ausreicht und daß die Streuung eine das Material kennzeichnende Größe ist, bei der Festigkeitsprüfung mit dem USTER- und dem Zwickgerät berücksichtigt worden. Bei der Zugfestigkeit z. B. haben wir nach unseren heutigen Gesichtspunkten keine Materialkonstante vor uns, sondern eine statistische Eigenschaft des Materials in dem Sinne, daß auch nach Eliminierung der apparativen und subjektiven Meßfehler noch eine statistische, dem Materialeigentümliche Verteilung übrigbleibt. Diese statistische Verteilung muß daher durch entsprechende Prüfgeräte festgestellt werden, um auch Aussagen über den störungsfreien Ablauf der Verarbeitungsprozesse zu erhalten. Zu diesen Prüfgeräten (vgl. § 21 c 2) gehört die Universal-Garnprüfmaschine, System FRENZEL-HAHN, die zur Messung der Gleichmäßigkeit der Spannung bei konstanter Dehnungsgeschwindigkeit am laufenden Faden dient.

Das Gerät ist in der Abb. IV, 25 schematisch dargestellt[1], arbeitet mit

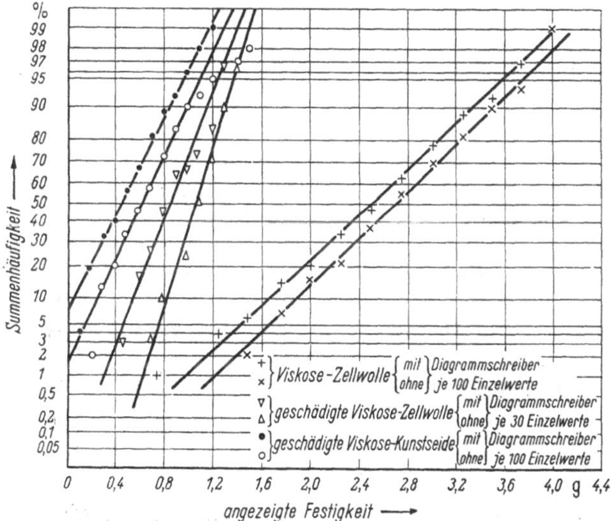

Abb. IV, 24. Summenhäufigkeitskurven der Versuchsergebnisse über den Einfluß der Schreibfederreibung bei der Faserfestigkeitsprüfung. (Nach F. WINKLER.)

[1] Eine Abart dieses Gerätes ist die automatische Prüfmaschine des Institut Textile de France (ITF) zur Bestimmung der Verwebbarkeit der Schußgarne, wie sie von A. MARTI: Bulletin de l'Institut Textile de France **6**, 7 (1954) beschrieben worden ist, sowie das von P. BRUER in L'Industrie Textile 865 (1954) erläuterte Gerät. Über die Anwendungen des Gerätes nach Abb. IV, 25 vgl. auch W. FRENZEL u. H. HESSE: Faserforsch. u. Textiltechnik **6**, 189 (1955).

konstantem Dehnungsverzug, der durch die Geschwindigkeitsdifferenz von Einzugs- und Abzugswalzen erzeugt wird. Die Abzugsgeschwindigkeit ist größer als der Fadenzulauf. Gemessen werden die beim konstant gedehnten Faden auftretenden kleinen Änderungen der Fadenspannung

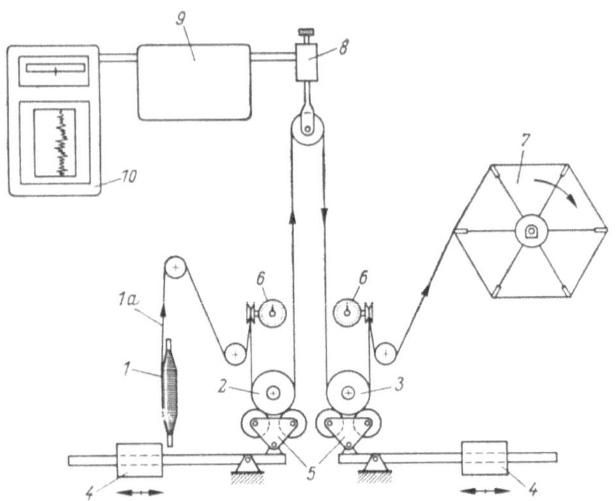

Abb. IV,25. Schematische Darstellung der Universal-Garnprüfmaschine. (System W. FRENZEL – G. HAHN mit Meßkopf nach H. STEIN).

1 Kops, *1a* Fadenlauf, *2* Einzugswalze, *3* Abzugswalze, *4* verschiebbares Andrückgewicht, *5* Andrückwalzenpaare, *6* Fadenspannungsmeßuhren, *7* Aufwickelhaspel, *8* Meßkopf nach H. STEIN, *9* Verstärker mit Meßbrücke, *10* Registriergerät.

bzw. der Belastungszunahme. Als Dynamometer wird die elektromagnetische Feinmeßeinrichtung nach H. STEIN (§ 22c) verwendet. Mit dem Gerät sind auch Elastizitätsmessungen möglich.

Vollständigkeitshalber sei auch ein schrittweise arbeitendes Gerät, der sog. BARMAG-Dehnungsprüfer, erwähnt, bei dem unter einer konstant eingestellten und weit unterhalb der Bruchgrenze des Fadens liegenden Belastung die Dehnung und deren Schwankungen gemessen und laufend aufgezeichnet werden. Die beiden Geräte sind besonders geeignet, um gefährliche Überstreckungen bzw. Überdehnungen, die sich besonders bei Reyongarnen, z. B. durch Glanzschüsse oder in Form von Glanzstreifen auswirken, festzustellen.

c) Weglose[1] Kraftmessung bei statischen Prüfmethoden.

Von den bisher geschilderten statischen Prüfgeräten arbeitet nur die Pendelwaage nach dem Prinzip der konstanten Dehnungsgeschwindigkeit. Sämtliche anderen Gerätetypen beruhen auf dem Prinzip der kon-

[1] Wir verstehen unter *weglos*, daß die Bewegung einer Einspannklemme bei der Kraftmessung vernachlässigbar gering ist. Vgl. auch § 22a: Nicht weglose Kraftmessung.

stanten Belastungsgeschwindigkeit und unterliegen daher der Einschränkung, daß die Belastungsgeschwindigkeit auf den Querschnitt bzw. auf den Titer bezogen werden muß. Die bisher bekannten Ausführungen dieser Gerätetypen haben jedoch den Nachteil, daß das Prinzip der konstanten Dehnungsgeschwindigkeit infolge der Lageveränderlichkeit beider Klemmen nur angenähert erfüllt ist. Hinzu kommt, daß durch die direkte Verbindung einer größeren Masse mit dem elastischen Prüffaden den Meßvorgang beeinflussende dynamische Massenwirkungen auftreten.

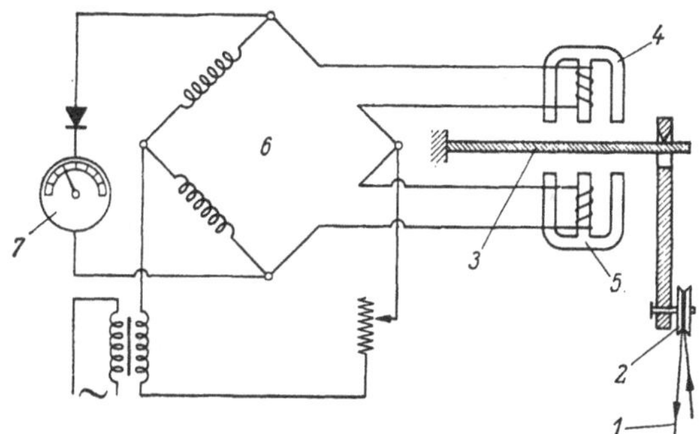

Abb. IV, 26. Induktive Kraftmessung. (Nach H. STEIN.)
1 Prüffaden, *2* Meßrolle, *3* Federmaßstab als Anker, *4*, *5* Magnetsysteme, *6* Meßbrücke, *7* Galvanometer.

Um diese grundsätzlichen Schwierigkeiten zu beheben, ist die Ausschaltung der Lageveränderlichkeit einer Einspannklemme und der dynamischen Massenwirkungen notwendig. Dieses Ziel kann durch die sog. weglose Kraftmessung erreicht werden, wobei einer der bekanntesten Meßköpfe für die statische Prüfung der *elektromagnetisch arbeitende Meßkopf nach* H. STEIN[1] ist (siehe Abb. IV, 26).

Die Kraft wird über den Faden *1* und über die Meßrolle *2* von einem Federstab *3*, der zur Erzielung verschiedener Meßbereiche auswechselbar ist, aufgenommen. Dieser befindet sich als Anker zwischen den beiden Magnetsystemen *4* und *5*, die in Brückenschaltung angeordnet sind und mit Wechselstrom gespeist werden. Beim Durchbiegen des Federstabes vergrößert sich der Luftspalt für das eine System, während sich gleichzeitig die Induktivität für das andere entsprechend verkleinert. Die hervorgerufene Meßspannung wird an der Brücke *6* abgegriffen, verstärkt und kann mit einem Galvanometer *7* gemessen oder mit einem Schreibgerät laufend registriert werden[2].

Eine weitere Möglichkeit der elektrischen weglosen Kraftmessung ergibt sich durch die Verwendung eines nach dem *elektrodynamischen Prinzip arbeitenden Verlagerungsaufnehmers* (siehe Abb. IV, 27), wie sie z. B. von den Firmen Philips, Eindhoven, Vibrometer u. a., in Verbindung mit einer direkt anzeigenden Meßbrücke hergestellt werden. Der Verlage-

[1] Aus E. WAGNER, Mech.-technologische Textilprüfungen, Wuppertal 1953, S. 81.

[2] Auch dieses System verlangt nicht nur eine sorgfältige Eichung, sondern auch eine laufende Überwachung, da der Federstab in seinen Deformationseigenschaften leicht der Veränderung unterliegt.

rungsaufnehmer ist für die Messung statischer und dynamischer relativer Verschiebungen entwickelt worden.

Der dünne Meßstift *1* wird an das schwingende Gebilde, dessen Verschiebung gemessen werden soll, herangeführt. In der Mitte ist auf diesem Meßstift ein kleiner „Ferroxcube"-Kern *2* befestigt und wird durch besonders ausgebildete Lagerfedern *3* unterstützt. Der kleine Kern ist in einem, an dem Gehäuse befestigten nicht-magnetischen Röhrchen, woran eine Primärspule *4* und zwei Sekundärspulen *5* angebracht sind, in der Längsrichtung beweglich. Die Wirkung dieses Aufnehmers beruht auf dem Prinzip der Kopplungsänderung, und zwar ändert sich die Kopplung der beiden Sekundärspulen durch die axiale Bewegung des Kerns und damit ändern sich die in den Sekundärspulen induzierten Spannungen in entgegengesetztem Sinne. Die Primärspule wird mit Wechselspannung gespeist. Der Spannungsunterschied kann an einer direkt anzeigenden Meßbrücke abgelesen werden. Durch eine kupferne Buchse *6* und das Stahlgehäuse *7* ist der Aufnehmer gegen Störfelder sowie gegenüber dem Auftreten von Fehlerquellen beim Einbau inMetallhalter geschützt. Da das bewegliche System eine nur sehr kleine Masse besitzt, kann es sehr großen Beschleunigungen folgen. Wird der Verlagerungsaufnehmer an eine je nach dem Kräftebereich verschieden starke Stahlfedermembran angesetzt, dann ist dieses System wieder ein induktiv arbeitender Meßkopf. In Verbindung mit einer direkt anzeigenden Meßbrücke können Durchbiegungen bis zu $0{,}05\,\mu$ abgelesen werden. Durch diese Eigenschaften können die Membranen entsprechend stark dimensioniert werden, so daß praktisch eine weglose Kraftmessung erreicht wird. Mit dem Aufnehmer können (statische) Kräfte bis zu einigen Milligramm herunter gemessen werden.

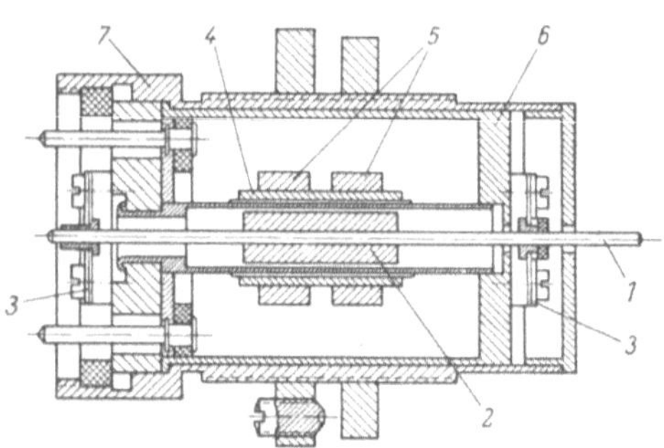

Abb. IV,27. Prinzip eines induktiven Verlagerungsaufnehmers (der Fa. Philips).
1 Meßstift, *2* „Ferroxcube"-Kern, *3* Lagerfedern, *4* Primärspule, *5* Sekundärspule, *6* kupferne Buchse, 7 Stahlgehäuse.

Eine besonders einfache Möglichkeit der weglosen elektrischen Kraftmessung besteht in der Anwendung von *Dehnungsmeßstreifen* (strain gauge)[1]. Nach diesem „OHMschen Prinzip" arbeiten Meßköpfe verschiedenen Fabrikats. Die Streifen werden zweckmäßigerweise in Brückenschaltung auf Federmembranen, die entweder einseitig oder beiderseitig eingespannt sind oder auf besonders ausgebildete Stahlbügel geklebt. Zu beachten ist dabei, ob in erster Linie statisch oder dynamisch gemessen werden soll. Wird statisch gemessen und als Anzeigegerät ein Galvano-

[1] Grundlagen und Anwendungen der Dehnungsmeßstreifen bearbeitet von K. FINK, Verlag Stahleisen, Düsseldorf 1952. — K. FINK u. CHR. ROHRBACH: Z. VDI 265 (1953). — CHR. ROHRBACH: Philips Techn. Inform. MA3, ATM. 135-7 (1955); ATM. 135-8 (1955); ATM. 135-9 (1955). — G. SCHULZ: Konstruktion 7, 299 (1955).

meter verwendet, dann ist es – um größte Empfindlichkeit zu erhalten – nach Berechnungen von A. U. HUGGENBERGER[1] zweckmäßig, den Galvanometerwiderstand gleich dem Widerstand des Dehnungsmeßstreifens zu wählen. Es ist bei statischen Messungen weiterhin zweckmäßig, den Widerstand der Meßstreifen so klein wie möglich zu halten (Herstellungsgrenze 100–120 Ohm).

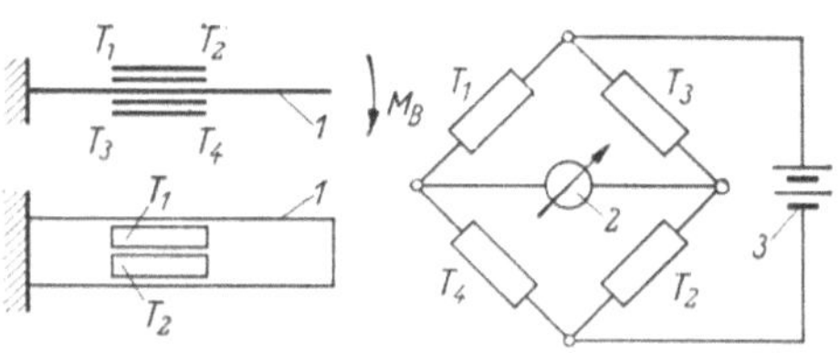

Abb. IV, 28. Anordnung von Dehnungsmeßstreifen auf einer Biegefeder. Schaltung mit Temperaturausgleich. (Nach A. U. HUGGENBERGER.) *1* Biegefeder, *2* Anzeigegerät, *3* Batterie, T_1, T_2, T_3, T_4 Dehnungsmeßstreifen.

Beim Einsatz der Dehnungsmeßstreifen ist der Einfluß der Temperatur stets zu beachten. Sämtliche Streifen werden deshalb in Brückenschaltung mit Temperaturkompensation angeordnet. Praktisch erreicht wird dies mit gleichzeitiger Empfindlichkeitssteigerung durch symmetrisches Aufkleben auf die Vor- und Rückseite der Federmembran.

Die Abb. IV, 28 zeigt ein Beispiel für die Anwendung der Dehnungsmeßstreifen mit Temperaturausgleich an einer Biegefeder. Die Meßstreifen werden gewöhnlich aus Batterien, in besonderen Fällen auch durch den Trägerfrequenzgenerator einer Meßbrücke gespeist[2].

d) Festigkeitsprüfgeräte nach dem Prinzip der weglosen Kraftmessung.

Der Statigraph nach H. STEIN (siehe Abb. IV, 29) arbeitet bei sämt-

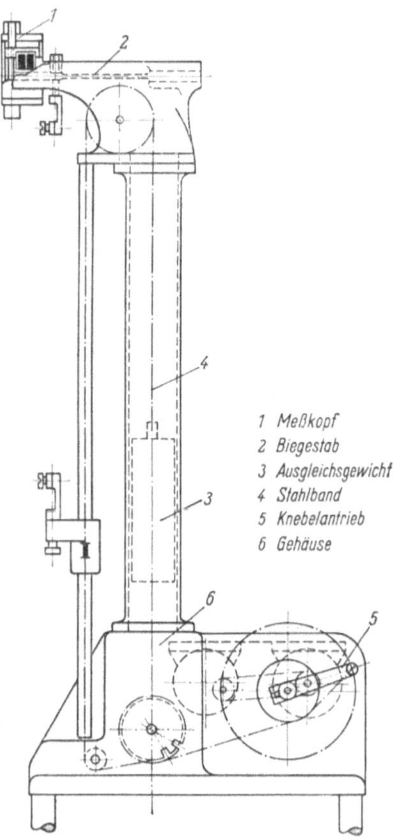

Abb. IV, 29. Schematische Darstellung des Statigraphen. (Nach H. STEIN.)

[1] HUGGENBERGER, A. U.: Schweiz. Arch. angew. Wiss. Techn. **18**, 4 (1952).

[2] THORSEN, W. J.: Textile Res. J. **24**, 5 (1954) hat, um auch die in § 21a geschilderten Relaxationserscheinungen an Einzelfasern experimentell erfassen zu können, folgenden Weg eingeschlagen: Die Messung kleinster Kräfte wurde mit der einfachsten Art von Dehnungsmeßstreifen, nämlich durch Spannen eines einzigen Drahtes aus besonderer Legierung durchgeführt. An die Stelle der vier Meßstreifen in Brückenschaltung und Batteriespeisung treten vier Einzeldrähte von je 20 cm Länge. An einem dieser Drähte greift die zu messende Kraft an. Mit dieser Anordnung werden Kräfte bis zu 3 mg über lange Zeit hindurch exakt gemessen.

lichen Versuchen mit konstanter, in bestimmten Stufen wählbarer Dehnungsgeschwindigkeit und besitzt zur Kraftanzeige den in § 22c geschilderten elektromagnetischen Meßkopf – nach H. STEIN – mit Meßbrückenanordnung, Röhrenverstärker und dem Netzanschlußteil mit Spannungskonstanthalter. Die Registrierung erfolgt durch Tintenschreiber. Das Diagrammpapier bewegt sich synchron mit den Bewegungen der Abzugsklemme. Durch die große Empfindlichkeit des elektrischen Meßkopfes und die Möglichkeit der Wahl sehr geringer Dehnungsgeschwindigkeiten ist zusätzlich zu den Möglichkeiten der früheren Geräte die Registrierung von Haft- und Gleitcharakteristiken an Faserbändern und Vorgarnen möglich[1]. Eine analoge Anordnung für die Einzelfaserprüfung stellt der Fafegraph dar. Dieses Gerät arbeitet mit stufenlos verstellbarer Dehnungsgeschwindigkeit und besitzt einen auswechselbaren Meßkopf bis herunter zu sehr kleinen Kräften von etwa 2 g.

Die beiden erwähnten Geräte erlauben nicht nur die Aufnahme des gewöhnlichen Kraft-Dehnungs-Diagrammes, sondern sind darüber hinaus in der Lage, mit den im statischen Zerreißversuch verwendeten Deformationsgeschwindigkeiten in Verbindung mit dem Steuergerät Robotex cyclische Be- und Entlastungen automatisch durchzuführen. Dadurch ist es möglich, ein bestimmtes Belastungs- oder Dehnungsspiel vorzugeben und den Prüffaden beliebig viele Male cyclische Beanspruchungen durchlaufen zu lassen. Dabei interessieren die auftretenden Änderungen im Dehnungsverhalten und die Ermittlung der zugelastischen Eigenschaften. Ebenfalls ist es mit dieser Anordnung möglich, dem Prüffaden langsam periodische Belastungen zwischen einer Vorbelastung und einem einstellbaren Belastungswert oder langsam wechselnde Dehnungen zwischen der Vorbelastung und einem einstellbaren Dehnungswert aufzuprägen. Der Lauf des Diagrammpapiers kann dabei entweder so geregelt werden, daß dieses bei der Aufhebung der Belastung jeweils dem gesamten Weg der Abzugsklemme oder daß es nur bis zum Erreichen einer eingestellten Vorbelastung, also nicht während der darauf einsetzenden Verschlappung des Fadens folgt. Weiterhin ist mit dieser Steuereinrichtung die automatische Einschaltung von Pausenzeiten vor der Entlastung bzw. der Rückwärtsbewegung durchführbar. Die obere Grenze der Frequenz dieser langsamen periodischen Vordehnung erreicht noch nicht 0,5 Hz.

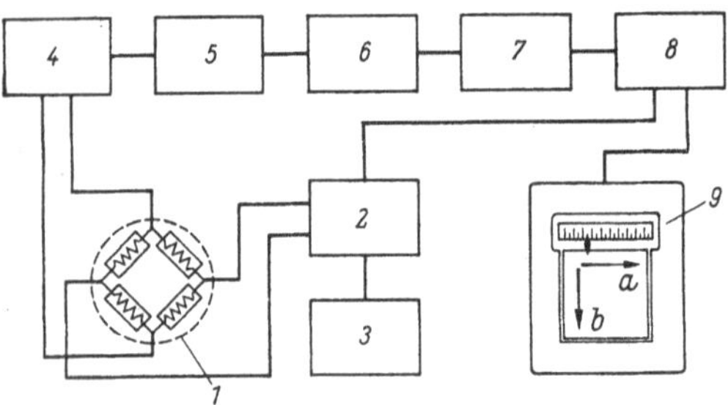

Abb. IV, 30. Kraftanzeige beim Instron-Tester.
1 Belastungszelle mit Dehnungsmeßstreifen in Brückenschaltung, *2* Trägerfrequenzgenerator (390 Hz), *3* Stabilisator von *2*, *4* Brückengleichgewichtsregler, *5* Verstärker I, *6* geeichter stufenweise regelbarer Abschwächer, *7* Verstärker II, *8* Gleichrichter, *9* Schreiber, *9a* Belastungsrichtung, *9b* Dehnungsrichtung.

Auch der Instron-Tensile-Tester[2] führt sämtliche Versuche mit konstanter Dehnungsgeschwindigkeit durch, die in Stufen eingestellt werden kann, und arbeitet gleichfalls nach der weglosen Kraftmessung, die mittels Dehnungsmeßstreifen (siehe § 22c) erfolgt und aus der Abb. IV,30 ersichtlich ist.

[1] STEIN, H.: Textilpraxis **10**, 527 (1955).
[2] HINDMAN, H. u. G. S. BURR: Transact. of A.S.M.E 789 (1949).

Das Gerät überträgt durch auswechselbare Kraftzellen einen Kräftebereich von einigen Gramm bis zu einigen 100 Kilogramm. In den kleinen Kräftebereichen erfolgt die Kraftanzeige durch Dehnungsmeßstreifen, die auf einen Federstab geklebt werden und dessen Durchbiegung gemessen wird. In den höheren Kräftebereichen wird die Deformation eines Stahlbügels direkt mit Meßstreifen gemessen.

Das Antriebssystem des Instron-Testers, bestehend aus dem Antriebsmotor, wird über Steuergeräte durch einen Steuergenerator so angetrieben, daß die vorgegebene Dehnungsgeschwindigkeit, die durch einen Synchronmotor über ein trägheitsarmes Wechselgetriebe erzeugt, genauestens eingehalten wird. Die Bewegung des Schlittens folgt mit der unteren Klemme praktisch trägheitsfrei jeder vorgegebenen Bewegung entweder von Hand aus oder automatisch. Größter Wert wurde bei dieser Konstruktion darauf gelegt, daß sowohl Kraft- wie Dehnungsanzeige völlig phasentreu von einem sich besonders schnell einstellenden Schreiber registriert werden. Das Diagrammpapier folgt den Bewegungen des Schlittens praktisch verzögerungsfrei, jedoch nur in ein und derselben Richtung. Ein Rücklauf des Diagrammpapiers bei Entlastung erfolgt also nicht.

Mit dem Instron-Tester ist es darüber hinaus möglich, jeweils zwei bestimmte Kraft- oder Dehnungswerte vorzugeben und die Probe von Hand aus oder automatisch beliebig viele Cyclen langsam durchlaufen zu lassen. Mit diesem Gerät sind daher besonders genaue Hysteresismessungen durchführbar, deren Auswertung jedoch durch den Gleichlauf des Diagrammpapiers etwas erschwert wird.

In der Abb. IV, 31 sind zwei Beispiele für die automatische periodische Beanspruchung des Prüffadens im Instron-Tester wiedergegeben, und zwar rechts bei konstantem Dehnungsspiel, bei dem die Belastungsspitzen mit der Anzahl der Belastungswechsel infolge der Relaxation absinken und die Kraftamplituden größer werden, was Zunahme der Steifheit des Prüffadens bedeutet. Im linken Diagramm bei konstanter oberer Lastgrenze und Einspannlänge sieht man das Fortschreiten der bleibenden Dehnung und die elastische Erholung.

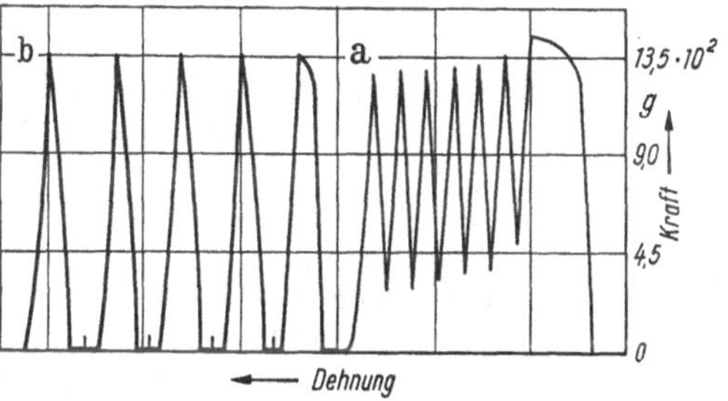

Abb. IV, 31. Automatische periodische Beanspruchung beim Instron-Tester: a) bei konstantem Dehnungsspiel; b) bei konstanter oberer Lastgrenze und Einspannlänge.

Auch beim Instron-Tester liegt die höchste Frequenz dieser langsamen periodischen Beanspruchungen noch $< 0,5$ Hz. Über die Fehlergrenzen auch dieses Geräts liegen nur wenige Angaben in der Literatur vor. Eine eingehende Beschreibung der Bau- und Arbeitsweise sowie der Vorteile und Fehlergrenzen des Geräts im Vergleich zu den Mängeln des Pendelgewichtsprüfgeräts wird von G. Broughton[1] gegeben.

[1] Broughton, G.: Wochenbl. f. Papierfabrikation 81, 158 (1953).

§ 23. Dynamische Prüfmethoden[1].

Die in den vorhergehenden Abschnitten beschriebenen Prüfgeräte nach H. STEIN und B. STEENBERG wie auch der Instron-Tester gehen bereits über die rein statisch arbeitenden Gerätetypen hinaus, da sie dem Prüffaden eine langsam wechselnde Beanspruchung aufzuprägen gestatten, wobei diese bis zum Bruch durchgeführt werden kann[2]. Der zeitliche Verlauf der Beanspruchung ist dabei in allen Fällen nicht sinusförmig, sondern linear. Die Frequenz dieses Be- und Entlastungsvorganges liegt in allen Fällen noch $< 0{,}5$ Hz[3].

a) Dynamisch arbeitende Prüfgeräte für Grundfrequenzen $< 0{,}5$ Hz.

Den Übergang zu den dynamischen Geräten bildet ein von W. WEGENER[4] angegebenes Meßgerät für dynamische Dauerstanduntersuchungen von Garnen, das auf der Grundlage der in § 22a geschilderten stetigen Kettenbelastung basiert und vollautomatisch nach dem Prinzip der periodisch wiederkehrenden Maximalbelastung und nach dem Prinzip der cyclisch wiederkehrenden Dehnungsbegrenzung bei jeweils konstanter Belastungsgeschwindigkeit arbeitet.

Die Belastung des Prüffadens erfolgt also wieder über einen Waagebalken, auf dessen einem Hebelarm das Gewicht des Teiles einer Kette wirkt. Auf optischem Wege registriert diese Apparatur Belastung und Dehnung als Funktion der Zeit. Die größte Frequenz der Belastungscyclen liegt bei etwa 0,075 Hz.

Zu den Geräten mit angenähert sinusförmiger Beanspruchung des Prüffadens gehört der Dehnungsspannungsmesser nach W. WEGENER[5], der nach dem Prinzip der proportional zur Dehnung abnehmenden Maximalbelastung arbeitet, wobei es möglich ist, sowohl den gesamten Dehnungs-Spannungs-Vorgang als auch den einzeln vorgegebenen Verlauf einer Be- und Entlastung zu registrieren.

Die bisher geschilderten Prüfgeräte arbeiteten in dem Frequenzgebiet $< 0{,}5$ Hz. Nunmehr kommen wir zu einem von W. WEGENER[6] vorgeschlagenen Gerätetyp, der diese langsame Grundfrequenz wohl noch beibehält, dabei jedoch eine Wechselbeanspruchung von einer Frequenz zwischen 5 bis 65 Hz überlagert.

Um den Prüffaden periodischen, praktisch sinusförmigen Be- und Entlastungen zu unterwerfen, wird folgendes Prinzip (vgl. Abb. IV, 32.)

[1] Vgl. Kap. I, § 3e, sowie § 6c und § 7d.

[2] Zu den dynamischen Prüfgeräten rechnet man z.B. auch das Pendelschlaggerät und andere Geräte, die für stoßweise Beanspruchungen gedacht sind; vgl. E. NÉMETH: Magyar Textiltechnika 155 (1954) und mit ausführlichen Literaturangaben: W. K. STONE, H. F. SCHIEFER u. G. FOX: Textilpraxis **10**, 1083 (1955) sowie A. SPRINGER: Faserforsch. u. Textiltechn. **6**, 76 (1955).

[3] Daß man mit diesen Frequenzen selbst bei der Untersuchung von Geweben nicht auskommt, zeigt J. S. MARGOLIN; er gibt daher für diese Prüfung eine neue Vorrichtung an: Textilprax. **8**, 475 (1953).

[4] WEGENER, W.: Melliand Textilber. **32**, 12 (1951).

[5] WEGENER, W.: Melliand Textilber. **31**, 10 (1950).

[6] Aus E. WAGNER: Mech.-technologische Textilprüfungen, Wuppertal 1953, S. 206.

angewandt. Die obere Klemme, in welcher der Faden eingespannt ist, befindet sich an dem einen Ende des Waagebalkens 1. Ein Motor mit Untersetzungsgetriebe treibt über das Ritzel *4* die beiden Scheibenräder *9* an, auf welchen die auswechselbaren Massen *3* als Unwuchten befestigt sind. Durch dieses System wird der Faden 22 zwischen Null und einem einstellbaren Höchstwert bei Frequenzen $< 0{,}5$ Hz periodisch be- und entlastet.

Gegenüber den in Abschnitt § 23a beschriebenen Geräten ist es bei diesem Gerät mit Hilfe eines Vibrators *23* möglich, diesem langsamen Belastungscyclus zusätzlich eine rasche periodische Beanspruchung im Frequenzbereich zwischen 5 bis 65 Hz zu überlagern. Die Amplitude des Vibrators ist jedoch gegenüber jener der Amplitude der Grundfrequenz klein und der zeitliche Kraftverlauf unbestimmt.

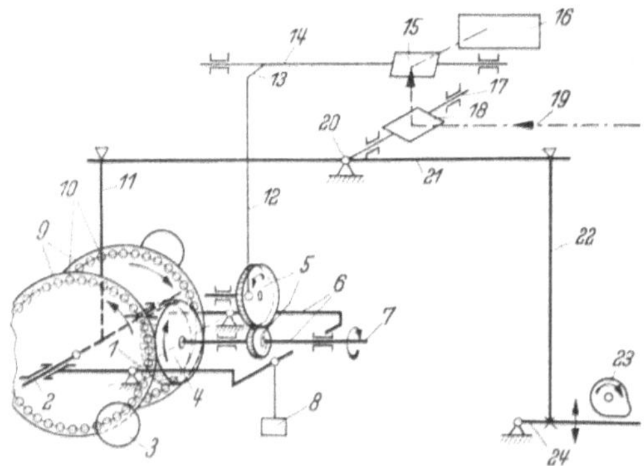

Abb. IV, 32. Schematische Darstellung eines Gerätes zur dynamischen Prüfung der Eigenschaften von Garnen. (System W. Wegener.)

1 Achse, *2* Welle, *3* Belastungsgewicht, *4* Ritzel, *5* Kurbeltrieb, *6* Waagebalken II, *7* Antriebswelle, *8* Tariergewicht, *9* Scheibenräder, *10* Punktverzahnung, *11* Übertragungshebel, *12* Kurbelstange, *13* Hebel, *14* Spiegelwelle *b*, *15* Belastungsspiegel, *16* Photopapier, *17* Spiegelwelle *d*, *18* Dehnungsspiegel, *19* Lichtstrahl, *20* Drehpunkt, *21* Waagebalken I, *22* Faden, *23* Vibrator, *24* Einspannhebel.

Der langsame periodische Be- und Entlastungsvorgang wird mit Hilfe eines Lichtstrahles *19* dem Dehnungsspiegel *18* und dem Belastungsspiegel *15* auf Photopapier festgehalten. Dabei werden die auftretenden Dehnungsänderungen des Fadens, also elastischer und bleibender Anteil, auf das Registriergerät übertragen. Der Einfluß der Vibration ist durch die verhältnismäßig kleine Amplitude gering. Für größere Amplituden des Vibrators bei kleineren Belastungsspielen und niedrigeren Frequenzen ist dieses Gerät nicht geeignet, da sonst dynamische Massenwirkungen durch das Umlaufsystem, welches über den Waagebalken mit dem Faden ein schwingendes System bildet, auftreten.

Bei den drei von W. Wegener[1] angegebenen Gerätetypen ist das Hauptaugenmerk auf die Spannungs-Dehnungs-Beziehung gerichtet.

b) Dynamisch arbeitende Prüfgeräte für Grundfrequenzen $> 0{,}5$ Hz.

Bei den Cordpulsatoren von Amsler und Zwick, die speziell zur Untersuchung von Reifencord entwickelt worden sind, wurde die Grundfrequenz, die bisher $< 0{,}5$ Hz lag, auf etwa 16 Hz und mehr erhöht.

[1] Wir halten die Bezeichnung dynamischer Dauerstandsprüfer nicht für glücklich gewählt, da es sich hier um eine Dauerwechselbeanspruchung handelt und nach F. Koerber[2] die Dauerstandfestigkeit diejenige Belastung bedeutet, bei der gerade noch kein dauerndes Weiterfließen eintritt.

[2] Koerber, F.: Z. techn. Physik 8, 421 (1927).

Das Amsler-Gerät (vgl. Abb. IV, 33) ist nach dem Schwingerprinzip gebaut und ermöglicht das Prüfen des dynamischen Verhaltens von Reifencord *1*, der in den Klemmen *2* eingespannt ist, bei vorgegebener statischer Vorspannlast und überlagerter periodischer Wechsellast. Die dadurch hervorgerufene periodische Dehnung (Amplitude) sowie die Gesamtdehnung werden gemessen. Infolge des Schwingerprinzips und des dadurch notwendigen abgestimmten Federsystems arbeitet dieses Gerät nur bei einer einzigen Frequenz und erfaßt nicht die Dämpfung des Materials. Es werden gleichzeitig zwei Proben der Prüfung unterzogen. Die Wechselkraft wird durch rotierende Scheiben *8* mit aufgesteckten Unwuchten *3* erzeugt, die an einem vertikal angeordneten Schwingarm *4*

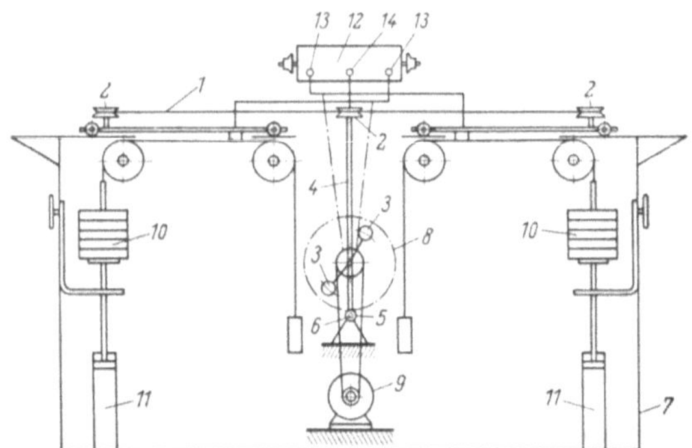

Abb. IV, 33. Schematische Darstellung des A. J. Amsler-Cord-Pulsers.
1 Cordfaden, *2* Klemmen, *3* verstellbare Unwuchten (1000 U/min), *4* Schwingarm, *5* Querachse, *6* Torsionsstabfeder, *7* Maschinengestell, *8* Umlaufscheibe, *9* Motor, *10* Gewichte, *11* Öldämpfung, *12* Schreibtrommel, *13* u. *14* Schreiber.

angebracht sind. Dieser ist an einer reibungslos gelagerten, hohlen Querachse *5* befestigt. Durch diese führt eine Torsionsstabfeder *6*, die einerseits am Schwingarm, andererseits am Maschinengestell *7* fest verankert ist. Die Eigenfrequenz des Schwingungssystems, Torsionsstab und Schwingarm (mit den daran befestigten Teilen), ist auf die Umlaufzahl der Unwuchtscheiben (1000 U/min) abgestimmt. Der Antrieb erfolgt durch einen Synchronmotor *9*. Die Wechsellast des Prüfkörpers wird daher durch die Zentrifugalkräfte der angebrachten Unwuchten erzeugt. Die eingestellte periodische Wechsellast bleibt während der Versuche konstant, unabhängig von der Änderung der Steifheit des Materials. Durch auflegbare Gewichte *10* wird die statische Vorlast dem Prüffaden aufgeprägt. Diese Last bleibt konstant, unabhängig vom Dehnungsverlauf des Materials. Die Gewichtsträger sind mit Dämpferkolben *11* ausgerüstet, die in mit Öl gefüllten Zylindern tauchen. Mit Hilfe zweier elektrischer Öfen können sämtliche Versuche bis zu Temperaturen von 120° C durchgeführt werden. Auf der Schreibtrommel *12* werden durch die beiden Schreiber *13*, die mit den beiden Wagen in Verbindung stehen, die Dehnungen der Proben und durch den Schreiber *14*, der mit dem Schwingarm in Verbindung steht, die gemeinsame Schwingamplitude registriert.

Die Wagen, die durch Gewichte gezogen werden, folgen bei größeren Vorspannlasten wohl der Dehnung des Materials; sie bilden aber mit dem Faden ein schwingendes System, dessen Trägheitskräfte trotz der Öldämpfung besonders bei kleineren Vorspannlasten in Erscheinung treten. Analog den auftretenden Trägheitskräften bei den nach dem Prinzip der Pendelwaage arbeitenden statischen Prüfgeräten sind auch sämtliche dynamischen Prüfgeräte, welche die Vorspannlast nach dem obigen Prinzip erzeugen, physikalisch nicht ganz einwandfrei. Bei kleineren Kräften und Frequenzen unter 10 Hz versagt sogar diese Methode. Mit dem Gerät kann daher nur ein starkes Material und dieses nur bei einer einzigen Frequenz geprüft werden. Der Vorteil liegt in der automatischen Konstanthaltung der statischen Vorlast wie der periodischen Wechsellast. Mit diesem Gerät ist es möglich, die Dauerwechselfestigkeit zu bestimmen, jedoch nur in sehr beschränktem Umfang die dauerelastischen Eigenschaften.

Im Prinzip völlig gleich ist der Cordpulsator nach ZWICK. An Stelle des Schwingungssystems – Torsionsstabfeder-Schwingarm – beim AMSLER-Pulser tritt hier eine Stabfeder. Die Wechsellast wird gleichfalls wieder durch eine verstellbare Unwucht erzeugt. Die Umdrehungszahl dieses Gerätes beträgt 2680 U/min. Alle bei der Beschreibung des AMSLER-Gerätes erwähnten Einzelheiten gelten auch für den ZWICK-Pulsator.

Auf Vorschlag von W. MESKAT wurde aus dem dynamischen Prüfgerät für Kautschuk von H. ROELIG[1] das dynamische Prüfgerät zur Bestimmung der Dauerwechselfestigkeit und der inneren Dämpfung entwickelt.

Dieses Gerät arbeitet ebenfalls mit konstanter Vorspannlast unabhängig von der Dehnung des Materials mit primär konstanter periodischer Dehnung. Periodische Dehnung und periodische Kraft werden durch ein optisches System auf einer Mattscheibe zu der Dämpfungsschleife zusammengesetzt.

Da bei diesem Gerät Prüffaden und Wagen mit den Zuggewichten ein schwingendes System bilden, das insbesondere bei kleinen Vorspannlasten und Frequenzen unter 10 Hz bereits kleine Schwingungen ausführt, ist eine physikalisch exakte Arbeitsweise nicht ohne weiteres erreichbar. Eine Verbesserung gegenüber den nach dem gleichen Prinzip arbeitenden Pulsatoren besteht in der Möglichkeit der Erhöhung der Trägheit des Wagens. Dadurch werden jedoch wieder die Reibungskräfte vergrößert und die Messungen bei kleiner Vorspannlast ungenau. Mit dem Gerät können noch Messungen bis zu einer unteren Grenze von etwa 800 g Zugkraft durchgeführt werden. Dieses Gerät ist von H. ROELIG und J. SCHMAHL[2] weiterentwickelt worden und erlaubt nunmehr auch bei wesentlich geringeren Zugkräften einwandfreie Messungen.

c) Weglose Kraftmessung bei dynamisch arbeitenden Prüfgeräten.

Mit der Heraufsetzung der Grundfrequenz über 0,5 Hz bis etwa 50 Hz bei den dynamischen Prüfgeräten sind dieselben Schwierigkeiten aufgetreten, wie wir sie bei den statischen Prüfgeräten in den sog. dynami-

[1] ROELIG, H.: Kautschuk **15**, 7 u. 32 (1939); Kunststoffe **30**, 164 (1940).
[2] ROELIG, H. u. J. SCHMAHL: Ausstellung des Gerätes auf der Achema XI (1955).

schen Massenwirkungen kennengelernt haben. Hinzu kommt, daß bei der Untersuchung des dynamischen Verhaltens bei Einzelfasern und Fäden besonders im Anfangsstadium bei kleinen periodischen Dehnungen kleine Kräfte auftreten, die mit den geschilderten mechanischen Dynamometern nur noch sehr schwer meßtechnisch erfaßt werden können. Um bei diesen Fasern und Fäden noch einwandfrei Messungen ihres Verhaltens bei dynamischer Beanspruchung zu erreichen, müssen die Dynamometer annähernd gleiche Empfindlichkeit in einem Kräftebereich von etwa 0 bis 10 g oder gar 0 bis 3 g (für feinste Fasern) bis etwa 5000 g (für starke Fäden, Cord usw.) aufweisen, außerdem strenge Linearität und Trägheitslosigkeit im gewünschten Frequenzbereich von 0,5 bis 50 Hz, sowie Nullpunktskonstanz, möglichst große Steifheit und keine Beeinflussung der Meßgenauigkeit bei Temperaturversuchen bis zu 120° C besitzen.

Die vollständige Erfüllung dieser Forderungen ist nur möglich durch die Einführung eines nicht mehr mechanisch arbeitenden, sondern auf elektrischen Prinzipien beruhenden Dynamometers.

Wir müssen nun nach dem Aufbau eines derartigen Dynamometers fragen, das den gestellten Anforderungen angenähert gerecht wird. Es ist naheliegend, wie bei den statischen Prüfgeräten, auch hier eine weglose Kraftmessung zu fordern, die nicht von der Frequenz der Wechselbeanspruchung beeinflußt wird. Von den in § 22c geschilderten Geräten entspricht der elektromagnetische Meßkopf von H. STEIN nicht diesen Anforderungen. Die induktiven Verlagerungsaufnehmer sind nur anwendbar, wenn eine phasentreue Kraftmessung bei der dynamischen Beanspruchung nicht erforderlich ist. Mit anderen Worten, für die Messung der inneren Dämpfung des Fasermaterials fallen auch diese induktiven Verlagerungsaufnehmer aus. Diese werden mit einer Trägerfrequenz gespeist und bei der Demodulation derselben findet eine Phasendrehung statt, die bei Frequenzen > 1 Hz merklich von der Frequenz abhängt und die Aussage ohne weitere Kompensationsglieder verfälscht.

Um auch die innere Dämpfung zu erfassen, bleiben bei dynamischen Messungen bisher nur die trägheitsfrei arbeitenden Dehnungsmeßstreifen übrig[1]. Diese Meßstreifen weisen gegenüber den Verlagerungsaufnehmern bei gleicher Verformung der Federmembran eine geringere Empfindlichkeit auf, erlauben jedoch dynamische Messungen bis zu den höchsten Frequenzen. Die Theorie hat ergeben, daß für derartige Untersuchungen Dehnungsmeßstreifen von besonders hohem Widerstand zweckmäßig sind (Herstellungsgrenze bis zu 1000 Ohm). Neben der völligen Träg-

---

[1] Von W. MESKAT und W. HOFFMANN[2] sind auch Versuche mit den „Statham Transducern Modell G 7", die ebenfalls auf dem Prinzip der Druck- bzw. Zug abhängigen Ohmschen Widerstandsänderung beruhen, unternommen worden. Außerdem ist von W. MESKAT und W. HOFFMANN[2] mit Erfolg der Weg beschritten worden, durch Aufkitten einer Fadenklemme auf die Schnittfläche eines piezoelektrischen Kristalles eine weglose Kraftmessung bei dem entwickelten dynamischen Prüfgerät durchzuführen. Während der Korrektur wurde den Verfassern eine Untersuchung von H. H. RUST und J. KROHN[3] über piezoelektrische Kraftmessung bekannt, die allerdings von ganz anderen Gesichtspunkten ausgehen.

[2] MESKAT, W. u. W. HOFFMANN: nicht veröffentlicht.

[3] RUST, H. H. u. J. KROHN: Z. angew. Physik 7, 61 (1955).

heitslosigkeit, allerdings nur bei Batteriespeisung, und der absoluten Linearität ist auch eine phasentreue Messung der Wechsellast bzw. der periodischen Dehnung möglich. Die Dehnungsmeßstreifen werden zu diesem Zweck auf Federmembranen in Brückenschaltung mit Temperaturausgleich aufgeklebt, wobei zur Erhöhung der Empfindlichkeit mehrere Meßstreifen hintereinander geschaltet werden können. Die Anwendungsgrenzen dieser Dynamometer sind gegeben einerseits durch die Dicke der Federmembran für den gewünschten Kräftebereich und andererseits durch die damit verbundenen Eigenfrequenzen dieser Membranen.

Anwendung der weglosen Kraftmessung bei dynamischen Prüfgeräten[1].

In der Abb. IV, 34 ist dieses Faden- und Faserprüfgerät schematisch dargestellt. Wir sehen, daß ein kleiner Elektromotor *1* ein Exzentersystem *2* über ein Zahnradgetriebe *3* so antreibt, daß mittels der Frequenzknöpfe *4* und *5* dem zu prüfenden Material Frequenzen von beispielsweise 0,5, 2, 5, 10, 25 oder 50 Hz aufgeprägt

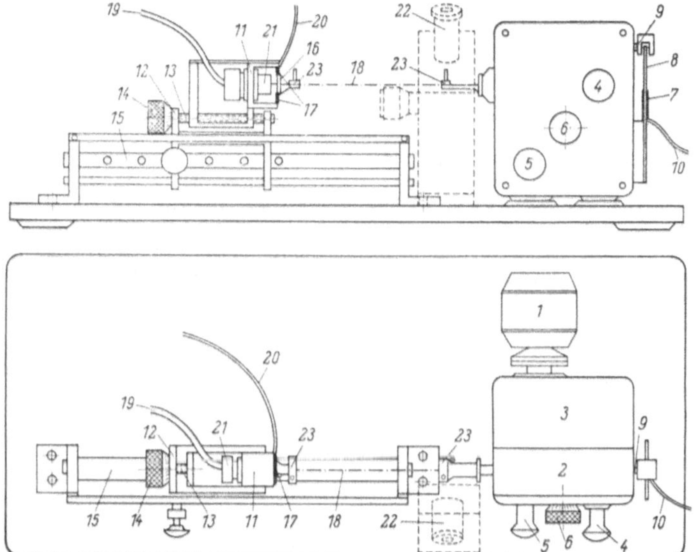

Abb. IV, 34. Schematische Darstellung eines Gerätes zur dynamischen Prüfung der Eigenschaften von Fasern und Fäden. (System W. MESKAT und O. ROSENBERG.)

1 Antriebsmotor, 2 Gehäuse mit Exzentersystem, 3 Gehäuse mit Zahnradgetriebe, 4 u. 5 Knöpfe zur Frequenzeinstellung, 6 Mikrometerschraube zur Amplitudeneinstellung, 7 Dehnungsmeßstreifen zur Amplitudenmessung, 8 Stahlfeder, 9 Kupplungsstift, 10 Kabel zum Verstärker des Phasenwinkelmeßgerätes, 11 Dynamometer, 12 Dynamometerschlitten, 13 Spindel, 14 Mikrometerschraube zum Messen der bleibenden Dehnung des Prüffadens, 15 Schlitten zur Einstellung der Einspannlänge, 16 Federmembran, 17 Dehnungsmeßstreifen zur Messung der periodischen Kraft, 18 zu untersuchende Probe, 19 Kabel zur Meßbrücke, 20 Kabel zum Verstärker des Phasenwinkelmeßgerätes, 21 Verlagerungsaufnehmer zur Messung der Vorspannlast, 22 Ablesefernrohr, 23 Fadenklemmen.

werden können. Die Amplitude, praktisch gleich der periodischen Dehnung, ist im Bereich von 0 bis ± 5 mm während der Bewegung mit Hilfe eines Schiebekeils kontinuierlich verstellbar, wobei die Einstellung mit Hilfe der Mikrometerschraube *6* erfolgt. Dadurch ist auch bei diesem Gerät ein Arbeiten bei konstanter Wechselbelastung durch entsprechende Kompensation der Amplitude möglich.

[1] MESKAT, W. u. O. ROSENBERG: Patentanmeldung „Textilprüfgerät", 11. Dez. 1953.

Die elektrische Messung der Amplitude erfolgt phasentreu und linear im gesamten Amplituden- und Frequenzbereich mittels der Dehnungsmeßstreifen *7*, die auf eine Blattfeder *8* geklebt werden, der durch den Stift *9* die jeweilige Amplitude aufgeprägt wird.

Die Meßstreifen *7* sind in Brückenschaltung mit Temperaturkompensation angeordnet, werden durch eine Batterie gespeist und die erzeugten Spannungsschwankungen werden mittels der Leitung *10* über einen Verstärker auf das eine Plattenpaar eines Kathodenstrahloszillographen gelegt.

Das Dynamometer *11* ist auf einem kleinen Schlitten *12* durch eine Spindel *13* und der Mikrometerschraube *14* verschiebbar angeordnet. Die Einspannlänge des zu prüfenden Fadens *18* kann auf einem großen Schlitten *15* zwischen 0 und 250 mm variiert werden.

Die im Fadenmaterial erzeugten Kräfte, also sowohl Vorspannlast wie überlagerte periodische Wechselbelastung werden von der Federmembran *16* aufgenommen. Um eine phasentreue Registrierung der periodischen sinusförmigen Wechselbelastung oder einer Wechselbelastung von beliebiger Kurvenform im genannten Frequenzbereich zu verwirklichen, wird diese gleichfalls mit Dehnungsmeßstreifen*17*, die auf die Federmembran *16* geklebt sind, elektrisch gemessen. Die erzeugte Wechselspannung kann wieder über einen Verstärker an das zweite Plattenpaar des Kathodenstrahloszillographen gelegt werden. Die Messung der Vorspannlast erfolgt mittels eines induktiven Verlagerungsaufnehmers *21* in Verbindung mit einer direkt anzeigenden Meßbrücke von hoher Nullpunktkonstanz. Die Vorspannlast kann direkt auf der geeichten Meßbrücke abgelesen und während des Versuchs nachgestellt werden. Auch Relaxations- und Kriechversuche hoher Genauigkeit sind mit dieser Kombination möglich.

Durch entsprechende Dimensionierung bzw. durch Auswechseln der Federmembran *16* ist es möglich, im gesamten geforderten Kräftebereich bis zu etwa 10 g herunter zu messen[1]. Die Gesamtdehnung des Materials wird mit der Mikrometerschraube *14* bestimmt. Durch Kenntnis der aufgeprägten periodischen Dehnung (elastische Dehnung) ist auch die bleibende Dehnung sofort zu bestimmen.

Mit Hilfe der rein mechanischen Ausführung ist eine Untersuchung des Materials bei vorgegebener Einspannlänge, Frequenz, periodischer Dehnung und Vorspannlast Null möglich. Tritt bei der dynamischen Beanspruchung des Materials Erschlaffung des Fadens ein, so ist diese im Ablesefernrohr bei einer eingetretenen bleibenden Dehnung von etwa 1 bis 2‰ (das entspricht stets $< 1\%$ der Zugfestigkeit) einwandfrei festzustellen.

Die phasentreuen Dämpfungsschleifen auf dem Schirm des Oszillographen werden photographiert, so daß auch das Anfangsstadium bei kleinen Frequenzen erfaßt werden kann.

Wesentlich zweckmäßiger und genauer ist es jedoch, an Stelle dieser Methode, die mit einer umständlichen und zeitraubenden Auswertung verbunden ist, die spezifische Dämpfung bzw. den mechanischen Verlustwinkel direkt mit Hilfe eines weiter unten beschriebenen Dämpfungsmeßgerätes zu bestimmen. Voraussetzung für diese Methode ist eine sinusförmige Form der im Faden erzeugten periodischen Wechselbelastung und Frequenzen $\geqq 2$ Hz.

Der Zusammenhang zwischen spezifischer Dämpfung und mechanischem Verlustwinkel bei sinusförmigem Verlauf von periodischer Kraft und periodischer Dehnung ist durch die Beziehung

$$D(\%) = \sin \varphi$$

gegeben.

---

[1] Um diese Grenze noch weiter nach geringeren Kräften zu verschieben, geht die Entwicklung dahin, die in dem induktiven Verlagerungsaufnehmer vorhandene Membran durch die Membran *16* zu ersetzen und diese unmittelbar mit dem Ferroxcub-Kern des induktiven Aufnehmers zu verbinden.

Unter der spezifischen Dämpfung D (%) verstehen wir das Verhältnis von irreversibler Formänderungsarbeit zu der in das Material hineingesteckten reversiblen elastischen Formänderungsarbeit je Belastungscyclus. Der mechanische Verlustwinkel φ ist der Phasenwinkel zwischen periodischer Kraft und periodischer Dehnung.

Die Bestimmung dieses Phasenwinkels aus den beiden Wechselspannungen ergibt sich aus der bekannten Vektorbeziehung (siehe Abb. IV, 35).

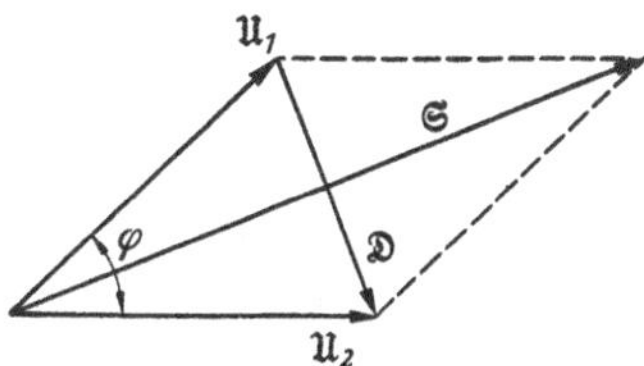

Abb. IV, 35. Vektorielle Darstellung von Summen- und Differenzspannung bei amplitudengleichen Vergleichsspannungen.

$$|\mathfrak{U}_1| = |\mathfrak{U}_2| = |\mathfrak{U}|$$
$$\mathfrak{U}_2 - \mathfrak{U}_1 = \mathfrak{D}; \quad \mathfrak{U}_1 + \mathfrak{U}_2 = \mathfrak{S}$$
$$|\mathfrak{D}| = 2 \cdot |\mathfrak{U}| \cdot \sin\varphi/2$$
$$|\mathfrak{S}| = 2 \cdot |\mathfrak{U}| \cdot \cos\varphi/2.$$

Bei amplitudengleichen Vergleichsspannungen folgt für die Differenzspannung die Beziehung

$$|\mathfrak{D}| = 2 \cdot |\mathfrak{U}| \cdot \sin\varphi/2.$$

Die Differenzspannung ist dann ein direktes Maß für den Phasenwinkel und ist unabhängig von der Frequenz.

Das Prinzip des Dämpfungsmeßgerätes besteht nun nach W. Denecke[1] darin, daß von den beiden zu vergleichenden Wechselspannungen zwei gleich große Vergleichsspannungen abgeleitet werden, deren Differenz einem Röhrenvoltmeter zugeführt wird. Die Anordnung ist in Abb. IV, 36 schematisch dargestellt.

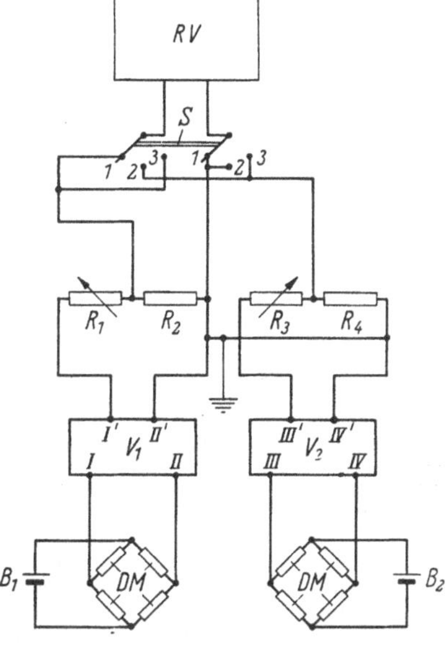

Abb. IV, 36. Schematischer Aufbau des Phasenwinkelmeßgerätes zum dynamischen Fadenprüfgerät.

B_1, B_2	Speisebatterien f. Dehnungsmeßstreifen,
DM	Dehnungsmeßstreifen,
V_1, V_2	Verstärker (etwa 20fach),
I–II, I'–II'	Eingangs- und Ausgangsspannung proportional der periodischen Dehnung,
III–IV, III'–IV'	Eingangs- und Ausgangsspannung proportional der periodischen Kraft,
R_1, R_3	Regelwiderstände,
R_2, R_4	Widerstände zur Abnahme der amplitudengleichen Vergleichsspannungen,
S	Schalter.

Schaltstellung 1: Vergleichsspannung $\mathfrak{U}_1$ (period. Dehnung)

Schaltstellung 2: Vergleichsspannung $\mathfrak{U}_2$ (period. Kraft)

Schaltstellung 3: Differenzspannung:

$$|\mathfrak{D}| = 2 \cdot |\mathfrak{U}| \cdot \sin\varphi/2.$$

RV Röhrenvoltmeter als Anzeigegerät, z. B. Philips GM 6017.

Wie bereits erwähnt, werden periodische Kraft und periodische Dehnung mit Dehnungsmeßstreifen in Brückenschaltung und Temperaturkompensation gemessen. Die Speisung erfolgt durch Batterien. Die erzeugten Primärspannungen aus den Dehnungsmeßstreifen DM werden durch zwei Verstärker V_1 und V_2 etwa 10- bis 20fach verstärkt; die Verstärkung erfolgt durch zwei gegengekuppelte zweistufige Verstärker. Die verstärkten Spannungen werden durch die beiden

[1] Denecke, W.: VDE-Fachberichte, 15. Bd. (1951).

Regelwiderstände R_1 und R_3 amplitudengleich abgestimmt. Die Differenzspannung wird durch Umschalten des Schalters S direkt am Röhrenvoltmeter RV sichtbar. Dieses besitzt eine lineare Skala, so daß der Verlustwinkel bzw. die spezifische Dämpfung sofort abgelesen werden kann[1].

Wie bereits in § 22b erwähnt, ist die Gleichmäßigkeit eine Materialeigenschaft bzw. die zugehörige Streuung eine Kenngröße von besonderer Bedeutung zur Beurteilung des Materials. Es wird daher genau so wie bei den statischen Prüfmethoden am laufenden Faden auch für die dynamische Prüfung unter denselben Bedingungen ein besonderes Gerät vorgeschlagen, das in der Abb. IV, 37 schematisch dargestellt

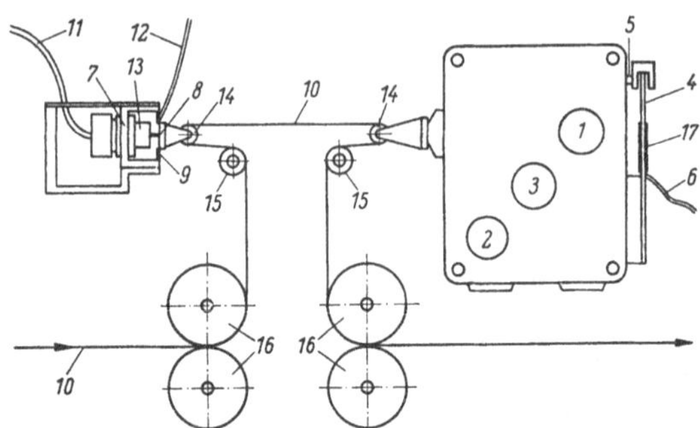

Abb. IV, 37. Schematische Darstellung eines Gerätes zur dynamischen Prüfung der Eigenschaften eines laufenden Fadens. (Nach W. MESKAT und O. ROSENBERG.)

1 u. *2* Knöpfe zur Frequenzeinstellung, *3* Mikrometerschraube zur Amplitudenmessung, *4* Stahlfeder, *5* Kupplungsstift, *6* Kabel zum Verstärker des Phasenwinkelmeßgerätes, *7* Dynamometer, *8* Federmembran, *9* Dehnungsmeßstreifen zur Messung der periodischen Kraft, *10* zu untersuchende Probe, *11* Kabel zur Meßbrücke, *12* Kabel zum Verstärker des Phasenwinkelmeßgerätes, *13* Verlagerungsaufnehmer zur Messung der Vorspannlast, *14* Rollenklemmen, *15* Umlenkrollen, *16* Rollenpaare zur Vordehnung der laufenden Fadenprobe, *17* Dehnungsmeßstreifen zur Amplitudenmessung.

ist. Während es mit den beschriebenen Geräten bisher nur möglich ist, entweder eine konstante Vordehnung oder konstante Belastung vorzugeben, wird bei dieser Apparatur einem konstant vorgedehnten Faden eine periodische Zugbeanspruchung überlagert. Die Anordnung ist im wesentlichen dieselbe wie beim dynamischen Faden- und Faserprüfgerät nach MESKAT und ROSENBERG. Die Fadenklemmen sind jedoch durch Rollenklemmen *14* ersetzt. Die Aufprägung der Vordehnung erfolgt wie vorhin durch Wahl des Verhältnisses der Drehzahlen der beiden Rollenpaare *16*. Die Fadengeschwindigkeit kann durch die Drehzahl der beiden Rollenpaare so eingestellt werden, daß eine vorgegebene Anzahl von Belastungswechseln bei der gewählten Frequenz der periodischen Beanspruchung während des Durchlaufens der Prüfstrecke dem laufenden Faden aufgeprägt werden kann. Die periodische Dehnung wird wie beim dynamischen Fadenprüfgerät durch ein verstellbares Exzentersystem erzeugt, die periodische Kraft wird mit dem elektrischen Dynamometer gemessen oder registriert. Mit einem kontinuierlich anzeigenden Phasenwinkelmeßgerät kann erstmalig die Dämpfung des Garns laufend gemessen und damit Veränderungen in dem dauerelastischen und im Ermüdungsverhalten verfolgt werden.

Dieses Gerät[2] verwirklicht die praktischen Gegebenheiten am Webstuhl noch mehr als die früher erwähnten Apparate, da bei den neuesten Schnellwebstühlen bereits periodische Dehnungen von 2 bis 2,5% bei Frequenzen bis zu 7 Hz auftreten und das Material als Faden einige tausend Belastungswechsel bis zum Gewebe auszuhalten hat.

---

[1] Einen anderen Weg schlägt K. HOMILIUS in ATM. V 3631–8 Lieferung 227, Okt. 1955 vor.

[2] Ein Normblatt für die Durchführung der Versuche mit derartigen dynamischen Prüfgeräten und für die Auswertung der Meßergebnisse existiert noch nicht. Es wurde bisher auf DIN 50100 und DIN 53513 zurückgegriffen und dabei die übertragbaren Begriffe und Angaben verwendet.

d) Dynamische Prüfgeräte mit einachsiger stetiger oder wechselnder Zugbeanspruchung unter gleichzeitiger Biegebeanspruchung.

Eine bereits seit langer Zeit durchgeführte dynamische Biegeprüfung von Fasern und Fäden ist in der Abb. IV,38 dargestellt[1]. Untersucht werden soll die Dauerbiegefestigkeit, man spricht hier auch von Knickbruchfestigkeit von Fasern, Garnen oder Geweben bei einer gleichzeitig einwirkenden Zugvorspannung unter einem vorgegebenen Biege- bzw. Knickwinkel bis zum Bruch. Wesentlich ist dabei, daß der Faserklemmpunkt genau mit der Drehachse der Biegeklemme übereinstimmt. Geringe Unterschiede im Klemmendruck und in der Klemmenbeschaffenheit, z. B. der Kantenabrundung, sind bereits von großem Einfluß auf die Streubreite der Meßwerte. Außerdem liegt es im Wesen dieser Methode, daß der Biegebeanspruchung des Prüffadens stets eine Scheuerwirkung an der Klemme, eine Art Kantenscheuerung, überlagert ist. Hinzu kommt, daß die Forderung der gleich großen Oberflächendehnung unabhängig vom Titer oder, mit anderen Worten, daß das Verhältnis von Biegeradius und Fadendurchmesser konstant sein muß, nicht erfüllt ist. Wie man sieht, ist die Gesamtbeanspruchung bei diesem Prüfgerät physikalisch undefiniert.

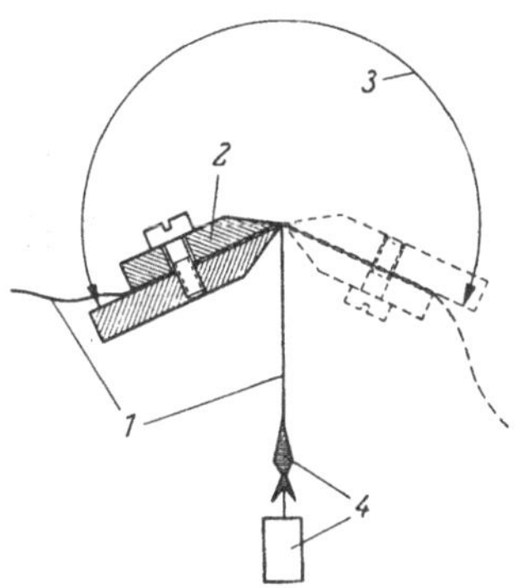

Abb. IV,38. Schematische Darstellung eines Dauerbiegeprüfers. (Nach E. FRANZ und H. J. HENNING.) *1* Faser, *2* Einspannklemme, *3* Ausschlagwinkel, *4* Vorspanngewicht mit Klemme.

Dieselben Schwierigkeiten ergeben sich bei der Untersuchung der Falzfestigkeit von Papier; nicht nur, daß der SCHOPPER-Falzer besonders hohe Streuungen der Einzelwerte zeigte (vgl. § 21c 3), sondern, wie W. BRECHT und L. KÖRNER[2] feststellten, beeinflußte die Zugfestigkeit des Papiers die Meßergebnisse. In dem Bestreben, die offensichtlichen Mängel des SCHOPPER-Falzers zu vermeiden, führte man Dauerbiegeprüfer, wie z. B. das KÖHLER-MOLIN-Gerät[3] ein. Man beachtete dabei jedoch nicht, daß bei allen diesen Geräten eine Zugwirkung auftritt, die bei der Erzeugung des Falzes gar nicht vorhanden ist; vielmehr wird der Falz durch Druck erzeugt. Aus dieser Erkenntnis entwickelten W. BRECHT und A. WESP[4] den sog. Druckfalzer. In der Abb. IV,39 sind für zahlreiche Papiere die Werte der Falzfestigkeit mit diesem Druckfalzer im Vergleich zu den SCHOPPER-Falzzahlen aufgetragen. Wir sehen, daß überhaupt kein Zusammenhang besteht. Zahlreiche Papiere lassen nach dem SCHOPPER-Falzer überhaupt keine Falzfestigkeit erkennen im Gegensatz

[1] Aus E. WAGNER: Mech.-technologische Textilprüfungen, Wuppertal 1953, S. 93. An den Methoden von J. C. GUTHRIE, D. H. MORTON u. P. H. OLIVER: J. Textile Inst. 45 T, 912 (1954) beschrieben.

[2] BRECHT, W. und L. KÖRNER, Das Papier **6**, 161 (1952).

[3] KÖHLER-MOLIN: Svensk Papperstidn. **54**, 710 (1951).

[4] BRECHT, W. u. A. WESP: Papier **6**, 443, 496 (1952).

zu den Ergebnissen der Praxis. Hier sieht man unmittelbar, daß bereits eine verhältnismäßig geringe Abweichung der Beanspruchung des zu prüfenden Materials im Prüfgerät von der Beanspruchung desselben Materials bei der Weiterverarbeitung oder im Gebrauch zu gegensätzlichen Ergebnissen führen kann. Dieses Ergebnis wird noch durch die Kurven der Abb. IV,40 unterstrichen, die noch einmal zeigt, wie wichtig die physikalische Analyse der auftretenden Beanspruchungsarten, unter denen das zu prüfende Material steht, ist und welchen Einfluß die relative Luftfeuchtigkeit auf die Meßergebnisse besitzt.

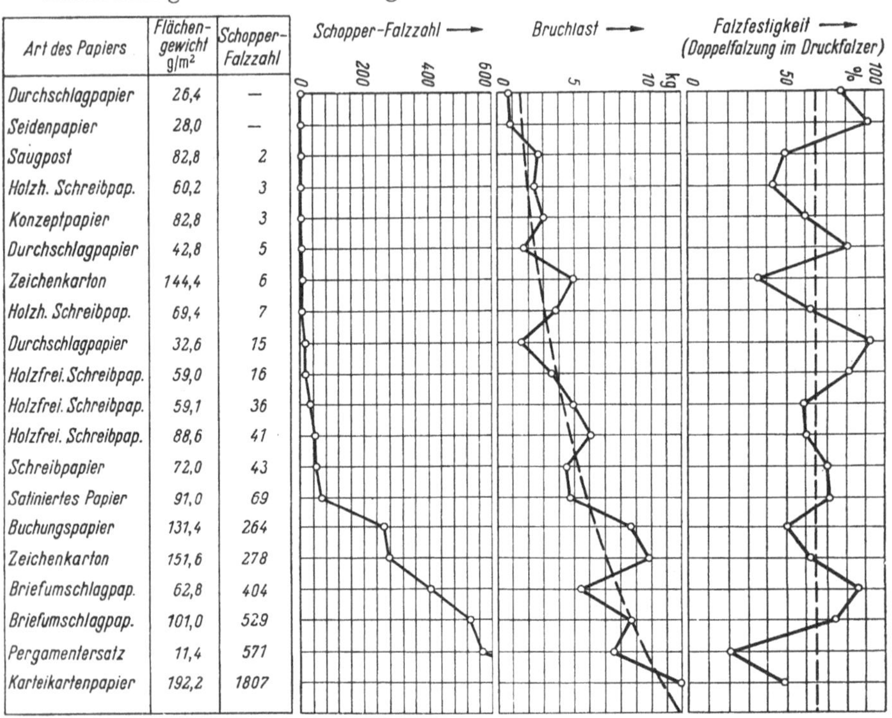

Abb. IV,39. Die Falzfestigkeit im Druckfalzer im Zusammenhang mit Bruchlast und SCHOPPER-Falzzahl für 20 verschiedene Papiere. (Nach W. BRECHT und A. WESP.)

Eine wechselnde Zugbeanspruchung unter gleichzeitiger Biegebeanspruchung kann unter Verringerung der im vorhergehenden Abschnitt bezeichneten undefinierten Beanspruchungsart mit dem Gerät von W. MESKAT und O. ROSENBERG durchgeführt werden. Es wird dabei die Schlingenfestigkeitsprüfung, wie in DIN 0053841 beschrieben wird, durch eine dynamische Methode ersetzt. Die Durchführung des Versuchs, insbesondere die Bildung einer Schlinge in Form zweier einfacher Kettenglieder kann dem genannten DIN-Normentwurf entnommen werden. Da nach neueren Untersuchungen zwischen der relativen Schlingenfestigkeit und der relativen Knotenfestigkeit Zusammenhänge bestehen, haben wir es hier mit zwei nicht voneinander unabhängigen Einflußgrößen zu tun, so daß die Messung dieser einen Größe zur Beschreibung des Verhaltens ausreicht.

§ 24. Prüfgeräte zur Messung zweiachsiger Beanspruchungen.

Die Grundtendenz der bisherigen Ausführung war, die Aufnahme der Spannungs-Dehnungs-Funktion bei einachsiger stetiger Zug- oder Zugwechselbeanspruchung und die Erweiterung dieser Betrachtungen durch zusätzliche Aufprägung einer Biegewechselbeanspruchung. Nun ist zu berücksichtigen, daß nach der Umwandlung der Fasern und Fäden zu einem Gewebe die Art der Beanspruchung der Gewebe und Gewirke nicht nur einachsig, sondern zweiachsig erfolgt. Nur auf diesem Wege kann unmittelbar der Einfluß der Bindung und anderer Größen, welche die Gewebestruktur bestimmen, in ihrem Einfluß auf das Deformationsverhalten geprüft werden[1].

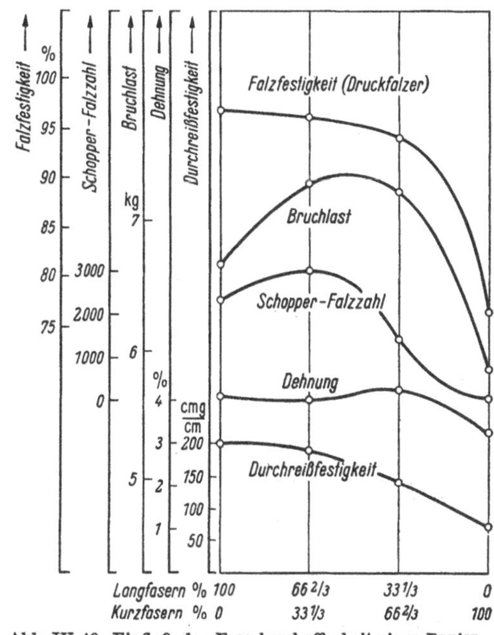

Abb. IV, 40. Einfluß der Faserbeschaffenheit eines Papierstoffes auf die Falzfestigkeit. (Nach W. Brecht u. A. Wesp.)

a) Die Wölb- und Berstwiderstandsprüfung.

In der Abb. IV, 41 ist der bekannte Berstdruckprüfer nach Schopper schematisch dargestellt. Die wichtigsten Anforderungen[2], die an ein Berstdruckprüfgerät gestellt werden müssen, sind folgende:

1. Arbeitet das Prüfgerät mit Druckluft, so muß es mit einem Haupt- und einem Regelventil ausgestattet sein, das die Strömungsgeschwindigkeit der Luft feinfühlig einzustellen gestattet.

2. Das Prüfgerät hat eine auswechselbare kreisförmige Prüffläche, zu der eine ebenfalls austauschbare Einspannvorrichtung gehört. Mit Hilfe einer Anpreßglocke wird die zu untersuchende Probe fest gegen einen Aufspannring, der im Unterteil eingelassen ist, gepreßt und die Prüffläche begrenzt. Die Anpreßglocke kann aus Plexiglas angefertigt sein, um die Vorgänge beim Wölb- bzw. Berstversuch sichtbar zu machen.

3. Das Gerät soll mit mehreren Druckmessern verschiedenen Meßbereiches, die wahlweise angeschlossen werden können, ausgerüstet sein. Der zur Messung jeweils verwendete Druckmesser ist so zu wählen, daß der mittlere Wölb- bzw. Berstdruck über dem ersten Fünftel des Meßbereichs liegt. Er muß mit einem Schleppzeiger ausgerüstet sein.

4. Das Prüfgerät muß mit einem Taster zum Messen der Wölbhöhe versehen sein, der die eingespannte Probe in der Mitte leicht berührt. Wölbt sich die Probe

[1] Abgesehen von der Möglichkeit, ein Gewebe wieder zu zerlegen und nach den bisher üblichen Methoden zu prüfen.

[2] Vgl. DIN-Norm 53860 – Entwurf Januar 1955.

unter der Wirkung des Überdruckes auf, so wird der Taster angehoben. Diese Bewegung wird auf ein Meßwerk übertragen und durch einen Zeiger auf einer Kreisskala angezeigt.

5. Als Membran soll beim Wölb- bzw. Berstversuch eine etwa 1 mm dicke füllstofffreie Naturgummischeibe verwendet werden. Membranen aus Buna sind hierfür nicht geeignet. Bei Proben mit sehr großer Dehnung (z.B. Strick- und Wirkwaren) hat sich die Verwendung von Kunststoff-Folien (z.B. Igelit-Folien) bewährt, es ist jedoch für jeden Versuch eine neue Membran notwendig.

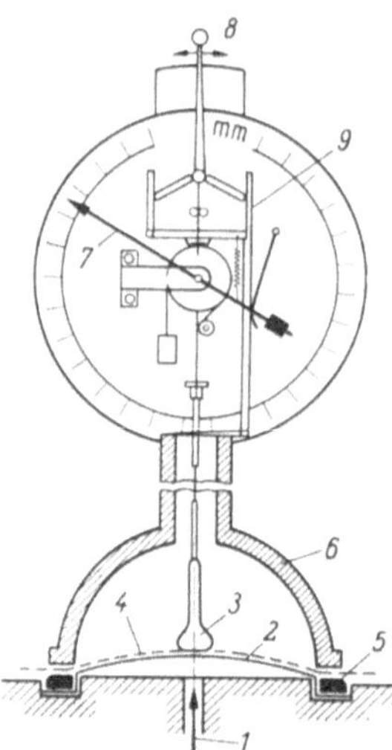

Abb. IV, 41. Einspannvorrichtung und Wölbhöhenmesser eines Berstdruckprüfers. (System L. SCHOPPER nach W. OESER.)

1 Preßlufteintritt, *2* Probe, *3* Taster, *4* Gummimembran, *5* Preßring, *6* Spannglocke, *7* Zeiger, *8* Bremsenauslösen, *9* Wölbhöhenmesser.

6. Stellt sich nach längerem Gebrauch an den Membranen eine bleibende Verformung ein, so ist diese durch eine neue zu ersetzen. Zur Ermittlung des Membranfehlers wird jede Membran für sich allein aufgewölbt und die Beziehung zwischen Druck und Wölbhöhe festgestellt (Membranfehler). Aus den Eichkurven wird für die beim Berstversuch ermittelte Wölbhöhe die entsprechende Berichtigung vorgenommen.

Gemessen werden der Berstdruck p und die Wölbhöhe h. Bei Berücksichtigung des Radius der Prüffläche r ergibt sich für die Berstfestigkeit K folgende Formel (nach SOMMER):

$$K = p \cdot \frac{r^2 + h^2}{4h} \, [\mathrm{kg/cm}] .$$

Voraussetzung für diese Ableitung ist, daß sich das Gewebe bei gleichen Festigkeits- und Dehnungseigenschaften in allen Richtungen als Kugelkappe aufwölbt.

Messungen der Wölbhöhen an verschiedenen Stellen der Berstprobe zeigten jedoch, daß die Aufwölbung nicht genau einer Kugelkalotte entspricht. Es ergibt sich die Frage, welchen Einfluß diese Abweichung auf die Spannungen und Dehnungen in der Probe haben.

Da die Berstdruckprüfung im allgemeinen bei Geweben und Folien angewendet wird, die als wenig biegungssteif anzusehen sind, wurde nach F. WINKLER[1] für die Berechnung der Berstfestigkeit die Theorie dünner Platten mit großer Ausbiegung verwendet. Die Untersuchung ergibt, daß die SOMMERsche Formel bis zu Wölbhöhen $h \approx 0{,}4\,r$ gut brauchbare Werte ergibt. Für größere Wölbhöhen, die bei Gewebe selten, bei Folien aber sehr häufig vorkommen, treten größere Abweichungen auf. Die Abhängigkeit des Berstdruckes von der Einspannfläche bei Papieren wurde besonders von W. VOLLMER[2] untersucht.

Die Einzelheiten bezüglich der geometrischen Größen sowie der Prüfdurchmesser, Wölbhöhe, Randwinkel, Wölbdehnung und Wölbdruck sind dem DIN-Entwurf 53860[3] zu entnehmen, bei dem auch die Versuchsausführung im einzelnen geschildert sowie die Auswertungen genau an-

[1] WINKLER, F.: Faserforsch. u. Textiltechn. 3, 449 (1952).
[2] VOLLMER, W.: Papier 8, 371 (1954).
[3] DIN 53860 – Entwurf Januar 1955.

gegeben worden sind, insbesondere auch die Genauigkeit, die bei der Messung bestimmter Einflußgrößen einzuhalten ist. In der Anlage zu diesem Normblatt sind auch die Umrechnungstabellen für die Berstfestigkeit sowie für die Wölbhöhe bzw. Berstdehnung enthalten. Die Eichfragen dieses Gerätetyps werden von O. BRAUNS, E. DANIELSSON und L. JORDANSSON[1] diskutiert und die Meßunsicherheit bei Berstversuchen mit Papier behandelt ausführlich H. L'HOMME[2].

b) Scheuerprüfgeräte.

Sowohl die Untersuchung des Deformationsverhaltens bei einachsiger Beanspruchung wie auch bei zweiachsiger Beanspruchung sowohl statischer wie dynamischer Art ermöglicht noch nicht die vollständige Beschreibung des Gebrauchswertes. Auch hier kann man den Ausführungen des § 21 b entnehmen, daß die statistische Theorie der Festigkeit, wie sie von W. WEIBULL und A. M. FREUDENTHAL entwickelt wurde, dazu führt, daß man zwischen der Schwachstellenverteilung an der Oberfläche und jener über das Volumen sehr wohl unterscheiden muß. Es sind daher die bisher geschilderten Methoden der Beanspruchung, die speziell auf die Schwachstellenverteilung über das Volumen ansprechen, nicht geeignet, um Aussagen über die Schwachstellenverteilung an der Oberfläche mit ausreichender Sicherheit zu machen. Da nun in der Praxis Fasern und Gewebe auf Scheuerung beansprucht werden und damit diese Oberflächenschwachstellenverteilungen besonders in Erscheinung treten, ist die Scheuerprüfung, die speziell auf diese Oberflächenschwachstellen anspricht, notwendig, um eine Vervollkommnung der Aussagen über den Gebrauchswert zu erreichen.

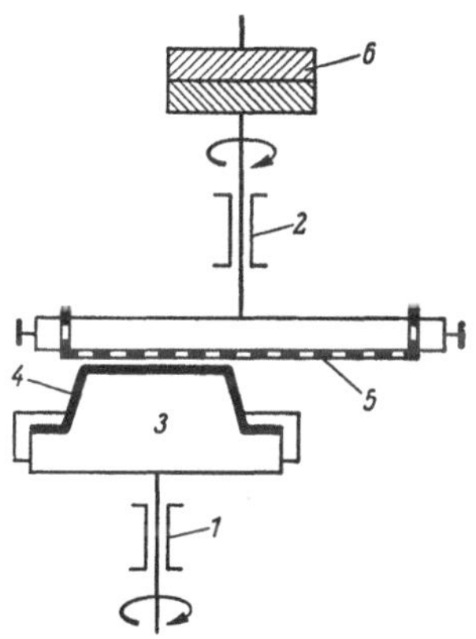

Abb. IV, 42. Rundscheuerprüfer nach H. F. SCHIEFER.
1 u. *2* Antriebe für *3* u. *5*, *3* Scheuerkopf, *4* Stoffprobe, *5* Reibkörper, *6* Belastungsgewicht.

Es werden Gewebe-, Garn- und Faserscheuerprüfung unterschieden. Nach SCHIEFER sind in der Scheuerprüfung drei grundlegende Probleme zu beachten[3]:

1. das theoretische Problem der idealen Scheuerbewegung,

2. das Problem des unveränderlichen Reib- oder Scheuermittels,

3. das Problem der Wertung der Scheuerung.

Das erste Problem wurde von H. F. SCHIEFER in der Weise gelöst (siehe Abb. IV, 42), daß sich der Prüfling und das Reibmittel mit gleicher Geschwindigkeit in der gleichen Drehrichtung, aber mit ihren Mittelpunkten in einer beliebigen Entfernung voneinander drehen. Da-

[1] BRAUNS, O., E. DANIELSSON u. L. JORDANSSON: Svensk Papperstidn. **57**, 867 (1954). — [2] L'HOMME, H.: Papeterie **75**, 731 (1953).

[3] Vgl. auch DIN-Neuentwurf 53 863, Grundlagen für die Scheuerprüfung von textilen Flächengebilden — Januar 1955.

bei muß das Reibmittel jederzeit die gesamte Fläche des Prüflings berühren.

Auch nach dem Prinzip des Flachscheuerprüfers von HAUSER-FRANK (siehe Abb. IV,43) durch zwei aufeinander senkrecht stehende Kreuzkurbelantriebe für Prüfling und Reibmittel ist es möglich, eine sehr gleichmäßige Scheuerung zu erzielen. Die Regelung der Stoffspannung erfolgt in der Regel durch Gewichte, neuerdings auch pneumatisch (siehe Abb. IV,44), unabhängig von der fortgeschrittenen Scheuerung der Probe.

Das zweite Problem ist bis heute noch nicht befriedigend gelöst. Zum Teil wird Schmirgelpapier verwendet, das in bestimmten Zeitabständen stets vom Faserstaub gereinigt wird. SCHIEFER verwendet eine Anzahl paralleler, gerader Federstahlstreifen. Hier fliegt der Faser- und Schlichtestaub wohl zwischen den Streifen heraus und verklebt das Reibmittel nicht, wie das bei Schmirgelpapier der Fall ist. Versuche zeigten, daß die Scheuerwirkung einer solchen Reibscheibe sich nur äußerst wenig ändert.

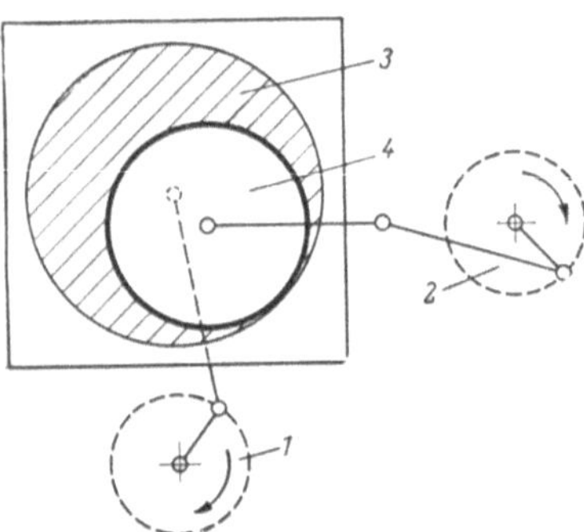

Abb. IV,43. Flachscheuerprüfer nach A. HAUSER-FRANK.
1 u. *2* Kreuzkurbelantrieb für *3* Scheuerplatte mit Reibkörper, *4* Probenkopf.

Beim Scheuern von Wolle wurde hingegen festgestellt, daß sich die Reibscheibe mit einer Materialschicht bedeckt, die die Scheuerwirkung sehr verstärkt. Schichtenbildung bei Nylon hingegen verringerte die Scheuerwirkung. Ein einheitliches Reibmittel für sämtliche Stoffe gibt es heute noch nicht.

Es sind Bestrebungen im Gang, das Reibmittel physikalisch zu definieren, z.B. mit dem SCHMALZschen Lichtschnittmikroskop, welches die Korngröße und Verteilung des verwendeten Schmirgelpapiers zu erfassen gestattet. Dagegen sind die Änderungen der physikalischen Eigenschaften durch Ablagerungen während des Scheuervorganges auch nach diesen Methoden nur sehr angenähert zu erfassen.

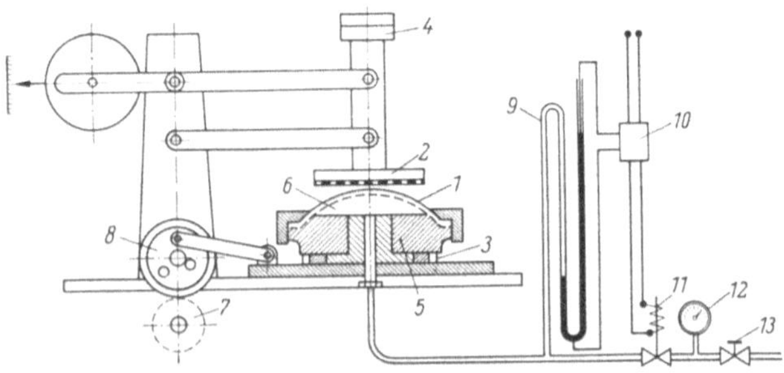

Abb. IV,44. Pneumatische Regelung der Stoffspannung bei neuzeitlichen Rund- und Flachscheuergeräten.
1 Stoffprobe, *2* Reibkörper, *3* Zahnrad, *4* Belastungsgewicht, *5* Probensockel, *6* Gummimembran, *7* Antriebsmotor, *8* Exzenter, *9* Kontaktmanometer, *10* Relais, *11* Steuerventil, *12* Manometer, *13* Reduzierventil.

Die Bewertung der Scheuerung erfolgt auf verschiedene Weise. Vorzuziehen sind Methoden, bei welchen der Prüfling nicht zerstört wird, so daß die Reibwirkung an demselben Prüfling wiederholt nach fortlaufend gesteigerter Scheuerung festgestellt werden kann. Während man bei der Prüfung von Teppichen z. B. mit einer Dickenmessung sehr gut auskommt, erlangte die Messung der elektrischen Kapazität vor allem bei Kleiderstoffen eine stets größere Bedeutung.

Für die Scheuerprüfung von Einzelfasern haben A. ZART[1] und D. WEIGEL ein Gerät entwickelt, bei dem eine ganze Anzahl von Fasern, die unter Vorspannung stehen, über eine mit Perlondraht bespannte und sich drehende Walze, die in eine mit Wasser und Netzmittelzusatz gefüllte Mulde eintaucht, bis zum Bruch gescheuert werden.

Während bei den eben genannten Prüfgeräten die Scheuerwirkung auf die Gewebe- bzw. Faserprobe ausschließlich beschränkt bleibt, gibt es im Gegensatz dazu eine Anzahl von Scheuerprüfgeräten, die kombinierte Beanspruchungen auf den Prüfling ausüben, so z. B. der Garnscheuerprüfer nach M. MATTHES, der den Faden gleichzeitig auf Zug, Biegung bzw. Knickung und Reibung bis zum Durchscheuern beansprucht. Ausgesprochen undefinierte Beanspruchungen erfolgen auch in dem in Abb. IV,45 dargestellten Scheuerprüfgerät für Faltenscheuerungen, sowie in dem in der Abb. IV,46 gezeigten Kantenscheuerprüfer.

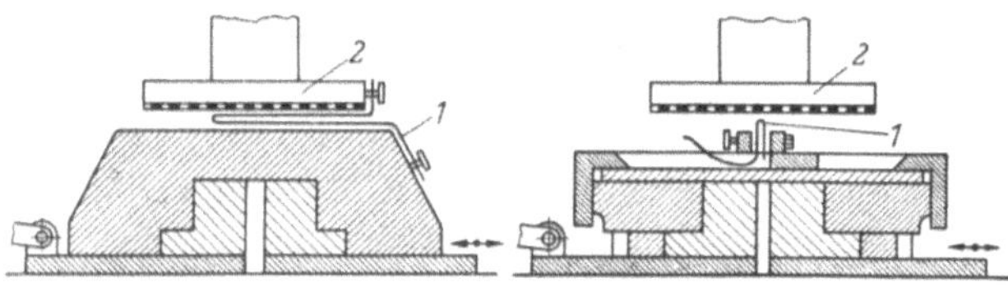

Abb. IV,45. Faltenscheuerprüfer. Abb. IV,46. Kantenscheuerprüfer.
1 Stoffprobe, *2* Reibkörper.

Eine sehr umfassende Arbeit, die den heutigen Stand der Prüfung der Scheuerfestigkeit von Textilien weitgehend darstellt, ist von H. SULSER[2] durchgeführt worden, wobei auch nahezu vollzählig 34 Prüfgerätetypen in der Art der Scheuerbeanspruchung, der Scheuerbewegung, der Scheuerelemente und des Prüfmaterials, dem Scheuerdruck usw., sowie der Messung des Verschleißes tabellarisch, kritisch zusammengefaßt, angegeben werden[3]. Die Arbeit führt u. a. zu dem Ergebnis, daß man aus dem Festigkeits-Scheuertourenzahl-Diagramm eine Scheuergütezahl ableiten kann.

Ausblick.

Mit der Bezeichnung „Scheuergütezahl“ ist bereits von H. SULSER zum Ausdruck gebracht worden, daß diese Größe nicht unmittelbar mit dem Gebrauchswert eines Faserstoffes in Beziehung steht. Es können daher auch Scheuerprüfgeräte, gleich welcher kombinierten Bauart, nicht als Gebrauchswertprüfapparate bezeichnet werden, wie dies noch

[1] Aus E. WAGNER: Mech.-technologische Textilprüfungen, Wuppertal 1953, S. 204. — [2] SULSER, H.: Diss. E. T. H. Zürich 1953.

[3] Eine Ergänzung der bisher bekannten Scheuerprüfer stellt der Appearence-Retention-Tester von der Fabric Development Tests, Brooklyn, New York, dar, der nicht nur die Scheuerung, sondern auch die Oberflächenabnutzung und das Pilling prüft.

durch H. BÖHRINGER[1] geschieht. Denn dieser Weg führt zwangsläufig zu der Entwicklung von Prüfgeräten, die sich immer weiter von der Übertragung eindeutiger physikalischer Beanspruchungen auf das Prüfstück entfernen und immer mehr einen speziellen praktisch komplexen Beanspruchungsvorgang nachahmen. Ein solches Gerät ist z. B. der „Spinntaxer" System LITTY zur Feststellung des technischen Spinnwertes eines Fasermaterials, der aus einer kleinen Präzisionskrempel (Taxkarde) und einer einspindeligen Perfektspinnmaschine mit Spezialstreckwerk besteht, und das Splittrigkeitsprüfgerät von H. BÖHRINGER[2], das die Vorgänge in einer Trommelwaschmaschine annähernd wiedergibt. Dieser erwähnte Splittrigkeitsprüfer erlaubt nur dann eine Aussage, wenn, wie in § 21d eingehend diskutiert, der spezifische Gebrauchswert nur von der Waschbeständigkeit des Faserstoffes bestimmt wird und alle anderen Einflußgrößen unterhalb des Störpegels liegen. Nur in diesem speziellen Fall wird die Waschbeständigkeitsgröße gleich dem spezifischen Gebrauchswert. Es darf nicht übersehen werden, daß dies Ausnahmefälle sind und daß im allgemeinen die Schadensendwerte bei den Faserstoffen sehr wohl durch mehrere definierte physikalische und chemische Einflußgrößen mit je nach dem Verwendungszweck verschiedenen statistischen Gewichten darstellbar sind, die mit Hilfe der Rangkorrelation bestimmt werden. Bis dieses Ziel wirklich erreicht ist, sollte man die laufend zu ergänzenden Gebrauchswertprüfungen, die nichts mit den vorstehend erwähnten Gebrauchswertprüfgeräten zu tun haben, z. B. Zellwolle einschließlich Verarbeitung der Fachgruppe chemische Herstellung von Fasern, anwenden, wobei ebenfalls die Rangkorrelation bzw. die von K. STANGE erläuterten statistischen Methoden[3] von Nutzen sein dürften[4].

[1] BÖHRINGER, H.: Faserforsch. u. Textiltechn. **5**, 1 (1954).

[2] BÖHRINGER, H.: Faserforsch. u. Textiltechn. **5**, 6 (1954).

[3] STANGE, K.: Mitt. f. math. Statistik **7**, 113 (1955).

[4] Prüfverfahren der Fachgruppe chemische Herstellung von Fasern, Z. Textil 1, Blatt 1–12, Neue Fassung, Januar 1950.

Fifth Chapter

The Structure and Mechanical Properties of Rubberlike Materials.

By

L. R. G. TRELOAR.

With 59 figures.

A. The thermodynamics of rubber elasticity.

§ 25. Elementary thermodynamic analysis.

a) The nature of rubber elasticity.

A proper understanding of the nature of rubber elasticity was hardly possible before evidence had become available concerning the macromolecular structure of rubber and other naturally occurring high polymers, e.g. wool, cotton, silk, etc. Our present conceptions are based on the view, originally put forward by MEYER, VON SUSICH and VALKO[1] and subsequently developed by KUHN[2,3] and GUTH and MARK[4] that the long-range extensibility arises from the thermal fluctuations of form of the long-chain molecules of which the material is composed. These fluctuations were assumed to take place as a result of unrestricted rotations about primary valence bonds in the chain. The elasticity of rubber, on this theory, is kinetic in origin, like the volume elasticity of a gas, and is not due to specific attractive forces, as is the elasticity of a normal solid.

The application of thermodynamics to MEYER's theory leads immediately to important conclusions, which can be subjected to experimental examination. The basis of the argument is that the deformation of a rubberlike material should involve no change in its internal energy. The entropy, on the other hand, being related to the number of configurations of the molecular chains, should *decrease* on deformation. From the first of these considerations it follows that the work done in a reversible deformation of the rubber should be converted quantitatively into heat, i.e. rubber should evolve heat on extension. This effect was

[1] MEYER, K. H., G. VON SUSICH u. E. VALKO: Kolloid-Z. 59, 208 (1932).
[2] KUHN, W.: Kolloid-Z. 68, 2 (1934).
[3] KUHN, W.: Kolloid-Z. 76, 258 (1936).
[4] GUTH, E. u. H. MARK: Mh. Chem. 65, 93 (1934).

studied by JOULE in 1859[1]. He observed a small fall in temperature up to about 14 per cent extension, followed by a much larger rise of temperature as the extension was progressively increased.

The second consideration (decrease of entropy), leads to the conclusion that the tension on a piece of rubber held at a constant stretched length should rise in direct proportion to the absolute temperature, or alternatively, that when subjected to a constant stretching force, the rubber should contract in length as the temperature is raised. This effect was first reported by GOUGH in 1805[2]. The two effects (contraction on heating and evolution of heat on stretching), which are together referred to as the GOUGH-JOULE effect, were shown by KELVIN to be thermodynamically related, but it was only with the rise of the kinetic theory that their significance became fully appreciated. Since this time however, these thermo-elastic effects have been the subject of intensive research. The question has proved to be considerably more complicated, both experimentally and theoretically, than was originally suspected, and it is only recently that a reasonably clear picture of the whole phenomenon has emerged.

b) Thermodynamic relations.

This section is concerned with the derivation of thermodynamic relations which will enable the changes in internal energy (U) and entropy (S) which accompany the deformation of an elastic body to be expressed in terms of the observed dependence of the elastic stress on length and temperature. The change in internal energy in any process is, by definition,

$$dU = dQ + dW \tag{V,1}$$

where dQ is the heat absorbed by the system and dW the work done on it. For a *reversible* process

$$dQ = T\,dS \tag{V,2}$$

and therefore

$$dU = T\,dS + dW. \tag{V,3}$$

Introducing the HELMHOLTZ free energy F, defined by

$$F = U - TS \tag{V,4}$$

we have for any reversible process

$$dF = dU - T\,dS - S\,dT. \tag{V,5}$$

For a process taking place at constant temperature $S\,dT = 0$, and therefore, from (V,3),

$$dF = dU - T\,dS = dW. \tag{V,6}$$

Let us consider a body of length l in equilibrium under the action of a

[1] JOULE, J. P.: Philos. Trans. Roy. Soc. London **149**, 91 (1859).
[2] GOUGH, J.: Mem. Lit. Philos. Soc. Manchester **1**, 288 (1805).

tensile force f. The work done in a small displacement being $f \cdot dl$, we have,

$$f = (\partial W/\partial l)_T = (\partial F/\partial l)_T \tag{V,7}$$

which, with (V,4) gives

$$f = (\partial U/\partial l)_T - T(\partial S/\partial l)_T. \tag{V,8}$$

Equation (V,8) expresses the tension as the sum of two terms representing respectively the internal energy and entropy. To relate these to experimentally observable quantities we write, from (V,5), (V,3) and (V,7)

$$dF = f dl - S dT \tag{V,9}$$

and therefore

$$f = (\partial F/\partial l)_T, \quad S = -(\partial F/\partial T)_l. \tag{V,10}$$

From the properties of partial differentials

$$\frac{\partial}{\partial l}\left(\frac{\partial F}{\partial T}\right)_l = \frac{\partial}{\partial T}\left(\frac{\partial F}{\partial l}\right)_T$$

and hence, from (V,10),

$$(\partial S/\partial l)_T = -(\partial f/\partial T)_l. \tag{V,11}$$

This is the required result. It states that the entropy change per unit extension is equal to the temperature coefficient of the tension, at constant length. The corresponding internal energy change is obtained by difference, using (V,8), i.e.

$$(\partial U/\partial l)_T = f - T(\partial f/\partial T)_l. \tag{V,12}$$

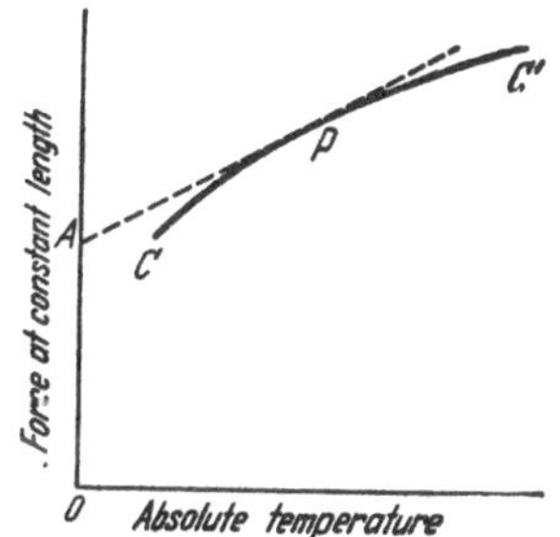

Fig. V,1. Slope and intercept of stress-temperature curve.

The meaning of equations (V,11) and (V,12) will be apparent from fig. V,1 in which the curve CC' represents the experimental variation of force, at constant length, with temperature. The slope of the tangent to this curve at any point P is $(\partial f/\partial T)_l$, which gives immediately the entropy change per unit extension [equ. (V,11)], while the intercept OA at $T=0$ is $f - T(\partial f/\partial T)_l$, which, from (V,12), is equal to the internal energy change per unit extension. If $(\partial U/\partial l)_T$ is zero, the tangent passes through the origin; the force is then proportional to the absolute temperature.

c) Experimental determination of stress-temperature relations.

In principle the experimental derivation of $(\partial U/\partial l)_T$ and $(\partial S/\partial l)_T$ is straightforward. It is only necessary to hold the material at a constant stretched length and to measure the tension as a function of temperature. In practice, the difficulties are considerable. These difficulties are associated with the imperfect nature of the elasticity of rubber, in consequence of which it is not possible to find a unique value of stress corresponding to specified values of the variables, length and temperature. This means that the concept of reversibility is not, in general, strictly applicable. The best that can be done is to establish an experimental procedure which will enable effectively reversible stress-temperature curves to be obtained over a limited range of the variables.

One of the earliest investigations of this kind was that carried out by MEYER and FERRI[1]. Their method was to allow the stretched rubber to relax, at constant length, at a relatively high temperature until the stress reached a steady value. Reproducible stress-temperature plots were then obtained so long as the initial relaxation temperature was not exceeded. One of their results, corresponding to an extension of 350 per cent, as shown in fig. V, 2, yields a linear increase of tension with temperature over a range of 120° C.

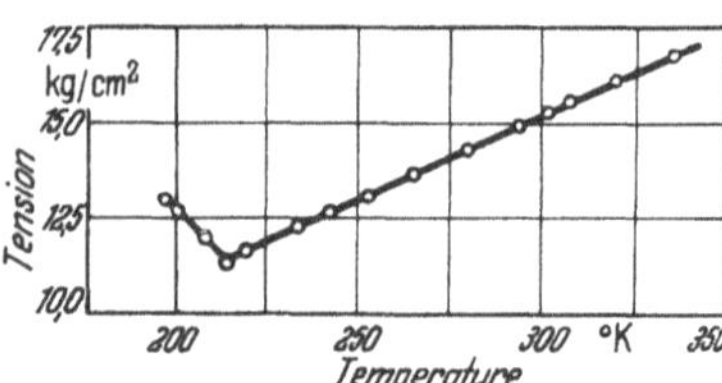

Fig. V, 2. Force at constant length as function of absolute temperature. Vulcanized rubber, 350 per cent extension. [MEYER and FERRI: Helv. Chim. Acta 18, 570 (1935).]

This variation, which is in the sense required by the kinetic theory, corresponds to a *reduction* of entropy on extension. [Equation (V, 11).]

MEYER and FERRI observed that at low extensions the stress decreased (instead of increasing) as the temperature was raised. This anomalous behaviour at low extensions is characteristic and appears, for example, in the more recent work of ANTHONY, CASTON and GUTH[2], reproduced in fig. V, 3. The extension at which the stress-temperature coefficient changes sign (ca. 10 per cent extension) is known as the *thermo-elastic inversion* point. It corresponds to the point at which the entropy of extension $(\partial S/\partial l)_T$ changes from a positive to a negative value.

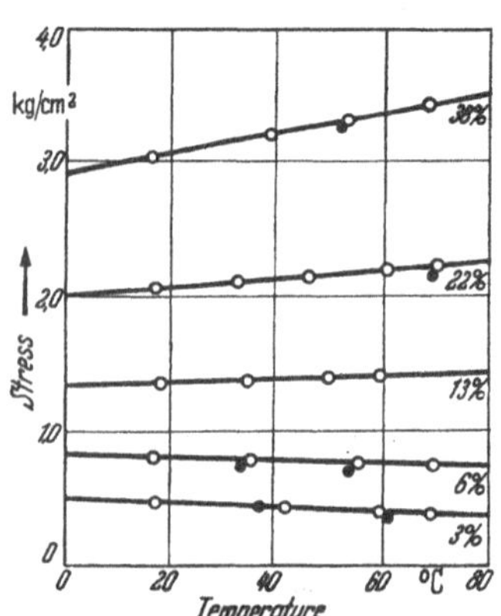

Fig. V, 3. Force at constant length as function of temperature for vulcanized rubber. Elongations as indicated. [ANTHONY, CASTON and GUTH: J. Phys. Chem. 46, 826 (1942).]

It was realized by MEYER and FERRI that the thermo-elastic inversion phenomenon was related to the thermal expansivity of the rubber, though the correct formulation of this connection was not given until later. The important fact is that if the variation of tension with temperature is measured at constant *extension ratio*, instead of at constant length, the thermo-elastic inversion phenomenon disappears. (The extension ratio is defined as the ratio of the stretched length to the unstretched length at the same temperature.) The results of GEE[3], for example, show that for extensions not exceeding 100 per

[1] MEYER, K. H. u. C. FERRI: Helv. Chim. Acta 18, 570 (1935).
[2] ANTHONY, R. L., R. H. CASTON u. E. GUTH: J. Physic. Chem. 46, 826 (1942).
[3] GEE, G.: Trans. Faraday Soc. 42, 585 (1946).

cent, the stress at constant extension ratio (λ) is strictly proportional to the absolute temperature, i.e.

$$f - T(\partial f/\partial T)_\lambda = 0. \qquad \text{(V,13)}$$

A similar conclusion may be drawn from the work of ANTHONY, CASTON and GUTH[1], and WOOD and ROTH[2] as well as from the original work of MEYER and FERRI. The full significance of this result, however, can only be understood on the basis of a more refined thermodynamic analysis than that given above.

§ 26. Advanced thermodynamic analysis.

a) Thermodynamic relations.

The requisite analysis has been carried out by ELLIOTT and LIPPMANN[3], and by GEE[4]. A distinction is made between extension at constant pressure and extension at constant volume. For extension at constant pressure we may write [analogously to (V,8)]

$$(\partial G/\partial l)_{p,T} = (\partial H/\partial l)_{p,T} - T(\partial S/\partial l)_{p,T} \qquad \text{(V,14)}$$

where H is the heat content (enthalpy) defined by $H = U + pV$, p and V are pressure and volume respectively, and G is the GIBBS free energy $(U - TS + pV)$. The quantity $(\partial H/\partial l)_{p,T}$ can be shown to be effectively equivalent under normal conditions (i.e. for $p = 1$ atmosphere) to $(\partial U/\partial l)_{p,T}$. It is related to experimentally observable quantities by the equation[4],

$$(\partial H/\partial l)_{p,T} = f - T(\partial f/\partial T)_{p,\lambda} + \beta l T(\partial f/\partial l)_{p,T} \qquad \text{(V,15)}$$

where β is the linear coefficient of expansion of the unstrained rubber. Using the experimental result (V,13), which is true for natural rubber up to 100 per cent strain, this gives

$$(\partial U/\partial l)_{p,T} \simeq (\partial H/\partial l)_{p,T} = \beta l T(\partial f/\partial l)_{p,T}. \qquad \text{(V,16)}$$

The corresponding expression for the entropy is

$$T(\partial S/\partial l)_{p,T} = \beta l T(\partial f/\partial l)_{p,T} - f. \qquad \text{(V,17)}$$

The analysis of GEE's experimental force-extension curve into internal energy and entropy components by means of equations (V,16) and (V,17) is shown in fig. V,4.

GEE next considers the connection between the internal energy term, represented by (V,16), and the change of volume due to the tensile stress. For this purpose he derives a relation between the internal energy change for extension at constant volume and the corresponding change in heat content for extension at constant pressure. This is

$$\left(\frac{\partial H}{\partial l}\right)_{p,T} - \left(\frac{\partial U}{\partial l}\right)_{V,T} = T\left(\frac{\partial f}{\partial l}\right)_{p,T}\left(\frac{\partial l}{\partial V}\right)_{f,T} \cdot \frac{(\partial V/\partial p)_{T,f}}{(\partial V/\partial p)_{T,l}}\left(\frac{\partial V}{\partial T}\right)_{p,l}. \qquad \text{(V,18)}$$

[1] S ehe S. 298, Fußnote 2.
[2] WOOD, L. A. u. F. L. ROTH: J. Appl. Physics **15**, 781 (1944).
[3] ELLIOTT, D. A. u. S. A. LIPPMANN: J. Appl. Physics **16**, 50 (1945).
[4] GEE, G.: Trans. Faraday Soc. **42**, 585 (1946).

This is as far as it is possible to go by purely thermodynamic arguments. To proceed further it is necessary to introduce certain approximations based on physical observation. These are

1. $(\partial l/\partial V)_{f,T} \simeq l/3V$. This assumes isotropic compressibility in the presence of tensile stress. It is strictly true only at zero stress.

2. $(\partial V/\partial p)_{T,f} \simeq (\partial V/\partial p)_{T,l}$ i.e. the compressibility at constant tension is equal to the compressibility at constant length.

3. $(\partial V/\partial T)_{p,l} \simeq 3\beta V$. This assumes the volume expansivity to be independent of strain. With these assumptions equation (V, 18) reduces to

$$(\partial H/\partial l)_{p,T} - (\partial U/\partial l)_{V,T} \simeq \beta l T (\partial f/\partial l)_{p,T} \qquad \text{(V, 19)}$$

which, with (V, 15), leads finally to

$$(\partial U/\partial l)_{V,T} \simeq f - T(\partial f/\partial T)_{p,\lambda}. \qquad \text{(V, 20)}$$

This approximate equation gives the required internal energy change *at constant volume* in terms of directly measurable quantities. Applying it to the data for natural rubber, as represented by equation (V, 13) we obtain

$$(\partial U/\partial l)_{V,T} = 0. \qquad \text{(V, 21)}$$

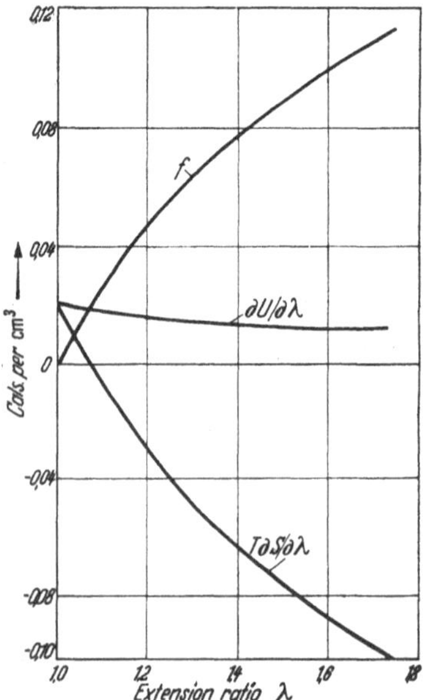

Fig. V, 4. Changes in internal energy, U, and entropy S accompanying extension of vulcanized rubber. [GEE: Trans. Faraday Soc. **42**, 585 (1946).] (f = force per unit unstrained area.)

b) Significance of volume changes.

The important result (V, 21) indicates that, if the extension could be carried out at constant volume (by suitably adjusting p) the change of internal energy would be zero. The natural inference is that the internal energy changes observed when the extension is carried out at constant pressure are associated entirely with the accompanying change in volume. This inference may be further developed and given quantitative form by relating the change of volume to the change of internal energy, assuming the relationship between these two quantities to be the same as when the change of volume is brought about by the application of a simple hydrostatic pressure, i.e.

$$(\partial U/\partial V)_T = 3\beta T/K \qquad \text{(V, 22)}$$

where K is the compressibility.

The volume change on stretching from length l_0 to length l thus becomes

$$\Delta V = \frac{K}{3\beta T} \int_{l_0}^{l} \left(\frac{\partial U}{\partial l}\right)_{p,T} dl. \tag{V, 23}$$

Using the experimental value (V, 16) for $(\partial U/\partial l)_{p,T}$ this gives,

$$\Delta V = \frac{K}{3} \int_{l_0}^{l} l \left(\frac{\partial f}{\partial l}\right)_{p,T} dl \tag{V, 24}$$

which enables the volume change to be calculated directly from the force-extension curve.

A direct test of this relation was made by GEE, STERN and TRELOAR[1], who measured the increase of volume and the force-extension curves for two different rubbers. On account of the smallness of the volume change (ca. one part in 10,000) the accuracy was not high, but the agreement with equation (V, 24) was considered statisfactory.

The mechanism responsible for the increase of volume on stretching is presumably simply that which is responsible for the ordinary bulk compressibility which rubber has in common with ordinary solids. The (negative) hydrostatic component of a tensile stress σ being $\sigma/3$, the corresponding volume change (assuming isotropic properties) for a sample of length l acted on by a tensile force f is

$$\Delta V = K f l/3. \tag{V, 24a}$$

This is the limiting form of (V, 24) as $l \to l_0$.

By way of confirmation, MEYER and VAN DER WYK[2] studied rubber in shear. Since a shear stress contains no hydrostatic component, there should be no change of volume in this type of deformation. In harmony with this consideration, MEYER and VAN DER WYK found the internal energy change to be zero.

c) General conclusion.

The foregoing analysis shows that the thermo elastic phenomena of rubber may be accounted for in terms of two mechanisms (1) a mechanism responsible for the essentially rubberlike properties, associated with a reduction of entropy without change in internal energy, and (2) a mechanism responsible for the volume compressibility, associated with an increase in *both internal energy and entropy*. These conclusions are true for natural rubber up to about 100 per cent extension. They are also approximately true (as the work of ROTH and WOOD[3] shows) for GR—S (butadiene-styrene) rubber at small strains. No mention has been made of the phenomena at larger strains, where, in the case of natural rubber, the picture is complicated by the effects of crystallization, which leads

[1] GEE, G., J. STERN u. L. R. G. TRELOAR: Trans. Faraday Soc. **46**, 1101 (1950).
[2] MEYER, K. H. u. A. J. A. VAN DER WYK: Helv. Chim. Acta **29**, 1842 (1946).
[3] ROTH, F. L. u. L. A. WOOD: J. Appl. Physics **15**, 749 (1944).

to a large *decrease* in volume, and probably also to corresponding *reductions* in internal energy.

The use of thermo-elastic data to characterize a material as rubberlike or otherwise would appear from this analysis to be a matter of unexpected difficulty. First there are the experimental difficulties, which are serious enough for rubber, and are likely to be even more serious in the case of other materials such as, for example, fibres, which are even less perfectly elastic, and are usually hygroscopic. Then, as we have seen, the direct interpretation of internal energy and entropy changes may be misleading. If we were to define rubberlike elasticity simply in terms of a negative entropy of extension we should be forced to conclude that rubber itself is not rubberlike below 10 per cent extension. This is because incidental effects, having essentially no connection with the primary mechanism of the elasticity, may nevertheless give rise to changes in internal energy and entropy sufficiently large to mask the primary phenomenon. In deciding whether a material should be classified as rubberlike, therefore, the evidence from thermodynamics alone is inadequate; in addition, all available evidence relating to the structure and physical properties (particularly the values of elastic constants, form of the stress-strain relations, etc.) should be taken into account.

B. The elasticity of long-chain molecules.

§ 27. Elementary chain statistics.

a) The random chain.

The development of the theory of rubberlike elasticity from the original concepts of MEYER may be thought of as taking place in two stages. In the first stage the statistical properties of the single molecule are examined mathematically, and expressions are derived for its entropy and hence for its free energy, as a function of its length. In the second stage the properties of a network formed by the cross-linking of an assembly of such chains, corresponding to a vulcanized rubber, are considered.

This section is concerned with the first stage — the statistical treatment of the single chain. In the general treatment of this problem it is convenient to represent the actual molecular structure by an idealized chain of equal links, such that the directions of successive links, in the absence of external restraints, are completely independent. A chain of this kind, which differs from any real molecular structure particularly in the omission of valence angle considerations, will be called a randomly-jointed, or randomly-kinked chain. Its relation to particular molecular structures will be considered later.

b) Distribution functions.

The statistics of the random chain have been studied from many different points of view. Formally, the problem is equivalent to the problem of the "random flight in three dimensions". The earliest applications to the theory of rubber elasticity were carried out by KUHN[1, 2] and by GUTH and MARK[3]. The problem is to calculate the relative number of configurations of the chain corresponding to a specified distance between its ends. For this purpose we imagine one end A to be fixed at the origin of coordinates (fig. V, 5) and calculate the probability of finding the other end B in the vicinity of the point (x, y, z). The approximate formula, as given, for example, by KUHN, is

$$p(x, y, z)\, dx\, dy\, dz = (b^3/\pi^{3/2})\, e^{-b^2(x^2+y^2+z^2)}\, dx\, dy\, dz. \qquad (\mathrm{V}, 25)$$

This expresses the fact that the required probability is proportional to the size of the volume element $(dx \cdot dy \cdot dz)$ within which the end B may lie.

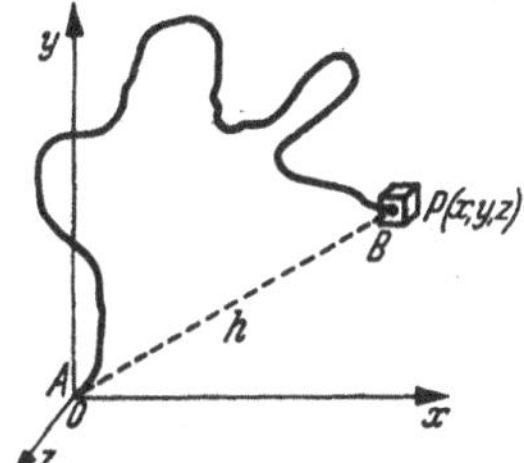

Fig. V, 5. Statistically-kinked chain.

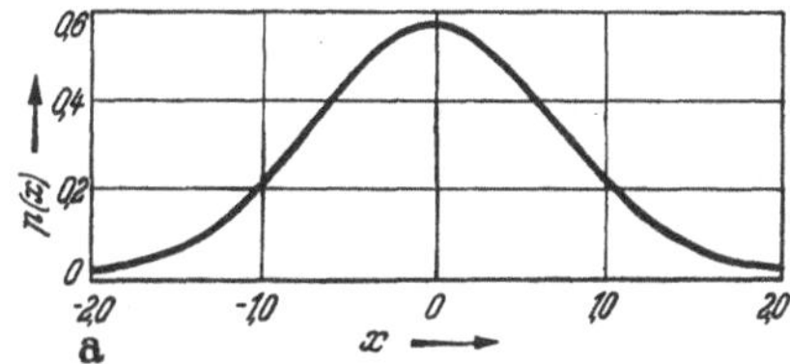

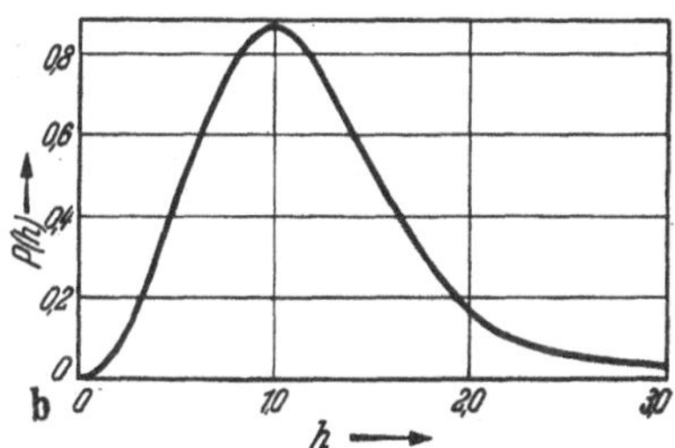

Fig. V, 6. a) Probability density for end B when A is fixed [equation (V, 25)]. b) Probability of vector length h [equation (V, 27)].

The function (V, 25) is the well-known "Gaussian" or normal error function. It contains only one adjustable parameter b, which is related to Z, the number of links and a, the length of the link, i. e.,

$$b^2 = 3/2\, Z a^2. \qquad (\mathrm{V}, 26)$$

The distribution (V, 25) is spherically symmetrical. Its form is shown in fig. V, 6a, which represents the variation of probability density $p(x, y, z)$ along a line passing through the origin. The maximum probability density occurs at the origin, i.e. when the two chain ends are coincident. This means that, if the end A is fixed at the origin, the probability of finding the other end B *within a volume element of specified size* is a maximum when that volume element is also at the origin.

It does not follow that the most probable length of the chain is zero. To find this, it is necessary to consider the probability $P(h)$ of a given distance h between the ends, i.e., the probability that the end B shall lie within a spherical shell of thickness dh. This is given by the product

[1] KUHN, W.: Kolloid-Z. **68**, 2 (1934).
[2] KUHN, W.: Kolloid-Z. **76**, 258 (1936).
[3] GUTH, E. u. H. MARK: Mh. Chem. **65**, 93 (1934).

of the probability density and the size of the volume element, namely $4\pi h^2 dh$, so that

$$\left.\begin{aligned} P(h)\,dh &= (b^3/\pi^{3/2})\,e^{-b^2(x^2+y^2+z^2)}\cdot 4\pi h^2\,dh \\ &= (4\,b^3/\pi^{1/2})\,h^2\,e^{-b^2h^2}\,dh\,. \end{aligned}\right\} \quad \text{(V, 27)}$$

This h-distribution function, which has the form shown in fig. V, 6b, gives a maximum, corresponding to the most probable value of h, or most probable length, at the point

$$h_{\text{most probable}} = 1/b = \sqrt{(2Z/3)}\cdot a\,. \quad \text{(V, 28)}$$

A more frequently required quantity is the root-mean-square value of h, which is given by

$$\sqrt{(\overline{h^2})} = \sqrt{(3/2\,b^2)} = a\sqrt{Z}\,. \quad \text{(V, 29)}$$

This, like the most probable value, is proportional to the square root of the number of links in the chain.

c) Elasticity of single chain.

The entropy of a chain is related to the number of configurations available to it, that is, to the probability function, by BOLTZMANN's relation

$$s = k\log p\,. \quad \text{(V, 30)}$$

s being the entropy, p the probability, and k BOLTZMANN's constant. For the Gaussian function (V, 25) this relation gives

$$\left.\begin{aligned} s(x,y,z) &= \text{const} - k\,b^2(x^2+y^2+z^2) \\ \text{or}\quad s(h) &= \text{const} - k\,b^2 h^2. \end{aligned}\right\} \quad \text{(V, 31)}$$

This means that if the ends of the chain are held at fixed points separated by the distance h, the corresponding entropy is proportional to h^2. [Note that $s(h)$ cannot be obtained from the h-distribution function (V, 27), since this implies that the end B is free to move over a spherical surface and is not confined to the neighbourhood of a fixed point.]

Assuming that all configurations have the same internal energy (Cf V, § 26), the HELMHOLTZ free energy $F\,(=U-TS)$ is therefore

$$F = \text{const} + k\,T\,b^2 h^2. \quad \text{(V, 32)}$$

The work required to move one end of the chain from the distance h to $h+dh$ with respect to the other end is equal to the corresponding change in HELMHOLTZ free energy. This is also equal to $\tau\cdot dh$, where τ is the mean force acting on the ends of the chain. Hence

$$\tau = dW/dh = dF/dh = 2k\,T\,b^2 h = 3k\,T\,h/Z a^2. \quad \text{(V, 33)}$$

This represents a mean tensile force acting along the line joining the ends of the chain, and proportional to the distance between the ends (fig. V, 7). The tension vanishes only when the two ends coincide.

It is important to distinguish between a free chain (corresponding to a molecule in solution) and a chain whole extremities are fixed in space. In the free chain there is no tension. But a chain whose ends are fixed exerts a force on its points of support whose mean value is given by (V,33); it behaves, therefore, like an extended spring.

The expressions (V,31), (V,32), (V,33) for the entropy, free energy, and tension, being based on the Gaussian probability function, are only approximate. They involve the assumptions (1) that the number of links is large and (2) that the distance h between the ends of the chain is small compared with its fully-extended length. The range of validity of these assumptions will be examined in the next section, where more accurate statistical treatments will be discussed.

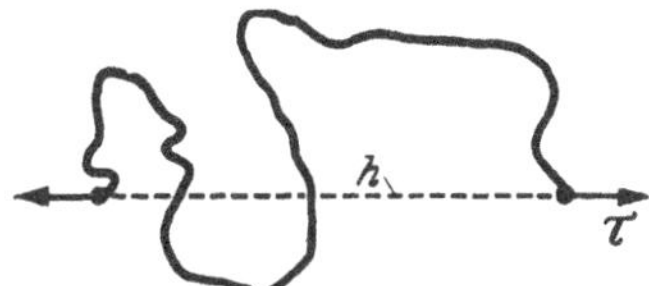

Fig. V,7. Tension on a chain whose ends are fixed.

§ 28. More accurate chain statistics.

a) The series distribution formula.

An exact solution to the random chain problem has been given by the author[1] who made use of a result derived by Hall[2] and Irwin[3] in connection with the theory of random sampling. It is expressed in the following series formula

$$P(h)\,dh = \frac{h}{2a^2}\,\frac{Z^{Z-2}}{(Z-2)!}\sum_{s=0}^{k}(-1)^s\binom{Z}{s}(m-s/Z)^{Z-2}\,dh \qquad (\text{V},34)$$

where $k/Z \leqslant m \leqslant (k+1)/Z$ and $m = (1-h/Za)/2$.

$\binom{Z}{s}$ represents the number of combinations of Z things taken s at a time.

This formula, which is valid for all values of Z and h, is unfortunately too complicated for general application. It has great value, however, in providing a standard by which other less accurate solutions may be judged. Examples of this application are given below.

b) The inverse Langevin approximation.

A function which is generally more useful, though less accurate, is that first derived by Kuhn and Grün[4] and independently by James and Guth[5]. Kuhn and Grün's method involves, as an intermediate step, the calculation of the angular distribution of the chain links. Assuming *a priori* that all link directions are equally probable, it is possible to write down the probability of any particular distribution of link orientations.

---

[1] Treloar, L. R. G.: Trans. Faraday Soc. **42**, 77 (1946).
[2] Hall, P.: Biometrika [London] **19**, 240 (1927).
[3] Irwin, J. O.: Biometrika [London] **19**, 225 (1927).
[4] Kuhn, W. u. F. Grün: Kolloid-Z. **101**, 248 (1942).
[5] James, H. M. and E. Guth: J. Chem. Physics **11**, 455 (1943).

The actual distribution will be that for which the corresponding probability has a maximum value.

If, in a chain of Z links, dZ represents the number in the angular range $d\theta$, referred to the line joining its ends, the resultant most probable distribution is

$$dZ = \frac{Z\beta}{\sinh\beta}\, e^{\beta\cos\theta} \cdot \frac{1}{2} \sin\theta \, d\theta \,. \tag{V,35}$$

This involves only one parameter β which is related to the fractional extension of the chain, h/Za, thus

$$h/Za = \coth\beta - 1/\beta = \mathfrak{L}(\beta) \tag{V,36}$$

where $\mathfrak{L}$ is the LANGEVIN function.

Alternatively we may write

$$\beta = \mathfrak{L}^{-1}(h/Za) \tag{V,36a}$$

where $\mathfrak{L}^{-1}$ is the *inverse* LANGEVIN function.

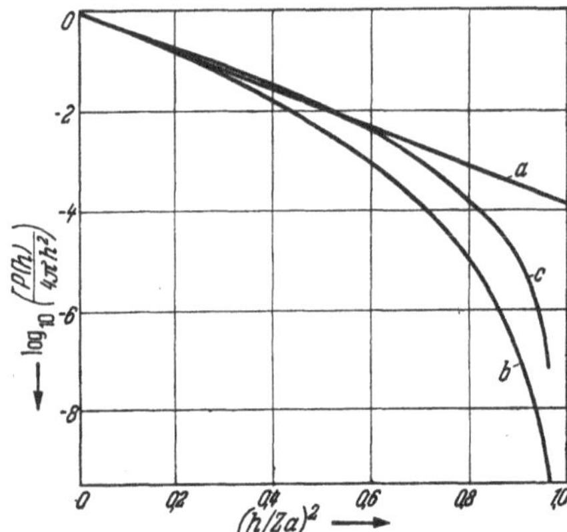

Fig. V,8. Distribution functions for random chain on logarithmic plot, $Z = 6$. a) Gaussian approximation [equ. (V,27)]; b) Inverse LANGEVIN approximation [equ. (V,37)]; c) Exact function (V,34).

It is interesting to note that the angular distribution function (V,35) involves only h/Za, the fractional extension of the chain. When $h/Za = 0$, i.e. when the two ends of the chain coincide, the distribution is spherically symmetrical.

The probability of the length h is readily obtained from the distribution of link angles; it is simply proportional to the probability of this distribution. The required probability density $p(h)$ is given by

$$\log p(h) = \text{const} - Z\left[\frac{h}{Za}\beta + \log\frac{\beta}{\sinh\beta}\right] \tag{V,37}$$

[The probability $P(h)$ of the length h is equal to $4\pi h^2 \cdot p(h)$.] Equ. (V,37) may be expanded into the following series,

$$\log p(h) = \text{const} - Z\left[\frac{3}{2}\left(\frac{h}{Za}\right)^2 + \frac{9}{20}\left(\frac{h}{Za}\right)^4 + \frac{99}{350}\left(\frac{h}{Za}\right)^6 + \cdots\right]. \tag{V,37a}$$

c) Relation to Gaussian distribution.

Comparison of (V,37a) with the Gaussian approximation (V,25) shows that the latter corresponds to the first term in the expansion (V,37a). The Gaussian approximation therefore becomes inadequate for values of h such that $(9/20)\,(h/Za)^4$ is no longer negligible compared with $(3/2)\,(h/Za)^2$, i.e., for fractional extensions of about 1/2 or greater.

The method of derivation of the inverse LANGEVIN approximation (V,37) rests on the assumption that Z, the number of links, is large. In order to examine the effect of this limitation, the distribution functions for chains of 6, 25 and 100 links calculated from the formula (V,37) have been compared with the corresponding exact distributions obtained from (V,34). These are shown in fig. V,8 and V,9 which include also the Gaussian forms. It is seen that while the inverse LANGEVIN approximation is substantially accurate for Z greater than 25, for $Z = 6$ it is quantitatively seriously in error, though even here it gives a better representation of the form of the distribution at high extensions than the Gaussian formula.

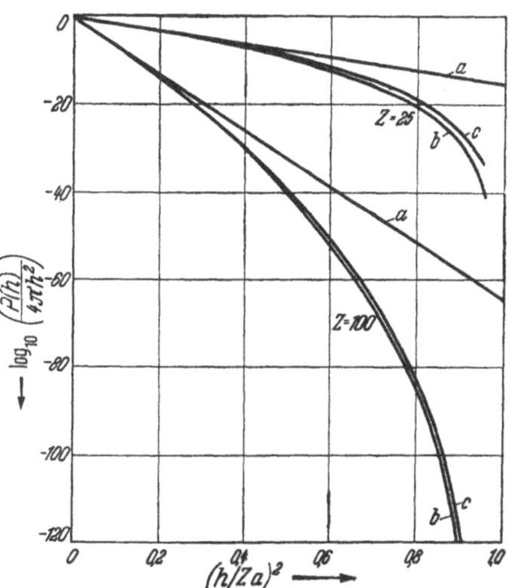

Fig. V,9. Distribution functions for random chain on logarithmic plot, $Z = 25$ and $Z = 100$. *a*) Gaussian approximation [equ. (V,27]; *b*) Inverse LANGEVIN approximation [equ. (V,37)]; *c*) Exact function (V,34).

It should be pointed out, however, that the logarithmic type of plot shown in fig. V,8 and V,9 places the emphasis on the high-extension region or "tail" of the distribution function. This is desirable for certain applications (e. g. for calculating the force-extension relation) but may be inappropriate for other purposes. If plotted on a linear scale, however, the significant differences between the respective formulae do not appear, and even for $Z = 6$ there is little obvious difference between either of the approximate formulae and the exact distribution. This will be seen from fig. V,10.

d) Force-extension curve for single chain.

As for the Gaussian chain, the entropy and hence the free energy corresponding to the inverse LANGEVIN approximation are obtained directly from the distribution function $p(h)$. Thus

$$F = -Ts = -kT\log p(h)$$

[from (V,30)] and therefore

$$f = kT\frac{d\log p(h)}{dh} \qquad (V,38)$$

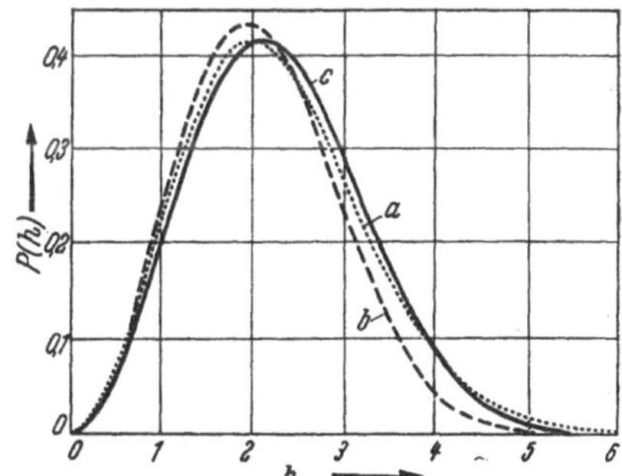

Fig. V,10. Distribution functions for random chain. Linear plot. $Z = 6$. *a*) Gaussian, *b*) Inverse LANGEVIN, *c*) Exact.

Hence, using the expression (V,37) for $p(h)$ we obtain

$$f = (kT/a)\,\mathfrak{L}^{-1}(h/Za) \qquad (V,39)$$

$$= \frac{kT}{a}\left[3\left(\frac{h}{Za}\right) + \frac{9}{5}\left(\frac{h}{Za}\right)^3 + \frac{297}{175}\left(\frac{h}{Za}\right)^5 + \cdots\right]. \qquad (V,39\,a)$$

This function, like the corresponding entropy function (V,37) is independent of the value of Z (except for a proportionality factor), but depends only on the fractional extension h/Za. Thus the force-extension relation for a single chain is independent of its length. The form of this function is shown in fig. V,11. The initial linear region corresponds to the Gaussian approximation (V,33). More direct derivations of the force-extension relation for a single chain have been given by JAMES and GUTH[1] and by FLORY[2]. FLORY considers the chain to be acted on by the tensile force τ and calculates the average extension by statistical thermodynamic methods. For a link at the angle θ_i the component of length parallel to τ is $x_i = a \cos \theta_i$, and the corresponding energy (work) is $-\tau x_i$. The probability of this particular link orientation, according to standard statistical thermodynamics, is

$$e^{\tau x_i/kT} d x_i .$$

The mean value of x_i is therefore

$$\bar{x}_i = \frac{\int_{-a}^{a} x_i e^{\tau x_i/kT} d x_i}{\int_{-a}^{a} e^{\tau x_i/kT} d x_i} = a \mathfrak{L}(\tau a/kT) . \qquad (V,40)$$

Putting $h = Z \bar{x}_i$ this leads immediately to (V,39).

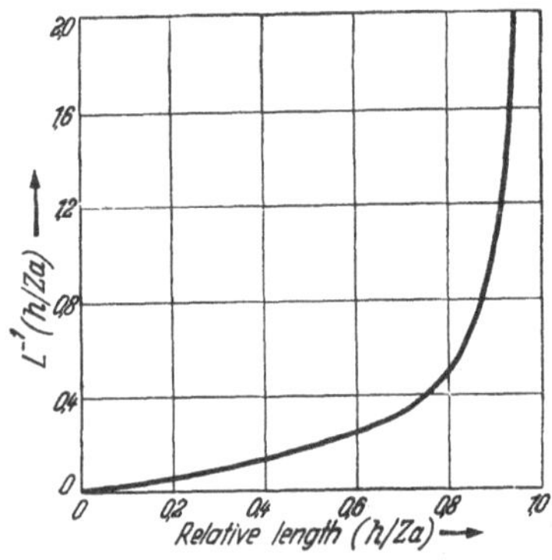

Fig. V, 11. Force-extension relation for single chain [equ. (V,39)].

e) Application to real molecular structures.

Any real molecular structure differs from the idealized random chain in a number of important properties. Certain aspects of these properties have been dealt with rather fully in Volume I. The most important considerations, so far as the statistical theory of elasticity is concerned, are the effects introduced by the presence of a valence angle, with either free or restricted rotation, and, more generally, by any particular sequence of bond lengths and bond angles in the structure. A further problem is the effect of the volume occupied by the atoms on the number of possible configurations of the chain. In this section we shall consider only the effects of chain geometry and valence angle, using the assumption of free (i. e. random) rotations about single bonds.

The simplest actual molecular structure is exemplified by polymethylene, $(-CH_2-)_Z$, which may be represented geometrically by a chain of links connected together at a constant angle, but otherwise unrestricted. The exact formula for the mean-square end-to-end distance for a chain of Z links of this type was first given by EYRING[3]. (Cf. Vol. 1 page 241.) For large values of Z EYRING's formula is

$$\overline{h^2} = Z a^2 (1 + \cos \theta)/(1 - \cos \theta) \qquad (V,41)$$

[1] JAMES, H. M. u. E. GUTH: J. Chem. Physics **11**, 455 (1943).
[2] FLORY, P. J.: *Principles of Polymer Chemistry*, Cornell 1953, p. 427.
[3] EYRING, H.: Physic. Rev. **39**, 746 (1932).

where θ is the supplement to the valence angle. For the tetrahedral valence angle $\theta = 70\,^1/_2{}^\circ$ (approximately) and $\cos\theta = 1/3$. With these values (V, 41) gives

$$\overline{h^2} = 2 Z a^2. \tag{V,42}$$

Comparing this with the value for a random chain (Za^2) it is seen that the polymethylene chain has $\sqrt{2}$ times the root-mean-square length of a random chain of the same number of links.

WALL[1] has given a method for the calculation of the R.M.S. length of a chain possessing an arbitrary sequence of bond lengths and valence angles, and has applied this to the *cis* and *trans* polyisoprene chains, corresponding to natural rubber and gutta-percha, respectively.

For small values of h the distribution function for any chain, of whatever structure, degenerates to the Gaussian form, provided that the number of (rotatable) links is sufficiently large. But the calculation of the complete distribution function, for all values of h, for a chain with valence angle is a formidable problem, which has not yet been solved in general terms. However, an approximate method which makes use of a graphical integration process has been devised by the author and applied to the paraffin[2] and polyisoprene[3] structures. The solutions obtained for the paraffin chain are illustrated in Vol. 1, page 244, figs. 92a and 92b.

The question now arises of the quantitative relationship between the random chain and the actual molecular structure. Is it possible to represent the actual chain (e.g. the paraffin chain) by means of a random chain with a suitably adjusted number of links? An examination of the solutions for paraffin chains of length ranging from 3 to 80 links throws light on this question. First, it is clear that for short chains, i.e. up to 5 links, the form of the distribution function is specific to the particular chain geometry. At about $Z = 10$ the distribution approximates to the normal or Gaussian form, and this approximation becomes closer with increasing chain length. So long as the Gaussian approximation holds (i.e. for h/Za small) the actual chain is obviously statistically equivalent to a random chain having a suitably chosen number of links. But the question of the form of the distribution function at high chain extensions remains to be examined. This examination has been carried out by fitting random chain distributions of the type represented by (V, 37) to the calculated paraffin and cis-polyisoprene distributions. The results are presented in fig. V, 12, from which it is seen that both these structures may be represented reasonably well by suitably chosen random chains over the whole range of extension, provided the number of chain links is sufficiently large. Thus, for example, the 80-link paraffin chain is effectively equivalent to a 34-link random chain having the same fully-stretched length. The random link is therefore statistically equivalent to about 2,35 C—C bonds. In the same way the polyisoprene chain may be re-

[1] WALL, F. T.: J. Chem. Physics **11**, 67 (1943).
[2] TRELOAR, L. R. G.: Proc. Physic. Soc. **55**, 345 (1943).
[3] TRELOAR, L. R. G.: Trans. Faraday Soc. **40**, 109 (1944).

presented by a random chain in which there are 1,4 links per isoprene unit.

A more critical examination of the figures[1], however, shows the equivalence to be not exact, particularly in the case of the paraffin chain. This may be seen by considering the Gaussian region. For the paraffin chain the mean square length [equ. (V,42)] is $2Za^2$, while the fully extended length is $Za\cos(\theta/2)$, or $\sqrt{(2/3)}Za$. The equivalent random chain is that for which both the mean square length and the fully extended length are the same. Thus, for a random chain of Z_1 links of length a_1, having the mean square length $Z_1a_1^2$, to be equivalent to a paraffin chain of Z links of length a, $Z_1a_1 = \sqrt{(2/3)}Za$ and $Z_1a_1^2 = 2Za^2$ giving

$$Z_1 = Z/3, \quad a_1 = \sqrt{6}a = 2\cdot245\,a.$$

In the Gaussian region, therefore, three paraffin links are equivalent to one random link, whereas the best fit to the curve (in which the high-extension region plays a predominant part) requires 2,35 paraffin links to be equated to one random link. A small part of this discrepancy may be attributed to systematic error in the paraffin-chain analysis, but the remainder represents a genuine difference and indicates that the paraffin-type chain is not strictly statistically equivalent to a random chain.

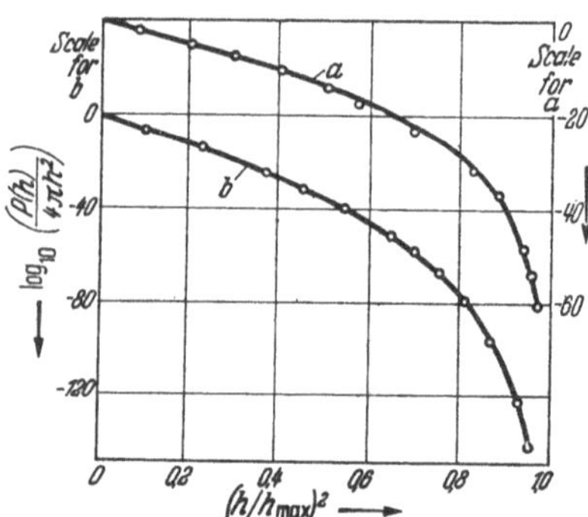

Fig. V,12. Comparison of paraffin and cis-polyisoprene chains with random chains. *a*) Circles, 80-link paraffin; continuous curve 34-link random; *b*) Circles, 64-isoprene; continuous curve, 90-link random.

For the polyisoprene chain, on the other hand, the distribution appears to agree almost perfectly with that for a random chain, though the errors involved in the calculation were here somewhat greater[2].

The conclusion to be drawn from these calculations is therefore that, for practical purposes, the statistical properties of any molecular chain, assuming free rotation, are not likely to differ in any significant way from the statistical properties of a randomly-jointed chain having the same extended length and a suitably chosen number of links.

This conclusion makes it possible to proceed to develop the statistical theory of the network on the basis of the random chain statistics, and to apply it to real molecular structures whose geometry is necessarily more complex than that of the random chain model.

[1] TRELOAR, L. R. G.: Trans. Faraday Soc. **42**, 77 (1946).
[2] TRELOAR, L. R. G.: Trans. Faraday Soc. **40**, 109 (1944).

C. Network theory of vulcanized rubber.

§ 29. General conditions for rubberlike elasticity.

The discussion in the last section has indicated that the phenomenon of rubberlike elasticity is fundamentally dependent on the statistical properties of long-chain molecules whose atoms are endowed with thermal agitation. This picture, as we have seen, enables us to draw certain general thermodynamic conclusions regarding the thermo-elastic behaviour of an ideal rubber, and also to understand the relation between rubberlike elasticity and certain specific features of the molecular structure, particularly the chain length and the freedom or otherwise of rotation about primary valence bonds. However, these molecular properties, though necessary, are not sufficient in themselves to ensure elasticity in the bulk material. This requires, in addition, the fulfilment of certain conditions relating to the manner in which the molecules are held together so as to form a coherent network. These essential conditions for the occurrence of rubberlike elasticity have been discussed by BUSSE[1]. They are

1. The presence of long-chain molecules, possessing freely rotating links.

2. Weak secondary forces between the molecules.

3. An interlocking of the molecules at a few places along their length to form a three-dimensional network.

The first condition has already been discussed. Condition (2) expresses the idea that the random thermal fluctuations of form of the individual chains is only possible if the intermolecular forces are sufficiently weak, comparable, in fact, with the intermolecular forces in a typical liquid. The third condition is introduced to meet the difficulty which would otherwise arise that with weak intermolecular forces the material would be expected to behave as a liquid, and not as a solid. The introduction of the cross-linking condition resolves the difficulty of reconciling the necessity for local molecular freedom of movement with the maintenance of a permanent macroscopic form.

These three conditions for elasticity are not necessarily realized in a strict and absolute sense in real materials. Rather, they serve as a guide to the interpretation of the actual properties of materials, and as an indication of the structural factors responsible for departures from ideal rubberlike properties in particular cases. This may be illustrated by considering unvulcanized natural rubber. This material is not chemically cross-linked; it does not, therefore, satisfy condition (3). The necessary structural coherence is brought about by local entanglements between chains, which, acting in conjunction with the VAN DER WAALS' forces, produce *effective* cross-linkages, which may perhaps be thought of as regions of local high viscosity. However, the elastic recovery of unvulcan-

[1] BUSSE, W. F.: J. Chem. Physics **36**, 2862 (1932).

ized rubber after straining is slower and less complete than that of vulcanized rubber. Moreover, a reduction of chain length, e.g. by milling, results in a very considerable loss of these already imperfect elastic properties. Vulcanized rubber, in which the chains are chemically cross-linked, comes much closer to the theoretical ideal, though in this material also the essential conditions for elasticity are statisfied only to a limited extent, and within a limited range of temperature. This is shown, for example, by the phenomena of crystallization (either with or without deformation) or by the transition to the glassy state which occurs at low temperatures (fig. V,2).

From the standpoint of the statistical theory we shall be concerned with the concept of an ideal rubber network in which the chains are cross-linked by unbreakable bonds and in which all other intermolecular forces are non-existent. The mechanical properties of such a network will first be developed theoretically with the restriction to the Gaussian chain statistics — the so-called Gaussian network. These properties will then be compared with the observed properties of vulcanized rubber under various types of strain. Following this, the extension of the theory by the incorporation of the non-Gaussian chain statistics will be considered. Finally in § 33 the analysis of certain departures of the experimental stress-strain characteristics of rubber from the forms deduced from the statistical theory in terms of certain more general phenomenological theories will be discussed.

§ 30. The Gaussian network.

a) Statistical treatment.

In terms of the foregoing picture of the ideal rubber as a cross-linked network of randomly-kinked chains, the problem is to calculate the entropy, and hence the free energy, as a function of the state of strain. From this the stress-strain relations corresponding to any particular type of deformation are readily derived.

Just as in the treatment of the single chain discussed in § 27, the essential problem is to compute the relative number of configurations under specified conditions of restraint. In the network problem this computation has to be carried out for the whole assembly of chains. Various methods of attack have been suggested, differing in the detailed simplifying assumptions introduced and in the degree of refinement of the statistical analysis[1, 2, 3, 4]. Fortunately these various treatments lead to substantially identical conclusions. The present discussion will therefore sufficiently illustrate the essential argument if it is limited to one particular model which displays all the principal features of the problem

[1] Wall, F. T.: J. Chem. Physics **10**, 485 (1942).
[2] Flory, P. J. u. J. Rehner: J. Chem. Physics **11**, 512 (1943).
[3] James, H. M. u. E. Guth: J. Chem. Physics **11**, 455 (1943).
[4] Treloar, L. R. G.: Trans. Faraday Soc. **39**, 36 (1943).

and is at the same time relatively simple and tangible. This is the model originally put forward by FLORY and REHNER[1].

In this model no attempt is made to deal directly with the whole assembly of chains. Instead, a small element, consisting of four chains radiating from a single point of cross-linkage, is taken to represent an average "cell" of the network, and the mathematical operations are performed on this elementary cell (fig. V, 13). Taking each of the four chains to have the same chain contour length (Z links each of length a) and to be terminated by junction points with neighbouring chains, it is assumed that the average positions of the four nearest neighbour junction points in the undeformed state will be at the corners of a regular tetrahedron. The effect of the deformation is assumed to be to displace the relative positions of the outer junction points in a manner corresponding to the macroscopic dimensional changes of the boundary surfaces of the network. FLORY and REHNER then calculate the entropy of formation of the elementary cell from its constituent chains, first in the undeformed state, then in the deformed state. The difference is the entropy of deformation.

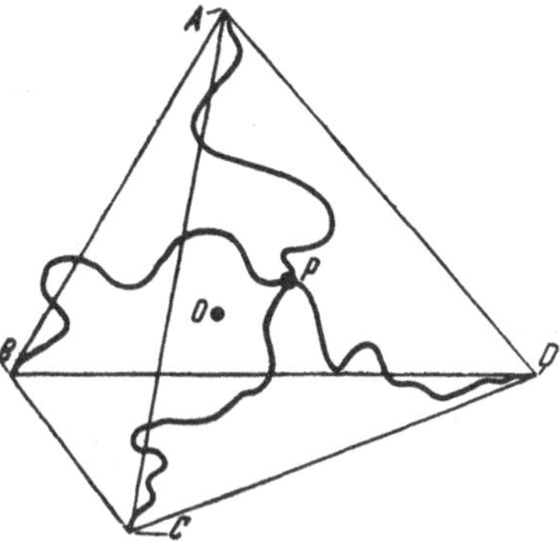

Fig. V, 13. The FLORY-REHNER model.

In the analysis outlined below the original treatment is somewhat modified so as to take account of the most general homogeneous deformation which can be applied. This type of strain is illustrated in fig. V, 14. It is such that a unit cube becomes transformed to a rectangular parallepiped of dimensions λ_1, λ_2 and λ_3, these being the three *principal extension ratios*. If, as will be assumed, the deformation takes place without change of volume (incompressibility condition) then,

$$\lambda_1 \lambda_2 \lambda_3 = 1 . \qquad (V, 43)$$

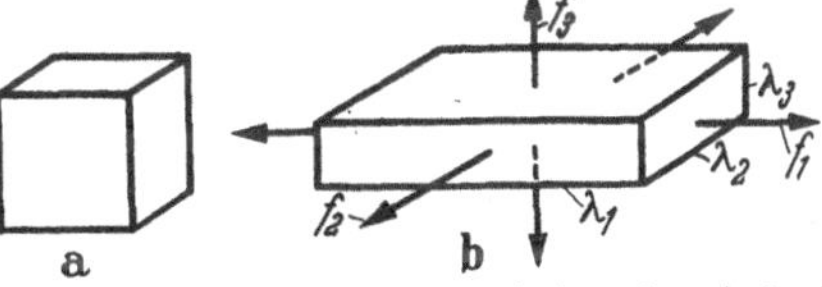

Fig. V, 14. Pure homogeneous strain. a) unstrained state, b) strained state.

As already stated, the calculation assumes an *affine* or *corresponding* displacement of the four corners of the tetrahedral cell; this fixes the position of the outermost junction points. The central junction point, on the other hand, is not fixed. The essential part of the treatment is concerned with the calculation of the probability that the four chains, whose outer extremities are fixed at A, B, C, D, shall meet at a particular point P. Integration over all positions of P then gives the total probability that these four chains shall meet at any point whatever.

The probability that the four chains shall meet within a small volume

[1] FLORY, P. J. u. J. REHNER: J. Chem. Physics **11**, 512 (1943).

element $\Delta\tau (= dx \cdot dy \cdot dz)$ in the neighbourhood of a given point P defined by coordinates (x, y, z) is [from (V,25)]

$$\omega(x, y, z)(dx . dy . dz)^4 = \prod_{i=1}^{4} p(x_i y_i z_i)\, dx_i\, dy_i\, dz_i \qquad \text{(V,44)}$$

where (x_i, y_i, z_i) are the components of the four respective chain vector lengths, i.e. the coordinates of P referred in turn to A, B, C and D, and $p(x_i, y_i, z_i)$ is the probability function for a single chain. For a Gaussian chain this has the form [equ. (V,25)]

$$p(x_i, y_i, z_i) = (b^3/\pi^{3/2})\, e^{-b^2(x_i^2 + y_i^2 + z_i^2)} = (b^3/\pi^{3/2})\, e^{-b^2 h_i^2}. \qquad \text{(V,45)}$$

Substitution in (V,44) gives, therefore,

$$\omega(x, y, z) = (b^{12}/\pi^6)\, e^{-b^2 \Sigma h_i^2}. \qquad \text{(V,46)}$$

The sum $\sum h_i^2$ is a function of the strain. If the centre point of the tetrahedron is fixed at the origin of coordinates while the corners A, B, C, D, are displaced in such a way that their y, x, and z coordinates become x', y', z', respectively, then

$$x' = \lambda_1 x, \quad y' = \lambda_2 y, \quad z' = \lambda_3 z. \qquad \text{(V,47)}$$

It may be shown that

$$\sum h_i^2 = 4[s^2 + h_0^2(\lambda_1^2 + \lambda_2^2 + \lambda_3^2)/3] \qquad \text{(V,48)}$$

where h_0 is the undeformed length $0A$, and $s = 0P$.

Integration over all positions of P gives, therefore, for the total probability of the four chains meeting at any point in space,

$$\Omega = (b^{12}/\pi^6) \int_0^\infty 4\pi s^2 e^{-4b^2[s^2 + h_0^2(\lambda_1^2 + \lambda_2^2 + \lambda_3^2)/3]}\, ds \qquad \text{(V,49)}$$

the volume element for the range ds being $4\pi s^2 ds$.

The integral (V,49) reduces to

$$\Omega = \text{const}\; e^{-4b^2 h_0^2(\lambda_1^2 + \lambda_2^2 + \lambda_3^2)/3}. \qquad \text{(V,50)}$$

The entropy, from Boltzmann's relation (V,30) is thus

$$S = k \log \Omega = \text{const} - (4/3)\, k b^2 h_0^2 (\lambda_1^2 + \lambda_2^2 + \lambda_3^2). \qquad \text{(V,51)}$$

For the undeformed state ($\lambda_1 = \lambda_2 = \lambda_3 = 1$) the corresponding entropy is

$$S_0 = k \log \Omega_0 = \text{const} - 4 k b^2 h_0^2$$

and hence the entropy of deformation becomes

$$S - S_0 = -(4/3)\, k b^2 h_0^2 (\lambda_1^2 + \lambda_2^2 + \lambda_3^2 - 3). \qquad \text{(V,52)}$$

The number of cells in the network being $N/4$, where N is the total number of chains, the total entropy of deformation for the network is therefore

$$\Delta S = -(1/3)\, N k b^2 h_0^2 (\lambda_1^2 + \lambda_2^2 + \lambda_3^2 - 3). \qquad \text{(V,53)}$$

If h_0 is given the root-mean-square value for a free chain, $h_0^2 = 3/2b^2$, this reduces to

$$\Delta S = -(1/2) N k (\lambda_1^2 + \lambda_2^2 + \lambda_3^2 - 3) \qquad (V,54)$$

which is the required solution to the problem.

b) Free energy of deformation.

Just as for the single chain, the HELMHOLTZ free energy is calculated on the assumption of no change in internal energy. The free energy is equal to the (isothermal) work of deformation W. Hence, from (V,54)

$$W = -T\Delta S = (1/2) N k T\, (\lambda_1^2 + \lambda_2^2 + \lambda_3^2 - 3) \qquad (V,55\,a)$$

$$= \frac{\varrho R T}{2 M_c} (\lambda_1^2 + \lambda_2^2 + \lambda_3^2 - 3) \qquad (V,55\,b)$$

where M_c is the "chain molecular weight", ϱ the density of the rubber, and R the gas constant.

Equation (V,55a) represents the elastically stored free energy for the most general type of strain. It is important to notice that it contains only one molecular parameter N, the number of chains (per unit volume) of the network. This may be related to the number of cross-linkages present, or alternatively, to the "chain molecular weight" M_c, i.e. the molecular weight between successive points of cross-linkage (equ. V,55b).

c) Stress-strain relations.

1. Pure homogeneous strain.

Equation (V,55a), [or the alternative form (V,55b)] contains all the information required for the calculation of the stresses necessary for the maintenance of the deformed state. Let us consider the state of pure homogeneous strain represented in fig. V,14. This state will be maintained by forces f_1, f_2 and f_3 acting in the directions corresponding to the principal axes of strain. To find the relation between the forces and the extension ratios the method adopted is to equate the change in stored (free) energy in any variation of λ_1, λ_2 and λ_3 to the work done by the external forces. By virtue of the incompressibility condition (V,43) only two of the three extension ratios are independently variable, hence, for any variation,

$$(\lambda_1 + d\lambda_1)(\lambda_2 + d\lambda_2)(\lambda_3 + d\lambda_3) - \lambda_1 \lambda_2 \lambda_3 = 0$$

giving

$$d\lambda_1/\lambda_1 + d\lambda_2/\lambda_2 + d\lambda_3/\lambda_3 = 0. \qquad (V,56)$$

Writing the stored-energy function (V,55a) in the form

$$W = (1/2) G (\lambda_1^2 + \lambda_2^2 + 1/\lambda_1^2\lambda_2^2 - 3) \qquad (V,55\,c)$$

(where $G = NkT$) the change in W for any variations in λ_1 and λ_2 is

$$dW = G[(\lambda_1 - 1/\lambda_1^3\lambda_2^2)\, d\lambda_1 + (\lambda_2 - 1/\lambda_2^3\lambda_1^2)\, d\lambda_2]$$

which, with (V,43), becomes

$$d\,W = G\left[(\lambda_1^2 - \lambda_3^2)\frac{d\,\lambda_1}{\lambda_1} + (\lambda_2^2 - \lambda_3^2)\frac{d\,\lambda_2}{\lambda_2}\right]. \tag{V,57}$$

The work done by the external forces is

$$d\,W = f_1\,d\,\lambda_1 + f_2\,d\,\lambda_2 + f_3\,d\,\lambda_3. \tag{V,58}$$

Making use of (V,56) to eliminate $d\,\lambda_3$, this may be transformed to

$$d\,W = (\lambda_1 f_1 - \lambda_3 f_3)\frac{d\,\lambda_1}{\lambda_1} + (\lambda_2 f_2 - \lambda_3 f_3)\frac{d\,\lambda_2}{\lambda_2}. \tag{V,59}$$

The identity of (V,57) and (V,59) for all variations $d\,\lambda_1$, $d\,\lambda_2$ leads to the result

$$\left.\begin{aligned}\lambda_1 f_1 - \lambda_3 f_3 &= G(\lambda_1^2 - \lambda_3^2)\\ \lambda_2 f_2 - \lambda_3 f_3 &= G(\lambda_2^2 - \lambda_3^2).\end{aligned}\right\} \tag{V,60}$$

The forces f_1, f_2, f_3 act on surfaces of unit area in the unstrained state. The principal *stresses* σ_1, σ_2, σ_3 are the forces per unit area *measured in the deformed state*. The area acted on by f_1 is $\lambda_2\lambda_3$, hence $\sigma_1 = f_1/\lambda_2\lambda_3 = \lambda_1 f_1$. In terms of the principal stresses, therefore, equations (V,60) become

$$\left.\begin{aligned}\sigma_1 - \sigma_3 &= G(\lambda_1^2 - \lambda_3^2)\\ \sigma_3 - \sigma_3 &= G(\lambda_2^2 - \lambda_3^2).\end{aligned}\right\} \tag{V,61}$$

The equations (V,61) are the general stress-strain relations for the Gaussian network. From their form it will be seen that a knowledge of the state of strain is not sufficient to enable us to determine the absolute magnitude of the stresses, but only the differences between any two of them. This fact, which is a direct consequence of the incompressibility condition, may be interpreted as meaning that the state of strain is unaffected by the superposition of an arbitrary hydrostatic pressure. Thus, if stresses σ_1, σ_2, σ_3, represent a particular solution for a given state of strain, then the stress system $\sigma_1 + p$, $\sigma_2 + p$, $\sigma_3 + p$, is also a solution. In practice, this indeterminacy does not usually introduce any difficulty, because one of the stresses is usually known, as, for example, when there is a free surface on which no external force acts.

2. *Simple elongation (or uni-directional compression).*

For an incompressible material a simple elongation is defined by the extension ratios

$$\lambda_1 = \lambda; \quad \lambda_2 = \lambda_3 = 1/\lambda^{1/2}. \tag{V,62}$$

Since $f_2 = f_3 = 0$, equations (V,60) then give, for the tensile force f, *per unit area of the unstrained section*, corresponding to an extension ratio λ

$$f = G(\lambda - 1/\lambda^2) = (\varrho\,R\,T/M_c)(\lambda - 1/\lambda^2). \tag{V,63}$$

This equation applies also to a compression in one direction (corresponding to $\lambda < 1$) with expansion in the other two directions (fig. V,15b).

3. Simple shear.

For this type of strain (fig. V.16) (which is not a *pure* strain, but involves rotation of the principal axes), it is preferable to revert to the stored energy function. The principal extension ratios corresponding to a simple shear are

$$\lambda_1 = \lambda; \quad \lambda_2 = 1; \quad \lambda_3 = 1/\lambda \tag{V,64}$$

the amount of the shear being[1]

$$\gamma = \tan\varphi = \lambda - 1/\lambda. \tag{V,65}$$

Inserting (V,64) in (V,55c) one obtains

$$W = (1/2)\,G\,(\lambda^2 + 1/\lambda^2 - 2) = (1/2)\,G\gamma^2. \tag{V,66}$$

The shear stress σ_{xy} is therefore

$$\sigma_{xy} = dW/d\gamma = G\gamma. \tag{V,67}$$

It is to be noted that, in addition to the shear stress σ_{xy}, there may be normal stresses present in simple shear[2]. These normal stresses, however, do no work, hence do not affect the stored energy, and will not be considered here.

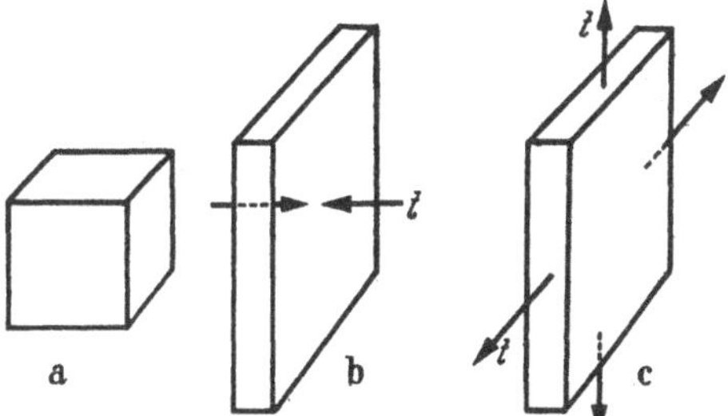

Fig. V,15. a) Unstrained state, b) uni-directional compression, and c) 2-dimensional extension.

It is seen from (V,67) that the constant G which occurs in the stored energy function (V,55c) and in the general stress-strain relations (V,61), is equal to the modulus of rigidity in simple shear.

d) Generalization of model.

1. Variation of strain axes.

We return now to the discussion in rather more detail of the properties of the Gaussian network, and to the question of the extent to which the particular model which has been chosen (i.e. the FLORY-REHNER model) as a basis for the formulation of these properties is unnecessarily restrictive. The first point to be noted about this model is that the calculated stored-energy function is independent of the directions of the principal axes of strain with respect to the undeformed

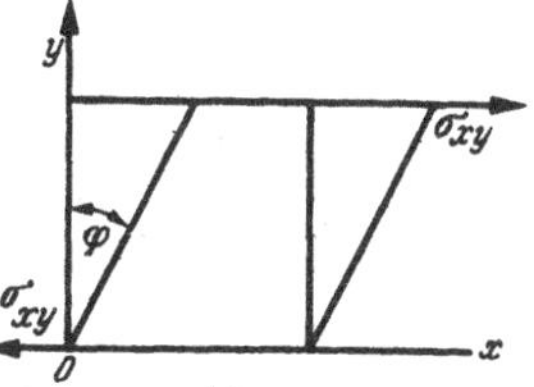

Fig. V,16. Simple shear.

[1] LOVE, A. E. H.: Mathematical Theory of Elasticity, Cambridge 1927, p. 33.
[2] RIVLIN, R. S., and D. W. SAUNDERS: Philos. Trans. Roy. Soc. London, A 240, 459 and 491 (1948); 241, 379 (1948).

tetrahedron[1]. The model is therefore *isotropic* in the undeformed state. (This property is restricted to the Gaussian region of strain.)

2. *Fluctuations of central junction point.*

With the outermost junction points fixed, the fluctuations of position of the central junction point are seen to be described by a Gaussian distribution function centred on the geometrical centre *0* of the tetrahedron (equ. V,49). The "breadth" of this distribution is one half that for a single free chain (equ. V,45). Further, it appears from (V,49) that these fluctuations are independent of the strain. Thus we arrive at the important result that the fluctuations of position of the central junction point, being unaffected by the strain, do not enter into the calculation of the entropy of deformation. This means that the calculated entropy would be unchanged if this junction point were considered to be fixed at its most probable position.

3. *Affine displacement of junction points.*

The above conclusion, that it makes no difference to the calculation of the entropy of deformation whether the central junction point is allowed to fluctuate or is fixed at its most probable position, may obviously be generalized to include all the junction points of the network. Further, it may be shown that the most probable positions of all the junction points, in the deformed state, correspond to an affine deformation of the chains. An alternative statement of this conclusion is that, on deformation, the junction points of the network move as if they were embedded in an elastic continuum. These properties of the Gaussian network have been established on a quite general basis by JAMES and GUTH[2]. It is because the network has these properties that the simplified treatments (e.g. the FLORY-REHNER model) lead to a correct solution.

4. *Distribution of chain lengths.*

In the normal vulcanization technique cross-links are introduced in a random manner. Consequently the assumption that all the chains have the same chain length is not justified. From equation (V,55b), however it is clear that the chain length does not affect the *form* of the strain-energy function. This will also be unaffected by the presence of a distribution of chain lengths. The same considerations apply also to the effect of the chain vector lengths (end-to-end distances) in the unstrained state, if these differ from the root-mean-square length. Both effects are included in the more general analysis of JAMES and GUTH[3], which is applicable for any distribution of chain contour lengths and chain vector lengths. This analysis requires the factor $(1/2)\,N k T$ in (V,55a) to be replaced by

$$(1/2)\,k\,T\sum Z_\tau\,\lambda_\tau^2 \qquad \text{(V, 68)}$$

where Z_τ is the number of links in the τth. chain, and λ_τ is its mean

[1] TRELOAR, L. R. G.: Trans. Faraday Soc. **42**, 83 (1946).
[2] JAMES, H. M. u. E. GUTH: J. Chem. Physics **11**, 455 (1943).
[3] JAMES, H. M. u. E. GUTH: J. Chem. Physics **15**, 669 (1947).

fractional extension in the unstrained state. If λ_τ corresponds to the R.M.S. length, then $\lambda_\tau^2 = 1/Z_\tau$ and (V,68) reduces to (1/2) NkT.

5. *Network defects.*

It is implicitly assumed in the foregoing theory that every chain in the network is connected at each end to three other chains. A consideration of the process of cross-linking, however, shows that this is an over-simplified representation of the structure, and that various types of imperfection, such as, for example, chains attached to the structure at one point only, cross-linkages between two points on the same chain forming closed loops, etc., will necessarily be present. Generally, the result of such defects will be that a certain number of the cross-linkages are rendered ineffective, since it is only those chains which form true network segments which take part in the elastic deformation.

In a quantitative treatment of the effect of chains with only one point of attachment, or "loose ends", FLORY[1] argues on the basis that, starting with N_0 original polymer molecules, it requires $N_0 - 1$ cross-linkages for these to be interconnected so as to form a single "giant" molecule. It is only after these are formed that the true network structure comes into existence. Since each subsequent cross-linkage produces two network segments, or chains, the effective number of chains, $\nu/2$, is given by

$$\nu = 2(\nu_0/2 - N_0) = \nu_0 - 2N_0 = \nu_0(1 - 2N_0/\nu_0)$$

where $\nu_0/2$ is the total number of cross-links. In terms of molecular weights, this may be written

$$\nu = \nu_0(1 - 2M_c/M) \tag{V,69}$$

where M is the molecular weight of the primary polymer, and M_c the mean molecular weight of the active chains. With this correction, the factor $(\varrho RT/2M_c)$ in (V,55b) becomes

$$\frac{\varrho RT}{2M_c}(1 - 2M_c/M). \tag{V,70}$$

§ 31. Comparison of observed properties with theory.

a) Experiments on elongation, compression and shear.

The analysis given above shows that the elastic properties of a rubber which conforms to the Gaussian statistics are describable in terms of a single elastic constant, which is related to the number of chains, or degree of cross-linking, in the network. The particular equations derived from the theory for the three simple types of deformation, elongation, unidirectional compression and shear, provide at once a basis for examining the quantitative validity of the theory as a means of representing the primary features of the elastic behaviour of an actual rubber. An experimental investigation on these lines, using vulcanized natural rubber,

[1] FLORY, P. J.: Chem. Reviews 35, 51 (1944).

has been carried out by the author[1]. The force-extension curve for simple elongation is shown in fig. V,17. Comparison with the theoretical form (V,63) reveals a somewhat greater curvature in the region of low strains, and a marked deviation at high strains; the value of G required to give agreement in slope at the origin is 4,0 kg/cm². These deviations from the theoretical form are discussed later; they are characteristic and significant. It is sufficient here to recall that the Gaussian statistics are not applicable in the region of very large strains.

The compression curve was obtained indirectly, by a two-dimensional extension of the sheet, a compression in the ratio λ in one dimension being equivalent to an extension in the ratio $1/\lambda^{1/2}$ in the other two dimensions (fig. V,15). The equivalent compressive stress is found by superposing on the deformed body a hydrostatic pressure equal to the tensile stress in the sheet; it is, in fact, equal to this tensile stress. Fig. V,18

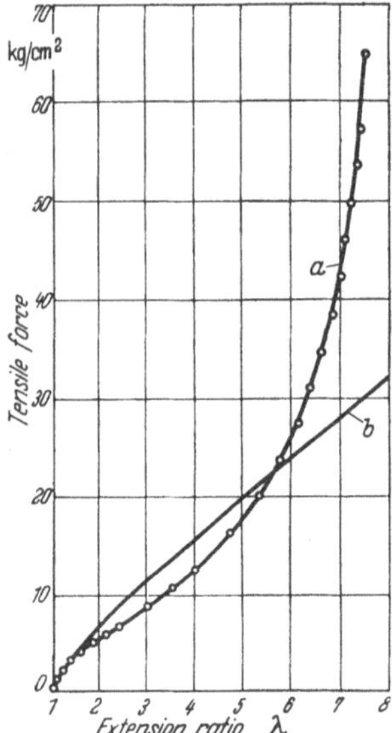

Fig. V,17. Force-extension curve for vulcanized rubber, simple elongation. a) Experimental, b) Equation (V,63), with $G = 4{,}0$.

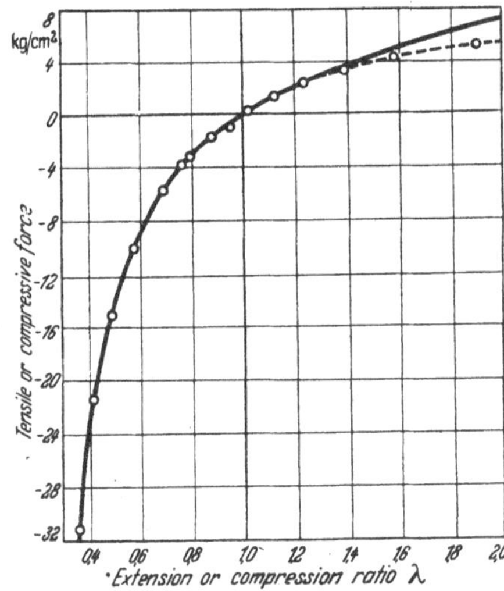

Fig. V,18. Compression data for vulcanized rubber, calculated from 2-dimensional extension. (Data for extension included.) ○ Experimental — Equation (V,63), with $G = 4{,}0$.

shows the compressive *force* (per unit area in the deformed state) as a function of the compression ratio for the same sheet of rubber as that represented in (V,17). The preceding data for simple elongation are included. It is seen that equation (V,63) adequately represents the behaviour under compression, and that the extension and compression data fall on a single continuous curve.

The data for *simple* shear, using the same material, were also obtained indirectly, from an experiment on *pure* shear. From (V,64) it is

[1] TRELOAR, L. R. G.: Trans. Faraday Soc. 40, 59 (1944).

seen that a shear is a type of deformation in which one dimension is unchanged. This may be effected by stretching a sheet in such a way that one dimension is held constant, for example by applying a tensile stress to a sheet whose "length" in the direction of stretch is small compared with its "width". If f is the tensile force, per unit area in the undeformed state, corresponding to a principal extension ratio λ, the equivalent shear stress σ_{xy} is

$$\sigma_{xy} = \frac{dW}{d\gamma} = \frac{dW}{d\lambda} \cdot \frac{d\lambda}{d\gamma} = f \cdot \frac{d\lambda}{d\gamma} \quad (\text{V}, 71)$$

where γ, the shear strain, is given by (V, 65), and $d\lambda/d\gamma$ is therefore $\lambda^2/(1+\lambda^2)$. The stress-strain relation for simple shear, calculated in this way, is shown in fig. V, 19; it is approximately linear, with an initial slope of 4,0 kg/cm².

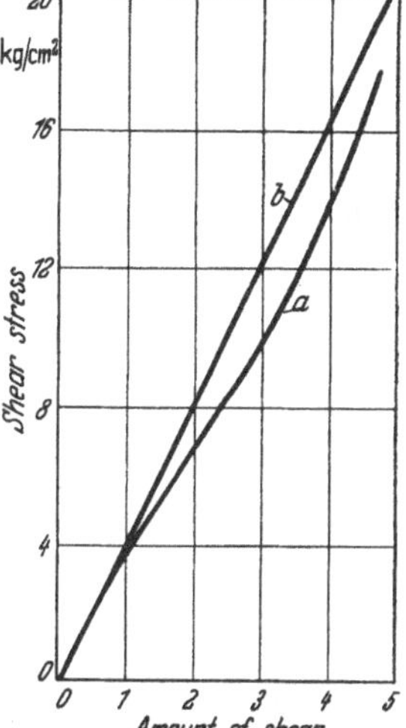

Fig. V, 19. Stress-strain relation for vulcanized rubber in simple shear, calculated from pure shear. a) Experimental, b) Equation (V, 67) with $G = 4{,}0$.

These experiments show the extent to which the Gaussian statistical theory is applicable to a particular vulcanized rubber. While not quantitatively accurate, the theory does correctly describe the general form of the curves in terms of a single adjustable parameter; it does, therefore, provide a sound basis for understanding, in a simple manner, the relations between the stress-strain curves corresponding to different types of strain. The significance of the deviations which occur will be considered in later sections.

b) Absolute value of the modulus.

1. Rubbers.

Equation (V, 63), or the modified form (V, 70), predicts a numerical relation between modulus and chain molecular weight, or degree of cross-linking. It is not easy to test this relation quantitatively because of the difficulty of providing an independent measure of the degree of cross-linking. The most hopeful line of attack is to look for a chemical reaction which produces inter-chain linkages in a quantitative manner. Ordinary sulphur vulcanizates, for example, are not satisfactory, because only a fraction of the combined sulphur is present in the form of simple monosulphide cross-linkages, the remainder being combined either in the form of polysulphide linkages or as ineffective attachments to the chain. The efficiency of cross-linking, moreover, is a function of the conditions of vulcanization, and is particularly affected by the presence of accelerators and other ingredients [1, 2].

The most extensive attempts to find a quantitative method of cross-linking have been made by Flory and co-workers. The most direct of

[1] Gee, G.: J. Polymer. Sci. **2**, 451 (1947).
[2] Flory, P. J.: *Principles of Polymer Chemistry,* Cornell 1953, p. 456.

these experiments were based on the use of bis-azo-dicarboxylates, having the structure typified by

$$R\langle\begin{matrix} O-CO-N=N-CO-O-CH_3 \\ O-CO-N=N-CO-O-CH_3 \end{matrix} \qquad (V,72)$$

in which reaction with a rubber chain occurs at each of the double bonds. The results obtained[1] in this way for natural rubber and GR—S (butadiene-styrene) respectively are shown in figs. V,20 and V,21, together with the theoretical lines representing the stress calculated on the basis of (V,63), assuming 100 per cent efficiency of cross-linking.

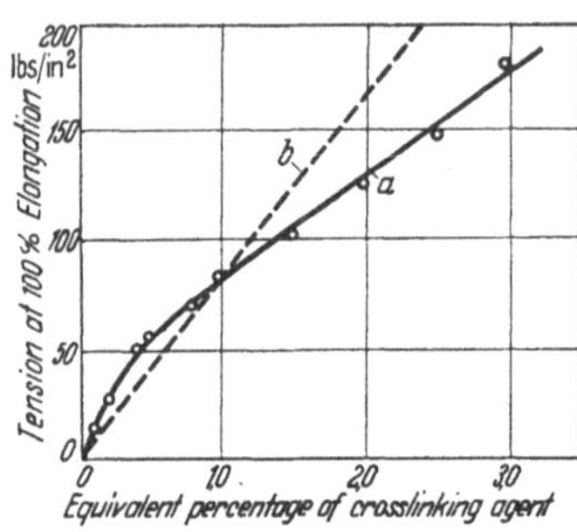

Fig. V,20. Dependence of force at 100% strain on degree of cross-linking. Natural rubber. *a*) Experimental, *b*) Theoretical. [FLORY, RABJOHN and SHAFFER: J. Polymer Sci. 4, 225 (1949).]

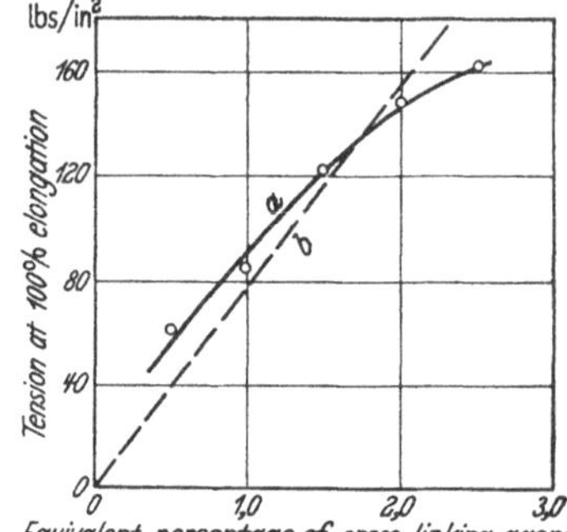

Fig. V,21. Dependence of force at 100% strain on degree of cross-linking. GR—S rubber. *a*) Experimental, *b*) Theoretical. [FLORY, RABJOHN and SHAFFER: J. Polymer Sci. 4, 225 (1949).]

2. *Multi-linked polyamides.*

An alternative attack, by SCHAEFGEN and FLORY[2] made use of synthetic polyamides of known structure which could be linked together with a diamine to form a network. The original polymer had the structure

$$R\{-CO[-NH(CH_2)_5CO-]_y OH\}_b \qquad (V,73)$$

in which b represents the number of chains proceeding from the group R, and y the mean chain length. Schematically the process may be represented, for the case when $b = 4$, as below

$$\begin{matrix} OH \\ | \\ OH-R-OH \\ | \\ OH \end{matrix} + H_2NR'NH_2 + \begin{matrix} OH \\ | \\ OH-R-OH \\ | \\ OH \end{matrix}$$

the groups R forming the "cross-linkages" in the final network. The number of chains was determined from a chemical estimation of the number of terminal groups prior to interlinking.

---

[1] FLORY, P. J., N. RABJOHN u. M. C. SHAFFER: J. Polymer Sci. 4, 225 (1949).
[2] SCHAEFGEN, J. R. and P. J. FLORY: J. Amer. Chem. Soc. 72, 689 (1950).

These polymers, like the nylons, to which they are closely related, are crystalline at room temperature, and assume rubberlike properties only on raising the temperature above the crystal melting point (ca. 200° C). Unfortunately at these high temperatures chemical instability leading to permanent "creep" or flow effects introduces complications, and makes it necessary to extrapolate the "creep" curves to zero time so as to obtain an estimate of the elastic strain for each applied load. The force-extension curves thus derived have the characteristic rubberlike form, and lie reasonably close to the theoretical curves.

Neither of these experiments is entirely conclusive. The use of the diazo-carboxylates introduces a form of "cross-link" which does not correspond to a simple junction point. In a typical compound the R in (V,73) was $C_{10}H_{20}$, and the total chain length between points of attachment (the double bonds) was not necessarily negligible. Moreover, there is a certain probability of the two reactive groups attaching themselves to two points on the same chain, with the formation of intra-molecular loops making no contribution to the network elasticity. Thus although the inter-linked polyamides provide a more satisfactory basis for estimating the degree of cross-linking, this advantage is to some extent offset by the uncertainty introduced by the poor elastic properties of these materials at the high temperature.

In addition to these experimental difficulties, the obvious oversimplification of the theoretical model should be borne in mind. For a theory of this generality, in which the finer points of the detailed network structure of a particular material are not included, it would be unrealistic to expect the absolute value of the modulus to be represented with strict accuracy. Taking into account all these factors therefore, the degree of agreement between theory and experiment revealed by the observations referred to above, as well as by a number of other investigations not specifically mentioned, may be regarded as satisfactory. Further quantitative illustrations pointing to the same conclusion are provided by the closely-related phenomena of swelling in solvents, which is also a function of the degree of cross-linking. This subject is discussed later.

§ 32. The non-Gaussian network.

a) Departures from Gaussian theory.

The departure of the forceextension curve from the Gaussian expression (V,63) becomes increasingly serious as the strain approaches the maximum extensibility of the network (fig. V,17). Similar departures are found also with other types of deformation, e. g. twodimensional extension or pure shear[1]. In this region the Gaussian approximation (V,45) to the chain probability function becomes inadequate, and one or other of the higher approximations discussed in § 28 have to be introduced.

The non-Gaussian network theory has neither the simplicity nor the generality of the Gaussian theory. The Gaussian theory gives an exact

[1] Treloar, L. R. G.: Trans. Faraday Soc. 40, 59 (1944).

solution to the problem of the idealized network in the region of small and moderate strains; this solution, as JAMES and GUTH[1, 2] have shown, is perfectly general, i.e. it is not dependent on the detailed structure of the network. There is no correspondingly exact general theory of the non-Gaussian network.

b) Simplified theory; three-chain model.

A simple treatment of the non-Gaussian network, which illustrates its essential properties, and at the same time provides a basis for discussion of the more sophisticated developments, is based on the three-chain model, the vectors of the chains lying along OX, OY, OZ. The inverse LANGEVIN expression (V, 36a) is used to represent the entropy for a random chain of Z links, each of length a, as a function of its length. Assuming the chains to be deformed in the same ratio as the bulk material (affine deformation) an extension λ in the direction OX changes the length of the corresponding chain from its initial value h_0 to

$$h_x = h_0 \lambda \qquad \text{(V, 74)}$$

while the other two become

$$h_y = h_z = h_0/\lambda^{1/2}. \qquad \text{(V, 75)}$$

The entropy for the three chains in the deformed state is therefore, from (V, 37)

$$\left.\begin{aligned} s_x + 2s_y = &-kZ\left[\frac{h_0\lambda}{Za}\mathfrak{L}^{-1}\left(\frac{h_0\lambda}{Za}\right) + \log\frac{\mathfrak{L}^{-1}(h_0\lambda/Za)}{\sinh\mathfrak{L}^{-1}(h_0\lambda/Za)}\right] \\ &-2kZ\left[\frac{h_0\lambda^{-1/2}}{Za}\mathfrak{L}^{-1}\left(\frac{h_0\lambda^{-1/2}}{Za}\right) + \log\frac{\mathfrak{L}^{-1}(h_0\lambda^{-1/2}/Za)}{\sinh\mathfrak{L}^{-1}(h_0\lambda^{-1/2}/Za)}\right]\end{aligned}\right\} \qquad \text{(V, 76)}$$

If there are N chains per unit volume, the entropy per unit volume will be $(N/3)(s_x + 2s_y)$; the tensile force is therefore

$$f = -\frac{NkT}{3}\cdot\frac{d(s_x + 2s_y)}{d\lambda}. \qquad \text{(V, 77)}$$

Making use of the result (V, 39) for the differential coefficient of (V, 37) with respect to h, one obtains in this way, from (V, 76)

$$f = \frac{NkT}{3}\cdot\frac{h_0}{a}\left[\mathfrak{L}^{-1}(h_0\lambda/Za) - \lambda^{-3/2}\mathfrak{L}^{-1}(h_0\lambda^{-1/2}/Za)\right]. \qquad \text{(V. 78)}$$

If the length h_0 in the unstrained state is given its root-mean-square value $aZ^{1/2}$ (V, 78) becomes

$$F = \frac{NkT}{3}\cdot Z^{1/2}\left[\mathfrak{L}^{-1}(\lambda Z^{-1/2}) - \lambda^{-3/2}\mathfrak{L}^{-1}(\lambda^{-1/2}Z^{-1/2})\right]. \qquad \text{(V, 79)}$$

In the Gaussian region $\mathfrak{L}^{-1}(t) = 3t$, (V, 39a), and (V, 79) reduces to the usual form (Cf. V, 63)

$$f = NkT(\lambda - 1/\lambda^2). \qquad \text{(V, 80)}$$

[1] JAMES, H. M. u. E. GUTH: J. Chem. Physics **11**, 455 (1943).
[2] JAMES, H. M. u. E. GUTH: J. Chem. Physics **15**, 669 (1947).

This form of the theory is equivalent to a treatment first given by JAMES and GUTH[1], though the method of derivation of their formula, which is essentially the same as (V,79), was by means of the expression (V,39) for the chain tension. They point out that the assumption of an affine deformation of the chains, in the non-Gaussian region of extension, is inexact, though it may be a useful approximation, except at extreme extensions.

The form of the force-extension curves, as calculated from equation (V,79), using three values of Z, is shown in fig. V,22. In the Gaussian region the quantity f/NkT is independent of Z; departures from this form occur at values of extension determined by the value of Z. There are thus two adjustable parameters, of which one, N, the number of chains per unit volume, or degree of cross-linking, determines the vertical scale, or modulus in the Gaussian region, while the other, Z, the number of random links in the chain, determines the horizontal scale, or maximum range of extension of the network, which is equal to $Z^{1/2}$.

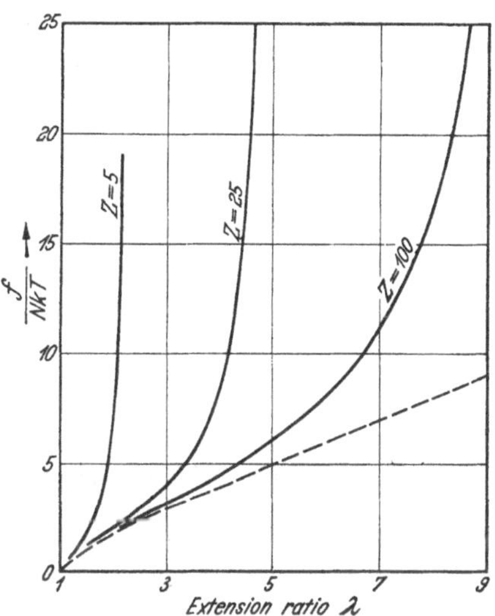

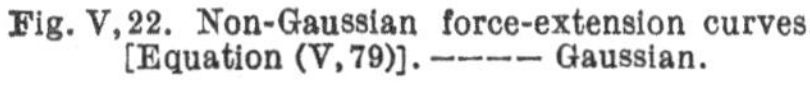
Fig. V,22. Non-Gaussian force-extension curves. [Equation (V,79)]. ----- Gaussian.

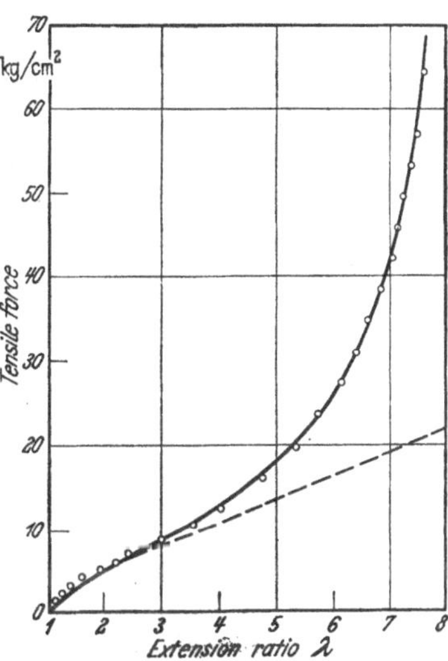

Fig. V,23. Non-Gaussian force-extension curve. ○ Experimental (fig. V,17) — Equation (V,79). $Z = 75$, $NkT = 2{,}78$ kg/cm². --- Gaussian.

For a given molecular structure the parameters N and Z are not, in fact, independently variable, but are both related to the degree of cross-linking. However, if they are regarded for the moment as unrelated, it is possible by means of equation (V,79) to obtain a very close approximation to the actual force-extension curve for a vulcanized rubber (fig. V,23).

c) Further developments of non-Gaussian theory.

The above simple model introduces the most important correction to the non-Gaussian theory, namely the more accurate chain-length distri-

[1] JAMES, H. M. u. E. GUTH: J. Chem. Physics **11**, 455 (1943).

bution function, and the very much better agreement with the observed behaviour of vulcanized rubber obtained with this modification justifies the belief that this comparatively simple solution contains the essential elements of the problem. This tentative conclusion seems to be substantiated by the various attempts at a more exact analysis of the non-Gaussian network, which yield (at a considerable cost in mathematical complexity) only minor refinements of the fundamental pattern already established. These further developments will therefore be considered only briefly.

The most serious error in the simplified treatment given above lies in the assumption of an affine deformation of the chain vector lengths. The author has attempted to examine the effect of this assumption by using the four-chain (tetrahedral) model of FLORY and REHNER (see page 313) and allowing the central junction point to take up its true equilibrium position under the action of the tensions in the four chains[1]. The effect of this extra freedom is to increase the maximum extensibility by a relatively small amount (roughly 25 per cent). Fig. V,24 shows the force-extension curve calculated on this model, with $Z = 25$, (*a*) assuming an affine deformation of the central junction and (*b*) with the central junction point fixed at its most probable (i.e. equilibrium) position. Unfortunately it is not possible to represent these curves by a general formula; each one has to be calculated numerically.

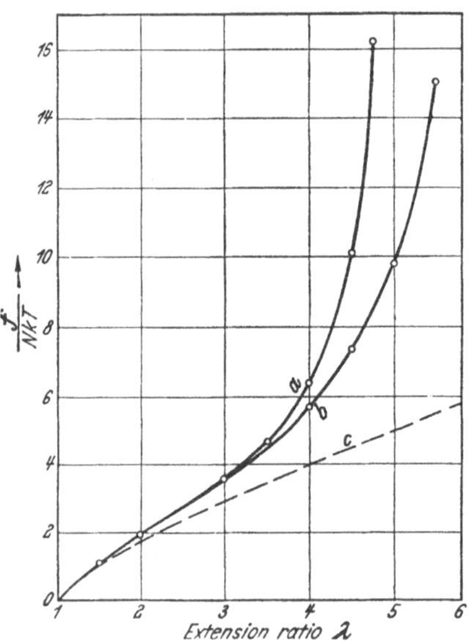

Fig. V,24. Non-Gaussian force-extension curve, 4-chain model, $Z = 25$. *a*) Affine displacement, *b*) Non-affine displacement of central junction point, *c*) Gaussian.

The next refinement is concerned with the effect of *fixing* the junction points at their equilibrium positions rather than allowing them to fluctuate in position. In the Gaussian case, as we saw, this restriction makes no difference to the entropy of deformation. Using the four chain model, the author[1] found that even in the non-Gaussian region it was of negligible importance, except possibly at values of Z as small as 5. A similar conclusion was arrived at by WANG and GUTH[2], on the basis of a more general argument.

WANG and GUTH[2] have discussed the properties of the non-Gaussian network in more general mathematical terms. However, on account of

[1] TRELOAR, L. R. G.: Trans. Faraday Soc. **50**, 881 (1954).
[2] WANG, M. C. u. E. GUTH: J. Chem. Physics **20**, 1144 (1952).

the complexity of the non-Gaussian distribution function (V, 37), progress can only be made by expanding it in the form of a series and utilizing a small number of terms (generally two). Clearly, this approach is of limited value; in particular, it tells us nothing about the limiting extensibility of the network.

In all these treatments all the chains are assumed to have the same contour length, represented by the parameter Z. This assumption, which is clearly at variance with the conditions of cross-linking in the vulcanization process, further limits the quantitative applicability of the non-Gaussian theory to an actual rubber.

d) The equivalent statistical link.

Returning now to the question of the evaluation of the parameters N and Z, which determine respectively the modulus and the extensibility of the network, it is necessary to relate these quantities to specific molecular properties. The quantity N, as we have seen [equations (V, 55 a) and (V, 55 b)] is related to the chain molecular weight M_c, i. e.

$$N k T = \varrho R T / M_c. \tag{V,81}$$

For a chain of given M_c, the value of Z, the number of links in the equivalent random chain, is a function of the chemical structure of the molecule. The reasonableness of the assumption of an equivalent random chain has already been discussed; it implies that the actual chain (with its specific valence angle structure) is statistically identical to a random chain having Z freely-jointed links each of length a. If M_s is the molecular weight of the submolecule or monomer unit of the structure, the chain of molecular weight M_c contains M_c/M_s monomers. Comparing this with a random chain of Z links, the number of monomer units per random link (A_m) is given by

$$A_m = \frac{M_c/M_s}{Z}. \tag{V,82}$$

Since A_m is specific to the molecular structure, the value of Z, and hence of the network extensibility, corresponding to a given value of modulus, will depend on the chemical structure of the rubber. But for a particular type of rubber the maximum theoretical extensibility should vary inversely as the square root of the degree of cross-linking, i. e. as the square root of the modulus.

In principle, it should be possible, by adjusting both N and Z to give the best fit to an experimental force-extension curve, to obtain an estimate of A_m. This may be illustrated by means of the curve in fig. V, 23 which has the parameters

$$N k T = \varrho R T / M_c = 2{,}78 \text{ kg/cm}^2; \quad Z = 75.$$

Taking $\varrho = 0{,}93$, $R = 8{,}314 \times 10^7$ ergs $= 84{,}75$ kg · cm, $T = 293°$ K, one obtains $M_c = 8307$. For isoprene $M_s = 68$, hence, from (V, 82), $A_m = 1{,}63$.

This is to be compared with the value $A_m = 0{,}71$ obtained from the geometry of the chain (page 310). While the above estimate is not quantitatively reliable, the difference between these two figures is probably significant, and may be taken as an indication of the effect of restricted rotation in reducing the flexibility of the chain. Further evidence pointing in the same direction will be discussed in the section dealing with the optical properties of the network.

§ 33. Phenomenological theory of large elastic deformations.

a) Introduction.

The theory of large elastic deformations has for its object the provision of a mathematical structure for the solution of problems involving a non-uniform distribution of stress and strain. In any such theory two types of consideration are involved. First, a method has to be found for representing the elastic properties of a given material in the most general way, and secondly, mathematical relations representing the variation of stress and strain from point to point of the deformed body have to be evolved. In this section we shall be concerned only with the first of these two considerations, for it is only this which involves a problem in physics; the remainder is pure mathematics.

It can be shown that for the purpose of the theory it is sufficient to define the elastic properties of the material in a *pure homogeneous* strain. The state of *pure* strain may be derived from the unstrained state by extensions along three mutually perpendicular axes, *without rotation.* A strain which is not pure involves, in addition, a rotation of the axes. This rotation requires no work to be done against the elastic stresses in the body, hence it does not enter into the description of the fundamental elastic properties. If the strain is not homogeneous, but varies from point to point, the body may be divided up into small elements of volume over each of which the state of strain is effectively homogeneous. The resultant stress distribution may then be related to the local stresses due to these homogeneously strained elements.

The fundamental elastic properties of the material may thus be completely defined with respect to the state of pure homogeneous strain. These properties may be expressed in terms of the elastically stored free energy per unit volume, or stored-energy function W. Alternatively, they may be expressed in terms of generalized stress-strain relations derivable from the stored-energy function. The stored-energy function is generally the more convenient method of representation, but for comparison with experiment the stress-strain relations are more direct. These are obtainable by partial differentiation of the stored-energy function. As we have seen, the Gaussian network theory leads to a stored-energy function of the form (page 315)

$$W = (1/2)\, G\,(\lambda_1^2 + \lambda_2^2 + \lambda_3^2 - 3) \qquad (\text{V},83)$$

where λ_1, λ_2 and λ_3 are the principal extension ratios.

The corresponding stress-strain relations are of the type

$$\sigma_1 - \sigma_2 = G\,(\lambda_1^2 - \lambda_2^2) \qquad (\text{V},84)$$

where σ_1, σ_2 and σ_3 are the principal stresses. Both the stored-energy function and the stress-strain relations involve only one physical constant of the material.

b) Pure homogeneous strain.

The application of a pure homogeneous strain of the most general type, i.e. one in which the principal extension ratios may be independently varied, under conditions in which the stresses may be measured, is not easy. The method adopted by the author[1] was to attach strings to a number of lugs projecting from the sides of a sheet of rubber cut to the form shown in fig. V, 25. By applying known forces to these strings the inner square *ABCD* could be deformed to a rectangle whose sides gave a measure of the two principal extension ratios λ_1 and λ_2. The third ratio, λ_3 (in the thickness direction) was obtained from the incompressibility condition

$$\lambda_1 \lambda_2 \lambda_3 = 1. \qquad \text{(V, 85)}$$

A plot of $\sigma_1 - \sigma_2$ against $\lambda_1^2 - \lambda_2^2$ (fig. V, 26) yielded an approximate straight line, in accordance with the theoretical form (V, 84). However,

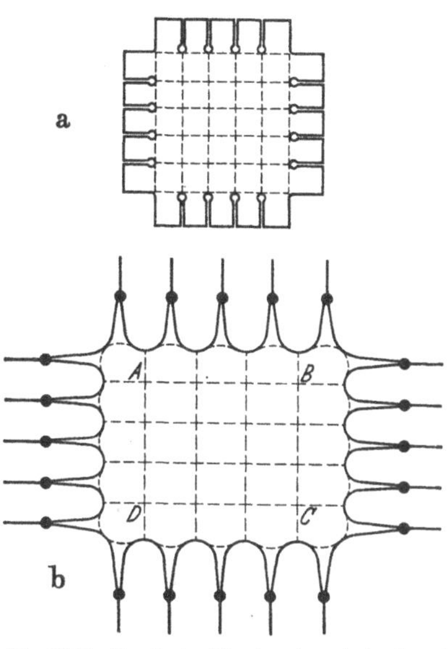

Fig. V, 25. Sheet of rubber in a) unstrained, and b) strained state. The middle area *ABCD* undergoes pure homogeneous strain.

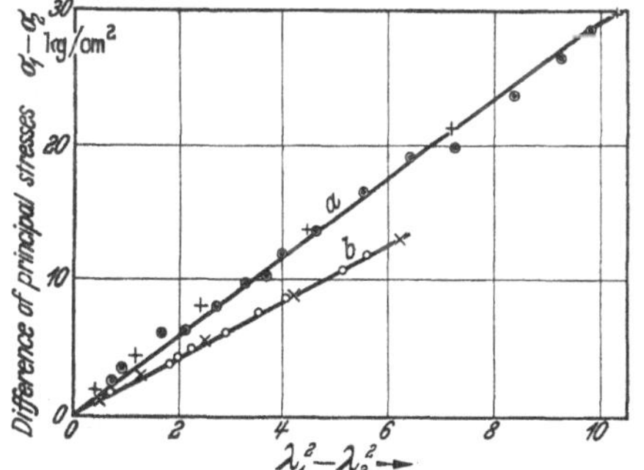

Fig. V, 26. Pure homogeneous strain. Difference of principal stresses in plane of sheet, $\sigma_1 - \sigma_2$, plotted against $\lambda_1^2 - \lambda_2^2$. a) dry, b) swollen in paraffin to $v/2 = 0{,}525$.

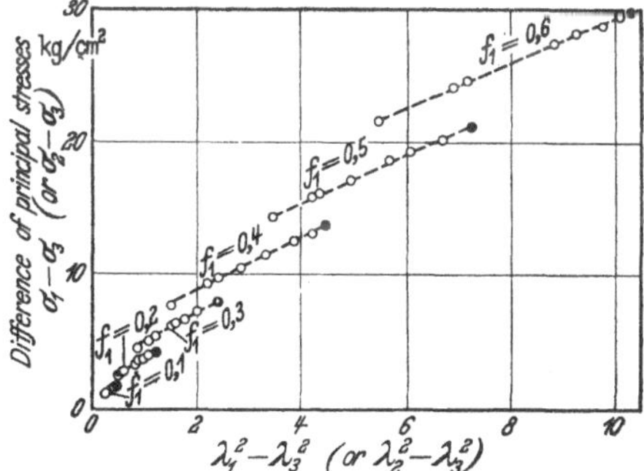

Fig. V, 27. Pure homogeneous strain. Difference of principal stresses $\sigma_1 - \sigma_3$ (or $\sigma_2 - \sigma_3$) plotted against $\lambda_1^2 - \lambda_3^2$ (or $\lambda_2^2 - \lambda_3^2$).

when $\sigma_1 - \sigma_3$ ($\sigma_3 = 0$) and $\sigma_2 - \sigma_3$ were plotted against $\lambda_1^2 - \lambda_3^2$ or $\lambda_2^2 - \lambda_3^2$ respectively this relation was found to be inapplicable (fig. V, 27).

[1] Treloar, L. R. G.: Proc. Physic. Soc. **60**, 135 (1948).

c) The Mooney stored-energy function.

From the assumption of a linear stress-strain relation in simple shear (or for a shear superposed on simple elongation), Mooney derived the 2-constant stored-energy function[1]

$$W = C_1(\lambda_1^2 + \lambda_2^2 + \lambda_3^2 - 3) + C_2(1/\lambda_1^2 + 1/\lambda_2^2 + 1/\lambda_3^2 - 3) \qquad \text{(V,86)}$$

from which the following stress-strain relations may be derived

$$\left.\begin{aligned} \sigma_1 - \sigma_3 &= 2(C_1 + C_2\lambda_3^2)(\lambda_1^2 - \lambda_2^2) \\ \sigma_1 - \sigma_3 &= 2(C_1 + C_2\lambda_2^2)(\lambda_1^2 - \lambda_3^2)\,. \end{aligned}\right\} \qquad \text{(V,87)}$$

Taking $\sigma_3 = 0$, and assuming a small value for C_2 (e.g. $C_2/C_1 = 0{,}1$ or $0{,}05$) these equations give a form of plot which reproduces some of the features of the experimental data (fig. V,28). Since λ_3 is less than 1 and

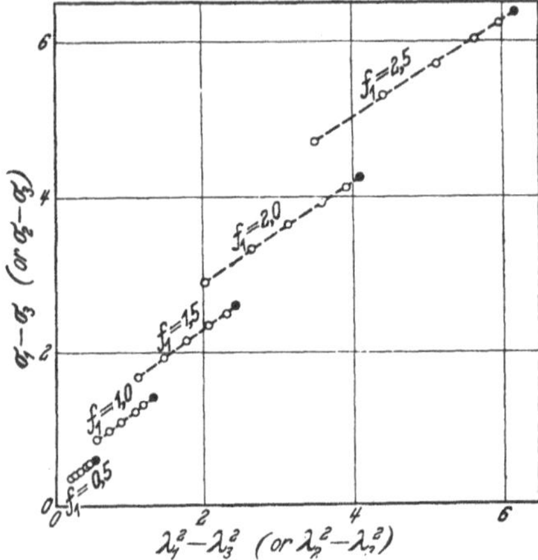

Fig. V,28. Mooney equation with $2C_1 = 1{,}0$; $2C_2 = 0{,}1$. Plot of $\sigma_1 - \sigma_2$ against $\lambda_1^2 - \lambda_2^2$.

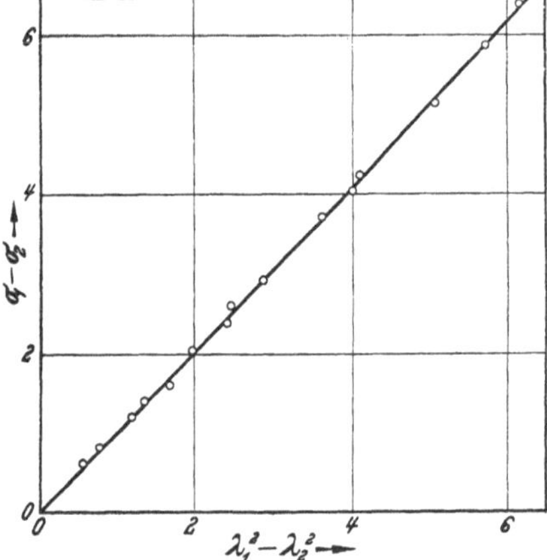

Fig. V,29. Mooney equation with $2C_1 = 1{,}0$; $2C_2 = 0{,}1$. Plot of $\sigma_1 - \sigma_3$ (or $\sigma_2 - \sigma_3$) against $\lambda_1^2 - \lambda_3^2$ (or $\lambda_2^2 - \lambda_3^2$).

diminishes rapidly as λ_1 and λ_2 increase, the term $C_2\lambda_3^2$ in the first of equations (V,87) is generally small compared with C_1, and the expression for $\sigma_1 - \sigma_2$ therefore approximates to the statistical form (V,84) (fig. V,29). The paradox of the apparent agreement with the statistical theory for the stress difference $\sigma_1 - \sigma_2$ coupled with disagreement for the stress difference $\sigma_1 - \sigma_3$ is thus resolved.

From equation (V,87) it follows that a plot of $\sigma_1 - \sigma_3$ against $[1 + (C_2/C_1)\lambda_2^2](\lambda_1^2 - \lambda_3^2)$ should yield a straight line. Using the best value of C_2/C_1, namely 1/20, the experimental points could thus be brought on to a continuous curve, which, however, differed appreciably from a straight line[2]. It must therefore be concluded that the Mooney equation, while giving a better representation of the experimental data than the statistical theory, is still not entirely adequate.

---

[1] Mooney, M.: J. Appl. Physics **11**, 582 (1940).

[2] Treloar, L. R. G.: Proc. Physic. Soc. **60**, 135 (1948).

For a better understanding of the mechanical properties of vulcanized rubber and of the significance of the MOONEY equation, it is necessary to consider the more general theory of RIVLIN[1], and the experimental data of RIVLIN and SAUNDERS[2].

d) General theory.

RIVLIN's theory[1, 2] is concerned with the possible forms which the stored-energy function for an isotropic incompressible elastic solid may assume. The basic postulate that the material is isotropic in the unstrained state requires that the stored-energy function shall be a symmetrical function of the three principal extension ratios. It may also be shown to be an even-powered function of these quantities. It follows that any possible form of stored-energy function can be represented as a function of the three symmetrical expressions (or *strain invariants*)

$$\left.\begin{aligned} I_1 &= \lambda_1^2 + \lambda_2^2 + \lambda_3^2 \\ I_2 &= \lambda_1^2\lambda_2^2 + \lambda_2^2\lambda_3^2 + \lambda_3^2\lambda_1^2 \\ I_3 &= \lambda_1^2\lambda_2^2\lambda_3^2 . \end{aligned}\right\} \qquad (V,88)$$

Introducing the incompressibility condition (V, 85) we have $I_3 = 1$, and hence W is a function of the two independent variables I_1 and I_2. These may be written in the form

$$I_1 = \lambda_1^2 + \lambda_2^2 + \lambda_3^2; \quad I_2 = 1/\lambda_1^2 + 1/\lambda_2^2 + 1/\lambda_3^2 . \qquad (V,89)$$

The most general functional relationship may be expressed as the sum of a number of terms, thus

$$W = \sum_{i=0,\, j=0}^{\infty} C_{ij} (I_1 - 3)^i (I_2 - 3)^j; \quad C_{00} = 0 \qquad (V,90)$$

the quantities $(I_1 - 3)$ and $(I_2 - 3)$ being inserted so that W vanishes automatically at zero strain ($I_1 = I_2 = 3$). On this basis the two simplest possible forms of W are

$$W = C_{10} (I_1 - 3) \qquad (V,91)$$

$$W = C_{01} (I_2 - 3) . \qquad (V,92)$$

Of these (V,91) is the form derived from the statistical theory for the Gaussian network. The form (V,92) does not appear to have acquired any physical significance, but the addition of (V,91) and (V,92) gives the MOONEY equation

$$W = C_1 (I_1 - 3) + C_2 (I_2 - 3) . \qquad (V,93)$$

This equation, which involves only the first powers of I_1 and I_2, is thus seen to be from the mathematical standpoint the simplest possible expression for the stored energy in terms of these invariants.

---

[1] RIVLIN, R. S.: Philos. Trans. Roy. Soc. London **A 240**, 459 u. 491 (1948); **241**, 379 (1948).

[2] RIVLIN, R. S. u. D. W. SAUNDERS: Philos. Trans. Roy. Soc. London **A 243**, 251 (1951).

It is to be noted that for either pure or simple shear

$$I_1 = I_2 = \lambda^2 + 1 + 1/\lambda^2,$$

and the MOONEY equation becomes indistinguishable from the statistical form (V,91). It is easy to show that either of these forms leads to a linear relation between shear stress and shear strain in simple shear.

The stress-strain relations may be derived from the stored-energy function by a method comparable with that set out on page 315 for the particular form (V,91). These may be expressed in terms of the partial derivatives of W with respect to I, and I_2 thus[1],

$$\sigma_1 - \sigma_2 = 2(\lambda_1^2 - \lambda_2^2)\left(\frac{\partial W}{\partial I_1} + \lambda_3^2 \frac{\partial W}{\partial I_2}\right) \qquad (V,94)$$

with similar expressions for $\sigma_2 - \sigma_3$ and $\sigma_3 - \sigma_1$. As in the statistical theory, the incompressibility condition requires that the stresses shall be indeterminate to the extent of an arbitrary hydrostatic pressure.

e) The experiments of RIVLIN and SAUNDERS.

Experimentally, the form of W may be obtained from measurements of the principal stresses corresponding to a given state of strain. These give directly the partial derivatives $\partial W/\partial I_1$ and $\partial W/\partial I_2$. A complete study would therefore aim at determining each of these partial derivatives as a function of I_1 and I_2.

An experiment along these lines, which is a modification of the author's original study of pure homogeneous strain, was carried out by RIVLIN and SAUNDERS[1]. Instead of varying λ_1 and λ_2, the principal extension ratios in the plane of the sheet, in an arbitrary manner, these authors arranged the strains so that I_1 was kept constant while I_2 was varied in steps, and conversely. Fig. V,30 (a) shows their values of $\partial W/\partial I_1$ plotted against I_1, while fig. V,30 (b) gives the corresponding plot for $\partial W/\partial I_2$. The data are consistent with a constant value of $\partial W/\partial I_1$ (independent of both I_1 and I_2) and a value of $\partial W/\partial I_2$ which is independent of I_1, but falls with increasing I_2. The stored-energy function is thus not precisely of the MOONEY form, ($\partial W/\partial I_1 =$ const., $\partial W/\partial I_2 =$ const.) but rather

$$W = C_1(I_1 - 3) + f(I_2 - 3) \qquad (V,95)$$

where $f(I_2 - 3)$ includes powers of higher order than the first.

Since the effect of higher-order terms in $I_2 - 3$ becomes negligible at small strains, it can be understood that the MOONEY formula should provide a valuable approximation so long as the strains are not too large. Moreover, since, over a limited range of I_2, $\partial W/\partial I_2$ may be regarded as approximately constant, it is possible that this equation may also give a useful approximation over a limited range of strain, or for a particular type of deformation, even when the strains are large. There is a danger, however, in applying it generally, without a careful examination of the

---

[1] RIVLIN, R. S. u. D. W. SAUNDERS: Philos. Trans. Roy. Soc. London **A 243**, 251 (1951).

type and magnitude of the strain and of the corresponding form of the stored-energy function for the particular rubber under investigation. This may be illustrated by the data given by RIVLIN and SAUNDERS for simple extension and compression. In terms of equation (V, 94) we have for this case,

$$\lambda_1 = \lambda; \quad \lambda_2 = \lambda_3 = 1/\lambda^{1/2}$$

$$\sigma_1 = \lambda f; \quad \sigma_2 = \sigma_3 = 0$$

and hence

$$\frac{\partial W}{\partial I_1} + \frac{1}{\lambda}\frac{\partial W}{\partial I_2} = \frac{f}{2(\lambda - 1/\lambda^2)}. \tag{V, 96}$$

A plot of $f/2(\lambda - 1/\lambda^2)$ against $1/\lambda$ therefore yields a curve whose slope gives $\partial W/\partial I_2$ and whose intercept at $1/\lambda = 0$ gives $\partial W/\partial I_1$. The Gaussian

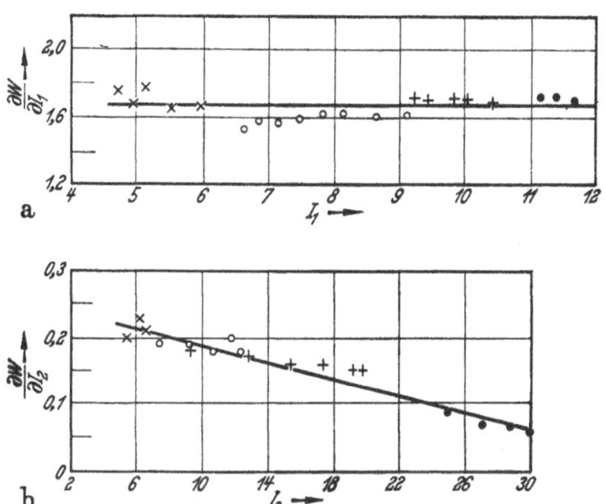

Fig. V, 30. a) Dependence of $\partial W/\partial I_1$ on I_1, b) Dependence of $\partial W/\partial I_2$ on I_2. Units: kg/cm². [RIVLIN and SAUNDERS: Phil. Trans. Roy. Soc. A 243, 251 (1951).]
a) × $I_2 = 5$, ○ $I_2 = 10$, + $I_2 = 20$, ● $I_2 = 30$.
b) × $I_1 = 5$, ○ $I_1 = 7$, + $I_1 = 9$, ● $I_1 = 11$.

network theory gives $\partial W/\partial I_1 = C_1$, $\partial W/\partial I_2 = 0$, and is represented by a horizontal straight line. This gives a fair representation of the compression data (fig. V, 31). The MOONEY equation, for which $\partial W/\partial I_1 = C_1$,

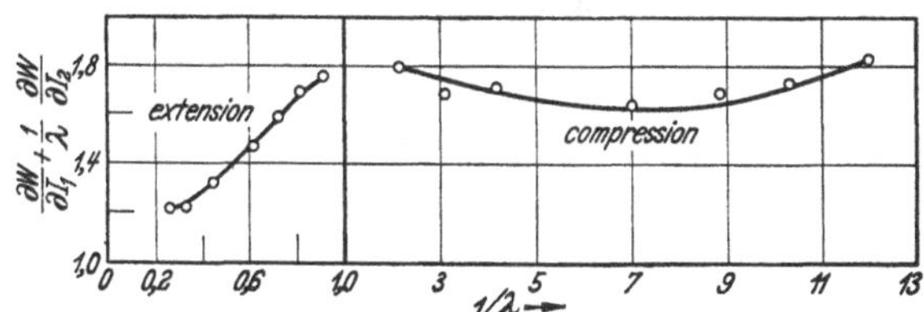

Fig. V, 31. Representation of data for extension and compression. Units: kg/cm². [RIVLIN and SAUNDERS: Phil. Trans. Roy. Soc. A 243, 251 (1951).] (Note change of scale at $1/\lambda = 1$.)

$\partial W/\partial I_2 = C_2$, is represented by a straight line whose inclination depends on C_2; the data for the extension region are consistent with this form.

(This would break down, however, at very large strains, due to the non-Gaussian effect.) It is important to note that it is not possible to obtain an accurate representation of the data for both extension and compression in terms of the MOONEY equation. This is consistent with the complex form of stored-energy function (V, 95) arrived at from the experiments on pure homogeneous strain.

f) General conclusion.

The general conclusion to be drawn from these studies may be set out in the following way. As a first approximation, particularly when simplicity of calculation is an important consideration, the stored-energy function derived from the Gaussian network theory may be used. For more accurate work the addition of a second term, of the MOONEY type, should give a very much closer approximation. It would not normally be profitable to pursue the investigation to include higher-order terms in I_1 and I_2, since these terms, being generally small, can only be evaluated on the basis of a rather extensive research. In any case, the properties of rubber are not usually sufficiently reproducible to justify this order of accuracy.

These remarks are only applicable so long as the strains are not too large. (The maximum values of λ in the experiments of RIVLIN and SAUNDERS were in the neighbourhood of 3,0.) In the region of very large strains (which are not usually encountered in engineering practice) the non-Gaussian effect discussed in § 31 predominates. This type of deviation can probably be most conveniently treated on the basis of the molecular theory, though it would also be possible to examine it from the purely phenomenological point of view.

Up to the present, no satisfactory molecular or structural basis for the MOONEY equation has been established. It is clear, however, that *any* departure from the Gaussian statistical theory will automatically introduce additional terms into the stored-energy function. The most likely additional term is the other first-order term (V, 92) which is *mathematically* on the same basis as (V, 91). These two together constitute the MOONEY equation. In order to establish a molecular basis for the MOONEY equation, therefore, it is not sufficient to show that a proposed model of the structure gives rise to a MOONEY term in the stored-energy function; it must also be shown that other terms, of higher order, are negligible. From this point of view, the attempt by ISIHARA, HASHITSUME and TATIBANA[1] to account for the MOONEY type of behaviour in terms of the non-Gaussian network statistics must be regarded as unsatisfactory. This has been shown by WANG and GUTH[2].

The phenomenological approach is essentially empirical, and the derivation of the form of the stored-energy function is the three-dimensional equivalent of curve-fitting. Since W is a function of the two independent variables I_1 and I_2, it may be represented by a surface in a rectangular

---

[1] ISIHARA, A., N. HASHITSUME u. M. TATIBANA: J. Physic. Soc. Japan **3**, 289 (1951). — [2] WANG, M. C. and E. GUTH: J. Chem. Phys. **20**, 1144 (1952).

system of coordinates. In this form of representation both the statistical form and the MOONEY equation correspond to plane surfaces. The MOONEY equation, being the more general, represents the first-order approximation to any curved surface (over a limited range of extension). The general problem is to determine the form of the representative surface over all values of the variables.

D. The photo-elastic properties of rubber.

§ 34. Optical anisotropy of long-chain molecules.

The phenomena of double refraction in strained rubber, considered as an amorphous cross-linked molecular network, find a simple explanation in terms of the statistical theory already developed to account for its mechanical properties. It is only necessary to assume that the links of the random chain (which is the statistical equivalent of the actual molecule) are themselves optically anisotropic, that is to say, that they may be characterized by two principal polarizabilities, one parallel to their length and the other in the transverse direction. On this basis, it may be shown that the random chain has optical properties which vary with the vector distance h between its ends. The first part of the problem of calculating the relation between double refraction and strain is to find the dependence of the principal polarizabilities of the single chain on h; the second part is to compute the principal polarizabilities for the whole network from the calculated chain polarizabilities and the distribution of vectors h. From these principal polarizabilities the corresponding principal refractive indices may be readily derived.

The solution to the single-chain problem was first given by KUHN and GRÜN[1]. They consider a random chain of Z equal links of length a. The principal polarizabilities of the single link being α_{01} and α_{02} for directions of electric vector respectively parallel and perpendicular to its length, it is required to find the mean principal polarizabilities of the whole chain as a function of the distance h between its ends.

Choosing a coordinate system such that h lies along the axis OX, the contribution of a link making the angle θ with OX to the total chain polarizability, for the direction of electric vector parallel to OX, is,

$$\alpha_{0x} = \alpha_{01}\cos^2\theta + \alpha_{02}\sin^2\theta . \qquad \text{(V, 97)}$$

If Φ is the angle which the plane containing the link and the axis OX makes with plane XOY, the polarizabilities parallel to OY and OZ respectively are

$$\alpha_{0y} = \alpha_{01}\cos^2\theta + \alpha_{02}\sin^2\theta$$

$$\alpha_{0z} = (\alpha_{01} - \alpha_{02})\sin^2\theta\sin^2\Phi + \alpha_{02} . \qquad \text{(V, 98)}$$

[1] KUHN, W. u. F. GRÜN: Kolloid-Z. **101**, 248 (1942).

The total polarizabilities in each of the three principal directions are obtained by integration over all the links, e.g., for the direction OX

$$\alpha_x = \int \alpha_{0x}\, dZ \tag{V,99}$$

where dZ is the number of links in the angular range $d\theta$. The derivation of the angular distribution has already been discussed (page 306); it is represented by

$$dZ_{\theta,\Phi} = (Z\beta/\sinh\beta)\, e^{-\beta\cos\theta} \cdot \frac{1}{2}\sin\theta\, d\theta \cdot \frac{d\Phi}{2\pi}\,;$$

$$\beta = \mathfrak{L}^{-1}(h/Za)\,. \tag{V,100}$$

Insertion of (V, 97) and (V, 100) in (V, 99) leads to an integral which when evaluated becomes

$$\alpha_x = Z\left[\alpha_{01} - (\alpha_{01} - \alpha_{02})\frac{2h/Za}{\mathfrak{L}^{-1}(h/Za)}\right]. \tag{V,101}$$

The corresponding polarizabilities along OY and OZ are

$$\alpha_y = \alpha_z = Z\left[\alpha_{02} + (\alpha_{01} - \alpha_{02})\frac{h/Za}{\mathfrak{L}^{-1}(h/Za)}\right]. \tag{V,102}$$

These two transverse polarizabilities are equal, as required by considerations of symmetry. The chain has therefore two principal polarizabilities, a longitudinal polarizability α_1 parallel to the line joining its ends, and a transverse polarizability α_2 at right angles to this direction. The optical anisotropy of the chain is represented by the difference of these two principal polarizabilities, i.e.

$$\alpha_1 - \alpha_2 = \alpha_x - \alpha_y = Z(\alpha_{01} - \alpha_{02})\left[1 - \frac{3h/Za}{\mathfrak{L}^{-1}(h/Za)}\right] \tag{V,103a}$$

$$= Z(\alpha_{01} - \alpha_{02})\left[\frac{3}{5}\left(\frac{h}{Za}\right)^2 + \frac{36}{175}\left(\frac{h}{Za}\right)^4 + \frac{108}{875}\left(\frac{h}{Za}\right)^6 + \cdots\right]. \tag{V,103b}$$

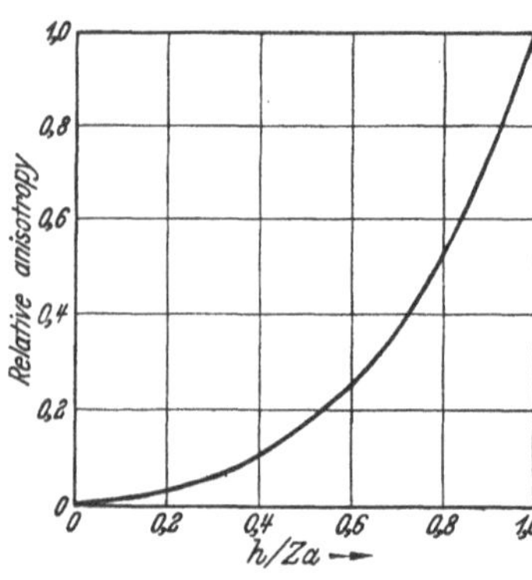

Fig. V, 32.
Optical anisotropy for single chain.

For sufficiently small extensions equation (V, 103b) reduces to

$$\alpha_1 - \alpha_2 = Z(\alpha_{01} - \alpha_{02})\frac{3}{5}\left(\frac{h}{Za}\right)^2 \tag{V,103c}$$

which corresponds to the Gaussian approximation in the corresponding expression for the entropy (page 306). For the limiting extension, when all the links are in line, ($h = Za$), the anisotropy becomes [from (V, 103b)]

$$\alpha_1 - \alpha_2 = Z(\alpha_{01} - \alpha_{02})$$

which is simply the sum of the anisotropies of the individual links. The function representing the relative anisotropy, $(\alpha_1 - \alpha_2)/Z(\alpha_{01} - \alpha_{02})$ is seen from (V, 103a) to involve only the *fractional* extension h/Za; it is therefore the same for any random chain. Its form is illustrated in fig. V, 32.

§ 35. Optical properties of molecular network.

a) The Gaussian network.

1. Derivation of strain-birefringence relation.

The optical properties of the whole network, in the Gaussian region, have been derived by KUHN and GRÜN[1] for the case of a simple extension (or uni-directional compression). It is assumed that the network is composed of N chains all of the same chain length (Z links), and that in the unstrained state all the chains have the same end-to-end distance (h_0) equal to the root-mean-square value for a free chain. It is further assumed that the components of length of the chains change in the same ratio as the corresponding dimensions of the bulk material (affine deformation) and that the volume is unchanged on deformation.

Taking an extension in the ratio λ along the direction OX, a chain whose h-vector, in the unstrained state, makes an angle θ_0 with OX will, in the deformed state, have a length h and angle θ defined by the equations

$$h\cos\theta = \lambda h_0 \cos\theta_0$$
$$h\sin\theta = \lambda^{-1/2} h_0 \sin\theta_0 . \qquad \text{(V,104)}$$

Whence

$$h_0^2 = h^2(\lambda^{-2}\cos^2\theta + \lambda\sin^2\theta)$$
$$\cos\theta_0 = \lambda^{-1}\cos\theta\,(\lambda^{-2}\cos^2\theta + \lambda\sin^2\theta)^{-1/2} . \qquad \text{(V,105)}$$

If Φ_0 is the angle made by the plane containing h_0 and OX with the plane XOY, and Φ the corresponding angle in the strained state, clearly $\Phi = \Phi_0$. The chain at (h, θ, Φ) contributes to the total polarizabilities along OX, OY, and OZ the respective amounts β_x, β_y, β_z, given by [Cf. equs. (V,97) and (V,98)],

$$\left.\begin{aligned} \beta_x &= \alpha_1\cos^2\theta + \alpha_2\sin^2\theta \\ \beta_y &= (\alpha_1-\alpha_2)\sin^2\theta\cos^2\Phi + \alpha_2 \\ \beta_z &= (\alpha_1-\alpha_2)\sin^2\theta\sin^2\Phi + \alpha_2 \end{aligned}\right\} \qquad \text{(V,106)}$$

in which, however, $\alpha_1 - \alpha_2$ is not constant, but is a function of h. Since the chains are distributed symmetrically around the axis OX, we shall be concerned only with the longitudinal and transverse polarizabilities, β_l and β_t, where

$$\beta_l = \beta_x , \quad \text{and} \quad \beta_t = (\beta_y + \beta_z)/2$$

and particularly with the difference between these two quantities, which, from (V,106) becomes

$$\beta_l - \beta_t = (\alpha_1 - \alpha_2)\left(\cos^2\theta - \frac{1}{2}\sin^2\theta\right) \qquad \text{(V,107)}$$

and is independent of Φ. The number of chains initially in the angular range θ_0 to $\theta_0 + d\theta_0$ is, by geometry

$$dN = (N/2)\sin\theta_0\, d\theta_0 . \qquad \text{(V,108)}$$

[1] KUHN, W. u. F. GRÜN: Kolloid-Z. **101**, 248 (1942).

Referred to the angle θ in the strained state this becomes, with (V, 105)

$$dN = (N/2)\sin\theta\,(\lambda^{-2}\cos^2\theta + \lambda\sin^2\theta)^{-3/2}\,d\theta\,. \qquad \text{(V, 109)}$$

The resultant difference of polarizabilities for the whole assembly is therefore

$$\left.\begin{aligned}\beta_1-\beta_2 &= \int(\beta_l-\beta_t)\,dN\\ &= N\int(\alpha_1-\alpha_2)\frac{\cos^2\theta-\frac{1}{2}\sin^2\theta}{(\lambda^{-2}\cos^2\theta+\lambda\sin^2\theta)^{3/2}}\cdot\frac{1}{2}\sin\theta\,d\theta\,.\end{aligned}\right\} \qquad \text{(V, 110)}$$

Insertion of the value (V, 103c) for the chain anisotropy, namely

$$\begin{aligned}\alpha_1-\alpha_2 &= Z(\alpha_{01}-\alpha_{02})\,\frac{3}{5}\cdot\left(\frac{h}{Za}\right)^2\\ &= Z(\alpha_{01}-\alpha_{02})\,\frac{3}{5}\cdot\left(\frac{h_0}{Za}\right)^2(\lambda^{-2}\cos^2\theta+\lambda\sin^2\theta)^{-1}\end{aligned}$$

yields finally

$$\left.\begin{aligned}\beta_1-\beta_2 &= \frac{3}{5}\,NZ(\alpha_{01}-\alpha_{02})\left(\frac{h_0}{Za}\right)^2\int_0^\pi\frac{\left(\cos^2\theta-\frac{1}{2}\sin^2\theta\right)\cdot\frac{1}{2}\sin\theta\,d\theta}{(\lambda^{-2}\cos^2\theta+\lambda\sin^2\theta)^{5/2}}\\ &= \frac{1}{5}\,N(\alpha_{01}-\alpha_{02})\,\frac{h_0^2}{Za^2}\,(\lambda^2-1/\lambda)\,.\end{aligned}\right\} \qquad \text{(V, 111)}$$

Putting $h_0 = a\,Z^{1/2}$ this becomes

$$\beta_1-\beta_2 = \frac{N}{5}(\alpha_{01}-\alpha_{02})(\lambda^2-1/\lambda)\,. \qquad \text{(V, 111a)}$$

The difference of polarizabilities has now to be related to the difference of refractive indices $n_1 - n_2$. This is obtained on the basis of the LORENTZ-LORENZ relation between refractive index n and polarizability per unit volume β,

$$\frac{n^2-1}{n^2+2} = \frac{4\pi}{3}\beta\,. \qquad \text{(V, 112)}$$

From its method of derivation this relation is strictly valid only for an isotropic medium. It is assumed, however, that it may still be applied without serious error when the medium is no longer isotropic. On this basis one obtains,

$$\frac{n_1^2-1}{n_1^2+2} - \frac{n_2^2-1}{n_2^2+2} = \frac{4\pi}{15}N(\alpha_{01}-\alpha_{02})\left(\lambda^2-\frac{1}{\lambda}\right). \qquad \text{(V, 113)}$$

In practice $n_1 - n_2$ is generally small, and (V, 113) may be reduced to the simpler expression

$$n_1-n_2 = \frac{(n^2+2)^2}{n}\cdot\frac{2\pi N}{45}(\alpha_{01}-\alpha_{02})\left(\lambda^2-\frac{1}{\lambda}\right) \qquad \text{(V, 114)}$$

in which n is the mean refractive index $(n_1 + 2n_2)/3$.

Distribution of chain contour lengths: — In deriving the foregoing equations, particularly (V,111) and (V,113), it was assumed that all the chains have the same contour length. From the method of derivation it is clear that this restriction may be removed by substituting the mean value of h_0/Za^2 for h_0/Za^2 in equation (V,111). But if we assume, as before, that $h_0/Za^2 = 1$ for all chain lengths this factor is unchanged. Equations (V,111) and (V,114) are therefore valid for any distribution of chain contour lengths.

2. *Stress-optical coefficient.*

Comparison of (V,114) with the stress-strain relation for the Gaussian network (page 316), namely

$$\sigma = \lambda f = N k T (\lambda^2 - 1/\lambda) \tag{V,115}$$

shows the dependence of birefringence on strain to be of the same form as the dependence of stress on strain. This means that the birefringence is proportional to the stress, i. e.

$$n_1 - n_2 = C\sigma. \tag{V,116}$$

This is Brewster's law. The constant C, the *stress-optical coefficient*, has the value

$$C = \frac{(n^2+2)^2}{n} \cdot \frac{2\pi}{45\,kT} (\alpha_{01} - \alpha_{02}). \tag{V,117}$$

3. *Pure homogeneous strain.*

In the general case of a pure homogeneous strain a medium which is mechanically and optically isotropic in the unstrained state acquires the optical properties of a biaxial crystal. This means that it may be characterized by a refractive index ellipsoid whose three principal axes correspond in direction to the principal axes of strain. This case is not so easy to deal with theoretically as is the simple extension, but by a suitable modification of the Kuhn-Grün model a solution may be obtained[1]. The difference between any two of the principal refractive indices, say $n_1 - n_2$, which represents the birefringence for light propagated along the direction of λ_3, is found to be related to the principal extension ratios λ_1 and λ_2 as follows:

$$n_1 - n_2 = \frac{(n^2+2)^2}{n} \cdot \frac{2\pi}{45} \cdot N (\alpha_{01} - \alpha_{02}) (\lambda_1^2 - \lambda_2^2) \tag{V,118}$$

where, as before, n is the mean refractive index, $1/3\,(n_1 + n_2 + n_3)$. The corresponding equation for the stresses is

$$\sigma_1 - \sigma_2 = N k T (\lambda_1^2 - \lambda_2^2). \tag{V,119}$$

Hence

$$n_1 - n_2 = C(\sigma_1 - \sigma_2) \tag{V,120}$$

which is the general form of Brewster's law. Equations (V,118) and (V,120) imply that the difference of refractive indices, in any principal

[1] Treloar, L. R. G.: Trans. Faraday Soc. **43**, 277 u. 284 (1947).

plane, is proportional to the difference of the corresponding principal stresses, or alternatively, to the difference of the squares of the corresponding extension ratios.

4. Significance of theoretical relations.

It is to be noted that whereas the birefringence-strain relations, like the corresponding stress-strain relations, involve N, the number of chains per unit volume, the stress-optical coefficient (V,117) is independent of N. The only *molecular* quantity involved in the expression for the stress-optical coefficient is $(\alpha_{01}-\alpha_{02})$, the difference of polarizabilities for the single link of the random chain. This is therefore a fundamental molecular parameter which is specific to a particular chain. It has considerable interest in connection with the estimation of the flexibility of particular types of chain molecule (see below).

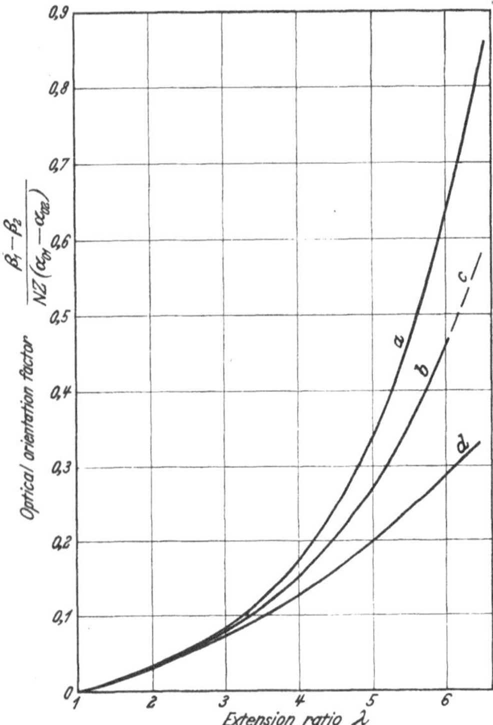

Fig. V,33. Theoretical optical anisotropy for network, $Z = 25$. a) Kuhn-Grün theory [equ. (V,121)]; b) Kuhn-Grün theory modified [equ. (V,122)]; c) Non-affine deformation theory (Treloar); d) Gaussian. (Note b) and c) are indistinguishable).

b) The non-Gaussian network.

The foregoing theory is applicable only in the Gaussian region, i.e. for strains which are not too large. Its extension to the non-Gaussian region requires the inclusion of the accurate expression (V,103a) [or its equivalent expansion (V,103b)], in place of the first-term approximation (V,103c), for the optical anisotropy of the single chain. A calculation on this basis has been carried out by Kuhn and Grün[1] who used the first three terms in the expansion (V,103b), and obtained the following formula for the optical anisotropy of the network

$$\left.\begin{aligned}\beta_1-\beta_2 = N(\alpha_{01}-\alpha_{02})\left[\frac{1}{5}\left(\lambda^2-\frac{1}{\lambda}\right)+\frac{2}{175Z}\left(6\lambda^4+2\lambda-\frac{8}{\lambda^2}\right)\right.\\ \left.+\frac{6}{875Z^2}\left(10\lambda^6+6\lambda^3-\frac{16}{\lambda^3}\right)\right].\end{aligned}\right\}\quad (V,121)$$

[1] Kuhn, W. u. F. Grün: Kolloid-Z. **101**, 248 (1942).

This shows that the optical anisotropy is no longer independent of the chain length (as represented by Z). The form of (V, 121) is shown in fig. V, 33 for the particular case $Z = 25$.

The author has discussed the problem of the optical properties of the non-Gaussian network in considerable detail[1]. By a minor modification of the KUHN-GRÜN treatment he obtained the result

$$\left.\begin{aligned} \beta_1 - \beta_2 = N(\alpha_{01} - \alpha_{02}) \Big[\frac{1}{5}\Big(\lambda^2 - \frac{1}{\lambda}\Big) + \frac{1}{150\,Z}\Big(6\lambda^4 + 2\lambda - \frac{8}{\lambda^2}\Big) \\ + \frac{1}{350\,Z^2}\Big(10\lambda^6 + 6\lambda^3 - \frac{16}{\lambda^3}\Big) \end{aligned}\right\} \qquad \text{(V, 122)}$$

which differs from their original expression only in the numerical values of the coefficients (fig. V, 33).

A more elaborate treatment was based on the FLORY-REHNER 4-chain model, and allowed for the non-affine displacement of the central junction point. These calculations involve numerical computation, and it is not possible to represent the results in general algebraic terms. An example is shown in fig. V, 33, which represents the case $Z = 25$.

A comparison of the three non-Gaussian solutions represented in fig. V,33 suggests that the precise form of the solution is not greatly dependent on the particular model chosen to represent the structure.

c) Experimental study of photo-elastic effects.

Experiments on the optical properties of strained rubber reveal complications due to crystallization. This is illustrated in the accompanying figures. In fig. V, 34 the birefringence is seen to follow the theoretical law (V, 114) up to a certain value of strain; thereafter the birefringence rises more rapidly than is required by the theory[2]. The range over which the theory applies becomes more extended as the temperature is raised, until at 100°C the agreement is very close right up to the breaking point. There is a good deal of evidence to suggest that these departures from the theory are due primarily to crystallization, which is reduced by raising the temperature. There is no definite evidence of a "non-Gaussian" effect of the type shown in fig. V, 33. The reason for its absence is obscure, but it may be that the magnitude of the extension was insufficient to reveal it.

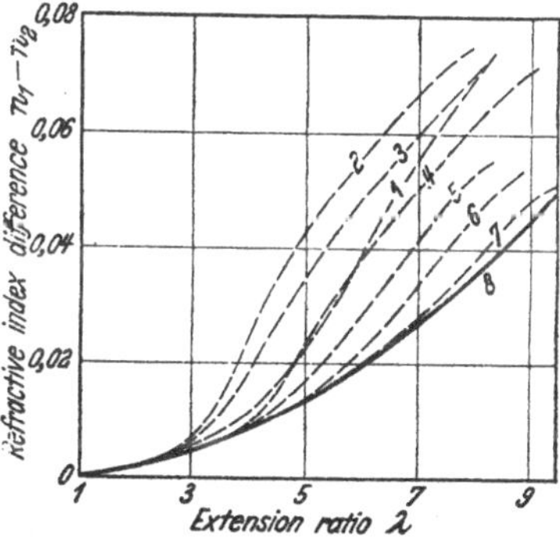

Fig. V, 34. Strain-birefringence for vulcanized rubber at various temperatures. (1) −50°C, (2) −25°C, (3) 0°C, (4) 25°C, (5) 50°C), (6) 75°C), (7) 100°C, (8) Gaussian theory.

Data for pure homogeneous strain[3] are shown in figs. V, 35 and V, 36. In these experiments the strains were insufficient to induce crystallization,

[1] TRELOAR, L. R. G.: Trans. Faraday Soc. **50**, 881 (1954).
[2] TRELOAR, L. R. G.: Trans. Faraday Soc. **43**, 277 u. 284 (1947).
[3] TRELOAR, L. R. G.: Proc. Physic. Soc. **60**, 135 (1948).

and the agreement with equations (V, 118) and (V, 120) respectively is good.

According to the theory, the value of the stress-optical coefficient should be independent of the degree of cross-linking or vulcanization of the rubber. This expectation is not borne out by the experiments of

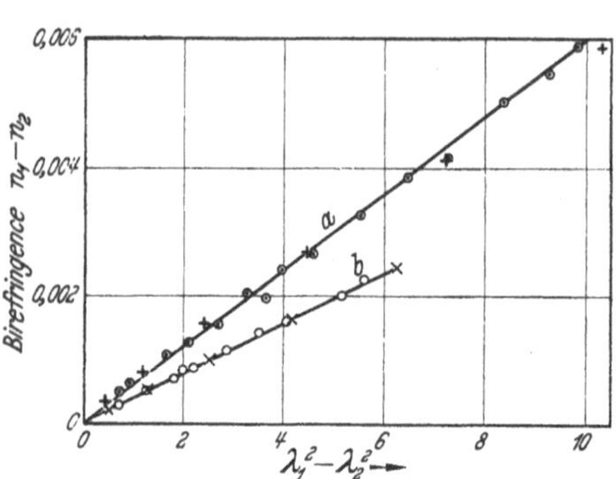

Fig. V, 35. Difference of principal refractive indices plotted against difference of squares of principal extension ratios. Vulcanized rubber a) dry, b) swollen in paraffin to $v_2 = 0{,}525$.

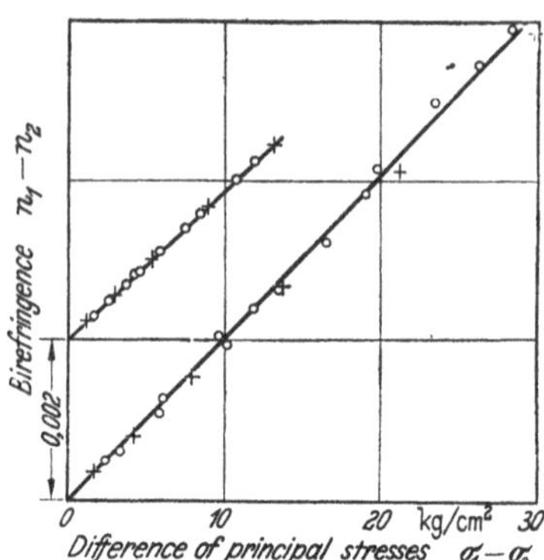

Fig. V, 36. Pure homogeneous strain. Birefringence plotted against difference of principal stresses. Vulcanized rubber a) dry, b) swollen in paraffin to $v_2 = 0{,}525$.

Thibodeau and McPherson[1], which indicate a linear increase of stress-optical coefficient with percentage of combined sulphur (fig. V, 37). However, sulphur is an inefficient cross-linking agent, and it is not unlikely that secondary reactions involving sulphur may effect either the chain polarizabilities or the length of the equivalent statistical link. In either case the quantity $\alpha_{01} - \alpha_{02}$, and hence the photo-elastic constant, would be a function of the amount of combined sulphur (equ. V, 118). For this reason the writer suggested that the photo-elastic constant relevant to the pure polyisoprene chain should be that obtained by extrapolation of Thibodeau and McPherson's data to zero sulphur content. In this way the figure $C = 1{,}96 \cdot 10^{-10}$ cm²/dyne was obtained.

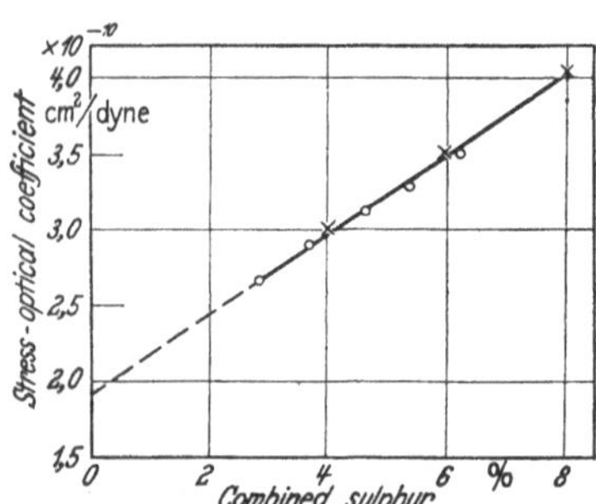

Fig. V, 37. Dependence of stress-optical coefficient on percentage of combined sulphur [Thibodeau and McPherson: Bur. Stds. J. Res. Washington **13**, 887 (1934).]
○ Compounds made with crude rubber.
× Compounds made with rubber hydrocarbon.

Alternative methods of cross-linking, not involving sulphur, have since been developed. One of these, depending on the use of a volatile peroxide, has been claimed to produce only C–C bonds, other reaction products being removed by evaporation. Using a series of these vulcaniza-

---

[1] Thibodeau, W. E. u. A. T. McPherson: Bur. Standards J. Res. Washington **13**, 887 (1934).

tes, SAUNDERS[1] obtained the values of stress-optical coefficient given in table V,1.

The first value in this table is the least reliable, on account of the shortness of the chain length and the possibility of a "non-Gaussian" effect. The variation with M_c is in any case very slight, and the mean value of C, $1{,}93 \cdot 10^{-10}$ cm²/dyne, is very close to the extrapolated value obtained from THIBODEAU and MCPHERSON's data for sulphur vulcanizates. This satisfactorily confirms the theoretical deduction.

d) Chain flexibility: the equivalent statistical link.

The question of the relation of any given molecular structure to the randomly-jointed chain has already been discussed. In the present context it will be sufficient to limit our consideration to the Gaussian range of extension, for it is only in this region that the linear relation between stress and birefringence, and hence a definite stress-optical coefficient, is to be expected. In this region, as we have seen, it is always possible to represent the actual structure by an equivalent random chain which has the same fully-extended length and the same Gaussian distribution of vector lengths; the value of Z for this equivalent chain defines the length of the "equivalent statistical link". For the rubber (polyisoprene) chain, treated geometrically, one random link is theoretically equivalent to 0,71 isoprene units (page 310).

Table V,1.
Stress-optical coefficients for peroxide-vulcanized rubbers (SAUNDERS).

M_c	Stress-optical coefficient (C)
2200	$1{,}83 \cdot 10^{-10}$ cm²/dyne
4460	$1{,}96 \cdot 10^{-10}$ cm²/dyne
5660	$1{,}91 \cdot 10^{-10}$ cm²/dyne
8390	$2{,}01 \cdot 10^{-10}$ cm²/dyne
Mean	$1{,}93 \cdot 10^{-10}$ cm²/dyne

The photo-elastic constant provides a basis for estimating the equivalent random link experimentally. For this constant gives $\alpha_{01} - \alpha_{02}$, the difference of polarizabilities for the "equivalent random link" [equation (V,117)]. Compare also Vol. III, § 28b if, therefore, the difference of polarizabilities for the monomer unit of the chain is known, it is immediately possible to derive the number of monomer units per random link.

The weak point of this procedure is the uncertainty with regard to the polarizabilities of the monomer unit. Values of longitudinal and transverse polarizabilities b_l and b_t for a number of bonds have been given by DENBIGH[2] but it is difficult to estimate their reliability. By making use of these values, the author[3] has calculated the anisotropy of the isoprene unit to be

$$b_l - b_t = 3{,}08 . 10^{-24}\ \text{cm}^3.$$

The value of $\alpha_{01} - \alpha_{02}$ for the equivalent random link, using the experimental figure of $1{,}96 \cdot 10^{-10}$ for the stress-optical coefficient, is

$$\alpha_{01} - \alpha_{02} = 4{,}67 . 10^{-24}\ \text{cm}^3.$$

[1] SAUNDERS, D. W.: Nature **165**, 360 (1950). Also unpublished work.
[2] DENBIGH, K. G.: Trans. Faraday Soc. **36**, 936 (1940).
[3] TRELOAR, L. R. G.: *Physics of Rubber Elasticity*, Oxford 1949, p. 149.

From these two figures one obtains

$$1 \text{ random link} = 1{,}52 \text{ isoprene units.}$$

This is rather more than twice the value obtained theoretically on the assumption of random rotation about each of the single bonds. It may be compared with the very rough figure 1,63 obtained from the non-Gaussian force-extension curve (page 327).

The difference between the experimental and theoretical equivalent random link is no doubt due to the effect of departures of the real chain from the ideal structure of freely-rotating bonds — departures which may be designated generally as "steric hindrances". While the above figures must be treated with reserve, on account of the uncertainty of the calculated monomer polarizabilities, the important point is that the photoelastic constant may be used to derive a fundamental molecular parameter which is directly related to the flexibility of the chain.

E. Phenomena of swelling in cross-linked polymers.

The theory of swelling, with particular reference to cross-linked systems, has been discussed in Vol. II, chap. 3. The present section is therefore limited to a more detailed examination of certain specific aspects of the subject, particularly the effect of stress on the swelling equilibrium, and the mechanical and optical properties of the swollen network.

§ 36. Equilibrium swelling and its dependence on strain.

a) General theory.

The phenomenon of swelling pressure is an example of the dependence of the amount of liquid absorbed by a gel in contact with a low-molecular liquid (or its vapour) on applied stress. It is thermodynamically equivalent to osmotic pressure. The more general case, in which any type of stress is applied, has received comparatively little attention, though the appropriate general thermodynamic relations have been worked out by BARKAS[1]. In the present context, however, we shall be concerned not with the general thermodynamic relations but with the particular problem of the swelling of a cross-linked polymer network, for which explicit expressions for the relevant free energy changes may be derived on the basis of the statistical theory.

Let us consider a piece of material which in the unstrained, unswollen state is in the form of a cube of unit edge length. Let this be placed in a swelling liquid and subjected to principal *stresses* (i.e. forces per unit area measured in the strained swollen state) σ_1, σ_2 and σ_3 acting normally to its surfaces, and let l_1, l_2 and l_3 be its dimensions when it has reached the equilibrium state. Let us further assume, for simplicity, that the

[1] BARKAS, W. W.: Swelling stresses in gels. H. M. Sationery Office 1945, p. 16.

volumes of polymer and liquid in the swollen gel are additive. The dimensions in the equilibrium state define the liquid content or swelling ratio. Thus if v_2 is the volume fraction of polymer,

$$1/v_2 = l_1 l_2 l_3 = 1 + n_0 V_0 \tag{V,123}$$

where n_0 is the number of moles of liquid of molar volume V_0 in the mixture.

The condition for equilibrium with respect to the liquid is that, for a small variation of the liquid content, the change in the HELMHOLTZ free energy of the system (swollen gel plus pure liquid) shall be equal to the work done by the externally applied forces[1]. In order to apply this condition, let us consider the absorption of a further quantity of liquid (δn moles) taking place in such a way that the dimension l_1 is increased to $l_1 + \delta l_1$, while l_2 and l_3 are held constant (by suitably adjusting σ_2 and σ_3). If F is the HELMHOLTZ free energy of the system, and δW the work done by the external forces, the equilibrium condition may be written

$$\left(\frac{\partial F}{\partial n_0}\right)_{l_2 l_3} \delta n_0 = \delta W. \tag{V,124}$$

The external work is equal to stress × area × distance, i.e.,

$$\delta W = \sigma_1 l_2 l_3 \, \delta l_1 . \tag{V,125}$$

From (V,123) we obtain, by differentiation, $\delta l_1 = (V_0/l_2 l_3)\,\delta n_0$, and therefore, for equilibrium [from (V,124) and (V,125)]

$$\left(\frac{\partial F}{\partial n_0}\right)_{l_2 l_3} = \sigma_1 V_0 . \tag{V,126}$$

This equation, with corresponding equations for σ_2 and σ_3, represents the general solution to the problem.

To proceed further it is necessary to introduce a specific expression for the free energy of the system. This may be regarded as the sum of two terms, i.e.,

$$F = F_m + F_e \quad \text{or} \quad \frac{\partial F}{\partial n_0} = \frac{\partial F_m}{\partial n_0} + \frac{\partial F_e}{\partial n_0} \tag{V,127}$$

of which the first, F_m is the free energy of mixing of the liquid with the polymer molecules, considered to be not cross-linked, while the second, F_e, is the free energy stored elastically in the cross-linked network. This method of analysis involves the assumption that the free energy of mixing can be considered to be unaffected by cross-linking.

The term $\partial F_m / \partial n_0$ in (V,127) is the free energy of dilution, i.e. the change in the free energy of the system per mole of liquid transferred from the liquid to the solid phase. We shall represent this by the FLORY-HUGGINS expression in the form[2]

$$\partial F_m / \partial n_0 = R T [\log (1 - v_2) + v_2 + \mu v_2^2] \tag{V,128}$$

[1] ROBERTS, J. K.: Heat and Thermodynamics. Blackie 1940, p. 317.
[2] FLORY, P. J.: J. Chem. Physics **10**, 51 (1942).

in which μ is a parameter whose value depends on the particular liquid and polymer. For the elastic network free energy we shall assume the Gaussian formula (V, 55b)[1],

$$F_e = \frac{\varrho R T}{2 M_c} (l_1^2 + l_2^2 + l_3^2 - 3). \tag{V,129}$$

Differentiation of (V, 129) gives, with (V, 123a)

$$\left(\frac{\partial F_e}{\partial n_0}\right)_{l_2 l_3} = \left(\frac{\partial F_e}{\partial l_1}\right)_{l_2 l_3} \left(\frac{\partial l_1}{\partial n_0}\right)_{l_2 l_3} = \frac{\varrho R T}{M_c} \cdot \frac{l_1 V_0}{l_2 l_3} = \frac{\varrho V_0 R T}{M_c} v_2 l_1^2. \tag{V,130}$$

The total free energy change in the process is therefore

$$\left(\frac{\partial F}{\partial n_0}\right)_{l_2 l_3} = R T \left[\log(1 - v_2) + v_2 + \mu v_2^2 + \frac{\varrho V_0}{M_c} v_2 l_1^2\right]. \tag{V,131}$$

The equilibrium condition (V, 126) thus becomes

$$\frac{R T}{V_0} \left[\log(1 - v_2) + v_2 + \mu v_2^2 + \frac{\varrho V_0}{M_c} v_2 l_1^2\right] = \sigma_1. \tag{V,132a}$$

The corresponding equations for σ_2 and σ_3 are

$$\frac{R T}{V_0} \left[\log(1 - v_2) + v_2 + \mu v_2^2 + \frac{\varrho V_0}{M_c} v_2 l_2^2\right] = \sigma_2 \tag{V,132b}$$

$$\frac{R T}{V_0} \left[\log(1 - v_2) + v_2 + \mu v_2^2 + \frac{\varrho V_0}{M_c} v_2 l_3^2\right] = \sigma_3. \tag{V,132c}$$

It is to be noted that v_2 is a function of l_1, l_2 and l_3. If l_1, l_2 and l_3 are taken as independent variables, the stresses σ_1, σ_2 and σ_3 are uniquely determined by equations (V, 132). Conversely, if σ_1, σ_2 and σ_3 are given, the state of strain, and the degree of swelling, *for equilibrium*, are uniquely determined. The application of these equations will now be illustrated by reference to particular types of strain.

b) Particular solutions.

1. Free swelling.

In the absence of an external stress, we have $l_1 = l_2 = l_3 = v_2^{-1/3}$ and $\sigma_1 = \sigma_2 = \sigma_3 = 0$. Equations (V, 132) then reduce to

$$\log(1 - v_2) + v_2 + \mu v_2^2 + (\varrho V_0/M_c) v_2^{1/3} = 0. \tag{V,133}$$

The solution of this equation in v_2 gives the equilibrium liquid concentration.

2. Hydrostatic pressure.

For a hydrostatic pressure p we have

$$\sigma_1 = \sigma_2 = \sigma_3 = p \quad \text{and} \quad l_1 = l_2 = l_3 = v_2^{-1/3}.$$

[1] The form (V, 129) follows from JAMES and GUTH's method [J. Chem. Physics **11**, 455 (1943)] of calculating the network entropy. WALL and FLORY [J. Chem. Physics **19**, 1435 (1951)] have given arguments in favour of a slightly different form, i.e. $\Delta S = -(\varrho R/2 M_c) [(l_1^2 + l_2^2 + l_3^2 - 3) - \log l_1 l_2 l_3]$ which is equivalent to (V, 129) when the volume is unchanged.

Hence, from (V,132), the equilibrium condition is

$$\frac{RT}{V_0}\left[\log(1-v_2)+v_2+\mu v_2^2+\frac{\varrho V_0}{M_c}v_2^{1/3}\right]=p. \qquad (V,134)$$

3. Simple extension (or uni-directional compression).

For a tensile stress acting in the direction l_1 we have

$$\sigma_2=\sigma_3=0;\quad l_2^2=l_3^2=1/l_1 v_2.$$

Hence from (V,132b)

$$\frac{\sigma_2 V_0}{RT}=\log(1-v_2)+v_2+\mu v_2^2+\frac{\varrho V_0}{M_c}\cdot\frac{1}{l_1}=0. \qquad (V,135)$$

This equation, originally derived by FLORY and REHNER[1], gives the liquid concentration in terms of the strained (swollen) length l_1. If l_1 and v_2 are known, the stress σ_1 is determined by (V,132a). If l_1 is greater than $v_2^{-1/3}$ the stress is tensile, while if l_1 is less than $v_2^{-1/3}$ the stress is compressive.

4. Uniform two-dimensional extension.

This state of strain may be produced by stretching a sheet by equal stresses in two directions at right angles. Hence

$$l_1=1/v_2 l_2^2=1/v_2 l_3^2;\quad \sigma_2=\sigma_3;\quad \sigma_1=0$$

and, from (V,132a)

$$\log(1-v_2)+v_2+\mu v_2^2+\frac{\varrho V_0}{M_c}\cdot\frac{1}{v_2 l_2^4}=0. \qquad (V,136)$$

This equation gives the liquid content in terms of the stretch ratio in the plane of the sheet. A knowledge of v_2 and l_2 enables σ_2 to be determined [equation (V,132b)].

5. Pure and simple shear.

The case of shear is rather more complicated. It has been discussed in a paper by the author[2] to which reference should be made for details. By definition, a shear is a state of strain involving no change of volume; hence, *by definition*, a shear strain cannot change the swelling ratio. One can, however, apply a shear strain and consider what stresses are required to maintain it. On the basis of equation (V,132) it is found that the state of *pure* shear (page 317) can be maintained by the following system of stresses

$$\left.\begin{aligned}\sigma_1&=0\\ \sigma_2&=\frac{\varrho RT}{M_c}v_2^{1/3}(\lambda^2-1)\\ \sigma_3&=\frac{\varrho RT}{M_c}v_2^{1/3}\left(\frac{1}{\lambda^2}-1\right)\end{aligned}\right\} \qquad (V,137)$$

[1] FLORY, P. J. u. J. REHNER: J. Chem. Physics **12**, 412 (1944).
[2] TRELOAR, L. R. G.: Proc. Roy. Soc. [London] **A 200**, 176 (1950).

where λ and l/λ are the principal extension ratios in the plane of the shear referred to the *swollen* unstrained state.

The stresses σ_2 and σ_3 are unequal in magnitude and of opposite sign. In the special case when the strain is small they become equal and opposite. The stress system then is equivalent to a simple shear stress. Since we have defined the system in such a way that the volume, and hence the degree of swelling, are constant, it follows that, so long as the strain is small, a shear stress does not change the state of swelling.

c) Experimental verification of theory.

The relation between the degree of swelling and the degree of cross-linking in rubbers, as represented by M_c in equation (V,133) has been examined experimentally by GEE[1] and by FLORY[2]. This work has already been discussed in Vol. II. (See figs. III,35 and (III,39.) The effect of simple extension on the equilibrium swelling has also been considered by GEE[1] for natural rubber, and by FLORY and REHNER[3] for butyl rubber. GEE found the theoretical relation (V,135) to apply quite well in the case of good swelling agents (e.g. benzene) but for poor swelling agents (e.g. ethyl acetate) significant departures from the theory were observed. In a more extensive investigation with vulcanized rubber the author considered three types of strain, simple extension, unidirectional compression and uniform two-dimensional extension[4]. Using good swelling agents (benzene and heptane), the dependence of liquid absorption on strain was in all cases in close agreement with the above theoretical equations. An example is given in fig. V,38.

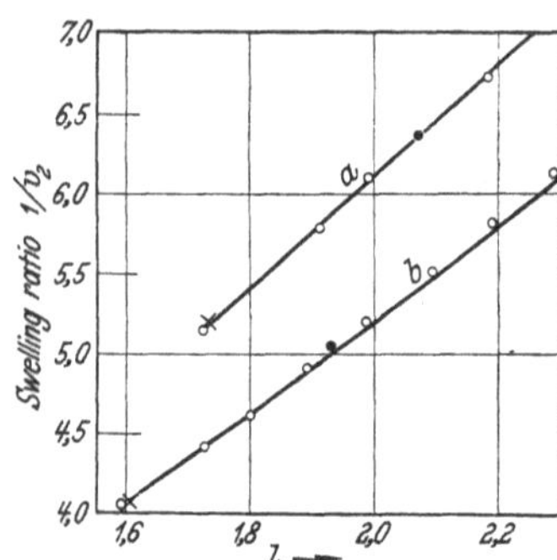

Fig. V,38. Effect of strain on equilibrium swelling of vulcanized rubber. 2-dimensional extension. *a*) benzene, *b*) heptane. l_2 is the ratio of the linear dimension in the strained swollen state to the corresponding dimension in the unstrained unswollen state.

d) Physical significance of stress-dependence.

As stated in (*a*) above, the dependence of the equilibrium swelling on stress is a generalization of the phenomenon of swelling pressure. The thermodynamic analysis is strictly valid only for a perfectly elastic material, and for such a material it makes no difference whether the change in the equlibrium swelling is related to the stress or to the strain. Physically, however, the change should be regarded as a direct consequence of the stress, and in particular, of the hydrostatic component of the stress. This is illustrated by the formulae derived above. For a tensile stress the hydrostatic component is negative; the swelling therefore increases. For

[1] GEE, G.: Trans. Faraday Soc. **42B**, 33 (1946).
[2] FLORY, P. J.: Chem. Review **35**, 51 (1944).
[3] FLORY, P. J. u. J. REHNER: J. Chem. Physics **12**, 412 (1944).
[4] TRELOAR, L. R. G.: Trans. Faraday Soc. **46**, 783 (1950).

a shear stress the hydrostatic component is zero; so long as the strain is small (i.e. so long as the material is substantially isotropic) the swelling equilibrium is unchanged.

The foregoing statistical theory is applicable only to an isotropic rubberlike material. But the general thermodynamic relations developed by BARKAS[1] are applicable to any material, whether isotropic or not. They have been applied by the author to interpret his observations on the reversible increase of water content in hair[2] and cellulose[3], when subjected to a tensile stress.

In the preceding discussion the stress has been considered to be homogeneous. When this condition is not satisfied, i.e. when the stress varies from point to point of the body, the state of swelling will, in general, also vary. Thus, in a polymer containing a definite quantity of imbibed liquid initially uniformly dispersed, the application of an inhomogeneous stress will lead to a redistribution of the liquid. This redistribution will operate in such a way that liquid tends to migrate from regions under pressure into regions under tension. The accompanying volume changes will obviously act in the direction required to reduce the local pressure or tension, that is, in such a way as to tend to reduce the inhomogeneity of the stress. This is a special case of the principle of LE CHATELIER and BRAUN.

§ 37. Physical properties of swollen rubbers.

From equation (V,129) the effect of swelling on the mechanical properties of a cross-linked polymer network may be readily obtained. For this purpose it is convenient to consider the deformation of the network as taking place in two stages, the first corresponding to an isotropic swelling, and leading to the free energy change

$$F_1 = \frac{\varrho R T}{2 M_c} (3 v_2^{-2/3} - 3) \qquad \text{(V,138)}$$

where $1/v_2$ is the volume swelling ratio, and the second corresponding to a deformation of the swollen network (without change in volume) leading to a free energy change F_2. Clearly

$$F_2 = F_e - F_1 \qquad \text{(V,139)}$$

where F_e is the total free energy change. Referring the deformation to the *swollen* volume by writing

$$\lambda_1 = l_1 v_2^{1/3}, \quad \lambda_2 = l_2 v_2^{1/3}, \quad \lambda_3 = l_3 v_2^{1/3}$$

this becomes [from (V,129)]

$$F_e = \frac{\varrho R T}{2 M_c} \left[v_2^{-2/3} (\lambda_1^2 + \lambda_2^2 + \lambda_3^2) - 3 \right].$$

[1] BARKAS, W. W.: Swelling stresses in gels. H. M. Sationery Office 1945, p. 16.
[2] TRELOAR, L. R. G.: Trans. Faraday Soc. 48, 567 (1952).
[3] TRELOAR, L. R. G.: Trans. Faraday Soc. 49, 816 (1953).

The free energy of deformation is therefore, from (V,138) and (V,139)

$$F_2 = \frac{\varrho R T}{2 M_c} v_2^{-2/3} (\lambda_1^2 + \lambda_2^2 + \lambda_3^2 - 3).$$

This refers to the volume $1/v_2$ measured in the swollen state. The work of deformation, W, per unit volume is therefore

$$W = F_2 v_2 = \frac{\varrho R T}{2 M_c} v_2^{1/3} (\lambda_1^2 + \lambda_2^2 + \lambda_3^2 - 3). \qquad (V,140)$$

This is the elastically stored free energy corresponding to the deformation of the swollen network, assuming the degree of swelling to be independent of the strain. It is applicable whether or not the swelling ratio corresponds to the equilibrium state. Putting $v_2 = 1$, equation (V,140) reduces to the standard form for the unswollen Gaussian network. The effect of swelling, therefore, is simply to reduce the modulus, and hence all the stresses, in the ratio $v_2^{1/3}$. The general stress-strain relations are therefore of the form [cf. (V,61)].

$$\sigma_1 - \sigma_2 = \frac{\varrho R T}{M_c} v_2^{1/3} (\lambda_1^2 - \lambda_2^2). \qquad (V,141)$$

This relation may also be derived directly from equations (V,132).

An examination of the effect of swelling on the force-extension curves for vulcanized rubbers has been made by GEE[1]. Some of his data are represented in fig. V,39 in terms of the quantity χ defined by

$$\chi = \frac{f_0}{T A_0 (l_1 - 1/l_1^2 v_2)} = \frac{f}{T v_2^{1/3} (\lambda - 1/\lambda^2)} \qquad (V,142)$$

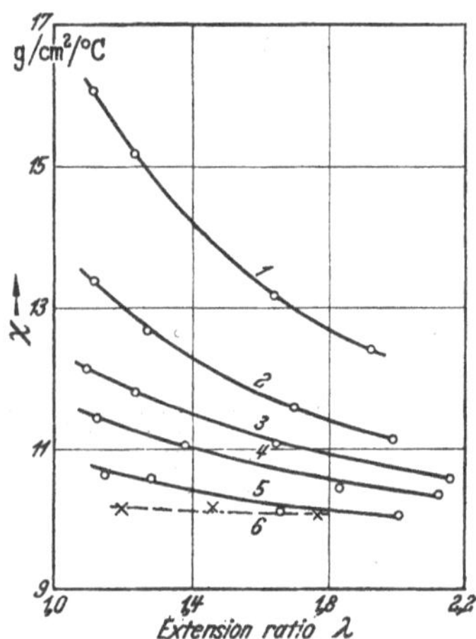

Fig. V,39. Variation of χ [equ. (V,142)] with extension of vulcanized rubber, for the following degrees of swelling in toluene. (1) $v_2 = 1{,}0$, (2) $v_2 = 0{,}774$, (3) $v_2 = 0{,}665$, (4) $v_2 = 0{,}563$, (5) $v_2 = 0{,}425$, (6) $v_2 = 0{,}331$. [GEE: Trans. Faraday Soc. 42, 585 (1946).]

where f_0 is the force acting on the unstrained unswollen area A_0, f is the force per unit unstrained *swollen* area, and λ is the extension ratio referred to the swollen dimensions. From equation (V,141) χ is seen to be equivalent to $\varrho R/2 M_c$ and should thus be independent of v_2 and λ. GEE found systematic departures from theory, the value of χ tending to fall both with increasing swelling and with increasing extension. The results shown in fig. V,39 are typical; similar effects were obtained with other rubbers and with a variety of swelling liquids. These departures indicate that the Gaussian theory gives a rather inexact representation of the free energy of network deformation for an actual rubber, but the evidence suggests that at high degrees of swelling the agreement between theory and experiment improves.

Data for the effect of swelling on the stresses, in a pure homogeneous strain, are given in fig. V,26.

[1] GEE, G.: Trans. Faraday Soc. 42, 585 (1946).

a) Photo-elastic properties of swollen rubber.

In this section we shall consider the modification of the photo-elastic theory discussed in § 35 required to take account of the swelling of the network. It will be assumed that the swelling liquid itself is optically neutral, that is to say, that it acts simply as a diluent (except in so far as it alters the mean refractive index).

In the calculation of the network polarizabilities the only alteration necessary is that h_0 in equation (V,111) shall be replaced by $h_0 v_2^{-1/3}$. This leads to the modified formula for the birefringence in simple extension

$$n_1 - n_2 = \frac{(n^2+2)^2}{n} \cdot \frac{2\pi N}{45} (\alpha_{01} - \alpha_{02}) v_2^{1/3} (\lambda^2 - 1/\lambda) \qquad \text{(V, 114 a)}$$

in place of (V,114).

A similar modification leads to the more general expression for the birefringence in a pure homogeneous strain defined by the extension ratios λ_1, λ_2, and λ_3,

$$n_1 - n_2 = \frac{(n^2+2)^2}{n} \cdot \frac{2\pi N}{45} \times (\alpha_{01} - \alpha_{02}) v_2^{1/3} (\lambda_1^2 - \lambda_2^2) \qquad \text{(V, 118 a)}$$

to replace (V, 118). In both (V, 114 a) and (V, 118 a) the effect of the swelling is thus to change the network anisotropy by the factor $v_2^{1/3}$. Comparison with (V, 141) shows that the stresses are changed in the same ratio. Thus, except to the extent that the swelling liquid may change the mean refractive index, the stress-optical coefficient is independent of the degree of swelling.

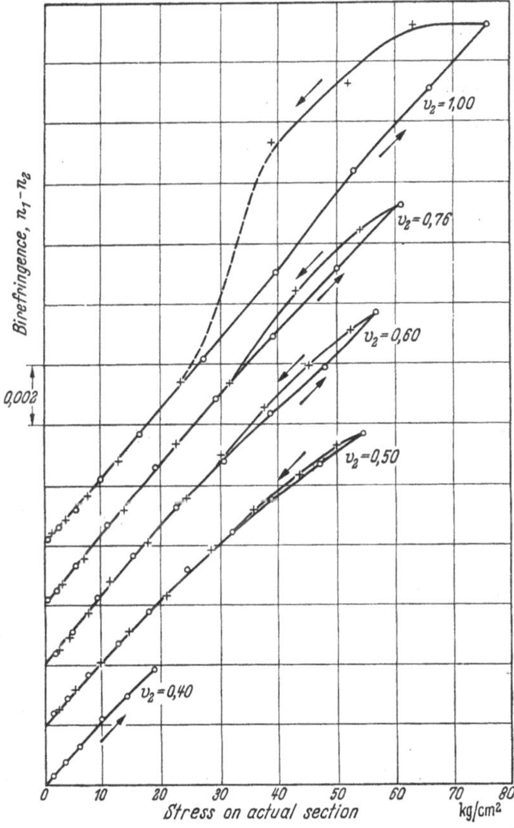

Fig. V, 40. Stress-birefringence relations for vulcanized rubber swollen in toluene. Simple elongation.

Data on vulcanized rubber[1] swollen in toluene and subjected to simple extension, reproduced in fig. V,40, show this theoretical prediction to be approximately fulfilled. This is further confirmed by the results of

[1] TRELOAR, L. R. G.: Trans. Faraday Soc. **43**, 277 u. 284 (1947).

an experiment on pure homogeneous strain[1]. The effect of swelling on the birefringence is shown if fig. V,35 and the effect on mechanical properties in fig. V,26. A plot of birefringence against difference of principal stresses, using the data represented in these two figures, is shown in fig. V,36. The slope of this line, which measures the stress-optical coefficient [equ. (V,120)], is seen to be only slightly affected by swelling the rubber to nearly twice its original volume.

F. The effect of crystallization on mechanical properties of rubbers.

§ 38. General.

The principal phenomena of crystallization in high polymers have been discussed in Vol. III, chapt. VIII. The present section is concerned with one particular aspect of this subject, namely the effect of crystallization on the mechanical properties of rubberlike materials.

In early years much difficulty was experienced in finding a basis for the reversible development of crystallization in rubber on extension. Gradually it became recognized that this remarkable, and at that time unique, property was only an extreme example of a wide range of comparable phenomena exhibited by a number of polymeric materials under a variety of conditions. Moreover, the phenomena of crystallization in rubber, and the contribution to its thermo-elastic properties (heat of extension, etc.) arising from crystallization confused the primary issue (at the time) of the nature and mechanism of the long-range extensibility of the material. The clarification of this issue was greatly facilitated by the quantitative development of the statistical theory of rubber elasticity and the consequent formulation of the essential mechanical and thermo-elastic properties of an ideal amorphous rubber.

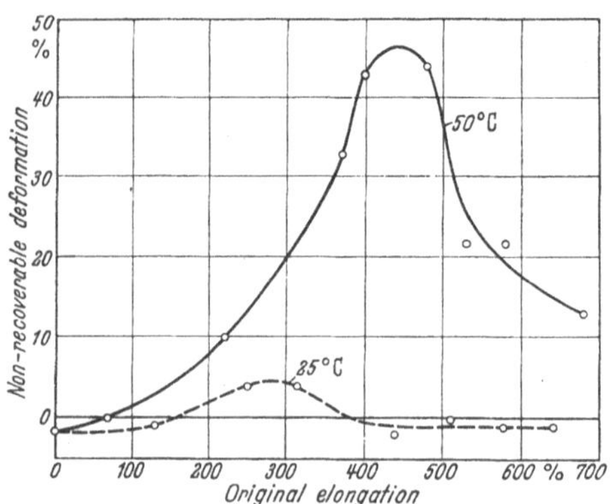

Fig. V,41. Plastic flow in raw rubber, as a function of extension.

The effects of crystallization are more marked in unvulcanized than in vulcanized rubber. In unvulcanized rubber the effect of crystallization in holding the molecules together and in suppressing plastic flow is well illustrated in fig. V,41. This refers to an experiment in which samples of natural rubber were subjected to varying degrees of extension for a fixed

[1] Treloar, L. R. G.: Proc. Physic. Soc. 60, 135 (1948).

time[1]. Thereafter the stress was removed and the samples were given temperature and swelling treatments designed to remove the elastic strain. The residual (plastic) strain, when plotted against the initial extension, showed a pronounced maximum at a region corresponding to the onset of crystallization.

The behaviour of raw rubber under strain is greatly dependent on the rate of application of the strain. If extended slowly, the processes of relaxation and flow predominate, the crystalline state is not developed, and the material eventually breaks under a comparatively low stress. If extended rapidly, on the other hand, the process of crystallization is in evidence, and the resultant strengthening is such that a tensile strength only slightly less than that for a vulcanized rubber may be attained.

Raw rubber which has been quickly stretched to something like its maximum extension retains this extension for an indefinite time after removal of the stretching force. In this condition it is highly anisotropic in mechanical and optical properties. The mechanical anisotropy is such that a characteristic fibrosity is developed. This was demonstrated by Hock[2] in 1925 by striking with a hammer at low temperatures, when the rubber splits "along the grain" into fibre-like bundles. Hock's explanation of these effects as due to the formation of an oriented crystalline state on extension was beautifully confirmed by Katz[3] in 1925 by X-ray diffraction methods. In fact, the state of stretched crystalline rubber is in all essentials comparable with that of a typical fibre.

In practice, of course, there is the important difference that the crystalline state in rubber has not the permanence required for a technically valuable fibre, since it is readily destroyed either by a moderate heating (30–40° C.) or by the action of swelling liquids. With such treatments the rubber reverts to the amorphous state, and at the same time retracts to its original form.

§ 39. Effect of crystallization on modulus of raw rubber.

Raw rubber may also be crystallized in the unstrained state. The rate of crystallization is governed by the temperature. At 0° C. the process requires about 10 days, but the rate increases as the temperature is lowered and is a maximum at −25° C., when it is completed within a few hours. Associated with the crystallization there is an increase in rigidity of the rubber, and also an increase in light scattering. The latter effect is due not to the individual crystallites (which are probably small compared with the wavelength of light) but to the so-called spherulitic clusters (compare Vol. III, § 45).

The increase in modulus of elasticity which accompanies crystallization has been examined by Leitner[4], who measured the Young's modulus for smooth-rolled smoked sheet during the progress of crystal-

[1] Treloar, L. R. G.: Trans. Faraday Soc. **36**, 538 (1940).
[2] Hock, L.: Z. Elektrochem. **31**, 104 (1925).
[3] Katz, J. R.: Chem. Z. **49**, 353 (1925).
[4] Leitner, M.: Trans. Faraday Soc. **51**, 1015 (1955).

lization at 0° C. (Fig. V, 42). The progress of crystallization was followed by simultaneously measuring the change in density. Her data show an increase by a factor of about 100 in passing from the amorphous state to the final state represented by a density change of about 2.2 per cent. The corresponding degree of crystallization may be estimated as 22 per cent[1] (compare for details also Vol. III, § 26).

§ 40. Crystallization in vulcanized rubber.

a) Effect on force-extension curve.

The phenomena of crystallization in vulcanized rubber in the unstrained state are qualitatively similar to those observed in raw rubber. The rates of crystallization are, however, generally very much lower, and are greatly affected by the state of vulcanization[2]. The associated change in mechanical properties is of considerable industrial importance, particularly when rubber is used in cold climates. It is true that the effects can be removed by heating; this, however, is not always convenient or practicable (e.g. in aeroplane tyres). However, precise quantitative measurements of the changes in mechanical properties during crystallization do not appear to be available.

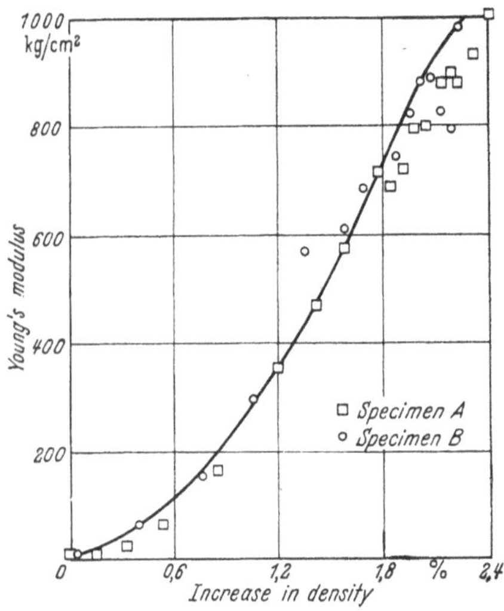

Fig. V, 42. Variation of modulus of raw rubber with crystallinity (represented by density change). LEITNER: Trans. Faraday Soc., **51**, 1015 (1955).

In the case of crystallization induced by extension the question arises of the effect of crystallization on the shape of the force-extension curve. The view has sometimes been expressed that the characteristic upward curvature of the force-extension curve for vulcanized rubber at high extensions (see, e.g. fig. V, 23) is due to crystallization, but there are good reasons for believing that this is not in fact the case. First, it has been shown that the general shape of the force-extension curve may be derived from the statistical theory of the amorphous network by taking into account the "non-Gaussian" effect. Secondly, the same type of upward curvature is evident in GR–S rubber, in which crystallization does not occur. Nevertheless, although crystallization may not be invoked for an explanation of the primary characteristics of the force-extension curve, it almost certainly does produce

[1] TRELOAR, L. R. G.: *Physics of Rubber Elasticity*, Oxford 1949, p. 172.
[2] RUSSELL, E. W.: Trans. Faraday Soc. **47**, 539 (1951).

effects of a secondary character. This may be illustrated by the curves of fig. V,43 taken at different temperatures[1]. It is seen that, as the temperature is raised from $-25°$ C. to $100°$ C. the position of the upward bend advances to higher extensions. Simultaneous measurements of the double refraction show a parallel shift of the extension at which crystallization

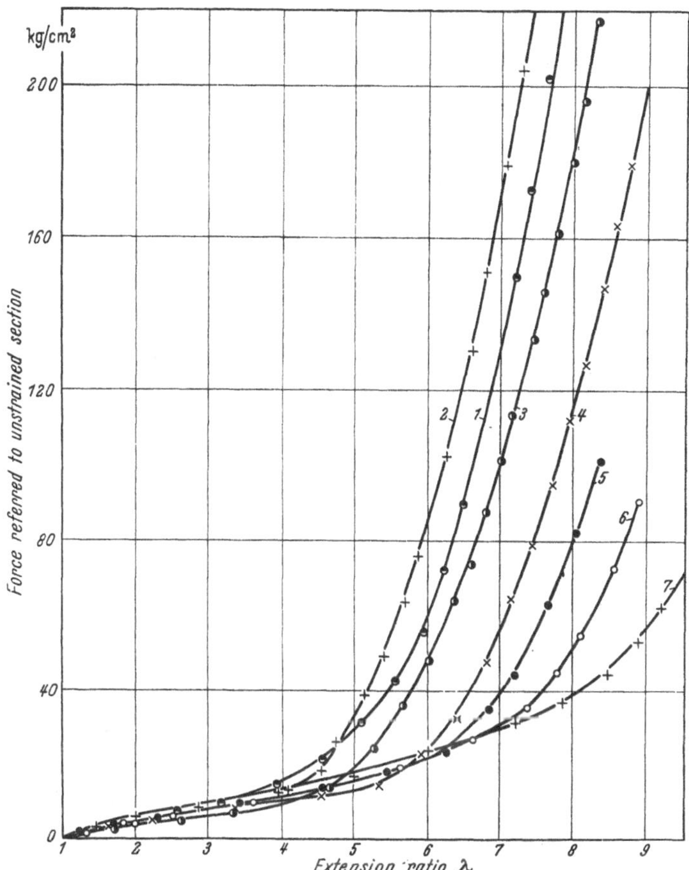

Fig. V,43. Force-extension curves for vulcanized rubber at various temperatures. (1) $-50°$ C, (2) $-25°$ C, (3) 0° C, (4) 25° C, (5) 50° C, (6) 75° C, (7) 100° C.

commences (fig. V,34). The natural inference is that the reduction of extensibility is due to the reduction of the effective average length of free chain as crystallization proceeds.

Any factor which increases the number of points of cohesion between molecules would be expected to increase the modulus and at the same time reduce the extensibility of the network (cf. p. 244), and it may well be that the effects just referred to are not entirely due to crystallization, but are connected with the general increase of intermole-

[1] Treloar, L. R. G.: Trans. Faraday Soc. **43**, 277 u. 284 (1947).

cular cohesion with reduction in temperature. That crystallization is an important factor is, however, revealed by the anomalous behaviour at −50° C. At this temperature, as shown in fig. V,43, the extensibility is higher than at −25° C., while the optical measurements indicate a reduction in the rate of crystallization. Thus, the general stiffening effect of the lower temperature is more than counterbalanced by the lower crystallinity.

b) Stress relaxation due to crystallization.

Crystallization of rubber, either raw or vulcanized, in the strained state, while increasing the rigidity, nevertheless leads to a *reduction* in the stress for a specimen maintained at constant length, or to an equivalent increase of length under constant stress. In raw rubber, at suitable (not too large) extensions the effect is so large that a sample clamped to a board and frozen may actually extend spontaneously by as much as 4 per cent of the unstrained length[1]. This effect appears to be due to the original nuclei for crystallization being preferentially oriented in the direction of the strain; further crystallization about the nuclei will then involve an increase in average molecular orientation which reveals itself in a reduction of stress, or in the extreme case, in a further extension.

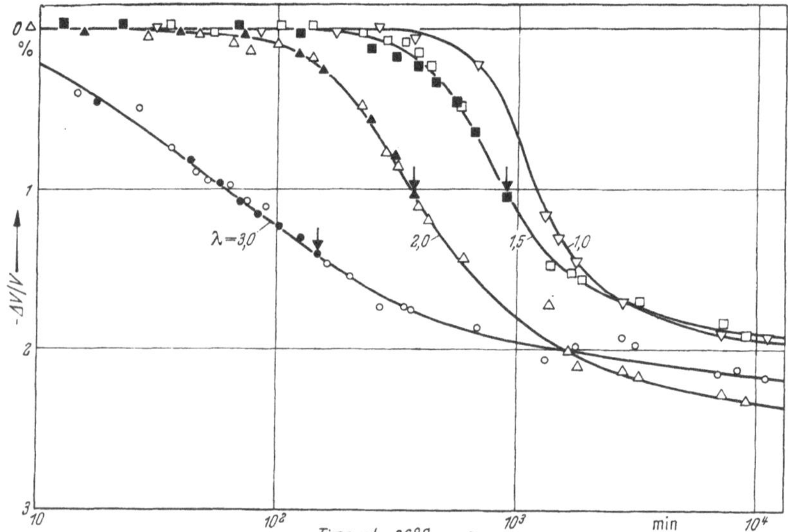

Fig. V,44. Stress relaxation and volume change in vulcanized rubber crystallizing in the extended state at −26° C. Volume change measurements ○ △ □ ▽. Stress change measurements (scaled appropriately) ● ▲ ■. GENT: Trans. Faraday Soc. **50**, 521 (1954).

In a study of stress relaxation in vulcanized rubber at low temperatures, GENT[2] demonstrated a direct proportionality between the amount of the stress decay and the degree of crystallization as estimated from simultaneous measurements of specific volume (fig. V,44). His data show

[1] SMITH, W. H. u. C. P. SAYLOR: Bur. Standards J. Res. Washington **21**, 257 (1938). — [2] GENT, A. N.: Trans. Faraday Soc. **50**, 521 (1954).

that the stress (at constant length) under these conditions falls to zero while the crystallization process is still far from complete; further crystallization then produces spontaneous extension.

The theory of crystallization in high polymers has been dealt with in Vol. III, chapt. VIII and will not be discussed here. Reference may, however, be made to the attempt by GENT to account for his stress-relaxation data on the basis of the statistical treatment of the problem of crystallization in a cross-linked polymer network given by FLORY[1]. In this theory it is assumed that all the crystal nuclei are oriented in the stretch direction, and that the crystallization proceeds to a thermodynamic equilibrium governed by the condition that the free energy of the system of crystallites plus "amorphous" chains shall be a minimum. For a given extension ratio λ there is an equilibrium fraction of crystalline material C_e and a corresponding equilibrium tensile force f_e. The relation between these quantities may be expressed in the form

$$C_e = \frac{f_0 - f_e}{f_0 (6Z/\pi)^{1/2} (\lambda - 1/\lambda^2)^{-1} - f_e} \tag{V,143}$$

where f_0 is the force in the wholly amorphous state at the same extension ratio λ and Z is the number of random links per chain. This form has the advantage that all molecular parameters except Z are eliminated; it should be remembered, however, that f_0 and f_e are not independent variables, but are determined by λ. GENT writes equation (V, 143) in the approximate form

$$C = (f_0 - f)(\lambda - 1/\lambda^2)(\pi/6Z)^{1/2} f_0^{-1} \tag{V,144}$$

and applies it under conditions of stress relaxation such that C and f (to be distinguished from C_e and f_e) are no longer equilibrium quantities. (This application is not strictly justifiable but is not unreasonable.) In a given experiment f_0 is constant; hence the reduction in stress should be proportional to the amount of crystallization. This is borne out experimentally. Furthermore, the stress should fall to zero when the crystallinity attains the value

$$C_{f=0} = (\lambda - 1/\lambda^2)(\pi/6Z)^{1/2} \tag{V,145}$$

obtained by putting $f = 0$ in equation (V, 144). To apply this result it is necessary to know Z, the number of random links per chain. The "chain molecular weight" is readily obtained from the value of f_0, but to estimate Z the number of monomer units per random link is required. GENT took the theoretical figure of 1,4 random links per isoprene unit (cf. page 310). Values of C thus derived were compared with the amounts of crystallinity obtained from the volume-change measurements, assuming 1% change in volume to be equivalent to 11.7% crystallinity. Some of his results are given below

Table V, 2.
Degree of crystallinity at point of zero stress, for different values of λ (GENT).

λ	3.0	2.5	2.0	1.5
C (from density)	0.152	—	0.117	0.122
C [from equ. (V, 145)]	0.170	0.138	0.103	0.062

[1] FLORY, P. J.: J. Chem. Physics **15**, 397 (1947).

The agreement appears reasonable, but in view of the uncertainties in the theory and in its application (e.g. in the choice of Z) its quantitative significance should not be over-emphasised.

c) Crystallization and tensile strength.

From the practical standpoint the most important of the mechanical properties which may be influenced by crystallization is the tensile strength. It is known that rubbers which do not crystallize have generally lower tensile strengths than crystallizable rubbers; GR–S for example, when vulcanized without fillers or reinforcing agents, has a tensile strength of only one tenth that for a pure-gum natural rubber vulcanizate. Moreover, the textile fibres, in which high tensile strength is a primary consideration are invariably highly crystalline.

However, the property of tensile strength is notoriously difficult to represent on a theoretical basis (compare chap. III.). The initiation of fracture is sensitive to minor features of the structure, or of the experimental method of test, and it is probably true to say that there is no material for which a satisfactory quantitative theory of fracture expressed in terms of fundamental molecular properties is available. In polymers particularly, there is the additional complication that the process of testing itself is liable to modify such relevant structural features as the amount of orientation of the crystallites. In these circumstances any interpretation of the relation between crystallinity and strength must necessarily be arrived at indirectly.

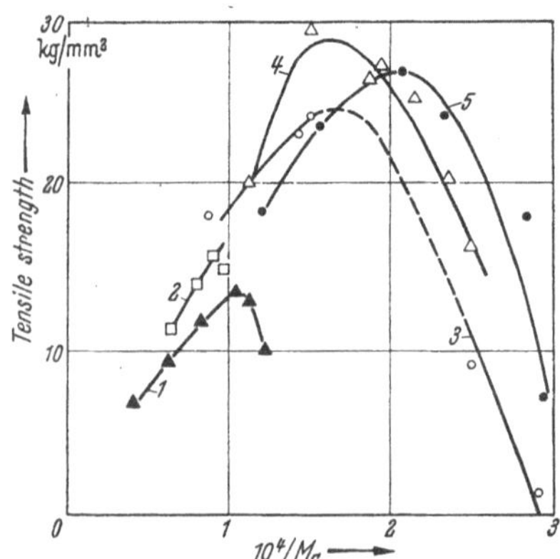

Fig. V,45. Dependence of tensile strength of sulphur-vulcanized rubbers on degree of cross-linking. (*1*) Rubber-sulphur, (*2*) T.M.T. Compounds, (*3*) Z.D.C. Compounds, (*4*, *5*) D.P.G. & M.B.T. with different proportions of sulphur. GEE: J. Polymer Sci 2, 451 (1947).

In the case of rubbers, it is possible to vary two parameters, the initial molecular weight, and the degree of cross-linking, without greatly affecting the lateral forces between chains. Both these parameters are found to exert a marked effect on tensile strength. Data by GEE[1] on a series of natural rubbers vulcanized with increasing amounts of sulphur showed the tensile strength to pass through a sharp maximum as the degree of cross-linking (determined from equilibrium swelling and modulus measurements) was increased (fig. V,45). However, experiments in which organic accelerators and other vulcanizing agents were employed showed that the degree of cross-linking corresponding to the maximum tensile strength depended on the type of compound introduced, due, no doubt, to the effects of secondary reactions in modifying the chemical structure of the chain. To overcome

[1] GEE, G.: J. Polymer Sci. 2, 451 (1947).

these difficulties MORRELL and STERN[1] employed a method of "vulcanization" involving the use of a volatile peroxide. This method is claimed to produce only C–C bonds, and to introduce no "impurity" into the system. Their results are reproduced in fig. V,46.

GEE has proposed the following explanation of the peculiar relation between tensile strength and degree of cross-linking. In the uncured rubber a sufficient degree of orientation for the initiation of crystallization is not attained, on account of intermolecular slippage. This is reduced by cross-linking, which thus promotes crystallization and enhances the tensile strength. In this way the initial rise in strength with progressive cross-linkage may be understood. To explain the subsequent fall, it is suggested that the degree of crystallinity at any given extension is substantially independent of the amount of cross-linking. If the tensile strength is further assumed to be determined by the crystallinity, there will then be a specific relation between tensile strength and extension at break which is valid for any degree of cross-linking. Such a hypothetical relation is indicated in fig. V,47, curve *4*. The actual breaking point will be the point at which the force-extension curve intersects this hypothetical curve. Clearly, the higher the modulus (degree of cross-linking) the lower will be the elongation at which this intersection occurs (fig. V,47). On this view the reduction of tensile strength with increasing cross-linking is due to the reduction in the amount of crystallinity at the breaking point.

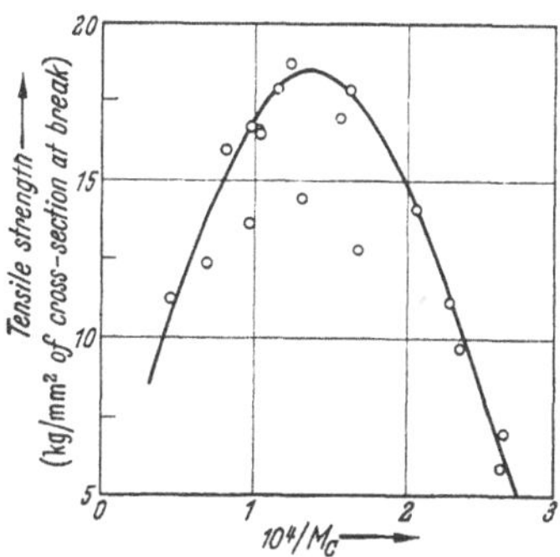

Fig. V,46. Dependence of tensile strength of peroxide-vulcanized rubbers on cross-linking. MORRELL and STERN: Trans. Inst. Rubber Ind. **28**, 269 (1953).

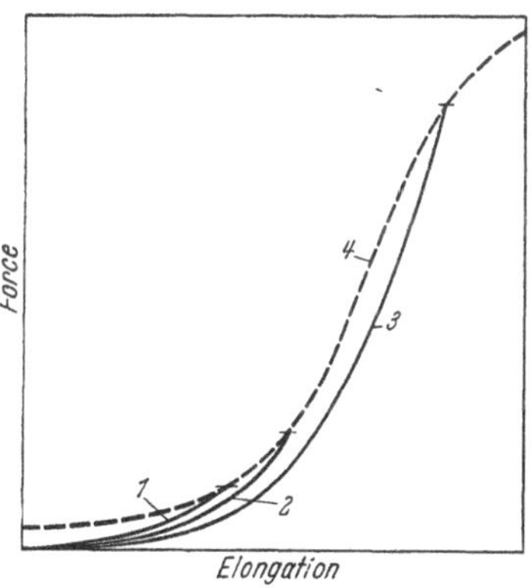

Fig. V,47. Force-extension curves for (*1*) high, (*2*) medium and (*3*) low degress of cross-linking. (Schematic.) (*4*) Hypothetical strength curve.

Support for this hypothesis is provided by the further work of MORRELL and STERN[1] on the reduction in volume on stretching for the range of peroxide-vulcanized rubbers referred to above. Fig. V,48 shows the volume change, extrapolated to the breaking point, as a function of degree of cross-linking. The maximum in this curve is found at about the same position as the maximum in the corresponding tensile strength curve (fig. V,46). Since the volume change may reasonably be taken as proportional to the amount of crystallization, the postulated relation between strength and crystallinity is substantiated.

[1] MORRELL, S. H. u. J. STERN: Trans. Inst. Rubber Ind. 28, 269 (1953).

Flory and coworkers[1] have examined the dependence of strength on degree of cross-linking, using natural rubber cross-linked with bis-azo compounds (cf. p. 322). Fig. V,49, which represents some of their data for two rubbers differing in molecular weight before cross-linking, shows the same general relation between tensile strength and degree of cross-linking as was found by Gee. Their explanation of the relation differs somewhat from that suggested by Gee.

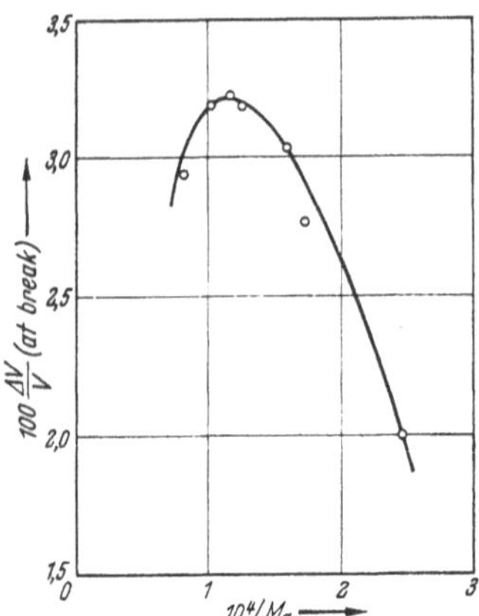

Fig. V,48. Volume change at break (extrapolated) for peroxide-vulcanized rubbers. [Morrell and Stern: Trans. Inst. Rubber Ind. **28**, 269 (1953).]

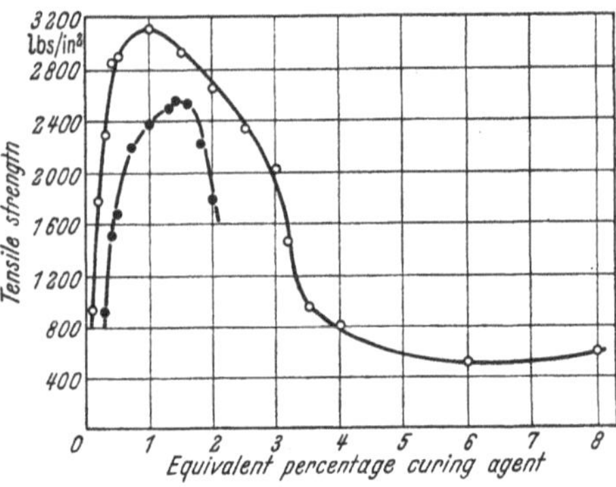

Fig. V,49. Dependence of tensile strength on cross-linking for natural rubber cross-linked with bisazo compound. (*1*) High molecular-weight rubber, (*2*) low molecular-weight rubber. [Flory, Rabjohn and Shaffer: J. Polymer Sci. **4**, 435 (1949).]

It is pointed out that the extensibility of a chain (i.e. the ratio of the fully extended length to the root-mean-square length) increases with its (contour) length; the mean link orientation, for a given extension ratio, is therefore higher, the shorter the chain length. If crystallization sets in when those chains which are most favourably oriented (i.e. parallel to the strain axis) are sufficiently straightened out, it follows that the overall network extension ratio at which crystallization commences will decrease as the degree of cross-linking increases. It is claimed that the formation of crystallites at a low value of extension ratio is unfavourable to the ultimate development of a high degree of oriented crystallinity.

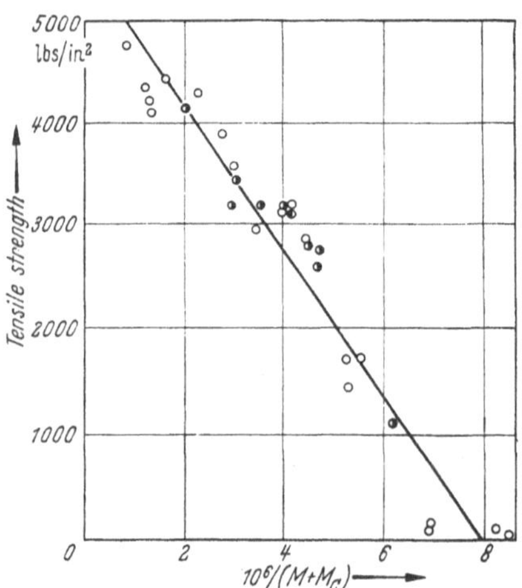

Fig. V,50. Tensile strength of butyl rubbers plotted against $1/(M + M_c)$ for homogeneous fractions or $1/(M_n + M_c)$ for mixtures. [Flory: Ind. Eng. Chem. **38**, 417 (1946).]

[1] Flory, P. J., N. Rabjohn u. M. C. Shaffer: J. Polymer Sci. **4**, 435 (1949).

Another aspect of the problem of tensile strength is brought out in the study by FLORY[1] of a series of butyl rubbers all cross-linked to the same extent but differing in molecular weight, and in molecular-weight distribution, before cross-linking. It is argued that it is only that fraction of the material which is actually in the form of network chains which contributes to the stress; the remainder, in the form of unconnected molecules or "loose ends", act merely as an unoriented diluent. The proportion of active network chains, w_a, is shown to be

$$w_a = 1 - 2M_c/(M + M_c). \qquad (V,146)$$

M being the initial molecular weight and M_c the mean molecular weight of network chains. A plot of tensile strength against $1/(M + M_c)$ is found to yield a straight line (fig. V, 50), in qualitative agreement with equation (V, 146). Quantitatively, however, there is a discrepancy, in that the tensile strength falls to zero at a value of M corresponding not to $w_a = 0$ but to $w_a = 0.41$. From this result it is concluded that a certain minimum amount of orientable material must be present for crystallization to start.

G. Stress relaxation phenomena.

By J. P. BERRY and L. R. G. TRELOAR.

§ 41. Introduction.

The ideal rubber, as considered by the statistical theory, is perfectly elastic in the sense that the stress is uniquely determined by the strain. As we know, such an ideal rubber does not exist, and in any real rubber the stress depends to a greater or less extent on the rate at which the strain is applied or the time for which it is maintained. The general nature of these stress-relaxation phenomena as they appear in a cross-linked rubberlike polymer is illustrated in fig. V, 51, taken from the work of TOBOLSKY and ANDREWS[2]. In these experiments the relaxation of stress at

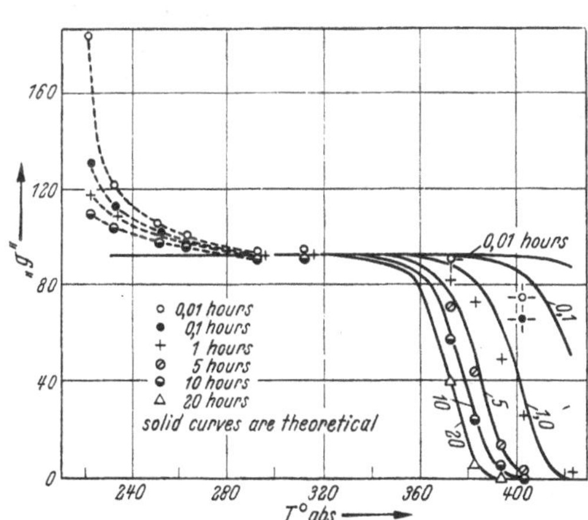

Fig. V, 51. "g" as a function of temperature for various times Hevea gum. [TOBOLSKY and ANDREWS: J. Chem. Phys. 13, 3 (1945).]

[1] FLORY, P. J.: Ind. Engng. Chem. 38, 417 (1946).
[2] TOBOLSKY, A. V. u. R. D. ANDREWS: J. Chem. Physics 13, 3 (1945).

constant strained length was observed at a number of temperatures; the curves show the variation of stress with temperature at particular times after the application of the strain. (Actually the "reduced stress" is plotted: this is the observed stress corrected for the kinetic elasticity temperature dependence by multiplying by T_0/T, where T is the working temperature and T_0 is a standard reference temperature.) Three fairly distinct regions are visible in the figure. First, there is the region of relaxations occurring at low temperatures and short times; then there is a middle region in which relaxation effects are negligible; and finally, there is a region in which further very pronounced relaxation sets in. The first region corresponds to the breakage of intermolecular bonds of a physical character; it represents the time required for the chains to rearrange themselves in their new equilibrium configurations and in this process no net change in the network occurs. The second region represents the state in which a vulcanized rubber is normally used, that is, where the equilibrium properties of the cross-linked network predominate. The high-temperature region is associated with irreversible chemical breakdown of the network structure and it is with this region only that we shall be concerned. The deterioration in the elastic properties of rubber in service (ageing) is due to this cause, and the elucidation of the chemical processes involved is of direct technological importance.

Contributions of TOBOLSKY *et. al:* A comprehensive survey of the field was made by TOBOLSKY and his co-workers[1, 2], who, in a pioneer investigation, established the main characteristics of the phenomenon. For vulcanized natural rubber at any temperature the stress decay in air at constant strained length was reported to follow an exponential law,

$$f = f_0 \exp. (-k' t).$$

The exponent k' was approximately independent of the amount of strain, over a wide range (10–100%) showing that the breakdown was not due to the stress, but occurred independently of it. It is presumably on the basis of this finding that it was assumed that the same reaction occurred in the unstrained state. The temperature dependence of k' could be described by the usual reaction rate equation.

$$k' = \frac{k\,T}{h} \exp \left\{ - \frac{\Delta F^{\ddagger}}{R\,T} \right\}.$$

$\Delta F^{\ddagger}$, the free energy of activation, was found to be 30,4 kg · cals/mole. Though other vulcanized polymers (Buna, Butaprene N, Neoprene, butyl rubber) gave relaxation curves which departed to some extent from the simple exponential form, a similar value for the activation energy was found in all cases, indicating an underlying similarity between the chemical reactions involved. The loss of elastic properties of vulcanized rubber has been recognised as an oxidative reaction for many

[1] TOBOLSKY, A. V., I. B. PRETTYMAN u. J. H. DILLON: J. Appl. Physics **15**, 380 (1944).

[2] TOBOLSKY, A. V., R. D. ANDREWS u. E. E. HANSON: J. Appl. Physics **17**, 352, (1946).

years. This fact was substantiated by carrying out stress relaxation experiments in nitrogen instead of air, when a thousand-fold reduction in relaxation rate was observed.

The results of the above investigation were interpreted in molecular terms by the application of the statistical theory of elasticity. Under the conditions of experiment, the observed force was considered directly proportional to the number of network chains per unit volume, and hence during relaxation this number decreased exponentially with time. If this were the only reaction occurring, the unstrained length of the sample should remain unchanged throughout the experiment; in practice, this length increased with increasing extents of degradation. The effect was considered due to a cross-linking reaction concomitant with the degradative process. It was assumed that the cross-links were formed between network elements which were in equilibrium configurations referred to the state of strain at which the relaxation was carried out. The network which was built up in this way was therefore considered to make no contribution to the stress during the relaxation, which measured only the breakdown of the original network. On removing the constraint, however, the unstressed length of the sample was determined by the equilibrium condition that the forces exerted by the two networks, considered separately, were equal and opposite. To assess the importance of the cross-linking reaction, changes in the modulus of a sample maintained at the relaxation temperature in the unstrained state were measured.

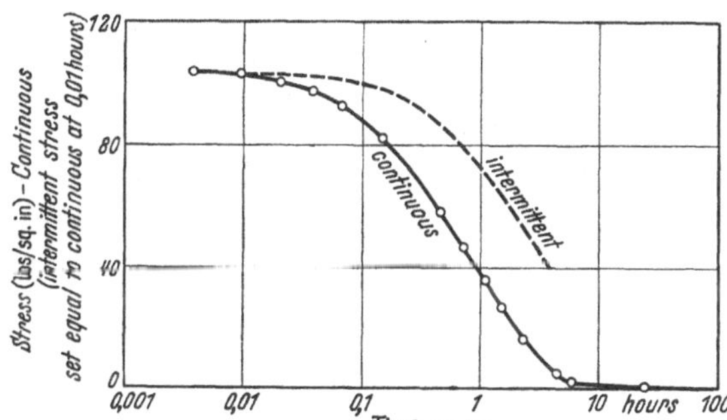

Fig. V, 52. Continuous and intermittent stress relaxation. Hevea gum stock, 130° C, 50 per cent elongation. [TOBOLSKY and ANDREWS: J. Chem. Phys. **13**, 3 (1945).]

On the assumption that both the scission and cross-linking reactions proceeded unchanged under these conditions the results gave the net change in the network, and since stress relaxation gives a measure of the scission reaction only, the two effects may be separated. Typical results obtained in this way are shown in figs. V, 52 and V, 53. It will be seen that while for natural rubber the rate of breakdown tended to exceed the rate of cross-linking, so that the material softened, the opposite effect was obtained with GR—S. Though the molecular interpretation probably represents a great over-simplification, the self-consistency of the results, and the successful extension of the theory to account for physical changes observed in vulcanized elastomers on ageing suggests that it is at least approximately valid.

Contributions of later workers: Following the early work of TOBOLSKY et. al. most of the investigations carried out in this field were directed towards establishing the technique of stress relaxation as an unequivocal

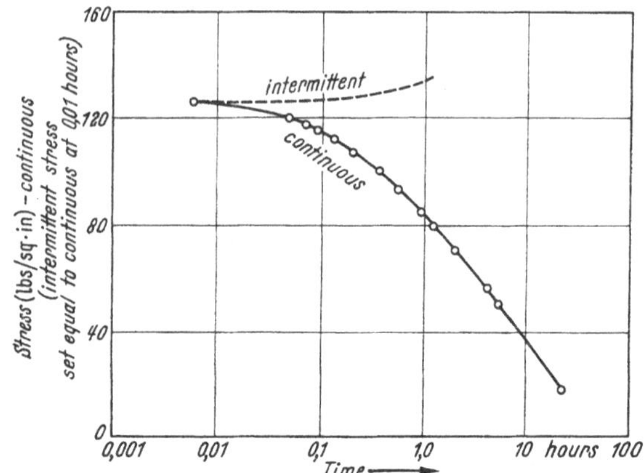

Fig. V, 53. Continuous and intermittent, stress relaxation. GR–S gum stock, 50 per cent elongation, 130° C stock. [TOBOLSKY and ANDREWS: J. Chem. Phys. **13**, 3 (1945).]

method of assessing accelerated ageing[1, 2, 3]. Unfortunately these efforts met with no real success, mainly due to the complexity of the chemical reactions responsible, which, though resulting in a fairly simple relaxation behaviour, make comparisons between results from rubbers compounded in different ways of doubtful validity.

LE BRAS[4] has suggested that agents which protect vulcanized rubber from the deleterious action of oxygen may be divided into two classes.

1. Antioxygens, which prevent the attack of oxygen on the rubber molecule.

2. Deactivators, which affect the further reaction of the oxidised molecule in such a way that scission does not result.

LE FOLL[5] has considered the action of such classes of compounds in stress relaxation. He concludes that antioxygens are effective in reducing the scission reaction rate, as measured by continuous relaxation methods, whereas deactivators promote the cross-linking reaction, and thus influence modulus changes in unstrained samples.

The stress relaxation curves which he obtains are not exponential (fig. V, 54), but may be represented by the sum of two exponential terms, viz.

$$f = (f_1)_0 \times 10^{-\alpha t} + (f_2)_0 \times 10^{-\beta t}.$$

[1] BERRY, D. S.: R. A. B. R. M. Research Bulletin R. 394.
[2] PEDERSEN, H. L.: Acta Chem. Scandinavica 4, 487. (1950).
[3] PEDERSEN, H. L. u. B. NIELSEN: J. Polymer Sci. 7, 97 (1951).
[4] LE BRAS, J.: Rev. gén. Caoutchouc **21**, 3 (1944).
[5] LE FOLL, J.: Rev. gén. Caoutchouc **30**, 559 (1953).

The first term, which is responsible for the deviation from linearity of the log force-time plot, is considered to represent the decay of secondary (VAN DER WAALS) forces, though it appears from his figures that these account for about 70% of the original force and persist at 100° C for some 12 hrs. Protective agents are found to reduce the rate (β) of the relaxation of the primary network forces, as represented by the second term in the above expression, but to have very little effect on the relaxation rate of secondary forces (α). This fact is put forward as confirmation of the general theory. An alternative explanation has been suggested by the authors (page 371).

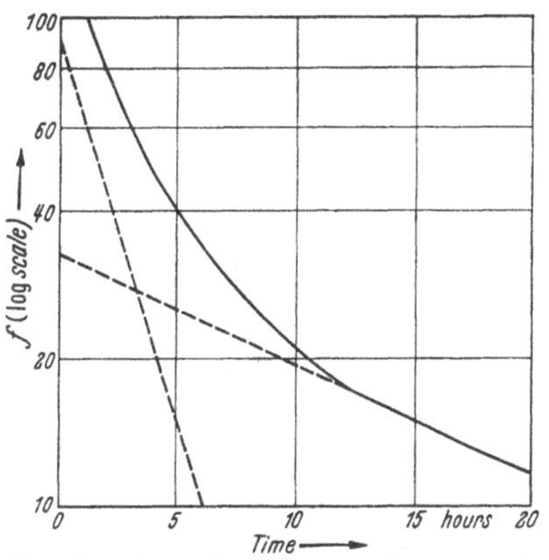

Fig. V, 54. Logarithm of tension f at constant elongation as a function of time during oxidative ageing. [LE FOLL: Rev. Gen. Caoutchouc **30**, 559 (1953).]

Recent developments. Recent investigations[1,2] though confirming the broad principles set down by TOBOLSKY, reveal deviations in relaxation behaviour which are of considerable practical and theoretical significance. Developments in technique which permit greater control over the experimental conditions have led to a clearer understanding of the fundamental processes involved.

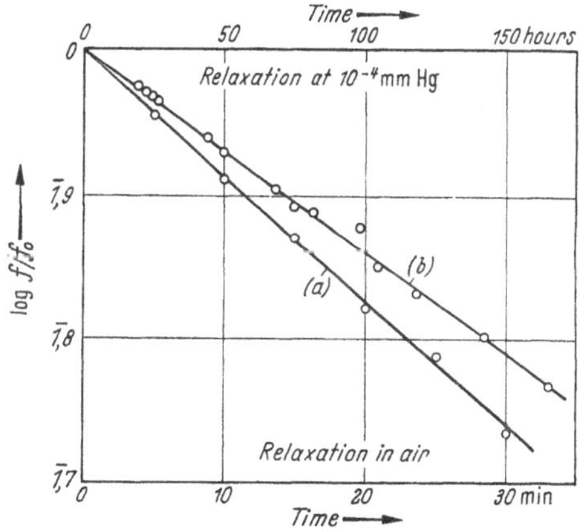

Fig. V, 55. Relaxation of a peroxide vulcanizate at 120° C. a) in air and b) in vacuo (10^{-4} mm. Hg).

In order to simplify the chemical picture as much as possible, a peroxide vulcanized purified natural rubber was first studied. By this process of vulcanization, the junction points are formed by direct linking of the carbon atoms in the rubber chains to produce a purely hydrocarbon network structure[3]. The relaxation of stress was strictly exponential with time for these rubbers, and the value of the relaxation constant k' was independent of oxygen pressure in the region 2–500 mm Hg. The removal of oxygen (10^{-4} mm Hg) reduced the relaxation rate by a factor of the same order (400) as found by TOBOLSKY in experiments carried out in nitrogen (fig. V, 55).

The more complex chemical structure of sulphur/accelerator vulcan-

[1] BERRY, J. P. u. W. F. WATSON: J. Polymer Sci., In press.
[2] BERRY, J. P.: Unpublished Work.
[3] FARMER, E. H. u. C. G. MOORE: J. Chem. Soc. London 142 (1951).

ized rubbers, where the rubber chains are linked through one or more sulphur atoms, produced a more complex type of relaxation behaviour. The vulcanizates studied were compounded according to the recipes

1. smoked sheet 100, zinc oxide 5, stearic acid 1, M.B.T. 1, sulphur 3.
2. smoked sheet 100, zinc oxide 5, stearic acid 1, Santocure 0,7, sulphur 2,5.

The decay was no longer strictly exponential, and the form of the curve depended on the oxygen pressure at which the relaxation was carried out (fig. V, 56). The value of k' obtained from the later exponential part of the decay curve was no simple function of oxygen pressure. In vacuo (10^{-4} mm Hg) stress relaxation occurred at a rate initially equal to the rate in oxygen (fig. V, 56) and proceeded to a considerable extent (fig. V, 57). These results do not confirm TOBOLSKY's observations on similar compounds[1]. The differences in the relaxation characteristics of these two types of vulcanizate and the deviations from the behaviour reported by TOBOLSKY in the case of

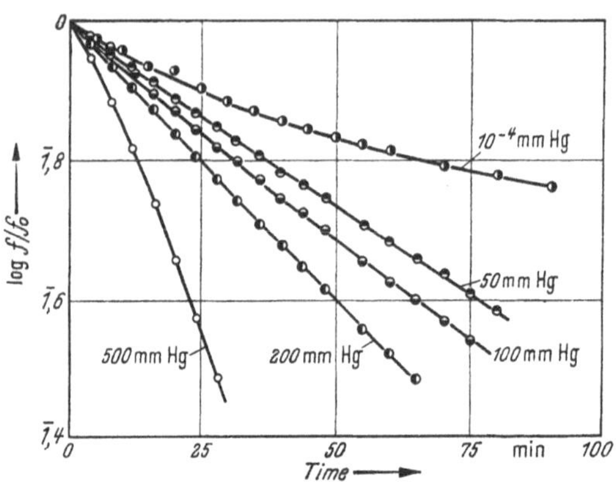

Fig. V, 56. Relaxation of sulphur-accelerator vulcanizate 1 at 120° C in oxygen at various pressures, and in vacuo (10^{-4} mm Hg).

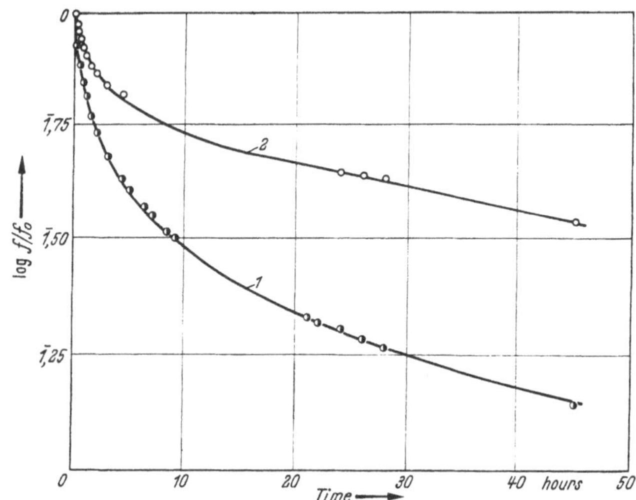

Fig. V, 57. Relaxation of sulphur vulcanizates 1 and 2 at 120° C in vacuo.

[1] TOBOLSKY, A. V., J. B. PRETTYMAN u. J. H. DILLON: J. Appl. Physics **15**, 380 (1944).

the second provide strong indications both of the site of the scission in the network, and the nature of the chemical reactions responsible. Degradation of the molecular network may occur either by random scission of the polymer chains or by scission at the cross-link. Experimental observations and theoretical predictions are consistent with the second mechanism of relaxation for both types of vulcanizate.

§ 42. Evidence for cross-link scission.

a) Stress relaxation results.

The differences in molecular structure of the two types of vulcanizate studied are confined to the cross-link. In peroxide vulcanizates this is formed by the direct intermolecular linking of carbon atoms[1], in sulphur/accelerator vulcanizates the linking takes place through a chain of one or more sulphur atoms, the detailed structure being determined by the vulcanization conditions. These structural differences at isolated points in the network would be expected to affect the chemical properties of the chains, which constitute its bulk, to only a very small degree. The wide differences in relaxation behaviour detailed above are therefore much easier to account for on the basis of reaction at the cross-link, rather than reaction in the chains. Further, the investigation of Tobolsky on polymers of widely different chemical structure and properties, but vulcanized in a similar way, indicated a common type of chemical reaction responsible for stress relaxation[2], a reaction probably common to the similar cross-links.

b) Influence of oxidation inhibitors.

The details of the reaction with oxygen of "model" compounds structurally related to rubber hydrocarbon have been thoroughly established[3,4]. The oxidation proceeds by a freeradical chain mechanism, and small proportions of certain compounds (inhibitors) retard the rate of this reaction by a considerable factor, since they react with one of the radical intermediates in the propagation steps. Experiments on unvulcanized rubber have revealed these same characteristics[5], and it has been shown that the scission reaction which occurs with this material probably also takes place in the propagation steps, due to the independent decomposition of the radical which is concerned in the inhibition[6]. Consequently the addition of inhibitors to unvulcanized rubber reduces both the rate of oxidation, as measured by oxygen absorption, and rate of scission, as measured by changes in molecular weight, by about the same factor. The same considerations would reasonably be expected to

[1] Farmer, E. H. and C. G. Moore: J. Chem. Soc. London 142 (1951).

[2] Tobolsky, A. V., J. B. Prettyman u. J. H. Dillon: J. appl. Physics **15**, 380 (1944). — [3] Bolland, J. L., Quart. Rev. **3**, 1 (1949).

[4] Bateman, L., Quart Rev. **8**, 147 (1954).

[5] Bolland, J. L. u. A. L. Morris: unpublished results.

[6] Morris, A. L.: unpublished results.

apply to the hydrocarbon chains of vulcanized rubber, though in this case the same exploratory techniques cannot be used, and direct verification is not possible. Incorporation of inhibitors into peroxidevulcanized rubbers produces a reduction in the scission reaction rate as measured by stress relaxation much less than would be expected if the above mechanism was involved. Hence, either scission of the hydrocarbon chain does not occur according to the normal oxidation scheme responsible for molecular scission in model compounds and unvulcanized rubber, or the site of the reaction is not in the hydrocarbon chains but at the cross-links. Of the two possibilities, the latter is the more plausible.

c) Theoretical treatments of network breakdown.

Some attempts have been made to predict the form of the stress-time relationships assuming either scission at random along the chains or at the cross-links. BUECHE[1] deduced that in a randomly cross-linked network, these two processes lead to essentially different types of stress decay. A more general treatment[2] considers both this factor, and also the effect of different chemical mechanisms for the reaction.

Random scission of chains. Rubber hydrocarbon is a long-chain polyisoprene, i. e. the rubber molecules consist of recurring isoprene monomer units. The action of vulcanization is to produce a non-uniform three-dimensional network, in which the distribution of lengths of chains, and consequently numbers of monomer units, between cross-links, is given by[3]

$$N_x = N\,\delta\, e^{-\delta x}. \qquad \text{(V, 147)}$$

N_x is the number of chains of x monomer units.

N is the total number of chains.

δ is the reciprocal of the average number of monomer units in a chain.

Thus $\delta = \frac{m}{M_c}$, where m is the molecular weight of the monomer unit, and M_c the molecular weight of the chain of average length.

If p is the probability that a monomer unit has undergone scission in time t, then the number of uncut chains of x monomer units is

$$N_{x,t} = N_{x,0}\,(1-p)^x. \qquad \text{(V, 148)}$$

Introducing (V, 147) in (V, 148) and summing for all values of x

$$\frac{N}{N_0} = \frac{\delta\,(1-p)}{e^{\delta} - (1-p)}. \qquad \text{(V, 149)}$$

The oxidation of model compounds and unvulcanized rubber referred to above is characterized by a continuously increasing rate (autocatalysis). An oxidative scission reaction in the network chains would reasonably

---

[1] BUECHE, A. M.: J. Chem. Physics **21**, 614 (1953).
[2] BERRY, J. P. u. W. F. WATSON: J. Polymer Sci. (In press).
[3] WATSON, W. F.: Trans. Faraday Soc. **49**, 1369 (1953).

be expected to exhibit the same characteristics and under these conditions the rate of scission of monomer units is

$$-\frac{dM}{dt} = r_0 + k_1 (M_0 - M) \qquad (V, 150)$$

r_0 is the reaction rate at zero experimental time.

M_0 is the number of unreacted monomer units at zero experimental time. p is equivalent to the ratio of the number of monomer units which have undergone scission in experimental time t to the initial number of monomer units.

$$\text{Hence } p = \frac{M_0 - M}{M_0} = \frac{r_0}{k_1 M_0} (e^{k_1 t} - 1). \qquad (V, 151)$$

Inserting (V, 151) in (V, 149), and assuming the relationship between force and number of network chains given by the statistical theory of elasticity,

$$\frac{f}{f_0} = \delta \left\{1 - \frac{r_0}{k_1 M_0} (e^{k_1 t} - 1)\right\} \Big/ \left\{e^{\delta} - 1 + \frac{r_0}{k_1 M_0} (e^{k_1 t} - 1)\right\}. \qquad (V, 152)$$

The initial rate of the reaction is usually determined by traces of catalyst, but it may be regarded as equivalent to an extent of reaction which has already taken place in a network which originally was perfect in the sense that no monomer units had undergone scission. The percentage (x_1) which does so by zero experimental time determines both r_0, the initial rate of the reaction, and M_0 the number of remaining unreacted monomer units. The relationship is

$$x_1 = \frac{r_0}{r_0 + k_1 M_0} \cdot 100\,\%. \qquad (V, 153)$$

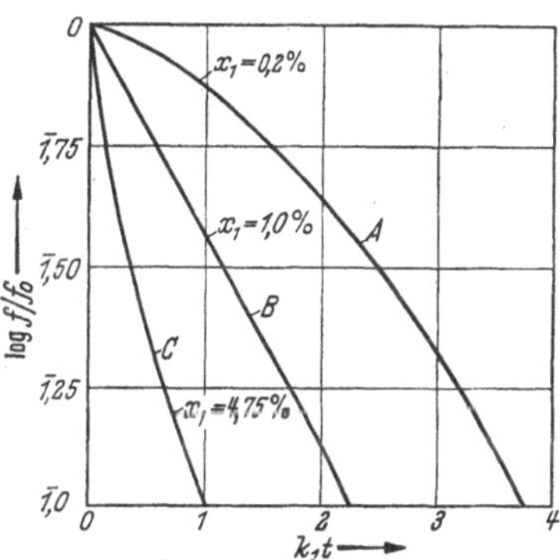

Fig. V, 58. Relaxation of stress by random scission of network chains assuming autocatalytic reaction of monomer units. [Equ. (V, 152) with $\delta = 0{,}01$.]

The shape of the predicted stress relaxation curve, adopting the usual mode of representation, depends to a large extent on the value of x_1 (fig. V, 58). For a particular value (1%) the stress relaxation is almost exactly exponential over the extent of degradation normally studied.

Scission at cross-links. Scission at the cross-links may take place either at one of the bonds by which the molecular chains are joined or in a monomer unit in a chain attached to the cross-link. Three types of plausible chemical reaction may be considered.

1. First order scission of the cross-link bond. The definitive rate expression is

$$-\frac{dX}{dt} = k_2 X. \qquad (V, 154)$$

Where X is the number of cross-links per unit volume. Hence

$$\frac{X}{X_0} = e^{-k_2 t}. \qquad (V, 155)$$

If end-effects may be neglected, and the conclusions of the statistical theory of elasticity assumed

$$\frac{f}{f_0} = e^{-k_2 t}. \qquad \text{(V, 156)}$$

On the basis of this mechanism therefore, the stress decay is strictly exponential (fig. V, 59, curve A).

2. *Autocatalytic scission of the cross-link bond.* Because of the nature of the molecular structure in the vicinity of the cross-link, the reaction may exhibit autocatalytic characteristics as discussed above.

Thus

$$-\frac{dX}{dt} = r_0 + k_3 (X_0 - X) \qquad \text{(V, 157)}$$

and

$$\frac{X}{X_0} = \frac{f}{f_0} = 1 - \frac{r_0}{k_3 X_0} (e^{k_3 t} - 1) \qquad \text{(V, 158)}$$

$$x_2 = \frac{r_0}{r_0 + k_3 X_0} \cdot 100\,\% . \qquad \text{(V, 159)}$$

The shape of the relaxation curve again depends on the value of x_2; those constructed assuming lower values than the one used in fig. V, 59, curve B, lie completely above this curve, with a lower initial and a higher final slope.

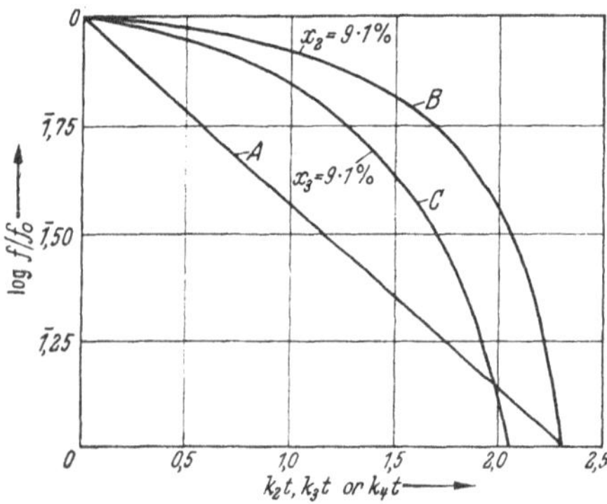

Fig. V, 59. Relaxation of stress by first order (A) and autocatalytic (B) scission of cross links and by autocatalytic scission of specific monomer units (C).

3. *Autocatalytic reaction of specific monomer units.* The process is considered to be non-random in the sense that the particular unit at which scission occurs is determined by its location with respect to the cross-link. It is clear however, that if the probability of undergoing scission is the same for the specific units equidistant from both ends of the same chain, this mechanism is equivalent to the random scission of a network of uniform chains of only two units in length. Though the treatment is the same as that employed for the more general case of random chain scission, the result is nearer in mathematical form to those given above. For this reason, it is included in its present context.

$$\left.\begin{aligned} \frac{f}{f_0} &= \left\{1 - \frac{r_0}{k_4 X_0} e^{k_4 t} - 1)\right\}^2 \\ x_3 &= \frac{r_0}{r_0 + k_4 M_0} \cdot 100\,\% . \end{aligned}\right\} \qquad \text{(V, 160)}$$

The curve corresponding to this equation is shown in fig. V, 59, curve C.

§ 43. Consideration of experimental results.

1. Peroxide vulcanizates. The strictly exponential stress decay of peroxidevulcanized rubber (fig. V, 55) may be accounted for by (1) the autocatalytic random scission for chains, with a particular initial rate (fig. V, 58, curve *B*) or (2) the first order scission of cross-link bonds (fig. V, 59, curve *A*). The pre-experimental history of the sample has no effect on the mode of decay, and no variation in the shape of the relaxation curve, corresponding to curves *A* and *C* (fig. V, 58) has been found. Theoretical predictions thus support the experimental evidence already put forward (page 367) for the second mechanism of stress decay in the case of peroxide vulcanized rubbers.

2. Sulphur/accelerator vulcanizate. The stress decay in these rubbers is only strictly exponential at a particular oxygen pressure. At higher pressures the relaxation rate (as defined by the slope of the log force-time graph) increases to a final constant value, at lower pressures it decreases to another constant value (fig. V, 56). The shapes of the curves cannot be identified with any of those predicted on theoretical grounds. They resemble superficially the curves illustrated in fig. V, 58, but differ from these in that after an initial interval, the experimental curves follow an exponential law (cf. fig. V, 54).

The experimental facts may be satisfactorily explained by assuming a two-stage reaction, the first resulting in an oxidised cross-link, which though retaining its physical function in that state, is ruptured by the subsequent firstorder reaction[1]

$$\begin{array}{lll} X + O_2 \rightarrow XO_2 & & k_1 \\ XO_2 & \rightarrow \text{scission products} & k_2 . \end{array}$$

The first reaction, though pseudomonomolecular for any particular value of oxygen pressure, is almost certainly complex. This complexity determines the form of the dependence on oxygen pressure of the rate constant k_1. It is assumed that a proportion of cross-links (x) have already been oxidised to XO_2 before relaxation starts. The form of the stress relaxation expression on the basis of these assumptions is

$$\frac{f}{f_0} = \frac{[X] + [XO_2]}{[X]_0 + [XO_2]_0} = \frac{1}{k_2 - k_1}\left\{(k_2 x - k_1)\, e^{-k_2 t} + k_2 (1 - x)\, e^{-k_1 t}\right\}$$

$$k_1 > k_2, \quad x = \frac{[XO_2]_0}{[X]_0 + [XO_2]_0}\,.$$

This equation accounts for the dependence of the shape of the relaxation curves, the slope of the final exponential part on oxygen pressure, the independence of the initial rate of relaxation on oxygen pressure and the form of the relaxation in vacuo (when $k_1 = 0$). The results of Le Foll (q. v.) may also be explained in terms of this hypothesis. The rate constant k_1, which determines the final slope of the

[1] Berry, J. P.: Unpublished results.

relaxation curve, is associated with the first, oxidative stage of the reaction, and any agent which affects the oxidation will likewise affect this slope. The first term (LE FOLL's "secondary force relaxation") is now identified with the decomposition of the oxidation product, a process which is not unexpectedly unaffected by the added protective agents.

3. The theoretical treatment of SCOTT *and* STEIN[1]. The results of TOBOLSKY have indicated that a cross-linking reaction takes place during relaxation and the assumption was made that this effect was not detectable under normal experimental conditions of continuous relaxation. SCOTT and STEIN have carried out a statistical treatment of the problem assuming first order scission of cross-links followed by their total recombination in instantaneous equilibrium configurations. Their analysis leads to the conclusion that the junction points so formed will contribute to the configurational entropy of the system, and hence to the force in the network. The resultant expression is of the form

$$f = f_0 e^{-\lambda k t}$$

where k is the true firstorder rate constant and λ is a factor which takes into account cross-linking effects. Its value depends on the initial extension of the sample. At low elongations ($\alpha_0 \simeq 1$), $\lambda = 0{,}5$, but this value falls rapidly to 0,43 at $\alpha_0 = 2$ and then decreases more gradually to 0,41 at $\alpha_0 = 7$. The shapes of the relaxation curves are unaffected by λ, and its dependence on extension introduces variations in slope which are only of the same order as experimental reproducibility. The value of the activation energy determined by the variation in slope of the relaxation curve with temperature will also be unaffected by λ, which however, should contribute to the entropy of activation of the reaction. In the usual form of relaxation experiments, according to this theory, the effect of the cross-linking reaction is undetectable. The self-consistency of the results on the basis of this assumption further justifies its adoption.

Books and general articles for Chapter V.

ALFREY, T.: Mechanical behaviour of high polymers, Interscience, New York 1948.

FLORY, P. J.: Principles of polymer chemistry, Cornell, New York 1953.

FRITH, E. M. and R. F. TUCKETT: Linear polymers, Longmans, Green, London 1951.

GUTH, E., H. M. JAMES and H. MARK: "Kinetic theory of rubber elasticity" in Advances in Colloid Science II, Interscience, New York 1946.

TRELOAR, L. R. G.: The thermodynamic study of rubberlike elasticity. Proc. Roy. Soc. [London] **B 139**, 506 (1952).

TRELOAR, L. R. G.: The physics of rubber elasticity, Oxford 1949.

[1] SCOTT, K. W. u. R. S. STEIN: J. Chem. Physics **21**, 1281 (1953).

Sixth Chapter

Structure and Mechanical Properties of Plastics.

By

John D. FERRY.

With 35 figures.

A. Viscoelastic Properties of Polymeric Liquids and Soft Solids.

§ 44. Introduction.

The characteristic structural feature of polymeric systems which leads to their remarkable mechanical properties is the existence of long chain molecules endowed with Brownian movement. The thermal motion may be thought of as translation of molecular segments with rotation around chain bonds. The translations and rotations of individual segments within a molecule are of course not independent. Some rearrangements are rapid, others slow, according to the degree of cooperation required among segments close together or far apart[1]. The overall motion is a complicated combination of different modes and covers a very wide range of time scale (§ 8).

Always, when an external time-dependent force is imposed on an internal Brownian movement, energy is both stored and dissipated, as most familiarly recognized in the theory of dielectric dispersion of DEBYE[2]. If the force is mechanical, the mechanical energies stored and dissipated correspond to elasticity and viscosity respectively, and the behaviour is described as viscoelastic. Because of the wide range of Brownian motions in polymers, these materials appear as viscoelastic under almost any kind of time-dependent stress.

This section deals with amorphous polymers above their glass transition temperatures, where some, at least, of the modes of thermal motion are rapid compared with the time scale of an ordinary experiment. Such materials have the superficial appearance of viscous liquids or soft solids, depending on the magnitude of the molecular weight. Their mechanical consistencies will, however, vary with the speed of measurement; as the time interval decreases (in a transient experiment), or the frequency

[1] ALFREY, T.: Mechanical Behaviour of High Polymers. New York: Interscience Publishers, 1948. — [2] DEBYE, P.: Polare Molekeln. Leipzig: Hirzel, 1929.

increases (in a periodic experiment), both the viscous liquids and the soft solids will approach the behaviour of hard elastic solids.

The experimental study of viscoelastic properties involves measuring the time dependence of the stress/strain ratio under various conditions. For small stresses, this ratio is a function of time alone and not of stress magnitude — the viscoelastic behaviour is linear. Then equivalent information can be obtained from many different types of experiments (§§ 3, 4). In practice, several different methods may be needed to provide a complete picture over a wide range of time scale. The effective range of time scale can be greatly increased by combining results at different temperatures[1, 2], provided the relaxation processes responsible for viscoelastic behaviour all have the same temperature dependence (§ 5). In terms of current molecular theories (§ 8), this is equivalent to saying that all the effective segmental frictional coefficients associated with thermal motions of various modes have the same temperature dependence.

For polymeric liquids and soft solids, it is usually possible to combine data at different temperatures by the method of reduced variables[3, 4] [§ 5] to separate the variables of time and temperature and provide two master functions. The *first* is a function of time alone and describes the viscoelastic behaviour at a single reference temperature. According to convenience it may appear in the form of the creep function, the stress relaxation function, the real and imaginary parts of the complex dynamic modulus, or the real and imaginary parts of the complex compliance; alternatively, it may be presented as a distribution function (spectrum) of relaxation times or of retardation times. The interrelations among these functions have been given in a previous Chapter (§ 4). The *second* is a function of temperature alone and represents the temperature dependence of the relaxation processes, corresponding on a molecular scale to the temperature dependence of the segmental friction coefficients. It may appear in the form of a factor a_T, the ratio of a relaxation time at temperature T to that at a standard temperature T_0; or the apparent activation energy of the relaxation processes, $\Delta H_a [= R d \ln a_T / d(1/T)]$. The latter is independent of the choice of an arbitrary standard temperature. The term "apparent" emphasizes the fact that ΔH_a is itself strongly temperature-dependent.

Examples of data will be shown both before and after analysis in terms of the two master functions. Most of the data refer to behaviour in shear, but some refer to behaviour in extension. For liquids and soft solids, where the bulk modulus greatly exceeds the shear modulus, data for shear and extension are interconvertible; the time-dependent YOUNG's modulus or the frequency-dependent complex YOUNG's modulus is greater than the corresponding shear modulus by a factor of 3.

There is a marked contrast between the viscoelastic behaviour of soft

---

[1] LEADERMAN, H.: Elastic and Creep Properties of Filamentous Materials and Other High Polymers. Washington: The Textile Foundation, 1943.

[2] TOBOLSKY, A. V. u. R. D. ANDREWS: J. Chem. Physics **13**, 3 (1945).

[3] FERRY, J. D.: J. Amer. Chem. Soc. **72**, 3746 (1950).

[4] FERRY, J. D. u. E. R. FITZGERALD: J. Colloid. Sci. **8**, 224 (1953).

linear polymers of molecular weights 5,000 and 1,000,000. For a polymer of low molecular weight, the relaxation and retardation spectra extend over only a moderately wide range of logarithmic time. There is a single broad peak in each spectrum corresponding to the transition (as a function of time, not temperature) from glasslike to soft consistency; at longer times there are no elastic mechanisms contributing to the spectrum at all. For a polymer of high molecular weight, a similar broad transition peak appears in the spectrum, but in addition other elastic mechanisms extend out to longer times and cover an enormously wide range of time scale. The contrast is illustrated in fig. VI, 1 where the retardation spectrum L is plotted for two samples of polyisobutylene. (The approximate coincidence of the maxima of two of the peaks is fortuitous.)

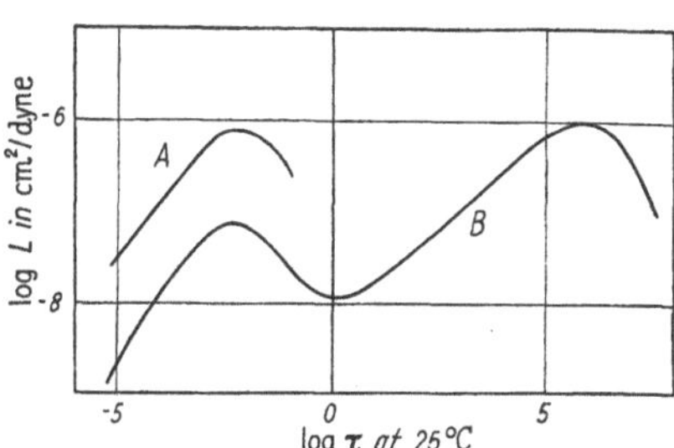

Fig. VI, 1. Retardation spectra of two polyisobutylenes, reduced to 25°C. A, M_w = 11,000 (LEADERMAN, SMITH and JONES); B, M_w = 1,560,000 (compilation by MARVIN).

The critical molecular weight above which the long-time mechanisms become perceptible and complicate the spectrum depends on the chemical structure, but it is of the order of 10,000 to 30,000. It corresponds to the point above which the average segmental friction coefficient becomes dependent on molecular weight[1] and the individual friction coefficients depend very much upon position along the polymer chain[2]. It also corresponds roughly to the distinction between liquids and soft solids in superficial appearance. The liquid polymers of low molecular weight are of course the easiest to describe; they are treated in § 45. For the soft solids of high molecular weight, it is convenient to divide the discussion into two parts. § 46, which deals with the transition range from rubberlike to glasslike consistency, corresponds to the region of time scale occupied by the first peak of curve B in fig. VI, 1. § 47, which deals with the range of rubberlike consistency, corresponds to the region of time scale occupied by the second peak.

§ 45. Polymeric Liquids of Low Molecular Weight.

The most extensive data on viscoelastic properties of polymer liquids are those for polyisobutylenes from measurements of creep by LEADERMAN[3] and dynamic rigidity and viscosity by DEWITT[4] and MASON and BAKER[5].

---

[1] BUECHE, F.: J. Chem. Physics **20**, 1959 (1952).

[2] HAMMERLE, W. G.: Ph. D. Thesis, Princeton, 1954.

[3] LEADERMAN, H., R. G. SMITH u. R. W. JONES: J. Polymer Sci. **14**, 47 (1954).

[4] HARPER, R. C. JR., H. MARKOVITZ u. T. W. DEWITT: J. Polymer Sci. 8, 435 (1952).

[5] MASON, W. P., W. O. BAKER, H. J. MCSKIMIN u. J. H. HEISS: Physic. Rev. **73**, 1074 (1948); **74**, 1873 (1949); **75**, 963 (1949).

a) Creep.

Creep in shear may be separated into instantaneous elastic, retarded elastic, and viscous contributions as follows (cf. § 3):

$$\gamma/\mathfrak{T} = J_0 + J\,\psi(t) + t/\eta \qquad \text{(VI, 1)}$$

where γ and $\mathfrak{T}$ are the shear strain and stress, t is the time of application of stress, J_0 the instantaneous compliance (difficult to determine but often negligible), J the steady state compliance, and η the steady flow viscosity. The normalized retarded elasticity function, ψ, as used by LEADERMAN, increases monotonically from 0 to 1 with increasing t.

The viscoelastic behaviour is described by the term $J\psi(t)$, which is plotted in fig. VI, 2 for a polyisobutylene of $M_w = 11{,}000$ at five different temperatures[1]. The strain/stress ratio increases gradually with time as the configurational distortions of the polymer chains approach their maximum values under steady-state flow. The effect of decreasing temperature is primarily to prolong the time scale within which the retarded elastic displacement occurs.

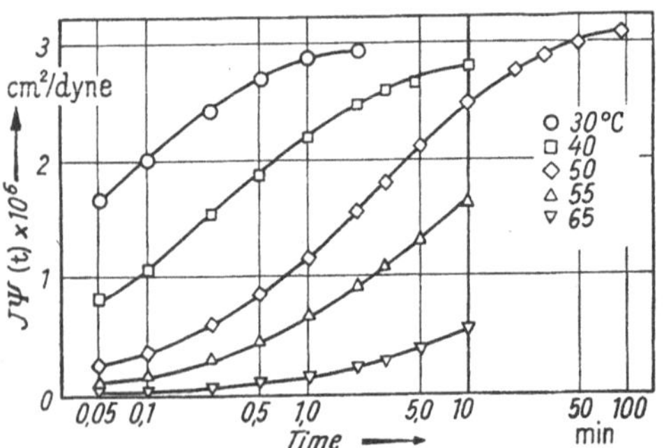

Fig. VI, 2. Retarded elastic component of creep of polyisobutylene, $M_w = 11{,}000$, at 5 (negative) temperatures as indicated (after LEADERMAN, SMITH and JONES).

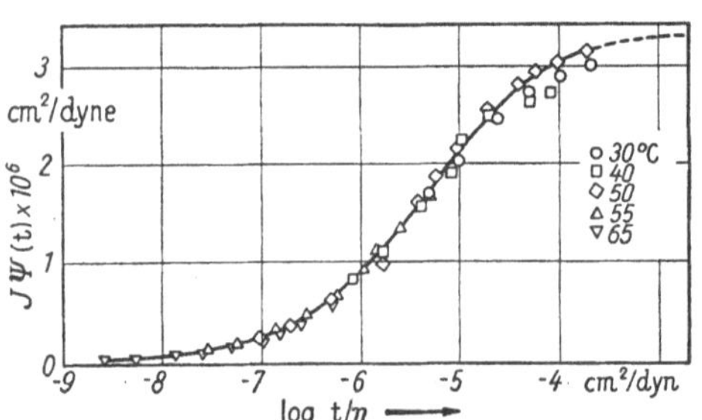

Fig. VI, 3. Creep data of Fig. VI, 2 reduced to a single temperature, using values of the steady flow viscosity, to give a single composite curve (LEADERMAN, SMITH and JONES).

According to the theory of reduced variables[2, 3], such data may be reduced to a standard temperature by plotting $J\psi(t)\,T\varrho/T_0\varrho_0$ against $t\eta_0 T\varrho/\eta T_0\varrho_0$. Here ϱ and η are the density and viscosity at the temperature of measurement, T, and ϱ_0 and η_0 are the corresponding values at the standard temperature T_0. In LEADERMAN's reduction, shown in fig. VI, 3, the factor $T\varrho/T_0\varrho_0$, which does not deviate greatly from unity, and the constant η_0 have been omitted for simplicity. The results superpose to form a single composite curve, showing that the relaxation times of all mechanisms involved do have the same temperature dependence. The composite curve represents the shape of the function $J\psi(t)$ over an extended time scale.

[1] LEADERMAN, H., R. G. SMITH u. R. W. JONES: J. Polymer Sci. 14, 47 (1954). — [2] FERRY, J. D.: J. Amer. Chem. Soc. 72, 3746 (1950).

[3] DAHLQUIST, C. A. u. M. R. HATFIELD: J. Colloid Sci. 7, 253 (1952).

b) Dynamic Measurements.

The response of a viscoelastic system to a periodic stress, varying sinusoidally with time, provides twice as much information as a transient measurement like creep; instead of one function of time, there are two functions of frequency. These may be chosen as the in-phase and out-of-phase stress/strain ratios, given by the real and imaginary parts of the complex shear modulus:

$$G^* = G' + iG'' = G' + i\omega\eta' \qquad \text{(VI,2)}$$

where ω is the circular frequency and η' is the real part of the complex dynamic viscosity (cf. § 3).

The polyisobutylene sample whose creep behaviour is shown in figs. VI,2 and VI,3 was subjected to dynamic measurements from 10 to 100 cycles/sec. by DEWITT[1]. The original values of G' and η' are not given here, but only the reduced values. If all relaxation mechanisms have the

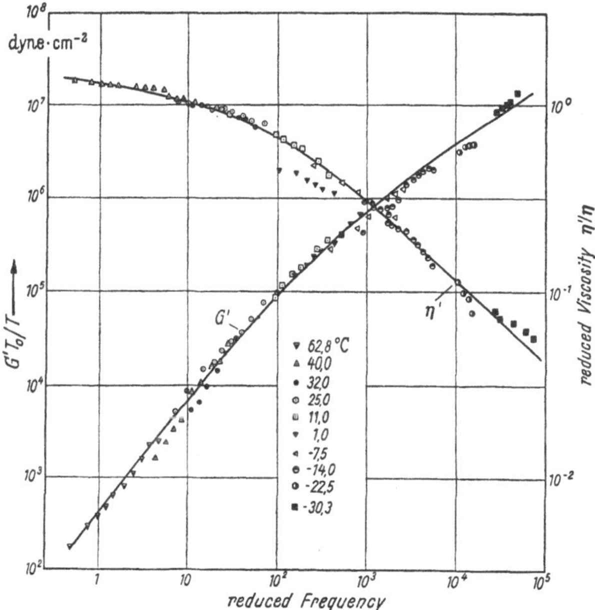

Fig. VI, 4. Real parts of dynamic rigidity and viscosity of polyisobutylene, $M_w = 11{,}000$, measured at various temperatures and reduced to 25° C. The reduced frequency used here is $\nu\eta T_0/\eta_0 T$ (HARPER, MARKOVITZ and DEWITT).

same temperature dependence[2], then $G' T_0 \varrho_0/T\varrho$ and η'/η plotted against $\omega\eta T_0\varrho_0/\eta_0 T\varrho$ should give composite curves reduced to T_0. In DEWITT's reduction, shown in fig. VI,4, the factor ϱ_0/ϱ is omitted and ω is replaced by $\nu = \omega/2\pi$; T_0 is chosen as 298° K. The precision is not sufficient to

[1] HARPER, R. C. JR., H. MARKOVITZ u. T. W. DEWITT: J. Polymer Sci. 8, 435 (1952).
[2] FERRY, J. D.: J. Amer. Chem. Soc. 72, 3746 (1950).

substantiate the reduction procedure, but the curves show the characteristic shapes of G' and η' over an extended frequency range. With increasing frequency, G' rises from a barely measurable value of 10^2 dyn/cm^2 to 10^7 and is still increasing; η' drops from the steady-flow viscosity η (which it must approach at zero frequency) to a considerably smaller value.

A polyisobutylene of similar viscosity-average molecular weight (as well as several others) was studied by MASON and BAKER[1] at far higher frequencies - from $2 \cdot 10^5$ to $2{,}4 \cdot 10^7$ cycles/sec. Though the data were not expressed in terms of G' and η', it appears that at 25° C. G' increases from $3 \cdot 10^7$ to $4 \cdot 10^9$ dyn/cm^2 and η'/η falls from 0,3 to 0,03 in this frequency range. The values at $2 \cdot 10^5$ cycles/sec. agree in order of magnitude with those in fig. VI,4 reduced from low frequency measurements at low temperatures. The values at $2{,}4 \cdot 10^7$ cycles sec. show that at very high frequencies this viscous liquid has the in-phase elastic response of a hard glass, but it still has a substantial out-of-phase response or elastic loss. There is thus a continuous transition from the consistency of a viscous liquid to that of a hard viscoelastic solid as a function of frequency.

c) Retardation and Relaxation Spectra.

That figs. VI,3 and VI,4 give essentially the same information can be most conveniently shown by calculating the retardation time spectrum from each. The creep data are first reduced to 298° K by multiplying the ordinates by 228/298 (the mid-point of the experimental temperature range having been about 228° K) and adding to the abscissa the logarithm of the steady-flow viscosity at 298° (= 2,93). The distribution $L d \ln\tau$ of contributions to shear compliance associated with retardation times whose logarithms lie between $\ln\tau$ and $\ln\tau + d\ln\tau$ is then obtained from SCHWARZL's second approximation formula[2] as used by LEADERMAN[3,4]:

$$L(\ln t/2) = d(J\psi)/d\ln t - d^2(J\psi)/d(\ln t)^2. \qquad \text{(VI,3)}$$

To calculate L from the dynamic properties it is necessary first to convert to the real and imaginary parts of the complex compliance, $J^* = J' - iJ''$, which are the in-phase and out-of-phase *strain/stress* ratios under sinusoidally varying shear stress:

$$\left.\begin{aligned} J' &= G'/(G'^2 + \omega^2\eta'^2) \\ J'' &= \omega\eta'/(G'^2 + \omega^2\eta'^2) \end{aligned}\right\} \qquad \text{(VI,4)}$$

Then, following the second approximation formulas of WILLIAMS and FERRY[5], we have

$$\left.\begin{aligned} L(-\ln\omega) &= -AJ'\,d\ln J'/d\ln\omega \\ L(-\ln\omega) &= BJ''\,(1 + d\ln J''/d\ln\omega) \end{aligned}\right\} \qquad \text{(VI,5)}$$

where A and B are correction factors which depend on $d\ln L/d\ln\tau$.

[1] MASON, W. P., W. O. BAKER, H. J. MCSKIMIN u. J. H. HEISS: Physic. Rev. **73**, 1074 (1948); **74**, 1873 (1949); **75**, 963 (1949).

[2] SCHWARZL, F. u. A. J. STAVERMAN: Physica **18**, 791 (1952).

[3] LEADERMAN, H., R. G. SMITH u. R. W. JONES: J. Polymer Sci. **14**, 47 (1954).

[4] The function denoted L by LEADERMAN is normalized and therefore differs from ours by a factor of $1/J$.

[5] WILLIAMS, M. L. u. J. D. FERRY: J. Polymer Sci. **11**, 169 (1953).

Values of L from creep and from the two dynamic moduli, J' and J'', are plotted in fig. VI,5. (The calculations from dynamic data for $\log \tau > -2$ are not shown; they are apparently unreliable because there is so little difference between η' and η). All the data are, of course, reduced to 25° C. The good agreement shows not only that this material follows the laws of linear viscoelasticity but also that the treatment of reduced variables is valid. Thus, in the region of overlap, the reduced data from creep come from a temperature range of $-65°$ to $-50°$ C, while the dynamic data come from a quite different range of $-30°$ to $+25°$ C.

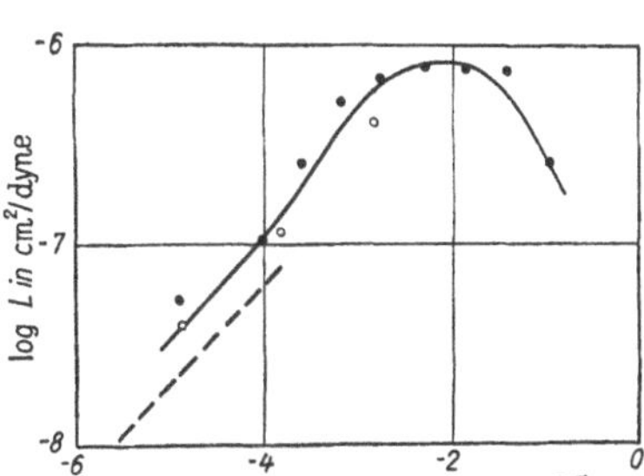

Fig. VI, 5. Retardation spectrum of polyisobutylene, $M_w = 11{,}000$, derived from creep (LEADERMAN), black circles; and from dynamic measurements (DEWITT), open circles. The dashed line is the prediction of the ROUSE theory (Equation I, 194).

Fig. VI,5 shows also the function L at short times as predicted by the ROUSE theory[1, 2] from the viscosity, density, and weight-average molecular weight of the polymer. Equation (I, 194) is used for this purpose with the solvent viscosity η_s omitted and nk replaced by its equivalent $\varrho R/M$. The calculated line lies close to the experimental curve in both position and slope, and there is little doubt that the retardation spectrum in this region reflects varying degrees of coordination of molecular segments as described by ROUSE. At longer times, the theory predicts a discrete rather than a continuous distribution, and if the sample were homogeneous with a molecular weight of 11,000 the longest retardation time would be about $10^{-4.0}$ sec. The existence of retarded elastic mechanisms at longer times may be attributed primarily to components of higher molecular weight; the shape of the function L near the maximum depends on molecular weight distribution.

Alternatively, the viscoelastic behaviour can be described by the relaxation spectrum. The distribution $H\,d\ln\tau$ of contributions to shear rigidity associated with relaxation times whose logarithms lie between $\ln\tau$ and $\ln\tau + d\ln\tau$ is obtained from dynamic data by the second approximation formulas

$$\left.\begin{aligned} H(-\ln\omega) &= A\,G'\,d\ln G'/d\ln\omega \\ H(-\ln\omega) &= B\,G''\,(1 - d\ln G''/d\ln\omega) \\ H(-\ln\omega) &= -\,B\,\omega\,\eta'\,d\ln\eta'/d\ln\omega\,. \end{aligned}\right\} \qquad \text{(VI, 6)}$$

Other methods for calculating H, some of which are more accurate, are discussed in § 4. It cannot be obtained conveniently from creep data.

The function H (reduced to 25° C) for the polyisobutylene studied by DEWITT is plotted in fig. VI, 6 together with the same function derived

[1] ROUSE, P. E. JR.: J. Chem. Physics **21**, 1272 (1953).

[2] FERRY, J. D., I. JORDAN, W. W. EVANS u. M. F. JOHNSON: J. Polymer Sci. **14**, 261 (1954).

from dynamic measurements on a sample of silicone polymer with viscosity-average molecular weight 60,000. Despite the difference in chemical structure of the two polymers, the two curves are very similar in shape, and at short times they lie close to the lines predicted by the ROUSE theory [equation (I, 193) with the same modifications introduced in equation (I, 194) above]. The longest relaxation time according to theory would be about $10^{-3.6}$ sec. for both (based on M_w for the polyisobutylene, M_η for the silicone). Actually, H extends over several logarithmic decades to the right of this point, reflecting again a distribution of molecular lengths, which is apparently of similar breadth in the two polymers.

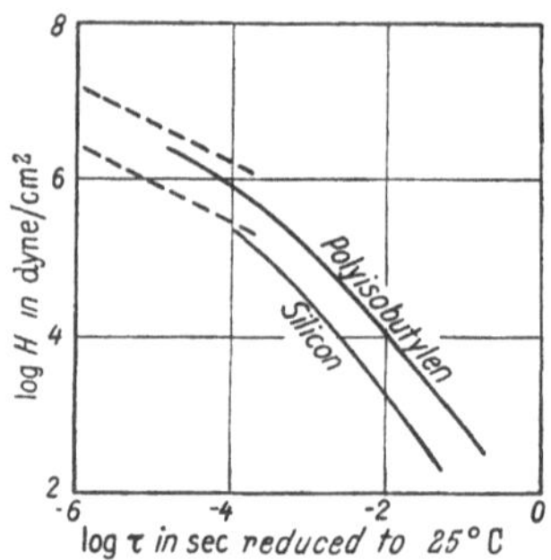

Fig. VI, 6. Relaxation spectra of polyisobutylene, $M_w = 11{,}000$, derived from the same data as fig. VI, 5, and of silicone, $M_\eta = 60{,}000$, derived from dynamic measuremetns of HARPER, MARKOVITZ and DEWITT. The dashed lines are the respective pre dictions of the ROUSE theory (Equation I, 193).

There must be a maximum in H as well as in L, but it lies at short times beyond the range of fig. VI, 6. The relaxation times in this region are probably influenced by local intramolecular interaction and steric effects which cannot be treated by any existing theory.

d) Influence of Molecular Weight and Chemical Structure on the Time Spectra.

Because of the scarcity of experimental data, discussion of the effect of molecular weight must rely on some guarded predictions from theory; conclusions from the calculations of ROUSE[1, 2] and BUECHE[3] are similar (§ 8). They are schematically illustrated in fig. VI, 7.

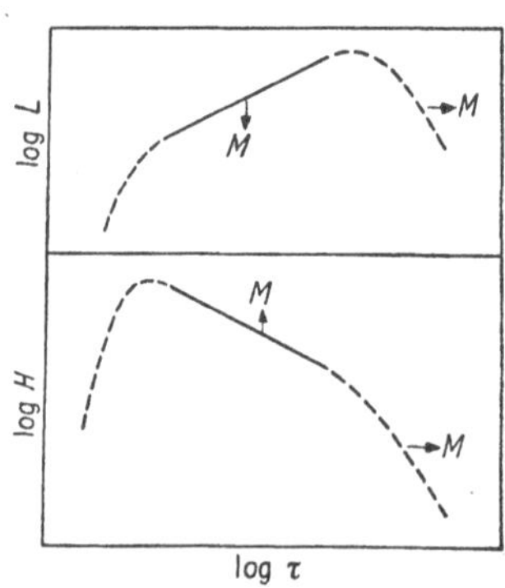

Fig. VI, 7. Effect of molecular weight on the retardation and relaxation spectra of a polymeric liquid, as predicted by the theory of ROUSE. The arrows denote the direction of shift with increasing molecular weight.

The portions of the L and H curves lying at very short times are beyond the scope of present theory, but their positions are probably almost independent of molecular weight. The remainder of either spectrum is predicted to be a linear portion on a log-log plot (slope of 1/2 for L, $-1/2$ for H) and a steeply falling portion at long times. The shape should be identical for all polymers regardless of chemical structure.

The location of either distribution function on the logarithmic time axis is specified, in these theories, by the segmental

[1] ROUSE, P. E. JR.: J. Chem. Physics **21**, 1272 (1953).

[2] FERRY, J. D., I. JORDAN, W. W. EVANS u. M. F. JOHNSON: J. Polymer Sci. **14**, 261 (1954). – [3] BUECHE, F.: J. Chem Physics **22**, 603 (1954).

mobility[1] of the polymer chains, or its reciprocal[2, 3], the segmental friction coefficient. The mobility depends on chemical structure and cannot be evaluated theoretically at present. However, the steady-flow viscosity η also depends on the mobility and the latter unknown quantity can therefore be eliminated to specify the time scale in terms of η. In this way, it is found that in the linear region H is proportional to $\sqrt{\eta/M}$, and L to $\sqrt{M/\eta}$ (cf. fig. I,41). Since η always increases with M faster than directly proportional, H increases with M. The steep drop at the right occurs (for a homogeneous polymer) at a time which is proportional to $M\eta$. For a heterogeneous polymer, of course, the shape at the right depends on molecular weight distribution, the drop becoming more gradual with increasing distribution breadth.

There are no detailed experimental data to test whether the shape of the central zone of the relaxation spectrum is indeed the same for liquid polymers of different chemical structure. The existence of viscoelastic behaviour is evident, however, in a variety of chemical types: in molten polyethylene from measurements of creep by DIENES[4] and dynamic viscosity by HOFF[5]; also in melts of vinyl chloride-vinyl acetate polymers[4] and low-molecular weight poly-α-methyl styrene[6].

e) Temperature Dependence of Viscoelastic Properties.

Figs. VI,3–VI,7 show various representations of viscoelastic behaviour as a function of time (or frequency) at a constant temperature T_0. The effect of changing to another temperature T is inherent in the method of reduced variables and is also predicted explicitly by the theories of ROUSE[1] and BUECHE[2]. There is a small change in the magnitude of H, L, J^*, G^*, etc., by the factor $T_0\varrho_0/T\varrho$ or its reciprocal (cf. the reduction equations above); and a larger displacement of the time scale such that t, τ, and ω become $a_T t$, $a_T\,\tau$, and ω/a_T. The temperature dependence of all the viscoelastic properties can then be simply described in terms of the temperature dependence of the factor a_T. Moreover, since[7] $a_T = \eta\,T_0\varrho_0/\eta_0 T\varrho$, it can be equally well described by the temperature dependence of the steady-flow viscosity. In terms of apparent activation energies,

$$\Delta H_a = \Delta H_\eta + R\,T - \alpha R\,T^2 \qquad \text{(VI, 7)}$$

where ΔH_a and ΔH_η are the activation energies for the relaxation (and retardation) processes and for viscous flow, respectively, and α is the thermal expansion coefficient. The last two terms are relatively small.

The magnitude and temperature dependence of ΔH_a are far more dependent on chemical structure than is the time function in its various

---

[1] ROUSE, P. E. JR.: J. Chem. Physics **21**, 1272 (1953).
[2] BUECHE, F.: J. Chem. Physics **22**, 603 (1954).
[3] BUECHE, F.: J. Chem. Physics **20**, 1959 (1952).
[4] DIENES, G. J.: J. Colloid Sci. **2**, 131 (1947).
[5] HOFF, E. A. W.: J. Polymer Sci. **9**, 41 (1952).
[6] MASON, W. P., W. O. BAKER, H. J. MCSKIMIN u. J. H. HEISS: Physic. Rev. **73**, 1074 (1948); **74**, 1873 (1949); **75**, 963 (1949).
[7] FERRY, J. D.: J. Amer. Chem. Soc. **72**, 3746 (1950).

forms of H, L, etc. Conversely, unlike the time function, ΔH_a is only slightly dependent on molecular weight.

For example, the polyisobutylene and silicone polymers whose relaxation spectra are shown in fig. VI, 6 have ΔH_a values of 16,2 and 4,7 kcal. respectively, which are roughly constant within the temperature ranges of the dynamic experiments. The marked difference must be correlated with the difference in chain structure. From viscosity measurements over wider temperature ranges by Fox and Flory[1], ΔH_η for polyisobutylene and polystyrene has been found to depend on temperature and molecular weight according to the following empirical equations:

$$\text{Polyisobutylene: } \Delta H_\eta = (5.1 \cdot 10^6/T)\, e^{-163/M} \tag{VI, 8}$$

$$\text{Polystyrene: } \Delta H_\eta = (8.1 \cdot 10^{17}/T^5)\, e^{-2530/M} \tag{VI, 9}$$

For molecular weights above 1600 and 25,000, respectively, these apparent activation energies are almost (within 10%) independent of molecular weight, as confirmed by measurements on polyisobutylene liquids by Leaderman[2] and Ferry and Parks[3]. But ΔH_η, and concomitantly ΔH_a, do depend on T; moderately for polyisobutylene, enormously for polystyrene. And, again, the magnitude of ΔH_η or ΔH_a depends on chemical structure, as illustrated in the discussion of soft solids in § 46 below.

The temperature dependence of a_T, or ΔH_a, reflects primarily the temperature dependence of the segmental friction coefficients[4,5]. This is determined by local chemical structure; the motion of a short chain segment requires a certain cooperation from segments in its immediate surroundings, which is made possible by a critical energy and/or free volume[6,4]. While the process has not been treated theoretically in detail, it will clearly be strongly influenced by intermolecular forces and steric demands of the polymer backbone and side chains.

f) Viscoelastic Constants.

Returning now to the time dependence of viscoelastic behaviour, we introduce some integrals of the distribution functions which are of particular interest:

$$\int_{-\infty}^{\infty} H\, d\ln\tau = \lim_{\omega\to\infty} G' = G_\infty \tag{VI, 10}$$

$$\int_{-\infty}^{\infty} \tau^2 H\, d\ln\tau \Big/ \left(\int_{-\infty}^{\infty} \tau H\, d\ln\tau\right)^2 = \int_{-\infty}^{\infty} L\, d\ln\tau = J \tag{VI, 11}$$

$$\int_{-\infty}^{\infty} \tau H\, d\ln\tau = \lim_{\omega\to 0} \eta' = \eta\,. \tag{VI, 12}$$

---

[1] Fox, T. G., jr. u. P. J. Flory: J. Appl. Physics 21, 581 (1950); J. Physical and Colloid Chem. 55, 221 (1951).
[2] Leaderman, H., R. G. Smith u. R. W. Jones: J. Polymer Sci. 14, 47 (1954).
[3] Ferry, J. D. u. G. S. Parks: Physics 6, 356 (1935).
[4] Bueche, F.: J. Chem. Physics 20, 1959 (1952).
[5] Bueche, F.: J. Chem. Physics 21, 1850 (1953).
[6] Fox, T. G., jr. u. P. J. Flory: J. Appl. Physics 21, 581 (1950).

The instantaneous rigidity, G_∞, represents the shear modulus under a deformation so rapid that no rearrangements of even the shortest polymer segments can occur. There is practically no information concerning its magnitude in polymeric liquids, because of the extremely high frequencies which would be necessary to measure it. However, it is probably[1] of the order of 10^{10} dyn/cm^2, as in molecular crystals and glasses of organic compounds of low molecular weight[2, 3].

The steady-state compliance, J, represents the maximum elastic response in steady-state flow. It can be determined from creep or creep recovery[4], and also from integration of the relaxation spectrum at the long time end[5]. For polyisobutylene liquids[4, 6] it is of the order of 10^{-5} cm^2/dyn. It appears to be independent of temperature within experimental error, although from theory[7, 5] it would be expected to be proportional to $1/T$. Its dependence on molecular weight can be predicted from a modification of the ROUSE theory[5]; for a homogeneous polymer it should be proportional to M, and for a heterogeneous polymer to $M_{z+1} M_z/M_w$ (§ 8). The marked dependence on breadth of molecular weight distribution which this relation implies is qualitatively in accord with experiment[4, 8]. The magnitude of J is important technically because the elastic energy stored in steady-state flow under a given stress is proportional to J. This energy must be dissipated after processing to avoid subsequent distortion.

Finally, the steady-flow viscosity η can be measured more easily than any other viscoelastic property, and a number of studies of low molecular weight polymeric liquids have been devoted to this property alone (for example, polyisobutylenes[9], silicone polymers[10], polyesters[11], polyamides[12], polyethylenes[13], and phenol-formaldehyde polymers[14]). A complete discussion of these results is beyond the scope of this chapter. The dependence of η on temperature has been mentioned above in connection with the temperature dependence of relaxation processes in polyisobutylene, polystyrene, and silicone polymer. In polyethylene, also, ΔH_η increases slightly with molecular weight[13]; near $M_n = 10{,}000$ and in the temperature range 110° to 160° C., it is 10 kcal. For various polyesters[11],

[1] MASON, W. P., W. O. BAKER, H. J. McSKIMIN u. J. H. HEISS: Physic. Rev. **73**, 1074 (1948); **74**, 1873 (1949); **75**, 963 (1949).

[2] KORNFELD, M. O. u. P. SHESTIKHIN: Compt. rend. Acad. Sci. U.R.S.S. **36**, 52 (1942). — [3] CRAWFORD, S. M.: Proc. physic. Soc. **B 66**, 953 (1953).

[4] LEADERMAN, H., R. G. SMITH u. R. W. JONES: J. Polymer Sci. **14**, 47 (1954).

[5] FERRY, J. D., M. L. WILLIAMS u. D. M. STERN: J. physic. Chem. **58**, 987 (1954).

[6] FERRY, J. D. u. G. S. PARKS: Physics **6**, 356 (1935).

[7] ROUSE, P. E., JR.: J. Chem. Physics **21**, 1272 (1953).

[8] VAN HOLDE, K. E. u. J. W. WILLIAMS: J. Polymer Sci. **11**, 243 (1953).

[9] FOX, T. G., JR. u. P. J. FLORY: J. Appl. Physics **21**, 581 (1950); J. Physic. and Colloid Chem. **55**, 221 (1951).

[10] HURD, C. B.: J. Amer. Chem. Soc. **68**, 364 (1946). — M. J. HUNTER, E. L. WARRICK, J. F. HYDE u. C. C. CURRIE: J. Amer. Chem. Soc. **68**, 2284 (1946).

[11] FLORY, P. J.: J. Amer. Chem. Soc. **62**, 1057 (1940).

[12] SCHAEFGEN, J. R. u. P. J. FLORY: J. Amer. Chem. Soc. **70**, 2709 (1948).

[13] UEBERREITER, K. u. H. J. ORTHMANN: Kolloid-Z. **126**, 140 (1952).

[14] DIENES, G. J.: J. Colloid Sci. **4**, 257 (1949).

ΔH_η appears to be independent of molecular weight up to 10,000 and is near 8 kcal. (between 80 and 120° C.). Crystallization prevents extension of these measurements[1,2] to low enough temperatures to observe a temperature dependence of ΔH_η such as is apparent for polyisobutylene and polystyrene in equations (VI,8) and (VI,9).

The dependence of η on molecular weight at constant temperature has been described by empirical equations such as

$$\log\eta = A + B M_w^{1/2} \qquad \text{(VI, 13)}^{1,2,3}$$

$$\log\eta = A + B\log M_w . \qquad \text{(VI, 14)}^{4}$$

The theories of ROUSE and BUECHE, assuming equal frictional coefficients for all molecular segments, would predict the form of equation (VI,14) with $B = 1$. Actually, for polyisobutylene ($M < 17{,}000$), $B = 1.75$, and for polystyrene ($M < 50{,}000$), $B = 2.34$. The discrepancy is attributed to the relative looseness of end groups, segments near chain ends actually having lower frictional coefficients than those near the middle.

g) Non-Newtonian Flow.

The above statements about viscosities all refer to values at vanishing shear stress, obtained either by measurements at very low stress or from extrapolation. At higher stresses, the apparent viscosity η_a (i.e., ratio of shear stress to rate of strain) is smaller, and decreases according to an equation of the form

$$\eta_a = \eta / f(\mathfrak{T}/G_a) \qquad \text{(VI, 15)}$$

in which G_a is a constant with dimensions of stress, often interpreted as an "internal shear modulus". The function f may be simply[5,6] $1 + \mathfrak{T}/G_a$, or a more complicated relation.

Qualitatively, non-Newtonian flow should give similar information to that of viscoelastic measurements, since both depend on the interaction of an applied stress with molecular Brownian motion. In fact, $1/G_a$ is usually found to be similar in magnitude to J, the steady-state elastic compliance[7,5]. More specifically, a recent macroscopic theory of DEWITT[8] shows that η_a as a function of $\dot\gamma\,(= \mathfrak{T}/\eta_a)$ should be equivalent to η' as a function of ω. From considerations on a molecular scale, BUECHE[9] has shown that this equivalence arises from the fact that in steady flow the individual molecules undergo periodic configurational changes as they rotate; however, the periodicity does not correspond exactly to a sinusoidal deformation, and the two functions $\eta_a(\dot\gamma)$ and $\eta'(\omega)$ should not be quite identical. Experimental values of η_a can be treated by the method of reduced variables[10] and the approximate form of the relaxation distribution H can be derived from them[8,11].

---

[1] Siehe S. 383, Fußnote 11. — [2] Siehe S. 383, Fußnote 13.
[3] Siehe S. 383, Fußnote 12. — [4] Siehe S. 383, Fußnote 9.
[5] LEADERMAN, H., R. G. SMITH u. R. W. JONES: J. Polymer Sci. **14,** 47 (1954).
[6] FERRY, J. D. u. G. S. PARKS: Physics **6,** 356 (1935).
[7] FERRY, J. D.: J. Amer. Chem. Soc. **64,** 1330 (1942).
[8] DEWITT, T. W.: J. Appl. Physics **26,** 889 (1955).
[9] BUECHE, F.: J. Chem. Physics, **22,** 1570 (1954).
[10] FERRY, J. D.: J. Amer. Chem. Soc. **72,** 3746 (1950).
[11] FERRY J. D., M. L. WILLIAMS u. D. M. STERN: J. Physic. Chem. **58,** 987 (1954).

§ 46. Soft Polymeric Solids in the Transition from Rubberlike to Glasslike Consistency.

The preceding section has dealt only with polymers of molecular weight below the order of 30,000. At higher molecular weights, linear polymers above their glass transition temperatures assume the superficial appearance of soft rubberlike solids. They still flow under steady stress, but within ordinary time intervals elastic deformation outweighs viscous flow. Moreover, the viscoelastic spectra become more complicated, as already illustrated by curve B in fig. VI, 1.

We now describe the properties of such soft solids in the region of time scale typified by the left peak in the above figure, where there is a transition (as a function of time, not temperature) from the consistency of a soft solid with a (real) shear modulus of the order of 10^6 to that of a hard solid with a modulus of the order of 10^{10} dyn/cm^2. Lightly cross-linked samples can be included in the discussion, because cross-linking has only a minor influence on the time-dependent mechanical properties in this region. The viscoelastic behaviour can be derived from measurements of creep, stress relaxation, and dynamic mechanical properties.

a) Creep.

The analysis of creep data is made according to equation (VI, 1), exactly as for polymeric liquids, the only difference being that the flow term t/η is relatively smaller so that it is easier to obtain the desired function $J\psi(t)$ by subtraction of t/η from $\gamma/\mathfrak{T}$. Creep of solids is more often measured in extension than in shear, in which case J and η must be replaced by the steady-state extension compliance and the so-called tensile viscosity, which (unless the consistency is very hard) are greater than the corresponding shear quantities by a factor of 3.

As an example, measurements of creep in extension were made by DAHLQUIST and HATFIELD[1] on a sample of GR—S (styrene-butadiene copolymer), viscosity-average molecular weight 306,000, at seven different temperatures. The behaviour was qualitatively similar to that of liquid polyisobutylene in fig. VI, 2 except that the compliance never reached its steady-state value within the time interval of the experiment. As before, the method of reduced variables can combine all the data in a single curve when $J\psi\, T/T_0$ is plotted against t/a_T, shown in fig. VI, 8. The factor ϱ/ϱ_0 is omitted here, being close to unity; a_T must be chosen empirically, rather than calculated from the steady flow viscosity, because the latter quantity is not available. The properties of a_T will be discussed subsequently. Similar composite curves were obtained for a high molecular weight polyisobutylene by DAHLQUIST and HATFIELD[1] from creep data in extension, and for six different polyisobutylenes by VAN HOLDE and WILLIAMS[2] from creep data in shear. None of these data

[1] DAHLQUIST, C. A. u. M. R. HATFIELD: J. Colloid Sci. **7**, 253 (1952).
[2] VAN HOLDE, K. E. u. J. W. WILLIAMS: J. Polymer Sci. **11**, 243 (1953).

extend to low enough values of reduced time, however, to encompass the entire transition from soft to glasslike consistency.

b) Stress Relaxation.

In stress relaxation measurements the sample is ordinarily subjected to a sudden strain and then, at constant strain, the stress/strain ratio is followed as a function of time. As an example, TOBOLSKY and associates[1] studied the stress/strain ratio in extension as a function of time at many different temperatures from $-33°$ to $-64°$C for a lightly cross-linked GR—S copolymer. The ratio decreased with time as the polymer chains, first responding only with the small molecular displacements of a hard glass, gradually assumed configurational distortions through Brownian motion. The lower the temperature, the slower the configurational changes, and below the glass transition temperature (about $-61°$) they scarcely took place at all within the experimental time interval of 1 hour.

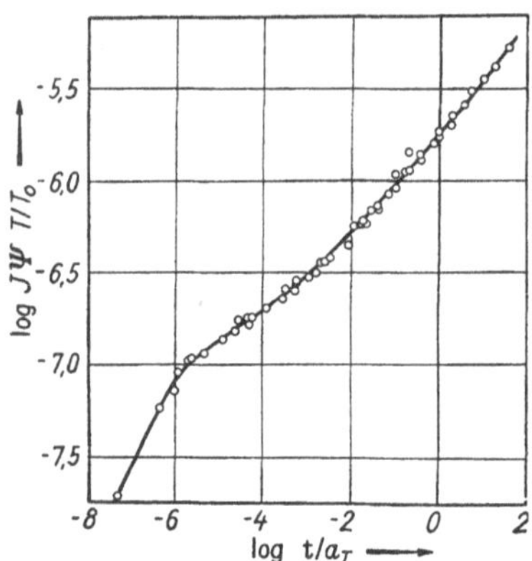

Fig. VI, 8. Creep of GR—S ($M_\eta = 306{,}000$) in the transition zone, measured at 7 temperatures by DAHLQUIST and HATFIELD and recalculated as creep in shear. Temperatures of measurement were 40°, 25°, 10°, $-5°$, $-20°$, $-35°$ and $-50°$ C, but the data are shown here reduced to 25° C, using empirical factors a_T.

Like other time-dependent data in the transition region, such data can be combined by reduction of the time scale, plotting the stress/strain ratio against t/a_T. The factor $T_0\varrho_0/T\varrho$ should be modified at low temperatures where the consistency becomes hard and configurational responses disappear; in TOBOLSKY's reduction, it has simply been omitted. The values of a_T must be chosen empirically. The resulting composite curve (fig. VI, 9) shows the stress/strain ratio falling from $10^{10.2}$ to $10^{7.4}$ dyn/cm² over 8 logarithmic decades of reduced time. Similar composite curves through the transition from soft to hard consistency have been obtained by TOBOLSKY and associates for other

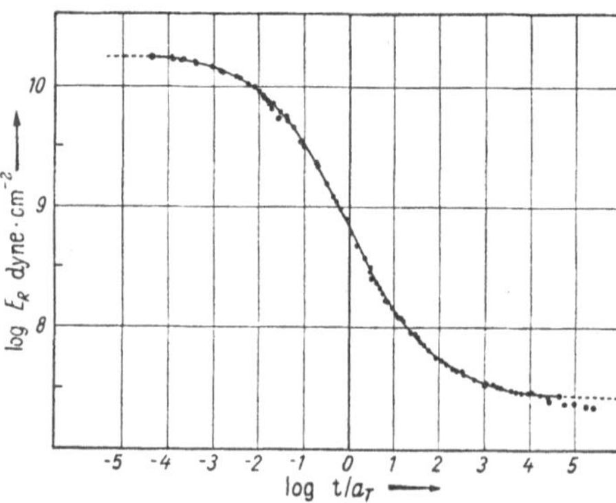

Fig. VI, 9. Stress relaxation of lightly cross-linked GR—S in the transition zone: stress/strain ratio in extension plotted against time, with logarithmic scales, measured at 11 temperatures and reduced to $-57°$, using empirical factors a_T (after BISCHOFF, CATSIFF and TOBOLSKY).

[1] BISCHOFF, J., E. CATSIFF u. A. V. TOBOLSKY: J. Amer. Chem. Soc. **74**, 3378 (1952).

styrene-butadiene copolymers of various compositions[1], polyisobutylenes of various molecular weights[2], and polymethyl methacrylate[3].

c) Dynamic Measurements.

The results of sinusoidal stress/strain measurements are given sometimes as the real and imaginary parts of a complex shear modulus [equation (VI, 2)] and sometimes as the parts of a complex shear compliance [equation (VI, 4)].

Examples of the real or in-phase compliance, J', are given in fig. VI, 10 for a polyisobutylene of high molecular weight (weight average 1,560,000)

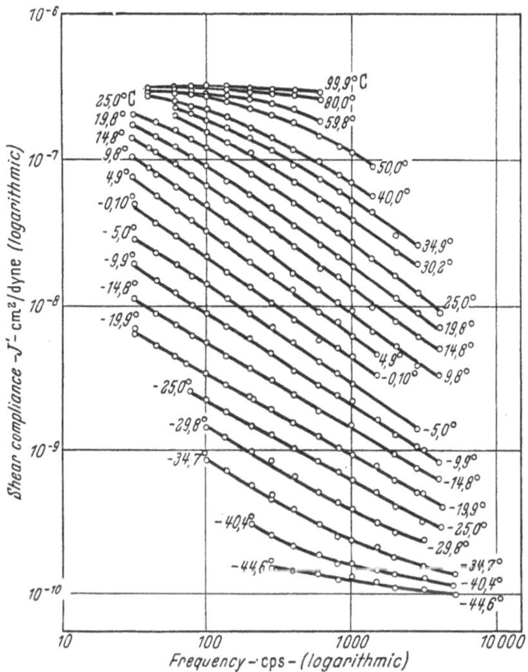

Fig. VI, 10. Real part of complex compliance of polyisobutylene ($M_w = 1{,}560{,}000$) in the transition zone, at 22 temperaturess a indicated (after FITZGERALD, GRANDINE and FERRY).

studied at twenty temperatures from 30 to 5100 cycles/sec. by FITZGERALD[4]. The compliance falls with increasing frequency from $10^{-6 \cdot 5}$ to 10^{-10} cm²/dyn, as configurational changes within the period of deformation become more and more restricted. The temperature has its usual effect on the time scale.

The method of reduced variables again produces a single composite curve from all these data and also from the associated values of the out-

[1] CATSIFF, E. u. A. V. TOBOLSKY: J. Appl. Physics, **25**, 1092 (1954).
[2] BROWN, G. M. u. A. V. TOBOLSKY: J. Polymer Sci. **6**, 165 (1951).
[3] MCLOUGHLIN, J. R. u. A. V. TOBOLSKY: J. Colloid Sci. **7**, 555 (1952).
[4] FITZGERALD, E. R., L. D. GRANDINE, JR. u. J. D. FERRY: J. appl. Physics **24**, 650 (1953).

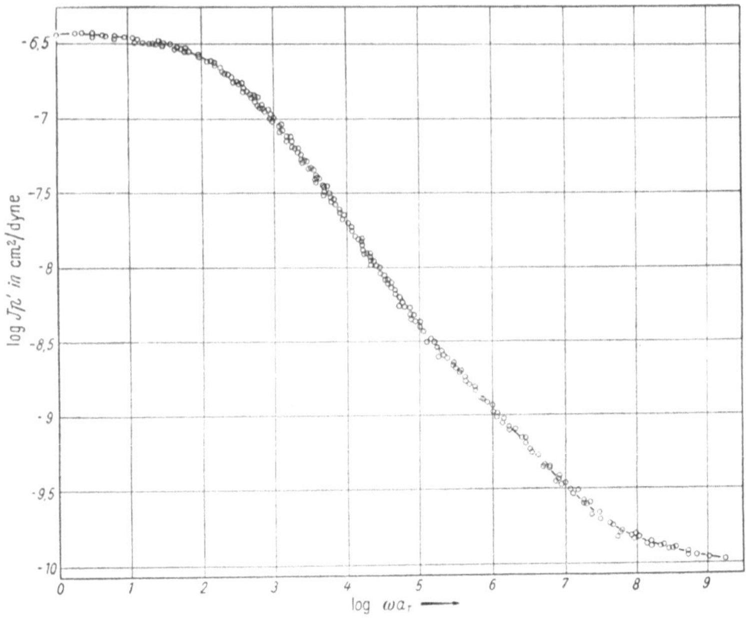

Fig. VI, 11. Real part of complex compliance, data of fig. VI, 10, reduced to 25° C., using empirical factors a_T (after FERRY, GRANDINE and FITZGERALD).

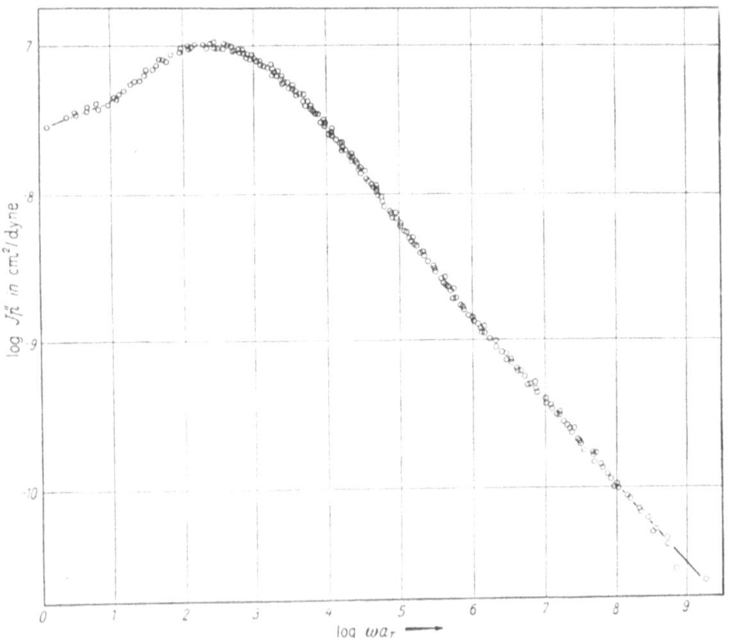

Fig. VI, 12. Imaginary part of complex compliance of polyisobutylene of fig. VI, 10, reduced to 25° C., using the same a_T factors as in fig. VI, 11.

of-phase compliance, J'', which is obtained from the same series of experiments. The corresponding reduced quantities are

$$J'_p = J'\,[T\varrho/T_0\varrho_0 + (J_0/J')\,(1 - T\varrho/T_0\varrho_0)] \qquad \text{(VI, 16)}$$

$$J''_p = J''\;T\varrho/T_0\varrho_0 \qquad \text{(VI, 17)}$$

where J_0, the instantaneous compliance, is estimated as the limiting value of J' at high frequency. In figs. VI, 11 and VI, 12, J'_p and J''_p are plotted logarithmically against ωa_T. The factor a_T must again be chosen empirically, but this is less arbitrary than in the case of creep (fig. VI, 8) or stress relaxation (fig. VI, 9) because the same value of a_T must make *both* J' and J'' data superpose. The temperature dependence of a_T as derived from all three types of measurements will be discussed in a subsequent section.

Similar data from 30 to 5100 cycles/sec. have been obtained by FERRY and associates for polystyrene[1] and polyvinyl acetate[2]. Polymethyl methacrylate has been studied in a lower frequency range (from 0,1 to 100 cycles/sec.) by SCHMIEDER and WOLF[3]. Working in a still lower frequency range, from $5 \cdot 10^{-4}$ to 3 cycles/sec., DEWITT and associates[4] have assembled dynamic data, also in shear, for a GR—S copolymer and natural rubber. Earlier measurements of NOLLE[5], of somewhat less precision but covering a very wide frequency range, provide dynamic data in extension for a Buna-N carbon-filled vulcanizate, as well as more fragmentary data for several other polymers; and those of GUTH and associates[6] and BLIZARD[7] data in extension for vulcanizates of several types of rubber[8].

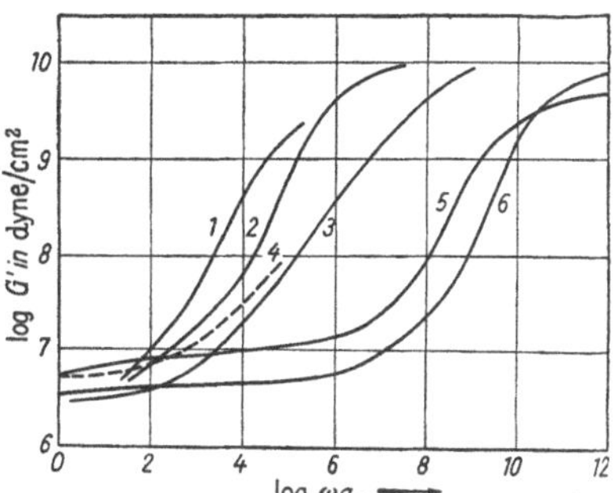

Fig. VI, 13. Real part of complex shear modulus of several polymers in the transition zone. *1*) polystyrene (GRANDINE and FERRY); *2*) polyvinylacetate (WILLIAMS and FERRY); *3*) polyisobutylene (FITZGERALD, GRANDINE and FERRY); *4*) butyl rubber (GUTH, BLIZARD); *5*) GR—S, 20% styrene (DEWITT); *6*) Hevea rubber (DEWITT). Reference temperature for reduction is 25° except for curves *1* (125°) and *2* (75°).

Composite reduced curves derived from some of these data are presented in figs. VI, 13 and VI, 14, where the in-phase shear modulus G' and the loss tangent $\omega\eta'/G' = J''/J'$ are plotted logarithmically against reduced frequency. Their positions on the abscissa axis are somewhat arbitrary, since not all have been reduced to the same temperature. Their

[1] GRANDINE, L. D., JR. u. J. D. FERRY: J. Appl. Physics **24**, 679 (1953).

[2] WILLIAMS, M. L. u. J. D. FERRY: J. Colloid Sci. **9**, 479 (1954).

[3] SCHMIEDER, K. u. K. WOLF: Kolloid-Z. **127**, 65 (1952).

[4] ZAPAS, L. J., S. L. SHUFLER u. T. W. DEWITT: J. Polymer Sci. **18**, 245, (1955).

[5] NOLLE, A. W.: J. Polymer Sci. **5**, 1 (1950).

[6] WITTE, R. S., B. A. MROWCA u. E. GUTH: J. Appl. Physics **20**, 481 (1949). — D. G. IVEY, B. A. MROWCA u. E. GUTH: J. Appl. Physics **20**, 486 (1949).

[7] BLIZARD, R. B.: J. Appl. Physics **22**, 730 (1951).

[8] See also BECKER, G. W.: Kolloid-Z. **140**, 1 (1955).

shapes are superficially similar, but the sharpness of the dispersion depends somewhat on chemical structure. In each case G' rises from a rubberlike to a glasslike magnitude, and the loss tangent goes through a maximum in the transition region. Curves for G' as in fig. VI, 13, and for stress relaxation as in fig. VI, 9 have been fitted empirically by TOBOLSKY to the GAUSS error function[1,2].

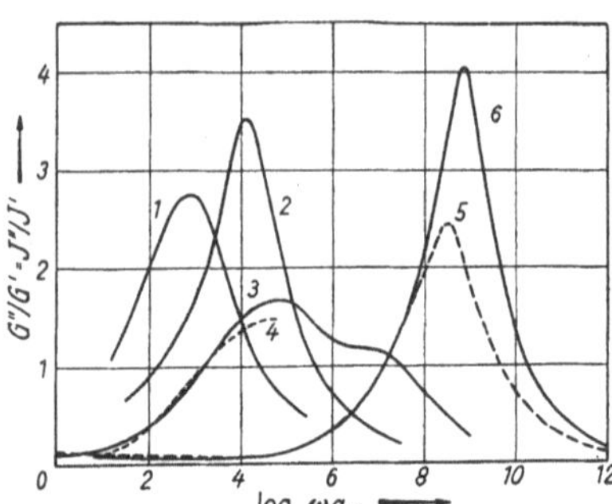

Fig. VI, 14. Loss tangent G''/G' for several polymers in the transition zone. Numerical key and reference temperatures for reduction same as in fig. VI, 13.

d) Relaxation Spectra.

Through the relaxation distribution function H, the dynamic and stress relaxation data can be brought together for comparison. The former have the advantage of providing double values [equation (VI, 6)] which test the internal consistency of the measurements. We use the second approximation formulas of equation (VI, 6) for calculations from G' and G'', and the corresponding approximation[3] from stress relaxation

$$H(\ln t) = -M(1/\gamma)\, d\mathfrak{T}/d\ln t \tag{VI, 18}$$

where M is another correction factor. For one polymer, polyisobutylene[4], the dynamic and stress relaxation data overlap over a considerable range and are in quite good agreement[5].

Relaxation distribution functions for several polymers, derived from both dynamic and relaxation data, are compared in fig. VI, 15. They are superficially similar in shape, though they differ in details of curvature. Each passes through a maximum in the neighborhood of 10^9 dyn/cm^2.

According to the theories of ROUSE[6] and BUECHE[7], this function should have a slope of $-1/2$ on a log-log plot at time intervals short compared with the longest relaxation times but long compared with relaxation times involving small segments [equation (I, 191)]. The theoretical slope is approached toward the lower end of the transition zone, where H is about 10^6 dyn/cm^2. At the upper end, many of the curves are considerably steeper; this discrepancy is not surprising, since the theories are unable to describe the motion of segments too short to be Gaussian and do not predict the occurrence of a maximum in the function.

The position of the curve on the logarithmic time scale is determined, according to theory (§ 8), by the friction coefficient per monomer unit,

[1] CATSIFF, E. u. A. V. TOBOLSKY: J. Appl. Physics **25**, 1092 (1954).

[2] TOBOLSKY, A. V.: J. Amer. Chem. Soc. **74**, 3786 (1952).

[3] FERRY, J. D. u. M. L. WILLIAMS: J. Colloid Sci. **7**, 347 (1952).

[4] MARVIN, R. S.: Proc. 2nd Intern. Congr. Rheology, p. 156, London: Butterworths Ltd. (1954).

[5] FERRY, J. D., L. D. GRANDINE, JR. u. E. R. FITZGERALD: J. Appl. Physics **24**, 911 (1953). — [6] ROUSE, P. E., JR.: J. Chem. Physics **21**, 1272 (1953).

[7] BUECHE, F.: J. Chem. Physics **22**, 603 (1954).

ζ_0, which measures the opposition to motion of chain segments. Since not all the polymers can be reduced to the same standard temperature, it is more instructive to compare the temperatures at which a given value of H (e.g., 10^6) corresponds to a standard point on the time scale (e.g., 10^{-2} sec.). This comparison will be made in the following section.

It is evident that, as a first approximation, the time-dependent behaviour in the transition region between rubberlike and glasslike consistency is independent of chemical structure and simply reflects the presence of large flexible molecules. The position of the region on the time scale depends, however, on local structure, as does the detailed shape of the relaxation spectrum, especially near its maximum. It is difficult at present to make any generalizations about the effects of chemical structure.

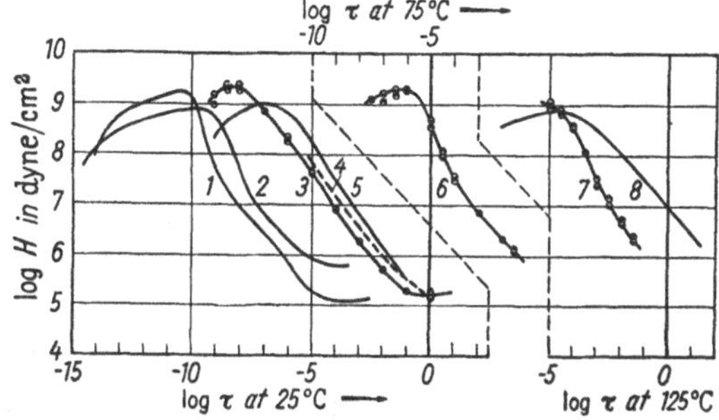

Fig. VI, 15. Relaxation spectra of several polymers in the transition zone, reduced to different reference temperatures as indicated. *1*) Hevea rubber (DEWITT); *2*) GR–S, 20% styrene (DEWITT); *3*) polyisobutylene (FITZGERALD, GRANDINE and FERRY); *4*) butyl rubber (GUTH, BLIZARD); *5*) GR–S, 25% styrene (BISCHOFF, CATSIFF and TOBOLSKY); *6*) polyvinyl acetate (WILLIAMS and FERRY); *7*) polystyrene (GRANDINE and FERRY); *8*) polymethyl methacrylate (MCLOUGHLIN and TOBOLSKY). Key to points: top black, calculated from real part of dynamic shear modulus; bottom black, calculated from imaginary part. Curves *5* and *8* are derived from stress relaxation, all others from dynamic measurements.

e) Temperature Dependence of Viscoelastic Properties.

The effect of temperature on the time-dependent mechanical properties in the transition from rubberlike to glasslike consistency may again be described by the factor a_T, obtained from application of reduced variables and representing the temperature dependence of the segmental mobility. The magnitude of a_T depends arbitrarily on the choice of a reference temperature; but the apparent activation energy ΔH_a calculated from it does not, so we choose ΔH_a as a basis for comparison.

In many cases steady-flow viscosity measurements are not available – obviously not for cross-linked materials – and equation (VI, 7) cannot be utilized nor tested. Moreover, the viscosity is dominated by configurational mechanisms with very long relaxation times which lie in a region of time scale far removed from the transition zone; it is not obvious that these should have the same temperature dependence. Nevertheless, for polyisobutylene[1] and polystyrene[2] ΔH_a in the transition zone is fairly close to ΔH_η as predicted by equation (VI, 7).

[1] FERRY, J. D., L. D. GRANDINE JR. u. E. R. FITZGERALD: J. appl. Physics **24**, 911 (1953).

[2] GRANDINE, L. D., JR. u. J. D. FERRY: J. appl. Physics **24**, 679 (1953).

Fig. VI, 16 shows ΔH_a plotted against temperature for a variety of polymers. The effects of chemical structure are much more apparent here than in fig. VI, 15. The steepness as well as the location on the temperature scale depends on intermolecular forces and steric relationships[1,2]. For each polymer, however, it is apparent that ΔH_a rises steeply with decreasing temperature as the glass transition temperature is approached. The approximate identification of the glass transition with the hardening or brittle temperature[3] reflects the extremely rapid shift of viscoelastic time scale with temperature in this region, such that the response to stress of any duration other than extremely long times will show a glasslike consistency.

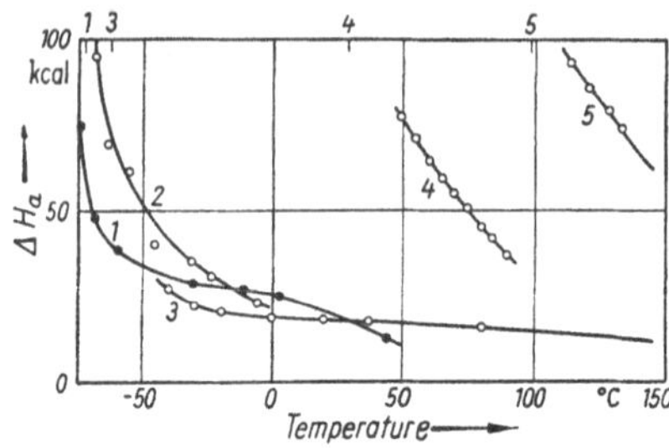

Fig. VI, 16. Apparent energy of activation for relaxation plotted against temperature. 1) Hevea rubber; 2) GR–S, 20% styrene; 3) polyisobutylene; 4) polyvinyl acetate; 5) polystyrene. Investigators same as in fig. VI, 15. Numbers at top denote glass transition temperatures.

The data for a_T from which fig. VI, 16 was derived can be used to determine the temperature at which H (from fig. VI, 15) attains a value of 10^6 dyn/cm² at 0,01 sec. These values are listed in table VI, 1. They correspond roughly to equal values of ζ_0, the friction coefficient per monomer unit. As would be expected, the bulkier and the more polar the structural unit, the higher the temperature at which the arbitrary characteristic mobility is achieved. The value of ΔH_a at the same temperature is also listed in each case. This appears to pass through a minimum near room temperature. It is difficult, again, to draw any detailed conclusions about the influence of local structure; but, as discussed under liquid polymers in the preceding section, the role of free volume is probably important[1,2].

Note added in proof. A recent analysis[4,5] has shown that when a_T is referred to a different reference temperature T_s for each polymer, suitably chosen, a_T is approximately a universal function of $T - T_s$ for many different polymer systems. It may be represented by the equation $\log a_T = -8{,}86\,(T - T_s)/(101{,}6 + T - T_s)$. As chosen, T_s lies about 50° above T_g, and with somewhat less accuracy the universal function for a_T reduced to T_g may be written $\log a_T = -17{,}44\,(T - T_g)/(51{,}6 + T - T_g)$. Comparison with a semiempirical equation of DOOLITTLE[6], and statistical calculations of the local fluctuations of free volume[7], show that the constants depend on the fractional free volume (f_g) at T_g and its thermal expansion coefficient (α_g). From the above numerical values, $f_g = 0{,}025$ and $\alpha_g = 4{,}8 \cdot 10^{-4}$ deg^{-1}. The latter agrees rather well with the increment in the macroscopic thermal expansion coefficient which occurs at T_g (see Volume III of this Series).

Sometimes the temperature dependence of viscoelastic properties of a polymer is expressed by plotting G' at constant frequency, or the stress/

[1] FOX, T. G., JR. u. P. J. FLORY: J. appl. Physics **21**, 581 (1950); J. Physical and Colloid Chem. **55**, 221 (1951).

[2] BUECHE, F.: J. chem. Physics **21**, 1850 (1953).

[3] BOYER, R. F. u. R. S. SPENCER: Advances in Colloid Science, Vol. II, p. 1, New York: Interscience Publishers (1946).

[4] WILLIAMS, M. L.: J. Physic. Chem. **59**, 95 (1955).

[5] WILLIAMS, M. L., R. F. LANDEL, u. J. D. FERRY: J. Amer. Chem. Soc. **77**, 3701 (1955).

[6] DOOLITTLE, A. K.: J. Appl. Physics **22**, 1471 (1951); **23**, 236 (1952).

[7] BUECHE, F.: J. Chem. Physics (in press).

Table VI, 1.
Comparison of Polymers at Temperatures of Coincident Relaxation Spectra.

Polymer	Temp. at which $H = 10^6$ at 0.01 sec. °C.	ΔH_a at same Temp. kcal.
Natural rubber[1]	− 30	29
GR–S (20% styrene)[1]	− 24	32
Polyisobutylene[2, 3]	12	19
Butyl rubber[4]	25	18
GR–S (25% styrene)[5]	31	14
Polyvinyl acetate[6]	85	41
Polystyrene[7]	137	69
Polymethyl methacrylate[8]	135	—

strain ratio at a given time interval after imposition of stress, directly against the temperature. This is the easiest type of experimental data to obtain, but it is difficult to interpret because the rate of change of the observed property with temperature will depend on the shapes of *both* the relaxation spectrum (or other time-dependent representation) *and* the apparent energy of activation (or other temperature-dependent representation).

For example, $(\partial \ln G'/\partial T)_\omega = (\partial \ln G'/\partial \ln \omega a_T)_T (\partial \ln \omega a_T/\partial T)_\omega$. The first of the latter two derivatives depends on H [equation (VI,6)], and the second depends on ΔH_a. Thus a rapid change in viscoelastic consistency with temperature may reflect a high value of H or a high value of ΔH_a or both[9, 10].

A plot of $\log G'$ against T at approximately constant frequency (order of 1 cycle/sec.) for unvulcanized natural rubber and several light vulcanizates, given by SCHMIEDER and WOLF[11], is shown in fig. VI, 17. Because of the rapid change of a_T with temperature, the increase of G' from 10^7 dyn/cm² (rubberlike) to 10^{10} (glasslike) is compressed into a very narrow temperature range; though the maximum slope is not quite so high as would be predicted from the product of the two derivatives cited above, based on the data for natural rubber of figs. VI, 13 and VI, 16. Close to the temperature where $\log G'$ is midway between its low and high limiting values, the loss tangent G''/G' goes through a very sharp maximum, as would be expected from figs. VI, 14 and VI, 16.

[1] ZAPAS, L. J., S. L. SHUFLER, u. T. W. DEWITT: J. Polymer Sci. **18**, 245 (1955).
[2] FITZGERALD, E. R., L. D. GRANDINE, JR. u. J. D. FERRY: J. appl. Physics **24**, 650 (1953).
[3] FERRY, J. D., L. D. GRANDINE, JR. u. E. R. FITZGERALD: J. Appl. Physics **24**, 911 (1953).
[4] BLIZARD, R. B.: J. Appl. Physics **22**, 730 (1951).
[5] BISCHOFF, J., E. CATSIFF u. A. V. TOBOLSKY: J. Amer. Chem. Soc. **74**, 3378 (1952).
[6] WILLIAMS, M. L. u. J. D. FERRY: J. Colloid Sci., **9**, 479 (1954).
[7] GRANDINE, L. D., JR. u. J. D. FERRY: J. Appl. Physics **24**, 679 (1953).
[8] MCLOUGHLIN, J. R. u. A. V. TOBOLSKY: J. Colloid Sci. **7**, 555 (1952).
[9] FERRY, J. D. u. E. R. FITZGERALD: Proc. 2nd Intern. Congr. Rheology, p. 140, London: Butterworths Ltd. (1954).
[10] MÜLLER, F. H.: Kolloid-Z. **134**, 77 (1953).
[11] SCHMIEDER, K. u. K. WOLF: Kolloid-Z. **134**, 149 (1953).

Plots similar to fig. VI,17, and the corresponding plots of G''/G' against temperature at approximately constant frequency, have been constructed for many different polymers by WOLF and collaborators[1, 2] and by NIELSEN and BUCHDAHL[3]. Since G' rises so steeply, it is easy to specify the temperature corresponding to the midpoint of this rise, and such temperatures, if compared always at closely similar frequencies,

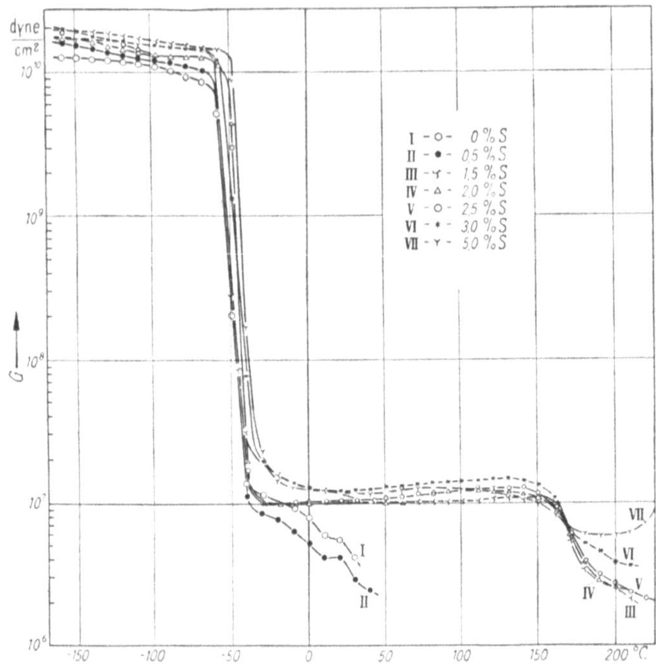

Fig. VI,17. Real part of complex shear modulus plotted against temperature at approximately constant frequency (order of 1 cycle/sec.) for unvulcanized natural rubber and vulcanizates there of with different amounts of sulfur as indicated (SCHMIEDER and WOLF).

provide a useful characterization of polymer types. An extensive compilation of these temperatures has been given by SCHMIEDER and WOLF[2] and is reproduced in part in table VI,2. They are in about the same order as the temperatures listed in table VI,1, but are lower, the differences ranging from 15° to 50°.

f) Influence of Molecular Weight and Cross-Linking.

In polymers of low molecular weight the viscoelastic properties, as illustrated by the spectra L and H, would be expected to depend considerably on molecular weight (fig. VI,7). In high molecular weight polymers, on the other hand, properties in the transition zone between rubber-

[1] SCHMIEDER, K. u. K. WOLF: Kolloid-Z. **127**, 65 (1952).

[2] SCHMIEDER, K. u. K. WOLF: Kolloid-Z. **134**, 149 (1953).

[3] NIELSEN, L. E., R. BUCHDAHL u. R. LEVREAULT: J. Appl. Physics **21**, 607 (1950).

Table VI, 2.
Comparison of Polymers at Temperatures Corresponding to Mid-Point of G' Transition at Frequencies of 0,1 to 3 Cycles/sec[1].

Polymer	Temp. of Mid-Transition	Freq. of Measurement cycles/sec.
Natural Rubber	− 50	1,2
Butyl Rubber	− 50	1,1
Polyisobutylene	− 48	1
Poly *n*-butyl acrylate	− 34	1,25
Polyvinyl *n*-butyl ether	− 32	0,8
Polyvinyl propyl ether	− 27	1,1
Polyvinyl ethyl ether	− 17	1,1
Polyvinyl methyl ether	− 10	2,7
Polyethyl acrylate	− 5	1,6
Polyvinyl *i*-butyl ether	− 1	1,2
Polyvinyl propionate	12	1,5
Polymethyl acrylate	25	1,2
Polyvinyl acetate	33	1,9
Polyvinyl *tert*-butyl ether	83	1,7
Polyvinyl chloride	90	0,7
Polystyrene	116	0,9
Polymethyl methacrylate	120	0,12

like and glasslike consistency should be rather insensitive to the magnitude of the molecular weight or the presence of a moderate degree of cross-linking[2]. Qualitatively, the appropriate configurational rearrangements involve cooperation only within relatively short chain lengths which are on the average far from chain ends, cross-links, or entanglements and hence are oblivious of the latter influences. Quantitatively, the theories of ROUSE[3,4] and BUECHE[5] predict that the magnitudes of H and L are proportional and inversely proportional, respectively, to the square root of the average friction coefficient per monomer unit, which is independent of the defined length of the mobile segment and also (at short times) independent of molecular weight.

The prediction that molecular weight is of minor importance at the short end of the time scale is borne out by comparison of polyisobutylenes[6,7,8] and polymethyl methacrylates[9] of different molecular weight; and the minor influence of cross-links is clearly evident in the close similarity of polyisobutylene and butyl rubber (curves *3* and *4*, fig. VI, 15) as well as the close juxtaposition of all the curves in fig. VI, 17 for natural rubber vulcanized to different extents.

[1] Siehe S. 394, Fußnote 2.
[2] ALFREY, T.: Mechanical Behaviour of High Polymers. New York: Interscience Publishers, 1948, pp. 145–148.
[3] ROUSE, P. E., JR.: J. Chem. Physics **21**, 1272 (1953).
[4] FERRY, J. D., R. F. LANDEL, u. M. L. WILLIAMS: J. Appl. Physics **26**, 359 (1955).
[5] BUECHE, F.: J. Chem. Physics **22**, 603 (1954).
[6] BROWN, G. M. u. A. V. TOBOLSKY: J. Polymer Sci. **6**, 165 (1951).
[7] TOBOLSKY, A. V. u. J. R. MCLOUGHLIN: J. Polymer Sci. **8**, 543 (1952).
[8] SCHMIEDER, K. u. K. WOLF: Kolloid-Z. **134**, 149 (1953).
[9] MCLOUGHLIN, J. R. u. A. V. TOBOLSKY: J. Colloid Sci. **7**, 555 (1952).

Nevertheless, Fox and LOSHAEK[1] have pointed out that cross-links should decrease the free volume and should therefore elevate the glass transition temperature, just as the introduction of free molecular ends (decrease of molecular weight) depresses it[2]. A concomitant small increase in ΔH_a with cross-linking would be expected, so that at lower temperatures the average effective friction coefficient would be somewhat greater in the presence of cross-linking. The small differences observed in figs. VI,15 and VI,17 are in the expected direction.

§ 47. Soft Polymeric Solids in the Range of Rubberlike Consistency.

The preceding section has described the mechanical behaviour of high molecular weight polymers in the range of time scale where the transition from rubberlike to glasslike consistency occurs. The essential consequence of the high molecular weight, differentiating these systems from the polymeric liquids discussed in § 45, is to place this transition far removed in time scale from phenomena involving long range chain cooperation and free ends; hence the relative unimportance of molecular weight and cross-linking as just demonstrated. Now we proceed to the long end of the time scale, typified by the right peak in curve B of fig. VI,1. Throughout this range a high molecular weight polymer has a soft, rubberlike consistency. For a linear polymer at sufficiently long times the elastic contributions vanish, but they persist to longer times than would be expected on the basis of molecular length alone. For a cross-linked polymer at long times the elastic contributions approach the equilibrium value prescribed by the theory of rubberlike elasticity.

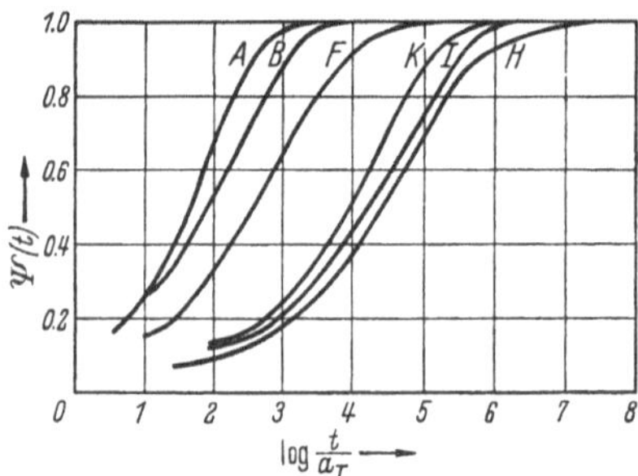

Fig. VI, 18. Normalized creep curves of 6 polyisobutylenes in the rubberlike range, reduced to 35° C. (after VAN HOLDE and WILLIAMS). M_w as follows: A) 110,000; B) 111,000; F) 200,000; K) 510,000; I) 640,000; H) 830,000.

a) Creep.

From creep data on six polyisobutylenes of weight-average molecular weight ranging from 110,000 to 830,000, VAN HOLDE and WILLIAMS[3] obtained the retarded elasticity function $\psi(t)$ in accordance with equation (VI,1) and they reduced data at various temperatures to 35° by the method of reduced variables. The results, shown in fig. VI,18, cover a range of reduced time scale lying far to the right of that of the creep experiments described in the preceding section (fig. VI,8), and in contrast to those experiments the compliance here reaches its steady-state value. However, for the highest weight-average

[1] Fox, T. G. u. S. LOSHAEK: J. Polymer Sci, **15**, 371 (1955).
[2] Fox, T. G., JR. u. P. J. FLORY: J. Appl. Physics **21**, 581 (1950).
[3] VAN HOLDE, K. E. u. J. W. WILLIAMS: J. Polymer Sci. **11**, 243 (1953).

molecular weight, this maximum value of unity for $\psi(t)$, corresponding to the maximum configurational distortions of the polymer chains in steady-state flow, is not achieved at 35° C. until 10^7 sec. This is about 12 decades of logarithmic time to the right of the transition zone between rubbery and glassy consistency (fig. VI, 15). Moreover, the time scale for these slowest configurational changes depends markedly on the molecular weight, in contrast to the behaviour of the more rapid mechanisms involved in the rubbery-glassy transition described above. The values of J were all of the order of 10^{-5} cm²/dyn, so the ordinates, if not normalized, would lie in the same range of magnitude as shown in fig. VI, 3 for a liquid polyisobutylene of much lower molecular weight.

b) Stress Relaxation.

The stresses portrayed in fig. VI, 9 do not relax to zero, because the polymer is lightly cross-linked. Actually, when the molecular weight is high, the behaviour in that transition region is much the same even in the absence of cross-links. The stress on an uncross-linked polymer can eventually decay to zero, but only after very long times, comparable to those required for establishment of steady-state flow in creep. Such times, as pointed out above, will be far in excess of the relaxation times involved in the rubbery-glassy transition. Fig. VI, 19 shows stress relaxation measurements of ANDREWS and TOBOLSKY[1] on six samples of polyisobutylene extended to very long times. The higher the molecular weight, the more slowly are the configurational changes imposed by the strain erased through coordinated motions of molecular segments. The temperature dependence of relaxation in these and other studies at very long times can be described by reduced variables, reflecting, again, identical temperature dependence of all the slow relaxation mechanisms within a given sample.

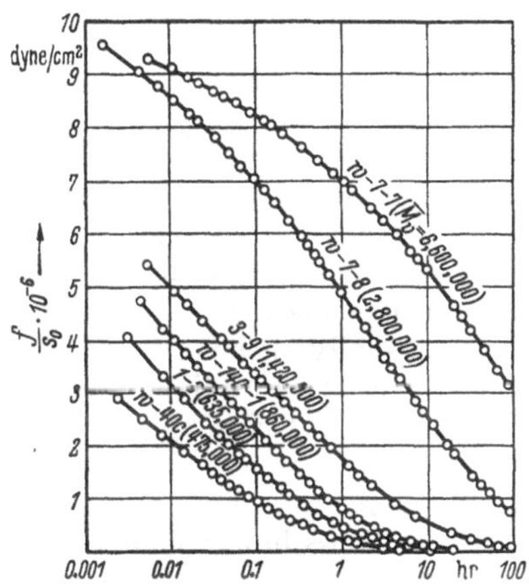

Fig. VI, 19. Stress relaxation of 6 polyisobutylenes in the rubberlike range, at 30° (after ANDREWS and TOBOLSKY). Viscosity-average molecular weights are indicated.

These stress relaxation and creep experiments on soft polymers of high molecular weight, as well as similar experiments on polystyrene[2] and polymethyl methacrylate[3], show the presence of some elastic mechanisms with extraordinarily long relaxation (or retardation) times. They have been qualitatively interpreted[2,4] as

[1] ANDREWS, R. D., HOFMAN-BANG u. A. V. TOBOLSKY: J. Polymer Sci. 3, 669 (1948); 7, 221 (1951).

[2] NIELSEN, L. E. u. R. BUCHDAHL: J. Colloid Sci 5, 282 (1950). – R. BUCHDAHL, L. E. NIELSEN u. E. H. MERZ: J. Polymer Sci. 6, 403 (1951).

[3] MCLOUGHLIN, J. R. u. A. V. TOBOLSKY: J. Colloid Sci. 7, 555 (1952).

[4] LONG, J. D., W. E. SINGER u. W. P. DAVEY: Ind. Engng. Chem. 26, 543 (1934).

due to occasional tight entanglements of the linear molecules, which act as temporary cross-links. The magnitude of the steady flow viscosity also reflects such entanglements[1], as will be discussed subsequently.

c) Dynamic Measurements.

Very low frequencies (or high temperatures) are required for measurements in the response range of the slow relaxation mechanisms. Such data have been obtained by PHILIPPOFF[2] and DEWITT[3], and fig. VI,20 shows an example[2] for polyisobutylene (the same sample whose behaviour at higher frequencies is depicted in figs.VI,10 to VI,15 inclusive). The real and imaginary parts of the shear modulus, G' and G'', are nearly constant over a very wide range of frequencies; the former rises slightly with increasing frequency, while the latter passes through a shallow minimum. Inevitably, at still lower frequencies, G' must fall off toward zero as it does for liquid polyisobutylene in fig. VI,4, and G'' must become directly proportional to frequency as the ratio $G''/\omega\,(=\eta')$ approaches the steady flow viscosity.

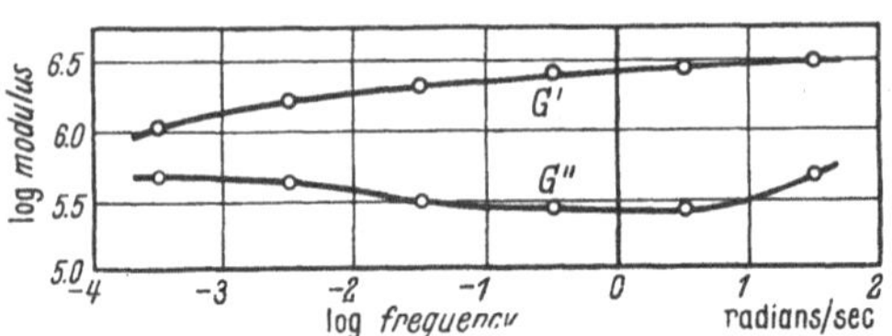

Fig.VI,20. Real and imaginary parts of the complex shear modulus of polyisobutylene (M_w = 1,560,000) in the rubberlike range, from dynamic measurements of PHILIPPOFF, reduced to 25° C.

Similar behaviour is also found for lightly vulcanized rubbers[2,4], where often G' increases slowly linearly with $\log\omega$ and G'' is practically independent of frequency. For a cross-linked sample, G' is only slightly higher than the equilibrium modulus in this region of time scale; this of course depends on the number of permanent cross-links per unit volume[5,6], but is usually of the order of 10^6 dyn/cm^2. It is a remarkable coincidence that many uncross-linked polymers show a modulus of about the same magnitude in this range of time scale (cf. fig. VI,20). Its existence may be qualitatively explained, again, by temporary cross-links due to tight entanglements[7], whose concentration happens to be similar to that of the chemical cross-links which are artificially introduced in light vulcanizates.

d) Relaxation and Retardation Spectra.

As usual, stress relaxation and dynamic measurements can be compared through the distribution function H. Fig. VI,21 shows excellent agreement between the respective data of CATSIFF and TOBOLSKY[7]

[1] BUECHE, F.: J. Chem. Physics **20**, 1959 (1952).
[2] PHILIPPOFF, W.: J. Appl. Physics **24**, 685 (1953).
[3] ZAPAS, L. J., S. L. SHUFLER, u. T. W. DEWITT: J. Polymer Sci. **18**, 245 (1955).
[4] KUHN, W. u. O. KÜNZLE: Helv. chim. Acta **30**, 839 (1947).
[5] TRELOAR, L. R. G.: The Physics of Rubberlike Elasticity. Oxford: Clarendon Press, 1949.
[6] FLORY, P. J., N. RABJOHN u. M. C. SHAFFER: J. Polymer Sci. **4**, 225 (1949).
[7] CATSIFF, E., u. A. V. TOBOLSKI: J. Colloid Sci. **10**, 375 (1955).

and PHILIPPOFF[1] on a standard polyisobutylene[2]. The distribution is nearly flat over several decades of logarithmic time for this and several other high molecular weight polymers. It has, in fact, been given an idealized representation of a horizontal line[3,4], described as a box[4] or plateau[5].

The plateau may finally be seen in its proper perspective in fig. VI, 22, where the data on polyisobutylene[2] (weight-average molecular weight 1,56 · 10^6) from figs. VI, 15 and VI, 21 are combined to show H and L over a time range of sixteen powers of ten. Comparison with figs. VI, 5 and VI, 7 shows that the effect of increasing the molecular weight is to split each of the simple spectra observed for polymer liquids into two limbs and to displace one limb several decades to the right. The left limb, representing the transition region between soft and glassy consistency, is essentially unchanged. The right limb, representing the drop in H and L associated with disappearance of the slowest relaxation mechanisms, is moved to longer times. In between, there appears a distribution of mechanisms which in H are represented by a plateau (actually a shallow minimum and secondary maximum) and in L are represented by a minimum and pronounced secondary maximum.

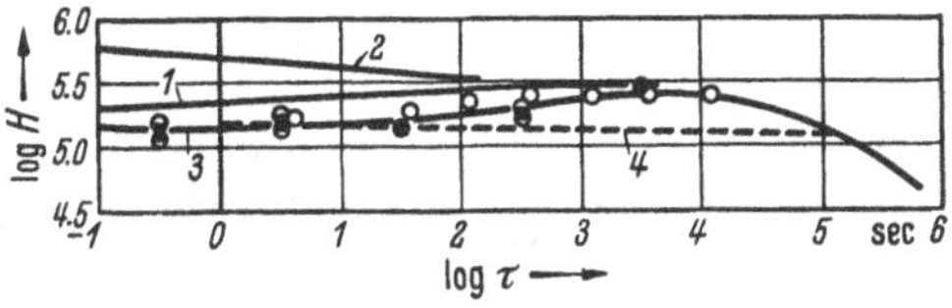

Fig. VI, 21. Relaxation spectra of several polymers in the rubberlike range, reduced to 25° C. *1)* Hevea rubber, raw (DEWITT); *2)* GR–S, raw, 20% styrene (DEWITT); *3)* polyisobutylene, M_w = 1,560,000, with points top black from real part of complex dynamic shear modulus (PHILIPPOFF), bottom black from imaginary part (PHILIPPOFF), and open circles from stress relaxation (CATSIFF and TOBOLSKY); *4)* Hevea rubber, lightly cross-linked (KUHN, KÜNZLE and PREISSMANN).

The new position of the right limb means that the slowest relaxation mechanisms have all been prolonged, to about the same extent, by a specific phenomenon associated with high molecular weight, undoubtedly the molecular

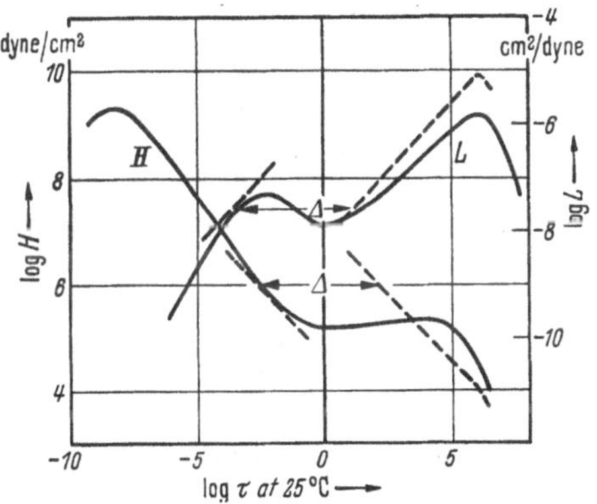

Fig. VI, 22. Retardation and relaxation spectra of polyisobutylene, M_w = 1,560,000, showing the transition, rubberlike, and terminal zones (compilation by MARVIN). Dashed lines at right show predictions of the ROUSE theory based on steady flow viscosity. Dashed lines at left indicate displacement of time scale necessary to make ROUSE theory coincide with lower part of relaxation spectrum in the transition zone.

[1] Siehe S. 398, Fußnote 2.

[2] MARVIN, R. S.: Proc. 2nd Intern. Congr. Rheology, p. 156, London: Butterworths Ltd. (1954).

[3] KUHN, W., O. KÜNZLE u. A. PREISSMANN: Helv. chim. Acta **30**, 307 (1947).

[4] ANDREWS, R. D., N. HOFMAN-BANG u. A. V. TOBOLSKY: J. Polymer Sci. **3**, 669 (1948); **7**, 221 (1951).

[5] FERRY, J. D., E. R. FITZGERALD, M. F. JOHNSON u. L. D. GRANDINE, JR.: J. Appl. Physics **22**, 717 (1951).

entanglement[1] or coupling[2] which has already been deduced from direct observation of the mechanical properties. In the plateau region, there is a partial effect of coupling which disappears at the left end.

Theoretical curves for H and L deduced from the ROUSE theory are also plotted in fig. VI, 22, based on the weight-average molecular weight and steady flow viscosity. The right limb is predicted rather successfully (not exactly, owing to broad molecular weight distribution in this sample), because the extreme prolongation of the slow relaxation times is reflected also in an extremely high viscosity, and the latter specifies the time scale of H in the theory [Equation (I, 193)]. However, the linear portions of the theoretical curves, corresponding to the transition from rubbery to glass-like consistency, lie too far to the right, because the theory cannot describe the gradual disappearance of the coupling phenomenon in the plateau region. The separation between theory and experiment at, e. g., $H = 10^6$ may be regarded as a measure of the magnitude of the coupling or entanglement effect of BUECHE[2]; symbolized in fig. VI, 22 by Δ, it amounts here to 4,7 logarithmic decades. Alternatively, if the time scale were specified by the friction coefficient per monomer unit ζ_0 in the absence of coupling, the left limb would agree approximately with theory and the right limb would be separated by 4,5 decades. However, ζ_0 cannot be determined at present by any independent method. The interpretation of the plateau as due to a gradual disappearance of intermolecular coupling is supported by approximate predictions of its shape with suitable modifications of the BUECHE[3] and ROUSE[4] theories.

e) Temperature Dependence of Viscoelastic Properties.

The effect of temperature on viscoelastic properties in the range of rubbery consistency — which now can be alternatively identified as the plateau region of the relaxation spectrum — may be expressed by the quantities a_T or ΔH_a, just as in the transition region described in § 46. Since the plateau region and the falling limb at the right involve the long relaxation times which are also reflected in the steady flow viscosity, equation (VI, 7) should hold, and for polyisobutylene in this region it has been found to hold exactly[5,6]. At a given temperature, ΔH_a in this region is independent of molecular weight[7], and the values for short and long relaxation times are closely similar[8,9]. Thus the coupling phenomenon to which the plateau is ascribed does not affect the temperature dependence

[1] NIELSEN, L. E. u. R. BUCHDAHL: J. Colloid Sci. **5**, 282 (1950). — R. BUCHDAHL, L. E. NIELSEN u. E. H. MERZ: J. Polymer Sci. **6**, 403 (1951).

[2] BUECHE, F.: J. Chem. Physics **20**, 1959 (1952).

[3] BUECHE, F.: J. Appl. Physics **26**, 738 (1955).

[4] FERRY, J. D., R. F. LANDEL u. M. L. WILLIAMS: J. Appl. Physics **26**, 359 (1955).

[5] FERRY, J. D., L. D. GRANDINE JR. u. E. R. FITZGERALD: J. Appl. Physics **24**, 911 (1953).

[6] VAN HOLDE, K. E. u. J. W. WILLIAMS: J. Polymer Sci. **11**, 243 (1953).

[7] ANDREWS, R. D., N. HOFMAN-BANG u. A. V. TOBOLSKY: J. Polymer Sci. **3**, 669 (1948); **7**, 221 (1951).

[8] FERRY, J. D., L. D. GRANDINE JR. u. E. R. FITZGERALD: J. Appl. Physics **24**, 911 (1953).

[9] GRANDINE, L. D., JR. u. J. D. FERRY: J. Appl. Physics **24**, 679 (1953).

of the average segmental friction coefficient characteristic of long-range cooperative motions, even though it increases the absolute magnitude of this friction coefficient greatly. The values of ΔH_a given in fig. VI,16 should, therefore, be applicable in the plateau region as well as the transition region of the relaxation spectrum.

f) Viscoelastic Constants.

For uncross-linked solid polymers, the characteristic constants defined in equations (VI,10–VI,12) can be derived from time-dependent mechanical properties. The value of G_∞ is close to 10^{10} dyn/cm² for polyisobutylene[1], natural rubber[2], GR–S[2], polyvinyl acetate[3], polyvinyl alcohol[4], etc. (See however § 52 below.) The value of J has been measured directly for various high molecular weight polyisobutylenes[5] and has been deduced from non-Newtonian flow measurements on high molecular weight polystyrenes[6]. In all cases it is of the order of 10^{-5} cm²/dyn, and its dependence on molecular weight and molecular weight distribution are qualitatively in agreement with the predictions of the ROUSE theory[7].

The steady flow viscosity, η, has been measured for high molecular weight polyisobutylenes and polystyrenes by FOX and FLORY[8] and others[6,9]. Its temperature dependence as expressed by $\Delta H_\eta(T)$ is practically independent of molecular weight, as already apparent from equations (VI,8 and VI,9) when M is high. Its molecular weight dependence is given by an equation of the form of (VI,14), except that the constant B is 3,4 for both polymers instead of the lower values obtained for low molecular weights. The transition from the low to the high coefficient occurs rather abruptly with increasing molecular weight, and is attributed by BUECHE[10] to the onset of coupling by entanglement. The exponent of 3,4 has also been obtained for polymethyl methacrylate[11].

If both the high value of B and the separation denoted by Δ in fig. VI,22 arise from this same cause, it should be possible to calculate Δ from the following formula[12]

$$\Delta = (B - 1) \log M_w/M_e) \qquad \text{(VI,19)}$$

where M_e is the minimum molecular weight for which the high value of B is applicable (17,000 for polyisobutylene, 50,000 for polystyrene[8]). The calculation is quite successful for the polyisobutylene described in fig. VI,22 and the polystyrene in fig. VI,15.

---

[1] FITZGERALD, E. R., L. D. GRANDINE JR. u. J. D. FERRY: J. Appl. Physics **24**, 650 (1953).
[2] ZAPAS, L. J., S. L. SHUFLER, u. T. W. DEWITT: J. Polymer Sci. **18**, 245 (1955).
[3] WILLIAMS, M. L. u. J. D. FERRY: J. Colloid Sci. **9**, 479 (1954).
[4] TOKITA, N. u. H. KAWAI: J. physic. Soc. Japan **6**, 367 (1951).
[5] VAN HOLDE, K. E. u. J. W. WILLIAMS: J. Polymer Sci. **11**, 243 (1953).
[6] SPENCER, R. S.: J. Polymer Sci. **5**, 591 (1950). – R. S. SPENCER u. R. E. DILLON: J. Colloid Sci. **4**, 241 (1949).
[7] FERRY, J. D., M. L. WILLIAMS u. D. M. STERN: J. Physic. Chem. **58**, 987 (1954). – [8] FOX, T. G., u. P. J. FLORY: J. Appl. Physics **21**, 581 (1950); J. Phys. u. Colloid Chem. **55**, 221 (1951).
[9] Siehe S. 400, Fußnote 8.
[10] BUECHE, F.: J. Chem. Physics **20**, 1959 (1952).
[11] BUECHE, F.: J. Appl. Physics **26**, 738 (1955).
[12] FERRY, J. D., R. F. LANDEL u. M. L. WILLIAMS: J. Appl. Physics **26**, 359 (1955).

For cross-linked soft polymers, some of the integrals in equations (VI,10–VI,12) must be interpreted differently. In equation (VI,10), G_∞ still has the same significance and about the same numerical value[1]. In equation (VI,11), the first member no longer has any simple significance, while the integral $\int_{-\infty}^{\infty} L\, d \ln \tau$ gives not the steady-state compliance but rather the equilibrium compliance as treated by the kinetic theory of rubberlike elasticity. Finally, the steady-flow viscosity is of course infinite; and the integral in (VI,12), if it converges, must be interpreted as a sort of internal viscosity damping motions of the permanently bonded network. No experimental data on cross-linked polymers appear to be available extending to long enough times to evaluate this integral.

g) Chemorheology.

When stress is applied to cross-linked polymers for time intervals longer than the longest configurational relaxation or retardation times, stress relaxation and creep can still occur if primary bonds are broken by chemical reactions, as investigated by Tobolsky[2]. The scission may occur at the cross-links alone or at random along the molecules comprising the network strands. In the former case, if the rate constant for scission is the same for all bonds, the stress relaxation at constant strain is approximately exponential[3]. The phenomenon corresponds to an additional discrete contribution to the relaxation spectrum with a single relaxation time equal to the reciprocal of the chemical rate constant. For chemical rupture at random along the molecules, however, the stress relaxation follows a more complicated course[3]. This subject is treatet in § 42.

B. Viscoelastic Properties of Polymer Solutions.

Introduction.

The mechanical properties of polymer solutions closely resemble in many respects those of the pure polymeric liquids and soft solids described in §§ 45–47 above. But the introduction of small solvent molecules into an assembly of flexible polymer chains has several effects which modify their viscoelastic behaviour. First, all modes of motion become more rapid because the frictional coefficients of all polymer segments are decreased. Second, the coupling or entanglement effect[4], characteristic of high molecular weight chains, is modified and at very low concentrations disappears entirely. This may be accompanied in some systems by more obvious changes in structure such as the disappearance of crystal-

[1] Nolle, A. W.: J. Polymer Sci. **5**, 1 (1950).

[2] Tobolsky, A. V., I. B. Prettyman u. J. H. Dillon: J. Appl. Physics **15**, 380 (1944). – M. Mochulsky u. A. V. Tobolsky: Ind. Engng. Chem. **40**, 2155 (1948). – [3] Bueche, A. M.: J. Chem. Physics **21**, 614 (1953).

[4] Bueche, F.: J. Chem. Physics **20**, 1959 (1952).

lites which provide cross-linkages[1] and the configurational limitations on network strands imposed by swelling[2]. Finally, at very low concentrations the response of the system to stress is dominated by the viscous flow of the solvent and the macroscopic effect of the polymer molecules is a relatively small viscoelastic perturbation of that flow. Nevertheless, the motions of the polymer molecules which are responsible for time-dependent mechanical properties appear to be much the same at a concentration of 0,1% as they are in 100% polymer.

If the only effect of solvent were to diminish the frictional coefficients, and if all frictional coefficients were equally affected, it should be possible to interrelate the effects of concentration and time or frequency by reduced variables[3], just as temperature has been repeatedly related to time or frequency in §§ 45–47. By assuming that all relaxation or retardation times change by the same factor with a change in concentration, and that the magnitudes of all elastic contributions are directly proportional to concentration, data at various concentrations can be reduced so a standard reference state[3]. This method is somewhat more limited in usefulness than the temperature reduction; the reference state usually chosen is a hypothetical one (of unit density and unit viscosity), and the reduction is successful only over limited ranges of concentration. However, it is found to be generally applicable at very low concentrations and also within a certain range at higher concentrations, the location and breadth of which depends on the molecular weight.

The discussion of polymer solutions can be conveniently divided into three regions of concentration. Very dilute solutions (less than 1% polymer), in which the polymer coils are more or less isolated and high molecular weight coupling is absent, are treated in § 48. Solutions in an intermediate range of concentrations are treated in § 49. Very concentrated polymer compositions, in the range of plasticized plastics, are treated in § 50.

§ 48. Dilute Solutions.

In dilute solution in a solvent of low viscosity, the frictional coefficients of polymer segments are very small and the relaxation spectrum is found in a region of short times which are accessible only by dynamic measurements at rather high frequencies. Because of the small magnitudes of stored elastic energy, measurements are difficult and have thus far been successfully carried out in only two laboratories, those of MASON and BAKER[4] and of ROUSE[5].

Measurements of G' by ROUSE and SITTEL[5] on a polystyrene fraction of high molecular weight ($6{,}2 \cdot 10^6$) in dilute solution (0,1% to 0,2%) are shown in fig. VI,23. The in-phase rigidity is of the order of 10 dyn/cm²

[1] FLORY, P. J.: J. Chem. Physics **17**, 223 (1949).
[2] FLORY, P. J. u. J. REHNER JR.: J. Chem. Physics **11**, 521 (1943).
[3] FERRY, J. D.: J. Amer. Chem. Soc. 72, 3746 (1950).
[4] BAKER, W. O., W. P. MASON u. J. H. HEISS: J. Polymer Sci. 8, 129 (1952).
[5] ROUSE, P. E., JR. u. K. SITTEL: J. Appl. Physics **24**, 690 (1953).

at 220 cycles/sec., and it increases proportional to the square root of the frequency. Concomitant measurements of the real part of the dynamic viscosity, $\eta'(=G''/\omega)$ revealed values which at the lowest frequencies were smaller than the steady flow viscosity by about a factor of two, and at higher frequencies fell progressively, approaching the viscosity of the pure solvent.

Both G' and η' for these and other dilute polymer solutions are in excellent agreement with values calculated from the theory of ROUSE[1], using the known molecular weight, concentration, and steady flow viscosity, indicating that the viscosity decrease and the appearance of rigidity are due to energy storage in the polymer molecules as they are subjected to sinusoidal deformations by the flow of solvent (§ I, 8). Even though the molecular weight is very high, the plateau region of the relaxation spectrum which is seen in undiluted polymers is entirely absent here; the viscoelastic time scale is correctly predicted from the steady flow viscosity, with no discrepancy Δ such as is apparent in fig. VI, 22. This means that an average segmental frictional coefficient, determined by the viscosity of the solvent medium and the openness of the individual polymer coil, can be used to describe all modes of molecular motion. There is no marked effect of intermolecular entanglement; the molecules are far enough apart to respond to the imposed stress as isolated individuals, in accordance with the conditions of the theory[1].

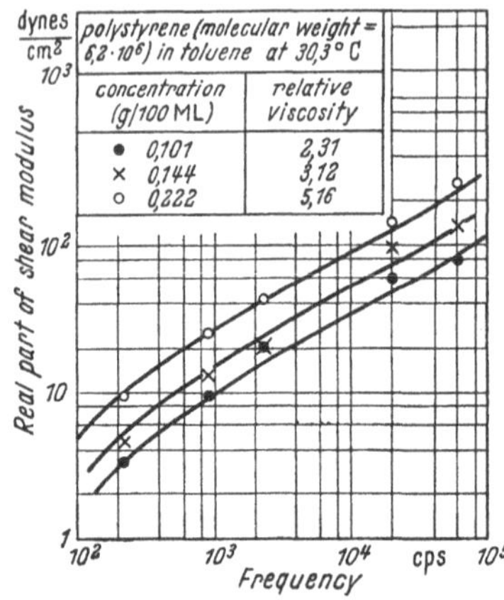

Fig. VI, 23. Real part of complex dynamic shear modulus in very dilute solutions of polystyrene (M_η = 6,200,000) in toluene, at 3 concentrations as indicated (after ROUSE and SITTEL). Points experimental measurements; curves, calculated from ROUSE theory.

This concordance with theory insures that the data of fig. VI, 23 and associated η' data would give single composite curves if plotted with reduced variables[2]. The theory specifies that the reduced relaxation spectrum H_r as a function of τ_r, as defined by equation (I, 195), will be the same for all concentrations and temperatures and even in different solvents within the range of very dilute solutions. The corresponding reductions for G' and η' as functions of ω are

$$G'_r = G' \, T_0/T \, c \tag{VI, 20}$$

$$\eta'_r = (\eta' - \eta_s)/(\eta - \eta_s) \tag{VI, 21}$$

$$\omega_r = \omega \, (\eta - \eta_s) \, T_0/T \, c \tag{VI, 22}$$

where c is the concentration in g. polymer per cc. solution and η_s is the viscosity of the pure solvent. A double logarithmic plot of G'_r or η'_r against ω_r provides a composite master curve, representing the properties of the system in a hypothetical reference state of unit viscosity and unit concentration. Then the effects of temperature, concentration, and change in solvent on the measured quantities G' and

[1] ROUSE, P. E., JR.: J. Chem. Physics 21, 1272 (1953).

[2] FERRY, J. D.: J. Amer. Chem. Soc. 72, 3746 (1950).

η' can be obtained simply by shifting the composite reduced curves by amounts which can be calculated from steady flow viscosity measurements. Thus, an increase in temperature or concentration affects the magnitude of G' through the factor Tc in equation (VI, 20) and shifts the frequency scale by a factor which depends on the concomitant change in steady-flow viscosity increment $(\eta - \eta_s)$. If $\eta - \eta_s$ were directly proportional to c (as it should be in extremely dilute solution) there would be no shift in frequency scale with concentration.

The effect of a change in solvent, at constant temperature and concentration, influences G' only through a shift in the frequency scale which is proportional to $\eta - \eta_s$. The increment $\eta - \eta_s$ is in turn approximately proportional to $[\eta]\,\eta_s$, where $[\eta]$ is the intrinsic viscosity. Thus the dispersion will be shifted to lower frequencies (relaxation spectrum shifted to longer times) by increasing the solvent viscosity and also by increasing the intrinsic viscosity, reflecting openess of coil configuration, through choice of a solvent with more favorable interaction. These predictions have been confirmed by comparisons of the dynamic properties of polyisobutylene in very dilute solution in various solvents[1].

To examine the effect of molecular weight, we turn to the relaxation spectrum. The discrete relaxation spectrum specified by the ROUSE theory can be expressed by a continuous distribution at short times, in the following forms before and after reduction to a standard reference state [cf. equations (I, 193) and (I, 196)]:

$$H = (\sqrt{6}/2\pi)\,[R\,T\,c\,(\eta - \eta_s)/M]^{1/2}\,\tau^{-1/2} \tag{VI, 23}$$

$$H_r = (\sqrt{6}/2\pi)\,(R\,T_0/M)^{1/2}\,\tau_r^{-1/2}. \tag{VI, 24}$$

These equations hold only for $\tau_r < M/2\pi^2 R\,T_0$. The effect of molecular weight on time-dependent mechanical properties is seen in equation (VI, 23). If $\eta - \eta_s$ were directly proportional to M (at constant concentration), in accordance with the STAUDINGER relation, then H (and G') would be independent of M. Actually, $\eta - \eta_s$ usually increases with a power of M less than unity, and H therefore decreases somewhat with increasing M, as shown by experiments of ROUSE and SITTEL[1] on polystyrenes.

The reduced relaxation spectrum, though independent of temperature, concentration, and choice of solvent, does depend on the molecular weight as seen in equation (VI, 24). With increasing M, the magnitude of H_r drops but the linear portion extends to longer times. The longest relaxation time present in a sample uniform with respect to molecular weight is, on the reduced scale, $\tau_{1r} = 6\,M/\pi^2 R\,T_0$.

Thus in very dilute solutions the time-dependent behaviour of homogeneous polymers can be expressed by a single function the form of which is the same for all concentrations, temperatures, and molecular weights, and does not depend on the chemical nature of either polymer or solvent; and this function is satisfactorily predicted by theory. The shifts in time scale accompanying changes in concentration, temperature, and molecular weight are always expressed in terms of a corresponding change in the steady-flow viscosity increment $\eta - \eta_s$. The dependence of $\eta - \eta_s$ itself on concentration, temperature, and molecular weight is of course a separate problem, treated in an earlier volume of this Series (Chapter V, Volume II).

The reduced variables and the ROUSE theory would not apply at times so short or frequencies so high that motions of very short molecular segments interact with applied stresses. In this case segmental mobilities would be influenced by local intramolecular interaction and steric hindrance as well as by the solvent viscosity; H would go through a maximum, as it does in fig. VI, 15, and its shape would depend on the chemical

[1] ROUSE, P. E., JR. u. K. SITTEL: J. Appl. Physics 24, 690 (1953).

structure of the polymer. But for very dilute solutions in low-viscosity solvents this region of the time scale lies beyond the reach of existing experimental methods.

§ 49. Moderately Concentrated Solutions.

As the concentration of a solution of a high molecular weight polymer is increased from 0,1% to 100%, the shape of the relaxation spectrum must change from the simple form predicted by theory as in fig. I, 41 to the complicated form shown in fig. VI, 22. The plateau in fig. VI, 22 will appear whenever molecular entanglements, arising from interpenetration of coils at higher concentrations, are tight enough to prolong the slowest relaxation mechanisms and displace the right limb of the relaxation spectrum to longer times. In solutions, the coupling is often insufficient to make a flat plateau, but a central region of smaller slope with a definite displacement Δ (cf. fig. VI, 22) is apparent for molecular weights of 10^5 at a concentration as low as 10%, and for higher molecular weights at still smaller concentrations.

The present section deals with solutions in the general concentration range from 5 to 60%, in which the effects of coupling by entanglement are present. Here, as in the case of undiluted polymers of high molecular weight, it is convenient to divide the range of time scale and discuss the transition region between soft and glasslike consistency separately from the region of soft or rubbery consistency.

a) The Transition Region: Relaxation Spectra.

Since complete studies of the transition region in concentrated solutions are few in number, each solution is treated here as a separate system with no attempt to evaluate the dependence of properties on concentration. Dynamic measurements have been made in the audiofrequency range over a wide range of temperatures on polyvinyl chloride at 10% and 40% by volume in dimethyl thianthrene[1]; polystyrene at 62% by weight in decahydronaphthalene[2]; polyvinyl acetate at 50% by volume in tricresyl phosphate[3]; and cellulose tributyrate at 20% by weight in dimethyl phthalate[4]. Data for each at various temperatures have been combined by reduction to a standard temperature of 25° C with a_T factors, in the manner used for undiluted polymers, with no attempt to include the concentration in the reduction. The distribution function of relaxation times has then been calculated from both G' and G'', and the results are compared in fig. VI, 24. The fact that the polyvinyl chloride and cellulose tributyrate solutions are actually gels, being effectively cross-linked by a few crystallites, does not interfere with the comparison because the proportion of crystalline material is undoubtedly small and the dynamic

[1] Fitzgerald, E. R. u. J. D. Ferry: J. Colloid Sci. **8**, 1, 224 (1953).
[2] Grandine, L. D., jr. u. J. D. Ferry: J. Appl. Physics **24**, 679 (1953).
[3] Williams, M. L. u. J. D. Ferry: J. Colloid Sci. **10**, 1 (1955).
[4] Landel, R. F. u. J. D. Ferry: J. Physical Chem. (in press).

properties reflect almost entirely the amorphous regions in this range of time scale[1].

The shape of H for these concentrated solutions is similar to those for the undiluted polymers in fig. VI,15. The function for 20% cellulose tributyrate follows the slope of $-1/2$ predicted by ROUSE[2] and BUECHE[3] over several decades of logarithmic time. The others (like those in fig. VI,15) have this slope at the rubbery end of the transition zone, near $H = 10^6$ dyn/cm², but at shorter times each rises more steeply and then passes through a maximum. Near the maximum the shape depends somewhat on the polymer concentration as well as on the chemical structure of the polymer, as can be seen by comparing the two concentrations of polyvinyl chloride.

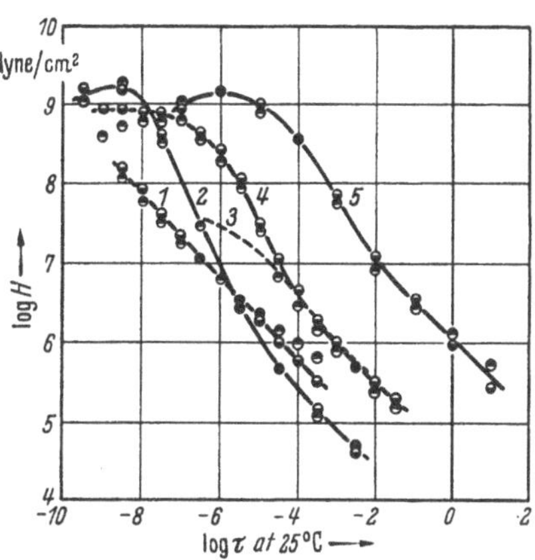

Fig. VI,24. Relaxation spectra of several concentrated polymer solutions in the transition zone, reduced to 25°C. 1) 20% (by weight) cellulose tributyrate in dimethyl phthalate (LANDEL and FERRY); 2) 10% (by volume) polyvinyl chloride in dimethyl thianthrene (FITZGERALD and FERRY); 3) 62% (by weight) polystyrene in decahydronaphthalene (GRANDINE and FERRY); 4) 50% (by volume) polyvinyl acetate in tricresyl phosphate (WILLIAMS and FERRY); 5) 40% (by volume) polyvinyl chloride in dimethyl thianthrene. Points top black, calculated from real part of complex dynamic modulus; bottom black, from imaginary part.

The primary effect of the added diluent has been to shift the H curves to shorter times by many decades of logarithmic time. It is impossible to determine the magnitude of the shift, even though two of the same polymers appear in diluted and undiluted form, respectively, in figs. VI,15 and VI,24, because they cannot be reduced to the same reference temperature[4]. The position on the time scale, again, depends on the segmental friction coefficient per monomer unit, ζ_0, representing now the force required to push short segments through a medium consisting of neighboring polymer chains interspersed with diluent molecules. The magnitude of ζ_0 reflects in a sense what the ratio of steady flow viscosity to molecular weight would be in the absence of effects of entanglement coupling and free ends[6].

b) The Transition Region: Temperature Dependence.

The apparent activation energies of the relaxation processes, ΔH_a, derived from the temperature reduction, are plotted in fig. VI,25 for the same five polymer-diluent systems. The activation energy increases with

[1] FITZGERALD, E. R. u. J. D. FERRY: J. Colloid Sci. 8, 1, 224 (1953).

[2] ROUSE, P. E., JR.: J. Chem. Physics **21**, 1272 (1953).

[3] BUECHE, F.: J. Chem. Physics **22**, 603 (1954).

[4] By utilizing the temperature dependence of dielectric relaxations, it has been calculated[5] that the diluent in the 50% solution of polyvinyl acetate diminishes ζ_0 by a factor of 10^7 at 40° C. — [5] Siehe S. 406, Fußnote 3.

[6] BUECHE, F.: J. Chem. Physics **20**, 1959 (1952).

decreasing temperature as it does in the undiluted polymers (fig. VI,16), though it does not attain quite such high values. Glass transition temperatures for the diluted systems are not available, but no doubt the effects are associated with changes in relative free volume as postulated for the undiluted polymers[1, 2], and depend on local molecular structure of both polymer and solvent.

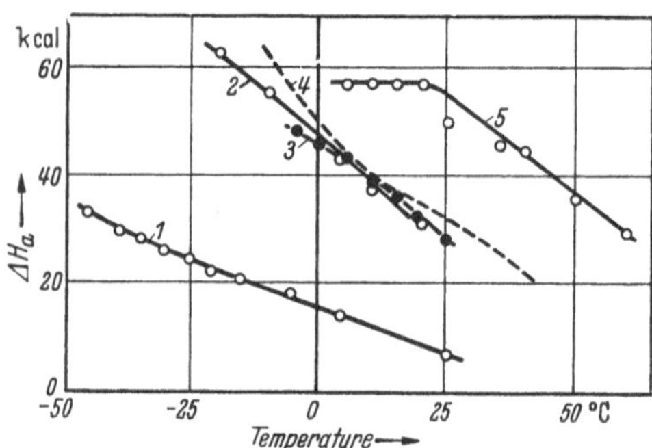

Fig. VI,25. Apparent energy of activation for relaxation plotted against temperature, for the polymer solutions referred to in fig. VI,24. Numerical key same as in fig. VI,24.

The values of a_T, the empirical relaxation temperature coefficient, from which fig. VI,25 was derived may be used to determine the temperature at which H attains a value of 10^6 dyn/cm² at 0,01 sec. These values are given in table VI,3 (cf. table VI,1). The arbitrary value of 10^6 is chosen, as before, because the logarithmic H plots all have the theoretical slope of $-1/2$ here, and ζ_0 can be calculated from their magnitudes[3]. The characteristic temperature depends of course on the concentration of polymer and the nature of the solvent, especially its viscosity. There is less variation in ΔH_a at corresponding temperatures than in the comparison of undiluted polymers in table VI,1.

Table VI,3.
Comparison of Polymer-Diluent Systems at Temperatures of Coincident Relaxation Spectra.

System	Temp. at which $H = 10^{-6}$ at 0,01 sec °C	ΔH_a at same temp. kcal.
10% Polyvinyl chloride in dimethyl thianthrene[4]	− 3	50
40% Polyvinyl chloride in dimethyl thianthrene[4]	45	41
50% Polyvinyl acetate in tricresyl phosphate[5]	13	39
62% Polystyrene in decalin[6]	13	38
20% Cellulose tributyrate in dimethyl phthalate[7]	−23	24

c) The Soft or Rubbery Region: Concentration Dependence.

The region of time scale where H exhibits a plateau, or at least a flatter portion, may be more often described as soft than as rubbery because at moderately low concentrations G' is considerably smaller than the values characteristic of rubberlike polymers. Unlike the transition region discussed above, the soft region has been extensively investigated and the dependence of viscoelastic properties on both temperature and concentration has been adduced for several systems. However, the pre-

[1] Fox, T. G., jr. u. P. J. Flory: J. appl. Physics **21**, 581 (1950).
[2] Bueche, F.: J. chem. Physics **21**, 1850 (1953).
[3] Ferry, J. D., R. F. Landel u. M. L. Williams: J. Appl. Physics **26**, 359 (1955). — [4] Fitzgerald, E. R. u. J. D. Ferry: J. Colloid Sci. **8**, 1, 224 (1953).
[5] Williams, M. L. u. J. D. Ferry: J. Colloid Sci. **10**, 1 (1955).
[6] Grandine, L. D., jr. u. J. D. Ferry: J. Appl. Physics **24**, 679 (1953).
[7] Landel, R. F. u. J. D. Ferry: J. Physical Chem. (in press).

cision of wave propagation studies[1] used for this purpose is inferior to that of some of the mechanical measurements described earlier. Somewhat better precision is obtained with direct dynamic stress-strain measurements[2].

Such data taken at different temperatures may first be reduced to a standard temperature as were the data in the transition region described in the preceding section. It has always been found that composite curves are obtained with a_T values calculated from the steady flow viscosity; thus, equation (VI,7) is generally applicable to concentrated solutions in the soft or rubbery range. Some examples of G' and η' for solutions of cellulose tributyrate in trichloropropane[3], reduced in this manner to 25° C., are shown in fig. VI,26. The individual temperatures of measurement ranged from −5° to 40° C. Both G' and η' increase with increasing concentration in the concentration range from 7% to 26%.

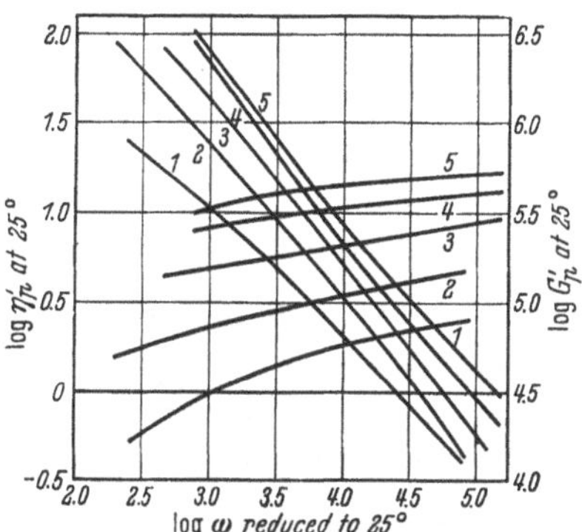

Fig. VI, 26. Real parts of dynamic rigidity (ascending curves) and viscosity (descending curves) reduced to 25°, for 5 solutions of cellulose tributyrate (M_η = 340,000) in 1, 2, 3-trichloropropane (LANDEL and FERRY).

$\eta'_p = \eta'\eta_0/\eta$; $G'_p = G'T_0c_0/Tc$;

ω reduced is $\omega\eta_0 Tc/\eta T_0 c_0$, where η_0 and c_0 are viscosity and concentration (g./cc.) at T_0 (298° K). Concentrations in weight per cent as follows: *1)* 7,2%; *2)* 10,4%; *3)* 15,6%; *4)* 21,0%; *5)* 25,6%.

Formulation of the concentration dependence at constant frequency would be difficult, however, because the primary effect of concentration change is a shift of the frequency scale, caused by changes in the friction coefficients and relaxation times. If all relaxation times are changed by the same factor, and all elastic contributions are proportional to the number of polymer molecules, then the treatment of reduced variables shows that reduction according to equations (VI,20–VI,22) should make data at different concentrations coincide, even though the shape of the resulting composite curve may be quite different from that prescribed by the ROUSE theory. Here η_s can be ignored as negligible in comparison with η. It is found that such reduction is indeed applicable in the general range from 5 to 50% polymer; when the data of fig. VI,26 are replotted in this manner, all five η' curves coincide to form a single composite curve, and similarly all five G' curves coincide. Similar tests on many other systems[4,5,6] of moderately high molecular weight have also been suc-

[1] ASHWORTH, J. N. u. J. D. FERRY: J. Amer. Chem. Soc. **71**, 622 (1949).

[2] MARKOVITZ, H., P. M. YAVORSKY, R. C. HARPER JR. L. J. ZAPAS u. T. W. DEWITT: Rev. sci. Instruments **23**, 430 (1952).

[3] LANDEL, R. F. u. J. D. FERRY: J. Physic. Chem. **59**, 658 (1955).

[4] GRANDINE, L. D., JR. u. J. D. FERRY: J. Appl. Physics **24**, 679 (1953).

[5] FERRY, J. D., E. R. FITZGERALD, M. F. JOHNSON u. L. D. GRANDINE, JR.: J. Appl. Physics **22**, 717 (1951).

[6] SAWYER, W. M. u. J. D. FERRY: J. Amer. Chem. Soc. **72**, 5030 (1950).

cessful. (The upper concentration limit of applicability depends on the molecular weight, and may be as low as 10% for a molecular weight of several million.) This result shows that in moderately concentrated solutions all cooperative motions contributing to soft viscoelastic properties are retarded to the same extent by increasing concentration and that the same retardation is reflected in the steady flow viscosity. Thus the concentration dependence of viscoelastic properties can be predicted if the concentration dependence of the steady flow viscosity is known. Similarly, the temperature dependence of viscoelastic properties has simply been absorbed, in this treatment, into the temperature dependence of the steady flow viscosity. The deviations from this principle which appear at higher concentrations and/or extremely high molecular weights will be discussed in § 50 below.

d) The Soft or Rubbery Region: Relaxation Spectra.

Application of the approximation equations (VI,6), or alternative approximation methods[1], to the reduced data for G' and η' yields the relaxation spectrum H_r reduced to unit viscosity and unit concentration at a standard temperature [cf. equation (I. 195)]. This calculation may be supplemented with data from stress relaxation measurements in concentrated solutions[2], as well as non-Newtonian flow[3], to provide H_r at the long end of the time scale. The resulting master curve can be shifted to represent the behaviour of a concentrated solution in any given solvent at a given temperature and concentration if the steady flow viscosity of the solution is known. Even in rather poor solvents[4,5], the shape of the function is apparently not seriously distorted in this range.

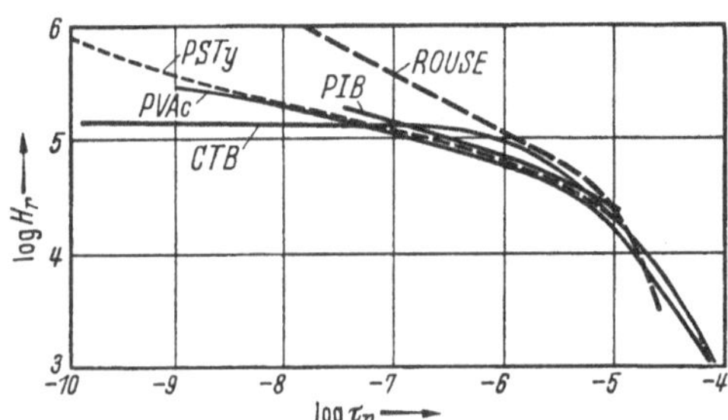

Fig. VI,27. Relaxation spectra of several polymer-solvent systems, representing concentrated solutions in the rubber-like region reduced to unit viscosity and concentration at 25° C. Polymers, solvents, and molecular weights as follows: CTB, cellulose tributyrate in 1, 2, 3-trichloropropane, M_η = 340,000 (data of fig. VI,26); PVAc, polyvinyl acetate in 1, 2, 3-trichloropropane, M_w = 420,000 (Ferry, Sawyer, Williams); PSty, polystyrene in decahydronaphthalene, M_w = 370,000 (Grandine, Ferry, Schremp); PIB, polyisobutylene in decahydronaphthalene, M_η = 320,00 (Ferry, Jordan, Evans and Johnson); Rouse, prediction of Rouse theory for molecular weight of 340,000.

The function H_r at 25° C. for concentrated solutions of cellulose tributyrate, determined in the soft range from the data of fig. VI,26, is plotted in fig. VI,27 together with the same spectrum for three vinyl

[1] Schwarzl, F. u. A. J. Staverman: Physica **18**, 791 (1952); see also § 4.

[2] Schremp, F. W., J. D. Ferry u. W. W. Evans: J. Appl. Physics **22**, 711 (1951).

[3] Ferry, J. D., M. L. Williams u. D. M. Stern: J. Physic. Chem. **58**, 987 (1955).

[4] Ferry, J. D., I. Jordan, W. W. Evans u. M. F. Johnson: J. Polymer Sci. **14**, 261 (1954).

[5] Grandine, L. D., jr. u. J. D. Ferry: J. Appl. Physics **24**, 679 (1953).

polymers of closely comparable molecular weights: polyvinyl acetate[1] (in trichloropropane), polystyrene[2] (in decahydronaphthalene), and polyisobutylene[3] (in decahydronaphthalene). The cellulose tributyrate shows a long flat plateau similar to that of soft high molecular weight polymers (fig. VI,21), and the height of the plateau is of the same order of magnitude. (The time scale is different because the reduction is based on a different reference state.) The more flexible vinyl polymers show spectra which are less flat but are remarkably like each other, apparently reflecting the properties of the flexible carbon chain backbone but revealing no difference attributable to chemical structure of either polymer or solvent.

The spectra of three of the four polymers are available from stress relaxation measurements at long times[4,5,6], and, as shown in fig. VI,27, they drop sharply past $H_r = 10^{-5}$, corresponding to the maximum relaxation time of $10^{-5 \cdot 2}$ predicted by the Rouse theory for a molecular weight of 340,000. Thus the Rouse theory correctly predicts the location of the terminal zone of the spectrum in these concentrated solutions, as it does in undiluted polymers of high molecular weight (cf. fig. VI,22). But, just as in undiluted polymers, the theoretical curve in fig. VI,27, with its time scale fixed in terms of the steady flow viscosity, diverges from the experimental results in the plateau region.

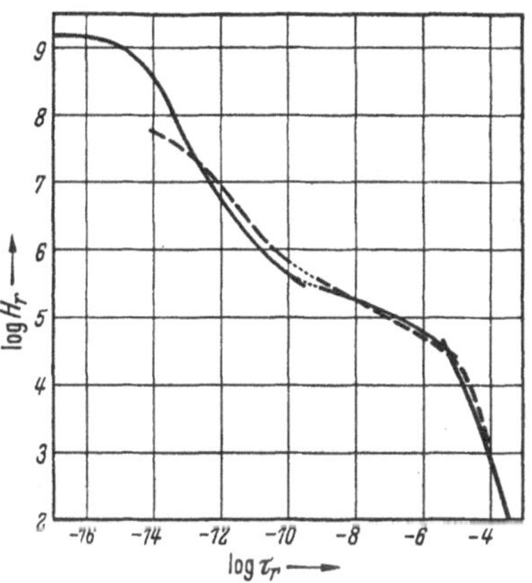

Fig. VI, 28. Relaxation spectra of two polymer-solvent systems in the range of moderately concentrated solutions, reduced to unit viscosity and concentration at 25° C., and encompassing transition, rubberlike, and terminal zones. Solid curve, polyvinyl acetate in tricresyl phosphate and 1, 2, 3-trichloropropane; dashed curve, polystyrene-decahydronaphthalene. (From data of figs. VI,24 and VI,27).

Through the plateau region, presumably the average friction coefficient decreases from a value f which depends on the entanglement characteristic of a given concentration and molecular weight to a value f_0 which is uninfluenced by entanglement. The latter fixes the location of the transition region (fig. VI,24) on the time scale. Data for both regions are available for only two polymers, polystyrene[2] and polyvinyl acetate[7], and for these only by superposing data reduced from rather widely different concentrations (fig. VI,28). The width of the plateau undoubtedly depends on molecular weight in a manner similar to that expressed by equation (VI,19) for undiluted polymers, although extensive data are

[1] Ferry, J. D., W. M. Sawyer, G. V. Browning u. A. H. Groth, jr.: J. Appl. Physics **21**, 513 (1950).
[2] Siehe S. 410, Fußnote 5. — [3] Siehe S. 410, Fußnote 4.
[4] Siehe S. 410, Fußnote 3.
[5] Landel, R. F. u. J. D. Ferry: J. Physic. Chem. **59**, 658 (1955).
[6] Siehe S. 409. Fußnote 2.
[7] Williams, M. L. u. J. D. Ferry: J. Colloid Sci. **10**, 1 (1955).

not available[1]. Moreover, the slope of the plateau depends on the molecular weight and the nature of the polymer. Thus, in fig. VI,27 the minimum (numerical) slope for the vinyl polymers is about 0,25. For polyisobutylene of molecular weight 1,000,000 or more, the plateau flattens down to a slope of zero. For cellulose tributyrate, however, the zero slope is observed at very much lower molecular weights, as low[2] as 55,000. These differences no doubt reflect the dependence of entanglements on concentration, molecular weight, and the structure of the central polymer chain.

e) The Soft or Rubbery Region: Temperature Dependence.

Equation (VI,7) usually applies to concentrated solutions, as stated above, so the value of ΔH_a which is needed to describe the temperature dependence of the viscoelastic properties can be easily obtained from temperature dependence of the steady flow viscosity. Estimates of the latter are available from a number of investigations on rubber[3], polyvinyl acetate[4], polyisobutylene[5], and polystyrene[6,7] in various solvents. The difference $\Delta H_\eta - \Delta H_{\eta 0}$, where the second term is the apparent activation energy for flow of pure solvent, is very roughly directly proportional to the polymer concentration c (expressed in g./cc.). At a temperature well above the glass transition temperature of the pure polymer, $\Delta H_\eta - \Delta H_{\eta 0}$ increases rather slowly with concentration, and it does not depend much on choice of solvent (even if the solvent is very poor) nor (as in the case of undiluted polymers) on molecular weight. At a temperature below the glass transition of pure polymer, however, there is a concentration region where ΔH_η increases much more rapidly, and the poorer the solvent the lower this region appears to be on the concentration scale. Qualitative predictions may be made from these facts concerning the effects of concentration and choice of solvent on the temperature dependence of viscoelastic properties, but again more work is needed to understand the relationships fully.

f) Cross-Linked Solutions.

In fig. VI,24 and table VI,3, it was unnecessary to treat the polyvinyl chloride and cellulose tributyrate solutions separately even though they are cross-linked by crystallites. The time-dependent properties in the transition region are largely independent of the number or disposition of the crystallites; and even if the latter are partly destroyed by increasing the temperature, for example, the viscoelastic properties involving short-range segment cooperation will be scarcely affected by such structural changes.

[1] Ferry, J. D., I. Jordan, W. W. Evans u. M. F. Johnson: J. Polymer Sci. **14**, 261 (1954).

[2] Landel, R. F. u. J. D. Ferry: J. Physic. Chem. **59**, 658 (1955).

[3] Cragg, L. H., L. M. Faichney u. H. F. Olds: Canad. J. Res. **26**, 551 (1948).

[4] Ferry, J. D., E. L. Foster, G. V. Browning u. W. M. Sawyer: J. Colloid Sci. **6**, 377 (1951).

[5] Johnson, M. F., W. W. Evans, I. Jordan u. J. D. Ferry: J. Colloid Sci. **7**, 498 (1952). — [6] Bueche, F.: J. Appl. Physics **24**, 423 (1953).

[7] Ferry, J. D., L. D. Grandine, jr. u. D. C. Udy: J. Colloid Sci. **8**, 529 (1953).

At longer times, however, viscoelastic properties are naturally profoundly affected by cross-links, either chemical (as would be present in swollen gels of vulcanized rubbers) or crystalline (as in the above-mentioned solutions and many other gelatinous or plastic compositions). The steady-flow viscosity is infinite; and the plateau and terminal regions of the relaxation spectrum involve cooperative motions of groups of network strands, as well as (in the case of crystallites) separation and rearrangement of individual molecules within crystallites. Although data on moderately concentrated solutions are rare[1], there is some information on more highly concentrated, "plasticized" systems which will be mentioned in § 50 below.

g) Viscoelastic Constants.

Of the characteristic constants defined by equations (VI, 10–VI, 12), very little is known about the first two for concentrated polymer solutions. The value of G_∞ appears to be between 10^9 and 10^{10} dyn/cm^2 for solutions of polyvinyl chloride[2] and polyvinyl acetate[3] of concentrations ranging from 10% to 50%. There are no reliable estimates, however, of its dependence on concentration or choice of solvent. The steady-state compliance J should, according to derivations[4] from the ROUSE and BUECHE theories, be proportional to M (for a heterogeneous polymer, $M_{z+1} M_z / M_w$) and to $1/c$ [equation (I. 197)]. There are a few isolated estimates of J from observation of creep recovery[5] and integration of H in accordance[6] with equation (VI, 11); they are of the order of magnitude predicted by theory.

In cross-linked systems, J is the equilibrium compliance, which depends on the number of cross-links per unit volume. In dilute polyvinyl chloride gels (10% in dioctyl phthalate), J has been found by WALTER[1] to be inversely proportional to T in accordance with the kinetic theory of rubberlike elasticity. Over the concentration range from about 4% to 20%, J is inversely proportional to a power of the concentration which ranges from 3,1 to 3,5 in most plasticizers. This means, presumably, that the number of effective cross-links increases rapidly with concentration, in a complicated manner which depends on the details of crystallite formation.

The steady-flow viscosity, on the other hand, has been the subject of many investigations. Its dependence on temperature has been discussed briefly above. Its dependence on concentration has been described by a wide variety of empirical equations[7, 8]. For polyisobutylenes, for

[1] WALTER, A. T.: J. Polymer Sci. **13**, 207 (1954).

[2] FITZGERALD, E. R. u. J. D. FERRY: J. Colloid Sci. 8, 1, 224 (1953).

[3] WILLIAMS, M. L. u. J. D. FERRY: J. Colloid Sci. **10**, 1 (1955).

[4] FERRY, J. D., M. L. WILLIAMS u. D. M. STERN: J. Physic. Chem. **58**, 987 (1954).

[5] FERRY, J. D.: J. Amer. chem. Soc. **64**, 1330 (1942).

[6] SCHREMP, F. W., J. D. FERRY u. W. W. EVANS: J. appl. Physics **22**, 711 (1951).

[7] DOOLITTLE, A. K., in: J. ALEXANDER: Colloid Chemistry, Vol. VII, pp. 161-162. New York: Reinhold, 1950.

[8] FERRY, J. D., L. D. GRANDINE JR. u. D. C. UDY: J. Colloid Sci. 8, 529 (1953).

example[1], η at 25° C. is proportional to c^5 over a considerable range and also proportional to $M^{3.4}$. The latter exponent was also found[2] for molecular weight dependence of η in polystyrene at a 44% concentration; and it applies, of course, to both these polymers in the undiluted state[3]. The critical value of M above which the exponent of 3,4 is applicable decreases with increasing concentration[2]. For polymers below their glass transition temperatures, η increases more rapidly with c than the fifth power[4].

The influence of choice of solvent on the viscosity of a concentrated solution depends primarily on the solvent viscosity η_s; if the polymer is above its glass transition temperature, η/η_s is roughly a function of c alone, being the same for different solvents compared at equal volume concentrations[1]. For polymers below their glass transitions, however, η/η_s may be higher in poor solvents than in good, because of the effect of solvent interaction on the apparent activation energy in this case[4]. Finally, the effect of the nature of the polymer on viscosity in concentrated solutions may be very roughly generalized by saying that vinyl polymers of equal chain lengths have approximately equal viscosities when c and η_s are comparable[4]. Solutions of cellulose derivatives, however, have far higher viscosities.

Further elucidation of the steady flow viscosity and its dependence on temperature, concentration, and the chemical composition of solvent and polymer will be important not only for its own sake but also because knowledge of η fixes the time scale of viscoelastic behaviour in many cases. Reduced plots of H or any other time-dependent mechanical property, as in figs. VI,27 and VI,28, can be shifted to predict viscoelastic behaviour under a wide variety of conditions provided the steady flow viscosities under these conditions are known.

§ 50. Very Concentrated Solutions.

Detailed information on viscoelastic properties of polymer solutions at concentrations above 60% is very sparse, despite the technical importance of such "plasticized" compositions. There are many studies of the stress/strain ratio at a fixed time interval after stressing, and the dependence of this ratio on temperature and concentration, as reviewed in earlier treatises[5,6]. However, as pointed out above for undiluted polymers, such measurements do not provide the shape of the relaxation or retardation spectrum, nor the apparent activation energies of relaxation processes.

---

1 JOHNSON, M. F., W. W. EVANS, I. JORDAN u. J. D. FERRY: J. Colloid Sci. 7, 498 (1952).

2 BUECHE, F.: J. Appl. Physics 24, 423 (1953).

3 FOX, T. G., JR. u. P. J. FLORY: J. Appl. Physics 21, 581 (1950).

4 Siehe S. 413, Fußnote 8.

5 ALFREY, T.: Mechanical Behaviour of High Polymers. New York: Interscience Publishers, 1948.

6 BOYER, R. F. u. R. S. SPENCER: J. Polymer Sci. 2, 157 (1947).

a) Behaviour Above Glass Transition Temperature of Undiluted Polymer.

In examining a series of solutions whose concentrations range from 60% to 100% at a temperature well above the glass transition of the undiluted polymer, a comparatively simple continuous gradation of properties may be expected. The apparent activation energy ΔH_a is not excessively high in the pure polymer (fig. VI, 16), and from its dependence on concentration below 60% (§ 49 above) ΔH_a should further increase gradually with concentration up to the value corresponding to pure polymer. Since (except for very low molecular weights) the plateau in the relaxation distribution function is present both at moderate concentrations (polyisobutylene in fig. VI, 27) and in pure polymer (fig. VI, 22), no profound changes in the shape of the relaxation spectrum are anticipated as the concentration is increased from 60% to 100%. It is true that H in the transition region is a little less steep in the presence of diluent (cf. figs. VI, 15 and VI, 24), but the general shapes are similar. The principal change with increasing concentration will be an enormous shift in the time scale as reflected approximately by the increase in steady flow viscosity.

In the transition region of time scale, dynamic measurements of SCHMIEDER and WOLF[1] have provided families of curves for G' and G''/G', plotted against frequency at different temperatures, for several plasticized compositions. Those for 80% polyvinyl chloride and 80% polystyrene, both in diethyl hexyl phthalate, resemble similar sets of curves for undiluted polymers[2] and for moderately concentrated solutions[3], and the data at different temperatures could probably be combined by the method of reduced variables.

The variation of G' and G''/G' with temperature at approximately constant frequency has been obtained for polyvinyl chloride plasticized with diethyl succinate and with dibutyl phthalate at various concentrations[1]. With decreasing polymer concentration, the fall in G' (cf. fig. VI, 17) is shifted rapidly to lower temperatures; its slope first decreases and then increases. The slope of this type of plot against temperature depends both on the slope of H and the magnitude of ΔH_a (cf. § 46 above), and the addition of diluent may increase the one (fig. VI, 24, curves 2 and 5) but decrease the other (fig. VI, 25); the net effect on the slope of G' *vs.* temperature (and the sharpness of the maximum of G''/G' *vs.* temperature) depends on the relative importance of these two opposing effects.

In the rubbery region of the time scale, an investigation of creep was made by HATFIELD and RATHMANN[4] on a polyisobutylene of molecular weight about 6,000,000 and two solutions in mineral oil at concentrations of 77% and 87% polymer (by volume). The composite creep curves for the three systems, each reduced to 25° C in the manner used for fig. VI, 8, are shown in fig. VI, 29. Though they are similar in shape, a profound

---

[1] SCHMIEDER, K. u. K. WOLF: Kolloid-Z. **127**, 65 (1952).

[2] FITZGERALD, E. R., L. D. GRANDINE, JR. u. J. D. FERRY: J. appl. Physics **24**, 650 (1953). — [3] LANDEL, R. F. u. J. D. FERRY: J. Physical Chem. (in press).

[4] HATFIELD, M. R. u. W. H. RATHMANN: J. Appl. Physics **25**, 1082 (1954).

change in magnitude is obviously caused by even a small proportion of diluent.

If the diluent were changing all friction coefficients by the same factor, and if each type of elastic contribution were proportional to the volume concentration of polymer, as postulated in the ordinary theory of reduced variables[1], then the reduction used in fig. VI, 28 should make these curves superpose. The appropriate plot for creep, corresponding to equations (VI, 20–VI, 22), is log $J\psi Tc/T_0$ *vs.* log $tTc/T_0\eta$. This plot is, however, unsuccessful; even though η is not available, it is found that no arbitrary shift on the abscissa axis can make $J\psi Tc/T_0$ coincide. No doubt the extent of entanglement is highly dependent on concentration in this region, so that the diluent not only alters the magnitude of ζ_0 but also modifies long-range structural features of the system.

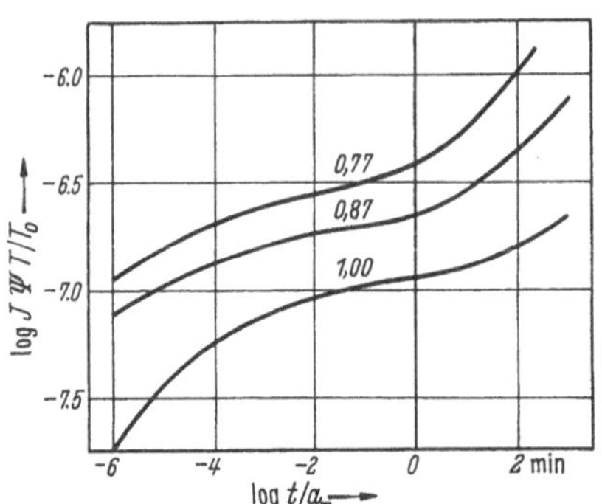

Fig. VI, 29. Creep of very concentrated solutions of polyisobutylene (molecular weight 6,000,000) in mineral oil, reduced to 25° C (data of HATFIELD and RATHMANN). Figures denote volume fraction of polymer.

HATFIELD and RATHMANN[2] found, however, that a change from c to c^2 in the expression for reduced compliance, calculated then as $J\psi Tc^2/T_0$, provided plots which could be brought to fairly close coincidence by arbitrary shifts along the logarithmic abscissa axis. Similar departures from the simple concentration reduction[1] have also appeared at lower concentrations for polymers of very high molecular weight. Thus, for polyisobutylenes with $M > 1{,}200{,}000$, the reduced variables defined by equations (VI, 21) and VI, 22) do not give superposition of data[3] at different concentrations in the range from 10 to 25%. However, if c is replaced by c^2 in equation (VI, 22) the data are brought into fairly close coincidence with each other and also with data similarly reduced for the undiluted polymer. Also, DEWITT[4] has noted in high molecular weight polyisobutylenes that superposition with equations (VI, 20–VI, 22) is obtained below a certain critical concentration, while above this concentration c must be replaced by c^2. The significance of this modification is not entirely clear, but it is no doubt related to a changing degree of entanglement.

On the other hand, wave propagation measurements at very high frequencies (2 to 10 megacycles) on swollen lightly vulcanized Buna-N rubber, by NOLLE and MIFSUD[5], indicate that the first power of c must

[1] FERRY, J. D.: J. Amer. chem. Soc. **72**, 3746 (1950).
[2] Siehe S. 415, Fußnote 4.
[3] FERRY, J. D., I. JORDAN, M. W. EVANS u. M. F. JOHNSON: J. Polymer Sci. **14**, 261 (1954).
[4] DEWITT, T. W., H. MARKOVITZ, F. J. PADDEN, JR. u. L. J. ZAPAS: J. Colloid Sci. **10**, 174 (1955).
[5] NOLLE, A. W. u. J. F. MIFSUD: J. Appl. Physics **24**, 5 (1953).

be used to achieve superposition with reduced variables. Though the concentration range is from 50 to 100%, the responses at these frequencies probably involve only short range cooperative motions which are oblivious of cross-links and entanglements.

b) Behaviour Below Glass Transition Temperature of Undiluted Polymer.

The effect of diluent on a polymer below its glass transition temperature must be much more complicated. Small amounts of diluent lower the glass transition temperature sharply[1], and hence must change ΔH_a rapidly (cf. fig. VI,16), so that the temperature dependence of viscoelastic properties is profoundly altered. The time scale is rapidly shifted from a region far to the left of the principal maximum of H shown in fig. VI,22 (see § 52 below) to the regions on the right of the principal maximum – the transition and plateau. The time dependences on the two sides of the maximum are entirely unrelated, and it is not easy to make any predictions about the changes in shape of H with concentration in this range. Some inkling of the complications may be gained from studies of diffusion of solvents into solid polymers[2–4].

c) Behaviour of Cross-Linked Solutions.

At long times, in the plateau and terminal regions of the relaxation spectrum, the viscoelastic properties of very concentrated solutions can be further modified by the presence of cross-links. Examples of such behaviour are provided by the extensive creep[5] and stress relaxation[6] measurements by Alfrey and Tobolsky and their associates on plasticized polyvinyl chloride, which is effectively cross-linked by crystallites. Since the number, size, and disposition of crystallites depend on temperature and concentration and even on previous thermal and mechanical history, the viscoelastic properties are influenced by all these variables in a complicated manner. The behaviour can be interpreted at least qualitatively, however, from the effects of temperature and concentration on the relaxation spectrum in simpler systems together with the features of microcrystalline structures as deduced from other physical properties.

§ 51. Conclusions.

In retrospect, the classification into three concentration ranges upon which this discussion has been based – very dilute, moderately concentrated, and very concentrated – may be replaced by a slightly more rational classification into three areas on a map of concentration and molecular weight, as shown in fig. VI,30. There is a characteristic visco-

[1] Boyer, R. F. u. R. S. Spencer: J. Polymer Sci. 2, 157 (1947).
[2] Kokes, R. J., F. A. Long u. J. L. Hoard: J. Chem. Physics 20, 1711 (1952).
[3] Park, G. S.: J. Polymer Sci. 11, 97 (1953).
[4] Crank, J.: J. Polymer Sci. 11, 151 (1953).
[5] Aiken, W., T. Alfrey, jr., A. Janssen u. H. Mark: J. Polymer Sci. 2, 178 (1947).
[6] Alfrey, T., jr., N. Wiederhorn, R. Stein u. A. V. Tobolsky: J. Colloid Sci. 4, 211 (1949).

elastic behaviour, as expressed by the shape of the relaxation spectrum and the manner in which the time scale changes with concentration, in each of these areas. It must be emphasized, however, that the boundaries are at present vaguely defined; those shown in the figure are very roughly the appropriate subdivisions for a flexible vinyl polymer — above its glass transition temperature, of course, and with no cross-links.

At very low concentrations or quite low molecular weights (area 1), the relaxation spectrum follows approximately the shape of the ROUSE theory. Changes in temperature, concentration, and molecular weight leave its shape unaltered, but shift the time scale and the magnitudes of mechanical properties to an extent which can be predicted from knowledge of the steady-flow viscosity (or, in the case of very dilute solutions, the viscosity increment over that of the solvent). Reduced variables can be used to obtain composite plots at different temperatures and concentrations [equations (I,195)]. There is no evidence of intermolecular coupling by entanglement.

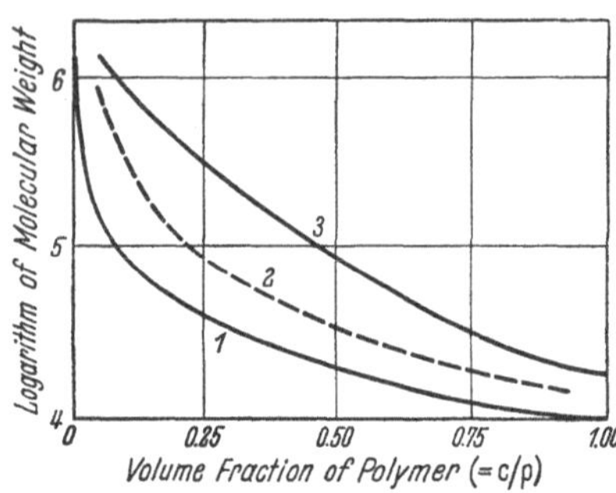

Fig. VI,30. Schematic classification of viscoelastic behaviour of solutions of a linear amorphous polymer above its glass transition temperature, on a map of molecular weight against volume fraction of polymer (see text for explanation).

At intermediate concentrations and/or molecular weights (area 2), the shape of the relaxation spectrum no longer follows the simple ROUSE theory, but shows a well-defined plateau; the slope here may not be zero, but is considerably flatter than the minimum (numerical) ROUSE value of 1/2 on a double logarithmic scale. Thus entanglement coupling is evident, according to the interpretations of the plateau region which have been given above. With increasing molecular weight, the plateau becomes longer and flatter.

Within area 2 one can imagine contours of equal entanglement, such as the dotted curve, along which the tendency to intermolecular coupling remains constant as we progress from moderately dilute solutions of high molecular weight to more concentrated solutions of lower molecular weight. Along such a contour the shape of the relaxation spectrum should remain constant. Actually, for a change of concentration at a given molecular weight (corresponding to a horizontal contour in fig. VI,30), the shape of the relaxation spectrum apparently does not change much over a considerable range, so that concentration reduction [equations (I, 195)] still provides composite curves and allows prediction of temperature and concentration dependence from the steady-flow viscosity.

At high concentrations and/or molecular weights (area 3), the relaxation spectrum is flat or passes through a minimum in the plateau region, showing strong entanglement coupling. Moreover, the ordinary concentration reduction does not bring data at different concentrations into coincidence, and it is concluded that structural features — specifically, the tendency toward coupling — change with concentration. However, the temperature reduction is still satisfactory, showing that at a single concentration all segmental mobilities or friction coefficients have approximately the same temperature dependence in these as in all other linear amorphous systems above their glass transition temperatures.

In the presence of cross-links or crystallites, or at temperatures below the glass transition point, complications appear which can be interpreted or predicted only qualitatively at the present time. However, the time is perhaps approaching when all aspects of viscoelastic behaviour may be explained on a semiquantitative basis and reasonable predictions of mechanical properties can be made from knowledge of chemical structure.

C. Viscoelastic Properties of Hard and Crystalline Plastics.

Introduction.

Below its glass transition temperature, any polymer or polymer solution, regardless of molecular weight or concentration, has the superficial appearance of a hard solid; the stress/strain ratio has a high value, 10^9 dyn/cm^2 or greater, for all practically attainable regions of time scale. Polymers with a high degree of crystallinity also exhibit high elastic moduli. The reason is, of course, that Brownian motion with long-range cooperation of molecular segments does not take place in these materials. The consequences of such Brownian motion, as reflected in the characteristic relaxation and retardation spectra described in §§ 44—51, are now absent. Instead, we find ourselves in effect on the left side of the principal maximum of the relaxation spectrum as seen in figs. VI, 15 and VI, 22. The generalizations which are so valuable in describing polymers above their glass transition temperatures — the concepts of segmental mobilities with equal temperature coefficients, of cooperative motions leading to a broad relaxation spectrum, and of intermolecular coupling by entanglement — are of little assistance here.

It should not be thought, however, that in the absence of extensive cooperative Brownian movements of the primary polymer chains the mechanical behaviour will become perfectly elastic. In highly crystallized systems there are very short segments of chains between crystallites which can still undergo configurational rearrangements at temperatures above the glass transition. Even below the glass transition, Brownian motion of side chains, and vibrations of side groups, are possible in both crystalline and amorphous systems. Elastic losses — i.e., dissipation of mechanical energy as evidenced for example by an out-of-phase component of the stress/strain ratio in dynamic measurements — may arise from these motions, as well as from other possible causes. Similarly, in polycrystalline metals elastic losses are attributed to slippage at grain boundaries, as well as several other phenomena[1].

[1] Zener, C.: Elasticity and Anelasticity of Metals. Chicago: Univ. of Chicago Press, 1948.

The viscoelastic properties of amorphous polymers below the glass transition temperature are discussed in § 52, and those of crystalline polymers in § 53.

§ 52. Amorphous Polymers.

The viscoelastic properties of hard, amorphous polymers, like those of the liquids and soft solids discussed previously, can be obtained from measurements of creep, stress relaxation, or response to dynamic stresses. They depend on the usual variables of temperature, time, and frequency, but in addition are profoundly influenced by certain other variables which are sometimes difficult to control.

First, the presence of traces of a diluent of low molecular weight has a far greater effect below the glass transition temperature than it does above, in loosening the structure and shifting the relaxation time scale to shorter times. This is apparent in the rapid drop of the glass transition temperature itself with small amounts of diluent[1]. The well-known difficulty of removing the last traces of solvent from a polymer makes it troublesome, therefore, to obtain reproducible results; samples must be carefully dried[2], and hygroscopic polymers must be protected from water vapor[3].

Second, the mechanical properties are influenced by thermal history. The very existence of the glass transition is usually attributed to failure to attain molecular-kinetic equilibrium at lower temperatures, and differences in rate of cooling through the transition temperature modify slightly the location of the transition temperature[4, 5], the specific volume[4, 5], and the heat capacity[6]. Specifically, quickly cooled samples have higher specific volumes below the transition temperature than do slowly cooled samples[4, 5]. In view of the role of free volume in determining the segmental mobilities of polymer molecules[7, 8], it is not surprising that stress relaxation occurs more rapidly in quickly cooled samples[3, 9].

Finally, the strains produced in hard polymers by moderate stresses are so small that rather large stresses are often used to obtain easily measurable deformations. Such large stresses may give departures from linear viscoelastic behaviour[10].

a) Creep and Stress Relaxation.

From fragmentary creep measurements on polystyrene[2, 11] well below the glass transition, it can be concluded that the deformations are extremely small (compliance of the order of 10^{-12} cm^2/dyn), and the rate of deformation decreases with time but never falls to zero. In these respects the creep resembles that of metals[12], rocks[13], and glass[14]. There appears to

---

[1] Boyer, R. F. u. R. S. Spencer: J. Polymer Sci. **2**, 157 (1947).
[2] Merz, E., L. Nielsen u. R. Buchdahl: J. Polymer Sci. **4**, 605 (1949).
[3] McLoughlin, J. R. u. A. V. Tobolsky: J. Colloid Sci. **7**, 555 (1952).
[4] Alfrey, T., G. Goldfinger u. H. Mark: J. Appl. Physics **14**, 700 (1943).
[5] Boyer, R. F. u. R. S. Spencer: J. Appl. Physics **17**, 398 (1946).
[6] Ferry, J. D. u. G. S. Parks: J. Chem. Physics **4**, 70 (1936).
[7] Fox, T. G., jr. u. P. J. Flory: J. Appl. Physics **21**, 581 (1950).
[8] Bueche, F.: J. Chem. Physics **21**, 1850 (1953).
[9] McLoughlin, J. R. u. A. V. Tobolsky: J. Polymer Sci. **7**, 658 (1951).
[10] Sauer, J. A., J. Marin u. C. C. Hsiao: J. Appl. Physics **20**, 507 (1949).
[11] Marin, J. u. G. Cuff: Proc. Amer. Soc. Test. Mater. **49**, 1158 (1949).
[12] Stanford, E. G.: The Creep of Metals and Alloys. Temple Press, 1949.
[13] Griggs, D.: J. Geol. **47**, 225 (1939).
[14] Taylor, N. W.: J. Appl. Physics **12**, 753 (1941).

be practically no information on the effects of temperature or molecular weight.

Measurements of stress relaxation of polymethyl methacrylate below its glass transition by TOBOLSKY[1] reveal a correspondingly high stress-strain ratio (of the order of 10^{10} dyn/cm^2) and slow relaxation — the stress falling by a factor of 2 in 100 hr., for example. Curves at different temperatures could be combined by reduced variables as in fig. VI, 9, indicating that the rearrangement processes responsible for disappearance of stress all have the same temperature dependence. The apparent activation energy for relaxation is of the order of 50 kcal.; this is higher than values characteristic of rubbery polymers, but lower than the values which some polymers achieve slightly above the glass transition temperature, where ΔH_a appears to pass through a maximum[2,3].

b) Dynamic Measurements.

The results of dynamic mechanical measurements below the glass transition temperature are often expressed in terms of the loss tangent E''/E' or G''/G' rather than the absolute values of the individual moduli, because of the ease with which it can be determined from various kinds of vibration experiments[3–5]. The contour of G''/G' plotted against frequency is very roughly similar[4] to that of H; a maximum in G''/G' is always associated with a change in the magnitude of G', but such changes below the glass transition temperature involve less than one power of ten, in contrast to the thousandfold variation in G' which occurs above the transition temperature (fig. VI, 13).

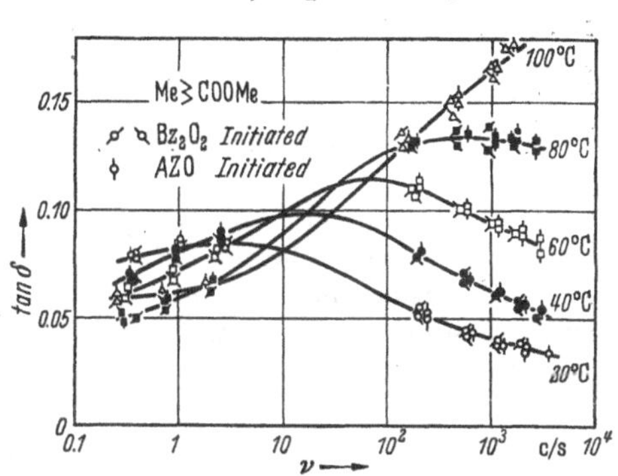

Fig. VI, 31. Loss tangent G''/G' or E''/E' of polymethyl methacrylate over a wide frequency range at several temperatures below the glass transition (after HEYBOER, DEKKING and STAVERMAN).

Examples of such data by HEYBOER, DEKKING and STAVERMAN[6] on polymethyl methacrylate are shown in fig. VI, 31. Although the gap in the frequency scale leaves the positions of some maxima in doubt, it is clear that a loss mechanism exists and that the frequency of maximum loss shifts upward with increasing temperature. The magnitude of the

[1] Siehe S. 420, Fußnote 8.
[2] McLOUGHLIN, J. R. u. A. V. TOBOLSKY: J. Colloid Sci. 7, 555 (1952).
[3] DEUTSCH, K., E. A. W. HOFF u. W. REDDISH: J. Polymer Sci. 13, 565 (1954).
[4] ZENER, C.: Elasticity and Anelasticity of Metals. Chicago: Univ. of Chicago Press, 1948.
[5] SACK, H. S., J. MOTZ, H. L. RAUB u. R. N. WORK: J. Appl. Physics 18, 450 (1947).
[6] HEYBOER, J., P. DEKKING u. A. J. STAVERMAN: Proc. 2nd Intern. Congr. Rheology, p. 123, London: Butterworths Ltd. (1954).

maximum is far smaller than that observed in the transition from rubberlike to glasslike consistency (cf. fig. VI,14), and the rate of shift with temperature is also relatively small. The latter corresponds to an apparent activation energy of only about 15 kcal., which is much less than values obtained for loss mechanisms involving relaxation of the polymer chain backbone (fig. VI,16). However, the latter values are strongly temperature-dependent, and it is impossible to make a direct comparison of results above and below the glass transition temperatures.

The loss mechanism of fig. VI,31 is seen schematically in perspective with the primary transition loss mechanism in fig. VI,32, synthesized by HEYBOER[1] for polymethyl methacrylate at 40° C. The separation of the two peaks on the time scale is uncertain, and may be temperature-dependent, the peaks drawing together with increasing temperature. The secondary maximum is attributed to the presence of the side chains, although it is not clear just what molecular motions are involved.

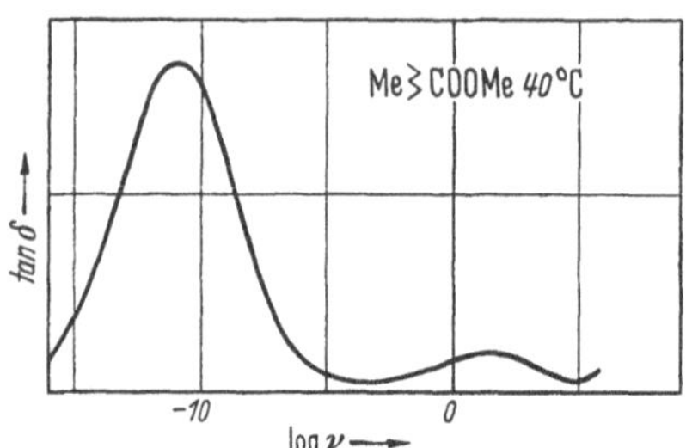

Fig. VI,32. Schematic plot of loss tangent of polymethyl methacrylate against frequency at 40° C. (HEYBOER, DEKKING and STAVERMAN). The primary maximum represents the transition from glass-like to rubberlike consistency; the secondary maximum is that evident in fig. VI,31.

Similar secondary loss maxima have been observed by SCHMIEDER and WOLF[2] and DEUTSCH, HOFF and REDDISH[3] in polymethyl methacrylate and many other polymers[4]. Moreover, when side chains of sufficient length and flexibility are present, a *third* loss maximum is evident[2,5]. In the absence of detailed studies over a range of frequencies, such loss mechanisms can be characterized by their mid-points on the temperature scale at some arbitrary and roughly constant frequency (cf. table VI,2). Several such temperatures are listed in table VI,4, taken from a compilation by SCHMIEDER and WOLF[2]. The locations of these loss mechanisms below the glass transition temperature, and the temperature dependence of their positions on the frequency scale, can be correlated with dielectric dispersion and loss[3,5].

c) Relaxation Spectra.

The dependence of mechanical properties on time or frequency is not yet known in sufficient detail to calculate relaxation spectra below the glass transition temperature. It is clear, however, that additional maxima must be present on the left of the principal maxima in figs. VI,15 and VI,22. It is conceivable that some of these may be so sharp that they can be better represented by discrete elastic contributions than by a

[1] Siehe S. 421, Fußnote 7.
[2] SCHMIEDER, K. u. K. WOLF: Kolloid-Z. **134**, 149 (1953).
[3] DEUTSCH, K., E. A. W. HOFF u. W. REDDISH: J. Polymer Sci. **13**, 565 (1954).
[4] See also IWAYANAGI, S., u. T. HIDESHIMA: J. phys. Soc. Japan **8**, 368 (1953).
[5] HOFF, E. A. W.: Private communication.

distribution function. Certainly they will reflect detailed chemical composition in a highly specific manner, involving structure of side chains as well as backbone, as evidenced by information already available[1-4].

d) Viscoelastic Constants.

In discussing polymers above their glass transitions, it was sufficient to cite G_∞ as about 10^{10} dyn/cm^2, representing the limiting value of G' at frequencies so high that chain backbone rearrangements do not occur; or, alternatively from equation (VI,10), the integral under H comprising the primary maximum in the transition region. Now, however, it is clear that each subsidiary maximum in H at very short times, apparent from measurements at low temperatures, must make an additional contribution to the integral of equation (VI,10). The final limiting value of G_∞ may thus be considerably higher than 10^{10}.

Table VI,4.
Temperatures of Secondary Loss Maxima at Frequencies of 5 to 12 Cycles/sec[1].

Polymer	Temperature of Loss Maximum	Freq. of Measurement Cycles/sec.
Natural rubber	−135	11
	− 85	10
Polyvinyl propyl ether	−160	5,5
Polyvinyl *n*-butyl ether	−150	10
	− 70	8
Polyvinyl *i*-butyl ether	−150	12
Polymethyl acrylate	− 80	9
Polyethyl acrylate	− 50	9
Poly *n*-butyl acrylate	−140	11
	− 80	9
Polyvinyl acetate	−100	12
	− 30	11
Polyvinyl propionate	− 40	5,4
Polymethyl methacrylate	36	10
Polyvinyl chloride	− 30	8,5
Polystyrene	−140	11

§ 53. Crystalline Polymers.

The gelatinous solutions of cellulose tributyrate and polyvinyl chloride whose properties are included in figs. VI,24 and VI,25 contain a small proportion of crystalline regions, but the viscoelastic properties appear to be determined largely by the amorphous regions; over wide ranges of temperature and time or frequency scale, there are no obvious anomalies attributable to the presence of crystallinity. The situation is not so simple for highly crystalline systems such as undiluted polyethylene, polyethylene terephthalate, polyesters, polyamides, etc., where the proportion of crystallinity approaches or exceeds 50%. Even though the compliance under stress is largely associated with the amorphous regions

[1-4] Siehe S. 422, Fußnoten 1–4.

(at least above the glass transition temperature), the chain segments in such regions are very short, and because of the restraints imposed during the crystallization process their freedom is restricted[1]; moreover, their length and distribution depend markedly on the extent of crystallinity. Since the latter is sharply influenced by temperature and the presence of small amounts of diluent, the latter variables alter the mechanical properties in a manner quite unrelated to the now familiar behaviour of amorphous systems.

a) Creep and Stress Relaxation.

Creep of polyethylene in very short time intervals, 10^{-3} to 1 sec., was measured by LETHERSICH[2]. In a typical case (Polythene grade 20 at 17°C.), the strain/stress ratio gradually increased from $0{,}7 \cdot 10^{-9}$ to $1{,}1 \cdot 10^{-9}$ cm²/dyn in this time interval, showing the presence of compliance mechanisms with retardation times in this range. Stress relaxation experiments in a range of much longer times by TOBOLSKY[3] and KOLSKY[4] show further gradual changes; the stress/strain ratio for the same type of Polythene at 20°C. fell[4] from $0{,}27 \cdot 10^9$ dyn/cm² at 15 sec. to $0{,}20 \cdot 10^9$ at 18 hr. At higher temperatures, the stress/strain ratios are smaller and relax somewhat more rapidly[3]. The primary effect of increasing temperature here is to decrease the proportion of crystalline material, and it is difficult to determine the temperature dependence of the relaxation mechanisms themselves.

The relaxation (and retardation) mechanisms may involve rearrangement of short chains caught between neighboring crystallites, though it is difficult to understand why such motions should be so slow; the effective friction coefficients would have to be enormously higher than in an amorphous polymer of similar composition at a similar temperature. Rearrangements involving removal of individual chains from crystallites may also be included. However, birefringence measurements[3] show that the orientation of whole crystallites remains fixed during the relaxation process.

b) Dynamic Measurements.

Wave propagation[4, 5] and torsional oscillation[2] experiments on polyethylene (Polythene 20) show a moderate increase of the in-phase shear modulus G' with frequency; for example[2], at 17°C., from $4 \cdot 10^8$ dyn/cm² at 10^{-4} cycles/sec. to $1{,}5 \cdot 10^9$ at 275 cycles/sec. Similarly, the in-phase YOUNG's modulus E' at 10°C rises from $7 \cdot 10^9$ dyn/cm² at 1000 cycles/sec. to $8{,}5 \cdot 10^9$ at 16,000 cycles/sec. These sets of values are quite consistent, since E' should exceed G' by a factor lying between 2,5 and 3, and both moduli increase with decreasing temperature. The loss tangents are compared directly in fig. VI,33; they are somewhat smaller than those attained by amorphous polymers above the glass transition (fig. VI,14) but larger than those attained by polymethyl methacrylate below the glass transition (fig. VI,31). The loss tangent changes more slowly with frequency than in either of the latter cases.

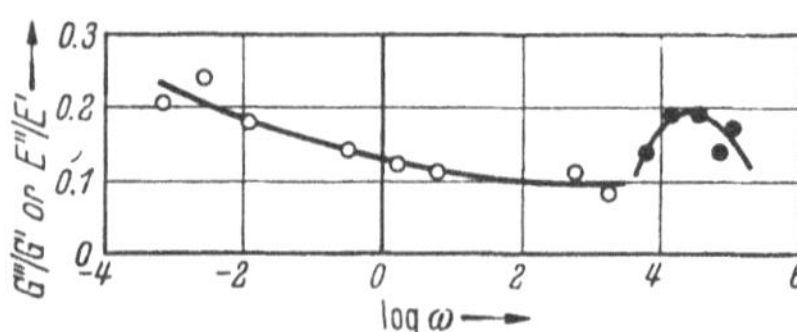

Fig. VI,33. Loss tangent of polyethylene over a wide frequency range. Open circles, G''/G' at 17°C. (LETHERSICH); black circles E''/E' at 10°C. (HILLIER).

With increase in temperature from 0° to 40°C the in-phase modulus E' falls by over a factor of 4. This is undoubtedly due to a structural change rather than

[1] SCHMIEDER, K. u. K. WOLF: Kolloid-Z. **134**, 149 (1953).
[2] LETHERSICH, W.: J. Sci. Instruments **27**, 303 (1950).
[3] STEIN, R. S., S. KRIMM u. A. V. TOBOLSKY: Textile Res. J. **29**, 8 (1949).
[4] HILLIER, K. W. u. H. KOLSKY: Proc. Physic. Soc. **B 62**, 111 (1949).
[5] BRYANT, W. M. D. u. R. C. VOTER: J. Amer. Chem. Soc. **75**, 6113 (1953).

primarily the shift in time scale which characterizes temperature dependence in amorphous polymers. Thus, although the maximum value of E''/E' depends on temperature[1], its location on the frequency axis is shifted relatively little with temperature change; an enormous shift would be required to explain the change in magnitude of E' on this basis.

When filaments of polyethylene are stretched to an elongation of over 100%, the in-phase modulus E' at 3000 cycles/sec. (in the longitudinal direction) decreases slightly for low elongations and then increases markedly, attaining three times the value for the unstretched material[2].

c) Relaxation and Retardation Spectra.

The transient and dynamic data can, as usual, be correlated by calculating the relaxation and retardation spectra, H and L. The spectrum L can be obtained from creep [equation (VI, 3)] and from complex compliance [equations (VI, 5)]; H can be obtained from the complex modulus of rigidity [equations (VI, 6)] and from stress relaxation [equation (VI, 18). Values for polyethylene are shown in fig. VI, 34. The agreement in L from LETHERSICH's creep and dynamic data is remarkably good. The black points are actually not the shear functions H and L, but the corresponding functions for extension, which should be larger by the factor $2(1+\mu)$, where μ is POISSON's ratio. The observed difference is in the right direction. The tagged points are derived from stress relaxation on filaments previously cold drawn to 100% elongation, and show a marked diminution of the contributions of relaxing mechanisms, associated no doubt with crystallite orientation and other structural changes.

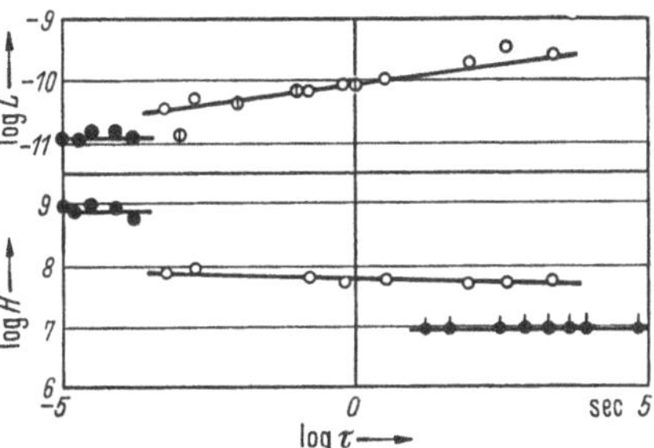

Fig. VI, 34. Retardation (L) and relaxation (H) spectra for polyethylene (Polythene 20). Open circles, dynamic torsion data of LETHERSICH, 17° C.; slotted circles, torsional creep of LETHERSICH, 17° C.; black circles, dynamic extension data of HILLIER, 10° C.; tagged circles, stress relaxation in extension, cold-drawn fibers, HILLIER and KOLSKY, 20° C.

The very broad plateau in the distribution functions in fig. VI, 34 far exceeds in width the plateau seen in soft amorphous polymers of high molecular weight, and presumably arises from an entirely different cause. It seems hardly possible that such a wide spread could be covered by motions of short chain segments caught between crystallites; the relaxation mechanisms may well involve separations of chains from the periphery of crystallites and subsequent ordering in other positions.

d) Temperature Dependence of Viscoelastic Properties.

To the left of the broad distribution for polyethylene seen in fig. VI, 34 lie more sharply concentrated relaxation mechanisms, which can be detected by measurements at moderate frequencies at lower temperatures. Studies over a range of frequencies are lacking, so the shape of the spec-

---

[1] BRYANT, W. M. D. u. R. C. VOTER: J. Amer. Chem. Soc. **75**, 6113 (1953).
[2] HILLIER, K. W. u. H. KOLSKY: Proc. Physic. Soc. **B 62**, 111 (1949).

trum cannot yet be specified, but dynamic measurements by SCHMIEDER and WOLF[1] at approximately constant frequency (order of 6 cycles/sec.) reveal a substantial loss mechanism at $-5°$ and another at $-107°$. These may involve motions of the short segments of the principal chains in amorphous regions, as well as motions of the short side branches present in polyethylene.

Similarly, dynamic measurements of G' and G''/G' for numerous crystalline polymers at roughly constant frequency reveal two or three loss mechanisms at low temperatures[1]. A typical plot, for three polyamides, is shown in fig. VI, 35. With increasing temperature, there are

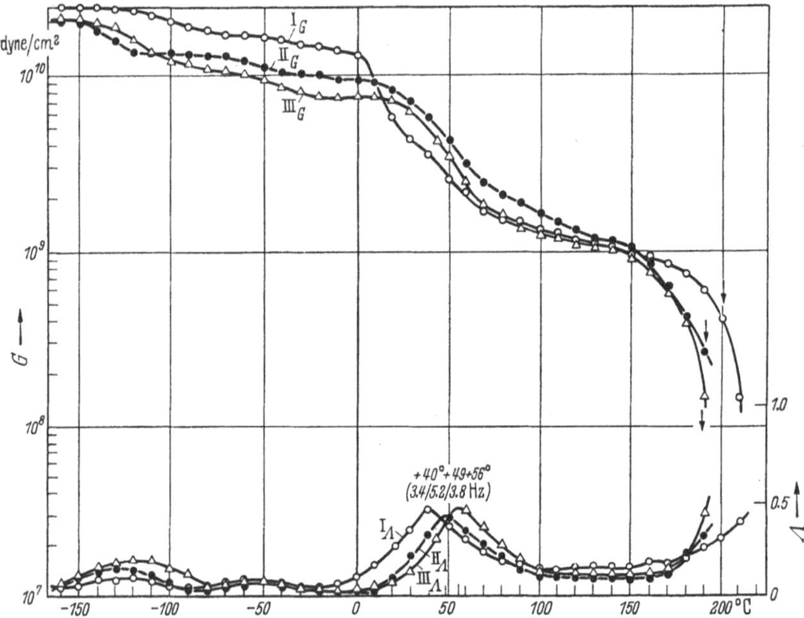

Fig. VI, 35. Real part of complex dynamic shear modulus (G') and logarithmic decrement in free oscillations ($\Lambda = \pi G''/G'$) plotted against temperature at approximately constant frequency (order of 3 to 10 cycles/sec.) for three poly-ω-amino carboxylic acids. Open circles, polycaprolactam; black circles, polycapryllactam; triangles, polyaminoundecanoic acid (SCHMIEDER and WOLF).

stepwise decreases in the real part of the modulus G' from $10^{10.5}$ down to 10^9 dyn/cm², each accompanied by a maximum in the loss tangent G''/G'. The effects of different numbers of CH_2 groups between the amide residues are apparent. The mechanism at the lowest temperature is tentatively identified with amorphous regions containing paraffin segments, and that at 50° with amorphous regions in which amide groups are concentrated. Above 180° to 200° C., G' drops sharply as the crystallites melt. Since the molecular weight is low in such polymers, the properties above the melting point are those of a polymeric liquid as described in § 45 at the beginning of this Chapter.

[1] SCHMIEDER, K. u. K. WOLF: Kolloid-Z. **134**, 149 (1953).

Siebentes Kapitel.

Struktur und mechanische Eigenschaften von Faserstoffen.

Von

W. Kast, W. Meskat, O. Rosenberg und A. K. van der Vegt.

A. Textur (Übermolekulare Ordnung und Eigenschaften).

Von W. Kast.

Mit 33 Abbildungen.

Die physikalischen Eigenschaften der hochpolymeren Faserstoffe werden letzten Endes natürlich durch die Eigenschaften der Einzelmoleküle bestimmt. Doch hängt es von der übermolekularen Struktur ab, wieweit die molekularen Größen, Länge und Beweglichkeit der Moleküle sowie Zahl und Verteilung der funktionellen Gruppen zur Auswirkung kommen. Hierfür nämlich ist die gegenseitige Lage und Vernetzung der Moleküle in den nichtkristallinen Gebieten und in den kristallinen Gebieten sowie Anteil, Größe, Form und gegenseitige Ordnung der letzteren maßgebend, übermolekulare Strukturgrößen also, die unter den Stichworten Orientierung und Kristallinität zusammengefaßt werden können.

Als kennzeichnende Eigenschaften werden meist Bruchspannung und Bruchdehnung genannt. Wenn diese auch hier speziell behandelt werden, so doch nur deshalb, weil für die Querfestigkeit, Abreibefestigkeit und Knitterfestigkeit, die für Textilien von nicht geringerer Bedeutung sind, wenn überhaupt, so doch nur roh qualitative Angaben über ihren Zusammenhang mit der Textur vorliegen. Bruchspannung und Bruchdehnung sollen aber erst als Endpunkte der Spannungs-Dehnungs-Kurve betrachtet werden, wenn zuvor deren gesamter Verlauf diskutiert worden ist. Nur dadurch kann nämlich ein hinreichend genaues Bild der mechanischen Eigenschaften gewonnen werden, zumal wenn man Bedingungen, wie Temperatur, Feuchtigkeit und Streckung (Orientierung) dabei variiert. Weitere Einblicke würden durch Einbeziehung der Kriech- und Erholungsvorgänge gewonnen werden. Doch ist allen Zeitabhängigkeiten ein eigener Abschnitt B vorbehalten.

§ 54. Spannungs-Dehnungs-Kurve.

Während das Gerät die Kraft über der Dehnung aufschreibt, rechnet man, um von der Dicke der Fasern unabhängig zu sein, die Kraft in die Spannung = Kraft/Querschnitt um. Dabei wird im allgemeinen die Abnahme des Querschnitts mit der Verstreckung nicht berücksichtigt, sondern auf den Anfangsquerschnitt bezogen. Ausnahmefälle, in denen der jeweilige Querschnitt eingesetzt ist, sind besonders gekennzeichnet.

Die Spannung (stress) wird hier stets in der Vertikalen aufgetragen und mit dem Buchstaben σ bezeichnet. Als Maß ist bei den Faserstoffen g/den üblich. Die *Denierzahl* mißt das Gewicht eines Fadens von 9000 m Länge in Gramm[1]; dieser sogenannte *legale Titer* ist also bei ein und demselben Fasermaterial ein Maß für den Querschnitt. Für verschiedene Materialien aber gelten, ihrer unterschiedlichen Dichte (ϱ g/cm^3) entsprechend, verschiedene Umrechnungsfaktoren in die physikalische Einheit kg/mm^2:

$$1\,\text{g/den} = 9\,\varrho\,\frac{\text{kg}}{\text{mm}^2}\,;$$

bei Celluloseregeneratfasern ($\varrho = 1{,}51\,\text{g/cm}^3$) also $1\,\text{g/den} = 13{,}6\,\text{kg/mm}^2$. Die Dehnung (strain) wird in der Regel mit γ bezeichnet und als prozentuale Verlängerung angegeben. Mitunter findet sich auch das Verhältnis der gestreckten Länge zur Anfangslänge als Maß für die Verstrekkung. Diese Zahl hat bei 50%iger Verstreckung den Wert 1,5.

a) Energie- und Entropiemechanismus der Deformation.

1. Elastische und blockiert-elastische Dehnung.

Wir legen unseren Betrachtungen zunächst die Verhältnisse bei den Cellulosefasern zugrunde; denn hier allein liegt ein umfangreiches und systematisches Material von Spannungs-Dehnungs-Kurven vor. Die synthetischen Fasern werden hierzu dann vergleichsweise betrachtet. In Abb. VII, 1 sind die Spannungs-Dehnungs-Kurven von Viscosereyon nach Messungen von HERMANNE[2] im trockenen und nassen Zustand wiedergegeben. Die Kurve für die trockenen Fäden läßt die typische Zusammensetzung aus zwei Kurvenästen verschiedener Neigung erkennen, die mit einem mehr oder weniger deutlichen Knickpunkt ineinander übergehen. Es ist vielfach gebräuchlich, den steilen Anfangsteil einem elastischen und den flacheren zweiten Teil einem plastischen Verhalten zuzuschreiben. Doch trifft diese Deutung bei den hochpolymeren Stoffen keineswegs zu.

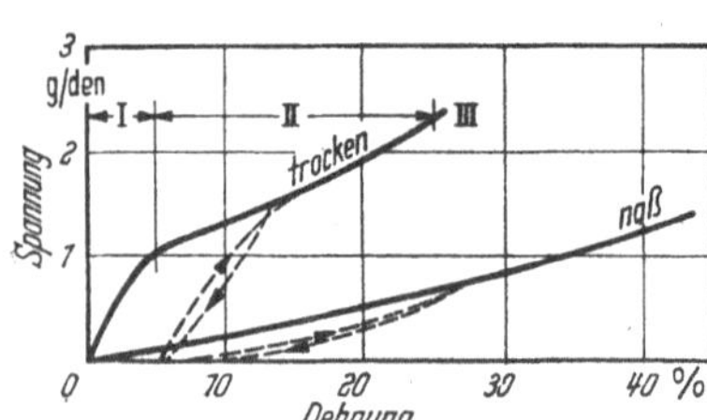

Abb. VII, 1. Spannungs-Dehnungs-Kurven von Viscosereyon trocken und naß. (Nach HERMANNE.)

Der erste Anstieg (Abschnitt I) entspricht wohl im wesentlichen der elastischen Deformation der kristallinen Gebiete, in denen, vor allem durch die Streckung der Valenzwinkel in den Molekülen, schnell eine hohe Spannung aufgebaut wird. Diese ist um so größer, je höher der Modul und je

---

[1] Ursprünglich, der alten französischen Gewichtseinheit 1 Denier = 0,05 g entsprechend, das Gewicht von 450 m Faden in Vielfachen von 0,05 g.

[2] HERMANNE, L.: Text. Res. J. **19**, 61 (1949).

größer der Anteil dieser elastischen Elemente ist. Diese Deformation ist mit einer Erhöhung der potentiellen Energie verbunden *(Energiemechanismus)* und kann eine Dehnung von 2—3% liefern. Die Höhe des Knickpunktes ist dann von der Spannung abhängig, die zur Einleitung des zweiten Mechanismus notwendig ist, der in der Regel den Hauptteil der Deformation liefert. Diese stellt nun keineswegs ein plastisches Fließen, also keine irreversible relative Schwerpunktsverschiebung der Moleküle dar. Das ist erst ziemlich am Ende der Spannungs-Dehnungs-Kurve (Abschnitt III) der Fall. Davor aber schiebt sich ein charakteristischer Deformationsmechanismus der hochpolymeren Faserstoffe ein, der ähnlich wie beim Kautschuk durch die Entknäuelung der Fadenmoleküle gekennzeichnet ist (*Entropiemechanismus*, Abschnitt II). Man darf den Übergang zwischen den Abschnitten I und II daher auch nicht als Fließgrenze bezeichnen.

Das Auftreten einer bleibenden Dehnung in diesem Abschnitt II ist nur scheinbar. Tatsächlich sind hier noch elastische Elemente wirksam, und es tritt auch eine weitere Verfestigung ein. Einmal nämlich wird kein Bruch beobachtet, wie lange auch die Faser in diesem Gebiet der Spannungs-Dehnungs-Kurve gehalten wird, und zum anderen führt ein Entlastungsversuch, wie er in Abb. VII, 1 für die trockenen Fäden z. B. nach 13% Dehnung eingezeichnet ist, nach neuerlicher Belastung erst oberhalb dieser Dehnung zu einer neuen Deformation. Die Deformationsgrenze ist durch die vorhergegangene Dehnung im Abschnitt II also erhöht worden. Aus der Steigung des zweiten Kurvenastes kann man auch einen Elastizitätsmodul für den Entropiemechanismus ableiten.

Weiter zeigt sich, daß die Entlastungskurve im trockenen Zustand zum Kurvenabschnitt I parallel läuft, wie es einem umkehrbaren Charakter von I und einer bleibenden Natur von II entspricht. Schon eine Benetzung des Fadens nach der Entlastung aber bringt die sogenannte bleibende Dehnung zum Verschwinden. Man spricht daher mit DE VRIES[1] besser von einer *quasipermanenten Deformation.*

HERMANNE[2] und HERMANS[3] bezeichnen diese Deformation als *blokkiert elastisch.* Das ist folgendermaßen zu verstehen: Im Gegensatz zum Kautschuk enthält die ungedehnte Cellulosefaser bereits kristalline Bereiche, die bei der Ausfällung der Cellulose aus der Spinnlösung entstanden sind. Dazwischen bleiben nichtkristalline Gebiete übrig, in denen die Ketten sich kreuzend und miteinander verschlaufend durcheinanderlaufen. Im Verlaufe der Umwandlung und Trocknung entstehen in diesen Gebieten Haftpunkte zwischen den Ketten, die im Gegensatz zu denen, die die kristallinen Bereiche zusammenhalten, als sekundäre Haftpunkte bezeichnet werden. Diese Haftpunkte müssen zunächst gelöst werden, ehe es zu einer Entknäuelung der Molekülteile in den nichtkristallinen Gebieten kommen kann. Das geht im ersten Teil der Kraft-Dehnungs-Kurve vor der Erreichung des Knickpunktes gleichzeitig mit der elastischen

---

[1] VRIES, H. DE: Appl. Sci. Res. **A 3**, 111 (1952).

[2] HERMANNE, L.: Text. Res. **J. 19**, 61 (1949).

[3] HERMANS, P. H.: Physics and Chemistry of Cellulose Fibres, Elsevier Publishing Co. Inc., Amsterdam 1948.

Deformation der kristallinen Gebiete vor sich. Der Vorgang im Abschnitt I wird dadurch kompliziert und verliert seinen rein momentanen Charakter. Je schneller die Belastung vor sich geht, um so steiler ist der Anstieg dieses ersten Teiles der Kraft-Dehnungs-Kurve, ein Verhalten, das nicht zu verstehen wäre, wenn es sich um einen reinen Energiemechanismus handeln würde, wie er in den Metallen vorliegt, solange das HOOKEsche Gesetz gilt. Nach der Überwindung der sekundären Haftpunkte, für deren Anzahl und Stärke also die Höhe der Zugspannung am Knickpunkt kennzeichnend ist, kann dann die Entknäuelung der Kettenteile in den nichtkristallinen Gebieten einsetzen.

HERMANNE[1] beschäftigt sich ausführlich mit der Frage der inneren Beweglichkeit der Celluloseketten. Hier ist die Valenz der Brückensauerstoffe zwischen den Glukoseringen drehbar, zwar unvollständig, aber doch so, daß man mit KUHN[2] längere statistische Kettenelemente einführen kann, die ihrerseits als frei beweglich gelten können. Dadurch erscheint auch hier eine Änderung der individuellen Form der Ketten und eine Modifikation ihrer gegenseitigen Anordnung ohne Änderung der potentiellen Energie möglich. Im nichtorientierten Material sind die Richtungen der statistischen Kettenelemente unter der Wirkung der Wärmebewegung gleichmäßig über alle Richtungen verteilt. Die Entropie als Funktion der Wahrscheinlichkeit W der Verteilung ($S = k \cdot \ln W$) hat dann ein Maximum. Die Verstreckung und Orientierung erfolgt dementgegen unter Verkleinerung der Entropie. Gleichzeitig wird durch die dichtere Packung der gestreckten Ketten, auch ohne daß es zu einer Kristallisation zu kommen braucht, eine Verfestigung hervorgerufen. Beim Kautschuk führt ein solcher Mechanismus zu Dehnungen bis 800%, bei Cellulose ist die Beweglichkeit wegen ihrer Vernetzung durch die Kristallisation der Ketten geringer, so daß nur sehr viel kleinere Dehnungen möglich sind. Bei orientierten Kunstseiden und ähnlich auch bei den orientierten synthetischen hochpolymeren Fasern liegt die Grenze unter 50%.

Diese Entknäuelung ist ihrerseits aber — der zunehmenden Verdichtung entsprechend — wieder mit der Bildung neuer sekundärer Haftpunkte verbunden und führt so einmal zu einer Verfestigung der Faser, zum anderen aber auch zu einer Blockierung der einmal herbeigeführten Dehnung. Die rücktreibende Kraft dieser Dehnung (II) wird durch die Mikro-BROWNsche Bewegung geliefert, die bei gestreckten Ketten im wesentlichen senkrecht zur Kettenrichtung erfolgt und unter Wiederherstellung der statistischen Verteilung der Kettenelemente die Entropie wieder zu erhöhen bestrebt ist. In der Cellulose sind diese Kräfte im trockenen Zustand aber zu schwach, um die allgemeine Anziehung der Ketten und die sekundären Haftpunkte zwischen ihnen zu überwinden. Die Dehnung ist daher quasipermanent. Wasser, das in die amorphen Gebiete eindringt, hebt die polaren Kräfte aber auf. Dadurch wird die Blockierung beseitigt und die Mikro-BROWNsche Bewegung in den Stand gesetzt, die ursprüngliche Verteilung der Kettenelemente in den nicht-

[1] Siehe S. 429, Fußnote 2.

[2] KUHN, W. u. H.: Helv. chim. Acta **26**, 1394 (1943).

kristallinen Gebieten und damit die Ausgangslänge wieder herzustellen. Bei den synthetischen Fasern Nylon 6 und Nylon 66 sind die gegenseitigen Anziehungen der zwischen die CO—NH-Gruppen eingeschobenen Methylenketten gering und ihre Beweglichkeit groß genug, um ohne Hilfsmittel eine Dehnung von 10% noch zu mehr als der Hälfte umkehrbar erscheinen zu lassen (s. unten Abb. VII, 17).

2. *Aufhebung der Blockierung.*

α) *Quellungsrückfederung und Thermorückfederung.* Bei der Cellulose ist zur Aufhebung der Blockierung allgemein ein Quellmittel erforderlich, das eine höhere Quellung hervorruft, als sie die Faser bei der Deformation besaß. Bei Fäden, die im trockenen Zustand gestreckt worden sind, genügt daher bereits Wasser. Sind sie im nassen Zustand gestreckt, so benötigt man eine 1n-NaOH-Lösung, bei Xanthogenatfäden, die in 2 n-$(NH_4)_2SO_4$-Lösung gestreckt wurden, eine verdünntere Salzlösung, etwa 0,5 n-Na_2SO_4-Lösung. Für den letzten Fall konnte HERMANS[1] nachweisen, daß die Umkehrbarkeit derart vollkommen ist, daß alle Eigenschaften, die von dem Grad der Dehnung abhängen, wie die Quellung, die Anisotropie der Quellung und die Doppelbrechung, exakt proportional mit der Kürzung der Faser zurückgehen. Eine Faser, die auf eine bestimmte Dehnung zurückgegangen ist, kann so nicht von einer anderen unterschieden werden, die nur einmal bis zu diesem Betrage gedehnt worden ist. Abb. VII, 2 zeigt, wie die Volumenabnahme, die ein Xanthogenatfaden bei der Streckung erfährt, bei seiner Zusammenziehung nach dem gleichen Gesetz, nach dem sie entstanden ist, auch wieder zurückgeht.

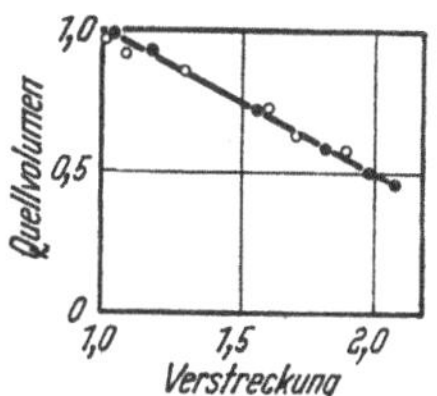

Abb. VII, 2. Abhängigkeit des Quellvolumens eines Xanthogenatfadens von der Verstreckung (Modellfaden, Ausgangszustand isotrop). Volle Kreise: Streckung in Ammoniumsulfatlösung; leere Kreise: Zusammenziehung in Natriumsulfatlösung; Messung in der originalen Ammoniumsulfatlösung. (Nach HERMANS.)

Der Rückgang der quasipermanenten Dehnung nach Aufhebung der Blockierung durch Lösung der bei der Verstreckung neu gebildeten Haftpunkte durch ein geeignetes Quellmittel wird einem Erholungsvermögen zugeschrieben, das im Gegensatz zu der *spontaneous recovery* als *latent recovery* bezeichnet wird; der Vorgang selbst wird nach HERMANS[2] und DE VRIES[3] *Quellungsrückfederung* oder *swelling retraction* genannt. Für die analoge Zusammenziehung von Kautschuk und Polystyrol nach kalter Streckung beim Erhitzen sagt man dann besser als *thermorecovery* (HOUWINK[4]): *thermoretraction oder Thermorückfederung* (HERMANS).

Von DE VRIES stammt auch das Recoverydiagramm (Abb. VII, 3), das die Zerlegung der Deformation auf experimentellem Wege zeigt. Nach der Entlastung tritt eine spontane Verkürzung ein, die dem momentanen

---

[1] HERMANS, P. H.: Cellulosechemie **19**, 117 (1942).

[2] HERMANS, P. H.: Physics and Chemistry of Cellulose Fibres, Elsevier Publishing Co. Inc., Amsterdam 1948.

[3] VRIES, H. DE: Appl. Sci. Res. **A 3**, 111 (1952).

[4] HOUWINK, R.: Physikalische Eigenschaften und Feinbau von Natur- und Kunstfasern, Leipzig 1934, S. 133ff.

und dem verzögerten elastischen Anteil (I) entspricht. Sie führt von der totalen auf die quasipermanente Deformation. Mittels Quellung wird dann die latente Verkürzung ausgelöst, die zu dem blockiert-elastischen Anteil (II) gehört. Danach bleibt schließlich die permanente Deformation (III) übrig. Wie man sieht, tritt ein geringer quasipermanenter Anteil schon von Anfang an auf; doch gibt es eine bestimmte Dehnung γ_c, von der an er stark zunimmt. Diese Dehnung γ_c liegt etwas höher als die zum Knick der Spannungs-Dehnungs-Kurve gehörige Dehnung γ_d. Sie findet sich aber als scharfer Knick im Verlaufe des dynamischen Elastizitätsmoduls mit der Dehnung (Abb. VII, 20). Die Spannungen jedoch, die zu den Dehnungen γ_c und γ_d gehören, stimmen praktisch überein, so daß σ_d weiter als die Spannung gelten kann, die zur Auslösung des Deformationsmechanismus II benötigt wird.

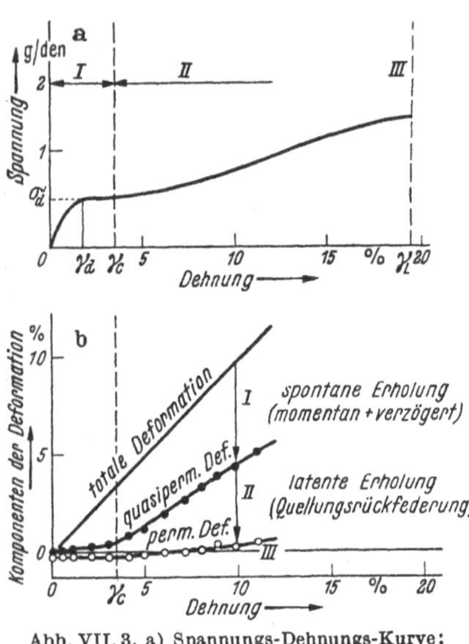

Abb. VII, 3. a) Spannungs-Dehnungs-Kurve; b) Recoverydiagramm von Viscosereyon. (Nach DE VRIES.)

In einer neuen Arbeit bestätigen BRYANT und WAKEHAM[1], daß die Streckung eines in konzentrierter Natronlauge ohne Spannung gequollenen Viscosegarnes im Wasser schon bei einer so niedrigen Temperatur wie 5° C nach dem Entropiemechanismus verläuft. Die Verfasser messen den Temperaturkoeffizienten der Kraft, die erforderlich ist, um die Fäden in Wasser auf einer bestimmten Länge zu halten. Diese Kraft läßt sich nach den thermodynamischen Gleichungen der Elastizität in eine Energie- und eine Entropiekomponente zerlegen:

$$f = (\partial E/\partial l)_{T,p} - T \cdot (\partial S/\partial l)_{T,p} = (\partial E/\partial l)_{T,p} + T \cdot (\partial f/\partial T)_{l,p} = f_E + f_S\,. \qquad \text{(VII, 1)}$$

Aus der Messung der Kraft f bei konstanter Länge und ihrer Abhängigkeit von der Temperatur, die zwischen 5 und 80° C variiert wird, können so Energie- und Entropieanteil getrennt bestimmt werden.

Im allgemeinen findet sich bei niedrigen Temperaturen ein Überwiegen des Energieprozesses, der durch einen negativen Krafttemperaturkoeffizienten charakterisiert ist, entsprechend einer normalen Wärmeausdehnung und Vermehrung der Entropie durch wachsende Unordnung. Bei mittleren Temperaturen, z. B. 45° C bei unbehandeltem Viscosegarn, durchläuft die Krafttemperaturkurve dann ein Minimum und zeigt anschließend einen positiven Temperaturkoeffizienten, der eine Abnahme

[1] BRYANT, G. M., u. H. WAKEHAM: Text. Res. J. 25, 224 (1955).

der Entropie oder Zunahme der Ordnung während der Streckung anzeigt. Der Gegenwirkung der Wärmebewegung wegen ist jetzt eine zunehmende Kraft erforderlich, um die Fäden auf einer gegebenen Länge zu halten. Unter der Annahme, daß mit wachsender Temperatur sich in erster Linie die Kraft zwischen den Ketten der nichtkristallinen Gebiete und weniger der Wassergehalt der Fäden ändert, kann die Temperatur, bei der das Minimum der Krafttemperaturkurve auftritt, als ein Maß für die Wechselwirkung zwischen den Ketten der nichtkristallinen Gebiete genommen werden. So sinkt die Übergangstemperatur und mit ihr die Energie der zwischenmolekularen Wechselwirkung durch die Quellung der Viscosefäden eben von 45 auf 5° C. Ein Gegenbeispiel ist die Erhöhung der Übergangstemperatur und der molekularen Wechselwirkung bei der Verseifung hochverstreckter Versuchsfäden aus Celluloseacetat von 50 auf etwa 100° C durch die Verkleinerung der Abstände zwischen den Ketten. Baumwolle zeigt einen durchgehend negativen Krafttemperaturkoeffizienten im ganzen Temperaturbereich, auch nach der Mercerisierung ohne Spannung. Selbst in diesem Zustande ist die zwischenmolekulare Wechselwirkung in den nichtkristallinen Gebieten bei der Baumwolle also noch merklich größer als in einem normalen Viskosegarn.

Bei den Viscosegarnen und den Acetatversuchsgarnen sind die Übergangstemperaturen unabhängig vom Betrage der Streckung. Bei kommerziellen Acetatgarnen aber treten kompliziertere Verhältnisse auf. Hier zeigen die Fäden bei kleinen Streckungen im ganzen Temperaturbereich einen negativen Krafttemperaturkoeffizienten, entsprechend einer Längung bei konstanter Kraft. Der Übergang zu einem positiven Koeffizienten (oder einer Verkürzung bei konstanter Kraft) tritt erst bei Streckungen über etwa 20% in Erscheinung, und dabei findet sich die Übergangstemperatur um so niedriger, je höher die Verstreckung ist. Die Autoren denken deshalb an eine Bildung von Ordnungsgebieten (rearrangements) bei niedrigen Verstreckungen und an ihre spätere Wiederzerstörung. Ähnliche Beobachtungen machte BRENSCHEDE[1] an Polyurethanfäden. Diese zeigen beim Tempern bei Dehnungsgraden unter 2,5 oder 150% eine Längung, die durch Kristallisation erklärt wird, bei der Kettenstücke in die Faserrichtung eingedreht werden. Bei größeren Dehnungen tritt dann eine Verkürzung oder *Thermorückfederung* ein[2].

β) Superkontraktion. Streckt man eine fertige Kunstfaser im trockenen Zustand, so geht in Wasser nur diese Dehnung wieder zurück. In einem stärkeren Quellmittel aber kann auch die Dehnung zum Verschwinden gebracht werden, die vorher beim Spinnen vorgenommen worden ist. Eine ähnliche Erscheinung ist als *Superkontraktion* bekannt.

Die erste Beobachtung und Benennung dieser Erscheinung geschah durch SPEAKMAN[3] an Keratinfasern. Normale ungedehnte Wollfasern

[1] BRENSCHEDE, E.: Z. Elektrochem. **54**, 191 (1950).

[2] Diese Längung wird auch bei gedehntem Kautschuk beobachtet, wenn er in gefrorenem Zustande weiterkristallisiert. Ihr entspricht die Spannungsrelaxation bei gedehntem, auf konstanter Länge gehaltenem Kautschuk, wenn dieser zusätzlich kristallisiert; vgl. Kap. V, Abschn. G.

[3] SPEAKMAN, J. E.: J. Textile Inst. **27**, P 231 (1936).

ziehen sich unter der Einwirkung von Chemikalien, wie Natriumsulfid oder Natriumbisulfit, Silbersulfat und Kaliumcyanid bis zu 40% zusammen. SPEAKMAN schrieb das der Aufspaltung der Cystinbrücken zu, nach der die Peptidketten sich zusammenfalten sollen, indem sie den Anziehungskräften, die von den verschieden geladenen Seitenketten ausgehen, nachgeben. ELÖD und ZAHN[1] wurden auf die wichtige Rolle aufmerksam, die die Temperatur dabei spielt, und konnten später röntgenographisch nachweisen, daß keine regelmäßige Faltung der Peptidketten, sondern eine statistische Desorientierung der Micellen und Ketten des Keratins stattfindet. Infolgedessen setzte ZAHN[2] die Erscheinung der Superkontraktion der Wolle in Analogie zu der Kontraktion der Muskeln oder der Schrumpfung des Kollagens. Danach ist es also die Möglichkeit der Ketten, eine wahrscheinlichere Konstellation anzunehmen, die zur Superkontraktion führt, wenn durch die Quellung die Voraussetzung hinreichender Beweglichkeit gegeben ist.

Daraus, daß diese Erscheinung auch in phenolischer Lösung nachgewiesen werden konnte, ging hervor, daß schon die Lösung der zwischen den Peptidgruppen und den Seitenkettengruppen betätigten Wasserstoffbindungen ausreicht. Wie danach zu erwarten war, gelang es ELÖD und ZAHN[3] in phenolischen Lösungen auch bei Polyamidborsten und -seiden Superkontraktion zu erreichen. Interessant ist das unterschiedliche Verhalten verschiedener Polyamide bzw. Polyurethane. So beträgt die Phenolkonzentration, die bei vierstündiger Einwirkung bei 18° C eine 15%ige Kontraktion hervorruft, bei Igamid 6 A-Seide, einem Mischkondensat aus Caprolactam und Hexamethylendiamin-Adipinsäure 0,9%, bei 6-Nylon 2,6%, bei 6,6-Nylon 3,1% und bei Polyurethan 4,4%. Auch bei Naturseide läßt sich eine Superkontraktion erreichen, hier allerdings erst in 50%iger Phenollösung.

In Abhängigkeit vom p_H-Wert findet sich in allen Fällen im sauren Gebiet eine Abnahme des Effektes, nach der basischen Seite dagegen bleibt er zunächst angenähert konstant. Erst oberhalb $p_H = 9$ wird der Effekt bei Polyamiden wieder kleiner. Bei tierischen Fasern, z. B. Roßhaar, dagegen findet oberhalb $p_H = 5$ ein starker Anstieg der Superkontraktion statt, den ELÖD und ZAHN mit der Simultanspaltung von Wasserstoffbrücken und Cystinbrücken durch das Phenol erklären. Unterhalb $p_H = 5$ ist die Superkontraktion auch bei Keratinfasern lediglich durch die Spaltung von Wasserstoffbrücken bedingt. Chemische Untersuchungen anderer Autoren[4] haben zu ähnlichen Konsequenzen geführt. Die Auffassung, daß die treibende Kraft, welche die durch die Spaltung der kovalenten Bindungen zwischen ihnen freigelegten Ketten verkürzt, in der Hauptsache auf Entropieeffekten beruht, findet sich dagegen nur bei ELÖD und ZAHN. Diese konnten auch nachweisen, daß der Wärmeausdehnungskoeffizient unter den Bedingungen der Superkontraktion in

[1] ELÖD, E. u. H. ZAHN: Kolloid-Z. **93**, 50 (1940); Melliand Textilber. **21**, 617 (1940); **28**, 2 (1947). — [2] ZAHN, H.: Diss. Karlsruhe 1940.

[3] ELÖD, E. u. H. ZAHN: Melliand Textilber. **30**, 17, 349 (1949).

[4] PHILLIPS, H.: J. Soc. Dyers Col. **62**, 203 (1946). — S. BLACKBURN u. H. LINDLEY: J. Soc. Dyers Col. **64**, 305 (1948).

negative Werte umschlägt, und die angenommene enge Beziehung zwischen Superkontraktion und Entropieeffekten dadurch entscheidend stützen.

b) Form der Spannungs-Dehnungs-Kurven.

1. Rolle der sekundären Bindungen.

Wenn man sich der Rolle erinnert, die die sekundären Haftpunkte in dem ersten Teil der im trockenen Zustand aufgenommenen Kraft-Dehnungs-Kurve spielen, und die Möglichkeit ihrer Lösung durch Quellmittel in Betracht zieht, so wird verständlich, warum bei einer Verstrekkung im nassen Zustand der anfängliche steilere Anstieg und der folgende Knickpunkt der Kraft-Dehnungs-Kurve nur schwach angedeutet sind. Hier bedarf es eben keiner besonderen Spannung mehr, um die Deformation durch Entknäuelung (Entropiemechanismus) einzuleiten. Man bezeichnet den Knickpunkt daher am besten als untere Grenze der quasipermanenten Dehnung. Die obere Grenze, bei der eine plastische Deformation an die Stelle der blockiert elastischen tritt, kommt im Verlaufe der Kraft-Dehnungs-Kurve nicht deutlich zum Ausdruck. Bei Belastungen aber, die von einer Faser zwar getragen werden, nach einer endlichen Zeit aber zu ihrem Bruch führen, ist die obere Grenze der quasipermanenten Dehnung bereits überschritten. Das ist nach ZART[1] bei Viscosefäden schon bei einer Dehnung der Fall, die erst 75% der Bruchdehnung erreicht hat. Eine ähnliche Variation des Spektrums der sekundären Haftpunkte, wie man sie bei der Cellulose durch Quellung und bei den vollsynthetischen Fasern durch Erwärmung erreichen kann, stellt sich bei den letzteren auch unter verschiedenen Bedingungen der Kristallisation, insbesondere also bei verschiedenen Abkühlungsgeschwindigkeiten, ein. Dadurch können bei diesen Fasern die mechanischen Eigenschaften gewollt oder auch ungewollt zusätzlich variiert werden.

Ein typisches Beispiel dafür scheint das unterschiedliche Verhalten dickerer und dünnerer Fäden aus 6,6-Nylon bei der Kaltverstreckung zu sein. Während isotrope Cellulosefäden, wie sie in den HERMANSschen Modellfäden vorliegen, mit Ausnahme des größeren Betrages ihrer Dehnbarkeit dieselben Spannungs-Dehnungs-Kurven zeigen wie die bei der Fabrikation verstreckten anisotropen Fäden, tritt bei unverstreckten synthetischen Fasern ein neuartiges Verhalten auf, das als Kaltverstreckung bezeichnet wird. Nach dem Überwinden des ersten Abschnittes führt eine geringe weitere Steigerung der Kraft hier zu einer Verlängerung um ein Vielfaches der Originallänge (Abb. VII, 4). Typisch für diesen Vorgang ist das Auftreten eines Halses. Die Verstreckung geht also nicht gleichmäßig über die ganze Fadenlänge vor sich, sondern setzt an einer bestimmten Stelle ein. Bei höheren Temperaturen, aber auch bei kleineren Titern, fehlt

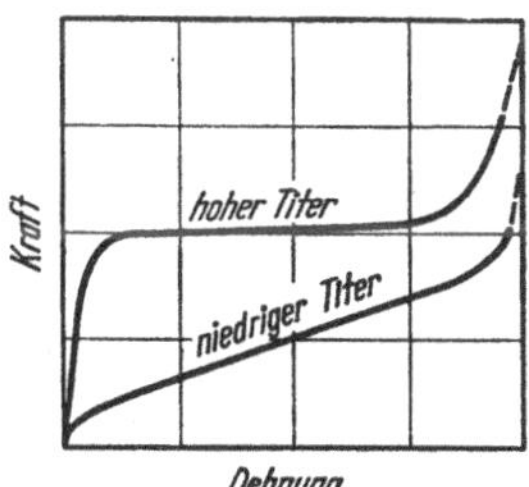

Abb. VII, 4. **Kraft-Dehnungs-Kurven unverstreckter Nylonfäden mit hohem und niedrigem Titer.** (Nach LODGE.)

[1] ZART, A.: Chemie **55**, 11 (1942).

der Hals vielfach. Für diese Fälle weist LODGE[1] auf eine Änderung der Kraft-Dehnungs-Kurve hin (Kurve „niedrig", Abb. VII, 4). Man muß es wohl mit einer Texturänderung in Verbindung bringen, die durch eine Verminderung der Zahl der Haftstellen gekennzeichnet ist – vielleicht in Form einer der höheren Abkühlungsgeschwindigkeit der dünnen Fäden entsprechend geringeren Kristallisation –, daß bei den Nylonfäden mit dem feineren Titer keine Stufe mit anschließendem Fließen bei konstanter Kraft, sondern eine gleichmäßige Deformation unter ständig wachsender Kraft auftritt. Der Vorgang der Entknäuelung und Verfestigung setzt erst am Ende der Kaltverstreckung ein. Hier schiebt sich also ein neuer Mechanismus zwischen die Gebiete der elastischen und der blockiert-elastischen Deformation ein.

Nach den vorstehenden Betrachtungen wird die Verstreckung aller hochmolekularen Fasern, bei den Naturfasern ebenso wie bei den Kunstfasern auf Cellulosebasis oder den vollsynthetischen Fasern, auf die Streckung der Kettenmoleküle zurückgeführt, und zwar zum geringen Teil auf die Streckung der Valenzwinkel der Kettenteile in den kristallinen Bereichen, hauptsächlich aber auf die Entknäuelung und Orientierung der Ketten in den nichtkristallinen Gebieten. Die Wechselwirkung der Ketten aber wurde nicht in Betracht gezogen. Auch dafür gibt es einen Energie- und einen Entropiemechanismus. Der erstere, der die sekundären Bindungen betrifft, kann nach MARK und PRESS[2] aber nur für 1% Dehnung aufkommen. Für den Entropieprozeß käme die Entropie der Zusammendrängung in Betracht; doch ist darüber praktisch nichts bekannt.

2. Einfluß von Temperatur, Quellung und Öffnung sekundärer Bindungen.

Wenn den sekundären Valenzen danach nur die Rolle der Einschränkung der Beweglichkeit zukommt, so ist gerade diese aber für den Zusammenhang von Textur und Eigenschaften von besonderer Bedeutung. Im folgenden soll deshalb die Änderung des Verlaufes der Spannungs-Dehnungs-Kurve gerade durch Vorgänge betrachtet werden, die die Beweglichkeit beeinflussen. Das ist die Erhöhung der Beweglichkeit durch Quellung und Temperaturerhöhung und ihre Erniedrigung durch Orientierung und Kristallisation. Dabei ist insbesondere die Lage des Knickpunktes zu beachten, der die untere Grenze der quasipermanenten Dehnung (Entknäuelung) charakterisiert, dazu die Steigung beider Äste, die gleichsinnig mit dem Elastizitätsmodul, also invers mit der Nachgiebigkeit des Materials verlaufen, und natürlich auch der Endpunkt, der den Bruchpunkt der Faser darstellt.

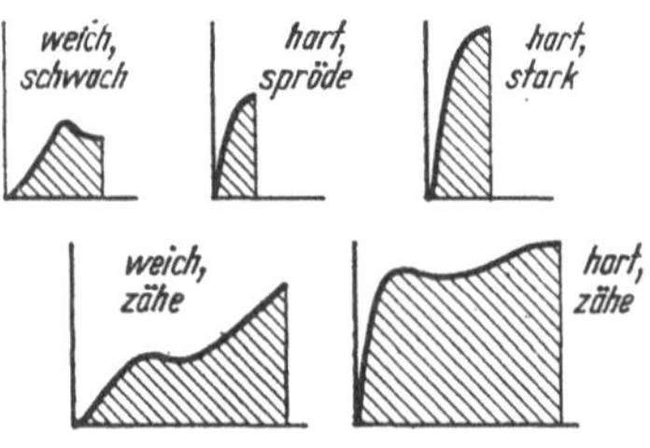

Abb. VII, 5. Spannungs-Dehnungs-Kurven für verschiedene Materialtypen. (Nach ALFREY.)

[1] LODGE, R.M. siehe R. HILL: Fibres from synthetic Polymers, Elsevier Publishing Co. Inc., Amsterdam 1953.

[2] MARK, H. u. J. PRESS: Rayon Text. Monthly **24**, 297, 339, 405 (1943).

Tabelle VII, 1. *Materialtypen* (aus ALFREY).

Typ		Elast.-Modul	Untere Grenze der quasipermanenten Dehnung	Bruch-dehnung	Bruch-spannung
soft weak	(weich schwach)	niedrig	niedrig	niedrig	niedrig
soft tough	(weich zähe)	niedrig	niedrig	hoch	hoch
hard brittle	(hart spröde)	hoch	undefiniert	niedrig	mäßig
hard strong	(hart stark)	hoch	hoch	mäßig	hoch
hard tough	(hart zäh)	hoch	hoch	hoch	hoch

ALFREY[1] unterscheidet in seinem Buch verschiedene Materialtypen auf Grund des Verlaufes ihrer Spannungs-Dehnungs-Kurven (Abb. VII, 5). Die aus diesen Kurven abzulesenden Eigenschaften sind in Tab. VII, 1 zusammengestellt. Die Haupttypen weich (soft) und hart (hard) unterscheiden sich durch die Anfangssteigung, also den Elastizitätsmodul. Spröde (brittle) und starke (strong) Materialien brechen beide schon an der unteren Grenze der quasipermanenten Dehnung, die im zweiten Falle nur höher liegt. Der weiche (weak) Typ besitzt dagegen schon einen quasipermanenten Dehnungsbereich. Im besonderen Maße ist das aber bei den zähen (tough) Typen der Fall, die eine lange Spannungs-Dehnungs-Kurve mit einem kräftigen Anstieg in diesem Gebiet zeigen. Die Einsattelung verschwindet bei genügend langsamer Verstreckung.

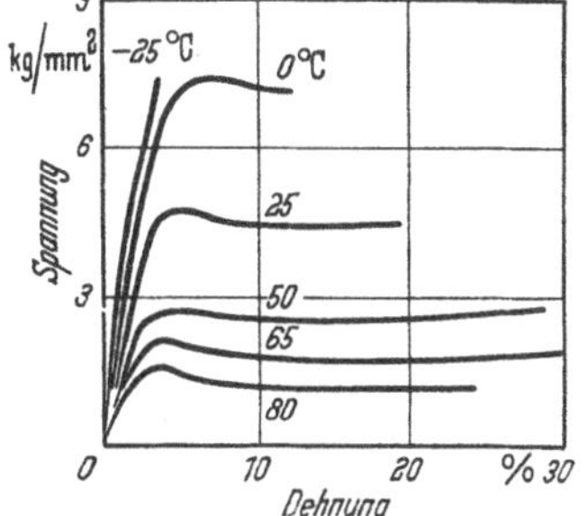

Abb. VII, 6. Einfluß der Temperatur auf die Spannungs-Dehnungs-Kurven von Celluloseacetatreyon. (Nach CARSWELL u. NASON[2].)

Daß diese Unterschiede im wesentlichen durch das verschiedene Maß der Beweglichkeit der Molekülketten hervorgerufen werden, zeigt der Vergleich mit der Abb. VII, 6, in der die Temperaturabhängigkeit der Spannungs-Dehnungs-Kurven von Celluloseacetatreyon dargestellt ist. Durch die Vergrößerung der Beweglichkeit mit wachsender Temperatur und ihre Verminderung bei Temperaturerniedrigung durchlaufen die Spannungs-Dehnungs-Kurven von Celluloseacetatreyon praktisch die ganze Reihe der in Abb. VII, 5 dargestellten Typen.

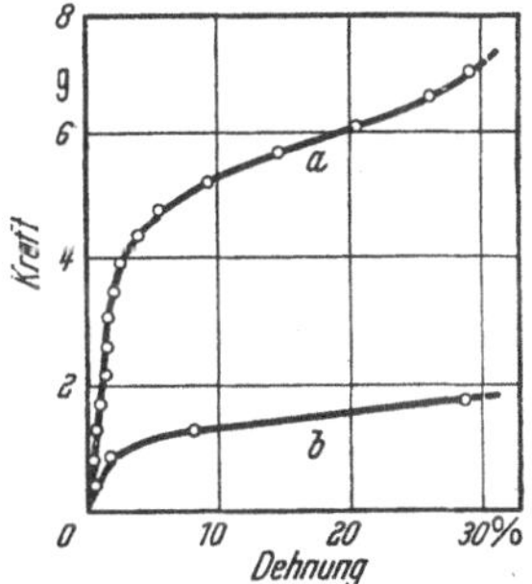

Abb. VII, 7. Kraft-Dehnungs-Kurven von Wolle. a) unbehandelt; b) nach Öffnung von 5/6 der Peptidbindungen durch Reduktion und Alkylation mit Methyljodid. (Nach HARRIS, MIZELL u. FOURT.)

Die Beweglichkeit der Ketten kann, soweit sie durch die Wirkung funktioneller Gruppen beschränkt ist, durch deren Öffnung und Blockierung vergrößert werden. Abb. VII, 7 zeigt die Änderung der Spannungs-Dehnungs-Kurve von

[1] ALFREY, T., jr.: Mechanical Behaviour of High Polymers, Intersc. Publ. Inc., New York 1948.

[2] CARSWELL, T. S. u. H. K. NASON: Amer. Soc. Test. Mater., Sympos. Plastics, Philadelphia 1944.

Wolle durch die Öffnung von 5/6 der Peptidbindungen durch Reduktion und Alkylation mit Methyljodid nach HARRIS, MIZELL und FOURT[1]. (Die Kurven sind hierbei nicht bis zum Bruch durchgeführt.) Eine ähnliche Wirkung zeigt auch die Wasseraufnahme von Wolle infolge der Solvatisierung der Peptidbindungen nach ASTBURY[2] (Abb. VII, 8).

Auf die die Beweglichkeit der Ketten erhöhende Wirkung des Quellwassers bei Cellulose wurde oben schon hingewiesen (Abb. VII, 1). Auf die Bedeutung des hyperbolischen Verlaufs der Spannungs-Dehnungs-Kurven der Regeneratfasern im nassen Zustand wird später noch zurückzukommen sein.

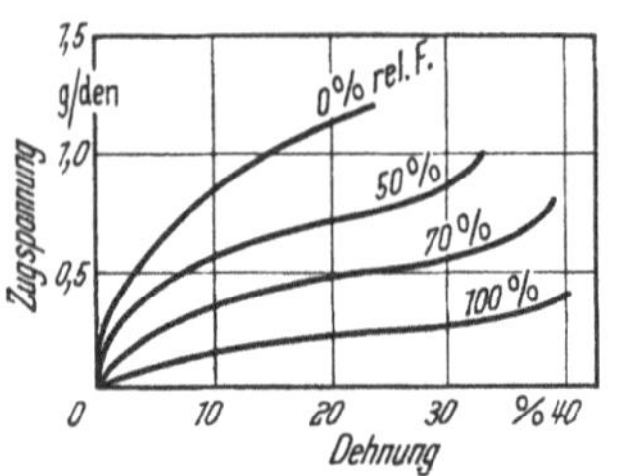

Abb. VII, 8. Spannungs-Dehnungs-Kurve von Wolle bei verschiedenen relativen Feuchtigkeiten. (Nach ASTBURY.)

3. *Einfluß von Orientierung und Kristallisation.*

Stark eingeschränkt wird die Beweglichkeit der Ketten durch die Orientierung.

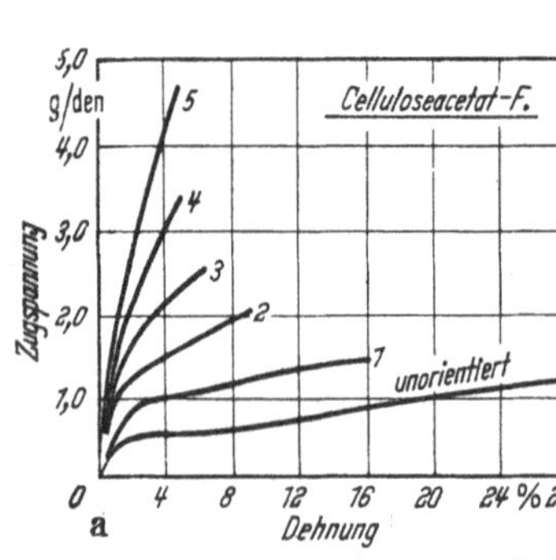

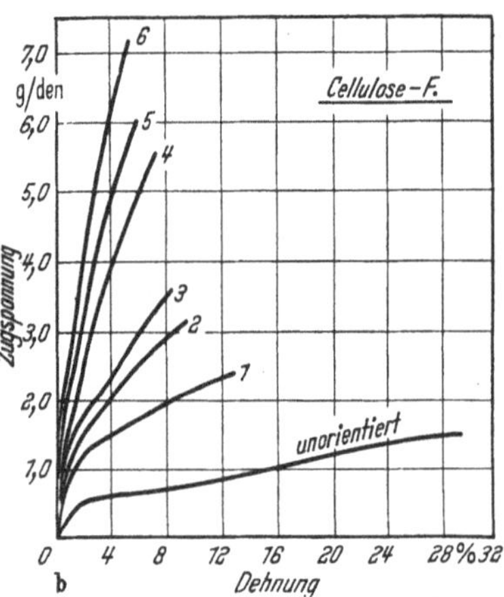

Abb. VII, 9. Einfluß der Verstreckung auf die Spannungs-Dehnungs-Kurven von Celluloseacetatreyon (a) sowie den daraus durch Verseifung erhaltenen Cellulosereyonproben (b). (Nach WORK.)

Mit wachsender Streckung durchläuft die Spannungs-Dehnungs-Kurve eines Materials daher ähnliche Formänderungen wie mit abnehmender Temperatur. Als Beispiel hierfür sind in Abb. VII, 9 die Spannungs-Dehnungs-Kurven von Proben von Acetatreyon nach Versuchen von WORK[3] wiedergegeben, die nach Quellung in einem Dioxan-Wasser-Gemisch in der Reihenfolge der angeschriebenen Zahlen in wachsendem Maße verstreckt wurden. Daneben sind auch die Spannungs- Dehnungs-Kurven der aus diesen Proben durch Verseifung erhaltenen Regeneratcellulosen eingetragen. Man sieht deutlich, wie bei höher orientierten Fäden die untere Grenze der quasipermanenten Dehnung immer höher

---

[1] HARRIS, M., L. R. MIZELL u. L. FOURT: Ind. Engng. Chem. **34**, 833 (1942).

[2] ASTBURY, W. T.: Foundamentals of Fibres Structure, Oxford Univ. Press. London 1933. — [3] WORK, R. W.: Text. Res. J. **19**, 381 (1949).

rückt und die Steigung beider Äste, sowohl des elastischen wie des blokkiert-elastischen, steiler wird. Auch die Gegenüberstellung der trocken und naß aufgenommenen Spannungs-Kurven von Viscosefäden verschiedener Streckung läßt die verminderte Beweglichkeit der Ketten im orientierten Zustand erkennen. Die Abb. VII, 10 ist dem Buche von HOUWINK[1] entnommen und zeigt, wie der Einfluß der Benetzung infolge verminderter Wasseraufnahme abnimmt, wenn die Orientierung erhöht ist. Sehr typisch ist der Einfluß der Orientierung auch bei der Polyesterfaser Terylen oder Dacron aus Terephthalsäure und Äthylenglykol. Es handelt sich hier um eine der wenigen synthetischen Fasern, bei denen man verschiedene Streckgrade fixieren kann. Das geschieht in diesem Falle durch Einleiten der Kristallisation. In Abb. VII, 11 sind nach RAY[2] die Spannungs-Dehnungs-Kurven einer frisch gesponnenen, einer mäßig und einer hochorientierten Dacronfaser zusammengestellt. Die Fasern wurden vor der Aufnahme der Kurven sämtlich in Wasser gekocht. Der Unterschied ist erheblich und zeigt einen völlig verschiedenartigen Charakter der verschieden stark gestreckten Dacronfasern an. Wir kommen darauf noch zurück.

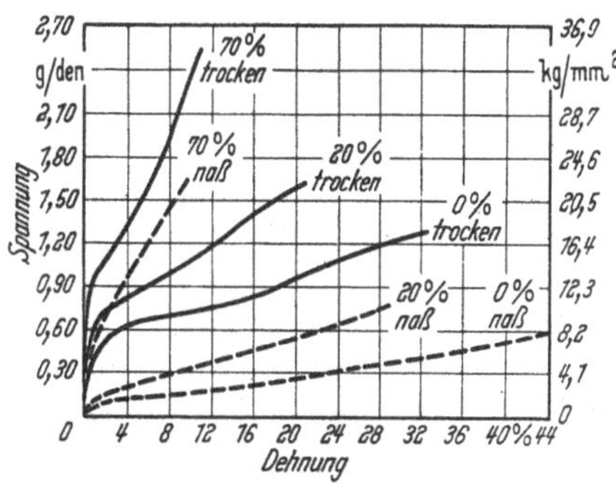

Abb. VII, 10. Abhängigkeit der mechanischen Eigenschaften von Viscosereyon vom Streckgrad. (Nach HOUWINK.)

Die Kristallisation bedeutet naturgemäß eine zusätzliche starke Herabsetzung der Beweglichkeit und damit ein weiteres Ansteigen des statischen Elastizitätsmoduls durch Erhöhung der gegenseitigen potentiellen Energie, ebenso aber auch des Elastizitätsmoduls der bleibenden Dehnung durch Verkürzung der Molekülketten zwischen den kristallinen Bereichen. Jedoch ist der Einfluß von Kristallisation und Orientierung in den meisten Fällen nicht zu trennen. Wie die Spannungs-Dehnungs-Kurve des Kautschuks zeigt, unterstützt die Kristallisation die Verlängerung, so daß die Kurve nach einsetzender Kristallisation fast horizontal verläuft. Erst wenn die Kristallisation nahezu vollendet ist, biegt die Kurve steil nach oben um und zeigt so die eingetretene Verfestigung an. Näheres darüber siehe Kap. V. Man darf aber nicht jede Zunahme der Steigung des blockiert-elastischen Kurvenastes als Kristal-

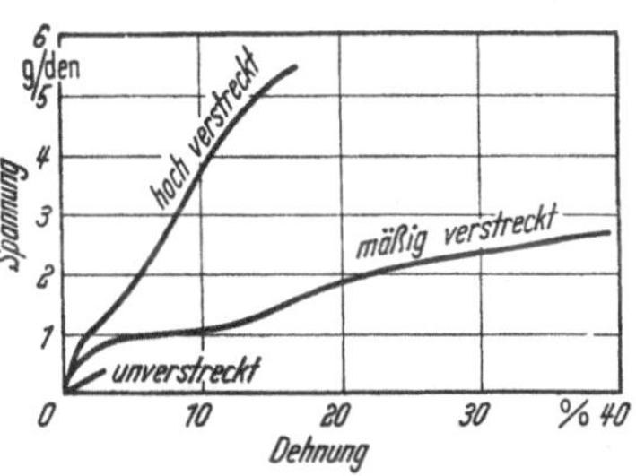

Abb. VII, 11. Einfluß der Orientierung auf die Spannungs-Dehnungs-Eigenschaften von Dacronfasern. (Nach RAY.)

[1] HOUWINK, R.: Grundriß der Technologie der synthetischen Hochmolekularen, Akad. Verlagsgesellsch., Leipzig 1952.
[2] RAY jun., L. G.: Text. Res. J. **22**, 144 (1952).

lisation deuten. Auch die Parallellagerung und Verdichtung der Ketten hat schon eine ähnliche Wirkung. Das wird in dem Fall der Cellulose klar (siehe oben Abb. VII,9 und Abb. VII,10), bei der die Verstreckung die Kristallisation nicht verändert.

Abb. VII,12. Spannungs-Dehnungs-Kurven von Terylen. (Nach MARSHALL u. WHINFIELD.)

Von den synthetischen Fasern ist bisher nur bei Terylen die Existenz sehr verschiedener Kristallinitäten durch Zahlen belegt. Diese reichen von 33–63% und geben, im Verein mit den unterschiedlichen Orientierungen, zu sehr verschiedenen Spannungs-Dehnungs-Kurven Anlaß. Die Abb. VII,12, die die Spannungs-Dehnungs-Kurven in der Darstellung von MARSHALL und WHINFIELD[1] wiedergibt, läßt dazu auf jeder Kurve drei mit scharfen Knicken gegeneinander abgesetzte Abschnitte erkennen. Die Knicke rühren daher, daß die Einfriertemperatur des Terylens +80°C beträgt, so daß die Spannungs-Dehnungs-Kurven im glasartigen Zustand aufgenommen wurden. Bis 1,25% erfolgt die Streckung durch Biegung der Valenzen in den kristallinen Gebieten ohne weiterreichende Gitterstörungen. Bis etwa 6% wirkt derselbe Mechanismus, jetzt aber von starken Gitterstörungen begleitet. Oberhalb 6% setzt dann das Nachgeben des Materials ein, das mit einer ständigen Zerstörung und Neubildung des molekularen Gefüges hier bezüglich der primären und sekundären Haftpunkte verbunden ist. In dem Charakter der Kurven zeigt sich die Wirkung eingefrorener Spannungen; nach dem Ausschrumpfen der Terylenfäden in kochendem Wasser sind die Kurven weicher geworden und gleichen in ihrem Charakter denen der anderen synthetischen Fasern, wie sie Abb. VII,11 und VII,13 zeigen.

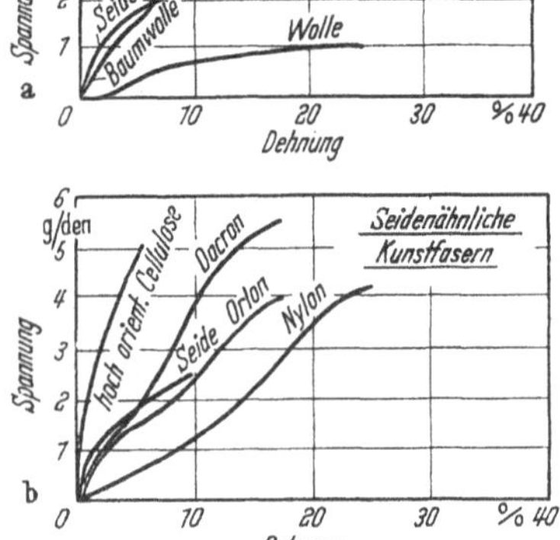

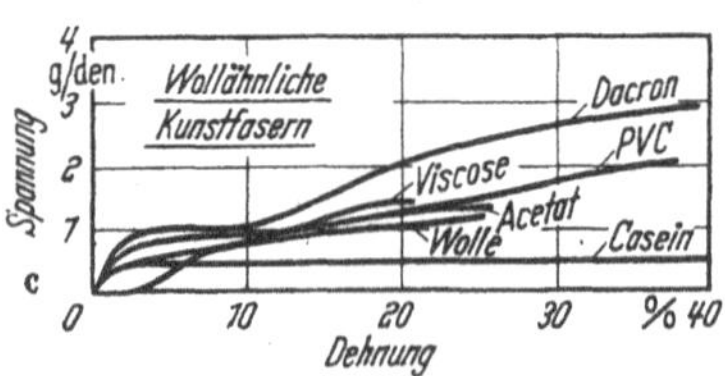

Abb. VII,13. Spannungs-Dehnungs-Kurven. (Nach RAY.) a) natürlicher Fasern; b) seidenähnlicher synthetischer Fasern; c) wollähnlicher synthetischer Fasern.

4. Seidenähnliche und wollähnliche Kurven.

In dem Teilbild 13a sind nach RAY die Spannungs-Dehnungs-Kurven der nativen Fasern zusammengestellt. Sie zerfallen in zwei Gruppen, von denen die eine die Seide und die Baumwolle, die andere die Wolle um-

[1] MARSHALL, I. u. J. R. WHINFIELD, siehe R. HILL: Fibres from Synthetic Polymers, Elsevier Publ. Co. Inc., Amsterdam 1953.

faßt. Die beiden ersten Kurven verlaufen ziemlich steil und endigen bei mäßiger Spannung und kleiner Dehnung. Die Kurve von Wolle verläuft mit wesentlich geringerer Steigung und erreicht eine viel höhere Bruchdehnung in Verbindung mit einer kleineren Bruchspannung. Entsprechend teilt Ray auch die synthetischen Fasern in seidenähnliche (silklike) und wollähnliche (wool-like) ein.

Zu den seidenähnlichen synthetischen Fasern (Abb. VII, 13 b) gehören 6,6-Nylon und 6-Nylon, ferner Orlon und Terylen bzw. Dacron. Hochorientierte Viscose hat eine steilere Kurve als Seide und Baumwolle und entspricht diesbezüglich etwa der nativen Ramie. Wollähnliche synthetische Fasern (Abb. VII, 13 c) sind die Caseinfasern[1], ferner Acetatreyon und Viscosereyon, weiter Polyvinylchlorid[1] und wieder Terylen. Das letztere findet sich also in beiden Gruppen, und auch Orlon kann mit verschiedenen Verstreckungen, mit niedrigen und hohen Graden von Orientierung und Kristallisation und je nach dem mit wollähnlichem oder seidenähnlichem Charakter erhalten werden.

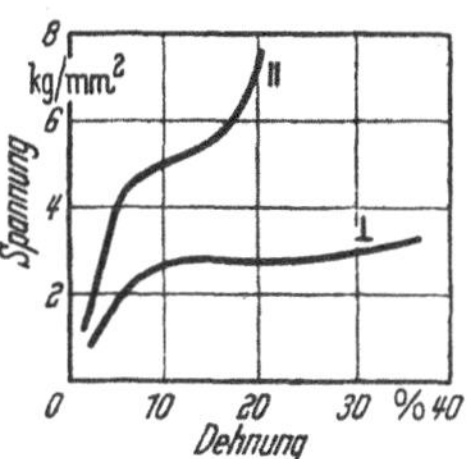

Abb. VII, 14. Spannungs-Dehnungs-Kurven von orientiertem Cellophan, parallel und senkrecht zur Streckrichtung aufgenommen. (Nach Houwink.)

Alle diese Kurven gelten für die Längsdehnung. Für die Querbeanspruchung ist das Verhalten ein anderes. Als Beispiel dafür mögen die Spannungs-Dehnungs-Kurven von verstrecktem Cellophan gelten, die parallel und senkrecht zur Richtung der Verstreckung aufgenommen worden sind. In dieser dem Buch von Houwink entnommenen Abb. VII, 14 ist die Zugspannung auf den jeweiligen Querschnitt bezogen. Daraus resultiert der steile Anstieg am Ende der Spannungs-Dehnungs-Kurve für die Längsbeanspruchung. Bei Querbeanspruchung liegt die untere Grenze der quasipermanenten Deformation niedriger; beide Kurvenabschnitte sind flacher, die Bruchspannung geringer, die Bruchdehnung größer.

§ 55. Elastizitätsmoduln.

Der Elastizitätsmodul ist definiert durch das Verhältnis zwischen der Zugspannung und der Dehnung, die sie hervorbringt, bzw. dem Spannungszuwachs und dem entsprechenden Dehnungszuwachs. Danach erlauben die Neigungen der beiden Abschnitte der Kraft-Dehnungs-Kurven eine Abschätzung der zugehörigen Elastizitätsmoduln für den Energiemechanismus und den Entropiemechanismus der Deformation. Die Spannungs-Dehnungs-Kurven der verschieden stark verstreckten Celluloseacetat- und Celluloseregeneratfasern (siehe oben Abb. VII, 9) und etwa auch der Terylenfasern (siehe oben Abb. VII, 11) lassen sich durch einen geknickten Linienzug annähern. Man kann danach hier von je einem *mittleren* Modul für die *elastische* und die *blockiert-elastische Deformation* sprechen. Beide Moduln sind um so höher, je besser die Orientierung der untersuchten Fasern ist. Dabei verschiebt sich auch die Grenze zwischen

[1] Die Kurven in Abb. 13 c für Casein- und Polyvinylchloridfasern sind dem genannten Buch von Houwink entnommen.

den beiden Deformationsmechanismen nach oben. Bei den Cellulosefasern ist das sicherlich eine Wirkung nur der Orientierung und Verdichtung, weil Unterschiede in der Kristallinität bei ihnen noch in keinem Falle nachgewiesen werden konnten. Beim Terylen dagegen kommt auch eine Erhöhung der Kristallisation und damit eine Neubildung primärer Haftpunkte neben den sekundären in Betracht.

a) Statischer und dynamischer Anfangselastizitätsmodul.

Zur zahlenmäßigen Bestimmung des Elastizitätsmoduls für den Energieprozeß zieht man die Tangente im Ursprung der Spannungs-Dehnungs-Kurve. Man muß dabei aber beachten, daß die Anfangssteigung mit wachsender Belastungsgeschwindigkeit zunimmt. Daran ist erkenntlich, daß schon hier nicht nur die reine Energieelastizität wirksam ist. Man kann daher nur Werte vergleichen, die mit gleicher Belastungsgeschwindigkeit aufgenommen worden sind, und gibt deshalb der *dynamischen* Messung mit Wechselbelastung bekannter Frequenz den Vorzug. Bei genügend hohen Frequenzen stellt sich dann ein oberer Grenzwert des *dynamischen* Anfangselastizitätsmoduls ein, der für alle gleichartigen Fasern, also Cellulosefasern einerseits oder Polyamidfasern andrerseits, in einem konstanten Verhältnis zum statischen Elastizitätsmodul steht. Dieses Verhältnis ist von DE VRIES[1] bei einer Frequenz von 8800 Hz für alle möglichen Cellulosefasern übereinstimmend zu 3/2 bestimmt worden. Abb. VII, 15 zeigt die Beziehung beider Moduln nach seinen Messungen. Dasselbe Verhältnis finden MESKAT und ROSENBERG bei vollsynthetischen Fasern schon bei 25 Hz (s. Abschnitt VII, C dieses Kapitels).

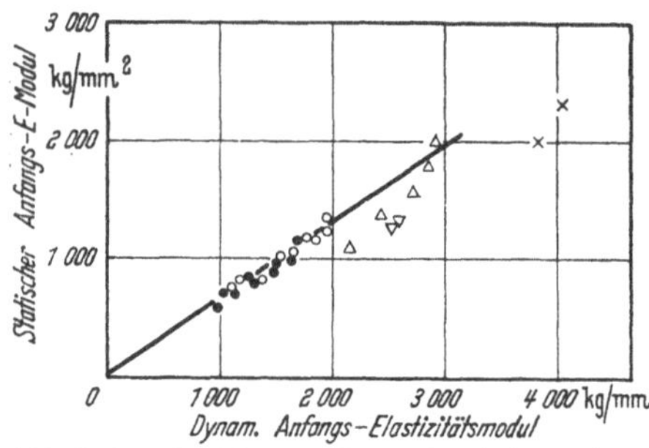

Abb. VII, 15. Beziehung zwischen den Anfangsneigungen der Spannungs-Dehnungs-Kurven und den dynamischen Anfangselastizitätsmoduln bei der Frequenz 8880 Hz. (Nach DE VRIES.) Kreise: normale Viscoseseide niedr. und mittl. Festigkeit; Dreiecke: Viscoseversuchsseide hoher Festigkeit; Kreuze: hochfeste verseifte Acetatseide Fortisan.

Die folgende Tab. VII, 2 gibt eine Zusammenstellung der dynamischen Anfangselastizitätsmoduln von Cellulosefasern; die Werte für die Regeneratfasern sind von DE VRIES[1] gemessen worden, die für die nativen Fasern angegebenen Zahlen sind Mittelwerte nach den Messungen von DE VRIES und von MEYER und LOTMAR[2]. Bei hochorientierten Flachs- und Hanffasern fanden LOTMAR, MEYER und DE VRIES dynamische Anfangsmoduln von 10200 und 10500 kg/mm². Den höchsten Wert verzeichnen MEYER und LOTMAR mit 11200 kg/mm² bei einer Flachsfaser, die im nassen Zustand gestreckt und unter Spannung getrocknet worden war.

Über die statischen Anfangselastizitätsmoduln der synthetischen

[1] VRIES, H. DE: Diss. Delft 1953.

[2] MEYER, K. H. u. W. LOTMAR: Helv. chim. Acta **19**, 68 (1936).

Tabelle VII, 2. *Dynamische Anfangselastizitätsmoduln bei 9000 Hz.*

Faser	Dynamischer Anfangselastizitätsmodul kg/mm²
Isotrope Viscose-Modellfäden	560
Gestreckte Viscose-Modellfäden	700—3400
Viscosereyon je nach Streckung	800—2000
Chemiekupferseide	2000
Hochfeste Viscose-Versuchsseide	2000—3100
Lilienfeld-Seide	3100—3600
Hochfeste verseifte Acetatstreckseide	4000
Baumwolle	5000
Ramie	5100
Hanf	7200
Flachs	7800

Fasern gibt die graphische Zusammenstellung von RAY[1] Auskunft (Abb. VII, 16). Die Breite der Bereiche ist durch die Variabilität von Orientierung und Kristallisation gegeben. Wieder fällt Terylen (Dacron) durch die Größe seiner Variationsbreite auf, aber auch Orlon und ebenso Viscosereyon können der Veränderlichkeit ihrer Orientierung entsprechend mit recht verschiedenen Elastizitätsmoduln hergestellt werden. (Bei den Acetatfäden ist nur die handelsübliche Variation berücksichtigt, nicht die durch Streckung in gequollenem Zustand erreichbare, die der Abb. VII, 9 zugrunde liegt).

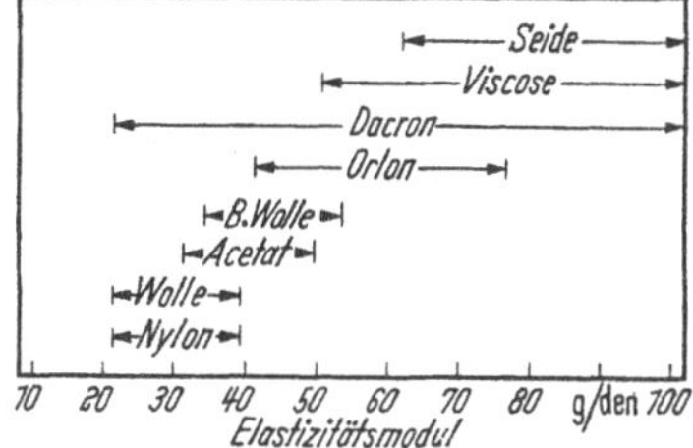

Abb. VII, 16. Bereich der Elastizitätsmoduln von natürlichen und künstlichen Fasern. (Nach RAY.)

Interessant ist auch die Reihenfolge der Elastizitätsmoduln. Ihr entspricht von links unten nach rechts oben eine abnehmende Beweglichkeit bzw. eine zunehmende Packungsdichte der Kettenmoleküle. Insbesondere kann auf den niedrigen Elastizitätsmodul von 6,6-Nylon hingewiesen werden, der völlig dem der Wolle entspricht.

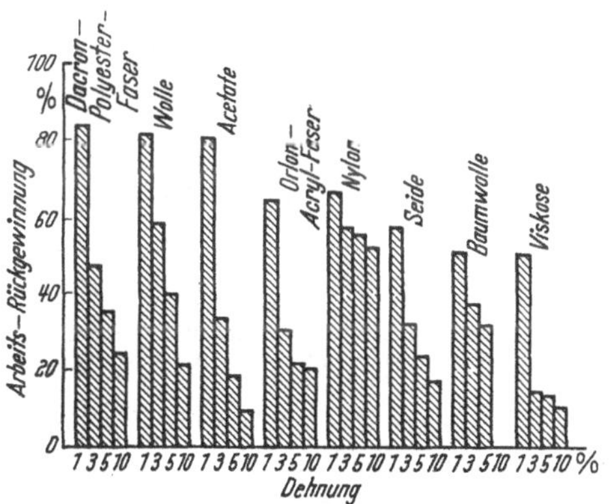

Abb. VII, 17. Arbeitsrückgaben verschiedener Faserstoffe bei 1-, 3-, 5- und 10%iger Dehnung. (Nach RAY.)

Andererseits ist die elastische Erholungsfähigkeit bei 6,6-Nylon wesentlich höher als bei Wolle und die höchste aller bekannten Fasern überhaupt. In Abb. VII, 17 sind nach RAY[1] die Arbeitsrückgaben (work recoveries) nach 1-, 3-, 5- und 10%iger Dehnung für eine Reihe von natürlichen und künstlichen Fasern aufgetragen. Wie man sieht, nimmt

[1] Siehe S. 439, Fußnote 2.

Nylon bezüglich des Erholungsvermögens eine einzigartige Stellung ein; bei 10%-iger Dehnung sind hier noch mehr als 50% umkehrbar gegenüber nur 20% bei Wolle. Wie schon oben ausgeführt wurde, haben die Molekülketten von Nylon eine weitgehende Mikrofreiheit, so daß die Entknäuelung beim Strecken (Entropiemechanismus) hier leicht und weitgehend umkehrbar ist. Darüber kann der Anfangselastizitätsmodul (Energiemechanismus) natürlich keine Auskunft geben.

Bei Terylen ist der Anfangsmodul wesentlich höher als bei Nylon. Das fällt besonders auf, wenn man unter entsprechender Einstellung von Kristallisation und Orientierung eine Terylenfaser herstellt, bei der der Endpunkt der Spannungs-Dehnungs-Kurve und damit die Werte von Bruchspannung und Bruchdehnung mit denen von Nylon praktisch zusammenfallen. In diesem Falle ist, wie Abb. VII, 18 zeigt, der anfängliche Verlauf beider Kurven völlig verschieden, die Anfangssteigung bei Terylen vielmals höher. Das liegt wieder daran, daß man bei Terylen bei Zimmertemperatur schon weit unter der Einfriertemperatur liegt, während man sich bei 6,6-Nylon offenbar noch darüber befindet.

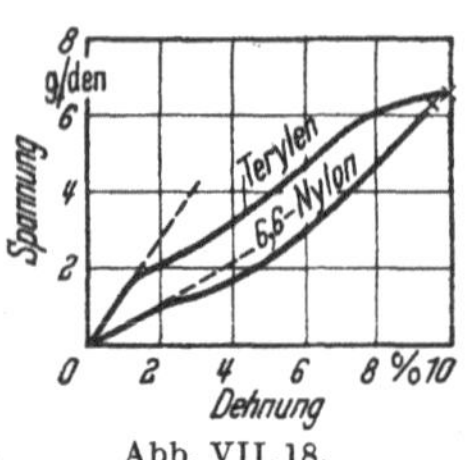

Abb. VII, 18. Spannungs-Dehnungs-Kurven von 6,6-Nylon und Terylen. (Aus HILL.)

Wenn der Elastizitätsmodul des Energiemechanismus und damit die Anfangssteigung der Kurve mit der Steifheit der Kettenmoleküle, mit der Zahl und der Verteilung der Anziehungskräfte zwischen den Ketten, mit dem Umfang der Kristallisation und mit dem der Streckung, die neue Haftpunkte zwischen den Ketten bildet, zunimmt, so wirkt die steigende Temperatur alledem entgegen. Der Anfangsmodul muß daher mit steigender Temperatur sinken und mit fallender Temperatur steigen.

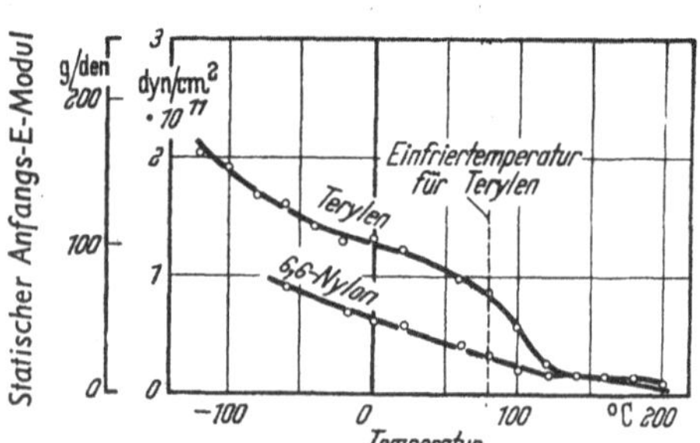

Abb. VII, 19. Temperaturabhängigkeit der Elastizitätsmoduln von Terylen und 6,6-Nylon. (Aus HILL.)

Ein eindrucksvolles Beispiel dafür liefert die Gegenüberstellung der Temperaturabhängigkeit der Anfangselastizitätsmoduln von Terylen und 6,6-Nylon (Abb. VII, 19). Oberhalb 150° C stimmen die Moduln beider Fasern überein. Sie nehmen mit fallender Temperatur zunächst auch gleichmäßig zu, doch zeigt Terylen zwischen 110 und 80° C einen steileren Anstieg, der dem Einfriervorgang entspricht. Unterhalb 80° C laufen beide Kurven wieder parallel, doch liegt der Modul von Terylen jetzt um etwa 50 g/den über dem von 6,6-Nylon.

b) Elastizitätsmodul des Entropiemechanismus.

1. Abhängigkeit von der Dehnung.

Die Erhöhung des E-Moduls des dreidimensionalen Netzwerkes der Kettenmoleküle durch die Dehnung kommt erst in dem Elastizitätsmodul des *Entropiemechanismus* zum Ausdruck. Die Abb. VII, 20 gibt den größten Teil des Materials über den Zusammenhang zwischen dem dynamischen Modul und der Verstreckung aus der Dissertation von DE VRIES wieder. Es handelt sich dabei um Viscosereyon, Lilienfeldreyon, Acetatreyon (verseift) und Modellfäden. Alle Kurven bilden eine Schar und schneiden sich nicht. Bis zu einem kritischen Wert γ_c der Dehnung ist der Modul annähernd konstant (E_c). Oberhalb dieses Wertes steigt er proportional mit der Dehnung merkbar an. Der kritische Dehnungswert γ_c liegt etwas höher als der Knickpunkt der Spannungs-Dehnungs-Kurve; die Spannung stimmt an beiden Stellen jedoch nahezu überein.

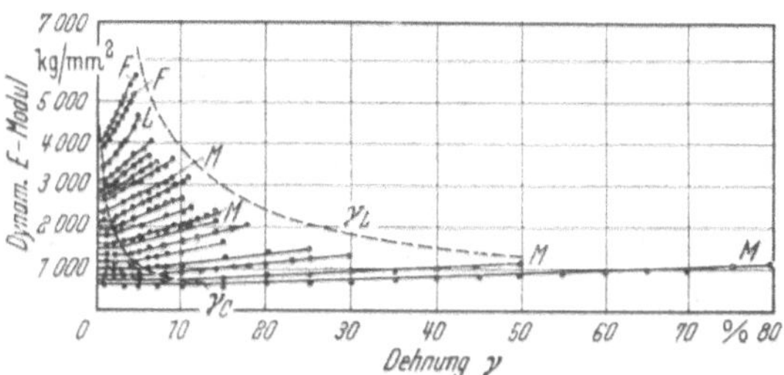

Abb. VII, 20. Dynamischer Elastizitätsmodul E verschiedener Celluloseregeneratfasern in Abhängigkeit von ihrer Dehnung γ. Unbezeichnet: Viscosereyon; M: Modellfäden, L: Lilienfeldreyon, F: Acetatreyon, verseift. (Nach DE VRIES.)

Durch seine eindeutige Markierung im Verlaufe des dynamischen Elastizitätsmoduls mit der Dehnung kann der Dehnungswert γ_c aber mit größerem Recht als die untere Grenze der quasipermanenten Dehnung angesehen werden.

Im zweiten Dehnungsabschnitt (Entropiemechanismus) besteht aber ein grundsätzlicher Unterschied im Verhalten gegen eine statische oder dynamische Beanspruchung; denn der Anstieg des dynamischen Moduls findet eine Parallele auf der statischen Seite nur im nassen Zustand mit seiner hyperbolisch ansteigenden Spannungs-Dehnungs-Kurve. Im trokkenen Zustand zeigt die Spannungs-Dehnungs-Kurve hier eine nur wenig veränderliche Steigung und damit einen nahezu konstanten Modul. Offenbar tritt durch die dynamische Belastung eine Lockerung des Gefüges ein, die dem Effekt der Benetzung ähnlich ist. Die Steigung des dynamischen Moduls ist um so steiler, je höher sein Anfangswert E_c liegt. In dem Maße, in dem die Fäden beim Spinnen verstreckt worden sind, findet sich also einmal der Anfangswert E_c, vor allen Dingen aber der folgende Anstieg des dynamischen Moduls beim Dehnungsversuch erhöht. Diese Beziehung zwischen dem dynamischen Modul und der Dehnung ist im Prinzip umkehrbar. Der Anfangswert kann jedoch, wenn die Faser erst einmal bis in das Gebiet der quasipermanenten Dehnung verstreckt worden ist, nicht vollständig wiederhergestellt werden; denn nach der Kontraktion durch Quellung, darauffolgender Rekonditionierung und neuerlicher Verstreckung tritt kein Bereich konstanten Anfangsmoduls wieder auf.

Zum richtigen Vergleich des Verhaltens der beim Spinnen in verschiedenem Maße vorgestreckten Regeneratfäden ist es notwendig, ein Dehnungsmaß einzuführen, das von dem gleichen Zustand ausgeht. DE VRIES geht dabei mit HERMANS[1] von dem isotropen Zustand aus, wie er in den HERMANSschen Modellfäden verwirklicht ist. Wird ein solcher Faden von der Ausgangslänge l_0 im Verhältnis v_0 bis zur Länge $l = v_0 \cdot l_0$ gedehnt, so steigt sein Modul auf $E_0 > E_{\text{iso}}$. Wird nun bei einem anisotropen Faden von der Ausgangslänge l' ein Elastizitätsmodul E_0 gefunden, so wird diesem entsprechend eine isotrope Ursprungslänge l'_0 zugrunde gelegt, die im Verhältnis v_0 kleiner ist als seine Ausgangslänge beim Dehnungsversuch: $l' = v_0 \cdot l'_0$. Eine Dehnung im Verhältnis $v_1 = 1 + \gamma$ auf die Länge $l_1 = v_1 \cdot l' = (1 + \gamma) \cdot l'$ bedeutet dann, bezogen auf den isotropen Zustand, ein Dehnungsverhältnis

$$v = \frac{l_1}{l'_0} = \frac{l'}{l'_0} \cdot \frac{l_1}{l'} = v_0 \cdot v_1 = v_0 (1 + \gamma) . \qquad \text{(VII, 2)}$$

Unter Benutzung des empirischen Zusammenhanges zwischen E und γ findet DE VRIES dann für den Zusammenhang zwischen dem auf den isotropen Zustand bezogenen Dehnungsgrad v und dem dynamischen Elastizitätsmodul E die Beziehung

$$\ln v = C \cdot (1/E_{\text{iso}} - 1/E) , \qquad \text{(VII, 3)}$$

wobei definitionsgemäß dem Werte $v = 1$ der Elastizitätsmodul $E = E_{\text{iso}}$ entspricht. In einem Diagramm, das die *Nachgiebigkeit* oder *compliance* $1/E$ als Abszisse und den sogenannten natürlichen Dehnungsgrad $\ln v$ als Ordinate enthält, wird also eine Gerade mit der Steigung $1/C$ gegen die Abszisse erhalten, die die Ordinate ($\ln v = 0$) im Werte $1/E_{\text{iso}}$ schneidet. Unter Ausnutzung der additiven Zusammensetzung des natürlichen Dehnungsgrades $\ln v = \ln v_0 + \ln (1 + \gamma)$ gemäß Gl. VII, 2 kann man auch

$$\ln (1 + \gamma) = C \cdot (1/E_0 - 1/E) \qquad \text{(VII, 3a)}$$

schreiben und die Meßwerte $1/E$ über $\ln (1 + \gamma)$ auftragen. E_0 bezeichnet den rückwärtigen Schnittpunkt der mit der Dehnung γ linear ansteigenden E-Kurvemit der Ordinatenachse ($\gamma = 0$). Mit dem differentiellen Elastizitätsmodul E läßt sich übrigens auch das HOOKEsche Gesetz in analoger Weise schreiben:

$$\ln (1 + \gamma) = (\sigma - \sigma_0)/E . \qquad \text{(VII, 4)}$$

Abb. VII, 21 zeigt das Material von DE VRIES in dieser Darstellung. Der lineare Zusammenhang zwischen der compliance $1/E$ und dem natürlichen Dehnungsgrad $\ln (1 + \gamma)$ findet sich danach für alle untersuchten Fäden gut bestätigt. Die abfallenden Kurven verlaufen sämtlich sehr nahe parallel zueinander und lassen so erkennen, daß die Konstante C bei allen Hydratcellulosen nur wenig variiert. Das ist ein ähnlicher Hinweis auf die weitgehend übereinstimmende Struktur ihrer Netzwerke wie der Befund von HERMANS und WEIDINGER[2], daß die Cellulosefasern auch für

[1] HERMANS, P. H.: Physics and Chemistry of Cellulose Fibres, Elsevier Publ. Co. Inc., Amsterdam 1948.

[2] HERMANS, P. H. u. A. WEIDINGER: J. Polymer Sci. 4, 135 (1949).

sehr verschiedene Orientierungen praktisch denselben kristallinen Anteil besitzen. Jedoch darf man C nicht als ein Maß für den kristallinen Anteil betrachten, sein Wert wird vielmehr maßgebend durch die Kohäsionskräfte bestimmt, die in den nichtkristallinen Gebieten der Fasern wirken.

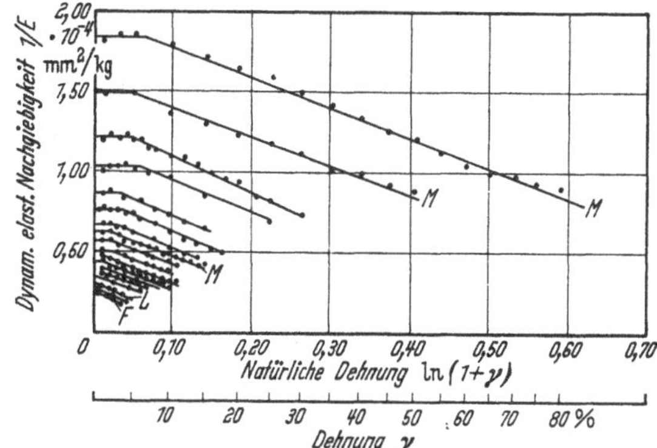

Abb. VII, 21. Dynamische elastische Nachgiebigkeit (compliance) $1/E$ derselben Celluloseregeneratfäden wie in Abb. VII,20 als Funktion ihrer natürlichen Dehnung $\ln(1+\gamma)$. (Nach DE VRIES.)

Im Mittel findet sich bei den Cellulosefasern $C = 580$ kg/mm² und $E_{iso} = 510$ kg/mm². Daraus ergibt sich unter Benutzung des von MEYER und LOTMAR[1] berechneten Maximalwertes des Elastizitätsmoduls für Cellulosefasern ($E_{max} = 12000$ kg/mm²) ein durchaus vernünftiger Wert für den maximalen natürlichen Dehnungsgrad, nämlich $\ln v_{max} = 1{,}09$ oder $v_{max} = (1+\gamma_{max}) = 3$ bzw. $\gamma_{max} = 200\%$.

2. *Abhängigkeit von der Spannung.*

Von WORK[2] existieren Messungen des dynamischen Elastizitätsmoduls an zwei Serien verschieden stark verstreckter Celluloseacetat- und Celluloseregeneratfäden. Da hier die Spannung angegeben wird, bei der die Fäden gemessen wurden, und nicht ihre Dehnung, zeigen die compliance-Kurven einen etwas anderen Verlauf. Sie lassen aber bei kleinen Spannungen d. h. kleinen Dehnungen ebenfalls ein Gebiet konstanter Nachgiebigkeit erkennen, das um so höher liegt, je größer die Orientierung der Fäden ist (siehe Abb. VII, 22). Die Zahlenwerte für das unverstreckte und das am stärksten verstreckte Material verhalten sich in beiden Serien wie 3,5 : 1.

Bei weiterwachsender Vorspannung nimmt die elastische Nachgiebigkeit dann ab, und zwar um so langsamer, je höher sie im ungespannten Zustand bereits war. Dabei fällt auf, daß in der Reihe der Celluloseacetatfäden der obere Teil jeder Kurve einer Probe mit niedrigerer Orientierung dazu neigt, die Kurve für die nächsthöhere Orientierung zu überlappen, während bei den regenerierten Cellulosen die benachbarten Kurven sich nur nähern und abbrechen, bevor eine Überlappung zustande kommt.

---

[1] K. H. MEYER u. W. LOTMAR, Helv. Chim. Acta **19**, 68 (1936).
[2] R. W. WORK, Text. Res. J. **19**, 381 (1949).

Das hängt wahrscheinlich mit der niedrigeren Kristallinität der Acetatfäden zusammen und stellt eine neue Technik zu ihrer Erfassung dar. Die Enden der Kurven, die die Nachgiebigkeiten am Bruchpunkt angeben, liegen in der Serie der Cellulosegarne durchweg bei kleineren Nachgiebigkeiten oder größeren Moduln als in der Reihe der Celluloseacetatgarne. Besonders groß ist dieser Unterschied bei dem nichtorientierten Material. Darin kommt wieder die geringere Kristallinität der Acetatfäden zum Ausdruck, die auf die störende Wirkung der Seitenketten zurückgeht. Entsprechende Verminderungen zeigen auch die Bruchspannungen (siehe unten).

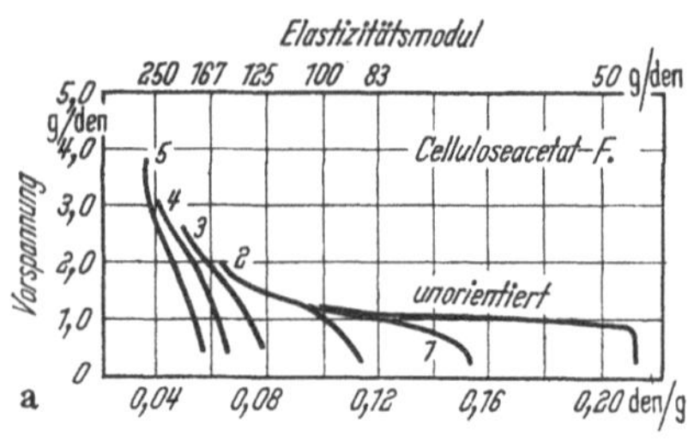

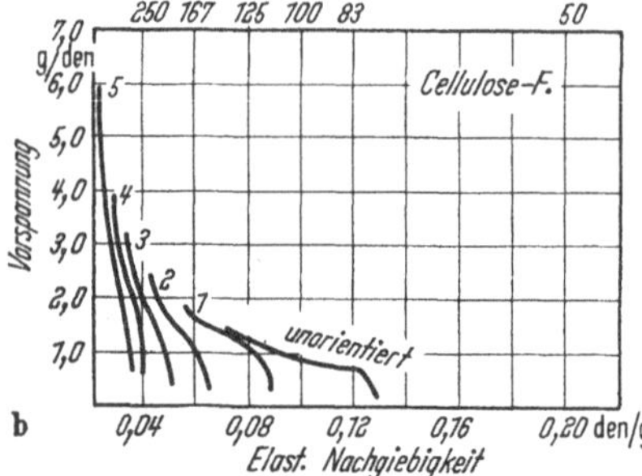

Abb. VII, 22. Zusammenhang zwischen der elastischen Nachgiebigkeit für eine Wechselbelastung mit Schallfrequenz und der Vorspannung bei verschieden stark verstreckten Celluloseacetatfäden (a) und den daraus durch Verseifung erhaltenen Cellulosehydratfäden (b). (Nach WORK.)

In dem linearen Zusammenhang zwischen der dynamischen elastischen Nachgiebigkeit und der im natürlichen Maße gemessenen Dehnung (siehe Abb. VII, 21) sieht DE VRIES eine für die Deformation eines dreidimensionalen Netzwerkes von Kettenmolekülen nach dem Entknäuelungsmechanismus charakteristische Beziehung. Er stützt diese Auffassung durch den Nachweis, daß die nach den Messungen von TRELOAR[1] an einem vulkanisierten Kautschuk gezeichnete Kurve, die die reziproke Steigung $1/S$ der im Maßstabe der natürlichen Dehnung $\ln(1+\gamma)$ gezeichneten Spannungs-Dehnungs-Kurve im Druck- und Zugbereich der Längendeformation in Abhängigkeit von $\ln(1+\gamma)$ darstellt (Abb. VII, 23), ebenfalls nach anfänglichem fast horizontalem Verlauf einen linearen Abfall zeigt, der sich über den Dehnungsbereich von -35 bis $+200\%$ erstreckt. Die Abweichung oberhalb 200% ist zu erwarten, weil hier die Kristallisation des vulkanisierten Kautschuks einsetzt.

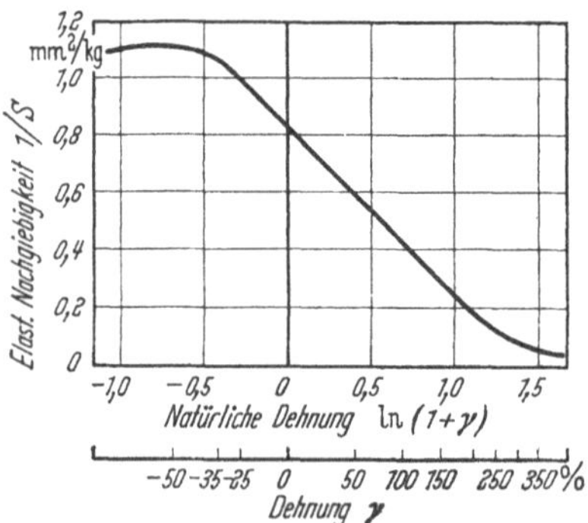

Abb. VII, 23. Zusammenhang zwischen der Nachgiebigkeit $1/E$ und der natürlichen Dehnung $\ln(1+\gamma)$ bei einem vulkanisiertem Kautschuk. (Nach Messungen von TRELOAR, dargestellt von DE VRIES.)

DE VRIES bringt weiter den Parameter C (Gl. VII, 4) mit dem Verhältnis von Zahl und Stärke der primären Valenzen, die ein Kettenmolekül

[1] TRELOAR, L. R. G.: The Physics of Rubber Elasticity, 1949, S. 84 u. 164ff.

zusammenhalten, und der sekundären Bindungen zwischen den benachbarten Ketten in Verbindung. Je höher dieses Verhältnis ist, um so schneller steigt der dynamische Elastizitätsmodul als Folge der Entknäuelung und Verdichtung der Ketten durch die Dehnung an und um so niedriger wird die Konstante C gefunden. E selbst hängt im gleichen Sinne von der Zahl der sekundären Bindungen ab, ist aber im Gegensatz zu C auch eine Funktion der Orientierung. Zieht man außer Cellulose auch andere Hochpolymere mit in Betrachtung ein, so ergibt sich eine erhebliche Variation von C im Bereiche von 11000 bis 500 kg/mm². Die Reihenfolge ist mit abnehmendem C:

Chemiekupferseide (regenerierte Cellulose)

Seide von Bombyx Mori (Kopolymer aus Alanin, Glycin und anderen Proteinen)

Viscosereyon und Acetatreyon, verseift (regenerierte Cellulose)

Orlon (Polyacrylnitril)

Terylen (oder Dacron, Polykondensat aus Terephthalsäure und Äthylenglykol)

Acetatreyon (Celluloseacetat)

Vinyon N (Kopolymer aus Vinylchlorid und Acrylnitril)

6,6-Nylon (Polykondensat aus Adipinsäure und Hexamethylendiamin)

6-Nylon (Perlon, Enkalon, Polykondensat aus ω-Caprolactam).

Es wäre interessant, so bemerkt DE VRIES zu dieser Reihe, die Beziehung der Konstanten C zu den molekularen Kohäsionen von MARK[1] zu untersuchen, die eine ähnliche Reihe ergeben (s. Band I dieses Werkes, Seite 67, Tabelle 27).

3. *Abhängigkeit von der Orientierung* (*Doppelbrechung*)

Quantitative Angaben über den Zusammenhang des dynamischen Elastizitätsmoduls und der Orientierung finden sich für Cellulosefasern ebenfalls bei DE VRIES[2]. Die Orientierung wurde dabei auf dem Wege über die Doppelbrechung bestimmt. Dabei wurde eine neue Methode benutzt, mit der man den Mittelwert der Brechungsindizes über eine Länge von mehreren Metern Faden auf einmal messen kann, so daß die Genauigkeit erheblich erhöht ist. Die Methode beruht auf der Spektralanalyse des Lichtes, das durch ein aus regelmäßigen Windungen der zu untersuchenden Fäden gebildetes Phasengitter abgebeugt wird. Dieses ist in eine Glaszelle mit einer Immersionsflüssigkeit eingetaucht. Wenn nun für eine bestimmte Wellenlänge der Brechungsexponent der Immersionsflüssigkeit mit einem der Hauptindizes der Fäden übereinstimmt, dann fehlt diese Wellenlänge in dem Beugungsspektrum. Da der konstante Wert des dynamischen Elastizitätsmoduls bei beginnender Dehnung keine Parallele im Verlauf der Doppelbrechung hat, wird die Doppelbrechung der ungedehnten Fäden zu dem Wert E_0 des dynamischen Moduls in Beziehung gesetzt, der sich durch rückwärtige Verlängerung der

[1] MARK, H.: Ind. Engng. Chem. **34**, 1343 (1942).

[2] VRIES, H. DE: Diss. Delft 1953.

im Dehnungsversuch gewonnenen E-Kurve ergibt. Der Doppelbrechung Null entspricht der Wert $E_0 = 500$ kg/mm², was gut mit dem oben angegebenen Modul der isotropen Fäden übereinstimmt. Abb. VII, 24 zeigt den Zusammenhang zwischen der Doppelbrechung und den E_0-Werten des dynamischen Elastizitätsmoduls für Viscosereyon und für Modellfäden. Die gestrichelte Kurve entspricht der aus den experimentell bestimmten Abhängigkeiten des Elastizitätsmoduls und der Doppelbrechung vom Dehnungsgrad v abgeleiteten Beziehung

$$\ln\left(1 + \frac{p}{m}\cdot[n_{\parallel} - n_{\perp}]\right) = p \cdot C \cdot (1/E_{\text{iso}} - 1/E) \qquad \text{(VII, 5)}$$

mit den Konstanten $p = 1{,}20$ und $m = 0{,}024$. In dem Falle, daß die lineare Beziehung zwischen der dynamischen Nachgiebigkeit $1/E$ und der natürlichen Dehnung $\ln(1 + \gamma)$ erfüllt ist, ist der dynamische Elastizitätsmodul danach zur Messung der mittleren Orientierung ebenso brauchbar wie die Doppelbrechung.

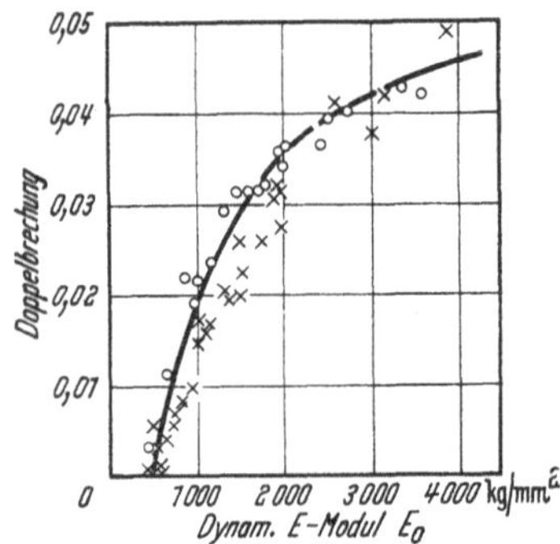

Abb. VII, 24. Zusammenhang zwischen dem dynamischen Elastizitätsmodul E_0 und der Doppelbrechung bei Viscosereyon (○) und Modellfäden (×). (Nach DE VRIES.)

DE VRIES gibt auch einen Ausdruck an, der das aus der Doppelbrechung abgeleitete mittlere Quadrat des Sinus des Orientierungswinkels $\overline{\sin^2\beta}$ mit dem dynamischen Elastizitätsmodul verbindet. Der Ausdruck ist jedoch so kompliziert, daß die vollkommene Interpretation noch offen ist. Von Bedeutung erscheint aber, daß der dynamische Elastizitätsmodul und die Doppelbrechung bzw. die Orientierung beide nur von dem Betrage der Deformation, nicht aber von den im Einzelfalle dazu notwendigen Spannungen abhängig gefunden werden. Nach STEIN und TOBOLSKY[1] weist dieses Verhalten auf einen inneren Ausgleich von örtlichen Spannungen in molekularen Dimensionen hin, den sie als „distortion relaxation" bezeichnen.

§ 56. Bruchspannung und Bruchdehnung.

Bruchspannung und Bruchdehnung sind jeweils durch den Endpunkt der Spannungs-Dehnungs-Kurven gegeben. Das bisher betrachtete Kurvenmaterial läßt also die Abhängigkeit dieser Größen von Temperatur und Feuchtigkeit sowie von Orientierung und Kristallinität bereits erkennen.

a) Rolle der sekundären Bindungen.

Zum Verständnis dieser Zusammenhänge muß man sich erinnern, daß es die seitlichen Nebenvalenzkräfte und vor allen Dingen die Wasserstoff-

[1] STEIN, R. S. u. A. V. TOBOLSKY: Text. Res. J. 18, 201 (1948).

brückenbindungen sind, die den Zusammenhalt der hochpolymeren Stoffe auch für Beanspruchungen in der Kettenrichtung bestimmen. Maßgebend für die VAN DER WAALSschen Kräfte sind Stärke und Dichte der funktionellen Gruppen, ferner der Abstand der Ketten voneinander und ihre gegenseitige Berührungsfläche, für die Wasserstoffbrückenbindungen in erster Linie die in Wechselwirkung stehenden Dipolmomente und ihr Abstand.

Auf die von MARK[1] berechneten Wechselwirkungsenergien einiger Fadenmoleküle wurde auf Seite 449 bereits hingewiesen. Diese Wechselwirkungsenergie nimmt mit der Entfernung schnell ab. Sie ist bei den VAN DER WAALSschen Kräften und den Wasserstoffbrückenbindungen umgekehrt proportional der 6. bzw. 3. Potenz der Entfernung. So kommt z. B. die Vergrößerung des Abstandes durch den Platzbedarf der Seitenketten in der starken Abnahme der Bruchspannung und Zunahme der Bruchdehnung von Fäden aus Celluloseestern mit wachsender C-Zahl deutlich zum Ausdruck (Abb. VII, 25).

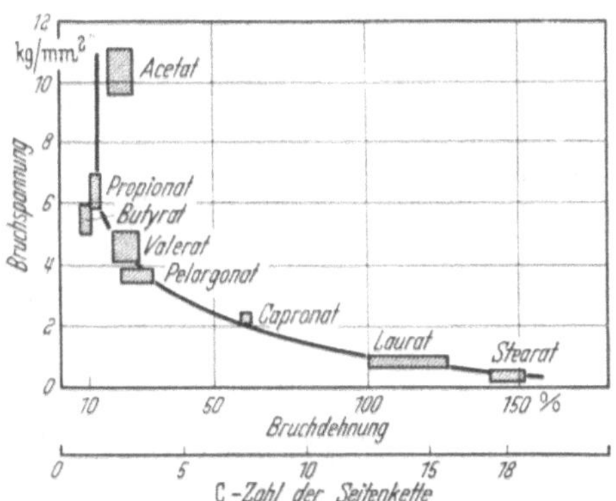

Abb. VII, 25. Abnahme der Bruchspannung und Zunahme der Bruchdehnung von Cellulosederivaten mit wachsender Länge der Seitenketten. (Nach HAGEDORN u. MÖLLER[2].)

Wichtig ist vor allem die Größe der Berührungsfläche benachbarter Kettenmoleküle, weil diese sich durch äußere Mittel, insbesondere durch die Streckung beeinflussen läßt. Sie bestimmt einmal die Stärke der VAN DER WAALSschen Kräfte, genauer der Dispersionskräfte, die direkt proportional mit der Berührungsfläche wächst, und zugleich auch die Zahl der Wasserstoffbrückenbindungen, indem um so mehr funktionelle Gruppen zur Wirkung kommen, je länger die nebeneinander parallel liegenden Kettenteile sind. Die Größe der Berührungsfläche hängt also vor allem von der Parallelorientierung der Kettenmoleküle ab, und das ist der Grund, warum alle hochpolymeren Faserstoffe während des Fabrikationsvorganges verstreckt werden müssen. Bei den nativen Faserstoffen war die Parallelorientierung ein so auffälliges Merkmal, daß sie zur Prägung der Bezeichnung „Faserdiagramm" für ihre charakteristischen Röntgendiagramme führte. MEYER und MARK[3] geben die in Abb. VII, 26 wiedergegebene schematische Skizze für die Bruch-

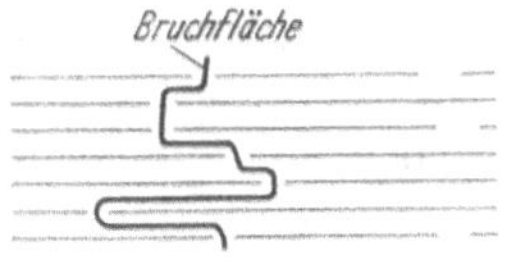

Abb. VII, 26. Schematische Darstellung der Bruchfläche einer hochpolymeren Faser. (Nach MEYER u. MARK.)

---

[1] MARK, H.: Ind. Engng. Chem. **34**, 1343 (1942).
[2] HAGEDORN, M. u. P. MÖLLER: Cellulosechemie **12**, 29 (1931).
[3] MEYER, K. H. u. H. MARK: Der Aufbau der hochpolymeren organischen Stoffe, Akad. Verlagsgesellsch. Leipzig, 1940.

fläche einer hochpolymeren Faser. MARK berechnet weiter die Bruchspannung einer Cellulosefaser für den Idealfall durch ihre ganze Länge durchlaufender Kettenmoleküle zu 40 g/den = 550 kg/mm². Für den Fall nicht durchgehender, aber zur Hälfte ihrer Länge miteinander verzahnter Ketten ergibt die Rechnung 8 g/den und liefert damit die richtige Größenordnung; denn in Ausnahmefällen sind schon 7 g/den erreicht worden.

Ein Beispiel für das Zusammenwirken von Abstand und Berührungsfläche bildet die Abhängigkeit der Bruchspannung und der Bruchdehnung HERMANSscher Modellfäden von der Cellulosekonzentration und dem mittleren Polymerisationsgrad der Viscosen, aus denen sie gesponnen waren (Abb. VII, 27). Es handelt sich dabei um vollkommen isotrope Cellulosefäden, und ihre Festigkeiten und Dehnungen steigen beide mit wachsender Konzentration und wachsendem Polymerisationsgrad an. Die Punkte liegen mit geringer Streuung auf einer Kurve, die die Form der Spannungs-Dehnungs-Kurven dieser Fäden hat. Diese haben für alle verschiedenen Viscosen also die gleiche Form und unterscheiden sich nur durch ihre Endpunkte.

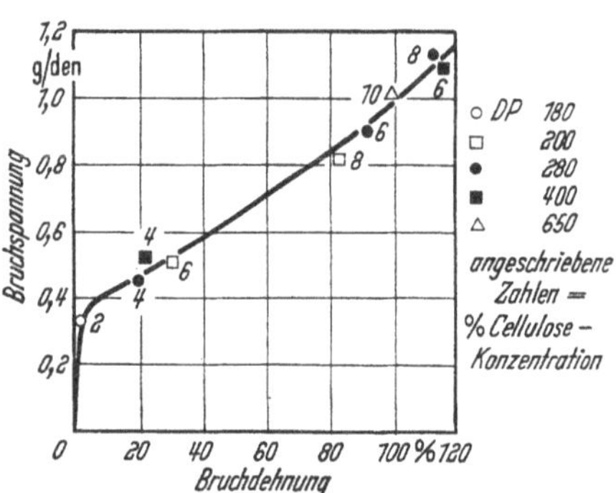

Abb. VII, 27. Zusammenhang zwischen Bruchspannung und Bruchdehnung isotroper HERMANSscher Modellfäden. (Nach HERMANS.) Die eingeschriebenen Zahlen geben die Cellulosekonzentration der Viscose in Gew.-% an.

Auch die Verringerung der Bruchspannung im nassen Zustand bei wasseraufnehmenden Fasern, wie Wolle und Regeneratfaser aus Cellulose (siehe oben Abb. VII, 8 und VII, 10), hat ihre Gründe in Änderungen der Abstände und der Berührungsflächen benachbarter Ketten in den nichtkristallinen Gebieten, die auf die Quellung und Lösung der VAN DER WAALSschen Kräfte bzw. Solvation der Peptidbindungen zurückgehen. Wenn im Gegensatz dazu die nativen Cellulosefasern durch Benetzung keine Verminderung ihrer Bruchspannung erfahren, so liegt das daran, daß die nicht-kristallinen Gebiete hier kürzer sind, und die Kettenmoleküle daher in ihnen nicht endigen. Die Verringerung der Bruchspannungen mit wachsender Temperatur (siehe oben Abb. VII, 6) geht ebenfalls auf die Vergrößerung der seitlichen Abstände und die Verminderung der Wirksamkeit der funktionellen Gruppen, in diesem Falle durch die Wärmebewegung, zurück.

b) Zusammenhang mit der Dehnung.

Die besonders starke Abhängigkeit der Bruchspannung von der Verstreckung geht aus der Abb. VII, 28 hervor, in der die Endpunkte der Spannungs-Dehnungs-Kurven der verschieden stark verstreckten Celluloseacetat- bzw. Celluloseregeneratfasern (aus Abb. VII, 9) aufgetragen worden sind. Vom unverstreckten Zustand (0) bis zur 5. Streckstufe (5) macht die Zunahme der Bruchspannung in beiden Fällen den Faktor 4 aus.

Einen ebensolchen hyperbolischen Zusammenhang erhält man nach DE VRIES[1], wenn man die auf den Bruchpunkt extrapolierten Werte der dynamischen Elastizitätsmoduln E_L verschiedener Celluloseregeneratfäden über der Dehnung aufträgt (gestrichelte Kurve in Abb. VII, 20). Auch der dynamische Anfangselastizitätsmodul E_c zeigt den gleichen Verlauf mit der Dehnung. DE VRIES schreibt dafür

$$E_c \cdot \gamma_c^n = a_c \quad \text{und} \quad E_L \cdot \gamma_L^n = a_L$$

mit $n \approx 2/3$ und $a_c = 150$ kg/mm², $a_L = 800$ kg/mm².

Weiter bestätigt DE VRIES auch den danach zu erwartenden engen Zusammenhang zwischen dem dynamischen Anfangselastizitätsmodul E_c und der Bruchspannung σ_L. Beide Größen finden sich im Bereiche der Bruchspannung 20–80 kg/mm² oder 1,5–6 g/den bei einem großen Untersuchungsmaterial einander proportional, und zwar beträgt die Bruchspannung stets ungefähr 2% des dynamischen Anfangselastizitätsmoduls.

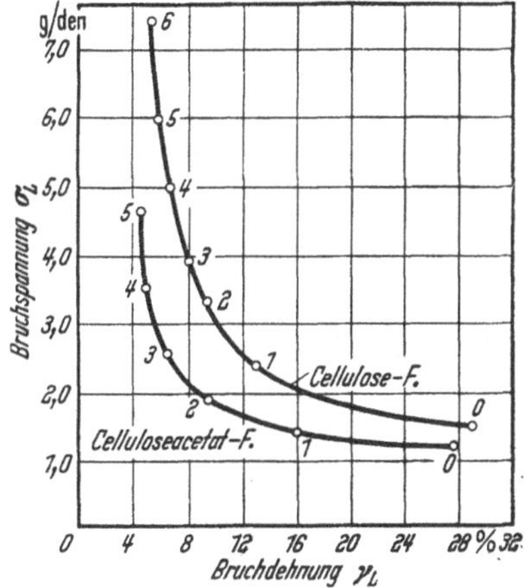

Abb. VII, 28. Bruchspannung und Bruchdehnung von Celluloseacetat- und Celluloseregeneratfäden verschiedener Nachverstreckung. (Nach WORK.)

Noch wirksamer als die bloße Parallellagerung durch die Dehnung sollte natürlich die zusätzliche Kristallisation von neuorientierten Ketten sein. Bei den Celluloseregeneratfasern ist aber als sicher bekannt, daß der kristalline Anteil sich mit der Verstreckung nicht ändert. Es findet vielmehr nur eine Streckung und Verdichtung der Ketten in den nichtkristallinen Gebieten und eine Eindrehung der kristallinen Gebiete in die Dehnungsrichtung statt. Über die Verkleinerung des Abstandes und die Vergrößerung der Berührungsfläche durch die Parallelisierung hinaus würde die Kristallisation den Zusammenhalt durch Einpassung der Seitengruppen in die gegenseitige Potentialmulde noch erhöhen.

Doch überwiegt auch bei gleichzeitiger Kristallisation der Einfluß der Orientierung. So liegt die Erhöhung der Bruchspannung, die beim Kautschuk durch eine Verstreckung von 600% und eine gleichzeitige Kristallisation zu 30% erreicht wird, mit dem Faktor 7 in der gleichen Größenordnung wie bei den nachverstreckten Cellulosefäden der Abb. VII, 28, und auch beim Saran verhalten sich die Bruchspannungen im unverstreckten und verstreckten Zustand wie 1 : 7. Eine gleichsinnige Wirkung wie die Kristallisation hat auch eine Vernetzung in den nichtkristallinen Bereichen, bei der ein Teil der sekundären Kräfte durch primäre ersetzt wird.

[1] H. DE VRIES: Diss. Delft 1953.

Mit der Erhöhung der Bruchspannung durch die Orientierung ist stets eine Verminderung der Bruchdehnung verbunden, (s. Abb. VII, 28) weil durch die Streckung und Orientierung bei der Herstellung ein Teil der Streckbarkeit des Materials vorweggenommen wird. Auch gilt die Erhöhung der Bruchspannung durch die Orientierung nur für die Faserachse. Für eine Belastung senkrecht dazu ist umgekehrt sogar eine Verminderung der Bruchspannung festzustellen.

In der folgenden Tabelle VII, 3 ist eine Zusammenstellung der Bruchspannungen und Bruchdehnungen von natürlichen und synthetischen hochpolymeren Fasern gegeben. Die auftretenden Variationsbreiten beider Größen sind durch Unterschiede in den Beträgen von Orientierung und Kristallinität bestimmt.

Tabelle VII, 3.
Bereiche von Bruchspannung und Bruchdehnung (nach RAY).

Material		Bruchspannung g/den	Bruchdehnung %
6,6-Nylon (normal)	Du Pont de Nemours & Co.	4,5—5,0	25—18
6,6-Nylon (hochfest)	Du Pont de Nemours & Co.	6,5—7,7	20—14
6,6-Nylon (Stapelfaser)	Du Pont de Nemours & Co.	3,8—4,5	37—25
Dacron (5600)	Du Pont de Nemours & Co.	4,4—5,0	22—18
Dacron (5400)	Du Pont de Nemours & Co.	3,0—3,9	40—25
Orlon (Type 81)	Du Pont de Nemours & Co.	4,7—5,2	17—25
Orlon (Type 41)	Du Pont de Nemours & Co.	2,0—2,5	45—20
Viscose (normal)	Du Pont de Nemours & Co.	1,5—2,4	30—15
Viscose (hochfest)	Du Pont de Nemours & Co.	2,4—4,6	20— 9
Acetatreyon	(Celanese Corp.)	1,3—1,5	30—23
Wolle		1,0—1,7	35—25
Baumwolle		2,0—5,0	7— 3
Seide		2,2—4,6	25—10

c) Zusammenhang mit der Orientierung.

1. Bedeutung der Vernetzung (Dehnbarkeit).

Für die Diskussion des Zusammenhanges zwischen Bruchspannung und Orientierung sind je nach dem Zustand, in dem die Verstreckung vor sich geht, zwei Fälle zu unterscheiden. Wenn noch keine Haftpunkte zwischen den Ketten bestehen, also kein zusammenhängendes Netzwerk existiert, so unterliegt das Material einer plastischen Deformation. Dabei kann die Ordnung und Orientierung durch das Strecken theoretisch bis zu einem Idealzustand erhöht und so die Festigkeit gesteigert werden. Diese Streckung erfordert nur kleine Kräfte und gleicht die Orientierung der Ketten aus, weil die am besten orientierten ihren Orientierungsgrad am langsamsten weiter verbessern und einfach gleiten.

Wird dagegen im vernetzten oder auch teilkristallinen Zustand verstreckt, so ist eine größere Kraft notwendig, um in dem zusammenhängenden Netzwerk eine höhere Orientierung zu erzwingen. Dabei werden insbesondere die gut orientierten Ketten noch weiter orientiert mit dem Ergebnis, daß die Spannung stark ansteigt. Die Belastung wird dann nicht mehr gleichmäßig von der Gesamtheit der Ketten getragen und ihre Verteilung ist um so ungünstiger, je mehr Streuung der Grad der

Orientierung zeigt. Der Bruch tritt jetzt bereits ein, ehe die Mehrzahl der Ketten ausgestreckt ist.

Beide Arten von Streckvorgängen können in den verschiedenen Zuständen, die ein Material während des Spinnvorganges durchläuft, nacheinander auftreten. Ihr Vergleich ist für den Fall des Spinnens aus einer Lösung dadurch möglich geworden, daß HERMANS[1] einen auf den trockenen Zustand bezogenen Dehnungsgrad (v_t) abgeleitet hat, der sowohl der Volumenänderung beim Entquellen als auch der Schrumpfung beim Trocknen Rechnung trägt.

Streckt man einen Faden von der Ausgangslänge l_1 auf die Länge l_2 und nimmt er, nachdem er freigegeben ist, die Länge l_3 an, so wird der Dehnungsgrad gemeinhin zu $v_3 = l_3/l_1$ definiert. Findet die Streckung im gequollenen Zustand mit dem Quellungsgrad q_1 statt, und hat der gestreckte Faden dann den kleineren Quellungsgrad q_3 angenommen, so berechnet sich unter Berücksichtigung des Quellungsunterschiedes der äquivalente Dehnungsgrad zu

$$v_a = \left(\frac{q_1}{q_3}\right)^{1/3} \cdot v_3 .$$

Dieser Ausdruck wurde schon von BAULE, KRATKY und TREER[2] in ihrer Theorie der affinen Deformation benutzt. HERMANS bezieht nun die Verstreckungen auf den trockenen Zustand ($q = 1$) und berücksichtigt zugleich das Verhältnis der nassen und trockenen Länge des verstreckten Fadens durch den longitudinalen Schrumpffaktor $\lambda = l_3/l_t$. Er erhält so für den auf den trockenen Zustand bezogenen Dehnungsgrad v_t den Ausdruck:

$$v_t = \frac{v_3}{\lambda} \cdot q_1^{1/3} = 1 + \gamma_t ,$$

wobei γ die prozentuale Dehnung bedeutet.

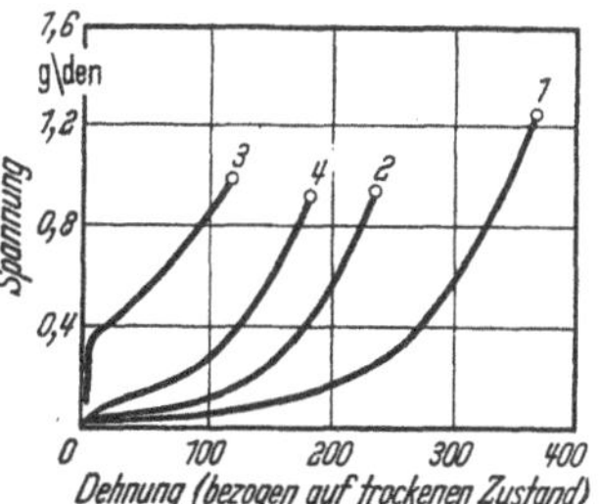

Abb. VII, 29. Spannungs-Dehnungs-Kurven von HERMANSschen Modellfäden, aufgenommen als Xanthogenatfaden (1), frischer Cellulosefaden (2), lufttrockener Faden (3), wiedergequollener Faden (4). (Nach HERMANS.)

Unter Benutzung dieses Dehnungsgrades können in Abb. VII, 29 die Spannungs-Dehnungs-Kurven von HERMANSschen Fäden, die im Xanthogenatzustand (1), im frischen Regeneratzustand (2), im trockenen Regeneratzustand (3) und im wiedergequollenen Zustand (4) aufgenommen wurden, in vergleichbarem Maßstab gegenübergestellt werden. Die Spannung ist dabei auf den jeweiligen Fadenquerschnitt bezogen worden. Da die Fäden sämtlich aus dem gleichen Cellulosegel hergestellt wurden, darf man annehmen, daß dem gleichen Dehnungsgrad (bezogen auf den trockenen Zustand) in jedem Falle nahe auch die gleiche Orientierung entspricht. Die Schnittpunkte der vier Spannungs-Dehnungs-Kurven mit einer bestimmten Abszisse zeigen also an, eine wieviel kleinere Spannung beim Verstrecken im Xanthogenatzustand im Verhältnis zum Verstrecken im vernetzten Regeneratzustand oder gar im trockenen Zustand nötig ist, um die gleiche Orientierung zu erhalten.

Die im trockenen Zustand aufgenommene Kurve unterscheidet sich von den drei anderen durch den von dem Energiemechanismus gelieferten

[1] HERMANS, P. H.: Physics and Chemistry of Cellulose Fibres, Elsevier Publ. Co. Inc., Amsterdam 1948.

[2] BAULE, B., O. KRATKY u. R. TREER: Z. physik. Chem. B **50**, 255 (1941).

steilen Anfangsteil, der mit einem Knick in die Kurve der quasipermanenten Dehnung übergeht. Die im nassen Zustand und insbesondere im Xanthogenatzustand aufgenommenen Kurven zeigen den typischen glatten und nach oben gekrümmten hyperbolischen Verlauf. Daraus leitet HERMANNE[1] an Hand der schematischen Abb. VII, 30 folgende Beziehung zwischen der Zugspannung σ, der Bruchdehnung γ_L und der Dehnung nach dem Entropiemechanismus γ_S ab:

$$\sigma \cdot (\gamma_L - \gamma_S) = \text{const} \cdot T\,.$$

Die Konstante auf der rechten Seite ist dem Charakter der Entropie entsprechend mit der absoluten Temperatur T multipliziert. Ihr Wert hängt von der Beweglichkeit der Kettenglieder, von ihrer Länge und von dem Verhältnis zwischen dem kristallinen und dem amorphen Material ab. HERMANNE schließt aus dieser Beziehung, daß die Deformationsgrenze γ_L eine physikalische Bedeutung haben und mit der Vollendung der Orientierung zusammenhängen muß.

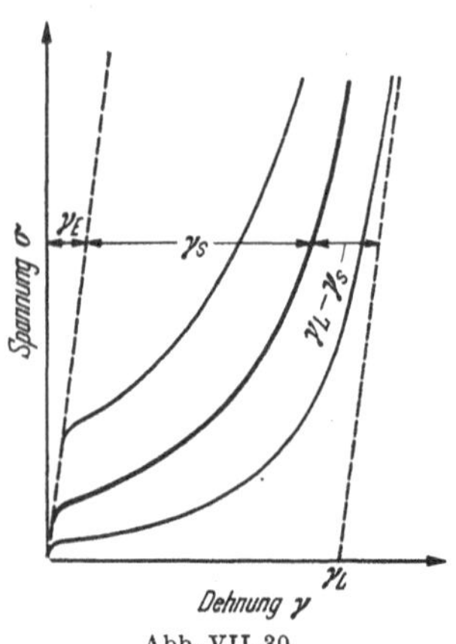

Abb. VII, 30. Schematische Spannungs-Dehnungs-Kurven für den Entropiemechanismus bei Viscosefäden. (Nach HERMANNE.)

Um den Energiemechanismus mit zu berücksichtigen, zieht man von dem Koordinatenursprung aus eine Gerade mit einer Neigung gegen die Ordinatenachse, die dem Elastizitätsmodul des Energieprozesses entspricht. Ihr Schnittpunkt mit der hyperbolischen Kurve des Energieprozesses liefert die untere Grenze σ_d der quasipermanenten Dehnung. Diese rückt mit den hyperbolischen Kurven nach oben, wenn die Konstante der Gleichung wächst, die Beweglichkeit oder Länge der Kettenglieder also abnimmt. So wird die Verschiebung der Elastizitätsgrenze nach oben mit wachsender Zahl der Vernetzungen einschließlich der sekundären Vernetzungen, die durch die Orientierung entstehen, verständlich.

2. *Zusammenhang mit der Endorientierung (Doppelbrechung).*

Durch ähnliche Überlegungen ist HERMANS gleichzeitig und unabhängig von HERMANNE zu Versuchen geführt worden, die den quantitativen Zusammenhang zwischen der Bruchspannung σ_L und der Vollendung der Orientierung an der Deformationsgrenze γ_L aufdeckten. Seine Untersuchungen erstreckten sich auf Modellfäden aus 9 verschiedenen Viscosen, die im Xanthogenatzustand jeweils in verschiedenem Maße vorgestreckt waren. Unter Benutzung des auf den trockenen Zustand bezogenen Dehnungsgrades kann nun der Gesamtbetrag der Dehnung (total v_t) ermittelt werden. Dieser berechnet sich als Produkt aus dem Trockenwerte $v_t = 1 + \gamma_t$ der Verstreckung im Xanthogenatzustand und der Streckung $v_L = 1 + \gamma_L$ im Reißversuch und liefert so wieder ein Maß für die Dehnung, das vom isotropen Zustand des Cellulosegels als Nullpunkt an zählt (s. auch oben S. 446). In diesem Maßstab wird deutlich (Abb. VII, 31), daß mit Hilfe der Vorverstreckung im Xanthogenat-

[1] HERMANNE, L.: Text. Res. J. **19**, 61 (1949).

zustand eine wesentlich höhere Gesamtverstreckung und damit eine größere Bruchspannung erreicht wird als bei der Streckung im trockenen Zustand, die jeweils schon auf der punktierten Kurve endet. Damit wird die technische Bedeutung der Verstreckung in einem frühen Koagulationszustand klar. Je höher die erreichbare Gesamtverstreckung ist, um so höher fällt auch die Bruchspannung aus.

Der Versuch, eine Beziehung zwischen der Bruchspannung und der Orientierung zu finden, schlug fehl, solange HERMANS die Doppelbrechung benutzte, die die Fäden vor dem Bruchversuch hatten. Beim Vergleich der Bruchspannung mit der Doppelbrechung, die der Faden am Bruchpunkt selbst aufweist, fand er aber ein weiteres interessantes Resultat. Die Bruchspannung zeigt sich jetzt in engen Grenzen durch diese Doppelbrechung, d.h. durch den am Bruchpunkt erreichten Orientierungszustand, bestimmt. Dieses Gesetz gilt für alle Viscosen unabhängig von

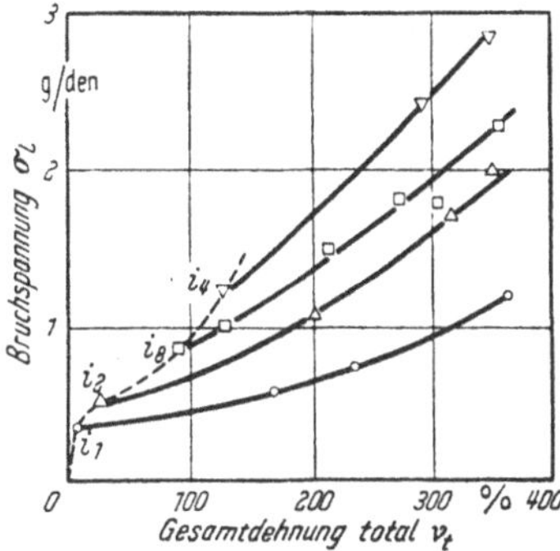

Abb. VII, 31. Zusammenhang zwischen der Bruchspannung σ_L und der Gesamtdehnung total ν_t am Bruchpunkt für Cellulosemodellfäden. (Nach HERMANS.)

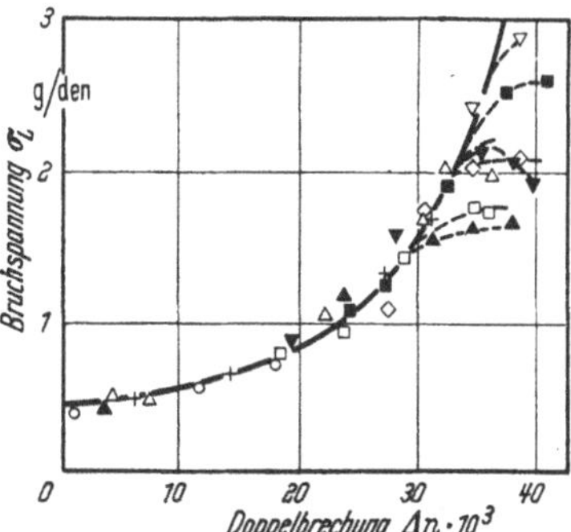

Abb. VII, 32. Spannung und Doppelbrechung am Bruchpunkt für verschieden stark vorgestreckte Modellfäden aus mehreren Viscosen. (Nach HERMANS.)

Konzentration und Polymerisationsgrad (Abb. VII, 32) (mit Ausnahme unangemessen hoher Bruchspannungen) und läßt sich durch die Formel

$$\sigma_L = \frac{\text{const}}{1 - f_0} = \frac{2}{3}\,\frac{\text{const}}{\sin^2 \beta_m}$$

ausdrücken. Darin bedeutet const eine Proportionalitätskonstante, deren Wert sich zu 0,45 g/den bestimmt und damit mit der unteren Grenze der Bruchdehnung der trockenen isotropen Modellfäden übereinstimmt. f_0 ist der optische Orientierungsfaktor, der durch das Verhältnis der gemessenen Doppelbrechung zu der für eine ideal orientierte Faser berechneten (0,050 bei Cellulose) gegeben ist, und $1-f_0$ die Abweichung von dem ideal vollkommenen Orientierungszustand. β_m schließlich bezeichnet den mittleren Winkel der Orientierung, den Orientierungswinkel also eines gedachten einheitlich orientierten Präparates, das dieselbe Doppelbrechung besitzt, wie das zur Untersuchung stehende.

Die aus der Abb. VII, 31 gezogene Folgerung, daß die Faser bis zum Bruchpunkt eine möglichst hohe Gesamtdehnung erreichen soll, wird also jetzt dahin präzisiert, daß ein möglichst hoher Orientierungszustand erreicht werden muß; denn die Bruchspannung findet sich umgekehrt proportional der Abweichung von dem ideal vollkommenen Orientierungszustand. Der Grad der Kristallisation geht in die Formel für die Bruchspannung nicht ein, weil sein Wert bei allen Celluloseregeneratfasern unabhängig von der Verstreckung übereinstimmt. Er beträgt etwa 40%. Näheres siehe Band III dieses Werkes, Kap. VB.

Die Bestimmung des mittleren Orientierungswinkels mit Hilfe der Doppelbrechung hat den Vorteil, daß die Richtungsverteilung der Ketten in beiden, den kristallinen und den nichtkristallinen Gebieten erfaßt wird. Die röntgenographische Messung liefert dagegen nur den mittleren Orientierungswinkel der kristallinen Bereiche und deren Orientierungsfaktor. Näheres darüber siehe Band III dieses Werkes, Kap. V, § 28 b. Im allgemeinen werden der unvollkommeneren Orientierung der Ketten in den nicht-kristallinen Gebieten wegen bei der röntgenographischen Bestimmung der mittlere Orientierungswinkel etwas kleiner und der Orientierungsfaktor entsprechend größer gefunden werden als bei der optischen Messung. Doch vermindert sich dieser Unterschied mit wachsender Orientierung.

3. Zusammenhang mit den röntgenographischen Orientierungsparametern.

Die obige Formel für den Zusammenhang von Bruchspannung und Orientierung enthält nur den Bruttowert der Orientierung in Form des Orientierungsfaktors oder des mittleren Orientierungswinkels. Nach den Feststellungen von KAST und PRIETZSCHK[1] sowie HERMANS und KAST[2] gibt es aber viele Fälle, in denen die Schwankung des Winkels der Kristallitachsen gegen die Faserachse sich gemäß der Beziehung

$$\overline{\sin^2}\beta = \overline{\sin^2}\alpha_0 + \overline{\sin^2}\alpha_3$$

in verschiedener Weise auf die Schwankungen der Winkel α_0 und α_3 der Lote auf den beiden praktisch aufeinander senkrecht stehenden achsenparallelen Flächen A_0 und A_3 gegen den Faserradius verteilt. Die Fläche A_0 zeigt im allgemeinen eine mehr oder weniger ausgeprägte Bevorzugung ihrer Parallelstellung zur Faserachse. Das hängt, wie schon KRATKY und Mitarbeiter[3,4] erkannt haben, mit der Ausbildung der kristallinen Bereiche als flache Blättchen mit A_0 als Blättchenfläche zusammen. Es hat sich aber gezeigt, daß man die auftretenden Unterschiede in dem Grade der Bevorzugung von A_0 nicht mit einer Formänderung in Verbindung bringen kann.

Ähnlich wie SISSON[5] die Parallelstellung der Blättchenflächen zu der Unterlage, auf der ein Cellulosefilm trocknet, auf die Querkräfte der Schrumpfung zurückführt, macht KAST[6] sich die Vorstellung, daß es die bei der Verstreckung eines vernetzten Systems auftretenden Querkräfte sind, die die kristallinen Bereiche mit ihren Blättchenflächen senkrecht zum Faserradius und damit parallel zur Faserachse stellen, während die

[1] KAST, W. u. A. PRIETZSCHK: Kolloid-Z. **114**, 23 (1949).
[2] HERMANS, P. H. u. W. KAST: Kolloid-Z. **121**, 21 (1951).
[3] BAULE, B., O. KRATKY u. A. TREER: Z. physik. Chem. B **50**, 280 (1941).
[4] Näheres siehe auch Band III dieses Werkes, Kap. V, § 27 d.
[5] SISSON, W.: J. physic. Chem. **40**, 343 (1936); **44**, 513 (1940).
[6] KAST, W.: Kolloid-Z. **125**, 45 (1952).

Längskräfte ihre Achsen in die Faserrichtung zwingen. Danach ist also eine bevorzugte Orientierung von A_0 charakteristisch für eine Verstrekkung, die im vernetzten Zustand erfolgt, während im plastischen Zustand die Flächen A_0 und A_3 des Fehlens der Querkräfte wegen beide gleichmäßig orientiert werden. Zur Erfassung dieses Blättcheneffektes[1,2] bzw. der für sie verantwortlichen Vernetzung kann man entweder aus den Halbbreiten α_{0h} und β_h der zugehörigen azimutalen Schwärzungskurven das „Orientierungsverhältnis" α_{0h}/β_h oder nach Integration der Schwärzungskurven das „paratrope Verhältnis" $\overline{\sin^2\alpha_0}/\overline{\sin^2\alpha_3}$ bilden. Die Tatsache, ob ein Streckvorgang mehr im plastischen Zustand oder mehr im vernetzten Zustand ablief, ist danach röntgenographisch noch am fertigen Faden festzustellen: Im ersteren Falle sind die Werte der beiden Orientierungsverhältnisse größer als im letzteren.

Dabei zeigt es sich wieder, daß eine um so größere Dehnbarkeit erhalten wird, je größer der Anteil der Verstreckung ist, der im plastischen Zustand vorgenommen wurde. Denn wie die Abb. VII, 33a für den Fall der Chemiekupferseiden und der verseiften Acetatstreckseiden erkennen läßt, fällt die Bruchdehnung um so höher aus, je größer das Orientierungsverhältnis gefunden wird. KAST und PRIETZSCHK bezeichnen das Orientierungsverhältnis daher als ein Maß für die Qualität des Orientierungszustandes und definieren entsprechend mit Hilfe des Produktes der Qualitätsgröße α_h/β_h und der Intensitätsgröße $1/\beta_h$ der Orientierung den Gütewert α_h/β_h^2 des Orientierungszustandes. Tatsächlich zeigt diese Orientierungsgüte entsprechend den Beziehungen zwischen dem Orientierungsverhältnis und der Bruchdehnung einerseits und zwischen dem Orientierungsbetrage und der Bruchspannung andererseits einen klaren Zusammenhang mit dem als Produkt von Bruchspannung und Bruchdehnung definierten sogenannten Textilfaktor (Abb. VII, 33b).

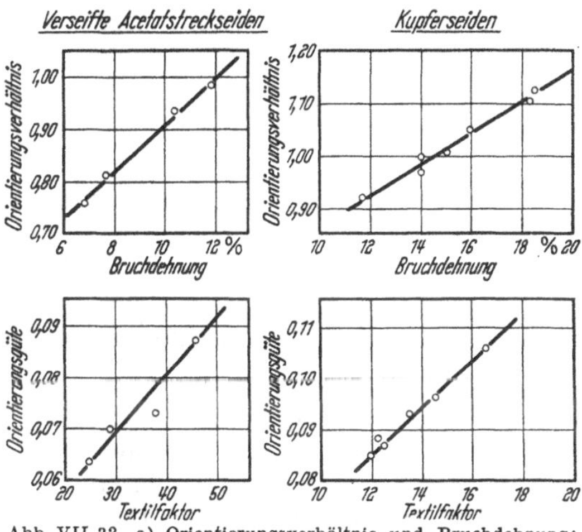

Abb. VII, 33. a) Orientierungsverhältnis und Bruchdehnung; b) Orientierungsgüte und Textilfaktor. (Nach KAST u. PRIETZSCHK.)

d) Schlußbemerkung.

Bei den synthetischen hochpolymeren Faserstoffen sind unsere Kenntnisse über den Verlauf und das Ergebnis der Verstreckung im Verhältnis

---

[1] Siehe S. 458, Fußnote 5. — [2] Siehe S. 458, Fußnote 6.

zu den Cellulosefasern noch sehr gering. Das liegt nicht nur daran, daß man bei den Polyamiden beispielsweise stets praktisch bis zu Ende verstrecken muß, also keine Zwischenstadien zwischen dem unverstreckten und dem völlig verstreckten Zustand herstellen kann, sondern vor allem daran, daß der Vorgang der Kaltverstreckung ein wesentlich komplizierterer ist. Bei der Cellulose wird die Verstreckung nach Möglichkeit schon vorgenommen, bevor eine Kristallisation stattgefunden hat; bei den synthetischen Hochpolymeren aber ist der umgekehrte Fall die Regel. Es handelt sich dann also nicht mehr um eine Orientierung von Ketten und die Eindrehung beweglicher Kettenpakete in die Dehnungsrichtung, sondern um eine durch Zerstörung und Wiederbildung der Kristallite gekennzeichnete und durch die beim Strecken auftretende Wärme begünstigte völlige Umbildung des kristallinen Gefüges (näheres siehe Band III dieses Werkes, § 32). Ein Blättcheneffekt, wie er beim Beginn der Kaltverstreckung mitunter nachgewiesen ist[1, 2], rührt dann von dem Abgleiten der Rostebenen aneinander her. Im Falle von Terylen bzw. Dacron und von Orlon gelingt es wohl, im heißen Zustand oder aus der Lösung im unkristallisierten Zustand zu verstrecken und die Verstreckung durch Einleitung der Kristallisation in verschiedenen Stadien abzubrechen, aber auch in diesen Fällen ist noch nichts Quantitatives über die Zusammenhänge von Textur und Eigenschaften bekannt.

B. Creep and Relaxation of Fibres.

By A. K. van der Vegt.

Introduction.

Textile fibres exhibit in their creep and relaxation behaviour some characteristic peculiarities.

Firstly there is a strong anisotropy, due to the molecular orientation. The ratio between Young's modulus and shear modulus is much higher than in unoriented polymers and the time-dependent behaviour in extension cannot be related in a simple way to the same in torsion[3]. Though in practice textile fibres undergo all kinds of deformations, nearly all measurements have only been performed in longitudinal direction.

Secondly most fibres show considerable deviation from linear viscoelastic behaviour. This phenomenon is revealed by creep tests under different stresses or by stress-relaxation at different extensions and will be discussed in more detail on the basis of the available results.

A considerable number of measurements on creep and relaxation have been performed, but only relatively few series of tests under systematic-

[1] Brill, R.: Z. physik. Chem. B **53**, 61 (1943).
[2] Fankuchen, I. u. H. Mark: J. appl. Physics **15**, 364 (1944).
[3] Hammerle, W. G. u. D. J. Montgomery: Text. Res. J. **23**, 595 (1953).

ally varied circumstances are reported. Moreover, comparison and connection of the data given by different authors is hardly possible owing to the great variation in materials and in the circumstances of measurements.

The properties of fibres depend strongly on the method of manufacturing (among other things on the degree of stretching) or, for natural fibres, on the geographical origin. Consequently even one and the same type of fibre shows a large variation in properties.

The circumstances under which experiments are performed by different investigators also show considerable differences. Some of the factors influencing the results are: temperature, relative humidity, the way of applying load or deformation[1], storing[2] or atmosferic conditioning[3] of the sample before the measurement, etc.

As a consequence of these facts only few general conclusions can be drawn. For that purpose, the available data will be discussed in the following order:

a general shape of the creep and relaxation curves,

b deviations from linearity,

c influence of temperature and humidity.

§ 57. General Shape of Creep- and Relaxation Curves.

Generally the creep can be divided into three components: γ_1, the immediate elastic strain; γ_2, the retarded elastic deformation or the primary creep; γ_3, the permanent flow, which is sometimes absent. Relaxation often proceeds unto a certain limit which is not zero. Consequently the tension can be separated into two fractions: a fraction σ_1 which is time dependent and tends to zero and a limiting value σ_2 being independent of time.

In many cases it is difficult to separate these different components. In a relaxation test for instance one has to wait unduly long in order to know the limit to which the stress approaches as a function of time. In a creep test a considerable amount of retarded response may have occurred before the first reading is made, which interferes with the measurement of γ_1. The separation of γ_2 and γ_3 would be easy if the rate of plastic flow were constant; in fibres, however, this condition is seldom fulfilled. For that reason O'Shaughnessy[2] carried out a number of creep tests for each load varying the total loading time. The amount of plastic flow was determined by means of a recovery cycle after each creep test. The results of these separation experiments will be discussed in the following section; among other things they appear to indicate that the non-recoverable extension as a function of time also seems to approach to a limiting value.

---

[1] Wegener, W.: Melliand Textilber. **30**, 90, 138, 184, 229, 282, 388, 443, 501, 558 (1949). — [2] O'Shaughnessy, M. T.: Text. Res. J. **18**, 263 (1948).

[3] Catsiff, E., T. Alfrey u. M. T. O'Shaughnessy: Text. Res. J. **23**, 808 (1953).

The determination of γ_2 becomes of course much simpler when the material shows no plastic deformation. This is the case with wet wool, and for small deformations, also with some other fibres. It has been shown by LEADERMAN[1] that the permanent flow can often be eliminated by means of mechanical conditioning. This procedure involves a repetition of creep and recovery cycles under identical load. After a few cycles the behaviour of the sample becomes reproducible and the deformation has become wholly reversible. Thus on one and the same sample a series of creep tests can be performed under different loads not exceeding the conditioning load, which only show immediate and delayed elastic deformation.

As was shown by SUSICH[2] the material is however rather seriously modified by this treatment. Presumably the elimination of γ_3 is not the only effect, but also the course of γ_2 with time may differ from the behaviour of the original fibre.

In relaxation experiments a similar mechanical conditioning procedure was performed by MEREDITH[3] in order to obtain reproducible stress-time curves.

Often swelling or heating restores the fibre into its original state.

When creep or relaxation is plotted against the logarithm of time most often a sigmoid shaped curve is obtained. For small values of time the curve runs approximately horizontally, for high values the relaxing tension or the primary creep (sometimes also the plastic deformation) approaches to a limiting value. Between these extremes the retarded elastic processes occur in a way characteristic for each material and for the circumstances.

Often the experimentally obtained data can be represented by means of an empirical mathematical expression, for instance in the form of a logarithmic, an exponential or a power law. In many cases these relationships are only valid within the time interval in which the masurements were carried out and thus depend on this interval.

The first part of the curve can in general be approximated by a power function of the time t.

WEBER[4] who was the first to report creep experiments, supposed the creep rate of silk to be proportional to some power of the deformation, which results in an equation of the form $\gamma = a(t + c)^n$.

WEGENER[5] reported an extensive investigation on the relaxation of several kinds of rayon and he expressed the relaxation during 300 seconds in the form $\sigma = a(t + c)^{-n}$. MEREDITH[3] who measured stress relaxation on the same materials but whose experiments were extended over larger time intervals (10^5 sec.), however showed that WEGENER's expression does not hold for higher values of t.

---

[1] LEADERMAN, H.: Elastic and creep properties of filamentous materials and other high polymers. The Textile Foundation, Washington D.C. (1943).

[2] SUSICH, G.: Text. Res. J. **23**, 545 (1953).

[3] MEREDITH, R.: J. Textile Inst. **45**, T438 (1954).

[4] WEBER, W.: Pogg. Ann. Physik **4**, 247 (1835).

[5] WEGENER, W.: Melliand Textilber. **30**, 90, 138, 184, 229, 282, 388, 443, 501, 558 (1949).

An expression by means of which the whole curve may be represented is the exponential law. Creep tests on wet wool made by RIPA and SPEAKMAN[1] revealed that the expression $\gamma = \gamma_0\left(1 - e^{-\frac{t}{\tau}}\right)$ was valid even for γ_0 values up to 80%. In these tests a constant load was applied to the fibre. Although the diminishing of cross sectional area becomes very important at these high strains, according to the authors however no compensation is needed, since from the absence of permanent flow the constancy of the number of molecular chains carrying the load can be concluded.

KUBU and MONTGOMERY[2] showed that wet wool in relaxation obeys the law $\sigma = \sigma_0\, e^{-\frac{t}{\tau}}$, τ being about 1000 minutes.

On cotton[3] and saran[4] relaxation tests have been reported by EYRING and collaborators. These are represented by the sum of two exponential functions with relaxation times in the order of 4 and 60 minutes for saran in water and 0,1 and 1000 minutes for cotton in water.

A more complicated case was described by PEIRCE[5] who derived from his torsion tests on a variety of textile and inorganic materials a generally valid expression for the relaxing torque $\sigma = \sigma_0 + a \exp\left(-\beta t^{1/3}\right)$.

When the measurements have been carried out neither at very small nor at very large times, only the central part of the sigmoidal curve may be found, which can often be approximated by a straight line. For this reason many results of creep and relaxation tests are expressed in terms of logarithmic laws $\gamma = a + b \log t$, and $\sigma = a - b \log t$[6–11]. The results of creep measurements on nylon made by ABBOTT[8] show even the validity of such a relation from 10 sec. up to $3 \cdot 10^7$ sec.

Sometimes combinations of different expressions are given. PRESS and MARK[12] expressed the creep of viscose and acetate rayon as a sum of four terms

$$\gamma = c_1 + c_2\left(1 + c^{-\frac{t}{\tau}}\right) + c_3 \log(t + 1) + c_4 t.$$

In this expression the first, second and fourth term are easily recognized as belonging to γ_1, γ_2 and γ_3 respectively. Recovery tests revealed that the third component is partially recoverable, partially permanent.

---

[1] RIPA, O. u. J. B. SPEAKMAN: Text. Res. J. **21**, 215 (1951); Nature **166**, 570 (1950).
[2] KUBU, E. T. u. D. J. MONTGOMERY: Text. Res. J. **22**, 778 (1952).
[3] LASATER, J. A., E. L. NIMER u. H. EYRING: Text. Res. J. **23**, 237 (1953).
[4] CHEN, M. C., T. REE u. H. EYRING: Text. Res. J. **22**, 416 (1952).
[5] PEIRCE, F. T.: J. Textile Inst. **14**, T390 (1923).
[6] STEINBERGER, R. L.: Text. Res. J. **7**, 83 (1936).
[7] KATZ, S. M. u. A. V. TOBOLSKY: Text. Res. J. **20**, 87 (1950).
[8] ABBOTT, N. J.: Text. Res. J. **21**, 227 (1951).
[9] DILLON, J. H. u. I. B. PRETTYMAN: J. appl. Physics **16**, 159 (1945).
[10] LYONS, W. J.: J. appl. Physics **17**, 472 (1946).
[11] HAMMERLE, W. G. u. D. J. MONTGOMERY: Text. Res. J. **23**, 595 (1953).
[12] PRESS, J. J. u. H. MARK: Rayon Text. Monthly **24**, 297, 339, 405 (1943).

The theoretical interpretation of the various measurements is simplest in the case of one or more exponential functions. Provided that the behaviour is linear the material can be represented by means of a simple mechanical model consisting of springs and dashpots.

In most cases, however, the curve extends over a larger region of the log time scale than is in a accordance with the exponential law. Mainly two different approaches have been made to account for this fact. The first is the introduction of a relaxation or retardation spectrum, the second is based on the absolute reaction rate theory of EYRING. Both treatments will be summarized.

HAMMERLE and MONTGOMERY[1] and MEREDITH[2] assumed a relaxation spectrum of the form

$$\left.\begin{aligned} g(\tau) &= \frac{g_0}{\tau} \quad \text{for} \quad \tau_1 < \tau < \tau_2 \\ g(\tau) &= 0 \quad \text{for} \quad \tau < \tau_1 \quad \text{and} \quad \tau > \tau_2 . \end{aligned}\right\} \qquad \text{(VII, 6)}$$

The stress σ as a function of time now becomes

$$\left.\begin{aligned} \sigma &= \gamma E \int_0^\infty g(\tau)\, e^{-\frac{t}{\tau}}\, d\tau \\ &= \gamma E g_0 \int_{\tau_1}^{\tau_2} \frac{e^{-\frac{t}{\tau}}}{\tau}\, d\tau \\ &= \gamma E g_0 \left[\mathrm{Ei}\left(-\frac{t}{\tau_1}\right) - \mathrm{Ei}\left(-\frac{t}{\tau_2}\right)\right]. \end{aligned}\right\} \qquad \text{(VII, 7)}$$

For $t \ll \tau_1 \ll \tau_2$: $\sigma = \gamma E g_0 \left(\ln\frac{t}{\tau_1} + c - \ln\frac{t}{\tau_2} - c\right) = \gamma E g_0 \ln\frac{\tau_2}{\tau_1}$ (VII, 8)

For $\tau_1 \ll t \ll \tau_2$: $\sigma = \gamma E g_0 \left(-\ln\frac{t}{\tau_2} + c\right) = \gamma E g_0 (\ln\tau_2 - c - \ln t)$ (VII, 9)

in which c = EULER's constant, being 0,5772.

For $\tau_1 \ll \tau_2 \ll t$: $\sigma = 0$. (VII, 10)

The central part [equation (VII, 9)] appears to be a linear function of $\log t$. An analogous approximation by means of three straight lines can be made for a retardation spectrum of the same form; in this case the term with $\ln t$ has the positive sign. From their relaxation experiments the authors mentioned calculated the dynamic viscosity, according to the theoretical treatment given by TOBOLSKY, DUNELL and ANDREWS[3].

An entirely different approach to account for the characteristic shape of creep and relaxation curves is made by HALSEY, WHITE and EYRING[4]

[1] HAMMERLE, W. G. u. D. J. MONTGOMERY: Text. Res. J. **23**, 595 (1953).
[2] MEREDITH, R.: J. Text. Inst. **39**, P 245 (1948).
[3] TOBOLSKY, A. V., B. A. DUNELL u. R. D. ANDREWS: Text. Res. J. **21**, 404 (1951).
[4] HALSEY, G., H. J. WHITE u. H. EYRING: Text. Res. J. **15**, 295 (1945).

who used the three element model of fig. VII, 34. The springs are Hookean, with stiffnesses k_1 and k_2, the dashpot does not obey the Newtonian law

$$\frac{d\gamma}{dt} = \frac{\sigma}{\eta} \tag{VII, 11}$$

but is characterized by the equation given by the theory of absolute reaction rates

$$\frac{d\gamma}{dt} = K \sinh \alpha \sigma. \tag{VII, 12}$$

This expression results from the following reasoning: When there is no force acting on the material, unbiased movements of flowing segments occur into empty places of the molecular configurations; they are supposed to be governed by symmetrical potential barriers. When a force is applied the barriers become asymmetric and this gives rise to a biased segment motion. The mathematical expression for the rate of flow, which will not be derived here, reads

$$\frac{d\gamma}{dt} = \frac{\lambda}{\lambda_1} \cdot k' \cdot 2 \sinh \frac{\sigma \lambda_2 \lambda_3 \lambda}{2kT}. \tag{VII, 13}$$

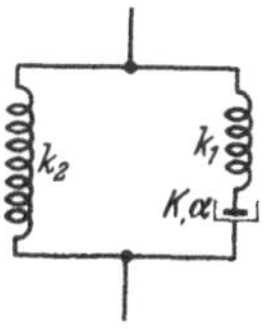

Fig. VII, 34. Three-element model with EYRING dashpot.

In this equation λ represents the distance between two equilibrium positions; λ_1 is the length and $\lambda_2 \lambda_3$ the effective cross-sectional area of the flow unit with respect to the flow direction; σ is the force per unit area, and k' the rate constant, indicating the frequency of movement of the flowing segments. According to the theory,

$$k' = \frac{kT}{h} e^{-\frac{\Delta F^*}{RT}},$$

ΔF^* being the free energy of activation for moving from one equilibrium position to the next. Substitution shows that in equation (VII, 12) the constants K and α are given by

$$K = \frac{\lambda}{\lambda_1} \frac{kT}{h} e^{-\frac{\Delta F^*}{RT}} \quad \text{and} \quad \alpha = \frac{\lambda_2 \lambda_3 \lambda}{2kT} = \frac{V_h}{2kT},$$

V_h being the volume of the flow hole.
When $\alpha\sigma \ll 1$, equation (VII, 12) merges into equation (VII, 11):

$$\frac{d\gamma}{dt} = K \sinh \alpha\sigma = K\alpha\sigma = \frac{\sigma}{\eta}; \quad \eta = \frac{1}{K\alpha}.$$

Returning again to the model represented in fig. VII, 34, the following equation can be derived for relaxation and creep:
Relaxation:

$$\sigma = \gamma \left\{ k_2 + \frac{2}{\alpha\gamma} \tanh^{-1} \left[\tanh \frac{\alpha k_1 \gamma}{2} \cdot e^{-\alpha K k_1 t} \right] \right\}. \tag{VII, 14}$$

Creep:

$$\gamma = \frac{\sigma}{k_2} \left\{ 1 - \frac{2}{\alpha\sigma} \tanh^{-1} \left[\tanh \left(\frac{\alpha\sigma}{2} \cdot \frac{k_1}{k_1 + k_2} \right) \cdot e^{-\frac{\alpha K k_1 k_2}{k_1 + k_2} \cdot t} \right] \right\}. \tag{VII, 15}$$

In order to obtain an impression of the course of these rather complicated functions it is useful to distinguish again three time intervals:

a. small values of t. The exponential factor is nearly equal to unity, and equation (VII, 14) reads approximately

$$\frac{\sigma}{\gamma} = k_2 + \frac{2}{\alpha\gamma} \cdot \frac{\alpha k_1 \gamma}{2} = k_2 + k_1 . \qquad \text{(VII, 16)}$$

b. intermediate values of t. Now the value of $\tanh \frac{\alpha k_1 \gamma}{2} \cdot e^{-\alpha K k_1 t}$ is determined by the second factor which differs in this region far more from unity than the tanh factor.

$$\frac{\sigma}{\gamma} = k_2 + \frac{2}{\alpha\gamma} \tanh^{-1} e^{-\alpha K k_1 t} = k_2 - \frac{1}{\alpha\gamma} \ln \frac{t}{\tau} \cdot \left(\tau = \frac{1}{\alpha K k_1}\right) \qquad \text{(VII, 17)}$$

c. large values of t. The exponential function disappears and

$$\frac{\sigma}{\gamma} = k_2 . \qquad \text{(VII, 18)}$$

Actually the approximations (VII, 16), (VII, 17) and (VII, 18) of equation (VII, 14) appear to be sufficiently good especially when $\frac{\alpha k_1 \gamma}{2} \gg 1$. Fig. VII, 35 shows how, according to equation (VII, 14), $\frac{\sigma}{\gamma}$ depends on $\log t$ for several values of $\frac{\alpha k_1 \gamma}{2}$. The higher the value of $\frac{\alpha k_1 \gamma}{2}$ the smaller the slope of the relative relaxation curve becomes. For very small values of this parameter however the curves approximate the steepest curve on the right which represents the behaviour of a linear three element model with a Newtonian dashpot and follows an exponential decay curve.

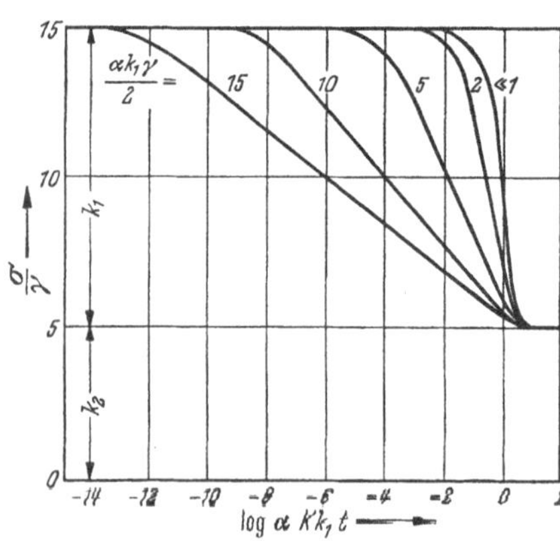

Fig. VII, 35. Reduced stress relaxation curves of EYRING three element model for different values of $\alpha k_1 \gamma$.

For creep, according to equation (VII, 15) a similar family of curves can be drawn, running in the opposite direction.

In this way the absolute reaction rate theory accounts for the wide log time interval over which creep or relaxation curves may extend on the basis of a rather simple three element model. From fig. VII, 35 it appears that the viscoelastic behaviour in this case will be essentially non-linear, for with constant α and k_1 the curves represent the relaxation at different initial deformation γ. Only in the case of very small $\frac{\alpha k_1 \gamma}{2}$, so only when an exponential law is valid, linearity occurs.

The use of a viscoelastic spectrum, on the contrary, always implies linearity.

Although in principle each of the two ways of treatment can be applicated to analyze a single experimental curve by proper choice of the characteristic parameters or functions, the validity of these descriptions can only be tested fully by analyzing the influence of stress and strain on creep and relaxation respectively. This point will be discussed in the next section.

§ 58. Deviations from linearity.

a) Stress dependence of creep.

One of the first thorough studies on the influence of stress on creep has been performed by LEADERMAN[1]. The materials investigated are mechanically conditioned nylon, silk, acetate and viscose rayon. In nearly all cases LEADERMAN finds an exact proportionality between spontaneous elastic deformation and applied stress. The delayed elastic component however is not proportional with stress. Still it appears possible to transform the curves for primary creep into each other by means of a deformation scale factor, which is a non-linear function of stress.

As LEADERMAN points out the existence of such a scale factor means that the creep can be represented by the product of a function of stress and a function of time:

$$\gamma = \frac{\sigma}{E} + J \cdot f(\sigma) \cdot \psi(t) \,. \qquad \text{(VII, 19)}$$

This equation in which $f(\sigma)$ replaces σ in the ordinary equation for linear creep, necessitates a modification of the BOLTZMANN superposition principle:

$$\gamma(t) = \frac{\sigma(t)}{E} + J \int_{\infty}^{0} \frac{d\sigma(t-\omega)}{d\omega} \psi(\omega)\, d\omega \qquad \text{(VII, 20)}$$

into

$$\gamma(t) = \frac{\sigma(t)}{E} + J \int_{\infty}^{0} \frac{df[\sigma(t-\omega)]}{d\omega} \psi(\omega)\, d\omega \,. \qquad \text{(VII, 21)}$$

When equation (VII, 21) is valid, the material has the following two properties in common with a material obeying the original superposition principle (VII, 20).

1. The recovery curve taken after a long time creep curve duplicates that creep curve.
2. The form of the long duration creep curve can be computed from a sequence of short duration creep and recovery curves.

In performing the described measurements LEADERMAN actually found that the materials tested showed this behaviour except nylon under higher loads, so that beside this exception the mechanical behaviour is not in disagreement with the modified Superposition principle.

---

[1] LEADERMAN, H.: Elastic and creep properties of filamentous materials and other high polymers. The Textile Foundation, Washington D. C. (1943).

The dependence of the scale factor in arbitrary units on stress is reproduced in fig. VII, 36. From this picture it is seen that in most cases the increase of creep with load is higher than in the linear case. For one of the nylon filaments the scale factor becomes more or less a constant at higher stresses. This means that the creep curves for high loads run parallel to each other at a distance equal to the difference in the immediate elastic deformation and in a $\left(\frac{\gamma}{\sigma}, \log t\right)$ plot they will run lower and less steeply the higher the load. From other measurements there are indications that this phenomenon is more general; LEADERMAN found that viscose rayon in a 100% relative humidity atmosphere behaves in the same way.

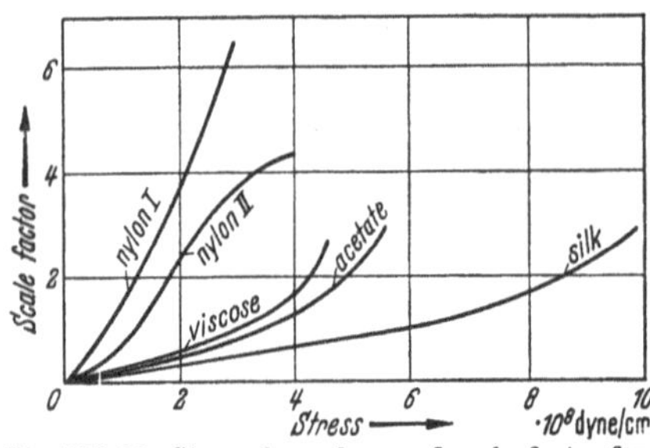

Fig. VII, 36. Stress dependence of scale factor for primary creep (after LEADERMAN).

The next investigation to be discussed is that of PRESS[1]. He measured creep and recovery of viscose rayon in the stress range from $1{,}3 \cdot 10^8 - 10{,}7 \cdot 10^8$ dyne/cm². The results are shown in fig. VII, 37.

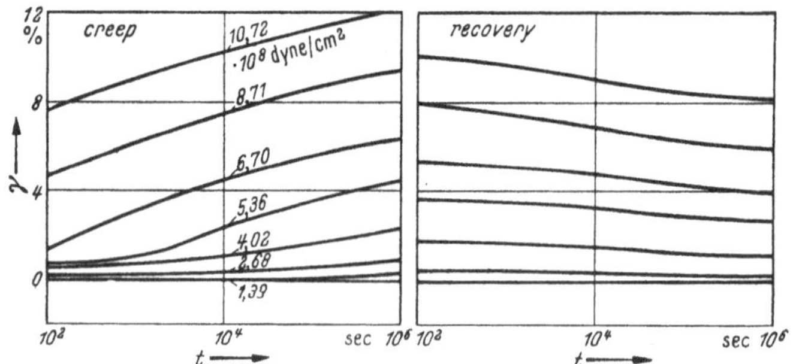

Fig. VII, 37. Creep and recovery of viscose rayon under various loads (after PRESS).

From this graph it appears that a time independent scale factor for each of the curves does not exist. This may be partly due to the higher stresses used, partly to the fact that the material was not mechanically conditioned.

A large amount of creep is non-recoverable especially at higher stress levels. The major part of the permanent deformation, however, can be removed by a special treatment consisting of wetting and drying, a fact which was also noted by LEADERMAN. On each of the stress levels used PRESS measured the deformation after recovery and after swelling successively and found two critical strain values governing this behaviour (fig. VII, 38). Extensions below $^3/_4$% are entirely recoverable; those between $^3/_4$% and 4% are only partly elastic but recover wholly after

[1] PRESS, J. J.: J. appl. Physics **14**, 224 (1943).

swelling; when the strain however has exceeded 4% a deformation remains even after recovery and swelling.

Concerning the creep curves, PRESS points out that the slope for high values of t as a function of load first increases proportionally with load; afterwards it increases more rapidly and finally, for stresses above $5 \cdot 10^8$ dyne/cm² the slope tends to become constant. From this phenomenon and from the fact that the elongation scale can be split up into the three regions mentioned, PRESS draws important conclusions concerning the deformation mechanism of viscose rayon from which it becomes clear that the partly amorphous, partly crystalline structure makes the viscoelastic behaviour very complicated. Shortly summarized these conclusions are as follows.

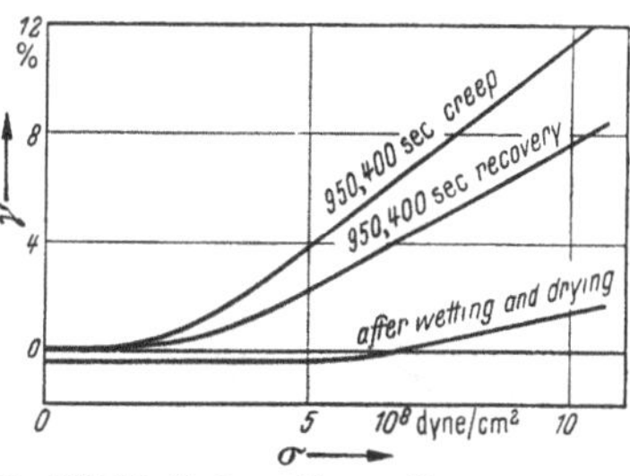

Fig. VII, 38. Deformations after creep, recovery and wetting out (after PRESS).

The behaviour at low strains up to $^3/_4$% is governed by the maximum movements of molecular segments before weaker secondary bonds in the relatively amorphous regions begin to break; in the second range ($^3/_4$–4%) these secondary bonds are broken allowing the molecular segments to move more freely. The latter however have not yet sufficient rotational and translational energy to come into new equilibrium positions of strong interaction. When the extension exceeds 4%, such positions can now be reached, which gives rise to an increase in viscosity. For short times the translational movement predominates (viscous flow), for long times the rotational movement (crystallisation) becomes more important.

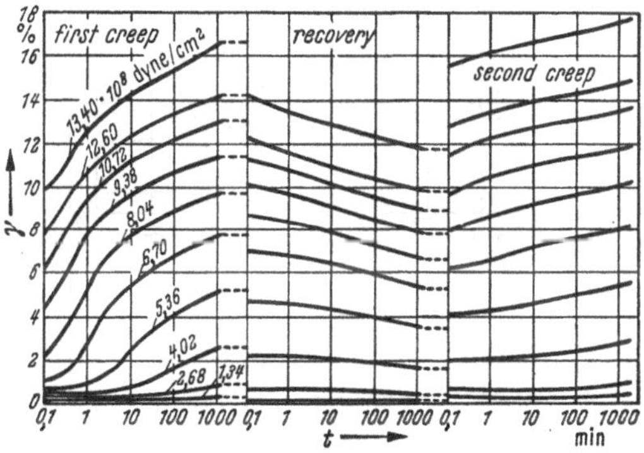

Fig. VII, 39. Creep, recovery and second creep of viscose rayon for various loads (after O'SHAUGHNESSY).

The creep curves reported by PRESS show a strong resemblance to those measured by O'SHAUGHNESSY[1] on rayon and cordura yarns. Some of his results are shown in fig. VII, 39 in which creep, recovery and second creep are plotted under different loads varying from $1{,}3 \cdot 10^8$ to $13{,}4 \cdot 10^8$ dyne/cm². The third set of curves shows the behaviour of mechanically conditioned rayon, the sample having gone through a 24 hours creep – 24 hours recovery cycle. Other experiments however reveal that the conditioning was not complete in this case.

---

[1] O'SHAUGHNESSY, M. T.: Text. Res. J. 18, 263 (1948).

Besides, O'SHAUGHNESSY performed separation tests, the results of which are reproduced in fig. VII, 40. The recoverable deformation curve derived in this way (curve b) exhibits in some cases a maximum. This indicates that the time of recovery increases with the deformation level reached during creep, so that the arbitrarily chosen recovery period was too short at higher loads. A complete analysis of the creep into its components is, therefore, not possible but nevertheless some conclusions can be drawn. The shape of the recoverable deformation curve is sigmoid (fig. VII, 40, a, b and c), the decrease after the maximum in fig. VII, 40, d and e will probably be replaced by a slow increase to a limiting value when longer recovery periods are applied. The quasi-permanent elongation is not proportional with time nor with log time but seems to exhibit the same general shape as the transient creep does. It must be kept in mind that, when the curves b in fig. VII, 40, c and d are adjusted to higher values in order to eliminate the effect of the insufficient recovery, then the slope of the curves c decreases. In view of this fact it appears to be justified to assume the existence of an upper limit for the secondary creep too.

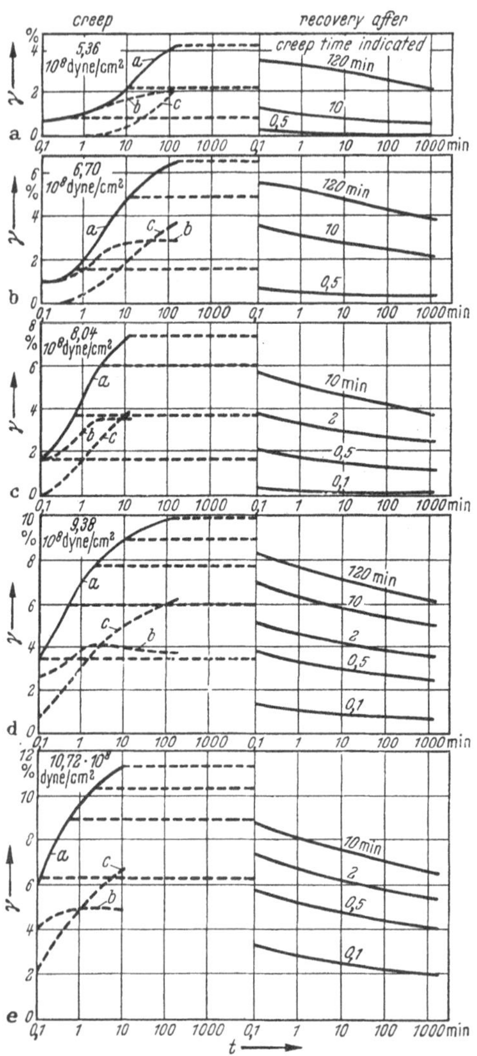

Fig. VII, 40. Creep (a) separated into recoverable deformation (b) and permanent flow (c) by means of recovery after different total loading times (after O'SHAUGHNESSY).

Concerning the load dependence of the upper limits of both components little can be derived from these data. Both components reveal a strong dependence on the stress in their position with respect to the time scale. With increasing load a considerable shift towards lower values of $\log t$ occurs. This shift is fairly equal for both components. At each load the non-recoverable flow makes its appearance when curve b bends upward. The occurrence of permanent deformation is, therefore, more or less dependent on the elongation reached. This confirms qualitatively PRESS' conclusions.

Returning again to O'SHAUGHNESSY's complete creep curves, fig. VII, 41 represents the same data as does fig. VII, 39 but now γ/σ is plotted against $\log t$. These reduced creep curves show approximate coincidence of their initial and final parts which means that with respect to the instantaneous and to the long term deformation the behaviour is nearly linear. Short time tests from $5 \cdot 10^{-3}$ min up to 1 min confirm the existence of the lower envelope for the creep curves at high loads too. The inversion in order of succession of the curves for large loads and long time is a phenomenon analogous to the tendency of LEADERMAN's scale factor to become constant. Further the picture of fig. VII, 41 is mainly characterised by the shift in the log time scale which indicates that the rate of the retarded processes is strongly load dependent. The experiments on cordura yarns show analogous results.

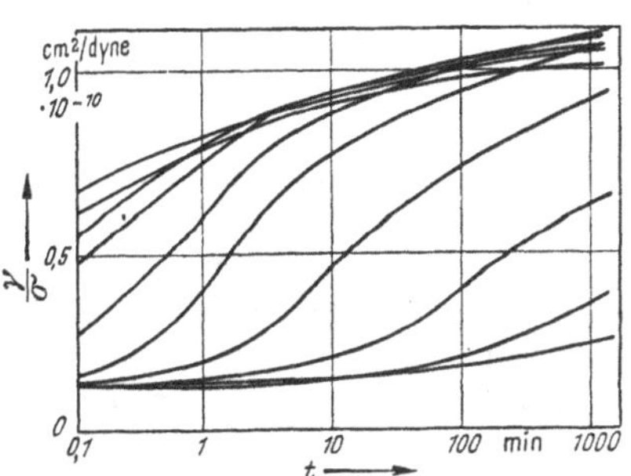

Fig. VII, 41. Reduced creep curves of viscose rayon (same data as fig. VII, 39) (after O'SHAUGHNESSY).

In order to arrive at a molecular interpretation of the creep data, O'SHAUGHNESSY supposes in the first flat region of the curves ordinary crystal like elasticity to be acting, which is rapid and completely recoverable. The rapidly increasing portion of the curves corresponds to configurational deformation which is to be seen as a stress-biased diffusion of the crystalline segments in the amorphous material, the restoring forces being attributed partly to a decrease of entropy. With increasing time this kind of deformation becomes more and more irreversible; the non-recoverable component however also seems to approach to a limiting value. As nearly all these deformations are removed by swelling, the author concludes that the structure of the material is characterized by the existence of a network in which the junction points formed by the crystallites are stable under the given conditions of stress, swelling and temperature.

Some investigators measured the creep of tyre cords. Most of the results can be expressed by a logarithmic function, $\gamma = a + b \log t$; in some cases a third term, ct, has to be added. DILLON and PRETTYMAN[1] mainly measured the elongation increment, so only the value of b could be determined. On rayon, fortisan, nylon and cotton under three different loads σ, at three temperatures, always a more or less rapid decrease of b/σ with increasing σ was found.

LYONS' data on cotton cords show the same result[2]. In one of his short time test series eight different loads were used, varying from $0,35 \cdot 10^8 - 20 \cdot 10^8$ dyne/cm². Other data of measurements over greater

[1] DILLON, J. H. u. I. B. PRETTYMAN: J. appl. Physics **16**, 159 (1945).
[2] LYONS, W. J.: J. appl. Physics **17**, 472 (1946).

time intervals contain a constant rate creep term, which is, however, too small to show a regular dependence on load. ABBOTT[1] measured the creep of nylon monofilaments. For the stress range used ($3 \cdot 10^8 - 30 \cdot 10^8$ dyne/cm²) the same logarithmic law as above was valid in which b was independent of stress up to $18 \cdot 10^8$ dyne/cm² but increased rapidly with stress above that value. The recovery curves ran parallel to each other in the same stress range in which b was constant.

These data do not allow conclusions about the load dependence of the separate components since the term a does not equal the immediate elastic deformation and the second term may include plastic flow.

For the interpretation of their creep measurements on rayon, cotton and wool HOLLAND, HALSEY and EYRING[2] used the non-linear model with the EYRING-dashpot already discussed, modified by the addition of an open dashpot. For different loads the characteristic parameters of the springs and dashpots, which were calculated from the measurements, all vary rather irregularly with load. This is illustrated by table VII, 4 which represents for wool the spring constant k and the dashpot parameters K and α in c.g.s. units derived from creep tests under different loads σ.

Table VII, 4.

$\sigma \cdot 10^{-9}$	$\alpha \cdot 10^{9}$	$k \cdot 10^{11}$	$K \cdot 10^{4}$
0,108	25	0,21	2,74
0,362	2,3	0,55	20,4
0,407	6,6	0,60	4,34

In a recent publication CATSIFF, ALFREY and O'SHAUGHNESSY[3] report extensive investigations on the stress dependence of creep in nylon. These authors arrived at a similar picture as the one PRESS and O'SHAUGHNESSY found for rayon and they performed a thorough graphical analysis of their results. It appeared possible to construct a "Master creep curve", using three different scale transformations for each curve. Firstly, the immediate elastic deformation can be separated from the total creep, secondly the curves are shifted along the $\log t$ axis until the time values for maximum slope of each curve coïncide and thirdly a scale factor for the creep-ordinate enables to a total coïncidence of the curves. In this way the stress dependence of creep is expressed

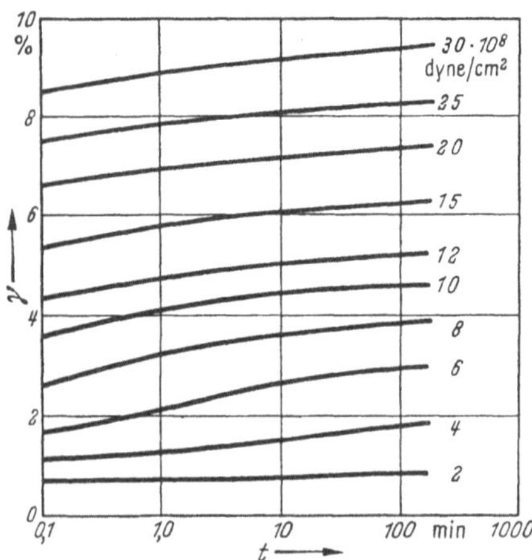

Fig. VII, 42. Creep curves of nylon under various loads (after CATSIFF et al.).

[1] ABBOTT, N. J.: Text. Res. J. **21**, 227 (1951).
[2] HOLLAND, H. P., G. HALSEY u. H. EYRING: Text. Res. J. **16**, 201 (1946).
[3] CATSIFF, E., T. ALFREY u. M. T. O'SHAUGHNESSY: Text. Res. J. **23**, 808 (1953).

by three functions of stress viz. the instantaneous elongation E_0, the characteristic retardation time τ and the creep scale factor F. In figs. VII, 42, VII, 43 and VII, 44 the results on nylon at 36° C and 30% R. H. are reproduced. For measurements under different conditions the curves show a similar shape.

It appears from fig. VII, 43, that E_0 shows deviations from linearity for small loads. In this region F is proportional to the load, but at higher loads F is constant. Except for the timeshift, the parameters F and E_0 may be compared to those found by LEADERMAN.

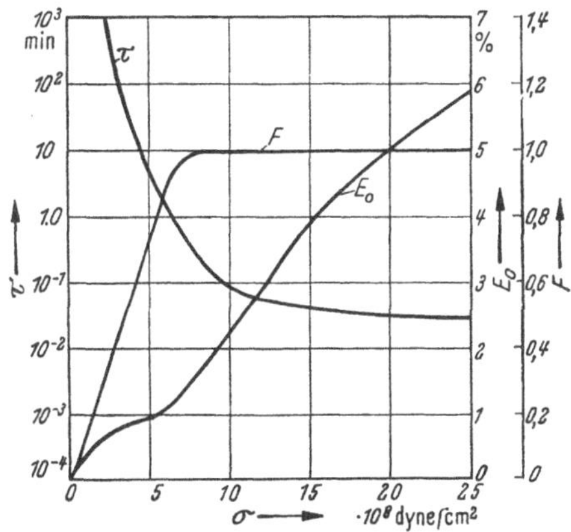

Fig. VII, 43. Characteristic retardation time, τ, instantaneous elongation, E_0, and creep scala factor, F, as a function of stress, from the data of fig. VII, 41 (after CATSIFF et al.).

The characteristic retardation time τ decreases strongly with increasing load when the load is small and varies only little for higher loads. This indicates, that even for very small stresses the behaviour is apparently not linear. It is useful to realize, that for a linear viscoelastic material E_0 and F would be represented by straight lines through the origin and τ by a horizontal straight line.

Concerning the relation between the characteristic retardation time and load, it is interesting to note, that from the creep data, reported by PRESS and O'SAUGHNESSY,

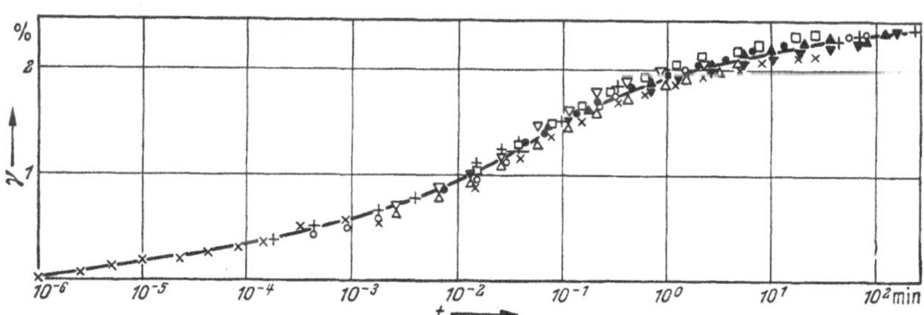

Fig. VII, 44. „Master creep curve", constructed from the curves of fig. VII, 41, by means of the transformation functions of fig. VII, 40 (after CATSIFF et al.).

similar results may be obtained. Furthermore, though LEADERMAN succeeded in describing his creep data by means of E_0 and F only, some of his measurements also suggest the existence of a timeshift (fig. VII, 45).

In fig. VII, 46 a comparison is given of the values of τ as a function of stress, derived from the reported data of the authors mentioned. In the same graph data of RIPA and SPEAKMAN[1] are plotted, who determin-

[1] RIPA, O. u. J. B. SPEAKMAN: Text. Res. J. 21, 215 (1951).

ed the influence of load on k, the rate constant in the creep equation for wool $\gamma = \gamma_0 (1 - e^{kt})$. These authors ascertained a linear relationship between log $(-k)$ and σ for two different temperatures.

b) Strain dependence of relaxation.

Several experiments are reported in which the relaxation function appears to be proportional to the deformation. On nylon monofilaments, HAMMERLE and MONTGOMERY[1] performed relaxation tests in torsion as well as in longitudinal direction. In both cases the material proved to behave as a linear system: the straight lines for stress/strain versus log t, coïncided for the deformations used, except for a longitudinal extension as high as 5%, which gave a greater slope. On the basis of these data the authors described the material by means of a broad rectangular distribution of relaxation times, but the spectra for torsion and elongation cannot be correlated quantitatively. On cotton, MEREDITH[2] made some measurements, from the graphical representation of which it may be concluded, that for the strains used (2, 4 and 6%) the behaviour is reasonably linear.

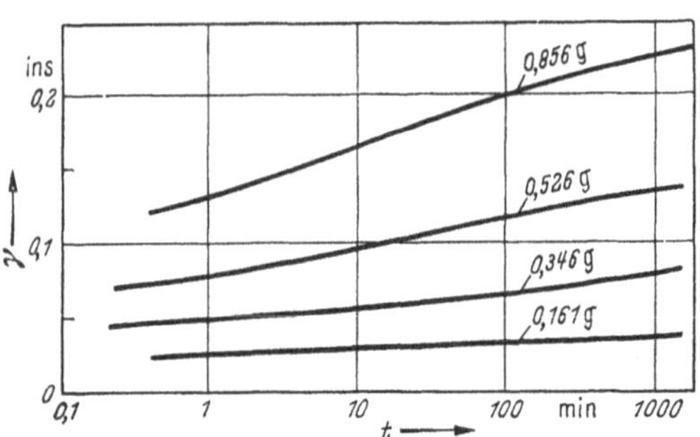

Fig. VII, 45. Creep of nylon under various loads (after LEADERMAN).

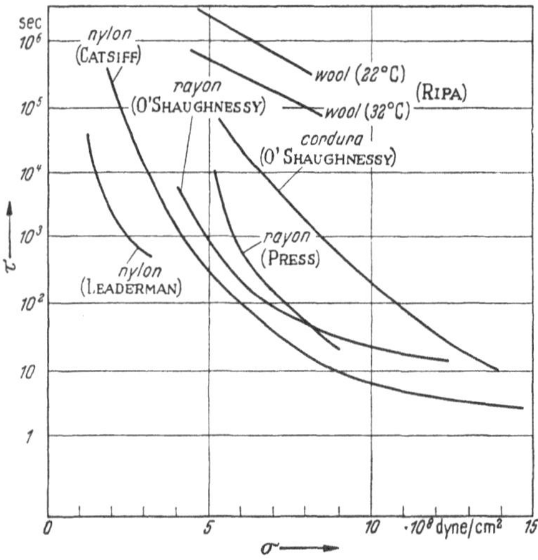

Fig. VII, 46. Characteristic retardation time, τ, versus stress, derived from data of various authors on different materials.

The measurements of SMITH and EISENSCHITZ[3] on viscose at low strains also indicate linearity. The authors established, that σ/σ_0 as a function of t was independent of σ_0 for $\sigma_0 = 2 \cdot 10^8$ to $6 \cdot 10^8$ dyne/cm². WEGENER's results[4] show strongly

[1] HAMMERLE, W. G. u. D. J. MONTGOMERY: Text. Res. J. **23**, 595 (1953).
[2] MEREDITH, R.: J. Textile Inst. **39**, P245 (1948).
[3] SMITH, H. DE WITT u. R. EISENSCHITZ: J. Textile Inst. **22**, T170 (1931).
[4] WEGENER, W.: Melliand Textilber. **30**, 90, 138, 184, 229, 282, 388, 443, 501, 558 (1949).

non-linear behaviour for three kinds of rayon, but the deformations applied varied from 2% to 16%, the smallest σ_0 value amounting to $8 \cdot 10^8$ dyne/cm². The constant-time sections through the graphs derived from these data all show the typical shape of the stress-strain curve above the apparent yield value.

Recent measurements on viscose and acetate rayon performed by MEREDITH[1] show some interesting facts. The strain levels applied vary from 0,5% to 20%. Even at low strains, there are deviations from linearity. The relaxation curves for acetate rayon are found to intersect, a phenomenon which is also shown by WEGENER's results; WEGENER's measurements, however, were carried out over a too small time interval to permit further analysis.

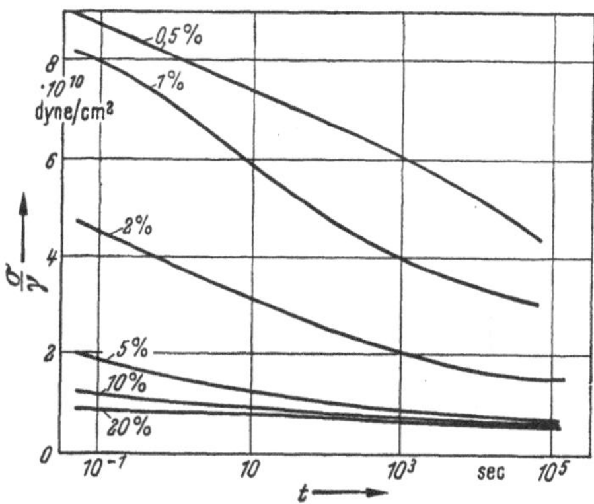

Fig. VII, 47. Reduced relaxation curves for viscose rayon at various elongations (derived from data of MEREDITH).

When from MEREDITH's data σ/γ is calculated and plotted against $\log t$, the pictures shown in figs. VII, 47 and VII, 48 are obtained. These indicate, that the major relaxation phenomenon is accelerated by higher strain. The graphs suggest a log time shift analogous to the one occuring in creep. It is clear, that in the case of a strong log time shift, the relaxation curves themselves may intersect.

WOOD[2] recently investigated the relaxation of human hair; he also observed this phenomenon. Some of his results, plotted in the same way, are shown in fig. VII, 49. This picture is much more complicated and suggests the existence of different relaxation mechanisms; the slowest of these mechanisms however seems to be subjected to a similar log time shift.

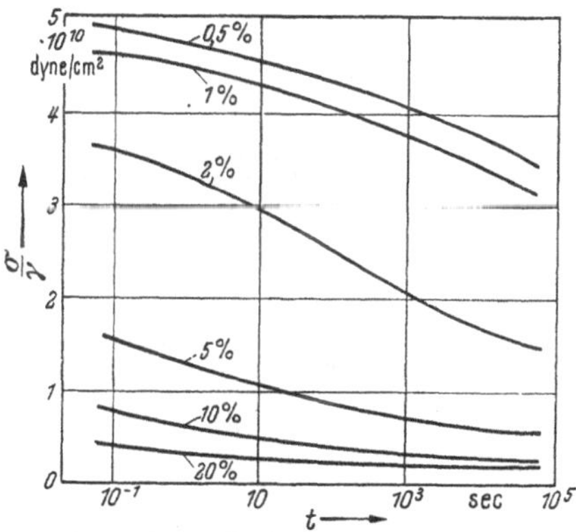

Fig. VII, 48. Reduced relaxation curves for acetate rayon at various elongations (derived from data of MEREDITH).

c) Conclusions.

Considering the data discussed in the foregoing paragraphs, the statement seems justified, that in general the region of linearity for fibres is,

[1] MEREDITH, R.: J. Textile Inst. **45**, T438 (1954).
[2] WOOD, G. C.: J. Textile Inst. **45**, T462 (1954).

if present at all, very small. As to creep, the data suggest a stress dependence of the relative creep rate even for small stresses (fig. VII, 46). In relaxation there may be a small strain region, in which the tension, though relaxing only little, is proportional to strain, but the major relaxation processes seem to exhibit a phenomenon analogous to that of creep. It is, therefore, not justified to describe the viscoelastic behaviour in terms of relaxation or retardation spectra, except in the few cases already mentioned. Instead of this treatment, many authors have applied the EYRING-model on the fibre properties to account for the non-linearity. In order to check the validity of this approach, it is useful to compare fig. VII, 35 in which the theoretical curves are shown, with for instance figs. VII, 41 and VII, 47, which represent some of the actual measurements. It appears, that the essential features of the theoretical and the experimental curves are strongly different. The latter show a more or less parallel shift, the former converge. Hence, the EYRING model can not be applied to account for the actual creep and relaxation of fibres.

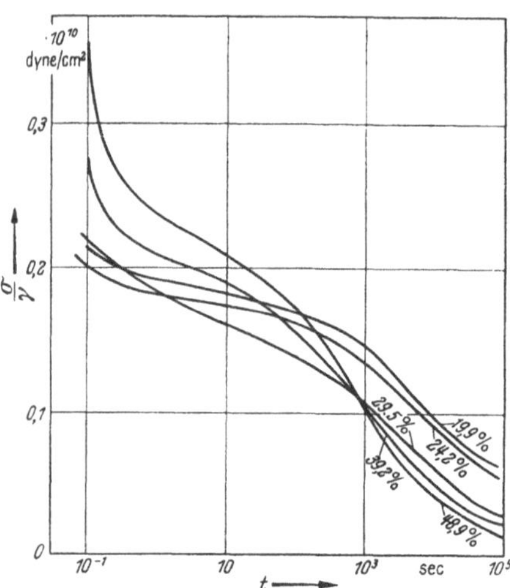

Fig. VII, 49. Reduced relaxation curves for human hair at various elongations (derived from data of WOOD).

If, notwithstanding this discrepancy, the viscoelastic properties are still interpreted in terms of this theory, the parameters of the three element-model must be assumed to be stress or strain dependent. When, for instance, experimentally, in relaxation an approximately linear behaviour is found, the value of the parameter α must be assumed to be inversely proportional to the strain γ applied, since the slope of the curve in the $\left(\frac{\sigma}{\gamma}, \log t\right)$ plot is $\frac{1}{\alpha\gamma}$. Some authors actually made that conclusion[1, 2]. BURLEIGH and WAKEHAM[1] interpreted this result as being caused by an increase of the number of flow units involved in the relaxation process, according to $\alpha = \frac{V_h}{2kT} = \frac{\lambda}{2NkT}$ where N denotes the number of parallel flowing units per unit area and λ the dimension of the unit in longitudinal direction. Starting from the assumption $\alpha\sigma_0$ = constant, these authors calculated the stress necessary to make $N = N_c$, the total number of chain molecules per unit area of the cross section. When this occurs, each molecule will flow as a separate unit. The authors interpreted this stress as being the theoretical tensile

[1] BURLEIGH, E. G. u. H. WAKEHAM: Text. Res. J. **17**, 245 (1947).
[2] TOBOLSKY, A. V. u. H. EYRING: J. chem. Physics **11**, 125 (1943).

strength and they found values, which in order of magnitude agree with those experimentally obtained.

Furthermore, α enters into the relaxation time $\tau = \frac{1}{\alpha K k_1}$, thus causing a shift of the whole curve towards higher values of log time with increasing deformation when K is constant. When this is not the case or when, on the contrary, a shift in opposite direction occurs, an increase of K has to be assumed, which results in a dependence of the free energy of activation ΔF^* on the deformation.

In this way any set of relaxation or creep experiments can be interpretated in terms of the EYRING theory, provided the curves exhibit qualitatively the shape of those in fig. VII, 35. The variations of the characteristic constants make, however, this treatment rather artificial; prediction of the behaviour under specified circumstances is only possible, if not only the four characteristic parameters are known, but also their dependence on stress or strain. Moreover, when the parameters are calculated from creep or relaxation experiments, during which respectively strain and stress vary, they are supposed to be independent of stress and strain. The assumption that in a relaxation test the parameters depend on the deformation (or on the initial stress) is in contradiction with this fact.

From these considerations it appears that the mechanical models with EYRING dashpots, though frequently used in the interpretation of viscoelastic behaviour of fibres, are not able to account for the real phenomena and must be rejected on the basis of the experimental data.

§ 59. Influence of atmospheric conditions.

a) Temperature effect.

LEADERMAN[1] performed a thorough study of the temperature dependence of creep. He measured the creep of mechanically conditioned filaments at different temperatures between 50° and 90° C. In fig. VII, 50 some of the results of the experiments on acetate rayon are represented. LEADERMAN plotted the log of the time necessary to reach specified deformations against the reciprocal absolute temperature and obtained in this way a series of parallel straight lines (fig. VII, 51). This indicates the possibility of shifting the creep curves into each other by a transformation of the time scale. For acetate rayon the temperature decrease, which causes a shift to higher values over one decade along the log time coordinate, amounts to 10,3° C. For nylon, which for small loads behaves in a similar way, the same shift is induced by a 12,6° C difference. From these data, the author calculated the energy of activation for the primary creep

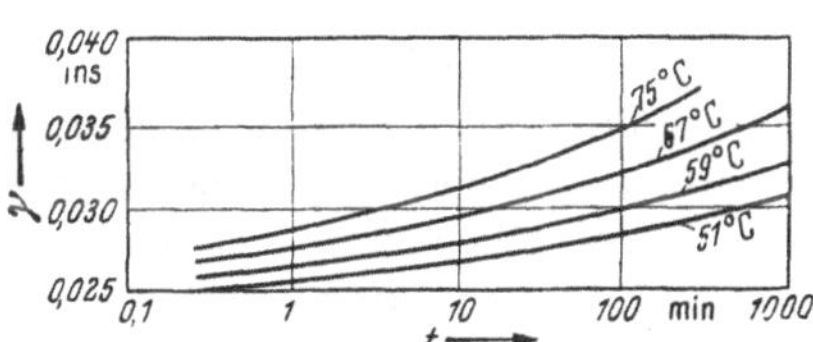

Fig. VII, 50. Creep of acetate rayon at different temperatures (after LEADERMAN).

[1] LEADERMAN, H.: Elastic and creep properties of filamentous materials and other high polymers. The Textile Foundation, Washington D. C. (1943).

and he arrived at values in the order of about $Q = 50$ kcal/mole and $Q = 40$ kcal/mole for acetate rayon and nylon respectively.

The magnitude of this energy appeared to be strongly dependent on the physical state of the filament; for acetate rayon for instance Q was reduced to about 30 kcal/mole after a small permanent deformation of the sample.

These results of LEADERMAN are consistent with the statement of SMITH and EISENSCHITZ[1] that changes in temperature (and also in relative humidity), leave the general form of the creep curve unaffected and only change the rate of creep. (Compare the discussion of the time temperature relationships given by SCHWARZL in § 5c). From the data of DILLON and PRETTYMAN[2] it appears, that in most cases the slope of the straight line representing the creep as a function of log time, increases with temperature; the results are, however, too fragmentary to permit a thorough analysis.

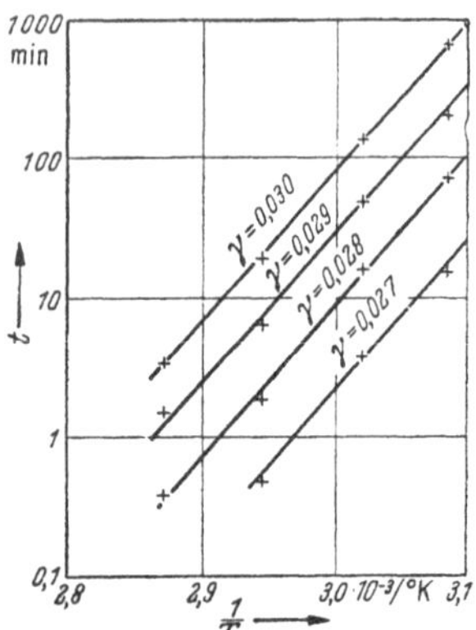

Fig. VII, 51. Time to attain specified deformations for acetate rayon (after LEADERMAN).

LASATER, NIMER and EYRING[3] who expressed the relaxation of cotton fibres by means of the formula $\sigma = C_1 e^{-r_f t} + C_2 e^{-r_s t}$, determined the temperature dependence of the relaxation frequencies r_f and r_s for two kinds of cotton in dry and in wet condition. The greater of the two, r_f, corresponding to the fast process, appears to be scarcely dependent on the temperature, whereas the smaller one, r_s, decreases with increasing T. Calculation of the activation energy from the data reported gives for the fast process in all cases small values below 1 kcal/mole, for the slow process values between 2 and 7 kcal/mole. The authors however treat the data in a different way. By means of the theory of absolute reaction rates, they calculate for each temperature the free energy of activation for the relaxation process. The obtained values appear to depend approximately linearly on T. When the formula $\Delta F = \Delta H - T\Delta S$ is applied, for the heat of activation ΔH, values are obtained, which are in the same order of magnitude as the ones mentioned above, the entropy of activation ΔS amounting to between -80 and -90 entropy units.

CHEN, REE and EYRING[4] measured the relaxation of saran at different temperatures and they expressed their results in the formula

$$\sigma = \sigma_0 \left[1 - \alpha_1 \left(1 - e^{-\frac{t}{\tau_1}}\right) - \alpha_2 \left(1 - e^{-\frac{t}{\tau_2}}\right)\right]$$

For the two exponential relaxation processes, the logarithms of the relaxation time are linearly related to the reciprocal of absolute temperature.

[1] SMITH, H., DE WITT u. R. EISENSCHITZ: J. Text. Inst. 22, T 170 (1931).
[2] DILLON, J. H. u. I. B. PRETTYMAN: J. appl. Physics 16, 159 (1945).
[3] LASATER, J. A., E. L. NIMER u. H. EYRING: Text. Res. J. 23, 237 (1953).
[4] CHEN, M. C., T. REE u. H. EYRING: Text. Res. J. 22, 416 (1952).

From the slope of the straight lines the heats of activation appear to be 2,9 and 4,4 kcal/mole. The total relaxing fraction of the initial tension, $\alpha_1 + \alpha_2$, increases with temperature. The components α_1 and α_2, however differ markedly in their temperature-dependence; in the region from 22° to 67° C, α_1 increases and α_2 decreases, both varying roughly by a factor three.

It may be remarked that the temperature dependence of creep and relaxation is narrowly connected with the thermoelastic behaviour. When the temperature of a fibre, suspended under a certain tension, is raised, the fibre may extend or shrink; which of the two phenomena occurs depends on the mechanism of elasticity. In the first case normal thermal expansion occurs, in the second case the change in entropy of the molecular chains in the amorphous parts predominates. Similar observations can be made when the elongation is held constant; here the tension will vary with the temperature.

Several investigators[1] have studied these phenomena; a detailed discussion of thermoelasticity, however, falls beyond the scope of this section.

b) Influence of relative humidity.

The influence of humidity on the viscoelastic properties is the more pronounced the higher the degree of swelling of the fibre. The fact has already been mentioned, that regenerated cellulosic fibres after creep and incomplete recovery are often restored to their original length by wetting and drying. From this phenomenon and also from the influence of water content on creep, it appears, that water acts qualitatively in the same way on water sensitive thermoplastic materials as heat.

Data of LEADERMAN[2] on creep of nylon under small load and different humidities are shown in fig. VII, 52. These curves indeed bear a certain resemblance to those obtained at various temperatures.

Largely the same result was found by CATSIFF and coworkers[3], who ascertained an important influence of humidity on the characteristic retardation time for nylon under small loads at small values of the rela-

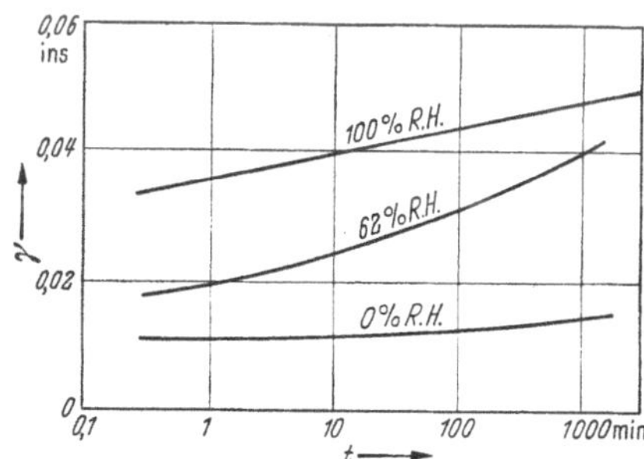

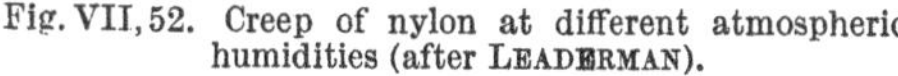
Fig. VII, 52. Creep of nylon at different atmospheric humidities (after LEADERMAN).

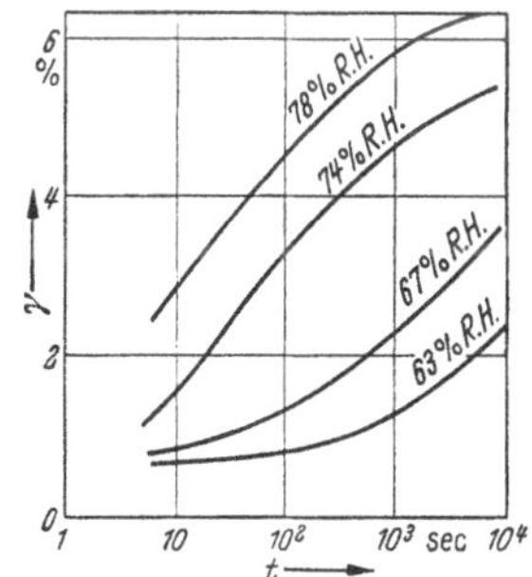

Fig. VII, 53. Creep of viscose rayon at different atmospheric humidities.

[1] BRENSCHEDE, W.: Kolloid-Z. **120**, 67 (1951). – G. M. BRYANT: Text. Res. J. **23**, 788 (1953). – J. F. CLARK u. J. M. PRESTON: J. Textile Inst. **44**, T596 (1953).

[2] LEADERMAN, H.: Elastic and creep properties of filamentous materials and other high polymers. The Textile Foundation, Washington D. C. (1943).

[3] CATSIFF, E., T. ALFREY u. M. T. O'SHAUGHNESSY: Text. Res. J. **23**, 808 (1953).

tive humidity; for higher R. H. values the quantities most affected were the immediate elastic deformation and the creep scale factor. Viscose rayon exhibits a still stronger dependence on the R. H., the material being able to absorb more water than nylon. Fig. VII, 53 represents measurements made by the author[1] on viscose at four different values of R. H. The main effect appears to consist of a log time shift, which roughly amounts to one decade for a 5% difference in R. H.

From these data it may be concluded, that swelling tends to decrease the viscosity and at the same time reduces the fibre stiffness.

C. Die dynamischen Eigenschaften und der Gebrauchswert von Faserstoffen.

Von **W. Meskat** und **O. Rosenberg**

Im Kapitel IV (§ 21 d) wurde der Weg zur Ermittlung des Gebrauchswertes von Faserstoffen aufgezeigt, wobei die Tabellen 4 und 5 dieses Kapitels die Zusammenhänge zwischen den verschiedenen Beanspruchungsarten, denen die Fasern und Fäden auf dem Verarbeitungswege ausgesetzt sind, und dem spezifischen Gebrauchswert schematisch darstellen. Hier wollen wir uns darauf beschränken, die Ergebnisse, die mit den geschilderten physikalischen Methoden – wie sie in dem Kapitel IV, §§ 22 und 23, beschrieben worden sind – gewonnen wurden, in Beziehung zu den Gebrauchseigenschaften zu bringen. Dabei gehen wir von den Zahlenwerten für das physikalische Verhalten sowie der relativen Werte bei undefinierter Beanspruchung der bekannten Faserstoffe aus, die in der folgenden Tabelle VII/5 von R. Domke[2] zusammengestellt worden sind. Die Tabelle umfaßt nur die Werte, die mit der sogenannten statischen Prüfgerätegruppe (vgl. Kapitel IV, § 22) erfaßt werden können. Sie enthält nicht die Werte, die uns die dynamische Prüfgerätegruppe (vgl. Kapitel IV, § 23) zu messen gestattet und auch nicht

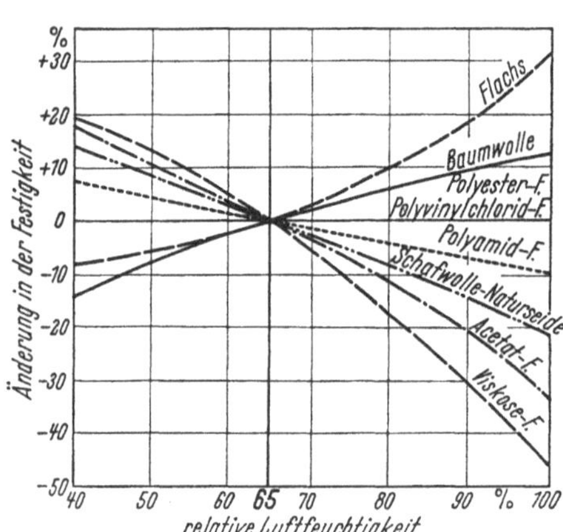

Abb. VII, 54a. Einfluß der relativen Luftfeuchtigkeit auf die Festigkeit der verschiedenen Faserstoffarten nach J. Pollitt.

[1] Vegt, A. K. van der: Unpublished results.

[2] Domke, R.: Vortrag vor dem Internationalen Chemiefaser-Kongreß, Paris Juli 1954. – Weiteres Zahlenmaterial ist den Beiträgen von P. A. Koch: Faserstoffe und P. Lagally: Papier, Zellstoff, Holzschliff, Landolt-Börnstein, Stoffwerte, 6. Auflage, Bd. IV, Teil 1, Springer-Verlag 1955, und Milton Harris: Handbook of Textile Fibers, First Ed. XII, Harris Research Laboratories, Inc., Washington, D. C., 1954, zu entnehmen, vgl. auch Chem. Eng. News **32**, 3950 (1954).

diejenigen Daten, die mittels der Prüfgerätegruppe zur Untersuchung der Oberflächenabnutzungen erfaßbar sind, weil insbesondere die dynamischen Prüfwerte in dem notwendigen Umfang bisher noch nicht vorlagen. Dagegen ist z. B. der starke Einfluß der relativen Luftfeuchtigkeit auf das Verhalten der Faserstoffe in der Tabelle mit angegeben. Anschaulicher zeigt uns den Einfluß der relativen Luftfeuchtigkeit, z. B. auf die Zugfestigkeit der verschiedenen Faserstoffe die graphische Darstellung der Abb. VII, 54a. Hinzu kommt, daß bei einer vorgegebenen Faserart, wir haben beispielsweise die Papierfaser gewählt, die mittels der verschiedenen Beanspruchungsarten erhaltenen Prüfdaten, wie die Dehnungswerte beim Zugversuch, die Werte des Falzwiderstandes, die Berstdruckwerte usw. eine unterschiedliche Abhängigkeit von der relativen Luftfeuchtigkeit erkennen lassen (vgl. Abb. VII, 54b). Daraus ergibt sich die Notwendigkeit, die allgemeinen Prüfbedingungen sehr sorgfältig einzuhalten, um überhaupt zu vergleichbaren Ergebnissen zukommen[1]. Trotzdem werden wir sehen, daß diese Kennzahlen, wie sie auch in der folgenden Tabelle VII/5 zusammengestellt, zur Beurteilung des Gebrauchswertes nicht ausreichend sind.

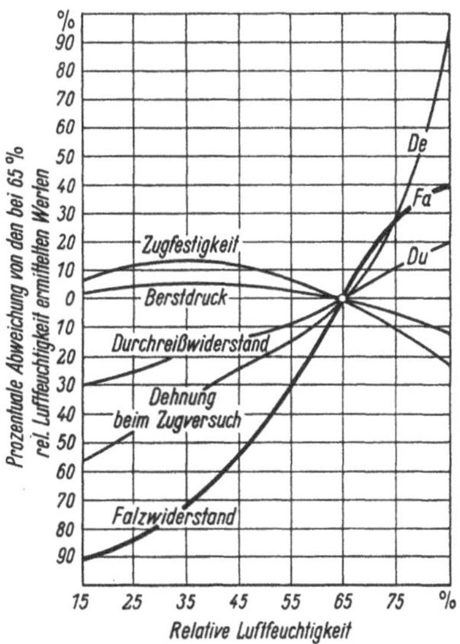

Abb. VII, 54b. Einfluß der relativen Luftfeuchtigkeit auf die Prüfwerte von Papierfasern bei verschiedenen Prüfverfahren. (Nach HOUSTON, CARSON u. KIRKWOOD[2].)

---

[1] Wir wollen in diesem Zusammenhang darauf verzichten, die bei der Ermittlung der physikalischen Eigenschaften von Faserstoffen zu beachtenden Normvorschriften zu erörtern und verweisen diesbezüglich auf DIN 50012, Entwurf August 1954 (Beschaffenheit des Prüfraumes, Messen der relativen Luftfeuchtigkeit), DIN 53802, Entwurf August 1954, Angleichen der Proben an das Normklima, DIN 53803, Richtlinien bei der Probenentnahme und DIN 53804, 1955, Auswertung der Meßergebnisse H. BISCHOFF: Melliand-Textil-Ber. **34**, 120 (1953), sowie auf die entsprechende A. S. T. M. Standards on Textile Materials vom A. S. T. M. Committee D–13 und S. N. V. 96441. Bei der Prüfung an Garnen und Zwirnen kommt noch die Nummernbestimmung hinzu, vgl. Kap. IV 13 § 21 c und DIN 53831. Diese allgemeinen Richtlinien werden ausführlich auch von E. WAGNER: Mech.-Techn. Textilprüfung, 6. Aufl. 1953, S. 3–52 diskutiert. Bezüglich der speziellen Richtlinien für die Bestimmung der Kennwerte mittels der sogenannten statischen Prüfgerätegruppe sei auf die Ausführungen und Literaturstellen des Kapitels IV, § 22 verwiesen. Diese Richtlinien sind notwendig, wie beispielsweise auch die Untersuchungen von R. MEREDITH: J. Text. Inst. **45**, T 30 (1954) sowie O. ANDERSSON u. L. SJOBERG: Svensk Papperstidn. **56**, 615 (1953), über den Einfluß der Dehnungsgeschwindigkeit auf das Festigkeitsverhalten von Viscose- und Acetat-Reyon sowie Nylon und Papier usw. zeigen; jedoch oft noch nicht hinreichend.

[2] HOUSTON, CARSON und KIRKWOOD: Papier Trade **7**, 51, 237 (1923). — P. LAGALLY in Landolt-Börnstein Stoffwerte, 6. Aufl., IV. Bd, Teil I, S. 304, Springer-Verlag (1955).

Tabelle VII, 5.

Faserstoff	Handelsbezeichnung	spez. Gewicht	Feinheit[1]		Breite	mittlere Stapellänge	Festigkeit und Dehnung		
							Reißfestigkeit		
							normalfeucht		spezifisch
		g/cm³	den	N_m	μ	mm	g/den	Rkm	kg/mm²
Chemiefasern									
Viscose-Reyon normal		1,52	1,5-5,0			endlos	(1,5-2,2)	(14-20)	(18-32)
Viscose-Reyon hochfest	Cordura, Tenasco, Glanzstoff R 100, Supercordura	1,52	2,3-2,4			endlos	(3,3-4,5)	(30-40)	(46-61)
Viscose-Zellwolle normal		1,52	1,0-60			20-230	1,5-3,0	14-26	21-40
Viscose-Zellwolle hochfest	Colvadur	1,52	1,2-3,0			30-60	2,6-3,3	24-30	37-46
Kupfer-Seide	Bemberg, Cupresa	1,52	0,5-1,5			endlos	(1,4-2,0)	(13-18)	(20-27)
Kupfer-Zellwolle	Cuprama	1,52	2,5-20			40-120	1,4-2,5	12-23	19-35
Acetat-Streckseide verseift	Fortisan	1,52	0,7-0,8			endlos	(6,5)	(59)	(90)
Acetat-Seide	Rhodiafil, Lonzona	1,31	2,8-6,6			endlos	(1,1-1,4)	(10-13)	(13-17)
Acetat-Zellwolle	Drawinella, Rhodiazellwolle, Fibroceta, Celafibre	1,31	3,0-28			40-140	1,1-1,8	10-16	13-21
Polyvinylchlorid-Seide	Rhovyl	1,40	1,4			endlos	(2,6-3,0)	(24-27)	(34-38)
Polyvinylchlorid-Faser	Fibravyl, PCU-Faser	1,40	2,0-3,0			40-120	2,6-3,5	24-32	34-44
Polyvinylchlorid nachchloriert	PeCe	1,44	3,4			40-120	1,7-2,2	15-20	22-29
Mischpolymerisat aus Vinylchlorid (85%)-Vinylacetat (15%)	Vinyon HH	1,35 bis 1,37	2,3-5,5			25-90	0,6-1,0	5-9	7-12
Mischpolymerisat aus Vinylchlorid (60%)-Acrylnitril (40%)	Dynel	1,28	1,5-24			40-150	3,0-4,2	27-38	35-49
Mischpolymerisat aus Vinylchlorid (15%)-Vinylidenchl. (85%)	Saran	1,72	70-3600			monofil endlos 20-70	(1,7-2,5)	(15-23)	(27-39)
Polyvinylalkohol	Vinylon, Kuralon	1,30	1,4-4,0			40-120	2,5-6,5	22-59	29-76
Polyacrylnitril-Seide	Orlon 81, PAN, Fibre D, Acrilan, X-51	1,18	1,0-5,0			endlos	(3,3-5,2)	(30-47)	(35-55)
Polyacrylnitril-Faser	Orlon 41/42, PAN, Fibre D, Dolan, Redon, Bayer-Acryl, X-51 Dralon	1,18	1,5-8,0			40	1,4-3,5	13-32	15-38

[1] Die Definition der metrischen Nummer N_m ist in Kap. IV, S. 248, Fußnote 2, erläutert. Es ist $N_m = \frac{L\,[\text{km}]}{G\,[\text{kg}]}$ die Anzahl der Längeneinheiten in km, je Gewichtseinheit in kg. Setzt man für G die Reiß- bzw. Bruchlast in kg ein, so folgt für die Lastlänge, nunmehr Reißlänge genannt und mit R bezeichnet, die Beziehung $R\,[\text{km}] = G\,[\text{kg}] \cdot N_m$.

Physikalisches Verhalten der Faserstoffe

Werte in Klammern: Garn 500 mm. Einspannlänge geprüft. Werte ohne Klammer: Einzelfasern 10 mm. Einspannlänge geprüft.					Elastisches Verhalten			Verhalten gegen Wasser			Verhalten gegen Hitze
Bruchdehnung	rel. Knotenfestigkeit	rel. Schlingenfestigkeit	Quersprödigkeitswinkel[1]	rel. Naßfestigkeit	Elast.-Grad bei 50%	Elast.-Grad bei 2/3 Bruchlast	Koeffiz. d. elast. Arbeit	Feuchtigkeitsaufnahme in % d. Trockengewichtes b. 65% r. L.	Feuchtigkeitsaufnahme in % d. Trockengewichtes b. 100% r. L.	Sättigungswert in % d. Trockengewichtes	
%	%	%	φ_D in	%	%	%	%	%	%	%	°C
(15-24)	45-48	62-67	50,5-54,5	(42-53)	42	37	33	13,5	39-42	66-125	Festigkeitsverl. bei 150 verkohlt bei 180
(16-23)	43-56	46-70	56,5-58,5	(55-59)	46	43	57	13,5	2	83-125	Festigkeitsverl. bei 150 verkohlt bei 180
11-30		28-32	50,5-54,5	45-63	37-55	34-47		13,5	44-48	95-110	Festigkeitsverl. bei 150 verkohlt bei 180
14-21		35	56,5-58,5	48-69	39	37		13,5		85-105	Festigkeitsverl. bei 150 verkohlt bei 180
(10-18)	65-70	70-75	55-56,5	(50-57)	66	52		12,5	38	85-105	zerfällt bei 150
17-24	85-90	55-70	48-50	60-65	67	47		12,5	36	110	zerfällt bei 150
(6-7)		51		(85-90)	50			10,5	20	47-63	erweicht bei 200—230 schmilzt bei 230
(20-25)	70-85	70-85	44-49,5	(64-75)	89	44	38	6,0	17-19	22-35	erweicht bei 200—230 schmilzt bei 230
16-39			44-49,5	58-70	76	31		6,0	15	22	erweicht bei 200—230 schmilzt bei 230
(14-20)		45	55	(100)			71	0			erweicht bei 72—75 schmilzt bei 200—210
25-35			50,5	100				0	1,5		erweicht bei 72—75 schmilzt bei 200—210
35-65			50,5	100	78	63		0			schrumpft ab 100 schmilzt bei 200—210
100-120		71-88		100				0	0,1	4-9	erweicht bei 54—60 schmilzt bei 135—149
15-40		71-88	44-45	100	32			0,5	bis zu 1,0	7-12	erweicht bei 70—115 schmilzt bei 150—160
(15-30)	43-65	50-65	50	(100)	53-56			0	0		erweicht bei 115 schmilzt bei 150—160 schrumpft ab 145
7-25	50-60	50	47,5-48	65-85				5,0	10	30	erweicht bei 220 schmilzt bei 232—238
(8-20)	63-65	70-81	55,5	(90-98)	30			1,2	2	17-19	erweicht bei 190—220 verkohlt vor Schmelzp.
22-33	75		52,5-56,5	90				2,0			erweicht bei 190—220 verkohlt vor Schmelzp.

Aus den Beziehungen auf S. 248 folgt weiter: 1 km = 1 g/Tex = 9 g/den, so daß 33,3 Rkm gleichbedeutend mit 3,70 g/den sind [vgl. auch G. SCHMALFUSS, Textilprax. **10**, 991 (1955)]. Diese und die weiteren Definitionen, wie z.B. die Erläuterung des Quersprödigkeitswinkels werden eingehend von P. A. KOCH in Landolt-Börnstein Stoffwerte, 6. Aufl., IV. Bd., 1. Teil, S. 385 usw. diskutiert.

Tabelle VII, 5.

Faserstoff	Handelsbezeichnung	spez. Gewicht	Feinheit		Breite	mittlere Stapellänge	Festigkeit und Dehnung: Reißfestigkeit normalfeucht		spezifisch
		g/cm³	den	N_m	μ	mm	g/den	Rkm	kg/mm²
Polyäthylen	Polythene, Courlene, Northylen	0,92	270-3600			monofil endlos	(0,6-3,0)	(5-27)	(4-25)
Tetrafluoräthylen	Teflon	2,2					0,06-0,09	0,5-0,8	1-1,75
6,6-Nylon	Nylon	1,15	1,5-15			endlos 38-115	(4,5-6,5)	(41-59)	(47-67)
6-Nylon	Perlon L, Grilon, Phrilon	1,15	1,4-20			endlos 33-120	(4,5-6,5)	(41-59)	(47-67)
Polyester-Seide	Terylen, Dacron Diolen	1,38	1,2-5,0			endlos	(4,0-6,9)	(36-63)	(50-88)
Polyester-Faser	Dacron	1,38	1,5-6,0			(Kabel)	3,0-4,7	27-42	37-58
Glas, Düsen-Ziehverfahren	Gerrix, Vetrotex	2,49			5-6,5	endlos	8,9-12,2	80-110	200-275
Glas, Düsen-Blasverfahren	Gerrix, Vetrotex	2,49			6-11	50-350	3,8-5,0	34-45	85-112
Glas, Stab-Ziehverfahren	Schuller	2,49			6-12	70-120	4,0-8,3	36-75	90-185
Naturfasern									
Baumwolle nordamerikanisch		1,55	2,4-1,5	3800-6000	14-17	16-30	2,3-3,3	21-30	33-46
Baumwolle ägyptisch		1,55	1,8-1,1	5000-8000	12-15	20-32	2,8-5,0	25-45	39-70
Flachs		1,49			11-31	20-39	4,4-7,6	40-69	60-102
Hanf		1,49			16-32	15-28	4,2-6,2	38-56	57-83
Ramie		1,55			40-50	120-140	5,2-7,5	47-67	73-103
Schafwolle		1,30	3,0-30	3000-300	18-60	40-250	1,1-2,1	10-19	13-24
Naturseide		1,37	1,3	7000	13-25		3,8-4,1	34-37	46-51
Hornblende-Asbest		2,9-3,0				60-70	2,9-7,9	26-71	75-225
Serpentin-Asbest		2,3-2,8			1-2 (0,02)	20-40	2,4-3,2	22-29	56-75

(Fortsetzung.)

Physikalisches Verhalten der Faserstoffe

Werte in Klammern: Garn 500 mm. Einspannlänge geprüft. Werte ohne Klammer: Einzelfasern 10 mm. Einspannlänge geprüft.					Elastisches Verhalten			Verhalten gegen Wasser			Verhalten gegen Hitze
Bruchdehnung	rel. Knotenfestigkeit	rel. Schlingenfestigkeit	Quersprödigkeitswinkel	rel. Naßfestigkeit	Elast.-Grad bei 50% Bruchlast	Elast.-Grad bei 2/3 Bruchlast	Koeffz. d. elast. Arbeit	Feuchtigkeitsaufnahme in % d. Trockengewichtes b. 65% r. L. %	Feuchtigkeitsaufnahme in % d. Trockengewichtes b. 100% r. L. %	Sättigungswert in % d. Trockengewichtes	
%	%	%	φ_D in	%	%		%	%	%	%	°C
(20-80)			11	(100)				0	0,5	0,5	erweicht bei 104—120 schmilzt bei 110 Umwandlungspunkt 327
(14-25)	87	85-86	33,5-42,5	(85-90)	91	89	83	4,0	8-9	13-17	erweicht bei 235 schmilzt bei 250
(14-25)		95	33,5-42,5	(85-90)				4,0	8-9	13-17	erweicht bei 160—170 schmilzt bei 212—219
(8-30)		73-97	43,5-47,5	(100)	33			0,5	0,6	5	erweicht bei 235—240 schmilzt bei 250
25-60			31-39					0,5	0,6		schmilzt bei 250
3,5-4,0	32-58	15-20	85-85,5							14	erweicht bei 676 schmilzt bei 843
2,1-2,9	32-58	15-20	86-87,5					0,5		14	erweicht bei 676 schmilzt bei 843
2,5-3,7			85-86					0,5		14	erweicht bei 676 schmilzt bei 843
6-10		95	53-56	100-120			29	8,0	16-27	42-53	zerfällt bei 150
6-10		95	53-56	100-120				8,0	16-27	42-53	zerfällt bei 150
3			68,5	102				8,5		46-55	
3-4				104-106				8,5		ca. 30	
2-4				116-125			36	7,5		ca. 30	
28-48		85	48,5-51,6	76-97	64		52	14,5	30	39-49	zerfällt bei 130 verkohlt bei 300
16-35			51	83	47		38	9,5	36-39	35-54	
	0	0		100							schmilzt bei 1150
	0	0		100				14,0			schmilzt bei 1550

§ 60. Die Ergebnisse der Gebrauchswertforschung.

Auf diesem Gebiet hat besonders H. BÖHRINGER[1] durch umfassende Arbeiten Wesentliches beigetragen. In zahlreichen Untersuchungen seines Institutes, und zwar durch Tragversuche mit den verschiedensten Bekleidungsstücken, wie Herrenoberhemden, Herrentrikotunterwäsche, Monteuranzügen, Herrensocken und Damenstrümpfen aus verschiedenen Textilmaterialien hat er den Gebrauchswert bzw. Schadensendwert nach den im Kapitel IV (§ 21d) geschilderten Bewertungsmethoden für den jeweiligen Verwendungszweck ermittelt. In der Abb. VII, 55a ist die Rangordnung einer Tragversuchsreihe mit Herrenoberhemden aus verschiedenartigen *B*-Zellwolltypen und zwei Baumwolltypen nach den Schadensendwerten wiedergegeben[2]. Hierbei bedeutet der niedrigste Schadensend-

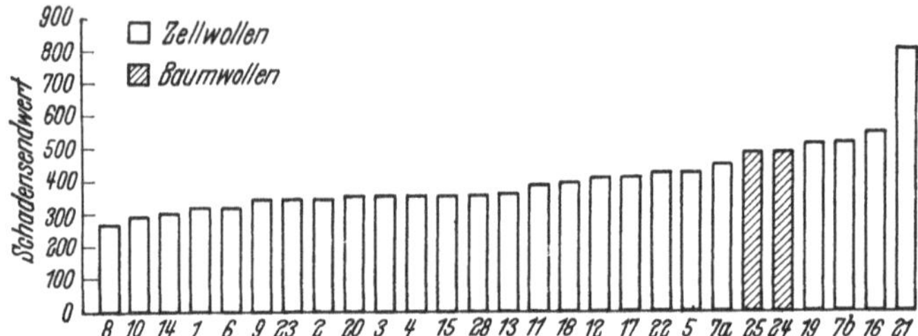

Abb. VII, 55a. Rangordnung einer Tragversuchsreihe mit Herrenoberhemden. (Nach H. BÖHRINGER.)

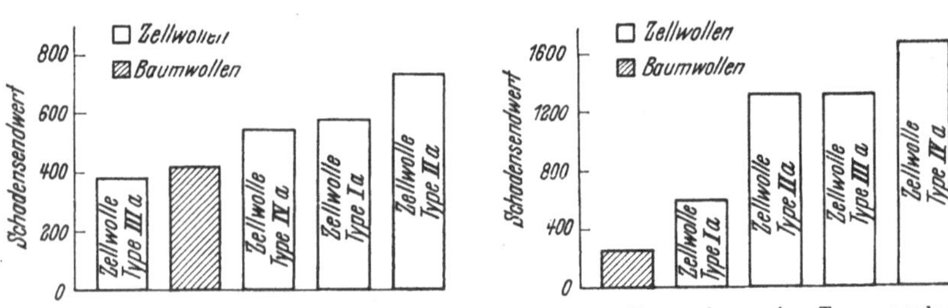

Abb. VII, 55b. Rangordnung einer Tragversuchsreihe mit Monteuranzügen. (Nach H. BÖHRINGER.)

Abb. VII, 55c. Rangordnung einer Tragversuchsreihe mit Herrentrikotunterwäsche. (Nach H. BÖHRINGER.)

---

[1] BÖHRINGER, H.: Textile Gebrauchswertprüfung I–VI. Faserforsch. u. Textiltechn. **5**, 9 (1954); **5**, 55 (1954); **5**, 91 (1954); **5**, 159 (1954); **5**, 193 (1954); **5**, 242 (1954); Textile Analyse der synthetischen Fasern **6**, 314 (1955); Textilprax. **10**, 751 (1955). — L. J. WEINER u. ST. J. KENNEDY: J. Text. Inst. **44**, 433 (1953). — W. SIMON: Rayon and Synthetic Textiles **33**, 31, 58, 61 (1952). Vgl. auch den kurzen Hinweis von F. FELBEL: Textil-Prax. **8**, 341 (1953) sowie A. M. KUSSNEZOW: Tekstiljnaja Promyschlenlostj **7**, 21 (1951). Für den speziellen technischen Sektor ist noch beispielsweise H. W. HOHLS: Melliand-Textilber. **34**, 121 (1953) zu erwähnen.

[2] Auf die Gebrauchswerteigenschaften, die nicht durch das mechanische Verhalten der Faserstoffe bedingt sind, wie z. B. die Gewebeverschmutzung und deren Messung (vgl. New York Section Amer. Dyestuff Reporter p. 322, 1952) oder der Voraussage des Wärmebehaglichkeitsgleichgewichtes aus den physikalischen Daten des Gewebes, wie sie von H. B. HARDY, J. W. BALLBOU u. O. C. WETMORE: Text. Res. J. **23**, 1 (1953) auf Grund sorgfältiger Messungen versucht worden ist, wollen wir trotz der unzweifelhaften Bedeutung für die Praxis nicht eingehen.

wert, wie in Kapitel IV (§ 21d) erläutert, das im Gebrauch am besten bewertete Material, während der höchste Schadensendwert auf das im Gebrauch am ungünstigsten sich verhaltende Textilmaterial schließen läßt. Nach dieser Rangordnung werden die Oberhemden aus Baumwolle durch die Oberhemden der verschiedenen Zellwolltypen in ihren Trageigenschaften erheblich übertroffen. Gehen wir zu der Abb. VII, 55b über, so stellen wir fest, daß die Verwendung von reiner Baumwolle bei Monteuranzügen erheblich günstiger ist und diese nur durch die Zellwolltype IIIa in der Haltbarkeit übertroffen wird. Noch krasser sieht man den Unterschied in der Abb. VII, 55c. Hier rückt Unterwäsche aus Baumwolle mit Abstand an die Spitze und läßt sämtliche Zellwolltypen ganz erheblich zurück. Daraus folgt bereits die Bestätigung der Ausführungen in § 21d, daß man nicht von einem allgemeinen Gebrauchswert sprechen kann, sondern daß dasselbe Material je nach dem Verwendungszweck in seinen Trag- bzw. Gebrauchseigenschaften sowohl an der ersten als auch an der letzten Stelle stehen kann.

Die mit den in Kapitel IV (§ 22a) geschilderten Gerätetypen ermittelten physikalischen Meßwerte wollen wir nunmehr in Beziehung setzen zu den Schadensendwerten, wie sie dem Gebrauchswert zugrunde gelegt werden. In der Abb. VII, 56a ist die Reißlänge bzw. die Zugfestigkeit der

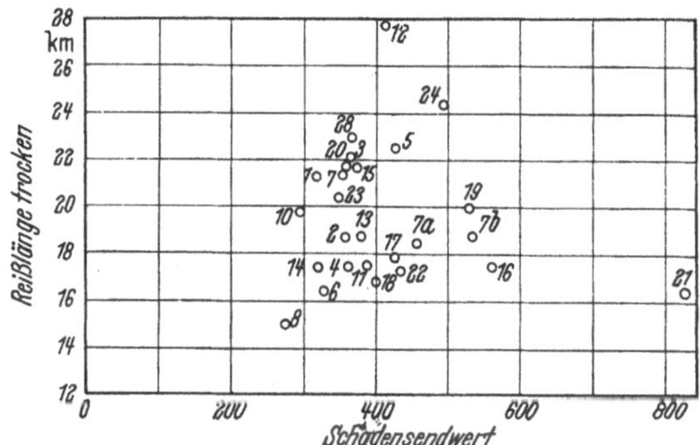

Abb. VII, 56a. Der Zusammenhang zwischen der Reißlänge der für die Tragversuchsreihe mit Herrenoberhemden verwendeten Textilmaterialien und den Schadensendwerten der Tragversuchsreihe. (Nach H. Böhringer.)

trockenen Fäden der verschiedenen Textilmaterialien, wie sie für die Tragversuchsreihe mit Herrenoberhemden (Abb. VII, 55a) verwendet wurden, in Abhängigkeit vom Schadensendwert aufgetragen[1]. Wir sehen, daß eine Relation zwischen dem Schadensendwert und der Zugfestigkeit des verwendeten Materials nicht besteht. Dasselbe Ergebnis finden wir, wenn wir die Bruchdehnung der trockenen Fäden dem Schadensendwert gegenüberstellen (vgl. Abb. VII, 56b).

Betrachten wir die Abb. VII, 56c, so wiederholt sich dieses Ergebnis auch bei der Prüfung des Zusammenhanges zwischen der relativen Schlingenfestigkeit und dem Schadensendwert. Ebenso führt die Gegen-

[1] Leider vermißt man bei den Untersuchungen von H. Böhringer die Angabe der Streuungen; die statistische Beurteilung bzw. Bewertung wird dadurch sehr erschwert.

überstellung der Zugfestigkeitswerte der nassen Fäden, sowie deren Bruchdehnung mit den Schadensendwerten zu keinem anderen Resultat. Damit kommt folgendes klar zum Ausdruck:

Die bei der üblichen statischen Beanspruchung der das Gewebe aufbauenden Fasern und Fäden ermittelten physikalischen Meßwerte stehen nach Überschreitung gewisser Mindestwerte in keiner Relation zu der Bewährung des Gewebes bei der Beanspruchung im Gebrauch.

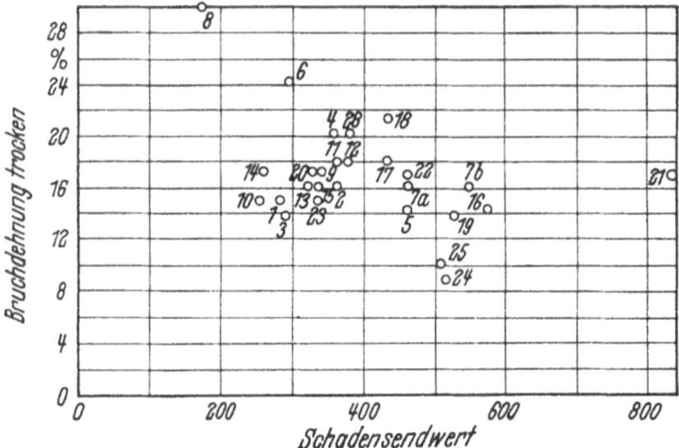

Abb. VII, 56b. Der Zusammenhang zwischen der Bruchdehnung der für die Tragversuchsreihe mit Herrenoberhemden verwendeten Textilmaterialien und den Schadensendwerten der Tragversuchsreihe. (Nach H. BÖHRINGER.)

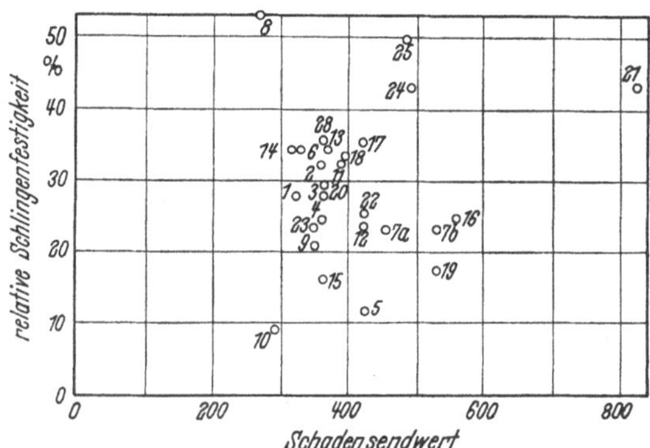

Abb. VII, 56c. Der Zusammenhang zwischen der relativen Schlingenfestigkeit der für die Tragversuchsreihe mit Herrenoberhemden verwendeten Textilmaterialien und den Schadensendwerten der Tragversuchsreihe. (Nach H. BÖHRINGER.)

In der Abb. VII, 57 sind die Meßwerte nach verschiedenen Behandlungsstufen einschließlich der Färbung und nach zehn Tragperioden bei Damenstrümpfen aus Viscose- und Kupferreyon wiedergegeben. Wir sehen bereits, daß grundsätzlich die Verarbeitung und Färbung auf die Reißlänge bzw. Zugfestigkeit, Gesamtdehnung, elastische Dehnung und elastisches Arbeitsvermögen[1] in allen Fällen sowohl bei Kupfer- als auch Viscosereyon einen wesentlich größeren Einfluß haben als die Beanspruchung durch zehn Tragperioden. Weiter ist zu sagen, daß die Drehung bei Kupferreyon sich wesentlich auf die physikalischen Eigenschaften

[1] Erläuterung auf S. 498/499.

und ihre Veränderungen auf dem Verarbeitungswege auswirkt, so daß man nur das Material gleicher Drehung von Kupfer- und Viscosereyon vergleichen darf. Hier kommt ebenfalls zum Ausdruck, daß die Zugfestigkeit am stärksten durch die Verarbeitung abnimmt[1] und, wie die Abb. VII, 57 erkennen läßt, an Viscose- und Kupferreyon im gleichen Umfang. Bei der Gesamtdehnung sehen wir bereits einen erheblichen Unterschied, wobei auch hier hervorgehoben werden muß, daß die Verarbeitung die stärkste Abnahme bedingt. Dagegen muß gesagt werden, daß der Färbeprozeß und die zehn Tragperioden sich wesentlich geringer bei Viscosereyon auswirken als bei Kupferreyon. Bei der elastischen Dehnung betragen die Werte 32% und 19%. Auch bei dem Gesamt- wie bei dem elastischen Arbeitsvermögen ist ein erheblicher Unterschied zwischen Kupfer- und Viscosereyon vorhanden, und zwar 53% und 44% bzw. 55% und 44%. Abgesehen von der starken Abnahme auf dem Verarbeitungsweg wirkt sich hier auch die Beanspruchung durch den Tragversuch wesentlich aus, und zwar erheblich stärker bei Kupferreyon als bei Viscosereyon[2]. Da das elastische Arbeitsvermögen einer Faser noch diejenige mit statisch arbeitenden Prüfgeräten zu ermittelnde Größe ist, die mit dem größten Gewicht in den Schadensendwert eingeht[3], so zieht H. BÖHRINGER auch den Schluß, diese Größe als Grundlage zur Erspinnung gebrauchswertoptimaler Chemiefasern zu benutzen.

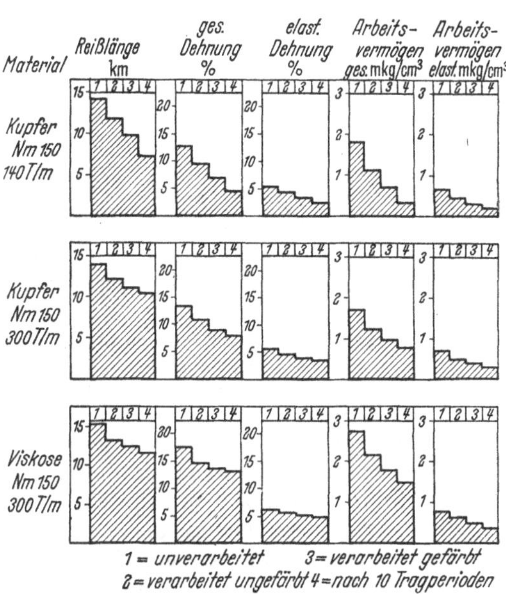

Abb. VII, 57. Tragversuch mit Damenstrümpfen aus Viscosereyon und Kupferreyon Nm 150. Physikalische Meßwerte nach verschiedenen Behandlungsstufen. (Nach H. BÖHRINGER.)

In der Abb. VII, 57 a ist dieses Schema dargestellt. Wir sehen daraus, daß das elastische Arbeitsvermögen bei dem Verstreckungsvorgang ein Optimum durchläuft, dem ein Dehnungs- und Verstreckungsoptimum

[1] Wie empfindlich auch die Chemiefaser auf Änderungen des Spinnverfahrens reagiert, läßt der Vortrag von H. ASHTON, gehalten vor dem Internationalen Chemiefaser-Kongreß, Paris, Juli 1954, erkennen.

[2] Dies ist nur ein praktisches Beispiel, wie sich die Verarbeitung auf die physikalischen Meßwerte auswirkt. Sowohl in der Zwirnerei wie Weberei und Wirkerei sind, wie in dem Kap. IV (§ 21 c) geschildert wird, naturgemäß noch eine Reihe weiterer Einflußfaktoren bei der Verarbeitung vorhanden (vgl. auch H. BÖHRINGER[4]). — [3] Ein Hinweis, das P—ΔL-Diagramm eingehend zu untersuchen.

[4] BÖHRINGER, H.: Faserforsch. u. Textiltechn. 5, 247–252 (1954).

zugeordnet ist. Die Festigkeit ist auch hier in keiner Weise durch einen besonderen Wert gekennzeichnet. Dieses Diagramm läßt ebenfalls erkennen, wie abwegig es wäre, optimale Festigkeitseigenschaften züchten zu wollen; sie könnten sich im Gebrauch katastrophal auswirken. Es darf allerdings nicht verschwiegen werden, daß diese Funktionen statistische Funktionen sind mit mehr oder weniger breiten Verteilungen, so daß nur auf statistischem Weg mit genügender Sicherheit derartige Aussagen gewonnen werden können und daher die notwendige Skepsis auch diesem Schema gegenüber angebracht erscheint.

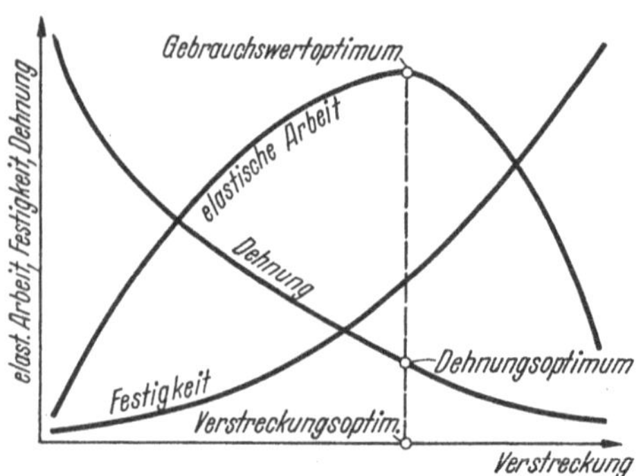

Abb. VII, 57a. Schema zur Erspinnung gebrauchswertoptimaler Chemiefasern. (Nach H. BÖHRINGER.)

Die bisherigen Ausführungen beschränkten sich auf die Untersuchungen der statischen Beanspruchung der Prüffasern und -Fäden sowie auf die Bewährung des aus diesen Fasern und Fäden hergestellten Materials und führte zu keinen gesicherten Relationen zwischen den ermittelten Prüfwerten und den Gebrauchseigenschaften.

Nun geht bereits aus dem Kapitel IV (§ 21c2 und § 21c3) hervor, daß sowohl bei der Verarbeitung[1] als auch im späteren Gebrauch eine dynamische Beanspruchung bzw. eine Wechselbeanspruchung auftritt, die es zweckmäßig erscheinen läßt, hier die Verbindung zu suchen, die zwischen dem Gebrauchswert und den auf statischem Wege ermittelten Meßwerten nicht festzustellen war. Eine bisher sehr oft angewandte Prüfmethode dynamischen Charakters ist die Bestimmung der Knickbruchfestigkeit. Es liegt jedoch eine zusammengesetzte komplexe Beanspruchung dabei vor, und das Prüfgerät ist eher mit einem Modellgerät zu vergleichen, da hier eine wirklich definierte physikalische Beanspruchung nicht realisiert wird (vgl. Kapitel IV, § 23d).

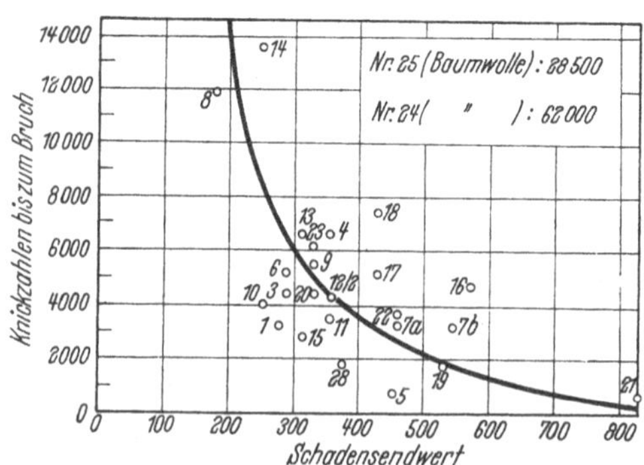

Abb. VII, 58. Der Zusammenhang zwischen den Knickzahlen bis zum Bruch der für die Tragversuchsreihe mit Herrenoberhemden verwendeten Textilmaterialien und den Schadensendwerten der Tragversuchsreihe. (Nach H. BÖHRINGER.)

Das Ergebnis dieser Prüfung (für Herrenoberhemden) zeigt die Abb. VII, 58.

---

[1] Welche Beanspruchungen bei der Verarbeitung auftreten können, zeigen insbesondere die Untersuchungen von W. FRENZEL und H. HESSE: Faserforschung u. Textiltechnik **6**, 189 (1955) an Garn verarbeitenden Maschinen.

Zwischen den Knickzahlen bis zum Bruch und dem Schadensendwert besteht eine gewisse Korrelation. Die Streuung ist außerordentlich groß und die Zuverlässigkeit der Aussage noch sehr gering. Das beweist auch, daß die Baumwolle, die für diese Herrenoberhemden verwendet wurde und einen großen Schadensendwert zeigte, hier außerordentlich günstige Werte erkennen läßt. Zu ähnlichen Resultaten kommt man auch bei der Scheuerprüfung, wie die ausführliche Erörterung in Kapitel IV, § 21d zeigt, wobei der Übergang zu Mischgespinsten, wie A. Wilhelm[1] und auch J. F. Sayre und A. J. Weldon[2] ausführen, noch den starken spezifischen Einfluß der Anteile bestimmter Fasermaterialien, wie z. B. Dacron (vgl. Abb. 58a) auf die Höhe der Prüfwerte erkennen läßt.

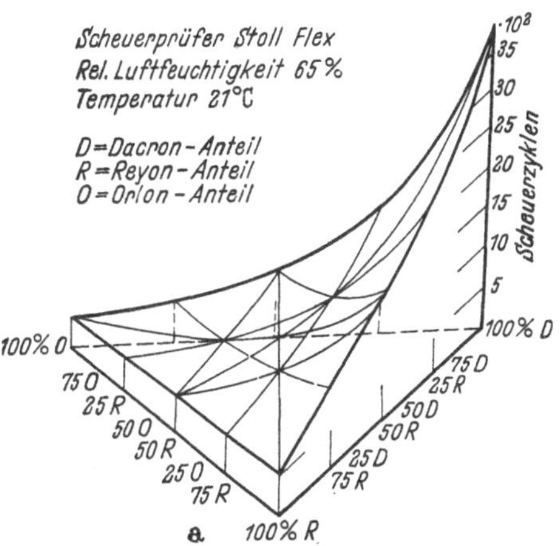

Abb. VII, 58a. Scheuerfestigkeit bei Dacron-, Reyon- und Orlonmischgeweben. (Nach J. F. Sayre und A. J. Weldon.)

Neben diesen Prüfgeräten wird von H. Böhringer das Splittrigkeitsprüfgerät besonders herausgehoben, mit dem eine annähernd lineare Beziehung zwischen den Prüfwerten und dem Schadensendwert festgestellt wurde. Die Ergebnisse sind nur sehr kritisch zu bewerten, da es sich hier um ein Gerät handelt, das den Waschvorgang in einer Trommelwaschmaschine nachahmt und daher die Beanspruchung durch den Waschvorgang anzeigt. Die gefundenen Relationen können also nur besagen, daß diesen Tragversuchen der Waschvorgang bzw. die Beeinflussung des Tragversuchsstückes durch die Wäsche so groß war, daß alle anderen Einflußfaktoren demgegenüber praktisch zu vernachlässigen waren[3]. Die Wäsche bedingt also hier den größten Verschleiß. In allen anderen Fällen wird man mit dem Splittrigkeitsprüfgerät keine Aussagen über den Gebrauchswert machen können.

Wir ersehen daraus, daß die bisherige Analyse der Vorgänge zumindestens zu grob gewesen ist, um die Zusammenhänge wirklich zu erkennen. Wir wollen daher jetzt die Beanspruchungen genauer analysieren und versuchen, auf diesem Wege weiterzukommen. Es ist naheliegend infolge des Hinweises Fußnote 3, S. 489, mit dem KD-Diagramm bzw. P–Δ L-Diagramm zu beginnen und diese Diagramme einer kritischen Betrachtung zu unterziehen.

---

[1] Wilhelm, A.: Textil-Prax. **7**, 590 (1952).

[2] Sayre, J. F. u. A. J. Weldon: Internat. Chemiefaser-Kongreß, Paris, Juli 1954.

[3] Vgl. auch S. Köhler: Text. Res. J. **24**, 173 (1954).

§ 61. Das KD (P⊿L)-Kraft-Längenänderungs-Diagramm.

Wie aus Kapitel IV (§ 22a) hervorgeht, müssen bei den nichtidealen elastischen Stoffen die Zeiteffekte mit berücksichtigt werden, da sie von wesentlichem Einfluß auf die Kraft-Längenänderungs-Kurven sind. Es wurde daher in diesem Abschnitt auch ausführlich darauf eingegangen und die zur einwandfreien Ermittelung der P⊿L-Kurven einzuhaltenden Bedingungen angegeben, insbesondere wurden die Voraussetzungen genannt, die die Prüfgeräte zur Aufnahme des K⊿L-Diagramms erfüllen müssen. Trotzdem müssen wir zwei Prüfmethoden und damit auch Prüfgerätetypen unterscheiden, und zwar einmal den Prüfgerätetyp, der mit einer konstanten Dehnungsgeschwindigkeit arbeitet, d.h., bei dem die Dehnung proportional mit der Zeit zunimmt, und der zweite Typ, bei dem die Belastungsgeschwindigkeit konstant ist, d.h., die Zugkraft proportional mit der Zeit zunimmt. Es ist ersichtlich und braucht nicht besonders erörtert zu werden, daß die beiden Gerätetypen bei demselben Material zu verschiedenen P⊿L-Diagrammen führt. Es sei daher besonders hervorgehoben, daß die in der Abb. VII,59 wiedergegebenen KD-

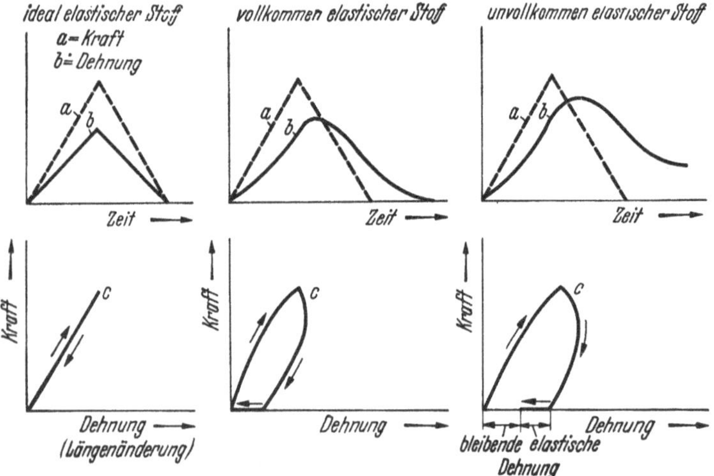

Abb. VII, 59. Typische Kraft-Längenänderungs- (Dehnungs-) Diagramme. (Nach DE VRIES.)

(P⊿L)-Diagramme nach DE VRIES[1] mit einer konstanten Belastungsgeschwindigkeit aufgenommen worden sind. Der verwendete Prüfgerätetyp ist die in dem Kapitel IV (§ 22a) beschriebene Neigungswaage. Die mit diesem Prüfgerätetyp aufgenommenen Diagramme veranschaulichen das Verhalten eines ideal-vollkommen und eines unvollkommen elastischen Stoffes bei dieser Beanspruchung, wobei *a* jeweils den Kraftverlauf in Abhängigkeit von der Zeit, *b* den Verlauf der Dehnung mit der Zeit

[1] VRIES, H. DE: Appl. Sci. Res. A **3**, 111 (1952). Eine graphische Analyse der Spannungs-Dehnungskurven speziell von Papier, nehmen O. ANDERSSON, B. IVARSSON, A. H. NISSAN u. B. STEENBERG: Proc. Techn. Sect. Paper makers. Assoc. **30**, 43 (1949) vor.

und *c* das Kraft-Dehnungs-Diagramm bzw. die Kraft-Längenänderungs-Kurve darstellen. Man sieht unmittelbar, daß die Zeiteffekte bei dem vollkommen und unvollkommen elastischen Stoff gegenüber dem ideal elastischen Stoff dadurch in Erscheinung treten, daß das Maximum der Dehnung nicht mit dem Maximum der Kraft zusammenfällt, sondern zeitlich verschoben ist. Der unvollkommen elastische Stoff unterscheidet sich von dem vollkommen elastischen dadurch, daß das PΔL-Diagramm bei dem unvollkommen elastischen Stoff nicht mehr geschlossen ist. Ein von Viscosereyon aufgenommenes PΔL-Diagramm zeigt die Abb. VII,60, und hier sehen wir bereits die Ähnlichkeit mit dem Kurvenzug des unvollkommen elastischen Stoffes der Abb. VII,59. Davon wesentlich verschieden ist das PΔL-Diagramm für Wolle trocken und naß, Abb 61. Vgl. auch W. Kast, Kapitel VII A, § 54 und die ausführliche Diskussion von Frode Andersen[1] über den Einfluß der relativen Luftfeuchtigkeit und der Quellung auf den Verlauf der Kraft-Längenänderungs-Diagramme.

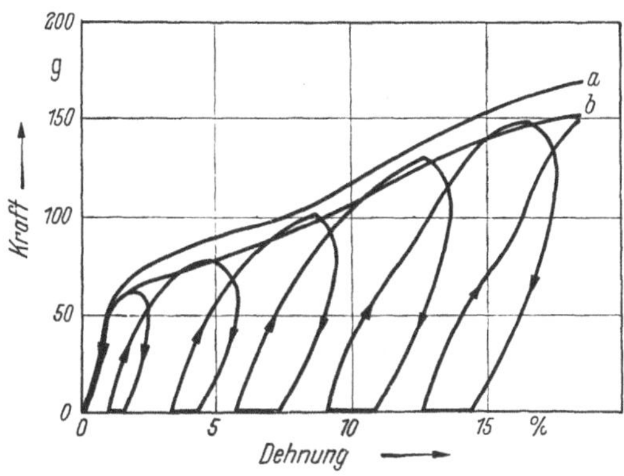

Abb. VII, 60. Kraft-Dehnungs-Kurve von Viscose-Reyon.

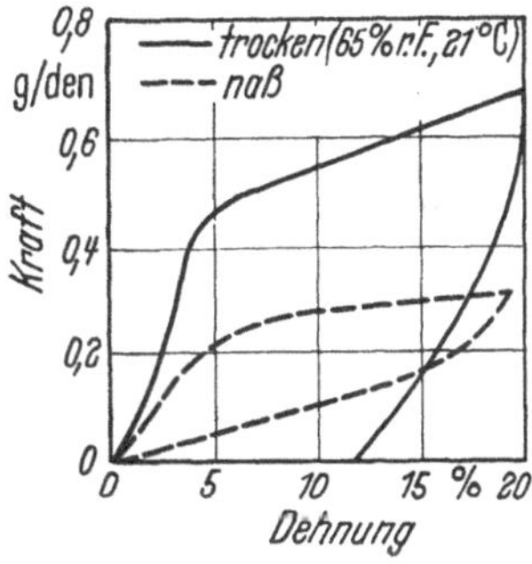

Abb. VII, 61. Kraft-Dehnungs-Diagramme von Wollfasern trocken und naß.

Beschränken wir uns zunächst auf den einmaligen Belastungsvorgang, wie er mit den üblichen statischen Festigkeitsprüfgeräten [vgl. Kapitel IV (§ 22a)] aufgenommen wird, so erhalten wir die in der Abb. VII, 60 wiedergegebenen Kurvenzüge *a* und *b* von Viscosereyon und die Diagramme der Abb. VII, 62a bis d sowie von verschiedenen PAN-Fasern (Abb. 62e[2]).

[1] Frode Andersen: Trans. Danish Acad. Techn. Sci. **3**, 1 (1950).

[2] Vergleicht man insbesondere die Bezeichnungen der Ordinaten bei den hier wiedergegebenen Diagrammen, so finden wir Kraft in g und Kraft in g/den sowie Reißlast in g/den, Belastung (Spannung) in kg/mm². Bei anderen Literaturstellen findet man genau so die Angabe Kraft in g/*g x* usw. Wir haben davon abgesehen, diese Ordinatenbezeichnungen zu ändern, da die Diagramme verschiedenen Literaturstellen entnommen worden sind; möchten jedoch, um Mißverständnisse zu vermeiden, auf die korrekten Definitionen, wie sie z. B. in dem Normentwurf DIN 53816 wie auch in dem Beitrag von P. A. Koch: Landolt-Börnstein Stoffwerte, 6. Auflage, Springer-Verlag 1955, Bd. IV, Teil 1, 4137, S. 385ff. und W. Kast, Kapitel VII, A, § 54, angegeben worden sind, verweisen.

In diesem Zusammenhang sei auch noch ein Hinweis auf die Auswertung von Kraft-Längenänderungs-Diagrammen gestattet, da immer noch die Mittelung der

Dieses umfangreiche Kurvenmaterial, das vorwiegend dem Bericht von R. DOMKE[1] entnommen ist, bildet eine notwendige Ergänzung zu den Zahlenwerten der Tabelle VII, 5 und gibt uns bereits ein anschauliches Bild, daß dieser mehr oder weniger komplizierte Kurvenverlauf im allgemeinen nicht durch wenige Kennzahlen beschrieben werden kann. Will man etwas über den Gebrauchswert aussagen, so wird man zumindest das gesamte Kraft-Längenänderungs-Diagramm in seinem Verlauf in Relation zu dem Verhalten bei der Verarbeitung oder dem Gebrauch dieser Fasern setzen müssen. Nun zeigt jedoch der Versuch, die üblichen Kenngrößen der Kurven der Abb. VII, 62e, die eine Reihe von PΔL-Diagrammen von PAN-Fasern darstellen, in Beziehung zu den Werten der Tabelle VII, 6, die sich aus der dynamischen Zugwechselbeanspruchung derselben PAN-Fasertypen ergeben, zusetzen, ein negatives Resultat. Es besteht keine Korrelation zwischen den üblichen Kennwerten der PΔL-Diagramme und dem Verhalten bei der dynamischen Zug-

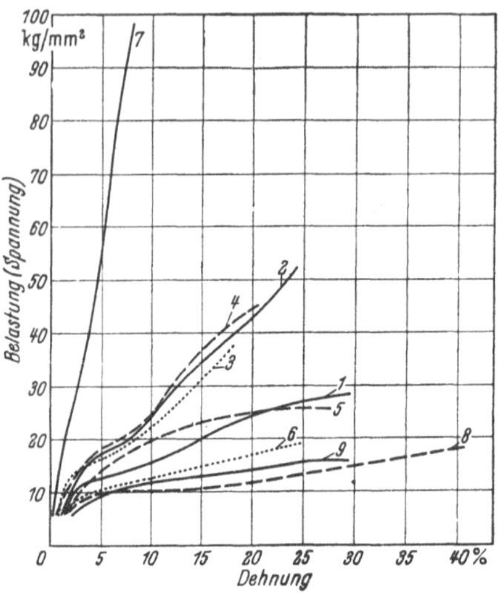

Abb. VII, 62 a. Spannungs-Dehnungs-Kurven von Chemiefasern auf Zellulosebasis. (Nach R. DOMKE.) *1* Viscose-Reyon, normal; *2* Viscose-Reyon, hochfest; *3* Viscose-Zellwolle, normal; *4* Viscose-Zellwolle, hochfest; *5* Kupfer-Seide; *6* Kupfer-Zellwolle; *7* Acetat-Streckseide, Fortisan; *8* Arcetat-Seide; *9* Acetat-Zellwolle.

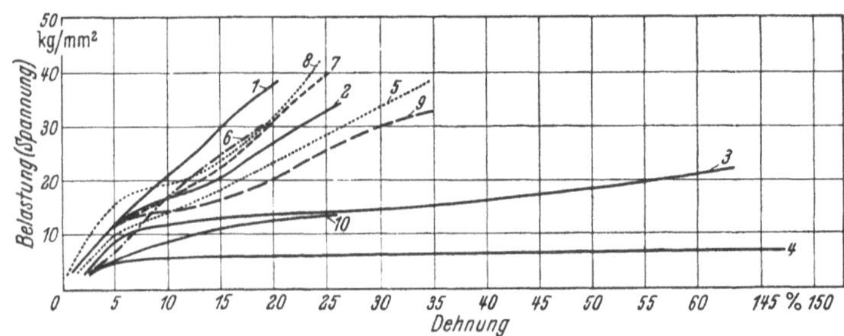

Abb. VII, 62b. Spannungs-Dehnungs-Kurven von Polymerisatfasern. (Nach R. DOMKE.) *1* Polyvinylchlorid, Rhovyl, endlos; *2* Polyvinylchlorid, Rhovyl, Faser; *3* Polyvinylchlorid, nachchloriert, Faser, PeCe; *4* Mischpolymerisat aus Vinylchlorid und Vinylacetat, Vinyon HH, Faser; *5* Mischpolymerisat aus Vinylchlorid und Acrylnitril, Dynel, Faser; *6* Mischpolymerisat aus Vinylchlorid und Vinylidenchlorid, Saran, endlos; *7* Polyvinylalkohol, Kuralon, Faser; *8* Polyacrylnitril, X-51, endlos; *9* Polyacrylnitril, X-51, Faser; *10* Polyäthylen, Courlene, monofil, endlos.

Dehnungswerte bei bestimmten Kraftstufen vorgenommen wird, die unzulässig ist. Ebenso ist die Mittelung der Kraftwerte bei bestimmten Dehnungsstufen unzweckmäßig, da z. B. bei vollsynthetischen Fasern mit sehr steilen Kraft-Längenänderungskurven in keiner Weise die Charakteristik des jeweiligen Schaubildes zum Ausdruck kommt. Es ist also nur die grafische Mittelung zu empfehlen.

[1] DOMKE, R.: Vortrag vor dem Internat. Chemiefaser-Kongreß D. T. 17, Paris 1954.

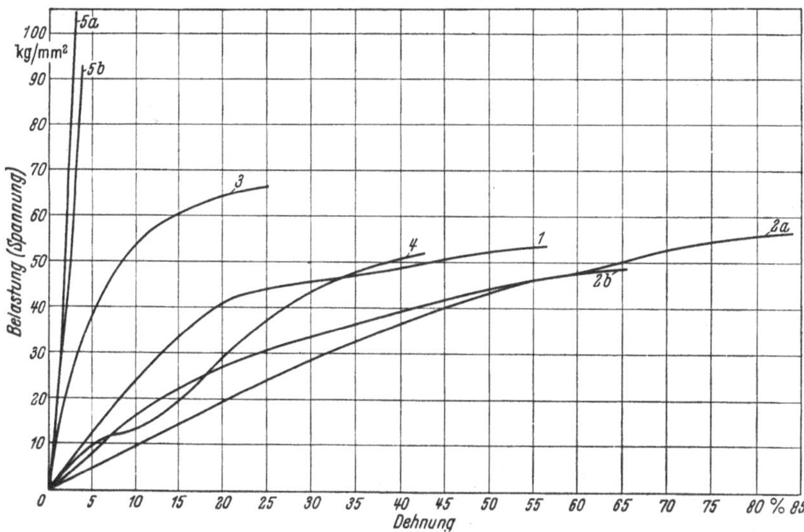

Abb. VII, 62c. Spannungs-Dehnungs-Kurven von Polykondensat- und Glasfasern. (Nach R. DOMKE.) *1* Polyamid 6,6, Nylon, endlos; *2a* Polyamid 6, Perlon, endlos; *2b* Polyamid 6, Perlon, Faser; *3* Polyester, Terylen, endlos; *4* Polyester, Dacron, Faser; *5a* Glasfaser, Düsenblasverfahren, Vetrotex; *5b* Glasfaser, Stab-Ziehverfahren, Schuller.

Tabelle VII,6.
Zugwechselfestigkeit von PAN-Fasern.
Einspannlänge: 25 mm,
Periodische Dehnung: 5%,
Frequenz: 25 Hz.

Probe	Belastungs-wechsel	Dehnung (%)
1	70000	30
2	>100000	43
3	7000	13
4	3000	49
5	400	18,5
6	300	30

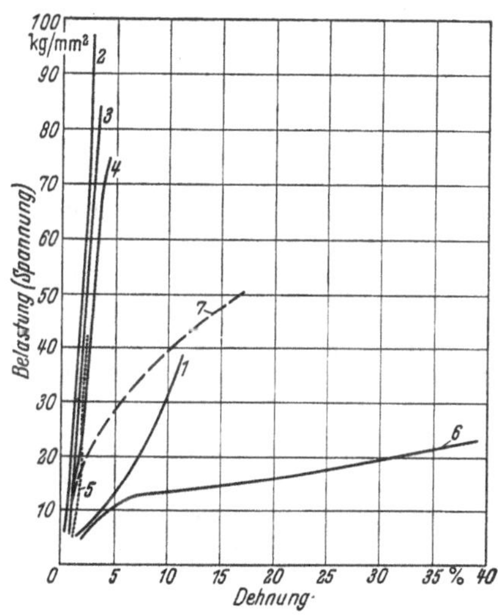

Abb. VII, 62d. Spannungs-Dehnungs-Kurven von Naturfasern. (Nach R. DOMKE.) *1* Baumwolle; *2* Flachs; *3* Hanf; *4* Ramie; *5* Jute; *6* Schafwolle; *7* Naturseide.

wechselbeanspruchung. Daher ist bei der einfachen Auswertung des PΔL-Diagramms auch prinzipiell keine Aussage über das Verhalten gegenüber Zugwechselbeanspruchungen, wie sie bei der Verarbeitung auftreten, möglich. Hieraus folgt bereits, daß man einerseits der Ansicht, aus dem PΔL-Diagramm abgelesene Eigenschaften als „Gesundheitszeugnis" des Materials zu werten, mit Skepsis begegnen muß und andererseits nicht übersehen darf, daß man immer die Veränderungen der *PΔL*-Diagramme zur Charakterisierung der Schädigung heranzieht[1].

[1] Vgl. auch W. SCHEFER: Textil-Rundschau **10**, 279 (1955).

Gehen wir nun wieder auf die Abb. VII, 60 zurück, so können wir uns die PΔL-Kurve auch als Enveloppe einer nicht geschlossenen Stufenkurve, wie sie durch eine wiederholte stufenweise Belastung kleiner Frequenz entsteht, auffassen. Die Aufnahme dieses stufenweisen PΔL-Diagrammes ist nun infolge der zu berücksichtigenden Zeiteffekte nach DIN 53835, Entwurf August 1954, genormt, und zwar sind sowohl die Höhe der einzelnen Kraft- oder Belastungsstufen wie die Belastungsgeschwindigkeit sowie die Erholungszeit nach völliger Entspannung genau festgelegt[1].

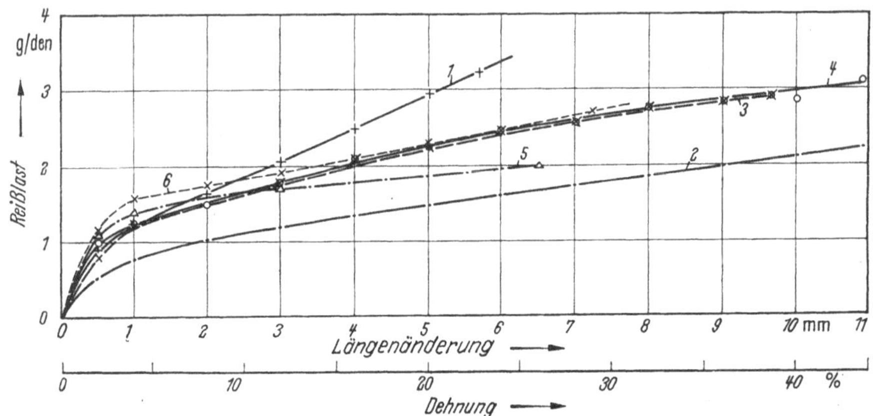

Abb. VII, 62e. Spannungs-Dehnungs-Kurven verschiedener PAN-Fasern.

In der Abb. VII, 63 ist das Schema wiedergegeben, nach dem die Probe stufenweise belastet wird. Die Höhe der einzelnen Kraftstufen ist nach DIN 53835 so zu wählen, daß zunehmend von 10% zu 10% der mittleren Reißlast beansprucht wird. Die Belastungsgeschwindigkeit soll so eingestellt werden, daß die vorgeschriebene Kraftstufe in etwa 20 Sekunden erreicht wird. Sofort nach Erreichen der jeweiligen Kraftstufe wird entlastet. Die Ruhezeit nach der völligen Entlastung auf die Vorlast P_v soll 1 Minute betragen. Bei Erreichen des Kraftwertes P_1 wird die Gesamtdehnung (ε_{ges1}) abgelesen, nach der sofort beginnenden Entlastung[2] sowie nach 1 Minute Ruhepause wird die Restdehnung ε_{r1} abgelesen. Dann beginnt das Belastungsspiel von neuem bis zur nächsthöheren Belastungsstufe P_2 usw. Um in die große Anzahl von abgeleiteten Begriffen bei der Auswertung von Elastizitätswerten Klarheit zu bringen, wollen wir nunmehr die einzelnen Begriffe näher erklären und an einem Beispiel (vgl. DIN 53835) deren Berechnung erläutern.

[1] Diese Normung bedeutet eine willkürliche Einschränkung der Variationsmöglichkeiten in der Richtung einer Einpunktsmethode.

[2] Diese DIN-Vorschrift berechnet die unmittelbar nach der Entlastung vorhandene Restdehnung $\varepsilon_{r\,s1}$ als Verhältnis der Restverlängerung zur ursprünglichen Einspannlänge und gibt dabei gar nicht an, wie diese Restverlängerung festgestellt wird. Dasselbe geschieht mit der elastischen Nachwirkung $\varepsilon_{el,\,n}$. Hieraus folgt, daß dieses Normblatt noch die üblichen Festigkeitsprüfgeräte, wie sie in Kap. IV (§ 22a) beschrieben worden sind, zugrunde legt, wobei die Pendelgeräte nach unserer Auffassung jedoch für derartige Untersuchungen gar nicht geeignet sind. Es sollte vielmehr ein Prüfgerätetyp mit wegloser Kraftmessung und automatischer Steuerung, wie z. B. der Statigraph – Kap. IV (§ 22d) – oder der Instrontester – Kap. IV (§ 22d) – vorgeschrieben werden.

Die Gesamtdehnung ε_{ges} ergibt sich demnach zu

$$\varepsilon_{ges} = \frac{L - L_0}{L_0} \cdot 100\,(\%)$$

als Verhältnis der Verlängerung zur ursprünglichen Meßlänge.

Beispiel: $L_0 = 500$ mm,
$L = 636$ mm,
folglich $\underline{\varepsilon_{ges} = 27{,}2\%}$.

Die sich nach der Entlastung und nach 1 Minute Ruhepause einstellende Restdehnung ε_r ist das Verhältnis der Restverlängerung zur Einspannlänge, also

$$\varepsilon_r = \frac{L_r - L_0}{L_0} \cdot 100\,(\%)\,.$$

Beispiel: $L_r = 546$ mm, daraus folgt

$$\underline{\varepsilon_r = 9{,}2\,\%}\,.$$

Die elastische Dehnung ε_{el} ist die Differenz der Gesamtdehnung ε_{ges} und der Restdehnung ε_r, also

$$\varepsilon_{el} = \varepsilon_{ges} - \varepsilon_r\,(\%)\,.$$

In unserem Beispiel finden wir für ε_{el} den Wert:

$$\underline{\varepsilon_{el} = 18{,}0\,\%}\,.$$

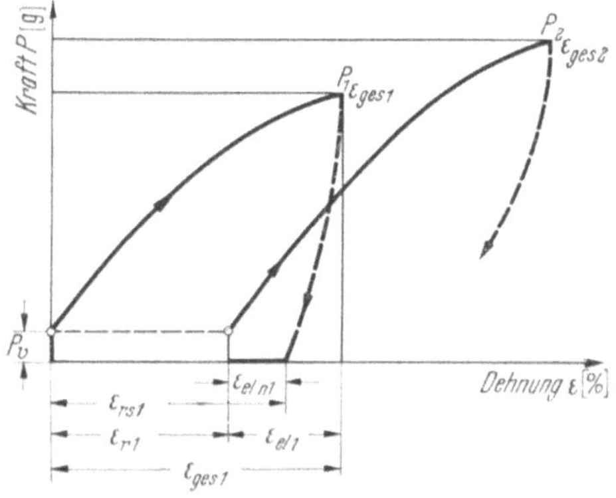

Abb. VII, 63. Schema des Elastizitätsversuchs nach DIN 53835.

Die unmittelbar nach der Entlastung vorhandene, experimentell mit nichtautomatischen Geräten nur schwierig zu bestimmende Restdehnung ε_{rs} wird als Verhältnis der Restverlängerung zur Einspannlänge errechnet.

$$\varepsilon_{rs} = \frac{L_{rs} - L_0}{L_0} \cdot 100\,(\%)\,.$$

In unserem Beispiel ist daher für $L_{rs} = 566$ mm der Wert für

$$\underline{\varepsilon_{rs} = 13{,}2\,\%}\,.$$

Die elastische Nachwirkung $\varepsilon_{el,n}$ ist die Differenz der unmittelbar nach der Entlastung vorhandenen Restdehnung ε_{rs} und der Restdehnung ε_r, so daß

$$\varepsilon_{el,n} = \varepsilon_{rs} - \varepsilon_r$$

ist und in unserem Beispiel den Wert

$$\underline{\varepsilon_{el,n} = 4{,}0\,\%}$$

annimmt.

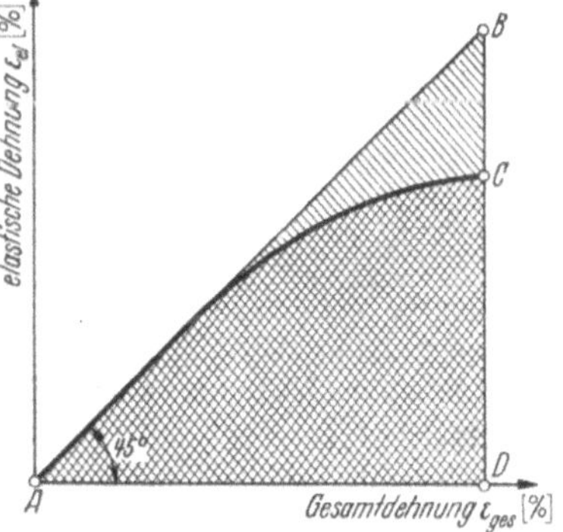

Abb. VII, 64. Elastizitäts-Schaulinie nach DIN 53835.

Durch mehrfaches, durch die Norm festgelegtes Wiederholen der Belastung bis zur nächsthöheren Belastungsstufe und darauffolgende Entlastung auf die Vorspannlast P_v erhält man zu jeder Kraftstufe P einen der eben angeführten und berechneten Werte für ε_{ges}, ε_r, ε_{el}, ε_{rs}, $\varepsilon_{el,n}$.

Werden über der Abszisse als Gesamtdehnung ε_{ges} die jeweils dazugehörigen Werte für die elastische Dehnung ε_{el} aufgetragen und die unter 45° geneigte Gerade für ε_{ges} eingezeichnet, wie dies in Abb. VII, 64 geschieht, so entstehen Punktreihen, die sich bei genügender Anzahl Kraftstufen zu Linien ergänzen lassen. Aus dieser Vorstellung heraus erhalten wir für jede Laststufe P_v den Elastizitätsgrad. Dieser ist definiert durch das Verhältnis von elastischer Dehnung ε_{el} zur Gesamtdehnung ε_{ges}, also

$$\varepsilon_{grad} = \frac{\varepsilon_{el}}{\varepsilon_{ges}} \cdot 100\,(\%)\,.$$

In unserem Beispiel finden wir für

$$\underline{\varepsilon_{grad} = 66\,\%\,.}$$

Der mittlere Elastizitätsgrad $\bar{\varepsilon}_{grad}$ ergibt sich aus der Abb. VII, 64 als Verhältnis der Schaulinienfläche $(A\,C\,D\,A)$ und der Elastizitätskurve $(A\,B\,D\,A)$. Bei der praktischen Auswertung werden diese Flächen ausplanimetriert

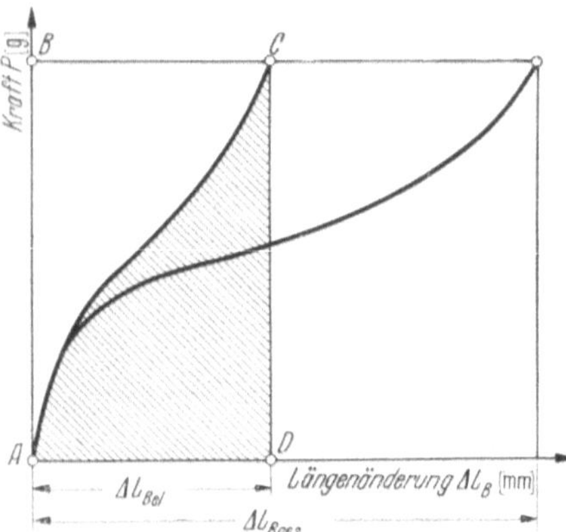

Abb. VII, 65. Diagramm der elastischen und der Gesamtarbeit nach DIN 53835.

$$\bar{\varepsilon}_{grad} = \frac{(A\,C\,D\,A)}{(A\,B\,D\,A)} \cdot 100\,(\%)\,.$$

Beispiel: $(A\,C\,D\,A) = 7530\,\text{mm}^2$,
$(A\,B\,D\,A) = 9248\,\text{mm}^2$.

Daraus folgt für

$$\underline{\bar{\varepsilon}_{grad} = 81\,\%\,.}$$

Die technische Elastizitätsgrenze σ_E ist diejenige auf den Querschnitt F_E der Probe bezogene Kraft P_E, die eine Restdehnung ε_r, das ist die Differenz zwischen Gesamtdehnung ε_{ges} und elastischer Dehnung ε_{el} von max. 0,1% nach einer Ruhezeit von 1 Minute zugeordnet ist, was aus der Abb. VII, 66 hervorgeht. Daher ist

$$\sigma_E = \frac{P_E}{F_E}\,.$$

Setzen wir für F_E den Ausdruck

$$F_E = \frac{1}{Nm \cdot \gamma_N\,(1 + \varepsilon_{E\,el})}\,,$$

so erhalten wir für

$$\sigma_E = P_E \cdot Nm \cdot \gamma_N \cdot (1 + \varepsilon_{E\,el})\,.$$

In unserem Beispiel:

P_E (abgelesen aus Abb. VII, 66) = 88 g,
metrische Nummer = 147 Nm (m/g),
$\gamma_N = 1{,}14\,\text{g/cm}^3$.
Elastische Dehnung, an der Elastizitätsgrenze abgelesen (aus Abb. VII, 66)
$\varepsilon_{E\,el} = 7{,}2\%$.

Daraus ergibt sich

$$\underline{\sigma_E = 15{,}8\,\text{kg/mm}^2\,.}$$

Der E-Modul E ist das Verhältnis von technischer Elastizitätsgrenze σ_E und der zugehörigen elastischen Dehnung $\varepsilon_{E\,el}$

$$E = \frac{\sigma_E}{\varepsilon_{E\,el}}\,(\text{kg/mm}^2)\,.$$

Gehen wir wieder zu unserem Beispiel über, so finden wir für

$$\underline{E = 219\,\text{kg/mm}^2\,.}$$

Wir kommen nun zur Ermittlung der elastischen Arbeit[1]. Diese ist ein Teil der Gesamtarbeit. In Abb. VII, 65 ist das Kraft-Längenänderungs-Diagramm des elastischen Anteils als Linie der Kraft über der elastischen Längenänderung gezeichnet.

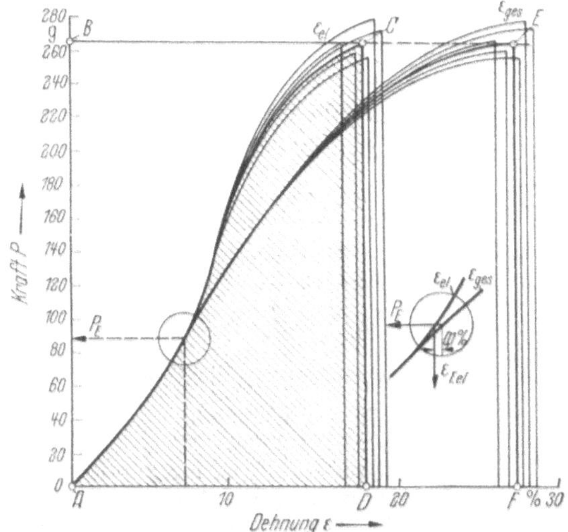

Abb. VII, 66. Schemazeichnung für die Auswertung von Elastizitätswerten nach DIN 53835.

Es dient zum Nachmessen der Kraft und der elastischen Längenänderung der Probe für jeden Zeitpunkt des Versuchs. Das Integral der Kraft über der elastischen Dehnung ε_{el} stellt die elastische Arbeit dar.

Die elastische Reißarbeit A_{el} ist der vom Beginn der Belastung bis zur Reißlast der Probe aufzuwendende elastische Anteil der Gesamtarbeit. Diese wird in der Abb. VII, 65 durch die Fläche $(ACDA)$ dargestellt. Ihre Größe ist gleich der Maßzahl dieser Fläche in den Einheiten der Koordinatenachsen. Für die praktische Auswertung wird vielfach folgende Formel benutzt:

$$A_{el} = q_{el} \cdot P_{max} \cdot \Delta L_{B\,el}.$$

[1] In neuester Zeit ist eine Größe in den Blickpunkt gerückt, die bereits im Jahre 1835 als ein Maß für die elastische Energie, die von einem Körper gespeichert wird, angesehen und mit ***Resilienz*** bezeichnet wurde. Dieser Begriff hat im Laufe der Zeit einen Bedeutungswandel durchgemacht und ist durch L. F. Beste und R. M. Hoffmann[2] zu einer Kenngröße geworden, die aus einem dreidimensionalen Kennlinienbild entnommen wird, dessen Achsen

1. die Steifheit (E-Modul),
2. die Abhängigkeit der Steifheit von der Dehnung,
3. die Erholung von der Deformation (Spannungs- und Arbeitserholung)

bedeuten. Diese Größe wurde dann in Beziehung zum Gebrauchswert gesetzt. Wir sind demgegenüber der Ansicht, daß dieser Weg nicht gangbar ist; denn erstens werden rein statische Größen zur Beurteilung des Materials herangezogen, zweitens kann man weder den Verlauf des E-Moduls noch die Nachgiebigkeit aus dem KΔL-Diagramm ermitteln. Außer L. F. Beste und R. M. Hoffmann haben sich auch W. J. Hamburger[3] und andere mit diesem Problem beschäftigt.

[2] Beste, L. F. u. R. M. Hoffmann: Text. Res. J. **20**, 441 (1950).

[3] Hamburger, W. J.: Text. Res. J. **18**, 102 (1948).

q_{el} ist das Verhältnis der Fläche $(A\,C\,D\,A)$ zu der Fläche des zugehörigen Rechtecks $(A\,B\,C\,D\,A)$, also

$$q_{el} = \frac{(A\,C\,D\,A)}{(A\,B\,C\,D\,A)}.$$

P_{max} ist die Reißlast, $\Delta L_{B\,el}$ ist die elastische Längenänderung für 100 cm Einspannlänge bis zum Bruch.

In unserem Beispiel: $q_{el} = 0{,}481$

P_{max} (abgelesen aus Abb. VII, 66) $= 264$ g

$\Delta L_{B\,el} = 18{,}0$ cm.

Daraus finden wir für

$$\underline{A_{el} = 2{,}29\,\text{kg/cm}.}$$

Unter dem elastischen Arbeitsmodul $a_{G\,el}$ verstehen wir nach DIN 53835 die auf die Masseneinheit umgerechnete elastische Reißarbeit A_{el}, also

$$a_{G\,el} = A_{el} \cdot Nm.$$

In dem Beispiel ergibt sich demnach für $a_{G\,el}$ der Wert

$$\underline{a_{G\,el} = 3{,}36\,\text{kgm/g}.}$$

Die spezifische elastische Reißarbeit $a_{V\,el}$ ist dagegen die auf die Volumeneinheit umgerechnete elastische Reißarbeit A_{el}. Sie wird aus dem elastischen Arbeitsmodul $a_{G\,el}$ mit Hilfe der Dichte im Normzustand γ_N berechnet, also

$$a_{V\,el} = a_{G\,el} \cdot \gamma_N.$$

Wir finden daher

$$\underline{a_{V\,el} = 3{,}83\,\text{kgm/cm}^3.}$$

Als letzte Größe wird der elastische Wirkungsgrad η_{el} als das Verhältnis der elastischen Arbeit A_{el} zur Gesamtarbeit A_{ges} bestimmt. Aus der Abb. VII, 66 ergibt sich als Verhältnis der ausplanimetrierten Fläche der elastischen Arbeit $(A\,C\,D\,A)$ A_{el} zu der ausplanimetrierten Fläche der Gesamtarbeit $(A\,E\,F\,A)$ A_{ges}

$$\eta_{el} = \frac{(A\,C\,D\,A)}{(A\,E\,F\,A)} \cdot 100\,(\%).$$

In unserem Zahlenbeispiel erhalten wir für

$$\underline{\eta_{el} = 52\,\%.}$$

Die ausführliche Diskussion der Untersuchung des elastischen Verhaltens führt zu einer größeren Anzahl von definierten Größen, von denen immer nur ein Teil ausgewertet worden ist.

So wird in der Abb. VII, 67[1] nach O. Eisenhut und W. Grether[2] das Dehnungsverhalten von Baumwolle, verschiedenen handelsüblichen Zellwollen sowie einer Viscose-Spezialzellwolle wiedergegeben, wobei nur

[1] In der Abb. VII, 67 sind ebenfalls Kraft-Dehnungs-Diagramme wiedergegeben. Hier findet man sogar die Kraft in km-Faserlänge ausgedrückt, eine Definition, die in der Textiltechnik leider gebräuchlich geworden ist. Es gilt auch hier der bereits in der Anm. 2 auf S. 493 gemachte Hinweis auf die korrekte Definition und außerdem auf eine klar definierte Maßeinheit. Es sollte insbesondere das Durcheinander in der Verwendung von Kraft und Spannung bei den Ordinatenbezeichnungen vermieden werden.

[2] Eisenhut, O. u. W. Grether: Melliand Textilber. **3**, 1 (1941).

zwischen der gesamtelastischen und bleibenden Dehnung unterschieden wird. Die Auswertung ist also nicht in dem Umfang erfolgt, wie es soeben an Hand von DIN 53835 geschildert worden ist. Hier setzt sich die angegebene elastische Dehnung (vgl. Abb. VII, 67b) aus zwei Anteilen zusammen, nämlich: der momentanen elastischen Dehnung und der elastischen Nachwirkung (verzögerte elastische Dehnung). Trotzdem erkennen wir bereits den Unterschied zwischen der Baumwolle und der Spezial-

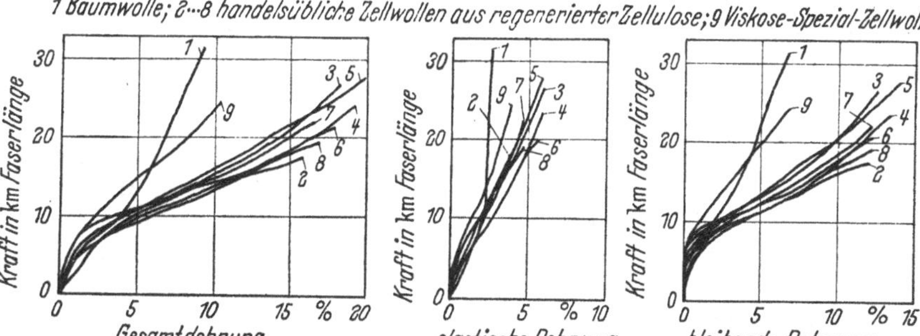

Abb. VII, 67. Kraft-Dehnungs-Kurven verschiedener Textilfasern. (Nach O. Eisenhut u. Grether.)

Viscosezellwolle einerseits und den handelsüblichen Zellwolltypen andererseits, wobei mit zunehmender Belastung diese Unterschiede sich immer stärker ausprägen. Hieraus ergeben sich bereits gewisse Schlüsse auf das spätere Verhalten.

Eine genauere Betrachtung der Abb. VII, 60a läßt erkennen, daß die Kurven bei der einmaligen Belastung bis zum Bruch, wie sie durch die Kurvenzüge *a* und *b* wiedergegeben werden, durch einen knieartigen Übergang nach kurzer Dehnung gekennzeichnet sind. Dieser Übergang scheint, wenn er scharf ausgeprägt ist, eine Proportionalitätsgrenze zu sein. Eine eingehende Analyse des Dehnungsvorganges kann nun auch vorgenommen werden, indem als Ordinate die Verformung und als Abszisse die Dehnung aufgetragen wird, wobei die Gesamtverformung gleich der Dehnung ist. Wir sehen dann, daß diese Proportionalitätsgrenze mit der sogenannten unteren Fließgrenze nach de Vries zusammenfällt. Außerdem beobachten wir nach dem Schema der Abb. VII, 62 nach de Vries[1] eine obere Fließgrenze und insbesondere eine Zerlegung der Verformung in einen spontanen Rückfederungsanteil und einen Anteil, der als latente bzw. blockierte Rückfederung bezeichnet wird, sowie der sogenannten permanenten Verformung. Dieses Dehnungsdiagramm gibt uns auch Auskunft über die Charakteristiken des Erholungsvorganges.

Vergleichen wir die Abb. VII, 62 mit der Abb. VII, 63, so sehen wir, daß die Gesamtverformung der Größe ε_{ges} und die sogenannte momentane (spontane) Rückfederung der Differenz aus ε_{ges} und ε_{rs} entspricht. Die latente (blockierte) Rückfederung setzt sich zusammen aus der elastischen Nachwirkung (verzögerte elastische Erholung) $\varepsilon_{el,n}$ und jenem Teil der Restdehnung ε_r, der durch Quellung oder Temperaturbehand-

[1] Siehe S. 492, Fußnote 1.

lung als elastischer Anteil in Erscheinung tritt. Die permanente Verformung ist jener verbleibenden Restdehnung ε_r gleichzusetzen, die auch durch diese Nachbehandlung noch erhalten bleibt. Diese Zerlegung der Dehnung in die verschiedenen Dehnungskomponenten legen G. SUSICH und ST. BAKER[1] ihren Untersuchungen mit dem Instrontester Modell TT-B 9 zugrunde, wobei sie allerdings Versuchsbedingungen wählen, wie z. B. eine Einspannlänge von 12,7 cm, eine Dehnungsgeschwindigkeit von 100%/min für den Belastungsvorgang und eine Entlastung, die mit derselben Geschwindigkeit bei rückwärts laufender Abzugsklemme vor sich geht, sowie eine Erholungszeit von 5 Minuten, die recht willkürlich anmuten, jedoch dadurch gegeben sind, daß man die praktisch höchste Abzugsgeschwindigkeit des Instron-Festigkeitsprüfers bei diesen Versuchen wählen wollte, um bei der Messung der elastischen Sofort-erholung bzw. der sogenannten momentanen Rückfederung Irrtümer auszuschalten. Weiter wollte man berücksichtigen, daß durch die Schreiberträgheit der Umkehrpunkt der Dehnung nicht exakt ausgezeichnet ist, weil infolge eines Weiterlaufens des Zugstangenkopfes bei der Änderung der Laufrichtung die vorgegebene Dehnung um einen zusätzlichen Betrag vergrößert wird. Zwischen der Forderung nach hoher Abzugsgeschwindigkeit zur Vermeidung der Einbeziehung eines erheblichen Teiles der sogenannten verzögerten Erholung in den Meßwert einerseits und der Forderung nach niedriger Geschwindigkeit, um die Schwierigkeiten auszuschalten, die aus der Trägheitsbewegung des Schreibers und des Zugstangenkopfes im Instron-Tester entstehen andererseits, ergibt sich als Kompromiß die obengenannte Abzugsgeschwindigkeit von 12,7 cm/min. Ebenso bestimmen praktische Erwägungen die Wahl der Erholungszeit von 5 Minuten. Sie zerlegen die Gesamtdehnung in eine momentan elastische Erholung (immediate elastic recovery), eine verzögerte elastische Erholung (delayed recovery, primary creep) sowie eine permanente Verformung. Diese verzögerte Erholung ist nach DE VRIES in der latenten Rückfederung enthalten und entspricht der elastischen Nachwirkung $\varepsilon_{\text{el},n}$ nach DIN 53835. Sie wird daher je nach der zur Verfügung stehenden Erholungszeit einen mehr oder weniger großen Anteil der latenten Rückfederung bzw. der Erholungsmöglichkeiten darstellen.

Eine Zusammenstellung des Erholungsverhaltens der verschiedensten Textilfasern[2] ist in der Abb. VII, 68 wiedergegeben, wobei zur besseren Veranschaulichung, insbesondere zum leichten Vergleich des Erholungsverhaltens verschiedener Fasertypen, die übliche Darstellung verlassen wird und die Dehnungskomponenten sowie die Spannung in $g/g\,x$ als Funktion der Gesamtdehnung aufgetragen werden. Diese spezielle Zerlegung der Dehnung in verschiedene Einzelkomponenten erlaubt, wie wir bei Betrachtung der Abb. VII, 68 sehen, eine wesentlich feinere Unterteilung der Eigenschaften der untersuchten Fasern und Fäden, die über die Bewertung von O. EISENHUT und GRETHER[3] wesentlich hinaus-

---

[1] SUSICH, G. u. ST. BAKER: Text. Res. J. **21**, 482 (1951).

[2] Erholungserscheinungen bei Papier untersucht J. KUBAT: Svensk Papperstidn. **56**, 670 (1953).

[3] EISENHUT, O. u. W. GRETHER: Melliand-Textilberichte **3**, 1 (1941).

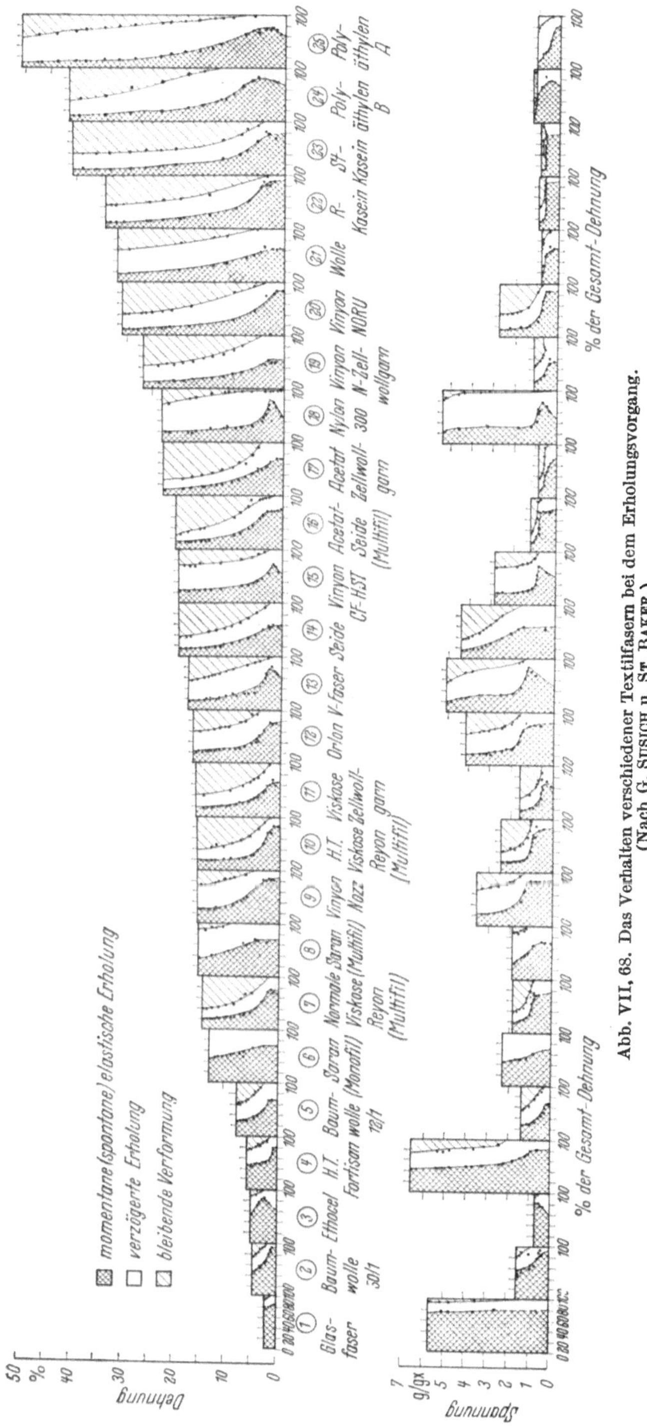

Abb. VII, 68. Das Verhalten verschiedener Textilfasern bei dem Erholungsvorgang. (Nach G. SUSICH u. ST. BAKER.)

geht. Wir sehen, daß z.B. Nylon 66 eine momentane Rückfederung von nur 4,2% zeigt, dagegen eine verzögerte Erholung von 12,6%, die erstaunlich hoch erscheint und von G. SUSICH und ST. BAKER dafür verantwortlich gemacht wird, daß beim Nähen mit Nylonfäden die welligen Säume entstehen. Die permanente Verformung beträgt bei Nylon nur 6,5%. Hieraus ergeben sich auch gewisse Folgerungen, warum bei der Bewertung von vollsynthetischen Fasern erhebliche Fehler gemacht werden. Man war der Ansicht, daß Nylon unter den Textilfasern (vgl. W. W. HECKERT[1]) eine einzigartige Stellung hinsichtlich seiner ungewöhnlichen elastischen Erholung nach starker Dehnung besitzt. Man hat daraus die Folgerung gezogen, daß Gewebe aus Nylon 66 auch eine besondere Knitterfestigkeit haben müßten. Wir wissen aber, daß sie von der Polyesterfaser Dacron in der Knitterfestigkeit wesentlich übertroffen wird, trotzdem deren elastische Erholung nicht einmal die Hälfte des Wertes von Nylon 66 beträgt. Umgekehrt glaubte man auf Grund der nachgewiesenen Knitterfestigkeit, daß für Teppiche[2] die Polyesterfaser Dacron am besten geeignet wäre. Hier zeigte sich jedoch, daß Nylon und Orlon dem Dacron überlegen sind.

Nun kann man zur Beurteilung des unterschiedlichen Verhaltens wieder von der soeben diskutierten Zerlegung des Erholungsvorganges in einen momentanen und verzögerten Anteil Gebrauch machen; denn diese verzögerte elastische Erholung hängt, wie wir gesehen haben, stark von der Zeit ab und damit auch von der Frequenz der wiederholten Beanspruchung. Aus der Größe dieser verzögerten elastischen Dehnung kann man daher gewisse Rückschlüsse auf das dynamische Verhalten

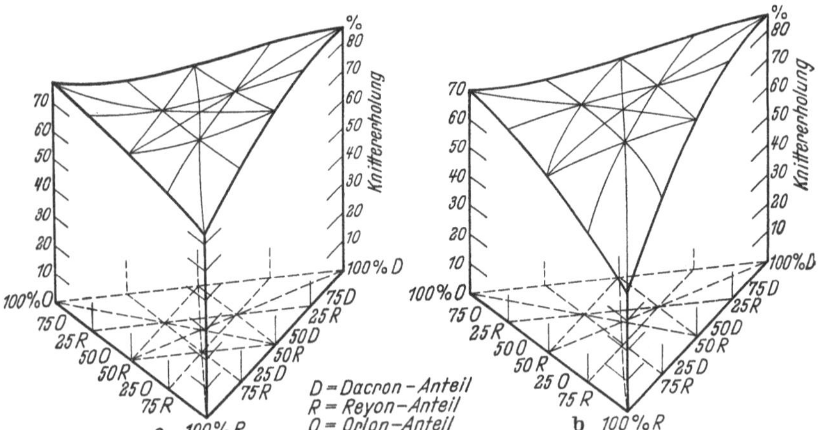

Abb. VII, 69a. Knittererholung bei Dacron-, Reyon- und Orlon-Mischgeweben. (Nach J. F. SAYRE und A. J. WELDON.) (Relative Luftfeuchtigkeit 65%, Temperatur 21°C.)

Abb. VII, 69b. Knittererholung bei Dacron-, Reyon- und Orlon-Mischgeweben. (Nach J. F. SAYRE und A. J. WELDON.) (Relative Luftfeuchtigkeit 90%, Temperatur 24°C.)

[1] HECKERT, W. W.: Ind. Engng. Chem. 44, 2103 (1952).

[2] Es sei darauf hingewiesen, daß speziell die elastische Erholung an Teppichfasern von J. L. BARACH: Text. Res. J. 19, 355 (1949) besonders untersucht worden ist.

ziehen, insbesondere auf die Ermüdung und Erscheinungen, die mit Relaxationsvorgängen in Verbindung stehen.

Diese Überlegungen sind nicht auf wiederholte Zugbeanspruchungen beschränkt, sondern können auf wiederholte Biegebeanspruchungen sinngemäß übertragen werden. Wir kommen damit zu den Begriffen des Knitteraufspringwinkels und des Knittererholungswinkels[1]. Es ist daher verständlich, daß man sowohl den Knitteraufspringwinkel wie auch den Knittererholungswinkel von Nylon und Dacron vergleichen muß. Der Knitteraufspringwinkel ist jedoch bei Nylon 66 durchaus nicht so ausgeprägt wie der Knittererholungswinkel, so daß hieraus auch das unterschiedliche Verhalten zu erklären ist. Wir sehen daraus, daß die Aufspaltung des Erholungsvorganges in diese Komponenten einen weitergehenden Einblick in die Knitterechtheit, Weichheit und Ermüdung gibt. Einen weiteren Einblick in diese Zusammenhänge erhalten wir durch die Einbeziehung von Fasergemischen in die Betrachtungen, wobei uns die dreidimensionalen Diagramme von J. F. SAYRE und A. J. WELDON[2] für kurzfristige 15 Sekunden und langfristige 5 Minuten Erholungsdauer vom Knittern bei relativen Luftfeuchtigkeiten von 65% und 90% der Dacron-, Reyon- und Orlon-Mischgewebe ein anschauliches Bild der Abhängigkeit der Erholung des Knitterwinkels von den Anteilen an Dacron, Reyon und Orlon in den Fasergemischen vermitteln. In den Abb. 69a und 69b sind zwei derartige Diagramme wiedergegeben; sie zeigen uns, daß der Wert der Knittererholung nicht immer dem Wert der Mischungsverhältnisse der Fasern proportional ist, sondern daß die Fasern in irgendeiner bisher noch nicht einwandfrei bekannten Weise aufeinander im Gewebe einwirken. Zum Beispiel weist ein 25%iges Reyon-Dacron-Mischgewebe einen unerwartet kleinen Verlust der Knittererholung gegenüber einem Gewebe aus 100%igem Dacron auf. Daß wir es bei der Knittererholung mit Eigenschaften der Gewebe zu tun haben, die zumindest den Gebrauchswert mitbestimmen, lassen die Beobachtungen erkennen, daß die 15 Sekunden Erholungswerte die Fähigkeit des Gewebes widerspiegeln, sein Aussehen beim Tragen in einem kurzen Zeitraum nicht zu ändern, während die Erholung nach 5 Minuten die Fähigkeit des Gewebes, sich über Nacht auszuhängen, annähernd wiedergibt.

Damit sind jedoch die Folgerungen, die wir aus dem K⊿L-Diagramm ziehen, noch nicht erschöpft. Es ist naheliegend, daß auch die Fließgrenze ein wesentliches Merkmal eines Materials darstellt und dem Verarbeiter

---

[1] SOMMER, H.: Faserforsch. u. Textiltechnik **2**, 468 (1951). Fa. Chem. Fa. Z. Tex. 8. DIN 53890, Entwurf Jan. 1954. Vgl. auch J. A. KALKMAN, Reyon Revue **9**, 49 (1955).

[2] SAYRE, J. F. u. A. J. WELDON: Vortrag vor dem Internat. Chemiefaser-Kongreß, Paris, Juli 1954. – Vgl. auch T. F. COOKE, J. H. DUSENBURY, R. H. KIENLE u. E. E. LINEKEN: Text. Res. J. **24**, 1015 (1954). — J. F. KRASNY u. A. M. SOOKNE: Text. Res. J. **25**, 493 (1955). — J. F. KRASNY, G. D. MALLORY, J. K. PHILLIPS u. A. M. SOOKNE: Text. Res. J. **25**, 499 (1955). Der Einfluß der sogenannten knitterfreien Ausrüstung auf den Knittererholungswinkel wird ausführlich von N. BARNABÉ diskutiert. Vortrag vor dem Internat. Chemiefaser-Kongreß, Paris, Juli 1954, und von E. PIEPER: Reyon, Zellwolle und andere Chemiefasern **31**, 239 (1953) eingehend behandelt.

gewisse Hinweise erlaubt. Leider zeigte sich, daß aus dem PΔL-Diagramm die obere Fließgrenze überhaupt nicht entnommen werden kann und die untere Fließgrenze bei nur wenigen Stoffen so ausgeprägt ist, daß sie daraus bestimmt werden kann[1,2].

Bisher überhaupt nicht erwähnt ist der sehr oft übersehene Einfluß der Temperatur auf den Verlauf des Kraft-Längen-Änderungs-Diagramms, insbesondere bei partiell kristallinen Substanzen mit ihren sekundären Erweichungsbereichen, wie es von K. Wolf[3] beschrieben wird. In diesem Zusammenhang sei auch auf das von J. F. Clark und J. M. Preston[4] bei höheren Temperaturen, und zwar bei Vinyon- und PC-Fasern beobachtete ausgeprägte gummielastische Verhalten aufmerksam gemacht.

§ 62. Veränderung der Dehnungseigenschaften durch auftretende Fadenspannungen in Abhängigkeit von der Zeit.

Die bisherigen Ausführungen lassen klar erkennen, welchen außerordentlichen Einfluß die Zeit auf den Verlauf des PΔL-Diagrammes ausübt und wie stark die aus der Analyse des PΔL-Diagrammes folgenden Dehnungs- und Erholungsgrößen von der Zeit abhängig sind[5]. Wir wollen diesen zeitlichen Vorgang der Veränderungen des PΔL-Diagrammes durch über kürzere oder längere Zeit aufgeprägte Spannungen hier erörtern, weil damit im Zusammenhang das Verhalten des Materials bei der Verarbeitung steht. In der Abb. VII, 70 sind die Veränderungen der Dehnungseigenschaften, die durch Vorbelastungen auftreten können, wiedergegeben. Sie wurden von H. Stein[6] mit dem Statigraph (vgl. Kapitel IV, § 22d) untersucht. Auf der rechten Seite sind die PΔL-Diagramme des ursprünglichen Materials aufgezeichnet, auf der linken Seite die entsprechenden PΔL-Diagramme, nachdem das Material 24 Stunden einer Vorlast, die etwa der halben Bruchlast entspricht, ausgesetzt wurde.

---

[1] Der Ausdruck Fließgrenze ist nicht glücklich gewählt. Es ist richtiger, im Sinne der vorstehenden Darlegungen von einer unteren und oberen Grenze der quasipermanenten Verformung zu sprechen.

[2] Dieser gesamte Fragenkomplex steht auch in engem Zusammenhang mit dem Problem der mechanischen Konditionierung (mechanical conditioning) von Textilfasern [vgl. G. Susich: Text. Res. J. **23**, 545 (1953)] und Papier [vgl. B. Ivarsson: Svensk Papperstidn. **51**, 383 (1948)], der die Analyse dieses Vorganges mit Hilfe des Eyringschen Drei-Elementen-Modells vornimmt (vgl. Kap. IV, § 21a).

[3] Wolf, K.: Vortrag vor dem DVM am 12. 10. 1955 in Stuttgart[4]. — J. F. Clark u. J. M. Preston: J. Text. Inst. **44**, T596 (1953). — R. Ecker: Schweizer Arch. 301 (1954); Gummi u. Asbest **8**, 111 (1955).

[4] Bueche, F.: J. appl. Physics **26**, 738 (1955); J. appl. Chem. **26**, 1133, (1955).

[5] Hier sei noch einmal auf den außerordentlich großen Einfluß der Feuchtigkeit auf die Gestalt des PΔL-Diagrammes hingewiesen, wie er z. B. von W. Wegener: Melliand Textilber. **33**, 37 (1952) und A. Webster: Text. Manufacturer **10**, 542 (1952) und anderen eingehend untersucht worden ist (vgl. Abb. VII, 61[7]).

[6] Stein, H.: Textil-Prax. **9**, 131 (1954).

[7] Bergen, W. von: Ind. Engng. Chem. **44**, 2157 (1952).

Wir sehen bereits daraus, daß nach der Verarbeitung, nachdem durch irgendeinen Prozeß eine Vorbelastung oder eine Fadenspannung auf das Material über längere Zeit hindurch ausgeübt wurde, dessen Dehnungseigenschaften sich wesentlich geändert haben und nicht mehr dem ursprünglichen Verhalten entsprechen. Hohe Fadenbeanspruchungen können, z. B. bei Spinn-, Schlicht- und Zwirnmaschinen und ebenso bei unzweckmäßig ausgebildeten Fadenbremsen, beim Scheren bzw. Zetteln und Schußspulen auftreten. Unterschiedliche Belastungen für die verwendeten Fadenbremsen an Schußspulmaschinen zeigen an der FRENZEL-HAHN-Garnprüfmaschine (vgl. Kapitel IV, § 22b) bei konstanter Dehnungseinstellung, wie stark die Dehnungseigenschaften verändert wurden. Haben die Fadenbeanspruchungen das Gebiet der elastischen Dehnung überschritten, dann entstehen bleibende Verformungen, die beim Weben störende Glanzeffekte verursachen können.

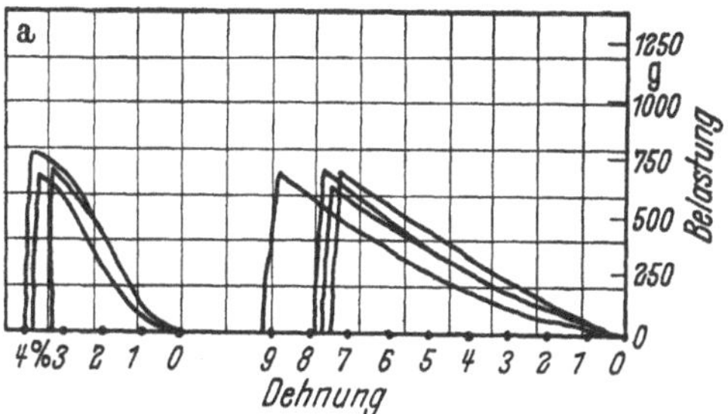

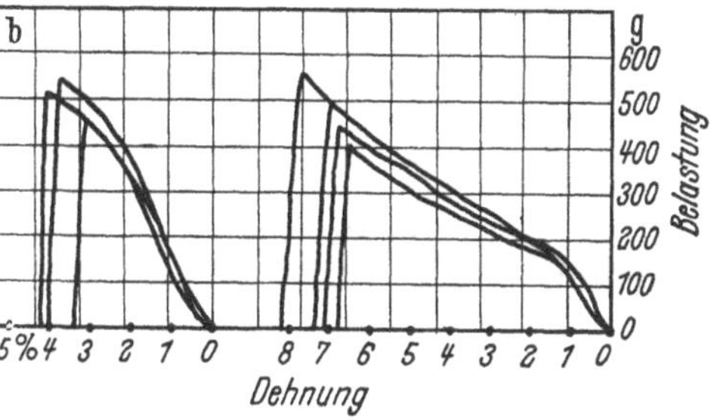

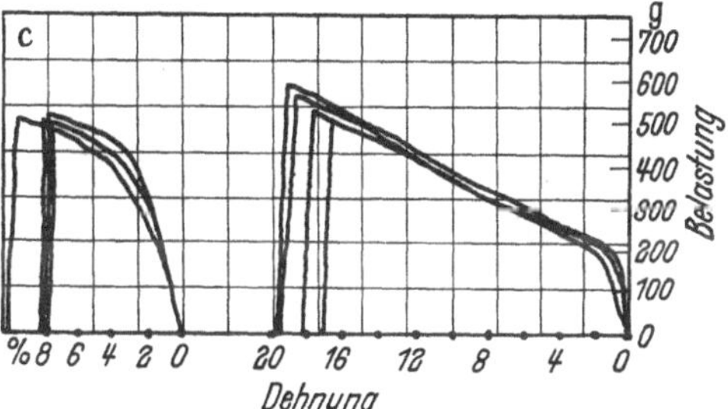

Abb. VII, 70. Veränderungen der Dehnungseigenschaften durch Vorbelastungen. (Nach H. STEIN.)

In der Abb. VII, 71 zeigt der obere Kurvenzug die Prüfung des Schußfadens von einem Gewebestück, das zu keinen Beanstandungen Anlaß gab. Der untere Kurvenzug stammt dagegen von einem Schußfaden aus einem streifigen Gewebestück. Man sieht zwei deutlich ausgeprägte periodisch wiederkehrende Belastungsschwankungen für die eingestellte Dehnung von 5%. Die Ursachen sind in Fehlern im Spulprozeß zu suchen (Störungen an der Fadenbremse), welche zu übermäßigen Beanspruchungen des Materials führten.

Derartige aufgezeigte Änderungen im Dehnungsverhalten durch die Verarbeitung machen sich vor allem beim Anfärben unangenehm bemerkbar. Hierbei ist zu betonen, daß solche Überdehnungen im praktischen

Gebrauch oft weniger durch gleichmäßige Fadenspannungen auftreten, als durch die Eigenarten der Bremsen und der Unregelmäßigkeit des Fadenmaterials, sowie durch im Herstellungsprozeß auftretende Spannungsspitzen (Stöße), die den Mittelwert stark übersteigen (oft um das Doppelte) und meist die eigentliche Ursache für das Auftreten von Überdehnungen sind. Solche Überdehnungen können aber auch durch Krumpfkräfte auftreten, die dann bei der Weiterverarbeitung des Gewebes wieder Glanzstellen und unterschiedliche Anfärbung verursachen. Krumpfkräfte lassen sich mit empfindlichen elektrischen Meßköpfen beim Einspannen des benetzten Materials bei Beginn der Trocknung messen (siehe Abb. VII, 72).

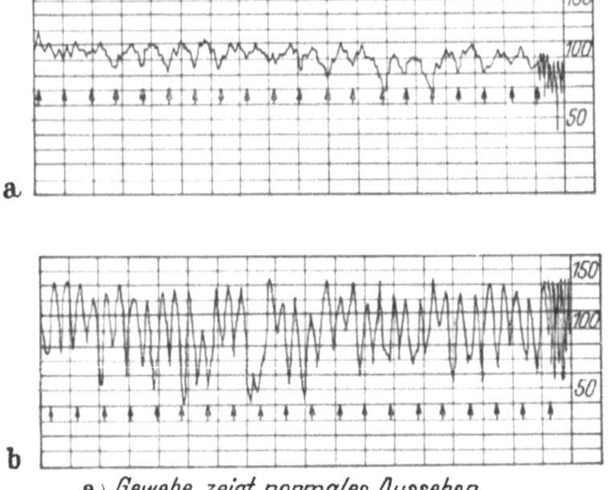

a) Gewebe zeigt normales Aussehen
b) Glanzfäden
↑) Folge der Schußeintragungen

Abb. VII, 71. Dehnungsprüfungen an Schußfäden. (Nach H. STEIN.)

Aus diesen Überlegungen folgt, daß durch Untersuchungen des Dehnungsverhaltens

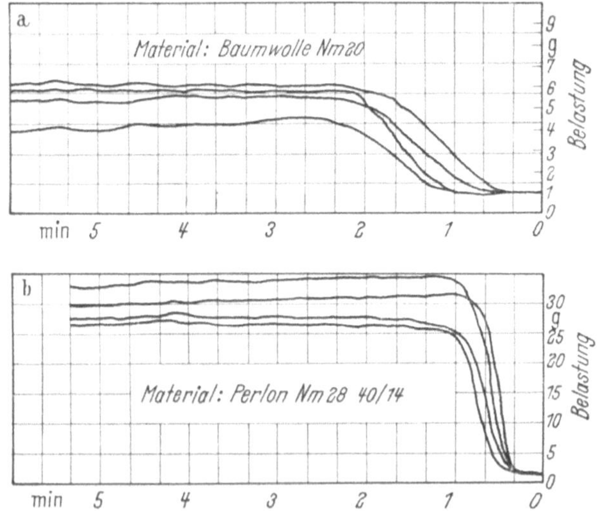

Abb. VII, 72. Krumpfkräfte von verschiedenem Fadenmaterial. (Nach H. STEIN.)

des Garns entweder in einem gewöhnlichen Festigkeitsprüfer einwandfreier Konstruktion oder am laufenden Faden an einer Garnprüfmaschine Zusammenhänge zwischen fehlerhaft arbeitenden Verarbeitungsmaschinen in Verbindung mit Unregelmäßigkeiten des Garnes feststellbar sind.

§ 63. Verhalten von Fasern und Fäden bei konstanter Dauerbelastung und Dauerwechselbeanspruchung.

Aus § 60 geht bereits hervor, daß es nicht nur eine Ermüdung des Materials durch wiederholte Beanspruchung, sondern auch durch eine langanhaltende konstante Beanspruchung gibt[1].

a) Verhalten von Fasern bei konstanter Dauerbelastung und wiederholter Belastung.

Von A. Zart[2] wird der Unterschied im Ermüdungsverhalten von Fasern durch dauernde und wiederholte Belastung aufgezeigt. Die vorliegenden Zellwollfasern wurden mit zwei Drittel der Reißlast einmal dauernd, das andere Mal wiederholt belastet, und zwar in jeweiligen Abständen

Tabelle VII, 7. *Ermüdungsversuche an Zellwollfasern nach* A. Zart.

	Bruchlast g	Rkm	Reißdauer bei 2/3 Rkm	
			Dauernde Belastung	Wiederholte Belastung
Faser A	9,85	23	56 Std. 1 Min.	64 Min.
Faser B	10,55	25	6 Std. 46 Min.	5 Min.
Faser C	8,75	26	33 Std. 45 Min.	18 Min.

von einer Minute. Die Zusammenstellung in Tabelle VII, 7 zeigt den großen Einfluß der wiederholten Beanspruchung auf das Material. Daraus ist ersichtlich, wie stark die Faser bei wiederholter kurzer Belastung leidet, im Gegensatz zur Dauerbelastung mit der gleichen Lasthöhe. Es ist zu betonen, daß sich die Proben in Bruchlast und Bruchdehnung nur sehr wenig unterscheiden. Vergleicht man außerdem noch die den jeweiligen Be- und Entlastungen entsprechenden bleibenden und elastischen Dehnungen, dann sieht man an dem Beispiel in Tabelle VII, 7a, daß schon bei der ersten Belastung mit zwei Drittel Bruchlast der Hauptteil der bleibenden Dehnung herausgestreckt wird. Bei den folgenden Belastungsspielen liegt die bleibende Dehnung im allgemeinen unter 1%. Es handelt sich bei diesen Untersuchungen demnach um Ermüdungsprüfungen.

Tabelle VII, 7a.
Dehnungsverhalten einer Zellwollfaser beim Ermüdungsversuch nach A. Zart.

Gesamte Dehnung %	Bleibende Dehnung %	Elastische Dehnung %
15,8	10,2	5,6
5,33	0,73	4,6
4,68	0,18	4,5
4,66	0,36	4,3
4,57	0,27	4,3
4,48	0,18	4,3
4,38	0,18	4,2
4,18	0,18	4,0
4,08	0,18	3,9
4,18	0,18	4,0
3,99	0,09	3,9
usw.	usw.	usw.

[1] Wegener, W.: Textil-Prax. **9**, 1115 (1954).
[2] Zart, A.: Melliand Textilber. **31**, 593 (1950).

Von T. HAJMASSY[1] wurde der Unterschied der Ermüdungseigenschaften für die Einzelfaser und das Garn bei wiederholter Belastung aufgedeckt. Die Lastwechselzahl beim Bruch und die im Laufe der Prüfung aufgetretene Gesamtdehnung bzw. bleibende Dehnung dienen als Kriterium für das Ermüdungsdiagramm. Es wurde gefunden, daß der Wollzwirn rascher ermüdet als die Einzelfaser. Außerdem wurde beobachtet, daß der Bruch der Einzelfaser bzw. des Zwirns unabhängig von der oberen Belastungsgrenze in jedem Fall nach Erreichen der gleichen Formänderung erfolgte. Für die Ermüdung ist die bleibende Dehnung bzw. die Veränderung derselben während des Dauerversuchs maßgebend.

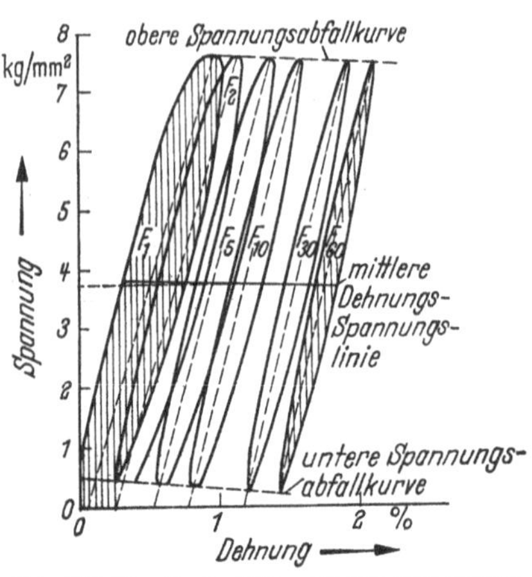

Abb. VII, 73. Hysteresiskurven bei dynamischer Dauerstanduntersuchung[2] an Viscose-Reyon. (Nach W. WEGENER.)

b) Messung der Dauerwechselfestigkeit bei Frequenzen <1 Hz.

Nach dem Prinzip der proportional zur Dehnung abnehmenden Maximalbelastung wurden von W. WEGENER[3] die Versuche bei sinusförmigem Verlauf der Belastung und 0,03 Hz an Viscosereyon durchgeführt. In der Abb. VII, 73 sind die Hysteresisschleifen nach den angegebenen Lastwechselzahlen wiedergegeben. Man sieht, wie die Maximalbelastung mit fortschreitender Dehnung abnimmt. Die im Faden bei vollkommener Entlastung noch vorhandene Dehnung ist die bleibende Dehnung. Die drei verschiedenen Dehnungsanteile sind in der Abb. VII, 73a in Abhängigkeit von der Zeit dargestellt. Aus der Abb. VII, 73 erkennt man, wie die Dehnungsanteile mit Zunahme der Dehnung immer geringer werdende Änderungen aufweisen. Die Breite der Hysteresisschleifen nimmt laufend ab, so daß nach der 60. Schwingung bei der Breite

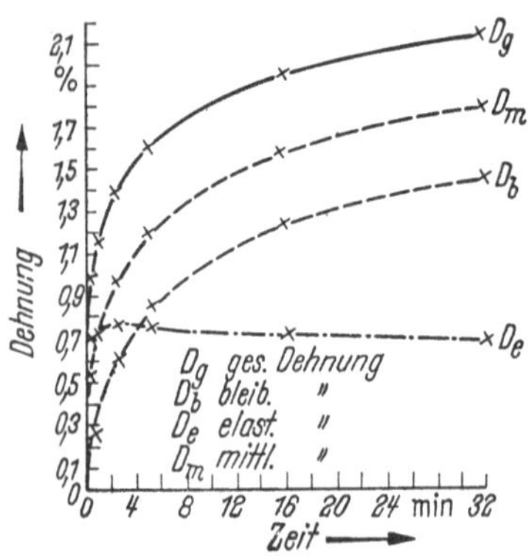

Abb. VII, 73a. Gesamte, bleibende, elastische und mittlere Dehnung in Abhängigkeit von der Zeit bei Viscose-Reyon. (Nach W. WEGENER.)

[1] HAJMASSY, T.: Magyar Textiltechnika **6**, 165 (1953).

[2] Wie bereits in Kap. IV (§ 23a) erwähnt, ist die Bezeichnung unglücklich gewählt, da es sich um eine Dauerwechselbeanspruchung handelt und nicht um eine Dauerbeanspruchung mit konstanter Last.

[3] WEGENER, W.: Z. ges. Textilind. **52**, 27, 55, 79, 108 (1950); Melliand Textilber. **34**, 7, 640 (1953).

des Lichtstrahles kein Unterschied zwischen dem hin- und rücklaufenden Strahl mehr festzustellen ist.

c) Messung der Dauerwechselfestigkeit bei Frequenzen > 1 Hz.

Analog dem statischen Zerreißversuch ist es möglich, bei der Untersuchung des dynamischen Verhaltens die Dauerwechselfestigkeit zu bestimmen. Dies ist eine Größe, die das Material während seiner Verarbeitung oder im praktischen Gebrauch hinsichtlich einer dauernden periodischen Zugbeanspruchung charakterisiert und für derartige Beanspruchungen den Gebrauchswert in hohem Maße mit bestimmt. H. P. ZÖPPRITZ[1] hat bei etwa 3 Hz Schaf- und Baumwollgespinste auf Dauerwechselfestigkeit untersucht und gefunden, daß ein Schafwollgespinst geringerer Reißfestigkeit infolge größerer Dauerwechselfestigkeit der Wolle bei dynamischer Beanspruchung prozentual eine höhere Spannung aushält als ein Baumwollgespinst mit höherer Reißfestigkeit. Diese Arbeit stellt auch die Verbindung zwischen den bis jetzt angeführten dynamischen Untersuchungen bei quasistatischen Frequenzen $\ll 1$ Hz und den analogen Untersuchungen bei höheren Frequenzen > 1 Hz her. Es handelt sich um ausgesprochene dynamische bzw. Ermüdungsprüfungen[2].

In der Autoreifenindustrie hat man sich schon frühzeitig mit diesem Problem beschäftigt, da der Cordfaden während des Gebrauchs dauernd wechselnden Beanspruchungen bei gleichzeitiger Temperatureinwirkung ausgesetzt ist.

Von den darüber veröffentlichten Ergebnissen sind die Arbeiten von P. KAINRADL und F. HÄNDLER[3] über die dynamischen Eigenschaften von Cord bekannt geworden. Sie arbeiten mit sinusförmiger Wechselbeanspruchung und verschiedenen Vorspannlasten nach einer Resonanzmethode bei Frequenzen > 50 Hz und untersuchen verschiedene Zusammenhänge zwischen den einzelnen dynamischen Größen. Es zeigt sich,

[1] ZÖPPRITZ, H. P.: Dauerprüfung und Ermüdung von Gespinsten. Diss. Aachen 1936.

[2] Die verwendeten dynamischen Prüfgeräte sind kritisch in Kapitel VII, § 23, diskutiert worden; dabei mußte eine Auswahl getroffen werden, die nach unserer Auffassung der neuesten Entwicklung gerecht wird. Wir wollen jedoch nicht versäumen, daß zahlreiche weitere Möglichkeiten, das zu prüfende Material dynamisch zu beanspruchen, gegeben sind. Soweit sie einer Kritik standhalten, ist gegen diese Prüfgeräte nichts einzuwenden. Die folgenden Literaturhinweise V. A. GORDEEV: Textilind. (russ.) **15**, 37 (1955); Z. BARTHA: Acta chim. Acad. Sci. hung. **5**, 481 (1955); Z. BARTHA: Magyar Textiltechnika **90** (1955); P. MAUVISSEAU: Bull. Inst. Textile France **51**, 47 (1955); W. WEGENER u. W. FALCH: Reyon, Zellwolle u. andere Chemiefasern **33**, 542 (1955); K. WOLF: Vortrag vor dem DVM am 12.10.1955 in Stuttgart; J. H. WAKELIN, E. T. L. VOONG, D. J. MONTGOMERY u. J. H. DUSENBURG: J. appl. Physics **26**, 786 (1955); H. BÖHRINGER: D. W. P. 7012; W. KERN (Mattiagerät) vgl. Fußnote 1 S. 521 und die Zusammenstellung von W. MESKAT u. O. ROSENBERG: Reyon, Zellwolle u. andere Chemiefasern **31**, 555, 617 (1955) können noch beliebig fortgesetzt werden, sollen jedoch genügen. Wegen der besonderen Ausführungen der Prüfgeräte bei mehrschichtigen Körpern mit Verbundcharakter, wie z. B. die Textilgummitransportbänder, sei auf A. MATTING: Vortrag vor dem DVM, 12. 10. 1955 in Stuttgart hingewiesen.

[3] KAINRADL, P. u. F. HÄNDLER: Kautschuk u. Gummi **5**, 1, WT 3 (1952); **5**, 2, WT 22 (1952); Deutsche Kautschuk-Gesellschaft, Vortragstagung 7.–9. 5. 1953 Goslar.

wie die dynamischen Eigenschaften von den jeweiligen Versuchsbedingungen abhängen. Um aus den Versuchsergebnissen auf den Gebrauchswert des jeweiligen Materials schließen zu können, ist es daher erforderlich, zuerst nach Möglichkeit die praktischen Verhältnisse zu übersehen, weil sich bei falscher Vorgabe der Prüfbedingungen oft falsche Rückschlüsse hinsichtlich der Bewährung des Materials im Gebrauch ergeben. Die Bedingungen, wie sie im Mittel an Autoreifen auftreten, sind jedoch noch umstritten, und die einzelnen Verfasser arbeiten folglich unter recht unterschiedlichen Prüfbedingungen. P. KAINRADL nimmt z.B. bei PKW-Reifen etwa folgende Verhältnisse an:

Vorspannung:	3 kg/mm[2]
Wechsellast:	0,3–3 kg/mm[2]
Frequenz:	etwa 40 Hz

sowie verhältnismäßig kleine periodische Dehnungen (< 0,5%) und betont, daß D. L. LOUGHBOROUGH, J. M. DAVIES und G. E. MONFORE[1] angeblich Spitzenbeanspruchungen gemessen haben, die diesen Wechsellasten entsprechen, so daß die Cordfäden kurzzeitig vollständig entlastet werden.

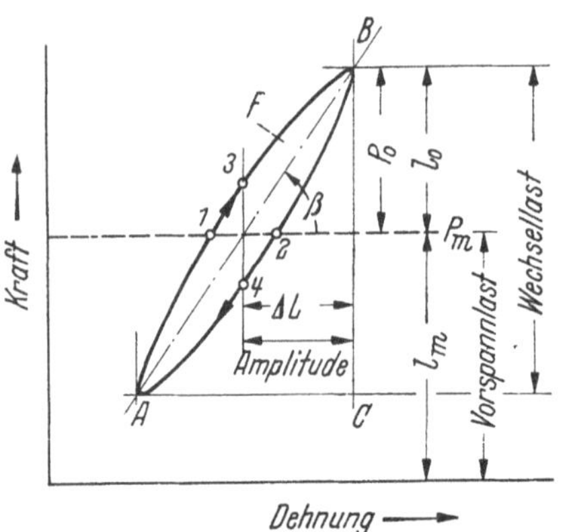

Abb. VII, 74. Auswertung der Hysteresis- bzw. Dämpfungskurve.

$$E_{\mathrm{dyn}} \sim St = \frac{L}{q} \cdot \mathrm{tg}\,\beta\,[\mathrm{kg} \cdot \mathrm{mm}^{-2}],$$

$$H = k\,\frac{C_F}{a^2} \cdot \frac{F}{V}\,[\mathrm{cal} \cdot \mathrm{cycl}^{-1} \cdot \mathrm{cm}^{-3}],$$

$$D = \frac{F}{2 l_0 \cdot \Delta L}\,[\%].$$

L Probenlänge; q Probenquerschnitt; k Maßsystemumrechnungsfaktor; C_F Apparatekonstante; a Vergrößerungsmaßstab; F Flächeninhalt der Dämpfungsschleife; V Volumen der Probe; St Steifheitsmodul

Trotz der Verschiedenheit der Prüfbedingungen der verschiedenen Autoren zeigt sich bei den Untersuchungen des grundsätzlichen dynamischen Verhaltens der Faserstoffe gute Übereinstimmung[2].

d) Abhängigkeit der dynamischen Größen von den Prüfbedingungen.

Der Auswertung der dynamischen Messungen legen wir die in Abb. VII, 74 wiedergegebene Hysteresisschleife (Dämpfungsschleife) zugrunde und benutzen die daraus abgeleiteten Begriffe und Bezeichnungen für die Darstellung der Ergebnisse in den folgenden Abbildungen.

Einfluß der Vorspannlast: Übereinstimmend mit den Messungen von P. KAINRADL fanden W. MESKAT und D. ROSENBERG Unabhängigkeit der dynamischen Zähigkeit bzw. der absoluten Hysteresisverluste von der Vorspannlast in einem weiten Bereich (siehe Abb. VII, 75). Sämtliche Verfasser finden ein Ansteigen des dynamischen E-Moduls bzw. der Steifheit St

[1] LOUGHBOROUGH, D. L., J. M. DAVIES u. G. E. MONFORE: Canad. J. Res. **28**, F 490 (1950).

[2] Es ist auch hier auf die Temperaturfunktion der mechanischen Dämpfung zu achten. Vgl. E. BUCHDAHL, N. E. NIELSEN: J. Polym. Sci. **15**, 1 (1955). W. MESKAT u. ROSENBERG, Tabelle VII/8.

mit der Vorspannlast. P. KAINRADL findet, daß die Cordkonstruktion von großem Einfluß ist. Höher gedrehte Zwirne ergeben den geringeren E-Modul.

Einfluß der Frequenz: P. KAINRADL findet für den dynamischen E-Modul höhere Werte als für den statisch gemessenen E-Modul. Bei Frequenzen > 50 Hz wird von ihm und anderen Autoren keine Abhängigkeit des E-Moduls mehr von der Frequenz festgestellt. W. MESKAT und O. ROSENBERG beobachteten in einem Frequenzbereich von 0,5 bis 25 Hz an Perlonmonofilen einen Anstieg des E-Moduls um einen Faktor $\sim 1{,}5$.

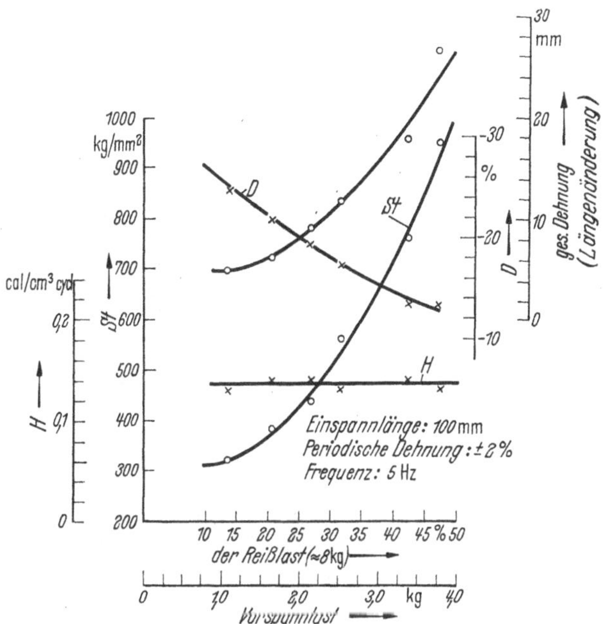

Abb. VII,75. Abhängigkeit der dynamischen Größen von der Vorspannlast nach 10^4 Belastungswechsel. (Nach W. MESKAT und O. ROSENBERG.)

Einfluß der Temperatur und der Feuchtigkeit: G. PALANDRI[1] verglich die Hysteresis bei konstanter Schwingungsamplitude von Baumwolle und Reyon. Er findet keinen oder nur geringen Einfluß. W. MESKAT und O. ROSENBERG stellen an Perlon mit zunehmender Temperatur und Feuchtigkeit starkes Absinken der Energieverluste und der Steifheit fest. Tabellen VII, 8 u. VII, 9. G. PALANDRI beobachtet an Reyon gleichfalls einen Abfall des E-Moduls mit zunehmender Feuchtigkeit. Dieses Ergebnis wird auch von H. WAKEHAM und E. HONOLD[2] bestätigt. Dagegen zeigt Baumwolle ein umgekehrtes Verhalten.

Einfluß der periodischen Dehnung und der periodischen Kraft: Allgemein wird quadratische Zunahme der Hysteresis mit der Schwingungsweite gefunden. Die Steifheit bzw. der E-Modul sinkt mit größer werdender

[1] PALANDRI, G.: Rubber Age **64**, 45 (1948).

[2] WAKEHAM, H. u. E. HONOLD: J. appl. Physics **17**, 698 (1946).

Tabelle VII, 8.

Abhängigkeit der dynamischen Größen von der Temperatur (nach 10^4 BW).

Versuchsbedingungen:

Einspannlänge: 100 mm,
Vordehnung: 10%,
Periodische Dehnung: ± 2%,
Frequenz: 5 Hz.

T °C	St (kg · mm⁻²)	D (%)	H[1] (cal · cycl⁻¹ · cm⁻³)	$\varepsilon_{ges\,dyn}$* (mm)
− 20	615	31	0,32	10,0
0	490	28	0,21	10,0
+ 25	421	21	0,126	10,0
40	380	19	0,12	10,0
50	360	16	0,12	10,0
60	363	17	0,11	10,0
80	362	16	0,10	10,0
100	315	14	0,08	10,0

* $\varepsilon_{elast\,dyn}$ konnte bei den Temperaturversuchen nicht ermittelt werden.

Tabelle VII, 9.

Abhängigkeit der dynamischen Größen von der relativen Feuchtigkeit (nach 10^3 BW).
Versuchsbedingungen wie Tabelle VII, 8.

rel F (%)	St	D	H	$\varepsilon_{ges\,dyn}$	$\varepsilon_{elast\,dyn}$
17	564	23	0,19	10,0	34
40	489	23	0,18	10,0	47
65	460	16	0,11	10,0	69
80	412	13	0,08	10,0	72
92	381	10	0,06	10,0	75

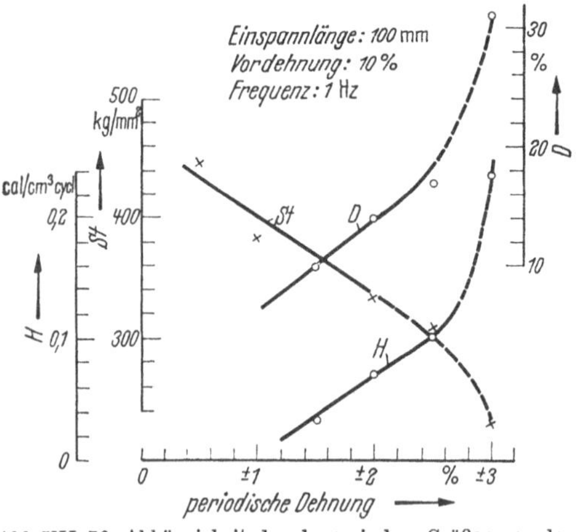

Abb. VII, 76. Abhängigkeit der dynamischen Größen von der periodischen Dehnung (Amplitude) nach 10^4 Belastungswechsel. (Nach W. MESKAT und O. ROSENBERG.)

Amplitude, wie auch P. KAINRADL in Übereinstimmung mit unseren Messungen (siehe Abb. VII, 76) feststellt. In der Abb. VII, 77a wird gezeigt, daß Nylon in diesem, den Reifenbauer besonders interessierenden Zusammenhang sich wesentlich besser verhält als Baumwolle, also viel geringere Verluste mit wachsender Amplitude zeigt.

Wird jedoch die periodische Wechselkraft vorgegeben, so ändert sich die Reihenfolge und Nylon zeigt eine besonders hohe Hysteresis, während Viscosereyon sich am besten verhält (Abb. VII, 77b). Aber nicht nur die Versuchsbedingungen, sondern auch die Konstruktion sind von großem

[1] Vgl. Kap. I, § 3e. Bezüglich der Bezeichnungen in den Tabellen VII, 8 u. 9, vgl. Abb. VII, 74, S. 512.

Einfluß. Versuche an Viscosereyon lassen erkennen, daß sich die Reihenfolge umkehrt, je nachdem, ob bei konstanter periodischer Dehnung oder konstanter periodischer Kraft gearbeitet wird[1] (siehe Abb. VII, 78a u. b).

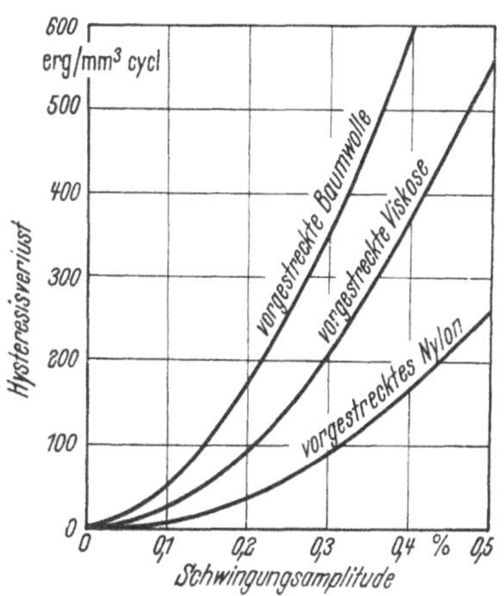

Abb. VII, 77a. Abhängigkeit der Hysteresisverluste von der Schwingungsamplitude bei verschiedenem Material. (Nach G. PALANDRI.)

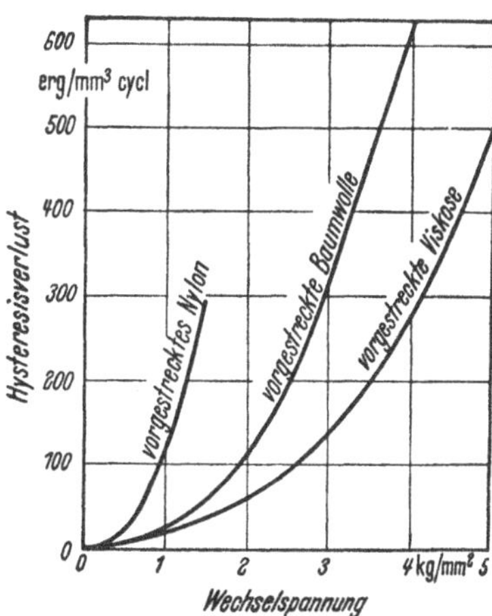

Abb. VII, 77b. Abhängigkeit der Hysteresisverluste von der Wechselspannung bei verschiedenem Material. (Nach G. PALANDRI.)

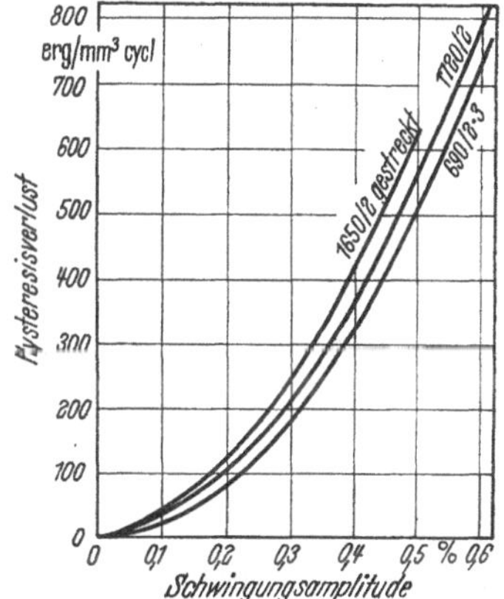

Abb. VII, 78a. Abhängigkeit der Hysteresisverluste von der Schwingungsamplitude bei Viscose-Cord verschiedener Konstruktionen. (Nach P. KAINRADL.[2])

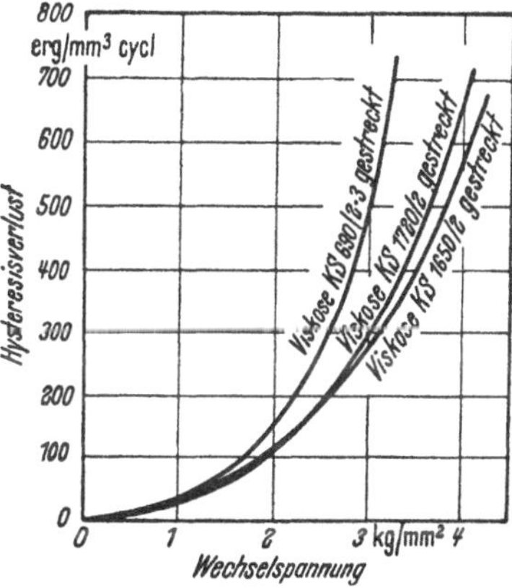

Abb. VII, 78b. Abhängigkeit der Hysteresisverluste von der Wechselspannung bei Viscose-Cord verschiedener Konstruktionen. (Nach P. KAINRADL.)

[1] Von W. WEGENER u. W. FALCH: Reyon, Zellwolle u. andere Chemiefasern **33**, 542 (1955), werden die bisher üblichen Prüfmethoden für Reifencord, wie die Schlagbiegebeanspruchung, Ermittlung des Biegewertes usw. eingehend betrachtet. Wir können diesen Ausführungen nicht immer folgen, da z.B. die Schlagbiegebeanspruchung heute nicht mehr angewandt wird und auch die Ermittlung des Biegewertes wie des Rollbiegewiderstandes nur in Verbindung mit einer Avivage-Prüfung brauchbar ist.

[2] Von P. KAINRADL u. F. HÄNDLER wird in einer neuen Arbeit auf den Einfluß des Zwirnens besonders eingegangen. Kautschuk u. Gummi **8**, W. T. 257 (1955).

Der Einfluß der Zeit: Die Tabelle VII, 10 zeigt die Abhängigkeit der dynamischen Größen von der Zeit bzw. der Anzahl der Belastungswechsel. Es tritt eine erhebliche Versteifung des Materials während der dynamischen Dauerbeanspruchung auf und die Energieverluste nehmen ab.

Tabelle VII, 10.
Abhängigkeit der dynamischen Größen von der Anzahl der Belastungswechsel.
(Nach W. MESKAT u. O. ROSENBERG.)

Einspannlänge: 100 mm,
Wechsellast: 0–25% der Reißlast,
Frequenz: 25 Hz.

BW	*St* (kg · mm⁻²)	*D* (%)	*H* (cal · cycl⁻¹ · cm⁻³)	$\varepsilon_{\text{ges dyn}}$ (mm)	$\varepsilon_{\text{elast dyn}}$ (%)
10^3	243	26	0,17	5,5	42
10^4	317	24	0,14	5,9	28
10^5	500	20	0,08	6,3	19
$5 \cdot 10^5$	595	16	0,05	6,4	18

Tabelle VII, 11. *Erholungsverlauf der Steifheit St.*

Die Steifheit des erholten Materials wurde an einer Probe bei gleichen Prüfbedingungen nach 10^3 BW gemessen.

10^3	10^4	$5 \cdot 10^5$ BW	Erholungszeit	$5m$[1]	$1h$	$3h$	$24h$	$3d$	$7d$	$14d$	$21d$
230	270	570		572	484	373	315	266	270	274	274

Die Gesamtdehnung wächst und die elastische Erholung geht mit der Anzahl der Lastwechsel zurück. Daraus folgt eine zunehmende Ermüdung des Materials. Aus der Tabelle VII, 11 ist der anschließende Erholungsverlauf der Steifheit nach der dynamischen Dauerwechselbeanspruchung zu entnehmen. Wir finden, daß der Wert bereits nach 3 Erholungstagen

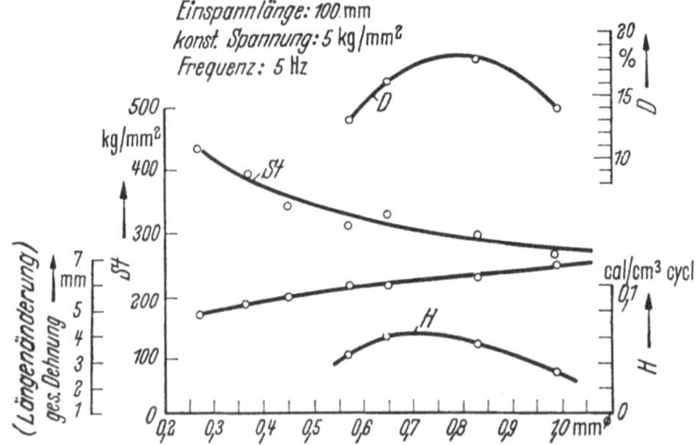

Abb. VII, 79. Abhängigkeit der dynamischen Größen vom Durchmesser des geprüften Monofils nach 10^4 Belastungswechsel. (Nach W. MESKAT und O. ROSENBERG.)

auf einen Betrag zurückgeht, der um etwa 15% gegenüber dem Anfangswert höher liegt und sich nicht mehr weiter verändert. Durch diesen Versuch ist klar bewiesen, welch großen Einfluß die Erholung nach dynami-

[1] m = Zeit in Minuten h = Zeit in Stunden d = Zeit in Tagen

scher Beanspruchung bei der Beurteilung des Materials besitzt und daß diese bei der Bewertung hinsichtlich des Gebrauchs mit einzubeziehen ist. Die Ergebnisse beziehen sich zunächst auf starkes Material (Perlonmonofile). An Perlonfasern fanden wir keine so beträchtliche Versteifung während der dynamischen Dauerbelastung. Wie die Abb. VII,79 zeigt, ist auch der Titer des Materials von großem Einfluß (vgl. die Untersuchung von W. MESKAT und O. ROSENBERG[1]).

e) Dauerwechselfestigkeit und WÖHLER-Kurve[2].

Während sich die vorhin beschriebenen Beanspruchungen gewöhnlich noch unterhalb der Bruchzone des Materials abspielen, kommen wir nun zu einer weiteren Charakterisierung des Materials. Dies ist die Bestimmung der Dauerwechselfestigkeit nach WÖHLER. Man versteht darunter die Beanspruchung des Materials mit bestimmter vorgegebener Amplitude oder Wechsellast bis zum eintretenden Dauerbruch. Abb. VII,80

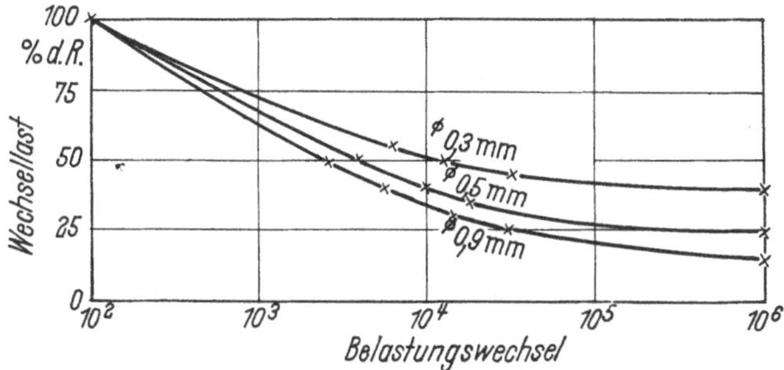

Abb. VII, 80. Wöhlerkurven von vollsynthetischen Monofilen bei verschiedenen Durchmessern. (Nach W. MESKAT und O. ROSENBERG.)

zeigt beispielsweise WÖHLER-Kurven an vollsynthetischem Material bei verschiedenen Titern. Jene Wechsellast, die das Material länger als 10^6mal erträgt, ist nach WÖHLER die Dauerwechselfestigkeit. Sowohl die Frequenz wie auch die Temperatur und Feuchtigkeit haben auf die Dauerwechselfestigkeit großen Einfluß.

Die Ursache des Dauerbruchs erkennt man aus den Zahlenwerten für $\varepsilon_{\text{elast dyn}}$ der Tabelle VII,12. Das Material versprödet mit zunehmender Wechsellast und zunehmender Anzahl der Belastungswechsel immer mehr, so daß schließlich $\varepsilon_{\text{elast dyn}}$ kleiner wird als die Amplitude, die zur

[1] MESKAT, W. u. O. ROSENBERG: Reyon-Zellwolle und andere Chemiefasern **31**, 555, 617 (1953).

[2] Im allgemeinen versteht man unter der WÖHLER-Kurve den funktionalen Zusammenhang zwischen der periodischen Kraft und der Anzahl der Lastwechsel. Bei Textilien kann man an Stelle der periodischen Kraft auch die periodische Dehnung in Abhängigkeit von der Anzahl der Belastungswechsel wiedergeben. Auch diese Darstellung wollen wir als WÖHLER-Kurve bezeichnen. Die besonderen Eigenschaften des textilen Materials verlangen darüber hinaus noch eine Erweiterung des Begriffs. Neben den Koordinaten der periodischen Dehnung und der Lastwechselzahl ist noch als dritte Koordinate das Fließen bzw. die bleibende Verformung des Materials hinzuzunehmen, so daß zur Darstellung des Verhaltens von Faserstoffen bei einer Dauerwechselbeanspruchung eine WÖHLER-Fläche im Raum benötigt wird.

Tabelle VII, 12.

Abhängigkeit der dynamischen Größen von der Wechsellast (nach 10^6 BW bei 25 Hz).

Belastung in % d. R.	St (kg · mm^{-2})	D (%)	H (cal · cycl^{-1} · cm^{-3})	$\varepsilon_{\text{ges dyn}}$ (mm)	$\varepsilon_{\text{elast dyn}}$ (%)	Bruchlast-wechselzahlen
0–20	400	25	0,05	3,6	49	—
0–25	426	25	0,06	5,6	27	—
0–30	480	25	0,08	6,0	22	20000–50000 vereinzelt noch 10^6
0–35	—	—	—	—	—	18000
0–40	—	—	—	—	—	10000
0–50	—	—	—	—	—	4000

Erzeugung der entsprechenden Wechsellast notwendig ist. In diesem Augenblick tritt der Bruch ein. Wird die Gesamtdehnung $\varepsilon_{\text{ges dyn}}$ kurz vor dem Eintritt des Dauerbruches beobachtet bzw. gemessen, so zeigt sich, daß in Übereinstimmung mit den oben angegebenen Ausführungen unmittelbar vor dem Bruch kein plötzlicher Dehnungszuwachs mehr erfolgt. Der Dauerbruch tritt bei der dynamischen Beanspruchung demnach nicht wie beim statischen Zerreißversuch beim Fließen des Materials auf und ist als ein spröder Bruch anzusehen. Die von uns geprüften Perlonmonofile zersplittern beim Dauerbruch wie dünne Holzstäbchen. Die bisherigen Überlegungen zeigen, daß durch die Einführung der Größe $\varepsilon_{\text{elast dyn}}$ die Möglichkeit der Beurteilung eines Materials in bezug auf seine Dauerwechselfestigkeit durch einen Kurzzeitversuch mit etwa 10^4 Belastungswechseln gegeben ist, da $\varepsilon_{\text{elast dyn}}$ sich als ein Maß für die Ermüdung des Materials herausgestellt hat.

Man kann das Problem, den Langzeit- durch einen Kurzzeitversuch zu ersetzen, auch so darstellen, daß wir als Maß für die eintretende Ermüdung das Fließen des Materials (bleibende Dehnung) – vgl. T. Hajmassy[1] – bei einer bestimmten vorgegebenen periodischen Dehnung oder periodischen Kraft betrachten. Es weist dann jenes Material die besseren dauerelastischen Eigenschaften auf, das bei einer vorgegebenen Beanspruchung am wenigsten fließt, also am wenigsten ermüdet. In der Tabelle VII, 13 ist die Untersuchung an verschiedenen Einzelfasern (Titer etwa 3–6 den) bei vorgegebener periodischer Dehnung gezeigt. Die Versuche wurden mit dem in Kapitel IV (§ 23b), Abb. IV, 34 dargestellten Faser- und Fadenprüfgerät durchgeführt. Mit der rein mechanischen Ausführung dieses Gerätes ist es möglich, wie angegeben, das Schlappwerden der Fasern bei bleibenden Dehnungen von etwa 1–2‰ durch ein Ablesefernrohr zu erkennen und jeweils so zu kompensieren, daß die Faser gerade noch gespannt ist. Zu der vorhin beschriebenen Wöhler-Kurve hat in der Textilindustrie als dritte Komponente noch das Fließen ganz besondere Bedeutung, wie wir bereits im Verlauf unserer Arbeit mehrfach festgestellt haben. Die bleibende Verformung ist für die Verarbeitung äußerst unerwünscht. Nach derartigen Tabellen ist es möglich, den Verarbeitern mitzuteilen, welche Amplituden bei deren Verarbei-

[1] Hajmassy, T.: Magyar Textiltechnika **6**, 165–169 (1953).

Tabelle VII, 13. *Dauerfestigkeit und Dauerelastizität von Fasern.*
(Nach W. MESKAT und O. ROSENBERG.)
Einspannlänge: 25 mm, Titer: 3-6 den, Frequenz: 5 Hz.

	1 %	2 %	3 %	4 %	5 %	6 %	7 %	8 %	10 %	12 %	14 %	15 %
Cupresa		> 20000 3 %	5000 8 %									
Viscose		> 20000 2 %	> 20000 5 %	500 10 %								
Cuprama		> 20000 8 %	300 10 %									
Naturseide		> 20000 1 %	> 20000 2 %	1100 5 %								
Hanf	1200 2 %											
Ramie		180 2 %										
Flachs		1300 4 %										
BAYER Acryl			> 20000 20 %	> 20000 30 %	5500 33 %							
Orlon 42			> 20000 23 %	500 20 %								
Schafwolle			> 20000 4 %		> 20000 15 %	1500 20 %						
Kuralon			> 20000 12 %	> 20000 18 %	> 20000 20 %	5000 28 %	150 20 %					
Terylen			> 20000 (11 %)		> 20000 24 %	> 20000 32 %	9000 38 %					
Perlon *U*					> 20000 5 %			> 20000 16 %	7500 36 %			
Perlon *L*					> 20000 4 %			> 20000 7 %	> 20000 10 %	> 20000 19 %	14800 29 %	770 29 %

tungsmaschinen höchstens auftreten dürfen, um das vorliegende Material durch periodische Beanspruchung (periodische Kraft bzw. periodische Dehnung) nicht gefährlich zu überdehnen. Die Kenntnis der Beanspruchung (periodische Kraft bzw. periodische Dehnung) durch Verarbeitungsmaschinen ist durch Einsatz elektrischer dynamischer Fadenspannungsmesser in den meisten Fällen zu ermitteln.

Bei Betrachtung der Tabelle VII, 13 fällt auf, daß bei sämtlichen Fasern die Anzahl der ertragenen Lastwechsel nach Überschreiten einer bestimmten periodischen Dehnung sehr rasch abfällt. Die Werte sind Mittel aus mindestens 10 Meßwerten. Um die Untersuchung zeitlich nicht zu sehr auszudehnen, wurde nach 20000 Belastungswechseln abgebrochen. Aus der zu diesem Zeitpunkt bereits eingetretenen bleibenden Verformung ist der Praxis in den meisten Fällen schon weitgehend gedient, wobei die hier auftretende größere Streuung zu beachten ist.

Andere Untersuchungen dieser Art ergaben an verschiedenen PAN-Proben, daß für das dynamische Verhalten nicht so sehr das Polymerisat, sondern in erster Linie die Verarbeitungsverfahren entscheidend sind. Gleiche Polymerisate, nach verschiedenen Spinnverfahren verarbeitet, zeigen wesentlich andere dauerelastische Eigenschaften. Auch der Einfluß von Mischpolymerisat auf die Dauerelastizität ist von erheblicher Bedeutung. Bei einigen Materialien stoßen wir auf einen starken Einfluß der Frequenz bei Versuchen mit > 10 Hz.

§ 64. Zusammenhang mit dem Gebrauchswert und Ausblick.

Wie wir bereits in Kapitel IV (§ 21d) und Kapitel VII (§ 60) eingehend erörtert haben, kann der Begriff des Gebrauchswertes nur auf den jeweiligen Verwendungszweck bezogen werden. Die starke Abhängigkeit der dynamischen Größen von den verschiedenen Einflußfaktoren zwingt uns dazu, die gestellten Forderungen noch zu erweitern, mit anderen Worten, auch dynamisch darf die Beanspruchung des zu prüfenden Materials im Prüfgerät sich nicht zu weit von den Beanspruchungen des Materials bei der Weiterverarbeitung und im Gebrauch entfernen. Das ist ein sehr wesentliches Ergebnis, das uns zeigt, warum die statisch ermittelten Kenngrößen infolge der dynamischen Beanspruchung des Materials auf dem Verarbeitungswege und im Gebrauch keine Aussagen über die Eignung des Materials erlauben und daß die statistischen Größen der Gleichmäßigkeit[1], wie sie diskutiert worden sind, in Verbindung mit dem elastischen Verhalten bei Dauerwechselbeanspruchung und der Scheuerfestigkeit sich als Kenngrößen von Textilmaterial herausheben.

Beispiele: Im Reifenbau ist bis heute noch kein befriedigender Zusammenhang zwischen den dynamischen Kenngrößen und der Lebensdauer gelaufener Reifen bekannt. K. R. Williams, J. W. Hannel und J. Swanson[2] machten es sich zur Aufgabe, den Zusammenhang zwischen

[1] Vgl. Kap. IV § 21, Abschn. 2 u. P. F. Grisshin: J. Textili Instit. 45 T 167, 179, 192, 209 (1954).

[2] Williams, K. R., J. W. Hannel u. J. M. Swanson: Ind. Engng. Chem. 45, 796 (1953).

den Ergebnissen von Labortestern, die auf Cordzerstörung prüfen, und der Lebensdauer gelaufener Reifen zu finden. Obwohl sämtliche Laborprüfungen Unterschiede zwischen den Corddaten aufdeckten, hat bisher keine von ihnen eine befriedigende quantitative Beziehung zu der Reifenstraßenprüfung gezeigt. Im Hinblick auf diesen Mangel ist der Wert jeder Ermüdungsprüfung zur Vorausbestimmung der Lebensdauer einer bekannten Cordtype im Reifen noch nicht ausreichend. Zunächst ist bekannt, daß Reifencord beim Fahren an Festigkeit verliert. Die Höhe dieses Festigkeitsverlustes des Cords hängt von der Drehung ab. Je niedriger die Drehung, desto höher der Festigkeitsverlust. Ferner wurde gefunden, daß die Beanspruchung des Cords im Reifen selbst verschieden ist. Cordfäden der inneren Lage zeigen den größten Festigkeitsverlust. Dies deckt sich mit der Praxis, die gleichfalls feststellt, daß im Reifen zuerst die innersten Cordlagen zerstört werden. Daraus wird ein Zusammenhang zwischen Cordfestigkeitsverlust und der Reifenermüdung abgeleitet. Weiter wurde beobachtet, daß bei der Prüfung von Einzelfasern die Unterschiede bei weitem nicht so ausgeprägt sind wie bei Cord. Daraus schloß man, daß der Cord in der Walkzone hauptsächlich durch die größere Zahl an gebrochenen Einzelfasern in diesem Reifenteil zerstört wird.

Außerdem zeigten Straßenversuche, daß der Cord im PKW-Reifen maximalen periodischen Dehnungen bis zu 2% und Stauchungen von 6–15% unterworfen ist, die sich grundsätzlich von den Beanspruchungen, wie sie in den Versuchsbedingungen von P. Kainradl (§ 63c) festgelegt worden sind, unterscheiden.

Erst mit Hilfe eines speziellen Prüfgerätes, dem Goodrich-Scheibentester[1], bei dem der in Gummi einvulkanisierte Cord nicht nur Dehnungen, sondern auch Stauchungen unterworfen wird, ist man zu Ergebnissen gekommen, die mit der Praxis vergleichbar sind.

Daraus folgt, daß das Prüfgerät die im Gebrauch auftretenden dynamischen Beanspruchungen angenähert auf das Prüfstück übertragen muß[2]. Es ist daher verständlich, daß auch die an Cordpulsern nach dem Typ Amsler und Zwick (vgl. Kapitel IV, § 23b) durchgeführten Versuche und die mit diesen Geräten ermittelte Dauerwechselfestigkeit im allgemeinen nicht mit den Ergebnissen der Praxis zur Deckung zu bringen sind, z.B. zeigen diese Versuche eine Abnahme der Dauerwechselfestigkeit mit stärker werdender Zwirnung, was den Erfahrungen der Reifenpraxis widerspricht.

Ebenso zeigen die durchgeführten dynamischen Versuche von W. Meskat und O. Rosenberg an PAN-Fasern, daß hohe Dauerwechselfestigkeit allein unabhängig von der bleibenden Dehnung betrachtet, nicht immer einen brauchbaren Hinweis auf das Verhalten in der Praxis gestattet (vgl. Tabelle VII, 6). Insbesondere verhalten sich PAN-Fasern mit hoher Dauerwechselfestigkeit bei der Verarbeitung durchaus nicht immer günstig. Diese

[1] Siehe M. W. Wilson: Text. Res. J. **21**, 47 (1951). — W. Kern (Mattiagerät): Kautschuk u. Gummi **8**, W. T. 195, W. T. 233 (1955) und auch W. Brötz: Papier **6**, 368 (1952).

[2] Man darf dabei nicht übersehen, daß diese Forderung die Gefahr der Entwicklung von Modellgeräten mit nicht definierter Beanspruchung heraufbeschwört.

Erkenntnis drängt uns zu der Annahme, daß es durchaus nicht entscheidend ist, wie viele Lastwechsel ein Material noch aushält, sobald eine bestimmte bleibende Dehnung überschritten ist. Wir beobachteten, daß Fasern, die oft hohe Lastwechselzahlen ertrugen, sich am Anfang genau so stark dehnten wie andere Fasern, die jedoch infolge einer geringeren Bruchdehnung früher zerstört wurden. Stationäre Verhältnisse bei Fasern großer Bruchdehnung stellten sich erst bei Dehnungen ein, die für die Praxis keine Bedeutung mehr haben. Auch hier bei der Dauerwechselfestigkeit in der Textilprüfung ist zu beachten, daß sie für die Beurteilung des Verhaltens der Fasern im Gebrauch oft von untergeordneter Bedeutung wird, weil die Prüfbedingungen gewöhnlich zu hart gewählt worden sind und zu wenig mit den Beanspruchungen in der Praxis übereinstimmen.

Bemerkenswert ist auch das wesentlich bessere dynamische Verhalten von Schafwolle gegenüber den PAN-Fasern (vgl. Tabelle VII, 5 und VII, 13). Wir sehen, daß die Schafwolle eine sehr hohe Elastizität bei langdauernder dynamischer Wechselbeanspruchung, sowie geringeres Fließen bzw. bleibende Verformung unter dieser Beanspruchung in Übereinstimmung mit den Ergebnissen der Praxis aufweist. Aus der Tabelle VII, 13 folgt auch, daß z. B. Viscosereyon eine periodische Dehnung von 3% noch ohne weiteres aushält, während Kupferreyon dabei bereits zerstört wird, ein Beispiel, warum bei den Tragversuchen an Damenstrümpfen nach H. Böhringer Viscosereyon besser als Kupferreyon abschneidet.

Die vorliegenden Versuche an Fasern (vgl. Tabelle VII, 13) wurden bei vorgegebener periodischer Dehnung durchgeführt. Dies ist einfacher, als bei konstanter periodischer Kraft zu arbeiten, weil man vom Titer bzw. von der statischen Reißlast unabhängig ist und nicht schon bei den Prüfbedingungen durch die starke Streuung der statischen Reißwerte einen Fehler bei der Vorgabe der periodischen Wechsellast hineinbekommt.

Unserer Ansicht nach ist die Meinungsverschiedenheit über die Vorgabe der periodischen Kraft oder der periodischen Dehnung von untergeordneter Bedeutung, wenn man nur darauf achtet, daß bei der Vorgabe der einen Größe gleichzeitig die andere gemessen wird. Wir haben bereits wiederholt darauf hingewiesen, daß die Beanspruchungen im dynamischen Prüfgerät mit den Beanspruchungen, denen das zu prüfende Material später ausgesetzt ist, angenähert übereinstimmen sollen. Zu dieser Ansicht kommen auch in neuester Zeit A. Bussmann[1] bei der Untersuchung der Treibriemenprüfung und A. Matting[1], der eingehend die Dauerprüfverfahren für Gummi-Textilförderbänder erläutert, die statische Prüfung ebenfalls als unzureichend ansieht und den dynamischen Prüfmethoden den Vorzug gibt und dabei noch besonders hervorhebt, daß sie so praxisnahe gestaltet werden müssen, daß sie den tatsächlichen Belastungsfall nachahmen. So weit allerdings können wir A. Matting nicht folgen, da wir der Ansicht sind, daß die Beanspruchung im Prüf-

[1] Bussmann, A.: Zum Problem der Riemenprüfung. A. Matting: Neue Dauerprüfverfahren für Gummitextilförderbänder. Vorträge vor dem DVM am 12. 10. 1955 in Stuttgart.

gerät immer noch physikalisch definiert bleiben muß, vgl. S. 255, Fußnote 2, mit anderen Worten das Prüfgerät darf nach unserer Auffassung nicht einfach zum Modellgerät werden. Daher müssen auch die das Material kennzeichnenden Größen aus dem Beanspruchungsvorgang, dem das Material bei der Weiterverarbeitung und im Gebrauch unterliegt, physikalisch abgeleitet werden, vgl. S 287. Diese kennzeichnenden Größen sind aber in sämtlichen Fällen, sowohl während der Verarbeitung als auch unter den besonderen Bedingungen des Gebrauchs weniger die periodische Kraft oder die periodische Dehnung, sondern Arbeitsgrößen[1]. Die Beispiele des § 63 bestätigen diese Auffassung, daß es aussichtsreich ist, das dynamisch-elastische Arbeitsvermögen oder, wie wir es vielleicht bezeichnen können, die dynamische Resilienz in die dynamische Prüfung einzuführen, um mit der Praxis übereinstimmende Aussagen über das spätere Verhalten des Materials zu gewinnen. Wir verstehen unter dynamischer Resilienz die elastische Energie des Materials, die bei periodischer Beanspruchung beliebig oft zwischen dem zu prüfenden Material und dem die Beanspruchung übertragenden Prüfgerät hin und her pendeln kann, ohne daß das Material überhaupt oder innerhalb für die Praxis annehmbarer Grenzen fließt bzw. sich bleibend verformt. Die Entwicklung der hier geschilderten Prüfmethoden und damit im Zusammenhang die einwandfreie Analyse physikalischer Beanspruchungen bei der Verarbeitung und im Gebrauch erhält laufend neue Impulse einerseits durch die Schaffung von Faserstoffen mit gelenkten physikalischen Eigenschaften mittels der Verfahren der Misch-, Block- und Pfropf-Polymerisation[2] und andererseits durch die chemische Modifikation von Naturfasern[3].

---

[1] Vgl. auch A. Springer: Faserforsch. u. Textiltechnik **6**, 76 (1955).

[2] Vgl. z.B. W. Griehl u. R. Hoffmeister: Faserforsch. u. Textiltechnik **6**, 504 (1955).

[3] Vgl. z.B. C. H. Fisher: Text. Res. J. **25**, 1 (1955).

Achtes Kapitel.

Struktur und elektrische Eigenschaften.

Von

F. WÜRSTLIN und H. THURN.

Mit 17 Abbildungen.

A. Grundlagen und Meßmethoden.

Einleitung.

An Hand der elektrischen Eigenschaften, wie sie sich nach den Leitsätzen zur elektrischen Prüfung von Isolierstoffen VDE 0303 bestimmen lassen, sind die hochmolekularen Substanzen durchweg zwischen Halbleiter und höchstwertige Isolierstoffe einzuordnen. Die Beurteilung und Einordnung geschieht im wesentlichen nach den Eigenschaften

Durchschlagsfestigkeit,
Kriechstromfestigkeit,
Widerstand,
Dielektrizitätskonstante,
Verlustfaktor

sowie deren funktioneller Abhängigkeit von Temperatur, Spannung, Frequenz, zeitlicher Einwirkung von Reagenzien u. ä.

Die beiden ersten Eigenschaftswerte lassen sich nur über Zerstörungsprüfungen bestimmen mit ihren bekannten Mängeln (unspezifische Werte, Abhängigkeit von vielen äußeren Bedingungen, große Streuung der Werte usw.). Sie spielen wohl für die Anwendungstechnik eine große Rolle, haben aber für die Grundlagenforschung bisher keine Bedeutung erlangt, da durch die komplexe Versuchsmethodik eine komplizierte Mischung der verschiedensten Einflüsse vorliegt und damit eine physikalische Charakteristik des Versuchsergebnisses meist nicht möglich ist. Der Widerstand ist dagegen physikalisch einwandfrei als spezifischer Wert bestimmbar und seine Bedeutung geht somit über die einer rein technischen Beurteilung einer Isoliereigenschaft hinaus, wobei allerdings festgestellt werden muß, daß man bis jetzt noch nicht viel Zusammenhänge zwischen dem elektrischen Widerstand und der Struktur hochmolekularer Substanzen kennt. Bei den dielektrischen Eigenschaften — Dielektrizitätskonstante und Verlustfaktor — handelt es sich um physikalisch definierte Eigenschaftswerte, die allein schon für die Anwendungstechnik von größerer Wichtigkeit sind als die Beurteilung und Kontrolle des Isoliervermögens. Im Laufe der letzten 1 bis 2 Jahrzehnte zeigte es sich jedoch immer mehr, daß die Bestimmung der dielektrischen Werte teilweise schon einer physikalischen Analyse des strukturellen Aufbaues gleichkommt. Von allen strukturell wichtigen elektrischen Untersuchungen hochmolekularer Substanzen nehmen die der dielektrischen Eigenschaften heute weitaus den größten Raum ein, und zwar nicht auf Grund der Beurteilung der Isoliereigenschaften, sondern wegen ihrer analytischen Aussagen über die Struktur. Polyvinylchlorid ist beispielsweise ein stark polarer Stoff, der wegen dieses polaren Aufbaues niemals als Isolierstoff im hoch- oder mittelfrequenten Bereich verwendet werden kann. Die Bestimmung der dielektrischen Eigenschaften erübrigte sich so von vornherein als Beurteilung seiner Isoliereigenschaften. Trotzdem wurde gerade diese Substanz sehr oft dielektrisch untersucht, da man mit diesen Werten sehr guten Einblick in die Struktur gewinnen kann. Es ist daher nicht verwunderlich, daß im Rahmen dieses Buches die Behandlung der dielektrischen Eigenschaften wesentlich umfangreicher ausfällt als die der übrigen elektrischen Eigenschaften.

§ 65. Durchschlagsfestigkeit, Kriechstromfestigkeit.

Jede technische Anwendung hochmolekularer Stoffe kann die Beantwortung der Frage nach den Grenzen dieser Anwendung nicht umgehen. Die Technik verlangt zahlenmäßige Angaben darüber, bis zu welchen Beanspruchungen ein Material unter ausreichender Lebensdauer brauchbar ist. Man sucht diese technisch notwendigen Zahlen mit Hilfe von Zerstörungsprüfungen zu erhalten, bei denen Beanspruchung und Dauer der Beanspruchung bis zur Zerstörung gesteigert werden.

Eine Anwendung der hochmolekularen Stoffe als Isolierstoffe in der Elektrotechnik verlangt die Kenntnis des Verhaltens solcher Stoffe unter hohen Feldstärken, da jedes im elektrischen Feld befindliche Dielektrikum bei sehr großen Feldstärken sein Isoliervermögen verliert. Die Technik verlangt aber im Grunde genommen keine Aufklärung über die Vorgänge, die sich beim Durchschlag abspielen, sondern sie will in erster Linie mit Hilfe der ermittelten Zahlenwerte ein Urteil abgeben können, ob das Material technisch brauchbar ist. Die zahllosen Versuche zur Bestimmung der Durchschlagsfestigkeit haben so neben den technischen Prüfunterlagen kaum einen Beitrag zur Strukturuntersuchung der Isolierstoffe und speziell der hochmolekularen Isolierstoffe gebracht. Ein derartiger Beitrag kann nach FRANZ[1] erst erwartet werden, wenn verschiedene — bei der technischen Anwendung und Prüfung immer vorhandene — Sekundäreffekte durch eine geeignete Versuchsmethode eliminiert werden können, so daß der eigentliche elektrische Durchschlag bei einer Spannung erfolgt, die weit über den Werten liegen muß, welche durch Sekundäreffekte bedingt werden. Als eine derartige Auswirkung eines Sekundäreffektes muß schon der Wärmedurchschlag angesehen werden.

[1] FRANZ, W.: Z. angew. Physik 3, 72 (1951).

Er kommt nach WAGNER[1,2] dadurch zustande, daß bei ungenügender Wärmeableitung der elektrische Strom eine Temperaturerhöhung und damit wieder eine Stromzunahme bewirkt, die über ein lawinenartiges Anwachsen des Ionenstromes bis zum Durchschlag führen kann. Die häufig beobachtete Abnahme der Durchschlagsfestigkeit bei Vergrößerung der Schichtdicke des Isoliermaterials ist oft durch die Verringerung der Wärmeableitung zu erklären, so daß also mit steigender Dicke der Wärmedurchschlag begünstigt wird. Der schon bei niederen Feldstärken sich abspielende Wärmedurchschlag kann durch kurzzeitige Beanspruchung unter Verwendung von Stoßspannung oder durch Messung bei genügend tiefen Temperaturen verhindert werden, wobei nach BIRNTHALER[3] aus der Temperaturabhängigkeit des Durchschlages die Abgrenzung zwischen elektrischem und Wärmedurchschlag erkennbar wird. Oft wird auch der Verlustfaktor $\operatorname{tg}\delta$ als Kriterium für einen zu erwartenden Durchschlag benutzt, da bei Messung der Durchschlagsfestigkeit in Abhängigkeit von der Beanspruchungszeit bei konstanter Feldstärke oder auch bei kontinuierlich ansteigender Feldstärke erfahrungsgemäß der Durchschlag durch einen plötzlichen Anstieg des $\operatorname{tg}\delta$ angekündigt wird. Bei diesen Messungen ist meist das Überschreiten der Ionisierungsgrenze des in Poren vorhandenen Gases Ursache des plötzlich ansteigenden Verlustfaktors und letzten Endes dann auch des Durchschlags[4-6].

Die Ausbildung von Randdurchschlägen ist ein weiteres Anzeichen von Sekundäreffekten. Man versucht diese Randdurchschläge durch besonders konstruierte Elektroden zu verhindern, die hohe Feldstärken an den Elektrodenbegrenzungen vermeiden[7]. Durch Verhinderung der angeführten und weiterer Sekundäreffekte ist es möglich, bis zu Durchschlagsfestigkeiten von etwa 10^6 V/cm zu kommen. Dieser im technischen Versuch nicht erreichbare rein elektrische Durchschlag fester Isolatoren wird nach heutigen Anschauungen von Elektronen getragen, die u.a. durch Stoßionisation frei gemacht werden können[8,9]. Erst hier kann man eigentlich von einer Physik der elektrischen Festigkeit sprechen, die aber bis heute nur wenige Erkenntnisse über die Einwirkung der Struktur geliefert hat.

Für die Anwendung von Isolierstoffen in der Elektrotechnik spielt weiter das Verhalten der Isolierstoffoberfläche eine große Rolle, wenn zwischen spannungsführenden Kontakten Feuchtigkeit, chemische Substanzen, Fremdkörper usw. einwirken. Auf der im trockenen, sauberen

---

[1] PERLICK, P.: Elektrotechn. Z. A **74**, 169 (1953).

[2] WAGNER, K. W.: Berl. Ber. 438 (1922); 101 (1934); Arch. Elektrotechn. **39**, 215 (1948). — [3] BIRNTHALER, W.: Kunststoffe **43**, 517 (1953).

[4] BRODHUN, C. G.: Physic. Rev. **86**, 653 (1952).

[5] MASON, J. H.: Nature [London] **165**, 932 (1950).

[6] RESS, H.: Papier **4**, 113 (1950).

[7] Von dieser Möglichkeit wird auch Gebrauch gemacht in dem neuen Entwurf der Leitsätze zur Bestimmung der elektrischen Durchschlagspannung, VDE 0303, Teil 2/53; Din 53481.

[8] HIPPEL, A. v.: Ergebn. exakt. Naturwiss. **14**, 104 (1935).

[9] STRIGEL, R.: Arch. techn. Mess. **23**, 831–833 (1935).

Zustand gut isolierenden Oberfläche kann sich unter der genannten Einwirkung ein Kriechstrom zwischen den elektrischen Kontakten ausbilden, der bis zum Ausbau eines Kriechweges mit sichtbaren Zerstörungen der Oberfläche und ihrer Isoliereigenschaft führen kann. Man ist seit Jahren bemüht, die Vorgänge versuchsmäßig so nachzubilden, daß man wenigstens auf Grund von konventionell getroffenen Versuchsausführungen vergleichbare Ergebnisse erhält[1]. Bei diesem Stand des technisch außerordentlich wichtigen Problems ist es verständlich, daß bis jetzt noch nicht über die Auswirkung der Ergebnisse auf die Physik und Struktur der Hochpolymere berichtet werden kann.

§ 66. Widerstand.

Eine Beurteilung eines elektrischen Isolierstoffes erfordert u. a. auch die Angabe eines Widerstandswertes. Ein Vergleich mit anderen Isolierstoffen geschieht dabei am besten unter Verwendung eines spezifischen Wertes, z. B. des spezifischen Widerstandes ϱ in $\Omega \cdot \mathrm{cm}$ oder der reziproken spezifischen Leitfähigkeit $\varkappa$ in $\Omega^{-1}\,\mathrm{cm}^{-1}$. Bei speziellen, meist anwendungstechnisch bedingten Problemen werden oft auch andere Widerstände bestimmt, wie nach VDE 0303 Teil 54 der Oberflächenwiderstand oder der Widerstand im Innern. Wir beschränken uns jedoch auf den spezifischen Widerstand ϱ, da hier noch am ehesten vergleichbare Zahlen vorliegen, wobei allerdings auch hier wieder festgestellt werden muß, daß im ganzen nur sehr wenig Arbeiten ausgeführt worden sind, die grundsätzliche Erkenntnisse über die elektrische Leitfähigkeit in hochmolekularen Substanzen in ihrer Abhängigkeit von der Struktur gebracht haben.

Alle im weiteren angeführten Messungen der elektrischen Leitfähigkeit hochmolekularer Substanzen kann man zwanglos unter der Annahme einer Ionenleitfähigkeit diskutieren. Die Zahl der Ladungsträger n, die Größe der Ladungen q und die Beweglichkeit der Ladungsträger v bestimmen diese Leitfähigkeit:

$$\varkappa = \sum n_i \cdot q_i \cdot v_i .$$

Die Zahl n_i und Ladung q_i der Ionen ist bei diesen hochmolekularen Stoffen praktisch nie bekannt, da es sich hier nicht um der Zahl und Art nach bekannte Ionen im Sinne einer Elektrolytleitfähigkeit handelt, sondern meistens um durch Unreinheiten verursachte Ionen und auch, besonders bei mehrphasigen hochmolekularen Systemen, um kolloidale Ionen, die ihre Ladung dem hochmolekularen Zustand verdanken und der Zahl und Art nach sehr unbestimmt sind. Auch über die Beweglichkeit v_i der Ionen ist meist keine Aussage zu machen mit Ausnahme der, daß die Beweglichkeit v_i in erster Annäherung als umgekehrt proportional der Viscosität η angenommen werden kann.

In reinen hochmolekularen Stoffen ist die Beweglichkeit der Ionen immer dann gering, wenn die Messung unterhalb der Einfriertemperatur

[1] Entwurf VDE 0303, Teil 1/52; Din 53480.

vorgenommen wird, da man schon bei der Einfriertemperatur eine Viscosität von etwa 10^{13} P annehmen muß, die mit weiterer Temperaturerniedrigung sehr rasch ansteigt. Harte hochmolekulare Substanzen haben auch bei polarem Aufbau somit einen hohen spezifischen Widerstand und bei Polystyrol sowohl als auch bei Polyvinylchlorid bestimmt man bei 20° C einen spezifischen Widerstand $\varrho > 10^{15}\,\Omega \cdot$ cm. Bei Abfall der Viscosität η, z. B. mit ansteigender Temperatur, erhöht sich die Ionenbeweglichkeit v_i und die Leitfähigkeit $\varkappa$ steigt an. Der Abfall der Viscosität η und des spezifischen Widerstandes ϱ über der ansteigenden Temperatur kann in parallelen Kurven vor sich gehen und deutet damit auf eine Erhaltung von Zahl und Art der Ionen, aber eine Erhöhung der Beweglichkeit v_i mit ansteigender Temperatur.

Bei zwei- und mehrphasigen hochmolekularen Systemen ist im allgemeinen eine höhere Leitfähigkeit zu erwarten allein schon deswegen, weil es sich um kolloidale Systeme handelt, bei denen mit hoher Wahrscheinlichkeit an den Grenzflächen Aufladungen entstehen, die einen zusätzlichen Ladungstransport verursachen können. Dies ist besonders dann zu erwarten, wenn die zweite Phase eine Substanz mit niederer Viscosität ist. Derartige zweiphasige hochmolekulare Systeme spielen technisch eine große Rolle in Form von hochmolekularen Substanzen, die mit äußerem Weichmacher gemischt werden, um Massen mit hoher Weichheit zu erzielen. Dielektrische Untersuchungen zeigten[1], daß bei der äußeren Weichmachung von Polyvinylchlorid der Weichmacher solvatmäßig an die hochmolekulare Substanz gebunden wird, wobei die Dipole des Polyvinylchlorids und des Weichmachers als Bindungsstellen für eine nebenvalente Dipolbindung dienen. Bei Erhöhung des Weichmacherzusatzes finden nach diesen Messungen schon bei geringen Weichmacherzusätzen nicht mehr alle Weichmachermoleküle eine entsprechende polare Gruppe der hochmolekularen Substanz, obwohl viele C—Cl-Dipole des Polyvinylchlorids noch nicht solvatgebunden sind. Dieses erste Auftreten von nicht solvatgebundenem Weichmacher ist gleichbedeutend mit dem ersten Auftreten eines zweiphasigen Systems, das eine starke Zunahme der Leitfähigkeit mit sich bringt. Die Abb. VIII, 1 läßt sich wahrscheinlich so deuten, daß etwa 20 Gew.-% des hier verwendeten Weichmachers vollständig solvatgebunden werden, wobei sich diese Weichmachermoleküle nicht in der Leitfähigkeit bemerkbar machen, da sie ihre Beweglichkeit durch die Bindung an die hoch-

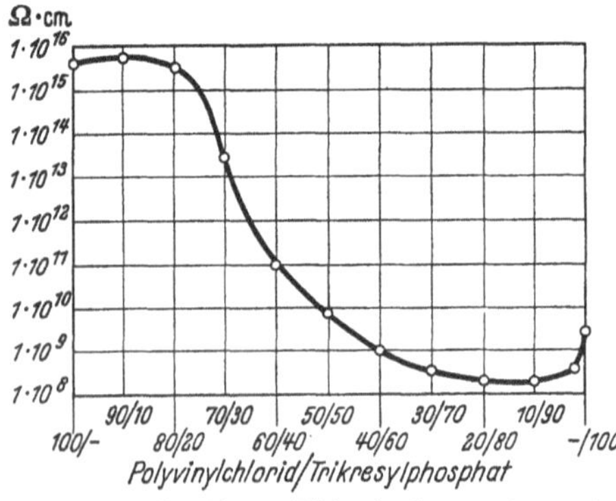

Abb. VIII, 1. Spezifischer Widerstand ϱ weichgemachter Polyvinylchloridmassen in Abhängigkeit von der Weichmacherkonzentration. (Nach WÜRSTLIN[2].)

[1] WÜRSTLIN, F.: Kolloid-Z. **113**, 18 (1949).
[2] WÜRSTLIN, F.: Kunststoff-Techn. **11**, 269 (1941).

molekulare Substanz verloren haben. Dagegen macht sich aber nach WÜRSTLIN[1] die Anwesenheit dieser solvatgebundenen Weichmacher sofort durch die Herabsetzung der inneren Viscosität bemerkbar, was sich an Hand der Verschiebung der DK-Kurven nach niederen Temperaturen zeigt. Übereinstimmend mit diesen dielektrischen Untersuchungen hat sich herausgestellt, daß das erste Auftreten von nicht solvatisiertem Weichmacher sich schon bei einer Weichmacherkonzentration anzeigt, bei der die Solvathülle bei weitem noch nicht ausgebaut ist. Bei weiterer Konzentrationserhöhung des Weichmachers wird dann sowohl die Solvathülle weiter ausgebaut, als auch der Anteil an nicht solvatisiertem Weichmacher erhöht. Der spezifische Widerstand fällt in Abb. VIII, 1 mit diesem ersten Auftreten von nicht solvatisiertem Weichmacher steil ab und es ist nach WÜRSTLIN[2] zu vermuten, daß diese beweglichen Weichmachermoleküle als Ionen wirken. Diese Auffassung wird von KNAPPE[3] bestätigt, der bei der Diffusionsmessung von Weichmachern in weichgemachtem Polyvinylchlorid eine Parallelität von Diffusion und Leitfähigkeit in ihrem Kurvenverlauf über die Konzentration feststellt.

Wasserhaltige Hochpolymere sind in gewissem Sinne auch zu diesem zweiphasigen System zu rechnen. Dabei kann das Wasser als Weichmacher wirken (wie z. B. bei den Polyamiden) oder es kann auch bei Substanzen, die an sich keine Tendenz zur Wasseraufnahme zeigen (wie z. B. Polystyrol oder Polyvinylchlorid) durch hygroskopische Beimengungen zu einer Wasseraufnahme kommen. Im letzteren Falle macht sich das Wasser auch in geringen Spuren oft durch eine zusätzliche elektrolytische Leitfähigkeit bemerkbar, da diese Beimengungen in Wasser dissoziieren können. Es ist bekannt, daß sich die Emulsionspolymerisate von den Suspensionspolymerisaten elektrisch dadurch unterscheiden, daß die meist hygroskopischen und wasserlöslichen Emulgatoren bei Wasserlagerung eine starke Erhöhung der Leitfähigkeit bewirken. In trockenem Zustand sind dagegen die Widerstandswerte meist nicht voneinander zu unterscheiden. Bei den Polyamiden wirkt sich die durch zahlreiche Amidgruppen verursachte Wasseraufnahme als äußere Weichmachung aus, so daß schon dadurch eine Erhöhung der Leitfähigkeit zu erwarten ist. Sie ist aber so außerordentlich stark — mit wenigen Gew.-% Wassergehalt steigt die Leitfähigkeit um mehrere Größenordnungen an —, daß dieser Effekt wohl nur mit dem Polyelektrolytverhalten der Polyamide erklärt werden kann[4].

Führt man nun vergleichende Untersuchungen des spezifischen Widerstandes an verschiedenen hochmolekularen Substanzen durch, so ist wieder daran zu erinnern, daß die Leitfähigkeit $\varkappa$ dem Produkt aus Zahl n_i, Ladung q_i und Beweglichkeit v_i über alle Ionen entspricht:

$$\varkappa = \sum_i n_i \cdot q_i \cdot v_i \,. \qquad \text{(VIII, 1)}$$

---

[1] Siehe S. 528, Fußnote 1.

[2] WÜRSTLIN, F.: Kunststoff-Techn. **11**, 269 (1941).

[3] KNAPPE, W.: Z. angew. Physik **6**, 97 (1954).

[4] SCHAEFGEN, J. R. u. C. F. TRIVISONNO: J. Amer. chem. Soc. **73**, 4580 (1951); **74**, 2715 (1952).

Da man eine bestimmte Leitfähigkeit mit einer großen Anzahl von Ionen hoher Ladung bei kleiner Beweglichkeit erhalten kann oder auch bei wenig Ionen mit hoher Beweglichkeit, wird man bei einem sinnvollen Vergleich einen Faktor dieses Produktes konstant halten oder eliminieren. Dies ist bei Zahl und Ladung der Ionen kaum möglich. Man kann dagegen mit Hilfe der Viscosität etwa auf gleiche Beweglichkeit der Ionen beziehen, besonders wenn es sich um Flüssigkeiten handelt, bei denen die Viscosität noch einwandfrei erfaßbar ist. Bei Isolierölen hat sich nach WALLRAFF[1] eine Beurteilung der Isoliereigenschaften nach dem Produkt $\varkappa \cdot \eta$ (Leitfähigkeit · Viscosität) bewährt, wobei gute Isolieröle einen kleinen und von der Temperatur nur sehr wenig abhängigen Wert zeigen.

Ein derartiger Vergleich ist natürlich bei Hochmolekularen nur schwer durchzuführen, da die Viscosität im allgemeinen nicht mehr direkt erfaßbar ist. Da jedoch die Temperaturlage der DK-Dispersion weitgehend durch die Viscosität der Substanz bestimmt ist, kann man diese Methode als indirekte Viscositätsmeßmethode benutzen. Hat man beispielsweise einige weichgemachte PVC-Massen mit verschieden hohem Weichmacheranteil, so findet man bei konstanter Frequenz die Dispersionsstelle in den verschiedenen Weichmassen bei verschiedenen Temperaturen. Mißt man bei diesen Temperaturen jeweils auch die Härte oder den spezifischen Widerstand, so ergeben sich nach DAVIES und Mitarbeitern[2] nahezu übereinstimmende Werte, d.h. man mißt tatsächlich etwa bei derselben Viscosität. Vergleicht man also zwei PVC-Weichmassen lediglich an Hand ihres spezifischen Widerstands, so kann man aus diesen Werten allein noch kein Urteil abgeben über ihr Isoliervermögen. Stellt man dagegen die beiden Weichmassen mit Hilfe der DK-Dispersion auf gleiche Viscosität ein und vergleicht bei diesen Temperaturen die spezifischen Widerstandswerte, so ist damit eine vergleichbare Beurteilung möglich.

Diese Methode ist selbstverständlich für eine laufende Untersuchung oder Betriebskontrolle zu umständlich. Im Grunde genommen verfährt man aber bei den technischen Messungen ähnlich, allerdings wesentlich gröber, wenn man statt der Viscositätskontrolle durch die DK-Dispersion die Kältefestigkeit heranzieht und sich bemüht, solche PVC-Weichmachermischungen herzustellen, die gerade eine bestimmte Kältefestigkeit erreichen. Bei gleicher Kältefestigkeit wird man dann der Weichmasse mit dem höheren spezifischen Widerstand als besser isolierender Substanz den Vorzug geben.

§ 67. Theoretische Deutung der Dielektrizitätskonstante und des Verlustfaktors; inneres Feld.

Die dielektrischen Eigenschaften eines Stoffes werden durch Angabe der Dielektrizitätskonstanten $\varepsilon^* = \varepsilon' + \varepsilon''$ und des Verlustfaktors $\operatorname{tg}\delta = \varepsilon''/\varepsilon'$ beschrieben. Diese Größen ändern sich, wie schon in Bd. I

[1] WALLRAFF, A.: Elektrotechn. Z. **63**, 539 (1942).

[2] DAVIES, J. M., R. F. MILLER u. W. F. BUSSE: J. Amer. chem. Soc. **63**, 361 (1941).

und II gezeigt, mit der Frequenz des angelegten Wechselfeldes und der Temperatur des Stoffes. In Bd. I, § 38 sind die Zusammenhänge zwischen ε' und ε'' ausführlich erläutert. Deshalb soll an dieser Stelle nur noch darauf hingewiesen werden, daß die Kurve $\varepsilon''/\varepsilon' = \operatorname{tg}\delta$ als Funktion der Frequenz nur näherungsweise eine symmetrische Glockenkurve ergibt[1]. Bei exakter Rechnung ist sie unsymmetrisch.

Verluste in Dielectrica entstehen im wesentlichen durch die Leitfähigkeit der Stoffe und durch die Einstellbewegungen von Dipolen mit ihren Nachbargruppen zu einem äußeren Wechselfeld. Bei guten Dielektrica sind bei Zimmertemperatur die Leitfähigkeitsverluste auch bei niedrigen Frequenzen klein. Bezeichnen $\varkappa$ die Leitfähigkeit in $\Omega^{-1}\,\mathrm{cm}^{-1}$, ε' die Dielektrizitätskonstante und $\omega = 2\pi\nu$ die Frequenz des Wechselfeldes, so beträgt der durch die Leitfähigkeit verursachte Verlustanteil[2]

$$\operatorname{tg}\delta = \frac{\gamma\cdot\varkappa}{\omega\cdot\varepsilon'}, \qquad \text{(VIII, 2)}$$

wobei der Faktor $\gamma = 4\pi\cdot 9\cdot 10^{11} = 1{,}13\cdot 10^{13}$ ist. Da sich $\varkappa$ und ε' nicht stark mit der Frequenz ändern, nehmen die Leitfähigkeitsverluste mit steigender Frequenz ab.

Bei den Einstellbewegungen von Dipolen mit ihren Nachbargruppen zu einem äußeren Wechselfeld wird dem elektromagnetischen Feld Energie entzogen und im Dielektricum in Wärme umgesetzt.

Die im Dielektricum umgesetzte Leistung beträgt[3]:

$$N = E^2\cdot\omega\cdot\varepsilon\cdot\operatorname{tg}\delta; \qquad \text{(VIII, 3)}$$

in Watt ausgedrückt beträgt der Leistungsverlust im Dielektricum

$$N = U^2\cdot\omega\cdot C\cos\varphi \approx U^2\cdot\omega\cdot C\cdot\operatorname{tg}\delta, \qquad \text{(VIII, 4)}$$

wenn U in Volt und C in Farad angegeben werden.

Da die in Wärme umgesetzte Energie von der Größe des $\operatorname{tg}\delta$, der Frequenz ω und dem angelegten Feld abhängt, läßt sich die Erwärmung von Hochpolymeren im Hochfrequenzfeld nur dann rationell durchführen, wenn der Stoff einen großen Verlustfaktor besitzt und das Wechselfeld eine möglichst hohe Frequenz mit großer Wechselspannung aufweist.

Zur Anregung der Einstellbewegungen von Dipolen mit ihren Nachbargruppen ist die Zufuhr einer Aktivierungsenergie notwendig. Die Aktivierungsgrößen sind in Bd. II dieses Werkes, § 95b, formelmäßig eingehend erläutert und abgeleitet. Wie aus den dort angegebenen Beziehungen hervorgeht, benötigt man zur Bestimmung der Aktivierungsenergie einer molekularen Einstellbewegung wenigstens zwei Kurven des $\operatorname{tg}\delta$, entweder als Funktion der Frequenz bei zwei Temperaturen oder als Funktion der Temperatur bei zwei Frequenzen, aus denen man die Lage des Verlustfaktormaximums entnehmen kann. Nach den in Bd. II angegebenen Formeln sollte man erwarten, daß die Aktivierungsenergie

---

[1] Müller, F. H. u. Chr. Schmelzer: Ergebn. exakt. Naturwiss. **25**, 359 (1951).
[2] Kohlrausch: Lehrbuch der Experimentalphysik.
[3] Kegel, R.: Z. Ver. dtsch. Ing. **91**, 25 (1949).

eine von der Frequenz unabhängige Konstante ist. Dies ist aber durchaus nicht der Fall. Das Experiment zeigt vielmehr, daß sich die Aktivierungsenergie mit der Frequenz teilweise recht beträchtlich ändert. Deshalb sollte die Angabe einer Aktivierungsenergie auch stets den Frequenzbereich enthalten, für den sie berechnet wurde.

Die Einstellbewegungen der Dipole zu einem äußeren Wechselfeld und ihr Zusammenhang mit den dielektrischen Größen wurden seit DEBYE eingehend untersucht. Dabei hat sich die Einführung eines *„inneren Feldes“* als nützlich erwiesen. Man beschreibt mit ihm die gegenseitige Beeinflussung der Dipole.

DEBYE hat 1912 die Beziehung zwischen der Dielektrizitätskonstanten und den molekularen elektrischen Konstanten für ideal verdünnte polare Gase und ideal verdünnte polar—unpolare Lösungen abgeleitet[1]. Er erhielt die bekannte Formel für die Polarisation:

$$P = \frac{\varepsilon - 1}{\varepsilon + 2} \cdot \frac{M}{\varrho} = \frac{4\pi}{3} N_L \left(\alpha + \frac{\mu^2}{3kT}\right) \qquad \text{(VIII, 5)}$$

(siehe Bd. II, Kap. 13). Hierbei bedeuten N_L die Anzahl der Dipolmoleküle je Mol der Lösung, α die (optische) Polarisierbarkeit, μ das Dipolmoment und kT die Energie der Wärmebewegung. Die DEBYEsche Ableitung setzt voraus, daß das ganze Molekül rotiert, daß ideale Verdünnung vorliegt und daß die Annahme von CLAUSIUS und MOSOTTI[2] gilt (siehe Bd. I dieses Werkes, § 36).

Der Ansatz vernachlässigt aber die Rückwirkung der umgebenden Moleküle auf das gerade betrachtete. Diese Voraussetzungen sind aber sicher nicht erfüllt. Einen wesentlichen Schritt zur Annäherung der tatsächlichen Verhältnisse stellt die Berücksichtigung der gegenseitigen Beeinflussung der Dipole durch die Einführung eines „inneren“ Feldes dar, da die gegenseitige Beeinflussung der Dipole auch ohne äußeres Feld bewirkt, daß in der Umgebung eines Bezugsdipols eine gewisse Ordnung herrscht.

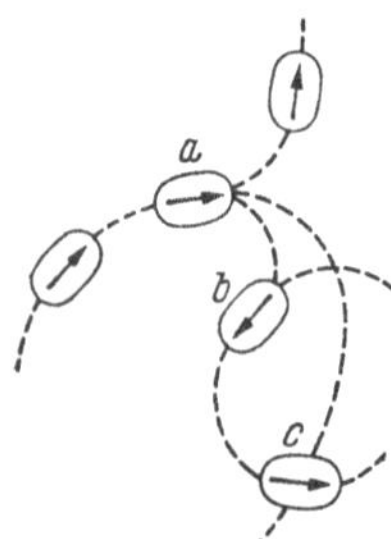

Abb. VIII, 2. Schema der Dipolorientierung. (Nach F. H. MÜLLER.)

Von dieser Ordnung macht man sich folgende Vorstellung: Jedem Dipol ist ein Feld zugeordnet, dessen Stärke mit einer gewissen, verhältnismäßig niedrigen Potenz der Entfernung vom Dipol abnimmt. Zwei Nachbardipole werden also durch ihre Felder aufeinander Richtwirkungen ausüben. Nach MÜLLER[3] übt ein Dipol a auf die benachbarten Dipole Richtungskräfte aus. Im Felde des Dipols a wird z. B. b zu a orientiert. Auf den Dipol c hat a jedoch kaum mehr Einfluß. Für c überwiegt vielmehr das Feld von b. Um jedes Molekül entsteht so eine Richtungssphäre von bevorzugt orientierten Molekülen. Diese Richtungssphäre betrifft aber nur die unmittelbare Nachbarschaft. Etwas weiter vom Zentrum abgelegene Moleküle sind in bezug auf dieses ungeordnet. Es herrscht somit Ordnung nur

[1] DEBYE, P.: Physik. Z. **13**, 97 (1912).
[2] MOSOTTI, O. F.: Mem. di matem. e ficsia in Modena **24**, 11, 19 (1850).
[3] MÜLLER, F. H. u. CHR. SCHMELZER: Ergebn. exakt. Naturwiss. **25**, 359 (1951).

im Kleinen und jedes beliebige Dipolmolekül kann als Zentrum aufgefaßt werden. Für diesen Ordnungstypus hat sich die Bezeichnung „Dipolsphäre“ eingebürgert.

Ein Dipol stellt zusammen mit dem Molekülrumpf ein polarisierbares Gebilde dar. Als weitere Dipolwirkung ist deshalb denkbar, daß das effektive Moment μ durch die Rückwirkung der „Dipolsphäre“ gegenüber dem Moment $\mu_{\text{gas}} = \mu_0$ des freien Moleküls geändert ist [1].

Die Koppelung der Dipole in einer Substanz ändert viele ihrer Eigenschaften. Verkoppelte Dipole sind in ihrer freien Rotation behindert und können sich im äußeren Feld nur vermindert orientieren. Dies bedeutet z.B. eine Verringerung der Dielektrizitätskonstanten gegenüber dem Wert, der sich aus Dipolmoment und Dichte bei einer als unbehindert angenommenen Drehbarkeit der Moleküle errechnen läßt [2].

Der erste Versuch, die Dipolwechselwirkung in der DEBYEschen Formel zu berücksichtigen, stammt von VAN ARKEL u. SNOEK [3]. Sie fanden halbempirisch eine Formel, die für viele Stoffe die Molekularpolarisation einer Dipolsubstanz in Mischungen mit unpolaren Lösungsmitteln in fast dem ganzen Konzentrationsbereich in guter Genauigkeit darstellt. VAN ARKEL u. SNOEK fügten in die DEBYEsche Gleichung noch ein Glied ein, das der Wechselwirkung der Dipole $\mu^2/v^3 \sim N_{\text{cm}^3} \cdot \mu^2$ proportional ist [4] und der durch das äußere Feld hervorgerufenen Orientierung ebenso wie die Wärmeenergie kT entgegenwirkt. Die Beziehung von VAN ARKEL u. SNOEK lautet:

$$p = \frac{\varepsilon - 1}{\varepsilon + 2} = \frac{4\pi}{3} N_{\text{cm}^3} \left(\alpha_0 + \frac{\mu^2}{3kT + c N_{\text{cm}^3} \mu^2} \right). \qquad \text{(VIII, 6)}$$

In die DEBYEsche Formel (VIII, 5) wurde also nur *eine* willkürliche Konstante c eingeführt, die zudem noch weitgehend von der beobachteten Substanz und der Temperatur unabhängig ist. Es wurde von verschiedener Seite versucht, eine Beziehung abzuleiten, in der die Dipol–Dipol-Wechselwirkung enthalten ist [5–10]. HARTMANN hat unter Umgehung des CLAUSIUS-MOSOTTI-Ansatzes eine Theorie der „Dipolschwarmbildung“ (= Dipolsphäre) für den Fall entwickelt, daß ein äußeres Feld vorhanden ist. Es läßt sich zeigen, daß die von HARTMANN [6] angegebene Formel für die Polarisation in erster Näherung mit der von VAN ARKEL u. SNOEK übereinstimmt. Die wichtigste Einschränkung der HARTMANNschen Rechnungen liegt darin, daß sie nur gelten, wenn die Dispersionskräfte zwischen den Dipolmolekülen und den Lösungsmittelmolekülen die gleiche Größenordnung haben wie die zwischen den Dipolmolekülen oder den Lösungsmittelmolekülen.

DEBYE hat eine Rotationskopplungsenergie E eingeführt, um so die gegenseitige, mit steigender Konzentration wachsende Kopplung der Dipolmoleküle zu berücksichtigen [7]. Er nimmt an, daß das Dipolmolekül aus seiner „Momentanrichtung“ nur unter Aufwendung einer gewissen Energie herausgedreht werden kann. F. H. MÜLLER [4] hat versucht, die Zusammenhänge der DEBYEschen Rotationskopplungsenergie mit anderen Größen zu finden. Er geht davon aus, daß die elektrischen Dipolfelder dafür sorgen, daß die Nachbardipole bevorzugt parallel zum Zentral-

---

[1] Siehe S. 532, Fußnote 3.
[2] FOWLER, R.: Proc. Roy. Soc. [London] A **149**, 1 (1935).
[3] ARKEL, A. E. VAN u. J. L. SNOEK: Physik. Z. **35**, 187 (1934).
[4] MÜLLER, F. H.: Physik. Z. **38**, 498 (1937).
[5] MÜLLER, F. H.: Z. Elektrochem. **43**, 863 (1937).
[6] HARTMANN, H.: Z. physik. Chem. **51**, 309 (1941).
[7] DEBYE, P.: Physik. Z. **36**, 100 (1935).
[8] HARTMANN, H.: Z. Elektrochem. **47**, 856 (1941).
[9] HARTMANN, H.: Z. physik. Chem. **53**, 37 (1942).
[10] HARTMANN, H.: Z. physik. Chem. **53**, 49 (1942).

dipol stehen. Am Orte des Zentraldipols herrscht dann ein inneres Feld F, das durch die umliegenden polaren Moleküle erzeugt wird. Ohne Dipolwechselwirkung wäre $F = 0$, weil dann die Dipole über alle Richtungen statistisch gleichmäßig verteilt wären.

DEBYE führt an Stelle des Gliedes $\frac{\mu^2}{3kT}$ seiner einfachen Gleichung die Rotationskopplungsenergie E ein und erweitert so seine ursprüngliche Gleichung um einen „Reduktionsfaktor". Es gelingt ihm dadurch sowohl die Bildung der Dipolsphäre wie auch die Rückwirkung der Sphäre auf den Bezugsdipol zu erfassen. BÖTTCHER[1] zeigte, daß sich die Formel (VIII,6) von VAN ARKEL u. SNOEK ableiten läßt, wenn man die im folgenden zu besprechenden ONSAGERschen Voraussetzungen[2] macht, daß nur ein Teil des inneren Feldes die Dipole ausrichtet. Die bloße Einführung des „Reaktionsfeldes" und der Umstand, daß dieses proportional dem Dipolmoment ist, genügen für die Ableitung.

Ein weiteres wichtiges Ergebnis der DEBYEschen Theorie ist die Aussage, daß durch die Dipolverkoppelung die Relaxationszeiten verkleinert werden[3]. Beim Übergang von unendlicher Verdünnung zu höheren Konzentrationen sollte theoretisch zunächst eine Aufspaltung in mehrere Relaxationszeiten erfolgen[4]. Experimentell wird eine Verschmierung der Relaxationszeiten beobachtet. DEBYE gibt außerdem den Zusammenhang der Relaxationszeit τ mit der Viscosität η an. Er findet:

$$\tau = \frac{4\pi\eta a^3}{kT} \quad (a = \text{Molekülradius}) \qquad \text{(VIII, 7)}$$

ONSAGER[4,5] hat die DEBYEsche Gleichung (VIII,5) auf andere Weise in Ordnung gebracht[2,4,5]. Er ging davon aus, daß der Mittelwert des inneren Feldes nicht mit dem des äußeren Feldes übereinstimmt. Das äußere Feld ist kleiner als das innere Feld. Die Differenz wird durch das „Reaktionsfeld" verursacht. Deshalb zerlegt ONSAGER das innere Feld in ein „Reaktionsfeld" und ein „Käfigfeld". Der permanente Dipol wird, wie bei LORENTZ, von seiner Umgebung durch eine kleine Umkugel getrennt gedacht.

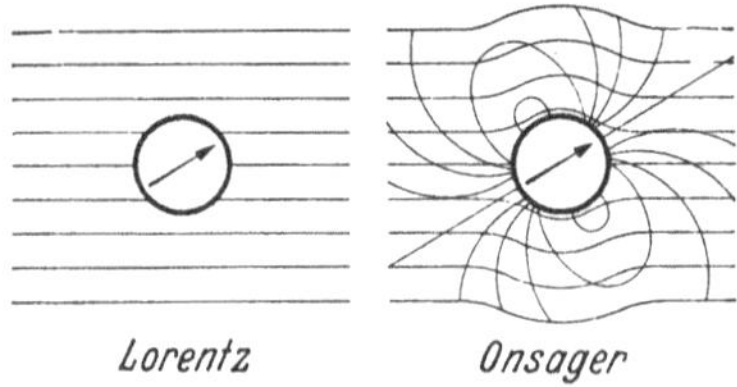

Abb. VIII, 3. Kraftlinienverlauf (a) bei der Berechnung nach LORENTZ, (b) bei der Berechnung nach ONSAGER. (Nach MÜLLER und SCHMELZER[4].)

Durch sein Dipolmoment polarisiert der permanente Dipol die Umgebung der Kugel, auch ohne daß ein äußeres Feld vorhanden ist. Die polarisierte Umgebung wirkt auf das Kugelinnere zurück und liefert im Kugelinnern ein homogenes Reaktionsfeld R. Es ist streng antiparallel zum permanenten Dipol (Dipolvektor), solange die Polarisation, die durch den permanenten Dipol verursacht wird, den Richtungsänderungen des Dipols ohne merkliche Verzögerung folgt, d.h. wenn keine Relaxation vorliegt. Durch das Reaktionsfeld wird die Größe des Moments im Kugelinnern geändert.

---

[1] BÖTTCHER, C. J. F.: Physica **5**, 635 (1938).
[2] ONSAGER, L.: J. Amer. chem. Soc. **58**, 1486 (1936).
[3] DEBYE, P. u. W. RAMM: Ann. Physik **28**, 28 (1937).
[4] MÜLLER, F. H. u. CHR. SCHMELZER: Ergebn. exakt. Naturwiss. **25**, 359 (1951).
[5] ONSAGER, L.: J. physic. Chem. **43**, 189 (1936).

Für das Käfigfeld, das allein orientierend auf den Dipol der Kugel wirkt, gilt der folgende Ausdruck[1]:

$$\mathfrak{F} = \frac{3\varepsilon}{2\varepsilon + 1} \cdot \mathfrak{E} + \frac{2}{3}\,\frac{\varepsilon - 1}{2\varepsilon + 1}\,4\pi\mathfrak{P}, \qquad \text{(VIII, 8)}$$

während nach LORENTZ gilt:

$$\mathfrak{F} = \mathfrak{E} + \frac{4\pi}{3}\mathfrak{P} \quad (\mathfrak{P} = \text{Polarisation}). \qquad \text{(VIII, 9)}$$

Da die Stärke des auf den Dipol im Kugelinnern rückwirkenden Reaktionsfeldes von der Dipolkonzentration in der Kugelumgebung abhängt, wird durch das Reaktionsfeld die Dipol—Dipol-Wechselwirkung erfaßt. Für rein polare Substanzen wird damit die entsprechende Reduktion[2]

$$R_{\text{ons}} = \frac{3\,\varepsilon_0\,(r^2 + 2)}{(2\,\varepsilon_0 + r^2)\,(\varepsilon_0 + 2)}\,; \quad \left[R_{\text{Deb}} = 1 - L^2\left(\frac{E}{kT}\right)\right]. \qquad \text{(VIII, 10)}$$

Die ONSAGERsche Theorie liefert wie die DEBYEsche Theorie eine Verkleinerung der Relaxationszeiten durch die Dipolkopplung. Die DEBYEsche Rotationskopplungstheorie und die ONSAGERsche Theorie führen aber nicht zum gleichen Ergebnis. Dies wird durch eine von MÜLLER und SCHMELZER[2] vorgenommene Gegenüberstellung der Reduktionsfaktoren besonders deutlich.

Die bisher besprochenen Theorien des inneren Feldes haben die Dipol—Dipol-Wechselwirkungen, also elektrostatische Kräfte, berücksichtigt. Es ist aber, wie DEBYE[3] und FOWLER[4] erkannt haben, in polaren Flüssigkeiten noch ein weiterer Effekt von Bedeutung. Außer der Dipol—Dipol-Wechselwirkung können nämlich auch noch zwischenmolekulare Kräfte kurzer Reichweite eine Molekülrotation oder Einstellung in auftretenden Feldern behindern. Solche Kräfte kurzer Reichweite bestimmen z. B. als VAN DER WAALSsche Kräfte und gegenseitige Abstoßungskräfte die Molekülformen. Die bisherigen Theorien vernachlässigten mit den molekularen Diskontinuitäten um das Dipolmolekül auch diese Nahwirkungskräfte.

KIRKWOOD[5] hat eine Theorie entwickelt, die nicht nur die elektrostatische Behinderung (Dipol—Dipol-Wechselwirkung), sondern auch den Einfluß der Nahwirkungskräfte berücksichtigt. Wenn man diese Theorie in Verbindung mit Modellen der Flüssigkeitsstruktur anwendet, läßt sie eine quantitative Beschreibung der dielektrischen Polarisation zu.

FRÖHLICH[6] hat das KIRKWOODsche Verfahren durch Zugrundelegen eines anderen Modells abgewandelt. Es treten jedoch bei der Berechnung der Wechselwirkung zwischen seiner nicht isolierten Probe und dem Feld Schwierigkeiten auf.

HARRIS u. ALDER[7] haben in Anlehnung an das Verfahren von KIRKWOOD eine Theorie auf Grund statistisch-mechanischer Überlegungen entwickelt, die sich von der KIRKWOODschen und FRÖHLICHschen Theorie im wesentlichen dadurch unterscheidet, daß eine genaue Berechnung des Verdrehungsterms (distortion) vorgenommen wird.

[1] Siehe S. 534, Fußnote 4.

[2] MÜLLER, F. H. u. CHR. SCHMELZER: Ergebn. exakt. Naturwiss. **25**, **359** (**1951**).

[3] DEBYE, P.: Physik. Z. **36**, 100 (1935).

[4] FOWLER, R.: Proc. Roy Soc. (A) **149**, **1** (1935).

[5] KIRKWOOD, J. G.: J. chem. Physics **7**, 911 (1939); Trans. Faraday Soc. **42**a, 7 (1946).

[6] FRÖHLICH, H.: Theorie of Dielectrics and Dielectric Loss, Oxford 1949.

[7] HARRIS, F. E. u. B. J. ALDER: J. chem. Physics **21**, 1031 (1953).

POWLES[1] wendet den Begriff des inneren Feldes auf relaxierende Dielektrica an mit dem Ziel, einen Ausdruck zu erhalten, der in seiner Gültigkeit der ONSAGERschen Näherung im statischen Falle entspricht. Eine modifizierte Form der DEBYEschen Theorie wurde unter Anwendung der ONSAGERschen Näherung von COLE[2] angegeben. POWLES erhält bei zeitabhängiger Rechnung einen Ausdruck von der gleichen Form wie COLE, jedoch gelten diese Beziehungen des inneren Feldes nur für einen Prozeß mit einer einzigen Relaxationszeit.

Im Gegensatz zu den bisher bekannten Theorien sieht LÖSCHE[3] die einzelnen Dipole zu einem beliebig herausgegriffenen Zeitpunkt nicht mehr als gleichberechtigte Teilchen an, er setzt vielmehr die Existenz von einzelnen Bereichen, in denen eine bestimmte Ordnung herrscht, voraus. Über die Struktur dieser Bereiche selbst kann nur unter speziellen Annahmen etwas ausgesagt werden. LÖSCHE zeigte, daß man mit dem Dipolmoment von Molekülgruppen ebenso rechnen kann wie bei Gasen mit dem einzelnen Dipol.

Eine Reihe von Arbeiten beschäftigt sich schließlich mit einer Verfeinerung des ONSAGERschen Modelles. So berechnet SCHOLTE[4] Ausdrücke für das Käfig- und Reaktionsfeld für ein ellipsoidförmiges Molekül. ABBOT u. BOLTON[5] zeigen, daß für einen Punktdipol zwischen den Brennpunkten auf der Achse eines Sphäroids der Mittelwert des Reaktionsfeldes der gleiche ist wie der, welchen SCHOLTE angibt. BUCKINGHAM[6] nimmt für seine Rechnungen ebenfalls ein sphäroidales Molekül an, geht aber von der einheitlichen Polarisierbarkeit ab und berechnet die Komponenten des inneren Feldes für die drei Hauptachsenrichtungen.

Während die bisher besprochenen Theorien Dielektrizitätskonstante und Verlustfaktor mit der Molekülgestalt, insbesondere mit polaren Gruppen verknüpfen, hat WAGNER[7] gezeigt, daß man bei manchen Substanzen das Verhalten des Dielektricums auch ohne spezielle Annahmen über den Molekülbau erklären kann. Schon MAXWELL hat darauf hingewiesen, daß man die an einem Dielektricum beobachteten Nachwirkungserscheinungen erklären kann, wenn es aus parallelen Schichten besteht. Ein solches Dielektricum weist Rückstandserscheinungen auf, wenn es von einem elektrischen Feld quer zur Schichtung durchsetzt wird, sofern das Verhältnis der DK zur Leitfähigkeit nicht für alle Schichten den gleichen Wert hat. WAGNER hat dieses grobe Modell verfeinert[8]. WAGNER zeigte, daß sich die dielektrischen Nachwirkungserscheinungen bei den meisten Dielektrica auch dann beschreiben lassen, wenn leitende Oberflächenschichten vermieden werden. Man hat also dann die Inhomogenität in der Struktur des Dielektricums selbst zu suchen („Inhomogenitätstheorie"). Bei der Verbesserung des Modells bettet WAGNER schließlich in eine nichtleitende Grundsubstanz leitende Kügelchen regellos verteilt ein. Beim Fehlen der Kugeln wäre das Feld im Kondensator homogen. Jede Kugel ruft eine gewisse Störung des Feldes hervor, die besonders innerhalb der Kugel und in ihrer näheren Umgebung ins Ge-

---

[1] POWLES, J. G.: J. chem. Physics **21**, 633 (1953).

[2] COLE, R. H.: J. chem. Physics **6**, 385 (1938).

[3] LÖSCHE, A.: Z. physik. Chem. **201**, 302 (1952); **202**, 95 (1953).

[4] SCHOLTE, J. W. A.: Physica **15**, 437 (1949).

[5] ABBOT, T. J. A. u. BOLTON, H. C.: Trans. Faraday Soc. **48**, 422 (1952).

[6] BUCKINGHAM, A. D.: Trans. Faraday Soc. **49**, 881 (1953).

[7] WAGNER, K. W.: Arch. Elektrotechn. **2**, 371 (1914).

[8] Es besteht im einfachsten Falle aus einer ebenen Schicht eines nichtleitenden vollkommenen Dielektricums, über die ein schlechter Leiter in einer Schicht von sehr geringer, aber von Ort zu Ort veränderlicher Dicke gelagert ist. Stellenweise kann diese zweite Schicht ganz fehlen.

wicht fällt, in großem Abstand von ihr aber unmerklich klein wird. Den Einfluß der Temperatur auf die dielektrischen Nachwirkungsvorgänge beschreibt WAGNER durch die Aussage, daß bei steigender Temperatur die Zeitkonstante T stark abnimmt[1]

$$T = \frac{3\varepsilon}{4\pi c^2 \lambda} \quad (\lambda = \text{Leitfähigkeit der Kugeln}). \qquad \text{(VIII, 11)}$$

Ein Anwachsen der Leitfähigkeit bedeutet also ein Abnehmen der Zeitkonstante.

Die von WAGNER berechnete scheinbare Änderung, welche die DK und die Leitfähigkeit des Grundstoffes durch die eingebetteten Kugeln erfährt, hängt lediglich von den Eigenschaften des Grundstoffes und der eingebetteten Substanz und ihrer Menge ab, dagegen stellt sie sich als unabhängig von der Größe der eingebetteten Kügelchen heraus. Die Kugelgestalt, die WAGNER der leichteren mathematischen Behandlung wegen eingeführt hat, geht in das Resultat nicht ein. Man kann deshalb annehmen, daß eine andere Form der Teilchen im wesentlichen zum gleichen Ergebnis führen würde.

Mit diesem Modell lassen sich die Nachwirkungsvorgänge sowie ihre Temperaturabhängigkeit beschreiben und ein Verteilungsgesetz für die Zeitkonstanten des Nachwirkungsvorganges gewinnen.

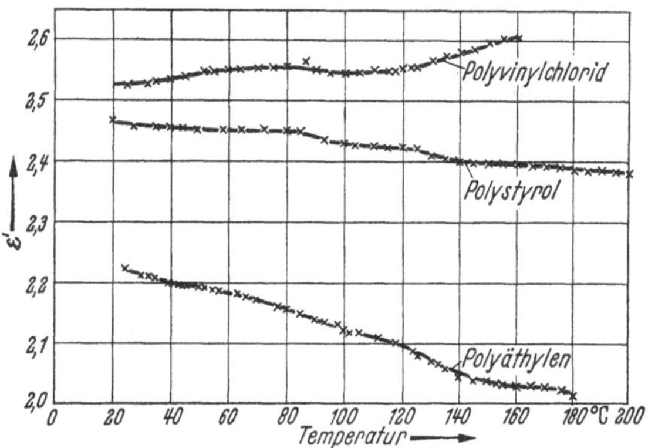

Abb. VIII, 4. Dielektrizitätskonstante von Polyvinylchlorid, Polystyrol und Polyäthylen bei der Frequenz $1 \cdot 10^{10}$ Hz als Funktion der Temperatur. (Nach THURN[3].)

Die angeführten Theorien beschreiben unter anderem die Frequenzabhängigkeit der dielektrischen Größen. In der Praxis kommt jedoch der Temperaturabhängigkeit von ε und $\operatorname{tg}\delta$ eine ebensogroße Bedeutung zu. GEVERS und DU PRE[2] haben den Temperaturkoeffizienten der DK an amorphen Dielektrica mit kleiner DK und kleinem $\operatorname{tg}\delta$ untersucht. Sie haben gezeigt, daß bei diesen Substanzen zwischen dem Temperaturkoeffizienten $\frac{1}{\varepsilon'}\frac{\partial \varepsilon'}{\partial T}$, dem $\operatorname{tg}\delta$ und dem linearen, thermischen Ausdehnungskoeffizienten α_l die Beziehung besteht:

$$\frac{1}{\varepsilon'}\frac{\partial \varepsilon'}{\partial T} = A \operatorname{tg}\delta - \alpha_l\left(1 + \varepsilon_\infty - \frac{2}{\varepsilon_\infty}\right), \qquad \text{(VIII, 12)}$$

wobei $A = \frac{2}{\pi T} \ln \frac{1}{\omega \tau_0}$ (τ_0 = Relaxationszeit).

GEVERS u. DU PRE haben festgestellt, daß fast bei allen verlustarmen Substanzen, die sie untersucht haben, der Faktor A eine Konstante mit dem Wert $A = 0{,}06$ grad^{-1} ist. Da bei den Hochpolymeren der thermische Ausdehnungskoeffizient α_l im allgemeinen groß ist, resultiert hier eine Abnahme der DK mit steigender Temperatur. Abb. VIII, 4 zeigt vergleichsweise die DK von Polyäthylen, Polystyrol und

[1] Wenn $T = R \cdot C$, entlädt sich ein Kondensator der Kapazität C_0 über einen Widerstand R wegen $C = C_0 e^{-\frac{T}{RC}}$ auf $C = C_0/e \approx 37\%$ von C_0 in der Zeit T.

[2] GEVERS, M. u. F. K. DU PRE: Dielectrics (Trans. Faraday Soc.) **42 a**, 47 (1946).

[3] Unveröffentlichte Messungen.

Polyvinylchlorid bei der Frequenz $1 \cdot 10^{10}$ Hz als Funktion der Temperatur. Bei den verlustarmen Hochpolymeren Polyäthylen ($\operatorname{tg} \delta < 10^{-3}$) und Polystyrol ($\operatorname{tg} \delta < 10^{-3}$) tritt die zu erwartende DK-Abnahme mit steigender Temperatur auf. Bei dem mit starken Verlusten behafteten Polyvinylchlorid ($\operatorname{tg} \delta > 10^{-2}$) steigt die DK mit der Temperatur an.

§ 68. Messung der Dielektrizitätskonstante und des Verlustfaktors.

In einem elektrischen Wechselfeld besteht in einer reinen Kapazität zwischen Strom und Spannung eine Phasenverschiebung von genau 90°. Bei allen Kondensatoren treten jedoch Verluste auf, die z.B. durch schlechte Isolation, Glimm- und Sprüherscheinungen, hauptsächlich jedoch durch die Einstellbewegungen von Dipolen des Dielektricums zu dem äußeren Feld hervorgerufen werden. Das äußere Feld liefert die Energie, die zur Durchführung der Einstellbewegungen verbraucht wird. Es entsteht so ein Energieverbrauch im Kondensator, den man sich so vorstellen kann, als wäre er durch einen im Kondensator vorhandenen Ohmschen Widerstand hervorgerufen worden. Die Verluste haben eine Abweichung um den Winkel δ von der Phasenverschiebung 90° zur Folge. Die hier hauptsächlich interessierenden Verluste durch Einstellbewegungen von Dipolen hängen im wesentlichen von der Art des Dielektricums, von der Frequenz und von der Temperatur ab.

$\operatorname{tg} \delta = R \omega C = \operatorname{cotg} \varphi$ wird als Maß für die Verluste des Kondensators verwendet und als *Verlustzahl* oder *Verlustfaktor* bezeichnet. Den Winkel δ, den man bei der Vektordarstellung der Widerstände erhält, bezeichnet man als *Verlustwinkel.* Da die Verlustfaktoren bei guten Dielektrica meist sehr klein sind ($\operatorname{tg} \delta \sim 10^{-4} - 10^{-3}$), gilt näherungsweise $\delta \approx \operatorname{tg} \delta$. Wenn deshalb manchmal der Verlustwinkel δ an Stelle des Verlustfaktors $\operatorname{tg} \delta$ angegeben wird, so ist der Unterschied zwischen den beiden Werten nur klein, solange δ klein ist. Vielfach werden die Verluste eines Dielektricums auch in ε'', dem Imaginärteil der komplexen Dielektrizitätskonstante $\varepsilon^* = \varepsilon' + j \varepsilon''$ angegeben. Es gilt hierbei die Beziehung $\varepsilon''/\varepsilon' = \operatorname{tg} \delta$.

Im allgemeinen werden Dielektrizitätskonstante und Verlustfaktor gleichzeitig gemessen. Die hierbei anwendbare Meßmethode ist im wesentlichen durch die Frequenz bestimmt, bei der die Messung durchgeführt werden soll. Außerdem spielt, insbesondere bei hohen Frequenzen, bei der Wahl der Meßmethode die Möglichkeit, die Temperatur des Prüflings zu variieren, eine gewisse Rolle.

Als Frequenzgeneratoren sind bis etwa 10^8 Hz Elektronenröhrenschaltungen normaler Ausführungen üblich. Im Dezimeterwellengebiet werden meist Rohrkreissender mit Scheibentrioden oder Magnetronsender verwendet, und im Zentimeter- oder Mikrowellengebiet haben sich Klystrons und Magnetrons als Frequenzgeneratoren bewährt[1].

---

[1] VILBIG, F.: Hochfrequenz-Meßtechnik, Carl Hanser Verlag, München 1953.

a) Messungen bei Frequenzen unter 10^6 Hz.

Für Messungen bei Frequenzen bis zu etwa 10^6 Hz werden im Handel eine Reihe von brauchbaren Meßinstrumenten angeboten. Bei diesen Frequenzen bereitet der schaltungsmäßige Aufbau von Meßbrücken keine besonderen Schwierigkeiten, denn hier sind die Abmessungen der Schaltelemente klein gegen die Wellenlänge.

Die wichtigsten Meßverfahren sind:

Brückenmessungen. Gebräuchlich sind Wien-Brücken, Schering-Brücken in reiner oder in abgewandelter Form. GIEBE und ZICKNER haben eine häufig benutzte Abwandlung der Schering-Brücke angegeben. Erdungs- und Symmetrisierungsschwierigkeiten umgeht man vielfach durch Kunstschaltungen, z.B. den WAGNERschen Hilfszweig.

Die Präparation der zu vermessenden Dielektrica bereitet oft Schwierigkeiten.

Bringt man feste Dielektrica zwischen die Kondensatorplatten, so muß man darauf achten, daß keine Lufträume zwischen den Platten und den Probenoberflächen auftreten. Vielfach bestreicht man deshalb die Kontaktflächen der Probe mit Leitsilber oder Aquadag oder man bedampft sie im Vakuum mit einem Metall. Dabei muß natürlich gewährleistet sein, daß durch das Aufbringen der leitenden Schicht keine die elektrischen Eigenschaften der Probe ändernden Effekte auftreten, wie z.B. Anquellen durch die Verdünnungsflüssigkeit des Leitsilbers[1].

Substitutionsverfahren[2]. Hierbei wird der zu vermessende Kondensator mit dem Dielektricum durch einen Luftkondensator gleicher Kapazität ersetzt.

Verstimmungsverfahren[2,3]. Es beruht auf der Feststellung der Änderung der Resonanzverhältnisse eines Schwingkreises beim Einbringen des zu untersuchenden Dielektricums.

Quotientenmeßverfahren[2]. Bei ihm wird die Güteänderung eines Resonanzkreises beim Einbringen eines Dielektricums bestimmt.

b) Messungen im Meter- und Dezimeterwellengebiet.

Im Frequenzbereich oberhalb etwa 10^6 Hz spielt die Induktivität der Leitungen und Schaltelemente eine so wichtige Rolle, daß man hier mit den für niedrige Frequenzen entwickelten Schaltungen nicht mehr arbeiten kann. Vielfach stehen die Schwierigkeiten der günstigsten Anordnung und des mechanischen Aufbaues, der meist große Präzision verlangt, den schaltungsmäßigen nicht nach.

Oft weichen die quasistationär gemessenen Werte von den tatsächlich bei hohen Frequenzen gemessenen erheblich ab.

Diese Fehler kann man bestimmen, indem man z.B. die Frequenzabhängigkeit des betreffenden Schaltelementes einmal in Luft, das andere Mal bei völliger oder teilweiser Einbettung in ein Dielektricum von be-

[1] KOHLRAUSCH: Lehrbuch der Experimentalphysik.
[2] VILBIG, F.: Hochfrequenz-Meßtechnik, Carl Hanser Verlag, München 1953.
[3] REIS, K. H.: Beitr. Fortschr. HF-Technik 1, 389 (1941).

kannten Eigenschaften (z. B. Tetrachlorkohlenstoff) bestimmt. Ein Beispiel für die Berechnung geben MÜLLER und SCHMELZER[1] an.

Leitmeßverfahren[2,3]. Das Leitmeßverfahren ist etwa im Bereich $5 \cdot 10^5$ bis $1 \cdot 10^8$ Hz anwendbar. An einen konstant erregten Schwingkreis sind zwei Belastungskreise angeschlossen. Der eine enthält das Prüfobjekt, der andere besteht aus einem Kondensator mit parallel geschalteter Triode. Das Meßverfahren beruht darauf, daß man die Verhältnisse am Prüfobjekt meßbar durch den Triodenkreis nachbildet.

Brückenmessungen. VOIGT[4] gibt eine Vierkapazitätenmeßbrücke für den Bereich $1 \cdot 10^5$ bis $1 \cdot 10^9$ Hz an. Die Brücke ist zur Erzielung der Frequenzunabhängigkeit so aufgebaut, daß die Brückenelemente möglichst symmetrisch angeordnet sind.

Für den Bereich von etwa $5 \cdot 10^5$ bis $1 \cdot 10^8$ Hz hat sich eine aus zwei parallel geschalteten T-Gliedern aufgebaute Brücke bewährt[5–7]. Die Brücke ist dann abgeglichen, wenn die Beziehungen

$$\left.\begin{aligned} w^2 L^{-1} &= C_0 + C_1 C_2 \left(\frac{1}{C_1} + \frac{1}{C_2} + \frac{1}{C_3}\right) \\ G_0 &= w^2 C_1 C_2 R \left(1 + \frac{C_G}{C_3}\right) \end{aligned}\right\} \qquad \text{(VIII, 13)}$$

erfüllt sind. Schaltet man an die Klemmen x, y eine unbekannte Admittanz $\mathfrak{G}_x = G_x + i Y_x$ hinzu, so kann durch Verändern von C_0 um ΔC_0 und von C_G um ΔC_G der gestörte Abgleich wieder hergestellt werden. Die Blind- und Wirkkomponente von $\mathfrak{G}_x$ sind dann $Y_x = w \Delta C_0$, $G_x = w^2 \frac{C_1 C_2}{C_3} \cdot R \cdot \Delta C_G$.

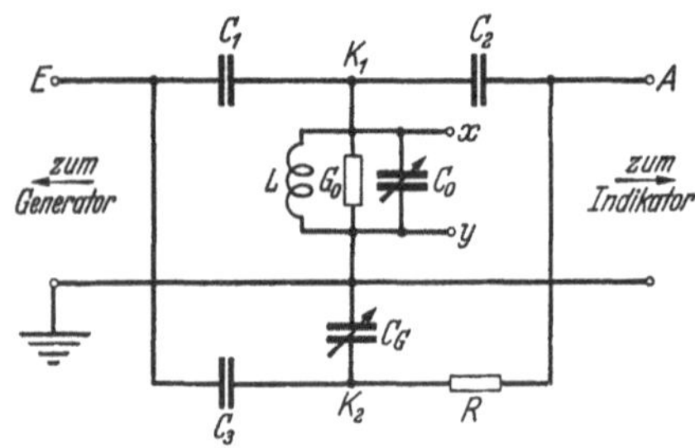

Abb. VIII, 5. Doppel-T-Brücke. (Nach MÜLLER und SCHMELZER[7].)

Messung mit einem Resonanztopfkreis. Für den Bereich von $5 \cdot 10^6$ Hz bis etwa $3 \cdot 10^9$ Hz haben WORKS und Mitarbeiter[8,9] Resonanztopfkreise zur Messung von DK und tg δ angegeben. Ein solcher Resonanztopfkreis besteht aus einem zylindrischen Hohlraum mit einem Innenleiter genau einstellbarer Länge. Ein Kurszchlußschieber erlaubt die Veränderung der Länge des Hohlraumes und damit der Frequenz. Die Endfläche des Innenleiters bildet mit der ihr gegenüberliegenden Fläche eine konzentrierte Kapazität. Das zu messende Dielektricum wird als kleine Scheibe auf eine dieser Flächen gelegt. Die Änderung der Kapazität und der Resonanzspannung

---

[1] MÜLLER, F. H. u. CHR. SCHMELZER: Ergebn. exakt. Naturw. **25**, 359 (1951).
[2] Siehe S. 539, Fußnote 2.
[3] RHODE, L. u. G. WEDEMEYER: E. T. Z. **61**, 577 (1940).
[4] VOIGT, H.: Arch. elektr. Übertr. **6**, 414 (1952).
[5] TUTTLE, W. N.: Proc. Instr. Radio Engr. **28**, 23 (1940).
[6] SINCLAIR, D. B.: Proc. Instr. Radio Engr. **28**, 310 (1940).
[7] MÜLLER, F. H. u. CHR. SCHMELZER: Ergebn. exakt. Naturw. **25**, 359 (1951).
[8] WORKS, C. N., T. W. DAKIN u. F. W. BOGGS: Proc. Instr. Radio Engr. **33**, 245 (1945).
[9] WORKS, C. N.: J. appl. Physics **18**, 605 (1947) u. **18**, 789 (1947).

ohne und mit Probe sind zusammen mit der Halbwertsbreite der Resonanzkurve ein Maß für DK und tg δ.

Von diesem Meßprinzip existieren eine Reihe von Abwandlungen.

c) Messungen im Zentimeterwellengebiet (Mikrowellengebiet).

Konzentrierte Schaltelemente lassen sich in diesem Bereich nicht mehr verwenden, weil die Versuchsanordnung nicht mehr klein gegen die Wellenlänge ausgeführt werden kann. Messungen, bei denen das zu untersuchende Material als Dielektricum in einen Kondensator eingebracht wird, sind deshalb nicht mehr möglich. Man benützt vielmehr ausschließlich Hohlleiter zur Fortleitung der elektromagnetischen Wellen und Resonanzhohlräume mit verteilter Induktivität und Kapazität. Da die Grenzwellenlänge von den Querschnittsabmessungen der Wellenleiter und Hohlräume abhängt, ist im allgemeinen für jeden Wellenlängenbereich eine eigene Apparatur erforderlich. Die gebräuchlichen Wellenlängen liegen um 10 cm, 3 cm und 1 cm. Die Zentimeterwellenmeßtechnik kennt eine große Zahl von genauen Meßmethoden für DK und tg δ. Sie beruhen alle darauf, daß man die Wellenlänge im Material mißt.

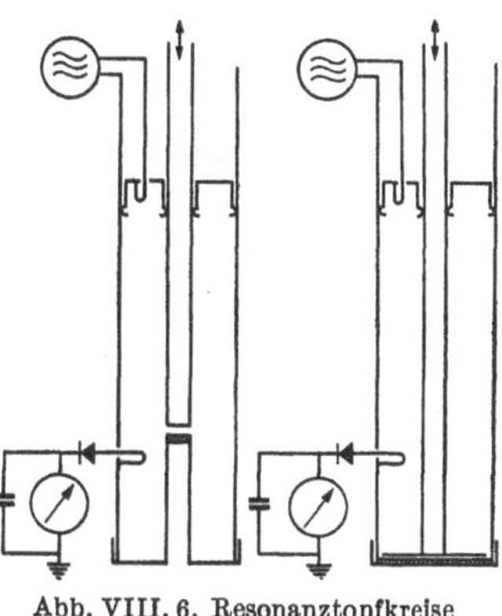

Abb. VIII, 6. Resonanztopfkreise (Schema).

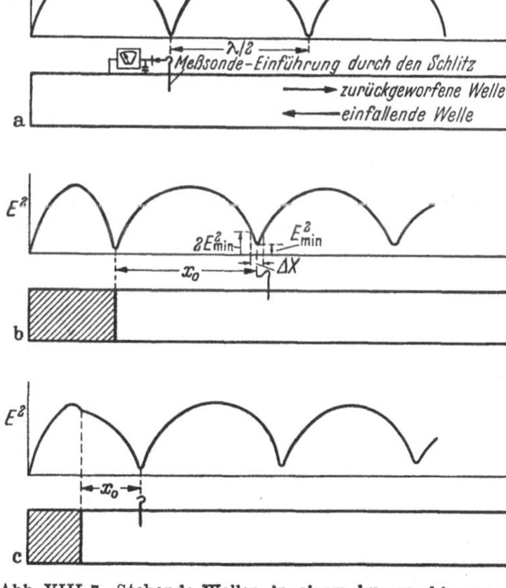

Abb. VIII, 7. Stehende Wellen in einem kurzgeschlossenen Wellenleiter. a) Dielektricum, b) Dielektricum der Lage $\lambda/2$, c) Dielektricum beliebiger Länge.

Bei der am häufigsten angewandten Methode wird ein rechteckiger Wellenleiter verwendet, der aus mehreren Gliedern zusammengesetzt ist. An dem einen Ende wird der Generator angekoppelt, das andere Ende ist durch eine Kurzschlußplatte abgeschlossen. Im Wellenleiter bildet sich bei einer senkrecht einfallenden Welle vor der vollständig reflektierenden Kurzschlußplatte ein System von stehenden Wellen aus. Die Lage der Maxima und Minima kann mit einer feinen, auf einem Schlitten beweglichen Sondenspitze, die durch einen Längsschlitz in den Wellenleiter hineinragt, abgetastet werden. Diese Sondenspitze, ein feiner Draht, ist über einen Gleichrichter mit einem Anzeigeinstrument verbunden. Bringt man ein Dielektricum vor die vollständig reflektierende Fläche, so tritt im Dielektricum eine Verkürzung der Wellenlänge auf. Der Abstand des ersten Minimums von der reflektierenden Fläche hängt von der Wellenlänge im Dielektrikum und von seiner Dicke ab. Das Einlegen des Dielektricums verschiebt das Minimum zum Ende hin. Der Abstand des ersten Minimums von der Dielektricumsfläche x_0 ist ein Maß für die DK. Wenn eine Dämpfung im Dielektricum auftritt, so reduziert sich das Verhältnis E_{max}/E_{min}, da nicht die ganze Intensität reflektiert wird. Die Änderung dieses Amplitudenverhältnisses ist ein Maß für den $\operatorname{tg}\delta$. Aus meßtechnischen Gründen wird vielfach nicht das Verhältnis E_{max}/E_{min} bestimmt, sondern der Abstand Δx der beiden Punkte der Amplitudenkurve, die den doppelten Minimumswert haben.

Eine ausführliche Beschreibung von 30 Methoden mit Literaturangaben findet sich bei MONTGOMERY[1] (siehe auch 2).

B. Meßergebnisse der dielektrischen Untersuchungen und ihre strukturelle Deutung.

Vorbemerkung.

Nach einem vorausgehenden Paragraphen, § 69, der dem Problem der Auswertung von Dispersionsmessungen gewidmet ist, werden in den folgenden Paragraphen 70—72 die in der Literatur bis heute vorliegenden Meßergebnisse dielektrischer Untersuchungen an hochmolekularen Substanzen diskutiert, wobei im wesentlichen nur die Messungen angeführt werden, bei denen mehr oder weniger ausgedehnte funktionelle Abhängigkeiten der dielektrischen Werte von Temperatur und Frequenz vorliegen und die so auch strukturelle Hinweise liefern. Der technischen Bedeutung der unpolaren hochmolekularen Substanzen entsprechend werden diese wichtigen Isolierstoffe der Hochfrequenztechnik zuerst behandelt, § 70.

In § 71 folgen dann die sehr zahlreichen polaren hochmolekularen Substanzen mit ihrem zum Teil recht komplizierten Verhalten der dielektrischen Werte als Funktion von Temperatur und Frequenz. Diese Stoffe sind nicht mehr allgemein als Isolierstoffe der Elektrotechnik zu verwenden. Ihr Einsatz fordert genaue Kenntnis der dielektrischen Werte, die in ihrer vielfachen Abhängigkeit von Temperatur und Frequenz meist gar nicht vorliegen. Die polaren Gruppen erweisen sich jedoch immer mehr als physikalisch-analytisch erfaßbare Gruppen, deren Beweglichkeit sich in der Dispersion der dielektrischen Werte über Temperatur und Frequenz widerspiegelt und ihrerseits wieder Hinweise gibt auf den

[1] MONTGOMERY, C. G.: Technique of Microwave Measurements, McGraw Hill Book Comp. 1947.

[2] SHAW, T. M. u. J. J. WINDLE: J. Appl. Physics 21, 956 (1950).

strukturellen Aufbau dieser hochmolekularen Substanzen. Die erwünschten dielektrischen Werte stammen so weniger von elektrotechnischer als vielmehr von physikalischer Seite und sie stehen oft in engem Zusammenhang mit anderen Äußerungen des strukturellen Aufbaus dieser Substanzen. Es wird daher im folgenden sehr oft verwiesen auf Bd. III, 11. Kapitel „Einfriererscheinungen", wo der strukturelle Aufbau an Hand der allgemeinen Einfriererscheinungen besprochen wird. In § 59 dieses 11. Kapitels wird auch die Abhängigkeit der Einfriererscheinungen vom Molekulargewicht behandelt, die hier insofern berücksichtigt wird, als man nach Möglichkeit für jede Substanz die Meßergebnisse von sehr hochmolekularen Proben angibt, die keine Abhängigkeit vom Molekulargewicht mehr zeigen. An manchen Stellen werden auch parallele Erscheinungen an niedermolekularen Substanzen erwähnt, wobei allerdings im wesentlichen auf die ausführliche Darstellung des „Dielektrischen Verhaltens im Zusammenhang mit dem polaren Aufbau der Materie" von F. H. MÜLLER und CHR. SCHMELZER hingewiesen werden muß.

Im letzten § 72 werden dann noch Meßergebnisse dielektrischer Untersuchungen an hochmolekularen Mischsubstanzen besprochen. Unter hochmolekularen Mischsubstanzen werden dabei im weiteren Sinne nicht nur durch Hauptvalenzen gebundene Mischungen verstanden, wie sie in Mischpolymerisaten und Mischkondensaten vorkommen, sondern auch Mischungen mit niedermolekularen Stoffen, Füllstoffen u. a., deren Hauptanteil hochmolekulare Substanzen sind. Die dielektrischen Messungen an Lösungen hochmolekularer Substanzen sind dabei schon in Bd. II, Kapitel 13, behandelt worden. Die Weichmachung hochmolekularer Substanzen durch Zumischung von äußerem Weichmacher und Lösungsmittel wird in Kapitel 9, zusammengefaßt, und lediglich einige elektrische Messungen an diesen weichgemachten Massen werden hier in diesem Kapitel erwähnt.

§ 69. Zur molekularen Auswertung von Dispersionsmessungen.

Überblickt man die dielektrischen Werte in ihrer Abhängigkeit von Frequenz und Temperatur, so findet man bei unpolaren hochmolekularen Substanzen in weitem Frequenzbereich nahezu konstante dielektrische Werte, wobei ε' etwa den Wert n^2 besitzt. Bei polaren Substanzen findet man dagegen in breitem Temperatur- und Frequenzbereich $\varepsilon' \gg n^2$, wenn sich Dipole im elektrischen Feld orientieren können und so einen Beitrag zur DK bringen. Mit fallender Temperatur oder steigender Frequenz muß jedoch dieser Orientierungsbeitrag der Dipole verschwinden, was in verschiedener Weise vor sich gehen kann (z. B. sprungartig oder Abfall in einem oder mehreren breiten Dispersionsbereichen, die sich jeweils durch eine mittlere Relaxationszeit kennzeichnen lassen). Die Lage dieses Abfalls über Temperatur und Frequenz und die Art des Abfalls ist weitgehend strukturbedingt. Dielektrische Untersuchungen sind daher ein wichtiger Teil der Strukturuntersuchungen von hochmolekularen Substanzen.

Bringt man einen polaren Stoff in ein elektrisches Wechselfeld, so orientieren sich seine polaren Gruppen nach dem Feld und folgen mehr oder weniger gut den Richtungswechseln des Feldes. Man kann diese Einstellbewegungen der Dipole mit ihren Nachbargruppen zu einem äußeren Feld nach EYRING[1] auch als Platzwechselvorgänge deuten. Diese Betrachtungsweise bringt insbesondere bei Hochpolymeren einige Vorteile. Es gelingt nämlich durch eine solche thermodynamische Beschreibung einen Teil der Schwierigkeiten zu umgehen, welche die Annahme einer speziellen Molekülgestalt bei der DEBYEschen Theorie mit sich bringt. Die Platzwechsel stellt man sich folgendermaßen vor (nach F. H. MÜLLER und CHR. SCHMELZER[2]): Das einzelne Molekül- oder Kettensegment schwingt infolge der Wärmebewegung mit ziemlich hoher Frequenz um seine momentane Lage. Die Fluktuation der thermischen Energie gibt ihm hin und wieder einen so hohen Energiegehalt, daß es die umgebende Potentialschwelle, die es an seine Lage kettet, überspringen kann, falls in der Nachbarschaft genügend freier Platz ist. Der Platzwechsel ist also dann möglich, wenn einerseits die Molekülgruppe genügend Aktivierungsenergie erhält, andererseits mindestens so viel Raum gegeben ist, daß zwei benachbarte Gruppen ihre Plätze tauschen können bzw. ein freier Platz in unmittelbarer Nachbarschaft vorhanden ist.

Da die Viscosität der Hochpolymeren stark von der Temperatur abhängt, ändert sich die Wahrscheinlichkeit für solche Platzwechsel ebenfalls mit der Temperatur. Man kann deshalb die Dispersion nicht nur als Funktion der Frequenz, sondern auch als Funktion der Temperatur messen. In den Temperaturkurven der dielektrischen Größen treten die gleichen Erscheinungen auf wie in den Frequenzkurven. Man beobachtet also z. B. Maxima in der Verlustfaktorkurve und Stufen in der DK-Kurve. Während aber die Maxima der Frequenzkurve eine breite Verteilung über wenigstens 1 Zehnerpotenz, gewöhnlich aber über 2 oder mehr Zehnerpotenzen zeigen, erstreckt sich das Maximum der Temperaturkurve über ein relativ kleines Temperaturintervall. Beide Meßmethoden sind äquivalent. Im allgemeinen werden aus meßtechnischen Gründen die Temperaturkurven bevorzugt, obwohl die Frequenzkurven theoretisch zur Zeit leichter zu erfassen sind.

Die Einstellbewegungen der polaren Gruppen bzw. die Platzwechsel folgen dem Wechsel des äußeren Feldes mit einer gewissen Verzögerung. Es tritt also eine Relaxation auf. Die Verzögerung wird durch die mittlere Relaxationszeit τ gekennzeichnet. Um diese mittlere Relaxationszeit existiert eine mehr oder weniger breite Relaxationszeitverteilung, die folgende Ursachen hat:

Bei den Hochpolymeren sind die Molekülketten unregelmäßig und verknäuelt angeordnet. Eine Folge davon ist, daß Schwankungen der Dichte und damit der Kraftfelder in kleinen Bereichen auftreten. Da die Orientierungsbewegungen eines Dipols von der Stärke der Kraftfelder in der Umgebung des Dipols abhängen, hat die Aktivierungsenergie für diese Bewegungen wegen der lokalen Unterschiede in der Stärke der Kraft-

[1] EYRING, H. K.: The Theorie of Rate Processes, New York 1941.

[2] MÜLLER, F. H. u. CHR. SCHMELZER: Ergebn. exakt. Naturwiss. **25**, 359 (1951).

felder keinen festen Wert, sondern ist um einen gewissen Mittelwert verteilt, wobei für die Orientierungsübergänge keine einheitliche Relaxationszeitverteilung vorliegt[1, 2].

Theoretische Überlegungen führen ebenfalls zur Einführung mehrerer Relaxationszeiten, wenn stark anisotrope Molekülformen[3] oder frei drehbare Gruppen im Molekül[4] vorhanden sind.

Der Verlauf von ε' und ε'' bei gleichzeitigem Auftreten mehrerer Relaxationszeiten τ muß sich, sofern die verschiedenen Mechanismen sich nicht gegenseitig beeinflussen, aus „DEBYE-Kurven" durch Überlagerung zusammensetzen lassen[5].

FUOSS und KIRKWOOD[6] haben ein Verfahren angegeben, bei dem die „DEBYE-Kurve" durch eine geeignete Funktion dargestellt und die Lösung in Integralform über die experimentelle Kurve angegeben wird.

Eine näherungsweise Berechnung der Relaxationszeitverteilungsfunktion ist nach einem Verfahren von WILLIAMS und FERRY[7] möglich. Die Methode wurde für mechanische Relaxationszeitverteilungen entwickelt und stellt ein Näherungsverfahren 2. Ordnung dar. Näherungsverfahren 1. Ordnung verwenden Zeit- oder Frequenzableitungen von experimentell gemessenen Größen. Das Näherungsverfahren 2. Ordnung führt zusätzlich eine zweite Ableitung nach der Zeit oder der Frequenz ein. Diese zweiten Ableitungen stellen tatsächlich die erste Ableitung der Verteilungsfunktion dar. Formale Analogien zwischen mechanischen und elektrischen Größen erlauben eine Übertragung des Verfahrens auf dielektrisch gewonnene Dispersionskurven.

Es gibt noch eine Reihe weiterer Methoden, die es erlauben, die Relaxationszeitverteilung oder wenigstens ein Vergleichsmaß für die Breite der Verteilung zu gewinnen.

Die wichtigste ist die von COLE und COLE[8]. Sie erfordert die Messung von ε' und ε'' über ein weites Frequenzgebiet. Trägt man die zusammengehörigen Werte ε'' gegen ε' auf, so liegen die Punkte auf einem Halbkreis, dessen Mittelpunkt dann auf der ε'-Achse liegt, wenn nur eine einzige Relaxationszeit auftritt. Liegt eine Verteilung von mehreren Relaxationszeiten vor, so findet man den Kreismittelpunkt unterhalb der ε'-Achse. Die Verschiebung des Mittelpunktes, ausgedrückt durch den Winkel α, ist ein Maß für die Relaxationszeitverteilung um einen häufigsten Mittelwert.

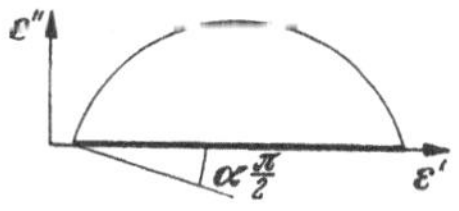

Abb. VIII, 8. Entnahme des Verteilungsparameters α aus der COLEschen Auftragung von ε'' gegen ε'. (Nach MÜLLER-SCHMELZER[5].)

FUOSS und KIRKWOOD[6] haben eine noch etwas andere Form errechnet.

Aus der Abweichung des COLE-Bogens von der Kreisbogenform kann

---

[1] Siehe S. 544, Fußnote 2.

[2] Das Relaxationsverhalten der Materie, 2. Marburger Diskussionstagung: Kolloid-Z. **134** (1953).

[3] FISCHER, E., A. BUDO u. S. MIYAMOTO: Physik. Z. **40**, 337 (1939).

[4] BUDO, A.: Physik. Z. **39**, 706 (1938).

[5] MÜLLER, F. H. u. CHR. SCHMELZER: Ergebn. exakt. Naturwiss. **25**, 359 (1951).

[6] FUOSS, R. M. u. J. G. KIRKWOOD: J. Amer. chem. Soc. **63**, 385 (1941).

[7] WILLIAMS, M. L. u. J. D. FERRY: J. Polymer Sci. **11**, 169 (1953).

[8] COLE, H. u. K. S. COLE: J. chem. Physics **9**, 341 (1941).

auf das Auftreten von weiteren Dispersionsstellen neben der Hauptdispersion geschlossen werden[1].

Während bei den bisher erwähnten Methoden die Messungen bei verschiedenen Frequenzen durchgeführt werden mußten, hat WALTHER[2] ein Verfahren angegeben, mit dem man, ohne die Relaxationszeiten selbst zu kennen, aus der Messung von $\operatorname{tg}\delta$ als Funktion der Temperatur Schlüsse auf das Auftreten von Relaxationszeitgruppen ziehen kann. Es wird hierbei eine Häufigkeitsanalyse für $\log \operatorname{tg}\delta$ als $f(T)$ durchgeführt. Das Auftreten von Relaxationszeitgruppen läßt sich so nachweisen, jedoch kann man ihre Verteilung nicht bestimmen.

Sowohl F. H. MÜLLER[3] als auch R. F. TUCKETT[4] wiesen vor Jahren darauf hin, daß den elektrischen Dispersionserscheinungen hochmolekularer Substanzen verschiedene Mechanismen zugrunde liegen können. Nach MÜLLER-SCHMELZER[5] sind wenigstens zwei verschiedene Dispersionsmechanismen möglich:

a) Orientierung von Partialdipolen, die starr mit der Hauptvalenzkette der hochmolekularen Substanz verbunden sind und somit die Beweglichkeit von Kettenstücken voraussetzen.

b) Umlagerungen eines einzelnen Atoms *(Protonensprung)*, die nicht die Kettenbeweglichkeit zur Voraussetzung haben.

Man muß diese beiden Fälle als häufig anzutreffende Grenzmechanismen ansehen, zwischen denen durchaus noch weitere Möglichkeiten der Dipolorientierung stehen. Zweifellos spielt der Fall a) bei linearen polaren hochmolekularen Substanzen eine große Rolle, bei denen längs der Kette die Dipole in sehr großer Konzentration durch Hauptvalenzbindungen eingebaut sind. Die Dipole können derart stark miteinander gekoppelt sein, daß wahrscheinlich keine Orientierungsbewegungen der einzelnen Dipole frei und unabhängig voneinander erfolgen, sondern Orientierungsbewegungen von einer Reihe von gekoppelten Dipolen in Form von Rotationsbewegungen von Kettenteilstücken um ihre Kettenachse.

Diese Rotationsbewegungen (Mikro-BROWNsche Bewegungen von Kettenteilstücken) werden thermisch erstmals ermöglicht bei der Einfriertemperatur, und es ist charakteristisch, daß im Falle a) die Orientierungspolarisation immer oberhalb der Einfriertemperatur erfolgt und daß man aus der Frequenzabhängigkeit dieser Dispersion nach KAUZMANN[6] auf diese Einfriertemperatur extrapolieren kann (siehe Bd. III, § 58, Abb. XI, 2). Es ist verständlich, daß im Falle a) weitgehende Parallelität der in § 58 aufgeführten Einfriererscheinungen besteht und daß alle strukturellen Elemente, die dort eine Erschwerung der Kettenbeweglichkeit mit sich bringen, hier eine Verschiebung der dielektrischen Dispersion nach höheren Relaxationszeiten verursachen.

In vielen polaren hochmolekularen Substanzen findet man mehrere Dispersionsstufen, deren strukturelle Zuordnungen eine möglichst weit-

[1] DAVIDSON, D. W. u. R. H. COLE: J. chem. Physics **19**, 1484 (1951).
[2] WALTHER, H.: Kolloid-Z. **117**, 75 (1950).
[3] MÜLLER, F. H.: Elektrotechn. Z. **59**, 1155, 1176 (1938).
[4] TUCKETT, R. F.: Trans. Faraday Soc. **40**, 448 (1944).
[5] MÜLLER, F. H. u. CHR. SCHMELZER: Ergebn. exakt. Naturwiss. **25**, 359 (1951).
[6] KAUZMANN, W.: Chem. Rev. **43**, 219 (1948).

gehende funktionelle Darstellung der dielektrischen Werte über Temperatur und Frequenz erfordert[1], mit der die Möglichkeit gegeben ist, die verschiedenen Orientierungsvorgänge u. a. durch ihre Aktivierungsenergie zu kennzeichnen. Es ist dabei zu erwarten, daß ein Protonensprungmechanismus mit einer wesentlich kleineren Aktivierungsenergie auskommt als eine Orientierungspolarisation von Kettenteilstücken. Die Aktivierungsenergie kann damit einen wesentlichen Beitrag liefern zu einer strukturellen Deutung einer Dispersionsstufe.

Schließlich sei noch vor der eigentlichen Diskussion der Meßergebnisse auf eine Komplikation hingewiesen, die durch die Kristallisationstendenz mancher hochmolekularen Substanz verursacht wird. Wie in Abb. VIII, 12 zu sehen ist, können bei der Abkühlung von Schmelzen derartiger Substanzen polare Gruppen im Kristallgitter festgelegt werden, womit dann ε' sprungartig abfällt. Dieser Abfall ist sehr stark versuchsbedingt und kann durch Abschrecken unter Umständen auch unterdrückt werden. Die Diskussion solcher Kurven kann gerade durch ihre Abhängigkeit von der Versuchsmethodik sehr erschwert werden.

Im folgenden wird an manchen Stellen von amorphen, kristallinen und gemischt amorph-kristallinen Hochmolekularen gesprochen. Um Mißverständnisse zu vermeiden, sei darauf hingewiesen, daß es einen völlig ungeordneten Zustand nur bei Gasen gibt und daß der andere Extremzustand einer dreidimensionalen Ordnung nur bei Einkristallen anzutreffen ist. Beide Extremzustände sind bei Hochmolekularen nie zu verwirklichen. Amorphe Hochmolekulare sind im Zustand einer Flüssigkeit, die zum mindesten Nahordnung zeigt. Die kristallinen Hochpolymeren besitzen ebensowenig eine dreidimensionale Ordnung ohne Begrenzung. Es sind vielmehr Substanzen, bei denen ein gewisser Volumenanteil in kleinen kristallinen Bezirken angeordnet ist, zwischen denen sich mit unscharfen Übergängen weniger geordnete Substanz befindet. Zwischen diesen beiden Vertretern von verschieden stark geordneten Hochmolekularen gibt es sehr viele Zwischenglieder, deren Ordnungsgrad immer von Vorbehandlung, thermischen Bedingungen u.ä. abhängt[2].

§ 70. Ergebnisse an unpolaren Stoffen.

a) Polyisobutylen.

$$\left[\begin{array}{c} CH_3 \\ | \\ -CH_2-C- \\ | \\ CH_3 \end{array} \right]_n$$

Polyisobutylen ist als reiner Kohlenwasserstoff eine unpolare Substanz mit hoher Kettenbeweglichkeit und niederer Temperaturlage der Einfriererscheinungen. Da die dielektrischen Werte niedrig und unabhängig von der Frequenz sind, wurde Polyisobutylen häufig als Isolier-

[1] Das Relaxationsverhalten der Materie, 2. Marburger Diskussionstagung: Kolloid-Z. **134** (1953).

[2] Stuart, H. A.: Kunststoffe **42**, 266 (1952).

stoff in Hochfrequenzkabeln benützt. Wegen mangelnder mechanischer Festigkeit, im besonderen wegen seiner Neigung zum „kalten Fluß", wurde in steigendem Maße Polyäthylen für diesen Zweck verwendet, so daß heute Polyisobutylen in seiner Bedeutung etwas zurückgetreten ist.

b) Polyäthylen. $[-CH_2-CH_2-]_n$

Polyäthylen als reiner Kohlenwasserstoff zeigt ebenfalls hervorragende dielektrische Eigenschaften. Nach FRÖHLICH[1] kann man wegen der Anordnung der Polyäthylenketten in Zickzackebenen erwarten, daß sich die an sich sehr schwachen CH_2-Dipole praktisch vollständig kompensieren. Diese Verhältnisse werden auch durch die Unsymmetrien, die bei Verzweigungen auftreten, nur unwesentlich gestört. Die Folge davon ist eine sehr niedrige DK von etwa $\varepsilon' = 2{,}25$ bis 2,30, die sich über den ganzen Frequenzbereich nicht ändert. Der sehr kleine Verlustfaktor $\operatorname{tg}\delta$ von etwa 1 bis $2 \cdot 10^{-4}$ steigt mit der Frequenz nur sehr wenig an und weist auch bei der Temperaturvariation nur geringe Änderungen auf. Allerdings tritt eine Erhöhung des $\operatorname{tg}\delta$ dann ein, wenn man Polyäthylen bei hohen Temperaturen ($> 150°$ C) lange Zeit der Einwirkung des Luftsauerstoffs aussetzt. Es wird dann Sauerstoff eingebaut, und die dabei entstehenden Dipole geben Anlaß zu Verlusten[2]. Polyäthylen nimmt praktisch kein Wasser auf, und seine guten dielektrischen Eigenschaften bleiben so auch bei Feuchtigkeitseinwirkung erhalten. Wegen dieser hervorragenden Eigenschaften findet Polyäthylen eine wachsende Anwendung in der gesamten Hochfrequenztechnik und vor allem in der Kabeltechnik.

c) Polystyrol.

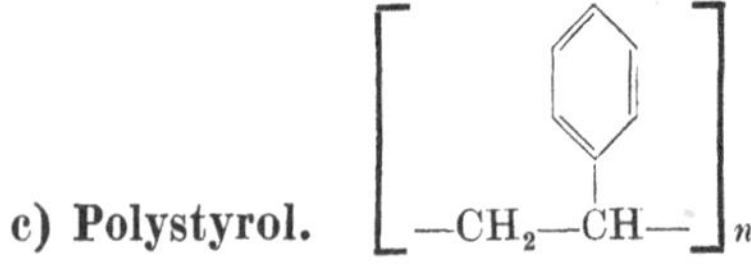

Polystyrol ist einer der wichtigsten synthetischen Isolierstoffe für die Hochfrequenztechnik, der neuerdings auch in Form von geschäumtem Polystyrol im Handel ist. Das Schaummaterial wird u.a. in der Kabeltechnik zur Herstellung von kapazitätsarmen Hochfrequenzkabeln verwendet. Dieses Styropor[3] läßt sich in den Dichten von 0,02 bis 0,3 herstellen. A. TELTSCHIK und SCHMELZER[4] bestimmten an derartigem geschäumtem Polystyrol bei 22°C und $1 \cdot 10^{10}$ Hz folgende Werte:

g/cm^3	0,05	0,10	0,15	0,20
ε'	1,05	1,105	1,16	1,21
$\operatorname{tg}\delta \cdot 10^4$	—	1,2	1,5	1,8

Bei Raumtemperatur zeigt reines Polystyrol im weiten Gebiet von niederen Frequenzen bis 10^9 Hz einen Wert $\varepsilon' = 2{,}5$ bis 2,6, der etwa n^2 entspricht ($n_D = 1{,}58$). $\operatorname{tg}\delta$ beträgt im selben Gebiet etwa 1 bis $3 \cdot 10^{-4}$.

[1] FRÖHLICH, H.: Theory of Dielectrics, Oxford 1949.
[2] JACKSON, W. u. J. A. S. FORSYTH: J. Instn. electr. Engr. **94**, 55 (1947).
[3] Handelsprodukt der Badischen Anilin- u. Soda-Fabrik AG., Ludwigshafen a. Rh. — [4] Persönliche Mitteilung.

Bei diesem verlustarmen Stoff ist nach GEVERS und DU PRE[1] ein schwacher Abfall von ε' mit der Temperatur zu erwarten, wie dies auch in

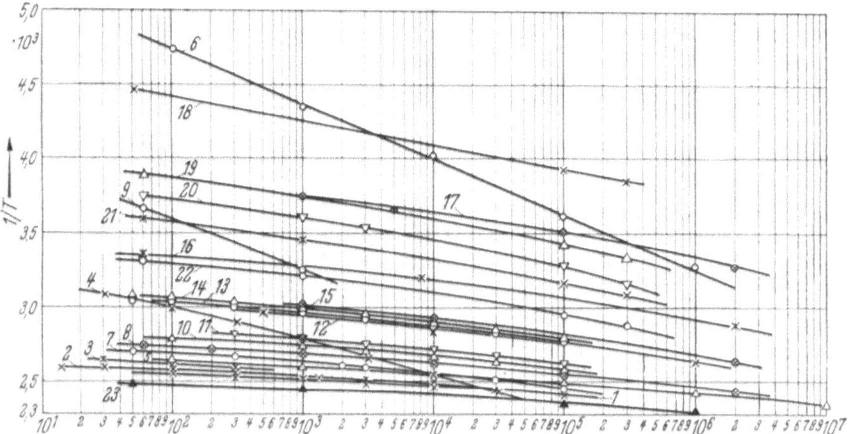

Abb. VIII, 9. Temperaturen der Verlustfaktormaxima in Abhängigkeit von der Frequenz [$1/T = f(\nu)$].

Tabelle VIII, 1.

Aktivierungsenergie für die in Abb. VIII, 9 eingezeichneten Hochpolymere für das Frequenzintervall $1 \cdot 10^3$ Hz bis $1 \cdot 10^5$ Hz berechnet.

Stoff	Aktivierungsenergie in kcal/Mol
1. Polyvinylformal	102
2. Polystyrol	101
3. Polymethacrylsäuremethylester Hauptmaximum	100 extrapoliert
4. Polymethacrylsäuremethylester Nebenmaximum	21
5. Terylen, Hauptmaximum	83
6. Terylen, Nebenmaximum	11
7. Polyacrylnitril	57
8. Polyvinylchlorid, Hauptmaximum	70
9. Polyvinylchlorid, Nebenmaximum	20 extrapoliert
10. Polyvinylacetal	61
11. Polyvinylbutyral	60
12. } Polyvinylacetat	48
13. } Polyvinylacetat	40
14. Polyvinylhexanal	47
15. Polyvinyl-2-äthylhexanal	46
16. Polyacrylsäuremethylester	35
17. Polyvinylmethyläther	35
18. Gummi + 0% Schwefel	30
19. Gummi + 8% Schwefel	30
20. Gummi + 10% Schwefel	30
21. Gummi + 12% Schwefel	31
22. Gummi + 16% Schwefel	35
23. Gummi + 32% Schwefel	114

[1] GEVERS, M. u. F. K. DU PRE: Disc. Faraday Soc. **42** a, 47 (1946).

Abb. VIII, 4 nach Messungen von THURN zum Ausdruck kommt. Oberhalb seiner Einfriertemperatur (und damit außerhalb seiner praktischen Verwendung als Isolierstoff!) wurde vor kurzem eine Dispersion gefunden, die dem schwachen Dipolmoment ($\sim 0{,}2\,DE$) der Phenylgruppe zugeordnet wird[1]. Aus der Temperaturverschiebung der Dispersion im Gebiet von 15 bis 10^5 Hz errechnet sich die Aktivierungsenergie zu 86 kcal/Mol. Diese Zahl im Zusammenhang mit der Temperaturlage der DK-Dispersion weist darauf hin, daß es sich nicht um eine Orientierung der Phenylgruppe allein handelt, sondern um eine Orientierung im Zusammenhang mit der Bewegung von Kettenteilstücken, die erst oberhalb der Einfriertemperatur frei wird. Die Auslegung wird neuerdings gestützt durch sehr genaue ε'-Messungen bei $2 \cdot 10^6$ Hz im Temperaturgebiet von -180 bis $+200°$ C, wobei von $+100°$ ab ein Anstieg von ε' in mehreren Sprüngen beobachtet wird, deren einzelne Stufen allerdings noch nicht geklärt sind[2]. Ein weiterer Hinweis für die durch die Phenylgruppe verursachte schwache Dispersion zeigt sich nach MILLANE[3] darin, daß nach vollständiger Hydrierung des Polystyrols zum Polyvinylcyclohexan bei 10^6 Hz die Werte $\varepsilon' = 2{,}25$ und $\operatorname{tg}\delta = 7 \cdot 10^{-5}$ erhalten werden.

Polystyrol hat als Hochfrequenzisolierstoff für manche technische Anwendung eine ungenügende Wärmefestigkeit. Man hat vielfach versucht, diese Wärmefestigkeit unter Erhalt der guten elektrischen Eigenschaften zu höheren Temperaturen auszudehnen. Dies gelingt beispielsweise durch erhöhte sterische Behinderung der Kettenbeweglichkeit im Poly-α-Methylstyrol oder in einem phenylsubstituierten Polystyrol[4]. Es ist aber auch möglich, eine Erhöhung der Wärmefestigkeit durch Einbau von Dipolen und damit Erhöhung der zwischenmolekularen Bindungen zu erzielen. Die nicht erwünschten dielektrischen Auswirkungen der Dipole können vermieden werden, wenn, wie im Poly-p-dichlorstyrol, die zwei Dipole sich kompensieren[5]. Eine Unsymmetrie wie im Poly-m-dichlorstyrol oder auch im Poly-α, β, β-trifluorstyrol macht sich sofort in den erhöhten Werten von ε' und $\operatorname{tg}\delta$ bemerkbar[6].

d) Polytetrafluoräthylen. $[-CF_2-CF_2-]_n$

Teflon ist eine aus Partialdipolen aufgebaute hochmolekulare Substanz, bei welcher der symmetrische Einbau der Dipole zu einer vollständigen Kompensation und somit zu einer unpolaren Substanz führt, die im ganzen Temperaturbereich von 20 bis 314° C und von Nieder- bis Hochfrequenz konstante und niedrige $\operatorname{tg}\delta$-Werte besitzt und deren ε'-Werte sich in diesem Temperaturbereich lediglich entsprechend der Dichteänderung verschieben[7,8]. Die vielen Dipole verursachen starke

[1] BAKER, E. B., R. P. AUTY u. G. J. RITENOUR: J. chem. Physics **21**, 159 (1953).
[2] THURN, H.: Z. angew. Physik **7**, 44 (1955).
[3] MILLANE, J. J.: Brit. Plastics **26**, 220 (1953).
[4] DIXON, J. K. u. K. W. SAUNDERS: Ind. Engng. Chem. **46**, 652 (1954).
[5] HIPPEL, A. v. u. L. G. WESSON: Ind. Engng. Chem. **38**, 1121 (1946).
[6] PROBER, M.: J. Amer. chem. Soc. **75**, 968 (1953).
[7] BEARDSLEY, J. H.: Rev. Sci. Instr. **24**, 180 (1953).
[8] EHRLICH, P.: J. Res. nat. Bur. Standards **51**, 185 (1953).

zwischenmolekulare Bindungen und einen ungewöhnlich hohen Schmelzpunkt der kristallinen Phase von 327° C. Der hohe Schmelzpunkt bringt Schwierigkeiten in der Verarbeitung mit sich und man hat so versucht, durch geringe Unsymmetrie im Aufbau den Schmelzpunkt zu drücken. Polytrifluormonochloräthylen $[-CF_2-CFCl-]_n$ ist eine solche Substanz mit geringer Unsymmetrie und einem Schmelzpunkt von 230—280° C. Elektrisch zeigt sich die Unsymmetrie durch eine Dispersion der dielektrischen Werte an[1].

§ 71. Ergebnisse an polaren Stoffen.

a) Polyisopren und synthetischer Kautschuk.

Polyisopren besitzt durch seine Doppelbindungen eine geringe Polarität, die sich in einer schwachen Dispersion bemerkbar macht. Die Orientierungspolarisation erfolgt mit einer geringen Aktivierungsenergie und in Zusammenhang mit der Kettenbeweglichkeit, die schon bei tiefen Temperaturen frei wird. Die für den technischen Gebrauch erforderliche Vulkanisation mit Schwefel bringt erheblich stärkere Polaritäten in die vernetzte Substanz, die nicht nur eine Erhöhung der dielektrischen Werte, sondern auch eine Verschiebung der Dispersion zu längeren Relaxationszeiten und zu höheren Aktivierungsenergien bewirken[2].

Für Polybutadien gilt etwa dasselbe wie für Polyisopren. Auch hier liegt eine schwache Dispersion durch die vorhandenen Doppelbindungen vor. Die Dispersion liegt bei hoher Frequenz bzw. tiefer Temperatur. Durch Mischpolymerisation mit Styrol (Buna S, GR—S) verschiebt sich diese schwache Dispersion zu längeren Relaxationszeiten[3–5], da die Mischpolymerisation mit Styrol eine Erschwerung der Kettenbeweglichkeit bedeutet. Auch hier wird die Polarität wesentlich erhöht durch die Vulkanisation. Ein technisch oft verwendetes weiteres Butadienmischpolymerisat ist der mit Acrylnitril als zweiter Komponente hergestellte Perbunan, der durch die Nitrilgruppe $-C{\equiv}N$ starke Polarität zeigt mit einer wesentlich ausgeprägteren Dispersion, die entsprechend niederfrequenter bzw. bei höheren Temperaturen liegt als die des Polybutadiens[3–6].

Ein ebenfalls nach seinen technisch-mechanischen Eigenschaften zum synthetischen Kautschuk zu rechnendes Produkt ist das Polychloropren (Neopren) $\left[-CH_2-\underset{\displaystyle Cl}{\underset{|}{C}}{=}CH-CH_2-\right]_n$. Der Einbau von C–Cl-Dipolen bedingt seine Polarität, die sich wieder an Hand der dielektrischen Werte

[1] Reynolds, St. J., V. G. Thomas, A. H. Sharbough u. R. M. Fuoss: J. Amer. chem. Soc. **73**, 3714 (1951).

[2] Scott, A. H., A. T. McPherson u. H. C. Curtiss: J. Res. nat. Bur. Standards **11**, 173 (1933).

[3] Carter, W. C., M. Magat, W. C. Schneider u. C. P. Smyth: Trans. Faraday Soc. **42**a, 213 (1946).

[4] Roelig, H.: Kautschuk **12**, 139 (1941).

[5] Roelig, H. u. W. Heidemann: Kunststoffe **38**, 125 (1948).

[6] Dalbert, R.: Rev. gén. Caoutchouc **29**, 515, 588, 649 (1952).

und einer Dispersion dieser Werte zeigt[1, 2]. Neopren steht in seinem chemischen Aufbau und seinen Eigenschaften etwa zwischen Polybutadien und Polyvinylchlorid, jedoch mehr in Richtung auf Polybutadien verschoben.

b) Polyvinyläther.

$$\left[-CH_2-\underset{\underset{\displaystyle R}{|}}{\underset{\displaystyle O}{\underset{|}{CH}}}- \right]_n$$

Das geringe Dipolmoment der Ätherbrücke verursacht, verglichen mit einem hochmolekularen Kohlenwasserstoff wie Polyisobutylen, eine kleine Verschiebung der Einfriererscheinungen nach höheren Temperaturen. Durch die Anwesenheit der Dipole wird hier im Gegensatz zum Polyisobutylen auch die elektrische Einfriererscheinung in der DK-Dispersion sichtbar. Die Aktivierungsenergie der DK-Dispersion eines Polyvinylmethyläthers ($R = -CH_3$) ist mit 33 kcal/Mol (10^3 bis $2 \cdot 10^6$ Hz)[3] erheblich kleiner als bei einer linearen hochmolekularen Substanz mit starker, elektrisch oder sterisch bedingter Behinderung der Kettenbeweglichkeit (siehe Polyvinylchlorid, Polystyrol, Polymethacrylsäuremethylester u. a.). Auch hier bei den Polyvinyläthern kann sich jedoch ebenfalls eine hohe Temperaturlage der Einfriererscheinungen und damit auch der DK-Dispersion einstellen, wenn in R sterisch wirkende Bauelemente verwendet werden. Vergleicht man die Temperaturlage der DK-Dispersion der drei isomeren Polyvinyläther der n-Butyl-, iso-Butyl- und tert. Butyl-Alkohole, so zeigt sich in dieser Richtung und vor allem beim Übergang vom iso-Butyl- zum tert. Butyläther ein sehr starker Anstieg der Temperaturlage der DK-Dispersion, der auch einen starken Anstieg in der Aktivierungsenergie vermuten läßt[4].

c) Polyvinylester.

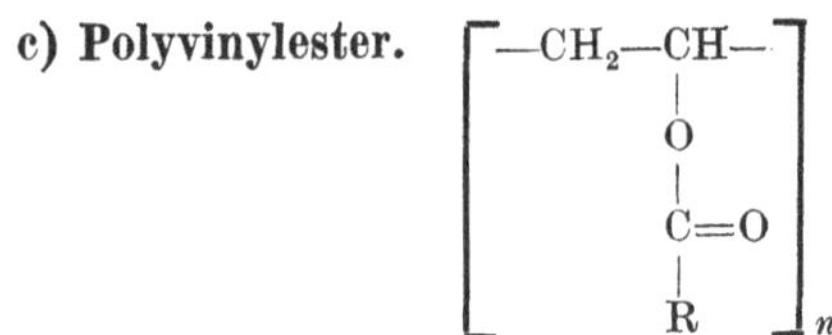

In dieser Reihe der Polyvinylester mit verschiedenen Alkylen R ist das Polyvinylacetat ($R = -CH_3$) dielektrisch am besten bekannt. Es zeigt nach Mead und Fuoss[5] eine ausgeprägte DK-Dispersion, die zusammen mit der Aktivierungsenergie auf eine gekoppelte Orientierungsbewegung der Estergruppen mit der Hauptkette hinweist. Nach Fuoss und Kirkwood[6] läßt sich aus diesen Werten das Dipolmoment μ der monomeren Einheit zu 2,3 *DE* errechnen. Die Temperaturlage der DK-Dispersion erhöht sich wie die Aktivierungsenergie mit steigendem Mole-

---

[1] Schneider, W. C., W. C. Carter, M. Magat u. C. P. Smyth: J. Amer. chem. Soc. **67**, 959 (1945). – [2] Reinisch, L.: Rev. gén. Caoutchouc **29**, 593 (1952).

[3] Unveröffentlichte eigene Messungen.

[4] Würstlin, F.: Z. angew. Physik **2**, 131 (1950).

[5] Mead, D. J. u. R. M. Fuoss: J. Amer. chem. Soc. **63**, 2832 (1941).

[6] Fuoss, R. M. u. J. G. Kirkwood: J. Amer. chem. Soc. **63**, 385 (1941).

kulargewicht und erreicht wie bei vielen hochpolymeren Substanzen einen Endwert beider Größen bei einem gewissen Mindestmolekulargewicht. Durch Zumischung von Weichmacher oder Lösungsmittel wird sowohl die Temperaturlage der DK-Dispersion als auch die Aktivierungsenergie erniedrigt[1].

Neben diesen Messungen an Polyvinylacetat liegen aus der Reihe der Polyvinylester nur noch eigene dielektrische Messungen an Polyvinylpropionat ($R = -CH_2-CH_3$) vor[2], die, verglichen mit Polyvinylacetat, eine niedrigere Temperaturlage und auch eine niedrigere Aktivierungsenergie für das Polyvinylpropionat ergeben. Auch hier erweist sich so wie bei den Einfriererscheinungen (Bd. III, Kap. 11 § 60) das Anwachsen des Alkyls als eine innere Erweichung der hochmolekularen Substanz, die sich sicher noch über das Polyvinyl-n-butyrat fortsetzen würde.

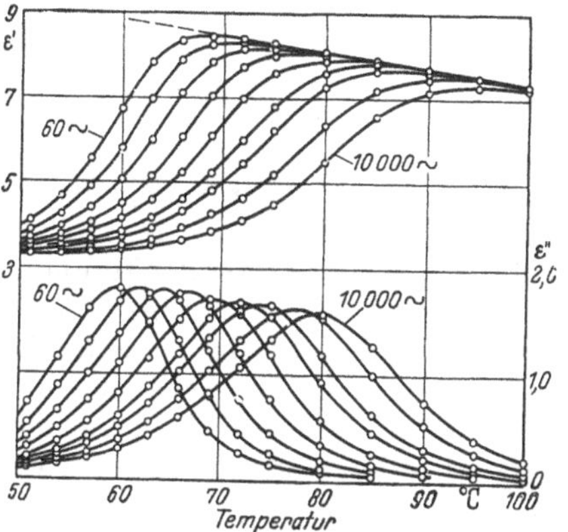

Abb. VIII, 10. Dielektrische Eigenschaften von Polyvinylacetat bei 60, 120, 240, 500, 1000, 2000, 3000, 6000 und 10000 Hz. (Nach MEAD u. FUOSS[3].)

MEAD und FUOSS[3] führten auch dielektrische Untersuchungen an Polyvinylchloracetat ($R = -CH_2Cl$) aus, wobei die dielektrischen Eigenschaften sehr ähnlich denen des Polyvinylacetats gefunden wurden mit Ausnahme dessen, daß die Kurven $\operatorname{tg}\delta = f(T)$ mit wesentlich höherem Maximum und geringerer Halbwertsbreite verliefen, was die beiden Verfasser als eine Orientierung der polaren Gruppen deuten, die nur noch wenig durch ihre Hauptvalenzbindung an die Kette behindert wird.

Sowohl beim Polyvinylacetat als auch beim Polyvinylchloracetat wurden Andeutungen für ein sekundäres Maximum mit hoher Relaxationszeit gefunden, das bei Zumischung von niedermolekularen Substanzen schnell verschwindet.

d) Polyacrylsäureester.

$$\left[\begin{array}{c} -CH_2-CH- \\ \quad\quad | \\ \quad\quad C=O \\ \quad\quad | \\ \quad\quad O \\ \quad\quad | \\ \quad\quad R \end{array}\right]_n$$

Auch hier wurden die ersten dielektrischen Messungen an Polyacrylsäuremethylester ($R = -CH_3$) wieder von MEAD und FUOSS[4] im Fre-

[1] WÜRSTLIN, F.: Kolloid-Z. **134**, 143 (1953).
[2] Unveröffentlichte Untersuchungen.
[3] MEAD, D. J. u. R. M. FUOSS: J. Amer. Soc. **63**, 2832 (1941).
[4] MEAD, D. J. u. R. M. FUOSS: J. Amer. chem. Soc. **64**, 2389 (1942).

quenzgebiet von 60 bis 8000 Hz durchgeführt. Aus diesen Messungen wurde das Dipolmoment der monomeren Einheit zu 2,0 *DE* bestimmt. Eigene Ergänzungen nach höheren Frequenzen[1] zeigten wieder das schon mehrfach gefundene Ergebnis, daß die Dispersionsstellen in der Darstellung $1/T = f(\log \nu)$ keine Gerade ergeben, sondern eine gekrümmte Kurve, aus der sich eine Zunahme der Aktivierungsenergie ergibt beim Übergang von höheren zu niederen Frequenzen. Auch hier beim Polyacrylsäuremethylester fanden MEAD und FUOSS bei 60 Hz ein schwaches Nebenmaximum im Bereich von -90 bis $-100°$ C.

Polyacrylsäuremethylester und Polyvinylacetat sind isomere hochpolymere Substanzen, die sich strukturell durch die Stellung der Gruppe $-\underset{\underset{O}{\|}}{C}-O-$ unterscheiden. Elektrisch zeigt der Polyacrylsäuremethylester die niedrigere Temperaturlage der DK-Dispersion mit niedrigerer Aktivierungsenergie, aber höherer Halbwertsbreite der tg δ-Kurve. Der starke Partialdipol $>C=O$, der den Hauptanteil an der Polarität der Estergruppe hat, ist im Polyacrylsäuremethylester durch seine unmittelbare Stellung neben der Hauptkette in seiner Beweglichkeit gehemmt, wodurch die breite Dispersionskurve zustande kommt. Die für die Dispersionslage und die Aktivierungsenergie hauptsächlich verantwortlichen zwischenmolekularen Dipolbindungen sind beim Polyacrylsäuremethylester kleiner als beim Polyvinylacetat, da im ersten Fall die zwischenmolekularen Abstände zwischen den $>C=O$-Dipolen benachbarter Ketten größer sind.

Dieser zwischenmolekulare Abstand wird innerhalb der Polyacrylsäureester vergrößert durch Verlängerung von $-R$, und auch hier ergibt sich so wieder das auch bei den Polyvinylestern, Polyvinylacetalen u. a. gefundene Bild der inneren Erweichung und Verschiebung der Dispersion nach niederen Temperaturen bei Verlängerung des *R*.

Tabelle VIII, 2.

Temperaturlage der tg δ-Maxima bei $2 \cdot 10^6$ Hz für Polyacrylsäuremethyl-, -äthyl- und n-butylester[2].

$R = -CH_3$	$-CH_2-CH_3$	$-(CH_2)_3CH_3$
74° C	36° C	10° C

In diesem Zusammenhang sei wieder auf das parallele Bild der Einfriererscheinungen hingewiesen (Bd. III, Kap. 11 § 66, Abb. XI, 9).

e) Polymethacrylsäureester.

$$\left[\begin{array}{c} CH_3 \\ | \\ -CH_2-C- \\ | \\ C=O \\ | \\ O \\ | \\ R \end{array}\right]_n$$

Die Polymethacrylsäureester zeigen gegen die Polyacrylsäureester wegen der zusätzlichen seitständigen CH_3-Gruppe eine erheblich verrin-

[1] Unveröffentlichte Untersuchungen.
[2] Unveröffentlichte eigene Untersuchungen.

gerte Kettenbeweglichkeit, die sich mechanisch durch um etwa 100° C erhöhte Werte der Einfriertemperatur, des „brittle point“ u.a. äußert. Die ersten dielektrischen Messungen von MEAD und FUOSS[1] an Polymethacrylsäuremethylester (PMAM) ergaben eine Dispersion, die nur um 23° höher lag als die des Polyacrylsäuremethylesters (PAM). Es wurde weiterhin festgestellt, daß beim PAM ein zweites bei tiefer Temperatur liegendes Maximum vorhanden ist, während beim PMAM sich dieses tiefliegende Maximum nicht zeigte. R. F. TUKKETT[2] wies darauf hin, daß die beobachtete Dispersion des PMAM wahrscheinlich der Orientierung der Estergruppe mit der Seitkette zuzuordnen ist und nicht einer gekoppelten Orientierungsbewegung mit der Hauptkette. Die Messungen selbst wurden nochmals von BROENS und MÜLLER[3] sowie von REDDISH[4] bestätigt. REDDISH und WÜRSTLIN[5] vertraten jedoch die Auffassung, daß alle diese dielektrischen Messungen eine sekundäre Dispersion des PMAM betreffen und die eigentliche bei höherer Temperatur liegende und eine höhere Aktivierungsenergie aufweisende Dispersion noch nicht erfaßt wurde.

Diese Auffassung wird in einer neuen Arbeit bestätigt durch ein bei niederen Frequenzen noch sichtbares Maximum des tg δ bei höherer Temperatur, das dieser Orientierung in Zusammenhang mit der Kettenbewegung zugeordnet werden muß[6]. Bei höherer Frequenz verschwindet dieses kleine Maximum im deutlicher ausgeprägten Maximum der eigen-

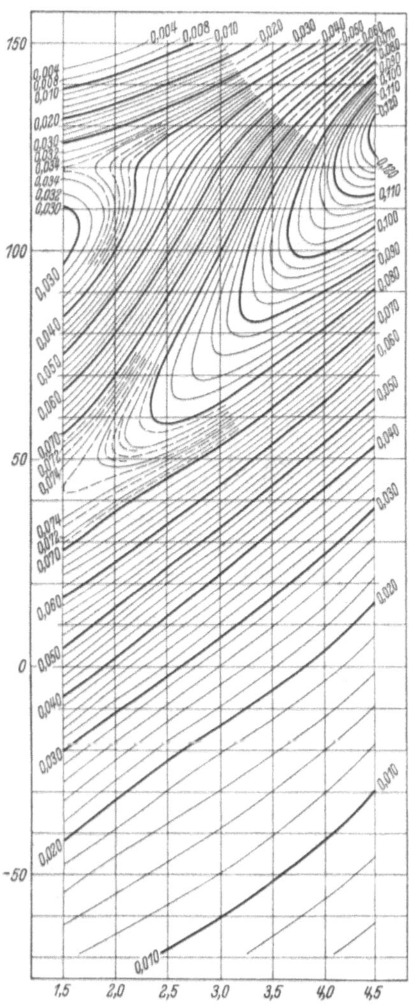

Abb. VIII, 11. tg δ als Funktion von Temperatur und Frequenz für Polymethacrylsäuremethylester. (Nach DEUTSCH, HOFF u. REDDISH[6].)

[1] MEAD, D. J. u. R. M. FUOSS: J. Amer. chem. Soc. **64**, 2389 (1942).
[2] TUCKETT, R. F.: Trans. Faraday Soc. **40**, 448 (1944).
[3] BROENS, O. u. F. H. MÜLLER: Kolloid-Z. **119**, 45 (1950).
[4] REDDISH, W.: Chem. and Ind. 1077 (1951).
[5] WÜRSTLIN, F.: Kolloid-Z. **134**, 135 (1953).
[6] DEUTSCH, K., E. A. W. HOFF u. W. REDDISH: J. Polymer Sci. **13**, 565 (1954).

beweglichen Estergruppen. Es verschwindet aber wohl deshalb, weil sich diese Hauptdispersion mit steigender Frequenz nicht so schnell nach höheren Temperaturen verschiebt wie das sekundäre Maximum. Das bedeutet eine höhere Aktivierungsenergie des Hauptmaximums gegen eine niedere Energie des Nebenmaximums. In eigenen Versuchen in Zusammenhang mit dem Problem der Weichmachung erwies sich das Hauptmaximum viel stärker reagierend auf den Weichmacherzusatz als das Nebenmaximum[1]. Es ist übrigens noch interessant festzustellen, daß bei der mechanischen Dispersion das Hauptmaximum wesentlich stärker ausgeprägt ist als das Nebenmaximum[2], daß elektrisch aber, wie schon betont, die Eigenbeweglichkeit der Estergruppen die gekoppelte gemeinsame Bewegung von Haupt- und Seitketten an Intensität weit übertrifft.

Wie von MEAD und FUOSS[3] wurde auch in der neuesten Arbeit von DEUTSCH, HOFF und REDDISH Poly-α-Chloracrylsäuremethylester

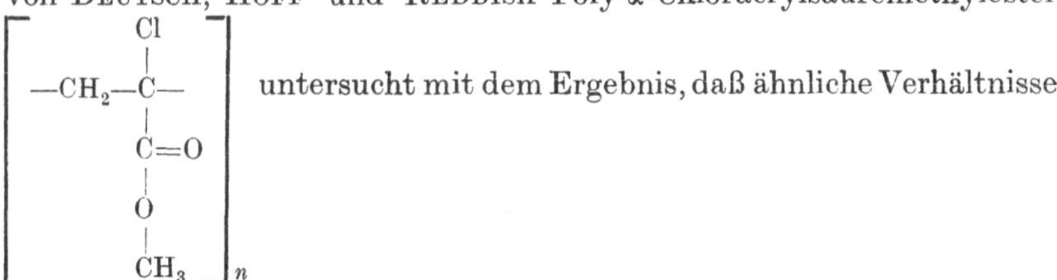

untersucht mit dem Ergebnis, daß ähnliche Verhältnisse wie beim PMAM vorliegen. Das an Stelle der CH_3-Gruppe stehende Cl-Atom bewirkt allerdings zusätzlich noch eine Verschiebung der beiden Dispersionen nach höheren Temperaturen und läßt das Hauptmaximum auch elektrisch wieder betonter werden gegen das Nebenmaximum.

f) Polyvinylacetale.

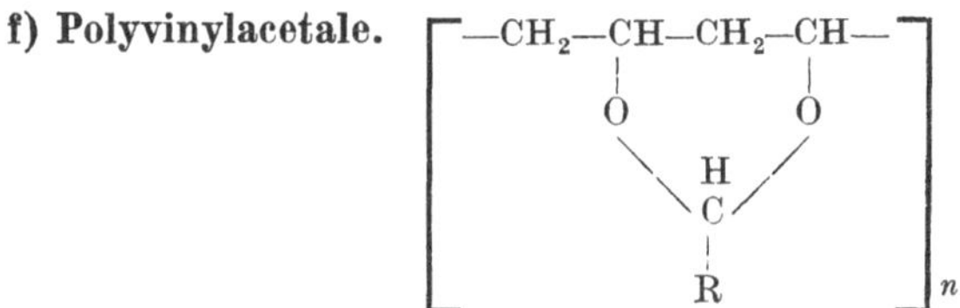

Polyvinylacetale mit verschiedenen Alkylen R wurden von FUNT und und SUTHERLAND[4–6] dielektrisch untersucht. Polyvinylformal ($R = H$) hat von allen vermessenen Polyvinylacetalen die niedrigsten Relaxationszeiten, die höchsten Temperaturlagen der DK-Dispersion und auch die höchste Aktivierungsenergie. Die Verlängerung der Seitenkette R über das Acetal ($R = -CH_3$) zum Butyral ($R = -CH_2-CH_2-CH_3$) und zum Hexanal ($R = -(CH_2)_4-CH_3$) erniedrigt die Temperaturen der DK-Dispersion und die Aktivierungsenergien stufenweise. Wie bei den Reihen der Polyvinyläther, Polyvinylester und Polyacrylsäureester erweist sich auch hier die Verlängerung der Seitenketten R als eine innere Erweichung,

[1] Unveröffentlichte Untersuchungen.
[2] SCHMIEDER, K. u. K. WOLF: Kolloid-Z. **127**, 65 (1952).
[3] MEAD, D. J. u. R. M. FUOSS: J. Amer. chem. Soc. **64**, 2389 (1942).
[4] FUNT, B. L.: Canad. J. Res. **30**, 84 (1952).
[5] FUNT, B. L. u. T. H. SUTHERLAND: Canad J. Res. **30**, 940 (1952).
[6] SUTHERLAND, T. H. u. B. L. FUNT: J. Polymer Sci. **11**, 177 (1953).

die sich sowohl in einer Verlagerung der Einfriererscheinungen nach niederen Temperaturen als auch in einem Abfall der Aktivierungsenergie zeigt.

g) Polyvinylchlorid. $[-CH_2-CHCl-]_n$

Polyvinylchlorid ist schon oft dielektrisch gemessen worden. Ein großer Teil dieser Messungen galt der Weichmachung von Polyvinylchlorid durch äußere Weichmacher und diese sollen deswegen im Abschnitt „Hochpolymere Mischungen“ diskutiert werden. Die frühesten ausgiebigen Messungen an Polyvinylchlorid stammen wieder von R. M. FUOSS[1], der bei Niederfrequenz in den Kurven $\operatorname{tg}\delta = f(T)$ zwei Dispersionen feststellte: Eine stark ausgeprägte Dispersion mit hoher Aktivierungsenergie, die mit der Einfriertemperatur und allen Einfriererscheinungen zusammenhängt und eine wesentlich schwächere Dispersion bei tieferer Temperatur, die einem Mechanismus mit geringer Aktivierungsenergie zugehört. Dieses letztere, sekundäre Maximum verschiebt sich mit steigender Frequenz sehr schnell nach höheren Temperaturen und verschwindet so schon bei relativ niederen Frequenzen im Hauptmaximum. Bei Raumtemperatur und der Funktion $\operatorname{tg}\delta = f(\nu)$ stellte YAGER[2] schon früher eine schwache Dispersion bei etwa $3 \cdot 10^4$ Hz fest, die offensichtlich identisch ist mit der eben erwähnten sekundären Dispersion.

Polyvinylchlorid ist ein typisches Beispiel für eine lineare hochmolekulare Substanz, bei der die vielen starken Dipole durch ihre inner- und zwischenmolekularen Dipolbindungen eine hohe Temperaturlage der Einfriererscheinungen verursachen sowie eine weit oberhalb Raumtemperatur sich abspielende DK-Dispersion mit einer sehr breiten Verteilung der Relaxationszeiten. Die breite Verteilung deutet darauf hin, daß die C—Cl-Dipole nicht frei orientierbar sind, sondern außerordentlich stark behindert werden in ihrer Beweglichkeit durch ihre Hauptvalenzbindung an die Kette.

Eine Verkleinerung des Dipolmoments der monomeren Einheit führt zu niedrigeren Temperaturen der Einfriererscheinungen und zu kürzeren Relaxationszeiten der Dispersion. Dies ist innerhalb der Reihe der Vinylpolymerisate beim Übergang z.B. zum Polyvinylmethyläther zu sehen. Man kann jedoch auch, wie die Reihe Methylchlorid CH_3Cl — Methylenchlorid CH_2Cl_2 — Chloroform $CHCl_3$ — Tetrachlorkohlenstoff CCl_4 zeigt, durch Dipolkompensation zu schwach polaren und zu unpolaren Stoffen kommen. Unter den hochmolekularen Substanzen entspricht das unpolare Polytetrafluoräthylen diesem aus Partialdipolen aufgebauten unpolaren CCl_4. In diesem Sinne verhält sich dann Polyvinylidenchlorid $[-CH_2-CCl_2-]_n$ zum Polyvinylchlorid wie CH_2Cl_2 zu CH_3Cl, d.h. das Dipolmoment der monomeren Einheit ist durch teilweise Kompensation der Partialdipole schwächer geworden. Dies zeigt sich wieder in der niedrigeren Temperaturlage der Einfriererscheinungen des Polyvinylidenchlorids. Es muß allerdings beachtet werden, daß es eine höhere Symmetrie des Aufbaus besitzt als Polyvinylchlorid und daß es deswegen im

[1] FUOSS, R. M.: J. Amer. chem. Soc. **63**, 369 (1941).

[2] YAGER, W. A.: Trans. Electrochem. Soc. **74**, 113 (1938).

Gegensatz zu Polyvinylchlorid eine sehr große Kristallisationstendenz aufweist und normalerweise als amorph-kristalline Substanz vorliegt mit einem Schmelzpunkt der kristallinen Phase von etwa 180° C und einer Einfriertemperatur der ungeordneten Phase von etwa −18° C[1].

OAKES und RICHARDS[2] stellten durch Nachchlorierung von Polyäthylen hochpolymere Substanzen mit bis zu 70 Gew.-% Cl her. Mit zunehmendem Einbau von Cl stiegen die *dielektrischen Werte an* und ebenso die mittleren Relaxationszeiten der Dispersionen. An diesen Substanzen wurden nach der Methode von FUOSS und KIRKWOOD die Dipolmomente der monomeren Einheiten bestimmt. Die Produkte gleichen Chlorgehalts mit Polychloropren und mit Polyvinylchlorid ergaben niedrigere Dipolmomente als Polychloropren und Polyvinylchlorid. Die niedrigen Dipolmomente in den nachchlorierten Substanzen können durch die Anwesenheit von $-CCl_2$- und auch $-CHCl-CHCl$-Gruppen erklärt werden.

h) Polyvinylcarbazol. $\left[-CH_2-CH(N\text{-Carbazolyl})-\right]_n$

Die im Vergleich zum Polystyrol wesentlich erhöhten Temperaturen der Einfriererscheinungen erklären sich rein sterisch mit der erheblich größeren seitständigen Carbazolgruppe gegenüber der Phenylgruppe des Polystyrols. In den dielektrischen Messungen von HOLZMÜLLER[3] ist ein Hauptmaximum oberhalb 160° C angedeutet, das wieder in Beziehung zu stehen scheint mit den übrigen Einfriererscheinungen. Ein sekundäres $\mathrm{tg}\,\delta$-Maximum ist dicht oberhalb Raumtemperatur sichtbar.

i) Polyacrylnitril. $\left[-CH_2-CH(C{\equiv}N)-\right]_n$

Nach MEAD und FUOSS[4] findet man bei 60 Hz ein Maximum von ε'' bei etwa 110° C. Aus diesen dielektrischen Messungen an der hochpolymeren Substanz läßt sich nach FUOSS und KIRKWOOD das Dipolmoment der monomeren Einheit bestimmen zu 4,6 *DE*, während bei Messungen an der monomeren Substanz das Dipolmoment der $-C{\equiv}N$-Gruppe mit 3,2 *DE* erhalten wird. Nach eigenen Messungen im Frequenzgebiet von 10^3 bis 10^5 Hz erhält man eine Aktivierungsenergie von 57 kcal/Mol[5]. Der Wert erscheint etwas niedrig bei einem Vergleich mit den Aktivierungsenergien anderer Hochpolymerer wie Polystyrol, Polymethacrylsäuremethylester, Polyvinylchlorid u. a. (siehe Tab. VIII, 1). Vielleicht ist auch hier wie beim Polymethacrylsäuremethylester das Nebenmaximum stär-

---

[1] BOYER, R. F., u. R. S. SPENCER: Second order transition effects in rubber and other high polymers. – Advances in colloid science. Vol. II. New York (1946).

[2] OAKES, W. G. u. R. B. RICHARDS: Disc. Faraday Soc. **42**a, 197 (1946).

[3] HOLZMÜLLER, W.: Physik. Z. **42**, 273 (1941).

[4] MEAD, D. J. u. R. M. FUOSS: J. Amer. chem. Soc. **65**, 2067 (1943).

[5] Unveröffentlichte Untersuchungen.

ker ausgeprägt und das Hauptmaximum dielektrisch bis jetzt noch nicht erfaßt worden, zumal durch die starke Leitfähigkeit dielektrische Messungen bei höheren Temperaturen nur noch begrenzt durchführbar sind.

k) Polyester.

$$\left[\begin{array}{c}-O-R'-O-\underset{\displaystyle O}{\underset{\|}{C}}-R''-\underset{\displaystyle O}{\underset{\|}{C}}-\end{array}\right]_n$$

Der lineare Bau der Polyester befähigt diese zu dreidimensionaler Anordnung in Kristallgittern, die gestört werden können, wenn seitständige Gruppen das enge Aneinanderlagern benachbarter Ketten verhindern. Der Einfluß dieser strukturellen Verschiedenheiten auf die dielektrischen Werte ist deutlich zu sehen in Abb. VIII,12, in der die dielektrischen Werte für 10^7 Hz in Abhängigkeit von der Temperatur aufgetragen sind für die zwei isomeren Polyester aus

1,4-Butandiol + Adipinsäure $\left[-O-(CH_2)_4-\overset{\displaystyle O}{\overset{\|}{C}}-(CH_2)_4-\overset{\displaystyle O}{\overset{\|}{C}}-\right]_n$

und 1,3-Butandiol + Adipinsäure $\left[-O-CH_2-\underset{\displaystyle CH_3}{\underset{|}{CH}}-CH_2-O\underset{\displaystyle O}{\underset{\|}{C}}-(CH_2)_4-\underset{\displaystyle O}{\underset{\|}{C}}-\right]_n$

Aus den Kurven ergibt sich, daß im kristallisationsfähigen Polyester ein großer Teil der polaren Estergruppen beim Abkühlen durch Kristallisation plötzlich eingefroren wird und nur noch ein in ungeordneten Bereichen befindlicher kleinerer Teil der Estergruppen unterhalb der Kristallisationstemperatur Orientierungsfähigkeit besitzt. In dem isomeren 1,3-Butandiolester verhindert die seitständige CH_3-Gruppe des Butandiols die Kristallisation und die DK-Dispersion spielt sich so in ausgeprägterer Form ab.

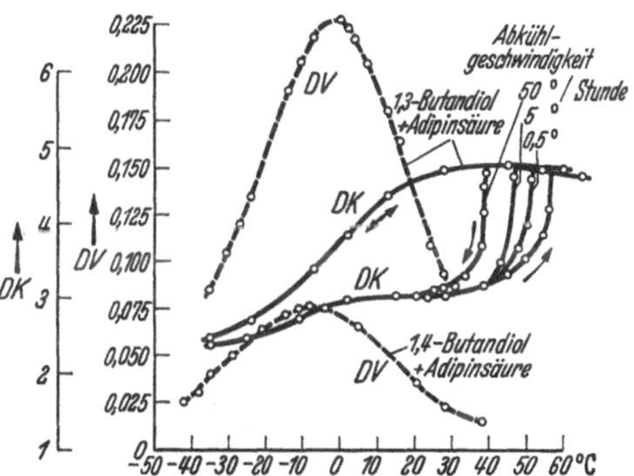

Abb. VIII,12. Anomale DK-Dispersion, gemessen bei 10^7 Hz an den Polyestern aus: 1,4-Butandiol-Adipinsäure (kristallin) und 1,3-Butandiol-Adipinsäure (nichtkristallin). (Nach WÜRSTLIN[1].)

Geht man von einem aliphatischen Polyester mit hoher Esterkonzentration aus und erhöht stufenweise die Anzahl der CH_2-Gruppen in R' und R'', so fällt die mittlere Relaxationszeit τ asymptotisch ab. Sie erreicht einen Endwert, wenn beide aliphatischen Kettenglieder etwa je 4 bis 6 CH_2-Gruppen enthalten[2]. Diese DK-Dispersion steht wieder in Zusammenhang mit der Einfriertemperatur und muß so als Orientierung der polaren Gruppen im Rahmen der Kettenbeweglichkeit gedeutet werden. An einem aliphati-

[1] WÜRSTLIN, F.: Kolloid-Z. **110**, 71 (1948).

[2] Unveröffentliche Messungen.

schen Polyester dieser Reihe (Polyester aus Äthylenglykol + Sebazinsäure, $R' = (CH_2)_2$, $R'' = (CH_2)_8$) wurde die Aktivierungsenergie für die DK-Dispersion zu 12,1 kcal/Mol bestimmt[1].

Bei kristallinen aliphatischen Polyestern kommt man nur zu Schmelztemperaturen von etwa 100°C. Bei Polyestern der Terephthalsäure werden um rund 100° höhere Schmelztemperaturen erreicht, was dieses „Terylen" zu einem sehr interessanten Textilrohstoff werden ließ. An einer trockenen kristallisierten Probe eines Polyesters aus Terephthalsäure und Äthylenglykol bestimmte W. REDDISH sehr ausführlich die

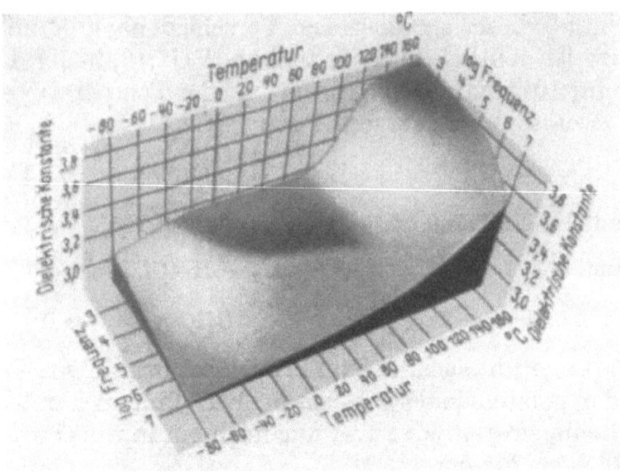

Abb. VIII, 13. Fläche der Dielektrizitätskonstante ε' für Terylen.

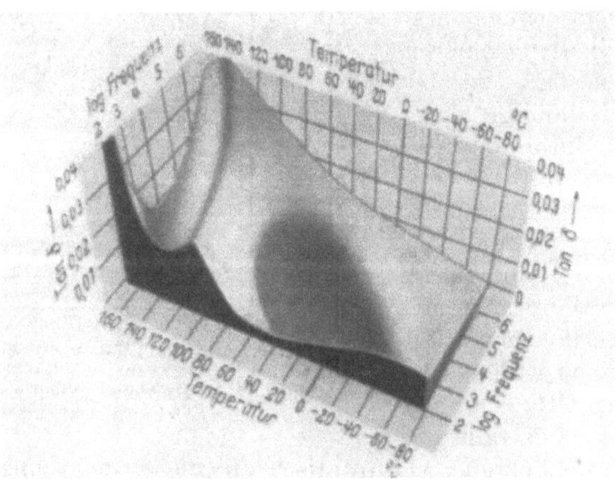

Abb. VIII, 14. Fläche des Verlustfaktors tg δ für Terylen.

dielektrischen Daten als Funktion von Temperatur und Frequenz[2]. Die in den Abb. VIII, 13 u. 14 wiedergegebene räumliche Darstellung von ε'

[1] BAKER, W. O. u. W. A. YAGER: J. Amer. chem. Soc. **64**, 2171 (1942).
[2] REDDISH, W.: Trans. Faraday Soc. **46**, 459 (1950).

und tgδ kennzeichnet neben dem starken Einfluß der Leitfähigkeit auf den tgδ bei hohen Temperaturen zwei Dispersionen, wobei dem Hochtemperaturprozeß eine hohe Aktivierungsenergie zukommt. Diese in den ungeordneten Bereichen des Polyesters sich abspielende Orientierungspolarisation steht in Zusammenhang mit der bei 80° bestimmten Einfriertemperatur des ungeordneten Volumenanteils. Die Einfriertemperatur ist um über 100° C gegenüber einem aliphatischen Polyester erhöht durch den Einbau der Terephthalsäuregruppe in die Kette, die eine sehr starke sterische Behinderung der Kette mit sich bringt. Die Bewegungsbehinderung gegenüber einem rein aliphatischen Polyester zeigt sich auch in der bis auf 90 kcal/Mol angewachsenen Aktivierungsenergie. Der Niedertemperaturprozeß mit 12 kcal/Mol Aktivierungsenergie wird der Orientierungspolarisation (Protonensprung) von nicht veresterten OH-Gruppen zugeordnet.

l) Polyamide.

$$\left[\begin{array}{l}-NH-R'-NH-\underset{\displaystyle O}{\underset{\|}{C}}-R''-\underset{\displaystyle O}{\underset{\|}{C}}-\end{array}\right]_n$$

Bei den Polyamiden liegen nur sehr wenige dielektrische Daten in Abhängigkeit von Temperatur und Frequenz vor. An einem Polyamid aus Hexamethylendiamin und Sebazinsäure [$R' = -(CH_2)_6-$, $R'' = -(CH_2)_8-$] wurde die DK-Dispersion bei verschiedenen konstanten Frequenzen in Abhängigkeit von der Temperatur bestimmt und daraus die Aktivierungsenergie zu 23,3 kcal/Mol errechnet[1]. Der im Vergleich zu einem aliphatischen Polyester wesentlich erhöhte Wert wird auf die im Polyamid wirksamen zwischenmolekularen Wasserstoffbindungen zurückgeführt, die die Orientierung der polaren Gruppen stark behindern. Bei Wasseraufnahme und Hydratation können sich weniger Wasserstoffbrücken ausbilden, die tgδ-Werte werden stark erhöht und die Dispersion wird nach tieferer Temperatur verschoben.

m) Cellulose.

Alle Cellulosen in den verschiedenen Verarbeitungsformen (Holz, Papier, Vulcanfiber usw.) zeigen bei Raumtemperatur ein breites Maximum des tgδ bei etwa 10^7 Hz. Sowohl F. H. Müller als auch Yager[2, 3] haben diese Dispersion als eine von der Hauptkette unabhängige Orientierungspolarisation (Protonensprung) der OH-Gruppen gedeutet und es ist wahrscheinlich, daß eine Orientierungspolarisation in Zusammenhang mit der Kettenbewegung bis jetzt noch nicht beobachtet wurde und wegen der thermischen Zersetzung bei hoher Temperatur wohl gar nicht erreichbar ist. Die Aktivierungsenergie dieses Protonensprunges ist praktisch dieselbe, wie sie beim Terephthalsäureglykolester von Reddish für das Nebenmaximum bestimmt wurde.

---

[1] Baker, W. O. u. W. A. Yager: J. Amer. chem. Soc. **64**, 2171 (1942).

[2] Müller, F. H.: Elektrotechn. Z. **59**, 1176 (1938) und in Werkstoffkunde der elektrotechnischen Isolierstoffe von H. Stäger. Berlin 1944.

[3] Yager, W. A.: Trans. Electrochem. Soc. **74**, 113 (1938).

§ 72. Ergebnisse an hochpolymeren Mischsubstanzen.

Unter hochpolymeren Mischsubstanzen sollen verstanden werden: Mischpolymerisate und -kondensate, dann auch hauptvalenzmäßige Verlängerungen und Vernetzungen durch intermolekulare Brückenatome oder Atomgruppen und schließlich auch Mischungen mit wenigstens einer hochmolekularen Substanz als Hauptkomponente.

Von großer technischer Wichtigkeit sind Mischpolymerisate und -kondensate, da sie eine breite Variationsmöglichkeit der Eigenschaften bringen, wobei aber durch die Hauptvalenzbindungen der hochmolekulare Charakter der Substanzen gewahrt ist und eine Trennung in die Mischkomponenten ohne Aufhebung der Hauptvalenzbindungen nicht möglich ist. Die Mischpclymerisation wird oft angewandt, um Änderungen im strukturellen Aufbau einer hochpolymeren Substanz zu erzielen. Polyacrylnitril läßt sich beispielsweise wegen seiner starken zwischenmolekularen Dipolbindungen, die bis zur Ausbildung von kristallinen Bereichen führen, nur in wenig Lösungsmitteln lösen. Eine Mischpolymerisation mit einer zweiten Komponenten stört diese Ausbildung kristalliner Bereiche und bringt so bessere Löslichkeit, was zusammen mit einer auch besseren Anfärbbarkeit vor allem für die Verwendung der PAN-Faser von Bedeutung ist. Der Abbau der kristallinen Bereiche durch Mischpolymerisation oder -kondensation — und zwar sowohl das Absinken des Schmelzpunktes als auch die Verringerung des Volumenverhältnisses kristallin/amorph — kann durch ε' und $\operatorname{tg}\delta$ als $f(T)$ verfolgt werden[1], da nach Abb. VIII, 12 die dielektrischen Werte über der Temperatur den gemischt amorph-kristallinen Aufbau deutlich zeigen.

Eine zweite wichtige Anwendung der Mischpolymerisation ergibt sich mit der Beeinflußbarkeit der inneren Beweglichkeit von Fadenmolekülen durch die Mischpolymerisation. Polymerisate mit hoher Temperaturlage der Einfriererscheinungen können durch Mischpolymerisation mit einer Komponente mit höherer innerer Beweglichkeit eine Verschiebung der Einfriererscheinungen nach tieferen Temperaturen erfahren. Die Mischpolymerisation spielt hier die Rolle einer inneren Weichmachung. Auch der umgekehrte Fall einer Erhöhung der Einfriertemperatur durch Mischpolymerisation mit einer Komponente mit stark polaren oder sterisch wirkenden Bauelementen wird oft beobachtet. Kennt man die beiden polymeren Endsubstanzen mit den Temperaturlagen der Einfriererscheinungen, so kann man — sofern bei der ganzen Mischreihe nur ungeordnete Substanzen auftreten — einen nahezu linearen Verlauf dieser Temperaturen über der Mischungszusammensetzung erwarten[2]. Bei polaren Mischpolymerisaten läßt sich diese im wesentlichen lineare Verschiebung der Einfriererscheinungen auch an Hand der Orientierungspolarisation verfolgen. Fuoss hat so die Mischpolymerisation von Acrylnitril und Acrylsäureäthylester dielektrisch untersucht[3], Roelig hat in den ver-

[1] Würstlin, F.: Kolloid-Z. **120**, 84 (1951).

[2] Nielsen, L. E., R. E. Pollard u. E. McIntyre: J. Polymer Sci. **6**, 661 (1951). — [3] Mead, D. J. u. R. M. Fuoss: J. Amer. chem. Soc. **65**, 2067 (1943).

schiedenen Bunasorten die Mischpolymerisation von Butadien mit Styrol dielektrisch gemessen[1].

Die vielseitige Anwendbarkeit der Mischpolymerisation zeigt sich auch in dem folgenden Beispiel. Polystyrol ist als unpolarer Stoff bekanntlich der Hochfrequenzerwärmung nicht zugänglich, da das Produkt $\varepsilon' \cdot \operatorname{tg}\delta$ keine genügend hohen Werte hat. Polystyrol kann jedoch im Hochfrequenzfeld verschweißt werden unter Verwendung einer dünnen Zwischenschicht eines Styrolmischpolymerisates, das einen für die Hochfrequenzerwärmung genügenden Anteil einer polaren Mischkomponenten enthält und sich andererseits im Ausdehnungskoeffizienten noch nicht sehr verschieden verhält von dem zu verschweißenden Polystyrol[2].

Eine Diskussion dielektrischer Messungen an vernetzten hochmolekularen Substanzen stützt sich hauptsächlich auf den klassischen Fall der Vulkanisation von Kautschuk mit Schwefel. Die Versuche wurden schon 1933 durchgeführt und es zeigte sich mit dem Einbau von Schwefel eine starke Dispersionsstufe in der Darstellung ε' und $\operatorname{tg}\delta = f(T)$, die mit zunehmender Menge an eingebautem Schwefel nach höheren Temperaturen rückte, wobei auch die Aktivierungsenergie dieser Polarisation in derselben Richtung anstieg[3]. Von F. H. Müller wurde darauf hingewiesen, daß man den vulkanisierten Kautschuk als eine Lösung von Schwefeldipolen in den unpolaren Isoprenketten auffassen kann. Dabei läßt sich die Debyesche Dipoltheorie auf die verdünnte Lösung der Schwefeldipole anwenden und F. H. Müller bestimmte so das Dipolmoment der Schwefelbrücke zum selben Wert wie das des niedermolekularen Thioäthers[4].

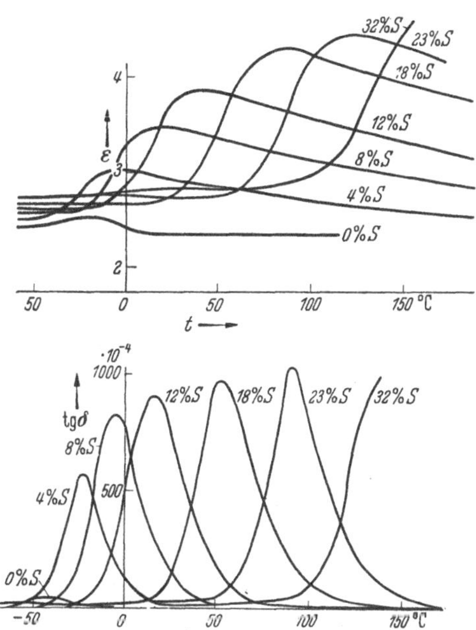

Abb. VIII, 15. DK und Verlustfaktor an vulkanisierten Kautschukproben als Funktion der Temperatur bei 1 kHz.

Leider sind dies die einzigen ausführlichen dielektrischen Untersuchungen an vernetzten hochmolekularen Substanzen, die sich in der Literatur finden. Es fehlen u.a. grundsätzliche dielektrische Untersuchungen über den Kondensationsvorgang und auch an fertigen dreidimensionalen Kondensationsharzen. Man kann wohl annehmen, daß in

[1] Roelig, H.: Kautschuk **12**, 139 (1941). — [2] DP 824388.
[3] Scott, A. H., A. T. McPherson u. H. C. Curtiss: J. Res. nat. Bur. Standards **11**, 173 (1933). — [4] Müller, F. H.: Kolloid-Z. **77**, 260 (1936).

derartigen polaren, sehr stark vernetzten Produkten keine Dispersion zu finden ist, die der Hauptdispersion des Polyvinylchlorids und ähnlicher linearer Hochpolymerer entspricht, welche der Orientierungspolarisation von nahezu starr mit Kettenteilstücken verbundenen Dipolen zugeordnet wird. Es findet sich jedoch oft bei diesen Substanzen in der Darstellung $\operatorname{tg}\delta = f(\nu)$ eine breite Dispersion[1], die der Orientierungspolarisation von Atomen (Protonensprung) oder auch von kleinen Atomgruppen zugeschrieben werden muß, die auch im dreidimensionalen Riesenmolekül noch Orientierungsfähigkeit besitzen. Bei Phenolformaldehydharzen kann diese Dipolorientierung eintreten, wenn noch freie OH-Gruppen vorhanden sind, was in jeder ausgehärteten Substanz praktisch noch der Fall ist. Diese Dispersion entspricht so dem hochfrequenten Nebenmaximum des Terylens (siehe Polyester). In Anilinformaldehydharzen sind auch nach der Vernetzungsreaktion noch N—H-Dipole vorhanden, so daß der Protonensprungmechanismus bei Raumtemperatur in einer Dispersion um 10^5 Hz zu sehen ist[2].

Da gerade die Kondensationsprodukte sehr häufig mit Füllstoffen beschwert werden, ist damit auch die Möglichkeit gegeben, daß die Füllstoffe unabhängig von der hochmolekularen Substanz eine Dispersion infolge ihres eigenen polaren Aufbaus verursachen. Unveresterte Cellulose als Füllstoff (in Form von Papier oder in textiler Form) bringt so immer die Dispersion des OH-Protonensprungs mit sich. Bei der Mischung hochmolekularer Substanzen mit Füllstoffen (Ruß, Mineralstoffen, Pigmenten u. a.) ist im allgemeinen zu erwarten, daß bei inerter Mischung von hochmolekularer Substanz und Füllstoff die dielektrischen Eigenschaften sich nach der Inhomogenitätstheorie von K. W. Wagner berechnen lassen aus den Eigenschaften der Komponenten. Ob sich ein Füllstoff als inerter Stoff verhält oder mit der hochpolymeren Substanz in Wechselwirkung steht, kann mit Hilfe der Relaxationszeit der Dipolorientierung entschieden werden, wobei eine Veränderung der Relaxationszeit des reinen Hochpolymeren durch Zumischung des Füllstoffes auf eine derartige Wechselwirkung deutet[3].

Unter den hochmolekularen Mischsubstanzen bedeuten die mit äußerem Weichmacher versetzten Mischungen eine vor allem technisch sehr interessante Gruppe von Mischsubstanzen. Der technische Weichmacher — im Grunde ein hochsiedendes Lösungsmittel — hat die Aufgabe, die Einfriererscheinungen der hochmolekularen Substanz nach tiefen Temperaturen zu rücken. Es handelt sich also bei diesen Weichmachern im Sinne von W. C. Schneider und Mitarbeitern[3] um Substanzen, die in Wechselwirkung stehen mit der hochmolekularen Substanz, was sich auch deutlich zeigt in Abb. VIII, 16, in der die Funktion $\operatorname{tg}\delta = f(T)$ für Polyvinylacetat und einige Mischungen mit zunehmendem Gehalt an Weichmacher bis zum reinen Weichmacher aufgetragen ist. Die Dispersion des Polyvinylacetats wird dabei durch Zumischung des Weichmachers zu niederen

---

[1] Yager, W. A.: Trans. Electrochem. Soc. **74**, 113 (1938).

[2] Müller, F. H.: Elektrotechn. Z. **59**, 1156, 1176 (1938).

[3] Schneider, W. C., W. C. Carter, M. Magat u. C. P. Smyth: J. Amer. chem. Soc. **67**, 959 (1945).

Temperaturen verschoben und ihre Aktivierungsenergie wird gleichzeitig verringert[1], wobei der Weichmacher selbst in den tg δ-Kurven erst bei höheren Konzentrationen sichtbar wird, wo neben den solvatisierten Weichmachermolekülen zum erstenmal auch nicht an die hochmolekulare Substanz gebundene Weichmachermoleküle auftreten. SCHALLAMACH und THIRION haben in gequollenem vulkanisiertem Kautschuk ähnliche dielektrische Kurven bestimmt[2].

Seit den ersten dielektrischen Untersuchungen von FUOSS an weichgemachtem Polyvinylchlorid[3] sind von vielen Autoren derartige Messungen veröffentlicht worden, wobei diese Arbeiten entweder einen Beitrag zum technischen Problem der Weichmachung oder zum physikalischen Problem der Orientierungspolarisation bringen sollten. Die Messungen der ersten Art sollen hier nicht erwähnt werden, da die Weichmachung in Kapitel 9 zusammenfassend beschrieben ist. Bei den Messungen der zweiten Art sind vor allem diejenigen anzuführen, die mit (technisch nicht interessanten!) unpolaren Weichmachungsmitteln durchgeführt wurden[5] zu dem Zweck, das Dipolmoment der monomeren Einheit der hochmolekularen Substanz zu bestimmen. Das nach FUOSS und KIRKWOOD[5] berechnete Dipolmoment der monomeren Einheit liegt für die reine hochpolymere Substanz höher als für das freie monomere Molekül. Mit steigender Verdünnung durch den unpolaren Weichmacher nähert sich aber der zu hohe Wert asymptotisch dem Wert für das monomere Molekül. Nach FUOSS bedeutet dies, daß sich die Dipole in der reinen hochpolymeren Substanz nicht so stark kompensieren können wie in einer niedermolekularen Flüssigkeit.

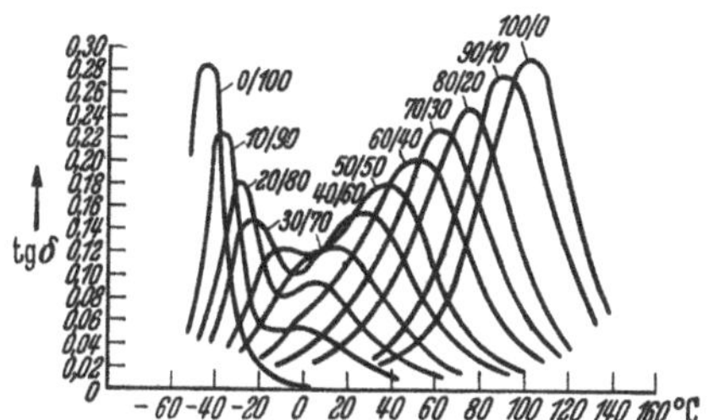

Abb. VIII, 16. tg δ als Funktion der Temperatur bei $2 \cdot 10^6$ Hz für Polyvinylacetat, weichgemacht mit Benzylbenzoat. (Nach WÜRSTLIN[4].)

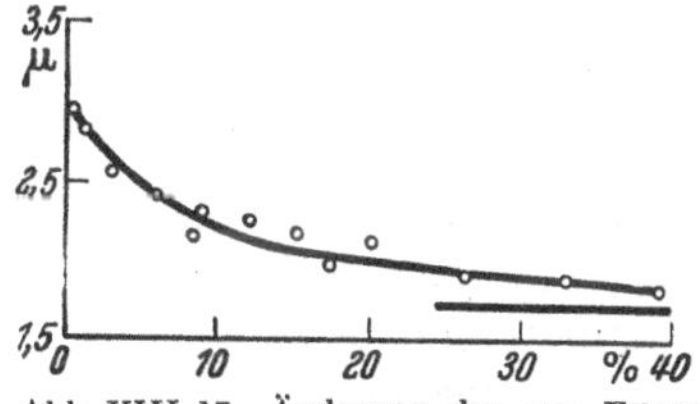

Abb. VIII, 17. Änderung des von FUOSS für den monomeren Rest bei Polyvinylchlorid errechneten Dipolmoments in Abhängigkeit vom Weichmachergehalt. (Nach FUOSS[3].)

Umgekehrt zu dem eben beschriebenen Versuch wurden auch niedermolekulare Substanzen mit hochmolekularen unpolaren Substanzen wie Polyisobutylen, Polyäthylen und Polystyrol gemischt. Die hochmolekularen Substanzen wirken in diesem Falle wie hochviscose Lösungsmittel und die hochfrequente Dispersion der reinen polaren Substanzen wird so durch die Lösung im hochviscosen Lösungsmittel zu niedrigeren Frequenzen verschoben[6].

[1] WÜRSTLIN, F.: Kolloid-Z. **134**, 135 (1953).
[2] SCHALLAMACH, A. u. P. THIRION: Trans. Faraday Soc. **45**, 605 (1949).
[3] FUOSS, R. M.: J. Amer. chem. Soc. **61**, 2334 (1939); **63**, 2401, 2410 (1941).
[4] WÜRSTLIN, F.: Kolloid-Z. **113**, 18 (1949).
[5] FUOSS, R. M. u. J. G. KIRKWOOD: J. Amer. chem. Soc. **63**, 385 (1941).
[6] PLESSNER, K. W. u. R. B. RICHARDS: Disc. Faraday Soc. **42** a, 206 (1946).

Neuntes Kapitel.

Die Wirkung von Weichmachern und ihre molekulare Deutung.

Von

Ernst Jenckel.

Mit 17 Abbildungen.

§ 73. Einleitung.

Viele Hochpolymere sind bei Zimmertemperatur mehr oder weniger hart und spröde. Sie teilen diese Eigenschaft mit wohl allen glasig erstarrten niedermolekularen Stoffen. Die technisch verwendbaren Stoffe dürfen jedoch meistens nicht spröde sein, manchmal wird sogar eine gewisse Geschmeidigkeit verlangt. Zu diesem Zwecke verdünnt man die Hochpolymeren zu etwa 30 bis 40% mit organischen Flüssigkeiten, sogenannten Weichmachern[1], deren Wirkung im einzelnen im folgenden besprochen werden soll. Im Prinzip kann jede organische Flüssigkeit als Weichmacher verwandt werden. Dementsprechend werden immer wieder neue Stoffe als Weichmacher empfohlen. Technisch durchgesetzt haben sich aber fast nur hochsiedende Ester, besonders der Phthalsäure und der Phosphorsäure, auch der Adipin- und Sebacinsäure. Gelegentlich werden auch Alkohole, Äther und Ketone verwandt, und für besondere Zwecke hochpolymere Ester, z.B. Polyacrylsäureester. Zahlreiche Weichmacher sind chemisch uneinheitlich und in ihrer Zusammensetzung nicht bekannt; sie kommen nur unter Phantasienamen auf den Markt. Die handelsüblichen Weichmacher sind mit ihren Eigenschaften bei Gnamm besprochen worden.

Die technische Verwendung von Weichmachern beschränkt sich auf eine verhältnismäßig kleine Gruppe von Hochpolymeren. Das weitaus wichtigste Polymere ist Polyvinylchlorid und seine Mischpolymerisate mit dem Acetat. Auch eine Reihe von Celluloseestern werden weichgemacht. Weiter wäre noch der Kautschuk zu nennen. Schließlich wirkt wahrscheinlich in den natürlichen und synthetischen Fasern das darin enthaltene Wasser als Weichmacher, worauf hier aber nur hingewiesen sei.

---

[1] Gnamm, H. u. R. Sommer: Lösungs- und Weichhaltungsmittel, 7. Auflage, Stuttgart 1956; vgl. auch H. Berger: Kunststoffe **42**, 324 (1952).

Die Systeme aus Weichmacher und Hochpolymeren sind im Sinne der physikalischen Chemie im allgemeinen als Lösungen zu bezeichnen. Man kann auf sie die Gesetze der Thermodynamik anwenden, worüber in Bd. II, Kap. 2 u. 3, berichtet wurde. Wenn die Löslichkeit sich über den gesamten Konzentrationsbereich oder wenigstens über den technisch interessierenden Bereich erstreckt, spricht man von einem *verträglichen* Weichmacher. Wenn dagegen die Löslichkeit nur beschränkt ist, und insbesondere im technisch wichtigen Bereich zwei Phasen auftreten, spricht man von einem *unverträglichen* Weichmacher. In beiden Fällen erreicht man eine Erweichung des Materials, jedoch dürfte bei verträglichen Weichmachern der Fließvorgang sich über das ganze im Sinne der Phasenlehre homogene Material erstrecken, bei unverträglichen Weichmachern dagegen im wesentlichen nur über die leicht bewegliche, weichmacherreiche Phase.

Es liegt nahe, das durch Zusatz von Weichmachern erreichte Weichwerden des Materials auf mechanischem Wege zu untersuchen, was auch für technische Zwecke vielfach geschieht. Das mechanische Verhalten der Hochpolymeren ist jedoch verhältnismäßig kompliziert, insbesondere überlagert sich dem elastischen Verhalten ein plastisches Verhalten, worauf in früheren Kapiteln dieses Bandes näher eingegangen ist. Daher eignet sich das mechanische Verhalten nicht sehr zur Beschreibung der Weichmacherwirkung. Nun geht offenbar die Erniedrigung der Einfriertemperatur durch einen Weichmacherzusatz dem Weichwerden weitgehend parallel. Die Erhöhung der inneren Beweglichkeit ist eben die gemeinsame Ursache sowohl für das mechanische Weichwerden wie für die Herabsetzung der Einfriertemperatur, was auch darin zum Ausdruck kommt, daß die Einfriertemperatur durch eine ganz bestimmte Viscosität von etwa 10^{13} abs. Einh. ausgezeichnet ist[1]. Die Messung der Einfriertemperatur stellt also letzten Endes eine Viscositätsmessung dar, die sich jedoch auf die Verschieblichkeit von Kettenstücken gegeneinander beschränkt, und andere Bindungsmechanismen, etwa diejenigen durch eine weitmaschige Vernetzung, nicht erfaßt. Es hat sich daher eingebürgert, die Wirkung der Weichmacher an ihrer Wirkung auf die Einfriertemperatur zu erfassen, worüber zunächst berichtet sei.

Die technische Anwendung bringt es weiterhin mit sich, daß ein Weichmacher nur möglichst langsam, sei es durch Abdampfen in die Atmosphäre oder durch Übergang in benachbarte Stoffe, verlorengehen soll. Diese Punkte werden weiter unten behandelt. Auf weitere Forderungen[2], etwa das Nichtverfärben im Lichte, überhaupt die Farblosigkeit, ferner die Geruchlosigkeit, Giftigkeit, Feuerfestigkeit und Netzbarkeit an Pigmenten kann hier nicht eingegangen werden.

---

[1] Vgl. E. Jenckel in Bd. III, § 54.

[2] Vgl. auch eine ausführliche Liste solcher Forderungen bei O. Fuchs in K. Winnacker u. E. Weingärtner, Chem. Technologie, München 1954, Bd. II, S. 624; weiter werden diese Fragen in dem schon zitierten Buch von Gnamm u. Sommer eingehend behandelt.

§ 74. Die Einfriertemperatur von Lösungen; Weichmacherwirkung.

a) Experimentelle Ergebnisse.

Für die Lösungen hochmolekularer Stoffe gelten hinsichtlich der Einfriertemperatur grundsätzlich die gleichen Betrachtungen wie für die chemisch einheitlichen Stoffe, und ebenso sind die experimentellen Methoden die gleichen. Sie wurden in Bd. III, § 54 besprochen. Es wurde die Einfriertemperatur nach mehr oder weniger großem Weichmacherzusatz in folgenden Lösungen bestimmt:

Polystyrol mit monomerem Styrol[1], Benzol[2], Toluol[2, 25], Äthylbenzol[2–4, 7, 25], sek. Butylbenzol[4], Nitrobenzol[2], Chlorbenzol[4, 25], Tetralin[5], CCl_4[2], $CHCl_3$[2], CH_2Cl_2[2], CS_2[2], C_2Cl_6[4]; versch. Phthalsäureestern[4, 7], Trikresylphosphat[2, 7], versch. Salicylsäureestern[2], versch. Essigsäureestern[2], Amylbutyrat[2], Methylhexylketon[4], Paraffinöl[6].

Polymethacrylester mit Benzol[2, 8], Toluol[2, 8], Nitrobenzol[2, 8], $CHCl_3$[2, 8], CBr_4[9], Aceton[2, 8, 9], Butylalkohol[2, 8], Benzylalkohol[9], versch. Essigsäureestern[2, 8, 9], Acetophenon[9].

Polyacrylsäureaethylester mit versch. Estern[21].

Polyvinylchlorid mit versch. Estern der Phthalsäure[10, 18a, 19, 20, 22], Sebacinsäure[10, 20, 22], Phosphorsäure[17, 19, 20], Trikresylphosphat[7, 10–13, 18a]; Diphenyl[7, 18], gesättigten und ungesättigten Fettsäuren[20, 22], weiteren Estern[21], Ketonen[14, 22], Amiden[14], versch. aromatische Kohlenwasserstoffe[22], versch. techn. Weichmacher[11, 13], Butylester zweibas. Säuren, z. T. schwefelhaltig[24].

Polyvinylacetat mit versch. Estern der Benzoesäure, Phthalsäure, ein- und zweibasischer Fettsäuren, Ketonen, Amiden[14].

Mischpolymerisat aus Vinylchlorid und Acetat mit zahlreichen Estern[19].

Äthylcellulose und Nitrocellulose mit Trikresylphosphat[15], versch. Estern der Phthalsäure, Zitronensäure und Glykolsäure[16].

Butadien-Acrylnitril-Mischpolymerisat mit Acrylnitril[9], Bromtoluol[9], CBr_4[9], Äthylacetat[9], Aceton[9].

Kautschuk mit *n*-Heptan, Paraffinöl, *i*-Amylbromid, Brombenzol, Dibutyläther, Digeranyläther[23], *Perbunan* mit Butylestern zweibas. Säuren, z. T. schwefelhaltig[24].

---

[1] ALEXANDROW, A. P. u. Y. S. LAZURKIN: C. R. Acad. Sci. URSS **43**, 376 (1944). — [2] JENCKEL, E. u. R. HEUSCH: Kolloid-Z. **130**, 89 (1953).

[3] ÜBERREITER, K.: Z. angew. Chem. **53**, 247 (1940).

[4] KARGIN, W. A. u. J. M. MALINSKI: Ber. Akad. Wiss. UdSSR, NS **73**, 967 (1950). — [5] JENCKEL, E.: Kunststoffe **31**, 209 (1941).

[6] JENCKEL, E. u. K. ÜBERREITER: Z. physik. Chem. A **182**, 361 (1938).

[7] BOYER, R. F. u. R. S. SPENCER: J. appl. Physics **16**, 594 (1945).

[8] JENCKEL, E. u. R. HEUSCH: Kolloid-Z. **118**, 56 (1950).

[9] ZHURKOW, S. N. u. N. R. LERMAN: C. R. Acad. Sci. USSR **47**, 106 (1945).

Wenn der Weichmacher selbst zur glasigen Erstarrung zu bringen ist (wenn er also nur geringe Neigung zur Kristallisation hat), ist es möglich (vollständige Löslichkeit vorausgesetzt), die Kurve der Einfriertemperatur über den ganzen Konzentrationsbereich vom reinen Polymerisat bis zum reinen Weichmacher zu verfolgen. Diese Weichmacher werden am ehesten Aufschluß über Gesetzmäßigkeiten ergeben.

Leider sind nur für einige wenige solcher Systeme die Einfriertemperaturen mitgeteilt, nämlich für Lösungen des Polystyrols mit Betol, Salol und Trikresylphosphat von JENCKEL u. HEUSCH[1] und weiteren Lösungsmitteln von KARGIN u. MALINSKI[2], für Lösungen der Äthylcellulose und der Nitrocellulose mit Trikresylphosphat von ÜBERREITER[3] und für Lösungen des Polyvinylchlorids mit Trikresylphosphat und zwei technischen Weichmachern von KNAPPE u. SCHULZ[4]. In Abb. IX, 1 u. IX, 2 sind einige Beispiele wiedergegeben. Alle diese Kurven, aufgetragen gegen den Gewichtsanteil, zeigen kennzeichnend eine Durchbiegung der Kurve nach unten, woran auch die Verwendung von Grundmolenbrüchen oder Volumenanteilen (vgl. auch Abb. IX, 6) nichts ändert. Die Einfriertemperatur der Lösung liegt also wesentlich tiefer als sich additiv aus den Einfriertemperaturen der Komponenten ergeben würde.

Folgende *niedermolekularen* Lösungen sind über den ganzen Konzentrationsbereich untersucht worden: Lösungen des Salicins mit Brucin und Piperin von TAMMANN u. KOHLHAAS[5], Lösungen der Glukose mit Glycerin von PARKS u. GILKEY[6] und solche des Phenolphthaleins mit Salicin, Betol und Trikresylphosphat von JENCKEL u. HEUSCH[1]. Diese Kurven sind — wie es scheint — weniger nach unten hin durchgebogen[7].

[1] JENCKEL, E. u. R. HEUSCH: s. S. 568, Anm. 2.
[2] KARGIN u. MALINSKI: s. S. 568, Anm. 4.
[3] ÜBERREITER, K.: s. unten, Anm. 15.
[4] KNAPPE, W. u. A. SCHULZ: s. unten, Anm. 13.
[5] TAMMANN, G. u. A. KOHLHAAS, Z. anorg. allg. Chem. **184**, 63 (1929).
[6] PARKS, G. S. u. W. A. GILKEY: J. physic. Chem. **33**, 1428 (1929).
[7] Vgl. auch H. HENDRICKS: Diss. München 1952. Hiernach ändert sich in einigen Systemen die Einfriertemperatur, bestimmt aus dem Auftreten von *Phosphorescenz*, merkwürdigerweise *nicht* mit der Zusammensetzung.

Fortsetzung der Fußnoten von S. 568.

[10] NIELSEN, L. E., R. BUCHDAHL u. R. LEVREAULT: J. appl. Physics **21**, 607 (1950). — [11] WÜRSTLIN, F.: Kolloid-Z. **105**, 9 (1943); **113**, 18 (1949).
[12] DAVIES, J. M., R. F. MILLER u. W. F. BUSSE: J. Amer. chem. Soc. **63**, 361 (1941). — [13] KNAPPE, W. u. A. SCHULZ: Kunststoffe **41**, 321 (1951).
[14] WÜRSTLIN, F. u. H. KLEIN: Kolloid-Z. **128**, 136 (1952); Kunstst. **42**, 445 (1952). — [15] ÜBERREITER, K.: Z. physik. Chem. B **48**, 197 (1941).
[16] MOELTER, G. M. u. E. SCHWEITZER: Ind. Engng. Chem. **41**, 685 (1949).
[17] AIKEN, W., T. ALFREY, A. JANSSEN u. H. MARK: J. Pol. Sci. **2**, 178 (1947).
[18] FUOSS, R. M.: J. Am. Chem. Soc. **63**, 369, 378 (1941).
[18a] FAULKNER, D.: Brit. Plast. **23**, 183 (1950).
[19] REED, M. C.: Ind. Engng. Chem. **35**, 896 (1943).
[20] LAWRENCE, R. R. u. E. B. MCINTYRE: Ind. Engng. Chem. **41**, 689 (1949).
[21] MAST, W. C. u. C. H. FISHER: Ind. Engng. Chem. **41**, 703 (1949).
[22] MEAD, D. J., R. L. TICHENOW u. R. M. FUOSS: J. Amer. chem. Soc. **64**, 283 (1942). — [23] SCHALLAMACH, A. u. P. THIRION: Trans. Faraday Soc. **45**, 605 (1949).
[24] STÖCKLIN, P.: Kautschuk u. Gummi **2**, 367 (1949); **3**, 45, 86, 199 (1950).
[25] JENCKEL, E. u. K. GORKE: Demnächst.

Mit den meisten übrigen Weichmachern, besonders mit denjenigen, die gewöhnlich als Lösungsmittel bezeichnet werden, läßt sich jedoch eine Untersuchung über alle Konzentrationen nicht durchführen, weil in den polymerisatarmen Lösungen das Lösungsmittel kristallisiert und daher die Bestimmung der Einfriertemperatur unmöglich wird. Es kann also nur ein Teil der ganzen Kurve bestimmt werden.

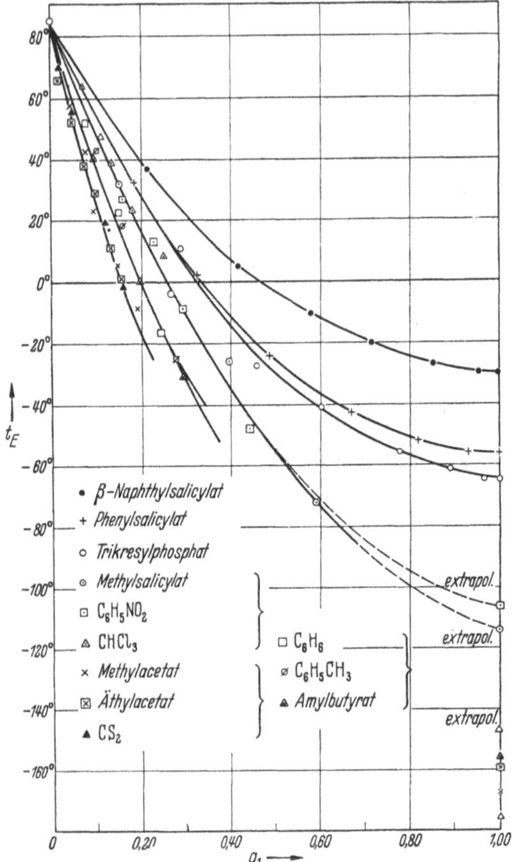

Abb. IX, 1. Erniedrigung der Einfriertemperatur des Polystyrols durch verschiedene Weichmacher. (Abscisse = Gewichtsanteil des Weichmachers.) (Nach JENCKEL u. HEUSCH.)

Wie zu erwarten und wie die Beobachtung bestätigt, wird durch nicht allzu große Zusätze an Weichmacher die Einfriertemperatur näherungsweise linear mit der Zusammensetzung erniedrigt

$$T_E = T_{E2} - c \cdot c_{g_1}. \quad \text{(IX, 1)}$$

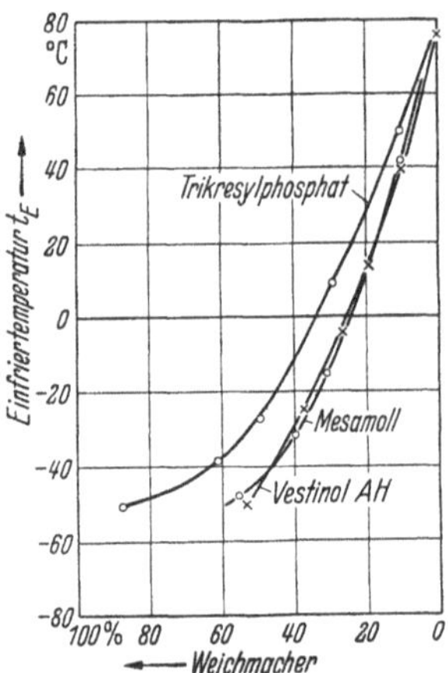

Abb. IX, 2. Erniedrigung der Einfriertemperatur d. Polyvinylchlorids. (Nach KNAPPE u. SCHULZ.)

(T_E und T_{E2} = Einfriertemperatur des weichgemachten Polymeren bzw. des reinen Polymeren, c_{g_1} = Gewichtsanteil des Weichmachers). Der Ausdruck

$$\left(\frac{d\,T_E}{d\,c_{g_1}}\right)_{c_{g_1}=0} \quad \text{(IX, 2)}$$

ist in seinem Zahlenwert für jedes Polymerisat-Weichmachersystem charakteristisch und sei als *Weichmacherwirkung* (*efficiency*) bezeichnet. Obwohl diese Kennzeichnung die Wirkung des Weichmachers nur unvollkommen beschreibt – dazu wäre die ganze Kurve nötig –, bleibt doch meistens gar keine andere Wahl.

Hingewiesen sei noch auf die ungewöhnliche Konzentrationsabhängigkeit bei weichgemachten, *teilkristallinen* Polymeren. Solche Stoffe darf man sicher näherungsweise als teilweise kristallin, teilweise amorph beschreiben. Der amorphe Anteil wird je nach der Temperatur als Gleichgewichtsschmelze oder als Glas vorliegen; die zugehörige Einfriertemperatur wird durch Weichmacherzusatz herabgedrückt. BOYER u. SPENCER[1] haben hierzu Beobachtungen an Saran (Mischpolymerisat aus Vinylchlorid und Vinylidenchlorid) mitgeteilt. Danach ändert sich die Einfriertemperatur des reinen, weichmacherfreien Polymeren *nicht* mit dem Ausmaß des kristallinen Anteils (zu erwarten wäre eine Zunahme von T_E, vgl. Bd. III, § 58), dagegen stark im weichgemachten Polymerisat. Im völlig amorphen Material wurde T_E durch einen nicht näher bezeichneten Weichmacher um 2° herabgesetzt, bei 38% kristallinem Anteil jedoch um 10°. Das ist wesentlich mehr, als sich aus der zu erwartenden Anreicherung des Weichmachers nur in den amorphen Bereichen ergeben würde — hinreichende Genauigkeit der experimentellen Angaben vorausgesetzt. — Auf die Erniedrigung des Schmelzbereichs der Kristalle durch den Weichmacher kann hier nur hingewiesen werden, vgl. Bd. III, § 47 a.

b) Zwei Bemerkungen zu den experimentellen Methoden.

1. Zur dielektrischen Messung[2].

Zur Bestimmung der Wirkung des zugesetzten Weichmachers braucht man nicht notwendigerweise die absoluten Werte der Einfriertemperatur zu bestimmen, es genügt vielmehr, nur ihre Erniedrigung gegenüber dem reinen Polymerisat, $\Delta T = T_{E_2} - T_E$, zu ermitteln. (T_{E_2} = Einfriertemperatur des reinen Polymerisats, T_E = Einfriertemperatur des Lösung.) Gerade für diese Fragestellung kann man sich vorteilhaft der Messung der Viscosität bedienen, welche, wie schon erwähnt, bei der Einfriertemperatur einen ungefähr konstanten Wert von 10^{13} abs. Einh. ausmacht. Es genügt auch, die Temperatur einer etwas kleineren Viscosität als 10^{13} (bei entsprechend höherer Temperatur) zu bestimmen, um den Wert ΔT zu erhalten. Das kann etwa in statischen Messungen, z.B. durch Aufnahme der Verformungskurve (Fließkurve) geschehen. Man kann aber auch die Temperatur des Maximums der Dämpfung langsamer mechanischer Schwingungen bestimmen; diese Messungen stellen im Grunde genommen ebenfalls eine Viscositätsmessung dar (vgl. Bd. III, § 58c u. 60). Da der Temperaturkoeffizient in den verschiedenen Lösungen einigermaßen gleich ist und man sich bei der Messung noch in nächster Nähe der Einfriertemperatur befindet, kann man mit guter Annäherung behaupten, daß diese durch den Zusatz des Lösungsmittels in der gleichen Weise zu tieferen Temperaturen verschoben wird wie die Temperatur einer bestimmten Verformungsgeschwindigkeit oder diejenige des Maximums der mechanischen Schwingungsdämpfung[3].

[1] BOYER, R. F. u. R. S. SPENCER: J. appl. Physics **15**, 398 (1944).

[2] Vgl. dazu auch F. WÜRSTLIN in Bd. III, § 58.

[3] Ausführlichere Angaben über das mechanische Verhalten finden sich z. B. in § 49.

Dagegen erscheint uns die Verwendung hochfrequenter elektrischer Schwingungen (10^6 Hz) nicht ganz unbedenklich. Auch hier beobachtet man bei einer bestimmten Temperatur ein Maximum der dielektrischen Verluste, welches durch Weichmacherzusatz zu einer tieferen Temperatur verschoben wird. Diese Messung stellt, wie oben schon ausgeführt wurde, in gewissem Sinne eine Viscositätsmessung dar, indem maximale Verluste gerade dann beobachtet werden, wenn die bei einer bestimmten Viscosität — auszudrücken durch eine Relaxationszeit τ — vorhandene Beweglichkeit der Dipole mit der Schwingungsdauer des aufgeprägten Wechselfeldes korrespondiert. Das Maximum der dielektrischen Verluste liegt erwartungsgemäß etwa 60 bis 80° höher als die Einfriertemperatur, weil die Schwingungsdauer von etwa 10^{-6} sec in solchen dielektrischen Versuchen viel kleiner ist als die Zeitdauer, die bei gewöhnlicher Abkühlung des Materials oder auch bei Torsionsschwingungen (Schwingungsdauer etwa 1 sec) zur Verfügung gestellt wird. Beobachtet man die dielektrischen Verluste in weichgemachten Polymerisaten, so bemerkt man eine Herabsetzung der Temperatur des Maximums, und zwar um so stärker, je mehr Weichmacher zugesetzt wurde. Diese Beobachtung und eine gewisse Parallelität zum Weichwerden des Polymerisats haben dazu geführt, die Erniedrigung der Temperatur maximaler dielektrischer Verluste durch Weichmacherzusatz der Erniedrigung der Einfriertemperatur ($T_{E_2} - T_E$) in Gl. (IX,1) gleichzusetzen[1].

Diese Übertragung der dielektrischen Ergebnisse auf die Erniedrigung der Einfriertemperatur setzt stillschweigend voraus, daß die Temperaturdifferenz zwischen der Temperatur der maximalen Verluste und der Einfriertemperatur beim reinen Polymerisat, beim reinen Weichmacher und auch in den binären Lösungen stets die gleiche sei. Das erscheint immerhin fraglich. Wichtiger ist vielleicht der folgende Umstand. WÜRSTLIN hat früher Messungen an Polyvinylchlorid[2] und Polyvinylacetat[1] mitgeteilt, nach denen in gewissen Konzentrationsbereichen zwei Maxima der dielektrischen Verluste nebeneinander existieren, die ihre Temperatur nicht oder nicht wesentlich ändern, wohl aber ihre Höhe. WÜRSTLIN hat hieraus, wohl mit Recht, auf die Existenz von Solvatmolekülen und freien Lösungsmittelmolekülen nebeneinander geschlossen. Man erhält hier ein Kurvenbild, wie es etwa in Abb. IX,3 schematisch wiedergegeben ist. Dielektrisch kann man eben die Beweglichkeit verschiedener Gruppen (solvatisierte und nicht solvatisierte Moleküle nach WÜRSTLIN) nebeneinander messen. Ein ähnlicher Kurvenverlauf, d.h. also die Beobachtung von zwei Einfriertemperaturen in der homogenen Lösung, ist aber bislang niemals

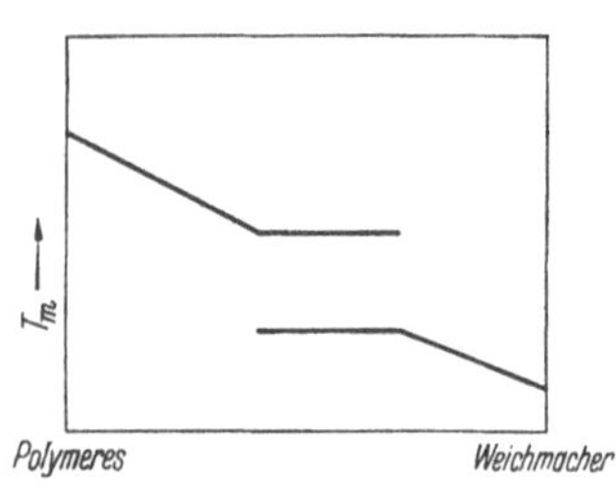

Abb. IX,3. Verschiebung des Maximums der dielektrischen Verluste durch zugesetzten Weichmacher. (Nach den Messungen von WÜRSTLIN.) Zugleich Erniedrigung der Einfriertemperatur in einem beschränkt löslichen Weichmacher. (Schematisch.)

[1] Vgl. F. WÜRSTLIN: Koll. Z. **120,** 84 (1951); **128,** 136 (1952).
[2] WÜRSTLIN, F.: Kolloid-Z. **113,** 18 (1949).

gefunden worden, weil die Einfriertemperatur von der gewöhnlichen, durch die Gesamtheit aller Gruppen erzeugten Viscosität abhängt.

Wenn wir nun auch zu dem Ergebnis kommen, daß die Messung der Verschiebung der Temperatur der maximalen dielektrischen Verluste nur eingeschränkt zur Messung der Erniedrigung der Einfriertemperatur geeignet ist, so soll doch nicht verkannt werden, daß diese Methode sicher angenähert richtig ist, und zwar um so besser, je weniger Weichmacher das Polymerisat enthält.

2. Zur „Brittle-Point"-Messung.

In vielen der oben skizzierten Untersuchungen, namentlich in den amerikanischen, ist die Einfriertemperatur nach der Methode des Brittle-Point oder doch nach einer mechanischen Methode bestimmt worden. Schon bei den weichmacherfreien Polymerisaten weicht die Temperatur des Brittle-Point manchmal erheblich von der richtigen Einfriertemperatur ab, welche etwa als Schnittpunkt der Volumenkurve bestimmt wird (vgl. hierzu Bd. III, § 54). In den weichgemachten Produkten können die Fehler erheblich größer werden, wenn der Weichmacher unverträglich ist. Auch unverträgliche Weichmacher lassen sich durch Kneten oder Walzen in gewissem Umfange in die Polymerisate einarbeiten. Auch hierdurch wird in mechanischer Hinsicht ein Weichwerden erzielt, welches den Brittle-Point herabdrückt. Das hat jedoch mit einer Erniedrigung der Einfriertemperatur in dem bisher besprochenen Sinne gar nichts mehr zu tun. Im Gegenteil: ein unverträglicher Weichmacher kann die wirkliche Einfriertemperatur nicht herabsetzen (vgl. weiter unten). Daher sind Bestimmungen der Weichmacherwirkung mit Hilfe des Brittle-Point nur mit großer Vorsicht zu verwenden.

§ 75. Weichmacherwirkung und chemische Konstitution.

a) Zur Wahl des Konzentrationsmaßes.

Will man zwei Weichmacher in ihrer Wirksamkeit hinsichtlich der Erniedrigung der Einfriertemperatur miteinander vergleichen, so tritt sogleich die Schwierigkeit auf, welches Konzentrationsmaß: Gewichtsanteil c_g, Volumenanteil c_v oder Grundmolenbruch x^* zugrunde gelegt werden soll.

Gewichtsanteile können keine theoretische Bedeutung haben. Vielleicht würde die Verwendung von Volumenanteilen gewisse Vorteile bieten, wenigstens vom Standpunkt der Viscositätsbetrachtung. Sehr bestechend erscheint dagegen eine molare Betrachtungsweise. Freilich darf man sicherlich nicht die richtigen Molenbrüche verwenden, denn bei hinreichend langen Ketten wird ebenso wie die Einfriertemperatur selbst, so auch ihre Erniedrigung durch Weichmacher von der Kettenlänge unabhängig sein. Man würde daher für verschieden lange Ketten verschiedene Werte der Weichmacherwirkung erhalten, was keinen rechten physikalischen Sinn hat. Geeigneter erscheint dann schon die Verwendung von Grundmolenbrüchen

$$x_1^* = \frac{n_1}{n_1 + n_2^*} = \frac{n_1}{n_1 + P\,n_2} \qquad \text{(IX, 3)}$$

[x_1^* = Grundmolenbruch des Weichmachers, n_1, n_2 und n_2^* = Anzahl der Mole des Weichmachers, des Polymerisats bzw. der Grundmole des Polymerisats, P = Polymerisationsgrad = Zahl der Grundmoleküle (monomeren Einheiten) im Kettenmolekül.]

Grundmolenbrüche sind zwar sehr leicht aus der Einwaage zu berechnen, wie weit sie jedoch wirklich eine molare Bedeutung haben, ist recht zweifelhaft. Beispielsweise könnte man im Polystyrol als Grundmolekül sowohl die Gruppe $C_6H_5CH \cdot CH_2$ ansehen, wie man es gewöhnlich tut, aber ebensogut auch das Doppelte oder Dreifache dieser Gruppe. Für das Verhalten des fertigen Polymerisates ist seine Entstehungsgschichte ganz gleichgültig; ein Polymerisat, welches aus dem Monomeren entstanden ist, wird sich ebenso verhalten wie ein solches, welches aus dem Dimeren oder Trimeren gebildet ist. Ähnlich kann man sich fragen, ob im Polyäthylen die Gruppe CH_2 oder $(CH_2)_2$ als Grundmolekül zu betrachten sei. Man erkennt an diesen Beispielen, daß die Verwendung der Grundmolenbrüche der molaren Betrachtung kaum näher steht als die von Gewichtsprozenten. Allenfalls könnte man als Grundmolekül die in einer Segmentlänge vereinigten Moleküle ansetzen. Man bezieht sich hierbei bekanntlich auf ein Modell des wirklichen Kettenmoleküls, welches aus starren, in Kugelgelenken miteinander verbundenen Stäben besteht und welches die gleichen Eigenschaften aufweist wie das wirkliche Kettenmolekül. Die Länge eines solchen Stabes, die eine mehr oder weniger große Anzahl der eigentlichen Grundmoleküle umfaßt, wird als Segmentlänge bezeichnet[1]. Die Segmentlänge eines Polymerisats ist in ihren Zahlenwerten jedoch zur Zeit noch so unbestimmt und außerdem von dem durch die angewandte Meßmethode beanspruchten Bewegungsmechanismus abhängig[2], daß hierauf praktisch eine Konzentrationsangabe nicht gegründet werden kann.

Diese Aussage muß jedoch eingeschränkt werden für den Fall der weichmacher*armen* Lösungen. Dann wird

$$x_1^* = \frac{n_1}{n_1 + n_2^*} \approx \frac{n_1}{n_2^*}\,. \qquad \text{(IX, 4)}$$

Betrachtet man jetzt Segmente als „Grundmoleküle", so ändert sich n_2^* und damit x_1^* um den Faktor $1/s$ bzw. s (s = Zahl der Grundmoleküle im Segment).

Obwohl sich die Segmentlänge s mit der Natur des Lösungsmittels (Weichmachers) ändern mag (vgl. Bd. II, § 29), muß doch offenbar in weichmacher*armen* Lösungen die Segmentlänge s_0 des reinen Polymerisats angestrebt werden. Damit aber wird der gewöhnliche Grundmolenbruch ein brauchbares „segment"molares Maß der Zusammensetzung.

---

[1] Siehe z.B. W. Kuhn: Experientia (Basel) **1**, 1 (1945); vgl. auch Bd. I, § 31.
[2] Vgl. H. A. Stuart: Kolloid-Z. **134**, 152 (1953); ferner Bd. III, S. 678.

Die Darstellung in Grundmolenbrüchen bzw. in Volumenanteilen hängt mit der Darstellung in Gewichtsanteilen *für hinreichend kleine Mengen* zugesetzten Weichmachers folgendermaßen zusammen:

$$\frac{d\,T_E}{d\,c_g} = \frac{d\,T_E}{d\,x_1^*} \cdot \frac{M_2^*}{M_1} = \frac{d\,T_E}{d\,c_{v_1}} \cdot \frac{\varrho_2}{\varrho_1} \qquad \text{(IX, 5)}$$

(c_{g_1} = Gewichtsanteil, c_v = Volumenanteil, M_1 und M_2^* = Molgewicht bzw. Grundmolgewicht, ϱ_1 und ϱ_2 = Dichte des Weichmachers bzw. des Polymerisats.)

b) Das Weichmacher-Polymerisatgemisch als Lösungsphase.

Polymerisat und Lösungsmittel sind entweder unbeschränkt ineinander löslich oder nur beschränkt. Im ersteren Fall spricht man von *Verträglichkeit* (*compatibility*) zwischen diesen beiden Stoffen, im letzteren Fall von *Unverträglichkeit*. Im Falle eines verträglichen Weichmachers, welcher also über den ganzen Konzentrationsbereich vom reinen Polymerisat bis zum reinen Weichmacher eine homogene Lösung bildet, liegt im Sinne der Phasenlehre nur eine einzige Phase vor, deren Eigenschaften sich kontinuierlich von der einen Komponente zur anderen hin ändern müssen. Eine solche Eigenschaft ist auch die Einfriertemperatur. Sie muß sich also kontinuierlich vom einen Grenzwert zum andern hin ändern. Über die Form dieser Kurve vermag die Phasenlehre keine Aussagen zu machen. Es scheint jedoch, daß Maxima und Minima auf dieser Kurve nicht vorkommen, sondern nur eine monotone Änderung.

Es möge hier noch der Fall der beschränkten Löslichkeit, also eines unverträglichen Weichmachers, behandelt werden. In Abb. IX, 12 ist die Löslichkeitskurve dargestellt, oberhalb der nur eine Phase, unterhalb der zwei Phasen koexistent sind. Dann wird die Einfriertemperatur im homogenen Gebiet stetig absinken. Im heterogenen Gebiet sollte man dagegen zwei unveränderliche Einfriertemperaturen beachten, von denen die eine der einen Phase und die andere der anderen entspricht. Abb. IX, 3 gibt hierzu eine schematische Darstellung. Ein Beispiel für diesen Fall scheint noch nicht untersucht zu sein[1].

c) Zur Konzentrationsabhängigkeit der Einfriertemperatur in Lösungen.

Wir wenden uns wieder den unbeschränkt löslichen Weichmachern zu. Wie Abb. IX, 4 lehrt, muß die Einfriertemperatur von zwei Lösungen mit verschiedenen Weichmachern auf dem weichmacherreichen Teil verschieden sein, weil die Einfriertemperaturen der reinen Weichmacher im allgemeinen verschieden sind. Hier muß also auf jeden Fall ein Einfluß der chemischen Konstitution zutage treten. Es fragt sich aber, ob auch bei

[1] Bei Materialien aus zwei unverträglichen *Polymeren* wurden schon zwei Einfriertemperaturen beobachtet; vgl. K. L. Floyd, Brit. Journ. Appl. Phys. **3**, 373 (1952); (Volumenmessungen an synthetischen Kautschuken) E. Jenckel u. H. U. Herwig, noch unveröffentlicht (Schwingungsdämpfung und Brechungsindex an Polymethacrylester – Polyvinylacetat und an Gemischen von zwei Styrol-Acrylester-Mischpolymerisaten), vgl. auch L. E. Nielsen, J. Am. Chem. Soc. **75**, 1535 (1953).

geringen Weichmacherzusätzen die Erniedrigung der Einfriertemperatur proportional der Erniedrigung der reinen Weichmacher gegenüber dem reinen Polymerisat wie in Abb. IX, 4a ist. Es wäre beispielsweise auch denkbar, daß die beiden Kurven in der Nähe des reinen Polymerisats das reine Polymerisat mit gleicher Tangente erreichen wie in Abb. IX, 4b. In diesem Bereich würde also die chemische Natur der Weichmacher völlig zurücktreten[1].

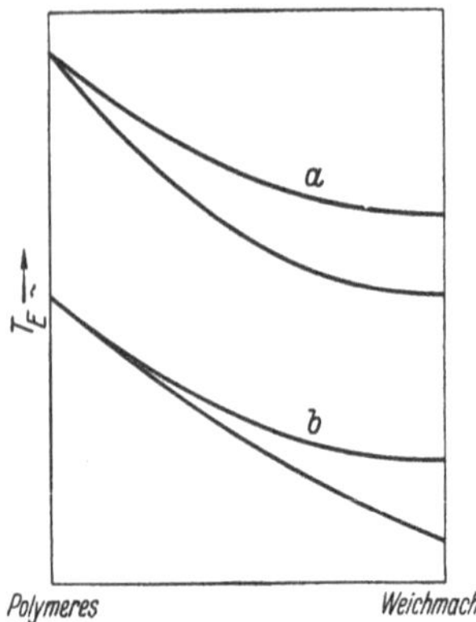

Abb. IX, 4. Zum Verhältnis der Kurve der Einfriertemperaturen von zwei Weichmachern.
a) Die Erniedrigung der Einfriertemperatur ist überall proportional dem Unterschied der Einfriertemperaturen der beiden Komponenten.
b) Die beiden Kurven fallen für kleine Weichmacherzusätze zusammen und trennen sich erst in den weichmacherreichen Lösungen.

In der Tat glauben ZHURKOW und LERMAN[2] gefunden zu haben, daß die chemische Natur des Weichmachers (in den weichmacherarmen Lösungen) keinen Einfluß ausübt, wenn man ein geeignetes Konzentrationsmaß, nämlich den Grundmolenbruch, wählt. Sie fanden an Polymethacrylester und einem Mischpolymerisat aus Butadien und Acrylnitril die gleiche Wirkung für verschiedene Weichmacher mit Molekulargewichten zwischen 60 und 380, wenn man Lösungen mit gleichem Grundmolenbruch untersucht. JENCKEL und GORKE[3] fanden das gleiche für Lösungen des Polystyrols mit Toluol, Äthylbenzol und Chlorbenzol[4, 5].

d) Der Einfluß der chemischen Konstitution[6].

Die Überprüfung der Regel von ZHURKOW und LERMAN[2] an einem größeren experimentellen Material ergibt aber doch recht erhebliche Abweichungen, d.h. es machen sich konstitutive Einflüsse geltend. An *Polymethycrylsäuremethylester* fanden JENCKEL und HEUSCH[7] mit 10 Weichmachern bei 10 Grundmolprozenten Erniedrigungen der Einfriertemperatur ΔT, die sich bis auf das $2^1/_2$fache ändern (vgl. Tab. IX, 1). An *Polyvinylacetat* kamen WÜRSTLIN und KLEIN[8] an Hand von umfangreichen

[1] Technische Produkte mit 30 bis 40 Gew.-% Weichmacher können im Sinne dieser Betrachtung höchstens dann als „weichmacherarm" angesprochen werden, wenn es sich um Weichmacher nicht allzu verschiedener Einfriertemperaturen handelt. — [2] ZHURKOW, S. N. u. N. R. LERMAN: C. R. Acad. Sci. USSR 47, 106 (1945).

[3] JENCKEL, E. u. K. GORKE: Demnächst.

[4] Vgl. auch Tabelle IX, 2.

[5] Nach W. A. KARGIN u. J. M. MALINSKY: Ber. Acad. Wiss. UdSSR, NS 73, 967 (1950), führt dagegen die Verwendung von Volumenanteilen zu ein und derselben Kurve bei verschiedenen Weichmachern (an Polystyrol). Dieser Befund ist, wie es scheint, anderweitig noch nicht bestätigt. Die Messungen von E. JENCKEL und R. HEUSCH: Koll. Z. **130**, 89 (1953), ebenfalls an Polystyrol, lassen sich auch mit Volumenbrüchen nicht einheitlich darstellen.

[6] Vgl. hierzu auch Bd. III, Kap. XI.

[7] JENCKEL, E. u. R. HEUSCH: Kolloid-Z. **118**, 56 (1950).

[8] WÜRSTLIN, F. u. H. KLEIN: Kolloid-Z. **128**, 136 (1952); Kunstst. **42**, 445 (1952).

Untersuchungen zu einem ähnlichen Ergebnis. 89 Weichmacher drückten in einer grundmolaren Konzentration von 10% das Maximum der dielektrischen Verluste um mindestens 24° und höchstens 48° herunter. Hierbei schwankten die Molekulargewichte in weiten Grenzen, etwa zwischen 50 und 500 und zeigten keinen systematischen Gang mit der Verschiebung des Maximums der dielektrischen Verluste. Was das *Polystyrol* anlangt, so lassen sich aus den Versuchen von JENCKEL und HEUSCH[1] die in Tab. IX,2 angegebenen Erniedrigungen der Einfriertemperaturen in der Lösung von 10 Grundmolprozent ermitteln. Auch hier variieren die Temperaturdifferenzen um mehr als das Doppelte. Immerhin haben chemisch ähnliche Stoffe — hier Ester der Salicylsäure — trotz verschiedener Einfriertemperatur der reinen Stoffe die gleiche Weichmacherwirksamkeit im Sinne von ZHURKOW.

Tabelle IX,1.

Erniedrigung der Einfriertemperatur ΔT an Polymethacrylsäuremethylester durch verschiedene Weichmacher bei einem Grundmolenbruch $x_1^* = 0{,}1$.

Weichmacher	ΔT	Molekülgestalt
$CHCl_3$	25°	gedrungen
C_6H_6	29°	
$C_6H_5CH_3$	33°	
$C_6H_5NO_2$	40°	
CH_3COCH_3	36°	
C_4H_9OH	39°	
Methylacetat	44°	
Äthylacetat	47°	
Propylacetat	54°	
Butylacetat	61°	gestreckt

Tabelle IX,2.

Erniedrigung der Einfriertemperatur ΔT an Polystyrol durch verschiedene Weichmacher bei einem Grundmolenbruch $x_1^* = 0{,}1$.

Weichmacher	ΔT	Molekülgestalt
C_6H_6	35°	gedrungen
$C_6H_5CH_3$	38°	
$C_6H_5NO_2$	33°	
CH_2Cl_2	39°	
$CHCL_3$	40°	
CCl_4	44°	
CS_2	45°	
Methylacetat	42°	
Äthylacetat	52°	
Butylacetat	53°	
Amylbutyrat	57°	gestreckt
Methylsalicylat	51°	gedrungen
Phenylsalicylat	53°	
β-Naphthylsalicylat	49°	gestreckt
Tricresylphosphat	73°	

Wie es scheint, kann man das experimentelle Material wohl am ersten übersehen, wenn man die Gestalt und damit die innere Beweglichkeit des Weichmachermoleküls betrachtet, worauf JENCKEL u. HEUSCH sowie WÜRSTLIN u. KLEIN zuerst hingewiesen haben. Gestreckte Moleküle, z. B. aliphatische Ketten mit hoher Beweglichkeit erniedrigen die Einfriertemperatur (bei gleicher grundmolarer Konzentration) stark, Moleküle mit gedrungener Gestalt, wie etwa starre Ringe oder kugelförmige Moleküle oder solche, die chlorsubstituiert sind, haben geringere Beweglichkeit und erniedrigen weniger[2].

[1] JENCKEL, E. u. R. HEUSCH: Kolloid-Z. **130**, 89 (1953).

[2] Nach dem gleichen Prinzip lassen sich im wesentlichen die Einfriertemperaturen der reinen Stoffe übersehen (vgl. Bd. III, § 60).

Sehr deutlich wird dieser Einfluß an Polymethacrylester. Er ist aber auch an Polystyrol noch gut sichtbar, wie Tab. IX,1 u. 2 ohne weitere Erläuterung erkennen lassen. Die ausgedehnte Untersuchung von WÜRSTLIN u. KLEIN geht noch weiter ins einzelne. Es fanden sich neun Weichmacher, deren polare Gruppen entweder in oder direkt neben einer aromatischen Gruppe eingebaut sind oder die nur eine polare Gruppe enthalten. Die polare Gruppe ist also in diesen Weichmachermolekülen behindert. Diese Weichmacher erniedrigen die Temperatur des Maximums der dielektrischen Verluste nur um etwa 28° bei 10 *Grundmol*prozenten Weichmacher. Andererseits fanden sie 15 Substanzen, die in der gleichen grundmolaren Konzentration die Temperatur des Maximums um etwa 46° herabsetzten. Es sind dies alles Substanzen, in denen die polaren Gruppen aliphatisch gebunden sind, so daß sie die größtmögliche Beweglichkeit haben. Die Verschiebung der Temperatur des Maximums der Verluste liegt bei den übrigen Substanzen zwischen diesen beiden Grenzwerten, was nach WÜRSTLIN u. KLEIN durch eine mittlere Beweglichkeit der polaren Gruppen leicht verständlich ist. Diese Regel gilt, wenn auch mit etwas größerer Streuung, auch noch bei grundmolaren Konzentrationen von 20 und 30% Weichmacher.

Schließlich haben LAWRENCE und MCINTYRE[1] verschiedene Weichmacher auf ihre Wirksamkeit an Polyvinylchlorid untersucht. Die Weichmacherwirksamkeit wird hier als die Menge Weichmacher angegeben, die vorhanden sein muß, um einen bestimmten Elastizitätsmodul bei 25° (1800 pounds/square inch) zu erzeugen. Diese Angabe kann nicht auf die übliche $\frac{d T_E}{d c_{g_1}}$ umgerechnet werden, dürfte dieser aber reziprok parallel laufen. Deutliche Effekte ergeben sich nur, wenn man im Weichmacher einen aromatischen oder noch besser einen hydroaromatischen Ring einführt, oder wenn man mit Chlor substituiert[2]. Durch diese Substitutionen wird die Weichmacherwirkung bedeutend geringer, d.h. man muß größere Mengen dieser Weichmacher hinzufügen, um den vorgegebenen E-Modul zu erreichen (Tab. IX,3), weil die innere Beweglichkeit, sei es infolge von Ringbildung, sei es durch Assoziation, vermindert ist.

Überraschend ist die starke Weichmacherwirkung der Salicylsäureester und besonders des Trikresylphosphats auf Polystyrol (Tab. IX,2), obwohl diese Stoffe starre Ringe enthalten. Auch LAWRENCE u. MCINTYRE fanden starke Wirksamkeit der Ester zweibasischer Säuren und der Phosphorsäure auf Polyvinylchlorid. Offenbar kommt es sehr auf die Anzahl der die Lösung (Solvatbildung) vermittelnden Gruppen und deren Beweglichkeit an, wie besonders WÜRSTLIN u. KLEIN betonen. Im Trikresylphosphat z.B. werden nicht nur die Estergruppen, sondern besonders auch die aromatischen Ringe die Lösung vermitteln, was freilich die Beurteilung der „Beweglichkeit" erschwert.

---

[1] LAWRENCE, R. R. u. E. B. MCINTYRE: Ind. Engng. Chem. **41**, 689 (1949). – Vgl. auch W. C. MAST u. C. H. FISHER: Ind. Engng. Chem. **41**, 703 (1949). – D. J. MEAD, R. L. TICHENOR u. R. M. FUOSS: J. Amer. chem. Soc. **64**, 283 (1942). – W. AIKEN, T. ALFREY JR., A. JANSSEN u. H. MARK: J. Polymer Sci. **2**, 178 (1947).

[2] Die Umrechnung von Gewichtsanteilen auf Grundmolenbrüche bringt keine wesentliche Änderung.

Tabelle IX, 3. *Einfluß verschiedener Weichmacher auf Polyvinylchlorid.*

Weichmacher	Reciproke Wirksamkeit (in Tl. Weichmacher auf 100 Tl. Polyvinylchlorid beim E-Modul 1800 $\frac{\text{pounds}}{\text{square inch}}$ und 25°)
Einfluß von Arylgruppen	
Di-n-butylphthalat	41
Butylbenzylphthalat	45
Dibenzylphthalat	60
Einfluß aliphatischer Ketten	
Di-n-hexylphthalat	50
Di-2-äthyl-butylphthalat	48
Di-cyclohexylphthalat	88
Einfluß der Chlor-Substitution	
Dibutyl-monochlorphthalat	46
Dibutyl-dichlorphthalat	52
Dibutyl-trichlorphthalat	60

Die ZHURKOWsche Forderung wird also nicht streng erfüllt, aber doch manchmal recht gut für Stoffe ähnlicher Molekülgestalt. Diese Stoffe können dennoch recht verschiedenen Schmelzpunkt haben, z.B. Benzol und Toluol. Wenn wirklich ihre Einfriertemperaturen – falls man sie messen könnte – sich verhalten wie die Schmelztemperaturen (vgl. Bd. III, § 51), müßte die Kurve so verlaufen wie in Abb. IX, 4b. Andere Weichmacher gehorchen jedoch der ZHURKOWschen Regel nicht; ihre Wirksamkeit wird beherrscht von der Einfriertemperatur des reinen Weichmachers und damit von seiner Schmelztemperatur. In der Tat findet die etwa im Verhältnis 1 : 2 unterschiedliche Wirksamkeit der Weichmacher ihr Gegenstück in dem Verhältnis 1 : 1,5 der Schmelztemperaturen.

§ 76. Weichmacherwirkung und Viscosität.

a) Die LEILICHsche Regel.

Bei der Einfriertemperatur beträgt die Viscosität ungefähr 10^{13} abs. Einh. Die Viscosität der Weichmacher dagegen ist bei Zimmertemperatur wesentlich kleiner, eben, weil sie tropfbar flüssig sind[1]. Ihre Einfriertemperatur liegt dementsprechend weit unter Zimmertemperatur, und zwar um so tiefer, je kleiner die Viscosität bei Zimmertemperatur ist. Neben dem Absolutwert der Viscosität kommt es hierbei auch auf den Temperaturkoeffizienten der Viscosität an. Je kleiner der Temperaturkoeffizient, bei um so tieferer Temperatur wird die Einfriertemperatur, d.h. die Viscosisät von 10^{13} erreicht sein. Die Lösungen aus Polymerisat und Lösungsmittel werden eine Einfriertemperatur haben, die zwischen der der beiden Komponenten liegt, wie oben dargelegt. Dementsprechend wird im allgemeinen die Einfriertemperatur eines bestimmten Polymerisats um so

[1] Feste Weichmacher sind im unterkühlten flüssigen Zustand zu denken.

stärker durch das zugesetzte Lösungsmittel erniedrigt werden, je tiefer dessen Einfriertemperatur liegt, d.h., je kleiner dessen Viscosität bei Zimmertemperatur und je kleiner ihr Temperaturkoeffizient ist.

Diese Regel ist zuerst von LEILICH[1] ausgesprochen worden. Sie bewährt sich im großen und ganzen, wie Abb. IX, 5, die einer Arbeit von WÜRSTLIN u. KLEIN[2] entnommen ist, zeigt. Die Abbildung zeigt zugleich, daß ein ungefähr linearer Zusammenhang besteht zwischen der Einfriertemperatur und dem Logarithmus der Viscosität, nicht der Viscosität selbst[3]. Das bedeutet also, daß man bei ziemlich hochviscosen Weichmachern mit einer weiteren Erhöhung der Viscosität nur noch eine geringe Veränderung der Einfriertemperatur erreicht. So ist auch die Beobachtung zu verstehen, daß zwei Polyester (Paraplex G25), die sich nur in ihrer Kettenlänge unterschieden und verschiedene Viscosität hatten, die gleiche Wirkung als Weichmacher hervorbrachten[4]. Ebenso gehört hierher die Feststellung von JONES[5], daß die Lösungsmittel mit steigendem Molekulargewicht in ihrer Viscosität zwar noch deutlich zunehmen, in ihrer Weichmacherwirkung dagegen nur sehr wenig, so daß ein Grenzwert des Molekulargewichts angegeben werden kann, oberhalb dessen die Weichmacherwirkung nicht mehr von der Viscosität des Lösungsmittels abhängt.

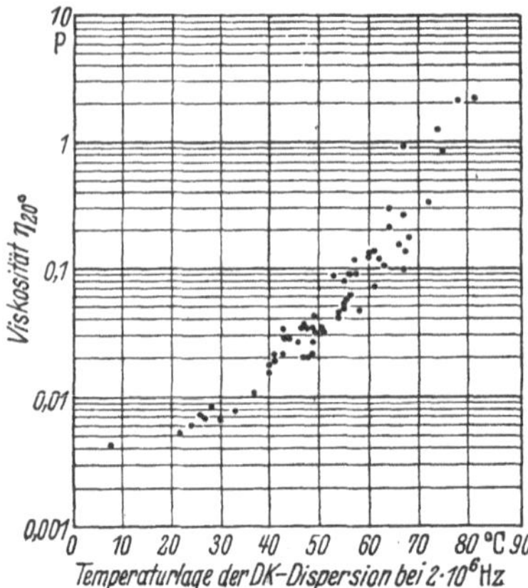

Abb. IX, 5. Der ungefähr lineare Zusammenhang zwischen der Erniedrigung der Einfriertemperatur bei gleicher gewichtsprozentischer Zusammensetzung und dem Logarithmus der Viscosität des reinen Weichmachers bei 20°. (An Polyvinylacetat nach WÜRSTLIN u. KLEIN.)

Dennoch gilt die Regel von LEILICH nur mit erheblicher Streuung. Das geht schon aus Abb. IX, 5 hervor und insbesondere auch aus den hier nicht wiedergegebenen Messungen von REED[6] und von JONES[5,7].

b) Ableitung der LEILICHschen Regel.

Hierzu sucht man die Viscosität der Lösung bei konstanter Temperatur aus der Viscosität der beiden reinen Komponenten zu ermitteln und

[1] LEILICH, K.: Kolloid-Z. **99**, 107 (1942).

[2] WÜRSTLIN, F. u. H. KLEIN: Kolloid-Z. **128**, 136 (1952).

[3] Vielleicht zurückzuführen auf die im wesentlichen doppelt logarithmische, nicht einfach logarithmische Abhängigkeit der Viscosität von der Temperatur, vgl. Bd. II, § 47, Bd. III, § 54.

[4] AIKEN, W., T. ALFREY, A. JANSSEN u. H. MARK: J. Pol. Sci. **2**, 178 (1949).

[5] JONES, H.: J. Soc. chem. Ind. **67**, 415 (1948).

[6] REED, M. C.: Ind. Engng. Chem. **35**, 896 (1943).

[7] Neuerdings haben F. WÜRSTLIN u. H. KLEIN, Z. Makromol. Chem. **16**, 1 (1955) gezeigt, daß die Weichmacherwirkung bei gleicher grundmolarer Konzentration bei den normalen (unverzweigten) o-Phthalsäuredialkylestern sich streng linear mit dem Logarithmus der Viscosität ändert. Bei den isomeren verzweigten Estern ist die Wirkung größer.

anschließend die Temperatur zu errechnen, bei der die Viscosität von 10^{13} abs. Einh. und damit die Einfriertemperatur erreicht wird. Man braucht also erstens eine Mischungsregel und muß zweitens die Temperaturabhängigkeit der Viscosität der Lösung kennen. Diese beiden Beziehungen sind aber nicht sicher bekannt. Man ist deshalb auf Ansätze angewiesen, die in mehr oder weniger großem Umfange gültig sind und dementsprechend auch nur Einfriertemperaturen liefern können, die der Beobachtung mehr oder weniger gut entsprechen.

Es liegt nahe, eine Mischungsregel von der Form

$$\ln \eta = c_{g_1} \ln \eta_1 + c_{g_2} \ln \eta_2 \tag{IX, 6}$$

zugrunde zu legen, welche die Viscosität von tropfbaren binären Flüssigkeitsgemischen einigermaßen wiedergibt[1]. (c_{g_1} und c_{g_2} = Gewichtsanteile des Weichmachers bzw. des Polymerisates, η, η_1 und η_2 Viscosität der Lösung bzw. des reinen Weichmachers bzw. des reinen Polymerisates bei konstanter Temperatur.)

Zweifellos werden Abweichungen von diesem Mischungsansatz auftreten, denen man in einfachen Fällen etwa durch den Ansatz[2]

$$\ln \eta = c_{g_1}^2 \ln \eta_1 + 2\, c_{g_1} c_{g_2} \ln \eta_{12} + c_{g_2}^2 \ln \eta_2 \tag{IX, 7}$$

oder[3]

$$\eta = c_{g_1}^2 \cdot \eta_1 + 2\, c_{g_1} \cdot c_{g_2} \cdot \eta_{12} + c_{g_2}^2\, \eta_2 \tag{IX,8}$$

Rechnung tragen kann. Dieser Ausdruck enthält jetzt jedoch die Größe η_{12}, die der unbekannten Wechselwirkung zwischen den beiden Komponenten entspricht und grundsätzlich nicht auf η_1 und η_2 zurückgeführt werden kann. Nun enthalten die gebräuchlichen Weichmacher meist ziemlich große Moleküle, in denen die spezifischen Eigenschaften einer charakteristischen Gruppe, etwa der Estergruppe, neben vielen wenig wirksamen Gruppen, etwa CH_2-Gruppen, etwas nivelliert erscheinen. Für diese Stoffe könnte gelten $\eta_{12} \approx \sqrt{\eta_1 \cdot \eta_2}$ bzw. $\eta_{12} \approx 1/2\,(\eta_1 + \eta_2)$, womit sich dann aus Gl. (IX,7) und (IX,8) die einfachen additiven Beziehungen (IX,6) bzw. (IX,16) ergeben. Nur mit dieser Voraussetzung läßt sich die LEILICHsche Regel in der vorliegenden Form, die eine spezifische Wechselwirkung nicht berücksichtigt, ableiten.

Für die Temperaturabhängigkeit wird gewöhnlich ein Ausdruck von der Form

$$\ln \eta = \frac{Q}{R\,T} + A \tag{IX, 9}$$

verwandt. Dieser gilt jedoch mit konstantem Q und A nur in einem verhältnismäßig kleinen Temperaturbereich. Verfolgt man nämlich an ein und demselben Stoff die Viscosität über möglichst große Temperaturbereiche, so erkennt man, daß die Viscosität viel stärker mit der Temperatur zunimmt, als sich aus Gl. (IX,9) ergibt. Man kann diesem Umstande Rechnung tragen, indem man die Aktivierungswärme zu $Q\left(1 + \varkappa \cdot e^{\frac{Q}{RT}}\right)$ ansetzt und von etwa 3000 bis 5000 cal bei hoher Temperatur auf etwa 50000 bis 100000 cal bei der Einfriertemperatur ansteigen läßt[4]. Das

[1] ARRHENIUS, S.: Z. physik. Chem. **1**, 285 (1887). – E. HATSCHEK: Die Viscosität der Lösungen, Dresden u. Leipzig 1929, S. 131.

[2] JENCKEL, E. u. G. REHAGE: Z. makromol. Chem. **6**, 243 (1951).

[3] STUART, H. A.: Z. Naturforsch. **3a**, 196 (1948); vgl. auch Bd. II, §§ 48 u. 50; wir wollen hier η_1, η_{12}, η_2 als konzentrationsunabhängig annehmen.

[4] JENCKEL, E.: Z. physik. Chem. A **184**, 309 (1939); vgl. auch Bd. II, § 47 und Bd. III, § 54.

wirkliche Verhalten der Viscosität dürfte sich jedoch ebensogut durch eine Veränderung von A mit der Temperatur bei konstantem Q darstellen lassen[1, 2].

Verwenden wir unbeschadet dieser sehr wesentlichen Einschränkungen zur Berechnung der Einfriertemperatur die Mischungsregel Gl. (IX,6) und die Temperaturabhängigkeit für die Viscosität nach Gl. (IX,9), so erhält man für die Viscosität einer bestimmten Lösung den folgenden Ausdruck

$$\ln \eta = c_{g_1}\left(A_1 + \frac{Q_1}{RT}\right) + c_{g_2}\left(A_2 + \frac{Q_2}{RT}\right). \qquad \text{(IX,10)}$$

Setzt man hier die für alle Stoffe gleiche Einfrierviscosität η_E ein[3], so wird die Temperatur T zur Einfriertemperatur T_E. Wir wollen weiter näherungsweise den Temperaturkoeffizienten im Polymerisat, im Lösungsmittel und in der Lösung einander gleichsetzen, $Q_1 = Q_2 = Q$, wozu wir schon deswegen veranlaßt werden, weil $Q = Q(T)$ eine komplizierte Funktion der Temperatur ist. Die Größe A_1 läßt sich leicht aus der Viscosität bei Zimmertemperatur ($T = 298°$) ermitteln. Die Größe A_2 ist ohnehin für ein und dasselbe Polymerisat konstant.

$$A_1 = (\ln \eta_1)_{298} - \frac{Q}{R \cdot 298}\,; \quad A_2 = (\ln \eta_2)_{298} - \frac{Q}{R \cdot 298} = \text{konst.} \qquad \text{(IX,11)}$$

Setzt man diese Werte ein, so erhält man schließlich:

$$\frac{R}{Q}\left[\ln \eta_E - c_{g_1}(\ln \eta_1)_{298} - c_{g_2}(\ln \eta_2)_{298}\right] + \frac{1}{298} = \frac{1}{T_E}. \qquad \text{(IX,12)}$$

Diese Gleichung läßt sich auch auf das reine Polymerisat ($c_{g_2} = 1$, $c_{g_1} = 0$) anwenden. Man erhält

$$\frac{R}{Q}\left[\ln \eta_E - (\ln \eta_2)_{298}\right] + \frac{1}{298} = \frac{1}{T_{E_2}}. \qquad \text{(IX,13)}$$

Zieht man beide Gleichungen voneinander ab, so ergibt sich

$$\frac{R}{Q}\, c_{g_1}\left[(\ln \eta_2)_{298} - (\ln \eta_1)_{298}\right] + \frac{1}{T_{E_2}} = \frac{1}{T_E}. \qquad \text{(IX,14)}$$

Danach ändert sich also die reziproke Einfriertemperatur linear mit der Zusammensetzung in Gewichtsanteilen und linear mit dem Logarithmus der Weichmacherviscosität[4], entsprechend den Angaben von Würstlin[5]. Schließlich kann man noch in Gl. (IX,14) die Viscositäten eliminieren,

---

[1] Siehe S. 581, Fußnote 3.

[2] Wyk, W. R. van, u. W. A. Seeder: Physica 4, 1073 (1937); 6, 129 (1939); 7, 45 (1940).

[3] $\eta_E = 10^{13}$ abs. Einh., vgl. Bd. III, § 54.

[4] Versucht man eine bessere Temperaturabhängigkeit einzuführen, so ist man aus mathematischen Gründen zu scharfen Näherungen gezwungen und erhält dann eine der Gl. (IX,27) fast gleiche Formel, die die Regel von Zhurkow wiedergibt. Damit begibt man sich jedoch des Hauptvorteils der Gl. (IX,14), die der unterschiedlichen Wirksamkeit der Weichmacher gerecht wird. Vgl. E. Jenckel, Kolloid-Z. 120, 10 (1951).

[5] Vgl. auch Abb. IX, 5.

indem man die Gl. (IX,14) für den reinen Weichmacher ($c_{g_1} = 1$) anwendet. Man erhält dann leicht

$$\frac{1}{T} = c_{g_1} \frac{1}{T_{E_1}} + c_{g_2} \frac{1}{T_{E_2}} \tag{IX,15}$$

d.h. die reziproke Einfriertemperatur ist additiv in den reziproken Einfriertemperaturen der beiden Komponenten.

Verwendet man den Ansatz nach STUART Gl. (IX,8) in linearer Form, indem man die Wechselwirkungsviscosität vernachlässigt,

$$\eta = c_{g_1} \eta_1 + c_{g_2} \eta_2 , \tag{IX,16}$$

so erhält man mit Gl. (IX,9) für die Einfriertemperatur entsprechend zu Gl. (IX,12, 13 und 14)

$$\frac{R}{Q} [\ln \eta_E - \ln (c_{g_1} \cdot \eta_{1\,298} + c_{g_2} \cdot \eta_{2\,298})] + \frac{1}{298} = \frac{1}{T_E} \tag{IX,17}$$

$$\frac{R}{Q} [\ln \eta_E - \ln \eta_{2\,298}] + \frac{1}{298} = \frac{1}{T_{E_2}} \tag{IX,18}$$

$$\frac{R}{Q} [\ln \eta_{2\,298} - \ln (c_{g_1} \cdot \eta_{1\,298} + c_{g_2} \cdot \eta_{2\,298})] + \frac{1}{T_{E_2}} = \frac{1}{T_E} , \tag{IX,19}$$

wobei wiederum eine von der Zusammensetzung unabhängige Aktivierungswärme angenommen ist.

Diese Gleichung läßt sich nicht in eine einfache Form bringen entsprechend Gl. (IX,15), insbesondere läßt sich der Wert Q *nicht* mehr eliminieren. Daher hängt die Einfriertemperatur nicht nur von den Einfriertemperaturen der beiden Komponenten, sondern außerdem von der Aktivierungswärme ab, wodurch der praktische Wert dieser Gleichung sehr beeinträchtigt wird. Im übrigen wird die LEILICHsche Regel nicht wiedergegeben, sondern eher die von ZHURKOW; vgl. S. 586.

Verwendet man in dem ja ganz willkürlichen Mischungsansatz nach Gl. (IX, 6) an Stelle der Gewichtsanteile die Grundmolenbrücke oder die Volumenanteile, so erhält man die Gl. (IX, 15) entsprechenden Formeln in diesen Konzentrationsmaßen.

Für einen Vergleich der abgeleiteten Formel mit der Erfahrung eignen sich im Grunde genommen nur Kurven der Einfriertemperatur, die über den ganzen Konzentrationsbereich aufgenommen sind. In Abb. IX, 6 gibt die Kurve *b* die nach Gl. (IX, 15) berechnete Einfriertemperatur im System Polystyrol ($t_E = 85{,}5°$) mit Betol ($t_E = -29{,}5°$) wieder. Ferner enthält die Abbildung die gemessenen Einfriertemperaturen, aufgetragen gegen die Gewichtsanteile $g_1 = c_{g_1}$, die Volumenanteile $w_1 = c_{v_1}$ und den Grundmolenbruch x_1^*. Die berechnete Kurve weicht immer noch erheblich von den beobachteten Kurven ab, selbst, wenn man die Konzentrationen in Gewichtsanteilen mißt. Dieses Verhalten ist offenbar typisch, jedenfalls für Polystyrollösungen[1]. Die Ursache für die Abweichung möchten wir in der zugrunde gelegten Formel für die Mischung (IX, 6) und die Temperaturabhängigkeit sehen, die schon das Verhalten der niedermolekularen Lösungen nur in grober Annäherung wiedergeben.

[1] JENCKEL, E. u. R. HEUSCH: Koll.-Z. **130**, 89 (1953).

c) Berechnung aus der Viscosität des Polymerisats.

Eine andere Mischungsregel hat FLORY[1] herzuleiten gesucht. An reinen geschmolzenen Polymerisaten verschiedener Kettenlänge fand er die Beziehung[2]

$$\log \eta = A' + \frac{Q}{RT} + C\sqrt{P}. \qquad \text{(IX, 20)}$$

Dabei bleiben in einer polymerhomologen Reihe die Werte A', Q und C konstant, während der Einfluß der unterschiedlichen Kettenlänge mit dem Ausdrucke $\sqrt{P}$ (P = Polymerisationsgrad) in die Rechnung eingeht. Gestützt auf diesen experimentellen Befund, setzt FLORY nun eine Mischungsregel an, indem er für die Lösung einen mittleren Polymerisationsgrad $\overline{P}$ einführt, der folgendermaßen definiert ist:

$$\overline{P} = c_{g_1} P_1 + c_{g_2} P_2 \qquad \text{(IX, 21)}$$

(c_{g_1}, c_{g_2} = Gewichtsanteil des Lösungsmittels bzw. Polymerisates). Da der Polymerisationsgrad P_1 des Lösungsmittels im allgemeinen gleich 1 anzusetzen ist, kann man an Stelle von Gl. (IX, 21) fast immer schreiben

$$\overline{P} \cong c_{g_2} P_2. \qquad \text{(IX, 22)}$$

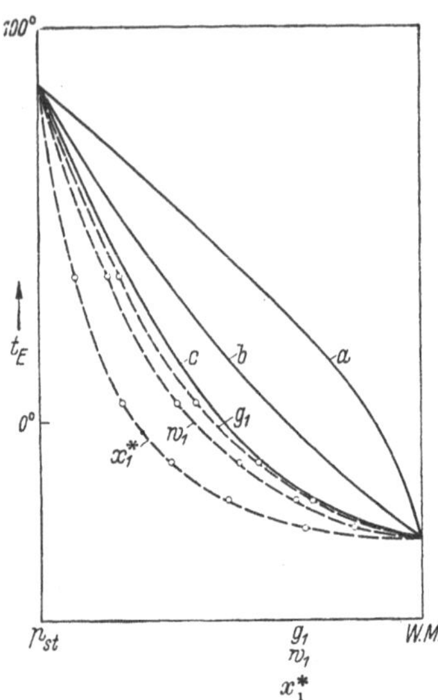

Abb. IX, 6. Vergleich berechneter und beobachteter Erniedrigungen der Einfriertemperatur durch einen Weichmacher in verschiedenen Konzentrationsmaßen. Kurve a) berechnet nach Gl. (IX, 28), Kurve b) berechnet nach Gl. (IX, 15), Kurve c) berechnet nach Gl. (IX, 53 u. 54); sämtlich wahlweise in Gew.-%, Volumenanteilen oder Grundmolenbrüchen. Gestrichelte Kurve: Beobachtete Werte im System Polystyrol-β-Naphthylsalicylat. In Gewichtsanteilen $g_1 = c_{g_1}$, in Volumenanteilen $w_1 = c_{v1}$, in Grundmolenbrüchen x_1^*,

Damit wird schließlich aus der ursprünglichen Viscositätsformel die Gleichung

$$\ln \eta = A' + \frac{Q}{RT} + C\sqrt{c_{g_2} P_2}. \qquad \text{(IX, 23)}$$

Wenn man nun mit BOYER u. SPENCER[3] den konstanten Wert η_E einführt, wird die zugehörige Temperatur die Einfriertemperatur. Damit wird aus Gl. (IX, 23)

$$\left(\ln \eta_E - A' - C\sqrt{c_{g_2} P_2}\right)\frac{R}{Q} = \frac{1}{T_E}. \qquad \text{(IX, 24)}$$

Wendet man diese Gleichung auf das reine Polymerisat an ($c_{g_2} = 1$), so erhält man

$$\left(\ln \eta_E - A' - C\sqrt{P_2}\right)\frac{R}{Q} = \frac{1}{T_{E_2}}, \qquad \text{(IX, 25)}$$

---

[1] FLORY, P. J.: J. Phys. Chem. **46**, 870 (1942).

[2] Gl. (IX, 20) enthält die gleiche Temperaturabhängigkeit wie Gl. (IX, 9): $A' + C \cdot \sqrt{P} = A$.

[3] BOYER, R. F. u. R. S. SPENCER: J. Pol. Sci. **2**, 157 (1947).

und zieht man die beiden letzten Gleichungen voneinander ab, so ergibt sich schließlich

$$\frac{1}{T_E} = \frac{1}{T_{E2}} + \left(1 - \sqrt{c_{g_2}}\right) \frac{R}{Q} C \cdot \sqrt{P_2} \qquad \text{(IX, 26)}$$

oder, weil $\left(1 - \sqrt{c_{g_2}}\right) \approx -\log c_{g_2}$ (BRIGGSscher Log.), wenn $1 > c_{g_2} > 0{,}4$,

$$\frac{1}{T_E} \approx \frac{1}{T_{E_2}} - (\log c_{g_2}) \cdot \frac{R}{Q} \cdot C \sqrt{P_2}\,. \qquad \text{(IX, 27)}$$

Die Konstanten dieser Gleichung sollten für ein und dasselbe Polymerisat und für alle Lösungsmittel entsprechend der Herleitung der Formel den gleichen Wert haben. Mit anderen Worten: ein und dasselbe Polymerisat sollte durch die gleichen Gewichtsanteile der verschiedensten Lösungsmittel in der gleichen Weise in seiner Einfriertemperatur erniedrigt werden. Das ähnelt der Regel von ZHURKOW (s. u.). Verzichtet man, wenn auch eigentlich etwas inkonsequent, auf diese Forderung, indem man in Gl. (IX, 26) für den reinen Weichmacher ($c_{g_2} = 0$) anwendet, so kann man den Ausdruck $R/Q \cdot C \sqrt{P_2}$ für jeden Weichmacher angeben. Die Gl. (IX, 26) läßt sich dann umformen in

$$\frac{1}{T_E} = \frac{1}{T_{E_1}} \left(1 - \sqrt{c_{g_2}}\right) + \frac{1}{T_{E_2}} \cdot \sqrt{c_{g_2}}\,. \qquad \text{(IX, 28)}$$

Danach ist die reziproke Einfriertemperatur der Lösung additiv aus den reziproken Einfriertemperaturen der beiden Komponenten zusammengesetzt, wobei jedoch die Gewichte der beiden Anteile in den merkwürdigen Größen $1 - \sqrt{c_{g_2}}$ und $\sqrt{c_{g_2}}$ ausgedrückt sind. Die mangelnde Symmetrie dieser Werte erscheint von vornherein sehr unbefriedigend. — Letzten Endes entstehen alle diese Schwierigkeiten, weil der Ausdruck $\sqrt{P}$, der für Hochpolymere sinnvoll mit ihrer Knäueleigenschaft zusammenhängt, in Gl. (IX, 21) in ganz formaler Weise auch auf den niedermolekularen Weichmacher angewandt wird.

Zu einem ähnlichen Ergebnis kommt man durch Anwendung des Ansatzes nach Gl. (IX, 8). Wenn man von den sehr verdünnten Lösungen absieht, kann man in Gl. (IX, 19) den ersten Summanden vernachlässigen und erhält

$$\frac{R}{Q} \left[\ln \eta_{2\,298} - \ln c_{g_2} \eta_{2\,298}\right] + \frac{1}{T_{E_2}} = \frac{1}{T_E} \qquad \text{(IX, 30)}$$

woraus sich sofort ergibt:

$$-\frac{R}{Q} \cdot \ln c_{g_2} + \frac{1}{T_{E_2}} = \frac{1}{T_E}\,, \qquad \text{(IX, 31)}$$

in der Form gleichwertig mit Gl. (IX, 27).

Entsprechende Betrachtungen gelten wieder, wenn man den Grundmolenbruch oder den Volumenbruch einführt.

Die Kurve *a* der Abb. IX, 6 wurde nach Gl. (IX, 28) berechnet; sie weicht stark von der Beobachtung ab. Gl. (IX, 26) wurde auch von BOYER u. SPENCER[1] an den Messungen im System von Nitrocellulose und Äthyl-

---

[1] BOYER, R. F. u. R. S. SPENCER: s. S. 584.

cellulose mit Trikresylphosphat, Glukose mit Glycerin und Polystyrol mit monomerem Styrol geprüft. Die Beobachtungen genügen der Gl. (IX, 26) jedoch nur bis höchstens 20% Weichmacher und weichen darüber hinaus stark ab. Bessere Übereinstimmung finden BOYER und SPENCER an weichgemachtem Polystyrol und Polyvinylchlorid[1].

d) Die Betrachtung von ZHURKOW[2].

Nach ZHURKOW ist die chemische Natur eines Weichmachers unwesentlich für seine Wirksamkeit, vgl. S. 576. Hierfür wird folgende Ableitung gegeben.

Zur Charakterisierung der Einfriertemperatur geht ZHURKOW auf ältere Vorstellungen zurück, nach denen diese Temperatur durch eine Aggregation der Moleküle in einem bestimmten Ausmaß gekennzeichnet ist. Diese Aggregation wird durch Bindungen veranlaßt, die mit sinkender Temperatur immer zahlreicher, mit zunehmendem Lösungsmittelgehalt infolge Solvatbildung immer geringer werden. Da nur die Bindungen zwischen den Kettenmolekülen von ZHURKOW als wesentlich für die Viscosität und damit die Einfriertemperatur betrachtet werden, nicht aber diejenigen zu den Lösungsmittelmolekülen und zwischen den letzteren untereinander, so ergibt sich am Ende der Betrachtung eine Erniedrigung der Einfriertemperatur, welche nicht von der Natur des Lösungsmittels abhängt. Wahrscheinlich ist die Vernachlässigung dieser Bindungen der Grund für die recht beschränkte Gültigkeit der ZHURKOWschen Regel.

Im einzelnen geht ZHURKOW wie folgt vor: Die Aggregation wird als Gleichgewicht beschrieben

$$\text{Kettenmolekül} + \text{Kettenmolekül} \rightleftharpoons (\text{Kettenmolekül})_2,$$

auf die das Massenwirkungsgesetz anwendbar ist. Die Gesamtzahl der Grundmoleküle werde mit n_2^*, die Zahl der Aggregationsbindungen im reinen Polymeren (bei der Temperatur T) mit Z_0 bezeichnet und daher die Zahl der „freien“ Grundmoleküle mit $n_2^* - 2Z_0$. Das Massenwirkungsgesetz ergibt dann

$$\frac{Z_0}{(n_2^* - 2Z_0)^2} = A \cdot e^{-\frac{\Delta H}{RT}}, \tag{IX, 32}$$

wobei ΔH die „Reaktionswärme“ bedeute. Im weichgemachten Polymerisat seien n_1 Weichmachermoleküle enthalten, die sämtlich an je ein Grundmolekül gebunden sein sollen. Jetzt betrage die Zahl der Verknüpfungen Z und die der „freien“ Grundmoleküle dementsprechend $n_2^* - n_1 - 2Z$. Hierauf das Massenwirkungsgesetz angewandt, ergibt

$$\frac{Z}{(n_2^* - n_1 - 2Z)^2} = A \cdot e^{-\frac{\Delta H}{RT}}. \tag{IX, 33}$$

Bei gleicher Viscosität, z.B. bei der Einfriertemperatur, soll nun die Zahl der Verknüpfungen einander gleich sein, $Z_0 = Z$. Damit wird in

---

[1] BOYER, R. F. u. R. S. SPENCER, J. Appl. Phys. **16**, 594 (1945); vgl. auch D. FAULKNER, Brit. Plast. **23**, 183 (1950).

[2] ZHURKOW, S. N.: C. R. Acad. Sci. URSS **47**, 475 (1945).

Gl. (IX, 32) $T = T_{E_2}$ und in Gl. (IX, 33) $T = T_E$. Man erhält durch Division beider Gleichungen

$$\left(\frac{1}{1-\frac{n_1}{n_2^* - 2Z}}\right)^2 \approx \frac{1}{x_2^{*2}} = e^{-\frac{\Delta H}{R}\left(\frac{1}{T_E} - \frac{1}{T_{E_2}}\right)} \qquad \text{(IX, 34)}$$

oder logarithmiert

$$\ln x_2^* = \frac{\Delta H}{R}\left(\frac{1}{T_E} - \frac{1}{T_{E_2}}\right) \approx \frac{\Delta H}{R} \cdot \frac{T_{E_2} - T_E}{T_{E_2}^2}. \qquad \text{(IX, 35)}$$

Danach sollte also die Erniedrigung der Einfriertemperatur bei allen Weichmachern die gleiche sein, gleicher Grundmolenbruch vorausgesetzt[1,4].

e) Weichmacherwirkung und Extrahierbarkeit.

Aiken, Alfrey, Janssen u. Mark[2] haben experimentell gezeigt, daß Weichmacher um so schneller mit Mineralöl aus dem weichgemachten Polymerisat zu extrahieren sind, je wirksamer sie sind. Nach diesen Autoren ist die Parallele zwischen Weichmacherwirksamkeit und Extrahierbarkeit viel besser erfüllt als diejenige zwischen Wirksamkeit und Viscosität des reinen Weichmachers im Sinne von Leilich. Ebenso scheinen wirksame Weichmacher aus dem Polymerisat leicht abzudunsten, schlecht wirksame nur langsam. In beiden Fällen wird der Verlust an Weichmacher durch dessen mehr oder weniger schnelle Diffusion (vgl. § 81) innerhalb des Polymerisats gesteuert.

Der hier beschriebene Zusammenhang ist jedoch eine Selbstverständlichkeit. Bei der Einfriertemperatur besteht etwa die gleiche Viscosität von 10^{13} abs. Einh. Auch im eingefrorenen Glas nimmt die Viscosität mit sinkender Temperatur noch zu, wenn auch mit kleinerem Temperaturkoeffizienten als in der Schmelze. Stark wirksame Weichmacher setzen die Einfriertemperatur stark herab; daher befindet sich das Material bei Zimmertemperatur in der Nähe der Einfriertemperatur und hat eine verhältnismäßig kleine Viscosität. Wenig wirksame Weichmacher setzen die Einfriertemperatur wenig herab, bei der Abkühlung bis auf Zimmertemperatur ist daher die Viscosität stark angestiegen. Die geringe Viscosität bewirkt hohe Diffusionsgeschwindigkeit und umgekehrt.

§ 77. Weichmacherwirkung und Güte des Lösungsmittels[3].

Als „gute“ oder „schlechte“ Lösungsmittel seien solche bezeichnet, deren Aktivität a_1 in der Lösung kleiner bzw. größer als ihr Molenbruch x_1 ist. (Hierbei ist die Aktivität a_1 beispielsweise durch das Dampfdruck-

[1] Zhurkow verlangt hier gleiches Grundmolverhältnis n_1/n_2^* infolge einer anderen Näherung.

[2] Aiken, W., T. Alfrey, A. Janssen u. H. Mark: J. Pol. Sci. **2**, 178 (1947).

[3] Vgl. dazu Bd. II, Kap. 2.

[4] Über eine Ableitung der Zhurkowschen Gleichung aus einer Viscositätsbetrachtung vgl. E. Jenckel. Koll. Z. **120**, 170 (1951).

verhältnis $a_1 = p_1/p_{01}$ definiert.) Es läßt sich nun zeigen, daß die Aktivität immer durch eine Reihe von der Form

$$\ln a_1 = \ln x_1 + \alpha\, x_2^2 + \cdots \tag{XI, 36}$$

beschrieben werden kann, in der das in x_2 lineare Glied fortfällt und α eine noch von der Temperatur abhängige Konstante bedeutet[1]. In α steckt die Lösungswärme und die über den idealen Wert hinausgehende Lösungsentropie.

Wir brechen für das folgende die Reihe mit dem quadratischen Glied ab.

Für hochmolekulare Lösungen verwendet man zweckmäßigerweise an Stelle des Molenbruches x_1 den Grundmolenbruch x_1^* definiert durch

$$x_1^* = \frac{x_1}{x_1 + P x_2} \quad \text{oder} \quad x_1 = \frac{x_1^* P}{x_1^* P + x_2^*}. \tag{IX, 37}$$

Man erhält

$$\ln a_1 = \ln x_1^* + \ln \frac{P}{x_2^* + P x_1^*} + \alpha \frac{x_2^{*2}}{(P x_1^* + x_2^*)^2} \tag{IX, 38}$$

oder mit einer bekannten Näherung[2] für verdünnte Lösungen

$$\ln a_1 = \ln x_1^* + \left(1 - \frac{1}{P}\right) \cdot x_2^* + \chi\, x_2^{*2}, \tag{IX, 39}$$

mit

$$x = \frac{1}{2}\left(1 - \frac{1}{P}\right)^2 + \frac{\alpha}{P^2}$$

wenn man die Konstanten in eine einzige Konstante χ zusammenfaßt. Diese Formel wurde von HUGGINS in etwas anderer Weise abgeleitet[3]. Werte von χ sind besonders mit Hilfe von Messungen des osmotischen Druckes und der Quellung bestimmt worden (vgl. auch Bd. II, § 30). Große positive χ-Werte charakterisieren schlechte Lösungsmittel bzw. Nichtlöser.

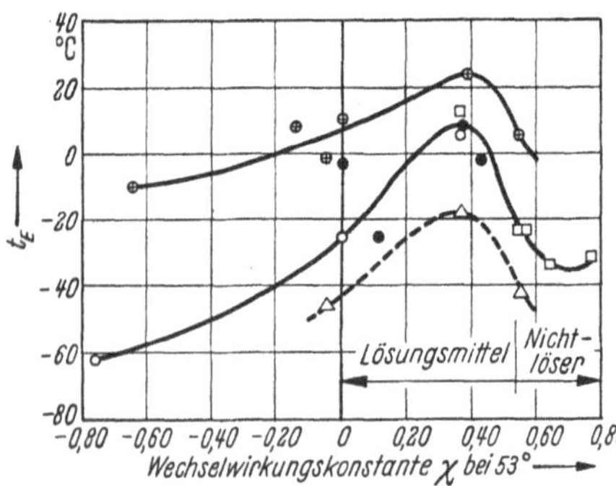

Abb. IX, 7. Zusammenhang der Wechselwirkungskonstanten mit der Weichmacherwirksamkeit. (Nach BOYER u. SPENCER.)

○ = Vinylite VYNW mit 30% Weichmacher, Brittle-Point-Messung
□ = desgl.
● = desgl., Volumen-Messung
△ = Polyvinylchlorid mit 40% Weichmacher, Brittle-Point-Messung.
⊕ = desgl. mit 33% Weichmacher, Messung der Plast. Deformation

Nach BOYER und SPENCER[4] soll nun ein Zusammenhang zwischen der Wechselwirkungskonstanten und der Weichmacherwirkung existieren. In Abb. IX, 7 ist die Einfriertemperatur von Lösungen des Polyvinylchlorids

---

[1] HAASE, R.: Z. Naturf. **8a**, 380 (1953).

[2] $$\ln \frac{P}{x_1^* P + x_2^*} = -\ln \frac{x_1^* P + x_2^*}{P} = -\ln\left(x_1^* + \frac{x_2^*}{P}\right)$$
$$= -\ln\left[1 - x_2^*\left(1 - \frac{1}{P}\right)\right] \approx + x_2^*\left(1 - \frac{1}{P}\right) + \frac{1}{2} x_2^{*2}\left(1 - \frac{1}{P}\right)^2 + \cdots.$$

[3] HUGGINS, M. L.: Ann. N. Y. Acad. Sci. **44**, 431 (1943).

[4] BOYER, R. F. u. R. S. SPENCER: J. Pol. Sci. **2**, 157 (1947).

mit konstantem Gehalt an Weichmacher (von 30 bis 40 Gew.-%) gegen die Wechselwirkungskonstante aufgetragen. Es ergibt sich, daß gute Lösungsmittel (große negative χ-Werte) die Einfriertemperatur stark erniedrigen, während schlechte Lösungsmittel mit einem Werte der Konstanten χ von $+ 0{,}4$ sie nur schwach erniedrigen[1].

Kritisch ist zu dieser Betrachtungsweise zu sagen: Da sich die Einfriertemperaturen monoton von der einen Komponente zur anderen ändern, so kann ein etwaiger Einfluß der Konstante χ, d.h. der Güte des Lösungsmittels, notwendigerweise nur einen Begleiteffekt neben demjenigen der Differenz in der Einfriertemperatur ΔT_E darstellen.

Um zu überzeugen, müßte die Regel jedenfalls an weiteren Lösungen geprüft werden. Im übrigen ist kürzlich ein Gegenbeispiel bekannt geworden. Toluol, Äthylbenzol und Chlorbenzol erniedrigen bei gleicher grundmolarer Konzentration die Einfriertemperatur des Polystyrols um fast genau den gleichen Betrag; nach Abzug der Einfrierwärme[2] verbleibt jedoch eine echte exotherme Lösungswärme für Chlorbenzol (gutes Lösungsmittel), während sie bei Toluol und Äthylbenzol verschwindet (athermische Lösung $\chi \approx 0{,}5$)[3].

§ 78. Die Einfriertemperatur der Lösung als Schnittkurve der Volumenflächen des Glases und der Schmelze[4].

Kennt man das Volumen der Gleichgewichtsschmelze in Abhängigkeit von Temperatur und Zusammensetzung und ebenso das Volumen des eingefrorenen Glases, so kann man leicht die Schnittkurve der beiden Flächen angeben, welche die Einfriertemperatur in Abhängigkeit von der Zusammensetzung darstellt.

a) Die zusätzliche Volumenkontraktion in den glasigen Lösungen.

Jenckel u. Heusch haben die Brechungsindizes von Schmelzen und Gläsern bestimmt, insbesondere von Lösungen des Polystyrols, und als niedermolekulares Gegenstück des Phenolphthaleins mit solchen Weichmachern, die selbst zur glasigen Erstarrung zu bringen sind, und die daher die Herstellung von Mischgläsern über den ganzen Konzentrationsbereich ermöglichen (vgl. Abb. IX, 1). Es wurde nun bei konstanter Temperatur der Brechungsindex in Abhängigkeit von der Zusammensetzung in Gewichtsanteilen betrachtet, und zwar aus Gründen der Zweckmäßigkeit,

---

[1] Merkwürdigerweise sollen die Lösungsmittel mit einem χ-Wert größer als 0,4 die Einfriertemperatur wieder stark erniedrigen. Wir möchten jedoch meinen, daß hier überhaupt keine Löslichkeit mehr stattgefunden hat, sondern daß das Material in Form zweier flüssiger Phasen vorliegt. An diesen kann man ohnehin nicht gut von der Einfriertemperatur *einer* Lösung sprechen (vgl. S. 575 u. 599). Zudem wurde zur Bestimmung der Einfriertemperatur die Methode des „Brittle-Point" verwandt, die für ein zweiphasiges Gemisch wahrscheinlich überhaupt keine Beziehung zu den Einfriertemperaturen der Phasen mehr hat.

[2] Jenckel, E. u. K. Gorke: Z. Naturforsch. 7a, 630 (1952).

[3] Jenckel, E. u. K. Gorke: Z. Naturforsch., demnächst.

[4] Jenckel, E. u. R. Heusch: Kolloid-Z. **130**, 89 (1953); E. Jenckel, Kunstst. **45**, 3 (1955).

für die Schmelzen bei 90° C und für die Gläser bei 0° C. Dabei wurde darauf geachtet, daß immer nur Werte für die Gleichgewichtsschmelze einerseits und für die eingefrorenen Gläser andererseits miteinander verglichen wurden. Diese scharfe Trennung wird bei KNAPPE u. SCHULZ[1] und bei BIRNTHALER[2] nicht beachtet. Die Werte wurden nötigenfalls durch eine Extrapolation erhalten, indem man den Brechungsindex des Glases zu höheren Temperaturen über die Einfriertemperatur hinaus extrapolierte oder umgekehrt das Volumen der Schmelze zu tieferen Temperaturen. Gegen dieses Verfahren können grundsätzliche Bedenken wohl nicht erhoben werden.

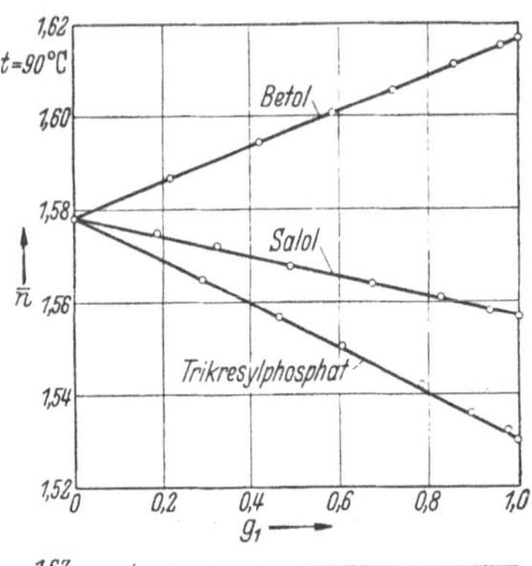

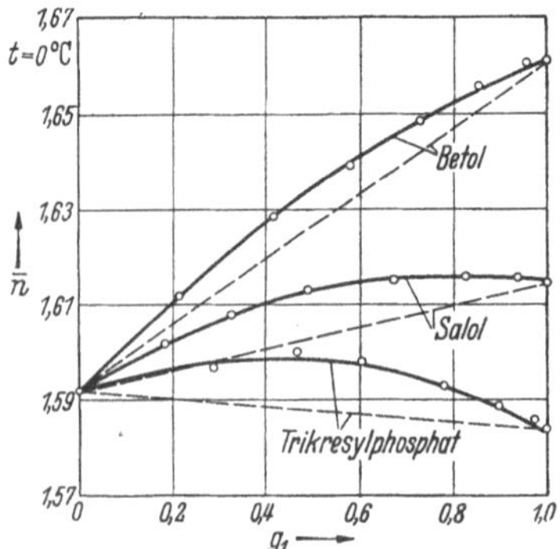

Abb. IX, 8. Der Brechungsindex in einigen Lösungen des Polystyrols. Obere Hälfte: Gleichgewichtsschmelzen. Untere Hälfte: Gläser. Die Schmelzen verhalten sich additiv, die Gläser zeigen eine zusätzliche Erhöhung des Brechungsindex, entsprechend einer Volumenkontraktion. (Nach JENCKEL u. HEUSCH.)

Der gemessene Brechungsindex n der Lösung läßt sich mit Hilfe der spezifischen Refraktion r in das spezifische Volumen v umrechnen

$$r = \frac{n^2 - 1}{n^2 + 2} \cdot \frac{1}{v}, \qquad \text{(IX, 40)}$$

für $n = 1{,}6$ ergibt sich hieraus leicht

$$\frac{\Delta v}{v} = 2{,}16 \frac{\Delta n}{n}. \qquad \text{(IX, 41)}$$

Es sei angenommen, daß sich die spezifische Refraktion r der Lösung additiv in Gewichtsanteilen aus spezifischen Refraktionen der beiden Komponenten r_1 und r_2 zusammensetzt

$$r = c_{g_1} r_1 + c_{g_2} r_2. \qquad \text{(IX, 42)}$$

Die Lösungen der Gleichgewichts*schmelzen* weisen in den untersuchten Beispielen einen Brechungsindex auf, der sich additiv aus dem der beiden Komponenten zusammensetzt, wie Abb. IX, 8 zeigt. (Grundsätzlich sind natürlich Abweichungen von der Additivität nach oben und nach unten möglich und auch beobachtet.) Die Berechnung des Volumens dieser additiven Brechungsindizes liefert zwar aus mathematischen Gründen keine additiven Volumina; die Abweichung vom additiven Verhalten ist jedoch gering, wie gezeigt werden konnte, so daß wir hier annehmen wollen, daß auch das Volumen sich additiv verhält. Für die Misch*gläser* dagegen beobachtet man für die gleichen Systeme eine Abweichung des Volumens von der Additivität nach unten; in den Mischgläsern beobachtet man also eine Volumenkontraktion. Die

[1] KNAPPE, W. u. A. SCHULZ: Kunststoffe **41**, 321 (1951).
[2] BIRNTHALER, W.: Kunststoffe **38**, 11 (1948).

Kontraktion in den Mischgläsern, die offenbar charakteristisch für den Glaszustand ist, ist besonders ausgeprägt in den polystyrolhaltigen Gläsern, wesentlich weniger ausgeprägt in den Gläsern, die an Stelle des Polymeren das niedermolekulare Phenolphthalein enthalten[1].

Gegen diese Schlußfolgerung ließe sich anführen, daß Gl. (IX, 42), obwohl aus der LORENZ-LORENTZ-Gleichung streng herzuleiten, doch nur eine Näherung an das wirkliche Verhalten darstellt. Ihre Gültigkeit kann zur Zeit nur aus Messungen an niedermolekularen Stoffen abgeschätzt werden. Nach TSCHAMMLER und Mitarbeitern[2] ist die Molrefraktion einer Lösung streng additiv, sogar in einer stark exothermen Lösung mit Volumenkontraktion und Ausbildung von Wasserstoffbrücken. Nach BÖTTCHER[3] betragen die Abweichungen der Refraktion von der Additivität 1 bis $2^0/_{00}$. Höchstens in diesem Ausmaß könnte bei additivem n das Volumen nicht additiv sein. Der Brechungsindex der Gläser in Abb. IX, 8 weicht um etwa $6{,}3^0/_{00}$ ab, entsprechend das Volumen um etwa $13{,}6^0/_{00}$, also um rund zehnmal mehr als man erwarten könnte. Wenn möglicherweise in hochmolekularen Lösungen die Refraktion stärker abweichen sollte (über $6^0/_{00}$ überadditiv in je einem guten, indifferenten und schlechten Lösungsmittel[4]), so bleibt immer noch zu erklären, warum in der Schmelze und im Glas sich n so verschieden verhält.

b) Volumenkontraktion und Einfriertemperatur.

Zur formelmäßigen Wiedergabe der Volumenkontraktion wurde ein Parabelansatz verwandt, so daß das Gesamtvolumen bzw. der Brechungsindex der Gläser folgendermaßen angegeben wird:

$$\left.\begin{aligned}\bar{v} &= c_{g_1}\bar{v}_1 + c_{g_2}\bar{v}_2 + \bar{k}_v c_{g_1}\cdot c_{g_2}\\ \bar{n} &= c_{g_1}\bar{n}_1 + c_{g_2}\bar{n}_2 + \bar{k}_n c_{g_1}\cdot c_{g_2}.\end{aligned}\right\}\qquad\text{(IX, 43)}$$

(überstrichene Werte beziehen sich auf den Glaszustand). Hierin ist die Größe $\bar{k}_v$ ein Maß für die Volumenkontraktion und $\bar{k}_n$ ein solches für die Abweichung von der Additivität des Brechungsindexes. Der Symmetrie des dritten Summanden folgen die experimentellen Kurven ziemlich gut.— Entsprechend gilt für die *Schmelze*

$$\left.\begin{aligned}v &= c_{g_1}v_1 + c_{g_2}v_2 + c_{g_1}\cdot c_{g_2}\cdot k_v\\ n &= c_{g_1}n_1 + c_{g_2}n_2 + c_{g_1}\cdot c_{g_2}\cdot k_n.\end{aligned}\right\}\qquad\text{(IX, 44)}$$

Wegen des besseren Anschlusses an die Versuche verwenden wir an Stelle des Volumens den Brechungsindex. Indem wir nun in Gl. (IX, 43) die Beziehungen

$$\begin{aligned}\bar{n}_1 &= \bar{\alpha}_1 + \bar{\beta}_1 T\\ \bar{n}_2 &= \bar{\alpha}_2 + \bar{\beta}_2 T\end{aligned}\quad\text{und}\quad \bar{\beta} = c_{g_1}\bar{\beta}_1 + c_{g_2}\bar{\beta}_2\qquad\text{(IX, 45)}$$

[1] Die bei BIRNTHALER (Fußnote 2, S. 590) mitgeteilte Volumenkontraktion bezieht sich auf das glasige Polymerisat und den flüssigen Weichmacher.

[2] TSCHAMMLER, H. u. Mitarbeiter: Mh. Chem. **79**, 394 (1948); **80**, 572 (1949);

[3] BÖTTCHER, C. F. J.: Theory of Electric Polarisation, Amsterdam 1952, S. 270.

[4] JENCKEL, E. u. G. REHAGE: Unveröffentlicht.

einführen, die nur besagen, daß der Temperaturkoeffizient β_1 bzw. β_2 der reinen Komponenten konstant ist und daß in der Lösung sich die Temperaturkoeffizienten additiv zusammensetzen, so erhält man

$$\bar{n} = c_{g_1}(\bar{\alpha}_1 + \bar{\beta}_1 T) + c_{g_2}(\bar{\alpha}_2 + \bar{\beta}_2 T) + \bar{k}_n c_{g_1} \cdot c_{g_2}. \qquad \text{(IX, 46)}$$

Wie erwähnt, ist $\bar{k}_n$ immer positiv, entsprechend einer Volumenkontraktion.

Eine analoge Gleichung gilt auch für die Gleichgewichtsschmelze

$$n = c_{g_1}(\alpha_1 + \beta_1 T) + c_{g_2}(\alpha_2 + \beta_2 T) + k_n c_{g_1} c_{g_2}. \qquad \text{(IX, 47)}$$

Die in den Schmelzen vorkommende Konstante k_n kann positiv oder negativ sein oder auch verschwinden wie in den Beispielen der Abb. IX, 8. Diese Konstante gibt also nur die gewöhnliche längst bekannte Volumenkontraktion oder Dilatation bei der Herstellung von Lösungen wieder. In Abb. IX, 9 sind die Volumenflächen von Schmelze und Glas in Abhängigkeit von Temperatur und Zusammensetzung schematisch wiedergegeben.

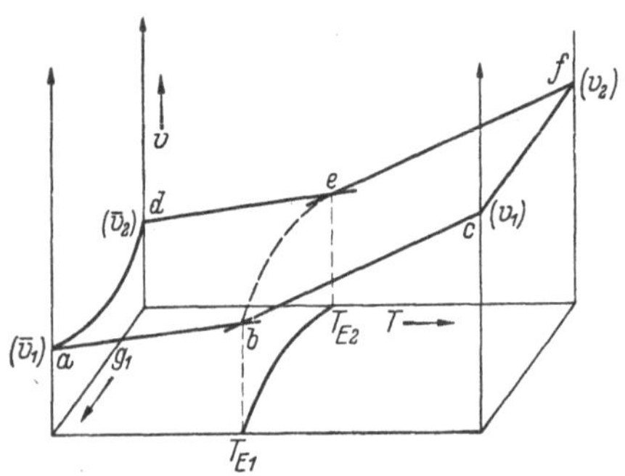

Abb. IX, 9. Volumen von Schmelzen und Gläsern über Zusammensetzung und Temperatur. Die fast ebene Fläche der Schmelze schneidet sich mit der gekrümmten Fläche der Gläser in einer Kurve. (Schematisch nach JENCKEL u. HEUSCH.)

Die Schnittkurve beider Flächen ergibt, projiziert in die T, c_g-Ebene, die Kurve der Einfriertemperaturen gegen die Zusammensetzung; sie ist durch die Bedingung $n = \bar{n} = n_E$ und $T = T_E$ gegeben. Man erhält

$$c_{g_1}(\Delta\alpha_1 + \Delta\beta_1 T_E) + c_{g_2}(\Delta\alpha_2 + \Delta\beta_2 T_E) + \Delta k_n \cdot c_{g_1} c_{g_2} = 0 \qquad \text{(IX, 48)}$$

mit den folgenden Abkürzungen:

$$\left.\begin{array}{ll} \Delta\alpha_1 = \bar{\alpha}_1 - \alpha_1; & \Delta\beta_1 = \bar{\beta}_1 - \beta_1 \\ \Delta\alpha_2 = \bar{\alpha}_2 - \alpha_2; & \Delta\beta_2 = \bar{\beta}_2 - \beta_2 \end{array} \quad \Delta k_n = \bar{k}_n - k_n. \right\} \qquad \text{(IX, 49)}$$

Aus den Grenzbedingungen für $c_{g_1} = 1$ und $c_{g_2} = 0$ folgt

$$\left.\begin{array}{l} \Delta\alpha_1 = -\Delta\beta_1 \cdot T_{E_1} \\ \Delta\alpha_2 = -\Delta\beta_2 \cdot T_{E_2}. \end{array}\right\} \qquad \text{(IX, 50)}$$

Hiermit wird aus Gl. (IX, 48)

$$T_E = \frac{1}{c_{g_1}\Delta\beta_1 + c_{g_2}\Delta\beta_2}\left[c_{g_1}\Delta\beta_1 T_{E_1} + c_{g_2}\Delta\beta_2 T_{E_2} - c_{g_1} c_{g_2} k_n\right]. \qquad \text{(IX, 51)}$$

Nimmt man noch an, daß

$$\Delta\beta_1 \approx \Delta\beta_2, \qquad \text{(IX, 52)}$$

eine Näherung, die meist hinreichend erfüllt ist, so wird aus Gl. (IX,51) der einfache Ausdruck

$$T_E = c_{g_1} T_{E_1} + c_{g_2} T_{E_2} - c_{g_1} c_{g_2} \cdot \frac{\Delta k_n}{\Delta \beta} . \tag{IX,53}$$

Danach setzt sich die Einfriertemperatur der Lösung additiv aus denjenigen der Komponenten zusammen, abzüglich eines Gliedes, in dem der Unterschied der Konzentrationskoeffizienten Δk und der Temperaturkoeffizienten $\Delta \beta$ steckt.

Die Beobachtungen gehorchen der Gl. (IX,53) in den hier betrachteten Systemen ohne Volumenänderung in der Schmelze ($k_n = 0$) recht weitgehend, wie Abb. IX, 6, Kurve c, zeigt. Bemerkenswerterweise findet man empirisch

$$\frac{\Delta k_n}{\Delta \beta} = T_{E_2} - T_{E_1} . \tag{IX,54}$$

Damit ist eine Formel gewonnen, die die Einfriertemperatur der Lösung lediglich aus denen der Komponenten zu berechnen gestattet. Es bleibt freilich zu prüfen, wie weit dieses Ergebnis allgemeingültig ist[1]. Unter dieser Voraussetzung jedoch würde in einer Lösung aus zwei Komponenten *gleicher* Einfriertemperatur diese nicht herabgesetzt werden und ferner würde auch die zusätzliche Volumenkontraktion der Gläser verschwinden, worauf wir sogleich noch einmal zurückkommen.

Aus Gl. (IX,53) erhält man die Weichmacherwirkung zu

$$\left(\frac{dT_E}{dc_{g_1}}\right)_{g_1=0} = -\left[(T_{E_2} - T_{E_1}) + \frac{\Delta k_n}{\Delta \beta}\right] \tag{IX,55}$$

oder in den hier betrachteten speziellen Fällen mit Gl. (IX, 54) zu

$$\left(\frac{dT_E}{dc_{g_1}}\right)_{g_1=0} = -2\,(T_{E_2} - T_{E_1}) . \tag{IX,56}$$

Auf der Seite des reinen Weichmachers läuft dann die Kurve waagerecht aus.

c) Zur molekularen Struktur weichgemachter Polymerisate.

Wie in Bd. III, § 55 für chemisch einheitliche Gläser ausführlich erläutert, müssen im Glase molekulare Hohlräume, „Löcher", entstehen, die dem zu großen Volumen eines Glases entsprechen. Die Kantenlänge dieser Löcher entspricht in hochmolekularen Gläsern ungefähr der Segmentlänge. Setzt man nun zu einer großen Menge eines Polymerisats mit verhältnismäßig langen Segmenten eine kleine Menge eines niedermolekularen Lösungsmittels, so werden dessen Moleküle in den Löchern verschwinden, ohne einen zusätzlichen Platz zu beanspruchen (Abb. IX,10). Das macht sich dann nach außen hin als eine Volumenkontraktion bemerkbar. Übrigens ist die beobachtete Kontraktion bei weitem nicht so

[1] Über eine Abschätzung von $\Delta k_n/\Delta \beta$ in anderen Systemen mit Volumenänderungen in der Schmelze ($k_n \neq 0$) vgl. E. JENCKEL u. R. HEUSCH: s. S. 489, Fußnote 4.

groß wie sich aus dem Schema der Abb. IX, 10 ergeben würde, in der nur die Tendenz angedeutet sein soll.

Diese Auffassung wird auch durch einige Messungen der spezifischen Wärme in weichgemachten Polymerisaten durch GAST[1] gestützt. Während in den reinen Polymerisaten die spezifische Wärme bei der Einfriertemperatur fast sprunghaft abfällt, erstreckt sich in den weichgemachten Polymerisaten der Abfall über einen viel größeren Temperaturbereich. Das weist unmittelbar darauf hin, daß in dem verhältnismäßig breiten Einfrierbereich der Lösungen sich zunächst noch bewegliche Anteile neben schon festgelegten Anteilen befinden, in völliger Übereinstimmung mit der eben skizzierten molekularen Vorstellung.

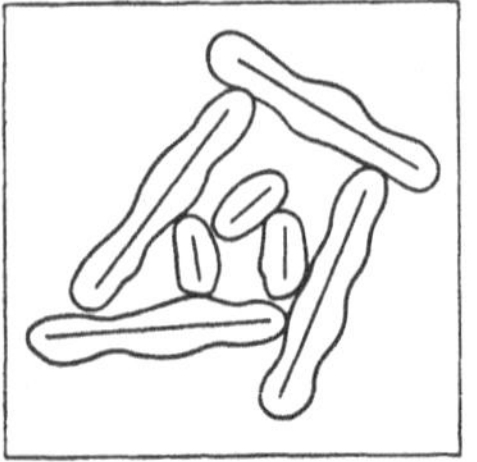

Abb. IX, 10. Struktur einer glasigen Lösung aus Hochpolymerem und niedermolekularem Weichmacher. Erklärung für die zusätzliche Volumenkontraktion in den weichgemachten Polymerisaten. (Schematisch nach JENCKEL u. HEUSCH.)

Schließlich sei auf eine elektronenoptische Aufnahme von SPURLIN, MARTIN u. TENNENT[2] hingewiesen, die BOYER u. SPENCER als Nachweis einer Zellenstruktur ansehen. Diese Auffassung stimmt mit der soeben vorgetragenen gut überein, wenn auch die Deutung solcher Aufnahmen immer unsicher bleibt.

Wir kommen noch einmal auf die Feststellung zurück, daß Stoffe gleicher Einfriertemperatur in ihrer Lösung keine Volumenkontraktion zeigen werden. Das ist nach dem Modell der Abb. IX, 10 nur möglich, wenn ihre Segmente gleich lang sind. Oder mit anderen Worten: In Stoffen gleicher Einfriertemperatur haben die Moleküle die gleiche Sperrigkeit[3].

§ 79. Der weichgemachte Stoff als kautschukelastisches Gel.

In der Technik wird häufig zwischen lösenden und quellenden Weichmachern unterschieden. Im Extremfall wird durch einen lösenden Weichmacher eine Lösung erhalten, die nur noch viscos fließt, dagegen kein elastisches Verhalten mehr zeigt. Durch einen quellenden Weichmacher dagegen erhält man ein mehr oder weniger kautschukähnliches Material, welches nach Fortnahme der wirkenden Spannung schneller oder langsamer in seine ursprüngliche Form zurückkehrt, also eine Gallerte, ein Gel. Gewöhnlich werden sogenannte quellende Weichmacher bevorzugt, weil die Einfriertemperatur der weichgemachten Stoffe unter Zimmertemperatur liegt und daher die eben erwähnten Unterschiede voll zur Auswirkung kommen.

Die Entstehung der Gallerte, die sogenannte Gelierung, hat tech-

[1] GAST: Kunststoffe **43**, 15 (1953). — E. JENCKEL u. H. KOPLIN: Noch unveröffentlicht.

[2] SPURLIN, H. M., A. F. MARTIN u. H. G. TENNENT: J. Polymer Sci. **1**, 63 (1946).

[3] Danach sollten kugelförmige Moleküle überhaupt nicht glasig einfrieren können, was allerdings experimentell nicht nachzuprüfen ist, da diese Stoffe leicht kristallisieren.

nisches Interesse. Zur Untersuchung wird in Abhängigkeit von der Zeit das Drehmoment eines kleinen Kneters, in dem sich z.B. Polyvinylchloridpulver und der Weichmacher befinden, registriert[1]. Anfänglich beansprucht die grobe Suspension von Pulver in Weichmacher beim Kneten nur ein geringes Drehmoment. Bei Weichmachern, welche nicht lösen bzw. quellen, bleibt dies kleine Drehmoment unverändert erhalten. Solche „Weichmacher" sind im Sinne dieses Kapitels eigentlich gar keine Weichmacher, da sie nicht lösen. Weichmacher, welche die Masse der Substanzen lösen, die Vernetzungen jedoch nicht angreifen – und das sind die wesentlichen und technisch brauchbaren Weichmacher –, führen nach einiger Zeit zur Quellung, erkennbar an einer erheblichen Steigerung des Drehmomentes. Nach vollendeter Quellung bleibt das Drehmoment konstant. Die unterschiedliche Geliergeschwindigkeit verschiedener Weichmacher, besonders auch bei verschiedenen Temperaturen, läßt sich so gut feststellen.

Eine sorgfältige Untersuchung haben EHLERS u. GOLDSTEIN[2] mitgeteilt, welche sich freilich auf sehr dünne Lösungen von 0,5 bis 2 Gew.-% Polymerisat, Polyvinylchlorid, beschränkt. Dennoch können wahrscheinlich ihre Ergebnisse auf die sehr viel höheren Konzentrationen an Polymerisat in den technischen Werkstücken übertragen werden. Sie erhitzen die Suspension von Pulver und Weichmacher und untersuchten dabei fortlaufend die Viscosität in einem Kapillarviscosimeter. Beim Erhitzen steigt die Viscosität zunächst nur sehr wenig (Kurvenstück *ab* in Abb. IX, 11), dann bei einer bestimmten von Weichmacher zu Weichmacher verschiedenen Temperatur stark an (*bc*), mit weiter steigender Temperatur nimmt die Viscosität zunächst stark (*cd*), dann langsam (*de*) wieder ab. Kühlt man jetzt wieder ab, so verläuft die Viscosität längs der Kurve *edf*. Der überraschende Kurvenverlauf längs *abcde* ist im wesentlichen durch die Temperaturen, nicht durch die Erhitzungsgeschwindigkeit bestimmt. Verfolgt man nämlich die Viscosität bei konstanter Temperatur mit der Zeit, so ändert sie sich nur verhältnismäßig wenig, und zwar nimmt auf dem Kurventeil *bc* die Viscosität ein wenig ab, auf dem Kurventeil *cde* ein wenig zu. Das gilt für Mono- und o-Dichlorbenzol und zwei technische Weichmacher, die

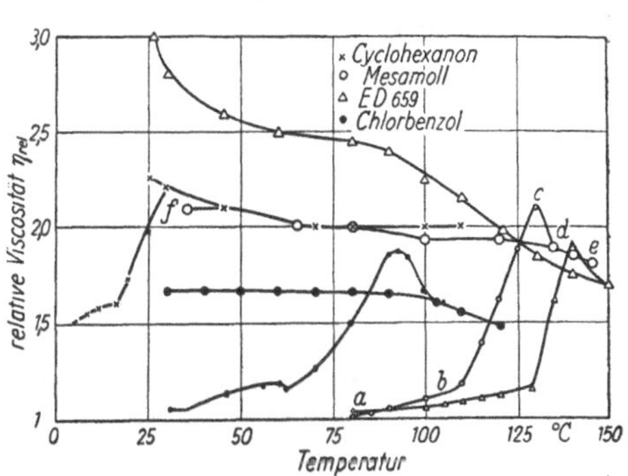

Abb. IX, 11. Scheinbare Viscosität von Aufschlemmungen des Polyvinylchlorids in Weichmacher in Abhängigkeit von der Temperatur. (Nach EHLERS u. GOLDSTEIN.)

[1] SCHMIDT, P.: Kunststoffe **41**, 23 (1951). – H. S. BERGEN u. J. R. DARBY: Ind. Engng. Chem. **43**, 2404 (1951).

[2] EHLERS, J. F. u. K. R. GOLDSTEIN: Kolloid-Z. **118**, 137, 151 (1950). Vgl. a. F. WÜRSTLIN u. H. KLEIN: Kunstst. **42**, 445 (1952); A. WESP, Kunstst. **41**, 213 (1951).

verhältnismäßig schlechte Lösungsmittel darstellen. Mit einem besseren Lösungsmittel, Cyclohexanon, wird einfach bei einer verhältnismäßig tiefen Temperatur, nämlich 30°, die Kurve der Lösung (*edf*) erreicht, ohne daß ein Maximum der Viscosität auftritt. Sehr bemerkenswert erscheint noch folgende Beobachtung. Kühlt man die Lösung aus dem Kurvenbereich (*bc*) ab, so scheiden sich beim Abkühlen nach einiger Zeit Flocken ab, die sedimentieren. Kühlt man dagegen aus dem Kurventeil *cde* ab, so bildet sich, je nach Weichmacher, ein klares oder trübes Gel, welch letzteres zum Ausschwitzen *(Synärese)* neigt.

Nach EHLERS u. GOLDSTEIN ist längs des Kurventeiles *bc* die Auflösung nur unvollkommen. Es bilden sich „Gelflocken". Diese führen einerseits zu der sehr hohen Viscosität beim Viscositätsmaximum, andererseits bewirken sie, daß beim Abkühlen sich wieder Flocken ausscheiden. Längs des Kurventeils *cd* ist die Auflösung schon fast vollkommen geworden, sie wird erreicht im Punkte *d*. Wenn jetzt beim Abkühlen wieder eine Ausscheidung stattfindet, bildet sich das Gel. Die letztere Annahme ist insofern nicht recht überzeugend, als sich geeignete Lösungen des Polystyrols[1], des Polymethacrylesters[2] und des Polyvinylchlorids[3] beim Abkühlen in jedem Fall durch die sich ausscheidende zweite flüssige Phase trüben, ohne daß eine Gallertbildung auftritt. Auch von EHLERS u. GOLDSTEIN wurden klare und trübe Gallerten beobachtet. Man kann daher, wie mir scheint, die Bildung der Gallerte nicht mit der besseren oder schlechteren Lösungsfähigkeit des Lösungsmittels in Verbindung bringen[4].

Hier sei vielmehr auf eine andere Deutung hingewiesen. Wie die Erfahrung lehrt, führen, im großen gesehen, der Zusatz von Weichmacher und die Erhöhung der Temperatur zu dem gleichen Ergebnis, nämlich zu einer Verbesserung der inneren Beweglichkeit. Untersucht man nun das reine weichmacherfreie Polyvinylchlorid, so findet man hier bereits das typische Kennzeichen des Gels, nämlich kautschukartiges Verhalten, sobald man über die Einfriertemperatur erhitzt hat, im Gegensatz etwa zum Polystyrol[5], bei dem dieses Verhalten kaum beobachtet wird und für das auch die Technik keinen quellenden Weichmacher entwickelt hat. Da, wie erwähnt, die weichgemachten Produkte im wesentlichen das gleiche Verhalten zeigen wie das reine Polymerisat, nur bei tieferer Temperatur, so muß man annehmen, daß auch im weichgemachten Polyvinylchlorid die Vernetzung nicht durch den Weichmacher erzeugt, sondern bereits vorher im Polymeren vorhanden war. Entsprechend diesem Gedankengange hat JENCKEL[6] schon früher bemerkt, daß man nicht von quellenden Weichmachern, sondern besser von quellbaren Polymerisaten sprechen sollte.

Kautschukelastisches Verhalten wird bekanntlich auf die Existenz

[1] JENCKEL, E.: Z. Naturforsch. **3**a, 290 (1948). – E. JENCKEL u. G. KELLER: Z. Naturforsch. **5**a, 317 (1950).

[2] JENCKEL, E. u. K. GORKE: Z. Naturforsch. **5**a, 556 (1950). – E. JENCKEL u. J. DELAHAYE: Z. Naturforsch. **7**a, 682 (1952).

[3] JENCKEL, E. u. K. H. SCHWERING: Unveröffentlicht.

[4] Vgl. auch F. WÜRSTLIN u. H. KLEIN: Z. Makromol. Chem. **16,** 1 (1955).

[5] JENCKEL, E. u. E. KLEIN: Z. Naturforsch. **7**a, 619 (1952).

[6] JENCKEL, E.: Kunststoffe **42**, 1 (1952).

von zwei „Bindungsmechanismen" im Material zurückgeführt, von denen der eine mit einem weitmaschigen Netzwerk verknüpft ist, der andere dagegen mit den (schwächeren) VAN DER WAALSschen Bindungen von Grundmolekül zu Grundmolekül (makro-BROWNsche und mikro-BROWNsche Beweglichkeit nach KUHN).

Die Entstehung des Netzwerkes kann offenbar nicht chemischer Natur sein, wie etwa im System Styrol/Divinylbenzol, denn bei Behandlung mit einem geeigneten[1] Lösungsmittel, etwa Tetrahydrofuran, erhält man beliebig verdünnte Lösungen.

Es bleibt eigentlich nur eine Vernetzung übrig durch gelegentliche[2] Assoziation über Dipole an Stellen des Kettenmoleküls, an denen der chemische Aufbau anomal und im übrigen ungeklärt ist, oder aber die Vernetzung durch kristalline Bereiche. Auch die letztere Möglichkeit kann nicht ausgeschlossen werden, ist es doch bekannt, daß Polyvinylchlorid im gereckten Zustand deutlich kristalline Struktur (Faserdiagramm) zeigt.

Nachdem man weiß, daß verschiedene Polymere je nach ihrer chemischen Konstitution in größerem oder kleinerem Ausmaße kristallisieren, ist es auch denkbar, daß andere Polymere, wie etwa das Polyvinylchlorid, in ungerecktem Zustande beispielsweise nur zu etwa 1% kristallisieren[3], wodurch bereits das kautschukelastische Verhalten einer Gallerte hervorgerufen werden dürfte. – Wie nun auch die Natur der Vernetzung sei, auf jeden Fall soll es sich nur um Vernetzungen zwischen den Kettenmolekülen handeln. Mit steigender Temperatur wird die Zahl der Vernetzungsstellen abnehmen und schließlich ganz verschwinden (es werden sich die Vernetzungen auflösen), während umgekehrt bei sinkender Temperatur wieder Vernetzungen auftreten werden, wenigstens, wenn man hinreichend lange wartet. (Über das Schmelzen und Kristallisieren hochmolekularer Lösungen vgl. Bd. III, Kap. VIII.) Auch die chemische Natur des Lösungsmittels wird wichtig sein, namentlich bei Dipolassoziation. Gute[4] Lösungsmittel werden schon bei ziemlich tiefer Temperatur die letzten Vernetzungen zum Verschwinden bringen, schlechte erst bei höherer Temperatur, ganz entsprechend den Beobachtungen von EHLERS u. GOLDSTEIN.

In ähnlicher Weise sind vielleicht Versuche von AICKEN, ALFREY, JANSSEN u. MARK[5] zu verstehen. Sie bestimmten sogenannte Kriechkurven, d.h., sie verfolgten die zeitliche Dehnung des Materials (in Form von Streifen) nach Belastung mit einem Gewicht. Die Versuche wurden durchgeführt am Polyvinylchlorid (Vinylit). Nach Fortnahme der Belastung schrumpfen im allgemeinen die gedehnten Kurven langsam, jedoch vollständig wieder zurück. Ohne an dieser Stelle auf das mecha-

[1] Die meisten Lösungsmittel werden nur beschränkt unter Quellung aufgenommen. — [2] Wegen der *weit*maschigen Vernetzung.

[3] Röntgenographischer Nachweis bei T. ALFREY JR., N. WIEDERHORN, R. STEIN u. A. V. TOBOLSKY: Ind. Engng. Chem. **41**, 701 (1949); J. Polymer Sci. **4**, 211 (1949).

[4] Vgl. auch über die „Güte" der Lösungsmittel: G. REHAGE: Z. Elektrochem. **59**, 78 (1955).

[5] AIKEN, W., T. ALFREY, A. JANSSEN u. H. MARK: J. Polymer Sci. **2**, 178 (1947). – Vgl. auch A. T. WALTER: J. Polymer Sci. **13**, 207 (1954).

nische Verhalten im einzelnen einzugehen, sei nur auf einen Unterschied der verschiedenen Weichmacher in der Form der Kriechkurve hingewiesen. Geringes zeitliches Fließen (Kriechen) wurde beobachtet bei Weichmachern mit aliphatischen Ketten, dagegen ausgeprägtes Kriechen bei Weichmachern mit Ringstrukturen. Das unterschiedliche Verhalten wurde besonders ausführlich an den Weichmachern Tri*oktyl*phosphat (flache Kriechkurven) und Tri*kresyl*phoshat (steile Kriechkurven) untersucht. Auf zwei von den Verfassern gebrachte Erklärungen sei hier nur hingewiesen, ohne sie im einzelnen zu diskutieren. JENCKEL[1] weist darauf hin, daß vermutlich von dem schlechten Lösungsmittel Trioktylphosphat die kristallinen Bereiche nicht gelöst werden und daher eine verhältnismäßig enge Vernetzung erhalten bleibt, somit hoher E-Modul, geringe makro-BROWNsche Beweglichkeit. Jedoch wird die Einfriertemperatur scharf herabgesetzt infolge der beweglichen aliphatischen Kohlenwasserstoffkette, also geringe Viscosität, hohe mikro-BROWNsche Beweglichkeit. Daher erreicht die Dehnung schnell einen niedrigen Endwert, ändert sich dann aber nicht weiter. Das ist das technisch erwünschte Verhalten. Mit dem guten Lösungsmittel Trikresylphosphat könnte dagegen wie in einer eutektisch schmelzenden Legierung die Zahl der kristallinen Bereiche (oder der Assoziate) geringer geworden sein und daher die Vernetzung viel weitmaschiger; das Material läßt sich daher über größere Beträge dehnen. Andererseits ist für diesen Weichmacher mit aromatischem Kern vielleicht eine höhere Einfriertemperatur zu erwarten, so daß die Dehnung sich erst in langer Zeit ausbildet. Daher dehnt das Material sich nur langsam, aber stetig, um erst nach sehr langer Zeit einen hohen Endwert der Dehnung zu erreichen. Das ist ein Weichmacher, der für die technische Anwendung unbequem ist.

Schließlich dürften sich auch die Gedankengänge von STOECKLIN[2] in den hier skizzierten Rahmen fügen. Nach seinen Vorstellungen sollen im Polymerisat zwei Mechanismen I und II vorhanden sein, auf die die Weichmacher in verschiedener Weise einwirken. Der Mechanismus I soll sich am reinsten in der elastischen Dehnung auswirken, in einem statischen Versuch, zu dem lange Zeiten und gewöhnlich erhöhte Temperatur erforderlich sind. Mechanismus II soll für das Verhalten bei kurzzeitiger Belastung bei tieferer Temperatur verantwortlich sein. Er wirkt sich am ausgeprägtesten in der Rückprallelastizität aus. Andere Prüfverfahren reagieren mehr oder weniger auf beide Mechanismen. Ordnet man nun die Weichmacher nach ihrer Wirksamkeit auf den Mechanismus I (beurteilt aus der elastischen Dehnung) in eine Reihe, so wird diese Reihe nicht dieselbe sein, wie wenn man sie beurteilt in der Wirksamkeit auf den Mechanismus II (beurteilt aus der Rückprallelastizität). Es ist klar, daß man so zu einer Beurteilung und Einteilung der Weichmacher kommt. Versuche hat STOECKLIN hauptsächlich mit Perbunan durchgeführt, z. T. auch mit Polyvinylchlorid. Sehr häufig ist man genötigt, die beiden Versuchsreihen zu variieren, um den Beobachtungen gerecht zu werden, wodurch sie jedoch an Überzeugungskraft nicht gewinnen.

---

[1] JENCKEL, E.: Kunststoffe **42**, 1 (1952).

[2] STOECKLIN, P.: Kautschuk u. Gummi **2**, 367 (1949); **3**, 45, 86, 199 (1950).

Vielleicht könnte man jedoch, worauf JENCKEL[1] hinweist, die STOECKLINschen Mechanismen I und II mit den zwei Bindungsmechanismen der Theorie des plastisch-elastischen Verhaltens identifizieren. Der Mechanismus I würde dann dem Netzwerk zuzuordnen sein; sein Elastizitätsmodul ist durch die Maschenweite des Netzwerkes und die dazugehörige plastische Verformung durch die Beständigkeit der vernetzenden Bindung gegeben. Der Mechanismus II würde den einzelnen Grundmolekülen zuzuordnen sein. Sein Elastizitätsmodul ist durch die zwischenmolekularen Kräfte gegeben, die dazugehörige plastische Verformung durch die Verschieblichkeit der Grundmoleküle gegeneinander. Die Versuche von STOECKLIN zeigen dann, daß ein Weichmacher auf die beiden Mechanismen verschieden stark einwirken kann.

§ 80. Zur Verträglichkeit der Weichmacher[2].

Wie oben sehr hervorgehoben wurde, müssen die Weichmacher, damit die Einfriertemperatur herabgesetzt wird, mit dem Polymerisat *Lösungen* bilden. Das ist nicht immer der Fall; man spricht von „Unverträglichkeit" des Weichmachers gegenüber einem bestimmten Polymeren. Vielmehr kann, ebenso wie bei den Systemen zweier niedermolekularer Flüssigkeiten, die Löslichkeit von Polymerisat und Weichmacher ineinander nur beschränkt sein. In diesem Falle, wie etwa im System Phenol—Wasser, erstreckt sich das Gebiet der homogenen Lösung nur über beschränkte Bereiche und bei mittleren Mischungsverhältnissen existieren zwei getrennte Phasen (vgl. auch Abb. IX, 12). Das Verhalten der hochmolekularen Lösung ist völlig analog. Es besteht lediglich der quantitative Unterschied, daß das Maximum der Löslichkeitskurve, der kritische Lösungspunkt, bei der Auftragung gegen Grundmolprozente[3] gewöhnlich ziemlich weit zur Seite des reinen Lösungsmittels verschoben ist, während bei niedermolekularen Lösungen der kritische Punkt im allgemeinen einigermaßen in der Mitte des Systems liegt (vgl. Bd. II, § 33).

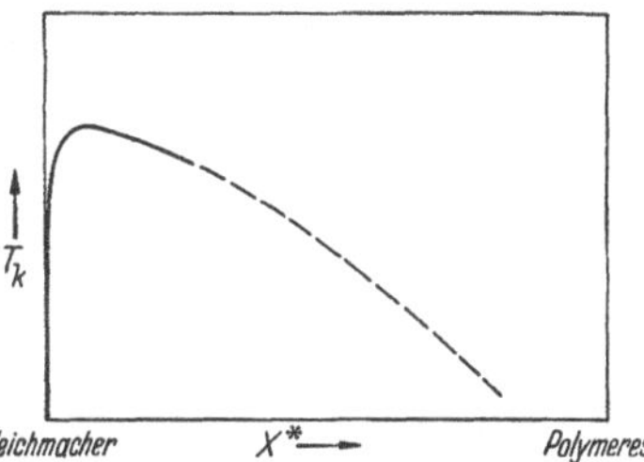

Abb. IX, 12. Darstellung der Löslichkeitskurve eines beschränkt löslichen Weichmachers; schematisch. Die ausgezogene Kurve ist der Messung zugänglich, die gestrichelte ist extrapoliert.

Die Bestimmung solcher Löslichkeitskurven bereitet keinerlei Schwierigkeiten, solange man sich im Konzentrationsgebiet zwischen 0 und 25 Gewichtsprozenten Polymeren bewegt. Man braucht hierzu die Mischung der beiden Komponenten von hinreichend hoher Temperatur, bei der alles gelöst ist und nur eine Phase existiert, nur langsam abzukühlen:

[1] Siehe S. 598, Fußnote 1. — [2] Vgl. auch S. 575 und Bd. II, § 33.
[3] Die Auftragung gegen richtige Molprozente würde die kritische Zusammensetzung noch sehr viel weiter gegen die Seite des reinen Lösungsmittels verschieben.

beim Unterschreiten der Löslichkeitskurve tritt infolge der Ausscheidung der zweiten Phase in Form feiner Tröpfchen eine deutlich sichtbare Trübung auf. Leider versagt diese Bestimmungsmethode der Löslichkeitskurve bei höheren Konzentrationen, weil auch die homogenen hochpolymeren Lösungen dann eine gewisse Trübung zeigen und daher die genaue Lage der Kurve nicht recht ermittelt werden kann.

Möglicherweise könnte man durch quantitative photometrische oder nephelometrische Messungen weiter kommen[1]. Dagegen scheint es manchmal schwierig zu sein, aus elektronenoptischen Bildern zwischen der Trennung in zwei Phasen und einer Assoziation in ein und derselben Phase zu unterscheiden[2]. In einem anderen Beispiel[3] ließ sich die Phasentrennung, wenn auch nur zu Mikrotröpfchen, elektronenoptisch deutlich erkennen, sobald eine gewisse Konzentration (Löslichkeitsgrenze) überschritten wird. Leider ist diese Untersuchung nur mit metallorganischen Weichmachern mit schweren Atomen möglich.

Man kann jedoch gar nicht daran zweifeln, daß die Löslichkeitskurve ungefähr in der in Abb. IX, 12 gestrichelten Weise weitergehen muß. Man gelangt dann in die Gebiete, die technisch interessant sind, mit etwa 60 bis 70 Gewichtsprozenten Polymerisat. Je nach der chemischen Natur des Lösungsmittels (Weichmachers) kann die Löslichkeitskurve bei ganz verschiedenen Temperaturen liegen. Vernetzung auch durch teilweise Kristallisation scheint die Löslichkeit herabzusetzen[4].

Die Unverträglichkeit des Weichmachers kann so ausgeprägt sein, daß es auch bei den hohen Temperaturen der Verarbeitung des Polymerisats nicht gelingt, eine homogene Lösung der beiden Komponenten zu erzielen, sondern nur eine mechanische Mischung aus zwei Phasen[5], die mechanisch weicher ist als das reine Polymerisat. Der „Weichmacher" hat dann den Charakter eines *Gleitmittels*, welches die Einfriertemperatur nicht herabsetzt, aber dennoch die Verarbeitung erleichtert. Nach kürzerer oder längerer Zeit sammelt sich die Weichmacherphase, die nach dem oben Gesagten aus fast reinem Weichmacher besteht, als Tröpfchen an der Oberfläche des Werkstücks: der Weichmacher *„schwitzt aus"*.

Es ist jedoch auch der Fall denkbar, daß bei der Verarbeitungstemperatur vollkommene Mischbarkeit besteht und, mit anderen Worten, die Löslichkeitskurve erst beim Abkühlen der Probe unterschritten wird. In diesem Falle scheidet sich die zweite Phase in Form sehr feiner Tröpfchen aus, welche eine opake Trübung des Werkstückes bedingen. Ob auch hier die Weichmacherphase sich an der Oberfläche des Werkstückes nach längerer Zeit sammelt, ist wohl nicht ganz sicher. Bemerkt werden soll jedoch, daß die Ausscheidung der zweiten Phase infolge Unterkühlung verzögert sein kann[6].

---

[1] Vgl. auch E. JENKEL u. G. BUTENUTH; unveröffentlicht (Brechungsindexmessungen).

[2] SPURLIN, H. M., A. F. MARTIN u. H. G. TENNENT: J. Polymer Sci. **1**, 63 (1946).

[3] RICHARD, W. R. u. P. A. S. SMITH: J. chem. Physics **18**, 230 (1950).

[4] BOYER, R. F.: J. appl. Physics **20**, 540 (1949).

[5] Die kritische Lösungstemperatur liegt hier also noch wesentlich über der Verarbeitungstemperatur.

[6] JENCKEL, E. u. G. KELLER: Z. Naturforsch. **5a**, 317 (1950).

§ 81. Diffusionsvorgänge in weichgemachten Kunststoffen.

Weichmacherverluste können entstehen:

1. infolge Abdampfens in die Luft,
2. infolge Wanderung in andere Kunststoffe.

Für beide Vorgänge ist die Diffusion die wesentliche Ursache.

a) Zur Theorie der Diffusion.

Die Diffusion führt zu einem Ausgleich zwischen Stellen unterschiedlicher Aktivität. Daher dampft Weichmacher ab, solange seine Aktivität a_1'' im Gasraum geringer ist als in der Lösung a_1', $a_1' > a_1''$, oder mit anderen Worten: solange über der Probe noch nicht der Sättigungsdruck des Weichmachers erreicht ist. Entsprechend wandert der Weichmacher aus einem weichgemachten Polymerisat mit der Aktivität a_1' in ein anderes, wenn dort seine Aktivität a_1'' kleiner ist.

Die Diffusion kann in Abhängigkeit von der Zeit verfolgt werden durch Beobachtung

1. der durch einen Querschnitt hindurchgeflossenen Substanzmenge und
2. der zeitlichen Änderung der Konzentration, insbesondere derjenigen an der Oberfläche.

Zur Berechnung von Diffusionsvorgängen wird stets das 1. und 2. FICKsche Gesetz zugrunde gelegt.

$$\frac{dm}{dt} = D \cdot F \cdot \frac{dc_v}{dx} \quad \text{und} \quad \left(\frac{dc_v}{dt}\right)_x = D \cdot \left(\frac{d^2 c_v}{dx^2}\right)_t . \qquad \text{(IX, 57)}$$

Hierin bedeutet m die diffundierende Substanzmenge, c_v die Volumenkonzentration, t die Zeit, x eine Ortskoordinate, D den Diffusionskoeffizienten und F den Querschnitt. Die Lösung der Differentialgleichung (IX, 57) gibt die Konzentration c_v als Funktion von Ort und Zeit und den unterschiedlichen Rand- und Grenzbedingungen der verschiedenen Versuche an. Spezielle Lösungen sind in verschiedenen Lehrbüchern angegeben. Außerdem sei auf den kurzen Abriß der Diffusion in Bd. II, § 44 dieses Werkes hingewiesen. Versuche sollten immer so angelegt werden, daß eine Lösung mathematisch möglich ist.

Verhältnismäßig einfache Lösungen ergeben sich für den halbseitig unendlich ausgedehnten Körper. Die hierfür errechneten Formeln haben die Form $m = A\sqrt{t}$ und gelten hinreichend, wenn wenigstens an einer Stelle des Körpers die ursprüngliche Konzentration durch die Diffusion noch nicht wesentlich verändert ist. Das letztere ist eine Frage der Versuchsdauer; daher gelten diese Lösungen für hinreichend kurze Zeiten stets.

Lösungen für Körper endlicher Ausdehnung im obigen Sinne lassen sich nur durch eine FOURIER-Reihe oder durch eine Reihe von GAUSSschen Integralen wiedergeben. Diese Lösungen müssen i. a. verwandt werden bei der Untersuchung von Folien.

Schließlich sind rechnerische Verfahren entwickelt worden, um der Änderung der Diffusionskonstante mit der Konzentration gerecht zu werden. Statt dessen kann man auch im Versuch die Konzentrationsunterschiede so klein halten, daß mit einem konstanten mittleren Koeffizienten gerechnet werden kann.

b) Das Abdampfen des Weichmachers.

Beim Lagern an der Luft ist der Gasdruck des Weichmachers über der Probe ganz undefiniert, so daß auf diese Weise die Diffusion nur sehr grob untersucht werden kann[1]. Zweckmäßig hält man vielmehr den Gasdruck auf Null, am besten durch ein angelegtes Hochvakuum, und erreicht dadurch einen maximalen Aktivitätsunterschied $a_1' - a_1''$. Es wurde auch vorgeschlagen, den Weichmacher aus dem Gasraum praktisch vollständig durch ein geeignetes Adsorptionsmittel, z.B. durch Silicagel, zu entfernen[2].

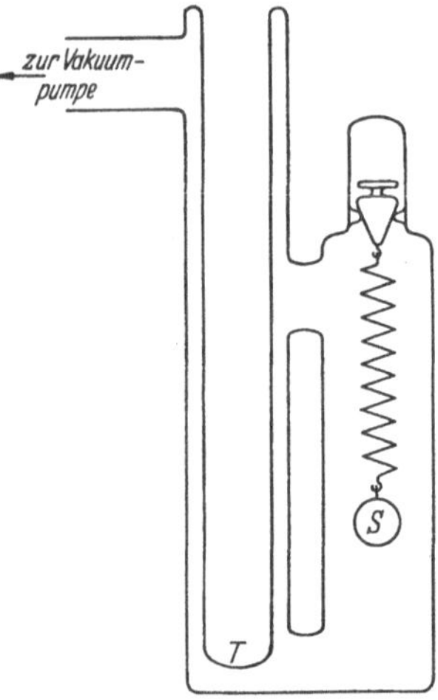

Abb. IX, 13. Gerät zur Bestimmung der Verdampfungs- und Aufnahmegeschwindigkeit eines Weichmachers aus der Gasphase. (Nach LIEBHAFSKY und Mitarbeitern.)

Zur Bestimmung des Gewichtsverlustes wird gewöhnlich eine Federwaage verwandt (Abb. IX, 13). Diese sehr bequeme Anordnung hat den Nachteil, daß nur Proben von geringer räumlicher Ausdehnung, etwa Folien, untersucht werden können. Der Verlust steigt im allgemeinen, wie zu erwarten, recht genau mit der Wurzel aus der Zeit, wie in Abb. IX, 14 und IX, 15 für die Verdampfung von Acetophenon aus einer Novolak-Lösung[3] mit 20 bis 50% Weichmacher belegt ist[4]. Weitere Versuche, bei denen auch gewöhnlich der umgekehrte Vorgang, die Aufnahme von Weichmacher aus dem Dampf, bestimmt wurde, wurden durchgeführt an: Polystyrol mit verschiedenen Lösungsmitteln[5], Polyvinylacetat und Aceton[6], Polyisobutylen und einer Reihe von chlorierten Kohlenwasserstoffen[7], Polyvinylchlorid und Trikresylphosphat, Sebacinsäuredibutylester, Phthalsäuredibutylester[8] und andern Phthalsäureestern[9].

Die Diffusionskonstanten nehmen mit der Weichmacherkonzentration zu. Das ist nicht verwunderlich, wenn man den engen Zusammenhang zwischen Diffusionskoeffizient und Viscosität beachtet und andererseits die Tatsache, daß durch Weichmacherzusatz die Einfriertemperatur herabgesetzt wird, also bei gleicher Temperatur auch die Viscosität herab-

[1] REED, M. C. u. L. CONNOR: Ind. Engng. Chem. **40**, 1414 (1948).
[2] GEENTY, J. R.: Ind. Rubber World **126**, 646 (1952).
[3] Novolak ist ein Phenol-Formaldehydkondensat.
[4] JENCKEL, E. u. J. KOMOR: Z. physik. Chem. A **187**, 335 (1941).
[5] PARK, G. S.: Trans. Faraday Soc. **46**, 684 (1950). – J. CRANK u. G. S. PARK: Trans. Faraday Soc. **45**, 240 (1949).
[6] KOKES, R. J., F. A. LONG u. J. L. HOARD: J. chem. Physics **20**, 1711 (1952).
[7] PRAGER, S. u. F. A. LONG: J. Amer. chem. Soc. **73**, 4072 (1951); hier auch Einfluß der Weichmacherkonstitution.
[8] LIEBHAFSKY, H. H., A. L. MARSHALL u. F. H. VERHOEK: Ind. Engng. Chem. **34**, 704 (1942).
[9] SMALL, P. A.: J. Soc. chem. Ind. **66**, 17 (1947).

gesetzt, d.h. die molekulare Beweglichkeit gefördert wird. Der zuerst von JENCKEL und KOMOR[1] angegebene Ausdruck

$$D = D_0 \cdot e^{\alpha c_v} \tag{IX,58}$$

ist später verschiedentlich bestätigt worden[2,3]. (Die Änderung des Diffussionkoeffizienten mit der Konzentration kann auch unmittelbar am Durchtritt von Weichmacherdampf durch eine Folie bestimmt werden[4].)

Die Temperaturabhängigkeit wird gewöhnlich durch einen Ausdruck von der Form

$$D = D_0 e^{-\frac{A}{RT}} \tag{IX,59}$$

beschrieben. Beachtet man, daß sich die Temperaturabhängigkeit der Viscosität nicht mit einer konstanten Aktivierungswärme wiedergeben läßt, sondern daß vielmehr mit sinkender Temperatur bei Annäherung an die Einfriertemperatur die Aktivierungswärme kräftig steigt (vgl. Bd. III, § 54), so wird man das gleiche auch hier sowohl bei wesentlich verschiedener Temperatur, als auch bei den verschiedenen Konzentrationen beobachten. In der Tat nimmt nach KNAPPE[5] die Aktivierungswärme und die Aktionskonstante D_0 an Polyvinylchlorid-Trioktylphosphat wie folgt ab:

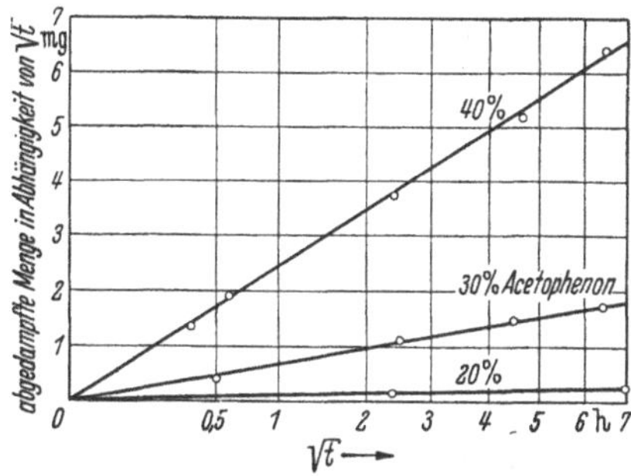

Abb. IX, 14. Verdampfungsverluste in Abhängigkeit von der Wurzel aus der Zeit im System Novolak–Acetophenon. Kleine Weichmacherkonzentrationen. (Nach JENCKEL u. KOMOR.)

Weichmacher	10	40	60	80	Gew.-%
Akt. Wärme	24,0	13,5	8,9	5,8	kcal
Akt. Konst.	$5{,}6 \cdot 10^4$	4,3	$2{,}0 \cdot 10^{-2}$	$7{,}3 \cdot 10^{-4}$	$cm^2 sec^{-1}$

Die absoluten Werte der gemessenen Diffusionskoeffizienten liegen etwa in der Größe von 10^{-8} bis 10^{-9} $cm^2 \cdot sec^{-1}$, bei KNAPPE[5] findet man Messungen zwischen 10^{-7} bis 10^{-11} $cm^2 \cdot sec^{-1}$. Eine Auswahl gemessener Werte sind in Tab. IX,4 verzeichnet.

JENCKEL und KOMOR[1] haben sich bemüht, die ungewissen Verhältnisse beim Lagern der Proben an der Luft in definierteren Versuchen zu erfassen. Zu diesem Zwecke war der Gasraum über der Lösung gegen das Hochvakuum durch eine Metallfolie

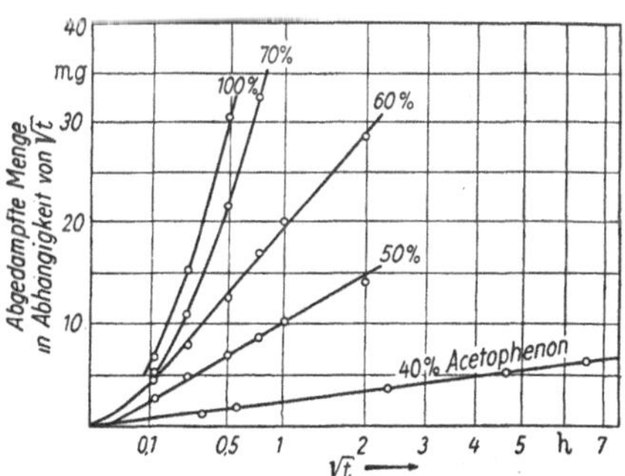

Abb. IX, 15. Verdampfungsverluste in Abhängigkeit von der Wurzel aus der Zeit im System Novolak–Acetophenon. Hohe Weichmachergehalte. (Nach JENCKEL u. KOMOR.)

[1] JENCKEL, E. u. J. KOMOR: s. S. 602, Fußnote 4.
[2] PRAGER, S. u. F. A. LONG: s. S. 602, Fußnote 7. — [3] PARK, G. S.: s. S. 602, Fußnote 5. — [4] PARK, G. S.: Trans. Faraday Soc. 48, 11 (1952).
[5] KNAPPE, W.: Z. angew. Physik 6, 97 (1954), dort auch weitere Systeme.

Tabelle IX, 4. *Auswahl einiger Diffusionskoeffizienten.*

Temp.	Gew.-% Weichmacher	Weichmacher	Polymerisat	Diff. Koeff. [$cm^2 \cdot sec^{-1}$] vgl. Gl. (IX, 58)	
				D_0	
25°	—	CH_2Cl_2	Polystyrol	$1950 \cdot 10^{-14}$	Nach PARK[1]
25°	—	CH_3J	Polystyrol	$460 \cdot 10^{-14}$	
25°	—	CH_2Br_2	Polystyrol	$250 \cdot 10^{-14}$	in Gl. (IX,58) bedeutet
25°	—	$CHCl_3$	Polystyrol	$58 \cdot 10^{-14}$	c = Volumenbruch
25°	—	$CHBr_3$	Polystyrol	$\sim 2{,}3 \cdot 10^{-14}$	$\alpha = 50$ bis 58, im Mittel 54
25°	—	CCl_4	Polystyrol	$\sim 0{,}9 \cdot 10^{-14}$	
				D_0	
35°	—	Propan	Polyisobutylen	$4{,}81 \cdot 10^{-9}$	Nach PRAGER u. LONG[2]
35°	—	n-Butan	Polyisobutylen	$3{,}24 \cdot 10^{-9}$	in Gl. (IX,58) bedeutet
35°	—	iso-Butan	Polyisobutylen	$1{,}45 \cdot 10^{-9}$	$c = \frac{\text{Gew.-Kohlenwasserstoff}}{\text{Gew.-Polymeres}}$
35°	—	n-Pentan	Polyisobutylen	$2{,}64 \cdot 10^{-9}$	
35°	—	iso-Pentan	Polyisobutylen	$1{,}32 \cdot 10^{-9}$	$\alpha = 21{,}1$ bis 25,5,
35°	—	Neopentan	Polyisobutylen	$0{,}62 \cdot 10^{-9}$	bei Neopentan 14,4
				$\overline{D}$	
Zimmer-temp.	80	Tri-Oktylphosphat	Polyvinylchlorid	$3{,}8 \cdot 10^{-8}$	Nach W. KNAPPE[3]
Zimmer-temp.	60	Tri-Oktylphosphat	Polyvinylchlorid	$3{,}0 \cdot 10^{-9}$	„wahre" Diffusionskoeffizienten
Zimmer-temp.	40	Tri-Oktylphosphat	Polyvinylchlorid	$5{,}0 \cdot 10^{-10}$	Konz. Unterschied = 20 Gew.-prozent
Zimmer-temp.	80	Di-Oktylphthalat	Polyvinylchlorid	$6{,}4 \cdot 10^{-9}$	Weichmacher-Wanderung
Zimmer-temp.	60	Di-Oktylphthalat	Polyvinylchlorid	$6{,}2 \cdot 10^{-10}$	

Zimmertemp.	80	Di-Butylphthalat	Polyvinylchlorid	$1{,}9 \cdot 10^{-8}$	
Zimmertemp.	60	Di-Butylphthalat	Polyvinylchlorid	$4{,}1 \cdot 10^{-9}$	
Zimmertemp.	40	Di-Butylphthalat	Polyvinylchlorid	$2{,}9 \cdot 10^{-10}$	
100°	80	Tri-Oktylphosphat	Polyvinylchlorid	$3{,}0 \cdot 10^{-7}$	
100°	60	Tri-Oktylphosphat	Polyvinylchlorid	$1{,}2 \cdot 10^{-7}$	
100°	40	Tri-Oktylphosphat	Polyvinylchlorid	$5{,}6 \cdot 10^{-8}$	
100°	10	Tri-Oktylphosphat	Polyvinylchlorid	$5{,}3 \cdot 10^{-10}$	
100°	80	Di-Oktylphthalat	Polyvinylchlorid	$1{,}4 \cdot 10^{-7}$	
100°	60	Di-Oktylphthalat	Polyvinylchlorid	$9{,}0 \cdot 10^{-8}$	
100°	40	Di-Oktylphthalat	Polyvinylchlorid	$2{,}1 \cdot 10^{-8}$	
100°	10	Di-Oktylphthalat	Polyvinylchlorid	$5{,}6 \cdot 10^{-10}$	
100°	80	Di-Butylphthalat	Polyvinylchlorid	$3{,}2 \cdot 10^{-7}$	
100°	60	Di-Butylphthalat	Polyvinylchlorid	$2{,}3 \cdot 10^{-7}$	
100°	40	Di-Butylphthalat	Polyvinylchlorid	$1{,}2 \cdot 10^{-7}$	
100°	10	Di-Butylphthalat	Polyvinylchlorid	$2{,}6 \cdot 10^{-9}$	
				$\overline{D}$ Foliendicke	Nach LIEBHAFSKY u. Mitarb.[4] Verdampfung ins Vakuum
145°	20	Tri-Kresylphosphat	Polyvinylchlorid	$2{,}6 \cdot 10^{-8}$ 1 mm	
145°	30	Tri-Kresylphosphat	Polyvinylchlorid	$5{,}9 \cdot 10^{-8}$ 1 mm	
145°	40	Tri-Kresylphosphat	Polyvinylchlorid	$7{,}1 \cdot 10^{-8}$ 1 mm	
145°	50	Tri-Kresylphosphat	Polyvinylchlorid	$13{,}5 \cdot 10^{-8}$ 1 mm	
145°	60	Tri-Kresylphosphat	Polyvinylchlorid	$25{,}4 \cdot 10^{-8}$ 1,087 mm	
145°	60	Tri-Kresylphosphat	Polyvinylchlorid	$18{,}5 \cdot 10^{-8}$ 0,498 mm	
145°	60	Tri-Kresylphosphat	Polyvinylchlorid	$35{,}5 \cdot 10^{-8}$ 2,282 mm	

[1] PARK, G. S.: Trans. Faraday Soc. **46**, 684 (1950); Werte auf Quellung korrigiert.
[2] PRAGER, S. u. F. A. LONG: J. Amer. chem. Soc. **73**, 4072 (1951).
[3] KNAPPE, W.: Z. angew. Physik **6**, 97 (1954).
[4] LIEBHAFSKI, H. H., A. L. MARSHALL u. F. H. VERHOEK: Ind. Engng. Chem. **34**, 704 (1942).

mit einem kleinen Loch abgeschlossen. Hierdurch wird eine gewisse Stauung des abdampfenden Weichmachers über der Probe erreicht, wie sie etwa der Stauung bei mangelhafter Luftströmung entsprechen dürfte. Die Gasströmung durch das Loch ist nach KNUDSEN berechenbar und proportional dem Druck. Die Diffusionsströmung läßt sich nach den Diffusionsgesetzen berechnen, wobei angenommen wird, daß zwischen der Oberfläche der Probe und dem Gasraum Gleichgewicht herrscht. Im stationären Zustand ist die Strömung durch das Loch und diejenige in der Oberfläche der Lösung gleich. Die langsamste dieser beiden Strömungen ist geschwindigkeitsbestimmend. Auf Einzelheiten der Rechnung kann hier nicht eingegangen werden. Die in den Abb. IX, 14 und IX, 15 schon gebrachten Versuche sind an einem einzigen Weichmacher bei 30°, jedoch für alle Konzentrationen von 0 bis 100% Weichmacher durchgeführt worden. Hierbei nimmt der Dampfdruck etwa linear, die Diffusionskonstante dagegen exponentiell mit der Weichmacherkonzentration zu. Dementsprechend ist bei Weichmachergehalten von 0 bis 40%, die etwa den technischen Produkten entsprechen, die im Laufe der Zeit abdampfende Menge diffusionsbestimmt, also proportional der Wurzel aus der Zeit, bei solchen von 70 bis 100% dampfdruckbestimmt, also proportional der Zeit selbst, während bei 50 und 60% beide Faktoren zusammenwirken. Das gleiche dürfte für niedriger siedende Weichmacher oder Lösungsmittel gelten, weil Dampfdruck und Diffusionskoeffizienten etwa parallel zunehmen. Mit steigender Temperatur scheinen jedoch die Diffusionskonstanten stärker zuzunehmen als die Dampfdrucke, so daß die Verdampfung jetzt eher dampfdruckbedingt sein könnte.

Die absoluten Mengen werden, gleichgültig, ob sie in ihrer Zeitabhängigkeit diffusions- oder dampfdruckbedingt sind, bei niedrig siedenden Lösungsmitteln größer sein, weil beide Faktoren größer werden. In der Tat, sind die technisch brauchbaren Weichmacher hochsiedende Flüssigkeiten.

c) Die Weichmacherwanderung.

Berühren sich zwei verschiedene organische Werkstoffe miteinander, wie es die technische Verwendung sehr oft mit sich bringt, so beobachtet man häufig die Wanderung des Weichmachers aus dem einen Werkstoff in den anderen hinein.

Zu einer vergleichenden Prüfung genügt es, weichmacherhaltige und weichmacherfreie Substanz, etwa in Form von Folien miteinander in Berührung zu bringen und nach hinreichender Zeit einzeln zu wiegen[1]. Eine genauere Untersuchung nach dem gleichen Verfahren an Polyvinylacetat-Aceton zur Ermittlung der Diffusionskonstanten haben KOKES, LONG u. HOARD[2] mitgeteilt. Schließlich hat W. KNAPPE[3] an Polyvinylchlorid mit Phthalsäureestern die Änderung der Konzentration über den Brechungsindex mit Hilfe eines ABBE-Refraktometers bestimmt. Hierzu wurde die Rückseite einer weichgemachten Folie von der Konzentration

[1] RÖSSIG, L.: Kunststoffe **44**, 250 (1954).
[2] KOKES, R. J., F. A. LONG u. J. L. HOARD: J. chem. Physics **20**, 1711 (1952).
[3] KNAPPE, W.: Z. angew. Physik **6**, 97 (1954).

c_1 mit einem anderen Weichmacher oder einer Weichmacherlösung von der Konzentration c_2 in Berührung gebracht. Dann wurde die Zeit bestimmt, nach der die Konzentration auf der Vorderseite der Folie sich auf den Wert $\frac{c_1 + c_2}{2}$ geändert hatte. Diese Messung gestattet, den Diffusionskoeffizienten zu berechnen. Die angeführte Arbeit gibt für Temperaturen zwischen Zimmertemperatur und 100° für einige Weichmacher die Diffusionskoeffizienten mit Werten zwischen 10^{-6} bis 10^{-12} $cm^2 \cdot sec^{-1}$ an (s. Tab. IX, 4).

Zur Unterdrückung der Weichmacherwanderung können hochpolymere Weichmacher verwandt werden, deren Diffusionskoeffizient verständlichweise sehr viel kleiner sein sollte. Versuche von J. R. GEENTY[1] ergaben allerdings nur geringfügig kleinere Werte.

d) „Anomale" Diffusion im Einfrierbereich.

Wie die Rechnung lehrt, sollte die Abgabe von Weichmachern an das Hochvakuum, und umgekehrt, die Aufnahme von Weichmacherdampf in den weichmacherfreien Kunststoff nach der gleichen Formel verlaufen, wenn man konstante, d. h. von der Konzentration unabhängige Diffusionskoeffizienten voraussetzt. In Wirklichkeit beobachtet man geringe Unterschiede, die jedoch einwandfrei zu berechnen sind, wenn man berücksichtigt, daß der Diffusionskoeffizient von der Konzentration abhängig ist[2]. Es wird also hier beachtet, daß innerhalb der der Diffusion unterliegenden Probe örtlich verschiedene Diffusionskoeffizienten einzusetzen sind. Es sind jedoch in einer ganzen Reihe von Beispielen weitere erhebliche Abweichungen sowohl für die Verdampfung wie für die Aufnahme, besonders auch im Verlauf dieser Kurven beobachtet worden[3-5].

Hierfür gibt Abb. IX, 16 zwei Beispiele. Während die Kurve der Verdampfung zwar von der $\sqrt{t}$-Beziehung abweicht, aber doch monoton verläuft, zeigt die Kurve der Weichmacheraufnahme sogar einen Wendepunkt. Außerdem überschneidet die letztere Kurve die erstere. Dieses Verhalten scheint typisch zu sein. Nun hat eine kürzliche Untersuchung von KOKES, LONG u. HOARD[3] ergeben, daß diese Anomalie immer nur dann auftritt, wenn im Probekörper Konzentrationen entstehen, deren Einfriertemperatur bei der Versuchstemperatur liegt. Mit anderen Worten: Es muß während des Diffusionsvorganges innerhalb der Probe das Einfriergebiet durchlaufen werden. Zum Nachweis wurden gesondert der Sättigungsdruck über der Lösung und die Einfriertemperatur als Funk-

---

[1] GEENTY, J. R.: Ind. Rubber World **126**, 646 (1952).

[2] CRANK, J. u. M. E. HENRY: Trans. Faraday Soc. **45**, 636, 1119 (1949). – J. CRANK: Trans. Faraday Soc. **47**, 450 (1951). – J. CRANK u. G. S. Park: Trans. Faraday Soc. **47**, 1072 (1951) sowie Resarch **4**, 515 (1951). Gleichwohl gelingt es, die einzelnen Kurven mit einem konstanten mittleren Wert $\bar{D}$ befriedigend wiederzugeben, etwa beim Abdampfen. $\bar{D}$ ist dann jedoch eine Funktion der Konzentration. – G. S. PARK: Trans. Faraday Soc. **46**, 684 (1950).

[3] KOKES, R. J., F. A. LONG u. J. L. HOARD: s. S. 606, Fußnote 2.

[4] KING, G. u. A. B. D. CASSIE: Trans. Faraday Soc. **36**, 445 (1940). – G. KING: Trans. Faraday Soc. **41**, 325 (1945).

[5] MANDELKERN, L. u. F. A. LONG: J. Polymer Sci. **6**, 457 (1951).

tion der Konzentration bestimmt. Die Anomalien werden nicht beobachtet, wenn nur Gleichgewichtsschmelzen auftreten, wie ausdrücklich betont wird; es scheint, daß sie auch nicht auftreten, wenn die Proben im Zustande des Glases vorliegen, mit anderen Worten: Wenn die Konzentration am Weichmacher hinreichend klein ist und die Versuchstemperatur unter der Einfriertemperatur liegt, wie etwa bei der Permeation von H_2 (vgl. Bd. II, § 45).

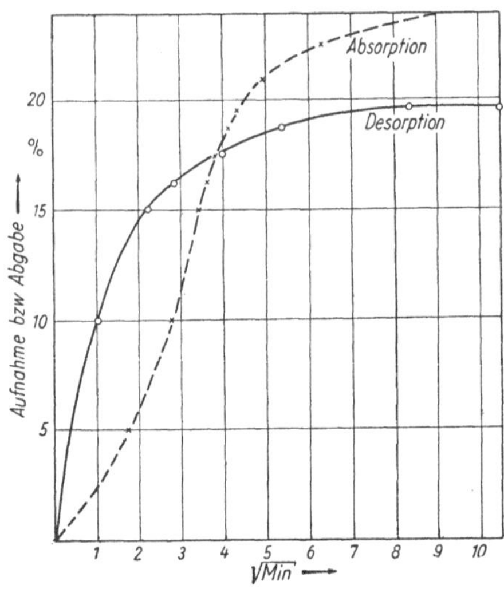

Abb. IX, 16. Anomale Diffusion. Unterschiedliche Abgabe und Aufnahme von Methylalkohol aus der Gasphase gegen die Wurzel aus der Zeit an Wolle. (Nach KING.)

Das Wesen des Einfriervorganges (vgl. Bd. III, § 55) bedingt, daß die Temperaturkoeffizienten der Eigenschaften — hier des Diffusionskoeffizienten — im glasigen eingefrorenen Zustande verhältnismäßig klein, näherungsweise gleich Null sind, dagegen in der Schmelze groß sind, und dasselbe dürfte für die Konzentrationskoeffizienten gelten. Dann ergeben sich für die Verdampfung und für die Aufnahme des Weichmachers die in Abb. IX, 17 schematisch skizzierten Werte für die Diffusionskoeffizienten in Abhängigkeit von dem Abstande von der Oberfläche. Diese Vorstellung ergibt in groben Zügen tatsächlich, daß die Verdampfung zunächst rasch verläuft, weil der Diffusionskoeffizient in der ganzen Probe groß ist, mit der Zeit aber sehr viel langsamer wird, weil sich eine oberflächliche, immer dicker werdende Schicht mit sehr kleinem Diffusionskoeffizienten bildet. Umgekehrt würde die Aufnahme von Weichmacher zunächst sehr langsam verlaufen, weil die Diffusionskonstante in der ganzen Probe klein ist; in dem Maße wie eine dicker werdende oberflächliche Schicht anwächst, in der der Diffusionskoeffizient groß ist, sollte die Aufnahme kräftig zunehmen, und schließlich sollte sie gegen Ende des Diffusionsvorganges wieder klein werden. Dieser oder ein ähnlicher Gedanke hat wohl auch den schon vorher veröffentlichten Untersuchungen von CRANK u. Mitarbeitern zugrunde gelegen. Es wurden hier u. a. extreme und diskontinuierliche Änderungen des Diffusionskoeffizienten mit der Konzentration

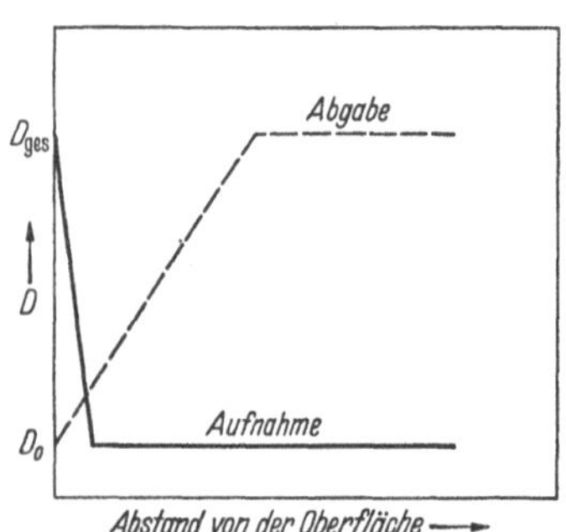

Abb. IX, 17. Die Diffusionskonstante in Abhängigkeit von der Entfernung von der Oberfläche. a) Aufnahme, b) Verdampfung. (Schematisch.)

untersucht. Das Ergebnis dieser Berechnung ist nun allerdings, daß zwar Abweichungen von der Proportionalität mit der Wurzel aus der Zeit und auch Unterschiede zwischen der Kurve der Verdampfung und der Aufnahme stattfinden können, nicht aber ein Wendepunkt auf der Aufnahmekurve. Nachdem auch eine unvollständige Einstellung des Gleichgewichts zwischen dem Gasraum und der Oberflächenschicht ziemlich ausgeschlossen werden konnte, werden bei CRANK u. PARK bereits zeitliche Änderungen der Diffusionskonstante (Nachwirkungen) diskutiert, wie sie für andere Eigenschaften in der Umgebung der Einfriertemperatur charakteristisch sind und die letzten Endes sämtlich darauf hinauslaufen, daß der eingefrorene glasige Zustand sich mehr oder weniger vollständig nach hinreichender Zeit dem Zustand der Schmelze nähert. Eine Berechnung im einzelnen, auch nur qualitativer Art, ist jedoch bis heute nicht möglich gewesen.

Zehntes Kapitel.

Vergleichende Betrachtungen zum Zusammenhang zwischen der molekularen Struktur und den makroskopischen Eigenschaften von Körpern aus Fadenmolekülen.

Von

H. MARK und H. A. STUART.

Mit 2 Abbildungen.

A. Allgemeine Betrachtungen.

Von H. A. STUART.

§ 82. Die verschiedenen Zustände bei hochpolymeren Stoffen aus Fadenmolekülen[1,2].

Bei Körpern, die aus langen Kettenmolekülen aufgebaut sind, finden wir eine im Gegensatz zu niedermolekularen Stoffen außerordentlich große Variationsfähigkeit und Temperaturabhängigkeit ihrer Eigenschaften.

Das liegt einmal daran, daß die Fadenmoleküle mehr oder weniger biegsam sind und daher ihre Form ständig ändern können. Ferner liegen die Moleküle nie durchweg so hoch geordnet und dicht gepackt, wie das bei niedermolekularen kristallinen Stoffen der Fall ist. So kommt es, daß die Molekülketten auch im festen Zustande in einem größeren Temperaturbereiche eine über die gewöhnliche thermische Schwingbewegung der Atome weit hinausgehende *mikro*-BROWNsche *Bewegungsmöglichkeit* besitzen (vgl. dazu auch § 5). Soweit Kettenmoleküle in größeren Stücken (Segmenten) beweglich sind, jedoch nicht völlig einander abgleiten können (freie mikro-BROWNsche und gehemmte *makro*-BROWNsche Bewegung) zeigt die Substanz kautschukartigen Charakter (vgl. Kap. V). Daher durchlaufen hochpolymere Körper aus Fadenmolekülen in einem bestimmten Temperaturbereich einen kautschukartigen Zustand, der um so ausgeprägter ist, je weniger die Substanz zur Kristallisation neigt und je höher das Molekulargewicht ist, vgl. Abschnitt C dieses Kapitels.

---

[1] Über die Definition von Faden- und Kornmolekülen vgl. Bd. II dieses Werkes, S. 371. — [2] Vgl. dazu z. B. H. MARK: Trans. Faraday Soc. **43**, 447 (1947).

Friert die mikro-BROWNsche Bewegung ein, so erhalten wir einen glasartigen Zustand, in dem nur noch thermische Schwingungen der Atome möglich sind[1]. Bei einer mechanischen Beanspruchung können nur noch die Atomabstände und die Valenzwinkel nachgeben. Ausweichmöglichkeiten durch Platzwechsel sind bei kurzzeitiger Beanspruchung (Stoß) nicht vorhanden. Der Körper wird spröde.

Während niedermolekulare Stoffe nur in zwei kondensierten Zuständen auftreten, ist bei hochmolekularen Körpern neben dem festen und flüssigen Zustande noch ein kautschukartiger Zustand möglich; vgl. hierzu auch § 5 und Kap. VI. Es ist nützlich, sich die molekularen Ordnungs- und Bewegungsverhältnisse in den drei Zuständen an einem Schema klarzumachen.

Tabelle X, 1 a. *Schema der Zustände in einem nichtkristallinen hochpolymeren Körper.*

fest	kautschukartig	flüssig
	(Einfriertemperatur ← → Fließtemperatur)	
glasartig mit Nahordnung	Nahordnung	Nahordnung
makro- und mikro-BROWNsche Bewegung erloschen	temporäre Fixpunkte, z.B. durch Kettenverschlingung	Kettenverschlingungen lösen sich, makro-BROWNsche Bewegung wird frei, der Körper kann fließen.
starrer Körper mit geringer Dehnbarkeit und hohem E-Modul spröde	nur mikro-BROWNsche Bewegung frei; fester Körper mit großer elastischer Dehnbarkeit, kleiner E-Modul	Elastoviscose Schmelze

Grenzen: Einfriertemperatur (zwischen fest und kautschukartig), Fließtemperatur (zwischen kautschukartig und flüssig).

In kristallinen Körpern liegen die Verhältnisse insofern anders, als die kristallinen Bereiche ein so dichtes Netzwerk bilden, daß größere Entknäuelungen der Moleküle in den amorphen Bereichen und damit auch größere Dehnungen unmöglich werden. So wird der Kautschukcharakter weitgehend zurückgedrängt; vgl. das Schema in der Tab. X, 1 b.

Tabelle X, 1 b. *Schema der Zustände in einem kristallinen hochpolymeren Körper.*

fest kristallin-amorph	fest kristallin-amorph	flüssig
in den amorphen Bezirken mikro-BROWNsche Bewegung erloschen, daher kleine Bruchdehnung, spröde	mikro-BROWNsche Bewegung in den amorphen Bezirken frei, daher zusätzliche elastische Reserve, Nachwirkungserscheinungen (vgl. z. B. § 40 b und Bd. III § 28 d), größere Bruchdehnung 20 und mehr Prozent, mechanisch zäh, faserbildend	Übergang fest—flüssig, meist ziemlich scharf, Schmelze elastoviscos

Grenzen: Einfriertemperatur (zwischen den beiden festen Zuständen), Schmelzpunkt (zwischen fest und flüssig).

[1] Alle Atome schwingen unter dem Einfluß des Kraftfeldes der umgebenden Atome um bestimmte Gleichgewichtslagen. Im Gegensatz zum Kristall sind diese aber (abgesehen von evtl. vorhandenen kristallinen Bezirken) nicht gitterartig geordnet, so daß die Kräfte, Frequenzen und Amplituden ziemlich schwanken.

Das Freiwerden der mikro-BROWNschen Bewegung oberhalb der Einfriertemperatur[1] erkennt man an der gegenüber dem Glaszustande erheblich größeren elastischen Dehnbarkeit und größeren Bruchdehnung[2] sowie den Kriech- und Relaxationserscheinungen. Der, auf die amorphen Bezirke beschränkte, Kautschukeffekt macht es möglich, bei kristallisierenden Hochpolymeren Werkstoffe zu entwickeln, die mit großer Festigkeit eine hohe Zähigkeit, Schlagbiegefestigkeit usw. verbinden. Hierbei ist es allerdings wichtig, bei der Verarbeitung innere Spannungen und Inhomogenitäten zu vermeiden, sowie eine geeignete kristalline Textur, unter anderem ein vernünftiges Verhältnis kristallin—amorph zu erzeugen. Ein zu hoher kristalliner Anteil verringert die elastische Reserve und macht das Material spröde, zum Teil als Folge der zusätzlichen inneren Spannungen.

Bei gut kristallisierenden, nicht zu langkettigen Molekülen ist der Übergang fest—flüssig oft erstaunlich scharf; er fällt dann auch sehr gut mit dem optischen Schmelzpunkt, siehe Bd. III, § 39, zusammen. Beispiele sind die Polyamide, Terylen und Polyäthylen. Bei genügend langen Ketten, die durch Verschlingung temporäre Fixpunkte bilden, würde die Schmelze kautschukartig werden. Außerdem beobachtet man eine überraschend hohe Standfestigkeit des geschmolzenen Materials[3]. Im gleichen Sinne dürften langkettige Verzweigungen wirken[4,5].

Dicht oberhalb des optischen Schmelzpunktes, wo noch kleinste, röntgenographisch und optisch nicht mehr faßbare Kristallkeime vorhanden sein mögen, müßte der Kautschukcharakter deutlicher werden, doch sind diesbezügliche Beobachtungen kaum bekannt.

Schon aus diesen Ausführungen erkennen wir, daß die Kautschukelastizität weniger die Eigenschaft einer bestimmten Gruppe von hochpolymeren Stoffen ist, als eine bei allen Stoffen aus Fadenmolekülen, auch bei Faserstoffen und plastischen Massen in einem bestimmten Temperaturgebiet je nach Konstitution, Kettenlänge usw. mehr oder weniger ausgeprägte Zustandseigenschaft ist. Es bestehen also zwischen faserbildenden, kautschukartigen und plastischen Massen keine grundsätzlichen Unterschiede in der chemischen Konstitution. So liefert aus der Schmelze abgeschrecktes Terylen ein amorphes, bei Zimmertemperatur glasartig sprödes Produkt. Erst wenn wir die Substanz etwas über die Einfrier-

[1] Häufig gibt es, vor allem bei kristallisierenden Hochpolymeren, neben der Einfriertemperatur im gewöhnlichen Sinne noch weitere, aber weniger ausgeprägte Übergangsgebiete, in denen andere molekulare Bewegungsmechanismen, wie die Bewegungen von Seitengruppen, einfrieren, vgl. die Ausführungen in Bd. III, Kap. XI, § 58c u. § 60c sowie diesen Bd., § 5.

[2] Beispiele zur Bruchdehnung unter- und oberhalb der Einfriertemperatur liefern Nylon und Perlon.

[3] In diesem Zusammenhange sei erwähnt, daß nach Beobachtungen von S. W. KANTOR u. R. C. OSTHOFF: J. Amer. chem. Soc. **75**, **931** (**1953**), eine Polymethylenprobe mit einem Molekulargewicht von rund 3 Millionen etwa 20° oberhalb des Schmelzpunktes von 132° unter dem Einfluß der eigenen Schwere auch nach 165 Stunden keine Verformung zeigte.

[4] Natürlich spielen noch die zwischenmolekularen Kräfte eine Rolle.

[5] Vielleicht erklärt sich so der zähflüssige Charakter einer Schmelze von Polytrifluorchloräthylen.

temperatur, nämlich auf etwa 80° erwärmen, kristallisiert sie. Verstreckt liefert sie ein ausgezeichnetes Fasermaterial. Oberhalb des Schmelzpunktes erhalten wir ein elastoviscoses Material, dessen Kautschukcharakter bzw. Formelastizität nur wegen des relativ niedrigen Molekulargewichts gering ist.

Dehnen wir Naturkautschuk bei Raumtemperatur genügend stark, so reicht die einsetzende Kristallisation aus, um bei Abkühlung auf mindestens —100° die gestreckte Form auch nach dem Entspannen zu stabilisieren. So erhält der Kautschuk in einem allerdings sehr tiefen und praktisch uninteressanten Temperaturbereich „Fasereigenschaften". Um einen praktisch brauchbaren Kautschuk oder eine wertvolle Faser herzustellen, muß allerdings, wie in den Abschnitten B und C ausgeführt wird, noch eine Vielzahl von Bedingungen erfüllt sein. So ist bei einem Kautschuk eine zusätzliche weitmaschige, aber dauerhafte Vernetzung erforderlich, die zwischenmolekularen Kräfte müssen schwach sein. Umgekehrt ist die Faserbildung und der dabei unerläßliche Prozeß der Verstreckung nur möglich, wenn die seitlichen Kräfte zwischen den Molekülen so stark sind, daß ein Abgleiten der Ketten bei stärkerer Beanspruchung in der Kettenrichtung verhindert wird, vgl. Bd. III, § 32, was aber nicht unbedingt starke Kristallisation mit gut ausgeprägten Gitterstrukturen voraussetzt (siehe Abschnitt B). Daneben ist eine gewisse kautschukartige Dehnbarkeit wichtig.

Die Zahl der Faktoren, die für die makroskopischen Eigenschaften eines hochpolymeren Stoffes von Bedeutung sind — wir bezeichnen sie im folgenden als die Elemente der *molekularen Struktur* — ist sehr groß und wird mit fortschreitender Erkenntnis immer größer. Wir werden im § 84 näher auf die Zusammenhänge zwischen diesen Parametern und den physikalischen und technologischen Eigenschaften eingehen.

Vorher müssen wir aber im nächsten Paragraphen noch die Frage erörtern, wie man die molekulare Ordnung und Beweglichkeit variieren und bestimmte Zustände in verschiedenen Temperaturbereichen verwirklichen kann.

§ 83. Die Beeinflussung des Ordnungs- und Bewegungszustandes, vor allem durch verfestigende und lockernde Zusätze.

In erster Linie werden die makroskopischen Eigenschaften eines Stoffes von den Eigenschaften des Einzelmoleküls, dem molekularen Ordnungszustand und der Beweglichkeit der Kettenstücke im kondensierten Zustande bestimmt. Ordnung und Beweglichkeit hängen natürlich ihrerseits wieder von der Konstitution der Kettenmoleküle, d.h. von den zwischenmolekularen Kräften und ihrer Geometrie (Regelmäßigkeit), genauer, von den potentiellen Energien im Wechselspiel mit der Energie der ungeordneten Temperaturbewegung ab. Das Ausmaß der Wärmebewegung wird außer von der Temperatur auch noch von der Biegsamkeit der Kette bestimmt, die ihrerseits wiederum von den innermolekularen, die Rotationsbehinderung bestimmenden Kräften, also letztlich wieder von der Konstitution des Moleküls abhängt. Ferner kann man durch Beimischungen den Ordnungs- und Bewegungszustand erheblich beeinflussen. Die Kettenlänge ist, wenn wir von ganz kurzen Ketten absehen, von untergeordneter Bedeutung. Dagegen kann man noch durch geeignete mechanische Verformung und durch die Art der Temperaturführung den Ordnungs- und Bewegungszustand in gewissen Grenzen variieren; vgl. dazu Bd. III dieses Werkes, vor allem die §§ 42ff.

Sehen wir von der letzten Möglichkeit ab, so sind es im wesentlichen die Eigenschaften des einzelnen Kettenmoleküls und diejenigen der (meist niedermolekularen) Zusätze sowie die Temperaturlage, welche den Ordnungs- und Bewegungszustand von vornherein bestimmen.

Bei den Eigenschaften des Einzelmoleküls handelt es sich um eine Reihe von Elementen, die in der Tab. X, 2 aufgeführt sind und die wir unter Hinweis auf die Ausführungen in Bd. III nur kurz nennen wollen.

Die chemische Konstitution der monomeren Einheiten, die regelmäßige oder unregelmäßige Folge derselben in der Kette, also die Ordnung innerhalb des Kettenmoleküls, bestimmen die zwischenmolekularen Kräfte und die geometrischen Verhältnisse (Symmetrie und Regelmäßigkeit des Kraftfeldes). Der Einfluß der Art der Kräfte (Dispersionskräfte, Kräfte bei stark polaren Gruppen etwa in Verbindung mit stark polarisierbaren Gruppen wie Benzolringen, Kräfte bei Wasserstoffbrückenpartnern), die Wirkung der Regelmäßigkeit im Bau, diejenige von seitlichen Gruppen oder von Verzweigungen, der Einfluß der Biegsamkeit auf die Kristallisationsneigung, Schmelzpunktlage usw., ist eingehend im Bd. III, Kap. IX „Kristallisation und Konstitution“ besprochen worden; über den Einfluß dieser Konstitutionseigenschaften auf die in Tab. X, 2 aufgeführten Elemente der Textur in einem hochpolymeren Körper ist allerdings bisher nur wenig bekannt. Ebenso ist die durch eine Copolymerisation hervorgerufene Herabsetzung der molekularen Ordnung und die damit verbundene Erhöhung der molekularen Beweglichkeit bereits in Bd. III, Kap. IX und Kap. XI eingehend besprochen worden. Über den Einfluß von Verzweigungen vgl. man § 84 b 1. Dagegen müssen wir jetzt etwas näher auf die verfestigende und lockere Wirkungen von Beimischungen eingehen.

Ein Körper fließt, sobald unter dem Einfluß äußerer Kräfte die Kettenmoleküle voneinander abgleiten können. Diese *makro*-Brownsche *Bewegung* kann man durch ein Netz von Fixpunkten unterbinden, wodurch der Körper formelastisch wird. Je dichter die Fixpunkte verteilt sind, desto geringer wird seine Dehnbarkeit und desto größer die Festigkeit.

Eine Vernetzung kann man auf verschiedene Weise bewerkstelligen. Schon durch die Verschlingung der einzelnen Ketten kann ein Abgleiten der Ketten, zumal bei kurzzeitiger Beanspruchung des Netzwerkes, weitgehend verhindert und zum mindesten immer erschwert werden. Daher besitzt auch eine Schmelze eine gewisse Formelastizität, die natürlich mit der Länge der Ketten immer ausgeprägter wird. Eine völlige Unterdrückung der makro-Brownschen Bewegung erreicht man bekanntlich durch eine Hauptvalenzvernetzung, also indem man mittels geeigneter chemischer Reaktionen die einzelnen Ketten durch S-, O- oder CH_2-Brücken miteinander verknüpft. Die Vernetzung durch starke zwischenmolekulare Kräfte, ausgeprägte Dipolkräfte, vor allem in Gegenwart von stark polarisierbaren Gruppen, wie aromatischen Ringen, oder durch Wasserstoffbrücken sowie durch kristalline Bereiche ist gegenüber länger dauernden oder stärkeren mechanischen Beanspruchungen oft nicht genügend stabil. Kristalline Bereiche verlieren überdies beim Aufschmelzen ihre Wirkung. Ferner kann man *aktive Füllstoffe* zusetzen, deren mikroskopische bzw. submikroskopische Partikelchen Stücke von Kettenmolekülen durch starke Kräfte adsorbieren und so sehr wirksame Fixpunkte bilden. Näheres über Vernetzungsmöglichkeiten siehe Abschnitt C.

Meist umgekehrt liegen die Verhältnisse bei einer molekularen Verteilung der zugesetzten Substanz. Bei starken Wechselwirkungskräften mit den Kettenmolekülen können bestimmte Gruppen der Ketten stark solvatisiert werden, so daß die unmittelbare Wechselwirkung der Kettenmoleküle stellenweise herabgesetzt, das molekulare Gefüge also gelockert und die mikro-BROWNsche Bewegung erhöht wird (*Weichmacherwirkung*), vgl. Bd. III, Kap. XI, sowie Bd. IV, Kap. IX, vor allem § 76 u. § 79. Werden die Ketten zu weitgehend solvatisiert, so verschwinden die Fixpunkte zwischen den Ketten, die makro-BROWNsche Bewegung wird frei und der Körper verliert seine Formfestigkeit.

Die *lokale Struktur* eines weichgemachten Thermoplasten und ihr Zusammenhang mit den mechanischen Eigenschaften ist von MARK u. Mitarbeitern[1] bei weichgemachtem Polyvinylchlorid näher untersucht worden. Bei Trioktylphosphat [$O{=}P(OC_8H_{17})_3$] als Weichmacher kann man annehmen, daß die Phosphatgruppe die Kettendipole solvatisiert, während die mit PVC unverträglichen Oktylschwänzchen sich zu submikroskopischen, die mikro-BROWNsche Bewegung zusätzlich fördernden „Klümpchen" zusammenballen.

Die Zahl der solvatisierten Gruppen ist in den praktisch vorkommenden Fällen relativ gering. So kommt bei einem Weichmachergehalt von 30 bis 40 Vol.-% eine Phosphatgruppe auf etwa 15 Grundeinheiten (CH_2—CHCl). So bleibt eine relativ große Zahl unabgeschirmter monomerer Einheiten zurück[2], die nach den Untersuchungen von ALFREY u. Mitarbeitern[3] ein kristallines Netzwerk bilden, das die Formbeständigkeit gewährleistet. Dabei ist der kristalline Anteil bei der hochpolymeren Komponente sicher sehr klein. Vgl. auch § 79, S. 598.

Es ist in diesem Zusammenhange bemerkenswert, daß Copolymere aus Vinylchlorid und Vinylacetat bereits bei 15% der letzteren Komponente nicht mehr die Langzeitformbeständigkeit besitzen wie reines Polyvinylchloridharz oder wie das Copolymerisat mit nur 5% Vinylacetat (VYNW). Der Grund ist der, daß mit steigendem Gehalt an Vinylacetat die regelmänige Folge der —CH_2—CHCl-Grundeinheiten immer mehr ge-

[1] AIKEN, W., T. ALFREY JR. A. JANSSEN u. H. MARK: J. Polymer Sci. **2**, 178 (1947).

[2] Man darf daraus natürlich nicht schließen, daß alle übrigen Grundeinheiten Kettenkontakte bilden, da die Trioktylphosphatmoleküle aus sterischen Gründen die Ketten erheblich auseinanderzwängen.

[3] ALFREY JR., T., N. WIEDERHORN, R. STEIN u. A. TOBOLSKY: Ind. Engng. Chem. **41**, 701 (1949). – Ferner J. Coll. Science **4**, 211 (1949).

stört, die Kristallisation also mehr und mehr zurückgedrängt wird, so daß das Netzwerk unvollkommener wird. Diese Störung der Kristallisation durch Copolymerisation ist auch der Grund, daß es z. B. bei PVC-Copolymeren nicht möglich ist, dieselben wertvollen mechanischen Eigenschaften zu erzielen wie durch eine äußere Weichmachung.

Die Frage, ob es sich um eine statische oder dynamische Solvatation handelt, d. h., ob die mittlere Lebensdauer eines Kontaktes von Ketten- und Weichmachermolekül groß oder klein gegenüber der Reaktionszeit der Segmentbewegung auf eine mechanische Beanspruchung ist, läßt sich noch nicht eindeutig beantworten. Nimmt man an, daß die kleinen Moleküle die bereits vorhandenen kristallinen Bereiche nur in geringem Umfange oder gar nicht sprengen, so würde man mit der statischen Vorstellung verstehen, weshalb man bei einer statistischen Copolymerisation nicht dieselbe Wirkung wie mit nachträglich zugesetzten kleinen Weichmachermolekülen erzielen kann. Die gewöhnliche Copolymerisation liefert notwendig eine statistische Unordnung längs der Ketten, welche die Kristallisation hemmt, so daß kein Netzwerk entstehen kann. Möglicherweise würde eine Copolymerisation von fertig polymerisierten Komponenten (blended polymers) günstigere Ergebnisse liefern.

Vgl. dazu auch die Diskussion bei MARK u. Mitarbeitern: siehe S. 615, Fußnote 1.

Vermögen die zugesetzten Moleküle gleichzeitig mit zwei Kettenmolekülen in Wechselwirkung zu treten, so erhalten wir eine Vernetzung durch starke VAN DER WAALSsche Kräfte bzw. Nebenvalenzkräfte. Ein Beispiel im Niedermolekularen ist die räumliche lockere Vernetzung von Alkoholmolekülen durch geringe Zusätze von Wasser[1], vgl. Bd. III, § 1.

Häufig wirken H_2O-Moleküle lockernd, indem sie, z. B. bei Polyamiden, H-Brücken sprengen und dann die einzelnen Partner absättigen können. So wird die Zahl der H-Brücken zwischen den Polyamidketten absolut verringert. Außerdem werden ständig H-Brücken gesprengt und an anderen Stellen neu gebildet (dynamisch statistisches Gleichgewicht zwischen abgesättigten und ungesättigten H-Brückenpartnern). So ergibt sich eine Lockerung des Gefüges mit erhöhter mikro-BROWNscher Bewegung, d. h. die Wassermoleküle wirken als *molekulares Schmiermittel*. Diese *innere molekulare Schmierung* äußert sich auch in den mechanischen Eigenschaften, vgl. § 85.

Wieder anders liegen die Verhältnisse bei nicht solvatisierenden Zusätzen begrenzter Verträglichkeit. Der überraschende Umstand, daß man auch mit begrenzt mischbaren Weichmachern praktisch brauchbare Produkte herstellen kann, läßt sich mit SPURLIN, MARTIN u. TENNENT[2] durch folgende Vorstellung erklären: Bei der Abkühlung des Stoffes von der Temperatur beim Mischvorgang kommt es zunächst zu einer submikroskopischen Phasentrennung, wobei die Weichmachermoleküle sich zu Mikrotröpfchen zusammenballen, die natürlich auch einige kurze, besser lösliche Ketten enthalten mögen. Diese submikroskopische Phasentrennung ist auch verschiedentlich elektronenmikroskopisch direkt nachgewiesen worden[3]. Eine solche Struktur ist natürlich thermodynamisch instabil. Doch kann in den zurückbleibenden weichmacherarmen Gebieten

[1] KAST, W. u. A. PRIETZSCHK: Z. Elektrochem. **47**, 112 (1941).

[2] SPURLIN, H. M., A. F. MARTIN u. H. G. TENNENT: J. Polymer Sci. **1**, 63 (1946).

[3] SPURLIN, H. M., A. F. MARTIN u. H. G. TENNENT: s. Fußnote 2. — W. R. RICHARD u. P. A. S. SMITH: J. chem. Physics **18**, 230 (1950).

die Beweglichkeit der Kettenstücke und damit auch die Diffusionsgeschwindigkeit der kleinen Moleküle stark absinken, wodurch die weitere Entmischung so langsam verläuft, daß wir einen weichgemachten, praktisch lange Zeit brauchbaren Thermoplasten erhalten[1]. Die submikroskopischen Tröpfchen haben sicher eine besonders gute weichmachende Wirkung, während das zurückbleibende Langkettengefüge die erforderliche Festigkeit besitzt.

§ 84. Die Elemente der molekularen Struktur und die Eigenschaften hochpolymerer Körper.

a) Allgemeines.

Von einer allgemeineren quantitativen oder auch nur qualitativen Theorie, mit deren Hilfe wir die makroskopischen Eigenschaften eines hochpolymeren Körpers aus seiner molekularen Struktur ableiten könnten, sind wir noch weit entfernt. Es gibt zwar eine Reihe von phänomenologischen Theorien, vor allem des mechanischen Verhaltens, die sich aber nur beschränkt mit der molekularen Struktur und mit molekularen Vorgängen in Verbindung bringen lassen. Am weitestgehenden sind noch die Theorien der elektrischen Eigenschaften auf molekularer Grundlage entwickelt worden.

Die Aufstellung einer allgemeinen Theorie ist schon deshalb erschwert, weil eine Grundvoraussetzung, nämlich die Kenntnis aller maßgebenden strukturellen Elemente und ihrer Bedeutung für bestimmte Eigenschaften noch gar nicht ausreichend erfüllt ist. Hier ist noch eine sehr große experimentelle Aufgabe zu erfüllen, nämlich die Durchführung systematischer Versuche an geeigneten und vor allem hinsichtlich ihrer chemischen Konstitution und ihrer molekularen Textur möglichst gut definierten Substanzen. Bekanntlich sind diese Vorbedingungen bei hochpolymeren Stoffen besonders schwer zu erfüllen. Die Erfahrungen der letzten Jahre, z.B. bei der Gewinnung einwandfreier Fraktionen oder bei der Bestimmung von Molekulargewichtsverteilungen, vor allem aber die Erkenntnis, daß Verzweigungen einen weit größeren Einfluß haben, als man bisher angenommen hatte, zeigen, daß die rein präparative Seite dieser Aufgabe einen um eine Größenordnung höheren Arbeitsaufwand, als bisher üblich, erfordert. Nur wenige Forschungsstellen und Wissenschaftler werden in der Lage sein, die hier nötige jahrelange und mühsame Zusatzarbeit zu leisten.

In den vergangenen Jahrzehnten ist ein riesenhaftes und immer noch anschwellendes Beobachtungsmaterial zusammengekommen, das, so nützlich es auch in anderer Hinsicht ist, doch nur zum geringsten Teile sichere Erkenntnisse geliefert und damit die wissenschaftliche Entwicklung wirklich gefördert hat. Wenn wir allmählich sichere Regeln und Grundsätze gewinnen wollen, um mit deren Hilfe Kunststoffe gewünschter Eigenschaften systematisch zu entwickeln, statt rein empirische und ebenso kostspielige wie zeitraubende Massenversuche anzustellen, müssen

[1] Vgl. auch Bd. III, § 80.

wir systematische Versuche an definierten und vergleichbaren Substanzen durchführen und den nötigen präparativen Aufwand in Kauf nehmen. Dieser mühsame Weg ist, auf lange Sicht, gesehen kürzer und billiger.

Die folgenden Betrachtungen dieses Kapitels mögen dem Wissenschaftler und Praktiker einige Hinweise geben, welche Faktoren nach dem heutigen Stande unseres Wissens für die physikalischen und technologischen Eigenschaften eines hochpolymeren Körpers von Bedeutung sind.

Die für die makroskopischen Eigenschaften wesentlichen molekularen Strukturelemente lassen sich vorläufig in fünf Gruppen einteilen. Die wichtigsten umfassen die Eigenschaften des einzelnen Makromoleküls, die Elemente der Textur sowie die Art und das Ausmaß der molekularen Beweglichkeit. Daneben spielen die Kettenlänge und ihre Verteilung sowie lockernd und verfestigend wirkende Beimischungen eine wichtige Rolle. Wir haben in der Tab. X,2 die genannten Gruppen und ihre zur

Tabelle X,2.

Die Elemente der molekularen Struktur eines hochpolymeren Körpers und ihre fünf Hauptgruppen.

I. Eigenschaften des Moleküls.
- a) Chemische Konstitution der monomeren Einheit, polare Gruppen, Wasserstoffbrücken-Partner, seitliche Substituenten.
- b) Zwischenmolekulare Kräfte und deren Geometrie: Kohäsionsenergie, Kraftfeld der monomeren Einheit, sterische Verhältnisse.
- c) Ordnung der monomeren Einheiten bei Copolymeren.
- d) Verzweigungen: kurze und lange Seitenketten, Ringbildung, innermolekulare Vernetzung.
- e) Statistische Formenmannigfaltigkeit und Biegsamkeit.

II. Kettenlänge und Kettenlängenverteilung.

III. Verfestigung und Lockerung des übermolekularen Gefüges, vor allem durch Zusätze.
- a) Vernetzung durch Haupt- und Nebenvalenzen.
- b) Vernetzung (Verfestigung) durch aktive Füllstoffe.
- c) Vernetzung durch Verschlingung langer Ketten (entanglements).
- d) Lockerung durch weichmachende Beimischungen einschließlich der Lockerung durch kurze Ketten und Monomere (innermolekulare Schmierung).

IV. Textur (alle Stufen der übermolekularen Ordnung).
- a) Ordnung benachbarter Kettenstücke (Kettenordnung).
  - Nahordnung in den „amorphen" Gebieten.
  - Mesomorphe Strukturen.
  - Gitterstrukturen.
- b) Kristallin–nichtkristallines Gefüge.
  - Verhältnis kristallin–nichtkristallin.
  - Form und Größe der kristallinen Bereiche.
  - Gegenseitige Ordnung der kristallinen Bereiche.
- c) Orientierung der Molekülketten in bezug auf äußere Richtungen.
  - In den kristallinen Gebieten.
  - In den nichtkristallinen Gebieten.
- d) Größere Ordnungseinheiten (morphologische Strukturen).
  - Fibrillen und Sphärolithstrukturen.
  - Gegenseitige Ordnung und Orientierung der Fibrillen.

V. Molekulare Beweglichkeit im Körper.
- Makro-Brownsche Bewegung (Bewegung der Kettenschwerpunkte).
- Mikro-Brownsche Bewegung (Bewegung von Kettenstücken).
- Drehbewegungen von kurzen Substituenten.

Zeit bekannten bzw. einer Untersuchung zugänglichen wichtigsten Elemente aufgeführt. Die Aufteilung ist wie jedes Schema natürlich nicht ganz willkürfrei. So könnte man unter III. auch die lockernde Wirkung durch innere Weichmachung (Copolymerisation) aufführen.

Der Begriff „*Textur*“ oder „*übermolekulare Ordnung*“ umfaßt sämtliche Stufen der übermolekularen Ordnung einschließlich der morphologischen Strukturen, ist also sehr weit gefaßt. Das ist schon deshalb zweckmäßig, weil die mit dem Elektronenmikroskop erfaßbaren morphologischen Formen bereits in die Abmessungen des kristallin—nichtkristallinen Gefüges hinabreichen.

Es ist ferner angebracht, bei der molekularen Beweglichkeit von der mikro-Brownschen Bewegung diejenige von kurzen Seitengruppen, z.B. OH-Gruppen, abzutrennen. Diese können ja erhebliche dielektrische Verluste und charakteristische Stufen in den mechanischen Konstanten, vgl. § 5, hervorrufen, ohne mit der mikro-Brownschen Bewegung, im üblichen Sinne, d.h. der Segmentbewegung, zusammenzuhängen.

Wir stellen nun der Tab. X, 2 mit den Elementen der molekularen Struktur jetzt eine Tab. X, 3 mit den wichtigsten makroskopischen Eigenschaften gegenüber. Es ist leider heute noch nicht allgemein möglich, anzugeben, ob und wie weit die einzelnen Eigenschaften von bestimmten Strukturelementen abhängen. Wir müssen uns daher begnügen, entsprechend unseren augenblicklichen Kenntnissen, wenigstens soweit möglich, die wichtigsten für die betreffende Eigenschaft maßgebenden Strukturelemente anzugeben (Spalte 2). Dabei haben wir den selbstverständlichen direkten Einfluß der Konstitution meist weggelassen. Die letzte Spalte enthält Hinweise auf die Stellen dieses Werkes, an denen die betreffenden Zusammenhänge besprochen werden.

In den folgenden Abschnitten B und C wird gezeigt werden, welche Strukturelemente und welche Kombinationen derselben für den faser- bzw. kautschukartigen Charakter eines Materials wesentlich sind. Wir werden daher in diesem Paragraphen nur noch die Bedeutung einzelner Elemente für bestimmte Eigenschaften herausstellen, und zwar vorwiegend solcher, die in den übrigen Kapiteln dieses Werkes gar nicht oder nur wenig behandelt worden sind.

b) Beispiele zu den Zusammenhängen zwischen den Strukturelementen und den physikalischen und technologischen Eigenschaften.

1. Verzweigung und innermolekulare Vernetzung.

In den letzten Jahren hat man immer deutlicher erkannt, daß es in vielen Fällen von großer Bedeutung ist, ob die Kettenmoleküle streng linear gebaut, oder ob sie verzweigt sind, welcher Art die Verzweigungen sind und ob Ringe bzw. innermolekulare Vernetzungen auftreten. Man hat ferner erkannt, daß Verzweigungen viel häufiger sind, als man früher, wenigstens stillschweigend, angenommen hatte, vor allem überall da, wo Übertragungsreaktionen in Frage kommen. Wahrscheinlich ist es überhaupt nur in Ausnahmefällen möglich, streng lineare Moleküle aufzubauen. Diese Situation ist recht unerfreulich, da damit sehr viele Unter-

Tabelle X, 3. *Makroskopische Eigenschaften von Körpern aus Fadenmolekülen und die wichtigsten maßgebenden Elemente der molekularen Struktur (Strukturparameter).*

Eigenschaft	Wichtigste Elemente der molekularen Struktur	Bemerkungen und Literaturhinweise
Festigkeit, längs und quer	Orientierung Kristallinität Mindestkettenlänge erforderlich bei amorphen Stoffen Vernetzung, Füllmittel s. Bd. IV, § 88 u. 89, 91 Morphologische Struktur	Statistisches Fehlstellenproblem, Bd. IV, § 16 u. 21 Abhängigkeit von Kettenlänge s. Bd. IV, § 86 Einfluß der chemischen Konstitution s. Bd. IV, § 56 u. 86 Faserstoffe § 54 u. 56
Bruchdehnung	Orientierung Kristallinität Vernetzung	Faserstoffe § 56
Dehnungs- und Torsionsmodul Dynamischer Modul	Orientierung, Kristallinität Vernetzung, Verzweigungen	Kautschuk Kap. V Faserstoffe § 54 u. 55
Zähigkeit	Kettenverteilung, sphärolithfreie Struktur	Anteil langer Ketten günstig; nicht zu hoch kristallin
Sprödigkeit	hohe Kristallinität, Sphärolithstruktur	s. § 84 b 3
Schlagbiegefestigkeit	geringe Mengen sehr niedermolekularer Anteile ungünstig	s. § 86 b
Knickbruchfestigkeit	Kristallinität, Orientierung, Vernetzung	
Plastizität	lockernde Zusätze	
Verformbarkeit bei höherer Temperatur	Kettenlänge und Verzweigungen	Mol.-Gewicht und Verzweigungen bestimmen die Schmelzviscosität, s. Bd. IV, § 86
Kaltverstreckbarkeit	kristalliner Anteil; Vernetzung	s. Bd. III, § 32; Bd. IV, § 13 c
Abrieb	Sphärolithgefüge günstig kleine Mengen sehr niedermolekularer Anteile ungünstig	s. § 84 b 3 s. § 86 b
Zeiteffekte Kriechen Elastische Erholung Relaxation d. Spannung wechselnde Beanspruchung	zwischenmolekulare Kräfte, Kristallisation Art der Vernetzung lockernde Zusätze; molekulare Beweglichkeit	s. Kap. I, V, VI, VII u. VIII, insbesondere § 5, 40, 47, 53, 54, 58
Ermüdung	geringe Mengen sehr niedermolekularer Anteile ungünstig	s. § 86 b
Verluste bei periodischer Beanspruchung	von M schließlich unabhängig	
Schmelzpunkt	chemische Konstitution, kristallines amorphes Gefüge, Orientierung	s. Bd. III, Kap. VIII u. IX

Tabelle X, 3. *(Fortsetzung).*

Eigenschaft	Wichtigste Elemente der molekularen Struktur	Bemerkungen und Literaturhinweise
Einfriertemperatur	chemische Konstituion, kristallin–nichtkristallin. Gefüge: innere Ordnung bei Copolymeren; lockernde Zusätze	s. Bd. III, Kap. X u. XI, ferner § 51 d. Ferner Bd. IV, Kap. IX
Löslichkeit, Quellbarkeit	chemische Konstitution, kristalliner Anteil, Kettenlänge, Verzweigung	s. Bd. IV, § 86; Bd. II, § 41
Wasserhaushalt	chemische Konstitution, kristalliner Anteil, Orientierung	s. Bd. IV, § 86 Bd. III, § 32 d, 44 b
Anfärbbarkeit	chemische Konstitution, kristalliner Anteil, Orientierung, Sphärolithgefüge	s. Bd. IV, § 86 s. Fußnote 1
Durchsichtigkeit	morphologische Struktur	vgl. Bd. VI, § 85; Bd. III, § 32
Dielektrische Verluste u. Dielektrizitätskonstante	chemische Konstitution, kristalliner Anteil lockernde Zusätze. Verunreinigungen z. B. durch Katalysatorreste	Bd. II, Kap. XIII; Bd. III, Kap. XI; Bd. IV, Kap. VIII
Elektrische Durchschlagsfestigkeit	Sphärolithgefüge ungünstig	s. § 84 b 3

suchungen der Zusammenhänge zwischen der Konstitution und Form des Einzelmoleküls und der Viskosität, Löslichkeit und den technischen Eigenschaften des Materials mit einer gewissen Unsicherheit behaftet sind.

Ist schon die Bestimmung der Molekulargewichtsverteilung und das Erkennen ihres Einflusses auf die makroskopischen Eigenschaften eines hochpolymeren Körpers eine sehr schwierige und meist noch nicht befriedigend gelöste Aufgabe, so trifft das in noch viel höherem Maße für das Studium des Einflusses von Verzweigungen auf diese Eigenschaften zu. Eine grundsätzliche Schwierigkeit ist dabei noch die, daß wir noch zu wenige chemische und physikalische Methoden haben, um die Größe und die Art der Verzweigung zu bestimmen. Es ist sinnlos, von Zusammenhängen zwischen Verzweigungen und irgendwelchen Eigenschaften eines festen Körpers oder einer Lösung zu sprechen, wenn man nicht die Verzweigung gleichzeitig näher definieren kann. So kann, wie schon in Bd. II, § 101 ausgeführt wurde, z. B. die Löslichkeit durch eine Verzweigung, je nach deren Charakter besser oder schlechter werden.

Auf Grund der bisherigen Erfahrungen scheint es zweckmäßig, vorläufig vereinfachend und schematisch folgende Typen einer Verzweigung zu unterscheiden und in Beziehung mit den makroskopischen Eigenschaften zu bringen (vgl. dazu auch Bd. II, § 101, Abb. XIV, 7):

a) Hauptkette mit Seitenketten ohne Folgeverzweigung, wobei wir ferner zwischen dem Fall kurzer und langer Seitenketten unterscheiden wollen.

[1] Forward, M. V. und S. C. Simmens: Journ. of the Textile Inst. **46**, 633 (1955).

b) Fadenmolekül mit so geringer Folgeverzweigung, daß man noch von einer Hauptkette sprechen kann.

c) Makromolekül mit sehr großer Folgeverzweigung und gelegentlicher innerer Vernetzung.

Am weitesten sind bisher die Verhältnisse beim Polyäthylen untersucht worden, wo man schon früh erkannt hat[1], daß die wertvollen Eigenschaften des technischen Produktes auf einer durch Verzweigung begrenzten Kristallisation beruhen. Streng linear gebautes Polyäthylen (Polymethylen), wie man es bei der Zersetzung von Diazomethan gewinnt[2], besitzt eine besonders hohe Dichte, ist hochkristallin und hat bei sehr langen Ketten einen Schmelzpunkt von 132° C[3], der im Gegensatz zu den technischen Produkten dicht an den Konvergenzpunkt für lineare Paraffine herankommt (vgl. dazu auch Bd. III, § 51 c).

Polyäthylen scheint wegen der Häufigkeit von Übertragungsreaktionen besonders stark zu Verzweigungen zu neigen. Nach den Untersuchungen von Nozaki[4], Beasley[5], und vor allem von Roedel[6] muß man beim Polyäthylen mit folgenden zu verschiedenen Verzweigungstypen führenden Übertragungsmechanismen rechnen[7]. Eine wachsende Radikalkette macht durch eine *Wasserstoff-Übertragungsreaktion* ein bereits fertiges Kettenmolekül von neuem nach dem Schema

$$R_1CH_2^* + R_2CH_2R_3 \rightarrow R_1CH_3 + R_2CH^*R_3$$

zu einem Radikal, so daß an dem betreffenden Kettenatom eine Verzweigung entsteht. Die neue Seitenkette wächst so lange, bis sie durch eine weitere Übertragungsreaktion auf ein anderes Molekül oder (häufiger) durch Absättigung mit einem anderen freien Radikal (bzw. durch Disproportionierung) abgebrochen wird. Man erhält so im allgemeinen eine lange Seitenkette.

Nimmt man diesen, offenbar auch thermodynamisch begünstigten Übertragungsmechanismus als richtig an, so muß man mit Roedel folgerichtig auch die Möglichkeit einer „*innermolekularen*“ *Übertragungsreaktion* zulassen.

```
                                             CH2 CH2
                                            /   \ /  \                CH2  CH2  CH2
  /CH2\    /CH2\    /CH2                   R    CH   CH2              /  \ /  \ /  \
R/     \CH2/    \CH2/    \CH2*  -->             |    |      -->    R     CH*  CH2  CH3
                                                H    CH2
                                                 ` . /
                                                   CH2*
```

¹ Vgl. R. B. Richards: J. appl. Chem. **1**, 370 (1951). – Vgl. ferner F. M. Rugg, J. J. Smith u. L. H. Wartmann: Annals of the New York Academy of Science **57**, 398 (1953).

² Buckley, G. D., L. H. Cross u. N. H. Ray: J. chem. Soc. [London] 1950, 2714. – Weitere Literatur bei R. B. Richards: Fußnote 1.

³ Kantor, S. W. u. R. C. Osthoff: J. Amer. chem. Soc. **75**, 931 (1953).

⁴ Nozaki, K.: Trans. Faraday Soc. Disc. **2**, 337 (1947).

⁵ Beasley, J. K.: J. Amer. chem. Soc. **75**, 6123 (1953).

⁶ Roedel, M. J.: J. Amer. chem. Soc. **75**, 6110 (1953).

⁷ Über einen weiteren, zu kurzen Verzweigungen führenden Reaktionsmechanismus s. F. M. Rugg, J. J. Smith u. L. H. Wartman, l. c., sowie J. Polymer Sci. **11**, 1 (1953).

An der Stelle, an der es zu einer vorübergehenden Ringbildung kommt, entsteht ein freies Radikal CH, so daß die Kette[1] hier weiterwächst, während die zurückbleibenden Ringglieder eine kurze abgesättigte Seitenkette bilden, deren C-Zahl am häufigsten bei 4 oder 5 liegen dürfte. Diese Reaktion muß sogar recht häufig sein, da für 5 oder 6 Glieder die sterischen Bedingungen für eine Wasserstoffübertragung besonders günstig werden. Natürlich kann das endständige Radikal, wenn auch viel seltener, auch weiter hinten aus der Kette ein H-Atom übernehmen, wodurch längere Seitenketten mit 8 bis 30 C-Atomen entstehen mögen. Eine Ringbildung bzw. innermolekulare Vernetzung erscheint auf Grund der obigen Übertragungsmechanismen sehr unwahrscheinlich.

Roedel diskutiert noch die Möglichkeit einer zweiten Verzweigung am gleichen C-Atom, indem nach der Addition von zwei weiteren Äthylenmolekülen die Bedingungen für eine weitere Ringbildung verbunden mit der Übertragung eines weiteren H-Atoms besonders günstig werden.

$$\underset{\substack{|\\C_4H_9}}{R{-}CH^*} + 2\,CH_2{=}CH_2 \longrightarrow \underset{\substack{|\\C_4H_9}}{R{-}CH}\overset{CH_2}{\diagup\;\diagdown}CH_2{-}CH_2{-}CH_2^* \longrightarrow \underset{\substack{|\\C_4H_9}}{\overset{\substack{C_4H_9\\|}}{R{-}C^*}}$$

Danach wären also das Auftreten von zwei kurzen Seitenketten am gleichen C-Atom besonders begünstigt.

Es ist schwer vorstellbar, daß eine Langkettenverzweigung ohne gleichzeitige Kurzkettenverzweigung auftritt. Umgekehrt können kurze Ketten ohne lange Ketten entstehen, nämlich dann, wenn die Konzentration an Kettenmolekülen bzw. das Verhältnis des polymeren zum monomeren Anteil genügend klein gehalten wird.

Durch Variation der Polymerisationstemperatur kann man den Anteil an kurzen Ketten in großen Grenzen variieren. Im groben Durchschnitt darf man mit etwa 3 Zweigstellen auf je 100 Kettenatome rechnen, was einem Zahlenmittelwert von etwa 60 Seitenketten je Molekül entsprechen würde[2]. Der Anteil an langkettigen Verzweigungen ist sehr viel kleiner, im Zahlenmittel liegt die obere Grenze praktisch bei einer Zweigstelle je Molekül. Da die langkettigen Verzweigungen aber bevorzugt bei höheren Molekulargewichten auftreten, erhält man als Gewichtsmittel gerechnet bis zu 10 Zweigstellen je Molekül. Dieser Umstand führt zu einer anomalen Verbreiterung der Verteilungskurve nach der Seite der hohen Molekulargewichte, so daß das Verhältnis $\overline{M}_w/\overline{M}_n$ bis zu 20 gehen kann[3].

---

[1] Bei nicht zu hoher Konzentration des polymeren Anteils ist die Wahrscheinlichkeit, daß das endständige Radikal ein Glied der eigenen Kette trifft, viel größer, als daß es auf ein fremdes Kettenmolekül stößt.

[2] Billmeyer, F. W.: J. Amer. chem. Soc. **75**, 6118 (1953). – J. K. Beasley: s. S. 622.

[3] Die Viscositätszahl eines sehr stark verzweigten Polyäthylens kann bis zu einem Zehntel des Wertes für eine unverzweigte Substanz vom gleichen Gewichtsmittelwert des Molekulargewichtes absinken. So kommt es, daß die Viscositätszahl ein ungefähres Maß des Zahlenmittelwertes unabhängig vom Verzweigungsgrade wird.

Dagegen hat die Kurzkettenbildung keinen direkten Einfluß auf die Verteilungskurve, da die weitere Polymerisation der Moleküle durch diesen Zwischenprozeß nicht beeinflußt wird[1].

Die Frage nach dem Einfluß der Verzweigungen auf die makroskopischen Eigenschaften einer Polyäthylensorte läßt sich vorläufig dahin beantworten — und diese Antwort gilt wahrscheinlich für viele andere hochpolymeren Stoffe —, daß es im wesentlichen auf die Kombination von drei Parametern ankommt, nämlich auf den Grad der kurzkettigen und der langkettigen Verzweigung, sowie auf das mittlere Molekulargewicht.

Da man diese drei Größen bei der Polymerisation unabhängig voneinander in ziemlich weiten Grenzen variieren kann, gelingt es, durch Veränderung des Molekulargewichts wachsartige bis zähfeste Massen herzustellen. Durch Variation des Kurzkettenanteils kann man biegsame bis steife, spröde Stoffe gewinnen, während mit wachsendem Langkettenanteil sowohl die Viscosität wie die Elastizität der Schmelze ansteigen.

Sicherlich spielt auch die Molekulargewichtsverteilung sowie die Verteilung der Längen der Seitenketten und der Zweigstellen eine Rolle, doch liegen darüber mangels geeigneter Meßmethoden noch keine näheren Erfahrungen vor.

Die kurzen Seitenketten hemmen vor allem die Kristallisation und setzen den kristallinen Anteil herab[2]. Lange Ketten sind hier von geringem Einfluß, einmal, weil sie relativ selten sind, und zum andern, weil lange Ketten wieder in andere Gitterbereiche eingebaut werden können.

Der Kurzkettenanteil beeinflußt also alle vom kristallinen Anteil abhängigen Eigenschaften. Mit seinem Ansteigen wird also die Dichte geringer, der Schmelzpunkt niedriger, das Material biegsamer und weicher sowie leichter löslich und durchlässiger gegen Dämpfe. Die Fließgrenze liegt tiefer. Die langkettigen Verzweigungen beeinflussen[3], wie schon gesagt, die viscoelastischen Eigenschaften der Schmelze, also auch die Verarbeitungsbedingungen. Die Reißfestigkeit des Materials sinkt mit steigendem Verzweigungsgrade bei sonst gleichen Bedingungen deutlich ab. Der Anteil der kurzen Ketten (Kriterium ist die Dichte) spielt dabei eine ganz untergeordnete Rolle. Die Bruchdehnung sowie die Härte und die Wärmefestigkeit (Vicat-Temperatur) werden dagegen vor allem von dem kurzkettigen Anteil und dem mittleren Molekulargewicht gesteuert.

Aus den Untersuchungen von SPERATI u. Mitarbeitern an Polyäthylen mit verschiedenen Verzweigungsgraden und mittleren Molekulargewichten folgt, daß manche Beobachtungen über die Abhängigkeit einer Eigenschaft, z. B. der Steifheit, vom Molekulargewicht in Wirklichkeit darauf

---

[1] Ein indirekter Einfluß mag, wie ROEDEL betont, dadurch hereinkommen, daß die Anwesenheit tertiärer C-Atome den Angriff durch freie Radikale begünstigt. M. J. ROEDEL: s. S. 622.

[2] Einzelne seitliche CH_3-Gruppen sollen nach R. B. RICHARDS: s. S. 622, die Kristallisation nicht sonderlich stören; allerdings sind CH_3-Substitutionen nach den besprochenen Übertragungsmechanismen kaum zu erwarten und tatsächlich nach Aussage der Ultrarotanalyse auch nicht in nennenswertem Umfange vorhanden.

[3] SPERATI, C. A., W. A. FRANTA u. H. W. STARKWEATHER JR.: J. Amer. chem. Soc. **75**, 6127 (1953).

beruhen[1], daß mit fortschreitender Polymerisation der Verzweigungsgrad zunimmt. Man erkennt auch an diesem Beispiele, wie vorsichtig man bei der Aufstellung von allgemeinen Regeln über die Abhängigkeit bestimmter Eigenschaften von bestimmten Strukturelementen sein muß.

Das Problem „Verzweigung – technologische Eigenschaften", dessen Schwierigkeiten erst in jüngster Zeit klarer erkannt worden sind, zeigt eindringlich die Notwendigkeit, alle systematischen Versuche über Struktur und Eigenschaften an möglichst definierten Substanzen durchzuführen.

2. *Kettenlänge und Kettenlängenverteilung.*

Mit wachsender Kettenlänge wird der Einfluß der Endgruppen immer geringer, so daß bei hinreichend langen Molekülen die Ordnung in der Umgebung irgendeines Kettenstückes sowie die mikro-BROWNsche Bewegung vom Polymerisationsgrad unabhängig werden müssen. Es müssen also *bei sonst gleichen Bedingungen*[2] im wirklich festen (formelastischen) Zustand alle mechanischen, elektrischen und optischen Eigenschaften sowie die Umwandlungspunkte gegen einen Grenzwert gehen. Nur da, wo die makro-BROWNsche Bewegung eine Rolle spielt, erhalten wir keinen Grenzwert, sondern eine dauernde Änderung. So nimmt die Schmelzviscosität wegen der mit der Kettenlänge ansteigenden gegenseitigen Verschlingung der Ketten ständig zu. Die Verformbarkeit bei höherer Temperatur unterhalb und oberhalb des Erweichungsintervalls bzw. des Schmelzpunktes wird immer schwieriger; der kautschukartige Charakter und damit die Störung durch elastische Nachwirkungen bei der Formgebung immer größer.

In sehr vielen Fällen, vor allem bei synthetischen kristallisierenden Hochpolymeren werden wegen der Kürze der Ketten die Grenzwerte noch lange nicht erreicht, so daß die mittlere Kettenlänge und die Kettenlängenverteilung eine ausschlaggebende Rolle spielen können. Vor allem ist für die Erreichung einer vernünftigen, mechanischen Festigkeit eine Minimallänge erforderlich; vgl. Abschnitt B.

Über den Einfluß der Verteilungskurve auf die mechanischen Eigenschaften läßt sich vorläufig nicht viel aussagen. Kleine Änderungen im Verlauf der Verteilungsfunktion scheinen keinen wesentlichen Einfluß auf die Festigkeit zu haben mit der Einschränkung, daß schon geringe Anteile an kurzen Ketten die Festigkeit entscheidend verschlechtern können. Dabei kommt es nicht auf den Gewichtsanteil, sondern auf die Zahl der kleinen Moleküle an. Offenbar spielt die Zahl der Endgruppen, die als „Schwachstellen" wirken, die maßgebliche Rolle[3].

---

[1] CAREY, R. H., E. F. SCHULZ u. G. J. DIENES: Ind. Engng. Chem. **42**, 842 (1950).

[2] Die Einschränkung „bei sonst gleichen Bedingungen" muß man bei der Beurteilung von vergleichenden Messungen immer im Auge behalten. So kann z. B. eine größere Kettenlänge das viscoelastische Verhalten so beeinflussen, daß bei der Verformung des Materials oder beim Abkühlen größere innere Spannungen entstehen und so das Material spröder wird statt zäher. Um zu einem einwandfreien Vergleich zu kommen, muß man also das Material etwa bei einer der größeren Viscosität angepaßten höheren Temperatur verformen.

[3] Vgl. dazu auch die Betrachtungen bei T. ALFREY JR. in: Mechanical Behaviour of High Polymers, New York 1948.

Umgekehrt kann schon ein geringer Anteil von anomal langen Ketten die Zähigkeit, Duktilität des Materials merklich verbessern.

3. *Morphologische Struktur.*

Die Bedeutung der morphologischen Struktur von nativen Faserstoffen für deren Eigenschaften ist schon lange bekannt. Dagegen weiß man bis heute über die bei synthetischen Stoffen (Fasern und massiven Formkörpern) auftretenden Strukturen und ihre technologische Bedeutung noch wenig. Die Sphärolith- und Fibrillenstrukturen sind in Bd. III, § 45, eingehend behandelt worden. Die exakte Führung der Schmelzbedingungen, vor allem die überall gleichmäßige Führung der Temperatur und ihres zeitlichen Verlaufs haben es ermöglicht, systematische Versuche über die technologischen Eigenschaften von Formkörpern aller Art, von der Seide bis zum massiven Körper, in Abhängigkeit von der morphologischen Struktur durchzuführen. Solche Ergebnisse sind von großer praktischer Bedeutung, aber bisher nur sehr spärlich bekannt geworden.

Es ist kein Zweifel, daß ein sphärolithaltiges Material, auch bei gleichem kristallinen Anteil, spröder als ein feinkristallines ist[1]. Die Art des Bruches, der entlang den radialen Fasern des Sphärolithen erfolgt[2], die verringerte elektrische Durchschlagsfestigkeit[3] gegenüber einem sphärolithfreien Material sowie die geringere Dichte weisen darauf hin, daß zwischen den Fasern radiale kapillare bzw. molekulare Risse vorhanden sind, die als Fehlstellen wirken[4]. In vielen Fällen ist also eine Sphärolithstruktur unerwünscht. Umgekehrt ist bei Beanspruchung auf Abrieb ein sphärolithisches Material günstig. Eine der wichtigsten Eigenschaften von Polyamiden als Werkstoff für Maschinenelemente, Zahnräder, Gleitlager, nämlich ihre hohe Verschleißfestigkeit, scheint wesentlich auf der Sphärolithstruktur[5] zu beruhen.

Schließlich bringen wir noch einige Hinweise auf weitere Zusammenhänge zwischen Strukturelementen und Eigenschaften, die an anderen Stellen dieses Werkes behandelt worden sind.

Kristallin-amorphes Gefüge: Kristalliner Anteil, Orientierung und mechanische Eigenschaften von Fasern, Kap. VII, § 54—56.

Einfluß der Kristallinität auf die Eigenschaften von Kautschuk, Kap. V, § 40.

Kristallinität, Orientierung und Verfestigung-Kaltverstreckung, Bd. III, § 32, und Bd. IV, § 13 c.

Vernetzung: Zwischenmolekulare Kräfte, Kriechen und Relaxation bei Faserstoffen, Kap. VII B.

---

[1] LANGKAMMERER, C. E. u. N. E. CATLIN: J. Polymer Sci. **3**, 305 (1948).

[2] Siehe die Aufnahmen von H. A. STUART, U. VEIEL u. M. HARTMANN-FAHNENBROCK: Naturwiss. **40**, 339 (1953); H. A. STUART u. B. KAHLE, J. Polymer Sci. **18**, 143 (1955), siehe Bd. III, § 45.

[3] REDING, F. P. u. A. BROWN: Ind. Engng. Chem. **46**, 1962 (1954).

[4] Daß man diese Risse durch Tempern zum Teil ausheilen kann, ergibt sich aus einer Beobachtung von BRENSCHEDE, wonach beim Erwärmen von Polyamiden (nach Bd. III S. 509) auf etwa positive Sphärolithe in negative übergehen, vgl. Bd. III, § 45, S. 509.

[5] JACOBI, H. R.: Z. Ver. dtsch. Ing. **96**, 1197 (1954).

Zwischenmolekulare Kräfte-Kraft-Dehnungsdiagramm von Fasern, Kap. VII A, § 54; Kap. X B, § 86.

Vernetzung-Kautschukelastizität, Kap. V; Kap. X C, § 89 und 91.

Aktive Füllmittel, Verfestigung von Kautschuk, Kap. X C.

Copolymerisation: Schmelzpunkte von Copolymeren, Bd. III, § 53.

Innere Weichmachung, Bd. III, Kap. XI, und Bd. IV, § 75[1].

Kautschukeigenschaften, § 90.

Dielektrische Eigenschaften, § 72.

Blended polymers (Block-Copolymere, Schmelzmischpolymerisate), Bd. III, § 53; Bd. IV, § 91.

B. Faserstoffe.

Von **H. Mark.**

§ 85. Allgemeines.

Das systematische Studium natürlicher Faserstoffe wie Hanf, Flachs, Ramie, Baumwolle, Seide und Wolle mit den Methoden moderner Forschung hat eine Reihe wichtiger, wenn auch in den meisten Fällen nur qualitativer Grundsätze aufgedeckt, die das endgültige Verhalten dieser Stoffe im praktischen Gebrauch und damit ihren Wert bestimmen, vgl. dazu auch Kap. VII. Diese Grundsätze sind in ihrer allgemeinsten Form bereits im Abschnitt A dieses Kapitels auseinandergesetzt worden, und es ist daher die Aufgabe des vorliegenden Abschnittes B, sie in besonderer Hinsicht auf die faserbildenden Eigenschaften natürlicher und synthetischer Hochpolymeren zu besprechen und ihren heuristischen Wert im Hinblick auf die Herstellung besserer und billigerer neuer Faserstoffe abzuschätzen[2].

Die technische „Güte" einer Faser ist nur in ganz wenigen Fällen (wie vielleicht bei Seilen und Nähgarnen) in überragender Weise von einer einzigen Eigenschaft (wie z.B. der Reißfestigkeit) bedingt. Im allgemeinen hängt sie vom sorgfältig herbeigeführten Ineinandergreifen mehrerer (manchmal sogar recht vieler) Eigenschaften ab. So wird man von einer guten Faser für die Herstellung von Hemdenstoffen verlangen, daß sie eine hohe Festigkeit hat und daher in sehr feine Garne mit großer Geschwindigkeit versponnen werden kann. Außerdem soll sie eine nicht zu geringe Bruchdehnung besitzen, denn sie muß hohe Abreibefestigkeit haben, sich den Bewegungen der Arme bequem anpassen und nicht leicht knittern. Man wird außerdem eine wohlbemessene Wasseraufnahmefähigkeit verlangen; nicht zu gering, damit sich das Gewebe nicht kalt und

[1] Vgl. auch W. O. Baker in Advancing Fronts in Chemistry, Vol. I „High Polymers" S. 105 ff.; Reinhold Publ. Corp. New York 1945.

[2] Zusammenfassende neuere Darstellungen sind zu finden bei: Rowland Hill: Fibres from Synthetic Polymers, Elsevier Publishing Company, Amsterdam 1953, oder H. R. Mauersberger: Textile Fibres, John Wiley, New York, 6. Edition 1954. Vgl. auch R. Pummerer: Die Chemie der Fasern und Folien, F. Encke, Stuttgart 1953. – J. M. Preston: Fibre Science, 2. Edition, Manchester 1953.

klebrig anfühlt und damit elektrostatische Aufladungen rasch genug abgeführt werden; nicht zu hoch, damit das Hemd nach dem Waschen rasch trocknet und in nassem Zustande nicht zuviel von seiner Festigkeit und von der ihm beim Zuschnitt gegebenen Form verliert. Darüber hinaus muß die Faser einen angenehmen Glanz besitzen, muß eine Oberflächenbeschaffenheit haben, die den Schmutz beim normalen Gebrauch nicht zu leicht aufnimmt und die, wenn er einmal an der Faser haftet, eine leichte Entfernung mit den gewöhnlichen Waschmitteln verspricht; sie muß sich in bequemer Weise anfärben und bedrucken lassen und muß schließlich eine hinreichende Widerstandsfähigkeit gegen Licht, Sauerstoff, Hitze und Kälte sowie gegen den Angriff von Säuren, Basen und organischen Lösungsmitteln aufweisen. Man sieht, daß eine „ideale" Faser selbst für ein relativ einfaches Gebrauchsgewebe einen recht komplizierten Kompromiß von Eigenschaften darstellt, deren gemeinsame Erreichung sicherlich nicht dem Zufall überlassen werden darf, sondern in wohlüberlegter Weise durch sorgfältige Wahl der geeignetsten Makromoleküle und ihrer besten Anordnung in der Faser herbeigeführt werden kann. Wir wollen versuchen, die hierfür maßgebenden Grundsätze in drei verschiedenen Stufen zu besprechen:

1. Im Hinblick auf die *isolierten Makromoleküle*, aus denen die Faser aufgebaut ist, ihr mittleres Molekulargewicht, ihre Molekulargewichtsverteilung, die charakteristischen Gruppen, welche sie tragen, sowie die molekularen Symmetrieeigenschaften und die Biegsamkeit der Ketten. Diese Besprechung ist im wesentlichen Gegenstand der organischen und physikalischen Chemie einzelner Moleküle und ihrer Teile und betrifft die Verhältnisse, die sich im Bereich von wenigen ÅNGSTRÖM-Einheiten abspielen.

2. Im Hinblick auf die Art und Weise, in der die Moleküle des Faserformers in der Faser zu axial und lateral geordneten Bereichen, den sogenannten kristallinen „Bereichen" oder „Micellen" zusammengefaßt sind, so daß eine *übermolekulare Ordnung der Molekülketten* entsteht, die den Größenordnungsbereich von 100 bis 1000 Å umfaßt, vgl. Tab. X,2. Diese Ordnung ist den Methoden der Phasenlehre in ähnlicher, wenn auch in etwas komplizierterer Weise zugänglich, wie bei der experimentellen und theoretischen Behandlung von Metallegierungen oder Salzgemischen.

3. Im Hinblick auf die im Mikroskop erkennbare *Grobstruktur* oder *Morphologie* der Faser, die den Bereich vom Mikron bis zum Zehntelmillimeter (10^4 bis 10^6 Å) umfaßt, und die bei natürlichen Fasern die Folge ihres Wachstums, bei synthetischen Gebilden das Ergebnis einer von außen nach innen fortschreitenden Koagulation, Koazervation oder Kristallisation ist. Diese Morphologie äußert sich im Vorhandensein von Kapillarsystemen, Fibrillen, Oberflächenhäuten, Rindensubstanzen und Schuppen, welche viele wichtige Eigenschaften der Fasern, Garne und Gewebe in erheblicher Weise beeinflussen.

Bei Punkt 2 und 3 können wir uns unter Hinweis auf frühere Abschnitte in Bd. III und IV dieses Werkes sehr kurz fassen.

Im Laufe der nun folgenden Besprechung wird es sich herausstellen, daß zwischen diesen drei Stufen ein enger Zusammenhang besteht in dem

Sinne, daß schließlich alle Eigenschaften einer Faser in irgendwelcher Weise von den molekularen Verhältnissen mitbestimmt sind. Gerade dieses Durchgreifen der Struktureinzelheiten des faserbildenden Makromoleküls in die Textur mit ihren verschiedenen Stufen der übermolekularen Ordnung ist es, welche es ermöglicht, allerlei interessante Kombinationen von praktisch wertvollen Eigenschaften zustande zu bringen.

Wir wollen daher versuchen, mit den molekularen Verhältnissen beginnend, ein Bild der für den Aufbau und die Eigenschaften von Faserstoffen maßgebenden Grundsätze zu entwerfen.

§ 86. Molekulare Eigenschaften.

a) Mittleres Molekulargewicht.

Es ist seit langem bekannt, daß die mechanischen Eigenschaften aller linearen Hochpolymeren vom Molekulargewicht in dem Sinne abhängen, daß bis zu einem kritischen unteren Grenzwert so gut wie kein Festigkeitsverhalten erzielt werden kann. Nach Erreichen dieser unteren Grenze nehmen die zu erwartenden mechanischen Eigenschaften mit zunehmendem Molekulargewicht relativ rasch zu, um in einem zweiten kritischen Bereich allmählich vom Molekulargewicht mehr oder weniger unabhängig zu werden. Graphisch dargestellt ergibt dieser Zusammenhang die in der Abb. X,1 wiedergegebene Kurve[1], welche die beiden technisch wichtigen kritischen Bereiche B_u und B_o deutlich zum Ausdruck bringt. Formelmäßig lassen sich die Verhältnisse angenähert durch die Beziehung

$$Z = A - \frac{B}{\overline{M}_n} \qquad (X,1)$$

darstellen[2], worin Z die gerade in Betracht gezogene mechanische Eigenschaft, z. B. die Zerreißfestigkeit, und $\overline{M}_n$ das mittlere Molekulargewicht (Zahlenmittel) darstellt. Die Konstanten A und B sind für ein gegebenes Material charakteristisch und hängen offensichtlich mit den kritischen Bereichen der Abb. X,1 zusammen. Jenes mittlere Molekulargewicht, unterhalb dessen die Festigkeit Null wird[3], ist durch

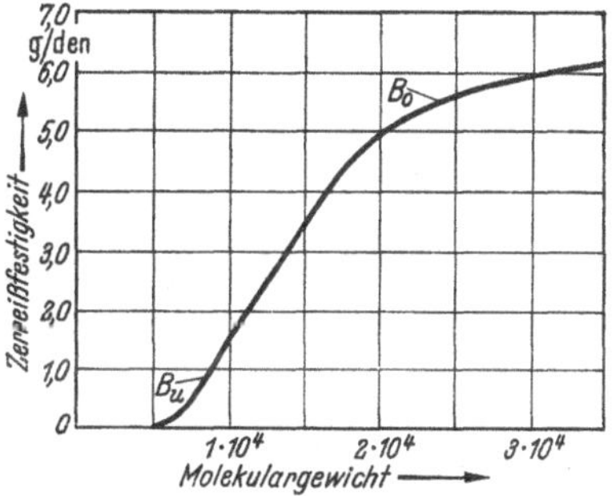

Abb. X, 1. Allgemeiner Zusammenhang zwischen mechanischer Festigkeit und Molekulargewicht am Beispiel von 6,6-Nylon.

$$\overline{M}_n = B/A \qquad (X,2)$$

[1] Vgl. H. MARK: Chemie und Physik der Cellulose, Springer, Berlin 1932. – H. MARK: Ind. Engng. Chem. **34**, 1343 (1942).

[2] FLORY, P. J.: J. Amer. Chem. Soc. **67**, 2048 (1945); auch A. M. SOOKNE u. M. HARRIS: Ind. Engng. Chem. **37**, 478 (1945).

[3] Die negativen Werte, zu denen man durch Gl. (X,1) geführt wird, sind offenbar durch den Wert Null zu ersetzen, was durchaus im Sinne der halbempirischen Natur dieser Gleichung liegt.

gegeben; der (asymptotisch sich einstellende) mögliche Höchstwert der Zerreißfestigkeit ist A.

Je stärker die zwischenmolekularen Kräfte — Wasserstoffbrücken, polare oder nicht polare VAN DER WAALSsche Anziehung — zwischen den Kettenmolekülen sind, um so größer ist A und um so kleiner ist B; doch hängen beide Größen nicht nur von der Natur und Zahl der anziehenden Gruppen je Längeneinheit der Kette ab, sondern auch von der genaueren Geometrie ihrer Verteilung, welche beim Parallellagern der Makromoleküle die Anzahl der ineinander passenden seitlichen Kontakte bestimmt und damit den Grad der inneren Kohäsion beeinflußt.

Obwohl noch sehr wenig Zahlenmaterial über A und B bzw. B_u und B_0 vorliegt, seien doch in Tab. X, 4 einige angenäherte Werte für die kritischen Bereiche der Abb. X,1 angedeutet, wie sie sich der gegenwärtig zugänglichen Literatur entnehmen lassen.

Tabelle X,4.
Kritische Bereiche B_u und B_0 für einige wichtige Hochpolymere.

Material	B_u			B_0		
	Mol.-Gew.	$\bar{P}_n$	Kettenlänge in Å	Mol.-Gew.	$\bar{P}_n$	Kettenlänge in Å
6,6-Nylon	6000	50	400	24000	200	1600
6-Nylon (Perlon)	6000	50	400	24000	200	1600
Terylen	8000	70	420	30000	250	1600
Cellulose	20000	130	650	75000	500	2500
Polyacrylnitril (Pan, Orlon)	15000	300	850	45000	900	2550
Polyvinylalkohol	15000	300	850	45000	900	2550
Polyvinylidenchlorid	25000	250	650	75000	750	2000
Polystyrol	60000	600	1500	300000	3000	7500

Wie zu erwarten, haben Polymere mit vielen, regelmäßig angeordneten Anziehungszentren (die daher leicht kristallisieren) niedrige B_u- und B_0-Werte, während typisch amorphe Materialien erheblich höhere kritische Bereiche zeigen. In der Tabelle sind auch die einzelnen Materialien nicht einfach nach ihrem Molekulargewicht, sondern sinngemäßer nach ihrem Polymerisationsgrad oder nach ihrer Kettenlänge verglichen. Solange man nun verschiedene Fraktionen ein und derselben Substanz vergleicht, ist es offenbar ganz richtig, in Gl. (X,1) das mittlere Molekulargewicht $\overline{M}_n$ als unabhängige Veränderliche zu verwenden; sobald man aber Substanzen von verschiedener chemischer Struktur miteinander vergleicht, ist es notwendig, den Polymerisationsgrad oder die ihm proportionale tatsächliche Kettenlänge einzuführen, da ein schwerer seitlicher Substituent, wie C_6H_5— in Polystyrol, $C_{10}H_7$ in Polyvinylnaphthalin oder $C_{11}H_{35}CO$— in Polyvinylstearat zwar das Molekulargewicht erhöht, aber die seitliche Anziehung nicht nur nicht fördert, sondern, im Gegenteil, sogar erheblich abschwächt.

In diesem Sinne zeigt die Tab. X,4 die Wichtigkeit von Wasserstoffbrücken oder stark polaren Bindungen für die Entwicklung hoher Festigkeitswerte.

b) Molekulargewichtsverteilung.

Bei der Herstellung einer Faser wird man sich demgemäß von dem unteren Grenzwert B_u geflissentlich fernhalten und würde vielleicht geneigt sein, der Sicherheit halber auch den oberen Grenzwert B_0 erheblich zu überschreiten, wenn man dadurch nicht in praktische Produktionsschwierigkeiten geraten würde. Diese werden dadurch hervorgebracht, daß mit zunehmendem Polymerisationsgrad die Viscosität der Spinnlösungen sowie auch der Spinnschmelzen sehr rasch ansteigt, was zu unüberwindlichen Hindernissen beim technischen Spinnen führt. Hierdurch gelangt man in ein Dilemma zwischen den besten erreichbaren Fasereigenschaften und der einfachsten (und daher billigsten) Faserherstellung, das zu den folgenden Schlüssen führt[1]:

Die Abhängigkeit der Viscosität η einer technischen Spinnflüssigkeit (konz. 20- bis 30-%ige Lösung oder Schmelze) vom Molekulargewicht, oder besser, wieder vom Polymerisationsgrad P ist gegeben durch

$$\eta = K \overline{P}_w^a, \tag{X,3}$$

wobei $\overline{P}_w$ das Gewichtsmittel des Polymerisationsgrades angibt. Dieses Gewichtsmittel ist in vielen Fällen (wenn nämlich die Molekulargewichtsverteilung nicht zu kompliziert ist) einfach ein Vielfaches des Zahlenmittels $\overline{P}_n$

$$\overline{P}_w = n \overline{P}_n, \tag{X,4}$$

wobei n für scharfe Fraktionen Werte zwischen 1,1 und 1,3 annimmt, im Normalfall in der Nähe von 2,0 liegt und nur im Falle größerer Breite der Verteilungskurve höhere Werte (bis zu 20)* annimmt.

Schreiben wir nun die Gl. (X,1) auch mit $\overline{P}_n$ anstatt mit $\overline{M}_n$ als

$$Z = Z_\infty - b/\overline{P}_n, \tag{X,5}$$

so haben wir in (X,3), (X,4) und (X,5) drei Gleichungen, die uns gestatten, das Problem in einer halbquantitativen Weise zu diskutieren.

Im allgemeinen wird von praktischer Seite her die Forderung erhoben werden, daß die Spinnviscosität einen gewissen kritischen Wert η_{kr} nicht überschreite. Daraus ergibt sich ein oberer Grenzwert $(\overline{P}_w)_0$ für das Gewichtsmittel

$$(\overline{P}_w)_0 = (\eta_{kr}/K)^{1/a}. \tag{X,6}$$

Dieser größte zulässige Wert für $\overline{P}_w$ ist um so kleiner, je größer K und a, wobei a als Exponent die wichtigere Rolle spielt. Im allgemeinen liegt a für konzentrierte Polymerlösungen zwischen 2 und 3 und für polymere Schmelzen zwischen 3 und 4, wobei sowohl die molekulare Biegsamkeit der Kette als auch das Vorhandensein anziehender Gruppen den genauen Wert des Exponenten in Gl. (X,3) bestimmen.

Die mit $(\overline{P}_w)_0$ erreichbare (beste praktische) Festigkeit ergibt sich dann als

$$Z = Z_\infty - n b/(\overline{P}_w)_0 \tag{X,7}$$

[1] Boyer, R. F.: J. Polymer Sci. 9, 289 (1953).

* Beim Polyäthylen; vgl. § 84 b 1.

Bei gegebenen Werten für Z_∞ und b erhält man offensichtlich die günstigsten Verhältnisse, wenn n den kleinstmöglichen Wert, nämlich $n = 1{,}0$ annimmt, d.h., wenn man eine möglichst homogene polymere Substanz verwendet. Man sieht also, daß die Molekulargewichtsverteilung einen erheblichen Einfluß auf die praktische Faserherstellung ausübt.

Abgesehen von diesen allgemeinen Erwägungen scheinen aber besonders sehr niedermolekulare Anteile schon in recht kleinen Mengen (5 bis 10%) für solche Eigenschaften wie Schlagbiegefestigkeit, Abrieb und Ermüdung besonders ungünstig zu sein, weil sie starre Bereiche (kleine Kriställchen aus Niedermolekularen) erzeugen, an denen sich starke Spitzenspannungen ausbilden, die zu einem vorzeitigen Nachgeben des Materials führen.

c) Funktionelle Gruppen.

Die zwischenmolekularen Kräfte in einem Hochpolymeren sind von ausschlaggebender Bedeutung für Löslichkeit, Schmelzpunkt, Umwandlungspunkte, für die endgültigen mechanischen Eigenschaften, Wasseraufnahme, Anfärbbarkeit und andere textile Eigenschaften des Materials. Von besonderer Wichtigkeit sind Gruppen, welche Wasserstoffbrücken bilden können, wie –CONH–, OH und NO_2, aber auch starke Dipole wie CN, CO und CCl. Wiederum ist nicht nur die Zahl dieser Gruppen je Gewichtseinheit des Materials wesentlich, sondern auch ihre Anordnung im einzelnen und die hieraus sich ergebenden gegenseitigen Bindungen zwischen den funktionellen Gruppen der einzelnen Ketten.

Je mehr funktionelle Gruppen zur van der Waalsschen Bindung zwischen den Makromolekülen verwendet werden, d.h., je besser die Orientierung und je höher die Kristallinität eines Präparates sind, desto weniger Gruppen bleiben für andere Funktionen wie das Binden von Wasser oder von Farbstoffmolekülen, übrig, so daß ein und dieselbe Substanz ganz verschiedene textile Eigenschaften haben kann, je nachdem welcher Prozentsatz der funktionellen Gruppen in einem bestimmten Fabrikationsstadium noch zugänglich ist.

Tab. X,5 gibt eine Reihe von Beispielen dafür, wie das Wasserbindungsvermögen von verschiedenen Fasern von der Anwesenheit funktioneller Gruppen abhängt und wie die Zugänglichkeit dieser Gruppen durch Quellung erhöht, durch Orientierung und Kristallisierung aber erniedrigt werden kann.

Tabelle X,5.
Wasserbindungsvermögen einiger Fasern in Gewichtsprozenten nach verschiedener Vorbehandlung.

Faser und Vorbehandlung	H_2O-Aufnahme in %	Faser und Vorbehandlung	H_2O-Aufnahme in %
Baumwolle	6—8	Nylon unverstreckt	4—5
Baumwolle, mercerisiert	8—10	Nylon verstreckt	3—4
Viscose Stapelfaser	8—10	Perlon verstreckt	∼4
Textilfaser Viscose	6—8	Dacron (Terylen)	0,5—1,0
Hochverstreckte Viscose	4—6	Orlon, Pan verstreckt	0,5—1,0
Fortisan	4—6	Acrilan verstreckt	0,5—1,0
Acetatseide	4—6	Dynel verstreckt	1,0—1,5

d) Einfluß der Seitenketten.

Von großem Interesse ist die Einführung gewisser Seitenketten in ein gegebenes Rückgrat-Makromolekül und der Einfluß dieser Ketten auf die mechanischen, thermischen und textilen Eigenschaften der aus dem substituierten Material gesponnenen Fasern. Je nach dem Charakter der Seitenketten oder Substituenten sowie auch in Abhängigkeit von ihrer Häufigkeit und Anordnung gelangt man zu ganz verschiedenen, aber in jedem Falle wichtigen Ergebnissen. Einige typische Fälle mögen hier angeführt werden.

1. *Unpolare, unregelmäßig verteilte* und *ungleichartige* Seitenketten wirken wohl stets als „Weichmacher", setzen den Erweichungspunkt des Materials herab, die Löslichkeit herauf und verhindern oder vermindern die Kristallisation selbst bei sehr starker Verstreckung. Als Beispiele nennen wir die deutlichen Unterschiede zwischen dem völlig unverzweigten Polymethylen, das ausgezeichnet kristallisiert und bei etwa 135° C schmilzt und dem normalen Polyäthylen (Lupolen, Alkathen, Alathon), das viele kurze (4 bis 6 C-Atome) und wenige lange Seitenketten besitzt, bei 112° C und weniger schmilzt, nur unvollständig kristallisiert und viel leichter löslich ist als die unverzweigte Modifikation. Zahlreiche Mischpolymerisate aus Äthylen, Propylen, Isobutylen und alkylierten Butadienen zeigen ebenfalls deutlich den weichmachenden Charakter unpolarer und ungleichartiger Substituenten an, welche daher hauptsächlich in der Chemie der synthetischen Kautschuke zur Herabsetzung der Sprödigkeitstemperatur verwendet werden, in der Herstellung vollsynthetischer Fasern aber eher vermieden als gesucht sind, da sie, abgesehen von ihrem ungünstigen Einfluß auf den Erweichungspunkt auch noch den Elastizitätsmodul herabsetzen, die Reißfestigkeit vermindern und die bleibende Dehnung sehr stark erhöhen.

2. *Unpolare (oder schwach polare) gleichartige* und *regelmäßig verteilte* Seitenketten bringen ein durchaus anderes Verhalten mit sich. Solange sie kurz sind, wie etwa im Cellulosetriacetat, -tripropionat oder im Polyvinylacetat und Polymethylmethacrylat wird durch ihre Einführung zunächst die Kristallinität der ursprünglichen Polymeren (Cellulose, Polyvinylalkohol oder Polyacrylsäure), wenn eine solche überhaupt vorhanden war, modifiziert und etwas herabgesetzt, aber nicht völlig zerstört. Die Wasserempfindlichkeit der Ausgangssysteme, die in vielen Fällen (Kunstseide, Cellophan, Polyvinylalkohol) sehr ausgeprägt ist, wird erheblich verbessert und an Stelle von Löslichkeit in wäßrigen Systemen tritt Löslichkeit in organischen Flüssigkeiten. Der Erweichungspunkt wird etwas herabgesetzt, behält aber, besonders bei den Cellulosederivaten, immer noch beachtlich hohe Werte. Die Verbesserung des Verhaltens im feuchten Zustand und die leichte Löslichkeit in gewöhnlichen organischen Lösungsmitteln hat zur weiten Verwendung der Celluloseacetate und -propionate auf dem Gebiete der Kunstfasern geführt. Macht man die unpolaren, gleichartigen und regelmäßig angeordneten Seitenketten länger und geht zu 5, 6 oder 7 Kohlenstoffatomen, so resultieren in allen Fällen nichtkristallisierende, tiefschmelzende und leicht lösliche,

plastische Produkte von großer Weichheit und sogar kautschukartigem Charakter. Geht man aber zu noch längeren Seitenketten, etwa 16, 18 oder mehr Kohlenstoffatomen, so beginnen diese wegen ihrer Gleichartigkeit und wegen ihrer dichten Packung selbst zu kristallisieren und man erhält mit Röntgenstrahlen ein typisches Paraffindiagramm, wie es von den freien Fettsäuren her wohlbekannt ist. Diese Erscheinung wurde bei Celluloseestern schon vor längerer Zeit von HAGEDORN[1] beobachtet und ist neuerdings von mehreren Autoren auch bei den höheren Fettsäureestern des Polyvinylalkohols und bei den höheren Fettalkoholestern der Polyacrylsäuren gefunden und eingehend untersucht worden. Die resultierenden Produkte haben wachsartigen Charakter mit niedrigen Erweichungspunkten (40 bis 60° C), völliger Wasserunempfindlichkeit und Öllöslichkeit. Im Textilgebiet können sie höchstens als Appreturmittel, aber nicht als Faserformer verwendet werden.

3. *Polare, kurze, regelmäßig verteilte* Seitenketten oder Substituenten sind für Faserformer von besonders vorteilhaftem Einfluß, denn sie erhöhen den Schmelzpunkt, vermindern die Löslichkeit, bewirken mehr oder weniger weitgehende Kristallisation und setzen die Zerreißfestigkeit hinauf. Polyacrylsäurenitril, Polyvinylalkohol, Polyvinylidenchlorid und Polyvinylchlorid sind Beispiele dafür, in welchem Maße polare, gleichartige und verhältnismäßig regelmäßig angeordnete Substituenten die faserbildenden Eigenschaften einer nur aus Kohlenstoffatomen bestehenden Rückgratkette beeinflussen. Bei natürlichen Fasern entsprechen die regelmäßig verteilten Hydroxylgruppen der Cellulose und die ebenfalls regelmäßig und in kurzen Abständen angeordneten Peptidbindungen der natürlichen Seide solchen verfestigenden, die Fasereigenschaften im allgemeinen günstig beeinflussenden Substituenten.

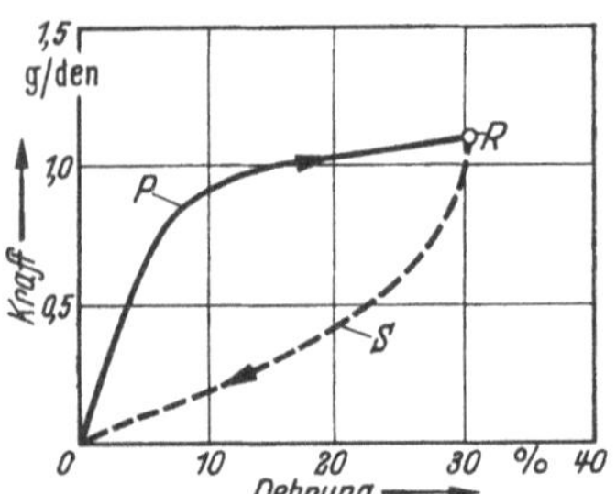

Abb. X,2a. Typisches Kraft-Dehnungs-Diagramm einer wollähnlichen Faser. Die gestrichelte Linie gibt das Verhalten bei der Erholung wieder.

4. *Polare, kurze, unregelmäßig verteilte* Seitenketten oder Substituenten sind hauptsächlich in Proteinfasern, besonders in der Wolle sowie in anderen Keratin- und Myosinfasern zu finden. Sie bewirken eine relativ hohe mittlere Molkohäsion und verhindern Kristallisation selbst im stark gedehnten Zustand, liefern aber eine sehr wirksame Vernetzung[2]. Bei Beginn der Dehnung müssen zunächst die polaren (eventuell Wasserstoffbrücken bildenden) Gruppen der Seitenketten aus ihren Gleichgewichtslagen herausgehoben werden, was eine relativ hohe Aktivierungsenergie erfordert und daher einen hohen, anfänglichen Elastizitätsmodul zur Folge hat (siehe Abb. X,2a, Strecke 0 P). Je mehr von diesen seitlichen Bindungen nun bei fortschreitender Verformung gelöst werden, desto geringer wird der Wider-

[1] HAGEDORN, M. u. P. MÜLLER: Cellulosechemie **12**, 29 (1930).
[2] MARK, H.: Ind. Engng. Chem. **44**, 2110 (1952).

stand gegen weitere Dehnung, was zu einer allmählichen Abflachung der Dehnungskurve führt. Beim Punkte R sind die meisten seitlichen polaren oder Wasserstoffbindungen gelöst und die Rückgratketten in einem gestreckten Zustand relativ niedriger Entropie. Läßt man an dieser Stelle die Faser sich ein wenig verkürzen, so werden sich infolge der erhöhten Beweglichkeit sogleich neue seitliche Bindungen bilden, die potentielle Energie sinkt stark ab. Durch diese Verfestigung des Gefüges sinkt die Spannung schon bei sehr kleiner Verkürzung stark ab und die Kontraktion der Faser vollzieht sich bei erheblich höherer innerer Viscosität und niedrigerer Spannung als die entsprechende Dehnung. Sie verläuft daher meist in ihrem letzten Teil sehr langsam und stellt die allmähliche Erholung wollähnlicher Fasern dar, was nach genügender Zeit unter Entropieerhöhung zur vollen Wiederherstellung des Ausgangszustandes der gestreckten und orientierten Rückgratketten führt. Die steilen Stücke des in Figur 2a dargestellten Kreisprozesses OP und RS entsprechen Verformungen, bei denen die Entropie im wesentlichen konstant bleibt, während die innere Energie von O nach P ansteigt und von R nach S abnimmt; die flachen Teile PR und SO beschreiben Vorgänge mit wesentlich konstanter innerer Energie und abnehmender bzw. zunehmender Entropie. Bei einem idealen Kautschuk würde sich die Schleife $OPRSO$ zu einer geraden Linie OR zusammenziehen und die Kontraktion entlang derselben Kurve wie die Dehnung erfolgen; siehe Abb. X, 2b. Je mehr die Elastizität einer Faser auf Änderungen der Entropie beruht, desto geringer wird der Elastizitätsmodul und die Hysterese. Doch liefert ein derartig schlaffes und nerviges Material keine gute Faser.

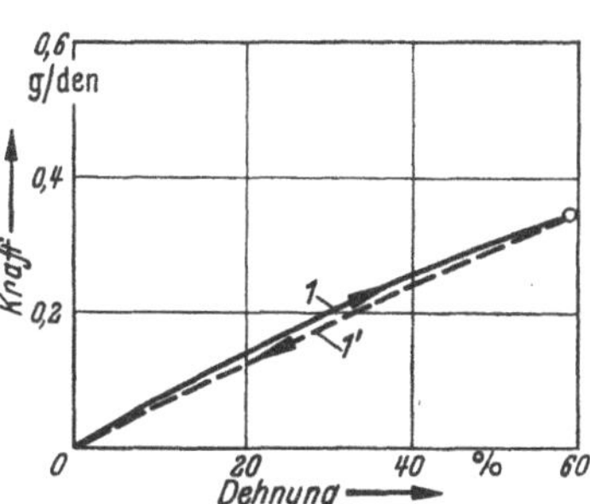

Abb. X, 2b. Kraft-Dehnungs- und Entspannungs-Kurve einer Faser. Kurve 1: Dehnungskurve; Kurve 1': Entspannungskurve.

Das Kraft-Dehnungs-Diagramm der Abb. X, 2a ist charakteristisch für Wolle und andere Fasern mit vielen kurzen polaren Substituenten wie Polyvinylalkohol und Polyacrylsäurenitril. Allerdings ist bei diesen beiden Systemen in gestrecktem Zustand eine erheblich höhere Kristallinität festzustellen als bei den Proteinfasern.

§ 87. Textur und Umwandlungspunkte.

a) Textur.

Wie bereits im Kap. VII, A eingehend gezeigt worden ist, hängen die Eigenschaften von Fasern wesentlich vom kristallinen Anteil und vom Orientierungsgrade ab. Die Kristallinität ihrerseits wird wiederum von der chemischen Konstitution des Kettenmoleküls sowie von seiner Symmetrie bestimmt, vgl. dazu das Kap. IX „Kristallisation und Konstitution“ in Bd. III dieses Werkes. Der kristalline Anteil von Faserstoffen

Tabelle X,6.
Kristallinität verschiedener Fasern in Prozenten.

Material und Behandlung	Prozent kristallin	Material und Behandlung	Prozent kristallin
Polyäthylen unverzweigt	95	Polyacrylnitril	niedrig
Polyäthylen etwas verzweigt	~ 70	Polyvinylalkohol	60—70
Polyäthylen stark verzweigt	~ 40	Ramie	75—85
Polyester	55—75	Fortisan	60—70
Nylon-66	50—60	Viscoseseide	30—40
Terylen, wollartig	30 →		
Terylen, seidenartig	→ 80		

liegt, wie Tab. X,6 zeigt, im allgemeinen zwischen 30 und 80%, nähere Angaben in Bd. III, § 26. Hohe Zerreißfestigkeit, hoher Elastizitätsmodul, Formbeständigkeit und Steifheit sind in den kristallinen Bereichen verankert, während die amorphen Bezirke die nötige Dehnbarkeit, Farbstoffaufnahme, Quellfähigkeit bestimmen. So erweist sich, wie bereits eingangs erwähnt, eine „gute" Faser als ein zusammengesetztes System, das eine Reihe von Eigenschaften in sich vereinigen muß.

Auf die Morphologie der nativen Fasern gehen wir hier nicht ein, da diese sehr eingehend in Bd. III, Kap. VI behandelt worden ist[1]. Dagegen seien noch kurz die morphologischen Strukturen von synthetischen Fasern besprochen. Auch hier treten ausgesprochene Fibrillenstrukturen auf[2]. Ihre Entstehung ist noch nicht klar, zum Teil dürfte sie einfach auf der geringen Querfestigkeit der Fasern beruhen, die beim Quellen, Auflösen oder mechanischen Zersplittern der Fasern Fibrillen liefert.

Oberflächenhäute, Randzonen sind zu erwarten, sobald die äußeren Schichten merklich schneller abkühlen als die inneren oder eine Koagulation in der Randzone so schnell verläuft, daß infolge der herabgesetzten Diffusion die Koagulation im Innern verzögert wird.

Der Übergang von dem seitlich und axial ungeordneten Zustand (amorph und unorientiert) eines polymeren Materials in den lateral und axial geordneten Zustand (kristallin und orientiert) kann sich grundsätzlich in zwei Weisen abspielen.

Man kann zuerst die Hauptvalenzketten parallel ausrichten, ohne sie seitlich in eine regelmäßige Anordnung zu bringen und erhält hierbei axial hochorientierte Fasern (im amorphen oder glasartigen Zustand), die sich nachher durch entsprechendes Tempern oder Quellen im hochverstreckten Zustand lateral ordnen lassen: Man orientiert *zuerst* und kristallisiert *nachher*. Die Herstellung hochorientierter Cellulosefasern im Wege über sekundäres Celluloseacetat stellt eine solche Arbeitsfolge dar. Zunächst werden die Kettenmoleküle des amorphen Acetats durch starke Verstreckung orientiert und dann durch Entfernung der Acetylreste in gut kristallisierende Cellulose übergeführt.

Man kann aber auch zuerst Makromoleküle der ursprünglich amorphen und nichtorientierten Substanz in gewissen Bereichen seitlich ord-

[1] Bei der Untersuchung der morphologischen Strukturen haben sich die Phasenkontrastmethode sowie das Interferenzmikroskop bewährt, vgl. zur letzteren Methode R. C. FAUST· Proc. physic. Soc. (B) **65**, 48 (1952).

[2] Siehe Band III, § 45.

nen (auskristallisieren) und hierdurch ein Agglomerat von winzigen Kristalliten oder Micellen erzeugen, die von amorpher Substanz umgeben sind. Dies geschieht meist beim Verspinnen kristallisierbarer Polymerer aus der Schmelze oder aus der Lösung, wenn ohne ausgeprägte Vorverstreckung gearbeitet wird. Nachher kann man dann durch Verstreckung die kristallinen Bereiche parallel zur Faserachse ausrichten, wodurch man auch hier schließlich zu einem Präparat gelangt, welches gleichzeitig hohe axiale und laterale Ordnung aufweist; bei diesem Verfahren kristallisiert man zuerst und orientiert nachher. In beiden Fällen gelangt man zu Endprodukten, die aus hochorientierten kristallinen Bereichen bestehen, welche in eine amorphe Masse eingelagert sind, und es entsteht die Frage, ob einem der beiden eben angedeuteten Wege der Vorzug im Hinblick auf bessere mechanische Eigenschaften gebührt. Überblickt man das gegenwärtig zur Verfügung stehende experimentelle Material, so gewinnt man den Eindruck, daß man in beiden Fällen Fasern von etwa gleicher Zerreißfestigkeit erhält, daß aber die Bruchdehnung, die Knickfestigkeit und der Widerstand gegen Abrieb und Zerfaserung bei jenen Präparaten besser ist, die nach dem zweiten Verfahren — erst kristallisieren und dann orientieren — gewonnen worden sind.

b) Umwandlungspunkte.

Selbstverständlich ist bei einer Faser ein hoher Schmelzpunkt erwünscht. T_F wird durch die Gleichung $T_F = \Delta H/\Delta S$ bestimmt, ΔH die Schmelzenthalpie und ΔS die Schmelzentropie. Wie diese Größen von der Konstitution des Einzelmoleküls, seiner Symmetrie und Biegsamkeit abhängen, ist in Bd. III, Kap. IX, eingehend besprochen worden. Letzten Endes sind also auch hier wiederum die Eigenschaften des Einzelmoleküls maßgebend.

Für die Gebrauchseigenschaften einer Faser ist neben dem Schmelzpunkt auch die Einfriertemperatur von Wichtigkeit, bei der die mikro-Brownsche Bewegung „einfriert". Da beide Umwandlungen wiederum von der Konstitution des Einzelmoleküls und anderen Kenngrößen desselben abhängen, vgl. Bd. III, § 51d, und Kap. XI, bestehen zwischen ihnen gewisse Beziehungen[1]. Es ist allerdings zu beachten, daß bei partiell kristallinen Stoffen die mikro-Brownsche Bewegung nicht bei einer einzigen Temperatur verschwindet, sondern daß die verschiedenen Bewegungsmechanismen nacheinander einfrieren bzw. auftauen. Das erkennt man deutlich aus den Torsionsschwingungsversuchen von Wolf u. Schmieder[2], die bei Polyamiden noch Einfrierbereiche bei −120° und −50° C zeigen, vgl. Abb. XI,7 in Bd. III.

Es leuchtet ein, daß die Temperaturlagen der Hauptstufe und die der Nebenstufen für die mechanischen Eigenschaften einer Textilfaser sehr wichtig sind. In der Tab. X,7 stellen wir die Umwandlungspunkte für eine Reihe von Faserstoffen zusammen.

---

[1] Bei vergleichbaren Stoffen besteht die Beziehung $T_E = a\,T_F$, wobei a häufig etwa = 2/3 ist, vgl. Bd. III, § 51 d.

[2] Wolf, K. u. K. Schmieder: Koll.-Z. **134**, 149 (1953).

Tabelle X,7.
Umwandlungspunkte einiger wichtiger Faserstoffe.

Material	Einfriertemperatur T_E (Hauptstufe)		Schmelzpunkt T_F	
	C°	K°	°C	°K
Polyäthylen, techn.	— 68	205	105—115	378—388
Polyvinylidenchlorid	— 17	256		
Polyacrylnitril (Orlon, Pan)	90	363		
Polyvinylalkohol	80	353	190	463
6-Nylon	45	318	220	493
6,6-Nylon	48	321	256	529
6,10-Nylon	45	318	213	486
Terylen (Dacron)	70	343	256	529

Wie man sieht, fallen die Hauptstufen, die Einfriertemperaturen, im bisher üblichen Sinne mitten in den Anwendungsbereich der betreffenden Fasern, was häufig unerwünschte Verformungen und mangelnde Widerstandsfähigkeit zur Folge haben wird. Eindeutige Zusammenhänge lassen sich allerdings heute noch nicht sicher angeben, weil ja unterhalb der Hauptstufe, wie z.B. bei 6- oder 6,6-Nylon noch Kristallisation möglich, d.h. noch eine gewisse Kettenbeweglichkeit vorhanden ist.

Wenn man daher im Interesse gewisser Eigenschaften nach einem Faserformer sucht, der einen Schmelzpunkt von 350°C (623°K) hat, so nicht so sehr deshalb, weil man diesen endgültigen Zusammenhalt des Materials für nötig erachtet, sondern vielmehr, weil man für die amorphen Bereiche einen Umwandlungspunkt von $\sim$132°C (415°K) $= 2/3\ T_F$ für wünschenswert erhält. Im Hinblick auf verschiedene Textilprozesse, wie Bügeln, Dämpfen oder Heißfixieren, ist dies sicherlich keine allzu übertriebene Anforderung.

C. Kautschukartige Stoffe.

Von **H. Mark.**

§ 88. Einleitung.

Lange Zeit war der aus dem Saft von Hevea Brasiliensis zu gewinnende polymere Kohlenwasserstoff — ein Poly-1,4-cis-isopren vom mittleren Molekulargewicht zwischen 100000 und 200000 — die einzige Substanz, die „Kautschukelastizität" zeigte, und man war daher geneigt anzunehmen, daß das Vorliegen dieser eigentümlich weichen, elastischen Verformbarkeit mit der *chemischen* Zusammensetzung des *Naturkautschuks* zusammenhängt. Tatsächlich bewegten sich die ersten Versuche zur Herstellung *künstlicher* Kautschuke durchweg in der Richtung, Polymere des Isoprens oder ihm nahestehender Kohlenwasserstoffe herzustellen und in ähnlicher Weise zu „füllen" und zu „vulkanisieren", wie dies in der Technologie des Naturkautschuks üblich ist. Obwohl diese Arbeiten zu sehr interessanten und praktisch überaus wichtigen Resultaten führten, gelangte man doch bald auf Grund theoretischer Überlegungen und experimenteller Ergebnisse zu der heute allgemein angenommenen Überzeugung, daß Kautschukelastizität eine viel allgemeinere Eigenschaft hochpolymerer Stoffe ist, die sich keineswegs auf polymere Kohlenwasserstoffe beschränkt, sondern im gleichen Maße auch bei ganz andersartigen Substanzen, wie Polyestern, Polyamiden oder Silikonen, vorkommt, wenn nur gewisse notwendige strukturelle Bedingungen erfüllt sind, deren Vorhandensein den Grad der weichen, reversiblen Dehnbarkeit und den Temperaturbereich bestimmt, in dem sich das Material als „nerviger" Kautschuk verhält[1].

Im Sinne dieser allgemeineren Auffassung kann man als „*Kautschuk*" oder „*Elastomer*" eine Substanz definieren, die

1. bei relativ geringen äußeren Zugkräften (10^6 bis 10^7 dyn/cm^2) bereits erhebliche Dehnungen (800 bis 1000%) annimmt und sich hierbei bis auf das 10- oder 100fache verfestigt,

2. diese Dehnung für längere Zeit (10^5 sec und mehr) ohne Abklingung der Endspannung beibehält und

3. beim Aufhören der äußeren Zugkräfte sehr rasch (innerhalb von Bruchteilen einer Sekunde) und so gut wie völlig (bis auf weniger als ein Prozent) in ihren ursprünglichen Zustand (sowohl was Form als auch was Eigenschaften betrifft) zurückkehrt.

Diese Kombination von Eigenschaften genügt zwar, um das grundsätzliche Vorhandensein von Kautschukelastizität zu garantieren, ist aber für einen praktisch brauchbaren Kautschuk nicht ausreichend. Ein solcher muß vielmehr noch eine Reihe anderer Qualitäten haben, wie etwa:

1. Aufnahmefähigkeit für verstärkende Füllstoffe, wie Ruß, Silicium- oder Aluminiumoxyd oder Zinkoxyd, deren Anwesenheit den Elastizitätsmodul und den Widerstand gegen Abrieb erhöht;

2. Reaktionsfähigkeit für die in der Kautschuktechnologie eingeführten Vulkanisationsmittel und deren Beschleuniger;

3. Widerstandsfähigkeit gegen Hitze und Licht in der Gegenwart von Sauerstoff und Wasser;

4. geringe Quellbarkeit in Wasser und organischen Substanzen, wie Benzin, Öl und Lösungsmittel.

Um einen „guten" Kautschuk zu erzeugen, muß man daher — genau wie dies in Abschnitt B für einen „guten" Faserstoff postuliert wurde — nach einem *Kompromiß* suchen, welcher *alle* oder doch die meisten wichtigen Eigenschaften in solcher Weise zusammenklingen läßt, daß sich

[1] Vgl. folgende zusammenfassende Darstellungen: C. E. H. Bawn: The Chemistry of High Polymers, Interscience Publishers, New York 1948. — H. Mark u. A. V. Tobolsky: Physical chemistry of high Polymeric Systems, Interscience Publishers, New York 1950. — P. J. Flory: Principles of Polymer Chemistry, Cornell University Press, Ithaca 1953. — K. H. Meyer: Natural and Synthetic High Polymers, Interscience Publishers, New York, Sec. Eol. 1951. — L. Zechmeister: Fortschritte der Chemie organischer Naturstoffe, Springer: Wien 1953, Bd. X, S. 119. — S. D. Gehman: Relationship between Molecular Structure and Physical Properties, Ind. Engng. Chem. 44, 730 (1952).

wertvolle neue Anwendungsmöglichkeiten ergeben. Die Bestrebungen für einen solchen *Ausgleich* der ineinandergreifenden Einflüsse werden offenbar um so erfolgreicher sein, je mehr sie sich auf ein grundsätzliches (womöglich quantitatives) Verständnis der maßgebenden Faktoren für die Existenz kautschukartiger Elastizität gründen. Wir wollen daher zunächst diese Grundsätze kurz beschreiben und nachher ihre Anwendbarkeit bei der Synthese neuer hochelastischer Stoffe skizzieren.

§ 89. Grundsätze der Kautschukelastizität.

Vom Standpunkt der Phasenlehre nehmen *Elastomere* eine Zwischenstellung zwischen den festen und flüssigen Substanzen ein; sie gleichen den Flüssigkeiten dadurch, daß kleine äußere Kräfte sehr erhebliche Formveränderungen bewirken, daß kautschukartige Substanzen im ungedehnten Zustand amorph sind, daß sie normale niedermolekulare organische Stoffe in erheblichem Ausmaße lösen und in sich diffundieren lassen und daß sie die Kompressibilitätsverhältnisse einer typischen Flüssigkeit zeigen. Den Festkörpern sind Elastomere dadurch verwandt, daß sie im unbeanspruchten Zustand eine wohldefinierte Gestalt haben, in welche sie selbst bei sehr starker Verformung rasch und vollständig zurückkehren, und daß viele von ihnen im gedehnten Zustand deutliche Kristallisationserscheinungen zeigen, die durch einen relativ scharfen Schmelzpunkt charakterisiert sind, vgl. Bd. III, § 43b.

Es ist offenbar eine leichte Beweglichkeit in kleinen Volumelementen notwendig, um die flüssigkeitsartigen Eigenschaften eines Kautschuks zu bewirken, während andererseits in größeren Bereichen gewisse Bindungen vorhanden sein müssen, die selbst unter dem längeren Einfluß relativ starker Kräfte nicht gelöst werden und für das festkörperähnliche Verhalten maßgebend sind. Ein Kautschuk kann als eine Flüssigkeit angesehen werden, in der ein System von festen Netzpunkten existiert; je geringer die Viscosität der Flüssigkeit, um so weicher ist der Kautschuk und um so rascher kehrt er in den unverformten Zustand zurück. Je stärker das Netzwerk, um so größere Spannungen kann das Material aushalten, ohne sich dauernd zu verformen.

Im Sinne dieser Anforderungen wird man daher bei der Herstellung eines synthetischen Kautschuks nach Makromolekülen suchen, die eine möglichst geringe mittlere Molkohäsion haben, damit die einzelnen Segmente der individuellen Ketten leicht aneinander vorbeigleiten können, wenn äußere Kräfte rasche Änderungen in der Konfiguration der miteinander verknäuelten Makromoleküle notwendig machen. Andererseits ist es notwendig, dieser Masse unregelmäßig angeordneter Ketten ein weitmaschiges Netz von erheblich stärkeren Bindungen zu überlagern, das ihren endgültigen Zusammenhalt und ihre Rückkehr in den Ausgangszustand gewährleistet. Dies kann in verschiedener Weise erreicht werden, z.B. durch starke chemische Bindungen, die zwischen den einzelnen Kettenmolekülen in größeren und im allgemeinen unregelmäßigen Abständen hergestellt werden (wie bei der Schwefelvulkanisierung), oder aber durch das Beimischen eines festen Füllstoffes (wie etwa Ruß, SiO_2 oder Al_2O_3), an dessen Oberfläche die Kautschukketten in ungewöhnlich starker Weise adsorbiert werden. Es ist scheinbar auch möglich, durch weitgehende Verzweigung der einzelnen Makromoleküle, besonders beim Vorliegen sehr hoher Molekulargewichte, so etwas wie eine sehr stark „*mechanische Verflechtung*" der Ketten herbeizuführen, so daß mehr oder weniger lokale Knoten entstehen, deren Entwirrung durch die BROWNsche Molekularbewegung so lange dauert, daß sie als praktisch permanente Vernetzungsbereiche angesehen werden können, ohne daß eine wirkliche chemische Bindung vorliegt. Diese Erscheinung der gegenseitigen mechanischen Verwicklung biegsamer Makromoleküle scheint hauptsächlich im Bereich sehr hoher Molekulargewichte vorzukommen und kann durch geeignete Verhältnisse bei der Polymerisation („Gel" und „Popkorn"-Polymerisation) offenbar begünstigt werden; sie scheint eine wichtige Rolle bei der Streckung gewisser synthetischer Kautschuke durch Schwerölfraktionen zu spielen. Schließlich können auch kristalline Bereiche eines hochpolymeren Stoffes die Rolle von vernetzenden Elementen übernehmen, besonders dann, wenn die Kristallite sehr klein und räumlich voneinander genügend entfernt sind, so daß sie eine im wesentlichen ungestörte Bewegung der dazwischenliegenden Kettensegmente gestatten. Man sieht aus diesen qualitativen Ausführungen, daß es eine ganze Reihe verschiedener Möglichkeiten gibt, die Forderung des gleichzeitigen Vorhandenseins von hoher Beweglichkeit in kleinen Bereichen und von festen Bindungen in größeren Abständen zu erfüllen. Dies hat zur Folge, daß es sehr viele polymere Systeme gibt, die in irgendeinem Temperaturbereich jenes Zusammenwirken dieser beiden Prinzipien in sich vereinigen, das zum Auftreten von Hochelastizität führt. Vom praktischen Standpunkt wird ein Elastomer naturgemäß um so brauchbarer sein, je weiter der Temperaturbereich ist, in dem die oben formulierten Bedingungen erfüllt sind. Tab. X,8 enthält einige kautschukartige Substanzen zusammen mit dem Temperaturbereich, innerhalb dessen sie sich als wirkliche Elastomere bewähren.

Tabelle X,8.
Temperaturbereich der Kautschukelastizität von verschiedenen Elastomeren.

Material	untere Grenze in °C	obere Grenze in °C
Silikonkautschuke	— 100	+ 150
Naturkautschuk	— 80	+ 130
Fluorierte Polyacrylate	— 70	+ 130
Butylkautschuk	— 70	+ 120
Buna-Kautschuk	— 55	+ 120
Neopren	— 35	+ 130
Vinylmischpolymere	— 10	+ 80

Die nächsten Fragen, die sich aus dem eben Gesagten offenbar ergeben, sind:

1. Welches sind die verschiedenen Möglichkeiten, hochpolymere Systeme mit hoher lokaler Beweglichkeit herzustellen und

2. auf welche Arten und Weisen kann das für den Nerv des Kautschuks notwendige System von Fixpunkten erzeugt werden?

§ 90. Charakteristisierung linearer Makromoleküle mit Kautschukelastizität.

Um eine lokale Beweglichkeit der einzelnen Ketten in einem hochelastischen Material zu begünstigen, wird man vor allem zunächst trachten, dem Prinzip der „freien" Drehbarkeit um die einfache C—C-, C—N- und CO-Bindung zur möglichsten Auswirkung zu verhelfen. Hierbei ist zu bedenken, daß eine wirklich „freie" Drehbarkeit nur dann besteht, wenn keine — anziehende oder abstoßende — Wirkung der Substituenten an den beiden einfach gebundenen Atomen in Frage kommt; so wurde gefunden, daß die beiden Methylgruppen in Dimethylacetylen

$$H_3C—C{\equiv}C—CH_3$$

und in Azomethan

$$H_3C—N{=}N—CH_3$$

gegeneinander völlig frei drehbar sind, weil ihr gegenseitiger Abstand so groß ist (etwa 4 Ångström), daß sie sich nicht mehr in merklicher Weise behindern. Hingegen besteht im Äthan, wo der gegenseitige Abstand der beiden CH_3-Gruppen nur etwa 1,5 Å beträgt, eine recht erhebliche Behinderung, die sich dadurch ausdrückt, daß bei der Rotation eine Potentialschwelle von etwa 2,7 kcal/Mol überwunden werden muß. Stärker behinderte Drehbarkeit liegt im 1,2-Dichloräthan (4,5 kcal) und besonders im Hexafluoräthan (5,5 kcal) vor, wie sich besonders aus Absorptionsmessungen im Infraroten ergeben hat, vgl. dazu Bd. I, § 28. Diese Zahlen lassen erkennen, daß normale Paraffinketten von großer Länge, wie sie im Polyäthylen vorliegen, und noch mehr chlorierte und fluorierte Paraffinketten, die im Polyvinylchlorid, Polyvinylidenchlorid, Polychlortrifluoräthylen und Polytetrafluoräthylen vorhanden sind, keine sehr große lokale Beweglichkeit besitzen können. In der Tat sind alle genannten Polymere typische plastische Massen und nehmen erst bei relativ hohen Temperaturen über einen ziemlich engen Temperaturbereich die Eigenschaften typischer Kautschukelastizität an. Auch haben sie, wegen der großen Regelmäßigkeit der Ketten (mit Ausnahme des Polyvinylchlorids), eine ziemlich ausgeprägte Tendenz, zu kristallisieren, wobei wegen der relativen Steifheit der Ketten und in einigen Fällen wegen der polaren Anziehungskräfte zwischen ihnen die Schmelzpunkte der kristallinen Bereiche ziemlich hoch liegen. Wenn man die Bereitwilligkeit zur Kristallisation durch Erhöhung der freien Drehbarkeit herabsetzen will, wird man versuchen, weniger stark substituierte Atome in die Kette einzuführen und diese in möglichst unregelmäßiger Weise über ihre Länge zu verteilen. Solche Kettenelemente sind z. B.:

$$—CH_2—S—CH_2—,\quad —CH_2—O—CH_2—,\quad —CH_2—NH—CH_2—,\quad —CH{=}CH—,\quad —C{\equiv}C—,$$

wobei die doppelte und dreifache Kohlenstoffbindung selbst zwar steif ist, aber wegen der geringeren Zahl der Wasserstoffatome (bzw. deren gänzlichen Fehlens) eine stark erhöhte Beweglichkeit der beiden benachbarten Einfachbindungen mit sich bringt. Eine besonders geringe Rotationsbehinderung weisen semipolare Bindungen, wie

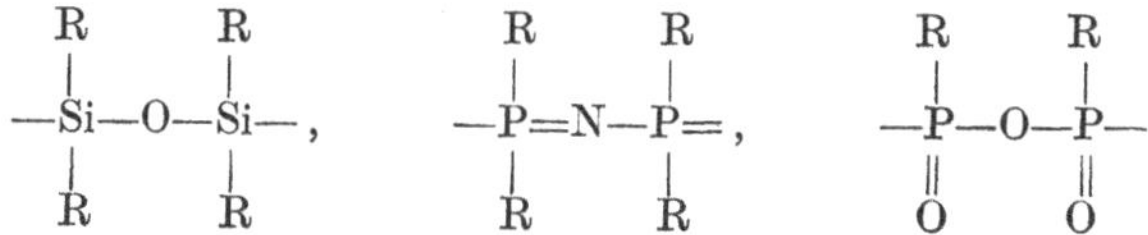

auf; sie führen in der Tat zu Polymeren von sehr ausgeprägter Kautschukelastizität, die sich über erstaunlich weite Temperaturbereiche erstreckt und im Beginn um so unerwarteter war, als diese Substanzen vom Naturkautschuk in chemischer Hinsicht völlig verschieden sind.

Ein anderer wichtiger Faktor für hohe lokale Beweglichkeit ist das Vermeiden von zahlreichen, gleichmäßig entlang der Ketten verteilter Atomgruppen, die entweder durch Wasserstoffbrücken, durch Polarität oder durch größere Polarisierbarkeit eine starke zwischenmolekulare Anziehung zur Folge haben. Um dieser Bedingung zu genügen, wählt man für Elastomere im wesentlichen entweder Kohlenwasserstoffe oder fluorsubstituierte Verbindungen der aliphatischen Reihen oder man schirmt etwa vorhandene polare Gruppen wie —CONH— oder —CO— durch Kohlenwasserstoff- oder Kohlenfluorsubstituenten mäßiger Länge (meist zwischen 4 und 6 C-Atome) ab. Es hat sich allerdings gezeigt, daß Kohlenwasserstoff- und Kohlenfluorketten immer noch zu leicht kristallisieren, um gute Kautschukelastizität zu geben, wenn die Makromoleküle unverzweigt sind und sich daher wegen ihrer einfachen, symmetrischen Gestalt leicht in ein dreidimensionales Gitter einfügen. Teflon und Polythene sind typische Beispiele für dieses Verhalten.

Daher nimmt man noch zu anderen Maßnahmen Zuflucht, um die Größe der zwischenmolekularen Anziehung herabzusetzen. Eine davon ist das Einführen voluminöser Gruppen von völlig unpolarem und möglichst unpolarisierbarem Charakter, welche die Hauptketten auseinanderdrängen und das Ausbilden eines dichtgepackten Gefüges verhindern. Da die VAN DER WAALSschen Anziehungskräfte sehr stark von dem gegenseitigen Abstand der zusammenwirkenden Gruppen abhängen, genügt schon eine relativ kleine Vergrößerung des mittleren seitlichen Kettenabstandes, um eine erhebliche Abschwächung der Molkohäsion herbeizuführen. Polypropylen und Polyisobutylen sind im Gegensatz zu Polyäthylen typische Kautschuke, weil die Methylgruppen abstandsvergrößernd wirken. Neben CH_3- scheinen Äthyl-, Isopropyl- und Isobutylgruppen besonders wirksam zu sein, doch darf man nicht außer acht lassen, daß diese Einführung typisch öllöslicher Kohlenwasserstoffradikale die Quellung des Systems in Benzin, Öl und aromatischen Lösungsmitteln stark erhöht und dadurch die praktische Verwendungsfähigkeit der Materialien ungünstig beeinflußt.

Eine andere Maßnahme ähnlicher Art ist die unregelmäßige Einführung gewisser Substituenten (Cl, $COCH_3$ usw.) in eine lineare Kohlenwasserstoffkette, welche die Gleichmäßigkeit des Moleküls vermindern und das Hineinpassen in ein kristallähnliches Gefüge erschweren. So haben Mischpolymerisate von Äthylen und Vinylchlorid oder Vinylacetat erheblich niedrigere Erweichungspunkte als Polyäthylen und sind viel kautschukähnlicher als dieses. Man spricht in diesem Sinne von einer *„chemischen" Weichmachung* durch Mischpolymerisation und hat dieses Prinzip im Gebiete der Kautschuke und plastischen Massen vielfach erfolgreich verwendet. Obwohl die Störung der Kristallisationsfähigkeit die gewünschte Weichmachung bewirkt, muß doch darauf hingewiesen werden, daß vom Standpunkt der praktischen Kautschukverwendung das Ausschalten der Kristallisation beim Dehnen auch nachteilige Folgen hat. Es ist eine der interessantesten und wichtigsten Eigenschaften des Naturkautschuks, daß er beim Dehnen nach etwa 200% Verlängerung zu kristallisieren beginnt. Auf diese Art erzeugt der Kautschuk im Verlaufe der mechanischen Beanspruchung seinen eigenen verfestigenden Füllstoff – nämlich Kautschukkristallite, die bei weitergehender Dehnung in steigender Menge entstehen und hierdurch ein rascheres Ansteigen des Elastizitätsmoduls bewirken. Allerdings – und das ist der entscheidende Unterschied zwischen Fasern und Kautschuken – haben diese Kristallite einen so niedrigen Schmelzpunkt, daß sie unter normalen Verhältnissen nur im gedehnten Zustand des Materials beständig sind und beim Nachlassen des äußeren Zwanges augenblicklich schmelzen. Die allgemeine technische Erfahrung geht nun dahin, daß in vieler Hinsicht (verfestigende Wirkung, Wärmeentwicklung bei raschperiodischer Verlängerung) die kristallisierenden Kautschuke, wie Naturkautschuk, Neoprene, Butylkautschuk und Vulkollane, bessere Eigenschaften als diejenigen Elastomere zeigen, die wegen mangelnder molekularer Regelmäßigkeit (wie Buna S, Buna N oder Perbunan) beim Dehnen nicht kristallisieren können. Man ist daher gegenwärtig hauptsächlich bemüht, die lokale Beweglichkeit künstlicher Elastomere in solcher Weise herbeizuführen, daß die Kristallisationsfähigkeit beim Dehnen nicht zu ungünstig beeinflußt wird. Im Butylkautschuk, in den bei tiefen Temperaturen oder nach dem MORTONschen Alfinverfahren[1] hergestellten Butadienkautschuken und bei den elastischen Polyestern und Polyamiden hat sich ein solcher Kompromiß mit gutem technischen Erfolg verwirklichen lassen.

Schließlich läßt sich hohe, lokale Beweglichkeit noch mit Hilfe von mechanischer Beimischung nieder- oder hochmolekularer Weichmacher erreichen. Besonders sehr hochmolekulare Systeme (Polybutadien und verschiedene Mischpolymere) haben die Eigenschaft, niedermolekulare Kohlenwasserstoffe (vom Charakter hochsiedender Öle) so weitgehend in dem Netzwerk ihrer verwickelten Ketten zu immobilisieren, daß nach geeigneter Vulkanisation äußerst brauchbare, hochelastische Materialien, die sogenannten *ölgefüllten* Kautschuke entstehen, die auf 100 Teile Kautschuk bis zu 100 Teile eines billigen Öls enthalten und bemerkenswerte technische Eigenschaften aufweisen.

[1] MORTON, M., P. P. SALATIELLO u. H. LANDSFIELD: Ind. Eng. Chem. **44**, 739 (1952).

§ 91. Herstellung eines dreidimensionalen Netzwerkes aus linearen Molekülen.

Das andere wesentliche Erfordernis für das Vorliegen von Kautschukelastizität ist die Existenz eines Systems von Fixpunkten, die an relativ weit voneinander gelegenen Stellen das Abgleiten von Molekülsegmenten verschiedener Ketten verhindern, hierdurch der Verformung ein Ende setzen und die Rückkehr in den Ausgangszustand ermöglichen.

Die älteste und noch immer am meisten gebrauchte Methode zur Herstellung solcher Vernetzungspunkte ist die Vulkanisation von Kautschuken mit Schwefel oder schwefelhaltigen Verbindungen. Obwohl alle Einzelheiten ihres (scheinbar recht verwickelten) Mechanismus durchaus noch nicht aufgeklärt sind, steht es doch fest, daß das Endergebnis eine Reaktion der Doppelbindungen des vorliegenden Kautschuks mit Schwefel und die Bildung von —S—- oder —S—S—-Bindungen zwischen den Ketten darstellt. Diese Bindungen sind im Verhältnis zu der geringen VAN DER WAALSschen Anziehung der Ketten sehr fest und werden jedenfalls bei normaler mechanischer Beanspruchung im Bereiche nicht zu hoher Temperaturen nicht gelöst. Diese Art der Vulkanisation verlangt offenbar das Vorliegen von Doppelbindungen in dem zu vernetzenden System und zugleich auch eine gewisse Reaktionsfähigkeit mit den chemischen Reagenzien, welche die Vernetzung bewirken. In den „klassischen" Kautschuken — wie Naturkautschuk, Buna S, Buna N und Zahlenbuna — sind aliphatische Doppelbindungen in großer Zahl vorhanden, so daß genügend Angriffspunkte für die Schwefelvulkanisation zur Verfügung stehen, sie sind sogar entlang der Kautschukketten so häufig, daß sie Abstände von nur wenigen ÅNGSTRÖM voneinander haben und daß daher nur ein sehr kleiner Prozentsatz von ihnen ausgenützt werden darf, wenn man ein Weichgummivulkanisat erhalten will. Im Butylkautschuk, der im wesentlichen aus Isobutylen besteht, werden die für die Vulkanisation nötigen Doppelbindungen in sehr geringer Zahl (etwa 1%) durch Mischpolymerisation mit Isopren oder Butadien eingeführt: sie haben — wegen ihrer geringen Zahl — erhebliche Abstände in den Molekülketten.

Das häufige Vorkommen von aliphatischen Doppelbindungen in den meistverwendeten Kautschuken hat zwar den Vorteil, daß diese rasch und leicht vulkanisiert werden können, es bewirkt aber auch eine starke, meist unerwünschte Reaktionsfähigkeit mit Sauerstoff, besonders bei höheren Temperaturen, im Verlaufe der Herstellung und beim späteren endgültigen Gebrauch. Dies ist der Grund für die merklich bessere Stabilität des Butylkautschuks beim Mastizieren und für seine größere Widerstandsfähigkeit gegen Luft und Licht im Gebrauche bei höheren Temperaturen. Die —S—- oder —S—S—-Brücken, die bei der normalen Schwefelvulkanisation zwischen den Hauptvalenzketten hergestellt werden, sind auch selbst gegen Sauerstoff und andere Oxydationsmittel nicht sehr beständig und tragen daher recht wesentlich zum „Altern" von vulkanisierten Kautschukartikeln bei. Man hat daher versucht, andersartige Querverbindungen zwischen den einzelnen Makromolekülen kautschukartiger ungesättigter Kohlenwasserstoffe herzustellen, was z. B. mit Hilfe von Peroxyden oder Nitroverbindungen möglich ist, doch ergeben sich nicht genügend Vorteile gegenüber der einfachen und hochentwickelten Schwefelvulkanisation, um diese Verfahren praktisch wichtig zu machen.

Viele verschiedene Reagenzien werden zum Vernetzen anderer Elastomeren benutzt. So werden z. B. die Querverbindungen in chlorsulfoniertem Polyäthylen (Hypalon) mit Polyaminen oder Bleiionen hergestellt, Silikone werden mit Poroxyden, fluorhaltige Polyacrylsäureester mit Natriumsilicat, hydroxylgruppenhaltige Polyester mit Diisocyanaten und hydroxylgruppenhaltige Polyäther mit Säureanhydriden vernetzt. In allen Fällen werden starke chemische Bindungen zwischen den Einzelketten erzeugt, die ein mechanisches Abgleiten an den Haftstellen verhindern. Will man einen Weichgummi von hoher Dehnbarkeit (800 bis 1000%), dann müssen diese Querverbindungen so weit voneinander entfernt sein (200 bis 300 Grundeinheiten), daß das ganze System noch eine erhebliche Beweglichkeit beibehält; soll andererseits das Vulkanisat eine geringere Dehnbarkeit und einen höheren Modulus besitzen, so werden die Querverbindungen in größere Nähe (50 bis 100 Grundeinheiten) gebracht, und will man schließlich Hartgummi, so erzeugt man ein starres, engmaschiges Netzwerk von Querverbindungen.

Haftstellen von größerer Festigkeit können auch durch das Beimischen aktiver Füllstoffe erzeugt werden. Am meisten verwendet man Ruß, SiO_2 und Al_2O_3 in solcher Form, daß sehr große (200 bis 400 m^2 je g) und gleichzeitig chemisch reaktive Oberflächen vorliegen, an denen die Ketten der Kautschukmoleküle durch starke Adsorptionskräfte gebunden sind, so daß das Rußteilchen die Rolle eines Verankerungspunktes mehrerer Kettenmoleküle spielt. Die Beimischung relativ geringer Mengen (1 bis 2%) eines feinverteilten aktiven Füllstoffes (Teilchendurchmesser von 2 bis 300 Å) wirkt in der Tat wie eine Vulkanisation, indem sie den anfänglichen Modulus und die Zerreißfestigkeit erhöht und die Bruchdehnung herabsetzt.

Schließlich scheint es möglich, daß das Vorhandensein weniger in dem System verteilter Kristallite, sowie auch die „mechanische" Verwicklung sehr hochmolekularer Ketten ein Fixpunktsystem erzeugen kann, welches an Stelle der chemischen Vulkanisation zu treten in der Lage ist. Versuche in der ersten Richtung werden hauptsächlich mit neuartiger Mischpolymerisation ausgeführt, die aus längeren Kettenstücken (1000 bis 1500 Å) schlecht kristallisierbarer monomerer Einheiten und aus kürzeren Stücken (200 bis 300 Å) besser kristallisierender Anteile bestehen (*Block-Copolymere*). Die Versuche zur „mechanischen Vernetzung" bewegen sich im wesentlichen in der Richtung, stark miteinander verfilzte Bereiche durch hohe Drucke oder andersartige mechanische Nachbehandlung zu erzielen.

Wir haben dem Sinne der Disposition dieses Kapitels folgend den Aufbau kautschukartiger Polymerer im wesentlichen vom Standpunkt ihrer molekularen Struktur dargestellt. Die Natur der kautschukelastischen Kräfte und die hier obwaltenden thermodynamischen Verhältnisse sind im Kap. V eingehend behandelt, so daß wir uns mit diesem Hinweis begnügen können.

Namenverzeichnis.

Sachverzeichnis.

Zeitfracht Medien GmbH
Ferdinand-Jühlke-Straße 7
99095 Erfurt, Deutschland
produktsicherheit@kolibri360.de

34,00-

ACCESO GRATIS *a la Lectura en la Nube*

Para visualizar el libro electrónico en la nube de lectura envíe junto a su nombre y apellidos una fotografía del código de barras situado en la contraportada del libro y otra del ticket de compra a la dirección:

ebooktirant@tirant.com

En un máximo de 72 horas laborales le enviaremos el código de acceso con sus instrucciones.

La visualización del libro en **NUBE DE LECTURA** excluye los usos bibliotecarios y públicos que puedan poner el archivo electrónico a disposición de una comunidad de lectores. Se permite tan solo un uso individual y privado

ESCOMBRERAS ILEGALES EN EXTREMADURA
Un análisis criminológico

ESCOMBRERAS ILEGALES EN EXTREMADURA

Un análisis criminológico

LORESA ARENAS GARCÍA
Directora

tirant lo blanch
Valencia, 2024

© TIRANT LO BLANCH
EDITA: TIRANT LO BLANCH
C/ Artes Gráficas, 14 - 46010 - Valencia
TELFS.: 96/361 00 48 - 50
FAX: 96/369 41 51
Email: tlb@tirant.com
www.tirant.com
Librería virtual: www.tirant.es
DEPÓSITO LEGAL: V-4369-2023
ISBN: 978-84-1197-574-2

Si tiene alguna queja o sugerencia, envíenos un mail a: *atencioncliente@tirant.com*. En caso de no ser atendida su sugerencia, por favor, lea en *www.tirant.net/index.php/empresa/politicas-de-empresa* nuestro procedimiento de quejas.

Responsabilidad Social Corporativa: http://www.tirant.net/Docs/RSCTirant.pdf

La autora agradece al Fondo Europeo de Desarrollo Regional
y a la Junta de Extremadura
(Consejería de Economía, Ciencia y Agenda Digital)
la cofinanciación de esta publicación
mediante la ayuda IB20050.

Índice

Capítulo 2

La regulación jurídico-administrativa de los RCD: de la normativa general de residuos a las ordenanzas municipales

PEDRO BRUFAO CURIEL

Capítulo 3

Fuentes de datos oficiales: calidad y márgenes de error

ALBERTO ALFONSO-TORREÑO

LOREA ARENAS GARCÍA

Capítulo 4

Análisis y caracterización de los focos de residuos

ALBERTO ALFONSO-TORREÑO

JOSÉ ANTONIO GUTIÉRREZ GALLEGO

Capítulo 5

Identificando características socioeconómicas de municipios con vertederos ilegales

JORDI ORTIZ GARCÍA

Capítulo 6

Detección de vertidos ilegales de residuos de construcción y demolición mediante fotointerpretación de modelos procedentes del DEM

MANUEL SÁNCHEZ FERNÁNDEZ

Capítulo 7

La percepción de las fuerzas y cuerpos de seguridad en la detección y control del vertido ilegal

LOREA ARENAS GARCÍA

Capitulo 8

El papel de los Agentes del Medio Natural

GUILLERMO AGUSTÍN EXPÓSITO PAULANO

Capítulo 9

El proceso de gestión del residuo sólido y relación con el vertido ilegal

SARA Mª MARCHENA GALÁN

Capítulo 10

Ayuntamientos y vertidos ilegales de residuos de la construcción: una aproximación a las percepciones de los responsables en las entidades municipales

JUANJO MEDINA ARIZA

Introducción

LOREA ARENAS GARCÍA[1]

En los últimos 50 años nuestro entorno natural ha sufrido las peores consecuencias de la acción humana debido a la explotación de los recursos naturales y la contaminación[2]. El desarrollo industrial, la expansión de los medios de producción extractivos, el consumo desenfrenado y el crecimiento demográfico, se identifican como las principales causas del problema. La primera conferencia mundial sobre el medioambiente celebrada en Estocolmo en 1972 supuso un punto de inflexión en la atención y respuesta internacional a esta problemática estableciendo un marco político estratégico y programático[3]. Este se plasmó en la adopción posterior de acuerdos, tratados, programas y acciones que han marcado la hoja de ruta a seguir por el conjunto de las naciones. A pesar de los esfuerzos realizados para frenar el deterioro ambiental, claves para visibilizar y sensibilizar a la población consciente de sus efectos más devastadores (olas de calor, inundaciones, escasez de agua, etc.), este problema continua *in crescendo* ocasionando pro-

---

1 Profesora Contratada Doctora en Criminología, Área de Derecho penal, Universidad de Extremadura.

2 Véanse informes del Grupo Intergubernamental de Expertos sobre Cambio Climático (conocido por sus siglas en inglés, IPCC). Es un ente científico creado en 1988 por la Organización Meteorológica Mundial y el Programa de Naciones Unidas para el Medio Ambiente. Su propósito es "proporcionar información objetiva, clara, equilibrada y neutral del estado actual de conocimientos sobre el cambio climático a los responsables políticos y otros sectores interesados". https://www.miteco.gob.es/es/ceneam/recursos/mini-portales-tematicos/Cclimatico/informe_ipcc.aspx [Consultado el 03.07.2023].

3 Naciones Unidas. Conferencias, medio ambiente y desarrollo sostenible. https://www.un.org/es/conferences/environment/stockholm1972 [Consultado el 03.07.2023].

blemas graves como el cambio climático, la destrucción de ecosistemas, la extinción de especies, la contaminación, entre otros.

Uno de los mayores desafíos para investigadores y agentes del mundo de la *praxis* son los daños y delitos que ocasionan los residuos ilegales, especialmente cuando son altamente contaminantes o peligrosos. Tienen lugar cuando se desecha cualquier tipo de sustancia u objeto en un lugar público o privado no autorizado impidiendo así su correcta recuperación y eliminación (Lu, 2018, 264). Las causas que precipitan y mantienen esta problemática son numerosas y de diversa naturaleza, a saber: la estrecha relación existente entre el crecimiento económico y la producción de residuos; la falta de información, atención y conciencia medioambiental por parte de la ciudadanía y los representantes políticos; la carencia histórica de una legislación integral, clara y coordinadora de los agentes sociales; la ineficaz gestión de las administraciones locales y nacionales para atajar el problema; la limitación de datos disponibles y su fiabilidad; las dificultades económicas y presupuestarias para implementar medidas preventivas; y la baja capacidad de detección y control de estas infracciones debido a su alta oportunidad delictiva, entre otras.

Esta problemática, de dimensiones globales y cuyos impactos afectan negativamente a la salud humana y medioambiental (Senior y Mazza, 2004; Triassi et al., 2015), fue situada en el centro del debate en la primera Estrategia Comunitaria de residuos de 1989 (resolución del Consejo de 7 de mayo de 1990). En esta se reconocía la necesidad de establecer un marco estratégico de gestión de los residuos para evitar su generación abogando por su reutilización, reciclado, valoración y, en última instancia, eliminación optimizada de aquellos no valorizados o recuperables. Tal jerarquía quedaba plasmada en las revisiones posteriores de la estrategia en 1996 (resolución del Consejo de 30 de julio de 1996) y 1997 (resolución del Consejo de 24 de febrero de 1997[4])

4 En este último se leía: “debería darse preferencia, siempre que sea una solución aceptable desde el punto de vista del medio ambiente, a la

estableciendo un objetivo primordial a seguir en posteriores políticas: reducir la creación del residuo y abogar por su valorización. La eliminación del residuo, al verterlo o incinerarlo, es contrario al compromiso del Horizonte 2020 (H2020)[5] y a la orientación del actual Programa Horizonte Europa[6], que nos orienta hacía una Europa eficiente en el uso de recursos, a la economía circular y la sociedad del reciclaje, por lo que la mera eliminación en vertederos debe ser la gestión de *ultima ratio.* A pesar del avance legislativo de las últimas décadas, materializados en nuestro país en la reciente Ley 7/2022, de 8 de abril, de residuos y suelos contaminados para una economía circular y en el Plan Estatal Marco de Residuos (PEMAR 2016-2022)[7], sigue sin alcanzarse una tasa adecuada de residuo reciclado[8], al tiempo que se desconoce la ubicación y magnitud de los focos de vertido.

Debido a ello, el Tribunal de Justicia de la Unión Europea (TJUE) condenó a España en los años 2016 y 2017[9] por no sellar y regenerar 61 vertederos ilegales de los 250 encausados una dé-

---

valorización de materiales sobre la valorización energética (…) Por otra parte, se considera que las estrategias energéticas que dependen del suministro de residuos no deberían perjudicar a los principios de prevención y valorización de material" (capítulo 3.1-5).

5 Es el Programa Marco de Investigación e Innovación (I+I) de la Unión Europea (UE) para el periodo 2014-2020. Véase: https://www.horizonteeuropa.es/anteriores-programas/h2020 [Consultado el 03.07.2023].

6 Es el Programa Marco de Investigación e Innovación (I+I) de la Unión Europea (UE) entre los años 2021 -2027. Véase: https://www.horizonteeuropa.es/que-es [Consultado el 03.07.2023].

7 El PEMAR contiene una estrategia general de la política de residuos, las orientaciones y la estructura a la que deben ajustarse los planes autonómicos.

8 Aunque en el año 2020 se recicló el 54,7% del residuo tratado, la media de los años anteriores si sitúa en torno al 38% (INE, 2022).

9 Sentencia de 25 de febrero de 2016, asunto C-454/14, que resuelve recurso por incumplimiento contra España de la Directiva 1999/31, relativa al vertido de residuos, y sentencia de 15 de marzo de 2017, que declara el incumplimiento del Reino de España de la Directiva 2008/98, sobre los residuos (arts. 13 y 15.1).

cada atrás. Por su parte, la Comisión Europea emitió un dictamen en 2018 denunciando la existencia de 1513 vertederos irregulares en territorio español e instaba a tomar acciones dirigidas a su sellado, clausura y posterior restauración del terreno[10].

La Fiscalía General del Estado también reseñó la situación descontrolada de un buen número de vertederos debido a incendios producidos en plantas e instalaciones de tratamiento y gestión de residuos, fenómeno que "se ha venido produciendo en los años precedentes y que, fundamentalmente en los dos últimos, ha alcanzado proporciones alarmantes" (2018, pp. 808-811). Dichas denuncias aluden a vertederos de grandes dimensiones, y mejor identificables, que almacenan residuos urbanos y cuya colapsada situación ha dado lugar a sucesos graves. Sirva de ejemplo el incidente de la planta vizcaína de Zaldívar[11]. En esta línea, organizaciones ecologistas como Greenpeace[12] y Ecologistas en Acción[13] llevan años denunciando la colapsada, irregular y cronificada situación de muchos de ellos, así como la falta de datos fiables sobre su ubicación y composición, sobre todo cuando se trata de vertidos más pequeños o en foco disperso.

Los datos publicados por el Ministerio del Interior no son más optimistas. Según los Anuarios Estadísticos del Ministerio del Interior (2009-2022) sobre las actuaciones del Servicio de Protección de la Naturaleza de la Guardia Civil (en adelante SEPRONA), el vertido de residuos, y sobre todo el residuo urbano, es la infracción administrativa contra el medioambiente más frecuente entre

---

10 Véase: https://ec.europa.eu/environment/legal/law/statistics.htm [Consultado el 03.07.2023].

11 Noticia de Público: https://www.publico.es/sociedad/zaldibar-vertedero-zaldibar-iban-residuos-nadie-queria.html [Consultado el 03.07.2023].

12 Véase: https://es.greenpeace.org/es/ [Consultado el 03.07.2023].

13 Véase: https://www.ecologistasenaccion.org/141691/la-comision-europea-reconoce-que-espana-podria-estar-incumpliendo-la-normativa-europea-de-tratamiento-de-residuos-domesticos/ [Consultado el 03.07.2023].

los años 2010 y 2019[14], representando en torno al 20% del total de las infracciones medioambientales (con una media anual de 28.258 infracciones). Estas son mucho más numerosas que las infracciones penales por depósito o vertido de residuos tóxicos o peligrosos (con una media de 8 delitos anuales entre 2014 y 2021).

Gracias a organizaciones ecologistas sabemos qué tipo de residuo se vierte en determinadas zonas y su ubicación exacta. Al respecto destaca el "Mapa interactivo de vertidos de la Comunidad en la Madrid"[15] desarrollado por Ecologistas en Acción en 2022 y en el que, únicamente en Alcorcón y Móstoles, se localizaron casi 300 puntos de vertido de residuos de enseres domésticos, residuos peligrosos sólidos como el amianto y residuos sólidos de construcción y demolición (en adelante RCD). Estos se hallan en zonas de fácil acceso por carretera y con escasa vigilancia (cunetas de caminos, zonas colindantes a polígonos industriales, en el extrarradio de las ciudades, en zonas poco frecuentadas de campo, cerca de bosques, ríos, etc.). A veces son esparcidos por los caminos, pero, si la práctica es reiterada, se depositan en pequeños montículos de forma piramidal y rectangular abarcando grandes extensiones.

Los estudios científicos desarrollados en nuestro país, principalmente desde el ámbito de la Geografía e Ingeniería, ponen de relieve la presencia de 1700 puntos de vertido descontrolados en Andalucía (Fernández et al., 1995; Jordá-Borrell et al., 2014; Lucendo-Monedero et al., 2015; Navarro et al., 2016) y 439 en las Islas Canarias (Quesada-Ruiz et al., 2018). Según Jordá-Borrell et al., (2014, 158), "los vertederos ilegales son visibles a menos de 500 metros de cualquier camino rural (93,18% de ellos) y el principal tipo de residuos proviene de la construcción y demolición (60% del total)". Los autores se refieren a los ya mencionados RCD, que son regulados por el Real Decreto 105/2008, de 1 de fe-

---

14 No obstante, y debido a la pandemia de la COVID, fueron superadas en 2020 y 2021 por infracciones a la normativa sobre sanidad pública y medicamentos.

15 Véase: https://www.ecologistasenaccion.org/193685/mapa-interactivo-de-vertidos-en-la-comunidad-de-madrid/ [Consultado el 03.07.2023].

brero, por el que se regula la producción y gestión de los residuos de construcción y demolición. En el artículo segundo se definen como: "cualquier sustancia u objeto que, cumpliendo la definición de «Residuo» incluida en la Ley 7/2022, de 8 de abril, se genere en una obra de construcción o demolición". Este decreto ya reconocía en su exposición de motivos que "el problema ambiental que plantean estos residuos se deriva no solo del creciente volumen de su generación, sino de su tratamiento, que todavía hoy es insatisfactorio en la mayor parte de los casos".

En la región extremeña, al igual que en otras zonas de España, su presencia es muy habitual en campos, caminos y cunetas. En 2015 la Comisión Europea remitió a España la infracción n.2015/2192 basada en una investigación preliminar sobre la presencia de vertederos ilegales de residuos inertes, ubicándose 133 en Extremadura. Precisamente su descontrol motivó la creación del Plan Integrado de Residuos de Extremadura o PIREX (2016-2022)[16] y su nuevo plan[17], ambos frutos del PEMAR, cuyo principal reto "es cumplir con los objetivos comunitarios y nacionales aplicables a la prevención de residuos y los específicos dirigidos al tratamiento de los residuos domésticos y a los residuos

16 Aprobado mediante Acuerdo del Consejo de Gobierno de 28 de diciembre de 2016 a fin de prevenir el residuo mediante una serie de medidas entre las que se encontraban la consolidación de la gestión adecuada de los residuos de construcción y demolición (RCD). Véase: http://extremambiente.juntaex.es/index.php?option=com_content&view=article&id=4428&Itemid=578 [Consultado el 03.07.2023].

17 Se trata del nuevo plan PIREX (2023-2030), cuya versión inicial fue publicada en septiembre de 2022 en el Diario Oficial de Extremadura, y que viene a proponer nuevas normas sobre residuos que formaban parte del denominado paquete de economía circular aprobadas por el Parlamento Europeo en abril de 2018. Persigue eliminar gradualmente el vertido de residuos y fomentar el uso de medios más económicos, reforzando la jerarquía de los residuos y priorizando aquellas acciones de prevención y tratamiento para la reutilización y reciclado frente al depósito en vertedero o incineración. Véase: http://extremambiente.juntaex.es/files/1_Version%20inicial%20PIREX%202023_2030(1).pdf [Consultado el 03.07.2023].

de construcción y demolición (RCD)" (PIREX, p.6). Alude especialmente a los residuos de construcción y demolición porque: "debe evitarse su vertido incontrolado, trasladándolos a instalaciones autorizadas para su adecuado tratamiento" (PIREX, p.7). Se refiere a aquellos residuos inertes generados en una obra de construcción o demolición, y los escombros procedentes de obras menores de construcción y reparación domiciliaria, que terminan siendo vertidos ilegalmente.

La respuesta a esta problemática ha sido diferente en la provincia de Cáceres que en Badajoz, apreciándose un mayor descontrol e impacto de vertidos ilegales en esta última. La Diputación provincial de Cáceres ha desarrollado dos planes en la zona norte y sur de la provincia gracias en parte a Fondos de Desarrollo Regional (FEDER) de la Unión Europea, lo que ha permitido la clausura de 49 vertederos ilegales en su territorio y la cofinanciación de 26 plantas de reciclaje (fijas y móviles) y de puntos limpios para la recogida de residuos procedentes de obras menores. Por su parte en Badajoz la gestión ha sido eminentemente privada con un menor número de instalaciones distribuidas de forma aleatoria. Según recoge *El Periódico de Extremadura*[18], en octubre de 2019, el Grupo Municipal Socialista de Badajoz criticó que "Badajoz está rodeada de vertederos ilegales, y que el consistorio culpa a la ciudadanía por ello, pero que la verdad es que no disponen de un lugar legal donde llevar los desechos". Esta situación es similar en otros municipios de la provincia, sobre todo en Don Benito, Villanueva de la Serena, Lobón y Guareña que, junto a Mérida, aglutinan buena parte de las denuncias del SEPRONA.

Dada la escasez de datos acerca de la localización de los vertidos inertes es difícil estimar la magnitud del problema, establecer sus causas y, por extensión, aplicar medidas preventivas. A lo ante-

---

18 Noticia de *El Periódico de Extremadura*: Disponible en: https://www.elperiodicoextremadura.com/noticias/badajoz/psoe-pide-ayuntamiento-badajoz-planta-recogida-residuos_1190197.html [Consultado el 03.07.2023].

rior se suma el hecho de que en nuestro país no existen trabajos criminológicos cuyo tema de análisis sean los vertederos ilegales, para ello es necesario recurrir a trabajos desarrollados desde otras disciplinas, siendo a su vez escasos, como la biología y la ingeniería (véanse: Gallardo et al., 2012; Navarro et al., 2016; Quesada Ruiz et al., 2018). A nivel comparado los estudios considerados de Criminología verde[19] han centrado su atención -tradicionalmente- en el cambio climático y el tráfico ilegal de especies. Los *desechos*, como línea central de investigación, integra los *delitos de vertido*, esto es, aquellos referidos "al comercio, tratamiento y eliminación de desechos en formas que violan las leyes ambientales internacionales o nacionales y que causan daños o producen riesgo para el ambiente y la salud humana" (Mol et al., 2017, 237). En torno a esta línea se integran múltiples temáticas: el crimen organizado y sus "crímenes de cuello sucio"; los desechos electrónicos en entornos legales transnacionales; las vulnerabilidades causadas por prácticas dañinas en el manejo de residuos en Europa; la relación del vertido con la ecovictimización; los estudios de casos particulares (por ejemplo, el caso Probo Koala[20]); el análisis de los delincuentes de desechos y su *modus operandi*; la detección, control y prevención del desecho; y los daños ambientales, sociales y económicos ocasionados.

Los estudios sobre vertederos ilegales han abordado transversalmente muchas de ellas pudiendo ser clasificados en seis grandes bloques. Trabajos que abordan los impactos sobre la salud humana y el medio ambiente utilizando indicadores epidemioló-

---

19 La Criminología verde se centra en el estudio de diferentes tipos de daño ambiental y describe su prevalencia temporal y geográfica, analiza críticamente las causas y consecuencias de tales daños, y reflexiona sobre cómo los cuerpos normativos, los sistemas de justicia penal, los individuos y los grupos responden, o deberían responder, a tales daños (Brisman y South, 2015; White y Graham, 2015).

20 En el año 2006 se vertieron 12 camiones de residuos tóxicos en 18 municipios de Abiyán (Costa de Marfil) descargados del buque Probo Koala. Debido a ello murieron 15 personas y 108.000 precisaron atención médica (Van Wingerde, 2015).

gicos (número de muertes y enfermedades, por ejemplo) y otros propios de las ciencias naturales (nivel de C02 en la atmósfera, hidrocarburos, metales pesados, etc). En estos últimos destacan los estudios de Senior y Mazza (2004) y Triassi et al. (2015).

Aquellos que pretender localizar o detectar vertederos empleando encuestas, autoinformes, bases de datos oficiales y diversas soluciones tecnológicas (Sistemas de Información Geográfica o SIG, *Machine Learning*, Apps, Drones, CCTV y GPS). En esta línea destacan los estudios de Romeo et al., 2003, Silvestri y Omri, 2007, Baugh y Kokaram, 2008; Biotto et al., 2009, Persechino et al., 2013, De Feo et al., 2014, Lega y Persechino, 2014, Glanville y Chang, 2015, Navarro et al., 2016, Zainun et al., 2016, Dabholkar et al., 2017, Jelks et al., 2018 y Lu, 2018. Un tercer grupo que centra su análisis en geoposicionarlos mediante SIG, fundamentalmente, para luego realizar mapas del delito y establecer *hop spots* (puntos calientes) e incluso efectuar predicciones (Tasaki et al., 2007; Uricchio et al., 2010; Kinobe et al., 2014; Seeboonruang, 2016; Seror y Portnov, 2018;). Estas técnicas se conocen como "Crime Mapping" o mapas del crimen y son características de la *Criminología ambiental* disciplina que, según Wortley y Mazerolle (2008), parte de tres premisas: la influencia del ambiente en la conducta delictiva, la no aleatoriedad de la distribución espacio-temporal del delito (no es azarosa), y la utilidad de los elementos anteriores en el control y prevención del delito. Los estudios referidos nos muestran que: el uso de algoritmos permite un seguimiento robusto de objetos generados a partir de la agrupación espacial de los mapas; es posible estimar la probabilidad de ubicación de un vertedero (por ejemplo: la distancia a la carretera principal más cercana, la profundidad del barranco y la proximidad al bosque determinan la probabilidad de un vertedero); sus metodologías permiten identificar y acotar zonas con mayor concentración de residuos; establecer áreas de riesgo facilita el seguimiento del residuo y reduce costes; nos ayudan a comprender los modelos de recolección y su influencia en la gestión de los residuos.

Un cuarto grupo que analiza la toma de decisiones o política en la gestión de residuos (el precio del reciclaje del residuo, las sanciones impuestas, las prácticas en la construcción, las fórmula reguladoras adoptadas, etc.) y su relación con la creación y mantenimiento del vertedero utilizando análisis sistémicos y dinámicos, teoría de juegos y modelos econométricos (Fraser y Choe, 1999; Allers y Hoeben, 2009; Hongping y Shen, 2010; Ichinose y Yamamoto, 2010; Yuan et al., 2010; Lu et al., 2011; Halvorsen, 2012; Hamilton et al., 2013; Kellenberg, 2012; Slavik y Pavel, 2012; Cossu y Masi, 2013; Ding et al., 2016; Chen et al., 2018). Estas investigaciones son útiles para facilitar una mejor comprensión de la dinámica de las actividades en la gestión de residuos entendiendo el fenómeno como un todo, realizar simulaciones y aplicar modelos que estudian las interacciones en estructuras formalizadas de incentivos. Por ejemplo, se sabe que: las herramientas que simulan el costo-beneficio de gestión de residuos son útiles para estimar el impacto de una política u otra; las tarifas impuestas a los usuarios pueden reducir o aumentar el reciclaje; la intensidad de la supervisión y sus costes asociados, las multas y el coste de la eliminación de desechos son los principales factores en la toma de decisiones de contratistas y agentes gubernamentales; aumentar la penalización sin garantizar un nivel adecuado de supervisión es ineficaz para controlar el vertido ilegal, mientras que la participación ciudadana en la supervisión podría ser un complemento importante; la prevención en la generación del residuo previene a su vez el vertido ilegal; el aumento de las opciones de reciclaje tiene un efecto significativo en el reciclaje de los hogares siendo los centros de recogida junto a la entrega puerta a puerta los dos métodos más efectivos; otorgar incentivos económicos por reciclar puede evitar el vertido ilegal; entre otros.

Un quinto grupo de trabajos que integra la perspectiva del crimen organizado a través del estudio de casos, en los que destacan los casos de EE. UU. e Italia por haber sido mejor documentados (Massari y Monzini, 2004; Liddick, 2009; Cherry y Sneirson, 2011; Crevron, 2015; D'Amato et al., 2015;); y un sexto que examina *el modus operandi*, el perfil criminal del delincuente (Crofts et al., 2010; Matos et al., 2012;) y las medidas preventivas. Estos dos últimos grupos de estudios recurren a las teorías criminológicas de la oportunidad delictiva desarrolladas por Felson y Clarke en 1998 para explicar las causas del delito y su prevención. De acuerdo con los autores, para que el crimen se cometa deben darse tres elementos: un delincuente predispuesto, una víctima propicia y una ausencia de control. Dichas teorías han sido empleadas tradicionalmente para explicar comportamientos delictivos de motivación económica (robos, hurtos, estafas, delincuencia de cuello blanco, etc.) en los que las variables espacio y tiempo modulan, junto al cálculo racional de costes y beneficios del delincuente, la oportunidad delictiva. No obstante, si en los delitos contra el patrimonio el beneficio es la sustracción de bienes materiales, en los delitos de vertido este se obtiene al ahorrar los costes del reciclaje. Los escasos estudios en la materia (Crofts et al., 2010; Matos et al., 2012) están en sintonía con estas teorías. Gracias a ellos sabemos que: el vertido tiene lugar por la noche y en lugares apartados con menor vigilancia (extrarradio de la ciudad); los residuos suelen aparecer cerca de polígonos industriales y alrededor de los nodos de actividad profesional y de las rutas entre ellos (teorías del patrón delictivo); siendo los escombros, materiales de obra y ciertos productos químicos los que más se vierten; los delincuentes suelen ser trabajadores de la construcción o intermediarios de estos (entre 35-55 años de edad) que deciden racionalmente desprenderse de los residuos para ahorrar costes; y que una prevención comunitaria basada en la conciencia del daño, la información sobre sanciones (con panfletos y charlas) junto a la vigilancia vecinal, podrían ser elementos clave para su prevención.

Tal y como se desprende de los estudios anteriores, son pues múltiples los aspectos a considerar y que explican la creación y mantenimiento del vertido ilegal. Por ello, el carácter multidisciplinar de la Criminología ofrece un marco ideal de análisis para abordar el objeto de estudio desde múltiples perspectivas, aportando así una visión integral y fundada del mismo. La perspectiva *jurídica*, mediante el análisis de la normativa autonómica y, en particular, de las ordenanzas municipales en materia de residuos a desarrollar por las entidades locales, así como de las sentencias emitidas sobre residuos; la perspectiva de *política y de gestión*, conociendo y evaluando el papel que juegan los diferentes agentes sociales (representantes políticos, empresas públicas y privadas gestoras de residuos, organizaciones no gubernamentales, etc.) en la toma de decisiones y ejecución de decisiones en la materia; la *tecnológica,* aplicando herramientas informáticas y soluciones tecnológicas para detectar y controlar los vertidos; y la propiamente *criminológica,* aplicando técnicas y enfoques teóricos de la disciplina para describir y explicar esta fenomenología criminal.

En base a la evidencia anterior, y considerando las mencionadas perspectivas de análisis, se creó el proyecto "Teledetección y análisis ambiental de vertederos ilegales" (IB20050-2021/2024), también denominado "Proyecto VIEX", para investigar los RCD ilegales gracias a fondos de la Junta de Extremadura y el Fondo de Desarrollo Regional (FEDER). La investigación se suma a los esfuerzos que otros organismos públicos y privados vienen desempeñando en la región para desarrollar el PIREX, dando a su vez respuesta al PEMAR y, por extensión, a los propios requerimientos de la Unión Europea.

El estudio se llevó a cabo en Extremadura, la cual es una región eminentemente rural y escasamente poblada que se dividide en dos provincias: Cáceres y Badajoz. Estas a su vez son las más extensas geográficamente de todo el territorio español. Extremadura abarca 41.633 km2, en la que poco más de 1 millón de habitantes están distribuidos en 388 municipios (Mapa 1). La densidad poblacional es de 27 habitantes km2 y los diez pueblos más poblados concentran el 46% de la población extremeña (Nieto y Cárdenas, 2017). A su vez, hay dos grandes factores que marcan las diferencias territoriales en la región: los ejes principales de comunicación y la actividad agroindustrial. La autovía A-66 (conocida como *Vía de la Plata*) y la autovía A-5 (Madrid - Badajoz) son las dos principales vías de comunicación en torno a las cuales se sitúan los pueblos más poblados de la comarca: Plasencia, Cáceres, Mérida, Almendralejo y Zafra en la A-66; y Navalmoral de la Mata, Mérida, Montijo y Badajoz por la A-5. A su vez destacan zonas con un sector agroindustrial competitivo ubicadas principalmente en zonas de regadío (por ejemplo, Vegas del Guadiana y Valle del Alagón) y de secano productivo, como Tierra de Barros.

Mapa 1

Ubicación geográfica y distribución de la población de Extremadura según censo de 2021.

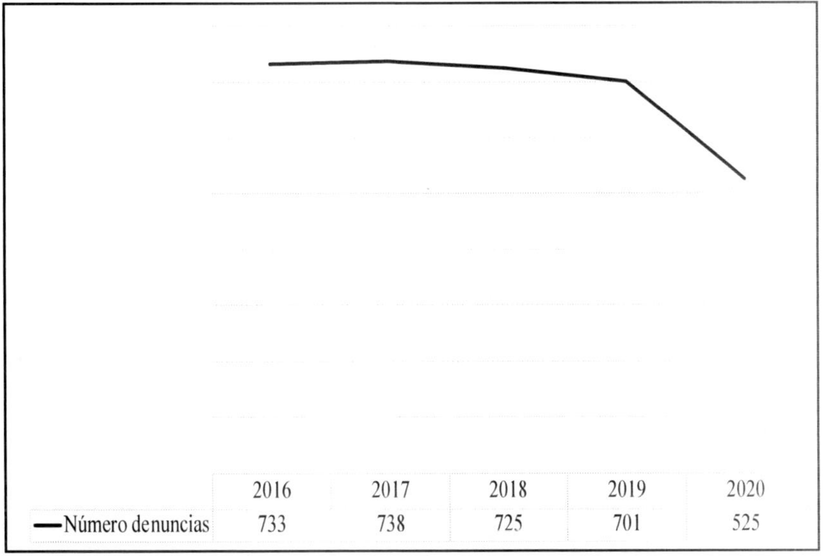

Nota: Se han utilizado los datos del censo poblacional del Instituto Nacional de Estadística (INE) para el año 2021 dada su renovación cada diez años.

La escasez de efectivos policiales, las vastas áreas de terrenos a examinar y la distancia que deben recorrer estos entre los pueblos para garantizar la seguridad ciudadana merman sus labores de control y vigilancia (Ortiz García, 2022). Obviamente, ello también dificulta la localización y control del vertido ilegal. A lo anterior hay que sumar que en nuestro país no existe ninguna herramienta capaz de localizar de forma rápida, sistemática y exhaustiva todos aquellos RCD esparcidos por diversas áreas urbanas y rurales. En este contexto, se plantearon tres objetivos generales investigación: *localizar los vertederos ilegales; realizar un análisis criminológico del fenómeno utilizando teorías ambientales y de la oportunidad delictiva; y conocer la toma de decisiones o política en la gestión de residuos a nivel local.*

Para la consecución de los objetivos referidos se han desarrollado diversos estudios de investigación materializados en los 10 capítulos que componen esta obra, siendo los capítulos 1 y 2 de carácter jurídico; los capítulos 3, 4, 5 y 6 de carácter tecnológico desarrollados fundamentalmente desde la Ingeniería; y aquellos de carácter eminentemente criminológico, como son los capítulos 7, 8, 9 y 10.

En el ***Capítulo 1*** se aborda la regulación penal en materia medioambiental con especial atención a los residuos y a las respuestas políticas-criminales en la materia. Se ha llevado a cabo un análisis jurídico de las sentencias e informes emitidos por diferentes organismos jurídicos.

En el ***Capítulo 2*** se profundiza en la regulación jurídico-administrativa del RCD a nivel local al tiempo que se analiza el nivel desarrollo de las ordenanzas municipales en la materia, esto es, qué municipios cuentan con ordenanzas locales en materia de residuos y cuáles no, o si tienen establecida la fianza obligatoria para obras menores y mayores.

En el ***Capítulo 3*** se examina la calidad de los datos proporcionados por el SEPRONA sobre las infracciones administrativas de RCD en la región entre los años 2016-2021. En particular, se analiza si las mismas están correctamente georreferenciadas y si son adecuadas para desarrollar modelos de teledetección del residuo. Antesala de la creación de una herramienta tecnológica capaz de localizar automáticamente todos los puntos de residuos en la región.

En el ***Capítulo 4*** se caracterizan los puntos del vertido ilegal para determinar las zonas con mayor probabilidad de presencia de RCD en función de la proximidad a determinados elementos geográficos (centros urbanos, vías de comunicación, hidrografía, etc.) y la relación entre la topografía del terreno con la presencia de este tipo de residuos.

En el ***Capítulo 5*** se analiza la asociación entre el número de denuncias de los municipios como variable dependiente y las variables sociodemográficas más características de estos (número de habitantes, número de efectivos policiales, existencia de puntos de acopio, etc.) para determinar cuáles de ellas están relacionadas con el vertido ilegal.

En el ***Capítulo 6*** se aborda la detección visual de acopios de RCD mediante fotointerpretación de modelos procedentes del DEM utilizando datos de campo del área urbana y periurbana de Mérida. En otras palabras, se desarrolla una herramienta piloto para teledetectar de forma rápida, sistemática y exhaustiva todos los residuos.

En el ***Capítulo 7*** se analizan las percepciones de los agentes y mandos del SEPRONA y de la Policía Local acerca de la frecuencia del vertido, su ubicación, el perfil del delincuente, *modus operandi*, los elementos facilitadores de la comisión de estas conductas y aspectos clave para la prevención.

En el ***Capítulo 8***, y siguiendo el mismo esquema de análisis que en capítulo anterior, se analizan las percepciones de los Agentes del Medio Natural, como operadores con competencias medioambientales y, por tanto, responsables de la vigilancia y control del residuo.

En el ***Capítulo 9*** se analizan las percepciones de los gestores autorizados de residuos que coordinan labores de gestión a nivel provincial, tanto en Cáceres como en Badajoz, a fin de comprender desde una perspectiva sistémica y dinámica la gestión efectuada y su relación con el vertido ilegal (sistema de incentivos propuestos, coste del reciclaje, licencias requeridas, cauces de entrega de residuos, etc.).

En el ***Capítulo 10*** se analizan las percepciones de operadores políticos, en particular, de alcaldes y concejales, como agentes clave en la toma de decisiones en la gestión de residuos a nivel local, ahondando en algunas cuestiones ya abordadas en otros capítulos, pero desde una óptica de la respuesta que se otorga desde la política de gestión local a los vertidos ilegales.

Finalmente, en el apartado de conclusiones se sintetizan los resultados más importantes de la obra para seguidamente aportar una serie de recomendaciones o buenas prácticas para eliminar o reducir los vertidos ilegales a fin de ser divulgadas entre diferentes organismos públicos y privados relacionados con el fenómeno.

REFERENCIAS BIBLIOGRÁFICAS

Allers, M.A. y Hoeben, C. (2009). Effects of Unit-Based Garbage Pricing: A Differences-in-Differences Approach. *Environ Resource Econ, Vol.* 45, 405–428. DOI: 10.1007/s10640-009-9320-6.

Baugh, G., y Kokaram, A. (2008). Feature-based object modelling for visual surveillance. *ICIP*, 352-1355. DOI: 10.1109/ICIP.2008.4712014.

Biotto, G., Silvestri, S., Gobbo, L., Furlan, L., Valenti, S. y Rosselli, R. (2009). GIS, multi-criteria and multi-factor spatial analysis for the probability assessment of the existence of illegal landfills. *International Journal of Geographical Information Science, Vol. 23* (10), 1233-1244.

Brisman, A. y South, N. (2015). An assessment of Tonry and Farrington's four major crime prevention strategies as applied to environmental crime and harm. *Varstvoslovje, Journal of Criminal Justice and Security, Vol.17*(2), 127-150.

Chen, J., Hua, C. y Liu, C. (2018). Considerations for better construction and demolition waste management: Identifying the decision behaviors of contractors and government departments through a game theory decision-making model. *Journal of Cleaner Production, Vol.* 212, 190-199. DOI: 10.1016/j.jclepro.2018.11.262.

Cherry, M.A., y Sneirson, J.F. (2011). Beyond profit: Rethinking corporate social responsability and greenwashing after the BP oil disaster. *Tulane Law Review, Vol.* 85, 4, 983-1039. DOI:10.2139/ssrn.1670149.

Cossu, R. y Masi, S. (2013). Re-thinking incentives and penalties: Economic aspects of waste management in Italy. *Waste Management, Vol.* 33, 2541–2547. DOI: 10.1016/j.wasman.2013.04.011.

Crevron, 2015. Ecuador lawsuit. The facts about Chevron and Texaco in Ecuador. Disponible en http://www.chevron.com/ecuador/#b2. [Consultado el 03.07.2023].

Crofts, P., Morris,T., Wells, K. y Powell, A. (2010). Illegal dumping and crime prevention: A case study of Ash Road, Liverpool Council. *Public Space: The Journal of Law and Social Justice, Vol.* 5 (4), 1-23. DOI: 10.5130/psjlsj.v5i0.1904.

D'Amato, A., Mazzanti, M. y Nicolli, F. (2015). Waste and organized crime in regional environments How waste tariffs and the mafia affect waste management and disposal. *Resource and Energy Economics, Vol.* 41, 185-201. DOI: 10.1016/j.reseneeco.2015.04.003

Dabholkar, A., Muthiyan, B., Srinivasan, S., Ravi, S., Jeon, H. y Gao, J. (2017). Smart illegal dumping detection, en 2017 IEEE Third International Conference on Big Data Computing Service and Applications (BigDataService), San Francisco, CA, USA, Apr. 6–9, 2017, 255–260.

De Feo, G., Cerrato, F., Siano, P. y Torretta, V. (2014). Definition of a multicriteria, web-based approach to managing the illegal dumping of solid waste in Italian villages. *Environmental Technology, 35*(1), 104-114. DOI: 10.1080/09593330.2013.816328.

Ding, Z., Yi, G., Tam, V.W.Y. y Huang, T. (2016). A system dynamics-based environmental performance simulation of construction waste reduction management in China. *Waste Management, Vol. 51,* 130–141. DOI: 10.1016/j.wasman.2016.03.001.

Fernández-Jurado, J.A., González-Almendros, C.M., & Sabariego-Rivero, S. (1995). Vertederos ilegales. *Boletín Criminológico, 1* (16), 1-4. DOI: 10.24310/Boletin-criminologico.1995.v1i.9070.

Fraser, I. y Choe, C. (1999). An Economic Analysis of Household Waste Management. *Journal of Environmental Economics and Management Vol. 38,* 234-246. DOI: 10.1006/jeem.1998.1079.

Gallardo, A., Bovea, Mª. D., Colomer, F.J., y Prades, M. (2012). Analysis of collection systems for sorted household waste in Spain. *Waste Management, Vol. 32,* 1623–1633. DOI: /10.1016/j.wasman.2012.04.006.

Glanville, K. y Chang, H.C. (2015). Mapping illegal domestic waste disposal potential to support waste management efforts in Queensland, Australia. *International Journal of Geographical Information Science, Vol.* 29(6), 1042-1058. DOI: 10.1080/13658816.2015.1008002.

Halvorsen, B. (2012). Effects of norms and policy incentives on household recycling: An international comparison. *Resources, Conservation and Recycling, Vol. 67,* 18– 26. DOI: 10.1016/j.resconrec.2012.06.008.

Hamilton, S.F., Sproul, T.W., Sunding, D. y Zilberman, D. (2013). Environmental policy with collective waste disposal. *Journal of Environmental Economics and Management, Vol. 66*, 37–346. DOI: 10.1016/j.jeem.2013.04.003.

Hongping, Y. y Shen, L. (2010). Trend of the Research on Construction and Demolition Waste Management. *Waste Management, Vol. 31*(4), 670-9. DOI: 10.1016/j.wasman.2010.10.030.

Ichinose, D. y Yamamoto, M. (2010). On the relationship between the provision of waste management service and illegal dumping. *Resource and Energy Economics, Vol. 33*, 79–93. DOI: 10.1016/j.reseneeco.2010.01.002

Jelks, N.O., Hawthorne, T.L., Dai, D., Fuller, C.H. y Stauber, C. (2018). Mapping the hidden hazards: Community-led spacial data collection of Street level environmental stressors in a degraded, urban watershed. *International Journal of Environmental Research and Public Health, Vol.15* (825). DOI: 10.3390/ijerph15040825

Jordá-Borrell, R., Ruiz-Rodríguez, F., y Lucendo-Monedero, Á.L. (2014). Factor analysis and geographic information system for determining probability areas of presence of illegal landfills. *Ecological Indicators, 37*(A), 151–160. DOI: 10.1016/j.ecolind.2013.10.001

Kellenberg, D. (2012). Trading wastes. *Journal of Environmental Economics and Management, Vol. 64*, 68–87.DOI: 10.1016/j.jeem.2012.02.003.

Kinobe, J.R., Nigawaga, C.B, Gebresenbet, G., Komakech, A.J. y Vinneras, B. (2014). Mapping out the solid waste generation and collection models: The case of Kampala City. *Journal of the Air y Waste Management Association, Vol. 65*(2), 197-205. DOI: 10.1080/10962247.2014.984818.

Lega, M. y Persechino, G. 2014. GIS and infrared aerial view: advanced tools for the early detection of environmental violations. *Waste Management and The Environment, Vol. 7*, 225-235. DOI:10.2495/WM140191.

Liddick, D. (2009). The traffic in garbage and hazardous wastes: An overview. *Trends in Organized Crime, Vol. 13*(2), 134-146. DOI: DOI:10.1007/s12117-009-9089-6.

Lu, W. (2018). Big data analytics to identify illegal construction waste dumping: A Hong Kong study. *Resources, Conservation y Recycling, Vol.141*, 264–272. DOI: 10.1016/j.resconrec.2018.10.039

Lu, W., Yuan, H., Li, J., Hao, J.J.L., Mi, X y Ding, Z. (2011). An empirical investigation of construction and demolition waste generation rates in Shenzhen city, South China. *Waste Management Vol. 31*, 680–687. DOI: 10.1016/j.wasman.2010.12.004.

Lucendo-Monedero, A.L., Jordá-Borrell, R., y Ruiz-Rodríguez, F. (2015). Predictive model for areas with illegal landfills using logistic regression.

Journal of Environmental Planning and Management, 58(7), 1309–1326. DOI: 10.1080/09640568.2014.993751

Massari, M. y Monzini, P. (2004). Dirty businesses in Italy: a case-study of illegal trafficking in hazardous waste. *Glob. Crime, Vol. 6* (3-4), 285-304. DOI: 10.1080/17440570500273416

Matos, J., Oštir, K. y Kranjc, J. (2012). Attractiveness of roads for illegal dumping with regard to regional differences in Slovenia. *Acta Geogr. Slovenica, Vol. 52*, 431–451. DOI: 10.3986/AGS52207

Ministerio de Agricultura, Alimentación y Medio Ambiente (2016). Plan Estatal Marco de Residuos 2016-2022 (PEMAR). Recuperado de: https://www.miteco.gob.es/es/calidad-y-evaluacion-ambiental/planes-y-estrategias/pemaraprobado6noviembrecondae_tcm30-170428.pdf [Consultado el 03.07.2023].

Ministerio Fiscal. (2018). Memorias de la Fiscalía General del Estado. https://www.fiscal.es [Consultado el 03.07.2023].

Mol, H., Rodríguez Goyes, D., South, N. y Brisman, A. (2017). *Introducción a la Criminología verde. Conceptos para nuevos horizontes y diálogos medioambientales.* Temis.

Navarro, J., Grémillet, D., Afán, I., Ramírez, F., Bouten, W. y Forero, M.G. (2016). Feathered detectives: real-time GPS tracking of scavenging gulls pinpoints illegal waste dumping. *PLoS One, Vol. 11* (7), e0159974. DOI: 10.1371/journal.pone.0159974.

Nieto, A. y Cárdenas, G. (2017). 25 años de políticas europeas en Extremadura. *Cuadernos de turismo, 39*, 389-416. DOI: 10.6018/turismo.39.290621.

Ortiz García, J. (2022). *Mito o realidad. Un estudio criminológico en las comunidades rurales de Extremadura.* Dykinson.

Persechino, G., Lega, M., Romano, G., Gargiulo, F. y Cicala, L. (2013). IDES project: An advanced tool to investigate illegal dumping. *WIT Transactions on Ecology and the Environment, Vol. 173*, 603-614. DOI: 10.2495/SDP130501.

Quesada Ruiz, L., Rodríguez-Galiano, V y Jordá Borrell, R. (2018). Identifying the main physical and socioeconomic drivers of illegal landfills in the Canary Islands. *Waste Management y Research, Vol.* 36(11) 1049–1060. DOI: 10.1177/0734242X18804

Romeo, V., Brown, S., Stuver, S. (2003). A GIS analysis of illegal dumping in the 78249 Zip Code of Bexar County, Texas, en Proceedings of Esri International User Conference, San Diego, CA, USA, 7–11 July 2003.

Seeboonruang, U. (2016). Geographic information system–based impact assessment for illegal dumping in borrow pits in Chachoengsao Province, Thailand. *The Geological Society of America, Special paper 520,* 393-406. DOI:10.1130/2016.2520(35).

Senior, K. y Mazza, A. (2004). Italian "Triangle of death" linked to waste crisis. *The Lancet Oncology, Vol.5,* 525-527. DOI: 0.1016/s1470-2045(04)01561-x

Seror, N. y Portnov, B.A. (2018). Identifying areas under potential risk of illegal construction and demolition waste dumping using GIS tools. *Waste Management, Vol.*75, 22-29. DOI: 10.1016/j.wasman.2018.01.027.

Silvestri, S. y Omri, M. (2007). A method for the remote sensing identification of uncontrolled landfills: formulation and validation. *International Journal of Remote Sensing, Vol.* 29(4), 975-989. DOI: 10.1080/01431160701311317.

Slavik, J. y Pavel, J. (2012). Do the variable charges really increase the effectiveness and economy of waste management? A case study of the Czech Republic. *Resources, Conservation and Recycling Vol.* 70, 68– 77. DOI: 10.1016/j.resconrec.2012.09.013.

Tasaki, T., Kawahata, T., Osako, M., Matsui, Y., Takagishi, S., Morita, A. y Akishima, S. (2007). A GIS-based zoning of illegal dumping potential for efficient surveillance. *Waste Manage, Vol. 27,* 256–267. DOI: 10.1016/j.wasman.2006.01.018.

Triassi, M., Alfano, R., Illario, M., Nardone, A., Caporale, O. y Montuori, P. (2015). Environmental Pollution from Illegal Waste Disposal and Health Effects: A Review on the "Triangle of Death". *Int. J. Environ. Res. Public Health, Vol. 12,* 1216-1236. DOI: 10.3390/ijerph120201216.

PEMAR (Plan Estatal Marco de Residuos 2016-2022). Ministerio para la Transición ecológica y el reto demográfico. Disponible en: https://www.miteco.gob.es/es/calidad-y-evaluacion-ambiental/planes-y-estrategias/Planes-y-Programas.aspx [Consultado el 03.07.2023].

PIREX (Plan Integrado de Residuos de Extremadura 2016-2022). Junta de Extremadura, Dirección General de Medio Ambiente. Recuperado de: http://extremambiente.juntaex.es/files/2017/P_AMBTAL/RESIDUOS/PIREX/PIREX_2016_2022.pdf [Consultado el 03.07.2023].

Uricchio, V.F., Palmisano, V. N., López, N. y Bruno, D.E. (2010). A centralized management data to prevention of environmental crimes fight the Altamura case, *2010 IEEE Workshop on Environmental Energy and Structural Monitoring Systems,* Taranto, 2010, 59-64.

Van Wingerde, K. (2015). The limits of environmental regulation in a globalized economy. *The Routedge handbook of White-collar and corporate crime in Europe,* 260.

White, R. y Graham, H. (2015). Greening justice: Examing the interfaces of criminal, social and ecological justice. *British Journal of Criminology, Vol.* 55(5), 845-865. DOI: doi.org/10.1093/bjc/azu117.

Wortley, R. y Mazerolle, L. (2008). *Environmental Criminology and Crime Analysis.* Willan.

Yuan, H.P., Shen, L.Y., Hao, J.J.L. y Lu, W.S. (2010). A model for cost–benefit analysis of construction and demolition waste management throughout the waste chain. *Resources, Conservation and Recycling Vol. 55*, 604–612. DOI: 10.1016/j.resconrec.2010.06.004.

Zainun, N.Y., Rahman, I.A. y Rothman, R.A. (2016). Mapping Of Construction Waste Illegal Dumping Using Geographical Information System (GIS), International Engineering Research and Innovation Symposium (IRIS), IOP Conf. Series: Materials Science and Engineering 160 (2016) 012049.

Capítulo 1

La regulación jurídico-penal de los delitos contra el medioambiente: claves político-criminales

GIORGIO DARIO MARIA CERINA[1]

1. LA PROTECCIÓN DEL MEDIOAMBIENTE COMO *CONDITIO SINE QUA NON*

Que desarrollo económico, vida humana y medioambiente formen una ecuación compleja es una obviedad. Y, si la búsqueda de un equilibrio lleva tiempo encima de la mesa, la expresión «desarrollo sostenible» propone una complicada síntesis que no siempre resulta fácil de aprehender[2].

---

1 Profesor Contratado Doctor, Área de Derecho penal, Universidad de Extremadura.

2 Sobre el origen y la evolución del concepto de desarrollo sostenible, vid., entre muchos, Bermejo Gómez de Segura, (2014). Ya en el año 1995, decía el Tribunal Constitucional (STC 102/1995, ECLI:ES:TC:1995:102) que "la calidad de vida como aspiración situada en primer plano por el Preámbulo de la Constitución, que en principio parece sustentarse sobre la cultura y la economía, aún cuando en el texto articulado se ligue por delante a la utilización racional de los recursos naturales y por detrás al medio ambiente, con el trasfondo de la solidaridad colectiva. En suma, se configura un derecho de todos a disfrutarlo y un deber de conservación que pesa sobre todos, más un mandato a los poderes públicos para la protección (art. 45 C.E.). En seguida, la conexión indicada se hace explícita cuando se encomienda a los poderes públicos la

La Resolución de la Asamblea General de las Naciones Unidas de 25 de septiembre de 2015 titulada "Transformar nuestro mundo. La Agenda 2030 para el Desarrollo Sostenible"[3] se construye alrededor de 5 ejes. Los primeros tres se rotulan significativamente "personas"; "planeta" y "prosperidad". El documento, firmado por la práctica totalidad de los países, no renuncia a un "crecimiento económico sostenido" pero lo acompaña con el adjetivo "sostenible"[4]. Parece reinar acuerdo sobre el hecho de que no puede plantearse seriamente el "fin" de "la pobreza y el hambre en todas sus formas", no se puede perseguir que "todos los seres humanos puedan disfrutar de una vida próspera y plena" y tampoco pueden propiciarse "sociedades pacíficas, justas e inclusivas que estén libres del temor de la violencia" si todo eso no se acompaña con la protección del "planeta contra la degradación".

---

función de impulsar y desarrollar se dice, la actividad económica y mejorar así el nivel de vida, ingrediente de la calidad si no sinónimo, con una referencia directa a ciertos recursos (la agricultura, la ganadería, la pesca) y a algunos espacios naturales (zonas de montaña) (art. 130 C.E.), lo que nos ha llevado a resaltar la necesidad de compatibilizar y armonizar ambos, el desarrollo con el medioambiente (STC 64/1982). Se trata en definitiva del "desarrollo sostenible", equilibrado y racional, que no olvida a las generaciones futuras, alumbrado el año 1987 en el llamado Informe Brundtland, con el título "Nuestro futuro común" encargado por la Asamblea General de las Naciones Unidas".

3 A/RES/70/1 disponible, en este momento en: https://www.google.com/url?sa=t&rct=j&q=&esrc=s&source=web&cd=&ved=2ahUKEwighLyumc3_AhVsTKQEHQ3jB2oQFnoECA4QAQ&url=https%3A%2F%2Functad.org%2Fsystem%2Ffiles%2Fofficial-document%2Fares70d1_es.pdf&usg=AOvVaw0mnysbnKrvA7OdbKhYYNaD&opi=89978449. [Consultado el 03.07.2023].

4 Véase el punto 9 de la Resolución en la que los firmantes declaran aspirar "a un mundo en el que cada país disfrute de un crecimiento económico sostenido, inclusivo y sostenible y de trabajo decente para todos; un mundo donde sean sostenibles las modalidades de consumo y producción y la utilización de todos los recursos naturales, desde el aire hasta las tierras, desde los ríos, los lagos y los acuíferos hasta los océanos y los mares".

Quizá tenga que ver con el concepto de "cascada de disponibilidad" o con la "aversión a la pérdida" (Kanheman, 2013) o simplemente con que los eventos negativos raros que "reciben menos atención de la que merecen según sus probabilidades objetivas" (Kanheman, 2013, p. 422) son cada vez menos raros (vid. también Silva Sánchez, 2001[5]); o quizá lo decisivo haya sido que, en materia de daños medioambientales, los supuestos responsables coinciden cada vez más con las víctimas reales. Sea como fuere, el diagnóstico del que parte la Agenda presenta "nuestro mundo actual" como "sumamente preocupante". Así, se dice que:

> El agotamiento de los recursos naturales y los efectos negativos de la degradación del medio ambiente, incluidas la desertificación, la sequía, la degradación de las tierras, la escasez de agua dulce y la pérdida de biodiversidad, aumentan y exacerban las dificultades a que se enfrenta la humanidad. El cambio climático es uno de los mayores retos de nuestra época y sus efectos adversos menoscaban la capacidad de todos los países para alcanzar el desarrollo sostenible. La subida de la temperatura global, la elevación del nivel del mar, la acidificación de los océanos y otros efectos del cambio climático están afectando gravemente a las zonas costeras y los países costeros de baja altitud, incluidos numerosos países menos adelantados y pequeños Estados insulares en desarrollo. Peligra la supervivencia de muchas sociedades y de los sistemas de sostén biológico del planeta.

Es bastante difusa la idea de que el propio concepto de medioambiente nazca a partir de una necesidad de defensa frente a un peligro acuciante[6]. Y lo cierto es que, hoy en día, basta ojear

---

5 Quien habla de "deterioro de *realidades tradicionalmente abundantes* y que en nuestros días empiezan a manifestarse como '*bienes escasos*', atribuyéndoseles ahora un valor que anteriormente no se les asignaba, al menos de modo expreso" (p. 25).

6 El propio Tribunal Constitucional (en la ya citada STC 102/1995), afirmó que el propio concepto de medioambiente nace "para reconducir a la unidad los diversos componentes de una realidad en peligro. Si éste no se hubiera presentado resultaría inimaginable su aparición por meras razones teóricas, científicas o filosóficas, ni por tanto jurídicas. (...). Diagnosticada como grave, además, la amenaza que suponen tales

cualquier medio de comunicación para percibir la actualidad de un problema cada vez más "cercano" a nuestro entorno más inmediato. En los tiempos que corren, entrar en cualquier confortablemente climatizado supermercado supone ciertamente un alivio toda vez que se acerca el verano que, año tras año, se presenta más asfixiante; pero, cuando llega el momento de pagar la compra y notamos el impacto sobre nuestro bolsillo del aumento de los precios de los productos frescos causado por la sequía, es difícil no reparar en cómo influye sobre nuestro bienestar el calor que producimos... para combatirlo. Sequía, desertificación, cambio climático, elevación del nivel del mar, acidificación de los océanos, ponen ciertamente en peligro la supervivencia de muchas especies... como la humana. Y esto es algo que sabemos desde hace tiempo: la novedad post-pandémica es la evidencia del impacto de ello sobre *nuestro* bienestar (y bolsillos).

2. EL MEDIOAMBIENTE COMO ALGO NECESITADO DE PROTECCIÓN PENAL

El espacio del que dispongo no me permite ir más allá. Así que, a la luz de lo dicho hasta ahora, asumamos que existe cierto consenso acerca de que la degradación del medioambiente sitúa en serios aprietos a la vida humana de tal forma que su preservación es absolutamente necesaria para garantizar no solo la convivencia en sociedad, sino la propia supervivencia de la especie.

---

agresiones y frente al reto que implica, la reacción ha provocado inmediatamente una simétrica actitud defensiva que en todos los planos jurídicos constitucional, europeo y universal se identifica con la palabra 'protección', sustrato de una función cuya finalidad primera ha de ser la 'conservación' de lo existente, pero con una vertiente dinámica tendente al 'mejoramiento', ambas contempladas en el texto constitucional (art. 45.2 C.E.), como también en el Acta Única Europea (art. 130 R) y en las Declaraciones de Estocolmo y de Río (FJ 7)".

Y dicho ello, recordemos lo que comúnmente se enseña que debe hacer el derecho penal: proteger aquellos «bienes jurídicos» cuya preservación es *imprescindible* para la convivencia en sociedad (Jescheck, 1993).

De hecho, que la protección del medioambiente sea algo que ha de interesar al derecho penal parece estar fuera de discusión.

La propia Constitución española lo sugiere expresamente: el apartado tercero del artículo 45 expresamente al medioambiente, afirma que "para quienes violen lo dispuesto en el apartado anterior, en los términos que la ley fije se establecerán sanciones penales o, en su caso, administrativas". Cierto es que se trata de una *opción,* que no de un mandato (las sanciones podrán ser penales "o" administrativas)[7]; y también es cierto que el apartado 3 solo se refiere al "anterior" que, a su vez, solo contiene un mandato dirigido a "los poderes públicos" (de "velar por la utilización racional de todos los recursos naturales, con el fin de proteger y mejorar la calidad de la vida y defender y restaurar el medio ambiente, apoyándose en la indispensable solidaridad colectiva"). Pero no es menos cierto que el hecho de que el constituyente español haya "animado" el legislador a hacer uso del derecho penal es ciertamente excepcional. Y que la sugerencia se realice con referencia al medioambiente no es algo que pueda pasarse por alto a la hora de sopesar la oportunidad de que la *extrema ratio* del ordenamiento jurídico se interese del asunto, con independencia de la postura que se asuma respecto de la necesaria relevancia constitucional de los bienes jurídicos a protegerse penalmente (vid., entre muchos, Alonso Álamo, 2009; Bricola, 1973; Cerina, 2022).

De la misma manera, dirigiendo la mirada hacia el ordenamiento europeo, el propio título de la Directiva 2008/99/CE del Parlamento Europeo y del Consejo, de 19 de noviembre de 2008, afirma inequívocamente que la "protección penal el medio ambiente" ha de llevarse a cabo utilizando "el Derecho penal" (vid.

---

7 En sentido contrario, entre otros, Martínez-Buján Pérez (2019) quien habla de un "mandato expreso de criminalización" (p. 995).

Vercher Noguera, 2023). Y aquí también no debería olvidarse de que se trata de una Directiva que impone obligaciones a los Estados miembros relacionadas con el Derecho penal nacional que entró en vigor *antes* del Tratado de Lisboa, cuando todavía se enseñaba la estructura en Pilares del Derecho comunitario y se decía que lo que tenía que ver con el Derecho penal era cosa del Tercer Pilar (el de la cooperación intergubernamental) en el que se usaban Decisiones Marco, que no Directivas, lo que implicaba significativas diferencias (Carrera Hernández, 2007; Vercher Noguera, 2005). Y, de hecho, fue precisamente la protección penal del medioambiente el terreno en el que se libró la batalla jurídica entre la Comisión y el Consejo que desembocó en la STJCE de 13 de septiembre de 2005, *Comisión c. Consejo*, C-176/03[8] que terminó reconociendo que, "para garantizar la plena efectividad de las normas" comunitarias en materia ambiental, el legislador europeo podía adoptar "medidas relacionadas con el Derecho penal de los Estados miembros" toda vez que las estimara "necesarias" (vid. Vercher Noguera, 2005). Así, por un lado, la necesidad de que, incluso a nivel europeo, se movilizara a la *extrema ratio* en materia ambiental representó la ocasión para reinterpretar e indudablemente reforzar la intervención europea en materia penal; por otro, la sentencia representó "un reconocimiento de la importancia del medio ambiente" y aquilató, "en definitiva, al Dere-

8 En síntesis, a propuesta del Reino de Dinamarca, el Consejo consideró que, de acuerdo con lo establecido en los artículos 29 e), 31 y, sobre todo, 34 TUE, la "protección del medio ambiente mediante el Derecho penal" habría de llevarse a cabo en el marco del Tercer Pilar y aprobó la Decisión marco 2003/80/JAI; la Comisión, por su parte, consideraba que, puesto que, conforme al art. 175 TCE, la competencia en asunto medioambientales se enmarcaba en el Pilar comunitario, "la armonización de las legislaciones penales nacionales, especialmente de los elementos constitutivos de infracciones contra el medioambiente sancionables penalmente" debía concebirse "como un instrumento al servicio de la política comunitaria en cuestión" (así literalmente el argumento que figura en el apartado 19 de la Sentencia citada en el texto).

cho penal como instrumento básico para su protección" (Vercher Noguera, 2005, p. 7).

En este mismo sentido, parte de la doctrina ha observado que el vigor con el que la respuesta europea ha apelado al Derecho penal para la protección del medioambiente no solo ha resultado decisivo para la revisión de las propias competencias europeas en materia penal (cristalizadas después en el Tratado de Lisboa), sino que también ha sido el primer terreno sobre el que se ha ido asentando la idea de una responsabilidad *penal* de las personas jurídicas como hoy la conocemos: como se ha observado, los arts. 6[9] y 7[10] de la Directiva 2008/99/CE configuraron una obligación que, en el año 2010, el Legislador español ya *tenía* que atender (Vercher Noguera, 2018).

---

9 El art. 6, rotulado "Responsabilidad de las personas jurídicas", dispone que "1. Los Estados miembros se asegurarán de que las personas jurídicas pueden ser consideradas responsables por los delitos a los que se hace referencia en los artículos 3 y 4 cuando tales delitos hayan sido cometidos en su beneficio por cualquier persona, a título individual o como parte de un órgano de la persona jurídica, que tenga una posición directiva en la persona jurídica, basada en: a) un poder de representación de la persona jurídica: b) una autoridad para tomar decisiones en nombre de la persona jurídica, o c) una autoridad para ejercer control dentro de la persona jurídica. 2. Los Estados miembros se asegurarán también de que las personas jurídicas puedan ser consideradas responsables cuando la ausencia de supervisión o control por parte de una persona a que se refiere el apartado 1 haya hecho posible que una persona bajo su autoridad cometa, en beneficio de la persona jurídica, alguno de los delitos a los que se hace referencia en los artículos 3 y 4. 3. La responsabilidad de las personas jurídicas de conformidad con los apartados 1 y 2 no excluirá la adopción de medidas penales contra las personas físicas que sean autoras, incitadoras o cómplices de los delitos a los que se hace referencia en los artículos 3 y 4.

10 Rotulado "Sanciones a las personas jurídicas" y cuyo texto es el siguiente: "Sanciones a las personas jurídicas. Los Estados miembros adoptarán las medidas necesarias para garantizar que las personas jurídicas consideradas responsables en virtud del artículo 6 sean castigadas con sanciones efectivas, proporcionadas y disuasorias".

3. DERECHO PENAL, SÍ. PERO ¿CÓMO?

3.1. El medioambiente que hay que proteger

Así que, por lo visto, no hay discusión: la necesidad de proteger el medioambiente empleando (también) el Derecho penal parece un hecho. Y es aquí donde empiezan los problemas: ¿cómo llevar a cabo la tarea?

En primer lugar, el principio de taxatividad implicaría que, por lo menos, sepamos a lo que nos estamos refiriendo cuando hablamos de medioambiente (Bacigalupo, 1981). Dejemos ahora a un lado la cuestión inherente a la ortografía (aunque la RAE sugiera la utilización de una única palabra[11], no es infrecuente encontrarse con la redacción "medio ambiente" que, en efecto, es la que utilizan tanto el Código penal como la Constitución). Lo primero que un apócrifo en la materia (y además, como yo, con una lengua materna distinta al español) observa es cierta aparente redundancia implícita en el mismo término "medioambiente": el pleonasmo sale a la luz tras un simple recorrido por la versión *online* del DRAE. Si se busca "medioambiente", el diccionario remite a "medio" entendido como "conjunto de circunstancias exteriores a un ser vivo"; "ambiente" es lo "que rodea algo o a alguien como elemento de su entorno" o el "conjunto de condiciones o circunstancias físicas, sociales, económicas, etc. de un lugar, una colectividad o una época"[12]. Así que, según algunos autores (Ma-

---

[11] Véase https://www.rae.es/dpd/medioambiente. [Consultado el 03.07.2023].

[12] Según algunos, el pleonasmo se debería a una traducción errónea del inglés *environment* en la cumbre de Estocolmo de 1972: el traductor habría omitido una coma entre "medio" y "ambiente" (véase, por ejemplo, Ramírez Mora, 2023). Nótese que la redundancia de la expresión "medio ambiente" ha sido observada también por el Tribunal Constitucional (vid. la ya citada STC 102/1995).

teos Rodríguez-Arias, 1992; Bacigalupo, 1981), habría sido suficiente hablar de "ambiente"[13].

Dicho eso, una primera posible pregunta podría ser si el ambiente que ha de interesar al Derecho penal es solo el ambiente "natural", entendido como "suma de las bases naturales de la vida humana" (Bacigalupo, 1981, p. 200) o si también hay que incluir en la ecuación el llamado "ambiente artificial" que se refiere aquella parte del entorno resultado de la obra del hombre, o si incluso lo conveniente sería adoptar una perspectiva amplia que abarque "factores culturales, sociológicos e incluso económicos" (Alastuey Dobón, 2004, p. 55)[14]. Pues bien, con carácter general, la doctrina penal parece haber optado por una postura restrictiva, haciendo coincidir el "objeto de protección a que debe referirse el derecho penal del medio ambiente" con el

> Mantenimiento de las propiedades del suelo, el aire y el agua, así como de la fauna y la flora y las condiciones ambientales de desarrollo de estas especies, de tal forma que el sistema ecológico se mantenga con sus sistemas subordinados y no sufra alteraciones perjudiciales (Bacigalupo, 1981, p. 201)[15].

---

13 Nótense, en España, la Ley 38/1972, de 22 de diciembre, de protección del *ambiente* atmosférico y la Ley 22/1973, de 21 de julio de Minas que se refiere en diversas ocasiones a la "protección del *ambiente*" (artículos quinto, ciento dieciséis y en la exposición de motivos).

14 El Tribunal Constitucional ha señalado cómo, aparte de los recursos naturales, la flora, la fauna y los minerales (definidos como "los tres "reinos" clásicos de la Naturaleza con mayúsculas), en el escenario que suponen el suelo y el agua, el espacio natural", desde la primera aparición en el ordenamiento jurídico español, el concepto de medioambiente ha incorporado "otros elementos que nos son naturaleza sino Historia, los monumentos, así como el paisaje".

15 Sobre este particular, no obstante, téngase en cuenta la postura del Tribunal Constitucional (SsTC 102/1995, ECLI:ES:TC:1995:102 y 306/2000, ECLI:ES:TC:2000:306) que, a la hora de justificar las incursiones de la legislación básica del Estado sobre protección del medio ambiente (art. 149.1 23ª) en ámbitos aledaños que la Constitución reserva a las Comunidades Autónomas (art. 148) ha optado sin dudas por un concepto amplio de medioambiente. El Código penal, por su

A partir de aquí, el debate alrededor del bien jurídico a protegerse mediante el Derecho penal ha sido sustancialmente absorbido por la discrepancia entre quienes han defendido una idea "antropocéntrica" del concepto de medioambiente que ha de interesar a la *ultima ratio* del ordenamiento jurídico y quienes, en el lado opuesto, han considerado plausible que el Derecho penal asumiera una aproximación "ecocéntrica". La discusión es demasiado conocida y extensa como para que ahora sea posible y conveniente recorrerla en toda su amplitud.

Remitiendo el lector a la doctrina especializada, podemos conformarnos aquí con apuntar brevemente que, según la primera postura (aproximación antropocéntrica), lo que ha de interesar al legislador penal es la protección de bienes jurídicos individuales referidos al hombre (Ferrajoli, 1996). Por consiguiente, para adquirir relevancia penal, la preocupación para el medioambiente debe traducirse en vulneración de bienes jurídicos individuales entre los que ciertamente destaca la salud.

Al lado opuesto, se halla quien considera que el entorno ha de ser protegido en cuanto tal, como conjunto externo e indepen-

---

parte, dentro del Título XVI ("De los delitos relativos a la ordenación del territorio y el urbanismo y la protección del patrimonio histórico y el medio ambiente") ha utilizado la expresión "medio ambiente" tan solo en el capítulo III referido significativamente a los delitos contra el medioambiente y "los recursos naturales" distinguiendo en cambio la ordenación del territorio y el urbanismo (capítulo I) así como el "patrimonio histórico" (capítulo II). De ahí que la doctrina haya sustancialmente optado por una idea "natural" de medioambiente haciendo hincapié en la escasa utilidad que, desde la óptica de la configuración del bien jurídico protegido, aporta una noción amplia. Con todo, quizás podría apuntarse que el argumento basado en la sistemática del Código penal debería tener en cuenta que los delitos incluidos en el capítulo IV parecen poderse rubricar como medioambientales incluso teniendo en la mente la definición estricta apuntada en el texto (vid., entre otros, Gómez Rivero, 2022; Muñoz Conde, 2022; Muñoz Conde F., López Peregrín, C., García Álvarez, P., 2015).

diente del hombre, sin necesidad de que la intervención penal esté supeditada a la afectación de bienes jurídicos individuales.

En posiciones intermedias, se coloca la mayoría de la doctrina, ora adoptando una postura antropocéntrica "moderada", ora asumiendo una aproximación ecocéntrica con matices relacionados con la función que el entorno representa para el individuo (por ejemplo, Martínez-Buján Pérez, 2019). De hecho, según el planteamiento mayoritario, el art. 45 CE contiene elementos de cada una de las posturas:

1. En clave antropocéntrica[16]: el apartado 1 consagra un derecho (no fundamental[17]) a "disfrutar de un medio ambiente *adecuado para el desarrollo de la persona"*, mientras que el apartado 2 obliga a los poderes públicos a velar "por la utilización racional de todos los recursos naturales *con el fin de proteger y mejorar la calidad de la vida (...) apoyándose en la indispensable solidaridad colectiva"*.
2. En óptica ecocéntrica: el precepto consagra no solo un derecho al disfrute del medio ambiente, sino también un "deber de conservarlo" (art. 45.1) y los poderes públicos deben velar por la utilización racional de todos los recursos naturales también "con el fin de (...) defender y restaurar el medio ambiente".

De la misma manera, recientemente se ha subrayado que, a la hora de desglosar las metas de los Objetivos de Desarrollo Sos-

---

16 Véase, una vez más, la STC 102/1995 ya citada en la que se lee que "el ambiente (...) es un concepto esencialmente antropocéntrico y relativo. No hay ni puede haber una idea abstracta, intemporal y utópica del medio, fuera del tiempo y del espacio. Es siempre una concepción concreta, perteneciente al hoy y operante aquí".

17 Vid. las consideraciones de Mateos Rodríguez-Arias (1992). En la jurisprudencia constitucional, vid., entre muchas, SsTC 199/1996 (ECLI:ES:TC:1996:199) y 84/2013 (ECLI:ES:TC:2013:84) que aclaran que "el art. 45.1 (...) enuncia un principio rector, no un Derecho fundamental".

tenible directamente relacionados con el medioambiente (13[18], 14[19] y 15[20]), la propia Agenda 2030 "aúna elementos de vocación ecocéntrica (enfatizando el valor inherente e independiente de los océanos, mares, recursos marinos, ecosistemas terrestres y ecosistemas interiores de agua dulce)[21] y referencias netamente antropocéntricas" (Gómez Lanz, 2023, p. 361)[22].

18 "Adoptar medidas urgentes para combatir el cambio climático y sus efectos".

19 "Conservar y utilizar sosteniblemente los océanos, los mares y los recursos marinos para el desarrollo sostenible".

20 "Proteger, restablecer y promover el uso sostenible de los ecosistemas terrestres, gestionar sosteniblemente los bosques, luchar contra la desertificación, detener e invertir la degradación de las tierras y detener la pérdida de biodiversidad".

21 Como ejemplos de metas descritas de forma claramente ecocéntrica, señalo: la 14.1 ("De aquí a 2025, prevenir y reducir significativamente la contaminación marina de todo tipo, en particular la producida por actividades realizadas en tierra, incluidos los detritos marinos y la polución por nutrientes"); la 14.2 ("De aquí a 2020, gestionar y proteger sosteniblemente los ecosistemas marinos y costeros para evitar efectos adversos importantes, incluso fortaleciendo su resiliencia, y adoptar medidas para restaurarlos a fin de restablecer la salud y la productividad de los océanos"); la 14.5 ("De aquí a 2020, conservar al menos el 10% de las zonas costeras y marinas, de conformidad con las leyes nacionales y el derecho internacional y sobre la base de la mejor información científica disponible").

22 A diferencia del autor, como ejemplo de meta claramente descrita de forma antropocéntrica, señalo la 13.1 ("Fortalecer la resiliencia y la capacidad de adaptación a los riesgos relacionados con el clima y los desastres naturales en todos los países");14.7 ("De aquí a 2030, aumentar los beneficios económicos que los pequeños Estados insulares en desarrollo y los países menos adelantados obtienen del uso sostenible de los recursos marinos, en particular mediante la gestión sostenible de la pesca, la acuicultura y el turismo"); 14.b ("Facilitar el acceso de los pescadores artesanales a los recursos marinos y los mercados").

3.2. Implicaciones político-criminales de la controversia entre ecocentrismo y antropocentrismo

El trasfondo del debate es ciertamente profundo: detrás está la propia idea de Derecho penal que, incorporando preceptos a tutela de "nuevos" bienes jurídicos supraindividuales, según algunos, se va expandiendo peligrosamente (Silva Sánchez, 2001), mientras que, según otros, evoluciona y se mantiene al paso con los tiempos (vid., por ejemplo, Martínez-Buján Pérez, 2002 y 2022). La cuestión filosófica imbrica tan de cerca la inherente a la idea de medioambiente que habría de ser asumida por el legislador penal, que no ha faltado quien ha dicho que la que atañe a la dicotomía antropocentrismo vs. ecocentrismo es, en realidad, una controversia "ideológica, no técnica" (Gómez Lanz, 2023, p. 361).

El hecho, sin embargo, es que parece que una diferente aproximación conceptual al bien jurídico «medioambiente» puede terminar teniendo repercusiones sobre las posibilidades que se ofrecen al legislador penal, por lo menos, en una óptica *de lege ferenda.*

Así, quien mantenga una concepción anclada a la protección de bienes jurídicos individuales podrá contar con un derecho penal posible construido a partir del esquema del delito de resultado que implica una lesión del bien jurídico protegido que ha de ser comprobada por el juez. No obstante, así procediendo, terminará dejando fuera del ámbito de aplicación de la normativa penal a comportamientos que no *causen* muertes o daños a la salud de personas concretas, y la utilización del Derecho penal se convertirá en lluvia de absoluciones toda vez que esta *causalidad* no pueda probarse ante un tribunal más allá de toda duda razonable.

Al lado opuesto, quien defienda una aproximación al medioambiente como bien jurídico que ha de gozar de protección penal sin necesidad de traer a colación bienes jurídicos personales, se las tendrá que ver con la compleja y discutida cuestión de la lesionabilidad de los bienes jurídicos supraindividuales generales (vid., entre muchos, Matellanes Rodríguez, 2019; Martínez-Buján

Pérez, 2022). De hecho, ¿cómo puede una sola conducta de un único individuo *lesionar*[23] al ambiente que es de todos[24]? Y, si el medioambiente es *esencial* para la propia supervivencia de la especie humana, ¿puede siquiera imaginarse un juez (humano) que sobreviva a la *destrucción*[25] del medioambiente para juzgar a los responsables?[26] Simplificando mucho, buscar la respuesta en el peligro abstracto para el entorno, por lo menos según el planteamiento que lo resuelve en peligro presunto por el legislador y que no hace falta demostrar en el juicio, es camino cuyas criticidades se han descrito con cierto detalle (vid. entre muchos, Pérez-Sauquillo Muñoz, 2017 a.; Silva Sánchez, 1997; Torío López, 1981). De la misma manera, anclar la punibilidad de una conducta al peligro de que la misma se repita incluso por parte de otras personas, tampoco parece resistir a las múltiples objeciones que se han formulado de cara a la legitimidad de un planteamiento "acumulativo" (vid. entre muchos, Pérez-Sauquillo Muñoz, 2017 b.; Silva Sánchez, 2001).

Problemas parecidos tendrá que resolver quien plantee una reconstrucción pluriofensiva cuyas variantes pueden ser numerosas: según que se prime una visión "moderada" que parta del

---

23 Piénsese en la lesión del bien jurídico "vida" en el delito de homicidio.

24 Como se ha observado, "los contornos indeterminados y abstractos que supuestamente caracterizan a los bienes jurídicos supraindividuales imposibilitarían o dificultarían en extremo la tarea de perfilar conceptualmente su lesión o puesta en peligro concreta por medio de un curso causal. Aun en aquellos supuestos en que tal lesión pudiera imaginarse —se añade—, esta no se produciría por efecto de una única conducta aislada, sino por un conjunto o acumulación de ellas, aparte de presentarse de todas formas serios problemas de prueba" (Pérez-Sauquillo Muñoz, 2019, p. 242 quien, sin embargo, es de otra opinión).

25 Como acontece con la vida en el delito de homicidio.

26 En la jurisprudencia, véase, por ejemplo, la STS 52/2003 (ECLI:ES:TS:2003:1220) en la que se lee que "la técnica adecuada de protección del medio ambiente frente a las trasgresiones más graves, que puedan constituir infracciones penales es la de los delitos de peligro, pues la propia naturaleza del bien jurídico "ambiente" y la importancia de su protección exige adelantarla antes de que se ocasione el daño".

ecocentrismo (o del antropocentrismo), el medioambiente (o la salud humana) puede pasar de ser mera *ratio legis*, a representar el bien jurídico inmediatamente protegido, siendo, en este segundo caso, la salud humana (o el propio medioambiente) el bien jurídico mediatamente tutelado por la norma. Sea como fuere, aquí también habrá que plantearse qué criterios hay que proporcionar al juez para la comprobación de la afectación al bien jurídico inmediata y, en su caso, mediatamente afectado por la conducta individual.

Naturalmente, no estamos en condición de abordar ni todas las soluciones posibles ni de enfrentarnos de forma solvente con todos los problemas apuntados. Lo que sí podemos hacer es tratar de restar algo de dramatismo a la cuestión.

Así, aun sabiendo que se trata de planteamientos discutibles y discutidos, de acuerdo con parte de la doctrina, quizás pueda apuntarse que incluso los bienes jurídicos genuinamente supraindividuales (cuya titularidad debe adscribirse a la colectividad entera) pueden verse lesionados por una conducta individual, siempre que la lesión se entienda como "afectación", algo no necesariamente equivalente a la destrucción total del sustrato. Exactamente como la salud y el honor, la integridad del medioambiente puede resultar afectada (en términos de lesión) por una conducta individual que no implique (ni puede implicar) su aniquilación (Cerina, 2022; Pérez Sauquillo Muñoz, 2019[27]).

---

[27] Como dice la autora (pp. 102-103), "el mero hecho de que un bien jurídico sea considerado como supraindividual en lugar de como individual no constituye motivo *per se* alguno para que su protección penal —esto es, su consideración como bien jurídico-penal— sea considerada ilegítima, incluso si esta protección adquiere el carácter de autónoma y no simultánea a la lesión o puesta en concreto peligro de un bien individual (...). En definitiva, los bienes jurídicos colectivos (...) deben estar legitimados siempre que en última instancia —y de conformidad con los parámetros de un Estado social y democrático de Derecho— repercutan en beneficio del individuo y su dignidad, lo que es compatible con una protección autónoma de los mismos cuando ello sea preciso

Esta reconstrucción, si, por un lado, ofrece interesantes posibilidades que permiten esquivar por lo menos parte de las conocidas críticas a las que se enfrentan los tipos penales (aparentemente) acumulativos y de peligro abstracto, por otro, permite vislumbrar con cierta claridad el otro evidente problema con el que se mide el legislador que decida utilizar su *extrema ratio* para proteger al medioambiente. De hecho, una vez aceptado que un individuo, con una sola acción, puede *lesionar* el medioambiente, con los conocimientos que hoy tenemos, no será difícil darnos cuenta de que no bastan los dedos de unas cuantas manos para contar las veces que diariamente realizamos conductas *lesivas* para nuestro entorno: arrancar el coche, montarse en un avión, enchufar a la corriente cualquier electrodoméstico, tirar la basura...son actos que se producen con cierta regularidad en nuestro día a día y todos implican clara e indiscutiblemente consecuencias negativas para el medioambiente. Así las cosas, para un ordenamiento jurídico que pretenda resultar plausible, el reto estribará no tanto en prohibir toda conducta perjudicial para el medioambiente, sino en la búsqueda del umbral más allá del cual el daño al entorno se considera intolerable.

Pero, así planteada la cuestión, aunque el problema sea mucho más complejo de como aquí se puede presentar, me parece que, en un plano político-criminal, podría encontrarse, si no un mínimo común denominador, por lo menos, un reto común que, *mutatis mutandis*, las distintas opciones plantean al legislador que considere utilizar el Derecho penal para proteger al medioambiente y/o al intérprete de la norma penal: el del riesgo (no) permitido más allá del cual la conducta es objeto de sanción (sobre el riesgo permitido y la expansión del Derecho penal, vid. Silva Sánchez, 2001). Así, en el primer caso (delito de resultado material orientado hacia la protección de bienes jurídicos individuales), la cuestión del riesgo permitido entrará en escena como imputación objetiva y/o adecuación social; en el segundo (peligro abstracto

---

y, también, con la existencia de tipos penales pluriofensivos en otros casos".

o delito acumulativo), se tratará de acotar el peligro típico (sobre las distintas opciones, vid. Pérez-Sauquillo Muñoz, C., 2017 a.); en el tercero (lesionabilidad individual del bien jurídico supraindividual), como hemos visto, la cuestión estribará en diferenciar las conductas *lesivas* del medio ambiente irrelevantes para el Derecho (penal) de las que no lo son.

Y el problema así planteado no parece tener soluciones fáciles ni susceptibles de cristalizarse en pautas de comportamientos constantes: hay detrás innumerables cuestiones técnicas que atañen tanto a las acciones perjudiciales como al impacto de las mismas sobre el entorno que, aparte de no resultar siempre pacíficas en el propio ámbito científico sectorial, se hallan en constante movimiento en la medida en que la propia ciencia avanza.

Además, ya en un plano más estrictamente político-criminal, las conclusiones científicas (si es que las hay) se plantean al legislador como elemento a tener en cuenta que debe ponderarse, por ejemplo, con la sensibilidad social y el impacto socio-económico de toda medida restrictiva que, a su vez, puede resultar más o menos asumible en razón de las soluciones científicas existentes[28].

De hecho, como emerge de una lectura atenta de la Agenda 2030, los distintos Objetivos del Desarrollo Sostenible y sus metas "tienen carácter integrado e indivisibles y conjugan las tres dimensiones del desarrollo sostenible: económica, social y ambiental". O, dicho de otra forma: cuando, en el año 2015, la práctica totalidad de los países del mundo confeccionaron la Agenda que dibujaba el itinerario a lo largo del cual había que desplazarse para "transformar nuestro mundo", acordaron que la sostenibilidad ambiental no puede perseguirse al margen de la idea de "prosperidad" que, como hemos visto, no prescinde de un "crecimiento económico sostenido".

---

28 Piénsese, por ejemplo, en el futuro del coche de combustión que ciertamente afecta a la libertad de movimiento de las personas y que se debate entre soluciones técnicas en constante evolución y permanentemente discutidas.

> Por poner un ejemplo relacionado con el objeto del proyecto de investigación en cuyo marco se inserta esta contribución, toda vez que el coste de la "preparación para la reutilización, reciclado y valorización material" de los Residuos de Construcción y Demolición[29] se termine trasladando al "usuario final", las repercusiones sobre el "derecho a disfrutar de una vivienda digna y adecuada" que el artículo 47 de la Constitución atribuye a "todos los españoles" tendrán que tenerse en cuenta a la hora de plasmar deberes, obligaciones y...sanciones[30].

Así dibujado el panorama, resulta bastante evidente que el principal terreno en el que se juega la partida medioambiental es el Derecho administrativo. Esta rama del ordenamiento configura deberes y obligaciones (con las correspondientes sanciones)[31] teniendo en cuenta la complejidad y amplitud de la materia y los continuos hallazgos científicos que no solo inciden en la individualización de las conductas lesivas para el entorno y en la necesidad de protección del mismo, sino también en las soluciones tecnológicas que hacen viable el mantenimiento de un equilibrio "sostenible" entre "prosperidad" y "sostenibilidad ambiental". El resultado es un panorama normativo de una enorme complejidad, fragmentario, a veces confuso e inabarcable y en constante ebullición, enriquecido por aportaciones que proceden desde el contexto supra e internacional (destacando el Derecho de la Unión Europea), y, en España, de ámbito nacional, autonómico e incluso local[32].

---

29 Véase, por ejemplo, lo que prescribe el artículo 30 de la Ley 7/2022, de 8 de abril, de residuos y suelos contaminados para una economía circular.

30 De la misma manera, piénsese en el supuesto en el que los RCD procedan de una obra cuya finalidad sea aumentar la eficiencia energética.

31 Vid. Muñoz Conde F., López Peregrín, C., García Álvarez, P. (2015, p. 20) quienes señalan que la regulación medioambiental administrativa, en primer lugar, más que de sancionar, tiene como finalidad "armonizar intereses contrapuestos", subrayando que aquí se trata de compatibilizar "desarrollo" con "protección del medioambiente".

32 En sentido contrario, Martínez-Buján Pérez (2019, p. 1000) afirma que la administración local "carece de competencias en materia de medio

Ante esta realidad, parece bastante obvio que la intervención penal habrá de realizarse a la luz del sistema delineado administrativamente por lo que deviene ciertamente central el problema de las relaciones entre del Derecho penal y el Administrativo.

4. DERECHO PENAL Y DERECHO ADMINISTRATIVO

La cuestión cobra ahora interés desde dos distintos puntos de vista.

4.1. El problema de las normas penales en blanco

En primer lugar, como hemos visto, la complejidad técnica de la materia y su inestabilidad relacionada con el progreso científico (y social) sugiere recurrir al Derecho administrativo para una primera discriminación entre conductas lícitas e ilícitas para el or-

ambiente". En realidad, el artículo 25.1 de la Ley 7/1985, de 2 de abril, Reguladora de las Bases de Régimen Local, afirma claramente que "El Municipio ejercerá en todo caso como competencias propias, en los términos de la legislación del Estado y las Comunidades Autónomas" en materias como "urbanismo" (letra a), "medio ambiente urbano" (letra b), "protección de la salubridad pública" (letra j). Por lo que respecta a la Comunidad Autónoma de Extremadura, la Ley 3/2019, de 22 de enero, de garantía de la autonomía municipal de Extremadura, en su artículo 15 ("Competencias propias de los municipios"), enumera las competencias de las que "dispondrán" los municipios "en las materias o ámbitos de actuación territorial e infraestructuras" y, en el punto 3ª menciona literalmente la "defensa y protección del medio ambiente y desarrollo sostenible, incluida la protección contra la contaminación acústica, lumínica y atmosférica en las zonas urbanas". Sobre la relevancia de las ordenanzas municipales para "completar" el tipo del delito ecológico del artículo 325 CP, véase la STS 52/2003 (ECLI:ES:TS:2003:1220). En materia de residuos, conviene remitir al art. 12 de la Ley 7/2022, de 8 de abril, de residuos y suelos contaminados para una economía circular que también distribuye competencias en la materia al Estado, Comunidades Autónomas y entidades locales.

denamiento jurídico ambiental (vid., entre muchos, Muñoz Conde F., López Peregrín, C., García Álvarez, P., 2015 y, en tema de residuos, Fuentes Loureiro, 2021). De ahí que al legislador penal se le plantea el problema de cómo construir el tipo teniendo en cuenta el equilibrio construido administrativamente.

La opción más inmediata es configurar el precepto penal remitiendo a la regulación extrapenal, lo que nos coloca ante el conocido debate alrededor de las normas penales en blanco que describen el supuesto de hecho haciendo referencia a preceptos de otra rama del ordenamiento jurídico. El Derecho penal, se dice, se convierte entonces en "accesorio" al Derecho extrapenal (administrativo, en nuestro caso). Las opciones con las que cuenta el legislador se clasifican normalmente atendiendo a:

1. Un modelo basado en la idea de accesoriedad limitada o relativa en virtud del cual "el tipo penal mantiene su esencia de desvalor, ya que los injustos penales se construyen exigiendo, *entre otros elementos,* la violación de normas administrativas, que pasan a ser un elemento más del tipo penal" (Matellanes Rodríguez, 2019, p. 131).
2. Un modelo de accesoriedad absoluta o extrema en virtud del cual el delito se construye ni más ni menos que como una violación de una(s) norma(s) administrativa(s).
3. La "accesoriedad de acto o formal" que puede integrarse en cualquiera de las dos anteriores aproximaciones y que hace pivotar la tipicidad sobre la ausencia de una autorización administrativa.

Las ventajas de recurrir a estas técnicas de tipificación son evidentes: no es necesario insistir en la escasa plausibilidad de una regulación medioambiental que se lleve a cabo mediante complejas y continuas modificaciones del Código penal. Una vez "encargado" el problema técnico y relacionado con la volatilidad de los parámetros a una rama del Derecho más adecuada para el cometido, la remisión a esta por parte de preceptos penales es garantía de cierta coherencia (Brandariz García, 2011; Muñoz Conde F., López

Peregrín, C., García Álvarez, P., 2015) de un sistema que además se consigue mantener permeable al cambio con cierta flexibilidad.

Empero, dicha coherente flexibilidad se logra pagando un precio (vid. por ejemplo, Muñoz Conde F., López Peregrín, C., García Álvarez, P., 2015). Así, una vez que se acepta que lo penalmente típico se acota (en todo o en parte) fuera del Derecho penal, hay que preguntarse dónde re-situar los límites al *ius puniendi* que, conviene no olvidarlo, son garantías del individuo frente al Estado.

Y, de hecho, como ya hemos visto, el derecho al disfrute del medioambiente no ha sido configurado como un derecho fundamental, de tal forma que la reserva de ley orgánica procedente del art. 81 CE no tiene razón de postularse en la legislación administrativa sobre la materia. De la misma manera, también se ha apuntado que las fuentes del Derecho (administrativo) medioambiental no se agotan en la ley estatal, sino que también implican al derecho internacional, al ordenamiento europeo (que podrá intervenir utilizando reglamentos[33] y no solo directivas[34]) y a la legislación autonómica. Además, es relativamente frecuente que los detalles técnicos de la compleja reglamentación ambiental se dejen a la administración (estatal, autonómica y local) que, como es sabido, no siempre se limita a *ejecutar* el mandato legal, sino que, aprovechando las autopistas dejadas libres por el legislador, termina *creando* un derecho que cuesta cada vez más distinguir del que procede de los parlamentos (vid. Habermas, 2005).

---

[33] Véase, por ejemplo, el Reglamento (UE) 2021/783 del Parlamento Europeo y del Consejo, de 29 de abril de 2021, por el que se establece un Programa de Medio Ambiente y Acción por el Clima (LIFE) y se deroga el Reglamento (UE) n. 1293/2013. Sobre la relevancia de Directivas y Reglamentos para completar la descripción típica del art. 325, véase la STS 52/2003 (ECLI:ES:TS:2003:1220).

[34] Mientras que, de acuerdo con el artículo 83 TFUE, en principio, a la hora de establecer "normas mínimas relativas a la definición de las infracciones penales...", el Parlamento Europeo y el Consejo utilizarán "directivas adoptadas con arreglo al procedimiento legislativo ordinario".

Así las cosas, la construcción de normas penales en blanco plantea problemas de compatibilidad no solo con los principios de legalidad y reserva de ley (orgánica)[35], sino también con el artículo 149.1 6ª CE que atribuye al Estado la competencia exclusiva en materia penal, desentonando el que, por vía de una remisión a la normativa autonómica o local, un determinado supuesto de hecho pueda resultar típico en un lugar de España y atípico en otros (vid. Muñoz Conde F., López Peregrín, C., García Álvarez, P., 2015). De la misma manera, toda vez que el tipo penal incorpora remisiones a ámbitos del ordenamiento jurídico-administrativo especialmente complejos, dispersos e inestables (como el medioambiental) otro problema que no debería infravalorarse es la frecuencia con la que podrá alegarse error sobre la normativa extrapenal que, con carácter general (como error de tipo), tenderá a excluir la relevancia penal del comportamiento doloso. Finalmente, sin ánimo de agotar ahora todos los problemas posibles, también podrían señalarse las dificultades relacionadas con las posibles modificaciones *contra reo* del tipo penal que se introduzcan vía derecho administrativo[36].

Y si estos problemas son comunes a ambos modelos de accesoriedad, en el caso de que la tipicidad se halle condicionada a la presencia de un acto administrativo (autorización), a lo anterior

35 Y, como ha señalado el Tribunal Constitucional (STC 24/2004, ECLI:ES:TC:2004:24), "la reserva de Ley no es sólo una garantía formal, sino que implica garantías materiales que podrían verse vulneradas en determinados supuestos de colaboración normativa". De hecho, la reserva absoluta de Ley guarda relación con la "exigencia de predeterminación normativa de las conductas y sus correspondientes sanciones", taxatividad, irretroactividad.

36 A este respecto, no dejan de resultar interesante las posibles implicaciones del principio "*stand still*" o de "no retorno" en el estándar de protección ambiental, que se desprende del art. 37 de la Carta de Derechos Fundamentales de la Unión Europea (sobre la posible relevancia Constitucional del principio, véase la STC 233/2015 (ECLI:ES:TC:2015:233), FJ 12).

se añade la cuestión de la posible detección *ex post* de vicios del mismo.

Por su parte, la accesoriedad absoluta pone en entredicho las propias funciones del Derecho penal que termina convirtiéndose en mero refuerzo del sistema regulador. Una vez subordinado por completo el tipo penal al Derecho administrativo, todo intento de diferenciar la sanción penal de la administrativa termina convirtiéndose en la persecución de una quimera: el sugerente criterio diferenciador teleológico (Silva Sánchez, 2001) termina relegado a un elegante segundo plano y postular una distinción cuantitativa es solo aparentemente tranquilizador: a faltas de unas pautas que acoten la genérica mayor "gravedad" del supuesto de hecho que puede ser objeto de atención por parte del legislador penal, el criterio cuantitativo no parece encajar con el ordenamiento jurídico vigente (Matellanes Rodríguez, 2019) y, lo que es más alarmante, se arriesga a convertirse en un simple contenedor formal que envuelve al más puro decisionismo de un legislador que, como sabemos, es proclive a "vender" a la opinión pública nuevos tipos penales (Matellanes Rodríguez, 2019), importando poco si, a veces, la sanción administrativa termina siendo cuantitativamente más gravosa que la penal[37].

En el mismo sentido, convendría recordar que el principio de legalidad y el de intervención mínima se erigen como obstáculos difícilmente conciliables con una técnica legislativa basada en la accesoriedad absoluta que plantearía un serio problema de "proporcionalidad de la reacción penal, que afectaría tanto del derecho a la libertad personal (art. 17.1 CE) como al principio de legalidad penal (art. 25.1) en cuanto comprensivo de la prohi-

---

[37] Véase, por ejemplo, la STS 1197/2001 (ECLI:ES:TS:2001:5276), en la que se lee que "la doctrina moderna europea, por el contrario, es prácticamente unánime: entre el ilícito del delito (...) y de las infracciones administrativas no existe más que una diferencia externa, constituida por la especie de consecuencia jurídica que se prevé para tales ilícitos y que depende de una decisión del legislador. El derecho comparado ratifica y refleja claramente este punto de vista".

bición constitucional de penas desproporcionadas". Legislador e intérprete solo han de incluir en el tipo penal "las conductas más graves e intolerables, debiendo acudirse en los demás supuestos al Derecho administrativo sancionador, pues de lo contrario el recurso a la sanción penal resultaría innecesario y desproporcionado" (STC 24/2004, ECLI:ES:TC:2004:24).

Así las cosas, no es de extrañar que el Tribunal Constitucional haya sustancialmente limitado la posibilidad de recurrir al modelo de accesoriedad extrema y que uno de los terrenos en los que ha sido llamado a pronunciarse haya sido precisamente el medioambiental: así, por ejemplo, a propósito de la aplicación del antiguo artículo 347 bis CP, la STC 127/1990 de 5 de julio (ECLI:ES:TC:1990:127)[38] precisó que el recurso a la ley penal en blanco es admisible solo si se cumplen una serie de requisitos:

1. «que el reenvío normativo sea expreso y esté justificado en razón del bien jurídico protegido por la norma penal;
2. que la ley, además de señalar la pena, contenga el núcleo esencial de la prohibición y sea satisfecha la exigencia de certeza o (...)
3. se dé la suficiente concreción, para que la conducta calificada de delictiva quede suficientemente precisada con el complemento indispensable de la norma a la que la ley penal se remite, y resulte de esta forma salvaguardada la función de garantía de tipo con la posibilidad de conocimiento de la actuación penalmente conminada»[39].

En obsequio a estos criterios, el Tribunal Constitucional ha llegado a deshabilitar determinadas fuentes normativas para com-

---

[38] En términos parecidos, las SsTC 118/1992 (ECLI:ES:TC:1992:118), 120/1998 (ECLI:ES:TC:1998:120), 82/2005 (ECLI:ES:TC:2005:82), 283/2006 (ECLI:ES:TC:2006:283), 24/2004 (ECLI:ES:TC:2004:24).

[39] Sobre la relevancia de esta sentencia, vid. Muñoz Conde F., López Peregrín, C., García Álvarez, P. (2015); Matellanes Rodríguez (2019); Fuentes Loureiro (2015).

pletar un tipo penal: así, se ha dicho que admitir que una orden ministerial pueda completar una norma penal, "diluiría de tal modo la función de garantía de certeza y seguridad jurídica de los tipos penales, función esencial de la reserva de Ley en materia penal, que resultaría vulnerado el art. 25.1" (así literalmente, STC 24/2004, ECLI:ES:TC:2004:24). Incluso se ha llegado a excluir (STC 341/1993, ECLI:ES:TC:1993:341) que una infracción leve establecida en la Ley Orgánica de Seguridad Ciudadana pueda configurarse remitiendo a "normas infralegales" (la norma se refería a "las reglamentaciones específicas" o a las "normas de policía dictadas en ejecución de las mismas").

Como se ha observado, se trata de criterios que distan mucho de ser perfectos

> En el sentido de que al fin y al cabo la infracción de la normativa extrapenal es un elemento tan fundamental del delito como los que configuran su "núcleo esencial", con lo que aun cumplimiento los requisitos establecidos, la Administración sigue definiendo parte de la conducta delictiva. Pero teniendo en cuenta que la remisión parece inevitable, dichos criterios contribuyen al menos a que se respeten los principios de legalidad y de intervención mínima (Muñoz Conde F., López Peregrín, C., García Álvarez, P., 2015, p. 68).

4.2. El problema del bis in idem. Apuntes

El otro conocido escollo en el que hay que procurar no encallarse es representado por el principio de *non bis in idem* que el Tribunal Constitucional considera parte del principio de legalidad en materia penal y sancionadora y que "veda la imposición de una dualidad de sanciones en los casos en que se aprecie identidad del sujeto, hecho y fundamento" (entre otras muchas, véase STC 2/1981, ECLI:ES:TC:1981:2). La idea es evitar "una reacción punitiva desproporcionada en cuanto dicho exceso punitivo hace quebrar la garantía del ciudadano de previsibilidad de las sanciones, pues la suma de la pluralidad de sanciones crea una sanción ajena al juicio de proporcionalidad realizado por el legislador y materializa la imposición de una sanción no prevista legalmente" (entre otras muchas, vid. la reciente STC 25/2022, ECLI:ES:TC:2022:25).

Lo anterior se completa en clave procesal con la prioridad de la vía penal sobre la administrativa que, sin embargo, es solo parcialmente tranquilizadora visto que no siempre es atendida (Matellanes Rodríguez, 2019).

Naturalmente, no tenemos aquí espacio para adentrarnos en la problemática. Lo que sí podemos hacer es apuntar que la cuestión no siempre ha resultado de fácil solución y que el propio Tribunal Constitucional ha reaccionado de manera no unívoca frente al administrado que, ya sancionado por la Administración, se veía luego condenado en sede penal. Una vez más, el delito medioambiental fue el contexto de una de las vicisitudes normalmente citadas por la doctrina (Matellanes Rodríguez, 2019): en el supuesto analizado por la STC 177/1999 (ECLI:ES:TC:1999:177), el consejero delegado de una empresa había sido sancionado por la Junta de Aguas de la Generalidad de Cataluña por verter indirectamente aguas residuales contaminantes (con elevado índice de cianuros y, sobre todo, de níquel) al cauce de un río sin contar con la autorización ni depuración previa. Como la conducta se consideró subsumible en el antiguo art. 347 bis del Código penal, también recibió una condena penal. El Tribunal Constitucional concedió el amparo, no considerando suficiente para excluir el *bis in idem* el que la jurisdicción penal hubiera tenido en cuenta la sanción administrativa ya satisfecha a la hora de cuantificar la multa.

El *overruling* no tardó ni 4 años. Con la STC 2/2003 (ECLI:ES:TC:2003:2), el Tribunal termina diciendo justamente lo contrario: el hecho de que el juez penal compruebe "el error cometido por la Administración al proseguir el expediente sancionador a pesar de que existía un procedimiento penal abierto", tan solo ha de implicar que, "en ejecución de la sentencia" penal de condena, "se le descuente de la pena (...) aquellas cantidades que acredite haber satisfecho" a la Administración. Así procediendo, se dice, la sanción finalmente impuesta es solo una: la penal, ya que la administrativa queda absorbida en ella. Y, si lo que materialmente se pretendía evitar con el *non bis in idem* era la imposición de una sanción no decidida por el legislador sino resultado de la suma de dos distintas...¡problema resuelto!

Naturalmente, el nuevo parámetro no tardó en ser aplicado en materia medioambiental (STS 833/2002, ECLI:ES:TS:2003:3760) y de tutela del patrimonio histórico (STS 654/2004, ECLI:ES:TS:2004:3570).

5. BREVE OJEADA A LA SOLUCIÓN DEL LEGISLADOR ESPAÑOL, ESPECIALMENTE RESPECTO DEL TRATAMIENTO PENAL DE LOS ILÍCITOS RELACIONADOS CON LOS RESIDUOS DE CONSTRUCCIÓN Y DEMOLICIÓN

Naturalmente, es ahora imposible pretender llevar a cabo un análisis pormenorizado de los tipos penales relacionados con la protección del medioambiente. Así que optemos por delimitarlo en un doble sentido: por un lado, teniendo en cuenta el objeto del proyecto de investigación en cuyo marco se inserta esta contribución, los que interesan son los tipos penales más relevantes para la sanción de conductas ilícitas relacionadas con los residuos de construcción y demolición (RCD); en segundo lugar, contentémonos con una rápida ojeada a la solución que ha dado el legislador a las problemáticas que hemos apuntado en las páginas anteriores.

Empezando por el primer punto, puesto que hablamos de "residuos", una primera idea podría ser dirigir la atención hacia el artículo 326 del Código penal que expresamente se ocupa de la materia, siendo el único lugar del Código en el que aparece la palabra "residuos". Ahora, desde su introducción en el Código penal de 1995, se debatió acerca de la necesidad de yuxtaponer una norma *ad hoc* en tema residuos al delito ecológico tipificado por el artículo 325 (para los términos del debate, vid. Fuentes Loureiro, 2021); con independencia de quien tuviera la razón, como ha observado la doctrina, es francamente complicado imaginar que las conductas incriminadas por el artículo 326 "causen un resultado lesivo" sin poderse subsumir a la vez en el artículo 325, "en cuyo caso el precepto a aplicar debería ser el art. 325, ya que aunque prevé las mismas penas, es más específico" (Muñoz Conde F., López Peregrín, C., García Álvarez, P., 2015, p. 244). Y, de hecho, parece que los supuestos delictivos que conciernen a RCD que han llegado a conocimiento de los tribunales del orden penal españoles[40], con carácter general, se han subsumido en el artículo 325 que ha llegado a considerarse especial respecto del 326 (STSJ Castilla La Mancha, 17/2023, ECLI:ES:TSJCLM:2023:751).

5.1. Elementos comunes

Los delitos tipificados en los artículos 325 y 326 del Código penal comparten algunas soluciones técnicas. *Mutatis mutandis*, en ambos casos, el tipo penal se configura de la siguiente manera:

1. remisión a la normativa administrativa para acotar una *conditio sine qua non* de la tipicidad;
2. descripción del núcleo de la prohibición mediante la mención expresa de las acciones (y omisiones) típicas (Fuentes Loureiro, 2021);
3. subordinación de la sanción penal a un daño o peligro de daño expresamente requerido y descrito en el tipo penal.

---

40 Se ha realizado una búsqueda en la base de datos online del Consejo General del Poder Judicial en fecha 24 de junio de 2023. En el apartado texto libre, se han insertado la frase "construcción y demolición". Pues bien, si se limita la búsqueda al artículo 325, los resultados son 25. Si se repita la búsqueda limitándola al artículo 326, los resultados son 13.

5.2. La remisión a la normativa extrapenal

En ambos casos, para que asuma relevancia penal, es preciso que la conducta típica se realice quebrantando el ordenamiento administrativo. Así

1. en el artículo 325, se exige la infracción de "las leyes u otras disposiciones de carácter general protectoras del medio ambiente";
2. en el artículo 326.1, la relevancia penal está supeditada a que se contravengan "las leyes y otras disposiciones de carácter general", sin mayor limitación;
3. en el supuesto tipificado por el artículo 326.2, se requiere la conducta se produzca "en algunos de los supuestos a los que se refiere el Derecho de la Unión Europea relativo a los traslados de residuos".

La primera cuestión estribará entonces en acotar qué debe entenderse por "leyes y otras disposiciones de carácter general".

Descartado que el carácter "general" se oponga a "sectorial", "autonómico" o "local", la postura mayoritaria defiende que lo que pretende el legislador penal con el adjetivo "general" es excluir la relevancia típica de aquellos comportamientos que simplemente vulneran "una actuación administrativa singular, dictada para la resolución de un caso particular" (STS 52/2003, ECLI:ES:TS:2003:1220). Así, como ya hemos visto, no parecen existir problemas en admitir que el tipo se complete, por ejemplo, con reglamentos comunitarios, leyes ordinarias (nacionales y autonómicas), reglamentos (de carácter nacional y autonómico) y ordenanzas municipales.

Más problemáticos parecen los supuestos de

1. un tratado internacional publicado en el BOE, pero no (correctamente) implementado mediante la modificación expresa del ordenamiento jurídico interno;

2. una directiva comunitaria no (correctamente) transpuesta cuya posible eficacia directa parece deberse limitar a los supuestos en el que el precepto comunitario determina una restricción del espectro aplicativo del tipo penal[41];

3. puesto que el tipo penal se refiere "leyes u *otras* disposiciones de carácter general", cabría preguntarse si puede completarse con preceptos contenidos en una ley singular[42] (so-

---

41 Para la solución de esta problemática, habría que recordar: la posible naturaleza *self executing* de determinados preceptos (anclada a su carácter preciso e incondicional); la jurisprudencia del TJUE que reconoce el principio de interpretación conforme (ya con respecto a las decisiones marco) y de no discriminación por razón de la nacionalidad. Con todo, precisamente la protección del medioambiente ha sido teatro de algunas de las más significativas Sentencias del Tribunal europeo (SsTJCE 11 de junio de 1987, *Pretore di Saló,* 14/86; 8 de octubre de 1987, *Kolpinghuis Nijmegen "aguas minerales"*, 80/86; 26 de noviembre de 1996, Arcaro, C-168/95) que han excluido con carácter general que una directiva no transpuesta pueda generar obligaciones para los individuos frente al Estado, con la consiguiente imposibilidad de que uno de sus preceptos pueda ser esgrimido como razón para ampliar el espectro aplicativo de la norma penal. Sobre la cuestión, véase, con carácter general, Mangas Martín (2017). Ahora bien, como ya se ha dicho, sería ciertamente interesante explorar las posibles consecuencias de la cláusula *Stand Still.*

42 Vid. la STC 129/2013 (ECLI:ES:TC:2013:129) que precisamente analizaba el caso de una planta de tratamiento de residuos cerrada en virtud de una sentencia del Tribunal Supremo y que se reabrió como consecuencia de la Ley 6/2007 de Castilla y León que contenía una disposición *ad hoc.* La ley se componía de un único artículo por medio del cual se aprobaba el "Proyecto Regional Ciudad del Medio ambiente" y la disposición adicional rezaba: "(…) la aprobación del presente Proyecto Regional comporta la directa modificación de las Normas Subsidiarias de Planeamiento Municipal de Garray, aprobadas por acuerdo de la Comisión Territorial de Urbanismo de Soria de 11 de noviembre de 1993 (BOCyL 1 de julio de 1996) y del Plan General de Ordenación Urbana de Soria aprobado por Orden FOM/409/2006, de 10 de marzo (BOCyL 16 de marzo de 2006), de forma que en el ámbito del Proyecto Regional las determinaciones urbanísticas aplicables serán las previstas en el propio Proyecto Regional".

bre el concepto de ley singular, vid. Mantilla Martos, 2015 y, recientemente, sobre un conocido caso que ha afectado a la Comunidad Autónoma de Extremadura que guarda relación con la protección del medioambiente, STC 134/2019, ECLI:ES:TC:2019:134[43]);

4. órdenes ministeriales o de consejeros autonómicos (vid. *supra* apartado 4.1.).

De cualquier manera, estas últimas consideraciones podrían abrir las puertas a un juicio *material* de "generalidad" que no necesariamente se limite al tipo de fuente normativa, sino que excluya de la ecuación a toda norma (o precepto) "singular", teniendo en cuenta los criterios desglosados al respecto por el Tribunal Constitucional.

Aclarado lo anterior, respecto del art. 326.1, la doctrina ha sugerido entender la referencia a las "leyes" y "disposiciones generales" como limitada a las que guardan relación con la "gestión integral de residuos", ya que, en caso contrario, la norma incriminaría como delictivo, por ejemplo, "un transporte de residuos excediendo el límite de velocidad permitido o incumpliendo cualquier disposición de la Ley de tráfico" (Fuentes Loureiro, 2021, p. 179).

Finalmente, a la hora de acotar la referencia realizada por el art. 326.2 al "Derecho de la UE relativo a los traslados de residuos", la doctrina especializada ha acotado la remisión al art. 2.35 del reglamento 1013/2006, lo que permitiría configurar un

---

43 La Sentencia, que guarda relación con el asunto "Valdecañas", recapitulando la doctrina del Tribunal Constitucional sobre el particular, recuerda que son leyes singulares: 1. las que "contienen una actividad típicamente ejecutivas, de aplicación a un caso concreto"; 2. las que se definen como "de estructura singular en atención a los destinatarios a los que van dirigidas" (expropiación Rumasa); 3. las que se dictan en atención a un supuesto de hecho concreto y singular, que agotan su contenido y eficacia en la adopción y ejecución de la medida tomada por el legislador ante ese supuesto de hecho, aislado en la ley singular y no comunicable con ningún otro".

tipo penal de traslado ilícito cuya ilicitud se definiría "con vocación de permanencia" (Fuentes Loureiro, 2021, pp. 262-263). En realidad, aparte del desliz técnico del legislador que ha olvidado mencionar que el traslado es delito solo si es contrario a derecho, el artículo 326 no ha aludido a la contradicción de la conducta con las "disposiciones de carácter general" (como, en cambio, se hace en el delito ecológico) sino que se ha remitido de forma genérica a "los supuestos a que se refiere el derecho de la UE relativo a los traslados de residuos" que no tiene por qué implicar solo normas generales[44].

5.2.1. La descripción de las conductas típicas en el art. 326.1.

Para que una remisión del tipo penal al ordenamiento extrapenal resulte constitucional, como hemos visto, se exige que "la norma penal tenga el núcleo esencial de prohibición. Esto es, se exige que el precepto penal describa la conducta prohibida, cumpliendo con la garantía de certeza exigida a las normas penales. En este sentido", se añade, "el art. 326.1 describe taxativamente tanto las conductas típicas como las posibilidades de resultado, por lo que precisa sobradamente el núcleo de prohibición" (Fuentes Loureiro, 2021, p. 189).

Pues bien, el precepto que ahora interesa describe las conductas típicas con los términos: recoger, transportar, valorizar, transformar, eliminar o aprovechar residuos, lo que, según parece, sal-

44 Ya hemos visto que la jurisprudencia europea excluye la posibilidad de que una directiva no transpuesta o transpuesta defectuosamente modifique *in peius* una norma penal, aunque lo haga interviniendo sobre la normativa administrativa a la que esta remite. En contra de la opinión de Fuentes Loureiro (2015, pp. 262-263), intuyo que, si es más o menos pacífico que los reglamentos europeos resultan incluidos por la referencia a "disposiciones generales" utilizada en el art. 325 y 326.1, la remisión al Derecho de la UE del art. 326.2 puede desde luego abarcar mucho más.

va la exigencia constitucional inherente a la necesidad de que sea la norma penal la que contenga el "núcleo de la prohibición".

No obstante, convendría tener en cuenta que los términos "recogida", "transporte", "valorización", "eliminación" aparecen definidos en la legislación vigente en materia de residuos nacional y, a veces, autonómica (Fuentes Loureiro, 2015). Y, puesto que la doctrina parece sugerir que las definiciones administrativas se utilicen para acotar la conducta *penalmente* típica (Fuentes Loureiro, 2015), no resulta tan obvio atribuir a su expreso desglose en el precepto penal un logro en términos de taxativa acotación del núcleo de lesividad. Además, una técnica legislativa que sustancialmente repite en el Código penal conceptos ya definidos en la normativa administrativa, no solo no aporta mucho en términos de construcción de un núcleo de lesividad autónomo, sino que pierde todas las ventajas de la remisión al derecho extrapenal sobre las que ya hemos insistido: la norma penal que cristaliza una definición administrativa quita la posibilidad de que la evolución del ordenamiento extrapenal conlleve automáticamente la del precepto penal. De hecho, en nuestro ámbito específico, no solo hay definiciones que la nueva ley de residuos de 2022 ha cambiado o ha incluido *ex novo*, sino que, según parece y como se dirá en otra parte de esta obra, es la propia lógica de economía circular que introduce nuevos parámetros ausentes en la regulación anterior y que ahora quedarán posiblemente ajenos a la norma penal que seguirá anclada a la lógica precedente (piénsese, por ejemplo, en el art. 30.2 relativo a los deberes de "clasificación" de RCD o al 30.3 en materia de demolición selectiva).

Por último, sin que podamos ahora detenernos en la cuestión, objeciones todavía más cruciales podría suscitar el supuesto tipificado en el apartado 2 del artículo 326 en el que lo único que parece distinguir la ilicitud penal es el carácter "no desdeñable" de la cantidad de residuos trasladados (ilícitamente). De hecho, en el marco normativo que ahora consideramos (arts. 325 y 326 CP), se trata del único caso en el que el Código penal no requiere expresamente ni daño ni peligro ni para el medioambiente ni para la salud de las personas (elemento sobre el que se dirá en seguida).

5.3. El peligro y la lesión para el bien jurídico

El punto clave de la construcción del injusto penal es ciertamente representado por la descripción expresa por parte del Código de la afectación necesaria para el bien jurídico. Así, el supuesto será típico solo si,

1. en el art. 325.1 (pena de prisión de 6 meses a 2 años, multa de 10 a 14 meses e inhabilitación especial para profesión u oficio de 1 a 2 años), los resultados materiales[45] de las conductas típicas[46] "por sí mismos o conjuntamente con otros"[47],

a. causan o

b. pueden causar

"daños sustanciales a la calidad del aire, del suelo o de las aguas, o a animales o plantas";

2. en el art. 325.2 (prisión de 2 a 5 años, multa de 8 a 24 meses e inhabilitación especial para profesión u oficio de 1 a 3 años), "las anteriores conductas, por sí mismas o conjuntamente con otras, pudieran perjudicar gravemente el equilibrio de los sistemas naturales";

3. en el párrafo II del art. 352.2 (pena de prisión en su mitad superior, pudiéndose llegar hasta la superior en grado), "si

---

45 "Vertidos, radiaciones, extracciones o excavaciones, aterramientos, ruidos, vibraciones, inyecciones o depósitos, en la atmósfera, el suelo, el subsuelo o las aguas terrestres o marítimas, incluido el alta mar, con incidencia incluso en los espacios transfronterizos, así como las captaciones de aguas".

46 Provocar o realizar directa o indirectamente.

47 La incorporación al tipo penal de esta referencia también ha sido causa de fuertes (y justificadas) críticas que la han relacionado con planteamientos acumulativos (vid. Martínez-Buján Pérez, 2019 y Muñoz Conde F., López Peregrín, C., García Álvarez, P. 2015 quienes proponen soluciones interpretativas alternativas más aceptables).

se hubiera creado un riesgo de grave perjuicio para la salud de las personas";

4. en el art. 326.1 (misma pena que el artículo anterior), si las conductas típicas se realizan "de modo que

a. causen o puedan causar
   - daños sustanciales a la calidad del aire, del suelo o de las aguas, o a animales o plantes
   - muerte o lesiones graves a personas

b. puedan perjudicar gravemente el equilibrio de los sistemas naturales"

5. En el art. 326.2 (pena de 3 meses a 1 año de prisión o multa de 6 a 18 meses e inhabilitación especial para profesión u oficio por tiempo de 3 meses a 1 año), es suficiente con la realización de la conducta típica (trasladar "una cantidad no desdeñable de residuos, tanto en el caso de uno como en el de varios traslados que aparezcan vinculados, en algunos de los supuestos a que se refiere el Derecho de la Unión Europa relativo a los traslados de residuos")[48].

5.3.1. ¿Antropocentrismo o ecocentrismo?

De acuerdo con una postura mayoritaria que se conforma con una lectura más literal, los preceptos compatibilizan una aproximación ecocéntrica (art. 325.1, 325.2 I y 326.1 en las hipótesis correspondientes) con una antropocéntrica (art. 325.2 II y 326): lo que parece estar fuera de discusión es que, en la medida en que se aproxima por lo menos el peligro para la salud humana, la pena va subiendo[49].

---

[48] Lo que nos vuelve a situar ante la problemática señalada *supra* en el apartado precedente.

[49] Así que se ha admitido el mantenimiento de una idea sustancialmente antropocéntrica moderada. Vid., entre otros, Fuentes Loureiro (2021).

5.3.2. La cuestión del principio de intervención mínima

No es preciso insistir en que la idea de condicionar la respuesta penal a la causación de un peligro, por lo menos, "sustancial", guarda relación con la necesidad de concebir esta rama del ordenamiento jurídico como *ultima ratio* reservada a los ataques más importantes para los bienes jurídicos cuya preservación es imprescindible para el hombre. Pero ¿es eso lo que consigue el legislador? Veamos.

La escala que dibuja el Código empieza con el peligro de "daño sustancial a la calidad del aire, del suelo, de las aguas, o a animales o plantas", sigue con el peligro de perjuicio grave para el "equilibrio de los sistemas naturales" y culmina con el riesgo de "grave perjuicio para la salud de las personas" (que, en el art. 326.1, se sustituye con el peligro de "muerte o lesiones graves a personas").

En su artículo 108.2, la ley de residuos de 2022 tipifica las infracciones administrativas muy graves y el precepto comienza con dos cláusulas generales entre las que destaca la de la letra b). Ahí el legislador dice que "la actuación (…) contraria a lo establecido en esta ley y en sus normas de desarrollo" es "muy grave" si ha "supuesto un peligro grave o daño a la salud de las personas" o ha producido un "daño o deterioro grave para el medio ambiente o cuando la actividad tenga lugar en espacios protegidos"[50]. De hecho, el art. 108.3 b) del mismo texto normativo dice que la misma conducta, si no supone "un peligro grave o un daño a la salud de las personas" o "un daño o deterioro grave para el medio ambiente", con independencia del lugar en el que haya tenido lugar[51], no supone una infracción "muy" grave, sino solo "grave".

---

[50] De hecho, la diferencia entre infracciones "muy graves" y "graves" se cifra a menudo en la presencia de estos parámetros (véase por ejemplo las infracciones graves tipificadas en el art. 108.3 a), b), c).

[51] Con lo que cabe interrogarse acerca de la actividad no peligrosa para la salud ni dañina para el medio ambiente que se desarrolla en espacios protegidos: ¿Infracción muy grave o grave?

Ahora, sin ánimo de entrar en detalle, a la vista está que, mientras que, para que haya infracción administrativa "muy grave", si no hay un peligro para la salud humana, es preciso un *daño grave para el medio ambiente*, la norma penal, si falta el riesgo para el hombre, se conforma con un *peligro* para el entorno que además lo es de un daño *no "grave"* sino tan solo "*sustancial*": sea la que fuere la diferencia entre los "daños *sustanciales* a la calidad del aire, del suelo o de las aguas, o a animales o plantas" (art. 325.1 y 326.1, I supuesto) y el perjuicio grave para el "equilibrio de los sistemas naturales" (art. 325.2 y 326.1, III supuesto), está claro que lo segundo es algo más que lo primero (véase, de hecho, STS 373/2023, ECLI:ES:TS:2023:2136). Pero, aun así, el peligro sustancial para el entorno, insuficiente para la configuración de una infracción administrativa muy grave, sí lo es para accionar una respuesta penal que parece no cumplir con la exigencia básica de acuerdo con la cual, "el nivel de daño mínimo a partir del cual la conducta contaminante puede ser considerada delictiva ha de ser notablemente superior al límite mínimo a partir del cual se entiende que existe una infracción administrativa" (Fuentes Loureiro, 2021, p. 212 quien, no obstante, no parece compartir la crítica).

5.3.3. El problema de la proporcionalidad

Otro problema atañe a la conocida crítica de la doctrina que ha objetado que los artículos 325 y 326 vulneran el principio de proporcionalidad. Y ello es así, en primer lugar, porque causar un peligro para algo no puede merecer la misma sanción que causar un daño a lo mismo: y ello con independencia de que el peligro y el daño sean graves o sustanciales y de que su objeto sea el aire, el agua o el equilibrio de los sistemas. Y, en segundo lugar, porque tampoco pueden pretenderse equivalentes en cuanto a pena

1. crear un "riesgo de grave perjuicio para la salud de las personas" (art. 325.2 II)

2. crear un peligro de "muerte o lesiones graves a personas" (art. 326.1 II supuesto)
3. causar "muerte o lesiones graves a personas" (art. 326.1 II supuesto)[52].

5.3.4. El peligro...de daño "sustancial" o "grave"

Ya hemos apuntado cómo la cuestión de la comprobación del peligro para el medioambiente no es baladí. Si se es demasiado exigentes y se requiere una comprobación efectiva *ex post* (peligro concreto), se corre el riesgo de acercar el peligro al daño y restringir así de forma muy poco operativa el ámbito de protección de bienes jurídicos supraindividuales; si, al lado opuesto, el intérprete se conforma con el pronóstico de peligrosidad realizado por el legislador a la hora de describir la acción típica y exime al juez de toda comprobación concreta, se termina dando entrada a una *praesumptio contra reo* que, sea *iuris et de iure*, sea *iuris tantum*, no deja de chirriar con los principios básicos del derecho penal, *in primis*, de lesividad y de presunción de inocencia (que, conviene no olvidarlo, lleva consigo la idea de que *actore non probante, reus absolvitur*).

Si bien, en un primer momento, la jurisprudencia ha oscilado entre el peligro concreto y el peligro abstracto (SsTS 52/2003, ECLI:ES:TS:2003:1220; 1828/2002, ECLI:ES:TS:2002:7059), hoy en día, parece reinar cierto acuerdo sobre el hecho de que los delitos de que hablamos son delitos de aptitud o de peligro hipotéti-

---

52 El argumento de acuerdo con el cual la nueva dicción refleja de forma literal el tenor de la Directiva 2008/99/CE (que emplea la fórmula "causen o puedan causar") no parece convincente, ya que, como es bien sabido, las obligaciones que descienden de una Directiva no incluyen la traslación literal de su contenido. En el caso específico, de la fórmula empleada por el legislador comunitario, tan solo se deduce que "causar" y "poder causar" (es decir, poner en peligro), han de ser conductas delictivas. Nada sugiere que hayan de tener la misma pena ni situarse en una misma horquilla sancionadora.

co, en los que el juez tiene que comprobar "las propiedades de la acción que permiten considerarla apta o idónea para producir, en su caso, un peligro real para el objeto de protección" (Torío López, 1981, p. 833), esto es, "acreditar la peligrosidad de la acción (desvalor real de acción) y la posibilidad del resultado peligroso (desvalor *potencial* de resultado) como exigencias del tipo" (Torío López, 1981, p. 846). Así, la expresión "pueda causar" parece apoyar la idea de que "la potencialidad lesiva debe ser constatada por el juez en el caso concreto", mediante, en primer lugar, un pronóstico de previsibilidad objetiva *ex ante* (que determine si la acción es apta para producir un peligro) y, en segundo lugar, ya *ex post*, verificando si, en el caso concreto, habría sido posible un contacto entre la acción y el resultado lesivo (Fuentes Loureiro, 2021, p. 209; vid. también, entre muchos, Martínez-Buján Pérez, 2022; Pérez-Sauquillo Muñoz, 2019). Así las cosas:

> Si una conducta no ha causado daños efectivos alguno de los elementos del medio ambiente, únicamente se podrá decir que es apta para causarlos cuando se demuestre científicamente que puede provocarlos, siendo insuficiente que, estadísticamente, sea frecuente que esa conducta derive en daños medioambientales (Fuentes Loureiro, 2021, p. 210).

Respecto de este extremo, la jurisprudencia ha concluido que:

> Es necesario, en definitiva, individualizar el posible perjuicio para el equilibrio de los sistemas naturales o para la salud de las personas. Lo decisivo en este aspecto es que se trata de una conducta que crea un riesgo que puede concurrir o no con otras conductas diferentes. La existencia de un daño efectivo no es necesaria para la consumación del delito, pero es un dato que en ocasiones permite identificar la conducta que lo ha ocasionado a través del examen de la causalidad y someterla a valoración (SsTS 926/2016, ECLI:ES:TS:2016:5469; 224/2020, ECLI:ES:TS:2020:1327 y, en la jurisprudencia menor extremeña, SAP Badajoz, 17/2023 ECLI:ES:APBA:2023:335).

De forma especialmente clara, también se ha dicho que:

> Aunque partamos de que estamos ante un delito de peligro hipotético, la gravedad del riesgo debe ser demostrada de forma fehaciente, con la necesaria práctica de prueba pericial que pueda

> llevar a este tribunal a la convicción de que en efecto el riesgo para el (...) bien jurídico protegido, ha existido. No podemos entender que, aunque sea la idoneidad de la conducta típica para producir el riesgo el elemento típico requerido, pueda presumirse de la mera conducta desplegada, siendo necesario en todo caso acreditar esa gravedad. No estamos ante un delito de peligro abstracto (SAP Badajoz 17/2023 ya citada).

Ahora bien, como hemos visto, para ser penalmente relevante, el peligro ha de ser "sustancial" o "grave". Y qué quiera decir "grave" o "sustancial" es algo que el legislador no ha especificado. Con carácter general, la jurisprudencia ha utilizado los siguientes parámetros:

> Intensidad del acto contaminante, probabilidad de que el peligro se concrete en un resultado lesivo, en definitiva, la magnitud de la lesión en relación con el espacio en el que se desarrolla, la prolongación en el tiempo, la afectación directa o indirecta, la reiteración de la conducta o la dificultad para el restablecimiento del equilibrio de los sistemas, proximidad de las personas o de elementos de consumo (Entre muchas, SsTS 849/2004, ECLI:ES:TS:2004:4640; 481/2020, ECLI:ES:TS:2020:3157; 610/2021, ECLI:ES:TS:2021:2940; 682/2022, ECLI:ES:TS:2022:2826. En la jurisprudencia menor extremeña, aparte de la ya citada SAP Badajoz 17/2023, SAP Cáceres 242/2018, ECLI: ES:APCC:2018:573)

En la práctica, para la valoración de la gravedad (o del carácter "sustancial" del daño real o posible) será "necesaria" la asistencia de un perito. Así explícitamente lo ha venido reconociendo la jurisprudencia que ha llegado a afirmar que

> El peligro grave para el medio ambiente, es un elemento ambiguo para cuya concreción es necesario que el Juez sea asesorado pericialmente por expertos que expongan los criterios relacionados sobre la gravedad del perjuicio ecológico y sobre los que se establezca la necesaria contradicción evitando que las percepciones del Juez se conviertan en presupuesto inseguro en la aplicación del tipo penal (SAP Badajoz 17/2023 ya citada).

5.3.5. ¿Y los RCD?

Aunque de forma más bien puntual, en algunas ocasiones, jueces y magistrados del orden penal se han ocupado de comportamientos contrarios a la normativa administrativa relacionada con vertidos de RCD[53]. A menudo, la conducta ilícita relacionada con los RCD ha sido analizada en conjunto con otras acciones peligrosas o dañinas para el medioambiente[54].

Con carácter general, no es infrecuente que el resultado de las actuaciones termine en absoluciones y sobreseimientos. Y no porque el comportamiento considerado en el caso concreto no contradiga abiertamente la normativa administrativa; ni tampoco porque haya dificultades en subsumir en el tipo penal la conducta ni porque falte el consiguiente perjuicio para el medioambiente (por lo menos, en términos de peligro). De hecho, se reconoce normalmente que, entre otros efectos perjudiciales, "el impacto medioambiental negativo de los residuos de construcción y demolición es paisajístico y de aumento del riesgo de incendio" (SAP Zaragoza 269/2015, ECLI:ES:APZ:2015:2160), de derrumbe del suelo y de contaminación de las aguas subterráneas (SAP Madrid 657/2021, ECLI:ES:APM:2021:16577). Y, ni que decir tiene que no son raros los supuestos en los que los RCD terminan mezclados con residuos peligrosos[55].

---

53 Vid., por ejemplo, SAP Madrid 657/2021, ECLI:ES:APM:2021:16577 y SAP Palencia, 9/2017, ECLI:ES:APP:2017:226, ambas con éxito condenatorio.

54 Vid., por ejemplo, SAP Guadalajara 8/2022, ECLI:ES:APGU:2022:357; SAP Madrid, 314/2012, ECLI:ES:APM:2012:10939 en la que, junto con RCD, se acumulaban "chatarra, hierro, plástico, madera"; SAP Madrid, 698/2019, ECLI:ES:APM:2019:15772, todas con éxito absolutorio. Vid., en cambio SAP Almería, 178/2022, ECLI:ES:APAL:2022:261, que termina con condena y en la que se contesta también una extracción ilegal de áridos.

55 Véase, por ejemplo, SAP Santa Cruz de Tenerife, 82/2017, ECLI:ES:APTF:2017:93 y STS 373/2023, ECLI:ES:TS:2023:2136 sobre la que volveremos en breve.

Empero, el escollo con el que se han encontrado jueces y magistrados del orden penal ha sido la comprobación de que el peligro (real o potencial) tenía que ser "grave para el equilibrio de los sistemas naturales" (antes de la reforma de 2015 del Código penal que, como hemos visto, tipificó el supuesto en el que el peligro es tan solo de "daños sustanciales a la calidad del aire, del suelo o de las aguas, o a animales o plantas"). En la mayoría de los casos enjuiciados, la valoración de las pruebas periciales ha determinado su insuficiencia para enervar la presunción de inocencia (que, no se olvide, tiene como corolario el principio *in dubio pro reo*). Paradigmático es el caso analizado por la muy reciente STS 373/2023 ya citada. Aquí, junto con un vertedero ilegal de residuos de construcción que se trasladaban sin clasificar y se abandonaban "sin medida preventiva alguna, contribuyendo así a un vertedero incontrolado", se mezclaban residuos "combustibles como plásticos, neumáticos, telas asfálticas, envases aislantes de tejados y frigoríficos". En el vertedero no autorizado, llegó a propagarse un incendio que "tardó 17 días en ser sofocado y que emitió al aire gases como monóxido de carbono, ácido cianhídrico, ácido sulfhídrico y amoniaco, en nubes de humo que afectaron a los vecinos". Pues bien, tanto en apelación como en casación, se concluyó que la conducta "no creó un grave perjuicio para el equilibrio de los sistemas naturales". Y, como los hechos se produjeron antes de la entrada en vigor de la reforma de 2015, lo que procedió fue la absolución.

Cierto es que, en la jurisprudencia, es difícil encontrar aplicaciones de la normativa vigente después de 2015 (más todavía si se restringe la búsqueda a supuestos relacionados con los RCD). Y, como se sabe, la introducción del nuevo supuesto tipificado por el art. 325.1 que se contenta con el peligro de "daños sustanciales a la calidad del aire, del suelo o de las aguas, o a animales o plantas" (y la correspondiente modificación del art. 326) podría abrir las puertas a un número mayor de condenas (como parece sugerir la propia STS 373/2023 ya citada).

6. CONCLUYENDO. DERECHO PENAL, MEDIOAMBIENTE Y RCD: *MUCH ADO ABOUT NOTHING*?

El proyecto de investigación en el que se enmarca esta contribución mueve de una probablemente difusa consternación por el relativamente frecuente incumplimiento de los estándares más elementales en materia de gestión de RCD. El análisis empírico de la difusión de prácticas por lo menos "molestas" se acompaña por una interesante búsqueda de herramientas tecnológicas que permitan su detección. Y, naturalmente, todo ello no solo tiene efectos preventivos, sino también disuasorios en la medida en que facilita la sanción temprana de infracciones de la normativa vigente en la materia.

Personalmente, considero que esta aproximación a la problemática es correcta y, conociendo la pericia de mis compañeros de aventuras, seguramente proporcionará resultados satisfactorios. Ciertamente más de lo que se puede esperar del Derecho penal.

En las páginas anteriores, de hecho, hemos dado breve e incompleta cuenta de algunas de las dificultades que entraña la configuración de una protección penal del medioambiente que, a la vez que pretenda ser mínimamente plausible en términos de eficacia, no eche completamente por tierra las garantías básicas que han de seguir limitando el *ius puniendi* del Estado. Y, como hemos visto, el camino del legislador que decida utilizar el Derecho penal en materia medioambiental está repleto de insidias. La solución española, más allá de un excesivo allanamiento a la literalidad de instrumentos comunitarios que no pretendían seguramente una transposición literal y de una impericia técnica ya crónica, se ha plasmado en una respuesta del legislador penal discutible y discutida, que ciertamente resulta escasamente disuasoria para quien "trapichee" con RCD.

Pero este epílogo no puede sorprender. A fin de cuentas, no debe extrañar que lo que la normativa sectorial califica por debajo de lo "no peligroso" (como los RCD) malamente encaje con un ámbito del ordenamiento que, en materia medioambiental, se construye precisamente a partir del *peligro*.

Así las cosas, al igual que mi aportación al proyecto de investigación y al reto que se plantea, la del Derecho penal a la lucha contra los vertidos ilícitos de RCD es y probablemente seguirá siendo periférica y fragmentaria. Y, en clave político-criminal, parece más que comprensible.

7. REFERENCIAS BIBLIOGRÁFICAS

Alastuey Dobón, M. C. (2004). *El delito de contaminación ambiental*. Comares.

Alonso Álamo, M. (2009), "Bien jurídico material y bien jurídico procedimental...y discursivo", en Carbonell Mateu, J. C., González Cussac, J. L., Orts Berenguer, E. (DIRS.), *Constitución, Derechos fundamentales y sistema penal,* Tirant lo Blanch, 91-121.

Bermejo Gómez de Segura, R. (2014). *Del Desarrollo Sostenible según Burndtland a la sostenibilidad como biomimesis*. Hegoa.

Brandariz García, J A. (2011), "Relaciones entre el derecho penal y el derecho administrativo. La relación de accesoriedad con el derecho administrativo". En Faraldo Cabana, P. y Puente aba, L. M. (Coords.), *Ordenación del territorio, patrimonio histórico y medio ambiente en el Código penal y la legislación especial,* Tirant lo Blanch, 91-105.

Bacigalupo Zapater, E. (1980-1981). La instrumentación técnico-legislativa de la protección penal del medio ambiente. *Estudios penales y criminológicos*, 191-214.

Bricola, F. (1973). Teoria generale del reato. *Novissimo Digesto, Vol. XIX.*

Cerina, G. D. M. (2022). *El bien jurídico protegido en el delito de cohecho.* Tirant lo Blanch.

Carrera Hernández, F. J. (2007). Requiem por las decisiones marco. A propósito de la Orden de detención europea. *Revista electrónica de estudios internacionales.*

Ferrajoli, L. (1996). *Diritto e ragione: teoria del garantismo penale.* Laterza.

Fuentes Loureiro, M. A. (2021). *Los delitos de gestión ilegal y traslado ilícito de residuos.* Tirant lo Blanch.

Gómez Lanz, J. (2023), "El Derecho penal y la consecución de los objetivos de desarrollo sostenible conectados con la protección del medio ambiente". En Gómez Lanz, J. y Gil Nobajas, S. (Dirs.), *El sistema penal y los Objetivos de Desarrollo Sostenible de la Agenda 2030,* Tirant lo Blanch.

Gómez Rivero, C. (2022), "Lección XIX" y "Lección XX", en Gómez Rivero, C. (Dir.), *Fundamentos de Derecho penal. Parte especial. Vol I,* Tecnos, 349-383.

Habermas, J. (1998). *Facticidad y validez.* Trotta.

Jesheck, H. H. (1993). *Tratado de Derecho penal. Parte general.* Bosh.

Kanheman, D. (2013). *Pensar rápido, pensar despacio.* Lectulandia (versión epub).

Mangas Martín, A. (2017). *Instituciones del Derecho de la Unión Europea.* Tecnos.

Martínez-Buján pérez, C. (2022). *Derecho penal económico y de la empresa. Parte general.* Tirant lo Blanch.

Martínez-Buján Pérez, C. (2019). *Derecho penal económico y de la empresa. Parte especial.* Tirant lo Blanch.

Martínez-Buján Pérez, C. (2002), "Algunas reflexiones sobre la moderna teoría del Big Crunch en la selección de bienes jurídico-penales (especial referencia al ámbito económico)". En Diez Ripollés, J. L. (Coord.), *La ciencia del Derecho penal ante el nuevo siglo: libro homenaje al profesor doctor don José Cerezo Mir,* Tecnos, 395-432.

Matellanes Rodríguez, N. (2019), "Derecho penal económico y derecho administrativo sancionador", en Camacho Vizcaíno, A. (Dir.), *Tratado de Derecho penal económico,* Tirant lo Blanch.

Mateos Rodríguez-Arias, A. (1992). *Derecho penal y protección del medio ambiente,* Colex.

Montilla Martos, J. A. (2015). Las leyes singulares en la doctrina del Tribunal Constitucional. *Revista Española de Derecho Constitucional,* 269-295. DOI: 10.18042/cepc/redc.104.09

Muñoz Conde, F. (2022). *Derecho penal. Parte especial.* Tirant lo Blanch.

Muñoz Conde F., López Peregrín, C., García Álvarez, P. (2015). *Manual de Derecho penal medioambiental.* Tirant lo Blanch.

Pérez-Sauquillo Muñoz, C. (2019). *Legitimidad y técnicas de protección penal de bienes jurídicos supraindividuales.* Tirant lo Blanch.

Pérez-Sauquillo Muñoz, C. (2017 a.), "Delitos de peligro abstracto y bienes jurídicos colectivos", Ponencia presentada en el XVIII Seminario Interuniversitario Internacional de Derecho Penal, Facultad de Derecho, Universidad de Alcalá, 18-19 de junio de 2015.

Pérez-Sauquillo Muñoz, C. (2017b). Reflexiones y críticas sobre el pensamiento de la acumulación. *La Ley penal, n. 128,* 1 de septiembre de 2017.

Ramírez Mora, J. N. (2021). "¿Medio ambiente o un solo ambiente?", en *El Heraldo Austral,* 14 de abril de 2021, Accesible en este momento en el enlace https://www.eha.cl/noticia/opinion/opinion-medio-ambiente-o-un-solo-ambiente-10699 [Consultado el 03.07.2023].

Silva Sánchez, J. M. (1997). *Política criminal y nuevo Derecho penal. Libro Homenaje a Claus Roxin*. Bosh.

Silva Sánchez, J. M. (2001). *La expansión del Derecho penal. Aspectos de la política criminal en las sociedades postindustriales.* Civitas.

Torío López, A. (1981). Los delitos de peligro hipotético. *Anuario de derecho penal y ciencias penales,* 825-848.

Vercher Noguera, A. (2005), "Una nueva vuelta de tuerca en pro de la protección ambiental de la Unión Europea: la Sentencia de 13 de septiembre de 2005", en *Letras Jurídicas: revista electrónica de derecho.*

Vercher Noguera, A. (2016). Algunas notas sobre la responsabilidad penal de la persona jurídica en el contexto penal ambiental comunitario. *Diario La Ley, n. 8805.*

Vercher Noguera, A. (2023). La evolución de los delitos contra el medio ambiente en el contexto europeo: la Directiva 2008/99/CE". *Diario La Ley, n. 10047.*

Vercher Noguera, A. (2018). Algunas notas sobre la responsabilidad penal de la persona jurídica en el contexto penal ambiental comunitario. *Diario La Ley, n. 8805.*

Capítulo 2

La regulación jurídico-administrativa de los RCD: de la normativa general de residuos a las ordenanzas municipales

PEDRO BRUFAO CURIEL[1]

1. INTRODUCCIÓN

La gestión de los residuos de construcción y demolición (RCD) es objeto de una detallada regulación administrativa, muestra de un creciente intervencionismo público. Los residuos en general representan uno de los grandes problemas ambientales, ya que el abuso del empleo de recursos naturales tiene como resultado una ingente producción de residuos de todo tipo. En el ámbito de la construcción y las infraestructuras, España destaca por el deficiente tratamiento de un ingente volumen de escombros, cuyo depósito o abandono en el territorio jalonan el paisaje y que solo desde hace unos años ha merecido una respuesta por parte de los poderes públicos, aunque el grado de ejecución difiere en el tiempo y en el espacio. La producción de RCD que incumplen la normativa no resulta una tarea fácil. Fenómenos como la autoconstrucción no solamente conllevan una verdadera insubordinación ante la legalidad urbanística y de residuos, sino que también suponen en miles de casos la degradación del territorio y el desperdicio de materias primas.

---

[1] Profesor Titular de Derecho Administrativo. Universidad de Extremadura.

En este capítulo esbozaremos los rudimentos del régimen jurídico de los RCD y analizaremos la normativa extremeña, regional y local desde el punto de su eficacia a la hora de afrontar y poner coto a este problema (Palomar Olmeda, 2022). Los efectos de la normativa administrativa, con un elevado contenido técnico, no acaban en sus estrictos límites, igualmente condicionan una eventual intervención penal y los aspectos criminológicos de una actividad que ha de gozar de la mayor seguridad jurídica; estos efectos comprenden igualmente diversos tipos de responsabilidad con consecuencias económicas y profesionales. Se hace así inexcusable, por tanto, acudir a la normativa administrativa que colma el tipo penal en blanco de la protección ambiental ante los residuos de construcción y demolición (Vercher Noguera, 2022).

2. ELEMENTOS BÁSICOS DEL RÉGIMEN JURÍDICO GENERAL DE LOS RESIDUOS DE CONSTRUCCIÓN Y DEMOLICIÓN

2.1. La normativa europea aplicable a los RCD

El ordenamiento jurídico vigente sobre residuos se recoge, de forma general, en la Directiva 2008/98/CE del Parlamento Europeo y del Consejo, de 19 de noviembre de 2008, sobre los residuos y por la que se derogan determinadas Directivas (DMR). Dado el amplísimo objeto de esta norma, se ha acudido a la regulación sistemática de los residuos con vistas a un tratamiento integral, que va desde la mera conceptualización de qué es un "residuo" y su diferencia respecto de los subproductos, algo especialmente de interés para los residuos de construcción y demolición (RCD) y los áridos, a la responsabilidad de sus productores y el establecimiento de porcentajes de reducción, reutilización y reciclaje. Se añaden también las medidas de valorización y eliminación teniendo en cuenta el ciclo de vida de los productos (Soriano García y Brufao Curiel, 2010).

La DMR ha aprobado la interpretación auténtica de conceptos como los ya dichos de residuo y de subproductos y otros como los de poseedor, negociante, agente, productor, prevención, reutilización, tratamiento, valorización, recogida, reciclado y eliminación (art. 3). Respecto de los RCD, el art. 11 incluye unos objetivos concretos sobre la reducción y reciclado, por los que antes de 2020 hubo de aumentarse hasta un mínimo del 70 % de su peso la preparación para la reutilización, el reciclado y la valorización de materiales, incluidas las operaciones de relleno que utilicen residuos como sucedáneos de otros materiales, de los residuos no peligrosos procedentes de la construcción y de las demoliciones, con exclusión de los materiales presentes de modo natural definidos en la categoría 17 05 04 de la lista de residuos, que son las tierras y piedras.

Con esta última referencia, nos encontramos así con una muestra de la prolija y no poco clara normativa de los residuos, en la que destaca la Decisión de la Comisión, de 18 de diciembre de 2014, por la que se modifica la Decisión 2000/532/CE, sobre la lista de residuos, de conformidad con la DMR. Sobre los de construcción y demolición, que figuran en la tabla inferior, se atribuye la capacidad interventora de las Administraciones competentes y de los interesados en su generación, tratamiento y ciclo de vida.

Tabla 1

Lista de residuos de construcción y demolición según Decisión de la Comisión, de 18 de diciembre de 2014.

17	RESIDUOS DE LA CONSTRUCCIÓN Y DEMOLICIÓN (INCLUIDA LA TIERRA EXCAVADA DE ZONAS CONTAMINADAS)
17 01 Hormigón, ladrillos, tejas y materiales cerámicos 17 01 01 Hormigón 17 01 02 Ladrillos 17 01 03 Tejas y materiales cerámicos 17 01 06* Mezclas, o fracciones separadas, de hormigón, ladrillos, tejas y materiales cerámicos que contienen sustancias peligrosas 17 01 07 Mezclas de hormigón, ladrillos, tejas y materiales cerámicos, distintas de las especificadas en el código 17 01 06	
17 02 Madera, vidrio y plástico 17 02 01 Madera 17 02 02 Vidrio 17 02 03 Plástico 17 02 04* Vidrio, plástico y madera que contienen sustancias peligrosas o están contaminados por ellas	
17 03 Mezclas bituminosas, alquitrán de hulla y otros productos alquitranados 17 03 01* Mezclas bituminosas que contienen alquitrán de hulla 17 03 02 Mezclas bituminosas distintas de las especificadas en el código 17 03 01 17 03 03* Alquitrán de hulla y productos alquitranados	

17 04 Metales (incluidas sus aleaciones)
17 04 01 Cobre, bronce, latón
17 04 02 Aluminio
17 04 03 Plomo
17 04 04 Zinc
17 04 05 Hierro y acero
17 04 06 Estaño
17 04 07 Metales mezclados
17 04 09* Residuos metálicos contaminados con sustancias peligrosas
17 04 10* Cables que contienen hidrocarburos, alquitrán de hulla y otras sustancias peligrosas
17 04 11 Cables distintos de los especificados en el código 17 04 10
17 05 Tierra (incluida la tierra excavada de zonas contaminadas), piedras y lodos de drenaje
17 05 03* Tierra y piedras que contienen sustancias peligrosas
17 05 04 Tierra y piedras distintas de las especificadas en el código 17 05 03
17 05 05* Lodos de drenaje que contienen sustancias peligrosas
17 05 06 Lodos de drenaje distintos de los especificados en el código 17 05 05
17 05 07* Balasto de vías férreas que contiene sustancias peligrosas
17 05 08 Balasto de vías férreas distinto del especificado en el código 17 05 07
17 06 Materiales de aislamiento y materiales de construcción que contienen amianto
17 06 01* Materiales de aislamiento que contienen amianto
17 06 03* Otros materiales de aislamiento que consisten en sustancias peligrosas o contienen dichas sustancias
17 06 04 Materiales de aislamiento distintos de los especificados en los códigos 17 06 01 y 17 06 03
17 06 05* Materiales de construcción que contienen amianto

17 08 Materiales de construcción a base de yeso
17 08 01* Materiales de construcción a base de yeso contaminados con sustancias peligrosas
17 08 02 Materiales de construcción a base de yeso distintos de los especificados en el código 17 08 01
17 09 Otros residuos de construcción y demolición
17 09 02* Residuos de construcción y demolición que contienen PCB (por ejemplo, sellantes que contienen PCB, revestimientos de suelo a base de resinas que contienen PCB, acristalamientos dobles que contienen PCB, condensadores que contienen PCB)
17 09 03* Otros residuos de construcción y demolición (incluidos los residuos mezclados) que contienen sustancias peligrosas
17 09 04 Residuos mezclados de construcción y demolición distintos de los especificados en los códigos 17 09 01, 17 09 02 y 17 09 03

Como se ve, la panoplia de elementos que se incorporan es amplia y más que prolija, a lo que se le suma cuestiones más complejas como la relativa al tratamiento y recogida del amianto, regulado por el RD 396/2006, de 31 de marzo, por el que se establecen las disposiciones mínimas de seguridad y salud aplicables a los trabajos con riesgo de exposición al amianto, así como la Directiva 2009/148/CE del Parlamento Europeo y del Consejo, de 30 de noviembre de 2009, sobre la protección de los trabajadores contra los riesgos relacionados con la exposición al amianto durante el trabajo. Esta regulación condiciona en grado sumo el tratamiento de los RCD: pensemos simplemente en la ingente cantidad de edificios e infraestructuras que contienen amianto.

La DMR impone a los Estados miembros unos objetivos ambiciosos sobre la reutilización y reciclado de los residuos. En su art. 11 se acuerda que los Estados miembros fomenten, de forma general, la reutilización de los productos y las actividades de preparación para la reutilización, promoviendo el establecimiento y apoyo de redes de reutilización y reparación, el uso de instrumentos económicos, los requisitos de licitación, los objetivos cuantitativos u otras medidas. Este fomento ha de dirigirse a lograr un reciclado de alta calidad y, a este fin, los Estados han de establecer una recogida separada de residuos, cuando sea técnica, económica y medioambientalmente factible y adecuada, para cumplir los criterios de calidad necesarios para los sectores de reciclado correspondientes.

Esto último es de gran relevancia, dado que es muy frecuente que los RCD se abandonen o depositen mezclados entre sí o que incluyan residuos peligrosos o domésticos. En el plano temporal, antes de 2020, debería haber aumentado hasta un mínimo del 70 % de su peso[2] la preparación para la reutilización, el reciclado y otra valorización de materiales, incluidas las operaciones de relleno que utilicen residuos como sucedáneos de otros materiales, de los residuos no peligrosos procedentes de la construcción y de las demoliciones[3].

2.2. El ordenamiento estatal sobre RCD

En la legislación nacional, hay que estar a lo regulado en la Ley 7/2022, de 8 de abril, de residuos y suelos contaminados para una economía circular (LRSC). Sobre los RCD en concreto, el art. 18 de esta ley los incluye en el objeto de las medidas de prevención de generación de residuos "tomando en consideración las mejores técnicas disponibles y las buenas prácticas ambientales". En cuanto a la preparación para la reutilización, reciclado y valorización a través de planes y programas de gestión de residuos (art. 26), la cantidad de residuos no peligrosos de construcción y demolición destinados a la preparación para la reutilización, el reciclado y otra valorización de materiales, incluidas las operaciones de relleno, con exclusión de los materiales en estado natural definidos en la categoría 17 05 04 de la lista de residuos citada

2 De acuerdo con el sistema del Reglamento 2150/2002 del Parlamento Europeo y del Consejo, de 25 de noviembre de 2002, relativo a las estadísticas sobre residuos, en la categoría 12.1, sobre los RCD.

3 Con exclusión de los materiales presentes de modo natural definidos en la categoría 17 05 04 de la lista de residuos.

más arriba, deberá alcanzar como mínimo el 70% en peso de los producidos. Un objetivo ambicioso donde los haya.

De forma concreta, el art. 30 se dedica a los RCD. Se prevén aquí las medidas que en las obras de demolición eviten su mezcla con otros tipos de residuos y prevén el manejo de manera segura de las sustancias peligrosas, en particular, el amianto. Las tareas de clasificación se convierten en una herramienta indispensable para la correcta gestión de los RCD. A este respecto, la norma prevé que, a partir del 1 de julio de 2022, los residuos de la construcción y demolición no peligrosos deberán ser clasificados en, al menos, las siguientes fracciones: madera, fracciones de minerales (hormigón, ladrillos, azulejos, cerámica y piedra), metales, vidrio, plástico y yeso. Asimismo, se clasificarán aquellos elementos susceptibles de ser reutilizados tales como tejas, sanitarios o elementos estructurales. Esta clasificación se realizará de forma preferente en el lugar de generación de los residuos y sin perjuicio del resto de residuos que ya tienen establecida una recogida separada obligatoria (Moreu Carbonell, 2022).

Yendo al momento inicial de la generación de estos residuos, la demolición, ésta se llevará a cabo preferiblemente de forma selectiva, y con carácter obligatorio a partir del 1 de enero de 2024, garantizando la retirada de, al menos, las fracciones de materiales citadas, previo estudio que identifique las cantidades que se prevé generar de cada fracción, cuando no exista obligación de disponer de un estudio de gestión de residuos. Asimismo, se hará de acuerdo con una jerarquía de tratamiento (art. 8), que prevé esta prelación: prevención[4], preparación para la reutilización, reciclado, otro tipo de valorización incluida la valorización energética y, finalmente, la eliminación. En el anexo II de la LRSC, que recoge las operaciones de valorización, prevé para los RCD diversas opciones. Estas se contemplan en la tabla que sigue a continuación.

---

4 Con un listado de posibles medidas de prevención en el Anexo VI.

Tabla 2

Opciones de valoración para los RCD según anexo II de la LRSC.

R12 Intercambio de residuos para someterlos a cualquiera de las operaciones enumeradas de R1 a R11	
R1201 Clasificación de residuos.	Instalaciones de clasificación de otros tipos de residuos (plásticos, papel/cartón, RCD, neumáticos fuera de uso, etc.).
R1208 Acondicionamiento de residuos para la obtención de fracciones combustibles.	Instalaciones de pretratamiento de residuos destinadas a la obtención de fracciones combustibles: – Instalaciones de pretratamiento de residuos domésticos mezclados, RCD, aceites usados, residuos líquidos orgánicos, etc. para la obtención de fracciones combustibles.

R05 Reciclado o recuperación de otras materias inorgánicas	
R0505 Reciclado de residuos inorgánicos en sustitución de materias primas para la fabricación de cemento.	Cementeras que utilicen áridos de RCD o tierras de excavación, etc. para la fabricación de cemento.
R0506 Valorización de residuos inorgánicos para la producción de áridos.	Instalaciones de producción de áridos a partir de RCD, de escorias negras de acerías de hornos de arco eléctrico de otros residuos inorgánicos cuando el material obtenido alcance el fin de la condición de residuo.
R0507 Reciclado de residuos inorgánicos en sustitución de materias primas en otros procesos de fabricación.	Utilización de áridos de RCD, tierras de excavación, etc. en sustitución de materias primas en procesos de fabricación distintos de la fabricación de cemento.
R0508 Valorización de materiales inorgánicos en operaciones de relleno (*backfilling*).	Relleno con residuos no peligrosos adecuados en restauraciones de huecos mineros, con fines constructivos, de acondicionamiento, y en restauración e ingeniería paisajística.

R0509 Valorización de materiales inorgánicos en operaciones distintas a las de relleno.	Uso de residuos no peligrosos adecuados en acondicionamiento de vertederos.
R0511 Preparación para la reutilización de residuos inorgánicos.	Instalaciones de clasificación y limpieza de residuos obtenidos en la demolición selectiva tales como tejas, piedras, etc. para su reutilización.

El Anexo III recoge las distintas operaciones de eliminación, que en relación con los RCD pueden ser:

Tabla 3

Operaciones de eliminación para los RCD según anexo III de la LRSC.

D01 Depósito sobre el suelo o en su interior (por ejemplo, vertido, etc.).	
D0101 Depósito sobre el suelo.	Depósito de residuos sólidos (por ejemplo, residuos de roca) en pilas. Depósitos de tierras naturales cuya valorización no sea factible.
D0102 Depósito en el interior del suelo.	

D05 Depósito controlado en lugares especialmente diseñados (por ejemplo, colocación en celdas estancas separadas, recubiertas y aisladas entre sí y del medio ambiente).	Se incluyen en esta operación los vertederos construidos de acuerdo con el Real Decreto 646/2020, de 7 de julio.
D0501 Depósito en vertederos de residuos inertes.	Vertederos de residuos inertes.
D0502 Depósito en vertederos de residuos no peligrosos.	Vertederos de residuos no peligrosos.
D0503 Depósito en vertederos de residuos peligrosos.	Vertederos de residuos peligrosos.

D13 Combinación o mezcla previa a su eliminación mediante cualquiera de las operaciones numeradas D1 a D12.	
D1301 Clasificación de residuos.	Instalaciones de clasificación de residuos para su eliminación posterior.
D1302 Separación de los distintos componentes de los residuos, incluida la retirada de sustancias peligrosas.	Instalaciones de separación de componentes de residuos, incluida la retirada de sustancias (no componentes) para su eliminación posterior.
D1303 Tratamiento mecánico (trituración, fragmentación, corte, compactación, etc.).	Instalaciones de trituración de residuos para su eliminación posterior.

D15 Almacenamiento en espera de cualquiera de las operaciones numeradas D1 a D14 excluido el almacenamiento temporal en espera de recogida en el lugar en que se produjo el residuo.	
D1501 Almacenamiento, en el ámbito de la recogida.	Puntos limpios.
D1502 Almacenamiento, en el ámbito del tratamiento.	Instalaciones de almacenamiento de residuos previo a su eliminación, en el ámbito del tratamiento

La norma principal directamente aplicable en España es el RD 105/2008, de 1 de febrero, por el que se regula la producción y gestión de los residuos de construcción y demolición (RDRCD). Esta norma reglamentaria comprende el régimen jurídico de la producción y gestión de los residuos de construcción y demolición, con el fin de fomentar, por este orden, su prevención, reutilización, reciclado y otras formas de valorización, asegurando que los destinados a operaciones de eliminación reciban un tratamiento adecuado, y contribuir a un desarrollo sostenible de la actividad de construcción (art. 1).

Junto a la definición de los RCD, como el obvio de cualquier residuo que se genere en la construcción y demolición, que hay que entender modificada por las normas posteriores[5], se incluye los siguientes conceptos, para cuya definición nos remitimos al RDRCD: residuo inerte[6]; obra de construcción o demolición[7]; los trabajos que modifiquen la forma o sustancia del terreno o del subsuelo[8]; obra menor de construcción o reparación domiciliaria; productor de RCD; poseedor de RCD; y tratamiento previo, que resulta de gran importancia para reducir el impacto ambiental y los efectos económicos de los RCD.

El RDRCD incluye diversas obligaciones a los productores de RCD, como la de incluir en el proyecto de ejecución de la obra un estudio de gestión con estimación de la cantidad, expresada en

---

5 Se excluyen (art. 3): las tierras y piedras no contaminadas por sustancias peligrosas reutilizadas en la misma obra, en una obra distinta o en una actividad de restauración, acondicionamiento o relleno, siempre y cuando pueda acreditarse de forma fehaciente su destino a reutilización. Los residuos de industrias extractivas regulados por la Directiva 2006/21/CE, de 15 de marzo. Los lodos de dragado no peligrosos reubicados en el interior de las aguas superficiales derivados de las actividades de gestión de las aguas y de las vías navegables, de prevención de las inundaciones o de mitigación de los efectos de las inundaciones o las sequías, reguladas por el Texto Refundido de la Ley de Aguas, por la Ley de Puertos del Estado.

6 Vid. la Decisión de la Comisión, de 30 de abril de 2009, por la que se completa la definición de residuos inertes en aplicación del artículo 22, apartado 1, letra f), de la Directiva 2006/21/CE del Parlamento Europeo y del Consejo sobre la gestión de los residuos de industrias extractivas.

7 "Tal como un edificio, carretera, puerto, aeropuerto, ferrocarril, canal, presa, instalación deportiva o de ocio, así como cualquier otro análogo de ingeniería civil".

8 Se citan las "excavaciones, inyecciones, urbanizaciones u otros análogos, con exclusión de aquellas actividades a las que sea de aplicación la Directiva 2006/21/CE del Parlamento Europeo y del Consejo, de 15 de marzo, sobre la gestión de los residuos de industrias extractivas". Vid. la Decisión de la Comisión, de 30 de abril de 2009, por la que se completan los requisitos técnicos para la caracterización de los residuos establecidos en la Directiva 2006/21/CE del Parlamento Europeo y del Consejo sobre la gestión de los residuos de industrias extractivas.

toneladas y en metros cúbicos y con los códigos correspondientes. Se le añaden (art. 4) medidas para la prevención de residuos en la obra objeto del proyecto; las operaciones de reutilización, valorización o eliminación a que se destinarán los residuos que se generarán en la obra; las medidas para la separación de los residuos en obra[9]; los planos de las instalaciones previstas para el almacenamiento, manejo, separación y, en su caso, otras operaciones de gestión de los residuos de construcción y demolición dentro de la obra, que habrán de adaptarse a las características particulares de la obra y sus sistemas de ejecución, previo acuerdo de la dirección facultativa de la obra; las prescripciones del pliego de prescripciones técnicas particulares del proyecto, en relación con el almacenamiento, manejo, separación y, en su caso, otras operaciones de gestión de los residuos de construcción y demolición dentro de la obra; una valoración del coste previsto de la gestión, que formará parte del presupuesto del proyecto en capítulo independiente. En las obras de demolición, rehabilitación, reparación o reforma, se exige un inventario de los residuos peligrosos, así como la previsión de su retirada selectiva, con el fin de evitar la mezcla entre ellos o con otros residuos no peligrosos y asegurar su envío a gestores autorizados de residuos peligrosos.

En cuanto a los requisitos documentales, habrá de acreditarse por este medio que los RCD realmente producido se gestiona, en su caso, en una obra o entregado a una instalación de valorización o de eliminación para su tratamiento por gestor de residuos autorizado. Esta documentación ha de guardarse durante los cinco años siguientes.

---

9 En particular se prevé que los residuos de construcción y demolición deberán separarse en las siguientes fracciones, cuando, de forma individualizada para cada una de dichas fracciones, la cantidad prevista de generación para el total de la obra supere las siguientes cantidades, en toneladas: Hormigón: 80 t. Ladrillos, tejas, cerámicos: 40 t. Metal: 2 t. Madera: 1 t. Vidrio: 1 t. Plástico: 0,5 t. Papel y cartón: 0,5 t (art. 5.5 del RDRCD).

En el caso de obras sometidas a licencia urbanística, se habrá de constituir, cuando proceda, en los términos previstos en la legislación de las comunidades autónomas, la fianza o garantía financiera (art. 6) equivalente que asegure el cumplimiento de los requisitos establecidos en dicha licencia, en relación con los residuos de construcción y demolición de la obra. En la práctica y en muchas ocasiones, la fianza se presta y se entiende como si fuera una tasa encubierta para proceder a su eliminación sin seguir las exigencias normativas, fianza que se da por perdida desde el principio por el interesado.

Por lo que respecta a los poseedores de RCD, el art. 5 obliga a quien ejecute una obra a que presente al propietario un plan sobre la gestión de los residuos. Si no los gestionase el poseedor, entraría el papel del gestor (art. 7) con vistas a que se destinen preferentemente, y por este orden, a operaciones de reutilización, reciclado o a otras formas de valorización[10]. La fehaciencia de la entrega al gestor ha de constar documentalmente, ya sea el encargado solo de la recogida, almacenamiento, transferencia o transporte (art. 12) o el que, por separado se encargue de su valorización (art. 8 y 9), que podrá realizarse en plantas móviles (art. 10). El órgano competente en materia medioambiental de la comunidad autónoma en que se ubique la obra, de forma excepcional, y siempre que la separación de los residuos no haya sido especificada y presupuestada en el proyecto de obra, podrá eximir al poseedor de los residuos de construcción y demolición de la obligación de separación de alguna o de todas las citadas fracciones (Arenas Cabello, 2010). El RDRCD prevé su destino a vertedero (art. 11), siempre que hayan recibido un tratamiento previo o sea inviable este tratamiento. Los residuos inertes se podrán emplear en la restauración ambiental de espacios degradados mediante su acondicionamiento o relleno, como en explotaciones de canteras o minas a cielo abierto algo que se califica como operación

10 Orden APM/1007/2017, de 10 de octubre, sobre normas generales de valorización de materiales naturales excavados para su utilización en operaciones de relleno y obras distintas a aquéllas en las que se generaron.

de valorización y no de vertido. A este respecto, no son pocos los casos en los que se emplean los residuos, de todo tipo, para falsas operaciones de restauración de explotaciones mineras.

De gran importancia para la gestión ordinaria de los RCD y para los ciudadanos particulares, las exigencias de los arts. 4 y 5 no se aplican a las pequeñas obras menores de reparación domiciliaria, origen de decenas de miles de pequeños vertidos que salpican nuestro territorio, y que se regulan mediante ordenanzas municipales (disposición adicional primera). Este régimen general tampoco se aplica a los restos de las excavaciones originados por obras públicas sometidas a evaluación de impacto ambiental, quedando a lo que la oportuna declaración disponga.

El régimen sancionador del RDRCD se remite al de la vigente LRSC en sus arts. 107 y ss. Los responsables pueden ser tanto los productores como los poseedores y los gestores de residuos, siendo solidaria dicha responsabilidad en algunos supuestos. Las infracciones que más nos interesa subrayar son el abandono y el vertido de los residuos, la actuación sin licencia o comunicación, el contrabando de residuos, la comunicación incorrecta de datos y la mezcla de residuos. Nos remitimos *in totum* al exhaustivo listado de infracciones y sanciones para su mejor comprensión. En cuanto a los órganos competentes, serán las entidades locales las que sancionen el abandono, vertido o eliminación incontrolados de los residuos regulados en sus ordenanzas. Pero más efectivas que las sanciones son las medidas cautelares: medidas de corrección, seguridad o control que impidan la continuidad en la producción del daño; el precintado de aparatos, equipos o vehículos; la clausura temporal, parcial o total del establecimiento, y la suspensión temporal de la autorización para el ejercicio de la actividad por la empresa.

De gran importancia para los RCD, hay que estar a lo previsto por el RD 646/2020, de 7 de julio, por el que se regula la eliminación de residuos mediante depósito en vertedero o escombreras. Esta norma comprende un tratamiento previo (art. 2), definido como “los procesos físicos, térmicos, químicos o biológicos, incluida la clasificación, a los que son sometidos los residuos con carácter previo a su eliminación mediante depósito en vertedero, que cambian las características de los mismos para reducir su volumen o su peligrosidad, facilitar su manipulación o incrementar su potencial de valorización". Para los residuos de construcción y demolición el tratamiento previo comprenderá, dice esta norma reglamentaria, como mínimo la clasificación y separación de fracciones valorizables (madera, fracciones de minerales-hormigón, ladrillos, azulejos, cerámica y piedra-, metales, vidrio, plástico y yeso), así como el triturado y cribado de dichas fracciones.

Los vertederos, de todo tipo, han de seguir los criterios constructivos del Anexo I, por los que se han de tener en cuenta los efectos derivados de su ubicación[11], el control de las aguas y la gestión de lixiviados[12], la protección del suelo y las aguas, el control de gases, molestias y riesgos, la estabilidad de la masa de residuos y estructuras asociadas, así como los cerramientos que eviten vertidos ilegales.

El Anexo II del reglamento de vertederos regula a su vez los procedimientos y criterios de admisión de residuos[13]. Sobre los

---

11 Uno de los grandes problemas es la ilegalidad misma de los propios vertederos de escombros.

12 Con valores límites recogidos en el Anexo II según el elemento o compuesto químico, como el cobre, arsénico, cadmio, mercurio, níquel, antimonio, selenio, bario, fluoruros, sulfatos o fenoles, entre otros.

13 Que son: a) El control documental de los residuos recibidos. b) La comprobación de que, de acuerdo con sus características físico-químicas, los residuos pueden ser admitidos en el vertedero. c) La inspección visual de los residuos recibidos. d) Su pesaje y, e) la inscripción en el archivo cronológico del vertedero. Este criterio cronológico comprende (art. 14.1. b. 5º de este RD): registro de cantidades de residuos admitidos construido por partida doble e independientemente, tanto a partir de los documentos de identificación como de los registros de pesada de las partidas de residuos admitidas; origen de los mismos; codificación de los residuos; fecha de entrega de los mismos; identificación del productor o el gestor que realiza la recogida en el caso de los residuos municipales; ubicación exacta en el vertedero si se trata de residuos

residuos inertes, el apartado 2.1 de este anexo incluye ciertas prescripciones que evitan la realización de pruebas previas de caracterización, como excepción. De esta manera, el residuo deberá ser un flujo único (una única fuente) de un único tipo de residuo. Los residuos que figuran en la lista que incluimos más abajo podrán ser admitidos conjuntamente siempre que procedan de la misma fuente (véase tabla 4). En cualquier caso, se aclara que "sin que la excepción a la realización de pruebas de admisión suponga merma alguna en relación con lo señalado en el artículo 1, la admisión de residuos que sean objeto de la misma se realizará sin perjuicio de la aplicación del resto de los procedimientos de admisión, incluida la realización de inspecciones visuales de los residuos".

En el caso de que, como resultado de una inspección visual o por el origen del residuo, se sospeche la existencia de contaminación, deberá bien efectuarse una caracterización básica o bien rechazar el residuo. Si los residuos enumerados están contaminados o contienen otros materiales (tales como metales, amianto, plásticos, productos químicos, yeso, etc.) o sustancias en cantidades que aumenten el riesgo asociado al residuo que justifique su eliminación en otras clases de vertederos, los mismos no podrán ser admitidos en un vertedero para residuos inertes.

---

peligrosos; cuando proceda, resultados de los ensayos y determinaciones analíticas de caracterización básica o pruebas de cumplimiento de acuerdo con lo señalado en el anexo II; cuando proceda, tratamiento previo al que han sido sometidos los residuos y resultados de los parámetros de eficiencia de dicho tratamiento. El archivo cronológico se mantendrá hasta la clausura definitiva del vertedero y deberá estar a disposición de las autoridades competentes.

Tabla 4

Admisión de residuos en vertedero.

Código LER	Descripción	Restricciones
10 11 03	Residuos de materiales de fibra de vidrio.	Solamente sin aglutinantes orgánicos.
15 01 07	Envases de vidrio.	
17 01 01	Hormigón.	Solamente residuos seleccionados de construcción y demolición*.
17 01 02	Ladrillos.	Solamente residuos seleccionados de construcción y demolición*.
17 01 03	Tejas y materiales cerámicos.	Solamente residuos seleccionados de construcción y demolición*.
17 01 07	Mezclas de hormigón, ladrillos, tejas y materiales cerámicos.	Solamente residuos seleccionados de construcción y demolición procedentes de gestor autorizado*.
17 02 02	Vidrio.	
17 05 04	Tierras y piedras.	Excluidas la tierra vegetal, la turba y la tierra y piedras de terrenos contaminados.
19 12 05	Vidrio.	Solamente el vidrio procedente de tratamiento de residuos de recogida separada.
20 01 02	Vidrio.	De acuerdo con el artículo 6.1.e) estos residuos no pueden ser admitidos.
20 02 02	Tierras y piedras.	Solamente residuos de parques y jardines. Excluidas la tierra vegetal y la turba.

Este Anexo II prevé diversas medidas para evitar los conocidos daños ocasionados por el amianto. La abundante presencia de este material tan nocivo conlleva el que se trate según estos criterios, como el de la gestión separada y el evitar su dispersión y el contacto humano. De modo muy importante sobre la regulación de seguridad laboral, se ordena que todos estos requisitos anteriores se deberán cumplir sin perjuicio de que las operaciones o actividades que se desarrollen en el vertedero en las que los trabajadores estén expuestos o sean susceptibles de estar expuestos a fibras de amianto o de materiales que lo contengan, deban cumplir con los requisitos aplicables del Real Decreto 396/2006, de 31 de marzo, por el que se establecen las disposiciones mínimas de seguridad y salud aplicables a los trabajos con riesgo de exposición al amianto (Cabero Morán, 2023). El art. 3 de esta norma comprende "los trabajos de demolición de construcciones donde exista amianto o materiales que lo contengan", de acuerdo con ciertos planes de trabajo (art. 11). Siguiendo con esta norma sobre vertederos, el Anexo II ter recoge las distintas opciones de vertido de residuos y su separación en obra y tratamiento previo, que en relación con los residuos inertes[14].

2.3. El régimen autonómico extremeño de los RCD

La norma principal directamente aplicable es el Decreto 20/2011, de 25 de febrero, por el que se establece el régimen jurídico de la producción, posesión y gestión de los residuos de construcción y demolición[15].

El decreto regional sigue la normativa estatal y europea y, a los efectos de este trabajo, contiene diversas disposiciones sobre el productor, el gestor y el poseedor de los RCD, así como las

---

14 "La primera pregunta a responder podría ser si el residuo está clasificado como peligroso o no. Si, atendiendo a las disposiciones de la Ley 22/2011, de 28 de julio, y a la lista europea de residuos, no lo es, la siguiente pregunta sería si el residuo es inerte o no. Si cumple los criterios de admisión en un vertedero de residuos inertes (clase A, véase la figura 1 y el cuadro 1, ambos de este anexo), el residuo podrá eliminarse en un vertedero de residuos inertes.
Alternativamente, los residuos inertes podrán eliminarse en vertederos de residuos no peligrosos, siempre y cuando dichos residuos cumplan los criterios apropiados".

15 En relación con los residuos inertes y los vertederos de la Ley 16/2015, de 23 de abril, de protección ambiental de Extremadura. En su día se aprobó el PIREX o Plan Integrado de Residuos de Extremadura (2016-2022): http://extremambiente.juntaex.es/index.php?option=com_contentyview=articleyid=4428yItemid=578
[Consultado el 03.07.2023].

diversas fases de la gestión, transporte, valorización y depósito en vertedero (Fuentes Loureiro, 2021). El art. 16 se detiene en la fianza necesaria para la valorización y eliminación mediante su depósito en un vertedero. También recoge el uso de residuos inertes en obras de relleno o restauración. Desde el punto de vista administrativo, se prevé la emisión de certificados de gestión o la creación de un registro de gestores. Sobre el régimen de control e inspección, las licencias de obras municipales incorporarán la gestión de los RCD previa caución de una garantía financiera, que en ningún caso será inferior al 0,4% del presupuesto de la obra, un límite ciertamente poco disuasorio desde el punto de vista del análisis económico del Derecho.

Por lo demás, esta norma sigue lo establecido en la normativa estatal dictada al respecto, como lo dispuesto para las obras menores que no precisen licencia, cuyos residuos tendrán la consideración de residuos sólidos urbanos, de acuerdo con unas ordenanzas municipales que habrán de adaptarse a este decreto, cuestión que deja mucho que desear. Para ciertas pequeñas poblaciones se ha previsto la posibilidad de excepción del tratamiento previo. Subrayamos el que este decreto no establece ninguna particularidad sobre el régimen de infracciones y sanciones, por lo que se remite sin más a la normativa estatal de residuos de los arts. 107 y ss. de la LRSC, por cuya novedad habrán de adaptarse los municipios.

2.4. Examen de ordenanzas municipales extremeñas sobre RCD

A continuación, analizaremos una muestra de ordenanzas municipales de 31 localidades de las provincias de Cáceres y Badajoz seleccionados en base al número de habitantes y denuncias interpuestas por el Servicio de Protección de la Naturaleza de la Guardia Civil (SEPRONA) entre los años 2016 y 2021. A priori pueden ser representativas de las variables demográficas y económicas de la región.

Tabla 5

Muestra de municipios con y sin ordenanza municipal en materia de residuos.

Nº	Municipio	Habitantes	Nº Denuncias	Ordenanza (sí o no)	General o específica (G o E)
1	Badajoz	150.984	122	SÍ	E
2	Cáceres	96255	207	SÍ	G
3	Mérida	59.548	74	SÍ	E
4	Don Benito	37.284	114	SÍ	E
5	Almendralejo	33.855	29	NO	
6	Villanueva de la Serena	25.752	87	SÍ	E
7	Navalmoral de la Mata	17163	32	NO	
8	Zafra	16.810	23	SÍ	E
9	Villafranca de los Barros	12.673	20	SÍ	E
10	Coria	12366	56	SÍ	E
11	Trujillo	8912	38	SÍ	E
12	Moraleja	6696	44	SÍ	E
13	Calamonte	6.170	18	NO	
14	Llerena	5.743	14	SÍ	E
15	Alburquerque	5.293	9	NO	
16	Arroyo de San Serván	4.080	32	Sólo tasa	
17	Malpartida de Cáceres	4076	38	SÍ	E
18	Valdelacalzada	2.721	9	SÍ	E
19	Alcuéscar	2554	67	SÍ	E
20	Cabezuela del Valle	2129	22	NO	
21	Villar del Rey	2.106	7	SI	G
22	Alconchel	1.659	3	NO	
23	Torreorgaz	1656	25	NO	
24	Cheles	1.173	3	NO	
25	Valdefuentes	1139	20	SÍ	E
26	Cordobilla de Lácara	897	2	NO	
27	Escurial	865	6	NO	
28	Valencia del Mombuey	725	13	SÍ	E
29	Palomas	671	3	NO	
30	Salvatierra de Santiago	298	19	NO	
31	Talaveruela de la Vera	298	1	NO	

A efectos expositivos, expondremos los municipios cuyas ordenanzas específicas sobre RCD siguen lo estipulado en el Decreto extremeño 20/2011 anteriormente citado, pero adolecen de su adaptación a los criterios infractores y sancionadores de la LRSC cuyo régimen es mucho más amplio que el limitado a las conductas infractoras y las eventuales sanciones aplicables:

2.4.1. Municipios con ordenanza específica:

.- Badajoz: Su ordenanza de 2015 apenas detalla unas cuantas conductas infractoras, centradas en el vertido y en la falta de comunicación de las actividades que generen los RCD. Las sanciones se limitan a las pecuniarias y están totalmente desfasadas respecto de la LRSC.

.- Mérida: La ordenanza de 2012 se remite sin más a la legislación estatal, con una breve referencia a la ya derogada Ley de Residuos de 1998.

.- Don Benito: La ordenanza de 2012 se remite a la ley estatal ley extremeña de protección ambiental.

.- Villanueva de La Serena: Su ordenanza de 2006 se remite a la normativa estatal e incluye la adopción de medidas cautelares, como la suspensión del vertido.

.- Zafra: La ordenanza de 2017 aplica el régimen general estatal.

.- Villafranca de los Barros: Siguiendo un modelo muy común, la ordenanza de 2012 se remite sin más a la legislación estatal.

.- Coria: Su ordenanza de 2020 es más detallada en cuanto al ejercicio de las competencias del alcalde, remitiéndose a la normativa general de RCD. Establece un régimen sancionador que no casa con el vigente.

.- Trujillo: La ordenanza de 2006 muestra una regulación muy simple y obsoleta. Se remite a la legislación general.

.- Moraleja. Solamente cuenta con una ordenanza fiscal, de 2014.

.- Llerena: Se remite a la normativa general. Fue publicada en 2015.

.- Arroyo de San Serván: Solo cuenta con una ordenanza fiscal, de 2012.

.- Malpartida de Cáceres: Este municipio cuenta con un detallado régimen sancionador, que incluye la obligación de reposición del terreno a su estado anterior y la reparación del daño. Se remite, de forma subsidiaria al régimen general de los residuos. El Ayuntamiento no ofrece la fecha de publicación.

.- Valdelacalzada: Siguiendo un modelo común a varios ayuntamientos, la ordenanza, de fecha de publicación desconocida, regula las infracciones y sanciones remitiéndose a la normativa general.

.- Alcuéscar: Siguiendo un modelo común a varios ayuntamientos, la ordenanza de 2015 regula las infracciones y sanciones remitiéndose a la normativa general.

.- Valdefuentes: Tras remitirse al régimen general, este ayuntamiento cacereño regula con detalle las conductas infractoras y las medidas de reparación. La ordenanza de 2015 recoge una serie de sanciones muy alejada del sistema vigente.

.- Valencia del Mombuey: Siguiendo un modelo común a varios ayuntamientos, la ordenanza de 2018 regula las infracciones y sanciones remitiéndose a la normativa general.

2.4.2. Municipios con ordenanzas generales:

.- Cáceres: La ordenanza de 2021 sobre residuos asimila los RCD a los residuos industriales, residuos no municipales y domésticos (art. 29), la colocación de contenderos en la vía pública siempre que no sobrepasen los 25 l/día. El art. 66 y ss. regula los RCD, como la cesión de la gestión a la Diputación de Cáceres, en un detallado régimen sobre su gestión completa. El régimen infractor hace un especial hincapié en la gestión de la información y en los sistemas de retirada de los RCD. Sin embargo, el régimen sancionador queda muy lejos de lo previsto por la normativa general.

.- Villar del Rey: Cuenta con una ordenanza que regula el punto limpio, donde se regulan los escombros.

En cuanto a las fianzas, hay ayuntamientos como el de Cáceres, Villanueva de la Serena, Don Benito, Zafra, Valdefuentes, Valencia del Mombuey, Coria, Alcuéscar, Arroyo de San Serván, Trujillo, Llerena, Villafranca de los Barros, Moraleja y el de Mérida, que incluyen una regulación detallada de las garantías financieras. Otros, como el de Badajoz, las citan, pero sin acordar el cálculo de su cuantía, por lo que hay que estar a la normativa general y a las ordenanzas fiscales propias de ese municipio. En algunos casos, como el de Almendralejo, ni se citan las fianzas.

3. CONCLUSIONES

El estudio criminológico de los problemas ambientales derivados de los RCD ha de estar inexcusablemente a la exhaustiva y detallada regulación administrativa, que comprende normas internacionales, europeas, nacionales, autonómicas y municipales.

El llamado tipo penal en blanco se colma gracias a la remisión a esta normativa, que se caracteriza por una complejidad técnica muy importante. A esta complejidad se le suma el juego de las distintas competencias administrativas entre las Administraciones territoriales y la propia evolución de una normativa que se modifica a pasos agigantados.

No solamente hay que prever la estricta normativa de residuos, sino que igualmente hay que tener presente el lugar donde éstos se depositan, tratan o simplemente se abandonan. Unos ejemplos nos pueden ilustrar: espacios naturales protegidos, riberas del mar o de los ríos, la cuenta de una carretera, solares, vertederos de residuos municipales o canteras. Estos lugares sufren el evidente incumplimiento de este ordenamiento, para cuya garantía hay que avanzar en medidas tributarias y de fomento del empleo circular de los RCD, de lo contrario son la sociedad y el medio ambiente quienes, al externalizarse los costes, acarrean las consecuencias.

Por otra parte, no hay que olvidar que el intervencionismo administrativo tiene su correlato en la práctica empresarial: no puede dejarse que se perpetúe la competencia desleal que implica el que quienes infringen la ley se vean beneficiados económicamente en un mercado donde los RCD son comunes a múltiples agentes económicos ni tampoco puede darse ninguna comprensión ante la actitud incívica de cientos de ciudadanos. Uno de los elementos que podrían disuadir de las prácticas ilegales es la fijación y el cobro de una fianza suficiente que sirva de freno al vertido clandestino de los RCD como "ticket moderador", cuestión que exige un adecuado análisis económico.

4. REFERENCIAS BIBLIOGRÁFICAS

Arenas Cabello, F. J. (2010). Marco jurídico de las instalaciones de gestión de residuos de construcción y demolición. *Actualidad Jurídica Ambiental, 2,* 1-13.

Cabero Morán, E. (2023). La evitación de la exposición al amianto en el ámbito laboral y los nuevos derechos a la reparación íntegra de los daños. *Trabajo y Derecho, 97.*

Fuentes Loureiro, Mª Á. (2021). *Los delitos de gestión ilegal y traslado ilícito de residuos (Art. 326 CP).* Tirant lo Blanch.

Moreu Carbonell, E. (2022). Marco jurídico de la minería urbana. *Revista Aranzadi de Derecho Ambiental, 51,* 23-44.

Palomar Olmeda, A. (dir) (2022). *Estudios sobre la Ley de residuos y suelos contaminados para una economía circular.* Thomson-Reuters Aranzadi.

Soriano García, J. E. y Brufao Curiel, P. (2023). Claves de Derecho Ambiental I. *Iustel,* 73 y ss.

Vercher Noguera, A. (2022). *Delincuencia ambiental y empresas.* Marcial Pons.

Capítulo 3

Fuentes de datos oficiales: calidad y márgenes de error

ALBERTO ALFONSO-TORREÑO[1]
LOREA ARENAS GARCÍA[2]

1. INTRODUCCIÓN[3]

Las infracciones administrativas y penales denunciadas por el Servicio de Protección de la Naturaleza (SEPRONA) es la principal fuente de datos oficial en materia de residuos que tenemos en nuestro país. Los datos elaborados por la Dirección General de la Guardia Civil son publicados en los Anuarios Estadísticos del Ministerio del Interior en el apartado "protección o conservación de la naturaleza". En este figuran según la naturaleza de la infracción (administrativa o penal) las cifras de denuncias (en números absolutos) de diversos conceptos que corresponden al ámbito de la infracción, a saber: ordenación del territorio; vías pecuarias; minería; turismo, ocio y deporte; leyes sanitarias; flora, bosques y montes; incendios forestales; fauna salvaje; C.I.T.E.S;

---

1 Instituto de Investigación para el Desarrollo Territorial Sostenible (INTERRA), Universidad de Extremadura.

2 Profesora Contratada Doctora, Área de Derecho penal, Universidad de Extremadura.

3 Este trabajo es una adaptación y traducción de la obra: Alfonso-Torreño, A., Gutiérrez Gallego, A., Ortiz García, J. y Arenas García, L. (2023). Measurement error of police data to identify solid waste dumping sites. *GeoFocus. International Review of Geographical Information Science and Technology, 31,* 39-53. DOI: 10.21138/GF.798.

animales domésticos; aguas marítimas; costas; aguas continentales; residuos; atmósfera; patrimonio histórico; otros. En cuanto al concepto "residuos", este aparece desglosado en otros subconceptos, tales como: urbanos, peligrosos, radioactivos, industriales, sanitarios y/o clínicos y otros. En esta subcategoría se engloban los datos oficiales que, dependiendo del anuario en particular, puede contener información más o menos detallada de los mismos (por ejemplo: si se trata de una falta, delito o infracción, si están esclarecidas o no, etc.). Gracias a estos datos sabemos que el vertido de residuos, y sobre todo el residuo urbano, es la infracción administrativa contra el medioambiente más frecuente entre los años 2010 y 2019[4]. Dichas infracciones alcanzan una media anual de 28.258 infracciones y representa en torno al 20% del total de las infracciones medioambientales. En cuanto a las infracciones penales por depósito o vertido de residuos tóxicos o peligrosos, son mucho menos numerosas que las anteriores (en torno a una media de 8 delitos entre 2014 y 2021).

Los datos publicados por el Ministerio del Interior no aparecen desglosados por autonomías ni tampoco sabemos nada de las actuaciones y denuncias que interponen otras fuerzas de seguridad encargadas de vigilar y controlar el vertido ilegal, como pueden ser las policías locales y autonómicas, o los propios agentes medioambientales. No obstante, gracias a los datos proporcionados por el SEPRONA para esta investigación podemos completar la información anterior conociendo que en la región se denunciaron 3769 infracciones administrativas a la normativa de residuos de 2016 a 2021[5]. En la figura inferior se muestra la evolución de las denuncias para dicho periodo.

---

4 No obstante, y debido a la pandemia de la COVID, fueron superadas en 2020 y 2021 por infracciones a la normativa sobre sanidad pública y medicamentos.

5 No se incluyen los datos de 2021 en la comparativa interanual porque fueron recogidos únicamente hasta el mes de julio.

Figura 1

Evolución del número de infracciones a la normativa de residuos por RCD de 2016 a 2020.

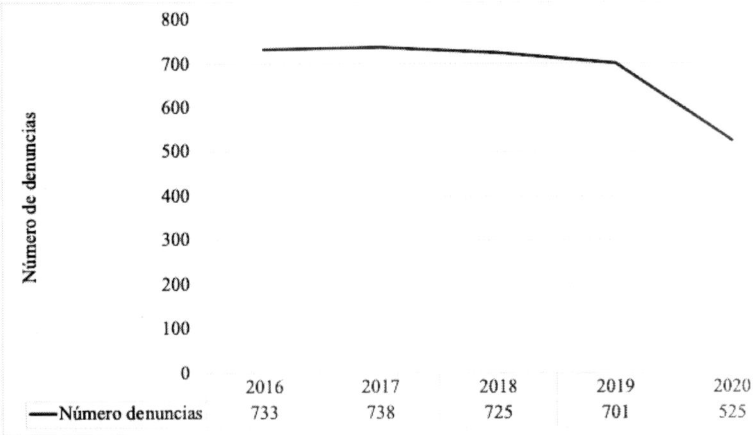

Tal y como se muestra en el gráfico, el número de denuncias permanece estable a lo largo de todo el periodo (rondando las 700 denuncias de media), si bien en el año 2020 estas disminuyen, muy probablemente debido al impacto de la COVID19 en las labores de vigilancia y control de los operativos policiales.

Adicionalmente, de las Memorias de la Fiscalía General del Estado se puede extraer información de las intervenciones realizadas en materia medioambiental en capítulo de "Fiscales coordinadores/as y delegados/as para materias específicas". En concreto, en su apartado tercero, se compendia toda la información estadística y sobre actividades en materia medioambiental. Si bien los datos son muy genéricos haciendo referencia, *grosso modo*, a: las diligencias investigadoras efectuadas, los delitos judiciales incoados, los escritos de acusación, las sentencias condenatorias y las sentencias absolutorias que han tenido lugar. Se trata de una fuente limitada de datos debido a su baja especificidad que poca información brinda para el caso de residuos inertes. La principal preocupación y alusión recurrente en las distintas memorias es acerca del potencial riesgo de incendio que tienen vertederos grandes de residuo urbanos debido a la mala gestión en las instalaciones.

Considerando la escasez de datos oficiales sobre residuos y de sus tipos, como puede ser el de construcción y demolición, es muy difícil determinar la calidad de estos, así como la cifra negra para este tipo de infracciones. Y es que, en nuestro país, tal y como se ha venido apuntando, se desconoce la cifra real de vertederos ilegales de residuo y, lo que es más importante, su ubicación exacta. La policía todavía no identifica ni localiza la mayoría de los vertederos ilegales, aunque necesita información precisa sobre sus características para desarrollar respuestas apropiadas. Por ejemplo, conocer la ubicación exacta y composición del foco de vertido enriquecería las estadísticas y facilitaría el análisis desde una perspectiva de la Criminología ambiental utilizando mapas del crimen.

Los mapas del crimen, también conocidos como *crime mapping*, son cada vez más utilizados en la labor policial. Se trata de una herramienta que aporta conocimientos geográficos e identifica los "puntos calientes" de la delincuencia. Ello posibilita el desarrollo de enfoques orientados a resolver problemas relacionados con la concentración espacio-temporal de determinadas formas de delincuencia. (Goldstein 1979; Braga, 2008; Braga and Weisburd, 2010). De hecho, en la literatura existente, hay muchos estudios sobre residuos utilizando Sistemas de Información Geográfica (en adelante SIG). Se trata de una herramienta valiosa en la investigación ambiental para: identificar y predecir donde se ubican los vertederos; comprender el complejo patrón de distribución espacial de los residuos ilegales y las variables específicas asociadas con su aparición y; analizar alternativas de gestión de residuos más inteligentes (Doak et al., 2007; Silvestri y Omri, 2008; Biotto et al., 2009; De Feo et al., 2014; Jordá-Borrell et al., 2014; Glanville y Chang, 2015; Zainun et al., 2016; Seeboonruang, 2016; Massarelli, 2018; Paz, Lafayette andSobral, 2018; Vijay et al., 2018; Seror y Portnov, 2018; AlZaghrini., et al, 2019; Hussen et al., 2019; Jakiel et al., 2019; Biluca et al., 2020). En consecuencia, los SIG son esenciales para estudiar la actividad delictiva, aunque gran parte del éxito de las técnicas espaciales radica en la calidad de los datos que se manejan. Y recordemos que, en el caso que nos ocupa, la ubicación de los vertederos de residuo de construcción y demolición o RCD es desconocida al tiempo que la investigación empírica es escasa y limitada. (Matsumoto y Takeuchi, 2011; Jordá-Borrell et al., 2014; Glanville y Chang, 2015).

Hay algunas investigaciones sobre la geocodificación policial para formas comunes de delincuencia en entornos urbanos, pero apenas sabemos nada sobre la georreferenciación de vertederos ilegales. Estos delitos a menudo ocurren en lugares no direccionables y geocodificarlos lo que entraña mayor dificultad (Chainey y Ratcliffe, 2005). Los criminólogos y geógrafos se han ocupado principalmente de estimar una tasa mínima para el proceso de geocodificación. Se trata de convertir una descripción textual, esto es, una dirección, en un conjunto de coordenadas geográficas (Brimicombe et al., 2007; Ratcliffe, 2004), Pero algunos estudios también han evaluado la precisión posicional (Hart y Zandbergen, 2012; Gerell, 2018) y el impacto del error geográfico en el análisis posterior (Harada y Shimada, 2006; Andresen et al., 2019; Briz-Redon et al., 2020), mientras que otros han tratado de estimar el impacto del sesgo de los datos policiales en pequeñas y grandes áreas geográficas (Buil-Gil et al., 2021).

Por todo ello, el componente espacial o territorial en un estudio criminológico comienza con una correcta georreferenciación o geocodificación de la localización del delito (Mazeika y Summerton, 2017). En este sentido, la mala praxis en la recopilación de datos policiales puede alterar y sesgar los análisis de la densidad de la delincuencia (e.g., Ivert et al., 2013). No obstante, cabe señalar que se espera una mayor fiabilidad de los datos cuando las coordenadas se han registrado en la escena del crimen utilizando dispositivos GPS que cuando son extraídas de las direcciones para ser geocodificadas a posteriori (Gerell, 2018).

2. OBJETIVOS

El objetivo de este trabajo es analizar la fiabilidad de los datos policiales proporcionados por SEPRONA para localizar vertidos sólidos en la región rural de Extremadura, así como proponer buenas prácticas a la policía para adaptar la tecnología SIG a los problemas ambientales y las necesidades de planificación local.

3. METODOLOGÍA

Los datos se basan en denuncias administrativas relacionadas con residuos de construcción y demolición que tuvieron lugar en Extremadura en 2018. El SEPRONA facilitó al equipo investigador un archivo vectorial (.shp)[6] con un total de 1238 geolocalizaciones sobre vertidos ilegales recogidos directamente en el lugar de la infracción mediante un dispositivo GPS. La distribución de los puntos se realizó utilizando el software QGIS. Se desconoce la instrumentación y el sistema de referencia de coordenadas utilizado para el monitoreo de residuos en los datos proporcionados.

Los datos de la Guardia Civil no incluían la fecha y hora del registro de la infracción. Sólo se ha tomado el año y no la hora o el día. Otra fuente de datos son las ortofotografías actualizadas del Instituto Nacional Geográfico Español (CNIG, 2019) visualizadas por el servicio de mapas web (WMS) a través de QGIS. Las ortofotografías tienen una resolución espacial de 0,25 m con un sistema de referencia de coordenadas del Sistema Europeo de Referencia Terrestre 1989 (ETRS 1989).

Se realizó una verificación detallada de los residuos de construcción y demolición a través de las ortofotografías proporciona-

6 La capa vectorial contenía la siguiente información: Id, año, ámbito penal, tipo de infracción, municipio, provincia, país y coordenadas geográficas (x, y). Las coordenadas se tomaron en el datum del Sistema Geodésico Mundial 1984 (WGS84) con proyección UTM Huso 29.

das por el WMS. Los datos georreferenciados se etiquetaron en base a tres tipos (Figura 2):

1. *Sí*: hay residuos de construcción y demolición donde se encuentra la infracción administrativa.
2. *Probablemente*: no hay vertido ilegal donde se encuentra la denuncia administrativa, pero sí hay escombros muy cerca de la geolocalización.
3. *No*: no hay residuos donde se encuentra la infracción administrativa (por ejemplo, la geolocalización está en un tejado)

Toda esta información se exportó a nuevos archivos Excel (.xls) para trabajar con ellos en una base de datos y añadiendo nuevos atributos, tales como: (1) la distancia real entre la ubicación de la denuncia y el vertido ilegal (en el caso de los datos definidos como "*probablemente*") y (2) el tipo de uso del suelo donde se ubicó la denuncia en los definidos como "*No*".

Figura 2

Ejemplos ilustrativos del enfoque metodológico, usando ortofotografías de 2019 de fondo.

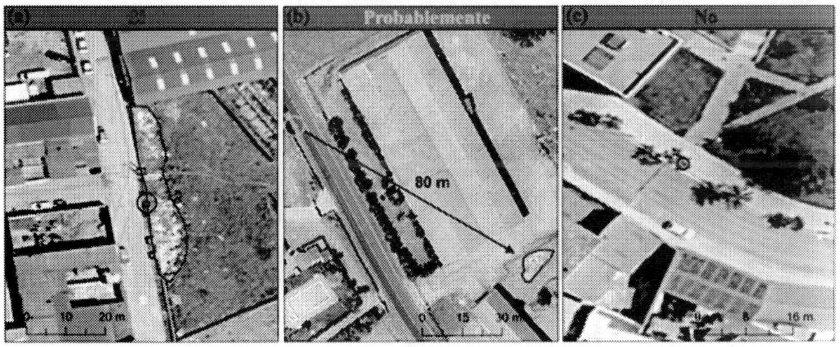

Nota: (a) realmente existe un vertido ilegal en el lugar de la queja, (b) la queja no se encuentra exactamente en los residuos de construcción y demolición y (c) no hay vertido ilegal.

La verificación de la presencia de escombros también se llevó a cabo mediante un estudio de campo (Figura 3). Un total de 58 lugares fueron verificados *in situ* para detectar la existencia o no de vertidos ilegales. Principalmente el trabajo de campo se realizó en Mérida, Cáceres, Badajoz y Don Benito.

Figura 3

Observaciones de campo que muestran residuos de construcción y demolición en dos localidades diferentes.

Nota: (a) en un suelo urbanizable en la ciudad de Cáceres y (b) en el terreno junto a una carretera sin pavimentar en el municipio de Mérida.

Finalmente, para comprender mejor los resultados obtenidos, se realizaron dos entrevistas abiertas a los agentes de la Guardia Civil que proporcionaron los datos.

4. RESULTADOS

A continuación, se presentan los principales resultados obtenidos en el análisis. De las 1.238 reclamaciones administrativas, 725 fueron denuncias con diferentes geolocalizaciones representando el 59% del total. En muchos casos (n = 513) más de una infracción coincidía con la misma ubicación eligiéndose, para estos casos, una sola ubicación para cada queja. Esto significa que un solo sitio geocodificado puede haber múltiples informes o registros policiales. Es común que una sola infracción conduzca al inicio de más de un procedimiento, principalmente porque se ha presentado más de una denuncia, o por cualquier otra razón que genere esta situación. (Serrano-Gómez, 2011). Por ejemplo, en una misma ubicación puede haber varios registros si el agente encuentra: un vertido ilegal, la posesión ilegal de alguna mercancía, un delito vial, etc. Weisburd (2015) define este escenario como la ley de concentración del delito en el lugar, que establece que para una medida definida del delito en una unidad microgeográfica específica, la concentración del delito tendrá lugar dentro de un rango espacial para una proporción acumulativa del delito.

Los municipios con al menos una queja administrativa por residuos de construcción y demolición representan el 61%. Aunque la distribución espacial de las quejas en la región es heterogénea, las quejas se concentraron principalmente en las poblaciones más pobladas (Cáceres, Badajoz, Mérida y Don Benito) y cerca de las principales vías de comunicación (Figura 4a).

El 90% de las coordenadas asociadas a las infracciones proporcionadas por el SEPRONA no correspondían a una ubicación con escombros tras su verificación en la ortofoto (Figura 4b). Solo en el 2% de los casos la coordenada de la infracción correspondía a un lugar con residuos de construcción y demolición observables.

Por otro lado, en el 9% de las denuncias los escombros se observaban muy cerca de la geolocalización registrada en los datos policiales. Analizando esta cuestión según provincia, 12 denuncias fueron correctamente geolocalizadas sobre residuos ilegales en Cáceres y solo 4 en Badajoz. Sin embargo, no se localizaron exactamente 24 denuncias sobre residuos ilegales en Cáceres y 42 en Badajoz, pero sí residuos muy cerca de la geolocalización (Figura 4c). Según Matsumoto y Takeuchi (2011) y Lucendo-Monedero et al. (2015), uno de los principales problemas en el estudio del vertido ilegal utilizando estadísticas oficiales es la falta de datos consistentes y fiables sobre la ubicación de los vertederos. Este parece ser el caso de nuestro estudio.

Figura 4

Distribución espacial de las denuncias de residuos y constatación de su existencia mediante verificación en la ortofoto.

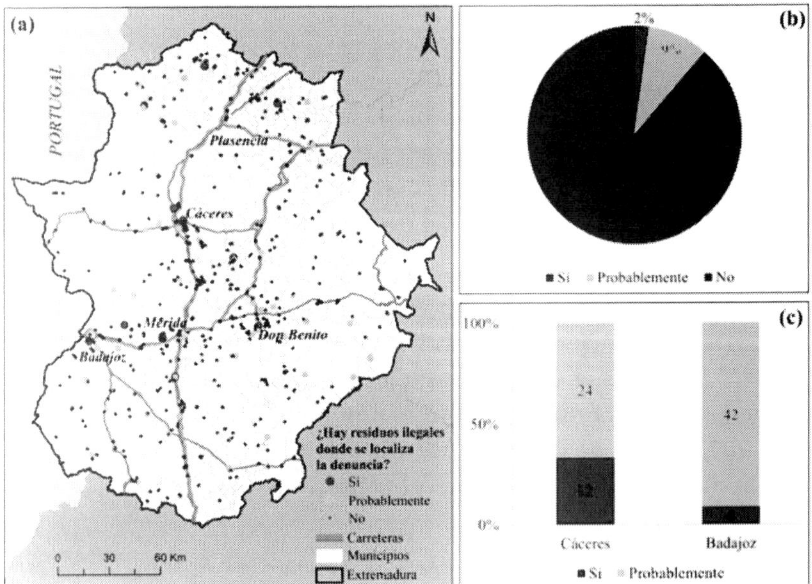

Nota: a) Distribución espacial de las denuncias administrativas por vertidos ilegales en Extremadura en función de la existencia o no de residuos en los que se localice la denuncia; b) porcentaje de infracciones administrativas en función de si se localizaron o no sobre residuos de construcción y demolición y; c) número de infracciones administrativas coincidentes con los residuos por provincias.

Según municipio, Cáceres y Badajoz registraron un gran número de denuncias administrativas con 43 y 24 respectivamente (Figura 5a). Don Benito es el municipio con mayor número de infracciones correctamente geolocalizadas respecto a residuos, representando el 45% de las mismas. Por el contrario, Cáceres y Badajoz fueron los municipios con menor número de denuncias correctamente geolocalizadas, con un 25% y un 19% respectivamente. La Figura 5b muestra la relación entre el número de infracciones administrativas y la población por municipios que se correlacionaron significativamente utilizando -para el análisis- funciones polinómicas de segundo grado.

Figura 5

Comparativa entre el número de denuncias y el número de denuncias geolocalizadas correctamente según municipio y volumen poblacional.

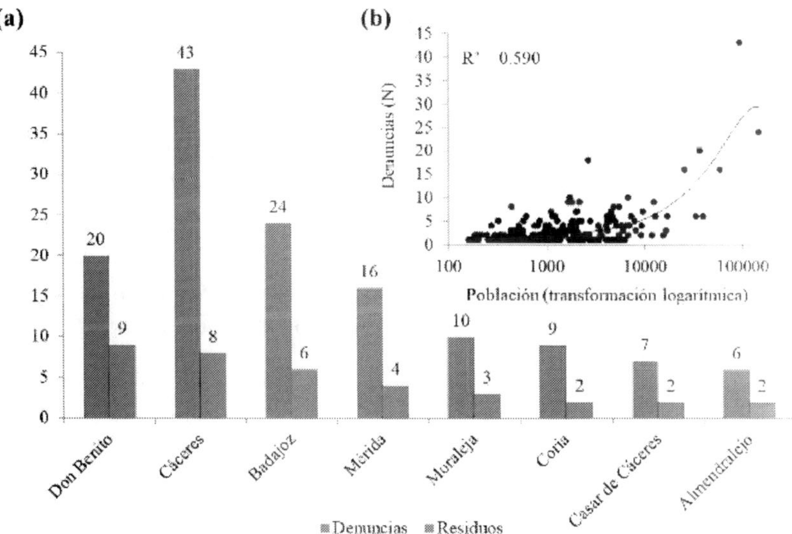

Nota: (a) Municipios con mayor número de denuncias administrativas y vertidos ilegales y (b) relación entre número de denuncias administrativas y población (transformación logarítmica) por municipios.

Para profundizar en el análisis, se consideró la distribución del error. En la figura 6 se muestra la diferencia de distancia entre el lugar de la coordenada de la denuncia administrativa y la ubicación real del vertido ilegal. La distancia media y mediana fue de 64 y 48 metros respectivamente. El 54% de las denuncias clasificadas como "probablemente" muestran una distancia a los residuos de menos de 50 metros. Una hipótesis razonable sería que la distancia inferior a 50 metros presumiblemente representa ubicaciones geocodificadas de manera inexacta pero próximas. Comparando las distancias encontradas en este estudio con el publicado por Ratcliffe (2001), se observa que la distancia media entre dos lugares diferentes fue menor en su caso, con una separación media de 47 metros. Por el contrario, Gerell (2018) encontró una distancia media de 83 metros, es decir, una distancia mayor que la estimada en este estudio.

Es probable que las distancias más grandes estén asociadas a lugares en los que no es posible determinar la ubicación exacta de los residuos porque se desconoce el lugar exacto del delito (Newton et al., 2014), lo que a veces resulta en una georreferenciación aproximada (ver Figura 2b, geocodificación en el centro de una carretera). También puede tratarse de lugares en los que no está permitido el acceso o, simplemente, se trata de datos mal grabados por una mala praxis del agente. Por ejemplo, si se registra la coordenada de la denuncia a posteriori en la estación de policía o desde la comodidad del coche policial en vez de en el lugar de los hechos. Como señaló Gerell (2018), las ubicaciones geográficas poco confiables pueden conducir a ubicaciones erróneas de un área completa con un impacto aún mayor en la efectividad de las operaciones policiales.

Figura 6

Distribución de distancias entre la ubicación de la denuncia administrativa y el vertido ilegal.

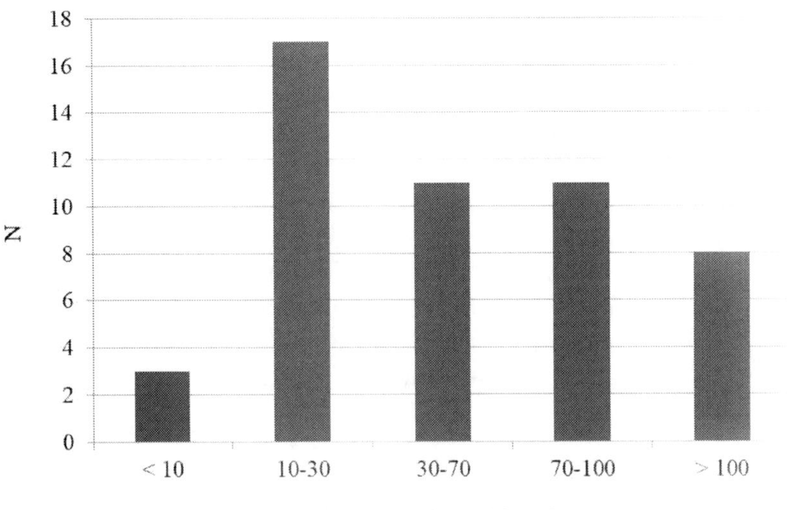

En la tabla 1 se muestra la ubicación de las denuncias donde no hubo residuos de construcción y demolición. Nuestro estudio reveló que el 40 % de las denuncias que no se ubicaron en residuos (es decir, las etiquetadas como "no") se ubicaron en suelo urbano, destacando las denuncias ubicadas en calles y azoteas con un 69 % y 27 %, respectivamente. En la literatura científica no existen estudios sobre la evaluación del error de medición para identificar vertederos de residuos sólidos que permitan comparar estos resultados. No obstante, existen varios estudios sobre áreas de probabilidad de presencia de vertidos ilegales que ayudan a comprender esta distribución de denuncias del tipo "no". Por ejemplo, Jordá-Borrell et al. (2014) identificaron, mediante herramientas de modelado geoestadístico, algunos factores relacionados con la presencia de vertederos ilegales en la región andaluza de España. Este estudio destacó que un gran porcentaje de los sitios se encontraban en áreas urbanas y en los pueblos más poblados (superiores a 10000 habitantes). Nuestro estudio reveló que el 38% de las denuncias de tipo "no" se ubicaron en terrenos rústicos y naturales, principalmente en zonas rurales y forestales. Estos hallazgos están en línea con los resultados de varios otros estudios, que revelaron la presencia de vertidos ilegales en áreas boscosas (Seror y Portnov, 2018; Silvestri y Omri, 2008). Este grupo de quejas (es decir, tipo "no") también se localizó en viales y suelo urbanizable, representando un 15% y un 7% respectivamente. Matos et al. (2012) y Jordá-Borrell et al. (2014) encontraron que la mayoría de los vertederos ilegales se encuentran cerca de las carreteras. Además, los residuos de construcción y demolición se pueden utilizar como material de relleno en caminos no pavimentados (Seror y Portnov, 2018). Estos resultados se visualizan con algunos ejemplos en la Figura 7.

Sin embargo, la no presencia de residuos donde se registraron denuncias y la geolocalización de residuos en elementos urbanos (e.g. cubiertas) puede deberse a varios motivos:

1. Probablemente, parte de los vertidos de RCD ubicados en espacios públicos (p. ej., carreteras, zonas urbanas y periurbanas) hayan podido ser eliminados por los sistemas de gestión local entre 2018 y julio de 2019. Se obtuvieron las ortofotografías del Instituto Geográfico Nacional de España. en julio de 2019 mientras que las denuncias administrativas del SEPRONA se registraron a lo largo de 2018.

2. Las infracciones administrativas geolocalizadas sobre cubiertas y embalses, por ejemplo, probablemente se atribuyan a una georreferenciación o geocodificación inexacta de los RCD.

3. Finalmente, las denuncias administrativas ubicadas en medio del campo y sin presencia de RCD pueden deberse a que el lugar exacto está en duda y se atribuye a lugares más generales (Gerell, 2018).

Tabla 1

Localización de denuncias administrativas donde no hay residuos de construcción y demolición.

Tipo	N	%	Subtipo	N	%
Suelo urbano	257	40%	Calle	177	69%
			Tejado	70	27%
			Zona industrial	4	2%
			Plaza	4	2%
			Parque	1	0%
			Aparcamiento	1	0%
Suelo urbanizable	43	7%			
Suelo rústico	248	38%	Campo	244	98%
			Embalse	2	1%
			Zona agraria	2	1%
Vías de comunicación	94	15%	Camino	34	36%
			Carretera	54	58%
			Rotonda	6	6%

Figura 7

Observaciones de ortofotografías que muestran lugares donde se ubicaron denuncias (puntos blancos) sin vertidos ilegales.

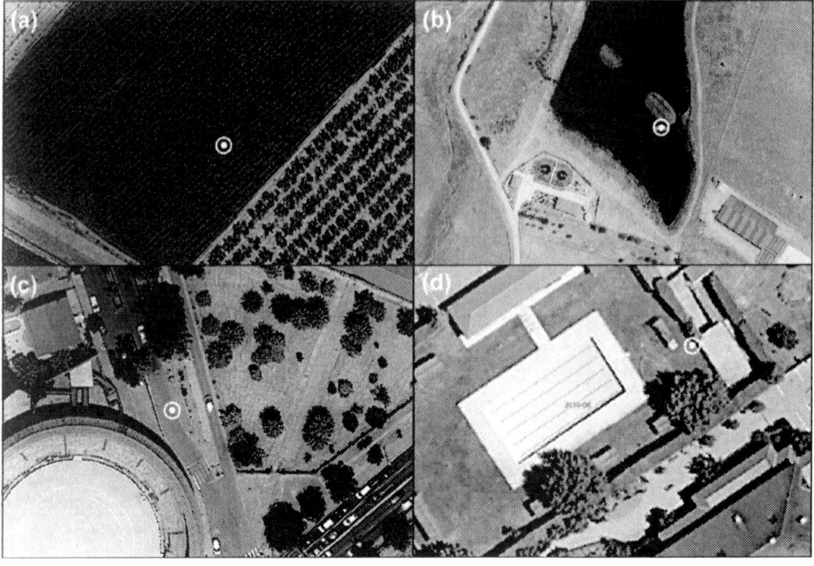

Nota: Estos son algunos ejemplos de ubicaciones: (a) campo, (b) embalse, (c) carretera y (d) parcela privada.

Teniendo en cuenta la gran cantidad de áreas donde no se observaron escombros, nos parecía poco probable que todos los puntos de desechos de RCD no estuvieran allí porque, en algún momento entre 2018 y 2019, habían sido limpiados. Para comprender mejor la ausencia de residuos en las coordenadas proporcionadas, mostramos los resultados obtenidos con los agentes de la Guardia Civil para encontrar respuestas. No se sorprendieron de los resultados. Para ellos, la razón del aparente volumen de error se debe al enfoque de los oficiales para codificar la información. Las coordenadas georreferenciadas se recogen como un campo más a rellenar en una aplicación informática, pero no (en la actualidad) con fines de explotación estadística o de inteligencia policial. Como señalaron Chainey y Ratcliffe (2013): "si bien es importante asegurarse de que los datos sobre delitos geocodificados sean de buena calidad, también es importante que, en la búsqueda de la perfección de los datos, el nivel de calidad de los datos que se requiere de los datos sobre delitos es proporcional a los fines a los que se aplicará". Tampoco se da formación a los agentes sobre cómo geoposicionar las infracciones, ni existen protocolos al respecto, ni medidas internas de rendición de cuentas. Esta sería probablemente la causa de la mala calidad de los datos. Por ejemplo, los agentes de campo no saben que deben permanecer en el lugar de la infracción durante un cierto tiempo antes de guardar la coordenada, ya que el GPS tarda en cuadrar correctamente su posición. Esto podría explicar, por ejemplo, por qué existe un desfase en la distancia entre el punto real y el georreferenciado en algunos casos, ya que la señal centra la posición del agente cuando está saliendo del lugar (e.g., Figura 2b). Otras veces se debe simplemente a malas prácticas intencionadas por parte de los agentes. Gerell (2018) señaló que una posible fuente de error sería la ubicación con un dispositivo GPS desde un lugar cómodo para el agente (dentro de un automóvil) en lugar de la ubicación exacta del incidente. En nuestro estudio, este hecho se ejemplifica en coordenadas que coinciden con bares o comisarías.

Además de una mala práctica en la recopilación de datos por parte de los agentes, otra posible fuente de error es el uso de métodos de recopilación de datos georreferenciados con grandes errores de medición. Por esa razón, se sugiere el uso de un receptor habilitado por el Servicio Europeo de Superposición de Navegación Geoestacionaria (EGNOS), que es el primer sistema europeo de navegación por satélite. Un receptor habilitado para EGNOS proporciona precisión de ubicación dentro de los 3 metros. Sin EGNOS, un receptor GPS estándar (por ejemplo, un teléfono móvil o un reloj inteligente) solo proporciona una precisión de 17 metros. Además, otra cuestión importante es que cada agente ha georreferenciado el delito utilizando diferentes dispositivos (un teléfono móvil, un reloj inteligente, etc.) induciendo un sesgo adicional. Los estudios criminológicos que incluyen el componente espacial, por medio de SIG, requieren que los datos geocodificados sean 100% precisos, precisos en la ubicación exacta del lugar donde ocurrió el delito, hayan sido registrados correctamente, estén completos y sean confiables (Chainey y Ratcliffe, 2013). En este sentido, parece que la falta de una cultura policial orientada a los SIG en España quizás explique la poca importancia que se da a estas prácticas y protocolos (Wang, 2012).

5. CONCLUSIONES

Los residuos de construcción y demolición forman parte del paisaje urbano y rural de España, siendo su localización y prevención un reto para investigadores y fuerzas de seguridad. En este artículo examinamos 1238 infracciones administrativas georreferenciadas de la Guardia Civil en la zona rural de Extremadura mediante una metodología de revisión caso por caso para comprobar si, efectivamente, existían escombros en las coordenadas proporcionadas por la policía. Encontramos que solo en el 2 % de las infracciones se confirmó la presencia de escombros en un lugar georreferenciado y que, en el 9 % de los casos, era probable su presencia cuando los escombros se ubicaban a una distancia menor de 100 metros. El hallazgo más relevante es que en el 89 % de los casos no hubo residuos en el lugar reportado. En estos casos, las coordenadas los ubicaban en calles y azoteas (en áreas urbanas) y en áreas de campo y bosque (en áreas naturales). La falta de precisión posicional en los datos policiales es un hallazgo común (Ratcliffe, 2004; Brimicombe et al., 2007) aunque rara vez es tan grave. De hecho, habíamos anticipado una mejor calidad de los datos porque los agentes los recolectaban directamente usando dispositivos GPS en lugar de tener que pasar por el proceso de geocodificación desde una dirección de texto. Para buscar explicaciones a estos hallazgos, los datos cuantitativos se complementaron con entrevistas a funcionarios clave del SEPRONA. Creen que los datos no fueron confiables para ubicar residuos porque no se han recopilado adecuadamente. Esto se debe fundamentalmente a tres razones: no existe una cultura policial de aprovechar los datos GIS para luego explotarlos para el análisis táctico y estratégico, es decir, no es un tema prioritario; que los agentes de campo no manejan técnicas precisas y adecuadas para la recolección de datos georreferenciados y que no cuentan con el entrenamiento y protocolos para geoposicionar adecuadamente las infracciones.

Las implicaciones de esta investigación son múltiples. En primer lugar, enriquece el campo de los estudios criminológicos sobre la calidad de los datos policiales para la geocodificación y georreferenciación de delitos recogidos por dispositivos GPS en delitos de vertidos ilegales. En segundo lugar, el hecho de no disponer de datos oficiales fiables para localizar los residuos dificulta: conocer su incidencia y las características del fenómeno, su asociación a distintas variables, patrones, etc.; predecir áreas vulnerables; desarrollar otras metodologías para localizar derrames que requieran una "verdad sobre el terreno", por ejemplo, en técnicas de teledetección; y que la policía pueda administrar mejor sus propios recursos para llevar a cabo una mejor prevención del delito. En tercer lugar, esto implica buscar fuentes de datos alternativas para contrastar la información proporcionada como forma de lograr una mayor confiabilidad geográfica (Gerell, 2018). En el contexto español esto solo sería posible desarrollando un trabajo de campo para verificar que los puntos GPS de las quejas son correctos; crear plataformas basadas en la web que involucren a los ciudadanos en la localización de vertederos; o desarrollar un procedimiento de detección remota capaz de reconocer automáticamente los vertederos mediante aprendizaje profundo.

En cuarto lugar, el uso de datos GIS fiables sería muy adecuado en zonas rurales y escasamente pobladas, como Extremadura, ya que los recursos policiales son escasos y se necesita un mayor esfuerzo para prevenir la delincuencia. Quinto, es muy importante concienciar a la policía sobre la calidad de los datos que genera y los beneficios que puede obtener de ellos. De acuerdo con (Ratcliffe, 2004), los profesionales y académicos en los campos del análisis del delito se han dado cuenta rápidamente de los requisitos de formación y han estado orientando a sus respectivas organizaciones para llevar a cabo, en primer lugar, una correcta recopilación de datos y, posteriormente, un análisis espacial completo del delito. En este sentido, consideramos que una buena formación de los agentes de campo, el desarrollo de protocolos de recogida de información, así como la elaboración de una guía de buenas prácticas pueden ser claves para paliar este problema.

6. REFERENCIAS BIBLIOGRÁFICAS

Alfonso-Torreño, A., Gutiérrez Gallego, A., Ortiz García, J. y Arenas García, L. (2023). Measurement error of police data to identify solid waste dumping sites. *GeoFocus. International Review of Geographical Information Science and Technology, 31,* 39-53. DOI: 10.21138/GF.798.

AlZaghrini, N., Srour, F.J, y Srour, I. (2019). Using GIS and optimization to manage construction and demolition waste: The case of abandoned quarries in Lebanon. *Waste Management, 95,* 139-149. DOI: 10.1016/j.wasman.2019.06.011.

Andresen, M., Malleson, N., Steenbeek, W., Townsley, M., y Vandeviver, C. (2020). Minimum geocoding match rates: an international study of the impact of data and areal unit sizes. *International Journal of Geographical Information Science, 34,* 1-17. DOI:10.1080/13658816.2020.1725015.

Biluca, J., Rodrigues de Aguilar, C.R., y Trojan, F. (2020). Sorting of suitable areas for disposal of construction and demolition waste using GIS and ELECTRE TRI. *Waste Management, Vol. 1*(114), 307-320. DOI: 10.1016/j.wasman.2020.07.007. Epub 2020 Jul 17. PMID: 32688063.

Biotto, G., Silvestri, S., Gobbo, L., Furlan, L., Valenti, S., y Rosselli, R. (2009). GIS, multi-criteria and multi-factor spatial analysis for the probability assessment of the existence of illegal landfills. *International Journal of Geographical Information Science*, Vol. *23* (10), 1233-1244. DOI: 10.1080/13658810802112128.

Braga, A.A., y Weisburd, D.L. (2010). *Policing Problem Places: Crime Hot Spots and Effective Prevention.* University Press. DOI:10.1093/acprof:oso/9780195341966.001.001.

Braga, A.A., (2008). *Problem-oriented policing and crime prevention, 2nd edn.* Lynne Rienner Publishers, Boulder.

Brimicombe, A., Brimicombe, L., y Yang, Li. (2007). Improving Geocoding Rates in Preparation for Crime Data Analysis. *International Journal of Police Science y Management, 9.* DOI: 10.1350/ijps.2007.9.1.80.

Briz-Redón, A., Martinez-Ruiz, F., y Montes, F. (2020). Reestimating a minimum acceptable geocoding hit rate for conducting a spatial analysis. *In-*

ternational Journal of Geographical Information Science, Vol. 34(7), 1283-1305. DOI: 10.1080/13658816.2019.1703994.

Buil-Gil, D., Moretti, A., y Langton, S.H. (2021). The accuracy of crime statistics: assessing the impact of police data bias on geographic crime analysis. *J Exp Criminol.* DOI: 10.1007/s11292-021-09457-y.

Chainey, S., y Ratcliffe, J. (2013). *Crime Map Cartography.* DOI: 10.1002/9781118685181.ch12.

Chainey, S., y Ratcliffe, J. (2005). *GIS and crime mapping.* John Wiley y Sons.

De Feo, G., Cerrato, F., Siano, P. y Torretta, V. (2014). Definition of a multicriteria, web-based approach to managing the illegal dumping of solid waste in Italian villages. *Environmental Technology, Vol. 35*(1), 104-114. DOI: 10.1080/09593330.2013.816328.

Doak, M., Khan, S., Kelly, G., y Silvestri, S. (2007). The use of remote sensing to map illegal landfills at the border of Ireland/ Northern Ireland. In: Paper presented at the international waste management and landfill symposium, 1–5 October, Sardinia.

Ministerio Fiscal. (2018). Memorias de la Fiscalía General del Estado. https://www.fiscal.es [Consultado el 03.07.2023].

EGNOS Service Definition Document (2013). Ref.: EGN-SSD os v. 2.3. Open Service. European commission. Directorate-Generale for Energy and Transport.

Gerell, M. (2018). Quantifying the geographical (un) reliability of police data. *Nordisk politiforskning, 5,* 157-171, DOI: 10.18261/issn.18948693-2018-02-05.

Glanville, K., Chang, H.C. (2015). Mapping illegal domestic waste disposal potential to support waste management efforts in Queensland, Australia. *International Journal of Geographical Information Science, Vol. 29*(6), 1042-1058. DOI:10.1080/13658816.2015.1008002

Goldstein H. (1979). Improving Policing: A Problem-Oriented Approach. *Crime y Delinquency, Vol. 25*(2), 236-258. DOI:10.1177/001112877902500207.

Harada, Y., y Shimada, T. (2006). Examining the impact of the precision of address geocoding on estimated density of crime locations. *Computers y Geosciences - COMPUT GEOSCI, 32,* 1096-1107. DOI: 10.1016/j.cageo.2006.02.014.

Hart, T., y Zandbergen, P. (2012). Effects of data quality on predictive hotspot mapping. Disponible en: https://www.ojp.gov/pdffiles1/nij/grants/239861.pdf [Consultado el 03.07.2023].

Hussen, N.U., Shimelis, G., y Ahmed, M. (2021). Spatial distribution of solid waste dumping sites and associated problems in Chiro town, Oromia regional state, Ethiopia. *Environ Dev Sustain, 23,* 389–397. DOI: 10.1007/s10668-019-00585-0

Ivert, A.-K., Chrysoulakis, A., Kronkvist, K., y Torstensson-Levander, M. (2013). Malmö områdesundersökning 2012: Lokala problem, brott och trygghet [Malmö neighborhood survey 2012: Local problems, crime and fear]. Malmö: Rapport från institutionen för kriminologi, Malmö högskola.

Jakiel, M., Bernatek-Jakiel, Anita., Maciej Filiks, AG., y Pufelska, M. (2019) Spatial and temporal distribution of illegal dumping sites in the nature protected area: the Ojców National Park, Poland. *Journal of Environmental Planning and Management, Vol. 62*(2), 286-305, DOI: 10.1080/09640568.2017.1412941

Jordá-Borrell, R., Ruiz-Rodríguez, F. y Lucendo-Monedero, A.L. (2014). Factor

Analysis and Geographic Information System for Determining Probability Areas of Presence of Illegal Landfills. *Ecological Indicators, 37,* 151–160. DOI: 10.1016/j.ecolind.2013.10.001.

Lucendo-Monedero, A.L., Jordá-Borrell, R., y Ruiz-Rodríguez, F. (2015). Predictive Model for Areas with Illegal Landfills Using Logistic Regression. *Journal of Environmental Planning and Management, Vol. 58*(7), 1309–1326. DOI:10.1080/09640568.2014.993751.

Massarelli, C. (2018) Fast detection of significantly transformed areas due to illegal waste burial with a procedure applicable to Landsat images. *International Journal of Remote Sensing, Vol. 39*(3), 754-769. DOI: 10.1080/01431161.2017.1390272.

Matos, J., Oštir, K., y Kranjc, J. (2012). Attractiveness of roads for illegal dumping with regard to regional differences in Slovenia. *Acta Geogr. Slovenica, 52,* 431–451. DOI: 10.3986/AGS52207.

Matsumoto, S., y Takeuchi, K. (2011). The Effect of Community Characteristics on the Frequency of Illegal Dumping. *Environmental Economics and Policy Studies, Vol. 13*(3), 177–193. DOI:10.1007/s10018-011-0011-5.

Mazeika, D., y Summerton, D. (2017). The impact of geocoding method on the positional accuracy of residential burglaries reported to police. *Policing: An International Journal of Police Strategies y Management, Vol. 40*(2), 459-470. DOI:10.1108/PIJPSM-03-2016-0048.

Ministerio del Interior (2021). Anuarios y estadísticas. Available on: https://www.interior.gob.es/opencms/es/archivos-y-documentacion/documentacion-y-publicaciones/anuarios-y-estadisticas/ [Consultado el 03.07.2023].

Newton, A. Partridge, H., y Gill, A. (2014). Above and below: Measuring crime risk in and around underground mass transit systems. *Crime Science, Vol.3* (1).

DOI: 10.1186/2193-7680-3-1.

Paz, DHFD., Lafayette, KPV., y Sobral, MDC. (2018). GIS-based planning system for managing the flow of construction and demolition waste in Brazil. *Waste Management, Vol. 36*(6), 541-549. DOI: 10.1177/0734242X18772096. Epub 2018 May 18. PMID: 29776320.

Ratcliffe, J. (2001). On the accuracy of TIGER-type geocoded address data in relation to cadastral and census areal units. *International Journal of Geographical Information Science, Vol. 15*(5), 473–486,

DOI: 10.1080/13658810110047221.

Ratcliffe, J. (2004) Geocoding crime and a first estimate of a minimum acceptable hit rate. *International Journal of Geographical Information Science, Vol. 18*(1), 61-72, DOI: 10.1080/13658810310001596076.

Seeboonruang, U. (2016). Geographic information system–based impact assessment for illegal dumping in borrow pits in Chachoengsao Province, Thailand. *The Geological Society of America, Special paper, 520,* 393-406. DOI:10.1130/2016.2520(35).

Seror, N., y Portnov, B.A. (2018). Identifying areas under potential risk of illegal construction and demolition waste dumping using GIS tools. *Waste Management, 75,* 22-29. DOI: 10.1016/j.wasman.2018.01.027.

Serrano-Gómez, A. (2011). Dudosa fiabilidad de las estadísticas policiales sobre criminalidad en España. *Revista de Derecho Penal y Criminología, 6,* 425-454.

Silvestri, S., y Omri, M. (2008). A method for the remote sensing identification of uncontrolled landfills: formulation and validation. *International Journal of Remote Sensing, Vol. 29* (4), 975-989. DOI: 10.1080/01431160701311317.

Vijay, R., Naik, K., Kushwaha, V., Gupta, I., y Kumar, R. (2018). Evaluation of pixel and object based image analysis for land use, land cover classification in the hilly region of Manali, India, Forest, Climate Change and Biodiversity. Klayani Publishers, Ludhiana, 3-13. ISBN 9789327289947.

Wang, F. (2012). Why police and policing need GIS: an overview. *Annals of GIS, Vol. 18*(3), 159-171. DOI: 10.1080/19475683.2012.691900.

Weisburd, D. (2015). The law of crime concentration and the criminology of place. *Criminology, Vol. 53* (2), 133-157. DOI: 10.1111/1745-9125.12070.

Zainun, N.Y., Rahman, I.A. y Rothman, R.A. (2016). Mapping Of Construction Waste Illegal Dumping Using Geographical Information System (GIS), International Engineering Research and Innovation Symposium (IRIS), IOP Conf. Series: Materials Science and Engineering 160 (2016) 012049.

Capítulo 4

Análisis y caracterización de los focos de residuos

ALBERTO ALFONSO-TORREÑO[1]
JOSÉ ANTONIO GUTIÉRREZ GALLEGO[2]

1. INTRODUCCIÓN

Los vertederos ilegales representan una amenaza mundial por la gestión ineficaz de los residuos en todo el mundo (Karak et al., 2012; Seeboonruang, 2016). La gestión sostenible de los residuos supone un reto para la sociedad por el gran volumen de desechos generados y por los peligros asociados al medio ambiente y a la salud humana (Silvestri y Omri, 2008; Quesada-Ruiz et al., 2018). La presencia de vertederos provoca diversos impactos ambientales como el deterioro de los paisajes locales (Matsumoto y Takeuchi, 2011) el aumento de la contaminación del aire (Bridges et al., 2000) y la contaminación de acuíferos (Hafeez et al., 2016).

Los materiales arrojados de manera ilegal a menudo contienen residuos de construcción y demolición (en adelante RCD), por ejemplo, paneles de yeso, tejas, madera, ladrillos, hormigón, electrodomésticos, neumáticos y basura doméstica (Seeboonruang, 2016). Las principales razones que favorecen el vertido ile-

1 Instituto de Investigación para el Desarrollo Territorial Sostenible (INTERRA), Universidad de Extremadura.

2 Instituto de Investigación para el Desarrollo Territorial Sostenible (INTERRA), Universidad de Extremadura.

gal por RCD incluyen la escasez de instalaciones de tratamiento de residuos, largas distancias y altas tarifas para transportar los residuos hasta los vertederos autorizados, desconocimiento en las consecuencias legales de no transportar los residuos a instalaciones habilitadas para su reciclado y falta de conciencia pública (Ichinose y Yamamoto, 2011; Matos et al., 2012). Algunos propietarios privados tienen la percepción equivocada de que pueden hacer lo que quieran en sus propiedades privadas (Seror y Portnov, 2018). Por tanto, la concienciación sobre los problemas ambientales causados por los escombros son condiciones previas necesarias para combatir el fenómeno de los vertederos ilegales (Biotto et al., 2009).

La existencia de vertederos ilegales sigue siendo un grave e incontrolado problema en países como Rumania (Apostol y Mihai, 2012), Serbia (Zelenović Vasiljević et al., 2012), Francia (Biotto et al., 2009), Austria (Allgaier y Stegmann, 2006), Italia (Silvestri y Omri, 2008) y España (Jordá-Borrell et al., 2014). Además, el interés por este fenómeno ambiental lo demuestra la gran cantidad de estudios científicos disponibles sobre vertederos ilegales en países de Asia (Moh y Abd Manaf, 2017), América del Norte (Jones, 2008), Australia (Glanville y Chang, 2015) y África (Hussen et al., 2021).

Los Sistemas de Información Geográfica (SIG) a través de herramientas geoestadísticas de cartografiado y modelado de fenómenos ambientales pueden ser de utilidad para identificar vertederos ilegales de RCD (Pelillo et al., 2014). Este tipo de herramientas permiten demostrar que los vertederos ilegales no se distribuyen aleatoriamente y que presentan un determinado patrón espacial (Biotto et al., 2009). Sin embargo, según Jakiel et al. (2019), los estudios sobre la distribución espacial de los vertederos ilegales son limitados. La localización exacta de escombreras de RCD suele ser una tarea ardua, principalmente en el medio rural, debido al gran tamaño de las áreas a georreferenciar y a la escasez de recursos humanos y financieros por parte de los organismos legislativos medioambientales (Seror y Portnov, 2018).

Este es precisamente el caso de Extremadura (España), región en la que se han desarrollado diferentes políticas para la prevención y gestión de este problema en los últimos años. A pesar de estos esfuerzos, se desconoce la cifra real de vertederos ilegales en esta comunidad autónoma. El presente estudio consiste en el análisis del comportamiento espacial de RCD ilegales localizados en Extremadura. En primer lugar, se identificó 244 residuos ilegales a través de fotografías aéreas y posteriormente se relacionó con variables topográficas, de accesibilidad y de usos y manejos del suelo. La comunidad autónoma de Extremadura situada al suroeste de Europa tiene una superficie de 41.633 km2 (el 8,3% de España) y una población de 1.051.738 habitantes (INE, 2022). De los 388 municipios, tan sólo 13 tienen una población superior a los 10.000 habitantes. Los diez municipios más poblados concentran el 46% de la población extremeña. Se trata de una región eminentemente rural, siendo una de las regiones más atrasadas de España y de Europa, con 19.072 € per cápita.

2. OBJETIVOS

El objetivo principal de este estudio es caracterizar las zonas donde se encuentran los residuos ilegales de construcción y demolición en una región periférica del suroeste europeo, así como la dimensión de estos.

Para alcanzar el objetivo propuesto se han desarrollado una serie de objetivos específicos: definir el tamaño tipo de los residuos, determinar las zonas con mayor probabilidad de presencia de RCD en función de la proximidad a determinados elementos geográficos (centros urbanos, vías de comunicación, hidrografía, etc.) y especificar la relación entre la topografía del terreno con la presencia de este tipo de residuos.

3. MATERIAL Y MÉTODOS

Los datos de partida tomados para iniciar el presente trabajo fueron proporcionados por el Servicio de Protección de la Naturaleza de la Guardia Civil (SEPRONA). Esta información estaba contenida en un libro de Excel que contenía las 6480 infracciones impuestas por el mencionado servicio desde el año 2016 a 2021. De cada infracción se tenía el año, el ámbito del hecho, el tipo de hecho, la provincia y la localización del hecho, definida a través de sus coordenadas de longitud y latitud.

A través de un estudio de campo se comprobó la fiabilidad de la información aportada por el SEPRONA, comprobando que había importantes errores en la localización de los hechos. Por tal motivo, se procedió a la identificación de las zonas de residuos a través de las imágenes del plan nacional de ortofotografías del Instituto Geográfico Nacional para el año 2019. Al mismo tiempo, a través de una nueva campaña de campo y la incorporación de otras fuentes, como la facilitada por la Asociación para la Defensa de la Naturaleza se identificaron un total de 244 de RCD.

Para la caracterización y el estudio de la probabilidad de presencia de RCD se elaboró una base de datos que contenía información sobre diversas variables topográficas y de usos y manejos del suelo, y las distancias entre los residuos ilegales identificados y diferentes elementos geográficos (Figura 1).

La altitud, pendiente y posición topográfica de cada residuo se obtuvo a partir del modelo digital del terreno del Instituto Geográfico Nacional (CNIG, 2010), con una resolución espacial de 5 m. La distancia a núcleos urbanos, vías de comunicación y red hidrográfica se estimó a través de la Base Cartográfica Nacional 1:200.000. Finalmente, los tipos de usos y manejos del suelo donde se ubican los RCD se identificó a partir del mapa de ocupación del suelo en España en 2018 (escala 1:100.000) correspondiente al proyecto europeo *Corine Land Cover*.

Los procedimientos y cálculos para estimar las diferentes variables topográficas, de accesibilidad y de tipos de usos del suelo se llevaron a cabo utilizando QGIS 3.22.

Figura 1

Variables topográficas, de accesibilidad y de tipos de usos del suelo seleccionadas para relacionar con los vertederos ilegales identificados.

Los RCD analizados se encuentran principalmente en la provincia de Badajoz. Concretamente, en la provincia pacense se han analizado 191 zonas de residuos, frente a las 53 zonas de Cáceres, es decir, la provincia de Badajoz tiene el 78% del total de zonas estudiadas y Cáceres solo el 22%. Las zonas de residuos analizadas se encuentran distribuidas en 43 municipios, 24 de la provincia de Badajoz y 19 de la provincia de Cáceres. Esta distribución más o menos equitativa desaparece si se consideran las zonas analizadas, como se ha indicado en el párrafo anterior. Para facilitar la comprensión de la distribución municipal de los residuos, se ha elaborado un histograma de los residuos en función de los municipios donde hay más de dos zonas analizadas (Figura 2). La capital pacense destaca en el conjunto de zonas consideradas, seguida de la capital autonómica (Mérida), estas dos capitales concentran el 57 % del total. Destaca en menor medida Cáceres con 15 zonas, así como Losar de la Vera que a pesar de ser un municipio turístico y pequeño tiene 9 zonas captadas.

Figura 2

Municipios con presencia de más de tres zonas de residuos de construcción y demolición.

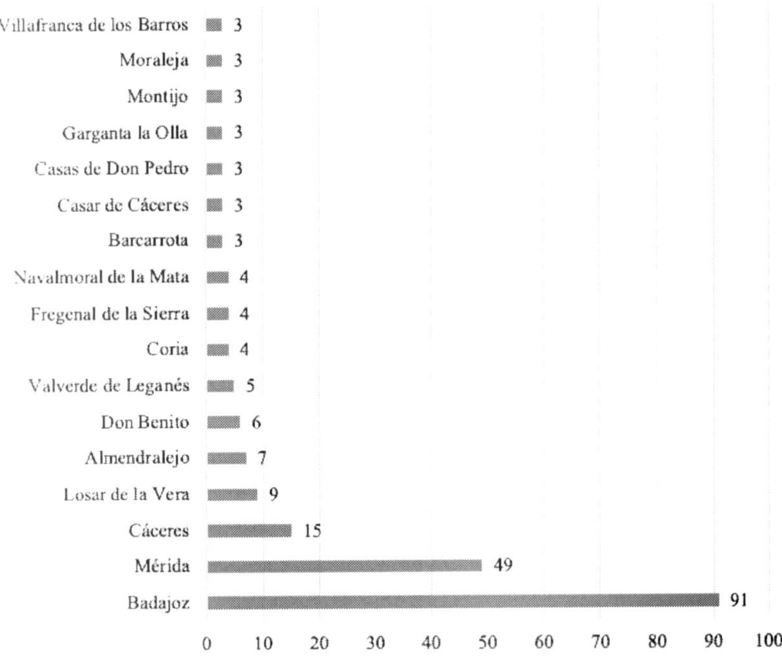

4. RESULTADOS

Un aspecto interesante en la caracterización de los residuos es su dimensión, pues esta variable puede condicionar los sistemas de detección de estos. En la figura 3 se muestra el histograma de frecuencias de la distribución de las zonas de RCD en función de su tamaño. En la figura se muestra el número de residuos en función de su dimensión en metros cuadrados. El 12,7% de los residuos analizados tiene menos de 10 metros cuadrados, esta dimensión es difícil de detectar si tenemos en cuenta la resolución de las ortofotografías empleadas en el estudio. El 80% de los residuos localizados tienen menos de 100 metros cuadrados, por tanto, se desaconseja el uso de imágenes espaciales con una resolución inferior a los 25 centímetros.

Figura 3

Histograma del tamaño de los residuos de construcción y demolición.

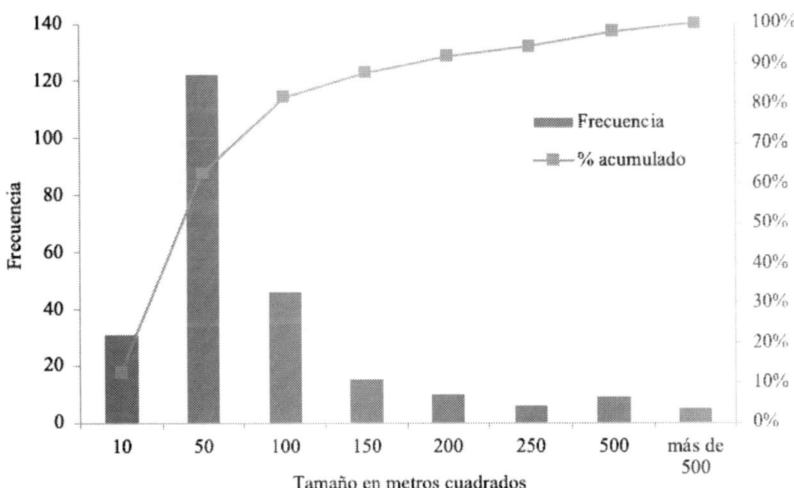

De las 244 zonas analizadas 63 se encuentran dentro del casco urbano, lo que supone un 18% del total de residuos (Figura 4).

La proximidad a los núcleos de población es muy relevante en la localización de los RCD, pues la mediana se sitúa en 186 metros, además más del 75% de los residuos se encuentran a menos de 500 metros. Porcentajes similares obtuvieron Jordá-Borrell et al. (2014) a través de un estudio geoestadístico para determinar áreas en las que existe probabilidad de presencia de vertederos ilegales, donde destacaron la proximidad de los vertederos ilegales a los núcleos urbanos. Menos del 8% de las localizaciones analizadas están a más de un kilómetro y medio.

Figura 4

Histograma de la distancia de los residuos de construcción y demolición a los núcleos de población.

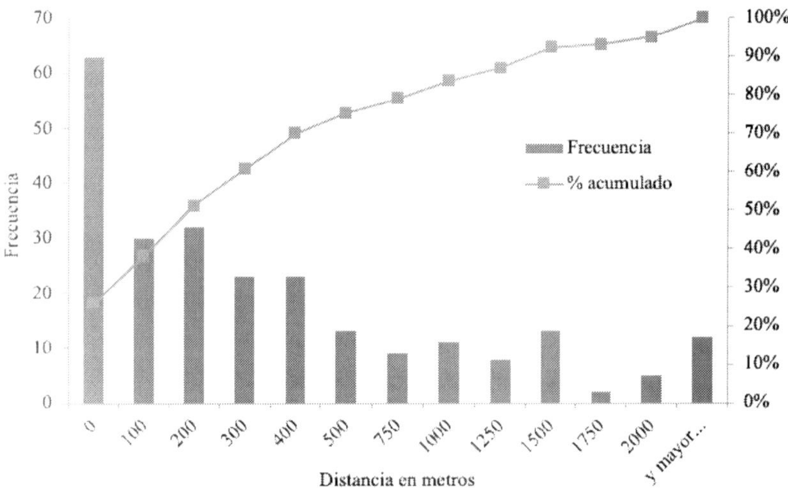

Sin embargo, la localización de los RCD con respecto a la red hidrográfica no es especialmente relevante. El 20% de los residuos se encuentran en torno al kilómetro y medio de la red hidrográfica. Como puede verse en la figura 5 hay una gran dispersión, así la desviación estándar supera 700 metros. Sin embargo, hay estudios que demuestran que más de un 75% de los vertederos ilegales se ubican a menos de 1 km de una vía fluvial (Jordá-Borrell et al., 2014).

Figura 5

Histograma de la distancia de los residuos de construcción y demolición a la red hidrográfica.

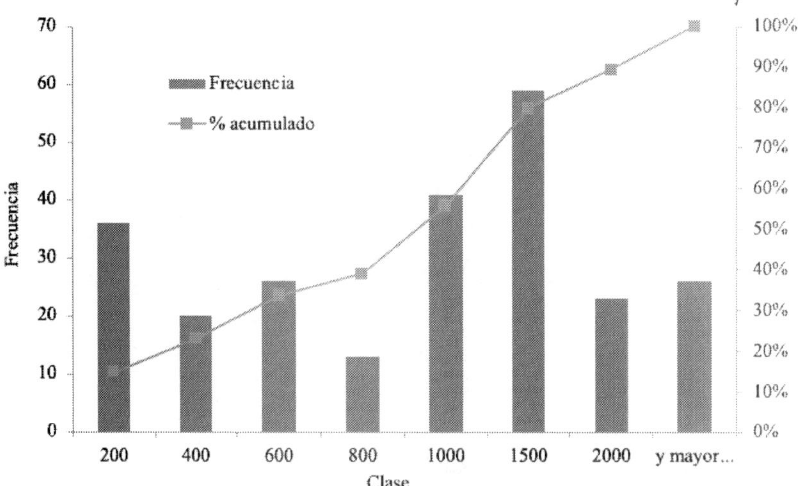

En cuanto a la distancia a las vías de comunicación, sólo se han considerado las zonas que no están dentro del casco urbano para no desvirtuar el resultado final. Por tanto, solo se han tenido en cuenta 181 localizaciones de los 244 totales. Como sucedía con la proximidad a los cascos urbanos, dos terceras partes de las zonas analizadas se encuentran a menos de medio kilómetro de las vías de comunicación (Figura 6). La mediana se sitúa en los 118 metros y la desviación estándar por debajo de los 300 metros. La mayoría de los vertederos ilegales son visibles a menos de 500 m de cualquier vía de comunicación, principalmente de caminos rurales con fácil acceso (Matos et al., 2012; Jordá-Borrell et al., 2014).

Figura 6

Histograma de la distancia de los residuos de construcción y demolición (fuera de los núcleos urbanos) a las vías de comunicación.

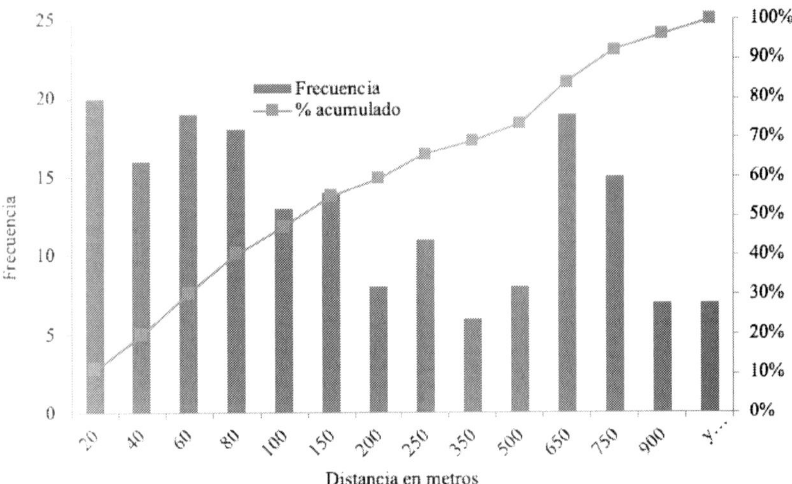

Las vías donde se localizan los residuos analizados se encuentran en las inmediaciones de las carreteras o caminos locales, como puede verse en la figura 7. El 50% de los residuos contemplados tienen como vía de comunicación más próxima una vía local, esta tipología de comunicación vial se corresponde con los caminos y carreteras vecinales, generalmente poco transitadas.

Figura 7

Vía de comunicación más cercana las zonas de residuos de construcción y demolición.

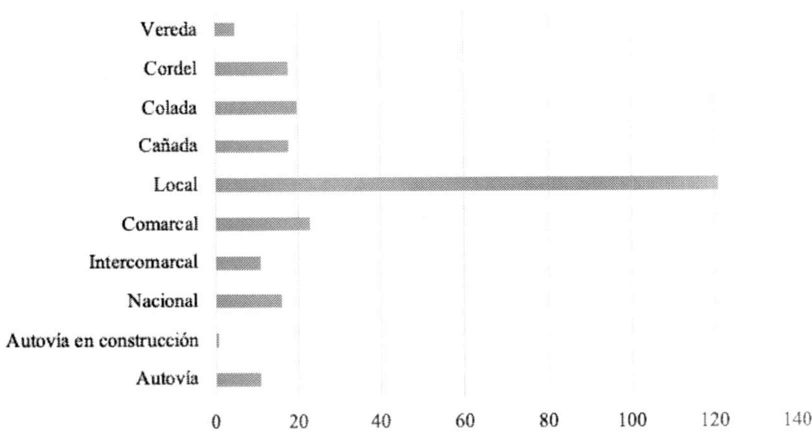

En cuanto a los usos del suelo, se debe destacar que las zonas agrícolas heterogéneas y las tierras de labor concentran más del 60 % de los residuos de RCD analizados, 140 zonas de las de las 244 están en estos dos tipos de usos de suelo (Figura 8). Nuestros resultados coinciden con los resultados obtenidos por Seror y Portnov (2018), que revelaron que la presencia de vertederos ilegales estaba significativamente asociada con áreas agrícolas y abiertas. En un segundo nivel se encuentran los espacios de vegetación arbustiva y/o herbácea y el tejido urbano que contienen el 28 % del total de las zonas analizadas.

Figura 8

Usos de suelos donde hay residuos de construcción y demolición.

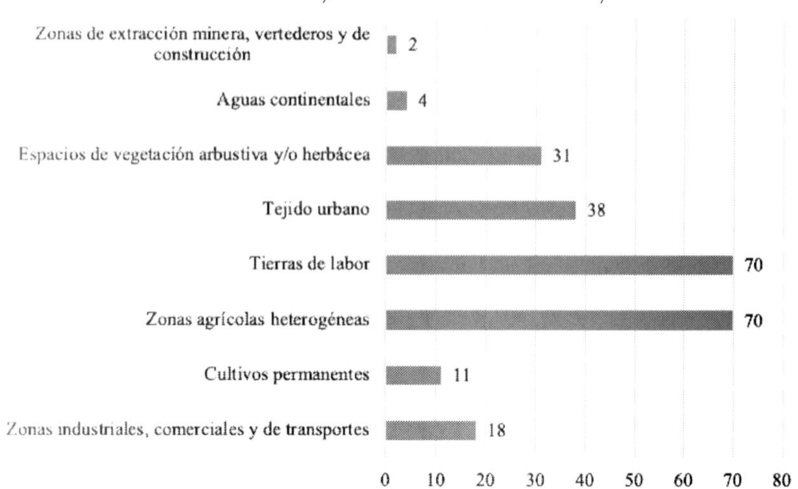

Atendiendo a la posición topográfica, destacan las zonas de vaguada y de media ladera contienen aproximadamente el 90% de las ubicaciones de residuos de RCD, con 127 y 89 de las localizaciones estudiadas respectivamente (Figura 9). Las zonas culminantes solo contienen el 11%, como cabría esperar por ser ubicaciones de mayor visibilidad. Desde un punto de vista geomorfológico, los vertederos ilegales se localizan sobre topografías onduladas y terrenos planos que faciliten el acceso a este tipo de ubicaciones (Jordá-Borrel et al., 2014).

Figura 9

Orografía donde hay residuos de construcción y demolición.

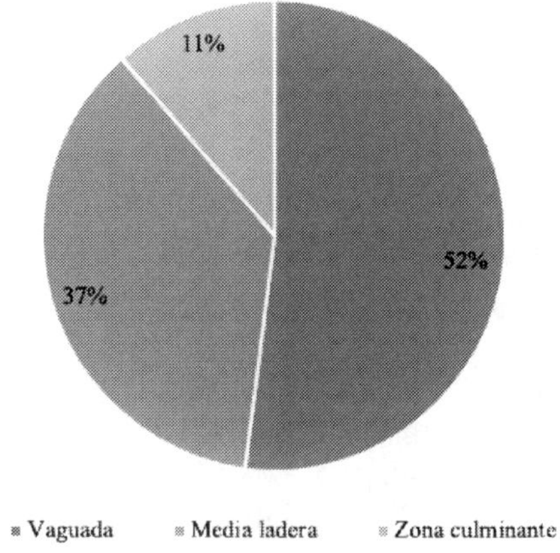

La localización de las zonas de RCD en las proximidades de los núcleos urbanos condicionan su altitud. Así, el 80% de los puntos de residuos analizados se encuentran por debajo de los 350 metros de altitud (Figura 10), siendo el valor predominante de altitud de las zonas de residuos los espacios que se encuentran entre los 200 y los 250 metros, cotas típicas de las zonas de vega.

Figura 10

Altitud de la ubicación de los residuos de construcción y demolición.

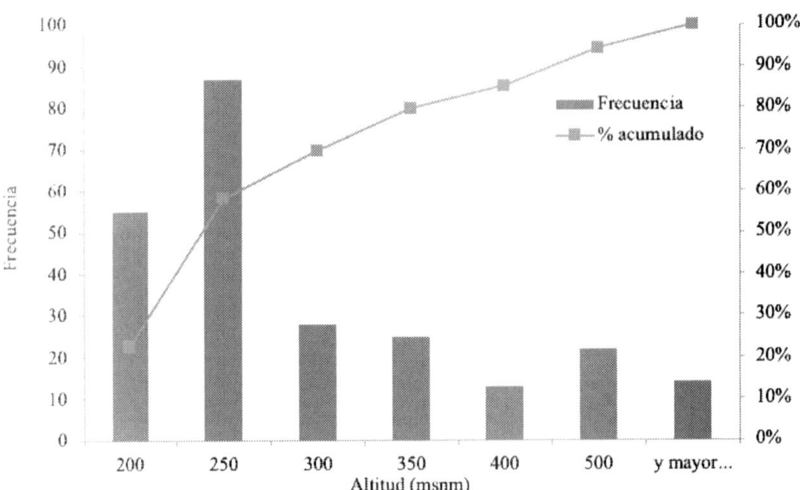

Por último, es necesario destacar que la pendiente no supone un elemento concluyente en cuanto a la localización de las zonas con presencia de RCD analizados. Como puede verse en la figura 11 solo destaca tímidamente la presencia de los RCD en terrenos con pendientes inferiores al 1%, cuestión esperada al abordar esta variable. Sin embargo, el resto de los valores tiene una distribución muy homogénea, lo que impide definir ningún límite efectivo, así como una clara prelación por la pendiente elegida.

Figura 11

Pendiente de la ubicación de los residuos de construcción y demolición.

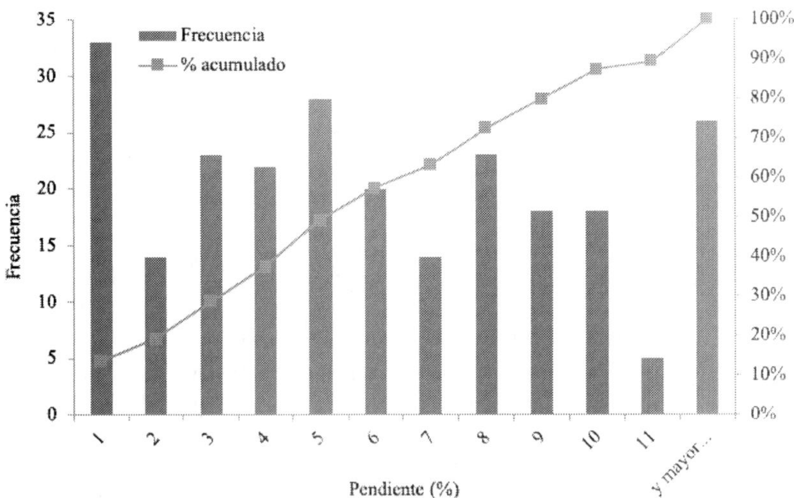

5. CONCLUSIONES

En este estudio, se caracterizan las zonas donde se ubican los RCD ilegales en Extremadura, y la dimensión de estos. Además, se analiza la probabilidad de presencia de residuos a partir de variables topográficas, de accesibilidad y de tipos de usos del suelo obtenidas mediante fuentes cartográficas del Instituto Geográfico Nacional. Se identificaron 244 RCD mediante una base de datos proporcionada por el SEPRONA y por la Asociación para la Defensa de la Naturaleza. Además, se llevó a cabo trabajos de campo para comprobar *in situ* la veracidad de las geolocalizaciones ya que el 80% de los residuos tenían menos de 100 m2.

De acuerdo con el análisis cartográfico desarrollado, los siguientes resultados determinan áreas con riesgo elevado de RCD:

1. El 75% de los residuos se localizan a menos de 500 m de un núcleo urbano, por tanto, la proximidad a los núcleos urbanos es muy relevante.
2. La localización de los RCD con respecto a la red hidrográfica no es especialmente determinante.
3. La mayoría de los residuos ilegales se sitúan a menos de 500 m de las vías de comunicación, principalmente de caminos rurales con fácil acceso y poco transitados.
4. Las zonas agrícolas heterogéneas y las tierras de labor concentran más del 60 % de los RCD analizados.
5. El 80% de las ubicaciones con residuos se encuentran por debajo de los 350 msnm, cotas pertenecientes a las zonas de las Vegas del Guadiana. Comarca con mayor densidad de población y con gran actividad agroindustrial.
6. Desde un punto de vista geomorfológico, los vertederos ilegales se localizan principalmente sobre topografías onduladas y terrenos planos que faciliten su acceso. Las zonas culminantes o cumbres no son áreas propensas a acumular residuos por ser ubicaciones de mayor visibilidad.

La aplicación de esta metodología basada en la identificación potencial de residuos ilegales a partir de elementos geográficos ha brindado resultados satisfactorios a nivel regional, convirtiendo a los SIG en una excelente herramienta para el estudio del comportamiento espacial de los residuos ilegales. El enfoque propuesto de analizar las áreas con mayor probabilidad de presencia de residuos ilegales puede utilizarse como punto de partida para identificar puntos calientes en grandes superficies y posteriormente desarrollar modelos basados en la clasificación automática de imágenes mediante técnicas de teledetección y *Deep Learning*.

6. REFERENCIAS BIBLIOGRÁFICAS

Allgaier, G., y Stegmann, R. (2006). Preliminary assessment of old landfills. Proceedings of Seminario-Workshop Tecnologie per la riduzione degli impatti e la bonifica delle discariche, Montegrotto Terme, Padova, Italia June.

Apostol, L., y Mihai, F.-C. (2012). Rural waste management: challenges and issues in Romania. *Present Environment and Sustainable Development* (2), 105-114. DOI:10.5281/zenodo.19121.

Biotto, G., Silvestri, S., Gobbo, L., Furlan, E., Valenti, S., y Rosselli, R. (2009). GIS, multi-criteria and multi-factor spatial analysis for the probability assessment of the existence of illegal landfills. *International Journal of Geographical Information Science, Vol. 23*(10), 1233-1244. DOI: 10.1080/13658810802112128.

Bridges, O., Bridges, J. W., y Potter, J. F. (2000). A generic comparison of the airborne risks to human health from landfill and incinerator disposal of municipal solid waste. *Environmentalist, 20,* 325-334. DOI: 10.1023/A:1006725932558.

Glanville, K., y Chang, H.-C. (2015). Mapping illegal domestic waste disposal potential to support waste management efforts in Queensland, Australia. *International Journal of Geographical Information Science, Vol. 29*(6), 1042-1058. DOI: 10.1080/13658816.2015.1008002.

Hafeez, S., Mahmood, A., Syed, J. H., Li, J., Ali, U., Malik, R. N., y Zhang, G. (2016). Waste dumping sites as a potential source of POPs and associated health risks in perspective of current waste management practices in Lahore city, Pakistan. *Science of the Total Environment, 562,* 953-961. DOI: 10.1016/j.scitotenv.2016.01.120.

Hussen, N. U., Shimelis, G., y Ahmed, M. (2021). Spatial distribution of solid waste dumping sites and associated problems in Chiro town, Oromia regional state, Ethiopia. *Environment, Development and Sustainability, Vol. 23*(1), 389-397. DOI: 10.1007/s10668-019-00585-0.

Ichinose, D., y Yamamoto, M. (2011). On the relationship between the provision of waste management service and illegal dumping. *Resource and energy economics, Vol. 33*(1), 79-93. DOI: 10.1016/j.reseneeco.2010.01.002.

Jakiel, M., Bernatek-Jakiel, A., Gajda, A., Filiks, M., y Pufelska, M. (2019). Spatial and temporal distribution of illegal dumping sites in the nature protected area: The Ojców National Park, Poland. *Journal of Environmental Planning and Management, Vol. 62*(2), 286-305. DOI: 10.1080/09640568.2017.1412941.

Jones, BJ. (2008). *The Geography of Open Dumps in Rural Appalachia.* Theses, Dissertations and Capstones. 128. https://mds.marshall.edu/etd/128

Jordá-Borrell, R., Ruiz-Rodríguez, F., y Lucendo-Monedero, Á. L. (2014). Factor analysis and geographic information system for determining probability areas of presence of illegal landfills. *Ecological Indicators, 37,* 151-160. DOI: 10.1016/j.ecolind.2013.10.001.

Karak, T., Bhagat, R., y Bhattacharyya, P. (2012). Municipal solid waste generation, composition, and management: the world scenario. *Critical Reviews in Environmental Science and Technology, Vol. 42*(15), 1509-1630. DOI: 10.1080/10643389.2011.569871.

Matos, J., Oštir, K., y Kranjc, J. (2012). Attractiveness of roads for illegal dumping with regard to regional differences in Slovenia. *Acta geographica Slovenica, Vol. 52*(2), 431–451-431–451. DOI: 10.3986/AGS52207.

Matsumoto, S., y Takeuchi, K. (2011). The effect of community characteristics on the frequency of illegal dumping. *Environmental Economics and Policy Studies, 13,* 177-193. DOI: 10.1007/s10018-011-0011-5.

Moh, Y., y Abd Manaf, L. (2017). Solid waste management transformation and future challenges of source separation and recycling practice in Malaysia. *Resources, Conservation and Recycling, 116,* 1-14. DOI: 10.1016/j.resconrec.2016.09.012 .

Pelillo, V., Piper, L., Lay-Ekuakille, A., Lanzolla, A., Andria, G., y Morello, R. (2014). Geostatistical approach for validating contaminated soil measurements. *Measurement, 47,* 1016-1023. DOI: 10.1016/j.measurement.2013.09.016.

Quesada-Ruiz, L., Rodriguez-Galiano, V., y Jordá-Borrell, R. (2018). Identifying the main physical and socioeconomic drivers of illegal landfills in the Canary Islands. *Waste management y research, Vol. 36*(11), 1049-1060. DOI: 10.1177/0734242X18804031.

Seeboonruang, U. (2016). Geographic information system–based impact assessment for illegal dumping in borrow pits in Chachoengsao Province, Thailand. *Geological Society of America Special Papers, 520,* 393-405. DOI: 10.1080/01431160701311317.

Seror, N., y Portnov, B. A. (2018). Identifying areas under potential risk of illegal construction and demolition waste dumping using GIS tools. *Waste Management, 75,* 22-29. DOI: 10.1016/j.wasman.2018.01.027.

Silvestri, S., y Omri, M. (2008). A method for the remote sensing identification of uncontrolled landfills: formulation and validation. *International journal of remote sensing, Vol. 29*(4), 975-989. DOI: 10.1080/01431160701311317.

Zelenović Vasiljević, T., Srdjević, Z., Bajčetić, R., y Vojinović Miloradov, M. (2012). GIS and the analytic hierarchy process for regional landfill site selection in transitional countries: a case study from Serbia. *Environmental management, 49*, 445-458. DOI: 10.1007/s00267-011-9792-3.

Capítulo 5

Identificando características socioeconómicas de municipios con vertederos ilegales

JORDI ORTIZ GARCÍA[1]

1. INTRODUCCIÓN

Como viene apuntándose en toda la obra, existe una mayor preocupación por parte de operadores políticos y agencias medioambientales del impacto que distintas problemáticas causan a nuestro ecosistema y a la salud humana (Ministerio de Transición Ecológica y Reto Demográfico, 2022)[2]. Entre los principales problemas medioambientales que viene padeciendo España en estos últimos años están los vertidos ilegales de residuos (Rio et al., 2010), un problema que se ha identificado gracias al elevadísimo número de sanciones interpuestas desde las instituciones europeas a nuestro país, por la falta de control y vigilancia de este tipo de comportamientos (El País, 2022). Este escenario ha provocado que gobiernos y administraciones muestren una ma-

---

1 Profesor Contratado Doctor, Área de Derecho penal, Universidad de Extremadura.

2 Desde el año 2018, el Gobierno de España cuenta con un Ministerio de Transición Ecológica y el Reto Demográfico. Sin duda, se trata de una medida con el objetivo de mejorar las políticas en materia de lucha contra el cambio climático, protección de la biodiversidad o del patrimonio natural. Disponible en: https://www.miteco.gob.es/es/ [Consultado el 03.07.2023].

yor concienciación sobre este problema ambiental, obligándose a buscar estrategias para mitigar o prevenir estos comportamientos (PIREX, 2016; Ecologistas en Acción, 2020[3]; Ministerio de Sanidad, 2022[4]; Plan Europeo de Economía Circular, 2020[5]).

Junto a las respuestas gubernamentales, el temor sobre el impacto y las consecuencias que tienen los vertidos ilegales en los seres humanos ha motivado a personal investigador de distintos campos de conocimiento a diseñar metodologías y estrategias eficaces para minimizar las consecuencias que tienen estas conductas ilícitas y peligrosas sobre la comunidad. Una problemática que tiene su amenaza en la proximidad a los municipios y las consecuencias que provoca sobre la salud humana y el medio ambiente (Puerta Ortiz, 2019).

A nivel internacional, la preocupación por el impacto ambiental de los vertidos ilegales es muy similar al nuestro. El miedo a las consecuencias provocadas por los vertidos ilegales en áreas urbanas es la principal preocupación para la inmensa mayoría de los países. Y de entre todos ellos, los vertidos de residuos de construcción y demolición (en adelante, RCD), fruto del auge en la construcción, son los que tienen mayor interés a nivel mundial. Así, podemos encontrar una importante literatura en países como México, Colombia o Alemania, lugares en los que se analizan tanto las políticas de gestión de estos residuos hasta sus efectos en

---

3 También encontramos organizaciones no gubernamentales como Ecologistas en Acción, que abordan cuestiones relativas a los residuos. A modo de ejemplo, véase: https://www.ecologistasenaccion.org/areas-de-accion/residuos/ [Consultado el 03.07.2023].

4 En la web del Ministerio de Sanidad podemos encontrar varios documentos para la gestión de residuos. A modo de ejemplo, encontramos la gestión de residuos tras la pandemia. Disponible en: https://www.miteco.gob.es/es/ministerio/medidas-covid19/residuos/ [Consultado el 03.07.2023].

5 La Comisión Europea propuso en el mes de marzo de 2022, las primeras medidas para la transición a una economía circular. Disponible en: https://eur-lex.europa.eu/legal-content/EN/TXT/?qid=1583933814386yuri=COM:2020:98:FIN [Consultado el 03.07.2023].

las áreas más pobladas (Gómez et al., 2008; Sánchez Núñez et al, 2009; Marmolejo de Oro, 2012; Antonini et al., 2015; Chamolí et al., 2016; Nelles et al., 2016; Jianguo et al., 2018).

Por todo ello, la finalidad de este trabajo es abordar la problemática de los vertidos ilegales de RCD e identificar elementos comunes en aquellos municipios en los que se han encontrado estos tipos de vertidos. Concretamente, este trabajo analiza las características sociales, económicas y de gestión ambiental de los municipios de la Comunidad Autónoma de Extremadura en los que se interpusieron sanciones administrativas por vertidos ilegales de RCD entre 2016 a 2021 (ambos inclusive). La finalidad de esta investigación tiene el propósito de complementar a los trabajos presentados en esta obra, y que analizan eventos geográficos de las áreas dónde se han localizado vertidos ilegales de RCD en Extremadura, como la distancia o tipos de suelo. Cabe destacar que esta investigación examina la totalidad de los municipios en los que se interpuso al menos una sanción administrativa por vertidos ilegales de RCD.

Las características socioeconómicas y de gestión medioambiental que hemos analizado en este trabajo, han sido seleccionadas a partir de un análisis previo de revisión bibliografía nacional e internacional sobre los factores no espaciales para la elección de áreas de vertederos ilegales por parte de la ciudadanía en los municipios (Jordá - Borrell et al, 2013).

La información de las variables analizadas ha sido obtenida de fuentes oficiales como el Instituto Nacional de Estadística (INE), Junta de Extremadura, Guardia Civil, Ministerio para la Transición Ecológica y el Reto Demográfico o Ministerio de Hacienda.

Para la investigación hemos utilizado una metodología descriptiva-estadística, que busca describir y resumir la información de las denuncias por vertidos de RCD. Las herramientas metodológicas seleccionadas para lograr nuestros objetivos han sido el software estadístico IBM SPSS (v. 27)[6], un instrumento empleado de manera frecuente en ciencias sociales y que hemos utilizado en este trabajo para el análisis de los correlatos entre las denuncias administrativas por vertidos ilegales de RCD por parte de la Guardia Civil en la Comunidad Autónoma de Extremadura (VD) y las características locales seleccionadas a partir de estudios previos (VI). Para la segunda parte de este trabajo, hemos utilizado el software geoestadístico Qgis (versión 3.30)[7], un programa que nos ha permitido diseñar y representar los mapas temáticos que aparecen en este trabajo. Los mapas nos permiten tener una imagen más nítida de la distribución geográfica de estos vertidos ilegales en los municipios extremeños. En todo caso, todas estas cuestiones se desarrollan en profundidad en el apartado metodológico.

Antes de presentar los resultados del estudio, abordaremos la fase de conceptualización y metodología empleada para esta investigación.

2. MARCO TEÓRICO

El cambio climático o la lucha frente comportamientos delictivos que dañan nuestra flora y fauna, son algunos de las problemáticas que parecen despertar un mayor interés entre la ciudadanía y líderes políticos[8]. Desgraciadamente, un desastre más silencioso golpea a la salud humana y el medio ambiente: los vertidos ilega-

6 La Universidad de Extremadura cuenta con la licencia del programa informático *Statistical Package for the Social Sciences* (SPSS) para uso del personal investigador.

7 El uso del software Qgis es un sistema de información geográfica libre y de código abierto. Se trata de una herramienta que nos permite visualizar, analizar o datos diseñar mapas. Disponible en: https://www.qgis.org/es/site/ [Consultado el 03.07.2023].

8 A modo de ejemplo, en nuestro país se puede comprobar el trabajo que desarrolla el SEPRONA de la Guardia Civil en la protección de medio ambiente. Disponible: https://www.mapa.gob.es/es/actuaciones-seprona/default.aspx [Consultado el 03.07.2023].

les. La liberación de determinadas sustancias tóxicas en la calidad del aire o la contaminación del suelo y el agua, son algunas de los problemas que generan estos vertidos, suscitando un mayor interés en el campo de la investigación (Triassi et al., 2015).

Además, el impacto en la salud de los seres humanos (Ministerio de Sanidad, 2022)[9], su cumplimiento normativo o las políticas públicas de control y vigilancia de estos comportamientos, son otros de los asuntos que desde el mundo académico y científico se viene analizando en estos últimos años. Unas investigaciones que tienen como finalidad minimizar los costes individuales y colectivos de esta problemática (Guardia Civil, 2023) tienen en los sistemas de información geográfica su principal herramienta metodológica para explicar el impacto de estos vertidos ilegales y desarrollar modelos que identifiquen áreas con mayores probabilidades de presencia de estos vertidos ilegales (Seeboonruang, 2016) y han tenido como marco espacial las áreas urbanas en países con características geográficas muy distintas a nuestro país.

Ahora bien, nuevas investigaciones apuntan que los vertidos ilegales son un fenómeno multiescalar, que precisa de variables no espaciales como las características socioeconómicas o culturales para conocer las causas de estos comportamientos, lo que implicaría la necesidad de estudiar a nivel municipal la gestión de los residuos o el comportamiento de la población (Farinos, 2001).

Y es por ello que disciplinas científicas como la Criminología puede jugar un papel determinante en este campo de la investigación sobre los vertidos ilegales. Una ciencia que años atrás, ha

---

9 Durante estos últimos años podemos encontrar trabajos de carácter divulgativo por parte de Instituciones públicas, que tienen como objetivo trasladar información sobre el impacto de los residuos en la salud de la población. A modo de ejemplo, encontramos el trabajo elaborado por el Ministerio de Sanidad denominado "Salud y residuos". Residuos no peligrosos. Evidencias sobre los efectos en la salud y restos para su mejor caracterización. https://www.sanidad.gob.es/ciudadanos/saludAmbLaboral/docs/SALUD_Y_RESIDUOS_ACCESIBLE.pdf [Consultado el 03.07.2023].

tenido como objeto de sus investigaciones problemas de carácter medioambiente urbano como el ruido (García y South, 2019) o conductas delictivas más popularizadas como los incendios o la caza (Salafranca y Maldonado, 2018; Liñán, 2019), pero que gracias otras disciplinas como geografía, biología o ingeniería (Navarro et al., 2016; Quesada et al., 2018) o el renacimiento de disciplinas como la Criminología Rural (Donnermeyer y Dekeseredy, 2014)[10], Criminología del Sur (Carrington et al., 2018; Serrano, 2020)[11] o Criminología Verde (Mol et al., 2017), muestran día tras día un mayor interés científico por problemas menos visibles, pero que afectan de manera muy grave a la salud de los seres humanos, como es el caso de los vertidos ilegales[12].

Desgraciadamente, el estudio de los vertidos ilegales por parte de la literatura criminológica en España sigue siendo escasa. Se trata de un comportamiento ilegal poco analizado a pesar del grave impacto sobre la población (Senior y Mazza, 2004). El acompañamiento a disciplinas jurídicas, sociales o técnicas como Derecho, Geográfica, Ingeniería o Biología están logrando entrever importantes elementos criminógenos que tienen que ser analizados en esta problemática. Por ejemplo, el estudio de factores que pueden favorecer o contribuir los vertidos ilegales, como el factor

---

10 A modo de ejemplo, cabe destacar la revista internacional sobre Criminología Rural. Una revista dedicada a publicar trabajos teóricos y empíricos sobre el crimen rural, disponible en: https://ruralcriminology.org [Consultado el 03.07.2023]. Además, se han creado grupos específicos sobre esta materia como The International Society for the Study of Rural Crime https://issrc.net [Consultado el 03.07.2023]. Por último, destacar grupo de trabajo, cómo el creado por la European Society Criminology sobre Rural Criminology.

11 Criminología del Sur es un nuevo proyecto político, teórico y empírico que mejora el conocimiento de la Criminología Global.

12 Actualmente, la Sociedad Española de Investigación Criminológica cuenta con un grupo de trabajo sobre Criminología Verde, cuyo trabajo está orientado a la investigación y conocimiento en torno al estudio del daño ambiental. https://criminologia.net/grupos-de-trabajo/ [Consultado el 03.07.2023].

espacial, el estudio de iniciativas públicas o privadas que puedan reducirlos o la respuesta social frente a estos comportamientos, son algunos elementos que pueden ser abordados desde la Criminología.

Respecto al factor espacial o el elemento geográfico, debemos apuntar que desde la Criminología existe un renacimiento del lugar como elemento en el estudio del delito en estos últimos años (Serrano, 2017, 192-193). El desarrollo de nuevos *softwares* más potentes y el uso de sistemas de información geográfica para el estudio del delito nos permiten identificar áreas con un riesgo alto o sensible de contaminación ambiental. Diferentes investigaciones apuntan al elemento espacial como la principal herramienta para localizar o identificar estos residuos. Por ello, el estudio ambiental desde la Criminología como herramienta de prevención y predicción parece ser una de las más populares entre el personal investigador.

En cuanto a la prevención de estos comportamientos ilícitos encontramos distintas fórmulas. Por un lado, políticas de concienciación, educación o fomento de alternativas sostenibles; por otro lado, políticas de seguridad frente a comportamientos normalizados y poco solidarios de la población, teniendo en cuenta el alto número de denuncias interpuestas por las fuerzas y cuerpos de seguridad en nuestro país, factores que son clave y recurrentes a lo largo de toda esta obra. Siendo la falta de servicios policiales o la despoblación factores de riesgo en la prevención de estos comportamientos en determinadas zonas, como puede ser el caso de la Extremadura. Una región preeminentemente rural[13], con

---

[13] La Comunidad Autónoma de Extremadura cuenta con un 90% de municipios rurales. Art. 19 del Decreto 115/2020, de 14 de mayo, por el que se crean y establecen las funciones de los órganos de gobernanza para la aplicación de la Ley de Desarrollo Sostenible del Medio Rural y se determina la delimitación y calificación de las zonas rurales de Extremadura, donde se califica el entorno rural en Extremadura como el espacio geográfico formado por la agregación de todos los municipios que integran la región extremeña, a excepción de los correspondientes

importantes carencias en los servicios públicos que pueden ser esenciales en la prevención de determinados comportamientos delincuenciales, como son los vertidos ilegales (Gutiérrez Gallego, 2023), no podemos olvidar el papel fundamental de instituciones como la policía o la comunidad en el medio rural, para la vigilancia y control de este tipo de comportamientos y la conservación del medio ambiente.

En 2023, la región extremeña cuenta con un total de 168 municipios con servicios policiales municipales de las 388 localidades que tiene la región, de los que un 50% sólo tienen 1 o 2 efectivos, lo que provoca unos servicios deficientes, pues causa que no puedan prestarse en horarios nocturnos o fines de semana. Una situación que limita sustancialmente el control y vigilancia de este tipo de conductas, a pesar de contar con Guardia Civil o Policía Nacional (Ortiz, 2022)[14]. En el siguiente mapa temático podemos observar la distribución geográfica de los municipios que cuenta con servicios policiales de Guardia Civil o Policía Local.

a Almendralejo, Badajoz, Cáceres, Don Benito, Mérida, Plasencia y Villanueva de la Serena.

14 La Guardia Civil opera en 172 localidades de la región extremeña, mientras que el Cuerpo Nacional de Policía se encuentra en las 7 localidades como más de 20.000 habitantes: Badajoz, Mérida, Don Benito – Villanueva de la Serena, Almendralejo, Cáceres y Plasencia.

Mapa 1

Distribución geográfica de las FCS en la Comunidad Autónoma de Extremadura.

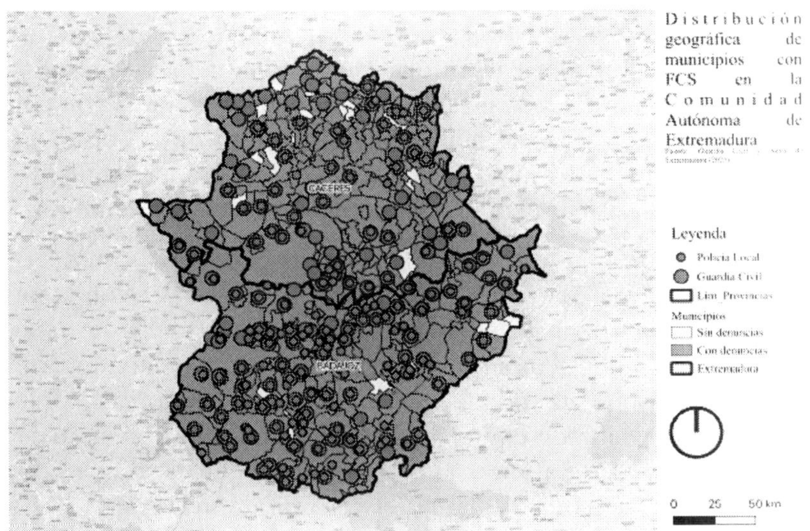

Nota: Elaboración propia a partir de datos obtenidos SEPRONA (Guardia Civil) y ASPEX (Junta de Extremadura).

La cultura del control es un elemento esencial para comprender los comportamientos desviados en nuestra sociedad (Medina, 2022). Conocer la respuesta social frente al delito resulta fundamental para entender ciertos comportamientos como los estudiados en esta investigación. Además, junto a las instituciones policiales, judiciales o penitenciarias, la respuesta de la sociedad civil es fundamental para luchar contra la inseguridad y el delito. En el caso de Extremadura, que viene sufriendo desde hace años un vertiginoso descenso de su población y de los servicios policiales en áreas rurales, es aún más significativo el papel de la comunidad, ya que junto al deterioro económico, descenso de la natalidad o envejecimiento de la población, la lenta caída de su población supone un grave riesgo para la vigilancia y control de comportamientos delincuenciales o antisociales, como pueden ser los comportamientos provocados por los vertidos ilegales próximos a los municipios. Sin duda, la pérdida de habitantes provoca un importante deterioro de los elementos que configuran el control informal en el medio rural. Es más, el fenómeno de la despoblación no sólo provoca un descenso de los elementos que configuran el control social informal, sino que también modifica modelos de vecindad y redes sociales que limitan esa relación con instituciones que se encargan de la vigilancia y control de determinadas conductas, como es la policía. Por ello, resulta esencial el papel de vecindario en la prevención de estos comportamientos.

Por lo tanto, para luchar contra los vertidos ilegales en determinados lugares de la región extremeña, no sólo basta con la implicación de las instituciones, también resulta esencial el papel de la población. Sin embargo, la respuesta de la ciudadanía ante este tipo de comportamientos no parece ser la más apropiada. La permisividad y falta de concienciación sobre este problema parece ser la tónica generalizada, atendiendo a los datos ofrecidos por el Centro de Atención de Urgencias y Emergencias de la Junta de Extremadura (112). En el año 2021, se registraron un total de 17.308 incidentes clasificadas como de Seguridad Pública, de las que sólo 3 llamadas hacían referencia a seguridad pública sobre el medio ambiente, y concretamente, sobre vertidos ilegales[15]. Una clara muestra de la baja conciencia social que tenemos todavía sobre esta problemática.

En el siguiente mapa temático podemos observar el tamaño de la población según el municipio (los colores más intensos son las áreas como una mayor densidad de población), mientras que los círculos en color claro muestran el número de denuncias por municipio.

---

15 Datos ofrecidos por la Subdirección Técnica del Centro de Atención de Urgencias y Emergencias 1.1.2. Extremadura. Dirección General de Emergencias, Protección Civil e Interior. Consejería de Agricultura, Desarrollo Rural, población y Territorio de la Junta de Extremadura.

Mapa 2

Distribución geográfica de tamaño de población por municipio en Extremadura.

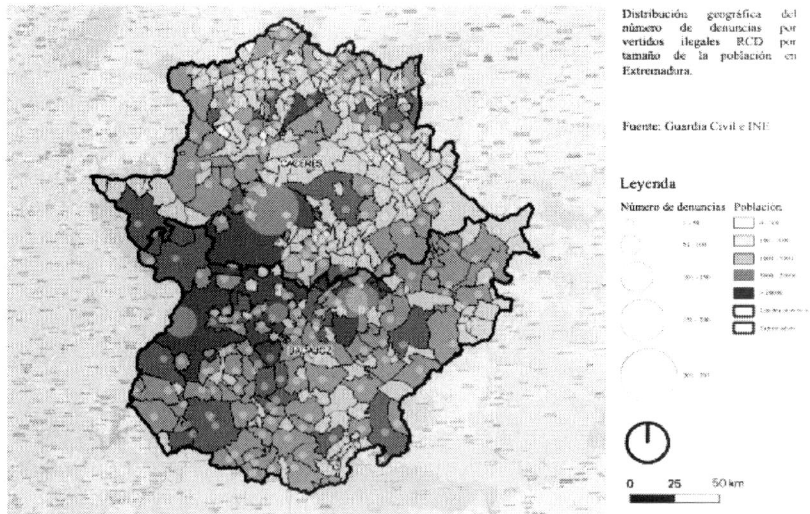

Nota: Elaboración propia a partir de datos obtenidos SEPRONA (Guardia Civil) y el Instituto Nacional de Estadística (INE).

Para finalizar, es preciso indicar que la aportación metodológica de los sistemas de información geográfica en el estudio de los vertidos ha resultado esencial para la planificación y gestión del medio ambiente (Seror y Portnov, 2018). Sin embargo, como apuntan distintas investigaciones, es necesario incrementar nuevas variables a escala local pueden mejorar la información para futuras investigaciones (Biotto et al., 2007).

3. METODOLOGÍA

Después de conocer la problemática de los vertidos ilegales en nuestro país y su relación con la Criminología, vamos a explicar la metodología utilizada para llevar a cabo esta investigación. Existe un objetivo común del grupo de investigación que consiste en la creación de un modelo predictivo para identificar vertidos ilegales de RCD que permita prevenir, diseñar y elaborar políticas públicas que eviten y puedan erradicar este tipo de comportamientos entre la ciudadanía. Para lograr nuestro objetivo general, debemos tener en cuenta otros procesos necesarios, como las características geográficas, demográficas, económicas, servicios, (presencia de un servicio policial permanente o plantas de reciclaje) o número de vivienda de los municipios que tienen sanciones administrativas a la normativa sobre vertidos y residuos RCD en la región extremeña, para identificar las áreas con mayor probabilidad de generar este tipo de vertidos ilegales.

La primera base de datos de la investigación estuvo compuesta por un total de 6.472 denuncias administrativas interpuestas por la Guardia Civil entre los años 2016 a 2021 (ambos inclusive). La información ofrecida por la Guardia Civil ha sido depurada debido a: errores en la localización, esto es, denuncias geolocalizadas fuera de la región extremeña (9 casos); por denuncias repetidas (2.672 casos); y la inclusión de denuncias no relativas a infracciones administrativas por RCD, tales como la infracción penal de recogida, transporte y eliminación de residuos peligrosos (1 caso) y la infracción a la normativa sobre centros gestores de residuos metálicos (13 casos). Como se observa, la mayor parte de las denuncias eliminadas eran denuncias repetidas. Las mismas fueron identificadas como tal porque compartían una misma coordenada X e Y en cada año. Este hecho fue clarificado con los agentes del SEPRONA (véase capítulo 3) pues asociaban -y añadían- una coordenada para cada hecho delictivo relacionado con una denuncia de RCD (por ejemplo, encontrar residuos en una finca abandona al tiempo que un pozo ilegal). Así, la muestra definitiva utilizada en esta investigación ha sido de 3.771 denuncias, lo que supone un 42% menos del total de denuncias recibidas por parte de la Guardia Civil. Considerando que nuestro trabajo está centrado en el estudio de características locales y la naturaleza de nuestra variable dependiente (número de denuncias) no fue preciso, tal y como se llevó a cabo en el capítulo 3, una verificación de las coordenadas de las denuncias muestrales en la ortofoto de 2019.

En otras palabras, no precisamos conocer la ubicación exacta de las denuncias, sino su inclusión en un determinado territorio (en esta línea: Buil et al., 2021).

El siguiente mapa muestra en color más oscuro los municipios que tienen una o más denuncias por infracción a la normativa sobre residuos y vertidos de RCD.

Mapa 3

Distribución geográfica de municipios con/sin denuncias de vertidos ilegales de RCD.

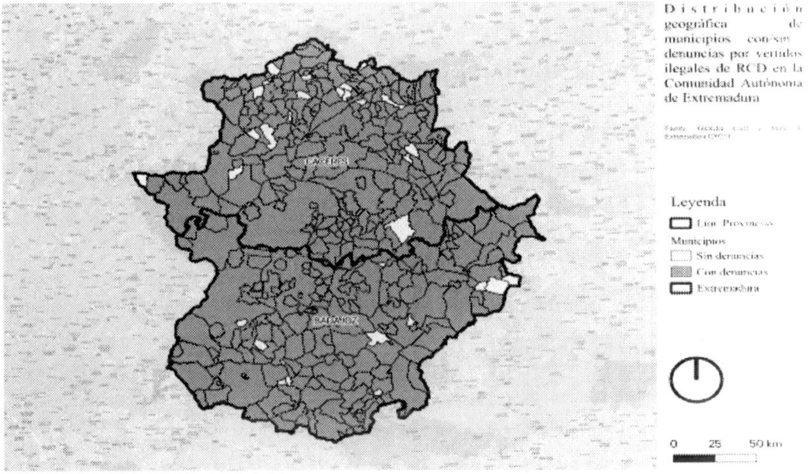

Nota: Elaboración propia a partir de datos obtenidos SEPRONA (Guardia Civil).

El marco espacial del estudio abarca, al igual que en el resto de los casos analizados en esta obra, la delimitación territorial de la Comunidad Autónoma de Extremadura. La región extremeña cuenta con un total de 388 municipios distribuidos en una superficie aproximada de 40.000 kilómetros. La población de la región extremeña en el año 2021 fue de 1.059.501 de habitantes (INE, 2021), distribuida entre sus dos provincias: Badajoz (669.943. habitantes) y Cáceres (389.558 habitantes).

Con respecto al ámbito temporal del estudio, la fecha de inicio de la investigación comienza a partir de las denuncias interpuestas por la Guardia Civil por vertidos ilegales en el 2016. La razón de analizar las denuncias desde ese año es poder contar un periodo más amplio de denuncias. En cuanto a la fecha de cierre del estudio, dado que la investigación se programa y comienza en el año 2020 hasta el año 2023, se opta por señalar el año 2022, como año de cierre para el análisis de los datos obtenidos, ya que se cuenta con denuncias hasta finales del año 2021.

Además de las denuncias interpuestas por vertidos ilegales de RCD, que se estableció como variable dependiente, se incluyeron como variables predictoras las siguientes: datos geográficos (superficie) datos demográficos (densidad de la población); datos económicos (renta media por persona; número de empresas; número de empresas de construcción e IBI) o datos sobre servicios (fuerzas y cuerpos de seguridad, instalaciones de residuos de construcción y demolición o puntos limpios), cómo se puede observar en la siguiente tabla.

Tabla 1

Operativización de las variables del estudio.

Bloque	Variable
Bloque de denuncias	Nº de denuncias (VD)
Bloque geográfico	Superficie
Bloque demográfico	Población del municipio.
Bloque económico	Renta Media por persona. Nº de empresas Nº de empresas de construcción IBI (Base imponible)

Bloque servicios	Parque de vehículos cada 100 hab. Servicios Policiales Planta Fija RCD Planta Móvil RCD Punto Limpio Piscinas cubiertas Piscinas no cubiertas
Bloque hogar	Nº de viviendas

Nota: Elaboración propia a partir de las variables analizadas en el estudio.

De las variables seleccionadas para esta investigación, el número de empresas de construcción sólo tenemos información del 57% de los municipios (n= 203), y respecto a las variables económicas, se seleccionó la base imponible del IBI porque supone una variable estática, que no depende de la autonomía financiera de los municipios.

La información de las diferentes variables seleccionadas para la investigación, han sido obtenidas de las siguientes fuentes de información: Servicio de Protección de la Naturaleza de la Guardia Civil (SEPRONA), Academia de Seguridad Pública de Extremadura (ASPEX), Junta de Extremadura, Instituto Nacional de Estadística (INE), Programa SIDAMUN (Ministerio para la Transición Ecológica y el Reto Demográfico) [16] y Ministerio de Hacienda y Función Pública.

Para el análisis espacial, hemos utilizado los datos cartográficos de Infraestructura de Datos Espaciales de Extremadura de la Dirección General de Urbanismo y Ordenación del Territorio de

16 El proyecto SIDAMUN ofrece una información detallada de los municipios españoles. Esta herramienta contiene datos demográficos, geográficos, económicos o de servicios a nivel local. Disponible en: https://www.miteco.gob.es/es/reto-demografico/temas/analisis-cartografia/ [Consultado el 03.07.2023].

la Junta de Extremadura (IDEEX, 2022)[17]. Los datos utilizados para este estudio ha sido la cartografía temática de las unidades administrativas de entidades de población y límites municipales, provinciales y autonómicos. Una vez recopilada la información cartográfica se ha realizado una unión con la base de datos descrita en párrafos anteriores, lo que nos ha permitido presentar los análisis descriptivos

Para evaluar cada una de las variables del estudio se creó una base de datos basada en estudios previos sobre modelos de prevención en las áreas con vertederos ilegales (Lucendo Monedero et al., 2015). La creación de la base de datos, así como la selección de los campos, constituyó un elemento fundamental para la consecución de los objetivos marcados para esta investigación. Los diferentes campos se estructuraron teniendo en cuenta las variables analizadas en el estudio (véase Tabla 1).

El diseño de la base de datos nos ha permitido realizar un estudio estadístico – descriptivo que busca someter a prueba las denuncias interpuestas por vertidos ilegales de RCD y la covariación entre las distintas variables.

Por último, como cualquier estudio de investigación, puede tener limitaciones en la muestra o la metodología utilizada que puede afectar a los resultados. En este caso, la falta de datos sociodemográficos y económicos de poblaciones de menos de 20.000 habitantes, que han sido los principales obstáculos para poder lograr datos más depurados y precisos en este trabajo.

4. RESULTADOS

Como ya se ha indicado durante toda la obra, los resultados de esta investigación forman parte de un trabajo más profundo

---

17 Existen otras webs de organismos públicos donde se puede obtener esta información. A modo de ejemplo, el Centro Nacional de Información Geográfica. Disponible en: www.ign.es [Consultado el 03.07.2023].

desarrollado por un grupo de investigación multidisciplinar de la Universidad de Extremadura. Por lo tanto, los resultados obtenidos del estudio deben ser analizados teniendo en cuenta otros elementos presentados otros capítulos de la obra. Sólo de este modo podremos tener una mirada amplia de esta problemática en la región extremeña. Los resultados se presentan en dos bloques separados: por un lado, los resultados obtenidos de un análisis descriptivo de las denuncias interpuestas por la Guardia Civil según el lugar, año, tipología y número de denuncias; por otro lado, los resultados del análisis de las relaciones bivariadas entre cada una de las variables y el número de denuncias.

4.1. Datos generales

Con respecto a los datos analizados en este trabajo, podemos indicar que el número total de denuncias administrativas interpuestas entre 2016 y 2021 (ambos incluidos) es de 3.771 en toda la Comunidad Autónoma de Extremadura, de las que 1.828 se interpusieron en la provincia de Badajoz (49%) y 1.949 en la provincia de Cáceres (51%). En cuanto a la evolución del número de denuncias en el periodo 2016-2021 (véanse datos absolutos en capítulo 2), cabe destacar que el número de denuncias en los primeros cuatro años mantiene un porcentaje anual cercano al 20%, mientras que en el último año[18] de análisis se ha producido un descenso estadísticamente significativo hasta un 9%.

Respecto a la tipología de la sanción, cabe indicar que todas las sanciones utilizadas para esta investigación fueron infracciones a la normativa sobre residuos y vertidos de RCD. Tal y como se apuntó, los datos relativos a denuncias en el ámbito penal, residuos metálicos o centros de recogida no se han tenido en cuenta en esta investigación.

---

18 Siendo este el año 2020, pues los datos de 2021 fueron proporcionados hasta el mes de septiembre y no permiten la comparativa interanual.

Para concluir, la distribución geográfica de las denuncias, de los 388 municipios que tiene la región, se han interpuesto denuncias en 356 localidades, lo que supone más de un 90 % de los municipios en la región. Los datos muestran la generalización y normalización de estos comportamientos entre la población. Además, debemos subrayar que más de un 80% de las denuncias se concentran en municipios de menos de 20.000 habitantes (N = 3.050), frente a las denuncias que se interpusieron en las 7 localidades de la región extremeña con más de 20.000 habitantes. Sin embargo, los municipios con el mayor número de denuncias se encuentran entre las localidades más pobladas: Badajoz (122), Cáceres (207), Don Benito (114). Mérida (74) y Villanueva de la Serena (87), como podemos observar en el siguiente mapa temático.

Mapa 4

Distribución geográfica de las denuncias por vertidos ilegales RCD en Extremadura.

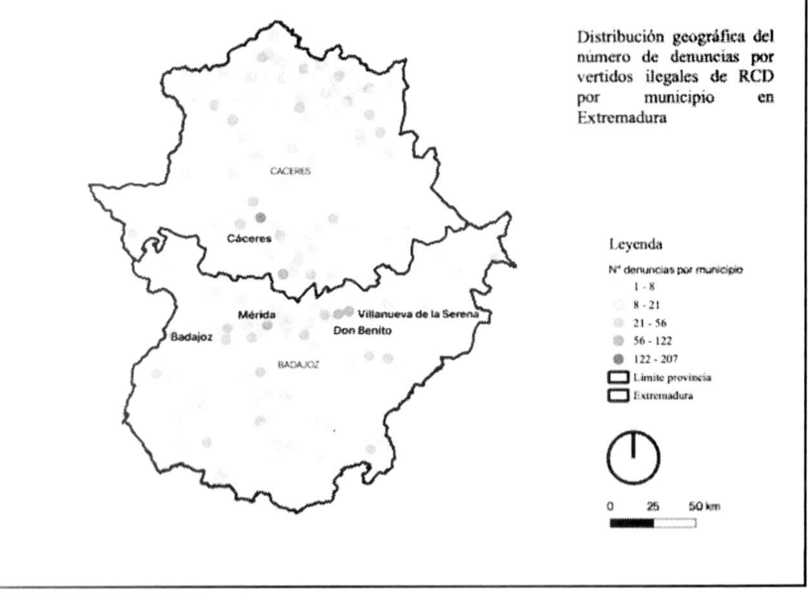

Nota: Elaboración propia a partir de datos obtenidos SEPRONA (Guardia Civil).

En el siguiente apartado, analizaremos los resultados obtenidos del análisis covariado entre las variables demográficas, socioeconómicas y de gestión ambiental con las denuncias por vertidos RCD.

4.2. Variables sociodemográficas, económicas y de gestión ambiental del estudio

En este apartado, analizaremos las variables locales que hemos seleccionado para esta investigación. Para una mejor compresión, lo presentaremos en cinco bloques: geográfico, demográfico, económico, servicios y hogar.

En el primer bloque, hemos seleccionado la variable *superficie*. La extensión del término municipal supone en algunos casos un territorio muy amplio que debe ser protegido y vigilado por un gobierno local, lo que provoca un problema para el cumplimiento normativo o la protección del medio ambiente. Los amplios límites administrativos que tienen muchos municipios en Extremadura, los convierten en una variable de enorme interés para nuestra investigación.

Así, en el análisis estadístico bivariado, la prueba muestra que existe una relación estadísticamente significativa fuerte entre el número de denuncias por vertidos RCD y el término municipal (,702**). Este resultado indicaría que el aumento de una variable conllevaría el aumento de la otra variable y, por lo tanto, una asociación entre ellas.

En cuanto al elemento demográfico, se ha seleccionado el tamaño de la población. Se trata de una variable que ha sido incorporada en otras investigaciones como un factor determinante en la aparición de vertidos ilegales, junto con los factores geográficos (Jordá et al., 2014). La prueba del análisis bivariado muestra que las dos variables analizadas están relativamente relacionadas (,787**). La relación entre denuncias y tamaño de la población es positiva y fuerte. Se trata de una correlación mayor que la variable superficie, lo que indicaría que no se trata de una relación al azar, sino que existe una asociación estadísticamente significativa entre ellas.

En el tercer bloque, hemos seleccionado algunas de las variables de carácter económico que pueden influir en los vertidos. Numerosos estudios apuntan a un modelo de construcción exagerado y excesivo en todo el mundo, y también encontramos aquellos que apuntan al valor económico social de los residuos RCD (Leigh y Patterson, 2006), resulta de enorme interés conocer el número de empresas que existen en las localidades analizadas. En el mismo sentido, sucede con la renta media de persona y el impuesto de bienes inmuebles, factores que pueden incidir de manera positiva o negativa a la aparición de estos vertidos, como apuntan algunas investigaciones (Tasaki et al. 2004, 2007; Gorsevski et al., 2012).

En este caso, del análisis bivariado de las cuatro variables seleccionadas, los resultados muestran una correlación estadísticamente significativa a la variable Empresa (,791**) y la variable IBI (,769**), mientras que las variables Renta Media por persona y Empresas de construcción es una correlación positiva pero muy débil, como se puede observar en la siguiente tabla.

Tabla 2

Análisis bivariado entre denuncias y características económicas del municipio.

Correlaciones						
		Denuncias	Renta Media Por persona	Empresas	Empresas Construcción (%)	IBI (base imponible)
Denuncias	Correlación de Pearson	1	,174**	,791**	,037	,769**
	Sig. (bilateral)		,001	<,001	,483	<,001
	N	356	356	356	356	356
Renta Media por persona	Correlación de Pearson	,174**	1	,173**	<,001	,171**
	Sig. (bilateral)	,001		,001	,998	,001
	N	356	356	356	356	356

Empresas	Correlación de Pearson	**,791****	,173**	1	,015	,996**
	Sig. (bilateral)	<,001	,001		,780	<,001
	N	356	356	356	356	356
Empresas Construcción (%)	Correlación de Pearson	,037	<,001	,015	1	,012
	Sig. (bilateral)	,483	,998	,780		,815
	N	356	356	356	356	356
IBI (base imponible)	Correlación de Pearson	**,769****	,171**	,996**	,012	1
	Sig. (bilateral)	<,001	,001	<,001	,815	
	N	356	356	356	356	356

** La correlación es significativa en el nivel 0,01 (bilateral).

En cuanto al análisis de la variable *servicios*, hemos selecciona-do un conjunto de elementos relacionados con las características del sistema municipal, tal y como apuntan distintas investigaciones. La selección de estos factores locales se debe a la incidencia que tienen los poderes públicos en la gestión de los residuos. Por ejemplo, la disponibilidad o no de instalaciones de reciclaje (Lucendo et. al, 2015) puede impactar en el comportamiento de la ciudadanía. De igual forma, el desarrollo urbanístico asociado al parque de vehículos, viviendas y otras instalaciones (como polideportivos y piscinas) puede tener relación con la mayor producción de RCD y, por extensión, con el vertido ilegal de los mismos. También la importancia de la presencia y mejora de los servicios policiales para la vigilancia y control de los vertidos ilegales con la creación de acciones específicas de patrullaje en determinadas zonas (Seeboonruang, 2016). En definitiva, un conjunto de factores relacionados con la gestión municipal.

Para ello, hemos realizado análisis bivariados entre la variable dependiente (denuncias) y el resto de las variables independientes. Este análisis nos ha permitido explorar las relaciones entre múltiples factores. Al igual que el resto de las variables, se intenta determinar si existe o no relación entre las variables.

Los resultados del análisis muestran que la relación de las denuncias con el número de vehículos en el municipio es moderada (,793**). En cambio, la correlación con los servicios policiales es más débil (,318**), lo que supone que la fuerza es menor a la variable anterior. En todo caso, la relación positiva de ambas variables nos permite observar cierta asociación entre ambas, como vemos en la siguiente tabla.

En cuanto a los resultados en el análisis de servicios de gestión ambiental muestran una relación positiva moderada entre las variables Planta Fija de RCD (254**,), Planta Móvil RCD (214**), Punto de Recogida (250**) y el punto de recogidas (343**), lo que indicaría una tendencia de aumento de una variable cuando las otras variables también aumenta, y. Sin embargo, en el caso de las piscinas cubiertas y no cubiertas, la correlación es muy débil, lo que indicaría una relación estadísticamente moderada y poco significativa (,399** y ,066), tal y como como puede observarse en la tabla inferior.

Tabla 3

Análisis bivariado entre denuncias y bloque servicios de los municipios.

Correlaciones

		Denuncias
Denuncias	Correlación de Pearson	1
	Sig. (bilateral)	
	N	356
Parque de vehículos por c/100 hab.	Correlación de Pearson	**,793****
	Sig. (bilateral)	<,001
	N	356
Servicios Policiales	Correlación de Pearson	,318**
	Sig. (bilateral)	<,001
	N	356
Planta fija RCD	Correlación de Pearson	,254**
	Sig. (bilateral)	<,001
	N	356
Planta móvil RCD	Correlación de Pearson	,214**
	Sig. (bilateral)	<,001
	N	356
Punto Limpio	Correlación de Pearson	,250**
	Sig. (bilateral)	<,001
	N	356
Punto Recogida	Correlación de Pearson	,343**
	Sig. (bilateral)	<,001
	N	356
Piscinas Cubiertas	Correlación de Pearson	,399**
	Sig. (bilateral)	<,001
	N	356
Piscinas No cubiertas	Correlación de Pearson	,066
	Sig. (bilateral)	,216
	N	356

** La correlación es significativa en el nivel 0,01 (bilateral).

Por último, hemos analizado los datos relativos al número de hogares del municipio. Para este apartado se ha seleccionado el total de las viviendas por municipio. Se trata de una variable relacionada con otras variables anteriormente analizadas y que muestran una relación estadísticamente fuerte con el número de denuncias: Población y Construcción. Cabe recordar el papel esencial que la construcción tiene en el aumento de este tipo de vertidos. Así, el análisis bivariado muestra que ambas variables tienen una relación estadísticamente significativa (,804**), siendo la variable con una relación mayor de las estudiadas en este trabajo.

Tabla 4

Análisis bivariado entre número de denuncias y número de viviendas en los municipios.

Correlaciones

		Denuncias	Nº viviendas
Denuncias	Correlación de Pearson	1	**,804****
	Sig. (bilateral)		<,001
	N	356	356
Nº viviendas	Correlación de Pearson	,804**	1
	Sig. (bilateral)	<,001	
	N	356	356

** La correlación es significativa en el nivel 0,01 (bilateral).

En resumen, los resultados del análisis de características socioeconómicas y de gestión ambiental para la identificación de áreas probables de vertederos ilegales de RCD, muestra que las variables superficie, población, construcción, IBI y número de viviendas son las características que tienen una relación estadísticamente significativa fuerte con la variable denuncias, mientras que el resto de variables, muestran diferentes valores de relación (baja o moderada), por lo que es necesario considerar tener en cuenta otros factores y realizar análisis más detallados.

5. CONCLUSIONES Y DISCUSIÓN

En base a todo lo argumentado en la obra y el análisis llevado a cabo en esta investigación, podemos concluir que los resultados obtenidos muestran la relación de factores no geográficos en la aparición de vertidos ilegales de RCD como superficie, población, construcción, IBI y número de viviendas, de acuerdo con Jordá et al., 2014. Y, al igual que ocurre en el estudio realizado en la Comunidad Autónoma de Andalucía, existen diferencias significativas si nuestra investigación se realizara a escala municipal o regional (Lucendo et al., 2015). El análisis descriptivo muestra tasas de denuncias por cada 1.000 habitantes muy significativas en municipios con poca población como Alcuéscar (2.554 hab.), Aldea del Cano (662 hab.) o Villanueva de la Sierra (441 hab.). Sin embargo, el análisis a nivel regional que hemos llevado a cabo en esta investigación sí muestra una correlación significativa con zonas de mayor concentración de población, tales como Badajoz, Cáceres, Mérida, Don Benito o Villanueva de la Serena, en la línea de lo apuntado por Lucendo et al. 2015. Sin embargo, hay un menor peso correlacional con los servicios de gestión ambiental como las plantas de reciclajes, a diferencia de otras investigaciones que asocian estos comportamientos a la falta de infraestructuras municipales de gestión ambiental o políticas punitivas (Sujuaddin et al., 2008; Ekere et al., 2009; Guerrero et al., 2013). Por último, hay que indicar que, a tenor de los resultados, variables menos significativas como los servicios policiales pueden ser factores que puedan confundir o distorsionar la relación entre las variables independientes más fuertes con las denuncias. No podemos olvidar que la falta de servicios policiales en muchos municipios de la región extremeña puede influir en la falta de control y vigilancia de estos comportamientos por parte de la ciudadanía. En todo caso, se trata de una cuestión que debe ser abordada con mayor profundidad en futuras investigaciones.

Para finalizar, es preciso insistir en que esta información puede ser útil para los responsables políticos y agencias medioambientales de cara a prevenir y mitigar los impactos negativos de los vertederos ilegales. Los resultados de este estudio contribuyen a aumentar la literatura que existe en este campo y pueden servir de base para nuevas políticas, medidas legales o fondos destinados a nuevos proyectos que permitan cambiar los modelos de construcción más ecológicos y sostenibles, frente a aquellos países que han decidido aplicar políticas más punitivistas como Alemania o Japón (Fujikura, 2011; Zou, 2011). Además, la representación de los mapas temáticos puede ser adoptada por organismos gubernamentales local y regional de Extremadura para incorporar medidas y estrategias de prevención para conservar las áreas rurales y el medio ambiente.

6. REFERENCIAS BIBLIOGRÁFICAS

Academia de Seguridad Pública de Extremadura (Junta de Extremadura).

http://aspex.juntaex.es/aspex/view/main/index/index.php [Consultado el 03.07.2023].

Alfonso-Torreño, A., Gutiérrez Gallego, A., Ortiz García, J. y Arenas García, L. (2023). Measurement error of police data to identify solid waste dumping sites. Geofocus.

Antonini, E., Tarantini, M. y Balázs, S. (2011) Application of life Cycle Assessment (LCA) methodology forvalorization of building demolition materials and products. *Proceedings of SPIE – The International Society for Optical Engineering*, Bologna, Italia.

Biotto, G., S. Silvestri, L. Gobbo, E. Furlan, S. Valenti., y R. Rosselli. 2009. GIS, Multi-Criteria and Multi-Factor Spatial Analysis for the Probability Assessment of the Existence of Illegal Landfills. *International Journal of Geographical Information Science, Vol. 2* (10), 1233-1244. DOI: 10.1080/13658810802112128.

Carrington, K, Hogg, R. y Sozzo, M. (2018). Criminología Sur. *Delito y Sociedad, 45*, 9-33.

Centro de Atención de Urgencias y Emergencias de Extremadura (112). http://instituciones.juntaex.es/112/ [Consultado el 03.07.2023].

Chamolí Canturín, W. (2016). Gestión de los residuos sólidos en la fase de construcción y demolición de las obras civiles en Huánuco y Amarilis. [Tesis para doctorado]. Universidad de Sevilla. Departamento de Construcciones Arquitectónicas I (ETSA) Sevilla. España.

Decreto 115/2010, de 14 de mayo, por el que se crean y establecen las funciones de los órganos de gobernanza para la aplicación de la Ley de Desarrollo Sostenible del Medio Rural y se determina la delimitación y calificación de las zonas rurales de Extremadura. http://doe.juntaex.es/pdfs/doe/2010/950o/10040125.pdf. [Consultado el 03.07.2023].

Donnermeyer, J. y Dekeseredy, W. (2014). *Rural Criminology*. Routledge. DOI: 10.4324/9780203094518

Ecologistas en acción. Disponible en: https://www.ecologistasenaccion.org [Consultado el 03.07.2023].

Ekere, W., Mugisha, J. y Drake, L. (2009) Factors influencing waste separation and utilization among households in the lake Victoria crescent, Uganda. *Waste Management, Vol. 29*, 12, 3047 – 3051. DOI: 10.1016/j.wasman.2009.08.001

Farinos, J. (2001). Reformulación y necesidad de una nueva geografía regional flexible. *Boletín de la Asociación de Geógrafos Españoles (AGE), 53*, 53-71.

Fujikura, M. (2011)-Japan's efforts against the illegal dumping of industrial waste. Environ. *Policy and Gov, 21*, 325 – 337.

Infraestructura de Datos Espaciales de Extremadura. Disponible en: http://ideextremadura.com/Geoportal/ [Consultado el 03.07.2023].

García Ruiz, A. y South, N. (2019). El ruido silenciado en la Criminología y en el medio ambiente. Apuntes preliminares para una criminología acústico – sensorial. *Revista Española de Investigación Criminológica, 17*, 1-27. DOI: 10.46381/reic.v17i0.297

Gorsevski, P.V., Donevska, K.R. , Mitrovsk, C.D. y. Frizado J.P (2012). Integrating Multi-Criteria Evaluation Techniques with Geographic Information Systems for Landfill Site Selection: A Case Study Using Ordered Weighted Average. *Waste Management, Vol. 32* (2), 287-296. DOI: 10.1016/j.wasman.2011.09.023

Gómez Parga, O., Nieto Beltrán, J.C. y Parada Suárez, O. (2008). Modelo de Gestión Ambiental participativo como instrumento para el manejo de los residuos de construcción y demolición RCD – Escombros – generados en Cartagena de Indias D.T.Y.C. [Tesis para maestría]. Universidad Tecnológica de Bolivar – UTB. Cartagena de Indias. Colombia

Grupo de Trabajo de Criminología Verde. Sociedad Española de Investigación Criminológica. https://criminologia.net/grupos-de-trabajo/ [Consultado el 03.07.2023].

Guerrero, L.A., G. Maas, and W. Hogland. (2013). Solid Waste Management Challenges for Cities in Developing Countries. *Waste Management, Vol. 33* (1), 220-232. DOI: 10.1016/j.wasman.2012.09.008 .

Gutiérrez Gallego, J. A. (2023). *Estudio y análisis de zonas prioritarias en materia de despoblación. El caso de la provincia de Badajoz.* Tirant lo Blanch.

Hanneke, M., Rodríguez Goyes, D. South, N. y Brisman, A. (2017). *Introducción a la Criminología Verde.* Temis.

Instituto Nacional de Estadística (INE). https://www.ine.es/dyngs/INEbase/es/categoria.htm?c=Estadistica_Pycid=1254735572981 [Consultado el 03.07.2023].

Jianguo, C., Yangyue, S., Hongyun, S. y Jindao, C. (2018). Managerial Areas of Construction and Demolition Waste: A Scientometrie Review. *US National Library of Medicine National Institutes of Health, 15,* 23-50.

Jordá- Borrell, R., Ruiz – Rodríguez, F. y Lucendo – Monedero. A. L. (2014). Factory analysis and geographic information system for determining probability areas of presence of illegal landfills. *Ecological Indicators, 37,* 151-160.

Leigh, N.G. y Patterson, L.M. (2006). Deconstructing to Redevelop: A Sustainable Alternative to Mechanical Demolition: The Economics of Density Development Finance and Pro Formas. *Journal Of the American Planning Association, 72,* 217-225. DOI: 10.1080/01944360608976740

Liñán Lafuente, A. (2019). EL delito de caza furtiva en tiempo de veda. Comentario a la SYS 3566/2020, de 3 de noviembre. *Revista de Derecho Penal y Criminología, 24,* 261-264. https://revistas.uned.es/index.php/RDPC/article/view/29070 [Consultado el 03.07.2023].

Lucendo – Monedero, A. L., Jordá – Borrell, R. y Ruiz – Rodríguez, F. (2015). Predictive model for areas with illegal landfills using logistic regression. *Journal of Environmental Planning and Management, 58.* DOI: 10.1080/09640568.2014.993751.

Marmolejo de Oro, G. A. (2012) Disposición Ilegal de Escombros y Localización Espacial de Conflictos Ambientales [Tesis para pregrado]. Departamento de Geografía, Facultad de Humanidades, Universidad del Valle, Santiago de Cali. Colombia.

Medina Ariza, J.J. (Coord.) (2022). *Instituciones de Control del Delito.* Dykinson.

Ministerio de Hacienda y Función Pública. https://www.hacienda.gob.es/es-ES/Paginas/Home.aspx [Consultado el 03.07.2023].

Ministerio de Transición Ecológica y Reto Demográfico. https://www.miteco.gob.es/es/ [Consultado el 03.07.2023].

Mol, Hanneke, Rodríguez, David, South Nigel y Brisman, Avi (2017). *Introducción a la Criminología Verde.* Temis.

Nelles, M., Grünes, J. y Morscheck, G. (2016). Waste Management in Germany – Development to a Sustainable Circular Economy? *Procedia Environmental Sciences, 35,* 6-14. DOI: 10.1016/j.proenv.2016.07.001.

Ortiz García, J. (2022). *Mito o Realidad: Un estudio criminológico de las áreas rurales de Extremadura.* Dykinson.

Planelles A. y Gómez M.V. (20 de febrero de 2022). De Doñana a las aguas fecales, los vertederos y los plásticos: España es el país europeo con más infracciones medioambientales abiertas. *El país.* https://elpais.com/clima-y-medio-ambiente/2022-02-20/de-donana-a-las-aguas-fecales-los-vertederos-y-los-plasticos-espana-es-el-pais-europeo-con-mas-infracciones-medioambientales-abiertas.html [Consultado el 03.07.2023].

PIREX (Plan Integral de Residuos de Extremadura 2016 – 2022). http://extremambiente.juntaex.es/files/2017/P_AMBTAL/RESIDUOS/PIREX/PIREX_2016_2022.pdf [Consultado el 03.07.2023].

Puerta Ortiz, I. M. (2019). *Impacto Ambiental en las escombreras. Revisión de la Literatura. 2008-2019.* [Trabajo Fin de Estudios]. https://repository.urosario.edu.co/items/6a0f065f-fc9a-470d-a596-17e92e77b8b7

Qgis (versión 3.28 Firenze). https://www.qgis.org/es/site/. [Consultado el 03.07.2023].

Rio Merino, M. Del, Izquierdo Gracia, P., Salto – Weis, A. Isabel and Santa Cruz Astorqui, J. (2010). La regulación jurídica de los residuos de construcción demolición (RCD) en España. El caso de la comunidad de Madrid. *Informes de Construcción, Vol. 62* (517), 81-86. DOI: 10.3989/ic.08.059.

Salafranca Barreda, D. y Maldonado Guzmán, D. (2018). Perfil geográfico de incendiarios urbanos. *Revista Española de Investigación Criminológica, 16,* 1-34. DOI: 10.46381/reic.v16i0.197

Sánchez Nuñez, J.M., Velázquez Serna, J., Serrano Flores, M.E., Ramírez Treviño, A. Balcázar Vázquez A. y Quitero Rodríguez, R. (2009) Criterios ambientales y geológicos básicos para la propuesta de un relleno sanitario en Zipnapécuaro, Michoacán, México. *Boletín de la Sociedad Geológica Mexicana, 61,* 305 324.

Seeboonruang, U. (2016). Geographic information system-based impact assessment for illegal dumping in borrow pits in Chachoengsao Province, Thailand. *The Geological Society of America, 520,* 393 – 405. DOI: 10.1080/01431160701311317.

Seror, N. y Portnov, B.A. (2018) Identifying areas under potential risk of illegal construction and demolition waste dumping using GIS tools. *Waste Management, 75,* 22-29. DOI: 10.1016/j.wasman.2018.01.027.

Serrano Maíllo, A. (2017). *Teoría Criminológica. La explicación del delito en la sociedad contemporánea.* Dykinson.

Serrano Maíllo A. (2021). *El resurgimiento de la criminología científica en América Latina.* Dykinson.

Sistema Integrado de Datos Municipales (SIDAMUN). Ministerio para la transición ecológica y el reto demográfico. Disponible en: https://public.tableau.com/views/SistemaIntegradodeDatosMunicipales/Portada?:language=es-ESy:display_count=ny:origin=viz_share_link?:showVizHome=no [Consultado el 03.07.2023].

Statistical Package for the Social Sciences (SPSS). Servidor Arquímedes. Universidad de Extremadura. Disponible en: www.arquimedes.unex.es. [Consultado el 03.07.2023].

Sujauddin, M., S.M.S. Huda, and A.T.M. Hoque. (2008). Household Solid Waste Characteristics and Management in Chittagong, Bangladesh. *Waste Management, Vol. 28* (9),1688-1695. DOI: 10.1016/j.wasman.2007.06.013

Tasaki, T, Kawahata, T., Osako, M, Matsui, Y, Takagishi, S. Morita, A. y Akishima, S. (2007). A GIS-Based Zoning of Illegal Dumping Potential for Efficient Surveillance. *Waste Management, Vol. 27*(2), 256 – 267. DOI: 10.1016/j.wasman.2006.01.018.

Triassi, M., Alfano, Rossella, Illario, Maddalena, Nardone, Antonio, Caporale, Oreste y Montuori, P. (2015). Environmental Pollution from Illegal Waste Disposal and Health Effects: A Review on the “Triangle of Death”.

Zou, X. (2011). Municipal Solid Waste Management in China with Focus on Waste Separation. Master’s Diss. Institute for Applied Material Flow Management (IfaS), Ritsumeikan Asia Pacific University, Japan. http://r-cube.ritsumei.ac.jp/bitstream/10367/3651/1/51209626.pdf

Capítulo 6

Detección de vertidos ilegales de residuos de construcción y demolición mediante fotointerpretación de modelos procedentes del DEM

MANUEL SÁNCHEZ FERNÁNDEZ[1]

1. INTRODUCCIÓN

En muchas regiones de nuestro país es habitual encontrar residuos sólidos derivados de la actividad de construcción y demolición esparcidos en campos y caminos. Lamentablemente se trata de un problema global que afecta a muchos países de nuestro entorno. Nos referimos a los ya referidos residuos RCD, cuyo acopio, almacenamiento y tratamiento queda regulado por el RD 105/2008 (Real Decreto 105/2008, de 1 de Febrero, por el que se regula la Producción y Gestión de los Residuos de Construcción y Demolición, 2008).

El trabajo realizado se centra en la detección visual de acopios de residuos de construcción y demolición, a partir de técnicas de teledetección, en el área urbana y periurbana de la ciudad de Mérida (España). En la actualidad, las contribuciones científicas en esta área se pueden dividir según diferentes campos de estudio.

---

1 Doctor, Graduado en Ingeniería Civil por la Universidad de Extremadura.

Du et al., (Du et al., 2021), realiza una revisión crítica de los diferentes aspectos que abarcan las contribuciones científicas de otros autores con relación a los vertidos de residuos. El trabajo abarca aspectos como la identificación de los contaminantes posibles del suelo, el seguimiento y control de los vertidos, detección de vertidos o mejoras en la gestión y toma de decisiones, ello analizado de forma general para las diferentes tipologías de vertidos posibles (Du et al., 2021).

En relación a la ubicación o emplazamiento del residuo, las contribuciones científicas se pueden separar en dos campos de estudio, los trabajos dirigidos hacia la predicción probabilística de la existencia de residuo en una ubicación determinada dentro de un marco geográfico (Biotto et al., 2009; Yadav, 2013; Jordá-Borrell et al., 2014; Quesada-Ruiz et al., 2019; Seror y Portnov, 2018;) y los trabajos encaminados a la detección de vertidos existentes (Glanville y Chang, 2015; Manzo et al., 2016; Angelino et al., 2018; Torres y Fraternali, 2021; Shahab y Anjum, 2022). De forma general los estudios citados se basan en datos espaciales y productos de teledetección analizados mediante el empleo de Sistema de Información Geográfica (GIS). Padubidri et al., (2022) proponen una metodología para identificación de vertidos ilegales a partir de ortofotografías aéreas empleando *Deep Learning*, los resultados muestran un déficit de información disponible en este sentido y la necesidad de entrenar el sistema de clasificación con imágenes sintéticas. Otros autores (Akinina et al., 2017; Parrilli et al., 2021) emplean sistemas de clasificación diferentes, integrado datos de fuentes de diferente naturaleza, como información espectral, ortofotografías aéreas, índices, etc. En esta línea Notamicola et al. (2004) realizan análisis directamente sobre los mapas generados a partir de información multiespectral procedente de sensores satelitales, ortofotografías y mapas de coberturas del suelo. En relación con los resultados y conclusiones de los trabajos revisados se observa que los vertidos detectados con frecuencia se ubican en zonas en que han existido vertidos históricos y zonas antrópicas relacionadas con las vías de comunicación y el desarrollo urbanístico. Emplean los datos de un mapa de cobertura del suelo para el análisis realizado poniendo de manifiesto la existencia de una relación entre el uso del suelo y la existencia o no de un vertido ilegal.

Este trabajo propone el empleo del Modelo Digital del Terreno (DEM) para la detección de vertidos de los residuos de construcción y demolición (RCD) o también llamados en inglés "Construction Demolition Waste" (CDW). El estudio parte de la singularidad topográfica de los vertidos de RCD, bien sean acopios individuales o zonas de vertederos con múltiples acopios. Para la mejora de la visualización e interpretación del DEM se han empleado las herramientas del software *Relief Visualization Toolbox (RVT)*. Estas herramientas han sido implementadas en otras disciplinas como estudios arqueológicos en los cuales permiten destacar estructuras desaparecidas o semi enterradas en el terreno existente. Dcihas estructuras provocan la modificación puntual del terreno generando túmulos o depresiones identificables por la topografía del terreno (Devereux et al., 2008; Hesse, 2010; Bennett et al., 2012; Guyot et al., 2021; Štular et al., 2021).

El estudio realizado evalúa el empleo de modelos de visualización del terreno obtenidos a partir del DEM para la detección visual de vertidos de residuos de construcción y demolición. De los siete modelos obtenidos para la mejora de la visualización del terreno mediante la herramienta RVT dos han permitido ubicar vertidos en condiciones determinadas y han permitido ubicar de forma sencilla emplazamientos naturales de antrópicos.

2. ÁREA DE ESTUDIO

La población, desarrollo urbanístico, orografía, influencia fluvial o costera entre otros son factores que afectan al comportamiento de la población y, por tanto, a los hábitos de vertidos de residuos. La zona de estudio se ha entrado en la hoja 777 de la distribución de hojas de escala 50.000 del IGN centrada en el municipio de Mérida (véase figura 1). El municipio de Mérida es una ciudad de pequeño tamaño (59.234 habitantes[2]) situado en un cruce de dos de las autovías principales de Extremadura (A-5 y A-66). La ciudad creció en población un 20% en la primera década del año 2000, y consecuentemente de forma simultánea se desarrolló urbanísticamente. La zona de estudio contiene más municipios de menor tamaño, influenciados en su desarrollo por la ciudad de Mérida. Parte de estos municipios, en la actualidad, funcionan como residencia dormitorio de la ciudad principal. Esta dependencia implica la existencia diaria tránsito de personas y profesionales entre Mérida y los municipios aledaños. La orografía de la zona de estudio es predominantemente llana, marcada por las vegas del río guadiana y por la existencia de varios embalses. Dentro del marco de estudio existen también algunas zonas de sierra como la zona de Valverde de Mérida, la sierra de las Cabrerizas próximas a Arroyo de San Serván y parte del Parque Natural de Cornalvo (en figura 1).

2 Según datos del padrón municipal del Instituto Nacional de Estadística. Disponible aquí: https://www.ine.es/jaxiT3/Datos.htm?t=2859 [Consultado el 03.07.2023].

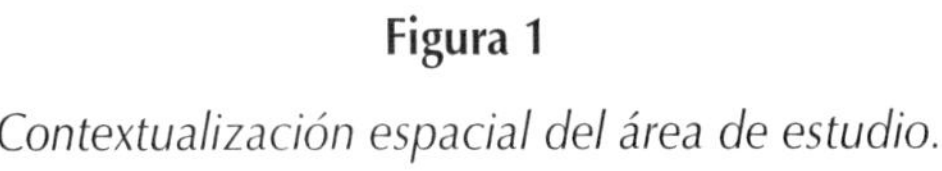

Figura 1

Contextualización espacial del área de estudio.

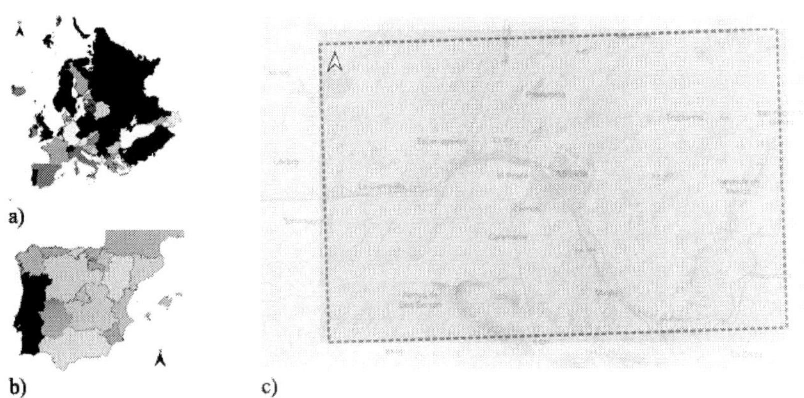

Nota: a) Emplazamiento de España (en azul) en Europa; b) emplazamiento de Extremadura (en verde) en España (gama de azules); c) Zona de estudio, (hoja de cincuenta mil 777 en rojo).

En una etapa inicial del estudio se ha procedido a la identificación de ubicaciones con vertidos de RCD. Esta fase se ha centrado en la zona periurbana de Mérida principalmente. 494 ubicaciones con vertidos han sido identificadas. Los vertidos se emplazan principalmente en parcelas no construidas, en urbanizaciones no consolidadas, en el margen de caminos de acceso a fincas y en parcelas no urbanas. A partir de la ubicación de los 494 vertidos se ha realizado una clasificación de la tipología de parcela en que se encuentra el vertido en relación con el uso del suelo. De la clasificación realizada se extraen 5 tipologías de suelo en función de su naturaleza geográfica y urbanística. Un total de 14 ubicaciones de dimensión 450 m x 450 m han sido extraídas (en figura 2). Las ubicaciones seleccionadas contienen el 84,4% de los vertidos localizados en los trabajos de inspección. Las tipologías de suelo establecidas son:

- Parcelas en suelo no urbano (3 emplazamientos). Esta tipología de suelo se refiere a parcelas rústicas en mitad de las cuales existen vertidos ilegales de RCD. En este caso puede hacerse una distinción entre las parcelas que fueron destinadas al uso de vertedero, anterior al 2011, que puedan haber generado hábito en la población de verter en ese punto los residuos y parcelas cuya topografía es el terreno natural.

- Suelo no urbano límite con suelo urbano (4 ubicaciones). Son las zonas de interfaz donde el final de un vial pavimentado de un municipio da comienzo a un camino rústico o zonas en las cuales existen viviendas directamente adosadas a parcelas rústicas.
- Caminos emplazados en suelo no urbano (3 ubicaciones). Caminos en suelo rústico que puedan conectar barrios, núcleos de población o fincas privadas.
- Suelo urbanizado (3 ubicaciones). Contempla toda aquella urbanización no consolidada en la cual existen una parte del terreno sin construir.
- Suelo urbano (1 ubicación). Únicamente se ha hallado una parcela de suelo urbano. Este es un caso singular ya que, aunque siendo urbano, es una parcela aislada y no edificada.

Figura 2

Parcelas de estudio empleadas en el trabajo

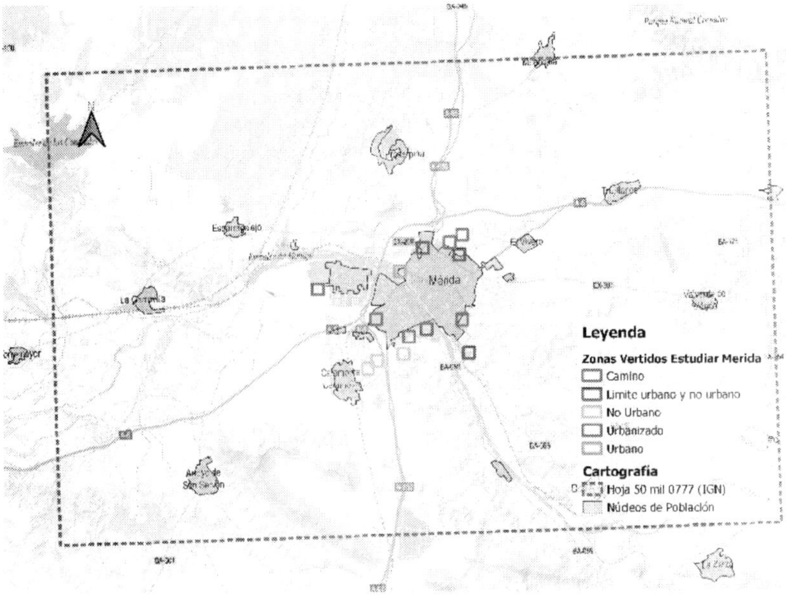

3. MATERIALES Y MÉTODOS

Este trabajo propone una metodología para la detección de los residuos mediante fotointerpretación de conjuntos de datos obtenidos a partir de modelo digital del terreno. Los conjuntos de datos calculados contribuyen a una mejor identificación geométrica del terreno realzando diferentes características a pequeña escala. Una de las consideraciones a tener en cuenta en el proceso de cálculo es el tamaño de los vertidos, estos pueden ir desde acopios de dimensiones, ancho, alto y largo, de 50 centímetros hasta varios metros. Este hecho implica la necesidad del empleo de modelos digitales del terreno de alta resolución espacial que permita identificar los acopios de menor dimensión. En la figura 3 se muestran diferentes acopios identificados en localizaciones en diferentes tipologías de suelos.

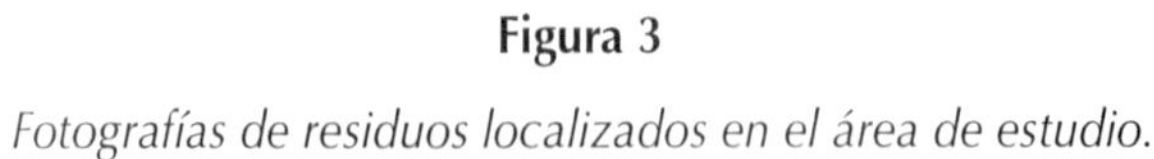

Figura 3

Fotografías de residuos localizados en el área de estudio.

Nota: a) y b) terreno no urbano; c) y d) en límite entre urbano y no urbano; e) urbano; f) margen de camino.

En la figura 4 se muestra el flujo de trabajo establecido en el estudio. Para la obtención del modelo digital del terreno se ha empleado la nube de puntos disponible por el proyecto PNOA LiDAR segunda cobertura del IGN. Según especificaciones técnicas del producto, la resolución del producto LiDAR empleado es de 1 punto por m2. La nube de puntos descargada se encuentra clasificada en 12 clases de los cuales la segunda clase se corresponde con los puntos de "suelo", (Martínez et al., 2015). A partir de los puntos clasificados como "suelo" se obtiene un DEM en formato ráster mediante una interpolación lineal. El tamaño de pixel del ráster de salida se establece en 0.25 m.

Figura 4

Flujo de trabajo implementado en el estudio.

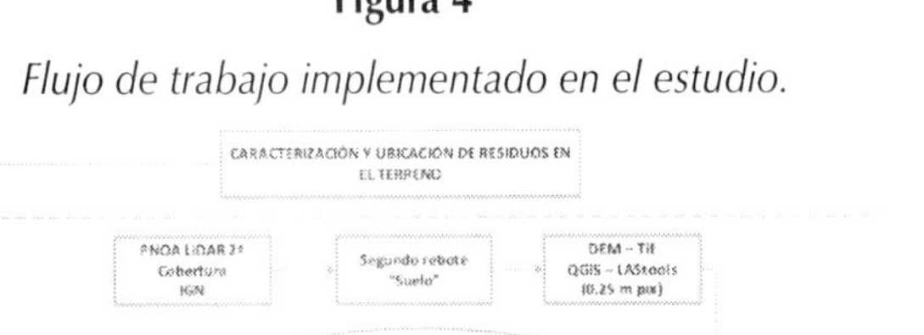

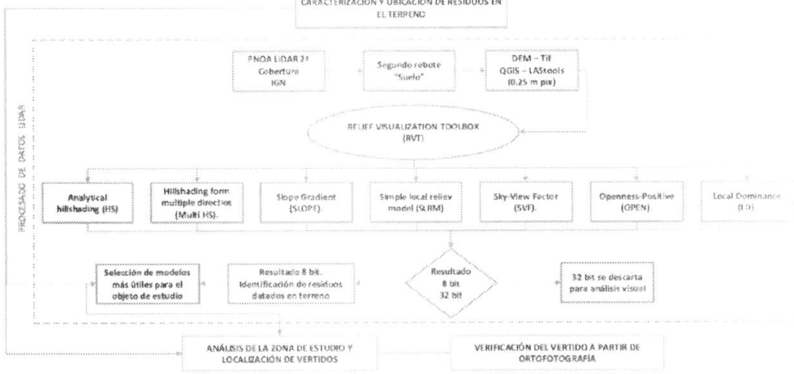

A partir del DEM generado se han obtenido siete modelos que muestran diferentes características del terreno basadas en la altimetría de la zona de estudio. Los modelos han sido calculados mediante el software *Relief Visualization Toolbox (RVT)*. Los resultados obtenidos resaltan pendientes, sombras y otras particularidades del terreno lo cual permite estudiar de forma visual si existen o no vertidos de residuos de construcción y demolición. Para el empleo del software RVT han sido tenidas en cuenta las consideraciones indicadas por otros autores y manual del software en cuanto a los parámetros empleados en los cálculos. Las herramientas empleadas han sido las siguientes (Zakšek et al., 2011; Kokalj et al., 2016; Kokalj y Hesse, 2017; Kokalj y Somrak, 2019):

1. *Analytical hillshading (HS)*. Permite observar el terreno a partir de las sombras generadas por un patrón de iluminación con una dirección y elevación determinada. Esta herramienta permite revelar características con poca luz en zonas llanas a partir de sombras oscuras y zonas muy iluminadas. Parámetros:
   - Sun azimuth 315 deg; Sun elevation angle 35 deg.

2. *Hillshading form multiple directios (Multi HS).* Parte del concepto de cálculo del "Analytical hillshading". Esta herramienta calcula las zonas iluminadas y de sombras en múltiples direcciones y realiza una representación en RGB de los patrones de sombras calculados. En este caso se han realizado los cálculos para 16 direcciones. Parámetros:
    - 16 direcciones; Sun elevation angle 35 deg.
3. *Slope Gradient (SLOPE).* Representa la tasa máxima de cambio entre cada celda y sus vecinas. Cuando se presenta en una escala de grises invertida (las pendientes pronunciadas son más oscuras). Esta herramienta permite observar con facilidad cambios bruscos de pendientes en el terreno.
4. *Simple local reliev model (SLRM).* Esta herramienta elimina los elementos morfológicos a gran escala (colinas, valles...) de los datos, de modo que sólo quedan los rasgos a pequeña escala. Parámetros:
    - *Radius for trend assessment* 17 pixeles.
5. *Sky-View Factor (SVF).* Es un indicador indirecto de la iluminación difusa y mide la proporción de cielo visible desde un punto determinado. En una imagen SVF las elevaciones se resaltan y aparecen en colores claros a blancos y las depresiones en colores oscuros. Parámetros:
    - Número de direcciones de búsqeuda 16; Radio de búsqueda 10 pixeles.
6. *Openness-Positive (OPEN).* El valor medio de todos los ángulos cenitales da una apertura positiva, mientras que el valor medio del nadir da una apertura negativa. Parámetros:
    - Parámetros igual a la herramienta Sky-View Factor
7. *Local Dominance (LD).* Esta herramienta ofrece buenos resultados para formaciones topográficas que presentan desniveles como puede ser una colina o un hueco en el terreno. Parámetros:

- Radio mínimo 15 pixeles y radio máximo 25 pixeles.

A partir de los siete modelos obtenidos se ha procedido a analizar el modo en que quedan representados los residuos registrados en el terreno en cada uno de los modelos obtenidos. El análisis abarca la identificación del propio vertido en los modelos y la caracterización del terreno en que se encuentra. Por último, una vez seleccionados los modelos en los cuales se identifican con mayor claridad los residuos se realiza una búsqueda de vertidos en el área de estudio. Para una optimización del análisis para este procedimiento se divide el área de estudio en áreas de 3695 x 4862 m obteniéndose 32 subzonas. La validación de la existencia de residuo se realiza a partir de la ortofotografía del PNOA del IGN de 0.25 m de resolución espacial.

4. RESULTADOS Y DISCUSIÓN

4.1. Formato de los resultados obtenidos del RVT

Las herramientas empleadas en el software RVT muestran los resultados en 32 y 8 bit. Las indicaciones del software muestran que los resultados expresados en 8 bit están optimizados para su visualización en software que no son herramientas GIS. En un primer análisis, se ha seleccionado para cada una de las herramientas empleadas el resultado que mejor permita identificar los cambios del terreno en relación con la naturaleza del estudio. La figura 5 muestra para una de las parcelas de estudio los resultados obtenidos para los modelos calculados en 8 bit y 32 bit. Los cálculos de Multi HS y LD no muestran resultados en 8 bit. La parcela mostrada en la figura 5 es una zona que previo a 2011 había sido empleada para el uso de vertidos de residuos de demolición y construcción, en la actualidad continúan realizándose vertidos ilegales, por tanto, el número de vertidos existentes es significativo. De forma general se observa que, el resultado en 8 bit presenta una escala de grises más amplia que el resultado expresado en 32 bit, por tanto, permite identificar mejor los cambios del terreno sin necesidad de ajustar la escala de grises (véase figura 5).

Figura 5

Comparativo entre resultados en 8 bit y 32 bit de los resultados obtenidos en una de las parcelas de estudio.

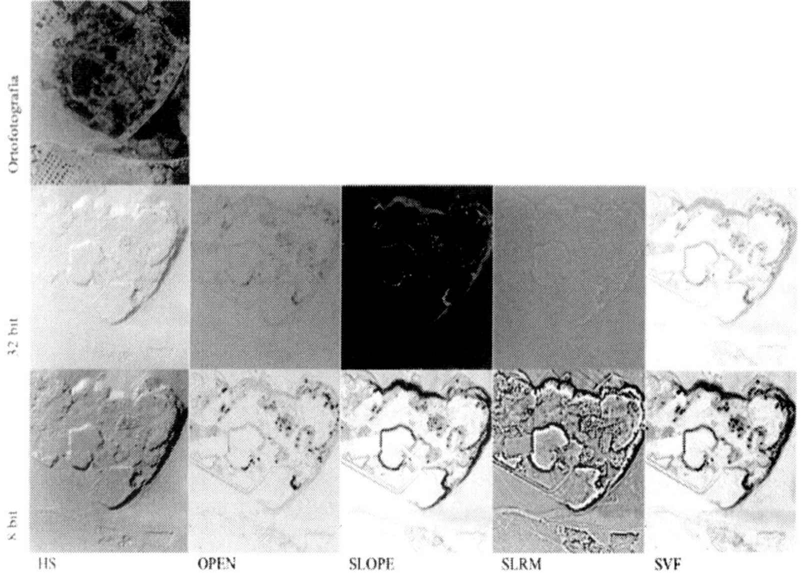

El resultado HS en 32 bit muestra una escala de grises a partir de la cual gran parte del terreno se representa en tonos de grises más claros, ello hace que los puntos más altos y las pendientes más pronunciadas no se signifiquen de igual modo al resultado de 8 bit, en el cual las pendientes pronunciadas y los "picos" de vertidos puntuales se representan en tonos más oscuros o más claros que el resto del terreno y resultan más sencillo su identificación (véase figura 5). Algo similar al resultado de HS ocurre con el resultado OPEN y de SVF. El ráster obtenido en 8 bit presenta una escala de grises más amplia y por tanto tiene mayor capacidad de representar pequeñas diferencias existentes en el modelo en los tonos más claros y oscuros. Al igual en los casos mencionados para las herramientas SLRM y SLOPE el resultado de 32 bit también muestra una escala de grises de menor rango de colores que el resultado de 8 bit, no obstante, en estos casos el resultado de 32 bit muestra en color significativamente más claro al resto de la imagen los puntos en los que existen sobreelevaciones puntuales en el terreno. Estas sobreelevaciones pueden ser vertidos de residuos, o puntos dispersos no filtrados de árboles, muros u otros elementos verticales. No obstante, la morfología del terreno continúa siendo más apreciable en el resultado en 8 bit (figura 5).

Del análisis preliminar realizado, en el que se comparan los dos resultados mostrados por el *software*, para cada una las herramientas empleadas (8 bit y 32 bit), se observa que, de forma generalizada los resultados expresados en 8 bit permiten distinguir con mayor claridad la morfología del terreno, en consecuencia, se descarta el empleo de los resultados expresados en 32 bit en los procesos posteriores.

4.2. Análisis de las zonas de vertidos localizados en campo

La morfología del terreno está condicionada en relación con la orografía y geología de la zona, el desarrollo urbano, la afección antrópica, etc. Ello implica, que, para cada tipo de suelo, la representación del terreno en los diferentes modelos calculados variará. En este caso la orografía y geología de la zona está marcada por el área de estudio en que se desarrolla la investigación. Dentro de esta área de estudio se han clasificado 5 tipologías de suelo en relación con la ubicación de los vertidos registrado en el terreno. En las figuras 6 – 11 se analizan los resultados de los modelos obtenidos para cada una de las tipologías de suelo establecidas en el trabajo. Con ello se pretende establecer cuál o cuáles de los modelos son los más adecuados para ser empleados según la naturaleza del trabajo. Para el análisis del tipo de suelo no urbano se ha estimado necesario dividir en dos subtipos, por un lado, parcelas que hayan sido modificadas de forma antrópica como son los vertederos históricos y terrenos con una topografía natural.

La figura 6 muestra el resultado obtenido para una parcela no urbana cuyo uso anterior al año 2011 fue el de vertedero de residuos de demolición y construcción. La morfología del terreno se muestra con claridad en los siete modelos. El modelo en que peor se representa, en este caso, es en el OPEN, debido a estrecha gama de grises que empela para la representación del terreno. Sin embargo, en el que mejor se observa la configuración del terreno es el Multi HS debido a la gama de colores RGB que resaltan los taludes. Los vertidos existentes, cuya particularidad topográfica es la de una sobreelevación puntual del terreno queda representada en el modelo LD como puntos blancos individuales, éstos pueden quedar ocultos a la vista ya que de igual modo taludes verticales y coronación de taludes se muestran de igual modo en tonos claros. Algo similar al LD ocurre con el modelo SLRM, en este caso los puntos blancos individuales se unen con los taludes y coronación de taludes de forma que resulta más complejo separar de éstos los acopios de vertidos.

Figura 6

Resultados obtenidos para el tipo de suelo no urbano. Emplazamiento de los vertidos en parcela de vertedero de residuos de demolición en una parcela.

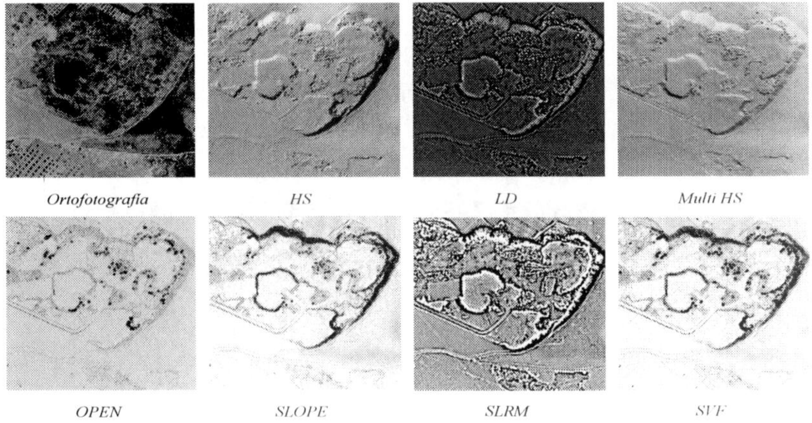

Nota: En rojo los puntos de vertido.

En la figura 7 se muestra una parcela no urbana con una topografía natural en gran parte de la escena. La orografía de la zona es principalmente llana, en este caso resulta complejo observar la morfología del terreno en una parte de los modelos calculados. Los modelos LD y SLRM son en los que se destaca mejor la orografía y los pequeños elementos contenidos en el terreno como caminos, lindes o cauces. El modelo Multi HS también permite observar la morfología del terreno, aunque en este caso los elementos de la imagen se encuentran menos marcados. En la zona analizada los vertidos están ubicados, bien en el margen de uno de los caminos o bien en el interior de las parcelas. Únicamente el modelo LD puede indicar de forma discreta la existencia de vertidos, en este caso, en el entorno de los vertidos, existen puntos con mayor altura y en consecuencia los puntos marcados en tonos de gris claro o blanco no se corresponden con vertidos. Del modelo de SLRM también indica de forma sutil la posibilidad de la existencia de los vertidos, pero queda eclipsado por la coronación de taludes de caminos y elementos verticales. El resto de los modelos calculados no permite determinar la existencia o no de posibles vertidos de demolición y construcción.

Figura 7

Resultados obtenidos para el tipo de suelo no urbano. Emplazamiento de los vertidos en caminos y parcelas rusticas.

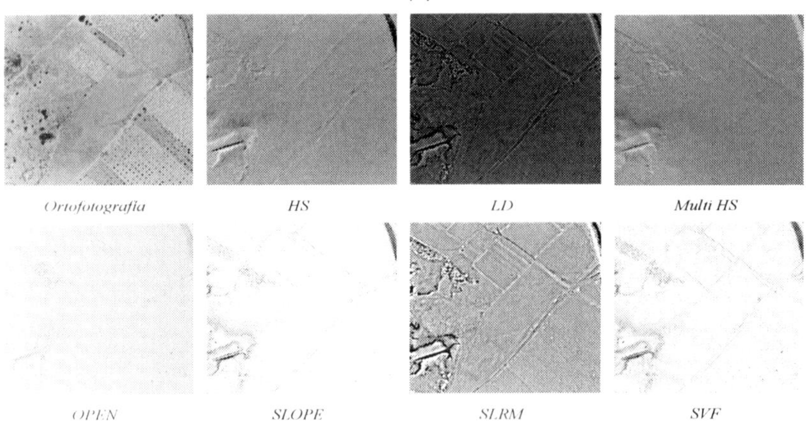

Nota: En rojo los puntos de vertido.

El límite entre el suelo urbano y no urbano es una zona donde pueden encontrarse parcelas no urbanas situadas entre dos parcelas construidas. En la figura 8 se muestran los resultados obtenidos en una parcela límite entre urbano y no urbano, en este caso la topografía natural de la parcela es característica de una zona llana. Al igual que ocurre en la figura 7 parte de los modelos calculados no representan con detalle la topografía de la parcela, estos son OPEN, SLOPE y SVF, el resto de los filtros (HS, LD Multi HS y SLRM) de forma general permiten conocer la topografía de la parcela y elementos contenidos en el terreno como caminos, lindes o cauces. La detección de residuos en este caso resulta compleja dada la orografía del terreno y la existencia de elementos verticales (construcciones) que representan puntos elevados en los modelos. El modelo LD es el único que representa puntos altos el color blanco que están ubicados en las zonas de los vertidos, no obstante, parte de estos puntos se corresponden con coronación de taludes de caminos, explanaciones o construcciones no siendo suficiente para determinar con exactitud la existencia de residuos. Otro de los modelos, el SLRM también permite identificar zonas sobreelevadas del terreno, pero en este caso los vertidos quedan camuflados por la coronación de taludes de caminos y construcciones existentes.

Figura 8

Resultados obtenidos para el tipo de suelo límite de urbano con no urbano.

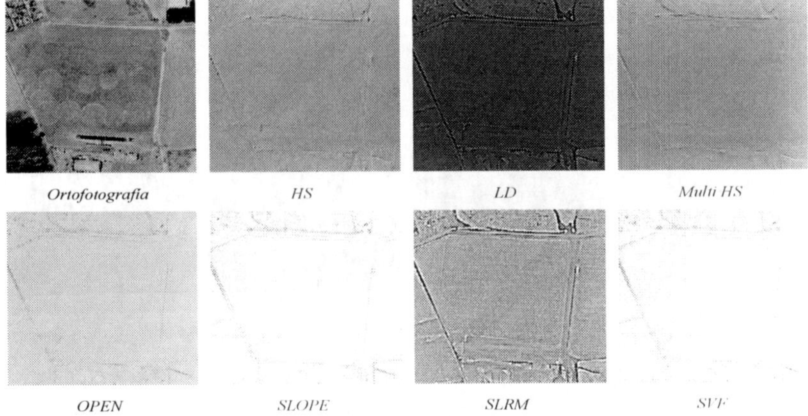

Nota: En rojo los puntos de vertido.

Los caminos rurales son un punto de depósito de vertidos de RCD habitualmente ligados a urbanizaciones próximas o itinerarios de conexión entre dos puntos de interés (Jordá-Borrell et al., 2014; Seror y Portnov, 2018; Quesada-Ruiz et al., 2019). La figura 9 muestra los modelos obtenidos para una zona en la cual existen vertidos en los márgenes de caminos, siendo además una zona de interfaz entre urbano y rústico, y el camino donde se encuentran los vertidos da acceso a una "urbanización" satélite distanciada del casco urbano. Por la orografía del terreno, los modelos OPEN, SLOPE y SVF, al igual que ocurre en los casos mostrados en las figuras 7 y 8, no reflejan en detalle los elementos contenidos en el terreno a diferencia del resto de modelos que con diferente grado de definición sí consiguen representarlos. En cuanto a la representación de los vertidos contenidos en la escena, es el modelo LD el único que permite distinguir en cierto grado la existencia de vertidos. En este caso, al encontrarnos en un escenario límite con suelo urbano en el que existen taludes verticales, se observa que la coronación de taludes queda reflejada de igual modo a un vertido, con la diferencia de que una coronación de talud se refleja de forma lineal y, de forma general, un vertido se representa de forma puntual e incluso múltiples vertidos juntos son representados como una sucesión de puntos.

Figura 9

Resultados obtenidos para el tipo de suelo "caminos".

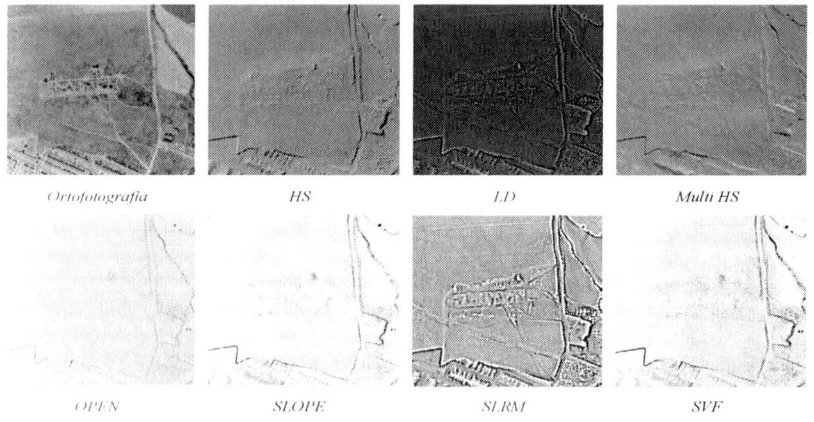

Nota: En rojo los puntos de vertido.

Las urbanizaciones no consolidadas son emplazamientos a menudo no vigilados ni transitados en los que existen parcelas vacías que en un futuro van a ser construidas. Por ello se puede tender a usarlas como zona de acopio puntual de tierras o vertidos o como zonas de préstamo de tierras adelantando en el tiempo la excavación de la cimentación de una futura edificación. En la figura 10 se muestra el resultado obtenido para una urbanización no consolidada. En este caso se observa que todos los modelos representan de forma adecuada la morfología del terreno destacando los viales construidos e incluso los movimientos de tierras realizados en parte de las parcelas existentes. Dentro de los modelos generados se destacan el LD y Multi HS, que reflejan un mayor nivel de detalle de la morfología del terreno, seguidos del HS y SLRM. Para la detección de vertidos es el modelo LD el que permite ubicar en cierto grado los vertidos existentes, aunque al igual que ocurre en los escenarios anteriores, este modelo también representa de igual modo a los vertidos las coronaciones de taludes y elementos verticales por lo que en ciertos casos es complejo saber si existe vertido o simplemente es un elemento antrópico como caminos, calles, muretes o terraplenes.

Figura 10

Resultados obtenidos para el tipo de suelo urbanizado no consolidado.

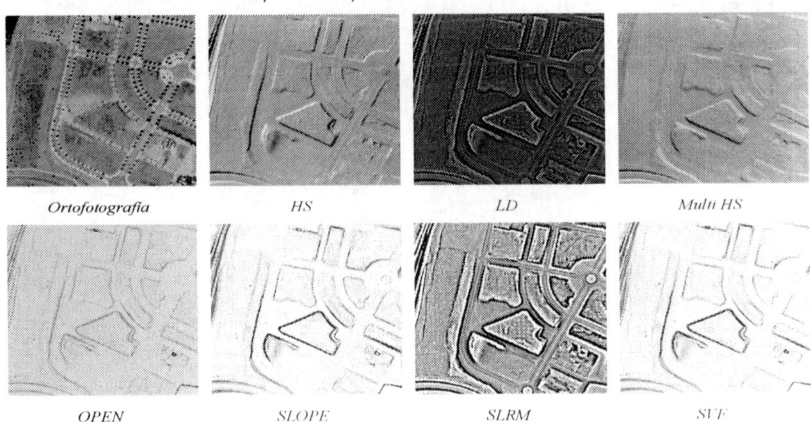

Nota: En rojo los puntos de vertido.

Por último, en la figura 11 se muestran los resultados para un escenario de una zona urbana. En este caso es una parcela urbana singular ya que no se encuentra construida, el uso es de recinto ferial y se encuentra aislada del terreno urbanizado en parte de sus márgenes por el cauce de un río. Este escenario podría ser integrado en el de suelo urbanizado no consolidado (figura 10) ya que comparte características con el suelo urbanizado, aislado y no edificado. Al igual que ocurre con parte de los escenarios estudiados, los modelos HS, LD y Multi HS son los que mejor han recogido la morfología del terreno, y LD el modelo que mejor representa la posibilidad de existencia de vertidos en la escena. Se ha de tener en cuenta que en este caso el talud del margen del río y los taludes de la explanada de construcción y viales próximos están representados de igual modo a los vertidos en el modelo LD y, por tanto, diluyen la capacidad de destacar estos.

Figura 11

Resultados obtenidos para el tipo de suelo urbano.

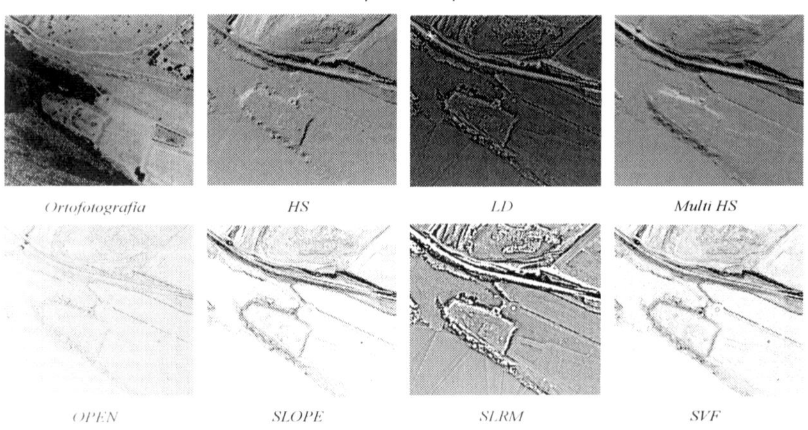

Nota: En rojo los puntos de vertido.

En las tablas 1 y 2, se muestra un resumen de la capacidad de representación de la morfología del terreno y la existencia de vertido por los modelos calculados según los casos estudiados. La escala empleada para la calificación ha sido mediante los valores enteros de 1 a 5, representando el valor 1 una muy mala capacidad de representación y 5 una buena capacidad de representación. El resultado mostrado en las tablas 1 y 2 es el promedio, redondeado a la unidad, de las 14 parcelas estudiadas en total en relación con la clase de suelo correspondiente. Como resultado general se puede observar que para la representación del terreno los modelos más adecuados son LD y Multi HS y para la representación de los vertidos únicamente el modelo LD.

Tabla 1

Resumen de adecuación de los modelos generados a las tipologías de suelo estudiadas en relación con la capacidad de representar la morfología del terreno.

	HS	*LD*	*Multi HS*	*OPEN*	*SLOPE*	*SLRM*	*SVF*
Suelo no urbano - 1	5	5	5	3	4	4	5
Suelo no urbano - 2	3	5	4	1	1	5	1
Limite urbano / no urbano	4	4	5	1	1	4	1
Caminos	4	4	5	1	1	5	1
Suelo urbanizado	5	5	5	3	4	4	5
Urbano	5	5	5	3	3	4	3
PROMEDIO	4	5	5	2	2	4	3

1	muy mal	2	mal	3	regular	4	mejorable	5	bien

Tabla 2

Resumen de adecuación de los modelos generados a las tipologías de suelo estudiadas en relación con la capacidad de representar la existencia de vertidos de demolición y construcción.

	HS	*LD*	*Multi HS*	*OPEN*	*SLOPE*	*SLRM*	*SVF*
Suelo no urbano - 1	1	5	1	4	3	3	3
Suelo no urbano - 2	1	3	1	1	1	2	1
Limite urbano / no urbano	1	4	1	2	2	3	1
Caminos	1	4	1	1	1	2	1
Suelo urbanizado	1	4	1	1	1	2	1
Urbano	1	4	1	1	1	2	1
PROMEDIO	1	4	1	2	2	2	1

1 muy mal 2 mal 3 regular 4 mejorable 5 bien

4.3. Aplicación de la metodología

Una vez establecida la metodología de trabajo se ha realizado un análisis en el marco de la hoja de cartografía 50 mil del IGN 0777 evaluando las posibles zonas de vertidos existentes. La zona de estudio, en parte, se encuentra centrada en área urbana de Mérida, conteniendo también otros núcleos urbanos como los de los municipios de Esparragalejo, Calamonte, Arroyo de San Serván, Trujillanos y Valverde de Mérida (figura 2). En la figura 12 se representan mediante puntos las zonas identificadas en el modelo LD con posibilidad de contener vertidos de RCD. A partir de las zonas detectadas, en un proceso posterior, se ha verificado mediante fotointerpretación, empleando la ortofotografía del PNOA del año 2019, la existencia de vertidos de RCD. En el proceso de verificación mediante fotointerpretación se ha reseñado la existencia afirmativa de vertidos (SI), si existen objetos que no se identifican con claridad y se asemejan a vertidos (PUEDE), si dadas las condiciones existe la posibilidad de que existan vertidos o hayan existido (PUEDE) y si son zonas antrópicas con otros usos y no existen vertidos (NO). El resultado obtenido está representado en la figura 12. La mayor parte de las zonas donde se han detectado parcelas antrópicas con posibilidad de contener vertidos están ubicadas en las filas 3 y 4, esto es debido a que en las filas 1 y 2 existe un mayor número de fincas de gran tamaño y menor superficie de parcelas de recreo además de un menor número de núcleos de población. La proximidad a un núcleo de población es un factor importante para determinar la posibilidad de la existencia de un vertido reflejado en publicaciones científicas de diferentes autores (Biotto et al., 2009; Jordá-Borrell et al., 2014; Quesada-Ruiz et al., 2019; Seror y Portnov, 2018; Yadav, 2013).

Figura 12

Zonas localizadas en el modelo LD con posibilidad de contener vertidos de RCD

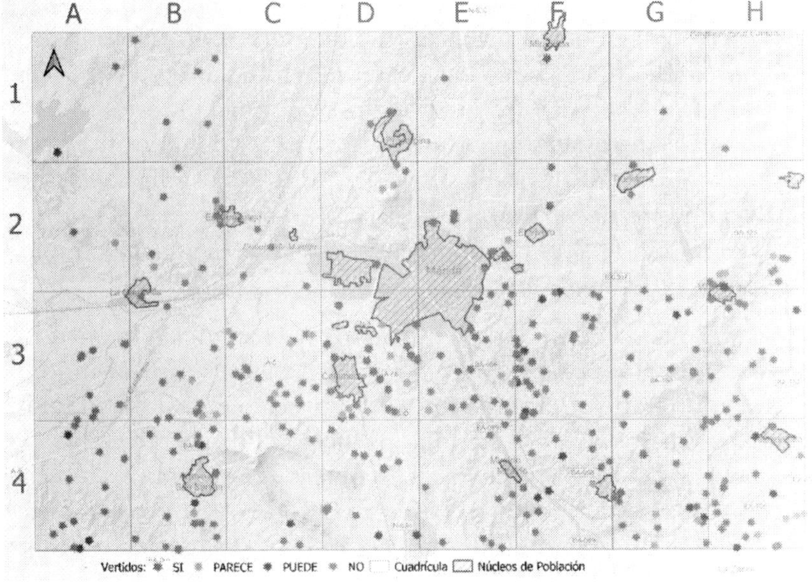

Un total de 322 ubicaciones fueron destacadas sobre el modelo LD como zonas antrópicas en que se pudiera esperar la existencia de vertidos en una superficie de estudio total de 574,66 Km2. En una posterior revisión realizada mediante fotointerpretación de ortofoto en color verdadero se determinó que 216 de las ubicaciones marcadas no contenían residuos (67%), estas ubicaciones representan en el modelo LD errores por vegetación, zonas construidas, charcas o rellenos de tierras de excavación de una misma parcela, entre otros elementos.

Del análisis realizado se observa que los emplazamientos de los vertidos ilegales están relacionados, en cierto grado, con terrenos antrópicos que pueden surgir de obras de viales no finalizadas, urbanizaciones no consolidadas o movimientos de tierras para préstamos o vertederos de tierras por obras.

5. CONCLUSIONES

Dos de los modelos empleados han permitido en todos los escenarios identificar con claridad la morfología del terreno, estos son LD y Multi HS. Estos modelos han permitido identificar morfologías antrópicas en el terreno permitiendo identificar con claridad taludes de terraplenes y desmontes existentes. Para el emplazamiento de vertidos de residuos de construcción y demolición el modelo LD es el que mejor resultados ha ofrecido en los escenarios objeto del estudio.

Ha resultado complejo determinar la existencia de vertidos en casos en que éstos no son verticales, en este caso la modificación topográfica del terreno no es apreciable en los modelos. Una parte de los vertidos estudiados han podido ser ubicados, de forma individualizada, en zonas antrópicas como caminos o urbanizaciones no consolidadas a partir del modelo LD. No obstante, ha resultado complejo la identificación concreta de los residuos individualizados, esto se debe a la existencia de otros elementos representados de igual modo, como es la coronación de un talud, o cualquier otra sobreelevación antrópica, así como los posibles errores existentes en el MDT, fruto de la existencia de puntos no deseados en el segundo rebote de la nube puntos LiDAR. Estos puntos no deseados son potencialmente puntos en copas de árboles, cubiertas, etc. o ruido propio de la nube de puntos que no han sido clasificado de forma adecuada.

La metodología empleada permite diferenciar de forma clara entre terrenos antrópicos y naturales, permitiendo también, según el caso, ubicar vertidos de construcción y demolición. A partir de la detección de terrenos antrópicos en el modelo LD derivado del DEM, en un proceso posterior mediante fotointerpretación usando como base ortofotografías en color verdadero ha sido posible ampliar la base datos de vertidos de partida del trabajo. Se ha analizado la hoja de 50 mil del IGN 0777 de 29 km de ancho x 19 km de alto obteniendo como resultado la localización de nuevas zonas de vertidos no conocidas. Esta metodología no ha dado resultados para la identificación de vertidos singulares o en caminos ya que la representación de los vertidos en el modelo LD es similar a cualquier sobreelevación como son las cabezas de terraplén y desmonte de las infraestructuras viarias.

6. REFERENCIAS BIBLIOGRÁFICAS

Akinina, N. V, Akinin, M. V, Taganov, A. I., y Nikiforov, M. B. (2017). Methods of detection in satellite images of illegal dumps by using a method based on tree classifier. *2017 6th Mediterranean Conference on Embedded Computing (MECO)*, 1–3. DOI: 10.1109/MECO.2017.7977179

Angelino, C. V, Focareta, M., Parrilli, S., Cicala, L., Piacquadio, G., Meoli, G., y Mizio, M. De. (2018). A case study on the detection of illegal dumps with GIS and remote sensing images. In U. Michel y K. Schulz (Eds.), *Earth Resources and Environmental Remote Sensing/GIS Applications IX, Vol. 10790*, 107900M, SPIE. DOI: 10.1117/12.2325557

Bennett, R., Welham, K., Hill, R. A., y Ford, A. (2012). A comparison of visualization techniques for models created from airborne laser scanned data. *Archaeological Prospection, Vol. 19*(1), 41–48. DOI:10.1002/ARP.1414

Biotto, G., Silvestri, S., Gobbo, L., Furlan, E., Valenti, S., y Rosselli, R. (2009). GIS, multi-criteria and multi-factor spatial analysis for the probability assessment of the existence of illegal landfills. *International Journal of Geographical Information Science, Vol. 23*(10), 1233–1244. DOI: 10.1080/13658810802112128

Devereux, B. J., Amable, G. S., y Crow, P. (2008). Visualisation of LiDAR terrain models for archaeological feature detection. *Antiquity, Vol. 82*(316), 470–479. DOI: 0.1017/S0003598X00096952

Du, L., Xu, H., y Zuo, J. (2021). Status quo of illegal dumping research: Way forward. *Journal of Environmental Management, 290,* 112601. DOI: 10.1016/J.JENVMAN.2021.112601

Glanville, K., y Chang, H.-C. (2015). Remote Sensing Analysis Techniques and Sensor Requirements to Support the Mapping of Illegal Domestic Waste Disposal Sites in Queensland, Australia. *Remote Sensing, Vol. 7*(10), 13053–13069. DOI: 10.3390/rs71013053

Guyot, A., Lennon, M., y Hubert-Moy, L. (2021). Objective comparison of relief visualization techniques with deep CNN for archaeology. *Journal of Archaeological Science: Reports, 38,* 103027. DOI: 10.1016/J.JASREP.2021.103027

Hesse, R. (2010). LiDAR-derived local relief models-a new tool for archaeological prospection. *Archaeological Prospection, Vol. 17*(2), 67–72. DOI: 10.1002/ARP.374

Jordá-Borrell, R., Ruiz-Rodríguez, F., y Lucendo-Monedero, Á. L. (2014). Factor analysis and geographic information system for determining probability areas of presence of illegal landfills. *Ecological Indicators, 37*(PART A), 151–160. DOI: 10.1016/J.ECOLIND.2013.10.001

Kokalj, Ž., y Hesse, R. (2017). *Airborne laser scanning raster data visualization: a guide to good practice, Vol. 14.* Založba ZRC.

Kokalj, Ž., y Somrak, M. (2019). Why Not a Single Image? Combining Visualizations to Facilitate Fieldwork and On-Screen Mapping. *Remote Sensing, Vol. 11*(7). DOI: 10.3390/rs11070747

Kokalj, Ž., Zakšek, K., Oštir, K., Pehani, P., Čotar, K., Somrak, M., Kokalj, Ž., y Kokalj, Ž. (2016). Relief Visualization Toolbox, ver. 2.2. 1 Manual. *Remote Sens, Vol. 3*(2), 398–415.

Lu, W. (2019). Big data analytics to identify illegal construction waste dumping: A Hong Kong study. *Resources, Conservation and Recycling, 141,* 264–272.

Manzo, C., Mei, A., Fontinovo, G., Allegrini, A., y Bassani, C. (2016). Integrated remote sensing for multi-temporal analysis of anthropic activities in the south-east of Mt. Vesuvius National Park. *Journal of African Earth Sciences, 122,* 63–78. DOI: 10.1016/J.JAFREARSCI.2015.12.021

Martínez, S. L., Carlos, J., Manrique, O., Rodríguez-cuenca, B., y González, E. (2015). Procesado y distribución de nubes de puntos en el proyecto PNOA-LiDAR. *Proceedings of the XVII Congreso de La Asociación Española de Teledetección, Murcia, Spain,* 3–7.

Notarnicola, C., Angiulli, M., y Giasi, C. I. (2004). Southern Italy illegal dumps detection based on spectral analysis of remotely sensed data and land-cover maps. In M. Ehlers, H. J. Kaufmann, y U. Michel (Eds.), *Remote Sensing for Environmental Monitoring, GIS Applications, and Geology III, Vol. 5239*, 483–493. SPIE. DOI: 10.1117/12.511668

Padubidri, C., Kamilaris, A., y Karatsiolis, S. (2022). Accurate Detection of Illegal Dumping Sites Using High Resolution Aerial Photography and Deep Learning. *2022 IEEE International Conference on Pervasive Computing and Communications Workshops and Other Affiliated Events (PerCom Workshops)*, 451–456. DOI: 10.1109/PerComWorkshops53856.2022.9767451

Parrilli, S., Cicala, L., VincenzoAngelino, C., y Amitrano, D. (2021). Illegal Micro-Dumps Monitoring: Pollution Sources and Targets Detection in Satellite Images with the Scattering Transform. *2021 IEEE International Geoscience and Remote Sensing Symposium IGARSS*, 4892–4895. DOI: 10.1109/IGARSS47720.2021.9555072

Quesada-Ruiz, L. C., Rodriguez-Galiano, V., y Jordá-Borrell, R. (2019). Characterization and mapping of illegal landfill potential occurrence in the Canary Islands. *Waste Management, 85*, 506–518. DOI: 10.1016/J.WASMAN.2019.01.015

Seror, N., y Portnov, B. A. (2018). Identifying areas under potential risk of illegal construction and demolition waste dumping using GIS tools. *Waste Management, 75*, 22–29. DOI: 10.1016/J.WASMAN.2018.01.027

Shahab, S., y Anjum, M. (2022). Solid Waste Management Scenario in India and Illegal Dump Detection Using Deep Learning: An AI Approach towards the Sustainable Waste Management. *Sustainability, Vol. 14*(23). DOI: 10.3390/su142315896

Real Decreto 105/2008, de 1 de febrero, por el que se regula la producción y gestión de los residuos de construcción y demolición., 13 (2008). https://www.boe.es/buscar/act.php?id=BOE-A-2008-2486 [Consultado el 03.07.2023].

Štular, B., Eichert, S., y Lozić, E. (2021). Airborne LiDAR Point Cloud Processing for Archaeology. Pipeline and QGIS Toolbox. *Remote Sensing, Vol. 13*(16). DOI: 10.3390/rs13163225

Torres, R. N., y Fraternali, P. (2021). Learning to Identify Illegal Landfills through Scene Classification in Aerial Images. *Remote Sensing, Vol. 13*(22). DOI: 10.3390/rs13224520

Yadav, S. K. (2013). GIS based approach for site selection in waste management. *International Journal of Environmental Engineering and Management, Vol. 4*(5), 507–514.

Zakšek, K., Oštir, K., y Kokalj, Ž. (2011). Sky-view factor as a relief visualization technique. *Remote Sensing, Vol. 3*(2), 398–415.

Capítulo 7

La percepción de las fuerzas y cuerpos de seguridad en la detección y control del vertido ilegal

LOREA ARENAS GARCÍA[1]

1. INTRODUCCIÓN

Los principales escenarios del vertido ilegal de residuos sólidos se ubican en el medio rural y en las zonas periurbanas, generalmente en campos y caminos de manera concentrada constituyendo *hot spots* o puntos calientes (Fernández et al., 1995; Jordá-Borrell et al., 2014; Lucendo-Monedero et al., 2015; Navarro et al., 2016; Quesada-Ruiz et al., 2018; Arenas García, 2023b). Debido a ello los agentes del Servicio de Protección de la Naturaleza de la Guardia Civil (en adelante SEPRONA), la Policía Local y el cuerpo de Agentes del Medio Natural son las principales fuerzas de seguridad observadoras de estas infracciones. Si el artículo 45 de la Carta Magna consagra que: "La conservación y disfrute de un medio ambiente adecuado para el desarrollo de la persona, es un derecho de todos los ciudadanos", el 104 encomienda a las fuerzas de seguridad su protección: "Las Fuerzas y Cuerpos de Seguridad, bajo la dependencia del Gobierno, tendrán como misión proteger el libre ejercicio de los derechos y libertades y garantizar la seguridad ciudadana". Para cumplir dichos preceptos, la Ley

---

1 Profesora Contratada Doctora en Criminología, Área de Derecho penal, Universidad de Extremadura.

Orgánica 2/1986, de Fuerzas y Cuerpos de Seguridad, asignó al Cuerpo de la Guardia Civil la competencia material específica de "velar por el cumplimiento de las disposiciones que tiendan a la conservación de la naturaleza y medio ambiente, de los recursos hidráulicos, así como de la riqueza cinegética, piscícola, forestal y de cualquier otra índole relacionada con la naturaleza". En consecuencia, en el año 1988 la Guardia Civil creó el SEPRONA, cuya regulación específica la hallamos en la actual Orden General número 29/2021 de 9 de septiembre, por la que se regula la especialidad de protección de la naturaleza y la estructura, organización y funciones de sus unidades. La misión de este organismo se incardina en torno a los siguientes objetivos: proteger el soporte físico natural (suelo, agua y atmósfera) y las especies vivas que lo pueblan (flora y fauna); preservar y proteger los espacios protegidos y el patrimonio histórico; velar por el cumplimiento de las normas sobre la ordenación del territorio; fomentar las conductas de respeto a la naturaleza y al medio ambiente; y prevenir, detectar, perseguir, denunciar e investigar los hechos y conductas ilícitos relacionados con estas materias, en colaboración con otras unidades y especialidades del cuerpo. Dichas competencias específicas son desarrolladas por distintas unidades: patrullas[2], destacamentos[3] y equipos[4] que operan en diferentes zonas y demarcaciones.

Las Policías Locales, cuyas funciones se recogen en el artículo 53 de la Ley Orgánica 2/1986, de Fuerzas y Cuerpos de Seguridad, realiza labores de policía administrativa en lo relativo a ordenanzas (como pueden ser aquellas específicas sobre residuos), la realización de diligencias de prevención "y cuantas actuaciones tiendan a evitar la comisión de actos delictivos en el marco de colaboración establecido en las Juntas de Seguridad". También les compete el: "Vigilar los espacios públicos y colaborar con las Fuerzas y Cuerpos de Seguridad del Estado y con la Policía de las

---

2 Se distinguen dos tipos: Las Patrullas de Protección de la Naturaleza (PAPRONA) y las Patrullas de Comandancia de Protección de la Naturaleza (PACPRONA). Las primeras constituyen la unidad operativa de vigilancia y protección del medio ambiente en un territorio determinado, son por ello las encargadas de vigilar e impedir, en su caso, las actividades que perturben el normal desarrollo de la fauna y la flora, particularmente el de las especies protegidas, vigilar el cumplimiento de la normativa que afecte a la protección de los recursos naturales y también denunciar cualquier tipo de agresión al medioambiente. En cuanto a las segundas, constituyen el instrumento especializado con el que el mando ejerce su acción, pudiendo desarrollar su labor en cualquier punto del ámbito geográfico de la Comandancia, por sí solas o en colaboración con otras unidades (Capítulo I, sección II, artículo 6 de la Orden General número 29/2021 de 9 de septiembre).

3 El Destacamento de Protección de la Naturaleza (DEPRONA) tiene como principal labor controlar, proteger y conservar los espacios naturales protegidos legalmente (por ejemplo, un parque natural). (Capítulo I, sección II, artículo 6 de la Orden General número 29/2021 de 9 de septiembre).

4 El Equipo de Protección de la Naturaleza (EPRONA) constituye el órgano técnico especializado para la detección, cuantificación e investigación de las agresiones graves al medio ambiente y de apoyo a las Patrullas y Destacamentos del Servicio, así como demás unidades de la Comandancia. Actúan como policía judicial en materia medioambiental, en la confección de atestados, denuncias, informes u otras diligencias relacionadas con el medio ambiente (Capítulo I, sección II, artículo 6 de la Orden General número 29/2021 de 9 de septiembre).

Comunidades Autónomas en la protección de las manifestaciones y el mantenimiento del orden en grandes concentraciones humanas, cuando sean requeridos para ello" (artículo 53, letra h). Aunque la norma no alude específicamente a funciones y competencias en materia medioambiental, la naturaleza de las funciones referidas conlleva *per se* la vigilancia y control de delitos medioambientales. Labor que sí aparece contemplada específicamente en el artículo 23, de la Ley 7/2017, de 1 de agosto, de Coordinación de Policías Locales de Extremadura, relativo a las funciones de la policía. Dice así: "Proteger el medio ambiente, velando por el cumplimiento de las disposiciones aplicables en la materia y denunciando su incumplimiento, de conformidad con el marco competencial atribuido a las respectivas Corporaciones Locales".

Extremadura, junto a las comunidades autónomas de Madrid, Cataluña, País Vasco y Canarias, contemplan funciones de policía medioambiental en sus respectivas leyes autonómicas de policía local. Si bien, es difícil estimar el número de policías locales en España que cuentan con "patrulla verde" o unidad especializada en la protección del medioambiente. Según estimaciones realizadas por la Unión Nacional de Jefes y Directivos de Policía Local[5], en nuestro país existen alrededor de 250 policías locales con dichas unidades, aunque el dato no aparece desglosado por municipio (FEMP, 2023). Si bien, en esta investigación se ha podido constatar que no es el caso de las policías de Badajoz y Cáceres.

En síntesis, las funciones que la ley atribuye a los agentes del SEPRONA y de las policías locales, sumado al hecho de que desempeñan sus labores de vigilancia y control en patrullas realizando rondas rutinarias en calles y caminos, siendo dichas batidas de campo uno de los métodos más eficaces en la detección del residuo (Arenas García, 2023a), les posiciona como actores clave en la percepción de esta forma específica de criminalidad.

---

5 Véase UNIJEPOL: https://unijepol.eu/quienes-somos/ [Consultado el 03.07.2023].

2. MARCO TEÓRICO

Gracias a los datos facilitados por el SEPRONA para esta investigación es posible conocer el número de infracciones administrativas y penales sobre residuos en la región extremeña de 2016 a 2021, pero poco se sabe sobre el perfil criminal del autor/a, su *modus operandi* y otras variables situacionales de interés para comprender la conducta criminal y adecuar la prevención. Para ello es necesario recurrir a la escasa evidencia empírica en la materia y, en particular, a los estudios de Crofts et al.,, 2010, Matos et al., 2012 y Arenas García, 2023b. En base a estos trabajos sabemos que: el vertido ilegal tiene lugar por la noche y en lugares apartados con menor vigilancia (extrarradio de la ciudad); los residuos suelen aparecer cerca de polígonos industriales y alrededor de los nodos de actividad profesional y de las rutas entre ellos; los escombros, materiales de obra o residuos de construcción y demolición (RCD), así como ciertos productos químicos son los que más se vierten; los delincuentes suelen ser trabajadores de la construcción o intermediarios de estos, hombres, con edades comprendidas entre 35-55 años, que deciden racionalmente desprenderse de los residuos para ahorrar costes; y que una prevención comunitaria basada en la conciencia del daño, la información sobre sanciones (con panfletos y charlas) y la vigilancia vecinal podrían ser clave.

Dichos resultados evidencian que el vertido se produce en escenarios cuyas variables situacionales aseguran una alta oportunidad delictiva. De ahí la importancia de analizar los principales enfoques teóricos del paradigma situacional, cuyo eje central es la noción de oportunidad en la aparición y mantenimiento del vertido ilegal. Desde la *teoría de las actividades rutinarias*, Cohen y Felson (1979) sostenían que el delito tenía lugar por la ocurrencia en un mismo espacio y tiempo de tres elementos: un sujeto motivado, un objetivo potencial o adecuado y la carencia de control o vigilancia -ausencia o deficiencia de control social formal-. En cuanto al primer elemento de la ecuación, se observa que se trata principalmente de trabajadores de la construcción cuya principal motivación es ahorrar costes al no pagar la tasa del reciclaje. En cuanto al objetivo potencial, este se identificaría con el lugar del vertido, pues se trata de un objetivo visible y accesible a la vez que valioso para cubrir las necesidades del agresor. Al ser el potencial objetivo el lugar del depósito, en zonas periurbanas y rurales, cabe concluir que los objetivos (o escenarios) son muy numerosos e intercambiables. En lo que respecta a la ausencia de control y la presencia de oportunidad, el postulado de la teoría clásica mantiene que los vigilantes no disuaden al actor de llevar a cabo la acción delictiva puesto que están ausentes o no son suficientes. En este sentido, es fácil anticipar que vigilar y controlar escenarios muy amplios y numerosos es una ardua tarea, sobre todo si tenemos en cuenta las características geográficas de Extremadura y la dispersión y baja dotación de efectivos policiales (Ortiz García, 2022).

Desde la *teoría de la elección racional*, Cornish y Clarke (1986) consideraban que la acción humana se definía por ser razonada por el individuo. Aquí, la racionalidad de la opción delictiva es utilitarista y aparece vinculada al factor oportunidad y al contexto situacional del autor (Burke, 2009; Chamard, 2010). Dicha racionalidad implicaba un fin y un significado racional de la acción. El elemento central de su teoría sostiene que el sujeto calcula, en términos de costes y beneficios, las acciones que elige cometer, que no atenderían más que a la consecución del placer, al igual que otras muchas acciones humanas no delictivas. Es decir, la conducta delictiva está instrumentalizada y el proceso de toma de decisiones que motiva su aparición está orientado a conseguir bienes que sean provechosos para el delincuente. De esta forma, si en los delitos contra el patrimonio el beneficio es la sustracción de bienes materiales, en los delitos de vertido este se obtiene al ahorrar diversos costes: las tasas del reciclaje, el coste del traslado (por ejemplo, el combustible) y el tiempo invertido. Y, en el caso de tratarse de un sujeto que trabaje en el sector de la construcción de manera clandestina, el beneficio sería ahorrarse todas las tasas requeridas para el trabajo legal. Además, el efecto disuasorio del castigo sería muy reducido debido a la elevada sensación de impunidad consecuencia del número bajo de detenciones (Arenas García, 2023b) y la escasa cuantía de las sanciones.

En tercer lugar, desde la *teoría del patrón delictivo*, heredada de la tradición ecológica y sustentada en el paradigma de la criminología ambiental, se subraya la importancia del medio o entorno físico en la etiología de la delincuencia y, sobre todo, en el particular atractivo criminógeno de determinados lugares para ciertos delincuentes (Felson y Clarke, 1998; Brantingham y Brantingham, 1991 y 2008; Vozmediano y San Juan, 2019). La literatura existente en la materia nos indica que los sujetos cometen delitos en lugares conocidos o que forman parte de su mapa geográfico del delito (Garrido Genovés, 2010) y que los patrones espacio-temporales de los delincuentes son iguales a los de cualquier persona (Vozmediano y San Juan, 2019). Ello se debe a que el conocimiento del medio les hace reconocer en mayor medida las oportunidades que brinda el mismo, así como las medidas de seguridad o guardianes que operan, por lo que la comisión del delito requiere un menor esfuerzo (Rengert, 2004). Es decir, los escenarios conocidos otorgan seguridad y se convierten en zonas de confort. En este orden de ideas, se explica que los vertidos se aglutinen en zonas concretas, generalmente periurbanas, a unas horas concretas, y también en aquellos caminos y carreteras con acceso a ellas, sin que los infractores recorran grandes distancias (Arenas García, 2023b).

3. OBJETIVOS

El objetivo general del estudio es analizar las percepciones de los agentes del SEPRONA y de la Policía Local acerca de los vertidos ilegales en el marco de sus labores de detección y control de estas conductas. De manera más concreta (objetivos específicos), se pretende:

- Obj. Esp. 1: Conocer los lugares más habituales del vertido ilegal.

- Obj. Esp. 2: Determinar el perfil del agresor/a.
- Obj. Esp. 3: Examinar su *modus operandi.*
- Obj. Esp. 4: Averiguar las causas del vertido ilegal.
- Obj. Esp. 5: Conocer si, en opinión de los agentes, se trata de una problemática que preocupa a la ciudadanía y a la clase política.
- Obj. Esp. 6: Determinar las medidas preventivas que proponen para reducir o eliminar la problemática.

4. METODOLOGÍA

Para conocer las percepciones de los agentes se ha aplicado una metodología mixta basada en cuestionarios y entrevistas personales semiestructuradas. Los cuestionarios fueron creados *ah hoc*[6] y administrados a los agentes utilizando la herramienta *Google Forms*[7], mientras que las entrevistas se efectuaron de manera presencial con los mandos policiales. En cuanto al modo de contacto y selección de los participantes, se utilizó una estrategia de investigación diferente para cada cuerpo policial dada su distinta naturaleza y organización.

En el caso de la Policía Local, se contactó a través de la dirección de la Academia de Seguridad Pública de Extremadura (en adelante ASPEX)[8] con todos los efectivos que operaban en la región. ASPEX, inaugurada en el año 1994 y adscrita a la Consejería

---

6 El cuestionario estaba compuesto por 20 ítems o variables que medían aspectos profesionales (demarcación de trabajo, cargo, años de experiencia, etc.) y otros relativos a la existencia, frecuencia, ubicación, características de los infractores, estrategias de seguridad y control, prevención del vertido, entre otros.

7 Véase: https://www.google.es/intl/es/forms/about/ [Consultado el 03.07.2023].

8 Véase: http://aspex.juntaex.es/aspex/view/main/index/index.php [Consultado el 03.07.2023].

de Agricultura, Desarrollo Rural, Población y Territorio, Dirección General de Emergencias, Protección Civil e Interior, tiene atribuidas las competencias de formación y coordinación de las Policías Locales, así como el desarrollo de actividades destinadas al estudio y divulgación en materia de seguridad pública[9]. Dadas sus competencias, la Academia posee el registro de las direcciones de correo electrónico de los efectivos policiales de toda la región, siendo una vía de difusión rápida y efectiva para transmitir información. Gracias a la colaboración de la dirección de la Academia fue posible realizar -hasta en dos ocasiones- un envío masivo de correos electrónicos a todos los agentes. En el correo se explicaba el proyecto de investigación, sus objetivos y avances, así como la manera de participar. Esta consistía en rellenar un cuestionario *online* para ser cumplimentado en una media de 10 minutos y cuyo enlace fue adjuntado en el propio correo de presentación del proyecto. Para incentivar la colaboración con la investigación se les expidió un certificado de participación tras finalizar el cuestionario, el cual fue enviado por correo electrónico junto a un texto de agradecimiento. Este correo fue aprovechado para solicitar de nuevo la colaboración de otros compañeros/as de la plantilla que no hubieran participado aplicando así la técnica de bola de nieve.

Con respecto a la Guardia Civil, se contactó directamente con los mandos que dirigen las plantillas del SEPRONA a nivel provincial (1 Sargento en Cáceres y 1 Teniente en Badajoz) para concertar una cita y explicar el proyecto de investigación, así como la forma de colaboración. Ellos también fueron entrevistados. Gracias a este primer contacto fue posible presentar el proyecto a todos los agentes y repetir el procedimiento efectuado con los policías locales. Si bien, los agentes del SEPRONA recibieron -además- dos jornadas formativas sobre residuos con posterioridad a la finalización de su participación a fin de mejorar sus conocimientos y competencias en la materia.

---

9 Según lo establecido en el artículo 59 de la Ley 7/2017, de 1 de agosto, de Coordinación de Policías Locales de Extremadura.

4.1. Muestra

En el estudio, de carácter exploratorio y no representativo, se aplicó un muestreo dirigido al total de los agentes de ambos cuerpos policiales. Según el Registro de las Policías Locales de Extremadura[10], en la región operan 1452 agentes de la Policía Local en 167 municipios: 990 efectivos en 119 municipios de Badajoz y 462 efectivos en 48 municipios de Cáceres. Ello significa que el 43% de los pueblos extremeños cuentan con este servicio policial (un 12,4% en municipios de Cáceres y un 30,6% en pueblos de Badajoz). Gracias a datos facilitados por el SEPRONA sabemos que en la provincia de Cáceres operan 36 agentes y en Badajoz 52 (estando operativos de estos últimos 43).

En el estudio se contó con la participación de 181 agentes (152 de la Policía Local y 29 de la Guardia Civil) lo que representa el 10% de los agentes de la Policía Local y el 36,7% de los agentes del SEPRONA. En cuanto a la Policía Local, se obtuvieron respuestas del 40,6% de los municipios que contaban con policías en toda la región. Desglosando el dato según provincia, se observa que participaron agentes del 41,6% de municipios de Cáceres y del 39,5% de Badajoz. Con respecto a la representatividad de los municipios muestrales según tamaño poblacional, cabe destacar que se obtuvieron cuestionarios de pueblos de todos los intervalos poblaciones[11]. No obstante, el 52,2% de las respuestas provenían de municipios de entre 1001 y 5000 habitantes, seguido de pueblos de entre 5001 y 10000 habitantes (24%). La menor representación de pueblos de menos 1000 habitantes (24% del total) es acorde a la menor existencia de los cuerpos policiales

---

10 Datos facilitados por la Consejería de Medio Ambiente y Rural, Políticas Agrarias y Territorio de la Junta de Extremadura en febrero de 2023.

11 Estos fueron establecidos a partir de datos del padrón municipal del Instituto Nacional de Estadística (INE) del año 2021 según número de habitantes: < 500, 501 a 1000, 1001 a 2000, 2001 a 5000, 5001 a 10000, 10001 a 20000, 20001 a 50000, 50001 a 100000, 100001 a 500000.

en estas poblaciones pues, según la normativa vigente[12], su creación es preceptiva en poblaciones superiores a 5000 habitantes. Por otro lado, se obtuvieron cuestionarios de los municipios más poblados de Extremadura, a saber: Badajoz (150984 habitantes), Cáceres (96255 habitantes), Mérida (59548 habitantes), Plasencia (39860 habitantes), Don Benito (37284 habitantes) y Almendralejo (33855 habitantes).

Con respecto al SEPRONA de Cáceres, no se pudo determinar el porcentaje de agentes que participaron en cada demarcación circunscrita al área de influencia de las Patrullas de Protección de la Naturaleza (PAPRONAS) al desconocerse el total de agentes en las demarcaciones debido a cuestiones de seguridad y confidencialidad. Dichas áreas de influencia se cubren con PAPRONAS en Cáceres, Trujillo, Guadalupe, Jarandilla, Plasencia, Camino Morisco, Coria y Valencia de Alcántara. También con aquellas dependientes directamente de la Jefatura de la Unidad de Protección de la Naturaleza (UPRONA), con las Patrullas de Protección de la Naturaleza situada en la capital de provincia (PACPRONA de Comandancia), el Destacamento de Protección de la Naturaleza (DEPRONA) del Parque Nacional de Monfragüe (sede en Casatejada) y el Equipo de Protección de la Naturaleza (EPRONA). Se obtuvieron respuestas de todas las unidades referidas a excepción de Guadalupe, Jarandilla y Coria.

Con relación al SEPRONA de la provincia de Badajoz, se organiza territorialmente en 10 demarcaciones en las que operan 10 PAPRONAS, abarcando así toda la provincia. Además, en la Comandancia de la Guardia Civil de Badajoz, hay ubicada 1 PACPRONA y 1 EPRONA, que sirven de apoyo a las patrullas territoriales. Nuevamente, debido a motivos de seguridad, no ha sido posible conocer el total de agentes que integra cada PAPRONA, ni los municipios que abarcan. Ahora bien, gracias a las respuestas de los cuestionarios, sabemos que participaron en la investigación:

---

12 Véase artículo 15 de la Ley 7/2017, de 1 de agosto, de Coordinación de Policías Locales de Extremadura.

el UPRONA de Badajoz, el EPRONA, las PAPRONAS de Fuente de Cantos (zona sureste de la provincia de Badajoz), Mérida (actúa en Tierra de Barros y Vegas Altas del Guadiana), Hornachos, Olivenza (que engloba los municipios de Olivenza, Almendral, Alconchel, Torre de Miguel Sesmero, Cheles y Valverde de Leganés) y Azuaga.

Para conocer las percepciones de los mandos se efectuaron entrevistas con el teniente de Badajoz, el sargento de Cáceres y el cabo de Mérida del SEPRONA. También con el jefe de grupo de la Policía Local de Badajoz, en la que no hay unidad o especialidad medioambiental como tal, pero en la práctica realiza funciones afines. Los cuatro tenían una experiencia media de 24 años en el cuerpo.

Finalmente, cabe destacar que los agentes de la Policía Local eran en su mayoría hombres (88,2%)[13], con una media de 16 años de experiencia en el cuerpo (desviación estándar de 10,3 años), mientras que los agentes del SEPRONA (todos hombres), contaban con una experiencia media de 25 años.

5. RESULTADOS

5.1. Incidencia, ubicación y frecuencia del vertido ilegal

La mayor parte de los agentes confirma que existen RCD ilegales en su localidad (88,2% de los policías locales y el 96,6% de la Guardia Civil). Ello pone de manifiesto que se trata de una infracción muy presente en buena parte de los municipios extremeños. Sin embargo, cuando se les pregunta si se trata de un problema en la región, los agentes del SEPRONA refieren en menor medida que los agentes municipales que esto sea así (76% frente a un

13 Dato acorde a la distribución según sexo del total de la plantilla policial extremeña, en la que el 91,1% son hombres y el 8,9% mujeres.

91%). Probablemente sea debido a que los agentes del SEPRONA conocen en mayor profundidad los problemas medioambientales de la región y señalan otras cuestiones más prioritarias para ellos (incendios, caza ilegal, venenos, etc.). Al respecto mencionaban que: "no creo que sea un problema ecológico, claro está, si solo son escombros, como restos de ladrillo o cemento, pero si un problema estético" (policía local), "no son residuos peligrosos que puedan tener una afección grave a fauna silvestre, hábitats o salud de las personas, si es cierto que generan un gran impacto visual o paisajístico propio de sociedades modernizadas" (agente del SEPRONA).

A la vista de la figura inferior, los campos y caminos son los lugares más habituales del depósito ilegal, seguido de arcenes o cunetas de los anteriores, carreteras y ríos. Se trata de zonas periurbanas, vías de salida de los municipios, zonas en construcción o expansión urbanística y, de forma más residual, polígonos o zonas industriales (Arenas García, 2023b). Los campos y caminos de zonas periurbanas se encuentran cerca del núcleo poblacional pero lo suficientemente lejos para no ser visto y emplear mucho tiempo y recorrido en llegar allí. En otras palabras, los delincuentes acuden a zonas conocidas y cercanas a sus domicilios sin que les suponga gran esfuerzo. Los campos, ríos y caminos ofrecen un escenario de oportunidad delictiva muy amplio e intercambiable. Lo mismo sucede con las zonas en expansión urbanística y las naves industriales abandonadas, en las que están creados los accesos a las parcelas a edificar (con calles y aceras) pero no cuentan con vigilancia vecinal. Los ríos, lugar de vertido referido en mayor medida por el SEPRONA, suelen ofrecer un contexto de oportunidad delictiva idóneo gracias a la pendiente de la ribera del río y la presencia habitual de bosque frondoso. Al mismo tiempo resulta claramente pernicioso su vertido en tales lugares debido a su alta biodiversidad y la contaminación de las aguas (Arenas García, 2023b).

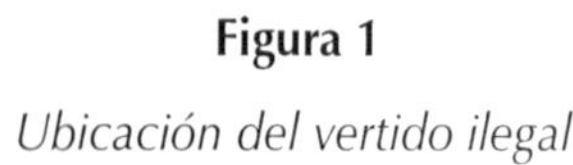

Figura 1

Ubicación del vertido ilegal.

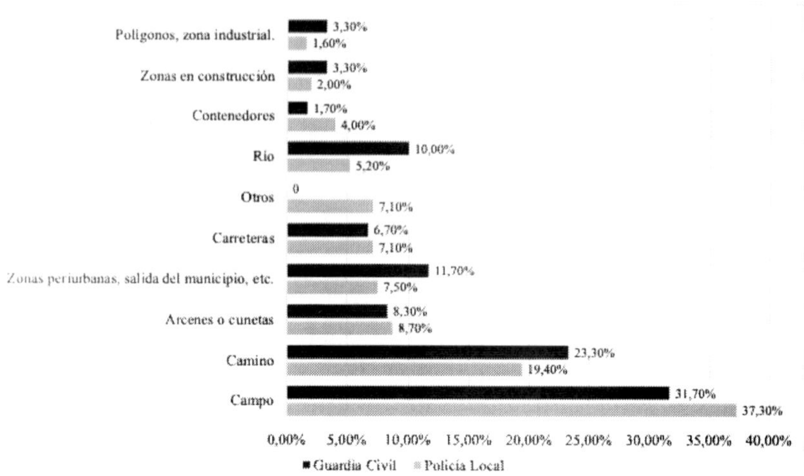

Con respecto a los contenedores urbanos, se trata de situaciones en las que se depositan dentro o cerca de estos los RCD, no siendo un punto limpio habilitado para ello. Por último, la categoría "otros" engloba lugares próximos al punto limpio, edificios que se quedaron en obras, parcelas privadas, canteras abandonadas y antiguos vertederos. Los edificios a medio construir y las naves industriales abandonadas son espacios aprovechados para depositar residuos en su interior o planta baja. Son lugares situados en el extrarradio, con poca vigilancia y que sirven de reclamo para depositar y camuflar los escombros con otros materiales de construcción ya existentes. Por otro lado, los agentes refieren que en algunas parcelas privadas los particulares vierten escombros para rellenar irregularidades del terreno, o simplemente acumulan pequeños montículos que son apreciables desde el exterior. Por último, en las afueras de los municipios se situaban vertederos (ilegales) que eran conocidos por todos los vecinos de la zona y aprovechados (por costumbre) para depositar residuos. Aunque la mayor parte de ellos fueron sellados y regenerados, sobre todo en la provincia de Cáceres, en muchos pueblos de Badajoz se siguen aprovechando estos lugares para verter escombros.

Con respecto a la frecuencia con la que se depositan vertidos en estos lugares, tanto las policías locales como los agentes de la Guardia Civil refieren en porcentajes similares (67,1% y 76% respectivamente) que se trata de una conducta ocasional seguido de todos los días (21% y 13% respectivamente). De manera más minoritaria consideran que se produce casi nunca y siempre (del 3% al 7%). Aunque se trate de una conducta ocasional, su gravedad estriba en que se viene produciendo desde hace muchos años, lo cual genera la acumulación y el crecimiento constante del residuo.

Figura 2

Frecuencia del vertido ilegal.

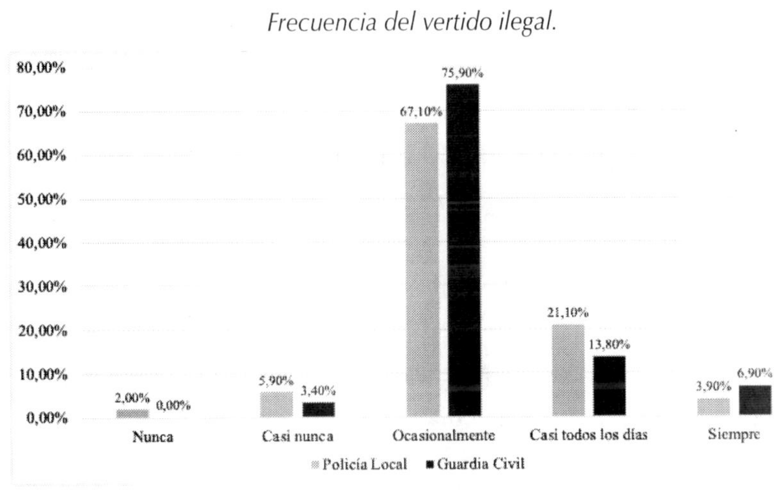

5.2. *Perfil del infractor y modus operandi*

Los infractores, tal y como son percibidos por los agentes, son en su mayoría hombres, de entre 30 y 55 años, que trabajan en el sector de la construcción, tanto de manera legal en pequeñas empresas (como autónomo o por cuenta ajena) como de forma ilegal. Con respecto al trabajador que declara su actividad, los agentes refieren que las motivaciones para verter ilegalmente los escombros son: ahorrar el pago de la tasa de reciclado, la lejanía del punto limpio (ahorro en combustible) y su poca flexibilidad de horarios. Respecto al trabajador ilegal, se distinguen a su vez dos perfiles: aquel que realiza pequeñas obras de manera ilegal como *modus vivendi*; y el que tiene un trabajo regularizado los días de entre semana, pero realiza obras los fines de semana para obtener un ingreso extra.

Los agentes denominan a estos actores como "el chapuzas que trabaja en *b*, sobre todo los fines de semana y luego tira los escombros por ahí". Al desempeñar su actividad en la economía sumergida no puede pedir licencia de obras, ni pagar fianzas o tasas de reciclaje, ni tampoco acudir a un gestor autorizado de residuos. Si puede trasladar los escombros al punto limpio (como ciudadano/a) pero el límite de escombros diarios a depositar es limitado, la matrícula del vehículo queda registrada y suele requerirse licencia de obras. Todo ello desincentiva la acción. Obviamente, para este actor, la motivación delictiva es el ahorro de todas las tasas que implican trabajar y gestionar los residuos de manera legal.

Con respecto al *modus operandi*, se identifican cinco acciones clave para llevar a cabo el vertido ilegal: la creación y almacenamiento, la recolección, el transporte, el depósito y el tratamiento del residuo. La creación y almacenamiento del residuo tiene lugar en la casa o local de un particular que quiere realizar una reforma de albañilería menor. Al tratarse de obras menores que no duran mucho tiempo, no suelen levantar sospechas ni molestias entre los vecinos. No obstante, según la policía, es habitual que se den estas obras en casas de campo y viviendas de zonas periurbanas, especialmente los fines de semana por lo que el nivel de vigilancia es mucho menor. En cuanto a los autores, aquí se distinguen nuevamente los dos perfiles referidos. El albañil dado de alta en autónomos que ha solicitado licencia de obras para dicha reforma, por lo tanto, estaríamos en el marco de una acción legal y, por otro lado, aquel que opera de forma clandestina en la economía sumergida. El albañil dado de alta que ha solicitado la licencia de obras deberá acudir a un gestor autorizado para obtener el certificado de depósito de residuos y así recuperar la fianza.

La recolección del escombro generado se lleva a cabo en la primera escena referida y en el vehículo que los infractores utilicen para transportarse. Suele ser una furgoneta de tamaño pequeño o mediano. Los escombros pueden cargarse en sacas, capazos o mantas desde el domicilio o local y transportarse hasta el vehículo. Se dirigirán a las zonas frecuentes del vertido ya mencionadas, al atardecer y/o anochecer, generalmente entre las 20:00 y 23:00 horas y también al mediodía (entre las 14:00 y 17:00 horas), cuando finaliza la jornada de trabajo. El vertido se realiza entre semana, pero sobre todo los fines de semana. Según los agentes, puede darse el caso de que, en algunos municipios extremeños con pocos habitantes y efectivos policiales, los infractores conozcan los horarios o momentos en los que la policía local esté ocupada y así aprovechar la oportunidad (por ejemplo, cuando se regula el tráfico en la entrada de un colegio). El vertido se realiza generalmente con el coche parado. Un sujeto conduce y el otro infractor baja del vehículo y vierte los escombros de las sacas, o vierte las propias sacas, o se sirve de una manta. En este último caso es habitual no bajar del vehículo e incluso no detenerlo del todo, esto es, en la parte de atrás de la furgoneta se abren las puertas, se acerca la manta que contiene los escombros en su interior a la puerta del furgón, se abre y se dejan caer al camino de campo, para luego cerrar la manta y las puertas traseras del vehículo. Este *modus operandi* es conocido como "manta" y reduce los riesgos de ser aprehendido al reducirse el tiempo de permanencia en la escena del delito. Tras realizar el vertido huyen de esta última y da comienzo la actividad post-delictiva, la cual consiste básicamente en regresar al núcleo urbano sin levantar sospechas. En el caso del albañil legal, no podrá recuperar la fianza sino obtiene el certificado de depósito de residuos en el gestor autorizado. En este sentido, caben dos opciones: renunciar a ella o falsear la cantidad de residuos reales que se generaron en la obra. Se trata de usuarios que no llevan la cantidad de residuos prevista en la licencia de obras, sino una inferior, para así obtener el certificado de depósito de residuos a menor precio (véase Arenas García, 2023b). Ello implica que los escombros restantes hasta alcanzar la cantidad prevista en la licencia de obras hayan sido vertidos ilegalmente.

5.3. Causas del vertido ilegal

La principal razón que indican los agentes como causa del vertido ilegal es el *ahorro de las tasas* que el ciudadano o empresario debe abonar para el reciclado del RCD, ya sea en el ayuntamiento correspondiente, en el punto limpio o en una planta de reciclaje (véase capítulo 9).

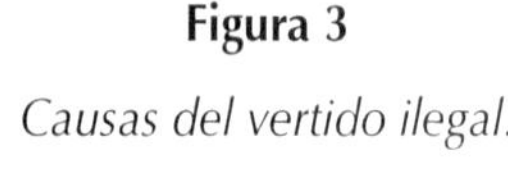

Figura 3

Causas del vertido ilegal.

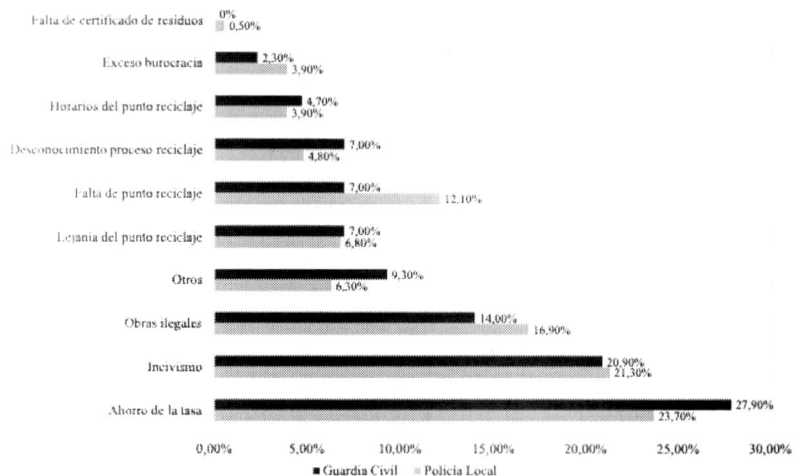

Claramente, el ahorro del dinero incentiva la comisión de la infracción. También la *falta de civismo* de la ciudadanía arraigada en la antigua costumbre, sobre todo en pequeños municipios, de no abonar tasas por depositar residuos, siendo estos trasladados a vertederos (ilegales) situados a las afueras de los pueblos. A este respecto un policía local reseñaba que: "La mayoría (de los ciudadanos) no lo ve como algo preocupante, lo ven como algo que ocurre desde siempre y a lo que están acostumbrados".

La tercera razón más referida es la *procedencia ilegal de las obras*. Son aquellas que se realizan al margen de todo el proceso administrativo, sin licencia de obras, fianzas o tasas. Los gestores autorizados suelen identificar al usuario[14] cuando deposita el residuo en sus instalaciones lo cual disuade que los infractores las utilicen -aunque sean gratuitas- con tal de no ser identificados. La *lejanía del punto de reciclaje*, sobre todo en aquellos municipios con puntos de recogida distantes (véase capítulo 9) o para usuarios sin vehículo, supone un mayor coste (en tiempo y combustible) para el depósito. De hecho, la distancia al punto de reciclaje es una variable predictiva del vertido ilegal, esto es: a mayor distancia, mayores posibilidades de vertido (Glanville y Chang, 2015). Casi en un porcentaje similar al anterior, aunque siendo más señalado por los agentes del SEPRONA, la *falta de un punto de reciclaje* en la localidad motiva su desecho en otros lugares. En este sentido, algunos autores (Halvorsen, 2010; Ichinose y Yamamoto, 2011), demostraron que cuántos más medios otorgas al ciudadano para reciclar más incentivas su utilización. Con respecto al *desconocimiento del punto de reciclaje*, se refiere a que no hay suficiente información entre la ciudadanía sobre el procedimiento para deshacerse de estos residuos, ni sobre los lugares de recogida y horarios. La *poca flexibilidad horaria* de estos lugares es otra de las causas incentivadoras del vertido ilegal. De forma más residual, parece que el *exceso de burocracia*, esto es, tener que tramitar la licencia de obras, la fianza, el abono de las tasas por el reciclaje, así como el reembolso de la fianza, desmotiva al usuario a seguir el procedimiento.

14 Solicitando datos personales, el número de matrícula del vehículo (en el que se trasladan y se pesan los ripios en una báscula habilitada para ello), la licencia de obras, entre otros.

El peso de la gestión la sufrirían en mayor medida los pequeños albañiles que realizan obras menores y que no cuentan con gestores. En cuanto al motivo *otros*, la falta de vigilancia, la deficiente y descoordinada actuación de las administraciones públicas, la falta de sanciones y la gran sensación de impunidad de los infractores que realizan estas conductas, son las causas más referidas. La alta cifra negra que acompaña a estas infracciones debido a la falta de vigilancia provoca que apenas existan casos en los que se pilla *in fraganti* al sujeto, siendo por tanto una conducta poco sancionada. Además, algunos agentes refieren que las cuantías de las sanciones son bajas y que no tienen un poder disuasorio.

Relacionado con el tema del incivismo de la ciudadanía y la descoordinación y deficiente actuación por parte de las administraciones públicas (en particular de los ayuntamientos), se ha preguntado a los agentes si consideran que los RCD ilegales preocupan a la ciudadanía y a la clase política. En la figura 4 se observa que los agentes de ambos cuerpos policiales tienen opiniones muy divididas acerca de la preocupación ciudadana. Los agentes del SEPRONA refieren positivamente y en mayor medida (58,9% frente a 41,4%) que el vertido ilegal preocupa a los ciudadanos. Señalan que: "afortunadamente la ciudadanía está cada vez más sensibilizada y concienciada con la conservación del medio natural que los rodea, ya que cada vez más interaccionan con el mismo, de una forma u otra".

Sin embargo, los agentes de la policía local se encuentran más igualados en sus opiniones (53,9% frente a 45,4%). Algunos señalan que: "cada vez más, son los avisos recibidos por los vecinos al observar tales hechos", "la vecindad si está muy en la labor de denunciar este tipo de actuaciones, ya que, desvirtúan el paisaje y las zonas más turísticas de sus localidades". Por el contrario, otros agentes indican que "no preocupan a la ciudadanía hasta que estos vertidos se producen cerca de sus viviendas" o "no se observa implicación directa por parte de la ciudadanía. Tener constancia sobre el hecho y no denunciarlo, es más habitual de lo que pensamos".

Figura 4

Percepción de la preocupación de la ciudadanía y la clase política sobre el vertido ilegal.

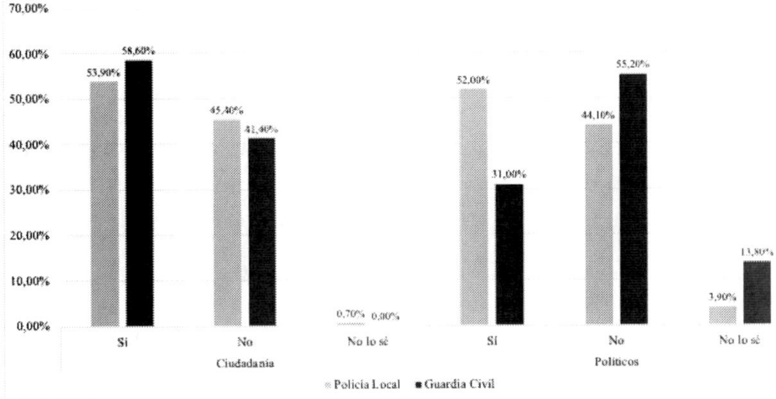

Si bien, ambos cuerpos policiales refieren que los vecinos preocupados son aquellos que sufren los escombros cerca de sus domicilios, parcelas o lugares que resultan muy visibles o molestos para ellos. Al respecto indicaron: "Solo (preocupan) cuando afectan a zonas cercanas de ocio al aire libre, pesca o caza", "no preocupan a la ciudadanía hasta que estos vertidos se producen cerca de sus viviendas" y "a los ciudadanos de las zonas colindantes a los vertidos, sí (les preocupa)". Por lo tanto, no se trata de una preocupación por el medioambiente fruto de una conciencia social sino más bien una respuesta fundamentaba en el interés propio.

Con respeto a los políticos, los agentes de policía local mantienen una opinión muy similar a la expresada para la ciudadanía, esto es, una visión muy dividida: el 52% opina que sí les preocupa y el 44%, no. Las opiniones de los agentes del SEPRONA son algo más nítidas. La mitad de ellos (el 55%) considera que el asunto no preocupa a los políticos. En este ítem los porcentajes de "no lo sé" se elevan, posiblemente porque se trate de un tema sensible y comprometido de contestar.

5.4. Propuesta de medidas preventivas

Conociendo las causas que generan el vertido, cabe preguntarse: ¿cómo podría reducirse o eliminar esta problemática?, ¿qué medidas preventivas se pueden adoptar? Las propuestas mencionadas por los agentes han sido recodificadas en diferentes temáticas, tal y como se muestra en la figura inferior. Los agentes municipales consideran el aumento del control, la vigilancia y las sanciones un aspecto clave de la prevención. Sin embargo, el SEPRONA apunta al mayor seguimiento y cumplimiento administrativo por parte de los responsables políticos, y a una mayor concienciación social y educación ciudadana.

El mayor control y vigilancia se traduce en aumentar el número de efectivos u equipos especializados en estas infracciones, así como al empleo de tecnologías con perspectiva situacional. Aunque son conscientes de que "no se puede tener un policía en cada esquina, sí se podrían colocar cámaras móviles ocultas en los puntos frecuentes donde se arrojan escombros dando publicidad a esta medida entre la ciudadanía". Las sanciones más severas se aplicarían a los dueños de los terrenos que no actúen sobre el vertido ilegal allí depositado. Refieren penalizar subsidiariamente al dueño del terreno (ciudadano o administración) si se abre el expediente (tanto en suelo público como privado). Por último, consideran que es importante la limpieza y regeneración constante de las zonas más afectadas para evitar el efecto llamada.

El mayor seguimiento y actuación de las administraciones se lograría: exigiendo la fianza obligatoria prevista en el ya mencionado Real Decreto 20/2011, de 1 de febrero, en todos los ayuntamientos; actuando con labores de limpieza o imponiendo sanciones en aquellos lugares en los que inició un expediente con motivo de un vertido ilegal; y realizando más inspecciones en las obras para verificar que, efectivamente, se exigen las fianzas y que las cantidades de residuos declaradas en las licencias de obras son las depositadas en los gestores autorizados.

Con respecto a la educación y concienciación social, señalan que es importante dar más información a la ciudadanía sobre el fenómeno y su impacto medioambiental, motivando incluso la colaboración ciudadana para detectar y denunciar el vertido. Es necesario "más educación por parte de los ciudadanos, con una concienciación sobre las consecuencias de tal hecho", "más cooperación vecinal, así como de las personas que transitan por los caminos, si ven algo extraño, alguna conducta extraña, avisarnos". Como medidas concretas, proponen la elaboración de campañas de comunicación en redes sociales, radio y televisión dirigidas a toda la ciudadanía, y campañas informativas para empresas de construcción y demolición. Proponen dar publicidad a los casos en los que se pilla *in fraganti* al agresor, con la foto del vertido, la sanción a imponer y las actuaciones policiales y judiciales desarrolladas. Se pretende buscar un efecto disuasorio y de concienciación visibilizando aquellos casos que no quedan impunes. En cuanto a las empresas, señalan que podrían facilitarse panfletos o boletines con los últimos avances legislativos en la materia, posibles sanciones, ubicación de los gestores autorizados de residuos e información sobre su gestión y teléfonos de interés. Es decir, otorgar más información a las empresas de construcción.

Figura 5

Propuesta de medidas preventivas.

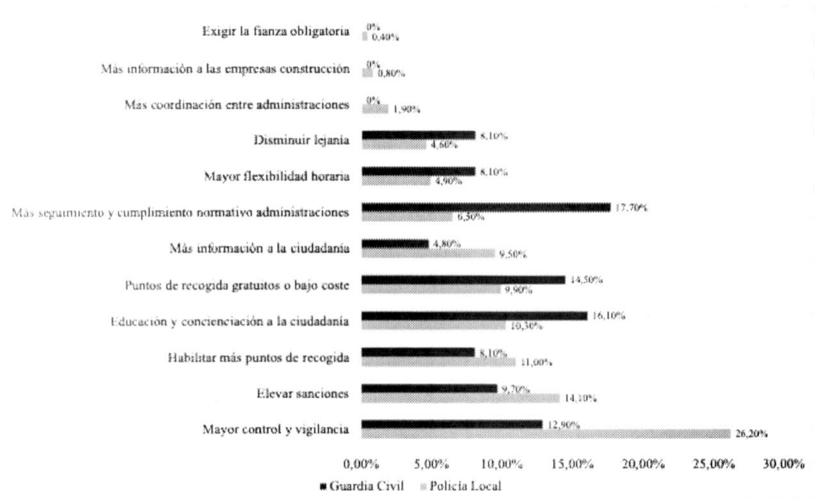

Para incentivar el depósito del residuo en lugares habilitados se reseñan diversas medidas, a veces contrapuestas, tales como: aumentar los puntos de recogida; facilitar a los usuarios la retirada de todo tipo de residuos sin coste o al mínimo coste; probar con contenedores urbanos para la recogida de pequeñas obras; habilitar zonas de entrada libre y no identificada a lugares de depósito para canalizar los vertidos procedentes de obras ilegales; dar más información a la ciudadanía sobre los horarios y funcionamiento de los puntos limpios; otorgar ayudas al sector de la construcción para el depósito y reciclaje de RCD; eliminar las tasas por reciclado al empresario; y elevar las cuantías mínimas de las fianzas en las obras menores para que no compense perderlas.

Por último, la Policía Local indica que es necesario una mayor coordinación entre las administraciones creando mecanismos entre los operadores encargados del control y vigilancia (policías locales, SEPRONA y agentes del medio natural), de estos con los ayuntamientos, y de los gestores autorizados para el depósito de los vertidos con todos los anteriores. Hoy día la coordinación efectiva es inexistente, pero muy necesaria para dar una respuesta integral al problema. A tal fin se propone: protocolizar las actuaciones de los agentes; compartir datos sobre las denuncias interpuestas; realizar encuestas a los agentes clave para recabar información sobre cuestiones diferentes a las denuncias (por ejemplo, inspecciones realizadas, contactos con ONG, etc.) y celebrar reuniones periódicas.

6. CONCLUSIÓN

La mayoría de los agentes participantes en la investigación afirman que existen RCD ilegales en aquellos municipios o demarcaciones en las que prestan sus servicios. Refieren que son frecuentes en zonas periurbanas y rurales, sobre todo en campos y caminos, lo que es acorde a la evidencia empírica en la materia (Fernández et al., 1995; Jordá-Borrell et al., 2014; Lucendo-Monedero et al., 2015; Navarro et al., 2016; Quesada-Ruiz et al., 2018; Arenas García, 2023b). Aunque se trate de una conducta ocasional, señalan que supone un problema en la región por el impacto paisajístico que ocasiona y el efecto llamada a nuevos residuos. El perfil del agresor, así como las variables situacionales presentes en estos casos y el *modus operandi,* son similares a los referidos en otros estudios (Crofts et al., 2010; Matos, Ostir et al., 2012). La profesión o sector profesional al que pertenece el agresor, los horarios y los lugares de depósito nos da una idea clara de la importancia de la oportunidad delictiva en este tipo de infracciones. Los albañiles, en sus actividades rutinarias, aprovechan aquellos lugares y horarios con menor presencia vecinal y policial para deshacerse de los vertidos acudiendo a zonas más vulnerables que concentran los focos más grandes del vertido. De ahí que propongan una mayor vigilancia y control de aquellos lugares poniendo en marcha medidas de prevención situacional, tales como: una mayor dotación de efectivos policiales o la instalación de cámaras de videovigilancia en los *hot spots.* Si bien, la naturaleza de estas medidas y sus efectos corto-placistas no ahondan en las causas de delincuencia pudiendo desplazarla, lo cual es muy probable que ocurra debido a la presencia de escenarios muy intercambiables. Por ello apuntan a la educación ciudadana y a promover una mayor conciencia social de la problemática medioambiental. En otras palabras, contar con una ciudadanía más respetuosa y concienciada con el medioambiente podría aumentar la eficacia colectiva (Sampson, 2014). Logrado esto, verter al margen de los puntos autorizados se concebiría como un mal hábito o costumbre al tiempo que incrementaría los casos de denuncia por parte de la ciudadanía. A lo anterior tendría que acompañar una actitud proactiva, coordinada y comprometida por parte de las administraciones, no permitiendo que los vertidos permanezcan abandonados durante décadas en los campos. Pues es habitual que, el residuo, una vez depositado de forma ilegal, permanezca abandonado y sin ser limpiado o recogido por las autoridades locales, salvo en circunstancias muy concretas (Arenas García, 2023b). En este sentido, es necesario "dar ejemplo" a la ciudadanía desde las administraciones para no crear un clima de permisividad e impunidad. Ello implica que las administraciones realicen un mayor seguimiento y control de las obras e infracciones a nivel administrativo. Al respecto el papel que juega la fianza es clave para incentivar o desincentivar el reciclado. Cuando se trata de obras menores -la mayoría- las cuantías bajas de la fianza, así como la falta de comprobación de las cantidades presentadas, son facilitadores del vertido ilegal. También que muchas de las zonas afectadas, ya sean de titularidad pública o privada, no se limpien generando un claro efecto llamada.

Todas las medidas referidas serían adecuadas para incidir en el fenómeno siempre y cuando se articulen de manera integral y dotadas de suficientes recursos humanos y materiales. En cualquier caso, parece evidente que responder al mismo requeriría de una mayor coordinación entre los distintos actores implicados, una coordinación que, a tenor de lo observado en el estudio, no parece existir en la actualidad.

7. REFERENCIAS BIBLIOGRÁFICAS

Arenas García, L. (2023a). ¿Qué funciona en la detección del vertido ilegal? En Muñoz Sánchez et al. (Eds), *Estudios jurídico-penales, criminológicos y políticos criminales. Libro Homenaje al Profesor José Luis Díez Ripollés*, 1695-1705. Tirant lo Blanch.

Arenas García, L. (2023b). El vertido ilegal de residuos sólidos: un estudio de caso. *Revista Española de Investigación Criminológica, Vol.21(2), 1-20. DOI:* 10.46381/reic.v21i2.820

Brantingham, P., y Brantingham, P. (1991). *Environmental Criminology*. Waveland Press.

Brantingham, P., Brantingham, P. (2008). "Crime pattern theory". En Wortley, R., Mazerolle, L. (eds.), Environmental Criminology and Crime analysis. Willan Pub.

Burke, R. H. (2009). *An introduction to criminological theory (3rd edition)*. Willan Pub.

Chamard, S. (2010). "Routine activities". En McLaughlin., y T Newburn (eds), The SAGE handbook of criminological theory. London.

Cohen, L.E., y Felson, M. (1979). Social change and crime rate trends: a routine activity approach. *American Sociological Review, Vol. 44*, 588–605. DOI: 10.2307/2094589.

Cornish, D., y Clarke, R. (1986). *The Reasoning Criminal: Rational Choice Perspectives on Offending*. Springer.

Crofts, P., Morris, T., Wells, K. y Powell, A. (2010). Illegal dumping and crime prevention: A case study of Ash Road, Liverpool Council. *Public Space: The Journal of Law and Social Justice, Vol. 5 (4)*, 1-23. DOI: 10.5130/psjlsj.v5i0.1904.

Felson, M., y Clarke, R. (1998). Opportunity Makes the Thief. Practical theory for crime prevention. *Police Research Series, 98*, 1-33. Disponible: https://popcenter.asu.edu/sites/default/files/opportunity_makes_the_thief.pdf [Consultado el 03.07.2023].

FEMP (Federación Española de Municipios y Provincias) (2023). Policías Locales y Protección Medioambiental. Material de José Francisco Cano de la Vega. Disponible en: http://femp.femp.es/files/3580-1832-fichero/JOSÉ%20F%20CANO.pdf [Consultado el 03.07.2023].

Garrido Genovés, V. (2010). *El rastro del asesino: el perfil psicológico de los criminales en la investigación policial*. Ariel.

Glanville, K. y Chang, H.C. (2015). Mapping illegal domestic waste disposal potential to support waste management efforts in Queensland, Australia.

International Journal of Geographical Information Science, Vol. 29(6), 1042-1058. DOI: 10.1080/13658816.2015.1008002.

Fernández-Jurado, J.A., González-Almendros, C.M., y Sabariego-Rivero, S. (1995). Vertederos ilegales. *Boletín Criminológico, Vol. 1* (16), 1-4. DOI: 10.24310/Boletin-criminologico.1995.v1i.9070.

Halvorsen, B. (2012). Effects of norms and policy incentives on household recycling: An international comparison. *Resources, Conservation and Recycling, Vol. 67*, 18– 26. DOI: 10.1016/j.resconrec.2012.06.008.

Ichinose, D., y Yamamoto, M. (2010). On the relationship between the provision of waste management service and illegal dumping. *Resource and Energy Economics, Vol. 33*, 79–93. DOI: 10.1016/j.reseneeco.2010.01.002.

Jordá-Borrell, R., Ruiz-Rodríguez, F., y Lucendo-Monedero, Á.L. (2014). Factor analysis and geographic information system for determining probability areas of presence of illegal landfills. *Ecological Indicators, 37*(A), 151–160. DOI: 10.1016/j.ecolind.2013.10.001.

Lucendo-Monedero, A.L., Jordá-Borrell, R., y Ruiz-Rodríguez, F. (2015). Predictive model for areas with illegal landfills using logistic regression. *Journal of Environmental Planning and Management, Vol. 58*(7), 1309–1326. DOI: 10.1080/09640568.2014.993751.

Matos, J., Oštir, K., y Kranjc, J. (2012). Attractiveness of roads for illegal dumping with regard to regional differences in Slovenia. *Acta Geográfica. Slovenica, Vol. 52*(2), 431–451. DOI: 10.3986/AGS52207

Ministerio del Interior. (2022). *Anuario estadístico.* Disponible en: https://www.interior.gob.es

Navarro, J., Grémillet, D., Afán, I., Ramírez, F., Bouten, W. y Forero, M.G. (2016). Feathered detectives: real-time GPS tracking of scavenging gulls pinpoints illegal waste dumping. *PLoS One, Vol. 11* (7), e0159974. DOI: 10.1371/journal.pone.0159974.

Ortiz García, J. (2022). *Mito o realidad. Un estudio criminológico en las comunidades rurales de Extremadura.* Dykinson.

Quesada Ruiz, L., Rodríguez-Galiano, V., y Jordá Borrell, R. (2018). Identifying the main physical and socioeconomic drivers of illegal landfills in the Canary Islands. *Waste Management y Research, Vol. 36*(11), 1049–1060. DOI: 10.1177/0734242X18804

Rengert, G. F. (2004). Journey to crime. En G. J. N. Bruinsma, H. Elffers., y J. de Keijser (Eds.), Punishment, Places, and Perpetrators: Developments in Criminology and Criminal Justice Research. Willan Publishing.

Sampson, R. J. (2014). Collective efficacy theory: Lessons learned and directions for future inquiry. In T. L. Anderson (Ed.), Understanding deviance: Connecting classical and contemporary perspectives, 128–139. Routledge/Taylor & Francis Group.

Vozmediano, L., y San Juan, C. (2019). *Criminología Ambiental: ecología del delito y de la seguridad.* Editorial UOC.

Capitulo 8

El papel de los Agentes del Medio Natural

GUILLERMO AGUSTÍN EXPÓSITO PAULANO[1]

1. INTRODUCCIÓN

Los agentes del medio natural de Extremadura son empleados públicos encargados de la custodia y labor de policía de los recursos naturales y del medio ambiente, controlando aquellas situaciones y procesos que, de forma natural o no, se originen en su ámbito de actuación. En España, pueden ser conocidos con otras nomenclaturas en función de la región: agentes de medio ambiente (Andalucía, Islas Baleares y Canarias), agentes medioambientales (Murcia, Castilla-La Mancha, Castilla y León y Comunidad Valenciana), guardas del medio natural (Principado de Asturias), agentes forestales (La Rioja y País Vasco) o agentes rurales (Cataluña).

Los agentes del medio natural (AMN, en adelante) poseen la condición legal de agentes de la autoridad, pudiendo acceder a cualquier terreno rural en el desarrollo de sus funciones. Entre sus deberes, también se encuentra el de informar a los ciudadanos con objeto de prevenir daños al medio natural o la comisión de infracciones. Su trabajo diario sobre el terreno les permite poseer un elevado conocimiento de este, estando actualizados sobre las actividades que se desarrollan en sus zonas de influencia, así

---

1 Personal Científico Investigador, Universidad de Extremadura.

como de las características medioambientales que son propias de estas. Por ello, su papel en la detección, control y prevención del vertido ilegal es de especial relevancia.

La especialidad de agente del medio natural tuvo su origen en el Decreto 268/2005, de 27 de diciembre, que reestructuró los colectivos de agentes forestales y agentes de medio ambiente para integrarlos en el cuerpo administrativo de la comunidad autónoma extremeña. Su organización y funcionamiento están descritos en el Decreto 269/2005, de 27 de diciembre, estructurándose de manera jerarquizada en las figuras del agente coordinador de unidad, agente coordinador adjunto y agente del medio natural.

En lo que respecta a sus funciones, se pueden señalar como principales las siguientes:

- Proteger y vigilar la riqueza forestal, cinegética y piscícolas de las aguas continentales, los espacios naturales, la flora, la fauna silvestre, y las vías pecuarias.
- Participar en la vigilancia, prevención y extinción de incendios forestales, así como colaborar en situaciones de emergencia cuando sean requeridos.
- Inspección y control de trabajos de conservación y mejora de los montes, repoblación y aprovechamiento, así como las actividades sujetas a la normativa sobre evaluaciones de impacto ambiental.
- Vigilancia y control en materia de actuaciones urbanísticas cuando el uso o construcción afecte a suelo no urbanizable, así como de vertidos de residuos y contaminación de las aguas, y de la atmósfera.
- Participación en la inspección, supervisión y control de actividades relativas a repoblaciones cinegéticas y piscícolas.
- Vigilancia, toma de datos y emisión de informes para el control y lucha contra enfermedades y plagas.
- Realizar censos, controles y seguimiento de especies de fauna silvestre cinegética, no cinegética y piscícolas.

- Colaborar y orientar a los ciudadanos en actividades a realizar en la naturaleza, así como informarles sobre cualquier asunto relacionado con el medio natural.

Para posibilitar el correcto desempeño de estas funciones, se redactó la Orden de 6 de marzo de 2007, por la que se determinaban las unidades territoriales de vigilancia que constituían el ámbito de funcionamiento de los Agentes del Medio Natural (modificada por la actual Orden de 4 de febrero de 2015). Estas *Unidades Territoriales de Vigilancia (UTV)* están formadas por municipios que guardan homogeneidad geográfica, forestal o ambiental.

Actualmente, la distribución de los AMN en Extremadura se divide en 10 zonas (ver figura 1): UTV-1 (aglutina las localidades de Coria, Hoyos y Villanueva de la Serena), UTV-2 (Hervás, Jaraíz de la Vera, Jarandilla de la Vera, Jerte, Navalmoral de la Mata, Plasencia, Villanueva de la Vera y Zarza de Granadilla), UTV-3 (Cañamero, Castañar de Ibor y Guadalupe), UTV-4 (Helechosa de los Montes, Herrera del Duque, Orellana la Vieja y Talarrubias), UTV-5 (Cabeza del Buey, Campillo de Llerena, Castuera, Don Benito, Guareña y Hornachos), UTV-6 (Azuaga, Fregenal de la Sierra, Jerez de los Caballeros, Llerena, Monesterio y Zafra), UTV-7 (Alburquerque, Badajoz, Barcarrota, Mérida y Montijo), UTV-8 (Alcántara, Alcuéscar, Brozas, Cáceres, Madrigalejo, Trujillo, Valencia de Alcántara y Zorita), UTV-9 (Jaraicejo, Serradilla, Torrejón el Rubio y Malpartida de Plasencia) y UTV-10 (Montehermoso, Pinofranqueado y Vegas de Coria).

Mapa 1

Distribución de las Unidades Territoriales de Vigilancia (UTV) de los AMN.

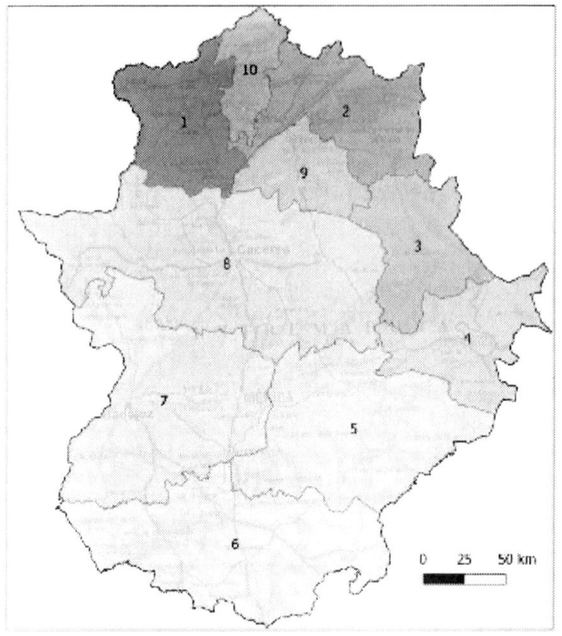

Nota: Consejería de Medio Ambiente y Rural, Políticas Agrarias y Territorio de la Junta de Extremadura (2022).

A tenor de todo lo expuesto resulta llamativo que, pese a las posibilidades que ofrece contar con un cuerpo como este en el ámbito de la *Criminología Verde*, son escasos los estudios sobre los AMN y, concretamente, aquellos que están ligados a vertidos ilegales como son los residuos de la construcción y la demolición (en adelante, RCD). Algunos estudios que se han encontrado sobre los agentes del medio natural son: Slatter, 2003; Lian et al., 2009; Sánchez Pinar, 2019; Curtis y Kaufman, 2020; Hertig y Palacios, 2020; y Rabasco Altamirano, 2021.

Debido a lo anterior, cabe destacar la originalidad de este estudio, cuyo objetivo principal es *analizar las percepciones de los AMN sobre el vertido ilegal de RCD para conocer el modus operandi, variables situacionales, las causas que lo generan y posibles medidas preventivas.* Los resultados que se expondrán más adelante invitan a pensar en un futuro abierto a nuevas líneas de investigación sobre la materia, pues existe una necesidad de mayor profundización al respecto.

2. MÉTODO

Conocer los detalles de la problemática del vertido ilegal de RCD desde la óptica de los AMN implica conocer sus percepciones, aplicando para ello una metodología cuantitativa basada en la técnica del cuestionario creado *ad hoc*.

Tal y como se ha descrito, la población que compone la muestra se encuentra dividida territorialmente en distintas UTV por toda Extremadura. Este hecho, sumado a que el cuerpo de AMN lo componen en torno a 300 personas podría suponer un proceso costoso en el momento de difundir el cuestionario preparado al efecto.

Para salvar esta situación se mantuvo una entrevista con el jefe de sección de coordinación de agentes, contacto clave para la difusión del cuestionario entre los AMN de la plantilla y a quien también se entrevistó de manera personal para conocer su percepción acerca de los RCD.

La herramienta escogida para diseñar el cuestionario fue *Microsoft Forms*, una aplicación web que permite crear y enviar formularios a múltiples participantes para su respuesta en línea.

En primer lugar, se dispusieron campos que permitieran conocer la demarcación geográfica en la que los AMN desempeñaban sus funciones y los años de experiencia en el puesto. Seguidamente, se abordaron aquellos aspectos relacionados con el fenómeno de los RCD: si tienen conocimiento de la existencia de vertidos ilegales sólidos en su demarcación; lugares en los que suelen ser encontrados y con qué frecuencia; motivos por los que se originan; perfil y modus operandi de las personas infractoras; acciones que se podrían desarrollar para prevenir esta conducta y; su percepción acerca de la preocupación que atribuyen ciudadanía y clase política a los RCD.

A lo largo del periodo de recopilación de información se enviaron tres recordatorios a las cuentas de correo electrónico de toda la plantilla. También se ofreció la posibilidad de obtener un certificado de participación al final de la encuesta para toda persona que lo solicitase, incentivando así la participación. Con todo, se recabó una muestra de 107 participantes, lo que supuso una tasa de respuesta del 35,6% del total de la plantilla y cuya distribución territorial se detalla a continuación:

Figura 2

Número de respuestas por UTV.

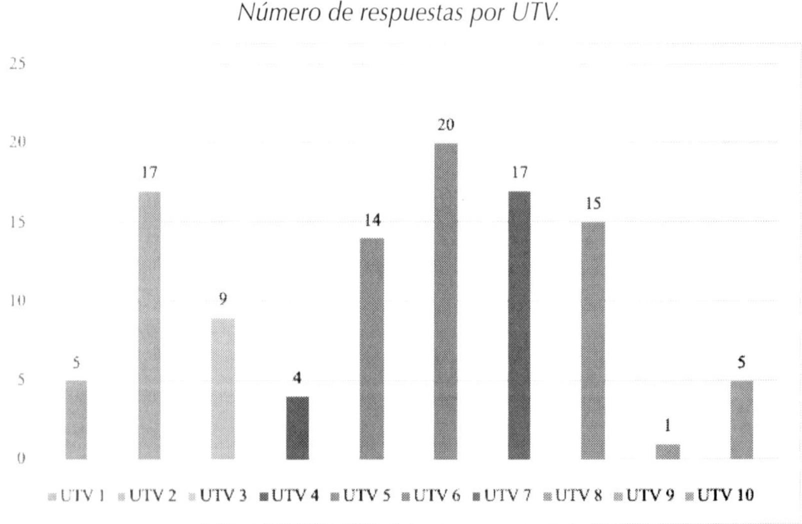

Atendiendo a lo apreciado en la figura anterior, el peso porcentual de respuestas de cada unidad territorial sobre el total sería del: 4,7% para la UTV 1; 15,9% para la UTV 2; 8,4% para la UTV 3; 3,7% para la UTV 4; 13% para la UTV 5; 18,7% para la

UTV 6; 15,9% para la UTV 7; 14% para la UTV 8; 1% para la UTV 9 y; 4,7% para la UTV 10.

3. RESULTADOS

Una vez vista la distribución de respuestas por UTV de las personas que participaron en el estudio cabe señalar que tanto la mediana como la media de años de experiencia en el puesto de los AMN fue de 17 años, siendo el mínimo de 2 años y el máximo de 37 años. Por otro lado, la desviación típica del total de la muestra fue de 10,6 años.

La UTV con mayor experiencia media fue la número 9, con 25 años; seguida por la UTV 8, con 24,4 años. Por otro lado, la UTV menos experimentada es la número 4, que posee una media de 3,5 años; por delante se encontraría la UTV 1 con 6,4 años de experiencia. A continuación, se pueden apreciar las medias de cada UTV:

Figura 3

Años de experiencia media por UTV.

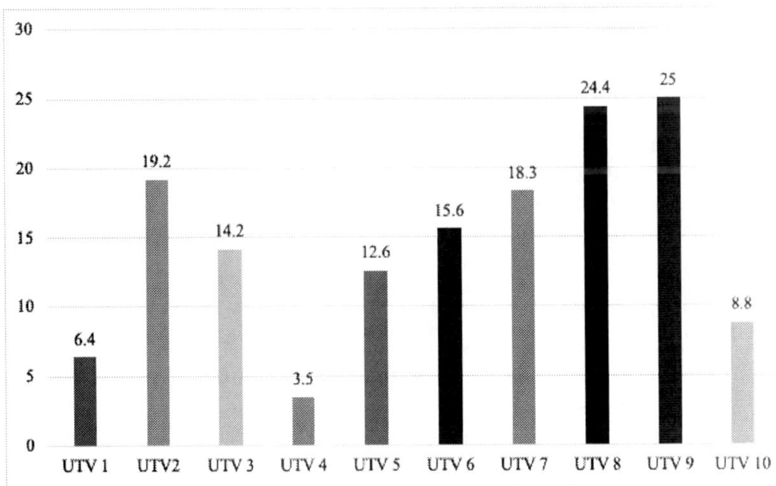

En lo concerniente a los RCD, el 14% de los participantes declaran la inexistencia de estos en sus demarcaciones y un 4% desconoce si hay vertidos ilegales de esta naturaleza en sus zonas de trabajo. En suma, se puede apreciar que se trata de un 18% de la muestra, un bajo porcentaje sobre el total si se compara con el restante 82% que advirtió la presencia de estos residuos en sus demarcaciones (véase figura 4). Sobre este asunto cabe señalar que se han comparado estas manifestaciones con las expresadas por otros AMN de las mismas demarcaciones y se han encontrado discrepancias en la mayoría de los casos.

Figura 4

Respuestas sobre existencia de RCD en sus demarcaciones.

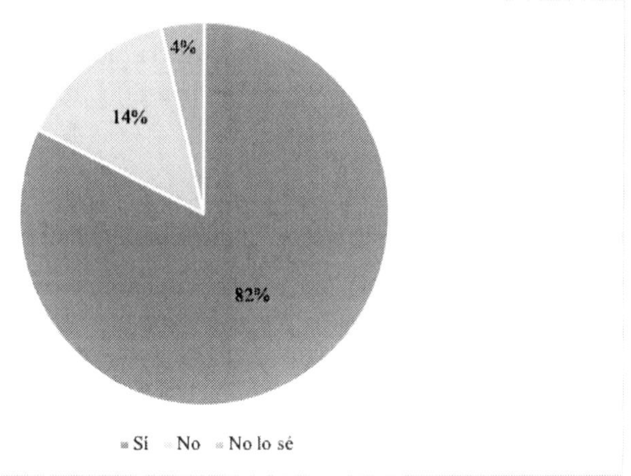

De forma similar se pronuncian respecto a la frecuencia con la que se producen estos vertidos. Solo el 8% de las personas han indicado que *casi nunca* se producen, mientras que otro 8% se posiciona en el otro extremo al marcar que se generan *casi todos los días*. El 84% restante sitúa la frecuencia como un hecho que se produce *ocasionalmente* (véase figura 5).

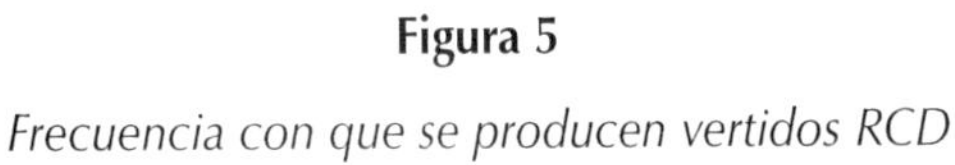

Figura 5

Frecuencia con que se producen vertidos RCD

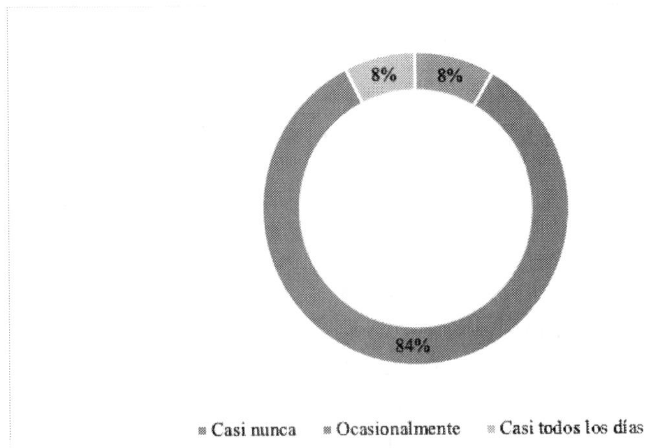

En cuanto a la ubicación en la que se suelen encontrar estos vertidos ilegales son múltiples los lugares detallados por los AMN, que coinciden mayoritariamente al destacar que los RCD aparecen en: vías pecuarias, caminos agrícolas, cunetas de carreteras poco transitadas y próximas al núcleo urbano, bordes de pistas forestales, antiguas escombreras, carreteras en desuso, márgenes de ríos y arroyos, fincas abandonadas, parcelas municipales y aquellas zonas perimetrales a los puntos limpios y plantas de reciclaje.

3.1. Motivos por los que se originan los RCD

Del mismo modo sucedió cuando fueron consultados por los motivos que consideraban que forman parte de la génesis de esta problemática, señalando una amplia variedad de elementos que se han aglutinado en tres bloques por similitud en los discursos:

A. *Desinformación*: en este grupo estarían las personas que desconocen la existencia de puntos específicos de acopio para estos residuos, los factores que se señalan atañen a la falta de información sobre su ubicación, funcionamiento, horarios o residuos que se pueden depositar en ellos. También se encontrarían en este apartado aquellos vecinos que creen que no están contaminando y que además utilizan este tipo de residuos para rellenar baches y caminos en los que tienden a atascarse los coches cuando hay lluvias. Los AMN destacaron que no hay suficientes campañas de información a la ciudadanía ni publicidad suficiente sobre los puntos de recogida.

B. *Incivismo*: En este bloque se encontrarían aquellas personas que quieren evitar el pago de tasas bien por ahorrar costes en los presupuestos de obras o bien porque no poseen licencia para la misma y estarían desempeñando esta actividad de manera ilegal. En esta línea se hallarían a su vez aquellos vecinos que por comodidad no quieren desplazarse hasta a los puntos de recogida o no les interesa acudir en los horarios establecidos por los municipios. El elemento que caracteriza a estas personas es que todas ellas llevarían a cabo estas conductas a sabiendas del daño medioambiental que están causando. Los motivos que alegarían los AMN son la falta de concienciación de estos vecinos y un sentimiento de impunidad al ejecutar estas acciones; este último debido a la escasez de personal que se dedique a la vigilancia y detección de quienes vierten estos residuos y a que, cuando se consigue, las sanciones no son lo suficientemente contundentes.

C. *Obstáculos en la gestión de los RCD*: Junto al punto anterior, la burocratización de la gestión de los RCD y la ausencia de facilidades por parte de la administración hacia el ciudadano son otros de los temas más señalados por los AMN como posibles causas de los vertidos ilegales. Aquí, se mostrarían aspectos como la gestión deficiente de las plantas de tratamiento y los puntos limpios, que se reflejaría en instalaciones que tienen difícil acceso para determinados vehículos o remolques, horarios y personal de atención bastante limitados (situación que se agravaría durante el fin de semana, que es cuando más demanda podría haber de estas instalaciones) y en unos requisitos administrativos para receptar los vertidos que resultarían excesivos y complejos a los ojos del ciudadano medio. Otra cuestión que se indicó al respecto serían los elevados costes económicos que supondría llevar a cabo la correcta gestión de estos residuos por parte de los ciudadanos ya que, además del pago de las tasas correspondientes por el tratamiento de los RCD, en ocasiones los contenedores destinados a tal efecto se encontrarían a las afueras de la población o en otros municipios, con el consiguiente gasto añadido en combustible al tener que desplazarse con los residuos hasta dichos puntos.

3.2. Perfil del infractor y modus operandi

En lo que concierne al perfil de la persona infractora la mayoría de los AMN indicaron que se trataría de varones de mediana edad que se dedican de forma autónoma a la construcción o albañilería y que llevan a cabo pequeñas reformas sin declarar de manera esporádica. En menor medida se señalaría a varones cercanos a la edad de jubilación dedicados a la ganadería y agricultura que realizan pequeñas obras en su finca particular.

Opinión casi unánime ha tenido la cuestión del *modus operandi* desarrollado para el depósito de estos vertidos ilegales. Por lo general, la forma de proceder consistiría en el desplazamiento de los RCD en pequeños sacos a través de un vehículo particular tipo furgoneta o bien mediante un pequeño remolque. Estos vertidos se llevarían a cabo en horarios de baja actividad como el amanecer, el anochecer o las horas de descanso del mediodía y durante los fines de semana o días festivos.

3.3. Percepción sobre la preocupación ciudadana y política sobre los RCD

Respecto a su percepción en cuanto a la preocupación mostrada por parte de la ciudadanía y los actores políticos sobre los RCD, la tendencia en las respuestas a ambas preguntas fue muy similar.

A nivel ciudadano, dos tercios de los AMN encuestados consideraron que este problema no suele preocupar a la población y, cuando existe esa sensibilización, se trata de colectivos relacionados con la protección del medioambiente. No obstante, apuntaron que está aumentando la concienciación en la defensa de la naturaleza por parte de la ciudadanía, algo que quedaría sustentado por aquellos AMN que sí consideraron que existía tal sensibilidad.

De otro lado, cuando fueron consultados por la preocupación que observaban en la clase política acerca de estos vertidos ilegales, 2/3 de los agentes señalaron que no es un asunto que parezca que les interese. De hecho, mostrarían menor preocupación que la ciudadanía, ya que solo atenderían a la cuestión si tuviera para ellos un coste mediático o electoral. En este sentido, algunas voces indicaron que perseguir y sancionar con mayor rigor los RCD podría conllevar enfrentamientos con los vecinos y ello, a la postre, una pérdida de votos. En cuanto a aquellos AMN que manifestaron que sí había una preocupación política por este asunto, subrayaron que no todos los municipios gozan de medios suficientes para llevar a cabo esta tarea.

3.4. Medidas preventivas contra el vertido ilegal de RCD

Finalmente, se consultó a los AMN cuáles serían las medidas que podrían implementarse para prevenir eficazmente la problemática de los vertidos ilegales de RCD. En general, muchas de las propuestas que se aportaron coincidieron entre las personas encuestadas, quedando reflejadas en los siguientes puntos:

- Desarrollar acciones de educación ambiental para concienciar a la ciudadanía sobre este problema, especialmente en colegios e institutos. Para ello, entienden que es necesaria una mayor implicación por parte de los colectivos que promueven la defensa del medioambiente.
- Mejorar la información sobre las zonas habilitadas y procedimientos para la recolección y gestión de RCD. Impulsando campañas de divulgación a través de los medios de comunicación y las redes sociales.
- Promover la accesibilidad a los puntos de recogida, ampliando los horarios de atención e impulsando alternativas menos burocratizadas que resulten en procedimientos claros, breves y sencillos para el ciudadano.
- Incentivar la gestión de los vertidos mediante bonificaciones. De un lado, permitiendo que las pequeñas empresas puedan depositar conjuntamente los RCD para abaratar los costes y; por otra parte, subvencionando las tasas de gestión de determinados materiales que tienen un alto coste, así como las pequeñas obras de particulares para que el proceso de gestión fuera gratuito.
- Optimizar el funcionamiento de las plantas de gestión de RCD por parte de los ayuntamientos y empresas privadas para poder garantizar el control de la trazabilidad del residuo. Según los AMN, para llevar a cabo lo anterior se necesitaría dotar a los municipios de más centros de gestión y valorización de escombros, así como disponer de contenedores que pudieran instalarse en puntos calientes de escombreras ilegales y en lugares próximos a las localidades.
- Flexibilizar la legislación en materia de RCD para que sea compatible con el coste real del tratamiento de estos residuos y con la infraestructura existente en los municipios.
- Endurecer las sanciones para que no sea rentable verter ilegalmente. Asimismo, los AMN entienden que es necesario que las administraciones ciertamente ejecuten los expedien-

tes sancionadores contra aquellas personas que infrinjan la ley. Para lograrlo, consideran fundamental que existe una red de colaboración entre administraciones, autoridades competentes y vecinos.

- Incrementar la vigilancia y control. Los AMN estiman que sería conveniente instalar cámaras de videovigilancia (CCTV) en aquellos lugares donde se ha generado costumbre social de verter escombros de manera ilegal para evitar el *efecto llamada* y posibilitar la identificación de las personas infractoras. De otro lado, señalan que es necesario llevar a cabo una supervisión *in situ* de las obras menores, así como examinar con detalle los procedimientos de concesión de licencias de obra municipales y de gestión de los RCD que se generan. Para lograrlo, manifiestan que es fundamental aumentar el número de efectivos en las plantillas de los AMN.

4. CONCLUSIONES

A la vista de la experiencia profesional de los AMN que participaron en este estudio (17 años de promedio), los resultados que se han expuesto en el apartado anterior gozan de bastante peso de cara a efectuar un análisis sobre la situación y variables desgranadas sobre el vertido ilegal de RCD.

Si bien es cierto que la participación podría haber sido mayor y que algunas UTV quedan descompensadas en representación respecto a otras, no es menos cierto que se aprecia bastante consenso entre los AMN en sus manifestaciones acerca de los RCD. A tenor de esto, cabe inferir que es ciertamente factible que la opinión del resto de los AMN concuerde con las voces que se han plasmado en este estudio.

Según indican los AMN los vertidos ilegales de RCD son un problema generalizado en toda la comunidad extremeña. La presencia ocasional de los mismos principalmente en zonas rurales y lugares de poco tránsito a las afueras de los núcleos urbanos guarda mucho sentido respecto al perfil de la persona infractora y su modo de proceder, variables que se desgranarán a tenor de los resultados obtenidos más adelante.

4.1. Sobre los motivos por los que se originan los RCD

Antes de comenzar, cabe precisar la interrelación que se vislumbra entre los elementos destacados como principales motivos que dan origen a la conducta del depósito ilegal de los RCD, *desinformación, incivismo y obstáculos en el proceso de gestión de los RCD.*

En este punto surgen dos tesis acerca de por qué se producirían estos comportamientos. La *primera* de ellas se plantea a un ciudadano con escasa información acerca de cómo proceder cuando llega la hora de gestionar estos residuos. Sabe que existe una planta de tratamiento en la periferia de su municipio, pero desconoce qué tipo de vertidos recogen y cuáles no. Tampoco sabe cuál es el horario de atención, si estará abierto o no cuando se acerque o si le cobrarán por receptar los residuos generados.

Consigue el número de teléfono de la planta en cuestión, contacta con un empleado que le explica que ha de pagar una tasa, que hay determinados residuos que ha generado que no puede depositar en esa instalación y que debe realizar unos trámites entre los que se encuentran: rellenar y entregar una serie de formularios (véanse la licencia de obra, resguardo de prestación de fianza, declaración responsable de gestión de residuos, etc.).

Nuestro protagonista ha pasado de no saber absolutamente nada sobre la cuestión a estar totalmente abrumado con todo *el entramado que le supone llevar a cabo la gestión de los RCD.* Ante tal panorama comenzaría a plantearse cómo proceder, decidiendo que sería mucho más fácil y barato verter los escombros en la cuneta que conecta un viejo camino con unas parcelas a las afueras de la localidad, cayendo en esa desidia *incívica* que señalaron los AMN.

El *segundo* postulado mostraría un escenario diametralmente opuesto al anterior. Se trataría de un ciudadano que *conoce la ubi-*

cación y funcionamiento de la planta de tratamiento o punto limpio del municipio. Esta persona *poseería la información sobre los requisitos administrativos que debe reunir para gestionar los RCD,* así como las tasas correspondientes para tal fin. Probablemente, este vecino habría tenido alguna experiencia previa al respecto y, por ello, *tendría decidido desde antes de que se generasen los residuos que no realizaría la gestión* de estos y, en su lugar, verterlos en cualquier zona que conociera y donde no pudiera ser descubierto.

Las razones que podrían alimentar esta conducta incívica pasarían desde ahorrase los costes de tasas y fianzas, hasta la comodidad de no querer desplazarse al punto de acopio ni adaptarse al horario estipulado por el mismo. En este caso se tendría la información, se conocería la burocracia de la gestión de los RCD, pero no existiría voluntad alguna de seguir los cauces legalmente establecidos.

Sin duda, ambos planteamientos han de ser tenidos en cuenta junto a otros factores con miras a ofrecer posibles medidas correctivas ante el vertido ilegal de RCD.

4.2. Al respecto del perfil del infractor y su modus operandi

Poco sorprende la fotografía mostrada por los AMN respecto al perfil del infractor en este tipo de vertidos. A tenor de lo expuesto, se sospecha que existirían dos retratos de personas que encajarían con este perfil: 1) aquellas que tendrían conocimientos básicos en albañilería y estarían llevando a cabo reformas en sus domicilios o parcelas, normalmente durante el fin de semana o días festivos y; 2) aquellas que se dedicarían profesionalmente al sector de la construcción y que buscarían unos ingresos extra a través de pequeños trabajos sin declarar a sus familiares o amigos.

Si esto fuera cierto, sería de justicia inferir que en ambas situaciones las personas involucradas no solicitarían licencia de obra, no prestarían fianza ni declaración para la gestión de los futuros RCD, por lo tanto, mucho menos se plantearían personarse en un punto de valorización y tratamiento abriendo la posibilidad a que puedan ser descubiertos en esas irregularidades. Entonces, ¿qué solución encuentran? *Lo que se ha hecho toda la vida,* sin importar las consecuencias medioambientales que puedan ocasionar.

Estos presupuestos ligarían perfectamente con el *modus operandi* descrito por los AMN. Tendría sentido así que, terminados los trabajos de construcción o reformas, estas personas buscaran las condiciones lumínicas y horarios de menor actividad social para cargar en sus vehículos los sacos de escombros y verterlos en lugares alejados de una mirada avizora.

4.3. Acerca de la percepción de la preocupación ciudadana y política sobre los RCD

Continuando el análisis con las percepciones que tienen los agentes encuestados respecto a la poca preocupación mostrada por parte de la ciudadanía y de los actores políticos en relación con los RCD interesa señalar que, sin dudar de las acciones de sensibilización ambiental que se han desarrollado a múltiples niveles hasta ahora, queda mucho trabajo por hacer.

A nivel macro es sencillo que cualquiera que lea estas letras visualice rápidamente algunos de los objetivos de la Agenda 2030 relacionados con el medioambiente o que recuerde el fenómeno que generó la aparición de Greta Thunberg en el panorama político y social alrededor de esta materia (Soto, 2022). Sin embargo, el mensaje que se lanza desde los AMN es que siguen siendo necesarias nuevas propuestas de concienciación a nivel regional y local que sean capaces de calar en toda la ciudadanía sin distinciones.

Pero no basta con estas iniciativas dirigidas a los vecinos, también se requiere acabar con la miopía electoralista de algunos gobernantes para poder desarrollar políticas públicas que afronten estos problemas con eficacia. Trayendo a colación algunas ideas que subrayan las personas encuestadas se hace necesario un compromiso serio por parte de las administraciones que procuren los medios materiales y humanos para garantizar que lo macro pueda hacerse realidad tomando en consideración las particularidades que presentan los diferentes municipios y comarcas extremeñas.

4.4. Tomando en cuenta las medidas preventivas contra el vertido ilegal de RCD

A tenor de todo lo expuesto, resultan oportunas muchas de las medidas que ofrecieron los AMN para atajar esta problemática. Enlazando con el párrafo anterior, los agentes evidenciaron la necesidad de promover tanto la sensibilización ambiental como la divulgación de información acerca de cómo gestionar los RCD. Especial mención tiene que los AMN depositen su confianza en los colectivos comprometidos con el medioambiente para llevar a cabo esta tarea, pues supone un reconocimiento a la labor que desarrollan y el potencial alcance que pueden llegar a tener en la población.

Por otro lado, se atisba en estas medidas una llamada de atención a las administraciones, que han de aceptar el reto de hacer más accesibles los puntos limpios y plantas de gestión de RCD. Para empezar, mejorando la información general disponible al ciudadano a través de los canales de que dispone y desarrollando campañas de divulgación de esta (como el buzoneo, por ejemplo). A continuación, podrían estudiarse alternativas que simplifiquen y agilicen los procesos relacionados con la gestión de los RCD con el objetivo de encontrar la manera de incentivar el tratamiento de estos entre los ciudadanos.

En el sentido de la propuesta anterior estaría la idea de subvencionar los costes del tratamiento de los RCD. No cabe duda de que a ningún vecino le resultará atractivo el pago de unas tasas por el depósito de los residuos, del mismo modo que la gratuidad no garantizaría *per se* la adecuada gestión de estos por el particular; pero puede ser un elemento más que aproxime a la consecución del fin pretendido. Por ello, se ha de repensar la legislación actual sobre la materia para encontrar una fórmula que permita bonificar al usuario y, por otro lado, optimizar los procesos de las empresas de valorización de RCD, especialmente en el momento dar salida al mercado del producto reciclado.

De la mano de lo que se ha comentado debe acompañar una inversión en infraestructuras. Como se ha señalado, la accesibilidad es muy relevante en este asunto, de manera que sería deseable disponer de puntos de acopio y gestión en todos los municipios. Sin embargo, ante la falta de medios puede ser interesante la instalación de contenedores en puntos cercanos a viejas escombreras con miras a atajar el depósito de RCD por costumbre. Así, el ciudadano predispuesto al vertido ilegal que conoce de estos puntos dispondría de una alternativa a menor coste (puede ir en cualquier momento, por lo que no se tiene que esconder; tampoco se le somete al pago de tasas o a la verificación de licencias, etc.).

En último lugar, se analizan las propuestas de vigilancia, control y castigo. En lo que refiere a las sanciones, parece contraproducente que dadas las circunstancias en que se materializan los vertidos ilegales se dé lugar a una gracia no llevando a término los expedientes sancionadores. Esto puede ayudar a mantener la percepción de impunidad de los infractores, que ven cómo los costes de oportunidad se rebajan aún más, aunque hayan sido denunciados. Se daría una suerte de retroalimentación pues se cometen irregularidades para ahorrar dinero y, cuando se tiene la oportunidad de identificar y sancionar al infractor, esto es, de hacer los costes mucho mayores por no haber seguido los cauces administrativos, se deja sin efecto la sanción. No hay razón para ello.

En cuanto a la vigilancia y control de los RCD, se sabe que la labor de detección se hace sumamente compleja principalmente por dos motivos: en primer lugar, porque el modo de proceder de los infractores persigue los resquicios en los que la vigilancia sea nula, por ejemplo, el final del turno del AMN de su localidad y, en segundo lugar, porque las plantillas de los agentes son insuficientes para llevar a cabo la persecución de estas actividades ilícitas en toda la comunidad extremeña. Ya se ha hablado de inversión en medios materiales, de modo que se piensa igual con respecto a los medios humanos. Un mayor número de efectivos en las plantillas permitiría un mayor control sobre las zonas que se consideran de mayor riesgo en cuanto a vertidos ilegales, así como de las pequeñas obras clandestinas que se llevan a cabo en los municipios.

Entendiendo las limitaciones existentes actualmente en cuanto a la dotación de las plantillas, resulta de gran interés exponer en este punto la posibilidad de instalar CCTV en los puntos calientes de vertidos ilegales. Esta es una propuesta más económica y podría provocar si no un efecto disuasivo en el efecto llamada de las escombreras en los infractores, si la identificación de estos para su propuesta de sanción por el vertido ilegal de RCD. Sin pecar de ilusorios, implementar esta medida debe llevar aparejado el estudio de la evolución de los RCD en las proximidades de estos puntos videovigilados, ya que se podría dar un desplazamiento de estas conductas ilegales a otros lugares del entorno.

Para terminar, hay que decir que se requiere un mayor estudio del fenómeno de los vertidos ilegales de RCD. En este texto se ha podido aproximar el estado de la cuestión y señalar aquellos puntos sobre los que se requiere empezar a trabajar para tratar de solventar el problema. Las propuestas de mejora aportadas por los AMN resultan ser interesantes con miras a reducir los problemas existentes que afectan a la materia y la posibilidad de aplicar algunas de ellas en conjunto implicaría cambios cuantificables.

En cualquier caso, toda medida correctiva que se lleve a cabo esté o no recogida en esta investigación, ha de ser evaluada para garantizar que se está avanzando adecuadamente hacia la reducción del vertido ilegal de RCD.

5. REFERENCIAS BIBLIOGRÁFICAS

Curtis, J. y Kaufman, S. (2020). "It's not what you see but what you hear...": Understanding environment protection officers' responsive decision making. *Journal of Environmental Management, 262,* 1-10. DOI: 10.1016/j.jenvman.2020.110336

Hertig, C.A. y Palacios, K.E. (2020). Role of the Professional Protection Officer. En S. J. Davies y L. J. Fennelly (Eds.), *The Professional Protection Officer* (2nd ed., 19-31). Elsevier Inc. DOI: 10.1016/C2018-0-01136-6

Lian, Y. L., Liu, J. W., Zhang, C. y Yuan, F. (2009). Comparison of strain of mental workers by characteristics. *Chinese journal of industrial hygiene and occupational diseases, Vol. 27*(12), 725-729.

Disponible en:

http://caod.oriprobe.com/articles/17669797/Comparison_of_strain_of_mental_workers_by_characteristics.htm [Consultado el 03.07.2023].

Rabasco Altamirano, F. J. (2021). *Perfil sociológico de los agentes de Medio ambiente de andalucía y análisis Sobre los problemas que afrontan.* [Trabajo Fin de Grado, Universidad Nacional de Educación a Distancia]. E-spacio. Disponible en: http://e-spacio.uned.es/fez/eserv/bibliuned:grado-CPyS-Sociologia-Fjrabasco/RabascoAltamirano_FranciscoJose_TFG.pdf [Consultado el 03.07.2023].

Sánchez Pinar, R. (2019). Agentes medioambientales de la Administración General del Estado. *Foresta, 74,* 32-33. Disponible en: https://www.forestales.net/Canales/Ficha.aspx?IdMenu=b6947309-987f-4bff-808d-4e7e974ccaf8yCod=502cd71d-a4c9-497c-854a-db4b609e4b24yIdioma=es-ES [Consultado el 03.07.2023].

Slatter, J. (2003). A Day in the Life of an Environment Protection Officer. *National Environmental Law Review, 3,* 45-56.

Disponible en: http://www.austlii.edu.au/au/journals/NatEnvLawRw/2003/25.pdf [Consultado el 03.07.2023].

Soto, R. (2022, noviembre 10). *El fenómeno Greta Thunberg: niños activistas contra el cambio climático en la COP27.* EFEverde. Disponible en: https://efeverde.com/ninos-activistas-cambio-climatico-cop27/ [Consultado el 03.07.2023].

Capítulo 9

El proceso de gestión del residuo sólido y relación con el vertido ilegal

SARA Mª MARCHENA GALÁN[1]

1. INTRODUCCIÓN

Los procesos de urbanización han provocado que la industria de la construcción sea uno de los principales productores de residuos sólidos y, por tanto, uno de los principales causantes de daños medioambientales. En España, la intensa actividad del sector de la construcción ha dado lugar a una intensa generación de residuos de construcción y demolición (en adelante, RCD), procedentes tanto de obras mayores como menores, siendo especialmente significativa en la zona de Levante[2] (Martínez Guirao, 2016). Aunque el desarrollo urbanístico en Extremadura no hay sido tan acuciante como en otras zonas de la geografía española, en los últimos 20 años el parque de viviendas ha aumentado consi-

---

1 Licenciada en Derecho y Personal Científico Investigador en Proyecto de Investigación VIEX "Teledetección y análisis ambiental de vertederos ilegales" en la Facultad de Extremadura.

2 En el periodo de 1987 a 2005 se construyó en la costa española una cuarta parte de todo lo construido, al menos hasta el año 2013 cuando este dato se recoge en el Informe "Destrucción a toda costa. Análisis del litoral a escala municipal", de Greenpeace España, del año 2013. Disponible para consulta en: http://archivo-es.greenpeace.org/espana/es/reports/Destruccion-a-toda-costa-2013/ [Consultado el 03.07.2023].

derablemente[3]. La significativa -y lógica- relación entre expansión demográfica y la producción de RCD evidenciada en esta obra (véase capítulo 5) no viene acompañada en muchas ocasiones de una adecuada gestión por parte de las administraciones públicas. La ausencia o la ineficaz gestión de los RCD puede dar lugar a su eliminación incontrolada ocasionando impactos muy negativos a nivel paisajístico, en los usos potenciales del suelo, en la ocupación de vías pecuniarias y en los sistemas fluviales, además de constituir un foco de incendios forestales[4].

Es por ello que, hoy en día, la gestión de los RCD para su adecuado tratamiento, reciclaje o eliminación es una de las claves en relación con el cuidado del medio ambiente al tiempo que centra nuestras preguntas de investigación: ¿cómo se gestionan los RCD?; ¿existen diferentes procesos o modelos de gestión?; ¿tiene lo anterior una relación con la aparición y mantenimiento del vertido ilegal?

En otras palabras, en este trabajo se aborda cómo las Administraciones Públicas extremeñas, en consonancia con el artículo 45 de la Constitución Española y las directrices europeas[5], están gestionando los residuos sólidos. Constituye, por tanto, una prioridad para las Administraciones Públicas el control del vertido ilegal, fenómeno que sigue persistiendo a pesar de las normas aprobadas en materia de gestión de residuos.

---

3 Pérez, Juan. (4 de febrero de 2022). El pasado año se construyeron en la región 107.518 viviendas nuevas, un 22,61% más. Canal Extremadura. Se alude en la noticia a un informe no disponible en línea. Disponible para consulta en: https://www.canalextremadura.es/noticias/extremadura/el-pasado-ano-se-construyeron-en-la-region-107518-viviendas-nuevas-un-2261-mas [Consultado el 03.07.2023].

4 Programa de vigilancia y control de la producción, gestión y vertido ilegal de residuos para el año 2022.

5 Véase la Directiva 2018/851 del Parlamento Europeo y del Consejo, de 30 de mayo de 2018, por la que se modifica la Directiva 2008/98/CE sobre los residuos.

Por *gestión de residuos* nos referimos a "la recogida, el transporte, la valorización y la eliminación de los residuos, incluida la clasificación y otras operaciones previas; así como la vigilancia de estas operaciones y el mantenimiento posterior al cierre de los vertederos", incluyendo también las actuaciones realizadas en calidad de negociante o agente, tal y como establece el artículo 2.n) de la Ley 7/2022, de 8 de abril, de residuos y suelos contaminados para una economía circular. Este proceso se desarrolla desde la previsión de la generación del residuo hasta su destino final en el lugar de reciclaje o tratamiento, interviniendo durante toda la secuencia diferentes actores y elementos que van a influir en el proceso de gestión y, por tanto, en la posible comisión del vertido ilegal.

2. ESTADO DE LA CUESTIÓN

La compleja relación entre el proceso de gestión y la aparición del vertido ilegal ha sido abordada ampliamente en la literatura existente en la materia desde una perspectiva jurídico-penal y criminológica.

Algunos de estos estudios indican que uno los principales problemas se encuentran en que, para los productores de residuos, el vertido sigue siendo el método más accesible y barato de eliminación de residuos (Slavik y Pavel, 2013). A su vez, señalan que el problema parece aún más grave–multiplicándose la casuística–en los municipios, donde las motivaciones individuales se ven influenciadas por la controversia entre quienes generan los residuos–principalmente particulares y empresas–y quien es el responsable de su recogida, fundamentalmente los Ayuntamientos. En este sentido, el Decreto 20/2011, de 25 de febrero, por el que establece el régimen jurídico de la producción, posesión y gestión de los residuos de construcción y demolición en la Comunidad Autónoma de Extremadura, dispone que es competencia de los entes locales la gestión de los RCD procedentes de obras menores de construcción y reparación domiciliaria, pudiendo delegar su ejercicio a las Mancomunidades Integrales[6]. Nos centramos, así, en el análisis del proceso del residuo generado en las obras menores[7], que representan a la gran mayoría, realizadas en locales o domicilios por parte de particulares o a cargo de empresas privadas de albañilería y sometidas a menor control urbanístico debido a sus características. Precisamente debido a su mayor número y menor control originan una multitud de vertidos ilegales en las zonas periurbanas de los pueblos extremeños. Prueba de ello es que la mayor parte de los puntos de residuo tienen un perímetro reducido (véase capítulo III). También proceden de obras menores las obras clandestinas. Por todo ello, resulta fundamental, de acuerdo con la literatura criminológica, rastrear las cadenas de contaminación desde el productor de los residuos hasta su destino final (Uricchio et al., 2010), para comprender en qué fallan estas medidas. Esto es así porque a lo largo del proceso de gestión de residuos se producen un conjunto de tomas de decisiones por parte de los sujetos implicados que influyen a la hora de actuar en relación a los mismos y, por tanto, en la producción o no de vertido ilegal (Chen et al., 2018).

2.1. Las medidas de gestión del residuo

El análisis de las medidas de prevención que se están tomando en diferentes partes del mundo para el control de vertido ilegal que puedan influir en el comportamiento y la toma de decisiones de los sujetos implicados resulta fundamental para aproximarnos

6 Artículo 3 Decreto 20/2011, de 25 de febrero, por el que establece el régimen jurídico de la producción, posesión y gestión de los residuos de construcción y demolición en la Comunidad Autónoma de Extremadura,

7 Por obra menor entendemos aquellas obras que se caracterizan por su escasa entidad, sencillez constructiva y pequeña cuantía y, por tanto, no requieren de proyecto técnico ni memoria habilitante, tal y como establece el artículo 1.c) del Decreto 205/2003, de 16 de diciembre, por el que se regula la memoria habilitante a efectos de la licencia de obras en Extremadura.

a entender qué puede estar fallando en un determinado modelo de gestión. Existen distintos tipos de medidas en este sentido, aunque las vamos a clasificar fundamentalmente en tres categorías: económicas, logísticas y de concienciación.

Las *medidas económicas* consisten en el establecimiento de una cantidad dineraria en concepto de tributo, garantía o sanción. Entre los tributos, por un lado, están las tasas, habiéndose apostado en los últimos años por el *sistema de tasas variables*, en sustitución de las tasas fijas por considerar que tienen una mayor eficacia (Slavik y Pavel, 2013). Consisten en el establecimiento de tasas diferentes en función del tipo de residuo y la cantidad que se deposita, entre otras características. Ello contribuye a que los hogares participen de la correcta gestión de los residuos separándolos y reduciendo su vertido en vertederos (Fullerton-Kinnaman, 1996; Miranda et al., 1996). En este sentido, en España, la Ley 7/2022, establece la obligación del productor inicial o del poseedor de residuos domésticos de separar en origen sus residuos y entregarlos en los términos establecidos por las ordenanzas municipales[8]. A su vez, el establecimiento de estas tasas o tarifas responden al principio de "quien contamina paga" que rige la actual normativa[9], de manera que el productor de los residuos debe hacerse cargo de los costes derivados de ellos[10]. En contraposición se encuentran, como hemos indicado, las tasas fijas que en algunos lugares se entienden como una "tasa solidaria" mediante la que todos contribuyen al sostenimiento del servicio. En España, la Ley 7/2022 establece que las entidades locales deben regular una tasa o prestación patrimonial de carácter público no tributario, específica, diferenciada y no deficitaria, que permita la implantación de sistemas de pago por generación, reflejando el coste real, directo o indirecto, de las operaciones de recogida, transporte, tra-

---

8 Artículo 20.3 de la Ley 7/2022, de 8 de abril, de residuos y suelos contaminados para una economía circular.

9 Artículo 11 de la Ley 7/2022, de 8 de abril, de residuos y suelos contaminados para una economía circular.

10 Plan Integrado de Residuos de Extremadura (PIREX) 2016-2022.

tamiento y vigilancia de los residuos, así como el mantenimiento y vigilancia posterior al cierre de los vertederos y las campañas de concienciación[11]. Por último, entre los tributos están los precios públicos que establecen los sectores privados cuando estos prestan el servicio.

Otra de las medidas de tipo económico es la *fianza,* es decir, una cantidad de dinero que se deposita de manera previa a la realización de la obra y se recupera una vez certificado que, efectivamente, los residuos se encuentran en el lugar habilitado para ello, actuando como una garantía. Por último, las *sanciones* también son un tipo de medidas económicas muy utilizadas, pero no se consideran lo suficientemente efectivas (Cossu y Masi, 2013; Chen, Hua y Liu, 2019). En España, la Ley 7/2022 ha establecido un régimen sancionador acorde a la realidad actual de las infracciones medioambientales, incluyendo infracciones como el abandono de basura dispersa o *littering,* a la vez que se ha definido el procedimiento sancionador en consonancia con la Ley 39/2015, de 1 de octubre, del Procedimiento Administrativo Común de las Administraciones Publicas. Clasifica las infracciones en muy graves, graves y leves[12], con sus consecuentes sanciones[13]. No obstante, las Comunidades Autónomas y los entes locales tienen potestad sancionadora en el ámbito de sus competencias. De esta manera, en relación a los RCD procedentes de obras menores de construcción y reparación domiciliaria, el Decreto 20/2011, de 25 de febrero, por el que se establece el régimen jurídico de la producción, posesión y gestión de los residuos de construcción y demolición en la Comunidad Autónoma de Extremadura, dispone la potestad sancionadora de los entes locales, que establecerán

---

[11] Artículo 11.3 de la Ley 7/2022, de 8 de abril, de residuos y suelos contaminados para una economía circular.

[12] Artículo 108 de la Ley 7/2022, de residuos y suelos contaminados para una economía circular

[13] Artículo 109 de la Ley 7/2022, de residuos y suelos contaminados para una economía circular.

sus régimen sancionador en las ordenanzas municipales[14], por ejemplo, en el caso de supuestos de abandono, vertido o eliminación incontrolada de los residuos cuya gestión corresponde a los entes locales.

Por *medidas logísticas* nos referimos al establecimiento de mecanismos y facilidades que se aplican al proceso de gestión para que los RCD vayan al destino legalmente establecido. Son medidas fundamentalmente relacionadas con la recogida y el traslado de los residuos. Por ejemplo, en algunos casos nos encontramos con el denominado *puerta a puerta*, consistente en la recogida de los residuos por parte del ayuntamiento en el propio domicilio del particular. Sin embargo, no es muy habitual dado que es difícil su sostenimiento económico y, en función del volumen y tipo de residuos, pueden producirse dificultades para tener la infraestructura necesaria para la recogida y transporte de los mismos. Se suele utilizar en casos de realización de una obra menor en un domicilio y a un bajo volumen de RCD, normalmente escombros. Otro tipo son los *puntos de recogida y de tratamiento* como una de las maneras más extendidas para el control de RCD donde la ciudadanía y las empresas van a poder acudir para depositar los residuos. Este elemento es fundamental ya que algunos estudios indican la falta de instalaciones adecuadas para el tratamiento de los residuos como uno de los factores principales en la producción del vertido ilegal (Ichinose y Yamamoto, 2011). Por tanto: a mayor número de instalaciones de tratamiento de residuos menor número de casos de vertidos ilegales.

En Extremadura existen cuatro puntos fundamentales para la recogida de RCD: los puntos limpios, las plantas de RCD, las plantas de transferencia y los puntos de acopio o bateas. Los *puntos limpios* son instalaciones de almacenamiento disponibles en las zonas

14 Artículo 32 del Decreto 20/2011, de 25 de febrero, por el que se establece el régimen jurídico de la producción, posesión y gestión de los residuos de construcción y demolición en la Comunidad Autónoma de Extremadura.

urbanas o periurbanas de los entes locales, en las que se ponen a disposición de la ciudadanía contenedores específicos para la recogida selectiva de residuos asimilables a residuos domésticos que requieran un tratamiento especializado[15]. Los RCD se consideran residuos admisibles ya que son residuos sólidos urbanos, requiriéndose a quien los quiera depositar la licencia de obra menor para permitir el vertido de los escombros, admitiendo la cantidad de hasta 150 kilogramos al día y estando sometido, en función de la normativa municipal, al pago de una tasa. Las *plantas de RCD*, son instalaciones en las que tanto la ciudadanía como las empresas pueden llevar RCD siendo el lugar donde se gestionan, valorizan y reciclan. Las plantas RCD son el centro neurálgico de la gestión del residuo, y en torno a ellas se disponen de puntos intermedios de recepción de residuos, que son las *plantas de transferencia* y los *puntos de acopio* o *bateas*. Las plantas de transferencia son plantas intermedias de almacenamiento entre el origen del residuo y el destino final en la planta de RCD. Los *puntos de acopio* o *bateas* son una especie de contenedores establecidos en diversos municipios para facilitar el depósito temporal de escombros hasta su traslado a la planta[16].

No obstante, ninguna de estas medidas funciona por sí sola, ya que todas están sometidas, de una forma u otra, a la voluntad o concienciación de los particulares y de las instituciones que deben velar por su buen funcionamiento, por tanto, son fundamentales las *medidas de concienciación* e información a todos los agentes implicados: desde la ciudadanía, hasta los técnicos municipales, pasando por las empresas privadas de construcción, albañilería y de gestión de los residuos. En este sentido, la Ley 7/2022, de 8 de abril, de residuos y suelos contaminados para una economía

---

15 Manual para la correcta gestión de un punto limpio de la Junta de Extremadura. Disponible en: http://extremambiente.juntaex.es/index.php?option=com_contentyview=articleyid=1573yItemid=578 [Consultado el 03.07.2023].

16 Esta cuestión se trata más detalladamente en el capítulo 5 de esta obra.

circular, prevé la elaboración de planes de prevención y de campañas de concienciación a nivel estatal, autonómico y municipal.

2.2. Las medidas del Decreto extremeño de gestión de RCD

Una de las herramientas fundamentales para la gestión de residuos está siendo la legislación referente a la misma, considerándose como la fuerza motriz de todo el sistema de gestión (Cossu y Masi, 2013). Por ello, debemos tener en cuenta la normativa básica en esta materia en Extremadura, que es el *Decreto 20/2011, de 25 de febrero, por el que se establece el régimen jurídico de la producción, posesión y gestión de los residuos de construcción y demolición en la Comunidad Autónoma de Extremadura.* Este Decreto señala que los RCD generados en Extremadura han alcanzado suficiente nivel como para ser objeto de correcta gestión y eliminación. Identifica en su preámbulo que persiste el vertido incontrolado, ocupando un gran volumen y funcionando como *efecto llamada* de otros residuos dificultando su gestión y su recuperación futura, elevando los costes medioambientales y suponiendo un factor económico importante. Es por ello que este Decreto 20/2011 establece una serie de mecanismos para la buena gestión de los RCD prohibiendo expresamente el abandono y vertido incontrolado, la eliminación total o parcial sin autorización y la mezcla que dificulte una correcta gestión[17]. En este sentido, clasifica los RCD en cuatro categorías[18] atendiendo a las dificultades que ocasionan a la hora de gestionarlos y con el objetivo de facilitar a los ayuntamientos el desarrollo de las ordenanzas municipales[19].

---

17 Artículo 4, Decreto 20/2011 de 25 de febrero, por el que se establece el régimen jurídico de la producción, posesión y gestión de los residuos de construcción y demolición en la Comunidad Autónoma de Extremadura

18 RCD que contienen sustancias peligrosas, RCD inertes sucios, RCD inertes limpios y RCD inertes adecuados para el uso en otras obras.

19 Artículo 5 Decreto de 25 de febrero, por el que se establece el régimen jurídico de la producción, posesión y gestión de los residuos de cons-

Así, uno de los aspectos fundamentales de este Decreto es la *fianza*. La fianza, descrita anteriormente como una medida económica de prevención, es uno de los elementos clave en el proceso de gestión de los RCD, ya que se concibe como un mecanismo de control vinculado a la posibilidad de realización de la obra, es decir, es requisito para obtener la licencia de obra, pero también lo es en el caso de que no se exija licencia urbanística[20]. La fianza se establece inicialmente como medida del régimen de control de la producción, posesión y gestión de RCD establecido en el Real Decreto 105/2008, de 1 de febrero, por el que se regula la producción y gestión de los residuos de construcción y demolición. De acuerdo con este Real Decreto, dicha fianza se debe incorporar a la legislación de las comunidades autónomas en cuantía suficiente para garantizar el cumplimiento de las obligaciones que le impone la ley[21]. Es así que el Decreto extremeño establece la constitución de una fianza como obligación del productor y del gestor de los RCD que debe depositarse en el ente local donde se vaya a realizar la obra[22]. Para determinar el coste de la fianza se tendrá en cuenta la clasificación de los residuos antes mencionada. Además, cada municipio debe incorporar a su ordenamiento interno la cuantía de la fianza, siendo el mínimo obligatorio un 0,4% del presupuesto de ejecución de la obra[23]. En el caso de las obras menores que nos ocupan, el Decreto dispone que los municipios deben establecer en sus respectivas ordenanzas las condiciones a

---

trucción y demolición en la Comunidad Autónoma de Extremadura.

20 Artículo 26 Decreto de 25 de febrero, por el que se establece el régimen jurídico de la producción, posesión y gestión de los residuos de construcción y demolición en la Comunidad Autónoma de Extremadura.

21 Artículo 6.2 Real Decreto 105/2008, de 1 de febrero, por el que se regula la producción y gestión de los residuos de construcción y demolición.

22 Artículo 7. c) Decreto de 25 de febrero, por el que se establece el régimen jurídico de la producción, posesión y gestión de los residuos de construcción y demolición en la Comunidad Autónoma de Extremadura.

23 Art. 25.1.d). Decreto de 25 de febrero, por el que se establece el régimen jurídico de la producción, posesión y gestión de los residuos de construcción y demolición en la Comunidad Autónoma de Extremadura.

las que deben someterse la producción, posesión y gestión de los RCD de obras menores (residuos sólidos), debiendo cumplir con la legislación establecida y los requisitos del Decreto, por tanto, también están sometidos a fianza[24]. De esta manera, forma parte de la tramitación de la obra el depósito de la correspondiente fianza, que se devolverá una vez quede certificado ante la propia administración local que se ha procedido al depósito de los RCD en el punto de recogida de RCD que sea gestor autorizado.

La figura del *gestor autorizado* o *gestor de residuos de construcción y demolición* también cobra un papel relevante en la gestión de residuos, ya que se define como la persona física o jurídica que, siendo o no el productor de los RCD, de naturaleza pública o privada, realice las operaciones de recogida, almacenamiento, transporte, valorización y/o eliminación de residuos, incluyéndose la vigilancia de estas actividades y de los lugares de depósito o vertido tras su cierre[25]. Los gestores deben cumplir con diferentes obligaciones. Una de ellas es la de obtener autorización para ser gestores, debiendo estar inscritos en el registro de la CCAA[26].

Una de las obligaciones clave de los gestores autorizados encargados de la recogida y transporte de los residuos es la de llevar un registro donde figure la identificación del productor, del poseedor y de la obra de la que proceden los RCD, así como la cantidad recogida o transportada, expresada en toneladas y metros cúbicos, y la identificación del gestor al que se destinan los residuos. Este registro debe conservarse durante 5 años, siendo puesto a

---

24 Disposición adicional segunda. Obras menores. Decreto de 25 de febrero, por el que se establece el régimen jurídico de la producción, posesión y gestión de los residuos de construcción y demolición en la Comunidad Autónoma de Extremadura.

25 Artículo 10 Decreto de 25 de febrero, por el que se establece el régimen jurídico de la producción, posesión y gestión de los residuos de construcción y demolición en la Comunidad Autónoma de Extremadura.

26 Listado de gestores autorizados disponible para consulta en: http://extremambiente.juntaex.es/index.php?option=com_contentyid=2563 [Consultado el 03.07.2023].

disposición del órgano competente en materia ambiental de la Junta de Extremadura, una vez sea requerido[27].

Finalmente, los gestores autorizados de recogida, al recibir los RCD y haber registrado los datos anteriores, se convierten en titular de los mismos, procediendo a formalizar un *certificado de gestión* donde consten los datos identificativos del productor o poseedor y, si procede, del gestor con el que se formalice la recepción con su correspondiente número de autorización, así como la indicación del tipo de residuo que se ha depositado, la cantidad y el proceso de gestión al que se van a someter.

Teniendo en cuenta la persistencia del fenómeno del vertido ilegal, a pesar de la obligación legal de la gestión del residuo, el establecimiento de sanciones, la concienciación medioambiental y la puesta a disposición de puntos limpios y plantas de RCD, nos planteamos qué puede estar disuadiendo o influyendo en que los sujetos generadores de residuos, o los responsables de depositarlos en el lugar correspondiente, estén incurriendo en un ilícito penal.

En este sentido, no hemos encontrado estudios en Extremadura que analicen la gestión de los RCD en relación con la oportunidad delictiva del vertido ilegal. Por ello, ante esta falta de estudios previos, nuestro objetivo general de investigación es analizar el modelo de gestión que se está aplicando y sus variables para identificar los factores, elementos o sujetos que pueden estar propiciando una mayor oportunidad delictiva de vertido ilegal.

3. METODOLOGÍA

Para conocer el modelo de gestión de RCD que se está aplicando en la práctica en la región y los elementos y/o sujetos que

27 Artículo 19 Decreto de 25 de febrero, por el que se establece el régimen jurídico de la producción, posesión y gestión de los residuos de construcción y demolición en la Comunidad Autónoma de Extremadura.

son claves en el proceso de la gestión de los residuos y, por tanto, en la generación de vertido ilegal, resultaba necesario identificar en primer lugar el ciclo material que siguen los RCD procedentes de obras menores. En estudio de carácter exploratorio, se llevó a cabo un método mixto consistente en un cuestionario semiestructurado creado *ad hoc*[28] vía telefónica a distintos ayuntamientos, puntos limpios municipales y empresas de RCD; entrevistas personales semiestructuradas a cuatro contactos clave; y una revisión de fuentes secundarias de información. Extrajimos la información de los puntos limpios existentes y de las empresas de RCD a través de listados públicos de la Junta de Extremadura. Una vez triangulada la información se obtuvo un total de 179 puntos conformados por puntos limpios, plantas RCD fijas y móviles repartidos entre 132 entes locales de toda Extremadura. De ese total se seleccionó una muestra de 65 municipios heterogéneos según población, existencia de puntos de recogida y mancomunidad de pertenencia, para obtener así una representatividad territorial.

De 132 municipios en los que se encuentra algún punto de destino (punto limpio o planta de reciclaje) se seleccionaron 65 (30 de Cáceres y 35 de Badajoz).

Tabla 1

Selección de municipios con puntos de destino RCD (plantas RCD, puntos limpios) por intervalos de población para realización de llamadas.

Intervalos de población (N º de habitantes)	Municipios con puntos de destino de RCD	Municipios seleccionados
+100000	1	1
50001-100000	2	1
20001-50000	4	4

[28] El cuestionario tiene 3 variables: el tipo de instalación, el número de habitantes del municipio de la instalación y la mancomunidad a la que pertenece el municipio.

10001-20000	5	5
5001-10000	14	11
2001-5000	36	13
1001-2000	37	13
501-1000	20	9
<500	13	8
TOTAL	132	65

Los 65 municipios representan y se clasifican, como puede comprobarse en la tabla, en distintos intervalos poblacionales obteniendo una representatividad demográfica. A su vez, se tuvo en cuenta que los municipios seleccionados se encontraran repartidos entre las 30 mancomunidades actualmente existentes en Extremadura, habiendo prácticamente 2 municipios por mancomunidad.

Para cumplimentar el cuestionario se realizaron llamadas telefónicas a empresas privadas de residuos, plantas de RCD, puntos limpios y Ayuntamientos a fin de conocer los servicios que prestaban. Para ello se utilizaron los teléfonos publicitados en sus correspondientes páginas web y de contacto para la ciudadanía. En cada llamada se solicitaba información sobre los trámites a seguir en una obra menor a la hora de depositar los escombros (tasas a pagar, obligatoriedad de la fianza, ubicación y distancia del punto de recogida, horarios, etc.).

De este primer cuestionario se consiguieron datos relativos al nivel de información que disponían los ayuntamientos en relación con la gestión de RCD. Se obtuvo información de los importes de las tasas de vertido y las tarifas privadas y se dedujo la existencia de las empresas con adjudicación del servicio público de gestión de los RCD en la provincia de Cáceres. Con base en la información recabada, se realizaron 4 entrevistas a los principales agentes autorizados en la gestión de los mismos, siendo las personas entrevistadas los encargados de las empresas de Araplasa de Residuos

S.A. (1 entrevista)[29], y Reciclados Cáceres Sur S.A. (1 entrevista)[30] en la provincia de Cáceres, y PROMEDIO (2 entrevistas)[31] en la provincia de Badajoz.

Araplasa S.A. es la empresa adjudicataria del concurso público denominado "Construcción de las infraestructuras y explotación del servicio de valorización y eliminación de los residuos de la construcción y demolición en la zona norte de la provincia de Cáceres"[32], por un periodo de 25 años iniciando su actividad en 2013. Por su parte, Reciclados Cáceres Sur S.A., se encarga del mismo servicio en la zona sur de la provincia de Cáceres, al ser adjudicataria del contrato "Construcción y explotación del servicio de gestión, valorización y eliminación de los residuos de construcción y demolición en la zona sur de la provincia de Cáceres", desde el año 2015. En la provincia de Badajoz nos encontramos con PROMEDIO, que en este caso se trata del Consorcio de Gestión Medioambiental de la Diputación de Badajoz. En estas entrevistas conocimos el modelo de gestión de RCD que se aplica por parte de estas empresas, completando la información adquirida en la primera ronda de llamadas. Esto nos permitió conocer las diferentes variables de ese modelo que influyen en la pérdida de trazabilidad de los RCD.

Posteriormente, una vez obtenida información por parte de estas empresas, se procedió a realizar una segunda ronda de llamadas telefónicas a los ayuntamientos en relación a la solicitud de la fianza para casos de obra menor, al entender este elemento como clave en la trazabilidad y prevención del vertido ilegal. De esta forma, se recabaron datos sobre la aplicación o no de la fian-

---

29 Más información disponible en: http://www.araplasaderesiduos.com/ [Consultado el 03.07.2023].

30 Más información disponible en: https://recicladoscaceressur.com/ [Consultado el 03.07.2023].

31 Más información disponible en: https://promedio.dip-badajoz.es/ [Consultado el 03.07.2023].

32 Disponible para consulta en: https://www.boe.es/diario_boe/txt.php?id=BOE-B-2011-9923 [Consultado el 03.07.2023].

za en la práctica de gestión de algunos ayuntamientos, pues se ha de recordar que el requerimiento de la fianza es obligatorio.

No obstante, hay que tener en cuenta ciertas limitaciones o márgenes de error en esta metodología. En primer lugar, el cuestionario inicial se debe entender como una fase de aproximación al papel de los distintos agentes en la gestión de los RCD, ya que no disponíamos de información alguna sobre la práctica de esta gestión. De ahí el carácter exploratorio de este estudio y la naturaleza semiestructurada del cuestionario. Esto es, se utilizó un método aproximativo a los diferentes agentes clave tratando de encontrar elementos comunes en la información recabada. Así, la información facilitada por los ayuntamientos se debe tomar con precaución, teniendo en cuenta que la conversación que se mantuvo con la persona encargada de atendernos telefónicamente a menudo era personal administrativo y, por tanto, podía desconocer cuestiones más técnicas. En el caso del segundo cuestionario sí se contactó principalmente con personal técnico en materia de urbanismo, preguntando exclusivamente por el tema de la fianza.

4. RESULTADOS

A raíz de los cuestionarios y las entrevistas efectuadas con los distintos agentes clave, hemos identificado el principal modelo de gestión y sus diferentes variables respecto a los RCD procedentes de obra menor en Extremadura.

4.1. Las empresas y las instalaciones de la gestión de RCD

En primer lugar, hay que tener en cuenta que, previamente al Decreto 20/2011, los municipios contaban con una zona conocida como *escombrera* donde se depositaban irregularmente los residuos. Con su aprobación, multitud de estas escombreras se han clausurado y se ha procedido a establecer medidas para dar una alternativa a la gestión de RCD. Sin embargo, no se ha logrado erradicar el problema debido a la escasez de instalaciones y a que estas están principalmente ubicadas en los núcleos más grandes de población al tiempo que son privadas. Por ello, los concursos públicos en materia de construcción de infraestructuras y de prestación del servicio de gestión de RCD se impulsaron, algunos gracias a fondos FEDER, tras la clausura de vertederos ilegales en los que la ciudadanía y empresas acostumbraban a depositar los RCD. Así, se pusieron en marcha una serie de políticas públicas para impulsar una gestión más controlada de los RCD y más facilidades a la ciudadanía y empresa. De esta manera, las primeras plantas de tratamiento de RCD comenzaron a funcionar en 2013 en la localidad de Plasencia y, en 2015, en el municipio de Cáceres. Las dos a cargo de Araplasa de Residuos. S.A., y Reciclados Cáceres Sur S.A, respectivamente. De esta manera, nos encontramos con que, en la provincia de Cáceres hay principalmente dos empresas privadas que cumplen con un servicio público mediante una concesión de la Diputación Provincial, siendo las que disponen de plantas y las que son gestores autorizados. Estas empresas se encargan de la puesta a disposición de puntos de acopio en distintos municipios, para el depósito temporal de RCD hasta su traslado a la planta de tratamiento y además ejercen una actividad de control activo de los vertidos ilegales, siendo a menudo denunciantes de la existencia de los mismos en los municipios. En total disponen de 151 puntos de acopio repartidos por toda la provincia. Cada uno de ellos pertenece a una planta de transferencia, que son un punto intermedio entre el punto de acopio y la planta de RCD final. Así, tenemos que, en la zona norte de Cáceres, cuya competencia está atribuida a Araplasa S.A., cada planta de transferencia cubre un radio de 25 kilómetros, tal y como se puede apreciar en el siguiente mapa.

Mapa 1

Distribución geográfica del área de influencia de las instalaciones de Araplasa en la provincia de Cáceres.

Por su parte, la zona sur de Cáceres, a cargo de la empresa Reciclados Cáceres SUR, S.A., dispone plantas de transferencia con un radio de 50 km.

Mapa 2

Distribución geográfica del área de influencia de las instalaciones de Reciclados Cáceres Sur en la provincia de Cáceres.

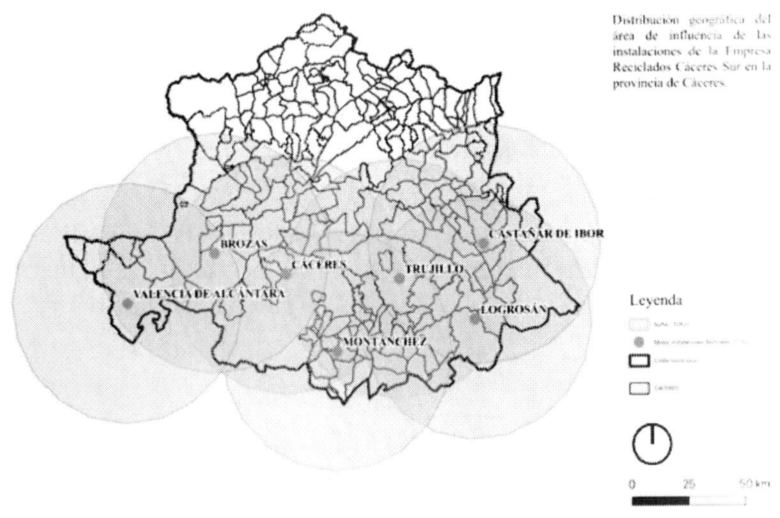

Como vemos, estas dos empresas disponen territorialmente estos puntos de recogida de RCD con el fin de llegar a la población más allá del municipio en que se encuentran las plantas de transferencia.

En Badajoz opera el servicio PROMEDIO, que es el consorcio para la gestión de los Servicios Ambientales de la Diputación de Badajoz, con la adjudicación a una empresa autorizada para el transporte y gestión de RCD del "Servicio de colocación, transporte de bateas y gestión de residuos de la construcción y demolición de competencia municipal en diversos municipios de la provincia de Badajoz"[33]. Mediante este servicio se puso a disposición de la

33 Toda la información sobre esta contratación se puede consultar aquí: https://contrataciondelestado.es/wps/portal/!ut/p/b0/04_Sj9CPyk-

ciudadanía y empresas un conjunto de bateas o contenedores en cada municipio adherido para el acopio de los restos de obras menores, facilitando así su retirada y tratamiento. La experiencia PROMEDIO se puso en marcha en 2018, consiguiendo un total de 73 municipios adheridos al servicio y gestionando más de 12.300 toneladas de RCD en cuatro años. Actualmente, dada la eficacia del servicio, se ha vuelto a licitar, esta vez con la adhesión de 72 municipios que agrupan un total de 101.644 habitantes, con la colocación de 77 bateas para el acopio de los RCD y el transporte hacia la planta de tratamiento[34].

Teniendo en cuenta estos datos recabados mediante los listados de la Junta de Extremadura, la región dispondría de las siguientes instalaciones en las que depositar RCD (véase tabla inferior).

Tabla 2

Número de instalaciones destinadas al depósito (temporal o no) de RCD en Extremadura y por provincias.

	Cáceres	Badajoz	Total
Puntos limpios	33	89	122
Plantas de RCD (fijas, móviles y de transferencia)	34	21	55
Puntos de acopio/bateas	151	77	228
Todas	218	187	405

Sin embargo, hay que tener en cuenta que no todas las referidas instalaciones se encuentran en funcionamiento[35]. En el caso de los puntos limpios no sólo no están activos algunos de ellos, sino que en muchos de ellos no se admiten RCD (pese a que están

ssy0xPLMnMz0vMAfIjU1JTC3Iy87KtUlJLEnNyUuNzMpMzSxKT-gQr0w_Wj9KMyU1zLcvQj3cMNzQrMvJxL3IPMMyxMVQ0KcnOL08pt-bfWBDEcAWzv7vg!!/ [Consultado el 03.07.2023].

34 Más información en: https://promedio.dip-badajoz.es/noticia.php?txt=157255 [Consultado el 03.07.2023].

35 Véase capítulo 5 en relación con los puntos limpios activos.

obligados a recibir un máximo de 150 kg/día). Ocurre que en muchas localidades puede existir un solapamiento entre el punto limpio y el punto de acopio, por lo que estaría justificado la no admisión de RCD en el punto limpio, ya que el depositarlo en el punto de acopio permite que la empresa responsable de ellos pueda acudir con la infraestructura necesaria[36] a recogerlos. Sin embargo, se produce también el hecho de que muchos ayuntamientos no hacen uso del punto de acopio/batea o lo tienen ubicado lejos del punto limpio[37].

Con esta información hemos determinado que hay puntos comunes en la gestión de los RCD y algunas diferencias entre las provincias de Cáceres y Badajoz. Entre los elementos comunes, se observa que las dos han apostado por la licitación de un servicio público a empresas privadas. También que han recurrido al establecimiento de puntos de acopio o bateas para llegar a multitud de municipios de distinto tamaño, y que en ambos casos el responsable de avisar para su recogida son los consistorios. Entre las diferencias nos encontramos con que en Cáceres las mismas empresas que ponen a disposición de los ayuntamientos los puntos de acopio son quienes tratan el residuo, ya que lo trasladan desde su punto de acopio a su planta de tratamiento. Sin embargo, en Badajoz, el servicio de PROMEDIO se encarga únicamente de la disposición de las bateas y de su recogida y traslado a plantas de tratamiento que pertenecen a diferentes empresas. De este modo, en la provincia de Cáceres hay 24 plantas fijas de RCD y cuatro empresas gestionándolas, teniendo las dos mencionadas la adjudicación del servicio público de la gestión de residuos en la zona norte y sur de Cáceres. Por su parte, en Badajoz, hay 13 plantas,

---

36 Hay que tener en cuenta que los puntos de acopio son un tipo de contenedor específico de la empresa de RCD de la que son propiedad, por ello, están adaptados a los camiones que dicha empresa utiliza. En el caso de los contenedores de los puntos limpios es habitual que no sean compatibles con los camiones, complicando así su recogida.

37 Véase capítulo 5 en relación con la inutilización de los puntos de acopio y las distancias.

cada una de diferentes empresas privadas, a las que PROMEDIO puede llevar los residuos. A su vez, vemos que, en términos territoriales, Cáceres presenta una estrategia de disposición de plantas de transferencia tratando de cubrir distancias considerables. La consecuencia es que Cáceres tiene un modelo territorial de gestión de RCD prácticamente homogéneo en toda la provincia con similar logística, tarifas, aplicados, procedimientos de registro, etc. Sin embargo, en Badajoz cada empresa tiene su propio funcionamiento promoviendo dicha diversidad la existencia de empresas no autorizadas que aplican sus propias tarifas. En esta línea, las empresas de la provincia de Cáceres

4.2. El modelo de gestión de RCD aplicado en Extremadura y sus fases

Una vez expuesto lo anterior, el modelo de gestión principal de RCD que hemos identificado consta del siguiente esquema en el que diferenciamos 3 fases fundamentales: la tramitación de la obra, la recogida y traslado de los residuos generados en la obra, y el depósito final de los residuos.

Tabla 3

Esquema del modelo de gestión de RCD identificado en Extremadura descrito por fases, pasos y sujetos que intervienen.

MODELO DE GESTIÓN DE RCD EN EXTREMADURA									
Fases	**TRAMITACIÓN DE LA OBRA**				**RECOGIDA Y TRASLADO**		**DEPÓSITO**		
Pasos	-Solicitud de la licencia.	-Emisión de la licencia/cobro de la tasa del residuo y establecimiento de la fianza. En Ayuntamiento/punto limpio/punto de acopio.		-Realización de la obra.	-Recogida en origen (en la obra).	-Recogida en intermediario (punto limpio, punto de acopio, planta de transferencia).	-Cobro de la tarifa del residuoen la planta RCD.	Emisión de certificado de gestión de RCD.	Recuperación de la fianza en el ayuntamiento.
Actores intervinientes	-Particular/ empresa privada de albañilería.	-Ayuntamiento/ Mancomunidad	-Particular/ empresa privada de albañilería. -Ayuntamiento. -Punto limpio.	-Particular/ empresa de albañilería.	-Particular. -Operario municipales (puerta a puerta) -Empresa privada del residuo. -Empresa privada de albañilería.	-Empresa privada del residuo. -Planta de RCD. -Ayuntamiento.	Gestor autorizado (planta RCD o empresa privada del residuo).		Particular o empresa privada de albañilería presentando el certificado de gestión.

De acuerdo con el modelo identificado, las principales fases de la gestión de los RCD son las que siguen a continuación.

4.2.1. Tramitación de la obra

La fase de *tramitación de la obra* comienza con la planificación de la obra hasta que efectivamente se realiza. Durante esta fase se realizan tres pasos fundamentales que debemos destacar:

- Solicitud de licencia de obra por parte del particular o de la empresa de albañilería en representación del particular ante el ayuntamiento del municipio en el que se realice la obra. En este momento el sujeto que solicite la licencia de obra debe presentar una previsión de los RCD que se van a generar en tipo y volumen. En función de ello se establecerá la fianza correspondiente aplicando el porcentaje establecido mediante ordenanza municipal o el mínimo de 0,4% establecido en el Decreto 20/2011. En el caso de que no sea necesaria licencia, se debe depositar igualmente la fianza.

- Emisión de la licencia: la emisión de licencia se produce por parte del ayuntamiento. En función del tipo de obra y de las ordenanzas municipales puede ocurrir que el particular deba esperar a la confirmación del consistorio para poder iniciar la obra, mientras que en otros casos se puede emprender la realización de la obra sin necesidad de esperar a la concesión de la licencia, mediante una comunicación previa. Este último caso suele suceder cuando la magnitud de la obra es muy pequeña. En cualquier caso, como decíamos anteriormente, de acuerdo con el Decreto 20/2011, la fianza es obligatoria.

- Cobro de la tasa del residuo: el depósito de RCD está sujeto al pago de una tasa que varía en función de la cantidad y el tipo de residuo, siendo la misma determinada por la ordenanza municipal. Esta exigencia responde al principio de "quien contamina paga", como decíamos anteriormente. El cobro de la tasa se produce al inicio del procedimiento en el punto limpio o punto de acopio que recibe el residuo, o bien directamente en el ayuntamiento, que es quien se encarga de certificar a la empresa autorizada que, en este caso, suele ser la planta de RCD que se encuentra en el mismo municipio o mancomunidad.

- Realización de la obra: es el momento en el cual se generan los RCD, ya sea a cargo de un particular o de una empresa de albañilería.

4.2.2. Recogida y traslado de los RCD

Una vez realizada la obra en un domicilio, los residuos pueden ser recogidos en origen, es decir, en la misma obra, por los siguientes sujetos:

a) El particular: el propio particular que ha generado los residuos puede recogerlos de su obra y trasladarlos al punto limpio, al punto de acopio/batea, o directamente a la planta de transferencia o planta de RCD final.

b) Los operarios municipales: a veces, los operarios municipales pueden encargarse de recoger los RCD generados en una obra pequeña, cuando son fundamentalmente escombros y poca cantidad, para trasladarlos al punto limpio del municipio en el caso de que esté en funcionamiento y admitan residuos, o bien al punto de acopio facilitado por las plantas de RCD constituidas como gestores autorizados (o las bateas en el caso de Badajoz).

c) La empresa de albañilería: la empresa contratada por el particular para realizar la obra se encarga habitualmente, como parte de sus servicios, de la gestión de los RCD debiendo trasladarlos fundamentalmente a la planta de transferencia o planta de RCD final, aunque también lo pueden llevar al punto de acopio.

d) La empresa privada de transporte de residuos: existen algunas empresas que se encargan de trasladar residuos sin ser necesariamente las empresas encargadas de las plantas RCD (o en el caso de PROMEDIO).

En el caso de que lleven los residuos a un punto de acopio o punto limpio, los usuarios solicitarán previamente al ayuntamiento el acceso a ese punto, debiendo llevar una declaración del volumen y del tipo de residuos a depositar. En el caso de los puntos limpios, en función de la información anterior, el consistorio exigirá el pago de una tasa.

Estos sujetos pueden llevar los residuos a un intermediario o a la planta final de RCD. Los puntos intermedios son, como hemos dicho, los puntos de acopio, los puntos limpios y las plantas de transferencia.

Teniendo en cuenta lo anterior, se puede producir que los residuos se dirijan desde el origen (la obra) al punto final (la planta RCD), sin pasar por intermediarios, pero es habitual, sobre todo en casos de obras menores, que pasen por un punto intermedio (punto limpio, punto de acopio o planta de transferencia). En todo caso, el transporte de los RCD desde el punto de acopio hasta la planta de reciclaje o de trasferencia es siempre responsabilidad del titular de la concesión, previo aviso del ayuntamiento.

4.2.3. El depósito

Por último, en esta fase se depositan los residuos en la planta principal de RCD. Allí, el sujeto que traslada los RCD debe rellenar una ficha en la que se solicitan un conjunto de datos que son el NIF/CIF, nombre, dirección, teléfono y personas de contactos del cliente a facturar, del productor de los residuos, del poseedor de los residuos, de la empresa de transportes, del conductor, y una ficha de la propia obra donde debe expresarse la denominación de la obra, la dirección, la referencia de la licencia, el titular de la obra y el teléfono[38]. En este momento, se realiza una comprobación de los residuos y se procede al cobro de la tarifa del residuo en función de la cantidad y el volumen de RCD que se hayan lleva-

38 Puede consultarse un ejemplo de ficha de registro en: http://www.araplasaderesiduos.com/documentacion.html [Consultado el 03.07.2023].

do[39]. Dicha tarifa responde a la aplicación de un precio privado, al estar la gestión del servicio público de valorización y eliminación de RCD en manos de una empresa concesionaria, en atención al régimen jurídico del artículo 2 de la Ley 588/2003, de 17 de diciembre, General Tributaria, vigente tras la aprobación de la Ley 2/2011, de 4 de marzo, de Economía Sostenible.

Una vez pagada esta tarifa, se emite un certificado denominado *certificado de gestión* que confirma que se han depositado los residuos. De esta manera, el particular o la empresa puede acudir al ayuntamiento con dicho certificado, presentarlo, y así recibir la devolución de la fianza inicialmente depositada como garantía de la correcta gestión del residuo. En el caso de que finalmente la cantidad entregada excediera la prevista, se tendría que pagar la tasa correspondiente a dicha cantidad en el Ayuntamiento.

Una vez expuesto el ciclo de gestión se indica en la tabla inferior las posibles variantes que nos podemos encontrar y tratar de identificar determinadas situaciones en las que se puede perder la trazabilidad de los RCD.

---

39 En el caso de las empresas de la provincia de Cáceres, de acuerdo con el Reglamento de Funcionamiento del Servicio de Gestión de los Residuos de Construcción, Demolición y Excavación de la Provincia de Cáceres, el importe es de 3,15 euros la tonelada, sin IVA, para RCD de categoría III limpio, con densidad superior a 1,2 Tn/m³; 9 euros la tonelada, sin IVA, para RCDs de categoría II sucio-mixto, con densidad entre 1,2 y 0,8 Tn/m³ y 13,50 euros la tonelada, sin IVA, los RCDs de categoría II sucio, con densidad inferior a 0,8 Tn/m³.

Tabla 4

Variables en tramitación de la obra.

Hipótesis de la gestión de RCD en la fase de la tramitación de la obra.			
A	B	C	D
El particular no solicita licencia de obra, por tanto, no deposita fianza.	La obra no requiere de licencia, pero sí de fianza. El particular la puede depositar o no.	El ayuntamiento no exige la fianza. No hay garantía financiera de gestión los RCD.	La cantidad de fianza exigida es exigua. Al particular no le compensa recuperarla.

En la *fase de la tramitación de la obra* puede ocurrir que el particular, al realizar una obra menor por su propia cuenta, no solicite licencia de obra y, por tanto, no deposite fianza. En el caso de una obra que no requiera de licencia, pero sí de fianza, tal y como establece el Decreto 20/2011, se reduce el mandato de depositarla a la voluntad del particular. Asimismo, se produce el hecho de que muchos ayuntamientos no exigen el pago de la fianza, de manera que no se encargan de que exista una garantía financiera de depósito de los RCD en su destino correcto. Es decir, se pierde la trazabilidad del residuo.

La fianza, se configura como la pieza clave que actúa en esta fase como garante de la correcta gestión del residuo. Sin embargo, dadas las posibles situaciones, la fianza no es una garantía por sí misma, sino que debe cumplirse con el mandato legal de ser exigida en cantidad suficiente como para que el particular encuentre una motivación en recuperarla. Las entrevistas realizadas nos han reportado multitud de casos en los que los particulares o empresas no solicitan en los gestores autorizados la certificación de haber depositado allí los RCD y, por tanto, la falta de depósito de fianza previa. Esta situación nos indica que, a pesar de ello, hay sujetos que depositan correctamente los RCD en el punto de destino habilitado. Sin embargo, la no exigencia de la fianza continúa siendo problemático dado que se constituyó como un elemento clave en el Decreto 20/2011 para la gestión del residuo. En el presente trabajo consultamos, de acuerdo con la metodología anteriormente expuesta, a un total de 65 ayuntamientos, por la solicitud o no de la fianza. De ellos nos encontramos con que al menos 19 no la solicitan en los casos de obras menores, es decir, un 29% de los ayuntamientos consultados. Parece existir una tendencia de no exigencia de la fianza en municipios de menos de 10.000 habitantes, siendo los núcleos poblaciones más grandes garantes de la misma. Sin embargo, no se puede determinar que cuanto más pequeños sean los municipios, demográficamente hablando, estos tiendan a no exigir fianza, ya que en la *Tabla 5* se observa que la mayoría sí lo exige. En la consulta también nos indicaron algunas de las cantidades aplicadas, variando desde la aplicación del 0,4% hasta la imposición de fianzas desde los 25 hasta los 120 euros. En cuanto a la diferencia provincial, nos encontramos con que, de la muestra consultada, 7 municipios de Cáceres no exigen fianza frente a 12 de Badajoz, por lo que se aprecia un cierto equilibrio, teniendo en cuenta que hemos consultado 30 municipios de Cáceres y 35 de Badajoz.

Tabla 5

Resultado de consulta de solicitud de fianza para obras menores en municipios de Extremadura.

Intervalos de población	Muestra	Exigen fianza para obra menor.	No exigen fianza para obra menor.
+100000	1	1	0
50001-100000	1	1	0
20001-50000	4	4	0
10001-20000	5	4	1
5001-10000	11	9	2
2001-5000	13	10	3
1001-2000	13	4	8
501-1000	9	6	3
<500	8	6	2
TOTAL	65	45	19

De acuerdo con estos resultados, no parece que esté generalizada la práctica de la no exigencia de la fianza. Sin embargo, hay que tener en cuenta las limitaciones expuestas en la metodología, ya que son resultados recabados de llamadas telefónicas que son cualitativamente diferentes respecto a un proceso de solicitud de licencia de obra en el que intervienen más factores. En este sentido, los agentes clave son coincidentes en la escasez de solicitudes de certificado de gestión para la recuperación de la fianza.

La *segunda fase* es clave desde el punto de vista logístico, ya que es el momento de la recogida y traslado de los RCD por lo que, en este punto se puede desviar el ciclo correcto del mismo hacia el vertido ilegal. Veamos pues, qué variables pueden intervenir:

Tabla 6

Variables en la fase de recogida y traslado de RCD.

Hipótesis de la gestión de RCD en la fase de recogida y traslado de RCD.		
A	B	C
Disponibilidad de puntos de destino. Horarios de puntos limpios y puntos de acopio. Distancias. Dificultad en el traslado.	Puntos de acopio inactivos.	Una ponderación de costes y falta de conciencia ambiental puede dar lugar a que la recogida en origen del residuo (por parte del ayuntamiento, la empresa de albañilería o la empresa privada) sea depositada en un lugar no habilitado para ello

En esta fase son fundamentales las medidas de tipo logístico, ya que la recogida y traslado de los RCD por parte de particulares se ve limitada en función de la puesta a disposición o no de instalaciones, de la distancia a las mismas, de las posibilidades y coste de transporte y de la información por parte de los entes públicos de la correcta gestión del residuo. En cuando al tipo de recogida, en algunos ayuntamientos se facilita la recogida de residuos mediante la intervención de personal operario que acude al domicilio del particular a recoger escombros, siempre que sea muy poca cantidad. No obstante, en la mayor parte de los casos es el particular quien los tiene que trasladar al punto de destino. En este sentido, el primer cuestionario telefónico a los ayuntamientos nos indicó que algunos puntos limpios municipales no recogen RCD, a pesar de la obligación de legal de ello. En otros casos, los puntos de acopio no están operativos o en activo, como ya hemos comentado anteriormente, y a las plantas de RCD solo se puede acceder en automóvil debido, fundamentalmente, a la distancia respecto del núcleo poblacional. En este punto, por tanto, se pierde la trazabilidad de la correcta gestión del residuo, pudiendo producirse el vertido ilegal por las dificultades de traslado por parte de particulares.

En caso de que los residuos sean depositados en un punto intermedio, pasa a intervenir el sujeto encargado de trasladarlos desde el punto intermedio al punto final, siendo los posibles intermediarios el ayuntamiento, la propia planta de RCD y la empresa privada del residuo, ya que cualquiera de estos tres actores puede realizar el traslado. No obstante, la responsabilidad de la gestión de los residuos es del Ayuntamiento, que debe encargarse de trasladar los residuos a la planta de RCD, contratar a una empresa que los traslade o, en el caso de que estén en los puntos de acopio o bateas, debe dar el aviso a la empresa de RCD propietaria de esos contenedores para que vayan a recoger los RCD.

Nos preguntamos, por tanto, qué ocurre en aquellos casos en los que ni el punto limpio ni el punto de acopio se están utilizando y, sin embargo, el ayuntamiento se encarga de los escombros. Algunos estudios señalan el alto vertido ilegal proveniente de los propios servicios públicos (Ichinose y Yamamoto, 2011), señalando una posible existencia de una mala práctica consistente en el almacenamiento de estos residuos en algunas parcelas municipales, con el objetivo de ahorrar los costes de transporte y gestión, e incluso para reutilizar los residuos generados que resulten de interés (como zahorra para caminos o desniveles de fincas).

En el caso de la realización de la obra por parte de una empresa de albañilería, dado que tienen disponibilidad de recursos para la recogida y traslado, los factores anteriores solo influyen en cuanto al coste que les supone (en tiempo y combustible, generalmente).

Tabla 7

Variables en la fase de depósito RCD.

Hipótesis de la gestión de RCD en la fase de depósito de la obra.			
A	B	C	D
Depósito de RCD obras sin licencia.	Discrepancia entre los datos sobre el volumen y tipo de RCD que figuran en la licencia de obras con respecto a lo que realmente se deposita.	Se da el caso de personas que no solicitan el certificado de gestión porque creen que el albarán es suficiente, sin embargo, no es un documento válido para la devolución de la fianza.	Competencia desleal de empresas no autorizadas.

El particular o empresa de albañilería acude con sus RCD, sin licencia, bien porque no se les ha requerido o bien porque no la han tramitado. En este caso puede darse la situación de que estén llevando todo el residuo generado, o solo una parte para ahorrar los costes que supondría pesar el total de los ripios generados. También puede darse el caso de que el particular o empresario, habiendo tramitado la licencia de obras, luego deposite en el gestor autorizado una cantidad menor de residuo a la prevista en la licencia. Esto es más difícil de detectar en las obras pequeñas, pues generalmente no se especifica el volumen de ripios a generar en términos cuantitativos, sino más bien el tipo de obra a realizar en clave cualitativa (ejemplo: reforma de un baño). Sin embargo, en las obras mayores, en las que es necesario un proyecto técnico y figuran necesariamente las cantidades de ripios a generar, es más fácil observar si existe un desfase entre dichas cantidades y las que se trasladan luego al gestor autorizado. En las obras menores resulta muy difícil saber de dónde proceden exactamente los RCD y si están todos los que se generaron realmente. No obstante, se considera beneficioso permitir la práctica de recibir residuos de obras sin licencia, aunque fuera requerida, dado que se entiende que los residuos están completando correctamente su ciclo final, con el matiz de que no sabemos si se han dejado algunos por el camino.

Otra situación es que acudan sin haber solicitado, en la fase de tramitación de la obra, licencia (por no ser requerida), pero sí habiendo pagado la fianza y, por lo tanto, soliciten el certificado de gestión para la devolución de esta.

Por todo ello, se plantea la hipótesis de que parte de los RCD se están desviando para su uso no autorizado o su tratamiento en plantas no autorizadas o que se estén eliminando parte de ellos para un ahorro en los costes. Ocurre asimismo que algunas personas no solicitan certificado de gestión, pero acuden con el albarán al ayuntamiento, el cual no es un documento válido. Aquí el problema se produce cuando los ayuntamientos aceptan estos albaranes, ya que pueden proceder de gestores no autorizados. En este sentido, se considera problemática la existencia de plantas de tratamiento no autorizadas, dada la competencia desleal que generan al cobrar a menudo tarifas más bajas, aunque realicen una mala gestión de los residuos.

4.3. Propuestas de mejora

Una de las pretensiones principales de las entrevistas a los agentes clave en la gestión de los RCD, fundamentalmente, las plantas encargadas de su gestión y reciclaje, ha sido la de obtener propuestas de mejora procedentes de la experiencia. Así, en materia de prevención, entre las principales preocupaciones se incide en la importancia de la fianza, tanto de su exigencia como del establecimiento de una cantidad adecuada que promueva la motivación de los particulares para su posterior recogida y, por tanto, garantice la efectiva gestión de los residuos. Para ello debe haber una mayor responsabilidad medioambiental por parte de los ayuntamientos y aumentar la formación técnica a los responsables de urbanismo.

En este sentido, se considera fundamental aumentar el control de las variaciones existentes entre la previsión de los RCD generados, los que se generan realmente y los que se llevan a la planta de RCD.

Desde el punto de vista del control del vertido ilegal, se considera fundamental aumentar el número de inspecciones por parte del SEPRONA a las plantas que no son gestores autorizado. Asimismo, indican la importancia del papel de las patrullas medioambientales de la policía local, que requieren de un aumento de personal, medios y horarios de atención.

Consideran que se debe dotar de más competencias y autoridad a las empresas de residuos que son gestores autorizados para identificar los desfases producidos entre lo que indica la licencia de obra y la cantidad que se deposita en la planta, ya que este es uno de los puntos clave en los que se detecta la posible desviación de RCD. Por ello también se requiere más control de los certificados que emiten los gestores autorizados, para comprobar que, efectivamente certifican lo que se recoge y los datos son correctos. Puede ser interesante también una mayor transparencia de las obras que se realizan. En algunos ayuntamientos se hacen públicas, mediante web o tablón de anuncios, las licencias de obras en vigor. De esta manera, se puede ejercer un seguimiento de si la obra efectivamente concluye con el ciclo correcto de gestión de los residuos una vez acuden al ayuntamiento en busca de la recuperación de la fianza.

Por último, resulta fundamental incidir en las medidas de concienciación ambiental a la ciudadanía y empresas. Explicar las consecuencias de los vertidos ilegales, y el beneficio social que supone una correcta gestión de los residuos por encima del coste económico que pueda suponer. En este sentido, hay que promover campañas a nivel municipal desde los ayuntamientos, puntos limpios y mancomunidades.

5. CONCLUSIONES

En primer lugar, es un hecho que las instituciones en Extremadura a nivel regional han sido conscientes del problema de la falta de instalaciones y de cómo ello influye en el vertido ilegal de RCD. Se puede decir que el impulso por medio de licitaciones públicas y ayudas europeas han ayudado a aumentar estas instalaciones, estando repartidas prácticamente por todo el territorio, y llegando a los municipios más pequeños. Sin embargo, a su vez nos encontramos con que aún muchas de estas instalaciones no están siendo debidamente utilizadas, a menudo por falta de voluntad, información o concienciación de algunos ayuntamientos.

En segundo lugar, la fianza como herramienta clave impulsada por el Decreto 20/2011, de 25 de febrero, por el que se establece el régimen jurídico de la producción, posesión y gestión de los residuos de construcción y demolición en la Comunidad Autónoma de Extremadura, sin duda es, de acuerdo con los agentes clave entrevistados, una buena medida. No obstante, en consonancia con el punto anterior, todavía falta mayor implicación por parte de los entes locales en cuanto a su exigencia y en cuanto a su importe. La falta de exigencia de la fianza es un factor disuasorio de la correcta gestión del residuo. Lo mismo ocurre con los bajos importes que se establecen en esta, pues a menudo no son suficientes para que compense su recuperación.

Por último, podemos afirmar que aunque existe un modelo más o menos homogéneo de gestión de RCD en toda Extremadura, presenta diferencias de carácter territorial, ya que en la provincia de Cáceres el seguimiento de este modelo lo realizan dos empresas que, además del traslado de RCD, son quienes los tratan en su destino final, teniendo las dos empresas la misma política de gestión en relación a organización territorial, control activo del vertido ilegal, registro de los RCD que se depositan, e importes de depósito, aunque se requiere una mayor implicación por parte de los Ayuntamientos en el uso y seguimiento de los puntos de acopio. Por su parte, la experiencia PROMEDIO en Badajoz también se considera muy positiva en relación con la recogida y traslado de RCD, suponiendo un mayor control en este sentido, pero quizás una mayor confusión para la ciudadanía y una mayor variación en la gestión final de los residuos al haber tantas empresas dedicadas a esta cuestión con sus diferentes procedimientos y tarifas.

6. REFERENCIAS BIBLIOGRÁFICAS

Allers, A. Maarten, Hoeben Corine. (2010). Effects of Unit-Based Garbage Pricing: A Differences-in-Differences Approach. *Environ Resource Econ, 45*, 405-428. DOI 10.1007/s10640-009-9320-6

Chen, Jianguo., Hua, Chunxiang., Li, Chenyu. (2018). Considerations for better construction and demolition waste management: Identifying the decision behaviors of contractors and government departments through a game theory decision-making model, School of Economics and Management.

Cossu, R., Masi, S. (2013). Re-thinking incentives and penalties: Economic aspects of waste management in Italy, *Waste Management, 33*, 2541-254. DOI: 10.1016/j.wasman.2013.04.011.

European Commision Directorate-General Environmen (2012). Preparing a Waste Prevention Programme. Guidance document.

Fullerton D, Kinnaman, TC. (1996). Household responses to pricing garbage by the bag. *American Economic Review, Vol. 86* (4), 971–984. http://www.jstor.org/stable/2118314

Greenpeace España (2013). Informe "Destrucción a toda costa. Análisis del litoral a escala municipal". Disponible para consulta en: http://archivo-es.greenpeace.org/espana/es/reports/Destruccion-a-toda-costa-2013/ [Consultado el 03.07.2023].

Ichinose, Daisuke, Yamamoto, Masashi (2010). On the relationship between the provisiono of waste management service and illegal dumping. *Resource and Energy Economics, Vol. 33* (1), 79-93. DOI: 10.1016/j.reseneeco.2010.01.002

Jofra Sora, M. (2016). "Metodología para la gestión ambiental de RCD en ciudades de América Latina". Disponible en:

https://residus.gencat.cat/web/.content/home/lagencia/accio_exterior/cooperacio/llibret-Bogota_baixa.pdf [Consultado el 03.07.2023].

Junta de Extremadura. *Plan Integrado de Residuos de Extremadura (PIREX) 2016-2022.*

Junta de Extremadura. *Programa de vigilancia y control de la producción, gestión y vertido ilegal de residuos para el año 2022.* Disponible en:

http://extremambiente.juntaex.es/files/Programa%202022%20Inspección%20Residuos_V4(F).pdf [Consultado el 03.07.2023].

Martínez Guirao, J.E. (2016). Implicaciones ecológicas y sociales del boom de la construcción. En Martínez Guirao, J.E., *et. al.* (Eds). *Perspectivas interdisciplinares en el estudio de la cultura y la sociedad.* Universidad Miguel Hernández de Elche.

Miranda, M.L, Bauer S.D., y Aldy, J.E. (1996). *Unit pricing programs for residential municipal solid waste: an assessment of the literature.* Duke University.

PROMEDIO. (2015). *Manual para la correcta gestión de un punto limpio.* Interreg. Disponible en: https://promedio.dip-badajoz.es/documentos/155524.pdf [Consultado el 03.07.2023].

Slavik, J., Pavel, J. (2013). Do the variable charges really increase the effectiveness and economy of waste management? A case strudy of the Czech Republic. *Resources, Conservation and Recycling, Vol. 70,* 68-77. DOI: 10.1016/j.resconrec.2012.09.013.

Uricchio, V.F., Palmisano, V. N., Lopez, N., y Bruno, D.E. (2010). "A centralized management data to prevention of environmental crimes fight the Altamura case," 2010 IEEE Workshop on Environmental Energy and Structural Monitoring Systems, Taranto, Italy, 2010, 59-64, DOI: 10.1109/EESMS.2010.5634178.

Capítulo 10

Ayuntamientos y vertidos ilegales de residuos de la construcción: una aproximación a las percepciones de los responsables en las entidades municipales

JUANJO MEDINA ARIZA[1]

1. INTRODUCCIÓN

Este capítulo versa sobre las percepciones que, a nivel municipal, los actores políticos (como alcaldes o concejales) o integrados en los órganos de gestión y gobernanza (como técnicos del ayuntamiento en el ámbito de urbanismo o medio ambiente) tienen sobre la problemática y las respuestas al fenómeno de los vertidos ilegales de residuos de construcción y demolición (en adelante RCD). Los ayuntamientos pueden y deben jugar un papel clave en la respuesta a este fenómeno. Que esto sea así se deriva claramente del marco normativo actual. La Ley 3/2019, de 22 de enero, de garantía de la autonomía municipal de Extremadura en su artículo 15,1, sobre competencias propias de los municipios, establece que los municipios extremeños podrán ejercer competencias propias municipales sobre, entre otras, "la ordenación, gestión prestación, y control de los servicios de recogida y tratamiento de residuos sólidos urbanos o municipales", así como la "planifica-

---

1 Investigador distinguido Talentia Senior. Departamento de Derecho Penal y Ciencias Criminales. Universidad de Sevilla.

ción, programación y disciplina de la reducción de la producción de residuos urbanos o municipales". En este mismo artículo, se reconoce también la competencia de los ayuntamientos para la "ordenación y gestión o ejecución de las relaciones de convivencia en el espacio público", sin que pueda caber duda de que el depósito ilegal de vertidos es un comportamiento que entra dentro de lo que se podrían considerar tales relaciones de convivencia.

La regulación en Extremadura hace que sean las entidades locales quienes, bien por gestión directa o indirecta, se hagan cargo del servicio de recogida, transporte y tratamiento de RCD generado en lo que tradicionalmente se denominaban obras menores. Los ciudadanos tienen la obligación de llevar estos materiales a los puntos limpios o a instalaciones de gestores autorizados. Mientras que, para obras mayores, son los titulares de la obra, la empresa que se encarga de la misma, quienes deben asumir la responsabilidad de la gestión adecuada de estos residuos y tienen la obligación de entregarlos a una planta de tratamiento, que son quienes emiten el certificado de gestión que permite al promotor recuperar la fianza depositada en el ayuntamiento. Es, por tanto, evidente que las entidades locales juegan un papel clave en el diseño para el control de estos residuos.

Tras una breve introducción que servirá para contextualizar el tema, procederemos a describir cómo se generaron y analizaron los datos, y sucintamente presentaremos los resultados de las entrevistas personales realizadas agrupando los mismos en torno a dos ejes: percepciones locales sobre el problema y percepciones sobre respuestas a este fenómeno, tanto las existentes como aquellas que estos actores consideran que podrían ser útiles de cara a una prevención más adecuada. El capítulo concluirá con unas consideraciones breves a modo de recomendaciones.

2. AYUNTAMIENTOS Y GESTIÓN DE LOS RESIDUOS ILEGALES

Como ya se ha indicado en capítulos previos, Extremadura cuenta con 388 municipios a lo largo de su territorio. Las características socioeconómicas, fundamentalmente rurales, de la región hacen que la gran mayoría de los municipios sean pequeños. De estos, 218 cuentan con menos de 1000 habitantes y 101 con poblaciones que oscilan entre los 1000 y los 2500 habitantes, si nos atenemos a los datos del último censo.

Esta realidad moldea de forma importante las capacidades de las que estamos hablando cuando nos referimos a las competencias de los ayuntamientos en la gestión de vertidos ilegales. Muchos de estos ayuntamientos cuentan con medios personales y materiales limitados, como se documentará más adelante en nuestros resultados, para poder desarrollar por sí solos funciones efectivas de gestión, vigilancia y control de este fenómeno. Son poblaciones, muchas de ellas, en las que el equipo de gobierno municipal tiene que estar para todo, en lugar de poder contar con un equipo grande y especializado por áreas de gestión de políticos liberados, con escasos técnicos, y en la gran mayoría de casos sin policía local o con un efectivo policial demasiado pequeño, como para poder prestar servicios policiales a todas horas y, por tanto, poder efectuar una vigilancia efectiva.

Aunque hay una amplia literatura sobre el creciente papel de los municipios y los órganos de gobierno local en la gestión y prevención del delito (ver, por ejemplo, Murria et al., 2022), este trabajo criminológico sufre de dos limitaciones a la hora de entender el problema al que nos enfrentamos en este capítulo. En primer lugar, esta literatura habla y se concentra en problemas más tradicionales de seguridad (ICPC, 2016), que solamente de forma indirecta están relacionados con el tema de los vertidos ilegales[2]. Las encuestas comparativas a agentes locales del *International Centre for the Prevention of Crime* claramente muestran que el tipo de comportamientos ilegales y delictivos contemplados por estos actores locales en el contexto de debates sobre seguridad tienen que ver con formas tradicionales de delincuencia callejera y comportamientos violentos (p.ej., delincuencia juvenil, robos, abusos de sustancias, seguridad vial, bandas callejeras, delitos de odios, prostitución y comercio sexual, etc.). No hay gran espacio para el análisis de formas delictivas o comportamientos ilegales tradicionalmente de interés a la Criminología verde por su afectación al medio ambiente.

En esta literatura, en la medida que se habla de residuos o de basura, generalmente se hace pensando en el papel que pueden tener como signos de desorden físico que pueden incentivar la aparición de otras formas delictivas, como se proponía por la teoría de las ventanas rotas (Wilson y Kelling, 1982). Este es un enfoque teórico que, por otro lado, en su trasposición a la práctica ha tendido a concentrarse en combatir el llamado desorden social, más que el desorden físico (Medina Ariza, 2023), por más que recientemente empiece a generarse una literatura que destaca, por medio de diseños experimentales, la importancia preventiva de atajar el desorden físico (Branas et al., 2018).

No obstante, hay que destacar que, en el contexto español, donde las ideas de la teoría de los cristales rotos más que tener un impacto en modelos de trabajo policial lo tuvo en la aparición de ordenanzas municipales de civismo y convivencia, la ordenanza tipo en este ámbito y desarrollada por la Federación Española de Municipios y Provincias (y ofrecida como posible modelo a adoptar) en su artículo 30 hace referencia a los residuos de obra, disponiendo que dichos residuos se "depositaran en los elementos de contención previstos en la ordenanza especifica en la ma-

---

2 Esto, por ejemplo, también se observa en la agenda del *European Forum for Urban Safety*, la mayor red europea de municipios organizados para pensar sobre problemas de seguridad y convivencia en el entorno local.

teria". Más adelante, en el art. 33, se prohíbe expresamente la realización de cualquier vertido en "la vía pública" y, en el art. 34, se consideran infracciones "la vulneración de las prohibiciones o mandatos contenidos en los artículos de esta ordenanza". Los artículos 124 y 127 regulan de forma más detallada el depósito de residuos, específicamente la prohibición de depositar RCD en contenedores de basura convencionales y la necesidad de llevar estos residuos a puntos limpios habilitados a tales efectos.

Un segundo problema con la literatura criminológica sobre gestión de los problemas de seguridad y convivencia es que fundamentalmente ha estado centrada en el análisis de problemas en contextos urbanos, sobre todo occidentales, y no ha prestado suficiente atención a las problemáticas y a la gestión de estas que se presentan a los municipios, más pequeños, en entornos más rurales. Es cierto, no obstante, que esta es una limitación histórica de la Criminología, que como proyecto surge en el siglo XIX en el contexto de entender y dar respuesta a fenómenos ligados al tránsito de una sociedad agraria a una sociedad urbana, que más recientemente está empezando a ser corregido por el desarrollo de lo que ha venido a denominarse como "Criminología rural" (Hollis and Hankhouse, 2019). La creación de una División de Criminología Rural en la Sociedad Americana de Criminología es un buen testamento de la creciente importancia de este ámbito de estudio, que ya cuenta con sus revistas especializadas, libros de texto y colecciones de artículos (como el *Routledge International Handbook of Rural Criminology*), y volúmenes especiales en revistas más genéricas.

Las peculiaridades y dificultades de responder al fenómeno de los residuos ilegales en el entorno rural son, al menos, reconocidos en los documentos políticos sobre el tema generados en el contexto extremeño. Así, por ejemplo, el Plan Integrado de Residuos de Extremadura 2016-2022 (en adelante PIREX) señalaba como la estrategia de la Junta de Extremadura de promover, desde el 2009, la instalación de puntos limpios mediante la concesión de subvenciones a entidades locales se ha visto condicionada por la falta de recursos económicos y las asociadas dificultades de ges-ción sobre todo en los municipios más pequeños. Como veremos este es un tema reflejado en nuestras entrevistas. Igualmente, el PIREX destaca como el control en la gestión de los RCD se ha visto condicionado por:

El retraso de numerosas Entidades Locales en la aprobación de la preceptiva ordenanza municipal que exija la constitución de una fianza de forma previa a la concesión de una licencia urbanística, con la cual el promotor de la obra responde del correcto tratamiento de los residuos generados... así como por la falta de personal cualificado, especialmente en los municipios de menor tamaño.

3. METODOLOGÍA

Nuestro estudio estaba orientado a aproximarnos a las percepciones que sobre esta problemática tienen los gestores locales. Como ya adelantamos, por gestores locales entendemos alcaldes, otros miembros del equipo de gobierno con responsabilidad sobre el tema, o técnicos especializados (generalmente jefes de servicio) a los que nos derivaban los responsables políticos. Para explorar estas percepciones utilizamos entrevistas semiestructuradas. En total se entrevistaron a 15 responsables políticos o técnicos en dicho número de ayuntamientos, con una cierta infrarrepresentación de municipios, sobre todo más grandes, en Cáceres. Los municipios incluidos en la muestra se presentan en la siguiente tabla.

Tabla 1

Participantes en la muestra.

Municipio y provincia	Población	Entrevistado	Partido político
Cáceres			
Acebo	542	Alcalde	PP
Acehúche	816	Alcalde	PSOE

Belvis de Monroy	737	Teniente alcalde	Extremeños
Guijo de Santa Barbara	387	Alcalde	PP
Badajoz			
Alburquerque	5144	Alcalde	PSOE
Alconchel	1644	Alcalde	PSOE
Almendralejo	33669	Técnico	No aplicable
Arroyo de San Servan	4054	Técnico	No aplicable
Don Benito	37310	Concejal	PSOE
Garbayuela	502	Alcalde	PSOE
La Codosera	2025	Alcalde	PSOE
Llerena	5670	Concejal	PSOE
Mérida	59324	Concejal	PSOE
Villagonzalo	1217	Alcalde	PSOE
Villanueva de la Serena	25873	Técnico	No aplicable

Conviene destacar que Cáceres y Badajoz presentan realidades diferentes en cuanto a la gestión de los RCD. La Diputación Provincial de Cáceres ha desarrollado planes para garantizar esta gestión con fondos FEDER de la Unión Europea. Se anticipó en la creación de plantas de reciclaje de RCD (en el PIREX se mencionan dos fijas y 20 con maquinaria móvil) y existen puntos limpios en la práctica totalidad de los municipios, mientras que en la provincia de Badajoz la administración ha delegado más en la iniciativa privada y se partía, como consecuencia, de una distribución menos planificada de las instalaciones de tratamiento, lo que históricamente ha generado que hubiera municipios alejados de las plantas de tratamiento, un factor que puede condicionar la aparición de vertederos ilegales. De ahí que para poder contextualizar debidamente este material también entrevistamos a 2 técnicos de la Dirección General de Medio Ambiente, Cambio Climático y Sostenibilidad de la Junta de Extremadura, 1 Director General de Urbanismo de la Delegación de Urbanismo y Medio Ambiente del Ayuntamiento de Mérida, al tiempo que concertamos cita con Diputación Provincial tanto de Cáceres como de Badajoz, aunque el diputado provincial de Badajoz canceló sin preaviso su entrevista cuando acudimos a la cita concertada.

Aunque en una primera etapa se estableció una muestra representativa de 31 ayuntamientos seleccionados según tamaño poblacional, provincia, número de infracciones del SEPRONA y desarrollo o no de ordenanza municipal sobre residuos, la baja tasa de respuesta para participar en la investigación motivó abarcar toda la población finita (N=388 municipios). El trabajo de campo se desarrolló entre febrero y mediados de mayo del 2023. El procedimiento de contacto consistió en localizar los teléfonos y correos electrónicos publicados en internet para cada consistorio. Algunos de ellos habían creado departamentos o unidades específicas sobre urbanismo y/o medioambiente, generalmente los más grandes, mientras que otros tenían un número o correo electrónico que centralizaba cualquier consulta. A dichos correos electrónicos se envió una carta de presentación del proyecto, invitando a participar en el mismo. Dicho requerimiento se envió hasta en dos ocasiones. Aquellos consistorios que contestaban al correo electrónico eran contactados posteriormente vía telefónica para concertar una entrevista (personal u *online*).

Muchos de los ayuntamientos que mostraron un primer interés (contestaron al correo de contacto) rehusaban posteriormente a mantener una entrevista, ya fuese por manifestar una negativa directa a no querer participar en el proyecto o bien por dejar de contestar a las llamadas o mensajes de teléfono. Desde que se enviaron los primeros correos electrónicos para participar en el proyecto hasta que finalizaron las entrevistas transcurrieron tres meses y medio. Posiblemente, la baja participación fuese parcialmente debida al periodo de elecciones municipales y autonómicas que transcurrían en paralelo a nuestro trabajo de campo y que desincentivan que los actores políticos quisieran significarse de alguna forma en un problema que afectaba a la mayoría de ellos. Aunque la participación fue baja y es imposible determinar el impacto que el sesgo de participación ha podido tener en nuestros resultados, es conveniente destacar que se observa una cierta saturación teórica cuando se analiza el material recogido.

En cuanto a la entrevista, la misma fue por videoconferencia con los responsables locales de los ayuntamientos, con la excepción del responsable local en Mérida que, al constituir esta ciudad un estudio de caso más profundo fruto de nuestro proyecto (véase Arenas García, 2023), pudo realizarse de manera presencial. Las entrevistas fueron transcritas y analizadas temáticamente, en duración oscilaron entre la media hora y los 45 minutos. Aunque se trataba de una entrevista exploratoria y abierta en muchas de sus preguntas, se utilizó una guía para conocer cuestiones clave ya preguntadas a otros agentes clave (de vigilancia y control), a saber: procedimiento administrativo de la gestión del residuo, establecimiento de la fianza, tasas de reciclado, respuesta al vertido ilegal, ejecución de las sanciones, regeneración y sellado de los focos de vertidos, características del vertido ilegal y de sus autores, medidas preventivas, etc. El guión se enviaba con carácter previo a la entrevista para permitir que los participantes pudieran reflexionar con antelación sobre los temas a tratar.

4. RESULTADOS

4.1. El problema de los vertidos

La percepción de la magnitud, y otras características del fenómeno, lógicamente varía entre los entrevistados, aunque existen más similitudes que diferencias y se podrían identificar una serie de temas comunes que a continuación enumeramos. En primer lugar, existe una percepción generalizada de que se trata de un problema municipal y regional con una incidencia recurrente (todos los años como mínimo), que se observa en casi todos los municipios en la muestra, tanto en Cáceres como en Badajoz, aunque la frecuencia con la que se repiten los vertidos y el carácter más o menos localizado o disperso de estos varía entre municipios. La única excepción era el Guijo de Santa Bárbara, donde el alcalde señalaba que no tienen constancia ni de problemas, ni quejas, lo que este atribuye al hecho de ser un pueblo muy pequeño con solo dos cuadrillas de albañiles, donde hay mucha conciencia vecinal al respecto, muy poca actividad de obra, y mucha vigilancia del SEPRONA y de guardias rurales al ser reserva regional de caza. Tampoco desde la Diputación de Cáceres confirmaron la presencia de RCD ilegal en la provincia, pues se aludía a que los focos habían sido limpiados y regenerados, aunque claramente esta una realidad que no se corresponde ni con los datos del SEPRONA, ni con las entrevistas con otros interlocutores, incluidos los que presentamos en este capítulo.

Existe, por tanto, diversidad de opiniones en cuanto a la frecuencia con la que aparecen estos vertidos ilegales en los territorios que los entrevistados gestionan (desde varias veces a la semana, sobre todo en los municipios más grandes, a al menos una vez al año), lo cual puede ser un reflejo de la diversa gravedad de este fenómeno en distintos municipios. De hecho, existe una clara percepción por parte de los responsables municipales que dicha variedad existe. Que hay municipios, algunos de los cuales ellos conocen bien por proximidad geográfica, en los que el problema está más o menos extendido que en el suyo. En las localidades más pequeñas se suele indicar que este es un fenómeno generalmente puntual que a veces repunta durante los meses primaverales y veraniegos, cuando se hacen más obras de reforma, sobre todo por parte de personas que viven en estos municipios de manera más esporádica.

Aquellos entrevistados que han tenido una responsabilidad política o técnica más extendida en el tiempo consideran que en los últimos años se ha podido observar una reducción en la prevalencia de la aparición de vertidos ilegales de RCD. Generalmente, estos entrevistados aluden a dos procesos paralelos que pueden explicar esta evolución, por un lado, la crisis inmobiliaria del 2008 que supuso un descenso considerable en la actividad generadora de dichos residuos y, por otra parte, las distintas actividades que, desde la Junta, Diputaciones, y ayuntamientos se han adoptado para ir generando una infraestructura de puntos limpios, centros de tratamiento de residuos y para limpiar escombreras ilegales. Algunos responsables locales en los municipios pequeños también aluden a factores generacionales y a una cultura extendida entre los más mayores a evitar el tributar por ninguna razón y como, el cambio generacional, unido a una mayor sensibilidad hacia el medio ambiente, quizá explique esta evolución. Recordemos en todo caso que estamos hablando de percepciones y que, al margen de no deber interpretar estas opiniones como inferencias causales, no contamos en Extremadura con estadísticas apropiadas para poder evaluar de forma debida la evolución de este fenómeno. En todo caso, estos entrevistados con una trayectoria de gestión más alargada en el tiempo son capaces de apreciar cambios en la gestión de los residuos y perciben o defienden un impacto positivo de las mismas.

No obstante, se sigue considerando que es un problema con un impacto medioambiental notable, que no ha desaparecido, y prácticamente todos los entrevistados consideran que seguirá existiendo sobre todo si no se toman medidas al respecto. En lo que tiene de degradación visual del paisaje, aquellos municipios que han hecho una apuesta por el turismo rural como una forma de energizar su economía son especialmente críticos con la degradación visual que implica la aparición de estos vertidos, sobre todo cuando lo hace en lugares particularmente escénicos o que forman parte de sus rutas de senderismo en sus términos municipales, como en ocasiones ocurre. Igualmente, existe una particular preocupación entre algunos de los entrevistados, sobre todo aquellos con un perfil más técnico, por aquellos residuos que incluyen amianto por su calidad de vertido peligroso y potencial efecto cancerígeno. Es más, algún entrevistado ha mostrado una cierta preocupación por lo que pueda deparar el futuro, observando que el ciclo de vida del amianto utilizado en muchas granjas como cubierta para corrales de ganado está llegando al final de su vida y que esto puede generar una avalancha de vertido ilegal de este tipo de material, sobre todo porque a diferencia de otros RCD, el procesamiento del amianto tiene un coste económico mayor para el ciudadano. También existe una cierta preocupación en torno al RCD que pueda resultar de futuras instalaciones mineras por las que parece que se está apostando, al menos desde ciertos sectores, en la región.

Aunque la mayor parte de los entrevistados reconocen que, al menos en su experiencia, nunca se ha pillado a nadie *in fraganti* ni se ha instigado expediente sancionador contra algún vecino o promotor por esta causa, el perfil descrito por estos entrevistados y aquellos que si conocen de casos en los que ha habido una identificación de personas responsables de vertidos en sus términos municipales coinciden bastante en la caracterización de estas personas: varones de entre 30 y 60 años.

La dificultad prácticamente insuperable de coger al infractor con las manos en la masa es considerada como muy relevante por varios de los entrevistados. Según los técnicos de la Junta y el contacto clave de Mérida, el problema no se corta de raíz porque "no se pilla al infractor". Las detenciones son anecdóticas. La ausencia de infractor motiva que el SEPRONA denuncie al propietario del suelo en que ha sido depositado el vertido, pues es responsable de garantizar que el suelo esté limpio. Se produce un requerimiento al titular del suelo otorgándosele un plazo para que retire los vertidos y, en caso de no hacerlo, se aplican multas. En ocasiones se recurre la denuncia al declararse perjudicado -y no culpable- al no ser el autor material. Según el responsable de urbanismo en Mérida, los vertidos se producen mayoritariamente en suelos privados de particulares de tipo rústico, no siendo obligatorio su vallado legalmente, lo que facilitaría el acceso a los mismos. No obstante, no es esa la percepción de los técnicos de la Junta que, por el contrario, señalan que, en general, los vertidos se abandonan en terrenos municipales abandonados de manera reiterada sin que los alcaldes actúen y que, cuando el residuo se encuentra en suelo privado, generalmente, el titular de la finca suele limpiar el residuo al primer requerimiento.

Generalmente, los entrevistados distinguen entre obra mayor y obra menor. Consideran que los controles burocráticos existentes en la actualidad para la obra mayor, y el hecho de que en este caso se trate de empresarios que operan con licencia de obra con un mayor grado de profesionalización, son factores que explicarían que en estos casos no se produzcan situaciones de vertido ilegal. Existe un consenso bastante generalizado en que el problema reside en la obra menor, en la típica reforma de interiores (por ejemplo, reacondicionar un baño) u otras chapuzas. Según uno de los entrevistados "puede que incluso el vecino no sabe lo que pasa con sus residuos y que el albañil le haya cobrado por el traslado y gestión del residuo, para luego dejarlos tirados por cualquier lado."

En estos casos de obra menor se considera que en numerosas ocasiones quien hace la obra es el propio vecino ("el típico manitas") o algún conocido que puede o no estar dado de alta como autónomo, y que puede incluso estar cobrando el paro ("el albañil que le gusta trabajar al margen"). Pueden ser también trabajadores del campo o de la construcción por cuenta ajena, que son manitas y que realizan obras sin licencias.

Algunos de los responsables entrevistados consideran que en los supuestos en que hay un participe que hace el trabajo dentro de la economía sumergida existe un mayor incentivo al vertido ilegal: "sin licencia, ni conciencia, cuando terminan la obra, cogen el residuo y lo tiran". El propio pago de la fianza o cualquier gestión con el ayuntamiento o la visita al punto limpio podría destapar esta forma de economía sumergida y, por tanto, el albañil o la persona que hace la obra podría hacer el vertido ilegal precisamente para evitar dejar registro de una actividad económica no declarada: "Hay un problema detrás, la economía sumergida que genera este problema, el tener miedo al que te quiten el paro, ese es el trasfondo del problema".

Al menos uno de los entrevistados reconoce que esta situación se puede dar no solamente con parados o autónomos no dados de alta, sino incluso con empresas de la construcción que en connivencia con el propietario (quizás para reformar o hacer obra en casas de campo -o más aisladas- en las que el control de estas obras es más difícil) pueden estar realizando la obra sin licencia alguna como una forma de abaratar costes: "Yo no creo que esto solo sea una cuestión de *chapucillas* como para explicar la cantidad de residuo que hay". En este caso, también habría un poderoso incentivo a la realización del vertido ilegal para evitar el descubrimiento de esta actividad económica. Lo mismo ocurre cuando es el propio vecino el que ejecuta la obra sin haber pedido licencia.

No obstante, incluso en casos menos extremos que estos, se considera que en los escenarios de obra menor al haber menos control también se da pie a que el vecino o la persona que ejecuta la obra disponga de los vertidos de forma ilegal por una mayor comodidad (por ejemplo, para evitar tener que esperar al día siguiente, porque el punto limpio esté cerrado, a depositar un residuo con el que ya se ha terminado) o por evitar cualquier tipo de tasa en el punto limpio, allí donde esta existe. Aunque conviene destacar que los vertidos ilegales también aparecen en municipios en los que, según sus responsables municipales, esta tasa es simbólica o no existe. En este caso se suele insistir en que es una cuestión de comodidad (de tipo temporal) y no tanto relacionada con la distancia al punto limpio, ya que esta es muy pequeña en los municipios de nuestra muestra. Además, se alude a que falta información sobre los requerimientos legales en materia de vertidos y aquellos recursos existentes para gestionarlos, o de socialización cultural en estas prácticas que incide en una falta de concienciación sobre el problema: "la gente ve que lo hace su primo, el cuñado, toda la vida, y no lo ven como problema, es que no se cortan, me lo dicen a mi como técnico que es lo que han hecho".

En cuanto al *modus operandi*, los responsables también ofrecen una caracterización común. Generalmente, al ser vertidos ligados a obra menor, se trata de pequeñas cantidades de RCD que pueden trasladarse al punto de vertido ilegal en un pequeño remolque ligado al coche, en una pequeña furgoneta, o incluso en el maletero. A veces esto puede implicar varios viajes si la cantidad es algo mayor. Aunque de forma singular puedan ser pequeños vertidos, la recurrencia y falta de recogida por las autoridades municipales puede hacer que el efecto acumulativo de estos vertidos, sobre en los núcleos de población mayores, pueda ser notable.

Para el vertido ilegal generalmente se buscan sitios más o menos aislados, donde haya escasa vigilancia formal o informal, pero que tengan fácil acceso por medio de vehículos motorizados. De ahí que muchas veces el residuo aparezca en las lindes o cunetas de caminos apartados y poco transitados, aunque también se describen otro tipo de zonas como antiguas graveras, antiguos vertederos ya limpiados, espacios entre fincas privadas, etc. Son espacios en los que no necesariamente hay vertido exclusivo de RCD, sino que también pueden aparecer en ocasiones otro tipo de enseres y mobiliario, latas de aceite, restos de poda, gomas de riego del sector agrícola, etc. En algunos municipios el RCD se deja en los contenedores ordinarios de basura ("el albañil se ve que coge los sacos y los va dejando en distintos contenedores por toda la zona"), lo cual genera problemas de calado para la recogida ordinaria de basura. Alguno de los entrevistados notaba como este residuo ha llegado a dañar los camiones de basura.

Por el momento en que se descubre el residuo o por los casos en que se ha pillado *in fraganti* a alguien, aunque también algunos entrevistados por pura especulación, se opina que generalmente el residuo se vierte en horas de menor tránsito (a primera hora de la mañana, durante la hora de la siesta y más calor en verano, y, sobre todo, a última hora del día cuando muchos puntos limpios pueden estar cerrados pero el constructor ha terminado su jornada). Suele ser también más común que este tipo de residuos aparezca durante los fines de semana, cuando muchas de estas chapuzas de orden menor se pueden realizar.

Como señalaba uno de los entrevistados:

No es por distancia. El punto limpio está a 300 metros del pueblo. Donde hacen los vertidos está más lejos... Tiene que ver con la organización personal de cada uno. Le da igual las horas que esté abierta o el precio. Tiene claro que va a hacer la obra el fin de semana, es cuando puede, y quiere tener el patio limpio el domingo, ya que el lunes tiene que estar de vuelta en el trabajo. Pues lo coge y lo tira en cualquier sitio.

4.2. ¿Cómo prevenir?

Algunas de las entrevistas más cortas y quizás menos informativas solían coincidir con ser aquellas en las que los responsables municipales ponían el énfasis, de forma un tanto superficial, en la necesidad de una mayor vigilancia policial, sanciones ejemplarizantes, e incluso la instalación de cámaras de vigilancia, al mismo tiempo que se reconocía que al ser pueblos, algunos sin policía por su tamaño o con una plantilla en todo caso limitada, esto no es una solución realista. Es más, aquellos que mencionaban la idea de cámaras eran conscientes que con tantas posibles opciones donde depositar el vertido ilegal, tan solo se conseguiría desplazar el punto de vertido.

Como señalaba otro entrevistado:

> Tenemos dos policías locales, pero no dan abasto. Entradas y salidas de colegios, control del tráfico, el jueves mercadillos, trámites administrativos, al final no da. Conocemos las rutinas de cuando se tira y donde, y por eso hemos cogido a alguna gente vertiendo, pero no tenemos, ni jamás tendremos, los recursos para más control. La solución no puede ser policial.

Casi todos los entrevistados consideraban que la formación y la sensibilización es fundamental, notando al mismo tiempo, algunos de los responsables, que mientras que en sus municipios se había hecho mucho hincapié por campañas de publicidad y concienciación sobre otros temas relacionados con vertidos (por ejemplo, la necesidad de separar plásticos, papeles y vidrios, usar los contenedores adecuados para el reciclaje, etc.), el tema de vertidos de la construcción era algo a lo que estas campañas no habían prestado suficiente atención. Algunos responsables notaban que las "charlas organizadas", organizadas sobre estos temas, habían tenido escasa participación ("solo vinieron 10 vecinos"). Mientras que otros mencionaban el uso de redes sociales como *WhatsApp* como un mecanismo útil de información al ciudadano y la necesidad de empezar con campañas de concienciación en los propios colegios de formación primaria y secundaria. Al margen de informar y sensibilizar a una ciudadanía que no ven como suficientemente concienciada o preocupada por este tema, lo cual probablemente contribuye a la tolerancia de este comportamiento, algunos responsables destacaban la necesidad de sensibilizar a los profesionales del sector. Un tema que surge en estas discusiones sobre sensibilización es la necesidad de poner una especial atención al tema del amianto:

> El tema del amianto, la gente no tiene conciencia de la gravedad de ese material, hay que informar mejor. No asustar a la gente por tenerlo en sus casas, que el problema es que es contagioso a la hora de manipularlo. Como toda la vida lo han manipulado ellos o familiares y no han visto, porque no es fácil verlo, efectos, pues piensan que lo pueden hacer ellos, y lo corto aquí, lo machaco, lo escondo porque reciclarlo es carísimo, creo que este es un tema muy importante para informar.

Y luego estaban los entrevistados que claramente habían reflexionado de forma más analítica sobre la naturaleza del problema antes de sugerir soluciones. Como indicaba uno de los responsables: "prohibir es muy fácil, pero no soluciona nada, lo que hay que hacer es facilitar el cumplimiento de la ley". En ese sentido, uno de los entrevistados destacaba:

> ¿Prevenir? Vamos a ver... ¿tú ves vertidos de metal en los caminos? ¿a qué no? ¿Por qué no lo ves? Porque todos sacan dinero, el que lo recoge, el que lo recicla, todos. Si eso pasara con los RCD, no verías un casquete por ningún lado. Si no sacamos el producto, si no lo valorizamos pasa lo que pasa. Mira, si tú tienes que pagar una persona para separar cosas. Otra que lo meta en un molino triturador, que se tire moliendo, con el coste de energía que eso tiene, dos o tres días, lo pasas por tamices con distintas granularidades, todo eso tiene un coste. Si a esa persona le das dos duros por un kilo de zahorra no le sale a cuenta. Aquí le das una sartén al chatarrero y le das una alegría, porque cuesta un dinero. Nosotros el material reciclado lo usamos de relleno de calle. No tiene sentido, desde un punto de vista ambiental dragar un rio para rellenar una calle. Hace años que hacemos esto y no se nos ha hundido ninguna calle.
>
> Si esas salidas del residuo estuvieran más generalizadas, si hubiera más usos permitidos y menos burocracia para gestionar la aprobación de esos usos, en una cadena como la del metal, no

> veríamos vertidos ilegales. Igual que antes veías cartoneros habría gente dedicada recoger los escombros porque les generaría un ingreso. Claro, si la Junta pone muchas pegas al uso del material reciclado, o le pone muchas pegas, si la Junta no considera el coste medioambiental de áridos no reciclados, eso dificulta la salida del material reciclado. Hay que ser más creativos y flexibles en ese sentido.

En un sentido parecido otro alcalde sugería la revalorización *in situ*:

> Una cosa que estamos planteando a nivel de la mancomunidad es hacer la revalorización *in situ* cuando tengamos prevista una gran obra. Ya hay empresas que nos ofrecen la posibilidad de hacer la revalorización en el propio punto de aprovisionamiento temporal. Les pagas por tonelaje. Nos permite que el material que se recicla lo podamos utilizar nosotros. Yo creo que va a ser el futuro. La última reunión que tuvimos en la mancomunidad la discusión iba por ahí. Si lo hacemos varios municipios abarata costes.

Valorizar el residuo para que desincentive su abandono es también un tema recurrente de los técnicos de la Junta y el contacto clave de la Diputación de Cáceres. Este último señala que en la provincia habían acopiadas 80.000 toneladas de RCD reciclado (zahorra) que "no tiene salida, no la compran". Ello se debe a que la zahorra reciclada no tiene "un código de calidad, un estándar de calidad tal, que motive su compra". Al respecto señala que desde la Diputación se llegó a desarrollar una guía técnica en 2018 para el uso de la zahorra reciclada a fin de incrementar su utilización. En esta línea, los técnicos de la Junta refieren que los áridos reciclados son más baratos que los naturales y tienen un menor impacto ecológico, debiéndose asegurar su calidad desarrollando un estándar obligatorio para su utilización. Refieren que para el año 2030 el 40-50% de los áridos tendrían que ser reciclados y que la ANEFA[3] (Asociación Nacional de Empresarios Fabricantes de Áridos) debe jugar un rol que lo incentive a pesar de los intereses económicos creados. A este respecto, y en paralelo al transcurso

---

3 Véase: https://www.aridos.org

de esta investigación, ha sido noticia un acuerdo suscrito entre la Junta de Extremadura, la Universidad de Extremadura, el Colegio de Ingenieros Técnicos de Obras Públicas de Extremadura, y varias empresas gestoras de residuos para tratar de homogeneizar el uso de áridos reciclados[4].

También estos responsables que daban respuestas más elaboradas sobre el tema del control identificaban fallos en el sistema actual de gestión y proponían otro tipo de medidas. Aunque casi todos los municipios tenían ordenanzas sobre el tema, algunas estaban anticuadas, y la *praxis* en torno a fianzas es todavía muy variable. Mientras se articulan soluciones reales, que pasan por la valorización del residuo, lo que algunos responsables municipales destacan son medidas orientadas a la limpieza del vertido, a mejorar la infraestructura de puntos limpios y centros de reciclaje, pero también a implicar de forma más activa la colaboración entre municipios y la intervención de la Junta.

Así, por ejemplo y como señalábamos anteriormente, parece existir un cierto grado de satisfacción con el sistema burocrático existente para obras mayores, pero se reconocen las limitaciones del sistema actual para obras menores:

> Si hay licencia de obra a nosotros nos tienen que traer una justificación del punto de reciclaje con lo metros cúbicos que han salido. El problema es que cuando es comunicación previa de obra pequeña, ahí se hace una estimación a ojo, y lo que a nosotros nos dicen, el técnico del ayuntamiento lo va a dar por bueno. ¿Cómo cotejas eso con lo que se ha generado en realidad y se ha llevado al punto limpio o de reciclaje?

Estas problemáticas de control con el sistema actual se experimentan de forma distinta en municipios pequeños y los grandes. Como nos decía un alcalde de un municipio pequeño sobre la po-

---

4 Véase la noticia de El Periódico Extremadura: https://www.elperiodicoextremadura.com/extremadura/2023/05/26/homogenizacion-aridos-reciclados-construccion-comienza-87753394.html

lítica de inversión en puntos limpios de la Junta de Extremadura (la cita es larga pero ilustrativa):

> La región ha procurado gestionar este problema por medio de la subvención a los ayuntamientos para que desarrollemos los puntos limpios. Pero hoy en día se ha demostrado que los puntos limpios en nuestros municipios han sido un fracaso. ¿Por qué ha sido un fracaso? Porque los municipios no podemos tener una persona en el punto limpio las 24 horas. Ni siquiera una jornada laboral o dos jornadas laborales al día, porque estos puntos limpios requieren que haya una persona controlando la entrada y salida de residuos fija durante la mañana y durante la tarde. Como eso los ayuntamientos no lo podemos mantener ¿Qué ha pasado en aquellos municipios pequeños que usaron la subvención para generar un punto limpio? Pues que en la mayoría de los casos los tienen cerrados. Si no lo tienen cerrado, lo tienen de pena. Es más, tenemos municipios en esta comarca donde se ha denunciado a varios ayuntamientos por la gestión del punto limpio. Ya no por el escombro que se deposita fuera, sino por la gestión del punto limpio. Nos ha generado otro problema. Hay municipios que decidimos no acogernos a esa subvención porque ya preveíamos que no íbamos a ser capaces de gestionarlo.
>
> Que intenta PROMEDIO, pues que sea diputación la que gestione esos puntos limpios de alguna manera. Claro, Diputación tampoco puede poner una persona en cada uno de los pueblos. Entonces, hay un verdadero problema. La Junta optó porque se crearán puntos limpios y los gestionaran los ayuntamientos, pero los ayuntamientos tienen ya bastante sobrecarga como para poder también hacerse cargo de eso. La Diputación ahí trabajando con los ayuntamientos insistió en la necesidad de hacer una valorización del residuo de la pequeña obra a través del contenedor gestionado por PROMEDIO.
>
> Ahí aportamos los ayuntamientos de acuerdo con los contenedores que tú recicles, y pagamos una tasa a Diputación dependiendo del número de contenedores que se hayan llevado. A nosotros nos supone dos mil, tres mil euros al año, algo que se puede mantener. Pero por parte de la Junta no se nos ha dado otro tipo de soluciones. Cuando tenía que ser la junta la que se encargara de que hubiera una red de puntos de acopio temporales y luego revalorizarlo en la central, las comarcales digamos.

Si bien, los municipios grandes no están exentos de problemas de gestión con el RCD. Sirva de ejemplo la localidad de Mérida, que ronda los 60.000 habitantes. Desde urbanismo indican que las fianzas depositadas en obras menores tienen una tasa muy pequeña, a veces representa un 0,4 del presupuesto de la obra, y que los usuarios prefieren perderla antes que tener que desplazarse y pagar las tasas del reciclado. De ahí que sugiera la necesidad de tener una cantidad mínima que sea más disuasoria. Además, se indica que el alto volumen de licencias a tramitar, sobre todo en las obras menores, al ser más numerosas (mencionan que por cada 3000 licencias de obra menor al año se tramitan unas 300 de obra mayor) y carecer de proyecto técnico sin especificarse las cantidades de RCD a generar, dificulta el control de su gestión. Esta carga de trabajo sumado a la falta de información precisa sobre las cantidades de ripios que se prevén generar motiva que los servicios municipales no cotejen las cantidades de la licencia de obras con las pesadas en el gestor autorizado y se ciñan únicamente a devolver la fianza si el usuario aparece con el certificado sellado por el gestor autorizado.

Incidiendo sobre el tema de la fianza, los técnicos de la Junta se quejan de su pobre implementación:

> El problema es que los ayuntamientos no establecen la fianza. Con la fianza identificas la obra y puedes personarte. El decreto 20/2011 (el de la fianza), se aplica en toda la región, desarrolla el artículo 105 (potestad fianza obligatoria), el decreto 20/2011 ya es obligatoria y cuantificada. Los ayuntamientos no piden esta fianza.

Igualmente, varios responsables insisten en la importancia de la recogida y limpiado rápido de los vertidos ilegales detectados para así evitar el efecto llamada observado por estos actores. Los mismos insisten en la eficacia de esta medida, en su experiencia, para evitar este efecto llamada que se produce cuando el vertido no se recoge con la suficiente celeridad.

Es necesario, según los responsables municipales, una mayor sensibilización de la propia clase política: "unos hacen las leyes desde la asamblea, pero luego somos nosotros quienes tenemos que aplicarla, y parece que no entienden la realidad". O como señalaba otro entrevistado:

> Sí. Es cierto que a la clase política no es algo que yo le haya escuchado en un debate plenario de la asamblea, o que en las reuniones del partido se hable mucho. Se habla de medioambiente, pero no tanto de residuos.

Desde estas posturas se reconoce la iniciativa, en Cáceres, adoptada por la Diputación Provincial y actores como PROMEDIO en los que observan concienciación sobre el problema, pero menos en la asamblea y en la Junta. Los responsables de estos municipios, como hemos insistido, con medios limitados por su tamaño, destacan la importancia de que estos actores se impliquen más en la búsqueda de soluciones y como no todo puede dejarse a Diputación, cuando es la Comunidad Autónoma la que tiene competencias en la materia. Sobre todo, en Badajoz seguimos observando una demanda de una infraestructura más adecuada y una red más completa de centros de tratamiento y reciclaje para lo cual hace falta ese tipo de iniciativa por encima del nivel municipal:

> Nosotros estamos pensando en comprar varias hectáreas al lado del pueblo para hacer una zona industrial, donde también haya un punto limpio en condiciones, lo que tenemos ahora es un corral. Pero debería también de haber plantas de reciclaje que no estén muy lejos. La más cercana está a más de 200 km. Es una distancia demasiado grande.

O como señala otro alcalde:

> En Cáceres está mejor que en Badajoz. Aquí no hay planta de aprovisionamiento temporal. Eso lo propusimos. Si nos facilitara la Junta esas plantas pequeñas, nosotros se la cedemos a la empresa de gestión de residuos, que se lo podrían llevar a sus plantas de revalorización y ya lo vendan ellos. Lo ideal sería estas plantas de aprovisionamiento temporal, donde venga la empresa de residuos a llevárselo, y nosotros gestionamos estas plantas. Así tienes una red comarcal, con varias plantas temporales y otras de revalorización ya más grande por comarca que podrían ir a una planta provincial más grande...
> Ahora estamos haciendo prácticamente lo mismo lo que pasa es que a pequeña escala, porque ahora lo que hacemos es con el contenedor de PROMEDIO tener un aprovisionamiento temporal, con un contenedor, si fuera de más metraje cubico se podría hacer directamente... La junta debería apostar más que por puntos lim-

pios, por puntos de acopio temporal que se gestionara por parte de un organismo central que abarate los costes y se valore como se valorizan estos materiales a nivel comarcal o provincial, si nos permitieran tener estos puntos de acopio temporal y pagar a una empresa para que haga la valorización *in situ*, que ahora hay máquinas que vienen y te lo hacen, yo creo que estaríamos en condiciones de hacerlo.

5. CONSIDERACIONES FINALES

Aunque este estudio exploratorio tiene sus limitaciones metodológicas, se pueden extraer conclusiones relevantes de cara a pensar sobre esta temática. Una primera de naturaleza metodológica y es que futuros estudios deberían incorporar otros actores que no incluimos en el nuestro por puro desconocimiento inicial de su relevancia, como es el caso de las mancomunidades comarcales. Es evidente que en un contexto de municipios pequeños en numerosas ocasiones la colaboración a nivel comarcal es particularmente relevante y futuros trabajos deberían incorporar este nivel. Aunque existen temas comunes en la caracterización del problema es evidente que no existen datos objetivos rigurosos que nos permitan valorar en qué medida esas percepciones se corresponden con la realidad. No existen estadísticas fiables que nos permitan entender la magnitud y la evolución de los vertidos ilegales de RCD en la comunidad. Quizá una forma de concienciar a la comunidad sobre este fenómeno sería subvencionar medidas comunitarias orientadas al mapeo de puntos de vertidos por medio de voluntarios, al menos en aquellos municipios de mayor tamaño.

En cuanto a la caracterización del problema, resulta evidente a la luz de estos resultados, que soluciones superficiales y simplistas de tipo policial son un parche en un queso gruyere. El problema es demasiado disperso y escondido en un contexto de limitada vigilancia formal e informal, y en el que existen incentivos económicos y de otro tipo, que hacen prever que la represión tiene poco recorrido. Evidentemente, eso no quiere decir que no haya formas de pensar una mejor gestión policial, pero ese no es el tema de este artículo.

Parece evidente también que el sistema actual de gestión presenta fisuras importantes en lo que se refiere a la obra menor o los trabajos que se realizan sin licencia de obra. El propio PIREX, igual que algunos de los responsables entrevistados, sugieren la necesidad de tener una red más completa de puntos limpios y de centros de reciclaje. Pero sobre todo es tema de la revalorización donde parece que deberíamos concentrar nuestros esfuerzos si queremos prevenir este fenómeno. Lo que debería pasar -como señala el PIREX- por la inclusión en "los costes de producción de árido natural, no solo los relativos a su explotación como hasta ahora, sino también el coste medioambiental" (PIREX, p. 31), así como la modificación del Decreto 20/2011, de 25 de febrero, por el que se establece un régimen jurídico de la producción, posesión y gestión de los residuos de construcción y demolición en la Comunidad Autónoma de Extremadura para incluir otras medidas de prevención y fomento de estos materiales.

6. REFERENCIAS BIBLIOGRÁFICAS

Arenas García, L. (2023). El vertido ilegal de residuos sólidos: un estudio de caso. *Revista Española de Investigación Criminológica, Vol.21(2),* 1-20. DOI: 10.46381/reic.v21i2.820

Branas, C., South, E., Kondo, M., y MacDonald, J. (2018). Citywide cluster randomized trial to restore blighted vacant land and its effect on violence, crime, and fear. *PNAS, 115,* 2946-2951. DOI: 10.1073/pnas.1718503115.

Hollis, M., y Hankhouse, S. (2019). The growth of rural criminology. *Crime Prevention and Community Safety, 21,* 177-180. DOI: 10.1057/s41300-019-00068-4.

International Centre for the Prevention of Crime (2016). 5th International Report. Crime Prevention and Community Safety: Cities and the New Urban Agenda.

Medina Ariza, Juan José (2023). Broken windows y la experiencia en Inglaterra y Gales. Francesc Guillen (ed.). 40 años de ventanas rotas: luces y sombras. J.M. Bosch.

Murria, M., Sobrino, C., González, C. (2022). Las políticas locales de seguridad (y prevención). En Medina Ariza, J. (ed.). Instituciones de control del delito. Dykinson.

PIREX (Plan Integrado de Residuos de Extremadura 2016-2022). Junta de Extremadura, Dirección General de Medio Ambiente. Recuperado de: http://extremambiente.juntaex.es/files/2017/P_AMBTAL/RESIDUOS/PIREX/PIREX_2016_2022.pdf [Consultado el 03.07.2023].

Wilson, J. Q. y Kelling, G. (1982). Broken windows: the police and neighbourhood safety. The Atlantic (March 1982), 29-38. Disponible en:

https://www.theatlantic.com/magazine/archive/1982/03/broken-windows/304465/ [Consultado el 03.07.2023].

Conclusiones

A lo largo de la obra se ha podido evidenciar la presencia de residuos de construcción y demolición en toda la región extremeña. Los datos policiales sobre infracciones administrativas mostraban lo que sería la punta del iceberg de una realidad más profunda y compleja caracterizada por una elevada cifra negra (Arenas García, 2023). Prueba de ello son la cantidad de focos de vertido localizados únicamente en el municipio de Mérida a través de batidas de campo, a los que se sumarán nuevos puntos teledetectados por los modelos de fotointerpretación cuando sean verificados *in situ*. Dicha metodología y futura herramienta de teledetección automatizada de puntos de vertido precisará de más fases de investigación y desarrollo para su correcto ajuste a entornos orográficamente diversos, garantizando así una mayor validez y fiabilidad en cualquier municipio extremeño. No obstante, los pasos ya andados conforman una buena base de conocimientos sobre la que seguir avanzando de cara a conseguir una adecuada teledetección. Si bien, el camino ya iniciado nos ha permito conocer que los datos sobre infracciones medioambientales proporcionados por el SEPRONA (al menos los relativos a residuos en Extremadura) presentan sesgos relativos a registros duplicados y georreferenciación. Esto es importante porque dificulta, y a veces impide, aprovechar dicha información para el análisis científico y, lo que es peor, para la inteligencia policial de agentes que operan en una región muy extensa y sin apenas recursos.

No obstante, gracias a las charlas formativas que se han impartido a los agentes a nivel provincial en Cáceres y Badajoz, ha sido posible mostrarles los resultados del análisis de sus propios datos a fin de mejorar su recogida: ya sea utilizando protocolos u otras herramientas tecnológicas (como *Apps*). El análisis de la información válida y fiable, esto es, aquella resultante de la depuración de los datos oficiales junto a la recabada por el equipo en el trabajo de campo, nos permite constatar que el residuo se ubica fundamentalmente en campos, caminos, cunetas, carreteras, etc. Siendo más numerosos los focos de vertido de perímetro pequeño y mayoritariamente situados a menos de 500 metros del núcleo urbano y de las vías de comunicación. Principalmente se trata de caminos rurales con fácil acceso y poco transitados, concentrando las zonas agrícolas heterogéneas y las tierras de labor más de la mitad ellos. Además, desde un punto de vista geomorfológico, los vertederos ilegales se localizan principalmente sobre topografías onduladas y terrenos planos que facilitan el acceso. Las zonas culminantes o cumbres no son áreas propensas a acumular residuos por ser ubicaciones de mayor visibilidad. Así mismo, existe una correlación significativa entre el número de denuncias de un municipio con respecto a su superficie, población, actividad de construcción, IBI y número de viviendas. Cuanto más aumentan las variables anteriores más lo hace el número de denuncias.

Muchas de las características anteriores son observadas y corroboradas por todos los agentes clave encuestados o entrevistados en el estudio. Ellos observan, en sus rondas rutinarias de vigilancia o bien por requerimiento vecinal, que las zonas periurbanas son las más vulnerables al vertido. Hablamos de las típicas parcelas o terrenos de campo que están sin vallar, próximas a carreteras y con poca o nula vigilancia informal. La alta oportunidad delictiva que presentan estas zonas y otras similares mencionadas en el estudio (zonas de extrarradio, en desarrollo urbanístico, polígonos industriales, etc.) facilitan que los infractores no recorran grandes distancias para desprenderse de los escombros. Además, existen lugares, sobre todo en municipios pequeños, en los que la gente vertía allí "por costumbre" sin pagar tasas y sin mayor control. Nos referimos sobre todo a los vertederos que fueron sellados y regenerados en la provincia de Cáceres para su sustitución por una red de puntos de acopio (bateas controladas, puntos limpios, plantas, etc.). Si bien, como el propio término costumbre implica, se trata de conductas muy arraigadas, sobre todo en municipios pequeños, en los que es necesario realizar una labor de pedagogía con el ciudadano para lograr que se dirija al punto habilitado y realice los trámites administrativos pertinentes, incluido el pago de una tasa. Aquí, las medidas preventivas que tendrían sentido pasarían por ofrecer más información a la ciudadanía sobre los puntos de depósito y los trámites a seguir, así como campañas de concienciación social (con campañas publicitarias, acciones educativas, difusión de material, entre otras). Y es que, el pagar por depositar, parece que es una de las principales causas del residuo ilegal. El beneficio económico que reporta su ahorro (en tasa, en tiempo burocrático y en combustible), junto a la alta presencia de escenarios (donde realizar un vertido ilegal) muy intercambiables y sin vigilancia, son factores clave que motivan la comisión de una infracción. Obviamente, tal y como apuntaba uno de los agentes "no se puede tener un policía en cada esquina", ni parece que la solución pase por ahí. La prevención situacional es importante para proteger zonas especialmente vulnerables, entiendo por estas: aquellas que están muy afectadas o saturadas de residuos y necesitamos frenar urgentemente su flujo, sobre todo porque a mayor cantidad de RCD, mayor probabilidad de amianto y de residuo sucio o mixto (maderas, plásticos, electrodomésticos), lo que aumenta la probabilidad de incendios y el mencionado "efecto llamada" (Arenas García, 2023); áreas próximas a reservas de biodiversidad o ríos, por el daño a la flora y fauna más presente en estos lugares; y las zonas en expansión urbanística y antiguos vertederos (ilegales), debido a la proximidad a las viviendas y, nuevamente, por el efecto llamada que generan debido a la "costumbre" de acudir "donde siempre". En estas zonas se podrían articular diversas medidas de corte preventivo-situacional, tales como: más vigilancia policial, cortar accesos (con cadenas, barreras, zanjas en los laterales de los caminos) e incluso instalar videocámaras en los antiguos vertederos con el correspondiente cartel informativo. Lo anterior podría complementarse con formación medioambiental a los agentes, especialmente al Cuerpo de la Policía Local, e incluso plantear la creación generalizada de patrullas o unidades "verdes" en este cuerpo. Todo ello junto a medidas pensadas para mejorar la recogida, análisis y puesta en común de los datos por ellos recabados. Dicha información, aparte de ser utilizada como inteligencia policial para optimizar sus actuaciones, debería ser compartida con el resto de los agentes clave (sobre todo consistorios y técnicos de medioambiente de la Junta) a fin de: visibilizar el problema, conocer su incidencia real y dar una respuesta coordinada y ajustada a cada municipio o mancomunidad.

Son precisamente los operadores políticos los que juegan un papel esencial en esta problemática, dada la naturaleza de sus funciones y la influencia que ejercen sobre el resto de los actores (ciudadanía, agentes del control y vigilancia, y gestores de residuos). De hecho, el resto de los actores apuntan en buena medida a la falta de proactividad y desidia de las administraciones públicas, especialmente de los ayuntamientos, para atajar el problema. Y es que, aunque en esta obra se hayan aportado datos más amplios y detallados acerca de la incidencia y características de los residuos, es una cuestión innegable que los representantes políticos que operan a distintos niveles (desde la Junta a los ayuntamientos) son muy conocedores de un problema visible a todas luces. No solo porque, como el lector ya sabe, Extremadura ha sido sancionada por la Comisión Europea debido a ello, sino porque es obligación de los poderes públicos de la región ejecutar planes y medidas que materialicen el mandato nacional y europeo. En su mano está, por tanto, desempeñar cuantas acciones sean necesarias en el marco de sus competencias para atajar el vertido, de ahí que la mayor parte de las recomendaciones compendiadas en el siguiente apartado vayan dirigidas a los responsables políticos.

Con respecto al nivel político cabe identificar varias problemáticas. En primer lugar, existe una opinión generalizada en torno a la idea de que los consistorios no responden de manera activa (persiguiendo, sancionando y limpiando) los vertidos, ni dando el trámite adecuado a los residuos situados en las fincas municipales, teniendo en estas, una responsabilidad subsidiaria de regeneración. No olvidemos que, los agentes de vigilancia y control detectan y denuncian las infracciones medioambientales abriendo el consecuente expediente sancionador, pero, en muy pocas ocasiones (ínfimas) detienen al responsable. Para que estos expedientes no mueran cuando se ubican en terrenos municipales sin autor localizado, los consistorios deben asumir responsabilidades y dar una respuesta pronta y eficaz al vertido. Ello tendría un efecto "ejemplarizante" en la población -e incluso disuasorio- al percibir la ciudadanía que se toman cartas en el asunto. A lo anterior habría que sumar la referida prevención general destinada a la población, pues ahí no solo incidimos en el infractor/a no identificado, sino en los potenciales infractores. En consecuencia, una mayor concienciación medioambiental podría aumentar la eficacia colectiva vecinal e ir paulatinamente reduciendo el incivismo en las generaciones futuras.

La segunda problemática, abordada en los capítulos de corte jurídico, es la falta de desarrollo, actualización y aplicación de las ordenanzas municipales en materia de residuos, así como la exigencia de la fianza obligatoria para conseguir la trazabilidad del residuo. La insuficiente implementación normativa junto a la no limpieza de los focos de vertido, pues sabemos que la mayor parte de las zonas afectadas llevan así años (décadas), genera un gran clima de permisividad a todos los niveles. Hoy día no se sabe a ciencia cierta cuántos municipios han desarrollado su ordenanza y/o exigen la fianza. Nótese que su obligatoriedad se introdujo en el año 2011, tiempo suficiente para haber articulado su implementación efectiva con garantías. Otra cuestión relevante tiene que ver con la cuantía de las fianzas, siendo a veces tan escasas en las obras menores que su pérdida no preocupa al particular o empresario. Ello explicaría que las fianzas se acumulasen en los ayuntamientos esperando a ser reembolsadas. Esta cuestión requiere de una reflexión y estudio econométrico sosegado, para examinar si el establecimiento de fianzas mínimas más altas incentivaría su recuperación o, por el contrario, fomentaría que se tramitasen menos licencias de obras. Por otro lado, que no se especifique en términos cuantitativos los ripios que se prevén generar en la obra menor dificulta que los operarios, ya sean de los servicios municipales o de empresas gestoras de residuos, puedan cotejar bien la cantidad reflejada en la licencia de obra con respecto a la pesada y tasada en el lugar de depósito. Este problema se acentúa en los municipios grandes con mayor volumen de obras menores y, por tanto, mayor carga de gestión. El desfase en las cantidades puede ser un claro indicador de vertido ilegal. En este sentido, es necesario implementar medidas destinadas a una mejor comprobación de la documentación presentada evitando la devolución de la fianza únicamente por la mera presentación de un certificado de pago de tasas.

En tercer lugar, la pluralidad de procesos o modelos gestión, la falta de evaluación de su eficacia, así como la carencia de una red de instalaciones amplia y accesible al ciudadano influyen en la proliferación del vertido ilegal siendo las autoridades conscientes de ello. Se puede decir que el impulso por medio de licitaciones públicas y ayudas europeas han ayudado a aumentar estas instalaciones, estando repartidas prácticamente por todo el territorio, y llegando a los municipios más pequeños. Sin embargo, a su vez, nos encontramos con que aún muchas de estas instalaciones no están siendo debidamente utilizadas, a menudo por falta de voluntad, información, medios para su gestión, o concienciación de algunos ayuntamientos. En este sentido, para mejorar el sistema de gestión existente se debería aumentar y optimizar la red existente de puntos de acopio y de centros de reciclaje, fomentar la reutilización *in situ* con plantas móviles, así como conocer y homogeneizar los modelos o procesos de gestión. Estos presentan un modelo más o menos homogéneo de gestión en toda Extremadura, pero con diferencias de carácter territorial, ya que en la provincia de Cáceres el seguimiento de este modelo lo realizan dos empresas que, además de trasladar el RCD y tratarlo en su destino final, han adoptado una misma política de gestión en relación con la organización territorial, el control activo del vertido ilegal, el registro de los RCD que se depositan y los importes del depósito. Sin embargo, en la provincia de Badajoz, hay una mayor confusión para la ciudadanía y variación en la gestión final de los residuos al haber tantas empresas dedicadas a esta cuestión con sus diferentes procedimientos y tarifas. De ahí que, podamos considerar experiencias como PROMEDIO muy positivas en relación con la recogida y traslado de RCD, ya que supone un mayor control del vertido de residuos.

En cuarto lugar, el actual modelo de gestión conlleva la realización de trámites administrativos para efectuar casi cualquier tipo obra, estando fuera de control (y detección) aquellos trabajos desempeñados sin licencia de obra. Ello hace que debamos plantearnos no solo mejorar el sistema gestión existente, con lo referido anteriormente y aumentado -por ejemplo- la red existente de puntos limpios y centros de reciclaje, sino también proponer un nuevo modelo de gestión que contemple puntos de acopio de acceso abierto, sin identificación, bonificados y con una política activa de valorización del residuo.

Se ha evidenciado suficientemente que el pago de la tasa y la burocracia asociada a la misma desincentiva el depósito, en consecuencia, plantearse un modelo de gestión (con respaldo normativo) basado en la bonificación del depósito no resulta descabellado. Dicha bonificación debería ser asumida en las partidas presupuestarias autonómicas estableciendo tarifas fijas o planas a los centros autorizados de recogida de residuos para sufragar su gestión. Obviamente, aquí estamos revirtiendo la lógica de la etiología delictiva al premiar el reciclaje. La motivación del infractor que opera desarrollando sus actividades dado de alta en la seguridad social deja de ser tal, quedando por encauzar en mayor medida la acción del albañil que trabaja en la economía sumergida, siendo este el más difícil de afrontar. Al respecto, y en el marco de una nueva política integral de gestión (y de prevención situacional) que descargue la tramitación administrativa, el facilitar el acceso a los puntos de acopio con tarjetas de crédito, sin requerir la presencia constante de operarios o, incluso, habilitando bateas sin vigilancia y sin requerir identificación personal, podrían ser medidas muy eficaces. Como se aprecia, se trata de crear un sistema más ágil y flexible que pudiese canalizar incluso el vertido proveniente incluso de las obras ilegales. Esto conlleva aceptar que "los chapuzas" siempre van a existir y que debemos procurar al menos que los restos de sus obras acaben en una parcela localizada.

Para lograr dicha bonificación sea viable en términos económicos debe ir acompañada de la valorización del residuo, aspecto clave para que el RCD deje de considerarse un "residuo basura" y pase a convertirse en "residuo preciado", tal y como sucede con algunos metales. En este sentido, el reciente acuerdo firmado por la Consejería de Movilidad, Transporte y Vivienda de la Junta de Extremadura, la Universidad de Extremadura, el Colegio de Ingenieros Técnicos de Obras Públicas de Extremadura y las empresas Gestoras de Residuos de Construcción y Demolición, supone un paso de gigante para la homogenización del uso de áridos reciclados en la construcción, posicionando a Extremadura como una región innovadora en el establecimiento de pliegos y normas básicas para el uso generalizado de áridos en la construcción. Gracias al desarrollo de un estándar de calidad es posible extender el uso del RCD a nuevas aplicaciones, pudiéndose además reciclar infinitamente. No obstante, se trata de una iniciativa reciente, cuya implantación de mecanismos referenciados y testados precisará de la correspondiente evaluación técnica para verificar que, efectivamente, se obtiene una guía del uso de áridos reciclados con condiciones técnicas generales y suficientes. De su éxito dependerá el exportar el conocimiento de la región al resto del territorio español dando así un gran impulso a los objetivos de la Agenda 2030.

Recomendaciones

Una de las principales finalidades de la Criminología es proponer estrategias y medidas encaminadas a intervenir sobre la realidad analizada para, en este caso, reducir los casos de vertido ilegal por RCD. A tal fin se han elaborado una serie de recomendaciones dirigidas a diferentes agentes clave en la gestión y vigilancia de los residuos, siendo estos: operadores políticos del ámbito local (como alcaldes, concejales de medioambiente, etc.), de las diputaciones provinciales y de la consejería de medioambiente de la Junta de Extremadura; operadores encargados de la vigilancia y control de los vertidos, como son las policías locales, el SEPRONA y los Agentes del Medio Natural; y los agentes autorizados en la recogida y reciclaje de los residuos. Dichas recomendaciones se efectúan a nivel macro, con medidas dirigidas a la educación para la ciudadanía y la concienciación social, y micro, con propuestas específicas ceñidas a la mejora de sus procedimientos. Algunas pretenden mejorar aspectos del modelo de gestión actual mientras que otras son características de un modelo nuevo. En cualquier caso, se trata de recomendaciones que deben explorarse, ajustarse y evaluarse en los diferentes contextos y necesidades de los municipios.

1. RECOMENDACIONES PARA LOS OPERADORES POLÍTICOS

- Establecer un plan regional que garantice la gestión de RCD a todos los niveles (infraestructura, destino final del residuo ya tratado, dotación y formación del personal) así como una mayor dotación de agentes que vigilen el cumplimiento de la legislación sobre la materia.
- Ante la incapacidad de los municipios pequeños para gestionar los RCD, la Junta u otros consorcios provinciales de-

ben asumir su gestión facilitando los mecanismos y/o servicios adecuados para ello.

- Crear mecanismos de coordinación entre municipios, mancomunidades, diputaciones, empresas de gestión de residuos y los agentes de vigilancia y control. A tal fin se deberían protocolizar sus actuaciones; establecer cauces para que compartan información relevante (datos sobre denuncias interpuestas, inspecciones realizadas, modelos de gestión, actividades, campañas etc.); y fomentar la celebración de reuniones periódicas entre ellos.
- Actualizar las ordenanzas municipales de acuerdo con la Ley 7/2022, de 8 de abril, de residuos y suelos contaminados para una economía circular.
- Crear y aplicar en los municipios una ordenanza modelo similar a la propuesta por entidades como PROMEDIO para favorecer el interés de los ayuntamientos.
- Crear un registro de municipios en el que conste si disponen de ordenanza sobre residuos, si efectivamente se aplica la fianza y las cuantías de las tasas del reciclaje.
- Revisar los importes de las fianzas solicitadas para la gestión de RCD de obras menores y establecer un importe o porcentaje en el caso de que no lo hubiera.
- Establecer fianzas lo suficientemente altas como para que se incentive su recuperación.
- Especificar las cantidades de ripios a generar (en términos cuantitativos) en las obras menores.
- Cotejar de manera más exhaustiva las cantidades de ripios que se prevén generar en las obras menores y mayores con respecto a las cantidades depositadas y certificadas por el gestor autorizado de residuos.
- Desburocratizar la gestión de los RCD, eliminando tasas y procedimientos que la dificulten tanto a particulares como a empresas.

- Bonificar la gestión de los RCD, especialmente de aquellos que son altamente contaminantes como el amianto.
- Habilitar contenedores urbanos para la recogida de pequeñas obras.
- Habilitar zonas de entrada libre y no identificada a lugares de depósito para canalizar los vertidos de obras ilegales.
- Aproximar los puntos de recogida y tratamiento de residuos a las poblaciones.
- Adaptar los horarios de apertura y cierre de los centros de recolección y tratamiento de RCD a las franjas en que se generan los mismos.
- Mejorar la información sobre los servicios que ofrecen los puntos limpios y las plantas de tratamiento.
- Mayor transparencia e información sobre recursos para la gestión de residuos desde los ayuntamientos y mancomunidades, a través de carteles informativos, redes, páginas web, etc.
- Creación de una página web desde la Junta de Extremadura y las Diputaciones con todos los recursos en materia de gestión de residuos, incluyendo puntos limpios, y mantenimiento de la información actualizada.
- Crear un *App* destinada a la ciudadanía para que registren (usando SIG) los impactos medioambientales cotidianos y de los que son testigos (contaminación de aguas, presencias de basura urbana, caza ilegal, maltrato animal, incendios, vertidos de RCD, entre otros).
- Realizar campañas informativas y de sensibilización hacía la ciudadanía en general (anuncios en radio, TV, redes sociales, etc.) y talleres y charlas en instituciones educativas, tales como colegios e institutos.
- Realización de talleres y charlas de concienciación medioambiental a nivel municipal en materia de residuos, tanto para ciudadanía como para personal técnico.

- Elaborar y divulgar panfletos o boletines a las empresas que operan en el sector de la construcción con los últimos avances legislativos en la materia, posibles sanciones, ubicación de los gestores autorizados de residuos e información sobre su gestión y teléfonos de interés.
- Aumentar la formación medioambiental y sobre residuos del personal municipal y mancomunal.
- Mejorar la inspección de las cantidades de residuo previstas y estipuladas en la licencia de obra con respecto al certificado de depósito del residuo.
- Aumentar las inspecciones a las instalaciones de gestores de residuos.
- Limpiar y regenerar las zonas más afectadas por residuos evitando así su acumulación y posible efecto llamada.
- Crear líneas de financiación a asociaciones preocupadas por el bienestar de la naturaleza a través de la Junta de Extremadura y las diputaciones (como futura línea FEDER) y líneas de ayudas a los ayuntamientos para la limpieza.
- Otorgar ayudas al sector de la construcción para el depósito y reciclaje de RCD.
- Otorgar ayudas a los gestores autorizados de residuos para la valorización del RCD.
- Desarrollar un estándar de calidad que permita la valorización y utilización del RCD para múltiples fines.
- Penalizar la extracción de los áridos naturales imponiendo el pago de tasas.
- Imponer la utilización obligatoria de áridos reciclados en obras públicas.

2. RECOMENDACIONES PARA LOS OPERADORES POLICIALES

- Georreferenciar los puntos de vertido ilegal a fin de utilizar la información como inteligencia policial focalizando los recursos en las zonas más vulnerables.
- Crear un protocolo para georreferenciar adecuadamente los puntos de vertido.
- Crear canales de comunicación y coordinación entre policías y agentes del medio natural.
- Establecer un canal de comunicación bidireccional entre asociaciones de vecinos y policías a fin de comunicar y compartir datos sobre vertidos ilegales de RCD en la zona logrando la implicación de los vecinos en la vigilancia informal de estas conductas.
- Realización de talleres y charlas de concienciación medioambiental a nivel municipal en materia de residuos, tanto para ciudadanía como para el personal técnico.
- En la medida que sea posible, crear unidades medioambientales o "patrullas verdes" en los cuerpos de la Policía Local.
- Dotar de mayor personal y recursos a las patrullas medioambientales de la policía local.
- Impartir formación continua en materia medioambiental y sobre residuos a los agentes.
- Aumentar la vigilancia y control de las zonas más afectadas por residuos mediante rondas rutinarias de los agentes, especialmente en las horas de mayor incidencia del vertido.
- Utilizar cadenas o zanjas que eviten el paso de vehículos a zonas más deterioradas por el vertido ilegal o enclaves vulnerables (ríos, reservas naturales, zonas con riesgo de incendio, etc.).

– Aumentar las inspecciones a las instalaciones de gestores de residuos.

3. RECOMENDACIONES PARA LOS AGENTES AUTORIZADOS EN LA RECOGIDA Y RECICLAJE DE LOS RESIDUOS

– Mayor control de la documentación presentada por los usuarios en las instalaciones a la hora de depositar y pesar los residuos. Todo ello para detectar y evitar desfases en las cantidades depositadas y pagadas de RCD.

– Ofertar servicios de recogida RCD a domicilio bajo cita previa en días de fin de semana y festivos.

– Comunicar a los consistorios los posibles indicios de fraude detectados a la vista de la documentación presentada o las cantidades depositadas de RCD.

– Elaborar y publicar informes sobre la actividad de sus instalaciones acerca del número de usuarios que atienden, registro de entradas y salidas, cantidades de ripios pesados y facturados, estado de sus instalaciones, número de recogidas de RCD que realizan fuera de sus instalaciones, composición de los residuos recogidos, incidencias relacionadas con el vertido ilegal, etc.

– Publicitar entre los empresarios y operarios del sector de la construcción la calidad, precio y aplicaciones del RCD reciclado en contraposición a los áridos naturales.

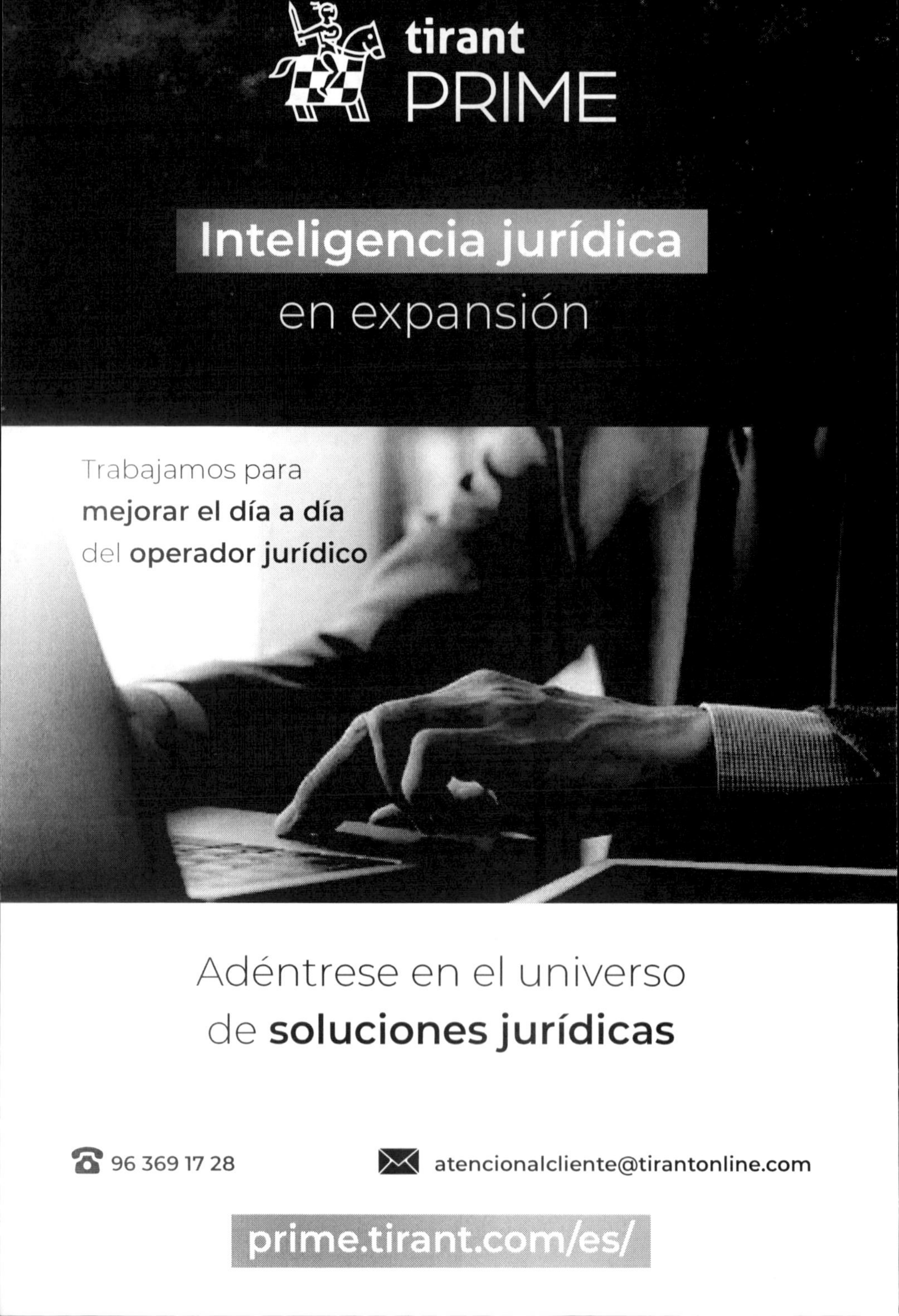
tirant PRIME
Inteligencia jurídica
en expansión
Trabajamos para
mejorar el día a día
del operador jurídico
Adéntrese en el universo
de soluciones jurídicas
96 369 17 28
atencionalcliente@tirantonline.com
prime.tirant.com/es/